T0191659

S. P. Venkateshan

Heat Transfer

Third Edition

Ane Books
Pvt. Ltd.

S. P. Venkateshan
Department of Mechanical Engineering
Indian Institute of Technology Madras
Chennai, Tamil Nadu, India

ISBN 978-3-030-58340-8 ISBN 978-3-030-58338-5 (eBook)
https://doi.org/10.1007/978-3-030-58338-5

Jointly published with ANE Books Pvt. Ltd. In addition to this printed edition, there is a local printed edition of this work available via Ane Books in South Asia (India, Pakistan, Sri Lanka, Bangladesh, Nepal and Bhutan) and Africa (all countries in the African subcontinent).
ISBN of the Co-Publisher's edition: 978-9-385-46207-8

This Springer imprint is published by the registered company Springer Nature Switzerland AG
The registered company address is: Gewerbestrasse 11, 6330 Cham, Switzerland

To My Parents with Respect

To the Shakkottai Family with Affection

Preface to the Springer Edition

The first edition of the book titled "A First Course in Heat Transfer" was published about 15 years ago. The second edition appeared in 2009. Time was ripe to revise the book further and the result was this third edition, which was published by Ane Books Pvt. Ltd. in 2017. With corrections, the same is now being brought out as Springer edition.

At one time it appeared that the field of 'heat transfer' had reached saturation and there was not much new in it. However, things have changed significantly in recent times. New applications in fields such as microelectronic devices, nuclear reactors, space propulsion systems, 3D printing made it necessary to move beyond what was possible only a few years ago. Improvements in computers and measuring instruments have made the field interesting once more and there is scope for much research in this area. The author has been involved in several of these developments and feels that it is time to look at the subject with renewed interest!

This edition is brought out with a complete overhaul of the book. Many new worked out examples are included in this edition. Also, many new topics have been added to bring the book, undoubtedly not only to a higher level, but also to a higher level of relevance. I have tried to intersperse the elementary aspects with several advanced topics so that the interested reader can explore more recent developments in heat transfer with a higher level of preparation.

The aim of the present edition remains the same as the earlier editions, viz., to move from elementary to advanced in a slow but steady progression. In order to keep the length of the book under check, I have tried to reformat the entire book using more advanced features available in "latex" along with the graphic environment "tikz". All plots have been redone using QtiPlot and the line drawings have been redone using "tikz". Problems at the end of each chapter are still the best way the reader will be challenged to test his learning of the material discussed in the chapters of the book.

I have tried to correct as far as possible all errors of various types in the earlier editions. However, I will be grateful if the reader would bring to my notice any errors still found in this edition.

Chennai, India
May 2020

S. P. Venkateshan

Preface to the Second Edition

The present book is an augmented and fully revised version of my earlier book *A First Course in Heat Transfer*. The book is now out of print and not available. Several typographical and factual errors that had crept into the book have now been corrected in the present book, which is titled "Heat Transfer". The change of the title seemed reasonable because the contents had been augmented to include many topics that were omitted in the earlier book. The number of chapters have grown to 17 as against 15 in the earlier book. Notable new material is to be found in all topics of heat transfer. Convection heat transfer and radiation heat transfer portions have been significantly improved with additional material that takes these closer to the current literature in these areas. I have also included the most relevant references as footnotes for the convenience of the reader. The present revised augmented book has taken over two years of my time.

I have improved the level of the book with additional materials included in almost all the topics. In view of the growing importance of numerical methods I have expanded the part that deals with numerical methods. Several appendices dealing with background material not easily accessible to the student are added to make it possible to deal with more advanced heat transfer topics in the book. The first time reader may want to skip some of these topics, without loss of continuity.

A few years ago I recorded a set of video lectures on Heat Transfer through the Educational Technology Cell of the Indian Institute of Technology, Madras. These are available in DVD form from the Educational Technology Cell, IIT Madras. These have subsequently been broadcast periodically over Eklavya, the Technology channel. As a part of the video effort I prepared notes on the various topics covered in the lecture series. The notes looked interesting and I felt that it would be worth while converting the Notes in to a book form and make it available to a wider audience. It seemed that, with many of the Regional Colleges of Engineering being upgraded to National Institutes of Technology, an introductory book directed towards students joining these institutes would be worth publishing and the present book is the result. With the general level of undergraduate programs undergoing qualitative change, the book should be relevant to undergraduate students studying in any of the many engineering colleges that have started functioning throughout the country. I believe the present book can be covered in a semester, say in the third year of the B.Tech. program as is done at the IIT's including IIT Madras. Some advanced sections may be omitted for this purpose.

Heat transfer as a discipline has grown over the past two centuries to a mature science. Rapid developments took place during the Second World War. Later the space age brought a new emphasis to the study of heat transfer under harsh conditions. The energy crisis focused the attention of heat transfer experts on solar energy applications. Developments in microelectronics have in recent times motivated heat transfer research. Heat Transfer in manufacturing processes like laser machining, electron beam welding, metal casting, to name a few, have also been major areas of recent research. However, it is fortunate that most of these have not required any more mathematics than that contained in a book like "Advanced Engineering Mathematics" by Kreyszig 1993. The knowledge of Physics required is more or less that covered in the Plus 2 followed by what is taught in the first year in most engineering colleges in the country. The background knowledge of Fluid Mechanics may be obtained from a book like "Introduction to Fluid Mechanics" by Fox and McDonald 1995. A good grounding in the fundamentals of thermodynamics, as is covered in the first year of engineering, is all that is needed to undertake a study of heat transfer. The present book assumes that the student has already had exposure to the above by the time he decides to use this book.

I do not have any pretensions regarding the originality of the material that forms the bulk of the book. These have been considered in one form or another by all the previous authors who have written books on Heat Transfer. I can only claim to a certain way the material has been presented in the present book. I use examples that are close to reality. A practicing Heat transfer engineer would probably think of similar examples when he is designing thermal systems. The problems are not the plug and play type. They do require some amount of "modeling" effort on the part of the student. Also, exercises at the end of each chapter require a fair amount of thinking on the part of the reader. I have deliberately avoided giving answers to problems. This will discourage the student from trying to get the provided answer by hook or by crook. We all make mistakes, in modeling or in calculations, and we learn more from mistakes than from perfectly executed solutions, the first time. This, I believe, is the best way of attaining some self sufficiency on the part of the student.

The writing of the book has involved support from several people. Dr. N. Ramesh, a former student of mine produced the first hand written version of the notes. Mrs. Lakshmi Suresh typed the first draft with a lot of care. I made all the figures and plots using the many software resources available on my PC. Particularly useful was MathCad 7 Professional and Microsoft EXCEL which helped in checking and rechecking the many solved examples presented in the book.

I have enjoyed writing the book since it gave me opportunity to learn a lot of new things. I hope the book also interests the students and other readers who may use it for learning heat transfer.

S. P. Venkateshan

Acknowledgements

Writing a book requires the help and support of many people. I have been fortunate in having a large number of students who have helped me in this effort. All my students are involved in heat transfer research and have provided support by carrying out my pet projects with dedication. At several places in the book, I have included data and results from theses of my students. The institution where I work—Indian Institute of Technology Madras——has provided an ambience conducive to academic pursuit and made this book revision a smooth affair. My past student Prasanna Swaminathan wrote a class file "bookspv.cls" which has helped me in improving the aesthetic quality of the book.

Most importantly I thank my wife for the wholehearted support she has extended by allowing me to work with the computer at home for long hours every day.

May 2016 S. P. Venkateshan

Contents

About the Author

Prof. S. P. Venkateshan obtained his Ph.D. from the Indian Institute of Science, Bangalore in 1977. After spending three years at Yale University, he joined Indian Institute of Technology Madras in 1982. He has been teaching subjects related to Thermal Engineering to both UG and PG students for the past 32 years. He has published extensively and has more than 100 publications to his credit. The areas of his interest are: (a) Interaction of natural convection with radiation, (b) Numerical and experimental heat transfer, (c) Heat transfer in space applications (d) Radiation heat transfer in participating media and (e) Instrumentation. Professor Venkateshan has been a consultant to ISRO, DRDO, and BHEL in India and NASA in the US. He has three patents to his credit in the area of instrumentation. He has also guided about 30 scholars towards the Ph.D. and a similar number of scholars towards the M.S. (by Research) degree at IIT Madras. The present book had its beginnings in the notes prepared for the course on Heat Transfer taught by him for several years at IIT Madras.

Nomenclature

Note:

- Many symbols have more than one meaning. The context will indicate the specific meaning.
- Symbols in limited use are not given here.

Latin Alphabet Symbols

A	Aspect ratio of a cavity, non-dimensional; Area (m^2)
a	Speed of sound (m/s)
B	Rotational constant ($1/s$)
Bi	Biot number (non-dimensional)
C_p or c	Specific heat ($J/kg \cdot K$ or $J/kg°C$)
C	Thermal capacity (W/°C)
c_0	Speed of light in vacuum (m/s)
C_f	Friction coefficient (non-dimensional)
d	Diameter (m)
D	Diameter (m)
D_H	Hydraulic diameter (m)
E	Electric field intensity (N/coul); Emissive power (W/m^2) (total) or ($W/m^2 \mu m$) (spectral)
E_n	Exponential integral function of order n
Ec	Eckert number (non-dimensional)
E_c	Total energy stored (J)
erf	Error function

erfc	Complementary error function
Eu	Euler number (non-dimensional)
F_{ij}	Diffuse view factor between surface i and surface j (non-dimensional)
f	Frequency (Hz); Friction factor (non-dimensional); Fraction of black body radiation between 0 and λT (non-dimensional)
Fo	Fourier number (non-dimensional)
g	Acceleration due to gravity (m/s^2)
G	Heat generation rate (W/m^3); Irradiation (W/m^2)
Gr	Grashof number (non-dimensional)
Gz	Graetz number (non-dimensional)
h	Heat transfer coefficient ($W/m^2 {}^\circ C$); Planck's constant (J – s); Enthalpy (J/kg)
H	Magnetic field intensity, Non-dimensional heat generation parameter
h_R	Radiation heat transfer coefficient ($W/m^2 {}^\circ C$)
h_{sf}	Latent heat of melting/solidification (J/kg)
I	Moment of inertia ($kg \cdot m^2$)
I	Radiation intensity ($W/m^2 \cdot sr$) (total); Radiation intensity ($W/m^2 \cdot \mu m \cdot sr$) (spectral)
J	Radiosity (W/m^2)
J	Rotational quantum number
k	Boltzmann constant ($kJ/kmol \cdot K$)
k	Thermal conductivity ($W/m^\circ C$ or $W/m \cdot K$)
L	Length (m)
$LMTD$	Logarithmic mean temperature difference (°C or K)
L_m	Mean beam length (m)
m	Complex index of refraction; Fin parameter for a uniform area fin (m^{-1}); Mass (kg)
M	Mach number (non-dimensional); Reduced mass (kg)
N	Environmental parameter (non-dimensional)
n	Refractive index
$\overrightarrow{n}$	Unit normal vector
N_{RC}	Radiation conduction interaction parameter (non-dimensional)
NTU	Number of transfer units (non-dimensional)
Nu	Nusselt number (non-dimensional)
p	Fin parameter for a variable area fin ($m^{-\frac{1}{2}}$)
P	Perimeter (m); Power (W)
p	Pressure (bar or Pa)
Pe	Peclet number (non-dimensional)
Pr	Prandtl number (non-dimensional)
q	Heat flux (W/m^2)
$\overrightarrow{q}$	Heat flux vector (W/m^2)
Q	Total heat transfer (W)
R	Electrical resistance (Ω)
r	Radial coordinate or radius (m); Recovery factor (non-dimensional)

R Radius (m); Ramp rate (°C/s); Capacity ratio (non-dimensional); Thermal resistance (°C/W or m^2°C/W)
Ra Rayleigh number (non-dimensional)
Re Reynolds number (non-dimensional)
R_f Fouling resistance (°C/W or m^2°C/W)
S Conduction shape factor (non-dimensional); Solar constant (W/m^2); Surface area (m^2)
S_D Diagonal pitch (m)
S_L Longitudinal pitch (m)
St Stanton number (non-dimensional)
S_T Transverse pitch (m)
Ste Stefan number (non-dimensional)
T Temperature (°C or K)
t Time (s)
t Transmittivity or transmittance
t^* Time lag (s); Charging time (s)
T_m Melting temperature (°C or K)
T_{ref} Reference temperature (°C or K)
U Free stream velocity (m/s); Overall heat transfer coefficient (W/m^2°C)
u x component of velocity, (m/s)
V Potential energy (J)
v y component of velocity (m/s)
v Vibrational quantum number
V Volume (m^3)
x x-coordinate (m)
y y-coordinate (m)
z z-coordinate (m)

Greek Symbols

α Absorptivity (no unit); Thermal diffusivity (m^2/s)
β Isobaric volumetric expansion coefficient (K^{-1}); Wedge angle (rad or °)
γ Ratio of specific heats (no unit); Surface tension (N/m)
δ Boundary layer thickness (m); Condensate layer thickness (m); Depth of penetration (m); Phase angle (rad)
ε Eddy viscosity (kg/m · s); Effectiveness of a fin array (no unit); Emissivity (no unit); Heat exchanger effectiveness (no unit)
ε_H Eddy diffusivity of heat (m^2/s)
ϕ Sub-cooling parameter (non-dimensional)
η Blasius similarity variable (non-dimensional); Fin efficiency (non-dimensional); Heat exchanger effectiveness (non-dimensional)

η_R	Radiating fin efficiency (non-dimensional)
κ	Absorption coefficient (m^{-1})
λ	Vacuum wavelength (m)
μ	Cosine of angle with respect to normal (no unit); Dynamic viscosity ($kg/m \cdot s$); Non-dimensional fin parameter (no unit)
ν	Kinematic viscosity (m^2/s); Photon frequency (Hz)
θ	Angle (° or rad); Characteristic rotational temperature (K); Non-dimensional temperature (no unit)
θ_{ref}	Environmental parameter (non dimensional)
ρ	Density (kg/m^3); Reflectivity (no unit); Resistivity ($\Omega - m$)
σ	Stefan–Boltzmann constant (W/m^2K^4); Surface tension (N/m)
τ	Characteristic time (s); Optical thickness (no unit); Shear stress (Pa); Time constant (s); Transmittivity (no unit)
ψ	Stream function (s^{-1}); Film resistance number (non-dimensional); Radiation number (non-dimensional)
ω	Circular frequency (rad/s)
Ω	Solid angle (sr)

Subscripts

amb	Pertaining to the ambient
c	Pertaining to convection
ch	Based on a characteristic length scale
f	Liquid
fg	Liquid-vapor
i	Pertaining to insulation layer
k	Pertaining to conduction
sat	Saturation condition
sf	Solid-liquid
R	Pertaining to radiation
r	Radial
w	At the wall
∞	Pertaining to free-stream or ambient
1,2, etc.	Pertaining to a specific position

Chapter 1
Introduction to the Study of Heat Transfer

Heat transfer is an important area of study since its applications are universal. Heat transfer plays a role in all sciences and technologies. It plays an important role in both living and non-living things. The scale at which heat transfer takes place spans the smallest scales at the atomic level to very large scales at the level of the universe. The present book is an introduction to the field of "Heat Transfer", not necessarily covering all its aspects. The present book aims to develop the subject ground up and brings it to a level sufficient to motivate the reader to look at current heat transfer literature.

1.1 Introduction

Heat transfer or the lack of it and its consequences are important in most day-to-day activities in industry and human life. Most engineering design involves, at some stage or the other, a consideration of heat transfer and its prediction for reliable operation of the designed system or product. Though a catalog of problems needing heat transfer prediction can be given, it is fortunate that, from a physical point of view, we can put them all under a small number of basic processes. The study of such processes forms the material of the course on "Heat Transfer".

Historically the birth of the discipline of heat transfer took place some two hundred and fifty years ago. The formulation of the basic principles of heat transfer paralleled developments in mechanics, thermodynamics, and the principles of fluid flow. The inclusion of the conservation principles and the laws of thermodynamics have brought the subject to a form, which is recognized today as "Heat Transfer". The mathematical foundation for the study of heat transfer was laid by the work of Newton, Fourier,

S. P. Venkateshan, *Heat Transfer*,
https://doi.org/10.1007/978-3-030-58338-5_1

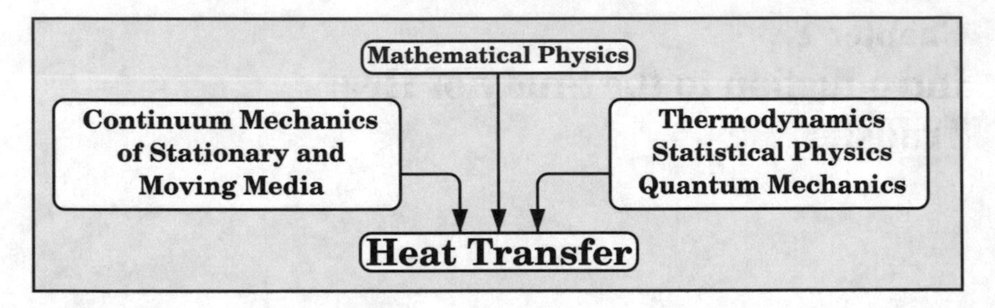

Fig. 1.1 Heat transfer—an interdisciplinary area

Laplace, and later by the work of Prandtl, Nusselt[1] and others. Developments in various branches of mathematics have also contributed to making the study of heat transfer a rigorous one.

Heat transfer is basically an interdisciplinary area. Figure 1.1 indicates why it is so. As indicated in the chart the discipline of heat transfer depends on the theoretical framework of the other disciplines indicated. With changing times, new situations requiring knowledge of heat transfer processes have become important. To this category belongs heat transfer for space applications, exploitation of solar energy, heat transfer in nuclear power applications, cooling of miniature electronic components, and so on. These applications have led to the development of new technologies solely for taking care of heat dissipation for the reliable operation of the components.

Basically, we recognize three modes of heat transfer—1. Conduction, 2. Convection, and 3. Radiation. However, it is seldom that these take place independently and in isolation, in a given application. It is a combination of these that takes place and hence the problems become mathematically very complex when we try to obtain the solution. These complications have led to the development of advanced analytical methods, approximate analytical methods, and numerical methods. With the availability of inexpensive computers, numerical methods of solution have also become attractive. Of course, experimental studies continue to be important in the study of heat transfer processes.

1.2 Basic Assumptions in the Study of Heat Transfer

Heat transfer takes place, in the presence of temperature differences within a system, or where there is a temperature difference across a boundary, which separates two systems. Whenever thermal energy leaves or enters a system, (except when there is a change of phase) its temperature changes. If this energy transfer is very rapid, the internal modes of energy may not all equilibrate and hence a *temperature*, under such

[1]These scientists were early contributors to the field of heat transfer; between 1600 and 1950 CE. Several *Non-dimensional numbers* in Fluid Mechanics and Heat Transfer are named after them.

conditions, does not characterize the system. We assume that such is not the case and hence assume that **L**ocal **T**hermodynamic **E**quilibrium or LTE prevails. Thus we associate a temperature to the smallest part of a system that may not, on the whole, be in thermodynamic equilibrium with the neighboring parts of the system.

The second assumption we make is that the matter within the system is a continuum. Hence, properties are defined as continuous functions of the space coordinates. Thus temperature, heat flux, pressure, enthalpy, etc., may be assigned to the smallest part of the system we may want to consider. The continuum is associated with a field—temperature, pressure, etc.—which are continuous functions of space and time. The local thermodynamic equilibrium assumption means that a macroscopic system may be defined as a collection of subsystems, each of which may be assigned properties such as density, temperature, pressure, etc.

Continuum and LTE may break down in case we encounter (a) Very low densities or (b) Very-high-speed flows (fluid velocity much much larger than the speed of sound in the fluid).

Continuum hypothesis

Matter is assumed to exist in the space-time continuum. However, matter is made up of discrete particles at the molecular or atomic level. The assumption that matter is distributed as a continuum breaks down as we reduce the size of the domain that contains matter. In the case of solids and liquids, the continuum assumption breaks down at the molecular level where the domain size is a few Angstroms ($1\,AA = 10^{-10}\,\text{m} = 0.1\,\text{nm}$). However, in the case of gases, the assumption may become invalid even at much larger domain sizes, depending on conditions that are discussed below.

Transport processes take place in gases due to collision between gas molecules as they keep moving around inside the gas volume. Two molecules will collide if the separation between them is less than or equal to the molecular diameter. The distance traveled by a molecule between successive collisions is known as the mean free path ℓ. The mean free path may be calculated using the formula

$$\boxed{\ell = \frac{\Re T}{\sqrt{2}\pi d^2 N_A P}} \tag{1.1}$$

where $\Re = 8.3145\,\text{J/K mol}$ is the universal gas constant, $N_A = 6.022 \times 10^{23}\,\text{/mol}$ is the Avogadro's number, T is temperature (K), P is pressure (Pa), and d is the molecular diameter (m). At mean sea level $P = 1.013 \times 10^5\,\text{Pa}$ and with $T = 288.15\,\text{K}$ the mean free path is $\ell = 9.82 \times 10^{-8} \approx 10^{-7}\,\text{m}$ assuming mean molecular diameter of $d = 3 \times 10^{-10}\,\text{m}$. Since the mean free path at sea level is of the order of a $0.1\,\mu\text{m}$ the continuum hypothesis is valid as long as the characteristic length of the gas volume is more than a μm. Frequent collisions between gas molecules and the presence of a large enough number of molecules within the smallest volume makes it possible to justify the continuum hypothesis.

As opposed to the above, the mean free path in the atmosphere at an altitude of 100 km is of the order of 0.1 m. The regime is one of slip flow. At an altitude of 300 km the mean free path is of the order of a 10 km. This regime corresponds to the free molecular flow regime.

The ratio of mean free path ℓ to a characteristic length scale L, $Kn = \frac{\ell}{L}$ the Knudsen number (after Martin Knudsen, 1871–1949, Danish physicist), is useful in classifying the flow regime.

Knudsen Number Range	Flow regime
$Kn < 0.01$:	Continuum flow regime
$0.01 < Kn < 0.1$:	Slip flow regime
$0.1 < Kn < 10$:	Transitional regime
$Kn > 10$:	Free molecular regime

When flow velocities are high such as in hypersonic flows (e.g.: reentry of a vehicle in to the atmosphere) shocks, regions of large gradients, are formed with a characteristic thickness smaller than the mean free path. Continuum hypothesis will break down in these cases also.

In **MEMS** (**M**icro **E**lectro **M**echanical **S**ystems) devices, the mean free path may again become larger than the characteristic dimension, say the diameter or width of passages. Continuum hypothesis may break down.

1.3 Basic Heat Transfer Processes and Examples

Heat transfer processes may be classified broadly as follows:

Diffusion Heat conduction, momentum transfer due to viscosity, mass diffusion[2]—generally, diffusion is a short mean free path process (for the molecules).

Radiation Thermal radiation—generally a long mean free path process (for the photons).

Convection Transport of mass, momentum, and energy by a moving fluid.

In engineering applications, these do not take place in isolation. A combination of two or more is common. The mathematical complexity is because of this. Typical examples are presented below.

[2]Mass transfer is not considered in the present book.

A Candle Flame

A candle flame is a fairly complex system. Air and fuel (hydrocarbon in the wax) react to form a flame. The reaction is exothermic and is dissipated by radiation (photons are released; we are able to see the flame since some of the photons are in the visible part of the electromagnetic spectrum) and is convectively transported and dissipated by the plume. Some heat is conducted/radiated into the candle and it melts. Figure 1.2 shows the state of affairs.

Heat Flow Across a Solid Wall

Figure 1.3 shows the situation schematically. The wall is a conducting element and heat transfer is by conduction across the solid. At the two boundaries, heat transfer is by convection and radiation. The temperature variation within the wall and the regions adjacent to the two boundaries of the wall, under steady conditions, is as shown in the figure. Note that the temperature variation adjacent to each boundary represents that in the ambient medium. Heat flux q is in the direction of decreasing temperature.

Cooling System for an Internal Combustion Engine

Figure 1.4 shows a schematic representation of the cooling system for an internal combustion engine used in automobiles. Cooling water is *forced* through the water jacket of the cylinder to maintain the cylinder wall at the desired temperature level.

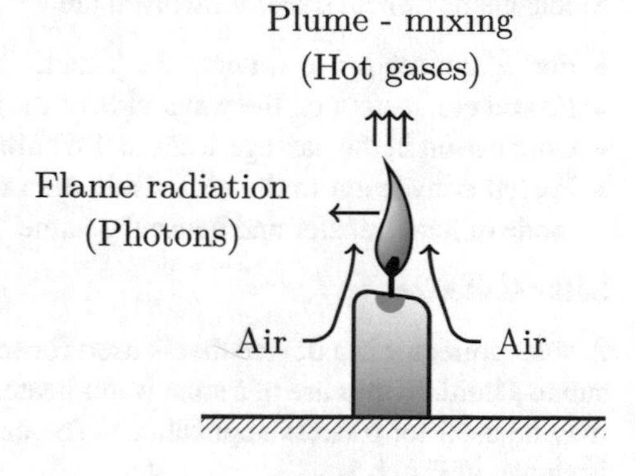

Fig. 1.2 Heat transfer processes in a burning candle

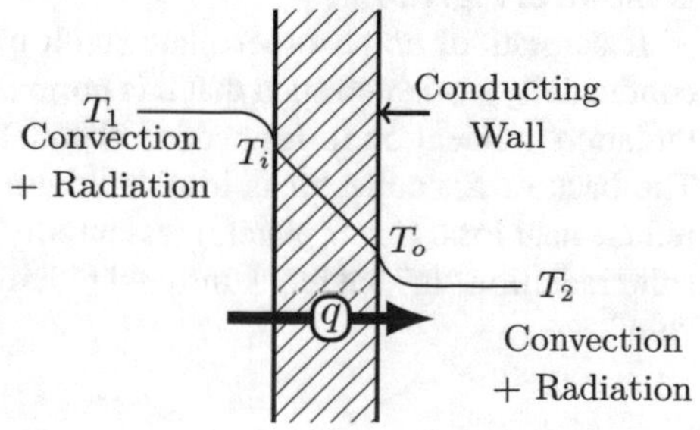

Fig. 1.3 Heat flow across a wall

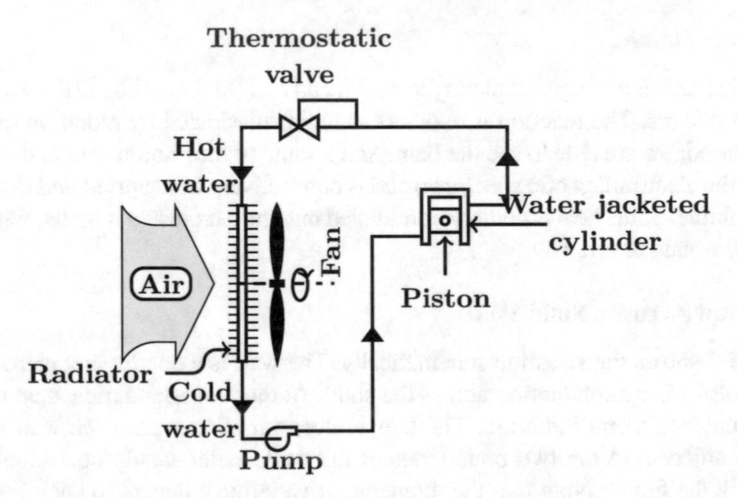

Fig. 1.4 Schematic of an IC Engine cooling system

The water picks up heat here and dissipates this heat to the atmosphere through the radiator. The radiator consists of passages through which the water flows and airflow takes place across the outside of the passages. Usually, fins (also referred to as extended surfaces—more on this later) are provided on the air-side to augment heat transfer.

The working of the engine cooling system is self-explanatory from the figure. Mechanisms of heat transfer involved are

- Forced convection to water in the jacket.
- Forced convection on the water-side of the radiator.
- Conduction in the passage walls and within the fins on the air-side of the radiator.
- Forced convection to the air outside the radiator—radiation is not the important mode of heat transfer and hence the name "radiator" is probably a misnomer.

Solar Collector

A solar collector is a device that is used for transferring the radiant energy from the sun to a fluid. In the case of a solar water heater, the fluid that gets heated is water that may be used for process applications. The cross section of a typical solar collector is shown in Fig. 1.5.

It consists of an absorber plate (with high solar absorptivity and high thermal conductivity), oriented such that it is normal to the solar radiation, that absorbs solar radiation incident on it. Embedded channels/tubes carry water that is to be heated. The back of the collector is insulated with a low thermal conductivity material to reduce heat loss. Cover plate(s) (essentially plane glass sheets) are provided to pass solar radiation "in" but block infrared radiation (emitted by the collector plate) going "out".

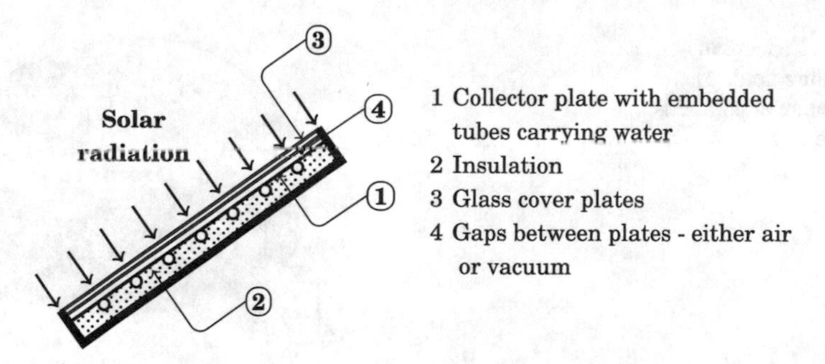

Fig. 1.5 Cross-sectional view of a typical solar collector

The following heat transfer processes take place in this application:

- Radiation from the sun is transmitted by the cover plate(s) and absorbed by the collector plate.
- Heat is conducted to the embedded tubes by conduction.
- Heat is transferred to water flowing in the tubes by convection.
- Some heat is lost from the back by conduction through the insulation.
- Some heat is convected/conducted across the air gap.
- Some heat is lost from the cover plate(s) by reflected solar radiation as well as by heat convection to the ambient.

1.3.1 Basic Definitions

Temperature field

The concept of the temperature field is that in a physical domain D, we associate temperature T to every point $P(x, y, z)$ at a given t (see Fig. 1.6). Time t is counted as positive from a conveniently chosen initial value. If $T(x, y, z, t) =$ constant, then D is an isothermal domain. In case T is a function of t alone, we have an unsteady *lumped* system. If T is a function of only (x, y, z), then the temperature field is steady. In such a case, $T =$ constant defines surfaces which are referred to as isothermal surfaces. An example of a two-dimensional steady temperature field is shown in Fig. 1.7. Several isotherms (solid lines) and several heat flux lines (dashed lines, assuming heat transfer is solely by conduction, tangent to flux line is the direction of heat flux) are shown thereon. Isotherm at T_2 intersects a heat flux line as indicated in the figure. These two intersect orthogonally and hence the angle between the tangents AB and CD is 90°. In general, isotherms and flux lines form a net of curvilinear rectangles.

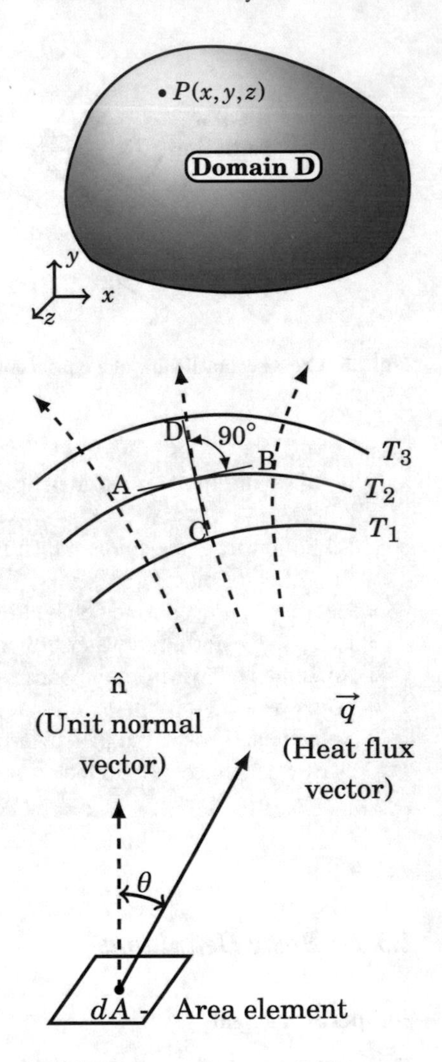

Fig. 1.6 Concept of temperature field. Temperature at point P is specified as $T(x, y, z, t)$

Fig. 1.7 Orthogonality of isotherms and heat flux lines—two-dimensional temperature field

Fig. 1.8 Heat flux vector

Heat Flux Vector

Imagine an area element dA with its normal along n̂ (see Fig. 1.8).
If $\vec{q}$ is the heat flux vector, we define the power crossing area element (i.e., energy crossing per unit time) dA as

$$\boxed{P = \vec{q} \cdot \hat{n}\, dA = q \cos\theta dA} \tag{1.2}$$

Note that P has units of power, i.e., W, $\vec{q}$ has units of W/m^2 and dA has units of m^2. Heat flux lines (see Fig. 1.7) are curves the tangents to which we define the direction of the heat flux vector.

Conduction Heat Transfer in a Homogeneous Medium

Conduction heat transfer in a homogeneous medium[3] at rest is the first mode of heat transfer we consider. Some general statements[4] that may be made are

1. Heat transfer takes place from a region of higher temperature to a region of lower temperature (i.e., no violation of Second law of thermodynamics).
2. No heat flows along an isotherm (definition of an isotherm).
3. Heat flows along the direction of largest temperature change and hence normal to an isotherm (flux lines and isotherms intersect at 90° as indicated in Fig. 1.7).
4. Heat flux is ***directly proportional*** to the temperature gradient (linear model).

The above observations led Fourier[5] to postulate a linear relation between conduction heat flux $\vec{q_k}$ and the temperature gradient vector as

$$\boxed{\vec{q_k} = -k\nabla T} \tag{1.3}$$

where, for a homogeneous medium, k is a constant and is known as the thermal conductivity. The negative sign makes sure that the heat flux is in the direction of *decreasing* temperature. Thermal conductivity k is assumed to be a property of the medium. Equation 1.3 represents Fourier law of heat conduction.

Heat Transfer by Radiation

Radiation is an independent mode of heat transfer. It takes place via electromagnetic waves or photons. It is a fact that bodies radiate solely due to their being at a temperature above 0 K. The most efficient radiator at a temperature T is called a black body. It radiates an amount of energy per unit time and area q_r (radiant heat flux) given by

$$\boxed{q_r = \sigma T^4} \tag{1.4}$$

[3] The medium is assumed to have the same property at all locations inside it.

[4] These are in the nature of postulates or axioms used in modeling conduction heat transfer in a homogeneous medium.

[5] Jean-Baptiste Joseph Fourier 1768–1830, French mathematician and physicist.

where $\sigma = 5.67 \times 10^{-8}$ W/m^{2}K^4 is the Stefan–Boltzmann constant.[6] If there are two black bodies that are at temperatures T_1 and T_2, respectively, the radiant heat transfer between them is given by

$$q_{1-2} = F\sigma\left(T_1^4 - T_2^4\right) \tag{1.5}$$

where "F" is a geometric factor. We note that the radiation heat transfer varies non-linearly with the temperatures. In the special case where $(T_1 - T_2) << T_1$ or T_2 we may approximate $T_1^4 - T_2^4$ as

$$\begin{aligned} T_1^4 - T_2^4 &= (T_1^2 - T_2^2)(T_1^2 + T_2^2) \\ &= (T_1 - T_2)\underbrace{(T_1 + T_2)}_{2T_{ref}}\underbrace{(T_1^2 + T_2^2)}_{\approx 2T_{ref}^2} \approx 4T_{ref}^3(T_1 - T_2) \end{aligned}$$

where $T_{ref} = \frac{T_1+T_2}{2}$. Equation 1.5 may then be rewritten as

$$\boxed{q_{1-2} = 4\sigma T_{ref}^3 F(T_1 - T_2) = h_r(T_1 - T_2)} \tag{1.6}$$

where $h_r = 4\sigma T_{ref}^3 F$ is referred to as the radiation heat transfer coefficient. Equation 1.6 represents what is called "linearized" radiation. It resembles Newton's law of cooling (to be presented below, in the case of convection heat transfer, see Eq. 1.8), in this special case.

Example 1.1

A surface that is maintained at 310 K is losing heat to a background at 300 K by radiation. What is the appropriate radiation heat transfer coefficient? Is it alright to use this concept? Justify.

Solution:
Figure 1.9 shows the geometrical arrangement along with the way radiation interaction takes place between the surface and the surroundings. The ambient receiving radiation is idealized as a black surface at the ambient temperature.

We assume that the radiation interchange is wholly between the surface and the surroundings, i.e., the geometric factor is $F = 1$. The reference temperature is taken as the mean of surface temperature and the temperature of the surroundings.

Step 1 Surface temperature is $T_s = 310$ K
Temperature of the surroundings is $T_{amb} = 300$ K
Reference temperature is

[6] Jožef Stefan, 1835–1893 and Ludwig Boltzmann—1844–1906, Austrian physicists.

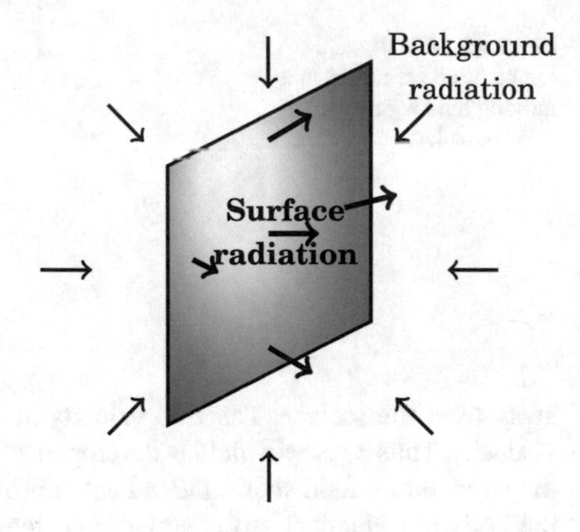

Fig. 1.9 Geometry for Example 1.1 showing radiation interaction

$$T_{ref} = \frac{T_s + T_{amb}}{2} = \frac{310 + 300}{2} = 305\,\mathrm{K}$$

Step 2 We use Eq. 1.6 to calculate the radiative heat transfer coefficient as

$$h_r = 4 \times 5.67 \times 10^{-8} \times 305^3 = 6.43\,\mathrm{W/m^2K}$$

Step 3 The heat transferred between the surface and the ambient is approximately equal to

$$q_{s-amb} \approx h_r(T_s - T_{amb}) = 6.43 \times (310 - 300) = 64.3\,\mathrm{W/m^2}$$

Step 4 We look at what will be the exact value obtained by the use of Eq. 1.5.

$$q_{s-amb} = 5.67 \times 10^{-8} \times (310^4 - 300^4) = 64.4\,\mathrm{W/m^2}$$

Step 5 Since the two values are very close to each other linear approximation is a valid approximation.

Heat Transfer by Convection

It is common knowledge (or experience, all you have to do is to sit under a ceiling fan on a hot day) that a flowing fluid increases the heat transfer from a surface. The reason for this is that the heated fluid near a surface moves away, carrying with it, the heat from the surface and cold fluid moves in to replace what has been swept away. The fluid motion thus promotes heat transfer. Figure 1.10 shows what is going on.

The fluid in contact with the surface is at rest if the body is standing still or otherwise, is moving with the same velocity as that of the surface (known as no-slip condition). The fluid velocity is the undisturbed fluid velocity U some distance

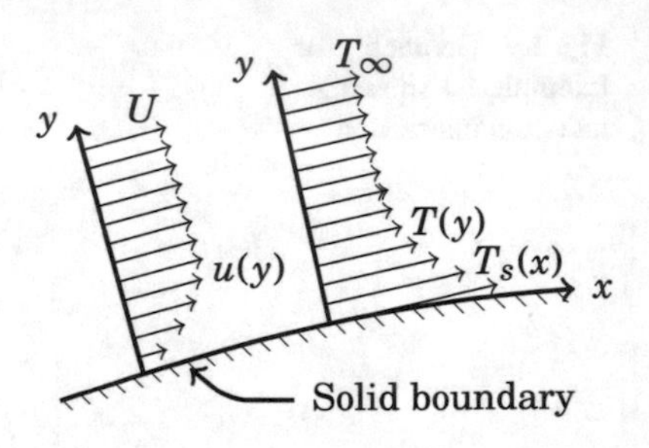

Fig. 1.10 Velocity and temperature variations in a moving fluid adjacent to a solid boundary

away from the surface. The zero velocity at the surface is a consequence of fluid viscosity. Thus a velocity field is developed as shown in the figure. Correspondingly the temperature field shows the indicated profile. The temperature of the fluid and the body are identical at the surface (no temperature slip), specified as $T_s(x)$. A short distance away the temperature of the fluid is the undisturbed value T_∞. The zero temperature *difference* between the solid surface and the fluid adjacent to it is a consequence of the thermal conductivity of the medium. Thus a temperature field is developed as shown in the figure. Since the medium is essentially at rest at the surface, the wall heat flux is given by the conductive heat flux in the fluid adjacent to the wall. The conductive heat flux component normal to the surface is written down using Fourier heat conduction law (Eq. 1.3) as

$$q_w(x) = -k_f \left.\frac{\partial T}{\partial y}\right|_{y=0} \tag{1.7}$$

The coordinate x is measured along the surface and the coordinate y is measured normal to the surface. k_f is the thermal conductivity of the fluid flowing past the surface. Equation 1.7 highlights the fact that the heat flux may vary with x. Also, the heat flux considered here is in the "local" y direction. The effect of movement of the fluid is an increase in the temperature gradient at the wall, $\left.\frac{\partial T}{\partial y}\right|_{y=0}$ as compared to the case where the fluid is at rest. This is shown schematically in Fig. 1.11.

The process of convection is difficult to model since

1. Velocity and temperature fields are, in general, interdependent;
2. The velocity problem is generally governed by a non-linear partial differential equation;
3. Flow and temperature fields exhibit different regimes such as laminar, transition and fully turbulent regimes.

Hence very few cases may be handled by analytical methods. One way of circumventing this problem is to rewrite Eq. 1.7 in the form

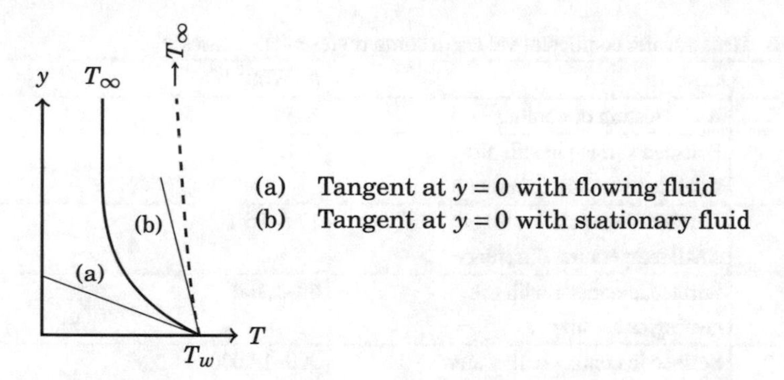

Fig. 1.11 Effect of fluid flow on the temperature variation adjacent to the wall

$$q_w(x) = h(x)\{T_w(x) - T_\infty\} \tag{1.8}$$

where we have postulated a linear relation between the surface heat flux and the temperature difference between the surface and the free stream. "$h(x)$" is referred to as the "local" heat transfer coefficient and Eq. 1.8 is referred to as Newton's[7] law of cooling. It is to be noted that Eq. 1.8 does not really represent a "law" and, at best, it is an expression that defines the heat transfer coefficient. Again the (x) indicates that the considered quantity may be a function of the position along the surface. The term "local" refers to this aspect. In applications, we also use a mean or an average value, in which case, the reference to (x) will have to be dropped. Generally the heat transfer coefficient "h" (units W/m^2°C) depends on many variables like the fluid velocity U, fluid viscosity ν, the geometry of the body through a characteristic dimension L, fluid thermal conductivity k_f, and so on. Most of the time the task of relating h to the flow variables is a challenging task. This will receive attention later on. The heat transfer coefficient values cover a wide range of values as indicated in Table 1.1.

It is to be noted that the entries in the table are to be interpreted in a general way, in that, the heat transfer mechanism *may not* be by convection alone. In fact, for an object in still air, "free" or "natural" convection heat transfer rate and radiation heat transfer rate, are of the same order of magnitude. The entry number 3 in Table 1.1 is the combined heat transfer due to natural convection and surface radiation obtained by using the concept of the radiation heat transfer coefficient introduced through Eq. 1.6.

Example 1.2

A flat plate of dimensions 100 × 50 mm and negligible thickness, maintained at 42 °C, is losing 0.85 W from both sides by natural convection to ambient air at

[7] Isaac Newton 1643–1727, English physicist and mathematician.

Table 1.1 Heat transfer coefficient values in some representative cases

No.	Case	h, W/m^2°C
1	Air - Heating or cooling	1–50
2	Polished surface in still air: small temperature difference	7–17.5
3	Blackened surface in still air: small temperature difference	17.5–25
4	Surface in contact with oil: heating or cooling	60–1,700
5	Surface in contact with water: heating or cooling	300–17,000
6	Surface in contact with boiling water:	1,700–50,000

30 °C. What is the heat transfer coefficient? Is it in agreement with the range shown in Table 1.1?

Solution:
The size of the surface is specified by $L = 100\,\text{mm} = 0.1\,\text{m}$ and $W = 50\,\text{m} = 0.05\,\text{m}$. Since the surface loses heat from both sides, the surface area for heat transfer is

$$A = 2L \times W = 2 \times 0.1 \times 0.05 = 0.01\,\text{m}^2$$

Surface temperature is $T_s = 42\,°\text{C}$
Ambient temperature is $T_{amb} = 30\,°\text{C}$
Total heat transferred from the two exposed surfaces of the plate is $Q = 0.85\,\text{W}$
By definition, the heat transfer coefficient is given by

$$h = \frac{Q}{A(T_s - T_{amb})} = \frac{0.85}{0.01 \times (42 - 30)} = 7.08\,\text{W/m}^2\,°\text{C}$$

This value is certainly in tune with the range shown in Table 1.1.

Concluding Remarks

In this chapter, we have given an overview of what constitutes the study of heat transfer. The succeeding chapters deal with a detailed study of each mode of heat transfer, in the order in which they have been presented above. The last chapter in the book will provide an introduction to multi-mode heat transfer, bringing all the modes of heat transfer together, in some simple cases amenable to an elementary treatment. That will also lead the reader beyond what is possible at the level of the present book.

1.4 Exercises

Ex 1.1 Answer the following in as much detail as you can. Some general reading is called for.

- Why does water kept in an earthen pot remain cool?
- Why does a metal chair feel cool while a wooden chair does not, on a cold day?
- Loose fitting clothes are known to be more comfortable in the tropics. What may be the reason for this?
- A healthy human being has the body temperature maintained at 37 °C. Explain how the human body accomplishes this.
- A friend tells you that keeping the door of a refrigerator open will cool the room. Do you agree with this? Explain giving reasons.
- Explain why a hot flat surface facing downwards in still air will lose heat at a lower rate than if it faces upwards.
- Why is the ventilator in a building at the roof level? Explain.

Ex 1.2 What a heat exchanger essentially does is to transfer a certain amount of heat per unit time from fluid 1 to fluid 2. In a certain heat exchanger, the state of affairs is as indicated in Fig. 1.12. By thermodynamic analysis determine the maximum possible heat exchange Q_{max} between the two fluids. What is the actual heat transfer Q_{act} between the two fluids? Can you attach any significance to the ratio of the latter to the former?

Ex 1.3 A steam power plant operates on the Rankine cycle. Make a schematic sketch of such a power plant and indicate the types of heat transfer equipment involved. Figure out what modes of heat transfer are involved in such equipment. It would help if you make a $T - s$ plot, i.e., temperature vs entropy plot of the cycle.

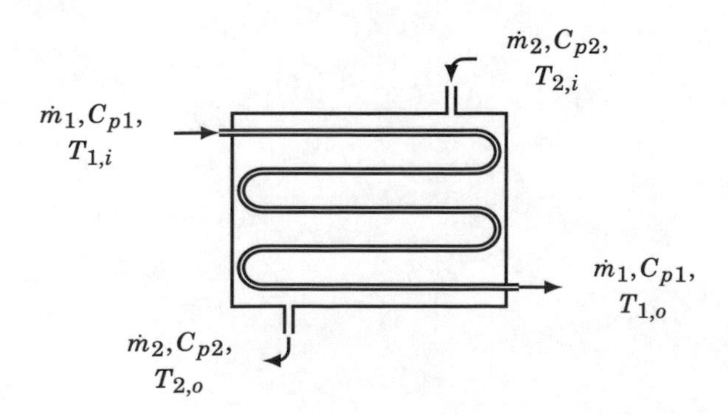

Fig. 1.12 Heat exchanger example in Exercise 1.2

Ex 1.4 An inventor claims that he has designed a heat exchanger that will cool 1 kg/s of air from 80 °C to 40 °C using 2 kg/s of cold water that heats up from 25 °C to 30 °C. Do you think the claim is genuine? Explain.

Ex 1.5 In a certain material of uniform thermal conductivity k, the temperature field is $T(x, y) = 50 \ln(x^2 + y^2)$. Determine x and y components of the heat flux at the point (3,2). What is the resultant heat flux and its direction? Make a sketch of isotherms and heat flux lines.

Ex 1.6 A fluid of thermal conductivity 0.5 W/m°C flows past a surface. The heat transfer coefficient has been measured to be 45 W/m^2°C when the surface temperature is 70 °C and the fluid temperature far away from the surface is 25 °C. What is the heat flux at the surface? What is the temperature gradient within the fluid normal to the surface?
Hint: Convection heat flux at the surface is equal to the conduction heat flux in the medium at the surface.

Ex 1.7 For free or natural convection in air the heat transfer coefficient from a horizontal surface is about. 6 W/m^2°C. Determine the largest heat flux that can be supported by a surface cooled by air at 30 °C if the surface temperature is not to exceed 100 °C.

Ex 1.8 A certain horizontal surface placed in air is losing heat by natural convection and radiation to an ambient at $T_\infty = 300$ K. The plate temperature is maintained at $T_s = 380$ K. The free convection heat transfer coefficient is $h = 10$ W/m^2°C. The plate surface has been coated with a paint that gives it an emissivity of $\varepsilon = 0.85$. Determine the heat loss from the plate by radiation and compare it with that due to free convection. Compare the heat loss by radiation computed above with the calculation based on linear radiation based on the radiation heat transfer coefficient $h_R = 4\varepsilon\sigma T_{ref}^3$, where σ is the Stefan–Boltzmann constant and T_{ref} is a suitably chosen reference temperature. Is the linear radiation model valid in this case? Explain giving reasons.

Chapter 2
Steady Conduction in One Dimension

It is traditional to consider simple conduction problems in one dimension to initiate the beginning student to the study of heat transfer. Steady conduction in one dimension leads to ordinary differential equations that are easy to solve. We present steady conduction in a slab, a cylinder, and a sphere to highlight the role of geometry in the temperature variations. Other modes of heat transfer such as convection and radiation make their appearance in the form of boundary conditions.

2.1 Preliminaries

This chapter begins a detailed study of conduction heat transfer. Preliminaries like units of quantities, thermal properties of various materials are considered before getting into the study of conduction heat transfer. In Chap. 1, we have already encountered the basic quantities that appear in conduction heat transfer. These are recapitulated here in Table 2.1.

2.1.1 On Thermal Conductivity Values

Thermal conductivity values of materials encountered in practice vary over some 6 orders of magnitude. Gases have the smallest thermal conductivity values of all materials. Liquids have thermal conductivity values in between those of gases and solids. Some fibrous (and hence porous) materials have values of thermal conductivity in between those of gases and liquids. Figure 2.1 shows the ranges of thermal conductivity values for gases, liquids, and solids. The thermal conductivity values

S. P. Venkateshan, *Heat Transfer*,
https://doi.org/10.1007/978-3-030-58338-5_2

Table 2.1 Units of physical quantities in heat transfer

Physical quantity	Unit
Temperature:	°C or K
Heat Flux:	W/m^2
Thermal Conductivity:	W/m°C or W/m K
Heat Transfer Coefficient:	W/m^2°C or W/m^2K

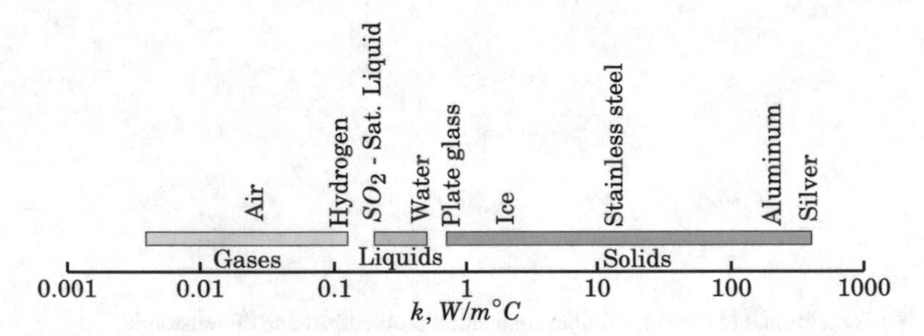

Fig. 2.1 Thermal conductivity of gases, liquids, and solids

are indicated on a logarithmic scale. Typical materials are shown on this line at locations that correspond to the respective thermal conductivity values. These values are typical since the thermal conductivity shows variation primarily with temperature and the presence of impurities in the sample.

The following points regarding some observed behavior of thermal conductivity of materials are of importance, in the study of heat transfer.

Some observed behavior of thermal conductivity of materials

1. Thermal conductivity of pure metals decreases with temperature.
2. Even small amounts of impurities reverse the above trend.
3. Thermal conductivity of most liquids decreases with temperature.
4. Thermal conductivity of gases increases with temperature. The thermal conductivity of gases decreases (in general) with increasing molecular weight (see Table 2.2).
5. Except at very high pressures, thermal conductivity is not influenced by pressure.

Table 2.2 Thermal conductivity of several gases at atmospheric pressure and 0 °C

Gas	Thermal Conductivity	Molecular Weight
Hydrogen	0.1672	2
Helium	0.1416	4
Nitrogen	0.0240	14
Air	0.0241	28
Carbon Monoxide	0.0232	28
Argon	0.0164	40
Carbon Dioxide	0.0154	44

2.1.2 Approaches to the Study of Conduction Heat Transfer

In general, there are two approaches to the study of conduction heat transfer. In the first, the general equation that governs the temperature field in a stationary medium is derived, from first principles, applying the law of conservation of energy to a volume element. The volume element may be represented using any one of the three orthogonal coordinate systems, viz., the Cartesian, the cylindrical, or the spherical coordinates. The material medium may be homogeneous with identical properties along all directions or a material with anisotropic properties. The thermo-physical properties may be allowed to vary with temperature. The resulting equation known as the "Heat Diffusion Equation" or simply the "Heat Equation" is a partial differential equation that may be solved either by analytical methods when amenable or otherwise by numerical methods. The simpler cases that correspond to, for example, steady conduction in one dimension, transient conduction in one dimension, etc., are then derived as special cases of the Heat Equation. This approach is more satisfying in terms of its elegance.

However, in the second approach, that is being followed in this book, we start directly with the special but simple cases and gradually build up the theoretical framework to culminate in the general Heat Equation. This approach is more satisfactory for the first time learner.

2.2 Steady One Dimensional Conduction

Steady one-dimensional conduction may take place in a plane wall (it is assumed to be infinite in extent or, at least, very much larger than the thickness of the plane wall so that edge effects are not significant), a long cylinder or a cylindrical annulus (in principle infinitely long but in practice the length of cylinder is very large compared to the outer diameter) and a sphere or a spherical shell. The plane wall represents the simplest geometry since the area for heat transfer does not vary with the independent space variable, say x. Usually, the calculations are based on a unit area perpendicular to the x-direction. When heat transfer takes place radially in a cylinder or a cylindrical annulus, the area available for conduction heat transfer increases linearly with the radial coordinate r. In the case of radial heat transfer in a sphere or a spherical shell, the area available for conduction heat transfer increases as the square of the radial coordinate r.

The governing equations in the three cases thus have significant differences and these lead to interesting consequences. The three cases are considered in what follows, in that order.

2.2.1 One-Dimensional Conduction in a Uniform Area Bar

Uniform Area Bar with Constant Thermal Conductivity

Consider a bar of material whose cross-sectional area is uniform. By uniform, we mean that the area does not change with x, as shown in Fig. 2.2. The temperature is assumed to vary only with x as shown in the figure that corresponds to the case of a bar with constant thermal conductivity (i.e., k is independent of T).

We write Fourier law for this case as

$$q_x = -k\frac{dT}{dx} = \text{constant} \tag{2.1}$$

where the heat flux has a non-zero component q_x along the x-direction. The conduction heat flux is constant since the temperature field is steady. Since the thermal conductivity is constant, we may conclude that the temperature gradient $\frac{dT}{dx}$ is itself a constant. Temperature variation along the rod is thus linear and is as indicated in Fig. 2.2. Based on the temperature variation in the bar, we have

$$\frac{dT}{dx} = \frac{CE}{AC} = \frac{T_L - T_0}{L} \tag{2.2}$$

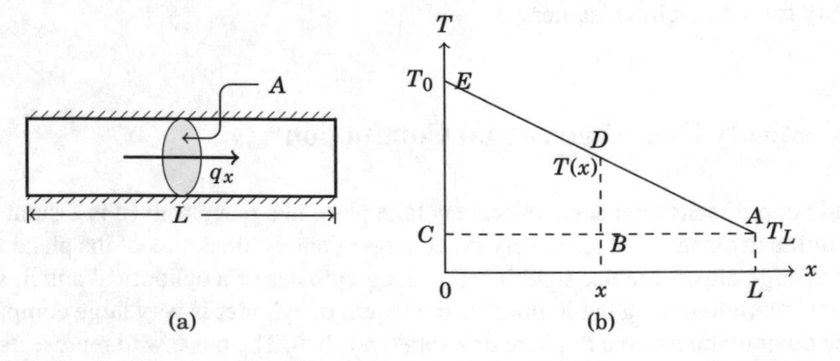

Fig. 2.2 Conduction of heat in a uniform area rod **a** Geometry **b** Temperature distribution

and

$$\frac{dT}{dx} = \frac{EF}{DF} = \frac{T - T_0}{x} \tag{2.3}$$

Equating these two alternate expressions for the temperature gradient, after some minor manipulation, we get

$$\boxed{\underbrace{\frac{T - T_0}{T_L - T_0}}_{\theta} = \underbrace{\frac{x}{L}}_{\xi}} \quad \text{or} \quad \boxed{\theta = \xi} \tag{2.4}$$

where $\theta = \frac{T-T_0}{T_L-T_0}$ is the non-dimensional temperature while $\xi = \frac{x}{L}$ is the non-dimensional axial distance from the left end. In this form, the temperature profile is independent of various temperatures, the length of the bar, and the constant thermal conductivity of the bar material. We may refer to this as the "universal temperature profile" for steady conduction in the bar. However, if one wants to calculate the heat transferred across the bar, all the physical variables will come in to the picture!

Uniform Area Bar with Variable Thermal Conductivity

When the thermal conductivity varies with temperature, the analysis becomes more complicated since the governing equation becomes non-linear. In order to get the flavor of this but not the complexity we look at the case of a bar of material whose thermal conductivity varies linearly with temperature according to the relation

$$k(T) = k_0(1 + bT) \tag{2.5}$$

where k_0 and b are specified constants. Equation 2.1 will have to be replaced by the equation

$$q_x = -k(T)\frac{dT}{dx} = -k_0(1 + bT)\frac{dT}{dx} = -A \tag{2.6}$$

where A is a constant. Equation 2.6 may be written in the expanded form

$$k_0\frac{dT}{dx} + k_0 bT\frac{dT}{dx} = k_0\frac{dT}{dx} + k_0 b\frac{d(T^2/2)}{dx} = A \tag{2.7}$$

Equation 2.7 may be integrated with respect to x to get

$$k_0 T + k_0 b\frac{T^2}{2} = Ax + B \tag{2.8}$$

where B is an integration constant. The two constants A and B are obtained by using the two boundary temperatures of the bar, viz., $T(x=0)=T_0$ and $T(x=L)=T_L$. These two give

$$\text{(a)}\quad k_0 T_0 + k_0 b\frac{T_0^2}{2} = B \quad \text{(b)}\quad k_0 T_L + k_0 b\frac{T_L^2}{2} = AL + B \tag{2.9}$$

Subtract Eq. 2.9(a) from 2.9(b) to get

$$A = k_0\frac{(T_L - T_0)}{L} + k_0 b\frac{(T_L^2 - T_0^2)}{2L} \tag{2.10}$$

This may be rearranged, noting that $T_L^2 - T_0^2 = (T_L - T_0)(T_L + T_0)$, as

$$A = \frac{(T_L - T_0)}{L}\underbrace{k_0\left(1 + b\frac{(T_L + T_0)}{2}\right)}_{k_m(T_0, T_L)} \tag{2.11}$$

We note that the term shown with underbracket in Eq. 2.11 is the thermal conductivity $k_m(T_0, T_L)$ of the material at the mean temperature $\frac{(T_L+T_0)}{2}$. Thus Eq. 2.11 may be rewritten in the compact form

$$A = k_m(T_0, T_L)\frac{(T_L - T_0)}{L} \tag{2.12}$$

Similarly Eq. 2.8 may be recast in the compact form

$$(T - T_0)\,k_m(T_0, T) = k_m(T_0, T_L)\frac{(T_L - T_0)\,x}{L} \tag{2.13}$$

Finally we may recast this in the non-dimensional form as

$$\text{(a)}\quad \frac{T - T_0}{T_L - T_0} = \frac{k_m(T_0, T_L)}{k_m(T_0, T)}\frac{x}{L} \quad \text{or} \quad \boxed{\text{(b)}\quad \theta = \frac{k_m(T_0, T_L)}{k_m(T_0, T)}\,\xi} \tag{2.14}$$

where $\theta = \frac{T - T_0}{T_L - T_0}$ and $\xi = \frac{x}{L}$. Eq. 2.14(b) will reduce to Eq. 2.4 in case $b = 0$, i.e., the thermal conductivity is independent of temperature. In this case the variation θ varies linearly with ξ.

Consider the case where $T_0 > T_L$ and $b > 0$, thermal conductivity increases with temperature. It is easily seen that $k_m(T_0, T_L)$ should be less than $k_m(T_0, T)$ since the mean temperature $(T_0 + T_L)/2$ is less than the mean temperature $(T_0 + T)/2$. Hence it is easily seen that $\theta < \xi$. We may conclude, by a similar argument, that for $T_0 > T_L$ and $b < 0, \theta > \xi$. Of course, at the two end points $\theta = \xi$ in all the cases.

Example 2.1

Thermal conductivity of a synthetic material varies linearly with temperature according to the relation $k(T) = 12.3\{1 - 0.007(T - 30)\}$ W/m °C. The validity of this relation is limited to temperature range given by $30 \leq T \leq 115$ °C. In a certain application, a 0.05 m thick layer of this material is used. One exposed surface of the slab is maintained at 95 °C while the other exposed surface is maintained at 35 °C. What is the heat transferred across a square meter of slab? What are the temperature gradients at the two surfaces of the slab?

Solution:

Step 1 Let $T_0 = 95$ °C be the temperature of the first surface and $T_L = 35$ °C the temperature of the second surface. Given:

$$k_0 = 12.3\,\text{W/m}\,^\circ\text{C};\; b = -0.007\,^\circ\text{C}^{-1};\; L = 0.05\,\text{m}$$

Step 2 The mean thermal conductivity $k_m(T_0, T_L)$ is calculated as

$$\begin{aligned} k_m(T_0, T_L) &= k_0\left[1 + b\left(\frac{T_0 + T_L}{2} - 30\right)\right] \\ &= 12.3 \times \left[1 - 0.007 \times \left(\frac{95 + 35}{2} - 30\right)\right] = 9.287\,\text{W/m}\,^\circ\text{C} \end{aligned}$$

Step 3 Using Eq. 2.12, the heat transfer across the slab per unit area is then given by

$$q = k_m(T_0, T_L)\frac{(T_0 - T_L)}{L} = 9.287 \times \frac{95 - 35}{0.05} = 11143.8\,\text{W/m}^2$$

Step 4 The gradients at the two surfaces of the slab are calculated based on Fourier law. At the surface at T_0, we have

$$\left.\frac{dT}{dx}\right|_0 = -\frac{q}{k(T_0)} = -\frac{11143.8}{12.3 \times [1 - 0.007(95 - 30)]} = -1662.4\,^\circ\text{C/m}$$

At the surface at T_L, we have

$$\left.\frac{dT}{dx}\right|_L = -\frac{q}{k(T_L)} = -\frac{11143.8}{12.3 \times [1 - 0.007(35 - 30)]} = -938.9\,^\circ\text{C/m}$$

Step 5 It is interesting to make temperature profile plots in two ways, using non-dimensional coordinates and using dimensional variables as shown in Fig. 2.3. The former makes the bounding values equal to 0 and 1 and the non-dimensional temperature increases with the non-dimensional dis-

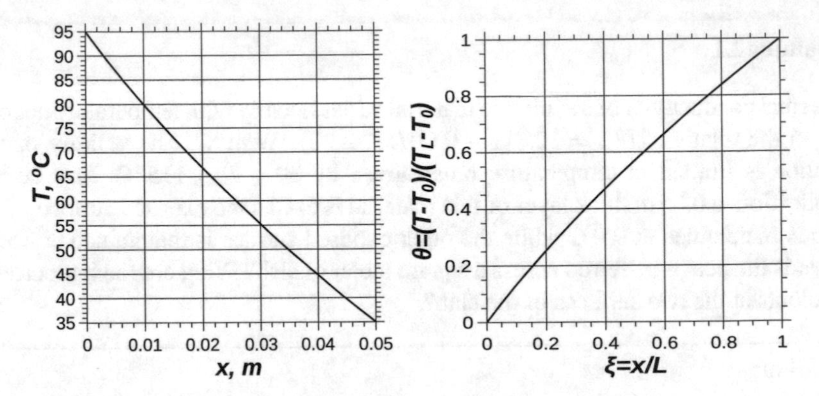

Fig. 2.3 Two types of plots of data in Example 2.1

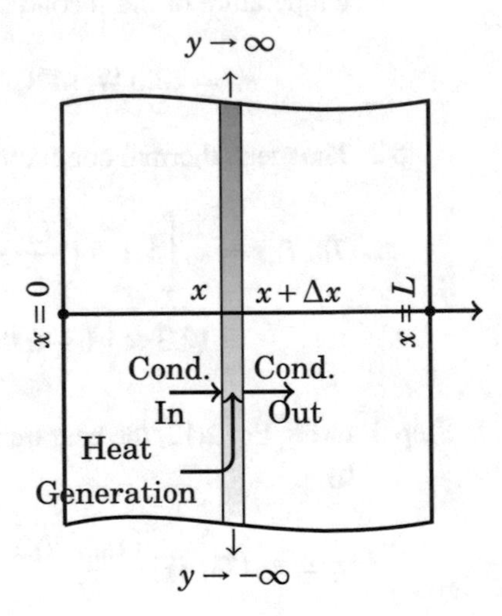

Fig. 2.4 Conduction in a slab with internal heat generation

tance. However the dimensional plot shows that the temperature *decreases* with increasing x. Both ways of plotting show that the temperature distribution is non-linear.

Plane Wall with Internal Heat Source

Consider a plane wall as shown in Fig. 2.4. Heat is generated internally (within the wall). Let the heat generation rate per unit volume be G W/m^3. This may, in general, be a function of x. Consider an element of volume $1 \times 1 \times \Delta x$ as indicated in the figure.

Under steady-state conditions, energy balance requires that the rate at which heat leaves the control volume by conduction be equal to the rate at which heat is conducted

in to the control volume and the rate at which heat is generated within the control volume.
Using Fourier law, we have

$$\text{Rate of heat conducted in} = -1 \times 1 \times k \left.\frac{dT}{dx}\right|_x \tag{2.15}$$

$$\text{Rate of heat generation within the control volume} = 1 \times 1 \times \Delta x \times G \tag{2.16}$$

$$\text{Rate of heat conducted out} = -1 \times 1 \times k \left.\frac{dT}{dx}\right|_{x+\Delta x} \tag{2.17}$$

We use Taylor expansion to recast Eq. 2.17 as

$$\text{Rate of heat conducted out} = -1 \times 1 \times k \left[\left.\frac{dT}{dx}\right|_x + \left.\frac{d^2T}{dx^2}\right|_x \Delta x \right] + O(\Delta x^2) \tag{2.18}$$

With these energy balance requires

$$-k\left.\frac{dT}{dx}\right|_x + \Delta x \times G = -k \left[\left.\frac{dT}{dx}\right|_x + \left.\frac{d^2T}{dx^2}\right|_x \Delta x \right] + O(\Delta x^2) \tag{2.19}$$

We now take the limit $\Delta x \to 0$ and get

$$\boxed{\frac{d^2T}{dx^2} + \frac{G}{k} = 0} \tag{2.20}$$

This is a non-homogeneous, second-order ordinary differential equation. Equation 2.20 is to be solved with the boundary conditions at the two faces of the slab, specified as follows.

$$x = 0, T = T_0; \;\; x = L, T = T_L \tag{2.21}$$

The reader should note that the boundary conditions are referred to as of first kind (or Cauchy condition)[1] since the function (dependent variable) is specified at the two boundaries. Other types of boundary conditions are also possible and will be considered later on. One integration of Eq. 2.21 yields

$$\frac{dT}{dx} + \frac{G}{k}x = A$$

[1]Named after Baron Augustin-Louis Cauchy 1789–1857, French mathematician.

assuming G to be a constant. Here A is a constant of integration. Second integration gives

$$T + \frac{G}{k}\frac{x^2}{2} = Ax + B \tag{2.22}$$

where B is a second integration constant. The integration constants are determined by the use of the boundary conditions. Using the boundary condition at $x = 0$ and $x = L$ in Eq. 2.22, we get

$$\text{(a)}\;\; T = T_0 = B \quad \text{(b)}\;\; T_L + \frac{G}{k}\frac{L^2}{2} = AL + B = AL + T_0 \tag{2.23}$$

From Eq. 2.23(b), we obtain A as

$$A = \frac{T_L - T_0}{L} + \frac{GL}{2k} \tag{2.24}$$

The constants A and B are substituted back in Eq. 2.22 and simplified to finally obtain

$$T - T_L = (T_0 - T_L)\left(1 - \frac{x}{L}\right) + \frac{GL^2}{2k}\left(\frac{x}{L} - \frac{x^2}{L^2}\right) \tag{2.25}$$

We introduce the following non-dimensional quantities:

$$\theta = \frac{T - T_L}{T_0 - T_L};\;\; \xi = \frac{x}{L};\;\; H = \frac{GL^2}{2k(T_0 - T_L)} \tag{2.26}$$

The quantity H is referred to as the heat generation parameter. Equation 2.25 is then recast as

$$\boxed{\theta = (1 - \xi)(1 + H\xi)} \tag{2.27}$$

The solution to the slab problem with internal heat generation, when cast in this non-dimensional form, depends on *only one* parameter H. The heat generation parameter combines the four quantities T_0, T_L, G, and k in to a single parameter that governs the problem.

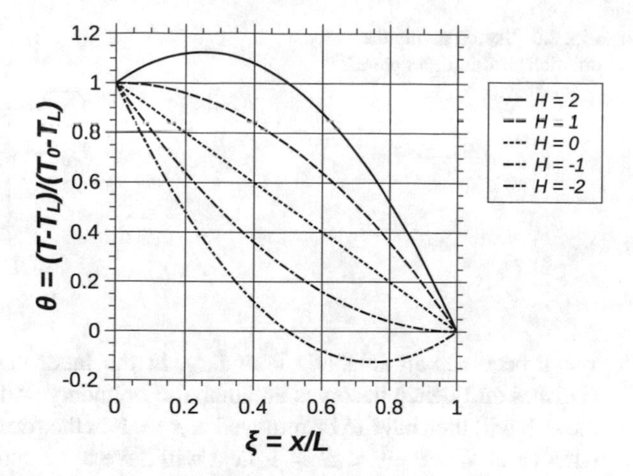

Fig. 2.5 Influence of H on temperature profiles

Heat generation parameter

This parameter may be rewritten as

$$H = \frac{GL^2}{2k(T_0 - T_L)} = \frac{GL}{\left(\frac{2k(T_0 - T_L)}{L}\right)} = \frac{\text{Heat generation rate in the slab}}{2 \times \text{Heat conduction rate across slab}}$$

In this form, the heat generation parameter is seen to be the ratio of heat generated per unit time inside the slab to twice the heat conducted by conduction across the slab per unit time, in the absence of heat generation. The magnitude of the temperature in the slab is affected by this ratio.

A plot of Eq. 2.27 is shown in Fig. 2.5 for various values of H. The profiles are non-linear for $H \neq 0$ and show the presence of an optimum (either a maximum or minimum in θ) under conditions to be derived below. The condition for optimum is given by

$$\frac{d\theta}{d\xi} = -(1 + H\xi) + H(1 - \xi) = 0 \quad \text{or} \quad \boxed{\xi = \frac{H-1}{2H}} \tag{2.28}$$

Noting that the value of ξ should be in the interval 0–1, an optimum exists only if $|H| > 1$. When $H > 1$, heat is to be removed at both the boundaries and when $H < 1$, heat is to be supplied at both the boundaries. When $H = 1$, the maximum temperature occurs at $\xi = 0$. When $H = -1$, the *minimum* temperature occurs at $\xi = 1$. The heat flux at $\xi = 0$ vanishes in the former case and hence the wall at

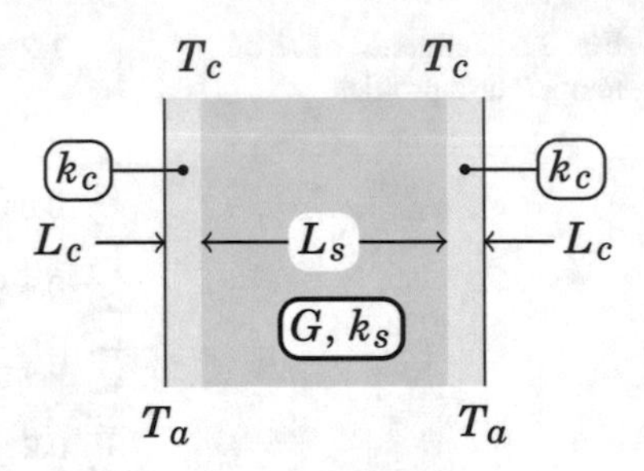

Fig. 2.6 Sketch giving the notation used in Example 2.2

$x = 0$ becomes an adiabatic boundary. In the latter case, the heat flux at $x = L$ vanishes and hence becomes an adiabatic boundary. All the heat generated within the slab will then have to be removed at $x = L$ in the former case and at $x = 0$ in the latter case. When $-1 < H < 1$, heat will have to be supplied at one boundary and removed from the other boundary, and the profile does not exhibit an optimum. The case $H = 0$, corresponding to the no heat generation case, gives a linear profile, as it should (Fig. 2.6).

Example 2.2

A large slab of thickness 0.1 m of a material of thermal conductivity equal to 10 W/m°C is generating heat at a constant volumetric heat generation rate of 10^5 W/m^3. The slab is covered on both sides by 0.025 m thick slabs of thermal conductivity 0.5 W/m °C. The two exposed surfaces are maintained at 85 °C by an external cooling arrangement. Determine the maximum temperature in the heat-generating slab and the temperature gradient in the covers.

Solution:

Step 1 The given data is written down first along with an explanatory sketch:

Step 2 Consider unit area (1 m^2) of the slab in a direction perpendicular to its thickness direction. The heat generated within the volume of the slab per unit surface area must equal the heat q going out per unit area through the two side covers. Because of symmetry half of this will pass through each cover.

$$\text{Thus heat conducted across each cover } q_c = \frac{q}{2}$$

$$= \frac{GL_S}{2} = \frac{10^5 \times 0.05}{2} = 5000\,\text{W/m}^2$$

Step 3 This will set up a temperature difference across each cover. Using Fourier law, the interface temperature T_c may be calculated as

$$T_c = T_a + \frac{q_c L_c}{k_c} = 85 + \frac{5000 \times 0.025}{0.5} = 335\,°\text{C}$$

Step 4 The temperature field is seen to be symmetric with respect to the mid-plane of the heat-generating slab since the two covers have identical temperatures on their outer surfaces. The mid-plane of the heat-generating slab is thus an adiabatic surface and the temperature in the slab is maximum there. This will thus correspond to $H = 1$. The characteristic dimension is $L = \frac{L_S}{2}$. We use Eq. 2.26 and identify the various quantities appearing in the definition of H as: (a) Maximum temperature in the slab $T_m = T_0$; (b) Interface temperature $T_c = T_L$. $H = 1$ may be recast as

$$H = 1 = \frac{GL^2}{2k_S(T_0 - T_L)} = \frac{G\left(\frac{L_S}{2}\right)^2}{2k_S(T_m - T_c)}$$

This may be solved for the mid-plane temperature T_m as

$$T_m = T_c + \frac{G\left(\frac{L_S}{2}\right)^2}{2k_S} = 335 + \frac{10^5 \times \left(\frac{0.05}{2}\right)^2}{2 \times 10} = 338.13\,°\text{C}$$

Step 5 Again, because of symmetry the magnitude of the temperature gradient in each of the covers is the same. While the gradient is negative in the right-side cover, it is positive in the left-side cover. The magnitude of the temperature gradient is obtained using Fourier law as

$$\left|\frac{dT}{dx}\right|_c = \frac{T_c - T_a}{L_c} = \frac{335 - 85}{0.025} = 10^4\,°\text{C/m}$$

Plane Wall Subject to Convection at Both Boundaries

In both cases considered above, the temperatures at the boundaries were specified and the boundary conditions were referred to as being of the first kind. In many engineering applications, it may not be possible to specify the boundary temperature because it is likely to be determined, by a balance between conduction within the solid and convection away from the boundary to an ambient fluid of specified temperature. Newton's law of cooling (see Chap. 1) provides a relationship that is to be satisfied at the boundary. The boundary temperature, indeed, floats at a value such that there is a balance between internal conduction and external convection. The boundary condition is referred to as boundary condition of the third kind or Robin condition.[2]

Refer to Fig. 2.7 now. At the two boundaries, neither the temperature nor the heat flux is specified, as mentioned above. For example, at $x = 0$ the temperature T_0 must take on such a value that the heat conducted into the solid is equal to the heat transfer by the fluid at T_1 to the surface at $x = 0$. A similar situation exists at $x = L$.

[2] After Victor Gustave Robin, 1855–1897, French mathematical analyst and applied mathematician.

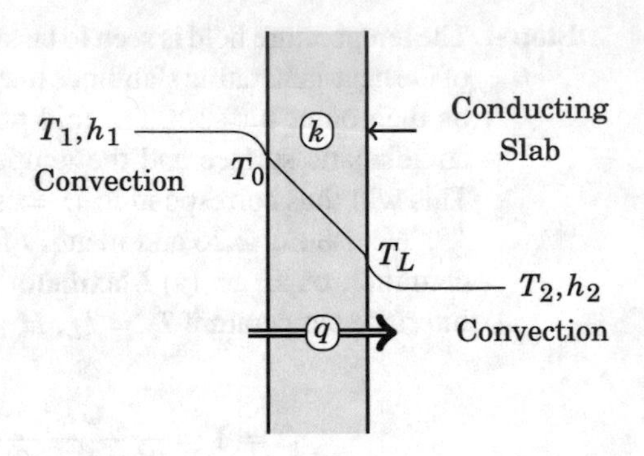

Fig. 2.7 Steady conduction in a slab subject to convection at the two boundaries

Assuming constant thermal conductivity and no heat generation ($G = 0$, $k =$ constant), and a convection heat transfer coefficient of h_1 we have the following relation at the left face of the slab:

$$h_1(T_1 - T_0) = q_c(x = 0) \tag{2.29}$$

Similarly we have the following relation at the right face of the slab:

$$q_c(x = L) = h_2(T_L - T_2) \tag{2.30}$$

In the present case of steady heat transfer across the plane wall, the two heat transfer rates $q_c(x = 0)$ and $q_c(x = L)$ are the same and are given by the relation

$$q_c(x = 0) = q_c(x = L) = \frac{k(T_0 - T_L)}{L} \tag{2.31}$$

Of course, T_0 and T_L are as yet unknown. These have to adjust to satisfy the two Eqs. 2.29 and 2.30. Using Eqs. 2.29 and 2.31, we have

$$T_1 - T_0 = \frac{kT_0}{h_1 L} - \frac{kT_L}{h_1 L}$$

This may be rearranged as

$$\left[1 + \frac{k}{h_1 L}\right] T_0 - \frac{k}{h_1 L} T_L = T_1 \tag{2.32}$$

Similarly, using Eqs. 2.30 and 2.31, we get

$$-\frac{k}{h_2 L} T_0 + \left[1 + \frac{k}{h_2 L}\right] T_L = T_2 \tag{2.33}$$

Subtract Eq. 2.33 from Eq. 2.32 and rearrange to get the following interesting result.

$$T_0 - T_L = \frac{T_1 - T_2}{\left(1 + \frac{k}{h_1 L} + \frac{k}{h_2 L}\right)} \tag{2.34}$$

At once we may calculate the heat transfer across the plane wall as (using Eq. 2.31)

$$q_c = \frac{k}{L} \frac{T_1 - T_2}{\left(1 + \frac{k}{h_1 L} + \frac{k}{h_2 L}\right)}$$

which may be simplified to

$$\boxed{q_c = \frac{T_1 - T_2}{\left(\frac{1}{h_1} + \frac{L}{k} + \frac{1}{h_2}\right)}} \tag{2.35}$$

Note that this is also the heat transferred across the wall from the fluid at T_1, to the fluid at T_2.

Electrical analogy

If we look at any one of the Eqs. 2.29–2.31, we see that heat transfer is analogous to a current, temperature difference is analogous to a potential difference, $\frac{1}{h_1}$ or $\frac{1}{h_2}$ is analogous to a resistance (referred to as Convective or film resistance R_f) and $\frac{L}{k}$ is also analogous to a resistance (referred to as the conduction resistance R_c). With these ideas in mind, Eq. 2.35 indicates that all the three resistances are in series and hence add up just as in the case of an electrical circuit containing three resistances in series (Fig. 2.8). The overall or equivalent *thermal* resistance is thus given by

$$\boxed{R_{EQ} = \frac{1}{h_1} + \frac{L}{k} + \frac{1}{h_2}} \tag{2.36}$$

These ideas find an interesting application in thermal engineering. The walls of a furnace are made of layers of different materials such as alumina brick layer facing the fire (k_1, L_1), low-cost bricks (k_2, L_2), and metallic outer layer (k_3, L_3), where the bracketed quantities represent the thermal conductivity values and thicknesses of the layers. One dimensional heat transfer assumption may be adequate in many

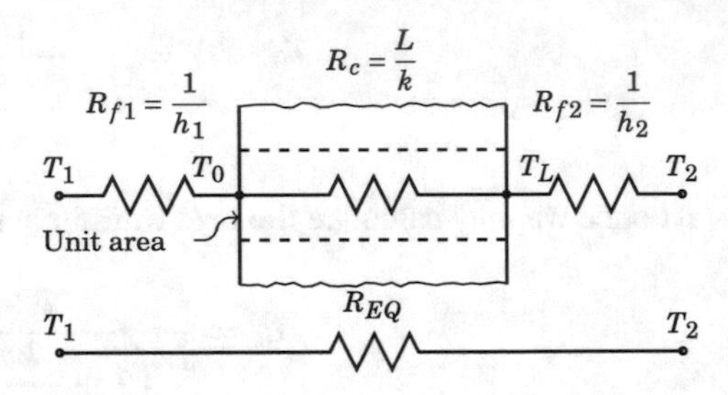

Fig. 2.8 Electrical analog of the slab problem subject to convection at the two boundaries

cases and the overall conductive resistance is given by

$$R_{c(overall)} = \frac{L_1}{k_1} + \frac{L_2}{k_2} + \frac{L_3}{k_3} \tag{2.37}$$

The brick layers are usually thick and provide a large conductive resistance because these materials have low thermal conductivities. The metal layer is usually thin (mostly provides structural rigidity) and provides a small thermal resistance because the thermal conductivity is normally very large. The brick layers act as thermal barriers and reduce the heat transfer from the fire side to the ambient. The application is one where the goal is to reduce heat transfer and the layered plane wall is an "insulating" system. Layering of the insulating materials is to reduce cost by using "high" temperature insulation facing the fire and "low" temperature insulation behind it. In case the layers do not form good "thermal" contact, additional conduction resistances are introduced and these are referred to as contact resistances. Contact resistance will introduce a significant temperature drop across any interface! Contact resistance at an interface depends on many factors such as the nature of the two surfaces that form the interface (smooth or rough), cleanliness of the two surfaces and the contact force, and so on. In applications where the contact resistance needs to be small one may use a thin layer of high conductivity grease at the interface.

Example 2.3

The walls of a large furnace are made of two layers of materials as shown in Fig. 2.9. The special brick surface facing the high-temperature environment receives heat from the high-temperature interior at 250 °C subject to a heat transfer coefficient of 45 W/m^2°C. The outside ambient is at 25 °C and removes heat via a heat transfer coefficient of 9 W/m^2°C. Determine the heat transfer per unit area across the wall and all the unknown temperatures indicated in the figure.

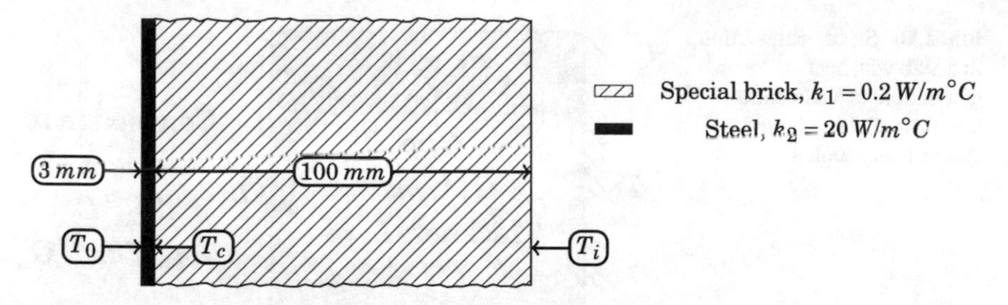

Fig. 2.9 Figure showing the geometry and nomenclature for Example 2.3

Solution:

Step 1 We make use of the concept of thermal resistance to solve the problem. The given data is written down first.

$$\begin{aligned}\text{High-temperature side:} T_1 &= 250\,°\text{C},\ h_1 = 45\,\text{W/m}^2°\text{C}\\ \text{Ambient side:} T_2 &= 25\,°\text{C},\ h_2 = 9\,\text{W/m}^2°\text{C}\\ \text{Special brick layer:} k_1 &= 0.2\,\text{W/m}\,°\text{C},\ L_1 = 100\,\text{mm} = 0.1\,\text{m}\\ \text{Steel layer:} k_2 &= 20\,\text{W/m}\,°\text{C},\ L_2 = 3\,\text{mm} = 0.003\,\text{m}\end{aligned}$$

Step 2 Equivalent thermal resistance between the inside and outside is calculated as

$$R_{EQ} = \frac{1}{h_1} + \frac{L_1}{k_1} + \frac{L_2}{k_2} + \frac{1}{h_2} = \frac{1}{45} + \frac{0.1}{0.2} + \frac{0.003}{20} + \frac{1}{9} = 0.63348\,\text{m}^2°\text{C/W}$$

Step 3 Heat transfer per unit area is then obtained as

$$q = \frac{T_1 - T_2}{R_{EQ}} = \frac{250 - 25}{0.63348} = 335.2\,\text{W/m}^2$$

Unknown temperatures are calculated in the next step.

Step 4 **Temperature** T_i: The above heat transfer should also equal the heat transfer by convection to the inner face. Hence, we have

$$q = h_1(T_1 - T_i) \quad \text{or} \quad \boxed{T_i = T_1 - \frac{q}{h_1} = 250 - \frac{335.2}{45} = 242.1\,°\text{C}}$$

Temperature T_c: The heat transfer may also be equated individually to the conduction heat transfer across the two layers of the wall. Thus:

$$q = k_1 \frac{T_i - T_c}{L_1} \quad \text{or} \quad \boxed{T_c = T_i - \frac{qL_1}{k_1} = 242.1 - \frac{335.2 \times 0.1}{0.2} = 64.5\,°\text{C}}$$

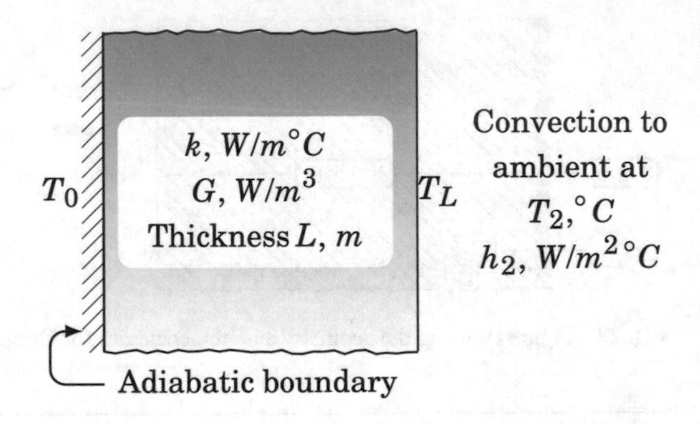

Fig. 2.10 Steady conduction in a slab with heat generation; one adiabatic boundary and one boundary convectively cooled

Temperature T_o:

$$q = k_2 \frac{T_c - T_o}{L_2} \quad \text{or} \quad \boxed{T_o = T_c - \frac{qL_2}{k_2} = 64.5 - \frac{335.2 \times 0.003}{20} = 64.45\,^\circ\text{C}}$$

Comment: It is noted that there is a negligibly small temperature drop across the metal layer. Most of the temperature drop occurs across the brick layer. If the heat transfer is considered to be excessive, it may be necessary to increase the brick layer thickness or back the special bricks with a layer of ordinary bricks of suitable thickness, thus introducing another conduction resistance.

Steady conduction in a slab with heat generation and convection at one boundary
Consider steady conduction in a slab as shown in Fig. 2.10. The slab has a thickness L and the uniform internal heat generation rate is G. The boundary of the slab at $x = 0$ is adiabatic while the boundary at $x = L$ communicates convectively with a fluid environment as indicated. Note that the boundary temperatures T_0 and T_L are as yet unknown. Adiabatic boundary condition is equivalent to specifying the heat flux, in this case, to be zero. This type of boundary condition is referred to as the boundary condition of the second kind or Neumann condition.[3] We note that the temperatures at both boundaries are unknown and hence need to be determined. We know, from the earlier discussion following Eq. 2.28, that the temperature within the slab is non-linear but corresponds to $H = 1$. Thus we have the important result from that analysis, viz.

$$T_0 - T_L = \frac{GL^2}{2k} \tag{2.38}$$

[3]After Karl Gottfried Neumann 1832–1925,German mathematician.

A second relation is obtained by noting that all the heat generated in the slab is removed by convection at the boundary at $x = L$. Thus

$$h_2(T_L - T_2) = GL \quad \text{or} \quad T_L = T_2 + \frac{GL}{h_2} \tag{2.39}$$

Substituting the value of T_L from Eq. 2.39 in Eq. 2.38 we get the value of T_0 as

$$T_0 = T_L + \frac{GL^2}{2k} = T_2 + \frac{GL}{h_2} + \frac{GL^2}{2k} \tag{2.40}$$

From Eqs. 2.38 and 2.40, we also get

$$\frac{T_0 - T_L}{T_0 - T_2} = \frac{\frac{GL^2}{2k}}{\left(\frac{GL}{h_2} + \frac{GL^2}{2k}\right)} = \frac{1}{\left(1 + \frac{2k}{h_2 L}\right)} = \frac{1}{\left(1 + \frac{2}{Bi_2}\right)} \tag{2.41}$$

where $Bi_2 = \frac{h_2 L}{k}$ is known as the Biot modulus or Biot number.

Biot number

This is a non-dimensional quantity and represents the ratio between a representative convective heat flux q_c and a representative conduction heat flux q_k. This may be verified by noting that

$$Bi_2 = \frac{h_2 L}{k} = \frac{h_2 \Delta T}{\left(\frac{k \Delta T}{L}\right)} = \frac{q_c}{q_k}$$

where ΔT is a characteristic temperature difference in the problem. We notice that when the Biot number tends to infinity, the convective heat transfer dominates and the boundary temperature is the same as the fluid temperature, i.e., $T_L = T_2$. This case corresponds to the one in which the boundary temperature is specified (case H = 1 in Fig. 2.5).

Example 2.4

Heat is generated in a slab according to the relation $G = G_0 \cos\left(\frac{\pi x}{L}\right)$ where x is measured from the left boundary of the slab that is perfectly insulated and L is the thickness of the slab. The right boundary is exposed to a convective environment at a temperature of T_{amb} with a heat transfer coefficient of h. Make a plot of the

temperature variation across the slab with the following data:
$L = 0.05\,m,\ k = 3\,\mathrm{W/m\,^\circ C},\ h = 15\,\mathrm{W/m^{2\circ}C},\ G_0 = 10^6\,\mathrm{W/m^3}$ and $T_{amb} = 30\,^\circ\mathrm{C}$.

Solution:
Equation governing the problem is given by Eq. 2.20, where G is taken as a function of x as specified in the problem. This equation is integrated with respect to x once to get

$$\frac{dT}{dx} + \frac{G_0}{k}\int \cos\left(\frac{\pi x}{2L}\right)dx = \frac{dT}{dx} + \frac{2LG_0}{\pi k}\sin\left(\frac{\pi x}{2L}\right) = A \tag{2.42}$$

where A is a constant of integration. A second integration yields the temperature given by

$$T - \frac{4L^2 G_0}{\pi^2 k}\cos\left(\frac{\pi x}{2L}\right) = Ax + B \tag{2.43}$$

where B is a second constant of integration. Since the left boundary is insulated—i.e., $\frac{dT}{dx} = 0$ at $x = 0$—constant of integration $A = 0$ in Eq. 2.42. Hence this equation gives, on putting $x = L$,

$$-k\frac{dT}{dx}\bigg|_{x=L} = \frac{2LG_0}{\pi} = q_c(x = L)$$

We equate this conductive flux inside the slab to the convective flux away from the slab to get

$$q_c(x = L) = \frac{2LG_0}{\pi} = h(T_L - T_{amb})$$

where $T_L = B$ (from Eq. 2.43 at $x = L$) is the temperature at the right surface of the slab. Solve the above for T_L and introduce in the equation for T to finally get the temperature profile as

$$T = T_{amb} + \frac{2LG_0}{h\pi} + \frac{4G_0L^2}{k\pi^2}\cos\left(\frac{\pi x}{2L}\right) \tag{2.44}$$

We make use of the data given in the problem to get the following temperature profile.

$$\begin{aligned} T &= 30 + \frac{2 \times 0.05 \times 10^5}{15 \times \pi} + \frac{4 \times 10^5 \times 0.05^2}{3 \times \pi^2}\cos\left(\frac{\pi x}{2 \times 0.05}\right) \\ &= 242.21 + 33.77\cos(10\pi x) \end{aligned} \tag{2.45}$$

Figure 2.11 shows a plot of temperature variation across the slab.

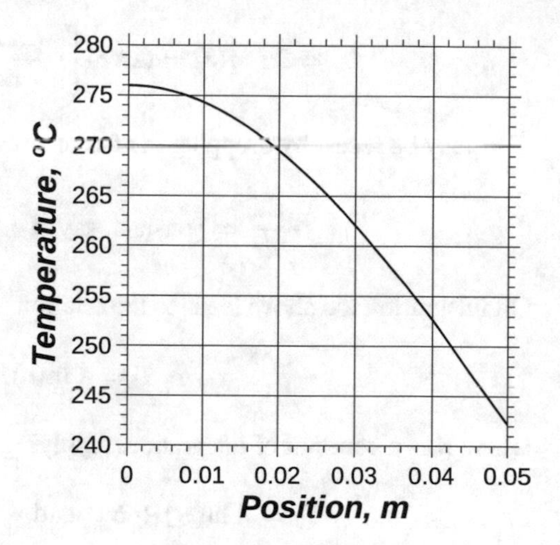

Fig. 2.11 Temperature profile across the the slab in Example 2.4

2.2.2 Steady One-Dimensional Conduction in Cylindrical Coordinates

Steady Conduction in a Long Cylindrical Annulus with Temperatures Specified at Boundaries

The difference between the slab geometry and the cylindrical geometry is mainly due to the increase of area available for conductive heat transfer in the radial direction with an increase in r. This makes the temperature variation non-linear in the radial direction even when i) there is no internal heat generation or ii) the thermal conductivity is constant. Figure 2.12 defines the geometry and introduces the nomenclature. Consider a unit thickness in a direction perpendicular to the plane of the figure. Steady condition requires that the total radial heat transfer at any r remain constant. Thus

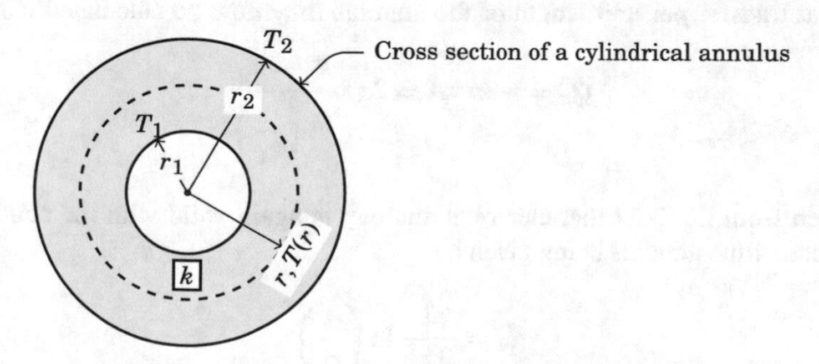

Fig. 2.12 Steady radial conduction in an annulus

$$Q_r = 2\pi r q(r) = (2\pi r)\left[-k\frac{dT}{dr}\right] = \text{constant, say } A' \tag{2.46}$$

This may be recast by dropping the factor $-2\pi k$ to get

$$r\frac{dT}{dr} = \text{constant, say } A \quad \text{or} \quad dT = A\frac{dr}{r} \tag{2.47}$$

On integration the above leads to the relation

$$T = A\ln(r) + B \tag{2.48}$$

where B is a constant of integration. Applying the boundary conditions, we have

$$T_1 = A\ln(r_1) + B \quad \text{and} \quad T_2 = A\ln(r_2) + B \tag{2.49}$$

By subtraction we eliminate B to get

$$T_1 - T_2 = A\ln(r_1) - A\ln(r_2) = A\ln\left(\frac{r_1}{r_2}\right) \quad \text{or} \quad A = \frac{T_1 - T_2}{\ln\left(\frac{r_1}{r_2}\right)} \tag{2.50}$$

From Eq. 2.49 we then have

$$\begin{aligned} B &= T_1 - A\ln(r_1) = T_1 - \frac{T_1 - T_2}{\ln\left(\frac{r_1}{r_2}\right)}\ln(r_1) \\ &= \frac{T_1\ln\left(\frac{r_1}{r_2}\right) - T_1\ln(r_1) + T_2\ln(r_1)}{\ln\left(\frac{r_1}{r_2}\right)} = \frac{T_2\ln(r_1) - T_1\ln(r_2)}{\ln\left(\frac{r_1}{r_2}\right)} \end{aligned} \tag{2.51}$$

The heat transfer per unit length of the annulus may now be calculated using Eq. 2.46 as

$$Q_r = -2\pi k A = 2\pi k\frac{T_1 - T_2}{\ln\left(\frac{r_2}{r_1}\right)} \tag{2.52}$$

It is seen from Eq. 2.52 that electrical analogy is again valid with the conduction resistance of the annulus being given by

$$R_a = \frac{1}{2\pi k}\ln\left(\frac{r_2}{r_1}\right) \tag{2.53}$$

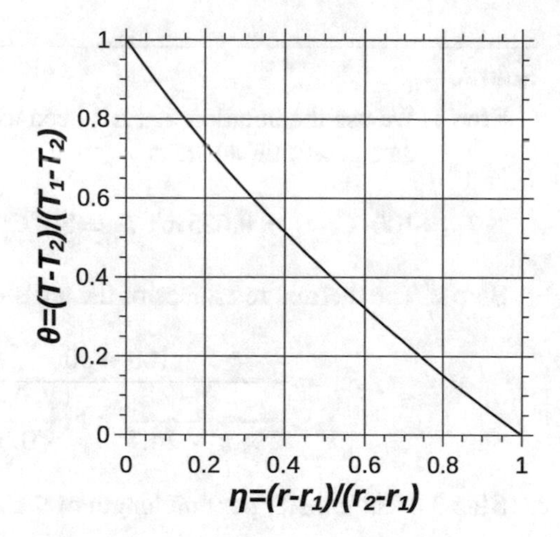

Fig. 2.13 Non-dimensional temperature profile in an annulus

The temperature profile in the annulus may be represented in the non-dimensional form by the use of the following normalized non-dimensional variables:

$$\eta = \frac{r - r_1}{r_2 - r_1}, \quad \theta = \frac{T - T_2}{T_1 - T_2}$$

We also define the non-dimensional radius ratio $\rho = \frac{r_2}{r_1}$. We may use Eqs. 2.48, 2.50 and 2.51 to show that

$$\boxed{\theta = 1 - \frac{\ln[1 + \eta(\rho - 1)]}{\ln(\rho)}} \tag{2.54}$$

It is easily verified that $\theta = 1$ at $\eta = 0$ and $\theta = 0$ at $\eta = 1$. Figure 2.13 shows a plot of the non-dimensional temperature variation with non-dimensional radial coordinate, for a specific value of the radius ratio $\rho = 2$. The temperature profile is dependent only on a single parameter, the radius ratio, and is independent of the boundary temperatures as well as the thermal conductivity of the annulus material.

Example 2.5

Consider a long cylindrical annulus of 50 mm ID and 100 mm OD. The material of the annulus has a thermal conductivity of 14.5 W/m°C. The inner surface is maintained at 100 °C while the outer surface is maintained at 30 °C. What is the heat transferred across the annulus in W/m? Make a plot of the temperature profile within the annulus. What are the temperature gradients at the inner and outer boundaries?

Solution:

Step 1 We use the notation that has been used in the text. Accordingly, the given data is written down as

$$T_1 = 100\,°\text{C};\ r_1 = 0.025\,\text{m};\ T_2 = 30\,°\text{C};\ r_2 = 0.05\,\text{m and } k = 14.5\,\text{W/m}\,°\text{C}$$

Step 2 The thermal resistance of the annulus is calculated using Eq. 2.53 as

$$R_a = \frac{100 - 30}{\frac{1}{2 \times \pi \times 14.7} \times \ln\left(\frac{0.05}{0.025}\right)} = 0.007608\,\text{m}\,°\text{C/W}$$

Step 3 Heat transfer per unit length of the annulus is then given by

$$Q_r = \frac{T_1 - T_2}{R_a} = \frac{100 - 30}{0.007608} = 9201\,\text{W/m}$$

Step 4 The dimensional temperature profile is given by $T(r) = A\ln(r) + B$. Using Eqs. 2.50 and 2.51 we have

$$A = \frac{100 - 30}{\ln\left(\frac{0.025}{0.05}\right)} = -100.99 \quad \text{and} \quad B = 100 + 100.99 \times \ln(0.025) = -272.54$$

Hence, we have $\boxed{T(r) = -100.99\ln(r) - 272.54}$. Note that r is in m and T in °C in this expression. Temperature variation across the annulus has been plotted in Fig. 2.14. Note that the radius ratio is 2 in this example also. Comparison may be made of Fig. 2.14 (plot uses dimensional variables) and Fig. 2.13 (plot uses non-dimensional variables). The shape is preserved in the two plots, if scales along the axes are chosen properly.

Step 5 The temperature gradient at any r is obtained by differentiating $T(r)$ with respect to r.

$$\frac{dT}{dr} = -\frac{100.99}{r}$$

Hence the gradients at the two boundaries are

$$\left.\frac{dT}{dr}\right|_{r_1} = -\frac{100.99}{0.025} = -4039.6\ °\text{C/m} \text{ and } \left.\frac{dT}{dr}\right|_{r_2} = -\frac{100.99}{0.05} = -2019.8\ °\text{C/m}$$

Cylinder with Convection at the Outer Boundary

This configuration is important because it is applicable to insulation design. Tubes carrying hot or cold fluids need to be insulated from the surroundings so as to reduce

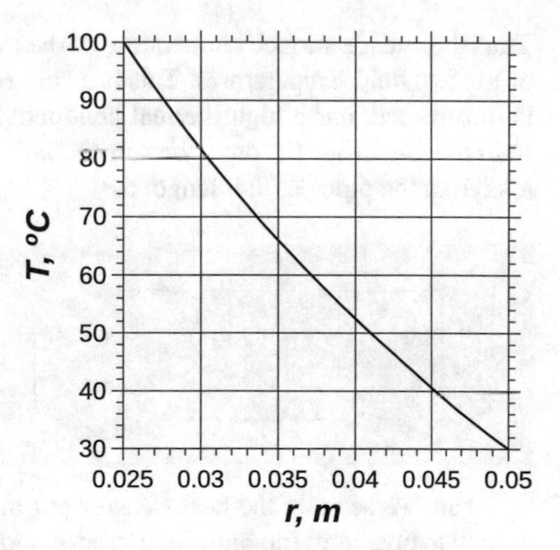

Fig. 2.14 Dimensional plot of temperature variation in an annulus of Example 2.5

heat loss form or heat gain by the fluid flowing inside. One can imagine the annulus in Sect. 2.2.2 to be an insulating material with a thin-walled pipe carrying a hot fluid placed inside it. The outer boundary at $r = r_2$ loses heat by convection to an ambient at T_∞ via a heat transfer coefficient h. Thus, we have

$$Q_c = 2\pi r_2 h(T_2 - T_\infty) \tag{2.55}$$

Under steady conditions, we should equate Eq. 2.55 and 2.52 so that temperature T_2 may be eliminated. Thus,

$$2\pi k \frac{T_1 - T_2}{\ln\left(\dfrac{r_2}{r_1}\right)} = 2\pi r_2 h(T_2 - T_\infty)$$

or solving for T_2

$$T_2 = \frac{T_1 + \dfrac{hr_2}{k} T_\infty \ln\left(\dfrac{r_2}{r_1}\right)}{1 + \dfrac{hr_2}{k} \ln\left(\dfrac{r_2}{r_1}\right)}$$

Thus the outside surface temperature of the insulation is seen to be a weighted mean of the two fluid temperatures. This is a consequence of the assumption that the pipe is thin walled, has a high thermal conductivity, and attains the temperature of the fluid flowing inside the pipe. We substitute this value of T_2 in Eq. 2.55 to get the heat loss from the pipe per unit length as

$$Q_r = Q_c = 2\pi r_2 h(T_2 - T_\infty) = \frac{T_1 - T_\infty}{\left[\dfrac{1}{2\pi r_2 h} + \dfrac{1}{2\pi k}\ln\left(\dfrac{r_2}{r_1}\right)\right]} \tag{2.56}$$

Again we see that the heat transfer per unit length of annulus is the ratio of an overall temperature (potential) difference and an overall thermal resistance given by the sum of a film resistance and a conductive resistance of the annulus. The denominator of equation is thus the sum of $R_f = \frac{1}{2\pi r_2 h}$ and R_a given by expression 2.53. We shall look at the overall resistance in more detail now. The overall resistance may be written in the form

$$R_{overall} = R_f + R_a = \frac{1}{2\pi r_2 h} + \frac{1}{2\pi k}\ln\left(\frac{r_2}{r_1}\right) = \frac{1}{2\pi k}\left[\frac{k}{hr_2} + \ln\left(\frac{r_2}{r_1}\right)\right] \tag{2.57}$$

Introduce the following non-dimensional quantities:

Non-dimensional resistance: $R = 2\pi k R_{overall}$
Biot number: $Bi = \frac{hr_1}{k}$
Radius ratio: $\rho = \frac{r_2}{r_1}$

Using Eq. 2.57, we then have

$$R = \frac{1}{\rho Bi} + \ln(\rho) \tag{2.58}$$

The non-dimensional resistance shows very interesting properties. Firstly it is valid only for $\rho > 1$. Secondly, it shows a monotonic behavior with ρ for $Bi > 1$. However,

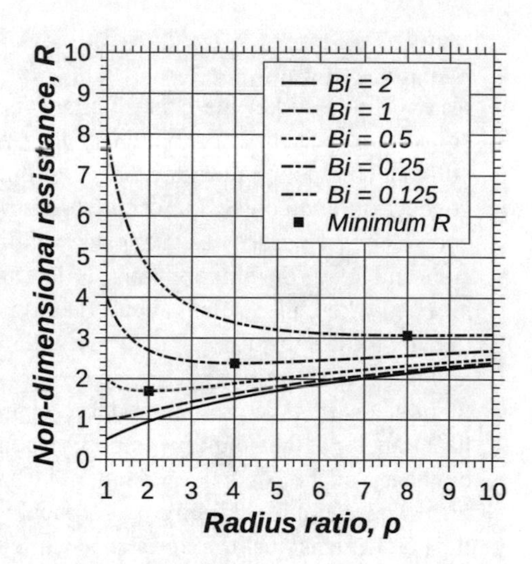

Fig. 2.15 Variation of overall resistance with radius ratio. Symbols show the location of the respective minima

for $Bi < 1$ the resistance decreases, and reaches a minimum value and then increases. This may be verified by differentiating Eq. 2.57 with respect to ρ and equating the derivative to zero, keeping Bi fixed. Thus,

$$\left.\frac{\partial R}{\partial \rho}\right|_{Bi-\text{fixed}} = -\frac{1}{Bi\rho^2} + \frac{1}{\rho} = 0 \tag{2.59}$$

This is satisfied for

$$\rho = \rho_{critical} = \frac{1}{Bi} \tag{2.60}$$

We have called the radius ratio that leads to an optimum as $\rho_{critical}$, the critical radius ratio. If $Bi > 1$ the critical radius ratio is less than one and this is physically not possible. For $Bi < 1$, the critical radius ratio is physically meaningful and gives the radius ratio for which the resistance is a minimum, as verified by the fact that the second derivative of R with respect to ρ is positive. For $\rho < \rho_{critical}$, the heat transfer actually increases with an increase in the insulation thickness. However for $Bi > 1$, increase in ρ (and hence an increase in insulation thickness) always decreases heat transfer. Figure 2.15 shows plot of non-dimensional resistance with radius ratio, for various values of the Biot number.

A note on Biot number

As explained earlier the Biot number represents the ratio of a representative convective heat flux at a boundary to a representative conduction heat flux within a conducting medium. This non-dimensional number occurs whenever conduction and convection occur together. In a typical practical problem, of course, the two processes take place simultaneously. The heat transfer coefficient and hence the film resistance is determined from a solution of a *conjugate problem* where the mathematical equations governing conduction and convection are to be solved simultaneously with the "boundary" condition becoming an interface condition.

It means that the present model is a simplified one wherein the convection heat transfer at the boundary imposes a constant heat transfer coefficient at the boundary of the conduction domain, i.e., at the outer surface of the annulus. Now we look at Eq. 2.58 from a physical view point. The first term represents the non-dimensional film resistance while the second term represents the non-dimensional conduction resistance. With an increase in the outer radius of the annulus, the convective resistance reduces while the conductive resistance increases. There is thus a competition between these two, the relative decrease and increase is determined by the Biot number. This competition, when the Biot number is less than one, leads to the appearance of an optimum or critical radius, as discussed already.

Example 2.6

A pipe (material thermal conductivity 45 W/m°C) carrying saturated steam at 180 °C has an one dimension of 50 mm and an OD of 55 mm. The outside surface is exposed to an ambient at 30 °C . The heat transfer coefficient from the outer surface of the pipe to the ambient has been estimated to be 25 W/m^2°C. What is the heat loss from the pipe per meter length? In a bid to reduce the heat loss the pipe is insulated by a 25.3 mm layer of an insulating material of thermal conductivity equal to 0.1 W/m°C. What is the heat loss per meter length of pipe now? Assume that the heat transfer coefficient remains the same in the two cases. Evaluate all the temperatures.

Solution:

Figure 2.16 shows the geometry and the nomenclature for the two cases, Case 1: without insulation and Case 2: with insulation.

The corresponding electrical equivalent shows that there are three resistances in series in Case 1 while there are four resistances in series in Case 2. In both cases, the film resistance R_{f1} is negligibly small since the steam side heat transfer coefficient is very large (see Table 1.1). Hence it is reasonable to ignore the film resistance on the steam side. Hence the pipe inner wall temperature is the same as the steam

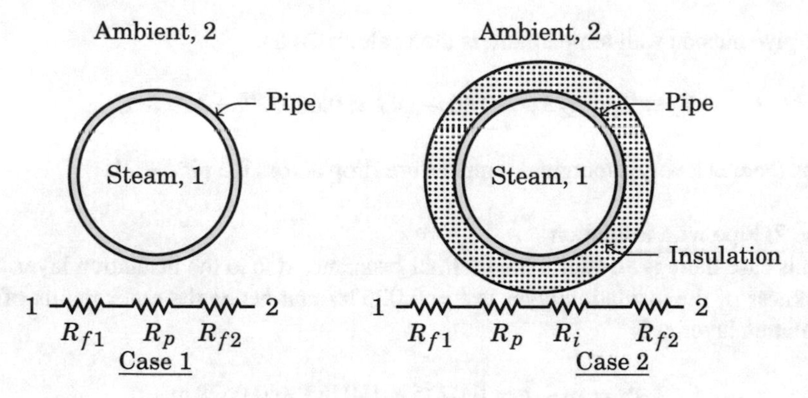

Fig. 2.16 Geometry and nomenclature for Example 2.6. Case 1 is without insulation, Case 2 is with insulation

temperature. The outside film coefficient R_{f2} is *different in the two cases* as will be seen later on. The given data is written down below:

Pipe inner wall temperature:	$T_1 = 180\,°\text{C}$
Pipe inner radius:	$r_1 = 0.025\,\text{m}$
Pipe outer radius:	$r_2 = 0.0275\,\text{m}$
Pipe material thermal conductivity:	$k_p = 45\ \text{W/m}\,°\text{C}$
Ambient fluid temperature:	$T_{amb} = 30\,°\text{C}$
External heat transfer coefficient:	$h = 25\,\text{W/m}^2°\text{C}$

Case 1: Bare pipe without insulation

The thermal resistances are calculated as follows:
Conduction resistance of the pipe:

$$R_p = \frac{1}{2\pi k_p} \ln\left(\frac{r_2}{r_1}\right) = \frac{1}{2\pi \times 45} \ln\left(\frac{0.0275}{0.025}\right) = 3.371 \times 10^{-4}\,\text{m}\,°\text{C/W}$$

Outside film resistance:

$$R_{f2} = \frac{1}{2\pi r_2 h} = \frac{1}{2\pi \times 0.0275 \times 25} = 0.232\,\text{m}\,°\text{C/W}$$

The overall resistance to heat transfer is then given by

$$R_{overall} = R_{f1} + R_p + R_{f2} = 0 + 3.371 \times 10^{-4} + 0.232 = 0.232\,\text{m}\,°\text{C/W}$$

The heat loss from steam per meter length of pipe is obtained as

$$Q = \frac{T_1 - T_{amb}}{R_{overall}} = \frac{180 - 30}{0.232} = 647\,\text{W/m}$$

The pipe outside wall temperature is then calculated as

$$T_2 = T_1 - QR_p = 180 - 647 \times 0.000337 = 179.8\,^\circ\text{C}$$

Thus there is a very a marginal temperature drop across the pipe wall.

Case 2: Pipe with insulation

In this case there is an additional thermal resistance due to the insulation layer. The thickness of the insulation layer is $t = 0.0253m$ and hence the outer radius of the insulation layer r_3 is

$$r_3 = r_2 + t = 0.0275 + 0.0253 = 0.0528\,\text{m}$$

Insulation layer conduction resistance:

$$R_i = \frac{1}{2\pi k_i} \ln\left(\frac{r_3}{r_2}\right) = \frac{1}{2\pi \times 0.1} \ln\left(\frac{0.0528}{0.0275}\right) = 1.038\,\text{m}\,^\circ\text{C/W}$$

Outside film resistance:

$$R_{f2} = \frac{1}{2\pi r_3 h} = \frac{1}{2\pi \times 0.0528 \times 25} = 0.1206\,\text{m}\,^\circ\text{C/W}$$

We see that the outside film resistance has changed to a lower value! The overall resistance to heat transfer is then given by

$$\begin{aligned} R_{overall} &= R_{f1} + R_p + R_i + R_{f2} \\ &= 0 + 3.371 \times 10^{-4} + 1.038 + 0.1206 = 1.159\,\text{m}\,^\circ\text{C/W} \end{aligned}$$

The heat loss in the presence of insulation is given by

$$Q = \frac{T_1 - T_{amb}}{R_{overall}} = \frac{180 - 30}{1.159} = 129.4\,\text{W/m}$$

The pipe outside wall temperature (this is also the temperature of the insulation layer in contact with the pipe wall) is

$$T_2 = T_1 - QR_p = 180 - 129.4 \times 0.000337 = 179.96\,^\circ\text{C}$$

The temperature drop across the pipe wall is indeed negligible. The insulation outside surface temperature (T_3) is given by

$$T_3 = T_2 - QR_i = 179.96 - 129.4 \times 1.038 = 45.6\,^\circ\text{C}$$

There is thus a substantial temperature drop across the insulation layer. We note in passing that the maximum allowable temperature of the outer surface of insulation, from safety considerations, is around 60 °C.

2.2.3 *Steady Radial Conduction in a Solid Cylinder with Internal Heat Generation*

Temperature Specified at the Boundary

A very-long-current-carrying conductor, very common in engineering applications, may be modeled as a solid cylinder with internal heat generation. Heat generated is by Joule heating and may be assumed uniform across the diameter of the conductor when the current is DC or when the frequency of the current is not very high. The formulation of the problem can be explained with reference to Fig. 2.17. Consider a cylindrical rod of radius R as shown. If the heat generation per unit volume is G, energy balance requires that all the heat generated within the cylinder of radius r (shown shaded in the figure) be conducted away at its boundary.

As usual we consider a unit length of the cylinder and hence we have

$$G\pi r^2 = 2\pi r q_r = -2\pi r k \frac{dT}{dr} \tag{2.61}$$

This may be simplified to

$$\boxed{\frac{dT}{dr} + \frac{Gr}{2k} = 0} \tag{2.62}$$

Only *one* boundary condition may be specified as $T = T_s$ at $r = R$. Integrate Eq. 2.62 with respect to r to get

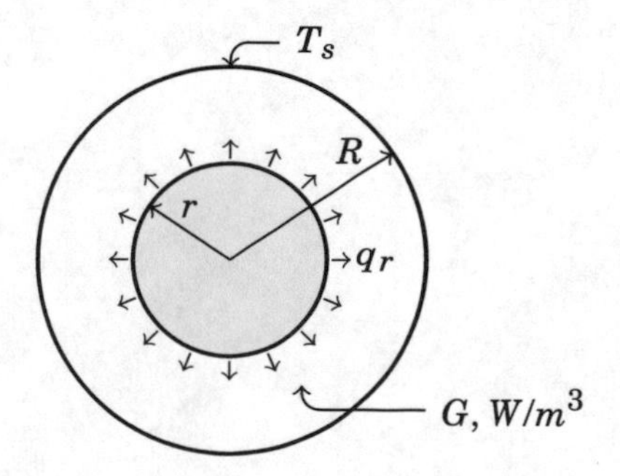

Fig. 2.17 Conduction in a cylindrical rod with internal heat generation

$$T + \frac{G}{2k}\frac{r^2}{2} = \text{constant, say } A \tag{2.63}$$

Applying the boundary condition at the surface of the cylinder, we get

$$T_s + \frac{G}{2k}\frac{R^2}{2} = A \tag{2.64}$$

Subtract 2.64 from 2.63 to get

$$T - T_s = \frac{G}{4k}(R^2 - r^2) \tag{2.65}$$

We easily see that the maximum temperature T_0 occurs at $r = 0$ and is given by $T_0 - T_s = \frac{GR^2}{4k}$. As usual we would like to represent the temperature variation with radius in the non-dimensional form since it leads to a single universal profile.

Introduce the following non-dimensional variables:

Non-dimensional temperature: $\phi = \frac{T-T_s}{\left[\frac{GR^2}{4k}\right]}$

Non-dimensional radius ratio: $\rho = \frac{r}{R}$

Equation 2.65 may then be written as

$$\boxed{\phi = 1 - \rho^2} \tag{2.66}$$

Temperature distribution given by Eq. 2.66 is shown plotted in Fig. 2.18.

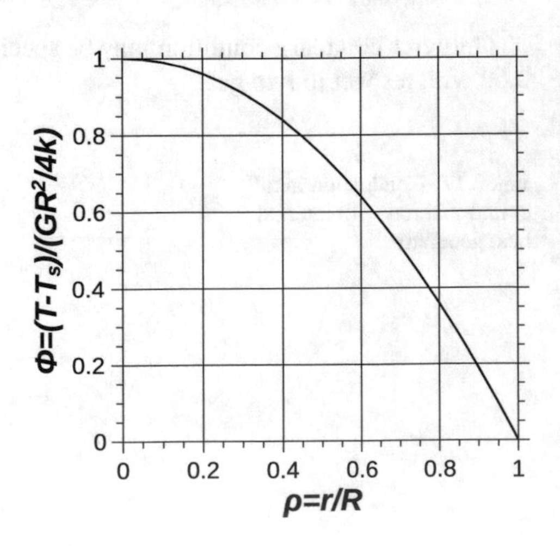

Fig. 2.18 Non-dimensional temperature profile in a heat-generating cylinder

Convection at the Surface to an Ambient at T_∞

This case is more appropriate in practice since the cylinder surface is exposed to an ambient fluid (usually the room air) and heat transfer from the surface is by convection, governed by a specified heat transfer coefficient h. The surface temperature T_s is eliminated by equating the conduction heat flux at $r = R$ within the cylinder to convection heat flux from the boundary to the ambient fluid. Alternately T_s may be eliminated by requiring that the total heat generated within the cylinder equal the heat removed by convection at the boundary. Thus

$$\pi R^2 G = 2\pi R h (T_s - T_\infty) \tag{2.67}$$

or

$$T_s - T_\infty = \frac{GR}{2h} \tag{2.68}$$

Equation 2.65 may be rewritten as

$$T - T_s = \frac{GR^2}{4k}\left[1 - \left(\frac{r}{R}\right)^2\right] \tag{2.69}$$

Adding Eqs. 2.68 and 2.69 we get the desired result.

$$T - T_\infty = \frac{GR}{2h} + \frac{GR^2}{4k}\left[1 - \left(\frac{r}{R}\right)^2\right] \tag{2.70}$$

Again, the maximum temperature occurs at $r = 0$ and is given by

$$T_0 - T_\infty = \frac{GR}{2h} + \frac{GR^2}{4k} = \frac{GR}{4h}[2 + Bi] \tag{2.71}$$

where $Bi = \frac{hR}{k}$, is the Biot number. The non-dimensional temperature profile may then be obtained as

$$\boxed{\phi = \frac{T - T_\infty}{T_0 - T_\infty} = \frac{2 + Bi(1 - \rho^2)}{2 + Bi}} \tag{2.72}$$

where $\rho = \frac{r}{R}$. The profile is a one parameter family of curves with Bi, the Biot number as the governing parameter. Figure 2.19 is a plot of Eq. 2.72 for several values of Bi. Note that the case $Bi = \infty$ corresponds to the case where the boundary temperature is specified ($T_s = T_\infty$). Note that for $\rho = 1$, i.e., at the cylinder surface, the non-dimensional temperature is given by

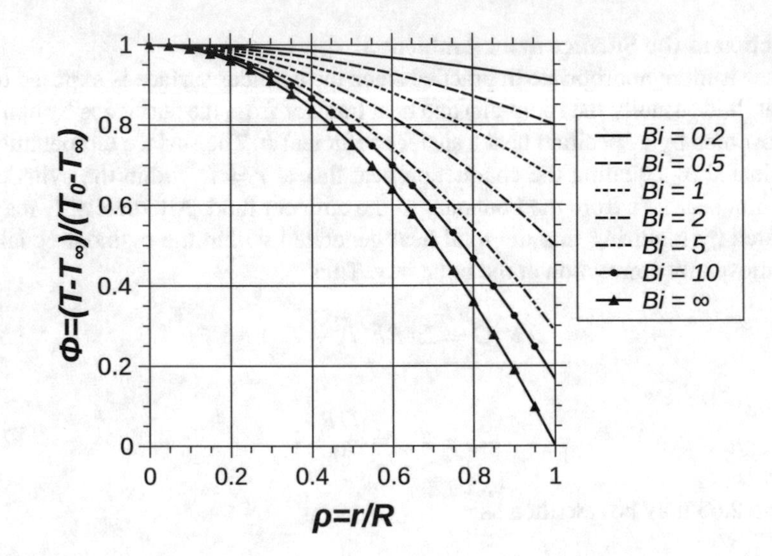

Fig. 2.19 Temperature profile in a heat-generating cylinder subject to convection at the boundary

$$\phi_s = \frac{T_s - T_\infty}{T_0 - T_\infty} = \frac{2}{2 + Bi} \tag{2.73}$$

For example, when $Bi = 5$, the non-dimensional cylinder surface temperature is $\phi_s = \frac{2}{7} \approx 0.286$.

Example 2.7

A round wire of 6 mm diameter is heated by a current passing through it. The heat dissipated per meter of the wire is 100 W. The wire is covered by a layer of insulation of outer diameter equal to 12 mm. The insulation has a thermal conductivity of 0.4 W/m °C. The outer surface of the insulation is exposed to an inert gas at 30 °C. The temperature of the outer surface of the insulation is measured and found to be 60 °C. Determine all the temperatures and the maximum temperature in the wire if the wire material has a thermal conductivity of 15 W/m °C.

Solution:

Step 1 Let h be the convection heat transfer coefficient to the inert gas from the outside surface of the insulation. This needs to be estimated from the data given in the problem.

Step 2 Given data is written down as:

Outside surface temperature of the insulation:	$T_s = 60\,°C$
Ambient temperature:	$T_\infty = 30\,°C$
Heat loss from the wire:	$q = 100$ W/m
Insulation layer outer radius:	$r_2 = \frac{0.012}{2} = 0.006\,m$

Using the above data we calculate the convective heat transfer coefficient as

$$h = \frac{q}{2\pi r_2(T_2 - T_\infty)} = \frac{100}{2 \times \pi \times 0.006 \times (60 - 30)} = 88.4\,W/m^2°C$$

Step 3 Now consider the insulation layer. The geometric and thermal parameters are:
Inner diameter of insulation layer r_1 is the same as the radius of the current carrying conductor $= \frac{0.006}{2} = 0.003\,m$.:
Thermal conductivity of the insulation material is $k_i = 0.4$ W/m°C.:
As far as the insulation and its interaction with the inert gas ambient is concerned, we may use the thermal resistance concept to model the process. Conduction resistance R_i of the insulation layer is calculated as

$$R_i = \frac{1}{2\pi k_i}\ln\left(\frac{r_2}{r_1}\right) = \frac{1}{2 \times \pi \times 0.4}\ln\left(\frac{0.006}{0.003}\right) = 0.276\,m\,°C/W$$

The wire outer temperature T_1 which is the same as the temperature of the inner surface of the insulation is given, using the thermal resistance concept as

$$T_1 = T_s + qR_i = 60 + 100 \times 0.276 = 87.6\,°C$$

Step 4 The volumetric heat generation rate G is such that the total heat generated in the wire equals the heat loss per meter length of wire. Thus

$$G = \frac{q}{\pi r_1^2} = \frac{100}{\pi \times 0.003^2} = 3.537 \times 10^6\,W/m^3$$

The maximum temperature in the wire occurs at its axis. This is calculated as

$$T_0 = T_1 + \frac{Gr_1^2}{4k_w} = 87.6 + \frac{3.537 \times 10^6 \times 0.003^2}{4 \times 15} = 88.1\,°C$$

where k_w is the thermal conductivity of the wire material. There is thus a mere 0.5 °C difference in temperature between the axis and the surface of the heat-generating wire.

Step 5 Alternately T_0 may be obtained using Eq. 2.73. The Biot number is based on the overall heat transfer coefficient based on the wire surface area. We have

$$R_{overall} = R_i + \frac{1}{2\pi r_2 h} = 0.276 + \frac{1}{2 \times \pi \times 0.006 \times 87.6} = 0.579m\ °\text{C/W}$$

The *overall* heat transfer coefficient h_w based on the wire surface area is then such that $R_{overall} = \frac{1}{2\pi r_1 h_w}$ or

$$h_w = \frac{1}{2\pi r_1 R_{overall}} = \frac{1}{2 \times \pi \times 0.003 \times 0.579} = 91.66\ \text{W/m}^2\text{°C}$$

The Biot number is then based on this heat transfer coefficient.

$$Bi = \frac{h_w r_1}{k_w} = \frac{91.66 \times 0.003}{15} = 0.0183$$

Using Eq. 2.73, we then get

$$\phi_s = \frac{T_1 - T_\infty}{T_0 - T_\infty} = \frac{2}{2 + Bi} = \frac{2}{2 + 0.0183} = 0.991$$

We then get the maximum temperature in the wire as

$$T_0 = T_\infty + \frac{T_1 - T_\infty}{\phi_s} = 30 + \frac{87.6 - 30}{0.991} = 88.1\ \text{°C}$$

Even though the alternate method is some what lengthy, it has been presented to point out that alternate methods are possible in solving the problem. The alternate method has also brought out the way one interprets the concept of overall thermal resistance.

2.2.4 *One-Dimensional Radial Conduction in Spherical Coordinates*

Conduction in spherical coordinates has interesting applications in (i) spherical pressure vessel, (ii) spherical dish ends of a boiler shell, (iii) nuclear reactor containment

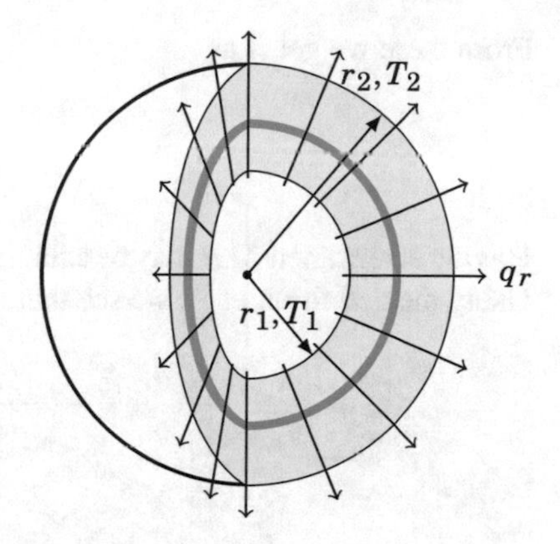

Fig. 2.20 Radial heat conduction across a spherical shell

vessel, and (iv) spherical tanks used to store natural gas and so on. In spherical coordinates, the area available for conduction in the radial direction varies as the square of the radius. Heat transfer in spherical food products may be modeled this way. Spherical particles are also used in many industrial process applications and heat conduction within the particles may be modeled using the analysis that will be developed below.

One-Dimensional Steady Conduction in a Spherical Shell

Consider the typical case of steady heat conduction across a shell of inner radius r_1 and outer radius r_2. Let the corresponding temperatures be specified as T_1 and T_2. The material has a constant thermal conductivity k. It is clear (see Fig. 2.20) that the total radial heat transfer Q given by

$$Q = -4\pi r^2 k \frac{dT}{dr} \tag{2.74}$$

remains constant across the shell, under steady heat transfer condition. The heat flux lines are radial lines and isotherms are spherical surfaces with a common center as indicated in Fig. 2.20.

Denoting $-\frac{Q}{4\pi k}$ as A, we get

$$\frac{dT}{dr} = \frac{A}{r^2} \tag{2.75}$$

Integrate Eq. 2.75 once with respect to r to get

$$T = -\frac{A}{r} + B$$

where B is a constant of integration. Using the boundary conditions specified in Fig. 2.20, we have

$$T_1 = B - \frac{A}{r_1} \quad \text{and} \quad T_2 = B - \frac{A}{r_2}$$

From these we get A as

$$A = \frac{T_1 - T_2}{\left(\frac{1}{r_2} - \frac{1}{r_1}\right)} \tag{2.76}$$

Having determined A, B may be calculated from either one of the boundary relations. Using the definition of A, we see that

$$Q = -4\pi k A = \frac{T_1 - T_2}{\underbrace{\frac{1}{4\pi k}\left(\frac{1}{r_1} - \frac{1}{r_2}\right)}_{R_k}} \tag{2.77}$$

Equation 2.77 shows that electrical analogy may be invoked in the case of radial heat conduction in spherical coordinates also. The thermal conduction resistance of the shell is simply the denominator of Eq. 2.78 (indicated as R_k).

One-Dimensional Steady Conduction in a Spherical Shell with Convection at Both Boundaries

It is not necessary to work out this case in detail. We invoke the thermal resistance concept. Let the convective heat transfer coefficients be h_1 and h_2 at the inner and outer boundary respectively. Let the corresponding ambient temperatures be T_1 and T_2. We recognize three resistances in series as follows:

Film resistance at inner boundary:	$R_{f1} = \frac{1}{4\pi r_1^2 h_1}$
Conduction resistance of the shell:	$R_k = \frac{1}{4\pi k}\left(\frac{1}{r_1} - \frac{1}{r_2}\right)$
Film resistance at outer boundary:	$R_{f2} = \frac{1}{4\pi r_2^2 h_2}$

As usual the total resistance is some of these three resistances in series. The heat transfer across the shell is then given by

$$Q = \frac{T_1 - T_2}{R_{f1} + R_k + R_{f2}} \tag{2.78}$$

Steady Conduction in a Sphere with Uniform Internal Heat Generation

Consider a sphere of radius R losing heat to an ambient as shown in Fig. 2.21. Let G be the constant volumetric heat generation rate. Under steady conditions, it is clear that the heat generated within the sphere of radius r must leave it by conduction. We thus have

$$Q_{generated} = Q_{conducted} \quad \text{or} \quad \frac{4}{3}\pi r^3 G = -4\pi r^2 k \frac{dT}{dr}$$

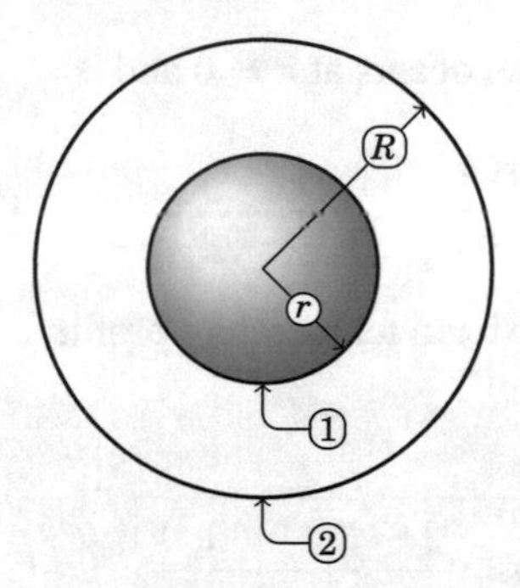

1 Heat generated within sphere of radius r leaves it by conduction
2 Heat generated within sphere of radius R leaves it by convection with heat transfer coefficient h to ambient at T_{amb}

Fig. 2.21 Heat removal by convection from a heat-generating sphere

Canceling common factor 4π , and simplifying, we get the following first-order ordinary differential equation.

$$\boxed{\frac{dT}{dr} + \frac{Gr}{3k} = 0} \tag{2.79}$$

Integration of Eq. 2.79 with respect to r gives

$$T + \frac{Gr^2}{6k} = B \tag{2.80}$$

where B is a constant of integration. At the surface of the sphere, $r = R, T = T_s$ (as yet unknown) and the total heat generated within the sphere is removed by convection from the surface. Hence, we have

$$\frac{4}{3}\pi R^3 G = 4\pi R^2 h(T_s - T_{amb}) \quad \text{or} \quad (T_s - T_{amb}) = \frac{RG}{3h} \tag{2.81}$$

From Eq. 2.80

$$T_s = B - \frac{GR^2}{6k}$$

Inserting this in Eq. 2.81 yields the appropriate value of B as

$$B = T_{amb} + \frac{GR^2}{6k} + \frac{RG}{3h} \tag{2.82}$$

The temperature variation in the sphere is obtained using Eqs. 2.82 and 2.80 as

$$T = T_{amb} + \frac{RG}{3h} + \frac{G}{6k}(R^2 - r^2) \tag{2.83}$$

The maximum temperature in the sphere, of course, occurs at $r = 0$ and is

$$T_0 = T_{amb} + \frac{RG}{3h} + \frac{GR^2}{6k} \tag{2.84}$$

From Eqs. 2.83 and 2.84, we have, the non-dimensional temperature profile

$$\frac{T - T_{amb}}{T_0 - T_{amb}} = \frac{\frac{RG}{3h} + \frac{G}{6k}(R^2 - r^2)}{\frac{RG}{3h} + \frac{GR^2}{6k}} \quad \text{or} \quad \boxed{\phi = \frac{2 + Bi(1 - \rho^2)}{2 + Bi}} \tag{2.85}$$

where $Bi = \frac{hR}{k}$ and $\rho = \frac{r}{R}$. Note that the non-dimensional temperature is identical to the expression 2.72 that was obtained for the case of a cylinder. Figure 2.19 also represents the present case. However, the expression for the maximum temperature T_0 is different and is given by Eq. 2.84. Non-dimensional temperature ϕ is a function of non-dimensional radial position ρ with a parametric dependence on Bi. The non-dimensional temperature varies between 1 at the center of the sphere to a value equal to $\frac{2}{2+Bi}$ at the surface. For $Bi << 1$, the temperature throughout the sphere is close to the maximum value. For $Bi >> 1$, however, the surface temperature is close to the ambient temperature. These observations are consistent with the observations made with respect to the heat-generating slab problem as well as the heat-generating cylinder problem. These observations are also consistent with the physical meaning of the Biot number.

Example 2.8

Consider a spherical shell of inner radius 5 cm and outer radius 10 cm. The inner and outer surfaces are maintained at 100 °C and 30 °C respectively. Make a plot of temperature variation with the radial position inside the shell. Compare this profile with those for a slab and annulus of the same thickness and the same surface temperatures.

Solution:

Inner radius and temperature of spherical shell: $r_1 = 5\,\text{cm},\ T_1 = 100\,°\text{C}$
Outer radius and temperature of spherical shell: $r_2 = 10\,\text{cm},\ T_2 = 30\,°\text{C}$
The temperature at any r is given by $T = -\frac{A}{r} + B$. From Eq. 2.76, we have

$$A = \frac{100 - 30}{\frac{1}{10} - \frac{1}{5}} = -700$$

Then B is obtained by using one of the boundary conditions, in this case, the inner boundary condition as

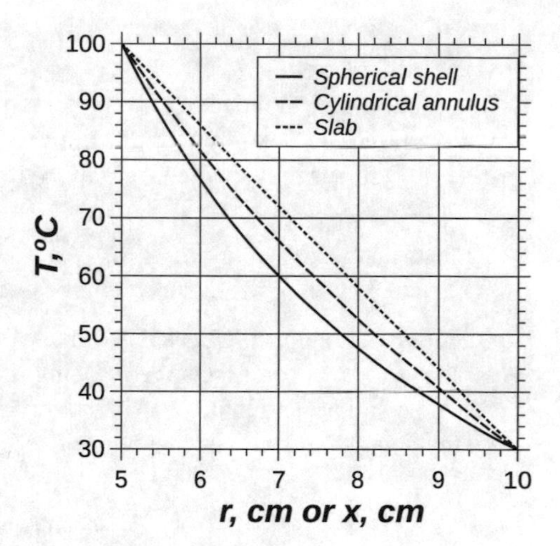

Fig. 2.22 Comparison of temperature profiles in a slab, a cylindrical annulus, and a spherical shell

$$B = T_1 + \frac{A}{r_1} = 100 - \frac{700}{5} = -40$$

The temperature profile is thus $T = \frac{700}{r} - 40$ where r is in cm and temperature is given in °C (shown in Fig. 2.22 by full line). Note that the constants A and B will be different if r is used in m (i.e., in SI unit). It is left as an exercise for the reader to derive the appropriate relations in the slab and cylindrical annulus.The plot shown in Fig. 2.22 shows all the three profiles. It is noticed that the profile is linear in the case of a slab. It is non-linear in both the cylindrical annulus as well as the spherical shell. The profiles do not depend on the thermal conductivity of the material of the spherical shell or the cylindrical annuls or the slab.

2.3 Generalization

The present chapter has dealt with heat conduction in one dimension—the dimension being a space dimension. All the problems were steady, with no changes with respect to time. The governing equations turn out to be ordinary differential equations, basically of second order, but cleverly written as first-order equations, using physical arguments!

We examine some general features of all the problems we have considered till now. The following development is for the more adventurous reader!

General features

In case of steady conduction in one dimension, the general feature is that the total conduction heat transfer does not change with x in the slab or r in the case of cylinder and sphere. We may write the total conduction heat transfer Q_k as

$$Q_k = -kA\frac{dT}{dx} = \text{constant} \tag{2.86}$$

where $A = 1$ in the case of slab geometry, $A = 2\pi r \times 1$ in the case of cylinder or $4\pi r^2$ in the case of the sphere. Using the symbol x to represent either x or r, and assimilating the factor k or $2\pi k$ or $4\pi k$ as appropriate in to the constant on the right hand side, we may write Eq. 2.86 as

$$x^n\frac{dT}{dx} = \text{constant} \tag{2.87}$$

where $n = 0$ in the case of slab, $n = 1$ in the case of the cylinder and $n = 2$ in the case of the sphere. If we differentiate Eq. 2.87 with respect to x, we get the following second-order differential equation.

$$\frac{d}{dx}\left(x^n\frac{dT}{dx}\right) = 0 \tag{2.88}$$

This equation is the general form of the steady one-dimensional heat conduction equation. In case of problem involving internal heat generation, we will have to modify the above equation in the following manner.

Heat generated $dQ_{generated}$ within an elemental volume dV can be written as

$$dQ_{generated} = GdV \tag{2.89}$$

where G is the heat generation rate per unit volume. The volume element may be written as

$$dV = cx^n dx \tag{2.90}$$

where $c = 1$, $n = 0$ for the slab, $c = 2\pi \times 1$, $n = 1$ for the cylinder and $c = 4\pi$, $n = 2$ in the case of the sphere. Figure 2.23 shows how these are obtained.

The net heat transferred across the volume element may be written as

$$dQ_k = -\frac{d}{dx}\left(kcx^n\frac{dT}{dx}\right) \tag{2.91}$$

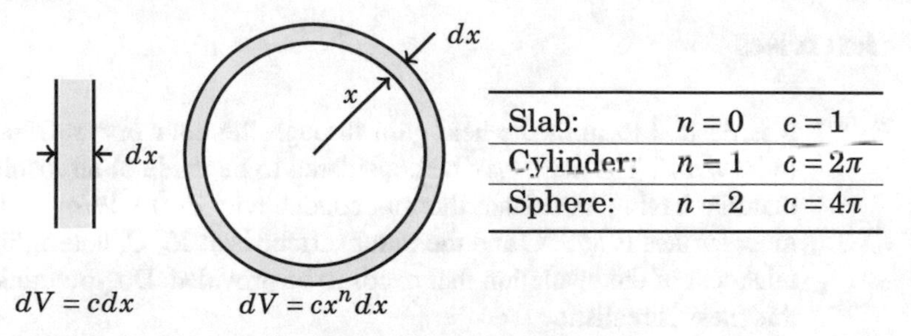

Fig. 2.23 Volume element in slab, cylinder, and sphere

Conservation of energy requires that the net heat transfer across the volume element must represent the heat generated within the volume element. We may remove the factor 1 or 2π or 4π as appropriate to get the following equation.

$$\frac{d}{dx}\left(x^n\frac{dT}{dx}\right)+\frac{Gx^n}{k}=0 \tag{2.92}$$

Of course Eq. 2.92 reduces to Eq. 2.88 when internal heat generation is zero. Equation 2.92 is the most general form of equation that represents steady conduction heat transfer in one dimension. The reader may verify that Eqs. 2.1, 2.46, 2.74, 2.20, 2.62, and 2.79 are obtained by performing one integration of the generalized equation presented here. We may refer to the equations cited above as representing an integral or integrated form of the general governing equation.

It is also clear that one integration constant is already present in the integral form of thc equation. However, the general form of the equation is of second order and hence supports two boundary conditions. It is rather unusual to specify two conditions at a single boundary (referred to as initial conditions) and hence all the problems considered in this chapter were boundary value problems.

Concluding Remarks

This chapter has dealt with steady conduction in one dimension. Problems of this type are the simplest to solve and occur in all three coordinate systems, viz., Cartesian, cylindrical, and spherical coordinate systems. Problems involving variable thermal conductivity, internal heat generation, convection at one or more boundaries have been considered. Toward the end of the chapter, we have given generalized governing equations that apply to steady one-dimensional conduction in all three coordinate systems.

2.4 Exercises

Ex 2.1 It is desired to limit the heat gain through the door of a refrigerator to 30 W/m^2. The door may be considered to be made of an insulating material having a constant thermal conductivity of 0.1 W/m°C. If the inner surface is at 3 °C and the outer surface is at 25 °C, determine the thickness of the insulation that needs to be provided. Do you think this thickness is realistic?

Ex 2.2 Thermal conductivity of a certain low carbon steel varies with temperature according to the formula $k(T) = 55 - 0.03T$, where T is in °C, k is in W/m °C and the formula is valid in the temperature range $0 \leq T \leq 400$ °C. In a certain application, a uniform area pin of diameter 10 mm and length 100 mm of this material is used to connect two structural members. The temperatures of these members are, respectively, 200 °C and 120 °C. Determine the heat transferred in W through the pin by conduction. What is the maximum temperature gradient in °C/m and where does it occur? Also determine the location and the magnitude of the largest departure of temperature from the linear profile.

Ex 2.3 A certain material has thermal conductivity varying linearly with temperature. The thermal conductivity of this material is specified as 15.5 W/m°C at 25 °C and 19.5 W/m°C at 60 °C. A slab of this material has a thickness of 0.15 m and its two surfaces are at 53 °C and 25 °C respectively. What is the heat transfer per unit area across the slab?

Ex 2.4 A brick partition 0.2 m thick and of thermal conductivity 0.72 W/m°C is covered with a 0.05 m thick fiberglass insulation of thermal conductivity 0.04 W/m°C as shown in Fig. 2.24. The exposed brick surface is maintained at 25 °C while the exposed surface of the insulation is maintained at 5 °C. Determine the heat transfer per unit area and the temperature at the brick fiberglass interface.

Ex 2.5 A composite wall is made of three layers of materials as shown in Fig. 2.25. The accompanying numbers indicate the data. Determine the heat transfer per unit area in this case. Also determine, using the electrical analogy, the temperatures at both the interfaces. Specify the conditions that are satisfied at an interface between the plaster and the brick.

Ex 2.6 A slab of thickness 0.05 m and of thermal conductivity 10 W/m°C is generating heat at a uniform rate of 3×10^5 W/m^3. The slab is sandwiched between two slabs (referred to as clad material) of equal thickness of 0.01 m and of thermal conductivity of 45 W/m°C. The external

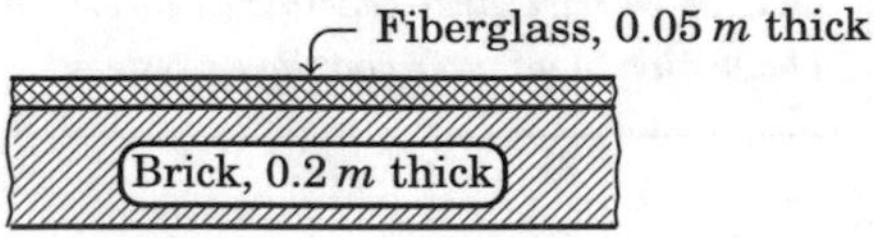

Fig. 2.24 Heat transfer across composite wall of Exercise 2.4

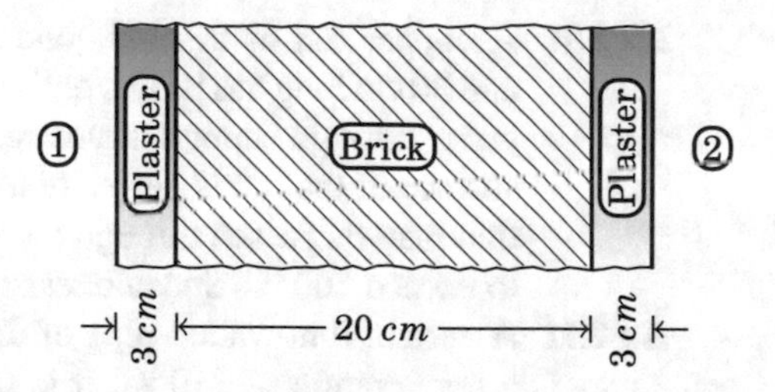

Fig. 2.25 Steady heat transfer across layered wall of Exercise 2.5

surfaces of the clad layers are exposed to an ambient at 30 °C subject to a heat transfer coefficient of 67 W/m^2°C. Determine the surface temperature of the material exposed to the convective environment as well as the maximum temperature. Assume that there is no contact resistance at the interface between the heat-generating slab and the surrounding clad layers.

If there is a thermal contact resistance between the heat-generating slab and the clad material what will happen to the clad surface temperature exposed to the ambient? Justify your answer.

If the heat transfer coefficient specified previously is increased what will happen to the temperature? Justify your answer.

Ex 2.7 A slab of material of constant thermal conductivity k and thickness L is generating heat at a uniform volumetric heat generation rate G. The surfaces at $x = 0$ and $x = L$ are in contact with a fluid at temperature T_∞ and are subject to convection to the fluid via a heat transfer coefficient h. Obtain an expression for the maximum temperature in the slab. Where does this maximum occur? What is the temperature at any one surface of the slab?

Ex 2.8 Consider a slab of material of thickness L and thermal conductivity k. Heat generation in the slab varies with x according to the relation $G(x) = G_0\left(1 - \frac{x}{L}\right)$ where G_0 is a constant. Both faces of the slab are maintained at a temperature equal to T_0. Derive an expression for the temperature T at any location inside the slab. Also, determine the location at which the temperature has the maximum value and the maximum value of the temperature. You are expected to present the analysis in non-dimensional form using suitably chosen non-dimensional variables and parameters.

Ex 2.9 A bare 2.5 cm diameter pipe has a surface temperature of 150 °C and is placed in air at 30 °C. The convection heat transfer coefficient between the surface and air is 6 W/m^2°C. It is desired to reduce the heat loss to 60% of its present value by the addition of a layer of insulation with $k = 0.1$ W/m°C. Assuming that the pipe surface temperature and the heat transfer coefficient remain the same determine the required insulation thickness.

Ex 2.10 A copper rod of thermal conductivity 380 W/m°C, 0.3 cm diameter and 30 cm long has its two ends maintained at 15 °C by circulating cold water through support structures in contact with the ends. The lateral surface of the rod is perfectly insulated. Determine the largest current that may be passed through the rod if the temperature in the rod is not to exceed 100 °C. Specific resistance of copper is $1.73 \times 10^{-6} \Omega \cdot \text{cm}$.

Ex 2.11 A certain thin-walled pipe of diameter 25 mm is conveying steam at a temperature of 110 °C. It is to be insulated such that the heat loss per meter of the pipe is at most equal to 15 W. The insulation material available has a thermal conductivity of 0.04 W/m°C. The heat transfer from the outside surface of the insulation is known to take place with a heat transfer coefficient of 15 W/m$^{2\circ}$C to room air at 35 °C. What is the insulation thickness? On a certain day, the room power fails and the heat transfer coefficient reduces to 5 W/m$^{2\circ}$C5. What will be the heat loss per meter on such a day? Comment on the result.

Ex 2.12 A cylinder of radioactive material of radius 50 mm, and thermal conductivity 42 W/m°C generates heat at the rate of 10^6 W/m^2. A 3 mm thick clad material of thermal conductivity equal to 1.5 W/m°C protects it. The composite cylinder is exposed to an environment that provides a heat transfer coefficient of 450 W/m$^{2\circ}$C to an ambient at 30 °C. Determine the maximum temperature, the outside surface temperature of the clad, and the value of heat flux at the clad outer surface.

Ex 2.13 A small diameter copper wire of diameter 3 mm is found to be at a temperature of 50 °C when a certain current is passed through it and it is exposed to ambient air at 20 °C. Assuming that copper has a resistivity of $1.73 \times 10^{-6} \Omega \cdot \text{cm}$, determine the current if the heat transfer coefficient to ambient air has been estimated to be 6 W/m$^{2\circ}$C. It was decided to insulate the wire by a 0.6 mm plastic insulation layer of thermal conductivity equal to 0.45 W/m°C. Assuming that the heat transfer coefficient remains unchanged and the current is held at the same value, determine the wire temperature. Would you advise an increase or decrease of insulation thickness if the goal is to increase heat transfer?

Ex 2.14 A material of thermal conductivity 200 W/m°C is in the form of a very long cylinder of diameter $D = 5$ mm. An air stream subjects the surface of the cylinder to cooling with a heat transfer coefficient of $h = 6$ W/m$^{2\circ}$C. Can we consider the cylinder to be isothermal when it is subject to uniform internal heat generation of $G = 10^5$ W/m^3? Whatever your answer is to the above, calculate the maximum temperature inside the cylinder.

Ex 2.15 An electrical heating element is shrunk in a hollow cylinder of carbon with a thermal conductivity of $k = 1.6$ W/m°C as shown in Fig. 2.26. The outer surface of carbon is in contact with air at 20 °C subject to a convection heat transfer coefficient of $h = 67$ W/m$^{2\circ}$C. Determine the maximum allowable heat generation rate per meter length if the maximum temperature of carbon is not to exceed 180 °C. The heating

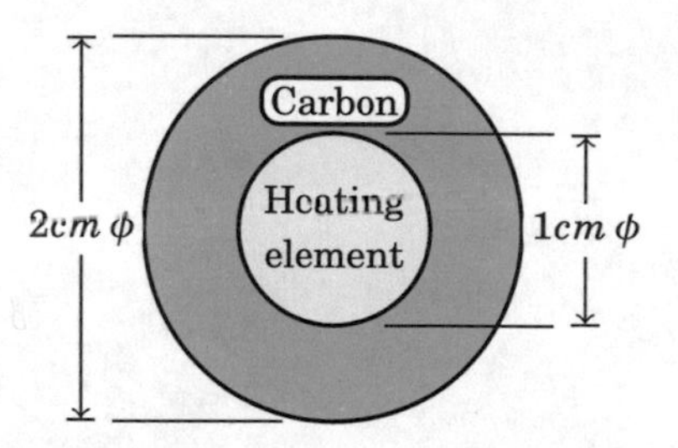

Fig. 2.26 Heat transfer in a composite cylinder: Exercise 2.15

element itself may be assumed to be isothermal. Refer to figure for other details

Ex 2.16 An insulating material of unknown thermal conductivity is sandwiched between two spherical shells of steel. The inner shell has an one dimension of 100 mm and a wall thickness of 5 mm. The outer shell has an one dimension of 150 mm and a wall thickness of 2 mm. In a certain experiment, it was found that the inner surface of the inner shell was at 110 °C when the outer surface of the outer shell was at 66 °C. The ambient fluid is air at 30 °C and the heat transfer coefficient from the spherical surface to the ambient has been estimated to be 6 W/m$^{2\circ}$C. What is the thermal conductivity of the insulating material? Take the thermal conductivity of steel as 45 W/m°C. The experiment was repeated another day when the ambient temperature had decreased to 14 °C. What will be the temperature of the outer surface of the outer shell this time if the inner surface of the inner shell indicated the same temperature?

Ex 2.17 A spherical shell of inner radius r_1, outer radius r_2 of a material of thermal conductivity k has the inner and outer boundaries at temperatures T_1 and T_2, respectively. Derive an expression for the heat transfer across the shell. Represent the solution in a non-dimensional form, after defining suitable non-dimensional variables.
If the outer surface of the shell is losing heat to an ambient at T_∞ subject to a convective heat transfer coefficient h, what is the heat transfer across the shell? Work out the appropriate electrical analogy for this problem?

Chapter 3
Unsteady Heat Transfer in Lumped Systems

When internal conduction resistance is much smaller than the thermal resistance for heat transfer from its surface, a solid body may be assumed to be at the same temperature throughout. The solid body is treatable as a lumped system. Similarly, a volume of liquid/gas may be considered to be a lumped system if it is well stirred. Cooling/heating of a lumped system is governed by ordinary differential equations with time t as the independent variable. Both first-order and second-order thermal systems are considered in this chapter.

3.1 Preliminaries

In Chap. 2 we have considered, in a simple way, the interaction of conduction within a solid and convection at its boundaries. The interaction manifests through the non-dimensional parameter, the Biot number. It was indicated that the magnitude of the Biot number determines the variation of temperature within the conducting medium. In the limit of $Bi \rightarrow 0$ (in practice, $Bi < 0.1$), the internal conductive resistance becomes very small in comparison with the film resistance at the surface. In this limit, the temperature gradients within the solid tend to zero. For example, in a solid cylinder with internal heat generation, the temperature difference between the center and the

S. P. Venkateshan, *Heat Transfer*,
https://doi.org/10.1007/978-3-030-58338-5_3

surface is less than 5% of the difference between the center and the temperature of the ambient medium, when $Bi < 0.1$. A lumped system approximation is a reasonable model, in such a case.

In the lumped system formulation, the temperature within the conducting medium is approximately the same throughout and hence a single temperature characterizes the system. Transient in a lumped system leads to a one-dimensional problem with time as the independent variable.

Examples of such systems are

- A thin shell cooling in air; L_{ch} = shell thickness (characteristic length scale).
- Thermometer bulb immersed in a hot/cold fluid.
- Well-stirred temperature-controlled bath or a well-stirred tank of hot water.

3.2 Governing Equation and the General Solution

Assumptions made in analyzing such a problem are explained with reference to Fig. 3.1. Note that the ambient temperature T_∞ may be either a constant or a specified function of time. For the following simplified analysis, the heat transfer coefficient h is assumed to be a constant.

If the temperature of the lumped system is *different* from the temperature of the ambient medium, heat transfer will take place between the system and the ambient. The temperature of the system will change with time and it is desired to determine this variation.

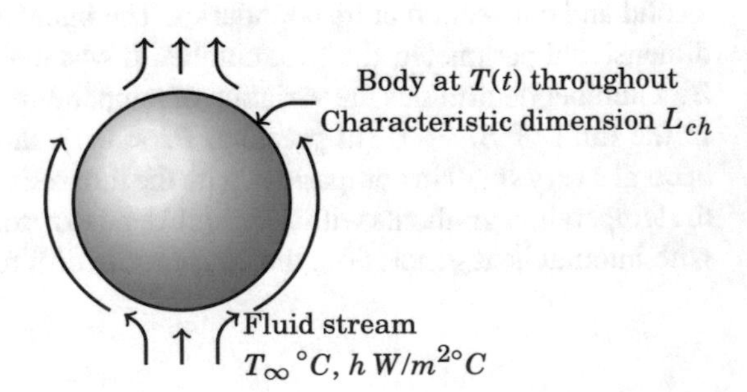

Fig. 3.1 Transient in a lumped system

3.2.1 Governing Equation

Consider the general case in which heat is generated within the system at a uniform volumetric rate of G, W/m^3. Let the following constant properties characterize the system:

Density of the conducting medium:	ρ, kg/m^3
Volume of the system:	V, m^3
Surface area exposed to the ambient:	S, m^2
Heat capacity of the conducting medium:	c, J/kg°C

Assume that the lumped system temperature is greater than the ambient temperature. Consider a time duration of Δt at t. In this time interval, the heat generated within the system is $Q_g = GV\Delta t$. The heat transferred from the surface to the ambient fluid is $Q_c = hS(T - T_\infty)\Delta t$. Since the temperature of the system changes in this interval, the change in the internal energy of the system is $Q_i = \rho Vc\frac{dT}{dt}\Delta t$. Energy balance applied to the lumped system requires that $Q_i = Q_g - Q_c$ or

$$\rho Vc\frac{dT}{dt} = GV - hS(T - T_\infty) \tag{3.1}$$

where we have dropped the time interval Δt that is common to all the terms. Equation 3.1 may be recast in the form

$$\boxed{\frac{dT}{dt} = \frac{G}{\rho c} - \frac{hS}{\rho Vc}(T - T_\infty)} \tag{3.2}$$

This is a first-order ordinary differential equation that can support an initial condition of the form $T = T_0$ at $t = 0$. Since the governing differential equation is of first order, the lumped system is referred to as a first-order system. Consider now the quantity $\frac{hS}{\rho Vc}$. This quantity has the units of reciprocal time (s^{-1}) and is denoted by the symbol $\frac{1}{\tau}$, where τ a characteristic time that governs the transient is referred to as the ***first-order time constant*** of the lumped system.

3.2.2 Electrical Analogy

A first-order system is analogous to an electrical circuit containing a resistance and a capacitance as shown in Fig. 3.2. The resistance R in the electrical circuit is analogous to the thermal resistance $\frac{1}{hS}$ while the capacitance C is analogous to the thermal

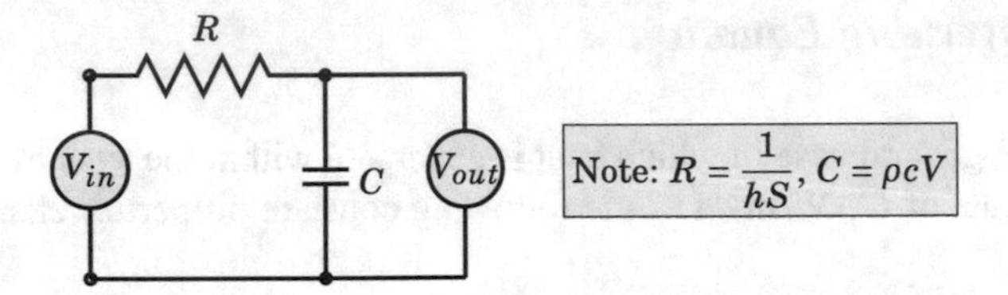

Fig. 3.2 Electrical analog of a first-order thermal system

capacity of the system given by ρVc or Mc where $M = \rho V$ is the mass of the lumped system. The input voltage to the electrical circuit is analogous to the impressed temperature while the output voltage is analogous to the system temperature. Note that the system temperature appears as output voltage across the capacitance in the electrical circuit. It is well known from electrical circuit theory that the time constant of the R-C network is $\tau = RC$. It is seen that the product of thermal resistance and the thermal capacity gives the time constant of the thermal system, i.e., $\tau = \frac{1}{hS} \times \rho Vc$.

3.2.3 *Characteristic Length Scale*

It is interesting to interpret the time constant in a different way. We may write it as

$$\boxed{\tau = \frac{\rho c}{h} \times \frac{V}{S} = \frac{\rho c}{h} \times L_{ch}} \tag{3.3}$$

where $L_{ch} = \frac{V}{S}$ is a characteristic geometric length scale while the other factor involves only thermal parameters. Thus the time constant of a first-order system is determined partly by the thermal parameters and partly by the characteristic length scale. Table 3.1 gives useful formulae for calculating L_{ch} in cases of practical importance.

Table 3.1 Characteristic length scale for a first-order system

Geometry	L_{ch}
Large slab of thickness δ losing heat from both sides	$\delta/2$
Long solid cylinder of diameter D	$D/4$
Solid sphere of diameter D	$D/6$
Long tube of small wall thickness $\delta << D$ losing heat from outside	δ
Spherical shell of small wall thickness $\delta << D$ losing heat from outside	δ

A Note on Time Constant

It is clear from the above that the time constant of a first-order thermal system is not a property of the system. It depends on parameters that relate to the system as well as the parameters that define the interaction between the system and the surrounding medium, specifically the thermal resistance to heat transfer between the system and the surroundings. Time constant also depends on the thermal mass of the system. The thermal mass depends on the physical size as well as the thermal properties of the material of the system. The way the time constant is written, as a product of thermal and geometrical factors, it is possible to manipulate the time constant of a system. Thermal mass reduction is one possibility. The other possibility is the reduction of thermal resistance. This may be achieved by increasing the interface area between the system and the medium. In general, this means a reduction in the characteristic dimension L_{ch} of the system. If time constant data is to be specified for a particular system it is necessary also to specify the thermal environment in which the time constant has been determined. For example, time constant is specified, for a particular temperature sensor (which may be treated as a first-order system) as follows:

Time constant = 0.14 s:	in water at room temperature
Time constant = 2.2 s:	in still air at 25 °C

3.2.4 General Solution

Equation 3.2 may be rewritten in the form

$$\frac{dT}{dt} + \frac{T}{\tau} = \frac{G}{\rho c} + \frac{T_\infty}{\tau} \tag{3.4}$$

This is a non-homogeneous equation that is amenable to solution by the use of an integrating factor given by $e^{\frac{t}{\tau}}$. On multiplication by the integrating factor, the left-hand side of Eq. 3.4 becomes a total differential given by $\frac{d\left(Te^{\frac{t}{\tau}}\right)}{dt}$. Hence we can integrate the governing equation with respect to time to get the general solution

$$Te^{\frac{t}{\tau}} = \frac{1}{\tau}\int_0^t \left(\frac{G\tau}{\rho c} + T_\infty\right) e^{\frac{t}{\tau}}\, dt + A \quad \text{or} \quad T = e^{-\frac{t}{\tau}}\left[\frac{1}{\tau}\int_0^t \left(\frac{G\tau}{\rho c} + T_\infty\right) e^{\frac{t}{\tau}}\, dt + A\right] \tag{3.5}$$

The constant of integration A is easily obtained by the use of the initial condition as $A = T_0$. Thus the general solution to the response of the first-order system is

$$\boxed{T = e^{-\frac{t}{\tau}}\left[\frac{1}{\tau}\int_0^t\left(\frac{G\tau}{\rho c}+T_\infty\right)e^{\frac{t}{\tau}}dt + T_0\right]} \quad (3.6)$$

The general solution given by Eq. 3.6 shows interesting features. The quantity $e^{-\frac{t}{\tau}}$ tends to zero as $t \to \infty$. Generally $t \to \infty$ may be interpreted as $t > 5\tau$ for which $e^{-\frac{t}{\tau}} < 0.01$. Terms on the right-hand side of Eq. 3.6 that contain the factor $e^{-\frac{t}{\tau}}$ become unimportant for large time, interpreted as mentioned above. Terms on the right-hand side that survive for large time constitute what is called the ***steady-state solution***. In practical problems, the interest may be in the steady state solution when the transients have died down.

3.2.5 Response of a First-Order System in Particular Cases

In this section, we are going to look at several interesting cases that are also of practical importance. The general idea is to study the response of a first-order system subject to specific inputs.

Response of a First-Order System to Step Input

Let a first-order system that is initially at temperature T_0 be exposed to an ambient at a constant (time invariant) temperature T_∞. Also, let the internal heat generation rate be $G = 0$. The problem is said to involve *step input* since the system, initially at a temperature different from the ambient temperature, is exposed to the ambient for $t > 0$. In practice, one may achieve this boundary condition, by quickly moving the system from one constant temperature environment with which it is in thermal equilibrium to another environment at a different temperature.

We may use the general solution given by Eq. 3.6, put $G = 0$ and take T_∞ outside the integral sign to get the solution to the present case. We note then that the integral on the right-hand side is

$$\int_0^t e^{\frac{t}{\tau}}dt = \left[\tau e^{\frac{t}{\tau}}\right]_0^t = \tau\left(e^{\frac{t}{\tau}} - 1\right)$$

Substitute this in Eq. 3.6 to get

$$T = e^{-\frac{t}{\tau}}\left[T_0 + \frac{T_\infty}{\tau}\tau(e^{\frac{t}{\tau}} - 1)\right] = (T_0 - T_\infty)e^{-\frac{t}{\tau}} + T_\infty$$

On rearrangement the non-dimensional response given by

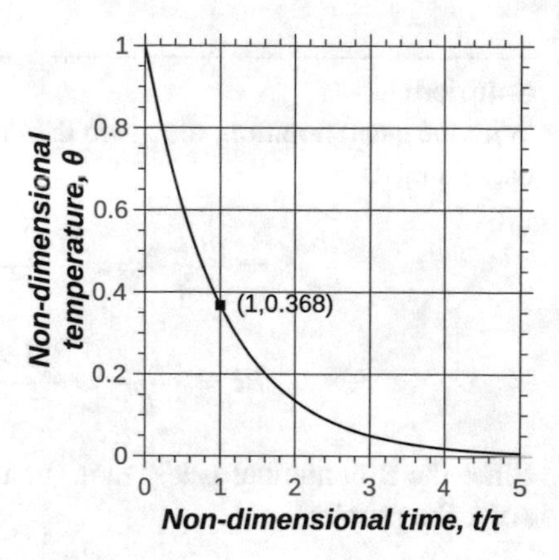

Fig. 3.3 Response of a first-order system to step input

$$\boxed{\underbrace{\frac{T - T_\infty}{T_0 - T_\infty}}_{\theta} = e^{-\frac{t}{\tau}}} \tag{3.7}$$

The response given by Eq. 3.7 is shown plotted in Fig. 3.3.

The response is an exponential decay of the initial temperature difference between the system and the ambient. At the end of one time constant, it is about 37% of the initial value. As $t \to \infty$ the temperature difference between the system and the ambient tends to zero, i.e., the system comes to equilibrium with the ambient. This represents the final steady state of the system and hence the steady-state response of the system is one in which the system is at T_∞, the ambient temperature.

Example 3.1

A thin shell of aluminum of diameter 5 mm and thickness 0.3 mm drops off a conveyor vertically down to the ground. The temperature of the shell is initially 75 °C. While it is falling to the ground that is some 15 m below, the shell cools by losing heat to the ambient air at 20 °C via a constant average heat transfer coefficient of 90 W/m^2°C. Determine the temperature of the shell as it hits the ground.

Solution:
With the usual notation, the given data is given in the table. The Biot number is then calculated as

$$\tau = \frac{\rho c L_{ch}}{h} = \frac{2700 \times 896 \times 0.0003}{90} = 8.064\ \text{s}$$

$$Bi = \frac{h L_{ch}}{k} = \frac{90 \times 0.0003}{236} = 1.14 \times 10^{-4}$$

Since the Biot number is less than 0.1 it is reasonable to use lumped formulation to solve the problem.

The first-order time constant may now be calculated using Eq. 3.3, to describe the temperature transient in the aluminum shell, as it falls to the ground.

Characteristic length(see Table 3.1): $L_{ch} = 0.3$ mm $= 0.0003$ m
Thermal conductivity of aluminum: $k = 236$ W/m°C
Density of aluminum: $\rho = 2700$ kg/m^3
Specific heat of aluminum: $c = 896$ J/kg°C
Heat transfer coefficient: $h = 90$ W/m^2 °C
Height of fall of shell: $s = 15$ m
Acceleration due to gravity: $g = 9.8$ m/s^2

Note: Properties of aluminum are taken from Table I.4

Time taken by the shell to fall to the ground is now calculated. Motion of the shell is under gravity and it is assumed that the shell starts to fall with zero initial velocity. The time t of travel before the shell hits the ground is

$$t = \sqrt{\frac{2s}{g}} = \sqrt{\frac{2 \times 15}{9.8}} = 1.75\ \text{s}$$

The temperature of the shell at the end of this time is determined using the response of a first-order system to a step input. With initial temperature of shell of $T_0 = 75\,°\text{C}$ and ambient temperature of $T_\infty = 20\,°\text{C}$, temperature of the shell as it hits the ground is obtained using Eq. 3.7 as

$$T = T_\infty + (T_0 - T_\infty)e^{-\frac{t}{\tau}} = 20 + (75 - 20)e^{-\frac{1.75}{8.064}} = 64.3\,°\text{C}$$

Example 3.2

A copper sheet of dimensions 50 × 100 × 2 mm is exposed on both sides to an ambient fluid at 25 °C with an average heat transfer coefficient of 7 W/m^2°C. Is it proper to consider the copper sheet as a lumped system for transient analysis? If the answer is yes, determine the time constant of the first-order system. If the initial temperature of the plate is 100 °C how long does one have to wait for the temperature to become 50 °C?

Solution:
Properties of copper are taken from Table I.4. They are

Density: $\rho = 8950$ kg/m^3
Specific heat: $c = 385$ J/kg°C
Thermal conductivity: $k = 386$ W/m°C

The characteristic length scale in the problem is semi plate thickness (see Table 3.1) given by $L_{ch} = 0.001m$. The Biot number is calculated as

$$Bi = \frac{hL_{ch}}{k} = \frac{7 \times 0.001}{386} = 1.813 \times 10^{-5}$$

Since the Biot number is smaller than 0.1 it is indeed reasonable to treat the cooling process as a lumped first-order process. The time constant for the cooling process is obtained as

$$\tau = \frac{\rho c L_{ch}}{h} = \frac{8950 \times 385 \times 0.001}{7} = 492.3 \text{ s}$$

Other appropriate data specified in the problem are given below:

Initial temperature of copper plate: $T_0 = 100$ °C
Ambient temperature: $T_\infty = 25$ °C
Final temperature of copper plate: $T_f = 50$ °C

We solve for t using Eq. 3.7 to get

$$t = -\tau \times \ln\left(\frac{T_f - T_\infty}{T_0 - T_\infty}\right) = -492.3 \times \ln\left(\frac{50 - 25}{100 - 25}\right) = 540.8\,s \approx 9 \text{ min}$$

Response of a First-Order System to a Periodic Input
If T_∞ is not a constant, but varies with time, interesting features result. For example, we may be interested in measuring a time-varying temperature of a fluid using a thermometer (system). A typical example is the variation of temperature inside

an internal combustion engine cylinder. The variation is certainly periodic, with the period dependent on the speed of the engine. The temperature variation is a complex wave with the shape dependent on the processes that take place within the cylinder. The analysis is made for a single sinusoidal temperature variation since *any* periodic function may be built up as a sum of its Fourier components (Fourier series representation will be considered in detail in Chap. 5). In case of an arbitrary periodic input, the output response will be a weighted sum of the response of the system to each of the Fourier components, since the system is governed by a linear differential equation. Let us assume that T_∞ is a periodic function of time, given by $T_\infty = T_a \cos(\omega t)$, where T_a is the amplitude and ω the circular frequency in rad/s ($\omega = 2\pi f$, where f is the frequency in Hz). The circular frequency of the Fourier components vary as integer multiples of a fundamental frequency. Assuming $G = 0$, the general solution has to be obtained by writing Eq. 3.6 as

$$T = e^{-\frac{t}{\tau}} \left[\frac{1}{\tau} \int_0^t T_a \cos(\omega t) e^{\frac{t}{\tau}} dt + T_0 \right] \tag{3.8}$$

The integral appearing on the right-hand side of Eq. 3.8 may be obtained by the repeated use of integration by parts.

$$\begin{aligned} I &= \int_0^t e^{\frac{t}{\tau}} \cos(\omega t) dt = \left(\tau e^{\frac{t}{\tau}} \cos(\omega t) \right)\Big|_0^t + \int_0^t \omega\tau e^{\frac{t}{\tau}} \sin(\omega t) dt \\ &= \tau \left[e^{\frac{t}{\tau}} \cos(\omega t) - 1 \right] + \omega\tau \left\{ \left[\tau e^{\frac{t}{\tau}} \sin(\omega t) \right] \Big|_0^t - \omega\tau \int_0^t e^{\frac{t}{\tau}} \cos(\omega t) dt \right\} \\ &= \tau \left[e^{\frac{t}{\tau}} \cos(\omega t) - 1 \right] + \omega\tau^2 e^{\frac{t}{\tau}} \sin(\omega t) - \omega^2\tau^2 I \end{aligned}$$

We see that the integral I has repeated itself after two integrations by parts. The above may be solved for I to get

$$I = \tau e^{\frac{t}{\tau}} \frac{\cos(\omega t) + \omega\tau \sin(\omega t)}{1 + \omega^2\tau^2} - \frac{\tau}{1 + \omega^2\tau^2} \tag{3.9}$$

The integral contributes the following to the solution.

$$\frac{e^{-\frac{t}{\tau}}}{\tau} I = \frac{\cos(\omega t) + \omega\tau \sin(\omega t)}{1 + \omega^2\tau^2} - \frac{e^{-\frac{t}{\tau}}}{1 + \omega^2\tau^2} \tag{3.10}$$

The first term on the right-hand side of Eq. 3.10 may be rearranged as

$$\frac{\cos(\omega t) + \omega\tau \sin(\omega t)}{1 + \omega^2\tau^2} = \frac{\frac{1}{\sqrt{1 + \omega^2\tau^2}} \cos(\omega t) + \frac{\omega\tau}{\sqrt{1 + \omega^2\tau^2}} \sin(\omega t)}{\sqrt{1 + \omega^2\tau^2}} \tag{3.11}$$

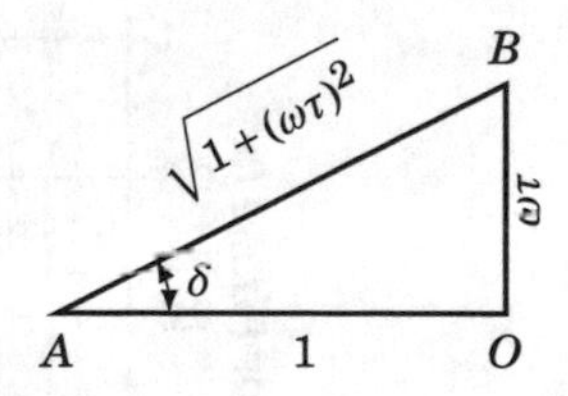

Fig. 3.4 Phase of a first-order system subject to periodic input

Refer now to Fig. 3.4. The ratios appearing as coefficients of 'cos' and 'sin' terms in the numerator of Eq. 3.11 may be interpreted as follows:

$$\frac{1}{\sqrt{1+\omega^2\tau^2}} = \cos\delta, \quad \frac{\omega\tau}{\sqrt{1+\omega^2\tau^2}} = \sin\delta$$

With these and familiar trigonometric identity $(\cos(A-B) = \cos(A)\cos(B) + \sin(A)\sin(B))$ Eq. 3.11 may be written as

$$\frac{\cos(\omega t) + \omega\tau\sin(\omega t)}{1+\omega^2\tau^2} = \frac{\cos(\omega t - \delta)}{\sqrt{1+\omega^2\tau^2}} \tag{3.12}$$

Substituting these in Eq. 3.8 we finally get the solution as

$$T = \underbrace{\left[T_0 - \frac{T_a}{(1+\omega^2\tau^2)}\right] e^{-\frac{t}{\tau}}}_{\text{Transient response}} + \underbrace{\frac{T_a}{\sqrt{1+\omega^2\tau^2}}\cos(\omega t - \delta)}_{\text{Steady state response}} \tag{3.13}$$

where $\delta = \tan^{-1}(\omega\tau)$.

Steady-State Response

A reference to Eq. 3.13 shows that the first term on the right-hand side decays to zero as $t \to \infty$ because of the exponentially decreasing factor. This term represents the transient set up due to the fact that there is an initial temperature difference between the mean of the imposed periodic input (in the present case, the mean is zero). However for $t \to \infty$, the transient dies down and the steady-state response persists, given by the last part on the right-hand side of Eq. 3.13. We see that the steady-state response is periodic with the same period or frequency as the input, but with a phase lag δ. The amplitude is also smaller by the factor $\frac{1}{\sqrt{1+\omega^2\tau^2}}$. Figure 3.5 shows the variation of the amplitude and phase with $\omega\tau$ product. It is seen that the amplitude reduction is marginal until about $\omega\tau = 0.1$. Beyond this value, the amplitude reduces rapidly. Similarly the phase lag is marginal up to $\omega\tau = 0.1$ and increases rapidly beyond that and reaches an angle of 90° or $\frac{\pi}{2}$ *rad* as $\omega\tau \to \infty$.

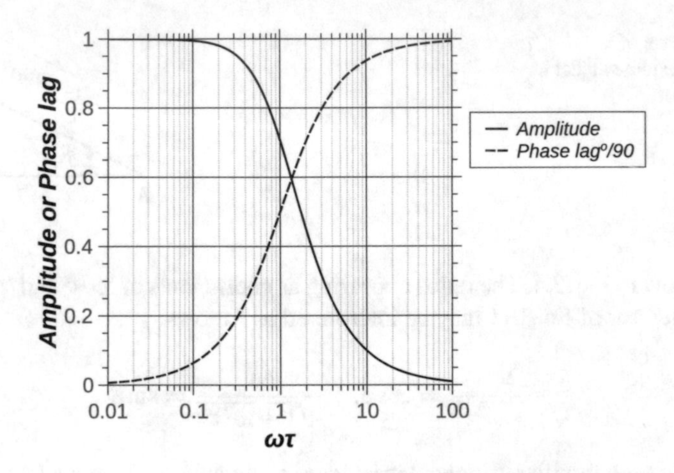

Fig. 3.5 Amplitude and phase response of a first-order system subject to periodic input

Example 3.3

A first-order system is characterized by a time constant of $\tau = 0.1$ s. The system is initially at a temperature of $T_0 = 30\,°\text{C}$. It is subject to a periodic exchange of heat at a circular frequency of 33 rad/s with an ambient that is given by $T_\infty = 50 + 20\cos(33t)$ where t is in s. Determine the time lag between the input and the system response when the steady state has been reached. Make a plot showing both the input and the response in the same plot.

Solution:

The problem will make use of the solution obtained above for a first-order system subject to periodic input.

1. The pertinent parameters in this problem are

Time constant: $\tau = 0.1$ s
Input wave circular frequency: $\omega = 33$ rad/s
Initial temperature of the system: $T_0 = 30\,°\text{C}$
Mean temperature of the input: $T_m = 50\,°\text{C}$
Amplitude of the input disturbance: $T_a = 20\,°\text{C}$

2. The phase lag is calculated referring to Fig. 3.4 as

$$\delta = \tan^{-1}(\omega\tau) = \tan^{-1}(33 \times 0.1) = \tan^{-1} 3.3 = 1.277\,\text{rad} \quad \text{or} \quad 73.14°$$

3. The time lag t_L is defined through the relation $\delta = \omega t_L$ and hence we have

$$t_L = \frac{\delta}{\omega} = \frac{1.277}{33} = 0.039\,\text{s}$$

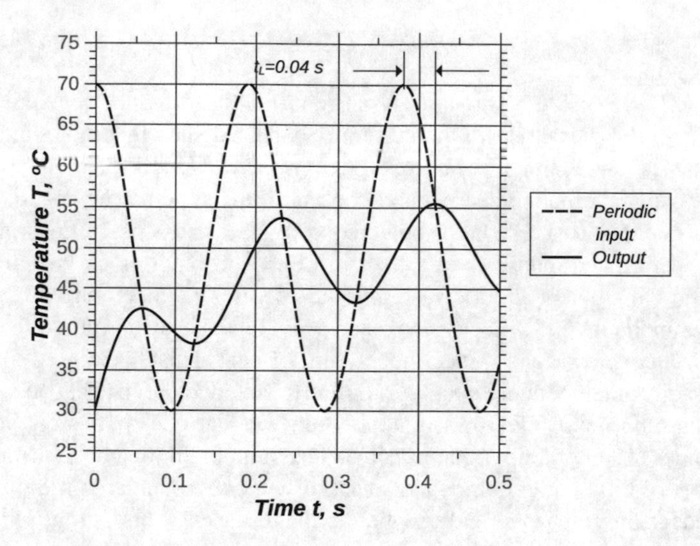

Fig. 3.6 Input and response for Example 3.3

4. The amplitude reduction factor ARF is given by

$$ARF = \frac{1}{\sqrt{1+\omega^2\tau^2}} = \frac{1}{\sqrt{1+33^2 \times 0.1^2}} = 0.290$$

5. Having completed the calculations, we make a plot as required in the statement of the problem. Since the mean temperature of the input is not zero, we need to combine (or superpose) the solutions obtained for the case of step input (Eq. 3.7) and the solution obtained for periodic input (Eq. 3.13). The solution turns out to be

$$T = T_m + \left[T_0 - T_m - \frac{T_a}{1+\omega^2\tau^2}\right] e^{-\frac{t}{\tau}} + \frac{T_a}{\sqrt{1+\omega^2\tau^2}} \cos(\omega t - \delta)$$

$$\text{or} \quad T = 50 - 21.57 e^{-\frac{t}{0.1}} + 5.8\cos(33t - 1.276)\ ^\circ\text{C}$$

We have made a plot of this in Fig. 3.6. Time lag $t_L \approx 0.04$ s is also shown in the figure.

Observations

Refer to Fig. 3.6. During the initial transient (that lasts about 5 time constants or 0.5 s in the present case) the system gets heated such that its temperature approaches the mean temperature. However, the oscillatory part asserts itself by imposing alternating heating and cooling of the system. The heat capacity effects manifest as a time lag, when the steady state is reached. The amplitude of the periodic response is less than the imposed input since heat transfer between the ambient and the system requires a finite temperature difference. This temperature difference is also determined by the heat capacity effects and the convective resistance since the amount of heat transfer required to bring about a given temperature change is directly related to the heat capacity.

When the input waveform is not a pure wave at one frequency we may decompose it in to Fourier components. The output will then be a linear combination of the Fourier components each of which will undergo a frequency-dependent phase change and amplitude change. Hence the output will be a modified wave with a change in shape.

As an example consider the input to be $T_a = 5\cos(33t) + 2\cos(66t)$ where the fundamental and the first harmonic alone are present in the input. Depending on the time constant of the system, the output will be modified since the phases and amplitudes in the output are dependent on τ and the individual frequencies. In the present case, these are as tabulated below for two systems with time constant of $\tau = 0.1$ s and $\tau = 0.01$ s, respectively.

τ, s	0.1	0.1	0.01	0.01
ω, rad/s:	33	66	33	66
δ, rad	1.277	1.420	0.319	0.583
ARF	0.290	0.150	0.950	0.835

The steady-state outputs of the two systems are obtained by the use of respective phases and amplitudes from the table. A plot is made and is shown in Fig. 3.7.

In the case of system with $\tau = 0.1$ s, the output does not resolve the minor peak present in the input. Both amplitude and phases change such that there is a significant reduction in amplitude as well as time delay between the input and the output. It is also clear from the figure that $\tau = 0.01$ s itself is not small enough for a faithful reproduction of the input wave. In fact, a value of $\tau = 0.003$ s or smaller will be required to give a fairly faithful response.

Response to a Ramp Input

A first-order system is subject to an ambient whose temperature varies linearly with respect to time given by $T_\infty = T_m + Rt$ where T_m is the initial temperature of the ambient and R is the constant rate of temperature increase. This type of input is met with in "thermal analysis systems" (instruments used for material characterization) where linear temperature rise with respect to time is quite common. The solution may easily be worked out by substituting the linearly varying temperature input in

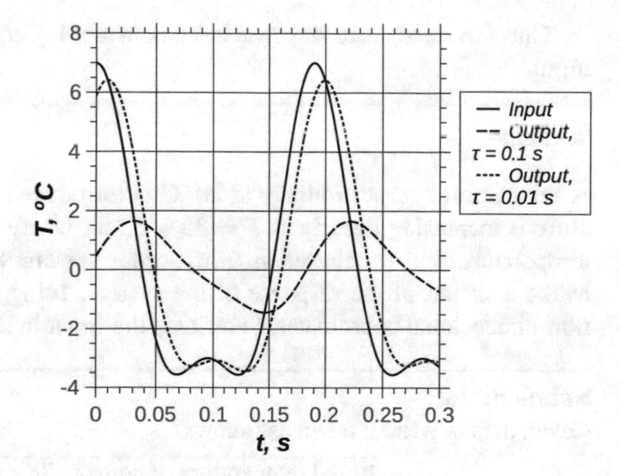

Fig. 3.7 Output of two systems with input containing two frequency components

the general solution given by Eq. 3.6. The constant part T_m leads to a response that is like the step input given by Eq. 3.7. The contribution to the general solution because of this part is written as

$$T_m + (T_0 - T_m)e^{-\frac{t}{\tau}} \tag{3.14}$$

The ramp part requires an integral that is obtained as follows, using integration by parts.

$$\begin{aligned} I &= \int_0^t t e^{\frac{t}{\tau}} dt = \left(t\tau e^{\frac{t}{\tau}} \right)\Big|_0^t - \tau \int_0^t e^{\frac{t}{\tau}} dt \\ &= t\tau e^{\frac{t}{\tau}} - \tau^2 e^{\frac{t}{\tau}} \Big|_0^t = t\tau e^{\frac{t}{\tau}} - \tau^2 \left[e^{\frac{t}{\tau}} - 1 \right] \end{aligned} \tag{3.15}$$

Introducing Eqs. 3.14 and 3.15 in to the general solution 3.6, and after some minor manipulations we get the response of the first-order system to ramp input as

$$\boxed{T = T_m + (T_0 - T_m + R\tau)e^{-\frac{t}{\tau}} + R(t - \tau)} \tag{3.16}$$

The steady-state response is easily obtained by dropping the exponential term to get

$$\boxed{T - (T_m + Rt) = T - T_\infty = -R\tau} \tag{3.17}$$

Thus the steady sate response has a constant lag equal to $R\tau$ with respect to the input.

Example 3.4

A temperature sensor initially at 20 °C is introduced into an oven whose temperature is increasing linearly as $T = 35 + 0.15\,t$ where T is in °C and t is in s. The temperature sensor behaves as a first-order system with a time constant of 10 s. Make a sketch of the response of the system, using both dimensional as well as non-dimensional coordinates. Show also the input in the dimensional plot.

Solution:
Given data is written down as follows:

Initial temperature of sensor:	$T_0 = 20\,°C$
Initial temperature of oven:	$T_m = 35\,°C$
Time constant of the sensor:	$\tau = 10$ s
Ramp rate:	$R = 0.15\,°C/s$

Introducing these numbers in to Eq. 3.16 we get the ramp response as

$$\begin{aligned} T &= 35 + (20 - 35 + 0.15 \times 10)e^{-\frac{t}{10}} + 0.15t - 0.15 \times 10 \\ &= 33.5 - 13.5e^{-\frac{t}{10}} + 0.15\,t \end{aligned}$$

The above relation is plotted in the form of a dimensional temperature—time plot in Fig. 3.8a. There is an initial "catching up" phase of roughly 60 s ($\approx$ 6 time constants) when the initial temperature difference between the sensor and the oven continuously reduces. Once the steady state is reached the temperature of the sensor increases linearly with the same rate as the input ramp but with a constant lag of 1.5 °C.

The non-dimensional plot requires the following definitions:

Non-dimensional temperature:	$\theta = \frac{T_\infty - T}{R\tau}$
Non-dimensional time:	$\eta = \frac{t}{\tau}$

Introducing the numerical values, we have

$$\theta = \frac{35 + 0.15t - T}{0.15 \times 10} = \frac{35 + 0.15t - T}{1.5}; \quad \eta = \frac{t}{10}$$

A plot of θ against η is shown in Fig. 3.8b. The numerical value of θ is 10 at $\eta = 0$ and asymptotically approaches unity as $\eta \to \infty$.

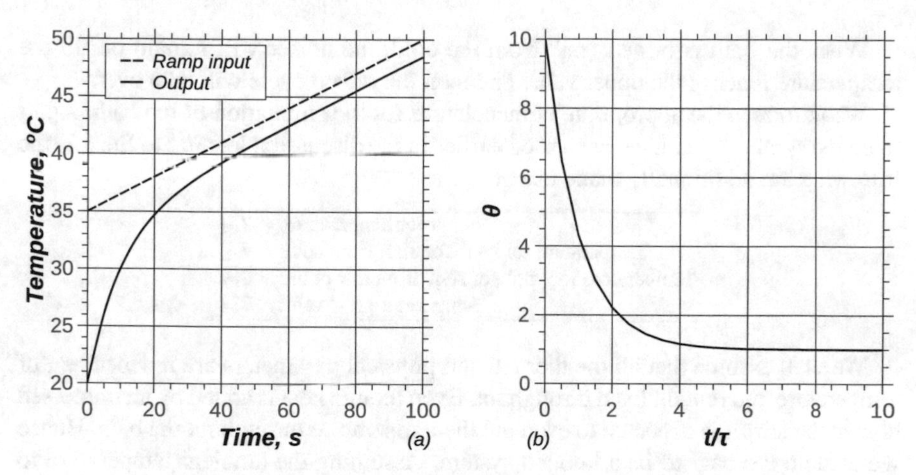

Fig. 3.8 **a** Dimensional response for Example 3.4 **b** Non-dimensional response for Example 3.4

Application to a Practical Problem: Constant Temperature Bath

In many laboratory applications, it is necessary to provide a constant temperature environment. A typical example is a water or oil bath that is used for the calibration of temperature sensors. Another example is a thermostat-controlled hot water system used in dwellings. The system is expected to maintain a certain mass of water or oil at a reasonably uniform temperature. Of course, since thermal losses are ubiquitous the temperature cannot be maintained at a desired value with perfect precision. The temperature of the system may be maintained around a narrow or acceptable band around the desired value, referred to as the set point. Different types of controllers are possible. However, in the present discussion, we consider the simplest of them, viz., an on-off controller. The controller is basically a switch that automatically turns on or off power to a heater that is immersed in the bath.

When the temperature of the bath goes below the set point by a small but measurable value the controller will switch on the heater. The heater will tend to increase the temperature back toward the set point. When the temperature goes above the set point by a small but measurable amount, the controller will switch off power to the heater. The bath will start cooling down. On-off cycles will go on perpetually or as long as the bath is in operation. The water bath problem may be simulated using the general equation for a first-order system that was presented as Eq. 3.2. During the "on" part of the cycle, G is non-zero while it is zero during the "off" part of the cycle. The temperature at the end of the "on" period provides the initial condition for the "off" period to follow. Similarly the temperature at the end of the "off" period provides the initial condition for the "on" period to follow.

When the bath is turned "on" from the cold, the heater will remain on till the temperature reaches the upper value and then the off-on cycle will take over.

We introduce the appropriate nomenclature for the simulation of the bath. Apart from the symbols that have been used earlier in the discussion leading to Eq. 3.2, the following are additionally made use of.

Set point temperature:	T_{sp}
Temperature for "on" for start from cold:	T_∞
Temperature for "on" for start from low point:	T_{lp}
Temperature for "off":	T_{hp}

We shall assume that all the thermal and physical parameters are independent of temperature and remain fixed throughout. Even though heat is added by an immersed heater, the stirrer is expected to even out the temperature throughout the bath. Hence we assume the bath to be a lumped system. Assuming the ambient temperature to remain fixed during the operation of the bath, we may use the temperature difference $\theta = T - T_\infty$ as the dependent variable to simplify the governing equation, during the "on" period as

$$\frac{d\theta}{dt} + \frac{\theta}{\tau} = \frac{G}{\rho c} \tag{3.18}$$

where G may be written as $G = \frac{Q}{V}$ where Q is the power dissipated by the heater and V is the volume of the bath. Equation 3.18 may be solved by using integration factor as before. The initial condition for the on period is $t = 0,\ T = T_\infty$ for cold start and $t = t_0,\ T = T_{lp}$ for subsequent operation. The solution may easily be shown to be (the reader should do this, as an exercise)

$$\text{Cold start:} \quad \theta = \frac{G\tau}{\rho c}\left[1 - e^{-\frac{t}{\tau}}\right]$$

$$\text{Subsequent operation:} \quad \theta = (T_{lp} - T_\infty)e^{-\frac{t-t_0}{\tau}} + \frac{G\tau}{\rho c}\left[1 - e^{-\frac{t-t_0}{\tau}}\right] \tag{3.19}$$

The solution given by Eq. 3.19 is useful till time t_{hp} such that the temperature reaches the high point of T_{hp}. At this time the power to the heater is switched off and the governing equation becomes

$$\frac{d\theta}{dt} + \frac{\theta}{\tau} = 0 \tag{3.20}$$

with the initial condition $t = t_{hp}, T = T_{hp}$. The solution during the cooling period is given by

$$\theta = (T_{hp} - T_\infty)e^{-\frac{t-t_{hp}}{\tau}} \tag{3.21}$$

This solution is valid till $t = t_{lp}$ at which time $T = T_{lp}$. At this time the heater will turn on and the on-off cycle will repeat.

Example 3.5

Consider a constant temperature bath that is in the form of a cylindrical vessel of 0.1 m diameter and 0.25 m height. The vessel contains water whose density and specific heat may be taken as 1000 kg/m^3 and 4200 J/kg°C respectively. The vessel is fairly well insulated and hence has an overall heat transfer coefficient of 5 W/m^2°C and interacts with an environment at 25 °C. The vessel is equipped with a small electric heater that provides 200 *W* when it is "on". The bath has an on-off controller that is adjusted to switch on when the temperature reduces to 53 °C and switches off when the temperature goes up to 57 °C. The bath is expected to have a set point temperature of 55 °C. Perform a simulation of the system.

Solution:

Step 1 The required background for the simulation has been presented above. The given data is written down first.

System related data:

Diameter of cylindrical vessel: $D = 0.1$ m
Height of cylindrical vessel: $H = 0.25$ m
Density of water: $\rho = 1000$ kg/m^3
Specific heat of water: $c = 4200$ J/kg°C
Heater power: $Q = 200$ W

Environment related data:

Ambient temperature: $T_\infty = 25\,°C$
Heat transfer coefficient: $h = 5$ W/m°C

Controller related data:

Set point temperature: $T_{sp} = 55\,°C$
Heater "on" temperature: $T_{lp} = 53\,°C$
Heater "off" temperature: $T_{hp} = 57\,°C$

The volume of the bath is calculated as

$$V = \frac{\pi D^2 H}{4} = \frac{\pi \times 0.1^2 \times 0.25}{4} = 0.00196\,\text{m}^3$$

The surface area of the bath interacting with the environment is

$$S = \pi D H + 2\frac{\pi D^2}{4} = \pi \times 0.1 \times 0.25 + 2 \times \frac{\pi \times 0.1^2}{4} = 0.0942\,\text{m}^2$$

The time constant of the bath treated as a first-order system is

$$\tau = \frac{\rho V c}{hS} = \frac{1000 \times 0.00196 \times 4200}{5 \times 0.0942} = 17500\,\text{s}$$

The internal heat generation rate is calculated assuming that the heater output is uniformly released throughout the volume of the bath.

$$G = \frac{Q}{V} = \frac{200}{0.00196} = 101859\,\mathrm{W/m^3}$$

Step 2 The simulation is done in two parts. In the first part, the bath starts from cold and the temperature is brought up to the operating temperature. For this part, the initial temperature is 25 °C and the heater is switched on at $t = 0$. The final temperature attained is 57 °C when the controller will switch off the heater. The bath temperature follows the first of relation 3.19.

$$T = 25 + \frac{101859 \times 17500}{1000 \times 4200}\left(1 - \mathrm{e}^{\frac{-t}{17500}}\right) = 25 + 424.41 \times \left(1 - \mathrm{e}^{\frac{-t}{17500}}\right)$$

Temperature reaches T_{hp} at $t = 1372$ s at which the heater is switched off by the controller. The cooling of the bath follows the Eq. 3.21 with the origin for time at 1372 s.

$$T = 25 + 32\mathrm{e}^{-\frac{t-1372}{14500}}$$

The temperature reaches a value of 53 °C at $t = 3714$ s. Thus the very first on-off cycle from cold start is of duration equal to 3714 s.

Step 3 Second and subsequent on-off cycles are now considered. During the on period the temperature variation with time is given by the second of Eq. 3.19 with $t_0 = 3714$ s. Hence we have

$$\begin{aligned} T &= 25 + (53 - 25)e^{-\frac{t-t_0}{17500}} + \frac{101859 \times 17500}{1000 \times 4200}\left[1 - e^{-\frac{t-t_0}{17500}}\right] \\ &= 25 + 28e^{-\frac{t-3714}{17500}} + 424.4\left[1 - e^{-\frac{t-3714}{17500}}\right] \end{aligned}$$

Calculations show that the temperature reaches the upper limit of 57 °C at $t_{hp} = 3891.5$ s. Subsequent cooling process is governed by Eq. 3.21 which translates to

$$\theta = (T_{hp} - T_\infty)e^{-\frac{t-t_{hp}}{\tau}}$$

or

$$T = 25 + (57 - 25)e^{-\frac{(t-3891.5)}{17500}} = 25 + 32e^{-\frac{(t-3891.5)}{17500}}$$

Calculations show that the temperature reduces to 53 °C at $t_{lp} = 6161.5$ s. Thus one on-off cycle takes $6161.5 - 3714 = 2447.5$ s. The temperature variation of the bath during one on-off cycle is shown in Fig. 3.9. The duration of an on-off cycle is decided by the choice of the temperatures

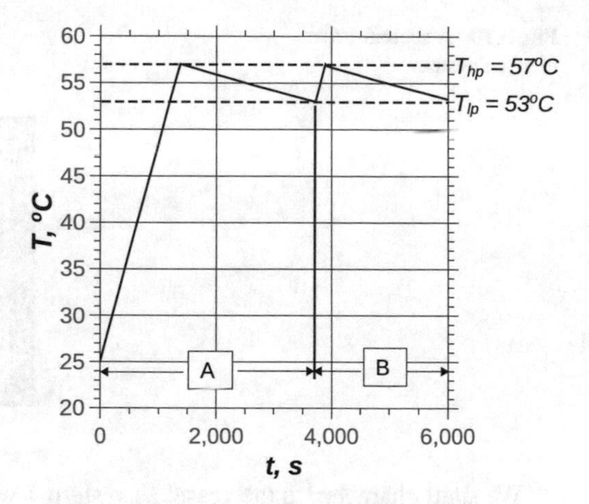

Fig. 3.9 Temperature variations during "on" and "off" periods in Example 3.5. **a**—First on-off period from cold start, **b**—second and subsequent on-off periods

T_{lp} and T_{hp}. If these are closer to the set point, the cycle time reduces. The on time is also determined by the heater dissipation. Larger the Q smaller the "on" time. If the range T_{lp}, T_{hp} is chosen very small the on-off periods will reduce and in the limit the switch may go in to rapid flutter!

3.3 Second-Order Thermal System: Response to Step Input

We have considered, in Example 3.5 a constant temperature bath as a first-order system. This is an acceptable model if the vessel itself is of negligible thermal mass. In many applications, this may not be a satisfactory assumption and the application is modeled as a second-order system. A second-order system is governed by a second-order ordinary differential equation (ODE). In mechanics, second-order systems are very common. For example, a spring–mass system is a second-order system since the governing equation is a second-order ODE. Mercury column in a U-tube manometer is also a second-order system when it is disturbed from the equilibrium position. However some types of pressure sensors are basically of first order![1] A simple example of a lumped system that is of second order is given in Fig. 3.10. The system consists of a thin-walled tank that contains a well-stirred liquid. The outside surface of the vessel at T_1 interacts with surroundings at T_∞ with a heat transfer coefficient h_1. The liquid in the container at T_2 interacts with the vessel from the inside with a heat transfer coefficient h_2.

Let us assume that $T_1 = T_2 = T_0$ at $t = 0$, $T_\infty \neq T_0$ and T_∞ is constant. Thus at any $t > 0$, we have, $T_1 \neq T_2 \neq T_\infty$.

[1] See, for example, S.P.Venkateshan, *Mechanical Measurements*, 2nd Edition, Ane Books and Athena Academic—Wiley, 2015.

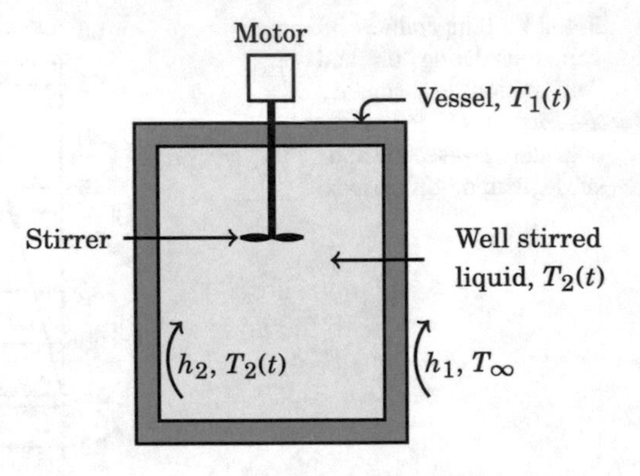

Fig. 3.10 A second-order thermal system

We shall characterize the vessel as system 1 with the following properties:

Density of vessel material:	ρ_1, kg/m^3
Volume of material of the vessel:	V_1, m^3
Outer surface area of the vessel:	S_1, m^2
Inner surface area of the vessel:	S_2, m^2
Specific heat of the vessel material:	c_1, J/kg°C

We shall characterize the liquid in the vessel as system 2 with the following properties:

Density of liquid:	ρ_2, kg/m^3
Volume of liquid in the vessel:	V_2, m^3
Surface area of contact with the vessel:	S_2, m^2
Specific heat of liquid:	c_2 J/kg°C

Energy balance for system 1 requires that the following hold:

$$\rho_1 c_1 V_1 \frac{dT_1}{dt} = h_1 S_1 (T_\infty - T_1) - h_2 S_2 (T_1 - T_2) \tag{3.22}$$

The above energy balance is made assuming that, at any time t, $T_\infty > T_1 > T_2$. The left-hand side represents the change in the energy stored in the vessel, the first term on the right-hand side represents the heat gained by the vessel from the ambient and the last term represents the heat lost by the vessel to the liquid contained in the vessel. Similarly we may perform energy balance for system 2 to get

$$\rho_2 c_2 V_2 \frac{dT_2}{dt} = h_2 S_2 (T_1 - T_2) \tag{3.23}$$

In this case the left-hand side represents the increase in the energy stored in the liquid while the right-hand side represents the heat transfer to the liquid from the vessel.

The reader will realize that there are two dependent variables T_1, T_2 in this problem. One would, however, be interested in the liquid bath and hence the time variation of temperature T_2 is of interest to us. It is thus obvious that we should eliminate temperature T_1 from the two Eqs. 3.22 and 3.23 to get a single equation that governs the variation of T_2 with time. We do this, by introducing the following three time scales:

$$\tau_1 = \frac{\rho_1 c_1 V_1}{h_1 S_1}, \quad \tau_2 = \frac{\rho_2 c_2 V_2}{h_2 S_2}, \quad \tau_3 = \frac{\rho_1 c_1 V_1}{h_2 S_2} \tag{3.24}$$

The time scales τ_1 and τ_3 are associated with system 1, the vessel, since it has thermal interactions with both the ambient and system 2, the liquid in the vessel. However, τ_2 is a single time scale that characterizes the thermal interaction between systems 2 and 1. Introducing the three time scales, the two governing equations may be recast in the form

$$\text{(a)} \quad \frac{dT_1}{dt} = \frac{T_\infty - T_1}{\tau_1} - \frac{T_1 - T_2}{\tau_3} \quad \text{(b)} \quad \frac{dT_2}{dt} = \frac{T_1 - T_2}{\tau_2} \tag{3.25}$$

Differentiate Eq. 3.25(b) once with respect to time to get

$$\frac{d^2 T_2}{dt^2} = \frac{1}{\tau_2}\left[\frac{dT_1}{dt} - \frac{dT_2}{dt}\right] \tag{3.26}$$

From Eq. 3.25(b) we also have

$$T_1 = T_2 + \tau_2 \frac{dT_2}{dt} \tag{3.27}$$

We substitute Eq. 3.27 in 3.25(a) to get

$$\frac{dT_1}{dt} = \frac{T_\infty - T_2 - \tau_2 \frac{dT_2}{dt}}{\tau_1} - \frac{T_2 + \tau_2 \frac{dT_2}{dt} - T_2}{\tau_3}$$

or

$$\frac{dT_1}{dt} = \frac{T_\infty - T_2}{\tau_1} - \tau_2\left(\frac{1}{\tau_1} + \frac{1}{\tau_3}\right)\frac{dT_2}{dt} \tag{3.28}$$

Substitute Eq. 3.28 in 3.26 to get

$$\begin{aligned} \frac{d^2 T_2}{dt^2} &= \frac{1}{\tau_2}\left[\frac{T_\infty - T_2}{\tau_1} - \tau_2\left(\frac{1}{\tau_2} + \frac{1}{\tau_3}\right)\frac{dT_2}{dt} - \frac{dT_2}{dt}\right] \\ &= \frac{T_\infty - T_2}{\tau_1 \tau_2} - \left(\frac{1}{\tau_1} + \frac{1}{\tau_2} + \frac{1}{\tau_3}\right)\frac{dT_2}{dt} \end{aligned} \tag{3.29}$$

We now introduce the following:

$$\theta = T_2 - T_\infty, \quad \frac{1}{\tau_1} + \frac{1}{\tau_2} + \frac{1}{\tau_3} = \frac{1}{\tau_s}, \quad \frac{1}{\tau_p} = \frac{1}{\tau_1 \tau_2}$$

With these the governing equation describing the variation of the bath temperature with time is

$$\boxed{\frac{d^2\theta}{dt^2} + \frac{1}{\tau_s}\frac{d\theta}{dt} + \frac{\theta}{\tau_p} = 0} \tag{3.30}$$

The equation has turned out to be a second-order equation and hence the system is referred to as a second-order system. This equation requires two initial conditions to be specified. Using the conditions $T_1 = T_2 = T_0$ at $t = 0$, we see that $\theta = (T_2 - T_\infty)|_{t=0} = (T_0 - T_\infty) = \theta_0$(say). Also from Eq. 3.25(b) the derivative of T_2 and hence the derivative of θ vanishes at $t = 0$. Thus the initial conditions that accompany Eq. 3.30 are

$$\theta = \theta_0 \quad \text{and} \quad \frac{d\theta}{dt} = 0 \text{ at } t = 0 \tag{3.31}$$

This equation may be solved easily using the two roots of the characteristic equation

$$m^2 + \frac{1}{\tau_s}m + \frac{1}{\tau_p} = 0$$

The roots are given by

$$m_1 = -\frac{\frac{1}{\tau_s} + \sqrt{\frac{1}{\tau_s^2} - \frac{4}{\tau_p}}}{2}, \quad m_2 = -\frac{\frac{1}{\tau_s} - \sqrt{\frac{1}{\tau_s^2} - \frac{4}{\tau_p}}}{2} \tag{3.32}$$

Both roots are real as long as $\frac{1}{\tau_s^2} \geq \frac{4}{\tau_p}$. Thus the solution is written down as

$$\theta = Ae^{m_1 t} + Be^{m_2 t} \tag{3.33}$$

where A and B are constants of integration. These are determined by the initial conditions. The first initial condition requires that

$$A + B = \theta_0 \tag{3.34}$$

The second initial condition requires that

$$Am_1 + Bm_2 = 0 \tag{3.35}$$

From Eq. 3.34 we have $B = \theta_0 - A$. Substituting this in Eq. 3.35, we get

$$Am_1 + (\theta_0 - A)m_2 = 0 \quad \text{or} \quad A = \frac{\theta_0 m_2}{m_2 - m_1}$$

But $m_2 - m_1 = \sqrt{\frac{1}{\tau_s^2} - \frac{4}{\tau_p}}$. Hence we have

$$A = -\frac{\theta_0}{2} \times \frac{\frac{1}{\tau_s} - \sqrt{\frac{1}{\tau_s^2} - \frac{4}{\tau_p}}}{\sqrt{\frac{1}{\tau_s^2} - \frac{4}{\tau_p}}} \tag{3.36}$$

Using this value of A we may find B as

$$B = \frac{\theta_0}{2} \times \frac{\frac{1}{\tau_s} + \sqrt{\frac{1}{\tau_s^2} - \frac{4}{\tau_p}}}{\sqrt{\frac{1}{\tau_s^2} - \frac{4}{\tau_p}}} \tag{3.37}$$

A worked example will bring out the nature of the solution obtained in the closed form above. Example 3.6 considers an interesting application where the response of a temperature sensor in the form of a thermocouple immersed in a liquid is to be described.

Example 3.6

A thermocouple is immersed in a liquid contained in a cylinder of inside diameter of 3 mm and an outside diameter of 3.5 mm. The cylinder material is copper with a specific heat of 383 J/kg°C and density 8900 kg/m^3. The heat transfer coefficient on the outer and inner surfaces, respectively, of the cylinder are 15 and 45 W/m^2°C. The density and specific heat of the liquid are known to be 1040 kg/m^3 and 3650 J/kg°C, respectively. The effective length of the cylinder is 10 mm. Obtain the step response of the thermocouple treating it as a second-order system. Make a plot of this response using suitable non-dimensional coordinates.

Solution:

Step 1 We follow the notation used in the text and specify the given data.

Outside heat transfer coefficients:	$h_1 = 15$ W/m^2 °C
Inside heat transfer coefficients:	$h_2 = 45$ W/m^2 °C
Outer diameter of cylinder:	$D_1 = 0.0035$ m
Inner diameter of cylinder:	$D_2 = 0.003$ m
Effective cylinder length:	$L = 0.01$ m
Density of cylinder material:	$\rho_1 = 8900$ kg/m^3
Specific heat of cylinder material:	$c_1 = 383$ J/kg°C
Density of liquid:	$\rho_2 = 1040$ kg/m^3
Specific heat of liquid:	$c_2 = 3650$ J/kg°C

Heat transfer areas are calculated as

$$S_1 = \pi D_1 L = \pi \times 0.0035 \times 0.01 = 1.1 \times 10^{-4}\,\text{m}^2$$
$$S_2 = \pi D_1 L = \pi \times 0.003 \times 0.01 = 9.425 \times 10^{-5}\,\text{m}^2$$

Volume of cylinder material is

$$V_1 = \frac{\pi(D_1^2 - D_2^2)L}{4} = \frac{\pi \times (0.0035^2 - 0.003^2) \times 0.01}{4} = 2.553 \times 10^{-8}\,\text{m}^3$$

Volume of the liquid is calculated as

$$V_2 = \frac{\pi D_2^2 L}{4} = \frac{\pi \times 0.003^2 \times 0.01}{4} = 7.069 \times 10^{-8}\,\text{m}^3$$

Step 2 Characteristic times:
Characteristic time τ_1 is calculated as

$$\tau_1 = \frac{\rho_1 V_1 c_1}{h_1 S_1} = \frac{8900 \times 2.553 \times 10^{-8} \times 383}{15 \times 1.1 \times 10^{-4}} = 52.75\,\text{s}$$

Characteristic time τ_2 is

$$\tau_2 = \frac{\rho_2 V_2 c_2}{h_2 S_2} = \frac{1040 \times 7.069 \times 10^{-8} \times 3650}{45 \times 9.425 \times 10^{-5}} = 63.27\,\text{s}$$

Characteristic time τ_3 is

$$\tau_3 = \frac{\rho_1 V_1 c_1}{h_2 S_2} = \frac{8900 \times 2.553 \times 10^{-8} \times 383}{45 \times 9.425 \times 10^{-5}} = 20.52\,\text{s}$$

With these, the two parameters that govern the problem are

$$\tau_p = \tau_1 \tau_2 = 52.75 \times 63.27 = 3337.5\,\text{s}^2$$
$$\tau_s = \left[\frac{1}{\tau_1} + \frac{1}{\tau_2} + \frac{1}{\tau_3}\right]^{-1} = \left[\frac{1}{52.75} + \frac{1}{63.27} + \frac{1}{20.52}\right]^{-1} = 11.98\,\text{s}$$

Step 3 Solution:
Using Eq. 3.32, we obtain the exponents m_1 and m_2.

$$m_1 = -\frac{\frac{1}{11.98} + \sqrt{\frac{1}{11.98^2} - \frac{4}{3338}}}{2} = -0.0797\,\text{s}^{-1}$$
$$m_2 = -\frac{\frac{1}{11.98} - \sqrt{\frac{1}{11.98^2} - \frac{4}{3338}}}{2} = -0.00376\,\text{s}^{-1}$$

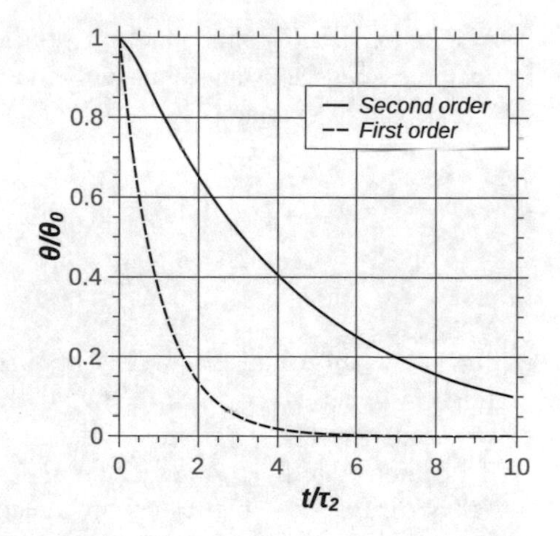

Fig. 3.11 Response of a second-order system to step input

The constants of integration are obtained now. Using Eq. 3.36

$$A = -\frac{\theta_0}{2}\frac{\frac{1}{11.98} - \sqrt{\frac{1}{11.98^2} - \frac{4}{3338}}}{\sqrt{\frac{1}{11.98^2} - \frac{4}{3338}}} = 1.0495\theta_0, \quad \text{and} \quad B = \theta_0 - A = -0.0495\theta_0$$

Introduce the following non-dimensional variables

$$\phi = \frac{\theta}{\theta_0}; \tau = \frac{t}{\tau_2}$$

to obtain the response as

$$\phi = 1.0495\mathrm{e}^{-0.0797t} - 0.0495\mathrm{e}^{-0.00376t}$$

Step 4 Plot of response of second-order system: The response of the second-order system shown in Fig. 3.11 uses the above non-dimensional variables along the two axes.

Comments

The second-order response is also compared with the first-order response with time constant equal to τ_2. The second-order system response is sluggish as compared to the first-order system. There is a time delay due to the wall that has to respond before the liquid starts responding to the input.

The two roots of the characteristic equations may be interpreted as two reciprocal time constants. One of them is labeled τ_h and the other is labeled τ_l. These may be obtained as

$$\tau_h = \frac{1}{0.00376} = 266.2\,\text{s}$$
$$\tau_l = \frac{1}{0.0797} = 12.54\,\text{s}$$

It is interesting to note that, when $h_2 \gg h_1$, the smaller of the two time constants tends to 0 and the larger of the two time constants tends to ∞. The response of the system becomes a first-order response with the bigger time constant. To demonstrate this, we redo Example 3.6 with $h_1 = 5\,\text{W/m}^{2\circ}\text{C}$ and $h_2 = 100\,\text{W/m}^{2\circ}\text{C}$. The coefficients now become $A = -0.0102$ and $B = 1.0102$. Also the two time constants become $\tau_l = 6.75$ s and $\tau_h = 668$ s. The temperature response becomes

$$\phi = 1.0102e^{-\frac{t}{668}} - 0.0102e^{-\frac{t}{6.75}}$$

Since the second term on the right-hand side is very small compared to the first term, the response is basically that of a first-order one with a time constant of 668 s.

Concluding Remarks

This chapter has dealt with transient response of lumped systems. When Biot number based on a characteristic dimension is small, typically $Bi < 0.1$, the system may be treated as lumped, i.e., the entire system is at a uniform temperature at any time t. The governing equation is an ODE. The response of a first-order system to step—periodic and ramp inputs are discussed. The chapter closes with the analysis of a second-order thermal system.

3.4 Exercises

Ex 3.1: It is proposed to manufacture glass beads of diameter 0.5 mm by spraying them into 20 °C air and allowing them to harden as they fall to the ground. Assume that the initial temperature of the glass bead is $T_i = 500\,°\text{C}$, the bead to air heat transfer coefficient is $h = 324\,\text{W/m}^{2\circ}\text{C}$ and the constant downward velocity is $U = 3.8\,\text{m}/s$. Glass properties are

$\rho = 2800\,\text{kg/m}^3$, $C = 810\,\text{J/kg°C}$ and $k = 0.81\,\text{W/m°C}$. The design calls for the production of beads whose temperature at the center does not exceed 40 °C at the instant it reaches the ground. From what height should the beads fall? Calculate the temperature of the bead surface when its center temperature reaches 40 °C. Compare these two temperatures and comment on whether the lumped capacitance method is applicable in the late stages of the cooling process.

Ex 3.2: A first-order system initially at T_0 is exposed to an ambient whose temperature varies linearly with time according to the relation $T_\infty(t) = T_1 + Ct$ where T_1 and C are specified constants. The heat transfer coefficient may be assumed to be constant and equal to h. Determine the temperature of the system at any time $t > 0$. Obtain the steady-state response of the system. If $T_0 = 20\,°\text{C}$, $T_1 = 30\,°\text{C}$, $C = 0.5°\text{C/s}$ and $\tau = 10\,\text{s}$ calculate the steady-state response of the system. Make a plot of system temperature as a function of time till the steady state is reached.

Ex 3.3: A thermometer initially at room temperature of 30 °C is immersed in boiling water at 100 °C. The thermometer indicates a temperature of 76 °C at the end of 10 s. Assuming the thermometer to be a first-order system, calculate its time constant in boiling water. The thermometer has a mass of $10^{-3}\,kg$, specific heat capacity of 7700 J/kg°C and a surface area of $10^{-4}\,\text{m}^2$. Determine the heat transfer coefficient in boiling water.
The thermometer in the above case reads 85 °C when it is removed and placed in an air stream at 30 °C. The thermometer shows a reading of 45 °C after a lapse of 30 min. Determine the heat transfer coefficient in air.

Ex 3.4: A first-order system with a time constant of 10 s in boiling water is exposed to boiling water at 100 °C for 15 s. The initial temperature of the system may be taken as 30 °C. The system, at the end of 15 s exposure to boiling water is transferred in to a bath of ice-cold water at 0 °C in which the time constant of the system is 25 s. How long should one wait for the temperature of the system to reach 5 °C?

Ex 3.5: A certain first-order system initially at a temperature of 100 °C is subject to convective cooling by a fluid at 30 °C with a time constant of 10 s. After the system has been cooling for 5 s the time constant suddenly changes to 5 s because of a change in the fluid velocity. The fluid temperature remains the same. Determine the temperature of the first-order system at 10 s and 30 s from the start.

Ex 3.6: Piston in an engine receives a thermal flux from a radiation source at the end of the cylinder. It loses energy by convection from the upper surface with a heat transfer coefficient of $h = 150\,\text{W/m}^2\text{°C}$. The piston is made of stainless steel with thermal conductivity $k = 14.5\,\text{W/m°C}$, specific heat capacity $C = 460\,\text{J/kg°C}$, density $\rho = 7810\,\text{kg/m}^3$. The radiation absorbed at the lower surface is a function of time and is given by $q(t) = q_0 \cos(\omega t)$ where $q_0 = 32000\,\text{W/m}^2$ and $\omega = 6\,\text{rad/s}$. Assume that the initial temperature of the piston is the same as the temperature of the fluid medium of 26 °C. Take the thickness of the piston as $L = 6\,\text{mm}$.

If the internal resistance due to conduction is neglected, show that the governing equation for transient temperature of the piston can be written as

$$\frac{d\theta}{dt} + m\theta = n \cos \omega t$$

where $m = \frac{h}{\rho C L}$ and $n = \frac{q_0}{\rho C L}$. Solve this equation and discuss the nature of the steady-state solution. Note that θ is the temperature excess with respect to the temperature of the medium.

Ex 3.7: The volume of the bulb of a thermometer is 0.25 ml and contains mercury just filling the bulb when the temperature is 0 °C. The bulb communicates with a capillary tube of 0.25 mm diameter. The bulb may be assumed to be perfectly rigid. The heat capacity of the bulb itself may be neglected in comparison with the heat capacity of the mercury contained within. The bulb is completely immersed in hot water at a temperature of 50 °C. The heat transfer coefficient between water and the bulb is known to be 45 W/m^2°C. Determine the time constant of the thermometer. The volume expansion coefficient of mercury is known to be 1.82×10^{-6}/°C. Calculate the length of the mercury column at a time equal to $\frac{\tau}{2}$. Also, determine the rate at which the mercury column is changing at the above time.
Other properties of mercury are
Specific heat $C = 140$ J/kg · °C and density $\rho = 13600$ kg/m^3
Make use of reasonable assumptions in solving the problem, after justifying them.

Ex 3.8: Two sensors experience the same heat transfer environment subject to the same heat transfer coefficient. The first sensor is in the form of a solid sphere of diameter $D = 0.003$ m while the second sensor is in the form of a square foil of thickness $\delta = 0.0002$ m and side of square of such length that the volume of the foil material is the same as that of the spherical sensor. Both the sensors are made from the same material. Determine the ratio of the time constant of spherical sensor to that of the foil-shaped sensor.

Ex 3.9: A first-order system has a time constant of 10 s in a particular environment. The environmental temperature has a sinusoidal variation with a frequency of 0.2 Hz. What is the phase lag and the amplitude of the temperature indicated by the system. What is the maximum frequency acceptable if the phase lag is to be less than 10° of angle? What is maximum frequency acceptable if the amplitude ratio is not less than 0.95 of the impressed value?

Ex 3.10: An engine runs at a maximum speed of 4500 rpm. It is desired to measure the temperature of the gases inside the engine cylinder, which may be assumed to vary periodically at a frequency corresponding to the engine speed. What should be the time constant of the sensor if the phase lag is to be limited to 1° of angle? What will then be the amplitude reduction factor?

Ex 3.11: A second-order system is characterized by $\tau_1 = 10\,\text{s}$, $\tau_2 = 15\,\text{s}$ and $\tau_3 = 5\,\text{s}$. Determine the two time constants that characterize the solution. Make a plot similar to Fig. 3.11, but using dimensional coordinates when $T_0 = 50\,°\text{C}$ and $T_\infty = 90\,°\text{C}$.

Chapter 4
Heat Transfer from Extended Surfaces

IMPROVING heat transfer from a base surface may be accomplished by increasing the surface area exposed to the ambient medium. This is accomplished by attaching fins or extended surfaces to the base surface. Fins are useful when the internal conduction resistance of the fin is less than the convective resistance at its surface. Both uniform and nonuniform area fins are considered in this chapter. Optimum profiled fins for minimum mass are also discussed. Analysis of an array of fins is given toward the end of the chapter.

4.1 Introduction

The present chapter looks at basically one-dimensional problems in heat conduction arising when there is an interaction between conduction along *one* direction and convection along a *different* direction. This model assumes that the temperature variation within the solid is significant along the former direction and insignificant along the latter direction. The conductive heat flux and convective heat flux are oriented along *perpendicular* directions. Just when this happens and when the model makes sense will be considered first. The analysis of such problems will follow afterward. The analysis assumes significance since there are many important applications where these conditions are satisfied. All these applications involve *extended surfaces* or *fins*.

Recall that convection heat transfer from a surface is given by

$$Q_c = hA\Delta T \tag{4.1}$$

where the symbols have the usual meanings. One will be able to increase convection heat transfer by increasing any one or all the three quantities that go on the right-hand side of Eq. 4.1. In practice, as we shall see later, increase of h is possible by aug-

S. P. Venkateshan, *Heat Transfer*,
https://doi.org/10.1007/978-3-030-58338-5_4

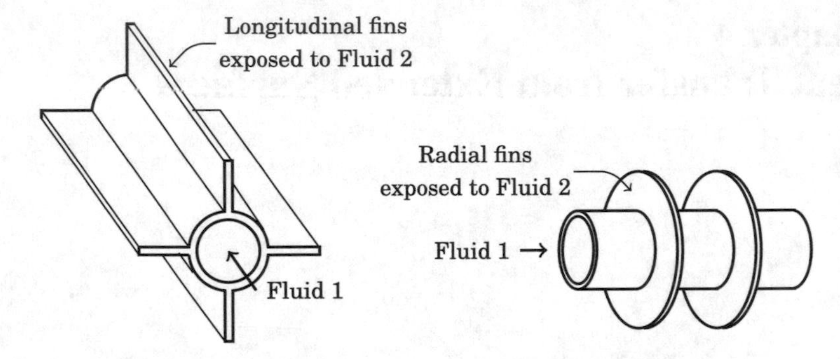

Fig. 4.1 Two types of extended surfaces

mentation methods. However, these are limited by pressure drop penalties. Increase in the temperature difference is certainly a possible solution. However, there may be constraints on ΔT because of safety or reliability considerations. Increase in the area available for heat transfer is certainly an option. An extended surface is essentially the use of this option. The intention is to increase the overall heat transfer from a base structure by providing an extra area attached to the base structure. Providing extra area will be beneficial only if the conductive resistance of the extended surface is not excessive. This may be assured by the choice of a high conductivity material and of proper geometric proportions. Typical examples of fins are shown in Fig. 4.1. In the first case shown in the figure, the fins are longitudinal (i.e., oriented parallel to the axis of the fluid-carrying tube), and are of uniform area for conduction. In the second case shown in the figure, the fins are placed normal to the tube axis and are of non-uniform area for conduction. The external surfaces of the fins are convectively cooled by a fluid flowing either parallel to the axis of the tube or normal to the axis of the tube. In either case, the present chapter will treat the heat transfer coefficient as being specified, and of constant value over the entire area of contact between the fluid and the extended surface.

Typical applications of extended surfaces

- Air-cooled Internal Combustion Engine Cylinder,
- Refrigerator Condenser,
- Automobile Radiator,
- Air compressor inter-cooler,
- Fins for cooling of electronic components—"heat sinks", and
- Cooling fins for radiation source in an optical instrument.

We have seen in Chap. 3 that lumping is justified when the Biot number is very small. Extended surfaces are normally made using a thin sheet of a material of high thermal conductivity. Typical materials are aluminum, copper, and brass. The thickness may be a fraction of a millimeter to a few millimeters. If there is significant temperature variation along the plate surface, we may consider the state of affairs over a small area element on the surface that is at a temperature T_s. The convective heat transfer to an ambient at T_{amb} will then be given by

$$Q_c = h\Delta A(T_s - T_{amb})$$

If the thickness of the sheet is $2t$, the conductive heat transfer within the sheet is given by

$$Q_k = k\Delta A\frac{(T_m - T_s)}{t}$$

where T_m is the mid-plane temperature within the sheet. At the surface, under steady state, there should be a balance between conduction and convection. Thus we have

$$\begin{aligned} Q_k = Q_c \text{ or } k\Delta A\frac{(T_m - T_s)}{t} &= h\Delta A(T_s - T_{amb}) \\ \text{or } T_m - T_s = \frac{ht}{k}(T_s - T_{amb}) &= Bi(T_s - T_{amb}) \end{aligned} \tag{4.2}$$

In the above, Bi represents the Biot number based on semi plate thickness. It is reasonable to expect that Bi will be small in most fin-related applications. The temperature variation across the thickness of the fin may be ignored, leading to the one-dimensional conduction along the length of the fin.

Typical example

Fin is 3 mm thick of aluminum alloy of thermal conductivity 200 W/m°C. The convection heat transfer coefficient h is 50 W/m^2°C (typical). The base structure temperature is 100 °C and the fluid temperature is 30 °C. On the right-hand side of Eq. 4.2, T_s may be replaced by the maximum possible value, i.e., 100 °C. Then

$$T_m - T_s = \frac{50 \times 0.0015}{200}(100 - 30) = 0.026\,^\circ\text{C}$$

This temperature difference is indeed very small and hence $T_m \approx T_s$ and the one-dimensional assumption holds. The Biot number in this case is

$$Bi = \frac{50 \times 0.0015}{200} = 0.000375$$

Indeed the Biot number is extremely small (much smaller than 0.1) in this case.

It will become clear as we develop the analysis that the temperature variation along the length of the fin will be significant and cannot be ignored. Also, the thermal performance of fins as augmenters of heat transfer will be determined by the longitudinal temperature distribution.

4.2 Fins of Uniform Area

4.2.1 *Analysis*

Figure 4.2 shows an example of a constant area fin. The fin is in the form of a flat plate attached at one end to the hot base structure at T_b. The lateral surfaces of the fin are cooled by a stream at T_{amb} via a constant heat transfer coefficient h. The length of the fin is L while its thickness is $2t$. The heat transfer analysis makes the following assumptions :

- The fin is in the steady state.
- Conduction heat transfer in the fin is one-dimensional and is along the x-direction.
- Heat loss from the sides is by convection with a constant heat transfer coefficient h to an ambient at uniform temperature T_{amb}.

In Fig. 4.2, the inset shows an elemental slice of fin for analysis. Consider a unit width of the fin in a direction perpendicular to the plane of the figure. Energy balance for the slice element requires that

$$Q_{k,x} = Q_{k,x+\Delta x} + Q_{c,x} \tag{4.3}$$

Conduction heat transfer takes place across the constant area of cross section of the element given by $A = 2 \times t \times 1 = 2t$. The surface area from which convection heat transfer takes place is $\Delta S = 2 \times \Delta x$ or $\Delta S = P\Delta x$, where P is the perimeter wetted by the ambient fluid. The reason we write the surface area in this form is that it is

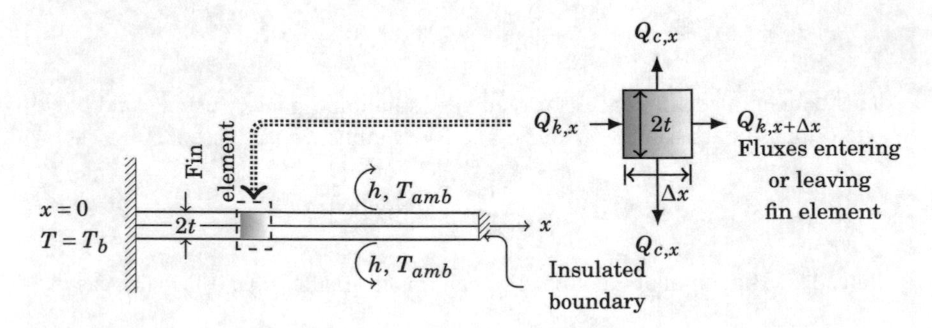

Fig. 4.2 Heat balance for an element of a uniform area fin

possible to generalize the analysis to a fin of any shape, as we shall see later. We use Fourier law to write Eq. 4.3 as

$$-kA\left.\frac{dT}{dx}\right|_x = -kA\left.\frac{dT}{dx}\right|_{x+\Delta x} + hP\Delta x(T - T_{amb}) \tag{4.4}$$

Taylor series expansion of the first term on the right-hand side of Eq. 4.4 gives

$$-kA\left.\frac{dT}{dx}\right|_{x+\Delta x} = -kA\left.\frac{dT}{dx}\right|_x - kA\left.\frac{d^2T}{dx^2}\right|_x \Delta x - \mathscr{O}(\Delta x)^2 \tag{4.5}$$

Introduce Eq. 4.5 in 4.4, cancel the $-kA\frac{dT}{dx}$ term on the two sides, remove common factor Δx, and take the limit as $\Delta x \to 0$, and rearrange to get the fin equation

$$\boxed{\frac{d^2T}{dx^2} - \frac{hP}{kA}(T - T_{amb}) = 0} \tag{4.6}$$

Let the temperature difference $T - T_{amb}$ be denoted by θ. Since T_{amb} is a constant $\frac{dT}{dx} = \frac{d\theta}{dx}, \frac{d^2T}{dx^2} = \frac{d^2\theta}{dx^2}$. Represent the composite parameter $\frac{hP}{kA}$ as m^2 to rewrite Eq. 4.6 in the standard form.

$$\boxed{\frac{d^2\theta}{dx^2} - m^2\theta = 0} \tag{4.7}$$

The fin equation, being a second-order ordinary differential equation, requires two boundary conditions. These are specified at the two ends, viz., $x = 0$ and $x = L$. It is customary to specify the temperature at $x = 0$ (the base of the fin) as $T = T_b$ (first kind of boundary condition) or $\theta = \theta_b$. Presumably this is the temperature (or the base temperature excess with respect to the temperature of the ambient, as the case may be) of the base structure to which the fins are attached. The location $x = L$ is referred to as the tip of the fin and any of three possible boundary conditions (BC) may be specified there. Even though the insulated tip condition will be used in what follows immediately, the other types of boundary conditions are discussed here, for the sake of completeness and use later on.

On Boundary Conditions

The three types of boundary conditions are

$$\text{I kind: } T(x = L) = T_t \text{ or } \theta = \theta_t$$

$$\text{II kind: } -k\frac{dT}{dx}\Big|_{x=L} = -k\frac{d\theta}{dx}\Big|_{x=L} = q_L$$

$$\text{III kind: } -k\frac{dT}{dx}\Big|_{x=L} = h_t(T_t - T_{amb}) \text{ or } -k\frac{d\theta}{dx}\Big|_{x=L} = h_t\theta_t$$

where h_t is the heat transfer coefficient for convection heat transfer from the tip surface. Note also that $\theta_t = T_t - T_{amb}$. In most applications, the boundary condition of the second kind is chosen by specifying the heat flux to be zero at the tip. This boundary condition is justified on the assumption that the heat loss from the small area of the tip surface is negligibly small. This amounts to specifying an insulated condition there, as indicated in Fig. 4.2. One may also, use a more general, third kind of boundary condition that will account for the heat loss from the tip as given by the boundary condition of the third kind.

In Eq. 4.7, parameter m is referred to as the fin parameter. It may easily be verified that m has the unit $\frac{1}{[L]}$, i.e., reciprocal length or m^{-1}. Now we consider typical examples to indicate how the fin parameter may be obtained.

Fin parameter in typical cases

1. **Flat fin of uniform thickness:** In the case of a fin in the form a flat plate the perimeter P is 2 (width—both top and bottom—of the fin in a direction perpendicular to the plane of Fig. 4.2) and the area A is $2t$. The fin parameter m then is

$$m = \sqrt{\frac{h \times 2}{k \times 2 \times t}} = \sqrt{\frac{h}{kt}}$$

2. **Pin fin of circular cross section:** In the case of a pin fin of uniform area and circular cross section the arrangement is as indicated in Fig. 4.3a. The perimeter P is πD and the area A is $\pi D^2/4$. Hence the fin parameter is

$$m = \sqrt{\frac{h \times \pi \times D}{k \times \pi \times \frac{D^2}{4}}} = \sqrt{\frac{4h}{kD}}$$

3. **Pin fin of square cross section of side** $2t$**:** In the case of a pin fin of uniform area and square cross section the arrangement is as indicated in Fig. 4.3b. The perimeter P is $8t$ and the area A is $4t^2$. Hence the fin parameter is

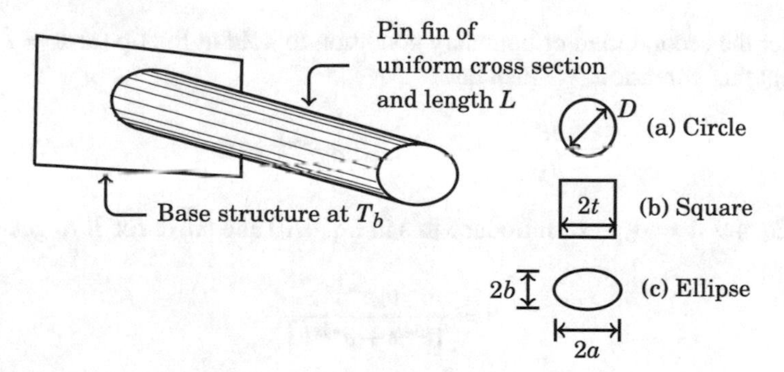

Fig. 4.3 Uniform area fins of various cross section shapes

Table 4.1 Fin parameter of pin fins of elliptic cross section

AR	m	AR	m	AR	m
2	$\sqrt{\frac{1.542h}{kb}}$	1.5	$\sqrt{\frac{1.683h}{kb}}$	1.25	$\sqrt{\frac{1.806h}{kb}}$
1.125	$\sqrt{\frac{1.891h}{kb}}$	$1^\dagger$	$\sqrt{\frac{2h}{kb}}$		

†Cross section is a circle with $b = D/2$

$$m = \sqrt{\frac{h \times 8 \times t}{k \times 4 \times t^2}} = \sqrt{\frac{2h}{kt}}$$

4. **Pin fin of elliptic cross section:** In the case of a pin fin of uniform area and elliptic cross section the arrangement is as indicated in Fig. 4.3c. The semi-major and semi-minor axis lengths are a and b, respectively. The ratio $\mathcal{P}/A$ depends on the aspect ratio $AR = a/b$. Hence the fin parameter values are given by the values shown in Table 4.1.

4.2.2 Solution to the Fin Equation

Insulated Tip Case

Equation 4.7 has the general solution given by

$$\theta = Ae^{mx} + Be^{-mx} \tag{4.8}$$

Using the base boundary condition, we have

$$\theta_b = A + B \tag{4.9}$$

Consider the second kind of boundary condition to hold at the tip (at $x = L$), with zero heat flux condition. We then have

$$\frac{d\theta}{dx} = Ame^{mL} - Bme^{-mL} = 0 \tag{4.10}$$

From Eq. 4.9 $A = \theta_b - B$. Introduce this in Eq. 4.10 and solve for B to get

$$B = \frac{\theta_b e^{mL}}{\left[e^{mL} + e^{-mL}\right]} \tag{4.11}$$

Then we also get

$$A = \theta_b - B = \theta_b - \frac{\theta_b e^{mL}}{\left[e^{mL} + e^{-mL}\right]} = \theta_b \frac{\left[e^{mL} + e^{-mL} - e^{mL}\right]}{\left[e^{mL} + e^{-mL}\right]} = \frac{\theta_b e^{-mL}}{\left[e^{mL} + e^{-mL}\right]} \tag{4.12}$$

Introduce these in Eq. 4.8 to get

$$\theta = \frac{\theta_b e^{-mL}}{\left[e^{mL} + e^{-mL}\right]} e^{mx} + \frac{\theta_b e^{mL}}{\left[e^{mL} + e^{-mL}\right]} e^{-mx} = \theta_b \frac{\left[e^{m(x-L)} + e^{-m(x-L)}\right]}{\left[e^{mL} + e^{-mL}\right]} \tag{4.13}$$

The above solution may be recast using hyperbolic functions as

$$\boxed{\theta = \theta_b \frac{\cosh\left[m(x - L)\right]}{\cosh(mL)}} \tag{4.14}$$

At this stage, it is instructive to recast the solution in non-dimensional form. For this purpose introduce the following non-dimensional quantities:

$$\phi = \frac{\theta}{\theta_b}; \quad \xi = \frac{x}{L} \quad \text{and} \quad \mu = mL \tag{4.15}$$

The non-dimensional independent variables are like the ones introduced earlier while studying other one-dimensional problems. The last one is the non-dimensional fin parameter that is the ratio of a characteristic length scale L in the problem to a characteristic internal length scale $\frac{1}{m}$—reciprocal of the fin parameter m—in the problem. More about this later.

The solution may now be rewritten, noting that the hyperbolic cosine is an even function, as

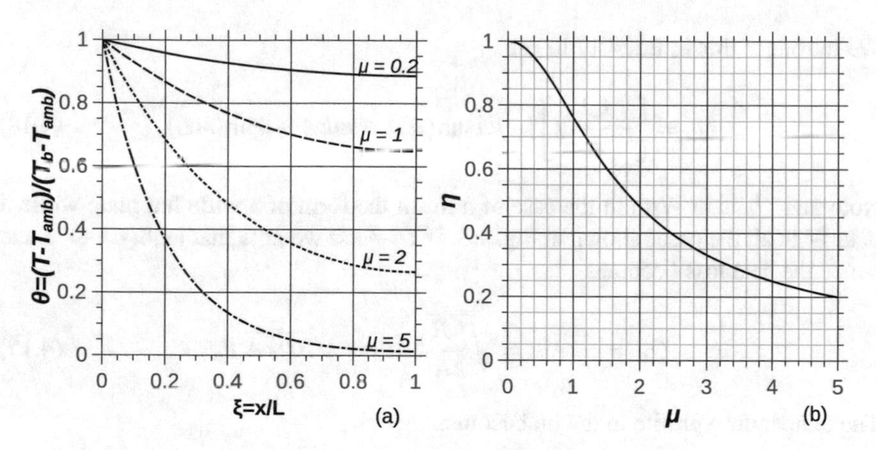

Fig. 4.4 **a** Fin temperature profile variation with fin parameter for a uniform area fin **b** Variation of efficiency of a uniform area fin with fin parameter

$$\boxed{\phi = \frac{\cosh[\mu(1-\xi)]}{\cosh(\mu)}} \tag{4.16}$$

ϕ represents the non-dimensional temperature distribution in terms of the non-dimensional x coordinate ξ. It is seen that the temperature variation depends on the non-dimensional fin parameter μ.

For typical values of h, P, A and L, μ depends inversely on k. The higher the k is, the smaller is the μ and more "full" is the temperature profile, as shown in Fig. 4.4a. The temperature at the fin tip approaches the base temperature, i.e., $\phi_t \to 1$ when $\mu \to 0$. When $\mu \to \infty$ the tip temperature tends to the ambient temperature ($\phi_t \to 0$).

Heat Transfer from Fin

The heat loss from the fin may be easily calculated by determining the heat Q_b entering the fin at the base, by conduction. This is so because, under steady conditions, the same amount of heat is lost from the surface of the fin to the ambient by convection as Q_c. By Fourier law, we have

$$Q_c = Q_b = -kA\frac{dT}{dx}\bigg|_{x=0} = -\frac{kA}{L}\theta_b\frac{d\phi}{d\xi}\bigg|_{\xi=0} \tag{4.17}$$

The derivative in Eq. 4.17 is obtained from Eq. 4.16 as

$$\frac{d\phi}{d\xi}\bigg|_{\xi=0} = \mu\frac{-\sinh[\mu(1-\xi)]\big|_{\xi=0}}{\cosh(\mu)} = -\mu\tanh(\mu)$$

We introduce this in Eq. 4.17 to get

$$Q_b = \left[-\frac{kA}{L}\theta_b\right][-\mu \tanh(\mu)] = mkA\theta_b \tanh(mL) \tag{4.18}$$

Note that Q_b is in W/m in the case of a fin in the form of a wide flat plate while it is in W in all the cases shown in Fig. 4.3. If $L \to \infty$ we note that $\tanh(\mu) \to 1$ and hence the heat loss becomes

$$Q_b = mkA\theta_b = \sqrt{\frac{hP}{kA}}kA\theta_b = \sqrt{hPkA}\,\theta_b \tag{4.19}$$

The temperature profile in the fin becomes

$$\theta = \theta_b e^{-mx} \tag{4.20}$$

A consequence of this is that the temperature of the fin approaches the ambient temperature for large x.

Fin Efficiency

The best that one may expect, or the maximum heat transfer one may expect occurs when the entire fin is at the base temperature. The heat loss in that case would be

$$\begin{aligned} Q_{max} &= \text{Fin surface area} \times \text{Heat transfer coefficient} \times \text{Temperature difference} \\ &= P \times L \times h \times (T_b - T_{amb}) = 2Lh\theta_b \end{aligned} \tag{4.21}$$

Obviously, the actual heat loss from the fin surface Q_b, given by Eq. 4.18 is less than or equal to Q_{max}. Hence one may refer to the ratio of the former to the latter as fin efficiency η to get

$$\boxed{\eta = \frac{Q_b}{Q_{max}} = \frac{mkA\theta_b \tanh(mL)}{PLh\theta_b} = \frac{\tanh(mL)}{mL} = \frac{\tanh(\mu)}{\mu}} \tag{4.22}$$

The last step follows from the defining expression for the fin parameter. A plot of fin efficiency η as a function of non-dimensional parameter μ is shown in Fig. 4.4b on page 105. It is seen that η decreases monotonically with increasing μ. The fin efficiency is unity in the limit $\mu \to 0$. In this limit, the actual heat transfer from the fin is zero! The fin parameter plays a very important role in determining the amount of heat transfer from the fin. We shall consider the fin parameter in more detail now.

The Fin Parameter

We may rearrange the square of the non-dimensional fin parameter as

$$\mu^2 = (mL)^2 = \frac{hP}{kA}L^2 = \frac{hL}{k} \times \frac{PL}{A} = Bi \times \left[\frac{PL}{A}\right] \tag{4.23}$$

where the now familiar Biot number appears as a factor. The second factor shown within braces is purely a geometric parameter that represents the ratio of two representative areas that characterize the fin geometry. In fact, the ratio is nothing but the ratio of surface area to the cross-sectional area of the fin. In practice, the Biot number is small but the area ratio is large. Recall that the Biot number represents the ratio of the two heat transfer rates, viz., the conduction within the fin over its length to convection from the surface to the ambient. In the limit $Bi \to 0$, the fin parameter will indeed tend to zero and the fin heat transfer tends to zero even though the fin is entirely at the base temperature (see temperature profile in Fig. 4.4). This limit is appropriate when the heat transfer coefficient is relatively small compared to the thermal conductivity and the fin length is also not too large. Naturally, the heat transfer from the fin is small. In the other limit of $Bi \to \infty$ the fin efficiency tends to zero. The fin, to a large extent of its length, will remain at the ambient temperature (refer Fig. 4.4) and does not really lose any heat to the ambient! This case actually represents the weak conduction limit. The reader is encouraged to figure out the details. The extended area is largely useless in the heat transfer process and hence *inefficient*. Thus very long fins are to be avoided in practice. Somewhere between these two limits lies the useful range of μ values. We shall look into this later when we discuss a simple optimum configuration.

The advantage of the graphical representation is that the fin heat loss may be easily calculated by reading off η from the graph and multiplying it by the expression for Q_{max} given by Eq. 4.21, thereby avoiding the cumbersome calculations involving hyperbolic functions.

Example 4.1

Consider a flat fin of thickness 1 mm, length 4 cm, and made of aluminum of thermal conductivity equal to 237 W/m°C. The fin is losing heat to an environment by convection subject to a heat transfer coefficient of 150 W/m^2°C. Calculate the fin heat loss per unit width and per unit temperature difference between the base and the ambient. Rework the problem with an iron fin with a thermal conductivity of 45 W/m°C, all other parameters being held fixed. Show the temperature profiles in the two cases on a common plot for a base temperature of 100 °C and an ambient temperature of 30 °C.

Solution:

Step 1 Given data is listed:

Fin thickness: $2t = 1$ mm $= 0.001$ m
and hence $t = 0.0005$ m
Fin length: $L = 4$ cm $= 0.04$ m
Heat transfer coefficient: $h = 150$ W/m^2 °C

Case 1: Aluminum fin

Step 2 With thermal conductivity of $k = 237\,\text{W/m°C}$ the fin parameter is calculated based on Table 4.1 as

$$\mu = L\sqrt{\frac{h}{kt}} = 0.04 \times \sqrt{\frac{150}{237 \times 0.0005}} = 1.423$$

The fin efficiency is calculated using Eq. 4.22 as

$$\eta = \frac{\tanh(1.423)}{1.423} = 0.626$$

The heat loss per unit width and unit temperature difference $\theta_b = 1\,°\text{C}$ is then calculated as

$$Q_b = 2hL\eta = 2 \times 150 \times 0.04 \times 0.626 = 7.512\,\text{W/m°C}$$

Step 3 We now generate the desired data for making a plot of temperature variation along the fin. The temperatures are taken as

$$T_b = 100\,°\text{C};\ T_{amb} = 30\,°\text{C}$$

We rewrite Eq. 4.15 in dimensional form as follows:

$$\begin{aligned} T(x) =& 30 + (100 - 30)\frac{\cosh\left[1.423\left(1 - \frac{x}{0.04}\right)\right]}{\cosh(1.423)} \\ =& 30 + 31.89\cosh\left[1.423\left(1 - \frac{x}{0.04}\right)\right] \end{aligned}$$

Case 2: Iron fin

Step 4 With thermal conductivity of $k = 45\,\text{W/m°C}$, the fin parameter is calculated based on Table 4.1 as

$$\mu = L\sqrt{\frac{h}{kt}} = 0.04 \times \sqrt{\frac{150}{45 \times 0.0005}} = 3.266$$

The fin efficiency is calculated using Eq. 4.22 as

$$\eta = \frac{\tanh(3.266)}{3.266} = 0.305$$

The heat loss per unit width and unit temperature difference $\theta_b = 1\,°\text{C}$ is then calculated as

$$Q_b = 2hL\eta = 2 \times 150 \times 0.04 \times 0.305 = 3.66\,\text{W/m°C}$$

We now generate the desired data for making a plot of temperature variation along the fin. The temperatures are taken as

$$T_b = 100\,°\text{C}; T_{amb} = 30\,°\text{C}$$

We rewrite Eq. 4.15 in dimensional form as follows:

$$\begin{aligned} T(x) &= 30 + (100 - 30)\frac{\cosh\left[3.266\left(1 - \frac{x}{0.04}\right)\right]}{\cosh(3.266)} \\ &= 30 + 5.33\cosh\left[3.266\left(1 - \frac{x}{0.04}\right)\right] \end{aligned}$$

Step 5 The temperature data generated above is plotted in Fig. 4.5. The figure compares the temperature variations along the fins of the two different materials. The figure indicates that the temperature variation is strongly affected by the thermal conductivity of the material. *Temperature variation is more pronounced in iron (lower thermal conductivity material) as compared to that in aluminum (higher thermal conductivity material).* It is as though the iron fin is thermally "longer" than the aluminum fin.

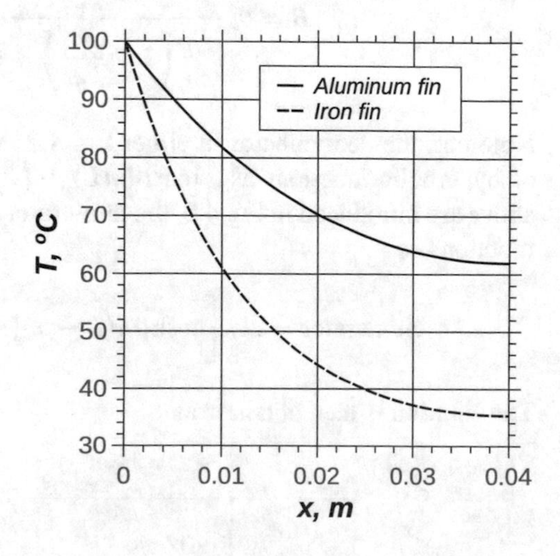

Fig. 4.5 Temperature variations in fins of two different materials in Example 4.1

4.2.3 *Uniform Area Fin Subject to Third Kind Boundary Condition at the Tip*

The second kind of boundary condition was justified, in Sect. 4.2.2, on the plea that the area available for heat transfer is small. We shall assess this assumption by changing it to the third kind of boundary condition. The general solution to the problem is still given by Eq. 4.8. The boundary condition at the base yields Eq. 4.9 as before. Now apply the tip boundary condition to get

$$-k\left(mAe^{mL} - mBe^{-mL}\right) = h_t\left(Ae^{mL} + Be^{-mL}\right)$$
$$\text{or } Ae^{mL}\left(1 + \frac{h_t}{km}\right) - Be^{-mL}\left(1 - \frac{h_t}{km}\right) = 0 \tag{4.24}$$

We use Eq. 4.9 in Eq. 4.24 to get A after some manipulations as

$$A = \theta_b \frac{e^{-mL}\left(1 - \dfrac{h_t}{km}\right)}{e^{mL}\left(1 + \dfrac{h_t}{km}\right) + e^{-mL}\left(1 - \dfrac{h_t}{km}\right)} \tag{4.25}$$

Substitute this back in Eq. 4.9 and get

$$B = \theta_b \frac{e^{-mL}\left(1 + \dfrac{h_t}{km}\right)}{e^{mL}\left(1 + \dfrac{h_t}{km}\right) + e^{-mL}\left(1 - \dfrac{h_t}{km}\right)} \tag{4.26}$$

Note that the denominator in either Eq. 4.25 or Eq. 4.26 may be written in terms of hyperbolic functions as $2\left[\cosh(mL) + \left(\frac{h_t}{km}\right)\sinh(mL)\right]$. When A and B given above are introduced in Eq. 4.8, the numerator may again be recast using hyperbolic functions as

$$\text{Numerator} = 2\theta_b\left[\cosh\{m(L-x)\} + \left(\frac{h_t}{km}\right)\sinh\{m(L-x)\}\right]$$

The solution is then obtained as

$$\boxed{\frac{\theta}{\theta_b} = \frac{\cosh\{m(L-x)\} + \left(\dfrac{h_t}{km}\right)\sinh\{m(L-x)\}}{\cosh(mL) + \left(\dfrac{h_t}{km}\right)\sinh(mL)}} \tag{4.27}$$

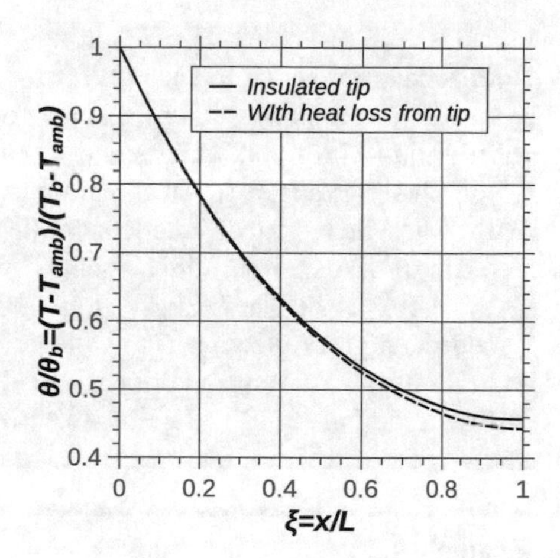

Fig. 4.6 Comparison of temperature profiles in a fin with and without tip heat loss

Note that if we set $h_t = 0$ we get back the solution given by Eq. 4.14, that corresponds to the insulated tip case. In the present case, an extra parameter given by $\left(\frac{h_t}{mk}\right)$ plays a role in addition to the fin parameter. In case we assume that the heat transfer coefficient h_t is equal to h, this second parameter turns out to be

$$\frac{h_t}{km} = \frac{h}{km} = \frac{h}{k\sqrt{\frac{hP}{kA}}} = \sqrt{\frac{hA}{kP}} \tag{4.28}$$

Notice that this parameter is non-dimensional. For a fin of uniform area in the form of a plate, this parameter is actually equal to

$$\sqrt{\frac{hA}{kP}} = \sqrt{\frac{h \times 2 \times t}{k \times 2}} = \sqrt{\frac{ht}{k}} = \sqrt{Bi_t} \tag{4.29}$$

where Bi_t ia a Biot number based on fin semi-thickness! We mentioned earlier that the one-dimensional assumption is based on a small value for this parameter! Hence the effect of the terms containing the hyperbolic sine in Eq. 4.27 is indeed small! The insulated tip condition would thus be a good assumption.

We consider the fin of Example 4.7 again, but by changing the boundary condition at the tip to that of third kind with $h_t = h = 67\,\text{W/m}^2\,^\circ\text{C}$. We make a plot of the temperature profile with this boundary condition and compare it with that obtained assuming insulated tip condition, in Fig. 4.6. It is seen that the two profiles are extremely close to each other and the difference does not exceed a maximum of around 0.3 °C!

Practical approach to account for tip heat loss
In practice, it is possible to account for heat loss from fin tip by extending it by **half the fin thickness** and **assuming adiabatic condition** at the fin tip. This was suggested in the pioneering work by D. R. Harper and W. B. Brown,"Mathematical equations for heat conduction in the fins of air cooled engines," NACA Rep. 158, National Committee on Aeronautics, Washington, DC, USA, 1922. The heat loss from the fin would be slightly enhanced by the added length. It is equivalent to assuming that the heat transfer coefficient for heat transfer from the additional lateral surface is the same as that along the rest of the fin.

Example 4.2

A pin fin of brass of diameter $D = 3\,\text{mm} = 0.003\,\text{m}$ and length $L = 60\,\text{mm} = 0.06\,\text{m}$ is used to remove heat from a parent surface maintained at $T_b = 70\,°\text{C}$. Fin loses heat to an ambient at a temperature of $T_{amb} = 20\,°\text{C}$ via convection with a uniform heat transfer coefficient of $h = 26\,\text{W/m}^2\,°\text{C}$. What is the heat transfer from the parent surface via the fin? What is the temperature at the fin tip? Take into account heat transfer from the fin tip assuming $h_t = h$.

Solution:

Step 1 Fin parameter is calculated as $m = \sqrt{\frac{4h}{kD}} = \sqrt{\frac{4\times 26}{116\times 0.003}} = 17.287$. Hence the non-dimensional fin parameter is $\mu = mL = 17.287 \times 0.06 = 1.037$. Parameter that accounts for tip heat loss is calculated as $\frac{h_t}{km} = \frac{h}{km} = \frac{26}{116\times 17.287} = 0.013$.

Step 2 Temperature gradient at the base of pin fin is obtained by differentiating with respect to x expression 4.27 and substituting $x = 0$ in the resulting expression. Thus we have

$$\frac{1}{\theta_b}\frac{\partial\theta}{\partial x}\bigg|_{x=0} = -m\left(\frac{\sinh(\mu) + \frac{h_t}{km}\cosh(\mu)}{\cosh(\mu) + \frac{h_t}{km}\sinh(\mu)}\right)$$

Substituting the numerical values we get

$$\frac{1}{\theta_b}\frac{\partial\theta}{\partial x}\bigg|_{x=0} = -m\left(\frac{\sinh(1.037) + 0.013\cosh(1.037)}{\cosh(1.037) + 0.013\sinh(1.037)}\right) = -13.603\,\text{m}^{-1}$$

Step 3 Heat loss from the fin may then be calculated as

$$Q = -k\left(\frac{\pi D^2}{4}\right)(T_b - T_{amb})\frac{1}{\theta_b}\frac{\partial \theta}{\partial x}\bigg|_{x=0}$$
$$= -116\left(\frac{\pi \times 0.003^2}{4}\right)(70-20)(-13.603) = 0.558\,\text{W}$$

Step 4 Tip temperature is determined by letting $x = L$ in Eq. 4.27 as

$$\frac{\theta_t}{\theta_b} = \frac{\cosh(0) + \frac{h_t}{km}\sinh(0)}{\cosh(\mu) + \frac{h_t}{km}\sinh(\mu)} = \frac{1}{\cosh(\mu) + \frac{h_t}{km}\sinh(\mu)}$$

We thus have

$$\frac{\theta_t}{\theta_b} = \frac{1}{\cosh(1.037) + 0.013\sinh(1.037)} = 0.617$$

Hence the fin tip temperature is

$$T_t = T_{amb} + \theta_t = 20 + (70-20) \times 0.617 = 50.86\,°\text{C}$$

4.3 Variable Area Fins

Material of a fin is effectively utilized when the longitudinal conduction heat flux within the fin is uniform. We return to this theme in Sect. 4.4. In the case of a uniform area fin, we have seen that the conduction flux reduces monotonically from the base as we move toward the tip. This may be easily appreciated by looking at the temperature profiles shown in Fig. 4.4. The longitudinal temperature gradient, i.e., $\left|\frac{dT}{dx}\right|$ is largest near the base and reduces to zero at the tip. Even when the tip heat loss is accounted for, as in Fig. 4.6, the state of affairs is not much different. Thus it is clear that if the temperature gradient were to be made more uniform, the longitudinal heat transfer may be made more uniform. This is possible by the use of a variable area fin wherein the area reduces with x, $\left|\frac{dT}{dx}\right|$ becomes more uniform and the material is used more efficiently in the heat transfer process. Thus a fin of triangular or trapezoidal profile represents a more efficient fin shape. In applications where radial fins of either uniform or non-uniform thickness are used, the area available for heat transfer, in fact, increases with the radial coordinate. Hence the fins are perforce of non-uniform area.

Hence the study of non-uniform or variable area fins is of importance from the viewpoint of applications.

4.3.1 General Analysis of Variable Area Fins

We start the analysis by formulating the governing equation for a pin fin of the variable area shown schematically in Fig. 4.7. The nomenclature used is also defined in this figure.

Consider an elemental length dx located at a distance x from the base plane. The lateral surface area of the fin from $x = 0$ to x is shown by the shaded surface. It is clear that the surface area of the element (that experiences convection to the ambient, shown in solid gray) is $dS = \frac{dS}{dx}dx$. Area of the section A is also a specified function of x. Energy balance for the element under steady state requires

$$\begin{bmatrix}\text{Conduction in} - \\ Q_{k,x}\end{bmatrix} = \begin{bmatrix}\text{Conduction out} - \\ Q_{k,x+dx}\end{bmatrix} + \begin{bmatrix}\text{Convection out} - \\ Q_c\end{bmatrix}$$

Convection leaving the element is shown as an elemental quantity since the surface area involved is an elemental area. Using Fourier law, we have

$$Q_{k,x} = -kA(x)\frac{dT}{dx}; \quad Q_{k,x+dx} - Q_{k,x} = \underbrace{-k\frac{d}{dx}\left[A(x)\frac{dT}{dx}\right]dx}_{\text{Based on Taylor expansion}}; \quad dQ_c = h\frac{dS}{dx}(T - T_{amb})dx$$

With these energy balance equation becomes

$$\frac{d}{dx}\left[A(x)\frac{d\theta}{dx}\right] - \frac{h}{k}\frac{dS}{dx}\theta = 0 \tag{4.30}$$

where, as usual, $\theta = T - T_{amb}$. Expanding the first term, we have

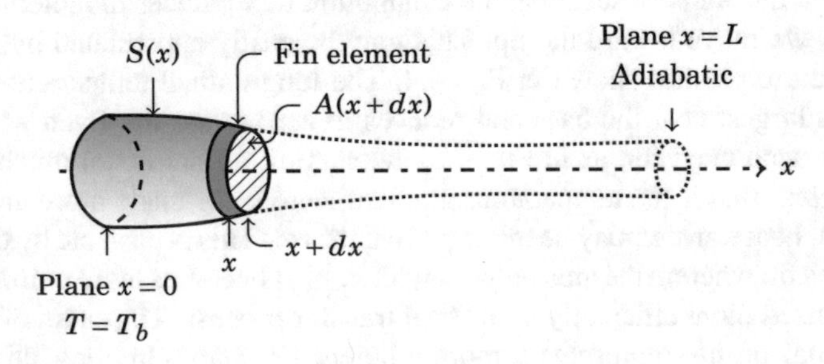

Fig. 4.7 Pin fin of variable area

$$\frac{d}{dx}\left[A(x)\frac{d\theta}{dx}\right] = A\frac{d^2\theta}{dx^2} + \frac{dA}{dx}\frac{d\theta}{dx} \tag{4.31}$$

Introducing this in Eq. 4.30 and dividing throughout by $A(x)$, we have

$$\boxed{\frac{d^2\theta}{dx^2} + \frac{1}{A}\frac{dA}{dx}\frac{d\theta}{dx} - \frac{1}{A}\frac{dS}{dx}\frac{h}{k}\theta = 0} \tag{4.32}$$

The coefficients $\frac{1}{A}\frac{dA}{dx}$ and $\frac{1}{A}\frac{dS}{dx}$ are, in general, functions of x and hence the governing equation is an ODE with variable coefficients. This is the essential difference between fins of constant area and fins of non-uniform area. The boundary conditions applicable to Eq. 4.32 are

$$\boxed{x = 0,\quad \theta = \theta_b;\quad \left.\frac{d\theta}{dx}\right|_{x=L} = 0 \text{ if } A(L) \neq 0 \text{ or } \theta_L \text{ is finite if } A(L) = 0} \tag{4.33}$$

4.3.2 *Particular Cases of Variable Area Fins*

Even though Eq. 4.32 was derived for a variable area fin in the form of a pin fin, it is very general and is valid for all variable area fins. These include fins of variable area made from a plate like structure or pin fins of variable area are usually referred to as "spines". By substituting expressions for S and A that characterize a particular fin shape, we get the appropriate fin equation, as particular/special cases.

Trapezoidal Fin

Analysis of a trapezoidal fin is made using the coordinate system and the nomenclature given in Fig. 4.8. Origin is located as indicated. Consider unit width in a direction normal to the plane of the figure.

Using the notation of Eq. 4.32, we have

$$A(x) = 1 \times 2y = 2y$$

From similar triangles OAB and OCD, we have

$$\frac{2y_b}{L} = \frac{2y}{x} \text{ or } y = \frac{y_b}{L}x$$

Hence $A(x) = 2\frac{y_b}{L}x$ and hence $\frac{dA}{dx} = 2\frac{y_b}{L}$. With this we get

$$\frac{1}{A}\frac{dA}{dx} = \frac{1}{2\frac{y_b}{L}x} \cdot 2\frac{y_b}{L} = \frac{1}{x} \tag{4.34}$$

The surface area is seen to be

$$S(x) = 2\ell \times 1 = 2\sqrt{(x - x_t)^2 + (y - y_t)^2} = 2(x - x_t)\sqrt{1 + \left(\frac{y - y_t}{x - x_t}\right)^2}$$

Since $\frac{y-y_t}{x-x_t} = \frac{y_b}{L}$ the above may be recast as

$$S(x) = 2(x - x_t)\sqrt{1 + \left(\frac{y_b}{L}\right)^2}$$

Hence

$$\frac{dS}{dx} = 2\sqrt{1 + \left(\frac{y_b}{L}\right)^2}$$

and

$$\frac{1}{A}\frac{dS}{dx} = 2\frac{\sqrt{1 + \left(\frac{y_b}{L}\right)^2}}{2\frac{y_b x}{L}} = \frac{L}{y_b x}\sqrt{1 + \left(\frac{y_b}{L}\right)^2} \tag{4.35}$$

On substituting Expressions given by 4.34 and 4.35 in Eq. 4.32, we get the governing equation for a trapezoidal fin as

$$\boxed{\frac{d^2\theta}{dx^2} + \frac{1}{x}\frac{d\theta}{dx} - \frac{p^2}{x}\theta = 0} \tag{4.36}$$

where the parameter p^2 is given by

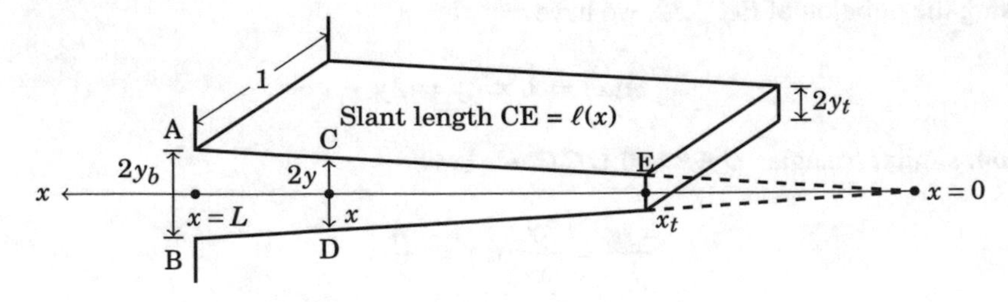

Fig. 4.8 Trapezoidal fin geometry nomenclature

$$\boxed{p^2 = \frac{hL}{ky_b}\sqrt{1+\left(\frac{y_b}{L}\right)^2}} \tag{4.37}$$

Solution to the Trapezoidal Fin Problem

Analytical solution to Eq. 4.37 is possible in terms of special functions called modified Bessel functions. These are tabulated functions available in handbooks of mathematics. An introductory discussion on these special functions, along with some useful formulae and tables of these functions is given in Appendix A.

Special Case of a Trapezoidal Fin: Triangular Fin

A fin of triangular profile is a special case of a trapezoidal fin when $x_t = 0$ and $y_t = 0$. In this case, the solution is very simple and is given by

$$\frac{\theta}{\theta_b} = \frac{I_0(2p\sqrt{x})}{I_0(2p\sqrt{L})} \tag{4.38}$$

where I_0 is the modified Bessel function of first kind and order zero. Note that in the case of triangular profiled fin, the length L is also the same as the physical length of fin. Using non-dimensional variables, as before, Eq. 4.38 may be rewritten as

$$\phi = \frac{I_0(2\mu\sqrt{\xi})}{I_0(2\mu)} \tag{4.39}$$

where $\phi = \frac{\theta}{\theta_b}$ is the non-dimensional temperature and $\mu = p\sqrt{L}$ is the non-dimensional fin parameter. Using formulae for derivatives of modified Bessel functions given by Eqs. A.55 in Appendix A it is possible to derive an expression for the fin efficiency as

$$\eta = \frac{I_1(2\mu)}{\mu I_0(2\mu)} \tag{4.40}$$

where I_1 is the modified Bessel function of first kind and order one.

The dependence of the temperature profile in a triangular fin on the fin parameter is brought out in Fig. 4.9. The most noticeable feature, when Fig. 4.9a is compared with Fig. 4.4(uniform area fin with insulated tip) is that the temperature gradient is non-zero at the tip. Since the area is zero at the tip for a triangular profiled fin, the temperature gradient need not vanish to make the heat flux zero. The second observation is that the temperature gradient is more uniform along a fin of triangular profile as compared with a fin of uniform area, even for a reasonably large value of μ such as $\mu = 1$. This signifies that a triangular fin uses the material more optimally as compared to a uniform area fin. Figure 4.9b shows the variation of efficiency of a fin of triangular profile with fin parameter μ.

Example 4.3

A fin of triangular section is made of an aluminum alloy of thermal conductivity equal to $k = 185\,\mathrm{W/m^{\circ}C}$. It has a base thickness of $2y_b = 3\,\mathrm{mm}$ and a length of $L = 56\,\mathrm{mm}$. The lateral surfaces of the fin are cooled by air at $T_{amb} = 25\,^{\circ}\mathrm{C}$ with a heat transfer coefficient of $h = 47\,\mathrm{W/m^{2\circ}C}$. How much heat does the fin dissipate if its width is $W = 0.3\,\mathrm{m}$ and the base is maintained at $T_b = 86\,^{\circ}\mathrm{C}$? Use fin efficiency from Fig. 4.9b to solve this problem.

Solution:

From the given data $y_b = 1.5\,\mathrm{mm} = 0.0015\,\mathrm{m}$. The fin parameter is calculated using definition 4.37 as

$$p = \sqrt{\frac{hL}{ky_b}\sqrt{1+\left(\frac{y_b}{L}\right)^2}} = \sqrt{\frac{47 \times 0.056}{185 \times 0.0015}\sqrt{1+\left(\frac{0.0015}{0.056}\right)^2}} = 3.08\,\mathrm{m}^{-\frac{1}{2}}$$

Hence the non-dimensional fin parameter is

$$\mu = p\sqrt{L} = 3.08 \times \sqrt{0.056} = 0.729$$

From Fig. 4.9b, the fin efficiency is read off as, $\eta = 0.77$. Hence the total heat loss from the fin is calculated as

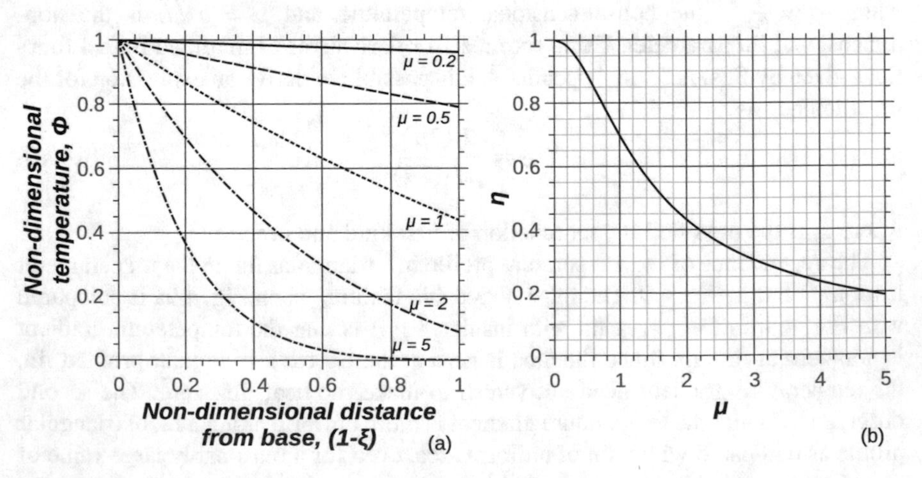

Fig. 4.9 **a** Fin temperature profile variation with fin parameter for a triangular fin **b** Variation of efficiency of a triangular fin with fin parameter

$$Q_b = 2L\sqrt{1+\left(\frac{y_b}{L}\right)^2}\,Wh\eta(T_b - T_{amb})$$

$$= 2 \times 0.056\sqrt{1+\left(\frac{0.0015}{0.056}\right)^2} \times 0.3 \times 47 \times 0.77(86-25) = 74.2\,\text{W}$$

The next example compares the relative thermal performance of uniform area fin with a triangular fin. We take the same length and base thicknesses in the two cases. This example looks at the effect of the longitudinal profile of fin on heat transfer.

Example 4.4

Compare the heat loss per unit mass for the following two configurations, viz., (a) Uniform area flat plate type fin, (b) Triangular fin. Both fins are 10 cm long. The thickness of the flat plate type fin and the base thickness of the triangular fin are both equal to 1.2 mm. Compare on the basis of unit width of fin in both the cases. The material of the fin is pure aluminum with thermal conductivity equal to 237 W/m°C and density of 2700 kg/m^3. The base and ambient temperatures are, respectively, 60 and 15 °C. The heat transfer coefficient has been determined to be 26 W/m^2°C in both cases.

Solution:
Data that is common to the two cases are

Fin length: $L = 0.1$ m
Thermal conductivity: $k = 237$ W/m°C
Heat transfer coefficient: $h = 26$ W/m^2°C
Base temperature: $T_b = 60$°C
Ambient temperature: $T_{amb} = 15$°C
Density of material of fin: $\rho = 2700$ kg/m^3

Case (a) Uniform area fin:

Step 1 The data specific to this case is

$$\text{Fin thickness: } 2t = 1.2\,\text{mm or } t = 0.0006\,\text{m}$$

The fin parameter may be calculated as

$$m_a = \sqrt{hkt} = \frac{26}{237 \times 0.0006} = 13.522\,\text{m}^{-1}$$

where the subscript "a" represents Case (a). The non-dimensional fin parameter is then calculated as

$$\mu_a = m_a L = 13.522 \times 0.1 = 1.352$$

Step 2 The fin efficiency is calculated as

$$\eta_a = \frac{\tanh(\mu_a)}{\mu_a} = \frac{\tanh(1.352)}{1.352} = 0.647$$

Alternately the efficiency may be read off Fig. 4.4b.

Step 3 The heat loss from the fin may be calculated as

$$Q_a = 2hL\eta_a(T_b - T_{amb}) = 2 \times 26 \times 0.1 \times 0.647 \times (60 - 15) = 151.4\,\text{W/m}$$

Step 4 Per unit width the volume of the fin is

$$V_a = 2tL = 2 \times 0.0006 \times 0.1 = 0.00012\,\text{m}^3$$

Mass of fin per unit width is

$$M_a = \rho V_a = 2700 \times 0.00012 = 0.324\,\text{kg/m}$$

Heat loss per unit mass of fin material is then given by

$$q_a = \frac{Q_a}{M_a} = \frac{151.4}{0.324} = 467.3\,\text{W/kg}$$

Case (b) Triangular fin:

Step 5 The data specific to this case is

$$\text{Base thickness: } 2y_b = 1.2\,\text{mm or } y_b = 0.0006\,\text{m}$$

The fin parameter may be calculated as

$$p_b = \sqrt{\frac{hL}{ky_b}\sqrt{1 + \left(\frac{y_b}{L}\right)^2}} = \sqrt{\frac{26 \times 0.1}{237 \times 0.0006}\sqrt{1 + \left(\frac{0.0006}{0.1}\right)^2}} = 4.276\,\text{m}^{-\frac{1}{2}}$$

where the subscript "b" represents Case (b). The non-dimensional fin parameter is then calculated as

$$\mu_b = p_b\sqrt{L} = 4.276 \times \sqrt{0.1} = 1.352$$

Step 6 The fin efficiency is read off Fig. 4.9b as $\eta_b = 0.58$. The heat loss from the fin may be calculated as

$$Q_b = 2hL\sqrt{1+\left(\frac{y_b}{L}\right)^2}\eta_b(T_b - T_{amb})$$
$$= 2 \times 26 \times 0.1\sqrt{1+\left(\frac{0.0006}{0.1}\right)^2} \times 0.58 \times (60-15) = 135.9\,\text{W/m}$$

Step 7 Per unit width the volume of the fin is

$$V_b = tL = 2 \times 0.0006 \times 0.1 = 0.00006\,\text{m}^3$$

Mass of fin per unit width is

$$M_a = \rho V_b = 2700 \times 0.00006 = 0.162\,\text{kg/m}$$

Heat loss per unit mass of fin material is then given by

$$q_b = \frac{Q_b}{M_b} = \frac{135.9}{0.162} = 839.1\,\text{W/kg}$$

Step 8 Even though the heat loss is a bit lower in the case of the triangular fin, it is seen that per unit mass it loses $839.1/467.2 = 1.796 \approx 1.8$ times the heat loss by the uniform area fin.

Conical Spine

Performance of a flat fin of uniform thickness could be improved by replacing it with a fin of triangular profile. Similarly the performance of a fin in the form of a rod of uniform diameter may be improved by replacing it by a conical spine. The resulting fin will look like what is shown in Fig. 4.10.

The governing equation for this case may be derived starting from the general fin Eq. 4.32. For this purpose, we identify the axial coordinate x as in Fig. 4.10. Area of the fin as a function of x is given by

$$A(x) = \pi r^2 = \pi\left(\frac{r_b x}{L}\right)^2 \tag{4.41}$$

The coefficient of the first derivative term is then given by

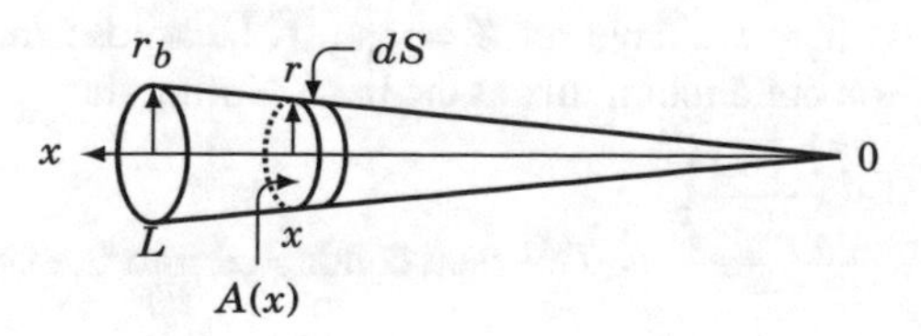

Fig. 4.10 Conical spine geometry and nomenclature

$$\frac{1}{A}\frac{dA}{dx} = \frac{2\pi x\left(\frac{r_b}{L}\right)^2}{\pi\left(\frac{r_b x}{L}\right)^2} = \frac{2}{x} \tag{4.42}$$

The surface area $S(x)$ is given by

$$S(x) = 2\pi\sqrt{1+\left(\frac{r_b}{L}\right)^2}\,r = 2\left(\pi\frac{r_b}{L}\right)\sqrt{1+\left(\frac{r_b}{L}\right)^2}\cdot x \tag{4.43}$$

The second coefficient that appears in the fin equation is then given by

$$\frac{h}{kA}\frac{dS}{dx} = \frac{2hL}{kr_b x}\sqrt{1+\left(\frac{r_b}{L}\right)^2} \tag{4.44}$$

On substitution in Eq. 4.32, the general fin equation simplifies as

$$\boxed{\frac{d^2\theta}{dx^2} + \frac{2}{x}\frac{d\theta}{dx} - \frac{p^2}{x}\theta = 0} \tag{4.45}$$

where $p^2 = \frac{2hL}{kr_b}\sqrt{1+\left(\frac{r_b}{L}\right)^2}$. The boundary conditions are specified as $\theta(x) = \theta_b$ at $x = L$ and $\theta =$ finite at $x = 0$. Equation 4.45 may be brought to standard form by multiplying throughout by x^2. It may be compared with the standard differential Eq. A.46 to write the general solution in terms of modified Bessel functions as

$$\theta = A\frac{I_1(2p\sqrt{x})}{\sqrt{x}} + B\frac{K_1(2p\sqrt{x})}{\sqrt{x}}$$

Since the latter term is unbounded at $x = 0$, we choose the integration constant $B = 0$. Setting $\theta(x = L) = \theta_b$ constant, A is obtained as $A = \frac{\sqrt{L}}{I_1(2p\sqrt{L})}$. Hence the solution may be written down as

$$\frac{\theta_x}{\theta_b} = \sqrt{\frac{L}{x}}\frac{I_1(2p\sqrt{x})}{I_1(2p\sqrt{L})} = \frac{2p\sqrt{L}}{X}\frac{I_1(X)}{I_1(2p\sqrt{L})} \tag{4.46}$$

where we have set $X = 2p\sqrt{x}$. Heat loss from the spine may be calculated based on conduction flux at the base. Noting that $\frac{d}{dx} = \frac{dX}{dx}\frac{d}{dX} = \frac{p}{\sqrt{x}}\frac{d}{dX}$, using the identity $\frac{d\left(\frac{I_1(X)}{X}\right)}{dX} = \frac{I_2(X)}{X}$ heat conducted into the spine at its base may be shown to be

$$Q_b = \frac{k\pi r_b^2 p\theta_b}{\sqrt{L}} \cdot \frac{I_2(2p\sqrt{L})}{I_1(2p\sqrt{L})}$$

The maximum possible heat loss Q_{max} occurs if the spine is at the base temperature throughout its length. It is given by

$$Q_{max} = h\pi r_b L\theta_b$$

Efficiency of the spine is given by the ratio of these two and hence we have

$$\eta = \frac{kr_b p}{hL^{\frac{3}{2}}} \cdot \frac{I_2(2p\sqrt{L})}{I_2(2p\sqrt{L})} \quad (4.47)$$

Example 4.5

A conical spine of a material of thermal conductivity $k = 45\,\text{W/m°C}$, base diameter of $d_b = 3\,\text{mm}$ and length $L = 100\,\text{mm}$ is used to remove heat from a parent surface maintained at $T_b = 120\,°\text{C}$. Fin loses heat to an ambient at a temperature of $T_{amb} = 20\,°\text{C}$ via convection with a uniform heat transfer coefficient of $h = 15\,\text{W/m}^2\,°\text{C}$. Make a plot of temperature variation along the fin. Determine the efficiency of the spine. What is the heat loss from the spine?

Solution:

Step 1 Radius of cross section of spine at its base is $r_b = \frac{d_b}{2} = \frac{0.003}{2} = 0.0015\,\text{m}$. Length of spine is $L = 100\,\text{mm} = 0.1\,\text{m}$. Base radius to length ratio is hence given by $\frac{r_b}{L} = \frac{0.0015}{0.1} = 0.015$. Hence parameter p may now be calculated, neglecting the factor involving this ratio as

$$p = \sqrt{\frac{2hL}{kr_b}} = \sqrt{\frac{2 \times 15 \times 0.1}{45 \times 0.0015}} = 44.4444$$

We then have $p\sqrt{L} = 2 \times 44.4444\sqrt{0.1} = 2.1082$.

Step 2 We shall introduce non-dimensional location along the spine by the relation $\xi = 1 - \frac{x}{L}$ such that $\xi = 0$ corresponds to the base of the spine and $\xi = 1$ corresponds to the tip of the spine. With $\phi = \frac{\theta}{\theta_b}$ where $\theta_b = T_b - T_{amb} = 120 - 20 = 100\,°\text{C}$ the non-dimensional temperature variation with ξ is given, using Eq. 4.46, as

$$\phi(\xi) = \frac{1}{\sqrt{1-\xi}} \frac{I_1\left(2p\sqrt{L(1-\xi)}\right)}{I_1\left(2p\sqrt{L}\right)} = \frac{1}{\sqrt{1-\xi}} \frac{I_1\left(4.2164\sqrt{(1-\xi)}\right)}{I_1(4.2164)}$$

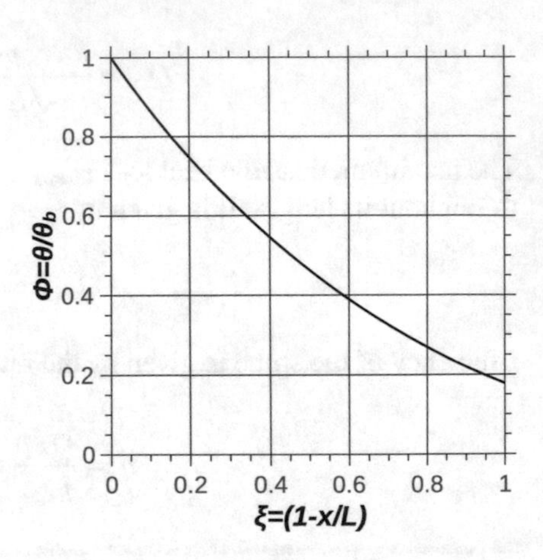

Fig. 4.11 Non-dimensional temperature profile in the conical spine of Example 4.5

Step 3 A plot of ϕ vs ξ is shown in Fig. 4.11.
The tip temperature is obtained by letting $\xi = 1$ in the expression for ϕ. It is noted that the ratio $\frac{I_1(4.2164\sqrt{t})}{\sqrt{t}}$ as $t \to 0$ is required. This ratio is given by $p\sqrt{L}$ and hence the tip temperature is obtained as

$$\phi(\xi = 1) = \frac{2.1082}{I_1(4.2164)} = 0.1774$$

Hence the dimensional tip temperature is $T(x = 0) = 0.1774 \times 100 = 17.74\,°\mathrm{C}$.

Step 4 We calculate the fin efficiency using expression 4.47 as

$$\eta = \frac{45 \times 0.0015 \times 44.4444}{15 \times 0.1^{3/2}} \cdot \frac{I_2(4.2164)}{I_2(4.2164)} = 0.6387$$

Maximum possible heat transfer by the spine is given by

$$Q_{max} = 15 \times \pi \times 0.0015 \times 0.1 \times 100 = 0.707\,\mathrm{W}$$

Hence the heat loss from the spine is given by

$$Q_b = Q_{max}\eta = 0.707 \times 0.6387 = 0.452\,\mathrm{W}$$

Radial Fin of Uniform Thickness

A second example is that of a radial flat fin of base radius r_1, tip radius r_2, and uniform thickness $2t$. Such fins are used to enhance heat transfer from the surface of a fluid-carrying tube. The nomenclature appropriate to this geometry is shown

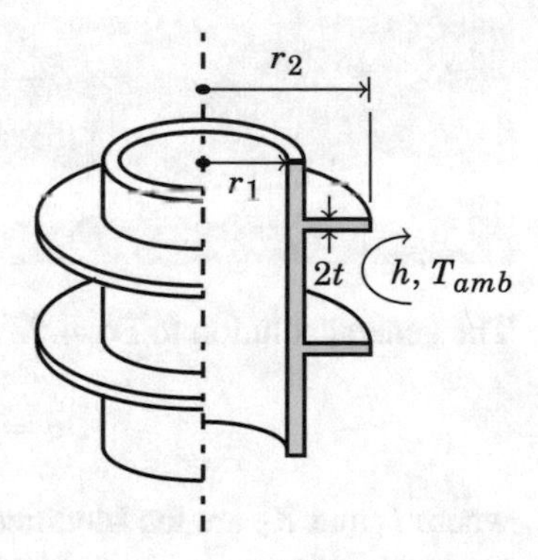

Fig. 4.12 Nomenclature for a radial fin of uniform thickness

in Fig. 4.12. The governing equation for this case may be derived starting from the general fin Eq. 4.32. For this purpose we identify the radial coordinate r as x in that equation. Area of the fin as a function of r is given by

$$A(r) = (2\pi r) \times (2t) = 4\pi rt \tag{4.48}$$

The coefficient of the first derivative term is then given by

$$\frac{1}{A}\frac{dA}{dr} = \frac{4\pi t}{4\pi rt} = \frac{1}{r} \tag{4.49}$$

The surface area $S(r)$ is given by

$$S(r) = 2 \times [\pi(r^2 - r_1^2)] \tag{4.50}$$

The second coefficient that appears in the fin equation is then given by

$$\frac{1}{A}\frac{dS}{dr} = \frac{1}{4\pi rt}(4\pi r) = \frac{1}{t} \tag{4.51}$$

Substituting Expressions 4.49 and 4.51 in Eq. 4.32, we get

$$\boxed{\frac{d^2\theta}{dr^2} + \frac{1}{r}\frac{d\theta}{dr} - p^2\theta = 0} \tag{4.52}$$

where $p = \sqrt{\frac{h}{kt}}$ represents the fin parameter appropriate to this case. Making use of insulated boundary condition at the tip, we also have

$$\theta|_{r=r_1} = \theta_b; \quad \left.\frac{d\theta}{dr}\right|_{r=r_2} = 0 \tag{4.53}$$

The general solution to Eq. 4.52 is given by (see Example 1 in Appendix A)

$$\theta = AI_0(pr) + BK_0(pr) \tag{4.54}$$

where I_0 and K_0 are the Modified Bessel functions of first and second kind, respectively, and of order 0. The constants of integration may be obtained using the boundary conditions. The resulting temperature profile is given by (see Example 3 in Appendix A)

$$\frac{\theta}{\theta_b} = \frac{K_0(pr_2)I_0(pr) + I_1(pr_2)K_0(pr)}{K_0(pr_2)I_0(pr_1) + I_1(pr_2)K_0(pr_1)} \tag{4.55}$$

In the above I_1 and K_1 represent the modified Bessel functions of first and second kind, respectively, and of order 1. An expression for fin efficiency may be derived as (see Example 3 in Appendix A)

$$\eta = \left(\frac{2\alpha}{\mu\left(1-\alpha^2\right)}\right)\left(\frac{K_1(\alpha\mu)I_1(\mu) - I_1(\alpha\mu)K_1(\mu)}{K_0(\alpha\mu)I_1(\mu) + I_0(\alpha\mu)K_1(\mu)}\right) \tag{4.56}$$

Thus there are two parameters that govern the problem, viz., the radius ratio, $\alpha = \frac{r_2}{r_1}$ and the non-dimensional fin parameter, $\mu = r_2\sqrt{\frac{h}{kt}}$. From an application point of view, it is useful to present a graph of efficiency versus fin parameter μ as shown in Fig. 4.13. A family of curves is obtained for different values of α. Useful or applicable range of μ and α are considered in making the plot.

Example 4.6

Consider an annular fin of uniform thickness equal to 1.6 mm. The base radius of the fin is 20 mm while the tip radius is 40 mm. The base structure to which the fin is attached is at 77 °C while the air flowing over the fin is at 15 °C. The average heat transfer coefficient is given to be 15 W/m^2°C while the fin has a thermal conductivity of 16 W/m°C. Determine the heat loss from a single fin.

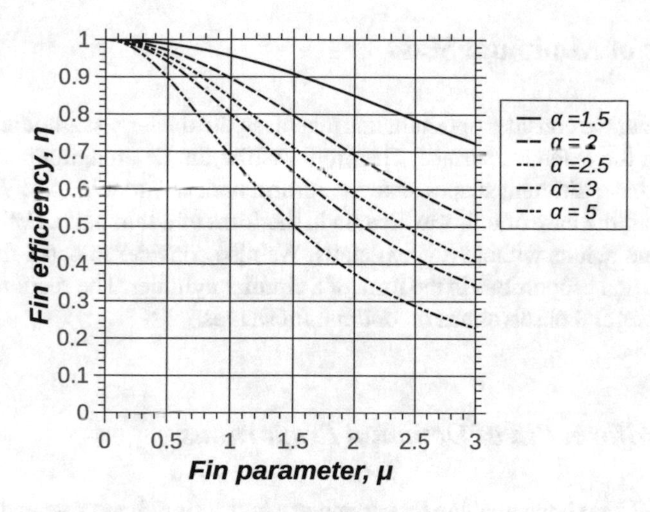

Fig. 4.13 Efficiency of a radial fin of uniform thickness

Solution:

The given data is recorded below, in the notation of the text.

$$r_1 = 0.02\text{ m};\ r_2 = 0.04\text{ m};\ t = 0.0008\text{ m};\ h = 15\text{ W/m}^2\,^\circ\text{C};$$
$$k = 16\text{ W/m}\,^\circ\text{C};\ T_b = 77\,^\circ\text{C};\quad T_{amb} = 15\,^\circ\text{C}$$

The two parameters that characterize the fin are calculated now.

$$\alpha = \frac{r_2}{r_1} = \frac{0.04}{0.02} = 2;\quad \mu = r_2\sqrt{\frac{h}{kt}} = 0.04 \times \sqrt{\frac{15}{16 \times 0.0008}} = 1.369$$

The fin efficiency from is read off Fig. 4.13 as $\eta = 0.82$. The heat loss from a single fin is then calculated as

$$\begin{aligned} Q_b &= 2\pi(r_2^2 - r_1^2)h\eta(T_b - T_{amb}) \\ &= 2 \times \pi \times (0.04^2 - 0.02^2) \times 15 \times 0.82 \times (77 - 15) \\ &= 5.75\text{ W} \end{aligned}$$

4.4 Fins of Minimum Mass

In this section, we consider optimum fins that dissipate the largest amount of heat per unit mass of the extended surface. The profile of the fin, i.e., longitudinal section of the fin may have different shapes such as uniform area or variable area. We consider in some detail the case of uniform area fin in the form of a thin plate. Following this, we give some results without giving details. We also consider in detail the case of a spine of optimal proportions in the form of a circular cylinder. The section concludes with some general observations on optimal fin shapes.

4.4.1 *Uniform Fin of Optimum Proportions*

Section 4.2.2 has shown that the fin parameter plays a significant role in determining the thermal performance of a fin. We have already dealt with the behavior of fin efficiency with asymptotic values of $\mu \to 0$ and $\mu \to \infty$. The pertinent question to ask now is whether a particular value of μ is the most desirable. For this, of course, we should set forth some criteria.

Consider a uniform area fin in the form of a flat plate made of a specific material. Let the heat transfer coefficient have a specified value. Thus the values of h and k are fixed. Let the base and ambient temperatures also have specified fixed values. It is desired to find the proportions of a fin of given mass per unit width that dissipates the largest amount of heat from the base structure to the ambient. Since the material of the fin is fixed and width may be taken as unity, the requirement may be rephrased by saying that the desire is to find the proportions of a fin of specified profile area $A_p = 2tL =$ constant that loses the most heat. The analysis should find a combination t^*, L^* that

$$\text{Maximizes } Q_b \text{ for } A_p = 2tL = \text{constant for fixed } h, k, T_b, T_{amb} \tag{4.57}$$

The optimization may be done by standard calculus method. The fin parameter may be represented as

$$\mu = \sqrt{\frac{h}{kt}}L = \sqrt{\frac{h}{k\left(\frac{A_p}{2L}\right)}}L = \underbrace{\sqrt{\frac{2h}{kA_p}}}_{=K}L^{\frac{3}{2}} = KL^{\frac{3}{2}}$$

where the constant K involves h, k and A_p which are all held fixed. The fin heat transfer per unit width and temperature difference may now be written using Eqs. 4.21 and 4.22 as

$$\frac{Q_b}{\theta_b} = 2hL\eta = 2h\frac{\tanh\left(KL^{3/2}\right)}{KL^{1/2}}$$

Defining $\frac{Q_b K}{2h\theta_b}$ as Q, we have to maximize

$$Q = \frac{\tanh\left(KL^{3/2}\right)}{L^{1/2}} \tag{4.58}$$

by the right choice of L. The condition that has to be satisfied is $\frac{\partial Q}{\partial L} = 0$. Thus, we have

$$\frac{\partial Q}{\partial L} = \frac{\text{sech}^2\left(KL^{3/2}\right) \times \frac{3}{2} \times KL^{1/2}}{L^{1/2}} - \frac{\tanh\left(KL^{3/2}\right)}{2L^{3/2}} = 0 \tag{4.59}$$

Equation 4.59 may be rearranged to read

$$3\mu\,\text{sech}^2(\mu) - \tanh(\mu) = 0 \tag{4.60}$$

This transcendental equation needs to be solved by a numerical method to find the desired root. Newton–Raphson method may be used as follows. Rewrite the hyperbolic functions in terms of the exponentials to recast Eq. 4.60 as

$$f(\mu) = e^{2\mu} - e^{-2\mu} - 12\mu = 0 \tag{4.61}$$

Ignoring the trivial root $\mu = 0$, the desired root may be determined by using the Newton–Raphson iterative scheme

$$\mu_{new} = \mu_{old} - \frac{f(\mu_{old})}{f'(\mu_{old})} = \mu_{old} - \frac{e^{2\mu_{old}} - e^{-2\mu_{old}} - 12\mu_{old}}{2\left(e^{2\mu_{old}} + e^{-2\mu_{old}}\right) - 12} \tag{4.62}$$

where f' is the derivative of $f(\mu)$ with respect to μ. The iteration starts with a guess value shown with the subscript "old" and yields a better value shown with the subscript "new". Convergence is very rapid as indicated in the following sequence:

μ_{old}	1.50000	1.42799	1.41934	1.41922
μ_{new}	1.42799	1.41934	1.41922	1.41922

It is sufficient to take the optimum value as $\mu^* = 1.4192$. The corresponding value of fin efficiency is $\eta^* = 0.6267$. We consider a typical example now.

Example 4.7

A sheet of aluminum alloy 1.5 mm thick is available for providing fins in a heat transfer application. Determine the length of the fin you would suggest such that it is an optimum fin. The thermal conductivity of the material of the sheet is 200 W/m°C while the heat transfer coefficient is 67 W/m^2°C. What is the heat loss per unit temperature difference between the fin base and the ambient? Discuss a few non-optimal cases.

Solution:

Step 1 Since the fin material is of specified thickness we assume that this corresponds to the optimum thickness. Thus

$$t = t^* = \frac{0.0015}{2} = 0.00075\,\text{m}$$

Step 2 The length of the fin is chosen such that $\mu = \mu_0 = 1.4192$. Since $\mu_0 = m^* L^*$, we have

$$L^* = \frac{\mu^*}{m^*} = \frac{\mu_0}{\sqrt{\frac{h}{kt^*}}} = \frac{1.4192}{\sqrt{\frac{67}{200 \times 0.00075}}} = 0.067\,\text{m}$$

Step 3 Figure 4.14 compares non-optimal proportioned fins with a fin that is proportioned to obtain optimal behavior. The figures are to scale. All the cases have a profile area A_p equal to $2t^* L^* = 2 \times 0.00075 \times 0.067 = 0.0001\,\text{m}^2$.

Step 4 Heat transfer per unit temperature difference may then be calculated as

$$\frac{Q_b}{\theta_b} = 2hL^*\eta^* = 2 \times 67 \times 0.067 \times 0.6267 = 5.639\,\text{W/m°C}$$

Heat transfer from the fin per unit volume and temperature difference is thus given by

$$\frac{Q_b}{\theta_b A_p} = \frac{5.639}{0.0001} = 56390\,\text{W/m}^3\text{°C}$$

Step 5 The calculations have been performed also for a number of non-optimal fins and the resulting data has been shown plotted, in Fig. 4.15, as a graph

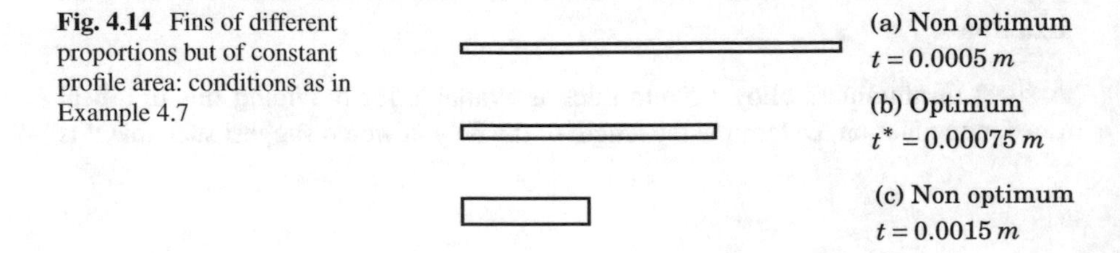

Fig. 4.14 Fins of different proportions but of constant profile area: conditions as in Example 4.7

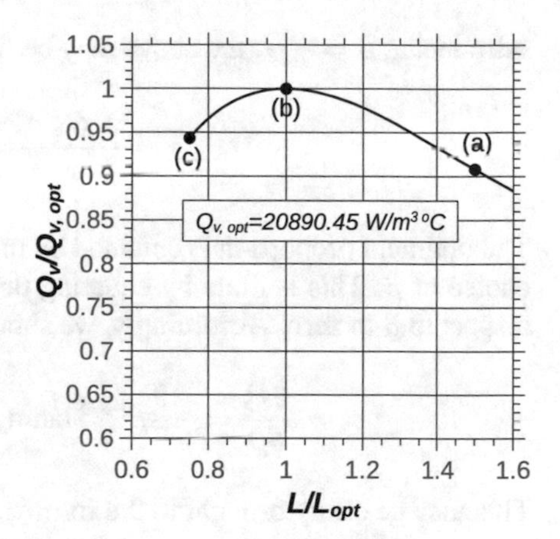

Fig. 4.15 Performance of fins of different proportions but of constant profile area: conditions as in Example 4.7

of ratio heat transfer from the fin per unit volume to the maximum possible heat transfer per unit volume and temperature difference, as a function of fin length. The optimum value is shown by the black circle labeled (b) in this figure.

4.4.2 Uniform Spine (pin Fin) of Optimum Proportions

We now consider a spine in the form of cylinder of circular cross section. It is desired to find the optimum diameter D^* and optimum length L^* of a cylindrical pin fin that loses the most heat per unit volume of fin, i.e., fixed $V = \frac{\pi D^2}{4} L$. The material thermal conductivity, heat transfer coefficient, base, and ambient temperatures are also held fixed, as before in the case of a fin in the form of a flat plate.

From Table 4.1, the fin parameter in this case is $m = \sqrt{\frac{4h}{kD}}$. Heat loss from the fin is given by Eq. 4.19. We thus have

$$Q_b = \sqrt{hPkA}\theta_b \tanh mL = \sqrt{h\pi Dk\pi \frac{D^2}{4}}\theta_b \tanh\left(\sqrt{\frac{4h}{kD}}L\right) = \theta_b \pi \sqrt{hk}\frac{D^{3/2}}{2} \tanh\left(\sqrt{\frac{4h}{kD}}L\right)$$

Define a volume parameter $V' = D^2 L$ such that $L = \frac{V'}{D^2}$. Then the argument under the hyperbolic tangent becomes $\sqrt{\frac{4h}{kD}}L = \sqrt{\frac{4h}{kD}}\frac{V'}{D^2} = 2V'\sqrt{\frac{h}{k}}\frac{1}{D^{5/2}}$. Introducing $K_1 = \theta_b \frac{\pi\sqrt{hk}}{2}$ and $K_2 = 2V'\sqrt{\frac{h}{k}}$ the above expression may be recast as

$$Q_b = K_1 D^{3/2} \tanh\left(\frac{K_2}{D^{5/2}}\right)$$

Introducing $\beta = \frac{K_2}{D^{5/2}}$, the above may be finally brought to the form

$$Q = \frac{Q_b}{K_1 K_2^{3/5}} = \frac{1}{\beta^{3/5}} \tanh(\beta) \tag{4.63}$$

The optimum proportion is obtained by maximizing the above expression by proper choice of β. This is done by equating the derivative of the above expression with respect to β to zero. Accordingly, we should have

$$\frac{\partial Q}{\partial \beta} = -\frac{3}{5}\beta^{-8/5} \tanh(\beta) + \beta^{-3/5} \text{sech}^2 \beta = 0$$

This may be easily brought to the form

$$\sinh(2\beta) = \frac{5}{3}(2\beta) \tag{4.64}$$

Transcendental Eq. 4.64 may be easily solved by Newton–Raphson method to get $\beta^* = 0.9193$. With this value of β the optimum diameter and length are given by

$$\text{(a)} \quad D^* = 1.3647 \left(\frac{V'^2 h}{k}\right)^{1/5} \quad \text{(b)} \quad L^* = 0.5370 \left(\frac{V' k^2}{h^2}\right)^{1/5} \tag{4.65}$$

Example 4.8

A spine in the form of a right circular cylinder, made of a material with thermal conductivity $k = 45\,\text{W/m}^\circ\text{C}$, is attached to a base structure at $T_b = 120\,^\circ\text{C}$. It loses heat to ambient air at $T_{amb} = 20\,^\circ\text{C}$ by convection via a heat transfer coefficient $h = 15\,\text{W/m}^{2\circ}\text{C}$. Determine the proportions of an optimum spine if the volume parameter is $V' = 10^{-6}\,\text{m}^3$. How much heat does it lose to the ambient? What is the heat loss per unit volume of the material of spine?

Solution:

Step 1 With the data specified in the problem, we calculate the parameters that appear in the optimum spine.

$$K_1 = \theta_b \frac{\pi\sqrt{hk}}{2} = (120 - 20)\frac{\pi\sqrt{15 \times 45}}{2} = 4081.0486$$

$$K_2 = 2V'\sqrt{\frac{h}{k}} = 2 \times 10^{-6} \times \sqrt{\frac{15}{45}} = 1.1547 \times 10^{-6}$$

Step 2 The diameter of optimum spine is obtained using Eq. 4.65a as

$$D^* = 1.3647\left(\frac{10^{-12}\times 15}{45}\right)^{1/5} = 0.0044\,\text{m}$$

The length of optimum spine is obtained using Eq. 4.65b as

$$L^* = 0.5370\left(\frac{10^{-6}\times 15^2}{45^2}\right)^{1/5} = 0.0526\,\text{m}$$

Step 3 Heat loss from the spine is then calculated as

$$Q_b = K_1 D^{3/2}\tanh\beta^* = 4081.0486\times 0.0044^{3/2}\tanh 0.9193 = 0.8529\,\text{W}$$

Step 4 Spine volume is given by

$$V = \pi\frac{D^{*2}L}{4} = \frac{\pi V'}{4} = \frac{\pi\times 10^{-6}}{4} = 7.854\times 10^{-7}\,\text{m}^3$$

Heat transfer per unit volume is then given by

$$Q = \frac{Q_b}{V} = \frac{0.8529}{7.854\times 10^{-7}} = 108590\,\text{W/m}^3 = 1.0859\,\text{MW/m}^3$$

Note on fin profile for optimum heat loss per unit mass:
Example 4.4 has indicated that per unit mass basis a triangular profile fin loses more heat than a rectangular profile fin. We have seen earlier that the temperature distribution in both cases is non-linear with distance from the fin base. The derivative of temperature decreases monotonically from the base to tip. This means that the conduction flux decreases monotonically from base to tip. It is conceivable that a fin with a profile that leads to a linear temperature variation with x and hence a constant temperature gradient will be the best profile. In order to elaborate on this, we work out the details in what follows.

Consider a fin profile that achieves a linear temperature variation with x given by $\theta = \theta_b\left(1-\frac{x}{L}\right)$ where L is the fin length. Then we have $\frac{d\theta}{dx} = -\frac{\theta_b}{L}$ and $\frac{d^2\theta}{dx^2} = 0$. In Eq. 4.32, we take $A = 2y(x)$—where$y(x)$ is semi-fin thickness at a location x along the fin, $\frac{dS}{dx} \approx 2$ (thin fin approximation), substitute the linearly varying θ to get

$$0 + \frac{1}{2y}\times 2\frac{dy}{dx}\times -\frac{\theta_b}{L} - \frac{1}{2y}\times 2\frac{h}{k}\theta_b\left(1-\frac{x}{L}\right) = 0$$

This simplifies to yield

$$\frac{dy}{dx} + \frac{hL}{k}\left(1 - \frac{x}{L}\right) = 0 \tag{4.66}$$

that defines the fin profile. We integrate the above once with respect to x, take $y = 0$ at $x = L$ to get

$$\boxed{y = \frac{hL}{k}\left(\frac{L}{2} - x + \frac{x^2}{2L}\right)} \tag{4.67}$$

The profile area A_p is related to the fin mass and may be easily calculated by the integral $\int_0^L 2y dx$. Thus

$$\boxed{A_p = \int_0^L 2\frac{hL}{k}\left(\frac{L}{2} - x + \frac{x^2}{2L}\right) dx = \frac{hL^3}{3k}} \tag{4.68}$$

Consider a typical example where $h = 15\,\text{W/m}^2\,^\circ\text{C}$, $k = 45\,\text{W/m}^\circ\text{C}$, $L = 0.1$ m and $\theta_b = 100\,^\circ\text{C}$. Fin profile is given by

$$y = 0.033(0.05 - x + 5x^2) \quad \text{where} \quad 0 \le x \le L$$

The profile is concave parabolic and becomes very thin as we approach the tip. The profile area is calculated as $A_p = 0.00011\,\text{m}^2$. Fin profile (top half) is shown plotted in Fig. 4.16 as "*Constant gradient profile*". The fin is symmetric with respect to the x axis and has the same profile for $-y$.

It is seen that the temperature gradient at $x = 0$, in this case, is $\frac{d\theta}{dx} = -\frac{\theta_b}{L} = -10\theta_b$ and hence the fin heat loss per meter width of fin is

$$q = -2ky(0)\frac{d\theta}{dx} = 2 \times 45 \times 0.00167 \times 10 \times 100\,\text{W/m} = \boxed{150\,\text{W/m}}$$

Figure 4.16 also shows half profiles of optimum proportioned rectangular and triangular fins. Both these are designed for maximum heat loss per profile area A_p, the same as that of the constant gradient fin. For a rectangular fin of minimum mass, we have $\mu^* = 1.4192$. By definition, we have $\mu = mL = \sqrt{\frac{h}{kt}}L$ and profile area $A_p = 2tL$. From the latter we have $L = \frac{A_p}{2t}$. This in the former gives

$$L = \left(\frac{kA_p\mu^2}{2h}\right)^{\frac{1}{3}}$$

Taking the optimum value of μ, i.e., μ^*, the corresponding fin length L^* is obtained as

$$L^* = \left(\frac{45 \times 0.00011 \times 1.4192^2}{2 \times 15}\right)^{\frac{1}{3}} = 0.0695\,\text{m}$$

The fin semi-thickness is then given by $t = \frac{0.00011}{2 \times 0.0695} = 0.0008\,\text{m}$. This profile is also shown in Fig. 4.16 labeled "*Optimum rectangular profile*". Heat loss from this fin may be calculated as under. Fin efficiency is given by

$$\eta^* = \frac{\tanh(\mu^*)}{\mu^*} = \frac{\tanh(1.4192)}{1.4192} = 0.627$$

Fin heat loss per unit width is then given by

$$q = 2L^* h\theta_b \eta^* = 2 \times 0.0695 \times 15 \times 100 \times 0.627 = \boxed{130.7\,\text{W/m}}$$

It may be shown by an analysis similar to that presented in the case of a fin of uniform area that an optimum fin of triangular profile satisfies the relation

$$y_b^* = \left(\frac{0.583 h A_p^2}{k}\right)^{\frac{1}{3}} = \left(\frac{0.583 \times 15 \times 0.00011^2}{45}\right)^{\frac{1}{3}} = 0.00134\,\text{m}$$

The fin length is then given by $L^* = \frac{A_p}{y_b^*} = 0.083\,\text{m}$. The corresponding half-profile is shown in Fig. 4.16 as "*Optimum triangular profile*". Heat loss from this fin may be calculated as under.

Fin parameter p may be calculated as

$$p = \sqrt{\frac{hL_{opt}}{k y_{b,opt}}} = \sqrt{\frac{15 \times 0.083}{45 \times 0.00134}} = 4.5460$$

Hence $\mu = p\sqrt{L_{opt}} = 4.5460\sqrt{0.083} = 1.3097$. Fin efficiency is given by

$$\eta^* = \frac{1}{\mu^*}\frac{I_1(2\mu^*)}{I_0(2\mu^*)} = \frac{1}{1.3097}\frac{I_1(2 \times 1.3097)}{I_0(2 \times 1.3097)} = 0.5936$$

Hence the fin heat loss per unit width is given by

$$q = 2L^* h\theta_b \eta^* = 2 \times 0.083 \times 15 \times 100 \times 0.5936 = \boxed{147.8\,\text{W/m}}$$

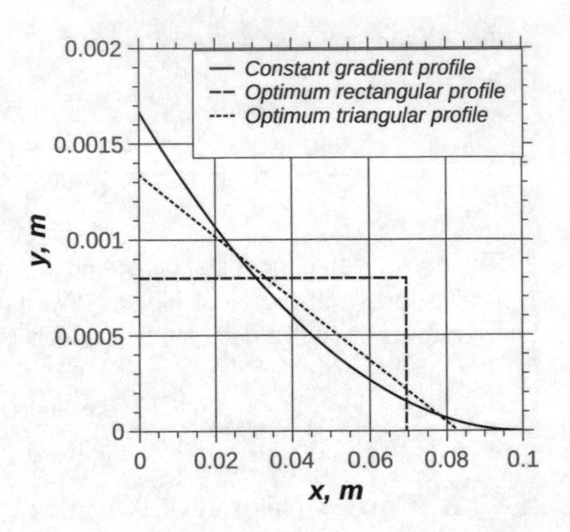

Fig. 4.16 Profile of a fin with linearly varying temperature compared with rectangular and triangular profiled fins

Comment: It is seen that the constant gradient profile is the best followed by the triangle profile. The least performance is shown by uniform area fin. However, the difference between the constant gradient profile fin and triangle profile fin is very small! Hence the triangle profile fin is a good choice in applications.

4.5 Heat Transfer from Fin Arrays

In applications, it is seldom that a single fin is made use of. It is common to have a parent surface covered with an array of fins. In such a case one may use either one of the following concepts:

(1) Overall surface efficiency (2) Effectiveness for the fin array

These two alternate concepts are described below in detail.

4.5.1 Overall Surface Efficiency of a fin array

Let A_t = total exposed area including fins, and A_f = total area of fins. Hence, the exposed parent area in the fin array is $A_e = A_t - A_f$. The total heat transferred from the fin array may be written as a sum of two parts: the first part is due to the exposed parent area given by $Q_e = A_e h\theta_b$ and the second part is due to the exposed area of fins given, using the concept of fin efficiency, as $Q_f = A_f h\theta_b\eta$. Thus the total heat transfer is given by

$$Q_t = Q_e + Q_f = A_e h\theta_b + A_f h\theta_b \eta$$
$$= (A_t - A_f)h\theta_b + A_f h\theta_b \eta = A_t h\theta_b \left[1 - \frac{A_f}{A_t}(1-\eta)\right] \quad (4.69)$$

In the above, η is the efficiency of each fin in the array, θ_b is the temperature excess of the base structure with respect to the ambient, and h is the constant heat transfer coefficient for heat transfer by convection from all the exposed surfaces. The reader will recognize that $A_t h\theta_b$ is the largest possible heat transfer rate Q_{max} that can take place from the fin array. Hence, analogous to the fin efficiency, an overall surface efficiency η_o is defined as

$$\eta_o = \frac{Q_t}{Q_{max}} = \left[1 - \frac{A_f}{A_t}(1-\eta)\right] \quad (4.70)$$

4.5.2 Effectiveness of a Fin Array

An alternate concept is that of effectiveness of a fin array, which is defined as the ratio of heat transfer in the presence of fins to the heat transfer, which would take place if fins were not provided. Let the parent area of the surface to which fins are attached be A_p. The heat transfer that would take place in the absence of fins is the heat transfer that would take place from this area by convection and is given by $Q_p = A_p h\theta_b$. The heat transfer that takes place when fins are mounted on the parent surface is given by Q_t given by Eq. 4.69. Hence, the effectiveness ε of the fin array is given by

$$\varepsilon = \frac{Q_t}{Q_p} = \frac{A_e + A_f \eta}{A_p} \quad (4.71)$$

Consider a typical fin array shown in Fig. 4.17. The parent surface is flat and the fins are attached normal to the parent surface. We consider a unit width in a direction normal to the plane of the figure. The fins may be of uniform thickness as shown in the figure, or in general, may be of non-uniform thickness. The parent surface is occupied by fins of total base area equal to $1 \times n \times 2t = 2nt$ and $n - 1$ gaps of area $1 \times (n-1) \times S = (n-1)S$. Hence, the area of parent surface is easily recognized to be $A_p = 2nt + (n-1)S$. The exposed fin array area is $A_f = 2nL$, since each fin has two heat losing sides. The total heat transfer area is given by $A_t = 2nL + (n-1)S$.

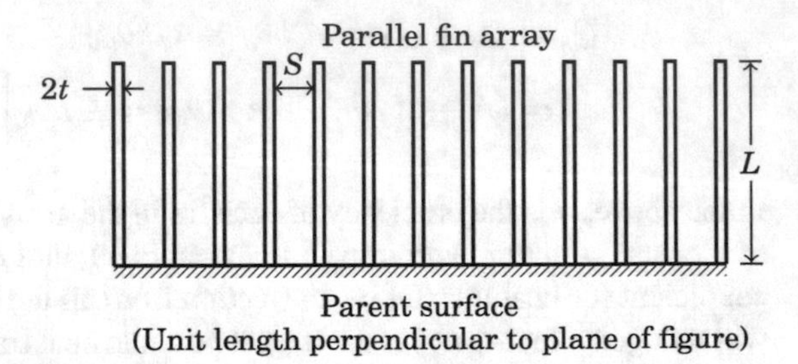

Fig. 4.17 Nomenclature for heat transfer from a fin array

The heat transfer when fins are not present is given by

$$Q_{\text{without fins}} = [2nt + (n-1)S]h\theta_b \tag{4.72}$$

The heat transfer when fins are present is given by (again the fins have an individual fin efficiency η)

$$Q_{\text{with fins}} = h\theta_b[2nL\eta + (n-1)S] \tag{4.73}$$

where we have assumed the fins to satisfy the insulated tip condition. Using the above two expressions, the effectiveness of the fin array is obtained as

$$\boxed{\varepsilon = \frac{Q_{\text{with fins}}}{Q_{\text{without fins}}} = \frac{2nL\eta + (n-1)S}{2nt + (n-1)S}} \tag{4.74}$$

4.5.3 Fin Array Applications

Fins are attached to a parent surface to enhance heat transfer. The parent surface may itself be subject to different environments over its exposed surfaces. For example, in the case of a wall, the inside may communicate convectively with one environment while the outside may communicate convectively with a different environment. The heat transfer coefficients on the two sides may have different values. In many applications two different fluids, such as water or a liquid on one side and air or a gas on the other side, may be involved. In such a case, the gas side heat transfer coefficient is normally the smaller of the two heat transfer coefficients. It is meaningful to attach fins on the side that experiences the lower heat transfer coefficient.

Fins Attached to a Plane Wall

A practical application is a plane wall with fins on one side as shown in Fig. 4.18. The figure also shows the electrical analog of the thermal problem. There are three thermal resistances in series given by the expressions in the figure. We have made

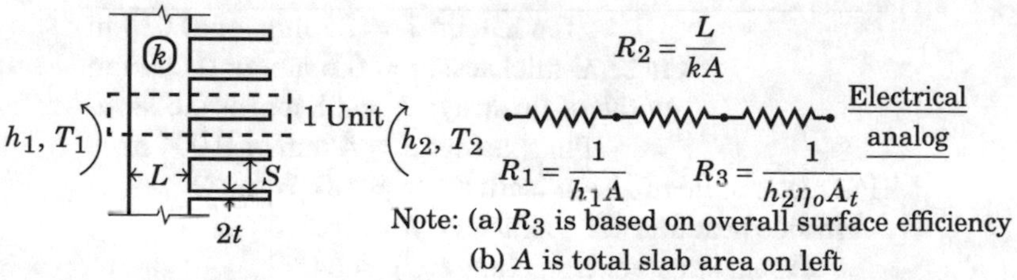

k 1 Unit h_1, T_1 L S $2t$ h_2, T_2

$R_2 = \frac{L}{kA}$

Electrical analog

$R_1 = \frac{1}{h_1 A}$ $R_3 = \frac{1}{h_2 \eta_o A_t}$

Note: (a) R_3 is based on overall surface efficiency

(b) A is total slab area on left

Fig. 4.18 Plane slab with fins on one side

use of the overall surface efficiency concept. Note that A is the area of the slab and A_t is the total exposed area on the finned side. The total thermal resistance may be written down as

$$R_{tot} = R_1 + R_2 + R_3 = \frac{1}{h_1 A} + \frac{L}{kA} + \frac{1}{h_2 \eta_o A_t} \tag{4.75}$$

Heat transfer across the wall may now be calculated as

$$\boxed{Q = \frac{T_1 - T_2}{\left(\frac{1}{h_1 A} + \frac{L}{kA} + \frac{1}{h_2 \eta_o A_t}\right)}} \tag{4.76}$$

Example 4.9

Eight uniform area fins of thickness 1.2 mm and length 35 mm are integral with a base in a heat sink used in electronic cooling application. The thermal conductivity of the material (an aluminum alloy) is 187 W/m°C. All the exposed surfaces are convectively cooled by ambient air at 22 °C subject to a heat transfer coefficient of 16.5 W/m ^{2}°C. The base temperature has been measured at 64 °C. Determine (a) the overall surface efficiency and (b) effectiveness of the fin array. Also determine the heat loss from the heat sink. The heat sink is 98 mm wide and the spacing between the fins is 4 mm.

Solution:

Step 1 The nomenclature follows that given in Fig. 4.18. The given data may be written down as under.

Fin length:	$L = 35$ mm or 0.035 m
Fin semi-thickness:	$t = 0.6$ mm or 0.0006 m
Width of fin array:	$W = 98$ mm or 0.098 m
Fin spacing:	$S = 4$ mm or 0.004 m
Fin material thermal conductivity:	$k = 187$ W/m°C
Fin side heat transfer coefficient:	$h = 16.5$ W/m^2 °C
Parent structure temperature:	$T_b = 64\,°C$
Ambient temperature:	$T_{amb} = 22$ °C

Step 2 The fin parameter is then calculated as

$$m = \sqrt{\frac{h}{kt}} = \sqrt{\frac{16.5}{187 \times 0.0006}} = 12.13\,\text{m}^{-1}$$

The non-dimensional fin parameter is then obtained as

$$\mu = mL = 12.13 \times 0.035 = 0.424$$

Step 3 Fin efficiency is then calculated using Eq. 4.22.

$$\eta = \frac{\tanh(0.424)}{0.424} = 0.944$$

Step 4 Various areas needed are calculated now. Number of fins is specified as $n = 8$. The breadth of the base is hence given by

$$B = 2nt + (n-1)S = 2 \times 8 \times 0.0006 + (8-1) \times 0.004 = 0.0376\,\text{m}$$

The total exposed area for convection may now be calculated as

$$A_t = 2LWn + SW(n-1) = 2 \times 0.035 \times 0.098 \times 8 + 0.004 \times 0.098 \times (8-1) = 0.0576\,\text{m}^2$$

Area of fins may be calculated as

$$A_f = 2LWn = 2 \times 0.035 \times 0.098 \times 8 = 0.0549\,\text{m}^2$$

Step 5 (a) **Overall surface efficiency** may now be calculated using Eq. 4.70.

$$\eta_o = \left[1 - \frac{0.0549}{0.0576}(1 - 0.944)\right] = 0.9466$$

Area of parent surface is calculated as

$$A_p = BW = 0.0376 \times 0.098 = 0.00369\,\text{m}^2$$

Exposed parent area is given by

$$A_e = (n-1)SW = (8-1) \times 0.004 \times 0.098 = 0.00274\,\text{m}^2$$

Step 6 (b) **The effectiveness** may now be calculated using Eq. 4.71 as

$$\varepsilon = \frac{0.00274 + 0.0549 \times 0.944}{0.00369} = 14.78$$

Step 7 Heat loss from the heat sink may now be calculated using **two different methods** as follows.
Based on the overall surface efficiency concept, we have

$$Q_b = A_t h(T_b - T_{amb})\eta_o = 0.0576 \times 16.5 \times (64-22) \times 0.9466 = 37.79\,\text{W}$$

Based on the effectiveness concept, we have

$$Q_b = A_p h(T_b - T_{amb})\varepsilon = 0.00369 \times 16.5 \times (64-22) \times 14.78 = 37.79\,\text{W}$$

Example 4.10

An extruded aluminum alloy heat sink consists of 4 parallel flat fins integral on a flat base. Air is blown parallel to fin flats and the convective heat transfer coefficient has been estimated from experiments to be $h = 28\,\text{W/m}^2\,°\text{C}$. Aluminum alloy has a thermal conductivity of $k = 175\,\text{W/m°C}$. The base of heat sink is $W = 30\,\text{mm}$, $H = 50\,\text{mm}$ with the fins being parallel to the 50 mm edge. All the fins as well as the base have a thickness of $2t = 0.4\,\text{mm}$. Choosing optimal proportioned fins determine the heat loss from the heat sink when the base is at a temperature of $T_b = 65\,°\text{C}$ and ambient air is at $T_{amb} = 30\,°\text{C}$.

Solution:

Step 1 With the data specified in the problem, we assume the material thickness to be $2t^* = 0.4\,\text{mm}$ or $t^* = 0.2\,\text{mm}$. We choose $\mu = 1.4192$ to obtain the fin length as

$$L^* = \frac{\mu^*}{m} = \frac{\mu^*}{\sqrt{\frac{h}{kt^*}}} = \frac{1.4192}{\sqrt{\frac{28}{175 \times 0.0002}}} = 0.05018\,\text{m} \approx 50.2\,\text{mm}$$

Step 2 With $n = 4$ and $W = 30\,\text{mm}$, fin spacing is given by

$$S = \frac{W - 2nt^*}{(n-1)} = \frac{30 - 2 \times 4 \times 0.2}{(4-1)} = 9.47\,\text{mm}$$

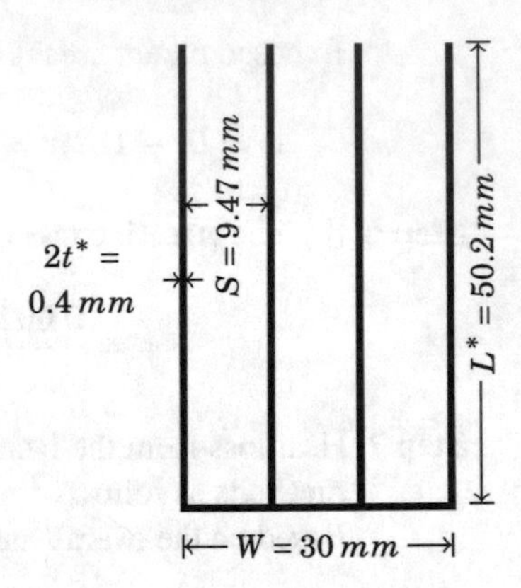

Fig. 4.19 Heat sink in Example 4.10. $H = 50\,\text{mm}$ is the dimension of the base and fins in the direction normal to the plane of the figure

Heat sink cross section will appear as shown in Fig. 4.19.

Step 3 Fin efficiency is then calculated using Eq. 4.22.

$$\eta = \frac{\tanh(1.4192)}{1.4192} = 0.627$$

The effectiveness of fin array may now be calculated using Eq. 4.71 as

$$\varepsilon = \frac{2 \times 4 \times 0.0502 + (4 - 1) \times 0.0947}{0.03} = 9.332$$

Step 4 Heat transfer in the absence of fins is given by

$$Q_{\text{no fins}} = W H h \theta_b = 0.03 \times 0.05 \times 28 \times (65 - 30) = 1.47\,\text{W}$$

Hence the heat transfer from heat sink is given by

$$Q_{\text{heat sink}} = \varepsilon \times Q_{\text{no fins}} = 9.332 \times 1.47 = 13.72\,\text{W}$$

Finned Tube

Another example that is used often is a tube with fins on one side (say the outside—automobile radiator and condenser coil in an air conditioner—are examples).

Figure 4.20 shows the details along with the applicable nomenclature. The corresponding electrical analog is also shown in the figure. The difference between the previous example and the present one is because of area variation due to cylindrical geometry. We may again use the concept of thermal resistance. The various thermal resistances are again in series. The concept of overall surface efficiency is used in writing down the appropriate thermal resistances. Expressions for the inner thermal

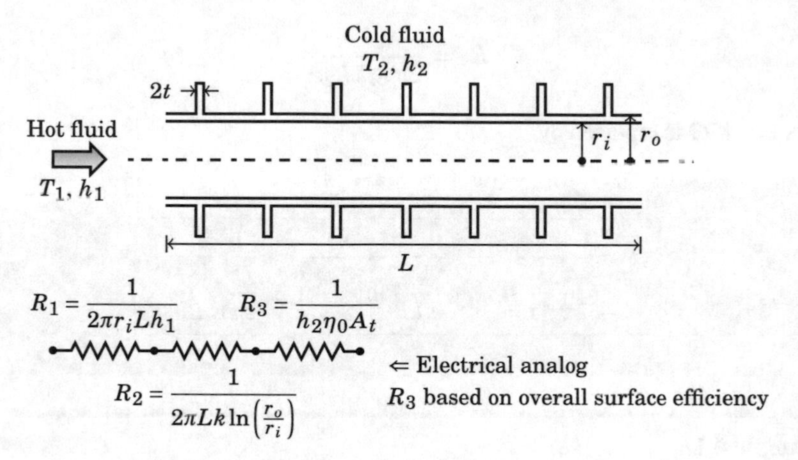

Fig. 4.20 Heat transfer from a fluid flowing in an externally finned tube

resistance R_1, tube wall conduction resistance R_2 and the fin side resistance R_3 are as indicated in Fig. 4.20. The total resistance is then given by

$$R_{tot} = R_1 + R_2 + R_3 = \frac{1}{2\pi r_i L h_1} + \frac{1}{2\pi L k} \ln\left(\frac{r_o}{r_i}\right) + \frac{1}{\eta_o h_2 A_t} \tag{4.77}$$

Note that the total area on the finned side is given by

$$A_t = 2\pi r_o (L - 2nt) + 2\pi (r_2^2 - r_o^2) n \tag{4.78}$$

where n represents the number of fins in length L of the tube and r_2 is the tip radius of the radial fin. The heat transfer from the externally finned tube is then given by

$$\boxed{Q = \frac{T_1 - T_2}{\frac{1}{2\pi r_i L h_1} + \frac{1}{2\pi L k} \ln\left(\frac{r_o}{r_i}\right) + \frac{1}{\eta_o h_2 A_t}}} \tag{4.79}$$

Alternately we may use the concept of fin array effectiveness in representing the thermal resistance R_3 on the outside. We note that the heat transfer from the outside is given by

$$Q_b = 2\pi r_o L \varepsilon h_2 \theta_b$$

The corresponding thermal resistance is

$$R_3 = \frac{1}{2\pi r_o L h_2 \varepsilon} \tag{4.80}$$

Thus Eq. 4.79 is replaced by

$$Q = \frac{T_1 - T_2}{\frac{1}{2\pi r_i L h_1} + \frac{1}{2\pi L k} \ln\left(\frac{r_o}{r_i}\right) + \frac{1}{2\pi r_o L h_2 \varepsilon}} \tag{4.81}$$

Example 4.11

A brass tube of 6 mm ID and 7 mm OD has short integral fins of height 7 mm and 0.3 mm thickness arranged with a spacing of 3 mm. The heat transfer coefficient to the hot fluid at 75 °C flowing inside the tube is 270 W/m^2°C. The fins and the external surface of the tube are exposed to air at 20 °C with a heat transfer coefficient of 45 W/m^2°C. Determine the heat transfer from a meter length of tube.

Solution:

Step 1 The nomenclature for this problem is based on Fig. 4.20. The given data is written down.

Tube inner radius:	$r_i = 0.003$ m
Tube outer radius:	$r_o = 0.0035$ m
Fin length:	$L = 0.007$ m
Fin tip radius:	$r_2 = r_o + L$
	$= 0.0035 + 0.007 = 0.0105$ m
Fin semi-thickness:	$t = 0.00015$ m
Fin spacing:	$S = 0.003$ m
Thermal conductivity of tube and fin material:	$k = 111$ W/m°C
Tube side heat transfer coefficient:	$h_1 = 270$ W/m^2 °C
Fin side heat transfer coefficient:	$h_2 = 45$ W/m^2 °C
Tube side fluid temperature:	$T_1 = 75$ °C
Fin side fluid temperature:	$T_2 = 20$ °C

Step 2 Non-dimensional fin parameter is calculated as

$$\mu = \sqrt{\frac{h_2}{kt}} \cdot r_2 = \sqrt{\frac{45}{111 \times 0.00015}} \cdot 0.0105 = 0.546$$

Radius ratio that characterizes the fin is

$$\alpha = \frac{r_2}{r_o} = \frac{0.0105}{0.0035} = 3$$

Step 3 Fin efficiency is read off in Fig. 4.13 as $\eta = 0.92$. The various areas needed are calculated per meter length of tube basis. Since the fin spacing is $S = 3$ mm or 0.003 m, the center to center distance between fins S' is

$$S' = S + 2t = 0.003 + 2 \times 0.00015 = 0.0033\,\text{m}$$

The number of fins per *m* length of tube is determined as

$$n = \frac{1}{S'} = \frac{1}{0.0033} \approx 303$$

The above is obtained by rounding the number to an integer. The parent area per *m* of tube length is obtained as

$$A_p = 2\pi r_o = 2 \times \pi \times 0.0035 = 0.02199\,\text{m}^2$$

Fin area per *m* of tube length is

$$A_f = 2\pi (r_2^2 - r_o^2)n = 2 \times \pi \times (0.0105^2 - 0.0035^2) \times 303 = 0.1866\,\text{m}^2$$

The total exposed area on the outside is

$$\begin{aligned} A_t &= A_f + 2\pi r_o (n-1)S \\ &= 0.1866 + 2 \times \pi \times 0.0035 \times (303 - 1) \times 0.003 = 0.2065\,\text{m}^2 \end{aligned}$$

Step 4 The overall surface efficiency may now be calculated, using Eq. 4.70 as

$$\eta_o = \left[1 - \frac{0.1866}{0.2065}(1 - 0.92)\right] = 0.928$$

The three resistances that correspond to those shown in Fig. 4.20 are

$$R_1 = \frac{1}{2\pi r_i h_1} = \frac{1}{2 \times \pi \times 0.003 \times 270} = 0.1965\,\text{m}^\circ\text{C/W}$$

$$R_2 = \frac{1}{2\pi k} \ln\left(\frac{r_o}{r_i}\right) = \frac{1}{2\pi \times 111} \ln\left(\frac{0.0035}{0.003}\right) = 2.2103 \times 10^{-4}\,\text{m}^\circ\text{C/W}$$

$$R_3 = \frac{1}{h_2 A_t \eta_o} = \frac{1}{45 \times 0.2065 \times 0.928} = 0.1160\,\text{m}^\circ\text{C/W}$$

The overall thermal resistance is then given by

$$R_{tot} = 0.1965 + 2.2103 \times 10^{-4} + 0.1160 = 0.3127\,\text{m}^\circ\text{C/W}$$

Step 5 Tube heat loss per m of tube length is then given by

$$Q = \frac{T_1 - T_2}{R_{tot}} = \frac{75 - 20}{0.3127} = 175.9\,\text{W/m}$$

Concluding Remarks

This chapter has presented an important application of conduction heat transfer, viz., enhancement of heat transfer by the use of extended surfaces. When the conductive resistance within the extended surface is smaller than the convective resistance from the surface of the extended surface, the addition of the extended surface enhances heat transfer. Analysis has been presented for both uniform and non-uniform area fins. Optimization of fin profile for minimum mass has been presented in the case of flat as well as pin fins. Toward the end, we have presented an analysis of fin arrays that are very useful in applications.

4.6 Exercises

Ex 4.1: Consider an aluminum fin of thermal conductivity 200 W/m°C. The fin is in the form of a plate, which is 3 mm thick and 150 mm long. It is exposed to an ambient at 30 °C with a heat transfer coefficient of 67 W/m^2°C. How much heat is dissipated per meter by the fin if the base is at 100 °C?

Ex 4.2: A straight fin of uniform circular cross section of area 0.2 cm^2, length 15 cm and thermal conductivity 200 W/m°C has its two ends maintained at 100 °C and 50 °C, respectively. The ambient air is at 30 °C and the convection heat transfer coefficient is 15 W/m^2°C. Determine the total heat loss from the fin to the ambient. What is the minimum temperature in the fin and where does it occur? Can you define fin efficiency for this case?

Ex 4.3: A metal cleaning brush with bristles of steel, which are 0.3 mm diameter and 10 mm long has the bristles mounted with center to center distance of 0.75 mm between adjacent bristles. The bristles are in an "in line" arrangement forming a square grid pattern. The bristles are mounted on a base block of the same material of size 75 × 25 × 10 mm. The bristles as well as the base block material have a thermal conductivity of 45 W/m°C.

The brush has been left, inadvertently, in an upward-facing position on a hot boiler accessory, which is at 80 °C, with the base of the brush in intimate contact with the accessory. The ambient temperature is 30 °C and all exposed surfaces are subject to a heat transfer coefficient of 15 W/m^2°C. Determine the total heat loss from the accessory due to the brush. Assume that there is a margin of 0.4 mm between the edges of the base and the nearest bristle. Mention any assumptions you make.

Ex 4.4: A uniform area pin fin of circular cross section has a diameter of 5 mm and a length equal to 125 mm. The base is at a temperature of 80 °C while the tip temperature (tip itself is treatable as insulated) has been measured to be 54 °C. The ambient temperature is 30 °C and the fin material thermal conductivity is known to be 45 W/m°C. Estimate the heat transfer coefficient between air and fin surface assuming it to be uniform. Based on the magnitude of the heat transfer coefficient identify a possible mode for heat exchange between the fin surface and air.

Ex 4.5: Water flows at 65 °C over the outside of a copper tube of 60 mm outside diameter, giving rise to a heat transfer coefficient of 1000 W/m ^{2} °C between the water and the tube. Air at 25 °C flows inside the tube whose inner diameter is 50 mm. The tube side heat transfer coefficient is 25 W/m ^{2} °C. To increase the heat transfer rate, four straight copper fins (thermal conductivity of copper may be taken as 400 W/m°C) of rectangular profile and 1.5 mm thickness are added to the tube side as shown in Fig. 4.21.

Calculate the heat transfer per meter of pipe with and without the internal fins. Assume that the tube side heat transfer coefficient does not change in the presence of the fins.

Ex 4.6: A heat sink used in electronic cooling is represented in its cross section as indicated in Fig. 4.22. The overall foot print of the heat sink is 45 × 45 mm. The fins shown are 0.5 mm thick and are 25.4 mm tall. The convection heat transfer coefficient may be assumed to be 8 W/m ^{2} °C from all the exposed surfaces. The material of the heat sink is an

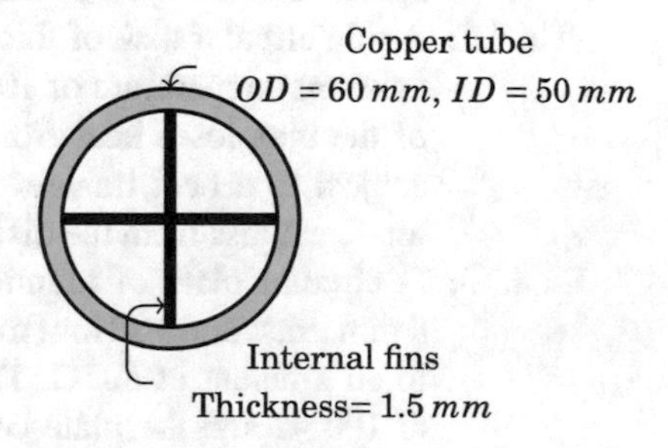

Fig. 4.21 Defining sketch for Exercise 4.5

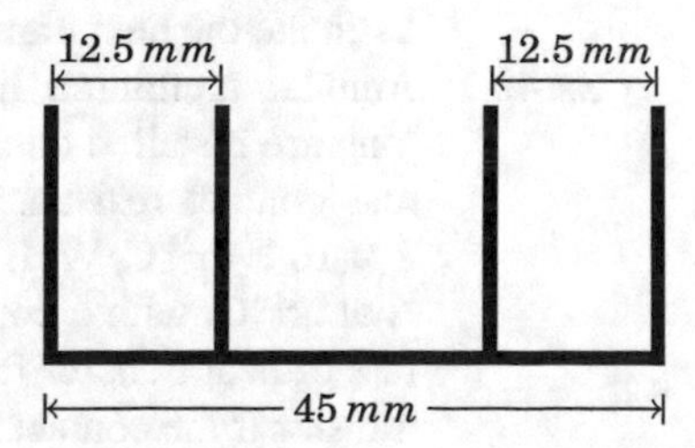

Fig. 4.22 Cross-sectional view of heat sink in Exercise 4.6. Thickness is uniform at 0.5 mm

alloy of aluminum and magnesium having a thermal conductivity of 175 W/m°C. What is the heat loss if the base is at a temperature of 60 °C when the ambient air temperature is 26 °C? Make suitable assumptions as necessary. What is the effectiveness of the heat sink?

A consultant has indicated that the heat transfer coefficient may be increased to 20 W/m ^{2}°C by the use of a small instrument fan mounted such that air is blown parallel to the fins and the base. What will be the base temperature if the same amount of heat is dissipated as above? What is the effectiveness of the heat sink in this case?

Ex 4.7: A circular disk of radius R, thickness δ, and thermal conductivity k is insulated on one side and absorbs a uniform heat flux of q on the other side. The circumference of the disk is maintained at a constant temperature of T_R. Derive the equation governing the variation of temperature in the disk. Specify suitable boundary conditions. Solve for the temperature. Obtain an expression for the maximum temperature in the disk.

A circular disk of thermal conductivity $k = 24$ W/m°C, thickness $\delta = 0.05$ mm and radius $R = 5.08$ mm is absorbing radiant heat at a constant rate of $q = 10000$ W/m^2 at one of the surfaces. The other surface is perfectly insulated. However, the periphery of the disk is cooled by chilled water to 10 °C. Determine the maximum temperature in the disk.

Ex 4.8: Derive the general fin equation for a variable area extended surface. Consider the particular case of a fin in the form of a cone with base diameter d_b and cone angle 2α. Deduce the particular fin equation governing heat transfer in this fin, starting from the general fin equation. Can you think of an appropriate boundary condition that should be specified at the apex of the cone?

Ex 4.9: A thin circular disk of thickness 2 mm and radius 100 mm is held at a constant temperature of 100 °C along its periphery. The lateral surface of the disk loses heat from one surface only to ambient air at 30 °C subject to a heat transfer coefficient of 15 W/m ^{2}°C. Determine the total heat loss from the disk to air and also the disk center temperature.

Ex 4.10: A circular plate of aluminum of thermal conductivity 205 W/m°C, 4 mm thick and 100 mm radius is insulated on one side and is exposed to an ambient at 30 °C. The circular edge of the plate is maintained at 100 °C and the plate center temperature is measured to be 70 °C. Estimate the heat transfer coefficient.

Ex 4.11: Annular aluminum fins ($k = 205$ W/m°C) 2 mm thick and 15 mm long are installed on an aluminum tube of 30 mm diameter. The thermal contact resistance between fins and the tube is known to be 2×10^{-5} m^2°C/W. If the tube wall is at 100 °C and the adjoining fluid is at 25 °C, with a convection coefficient of 75 W/m ^{2}°C, what is the rate of heat transfer from a single fin? What would be the rate of heat transfer if the contact resistance could be eliminated?

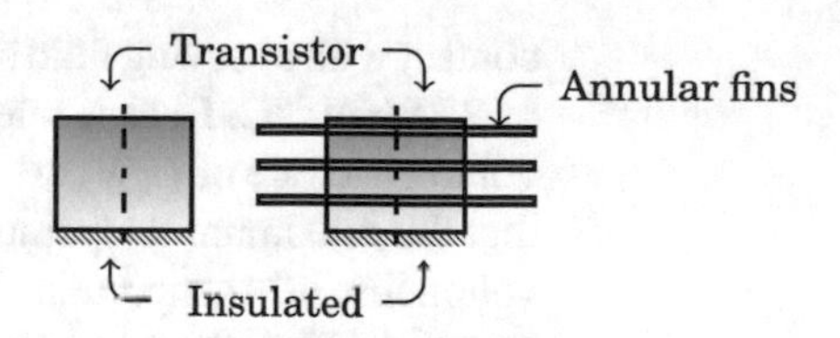

Fig. 4.23 Figure for transistor problem in Exercise 4.13

Ex 4.12: A refrigerator coil has a tube of 3 mm ID and 4 mm OD with radial fins of outer radius 10 mm and thickness 0.2 mm attached to the outside with a spacing of 2 mm. Other pertinent data are

- Temperature of fluid flowing inside the tube: 60 °C.
- Heat transfer coefficient on the tube side: 270 W/m^2°C.
- Temperature of the fluid on the outside of the tube: 30 °C.
- Heat transfer coefficient on the outside: 24 W/m^2°C.
- Tube and fin material thermal conductivity: 380 W/m°C.

Determine the heat transfer per meter length of the coil.

Ex 4.13: A transistor (see Fig. 4.23) may be considered as a short cylinder 4 mm long and 5 mm diameter. The surface of the cylinder may be assumed to be at a uniform temperature of 70 °C during its operation. If the heat transfer coefficient from the surface of the transistor is 5 W/m^2°C how much heat will be dissipated to an ambient at 30 °C. Assume that the bottom side of the transistor is essentially insulated.
In order to improve heat transfer, a heat sink is attached to the transistor. The heat sink essentially is equivalent to adding three 0.5 mm thick annular fins of 5 mm ID and 10 mm OD. The heat sink material has a thermal conductivity of 207 W/m°C. The heat transfer coefficient remains the same as in the previous case. What will be the operating temperature of the transistor in this case if the heat dissipation remains the same as in the previous case?

Ex 4.14: A pin fin is in the form of a rod of uniform diameter D, L long made of a material with a thermal conductivity k. The base of the fin is maintained at a temperature T_b greater than the ambient temperature T_∞. The heat transfer coefficient is h from all the exposed surfaces. Formulate the problem and specify the boundary conditions. Obtain an expression for the variation of temperature in the fin. Also obtain an expression for the efficiency of the fin.
What is the tip temperature if $T_b = 77$ °C, $k = 100$ W/m°C, and $h = 22$ W/m^2°C for a pin fin with $D = 6$ mm and $L = 115$ mm? What is the heat loss from the fin? What will these become if the heat loss from the fin tip is ignored?

Ex 4.15: A very wide 3 mm flat metal plate of thermal conductivity 15 W/m°C is 0.15 m long. The two edges of the plate are maintained at a uniform temperature of 40 °C. The lateral surfaces of the plate are in thermal contact with a moving fluid at 40 °C subject to a heat transfer coefficient of 27 W/m ^{2}°C. Heat is internally generated in the plate at a uniform volumetric rate of G W/m^3. Determine the maximum value of G such that the maximum temperature in the plate is limited to 100 °C. If the volumetric heat generation rate is maintained at this value and the lateral surfaces of the plate are perfectly insulated what would be the maximum temperature in the plate?

Ex 4.16: A metal rod with a thermal conductivity of 16.5 W/m°C is left immersed up to a depth of 10 cm in a pot of boiling water. The total length and the diameter of the rod are 15 cm and 6 mm respectively. The surface of the rod outside the water loses heat by convection to room air at 35 °C with a heat transfer coefficient of 8 W/m ^{2}°C. What is the lowest temperature in the rod? If this temperature should be less than 50 °C what should be the total length of the rod? If the rod length may not be changed, what will have to be the material thermal conductivity such that the minimum temperature is just equal to 50 °C?

Chapter 5
Multidimensional Conduction Part I

We commence a study of multidimensional heat conduction in this chapter. Two types of problems will be dealt with—(a) conduction heat transfer involving transient one-dimensional heat conduction, i.e., temperature varies with one space dimension and with respect to time, (b) steady conduction heat transfer in two space dimensions. Similarity analysis and the approximate integral method are useful in one-dimensional transient conduction. However, the basic approach is the use of the method of separation of variables to solve the governing equations in these two cases. We also present the use of complex variables for the solution of steady heat conduction in two dimensions. Conduction shape factors are introduced to analyze steady conduction in two dimensions.

5.1 Introduction

In Chaps. 2–4, we have been basically considering conduction in a single dimension (either one dimension in space or one dimension in time). We shall commence a study of multidimensional conduction now. Multidimensional conduction poses a challenge since the governing equation is a partial differential equation. The solution of such equations requires a mathematical background that will be built up as a part of the discussion here.

5.1.1 Integral Form of Governing Equation

Consider a volume of material of arbitrary shape shown in Fig. 5.1. We intend to make an energy balance for this volume now. Various quantities that affect the amount of energy within the volume are written down. The rate at which heat is generated

S. P. Venkateshan, *Heat Transfer*,
https://doi.org/10.1007/978-3-030-58338-5_5

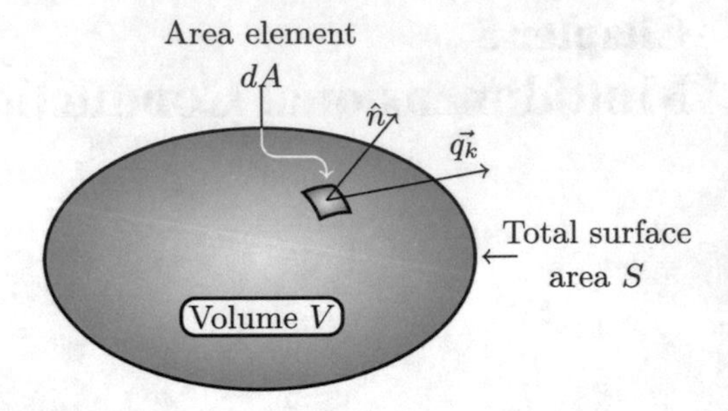

Fig. 5.1 Integral formulation of conduction in three dimensions

inside the volume V is given by integrating the volumetric heat generation rate over the volume as

$$\text{Rate at which heat is generated} = \iiint_V G\mathrm{d}V \tag{5.1}$$

The net heat transfer across the boundary S of the volume V is given by the following surface integral.

$$\text{Net rate of heat transfer across boundary} = \iint_S \vec{q}_k \cdot \vec{n}\mathrm{d}S \tag{5.2}$$

In the above, the integrand is the dot product of conduction heat flux vector $\vec{q}_k$ at the surface and the unit normal $\vec{n}$ to the surface. The rate at which energy is stored inside the volume V is given by

$$\text{Rate of energy storage inside} = \rho c \iiint_V \frac{\partial T}{\partial t}\mathrm{d}V \tag{5.3}$$

where ρ is the density and c the heat capacity of the medium inside V. We have assumed that the properties are constant independent of temperature. Energy balance for the volume requires that the rate of energy storage within volume V equals the difference between the rate at which heat is generated within the volume V and the rate at which heat is transferred across the boundary S. Hence, we have, using Eqs. 5.1–5.3 the integral form of the energy equation

$$\boxed{\rho c \iiint_V \frac{\partial T}{\partial t}\mathrm{d}V = \iiint_V G\mathrm{d}V - \iint_S \vec{q}_k \cdot \vec{n}\mathrm{d}S} \tag{5.4}$$

5.1.2 Differential Form of Governing Equation

In order to write the governing equation in the differential form, it is necessary to apply the integral equation to a differential volume. The differential volume may be represented in any of the three orthogonal systems of coordinates—Cartesian, cylindrical, or spherical. We begin the exercise by applying Eq. 5.4 to a volume element shown in Fig. 5.2. The volume element is easily recognized to be $dV = dxdydz$. The two volume integrals given by Eqs. 5.1 and 5.3 become very simply

$$\iiint_V G dV = G dx dy dz, \quad \rho c \iiint_V \frac{\partial T}{\partial t} dV = \rho c \frac{\partial T}{\partial t} dx dy dz \tag{5.5}$$

The interpretation of these two is that the volume integral is simply the product of the integrand, evaluated say at the center of the volume element, and the differential volume of the element. The surface integral requires some effort. We should like to write it in the form of a volume integral. This is done by the use of Fourier law and the application of Taylor expansion, retaining first-order terms, and taking the limits as the element shrinks to zero volume.

For this purpose, fluxes crossing boundaries are calculated as indicated below. The surface area S consists of six faces of a parallelepiped element as shown—a typical face is 1234. Flux crossing face 1234 is given by

$$\underbrace{q_{k,x}}_{\substack{x \text{ component} \\ \text{of conduction flux}}} \times \underbrace{dy dz}_{\substack{\text{Area} \\ \text{normal to } x}} \tag{5.6}$$

Flux crossing face 5678 is calculated by using Taylor expansion of the x component of the conduction heat flux

$$q_{k,x+dx} dy dz = \left[q_{k,x} + \frac{\partial q_{k,x}}{\partial x} dx \right] dy dz \tag{5.7}$$

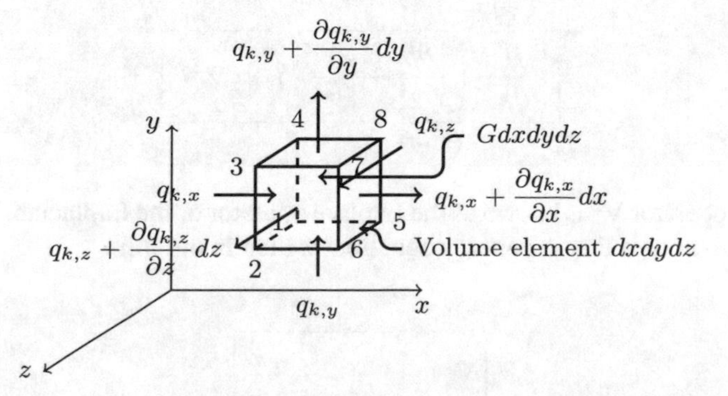

Fig. 5.2 Conduction in three dimensions—differential volume in Cartesian coordinates

Thus, the net flux crossing the volume element along the x-direction is given by the difference between the leaving and entering fluxes. Thus the net flux because of heat transfer across the two faces of the volume element normal to the x-direction is

$$\frac{\partial q_{k,x}}{\partial x}\mathrm{d}x\mathrm{d}y\mathrm{d}z = -k\frac{\partial^2 T}{\partial x^2}\mathrm{d}x\mathrm{d}y\mathrm{d}z \tag{5.8}$$

where the last part is based on Fourier law of heat conduction, assuming constant thermal conductivity. It is clear that the above has transformed the surface integral to a volume integral! Similarly the net heat transfer across the other four surfaces along with the above leads to

$$\iint_S \vec{q}_k \cdot \vec{n}\mathrm{d}S = -k\left[\frac{\partial^2 T}{\partial x^2} + \frac{\partial^2 T}{\partial y^2} + \frac{\partial^2 T}{\partial z^2}\right]\mathrm{d}x\mathrm{d}y\mathrm{d}z \tag{5.9}$$

Substitute Eqs. 5.5 and 5.9 in the integral form of energy Eq. 5.4 and cancel the common factor $dxdydz$ to get

$$\rho c\frac{\partial T}{\partial t} = G + k\left[\frac{\partial^2 T}{\partial x^2} + \frac{\partial^2 T}{\partial y^2} + \frac{\partial^2 T}{\partial z^2}\right] \tag{5.10}$$

or

$$\boxed{\left[\frac{\partial^2 T}{\partial x^2} + \frac{\partial^2 T}{\partial y^2} + \frac{\partial^2 T}{\partial z^2}\right] + \frac{G}{k} = \frac{1}{\alpha}\frac{\partial T}{\partial t}} \tag{5.11}$$

where $\alpha = \frac{k}{\rho c}$ is a property of the medium and is referred to as the thermal diffusivity. The terms in Eq. 5.11 that contain the second derivatives of temperature are represented in the operator form as

$$\boxed{\left[\frac{\partial^2}{\partial x^2} + \frac{\partial^2}{\partial y^2} + \frac{\partial^2}{\partial z^2}\right]T = \nabla^2 T} \tag{5.12}$$

where the operator ∇^2 is known as the Laplace operator or the Laplacian. The important thing to note is that the energy equation written in the form

$$\boxed{\nabla^2 T + \frac{G}{k} = \frac{1}{\alpha}\frac{\partial T}{\partial t}} \tag{5.13}$$

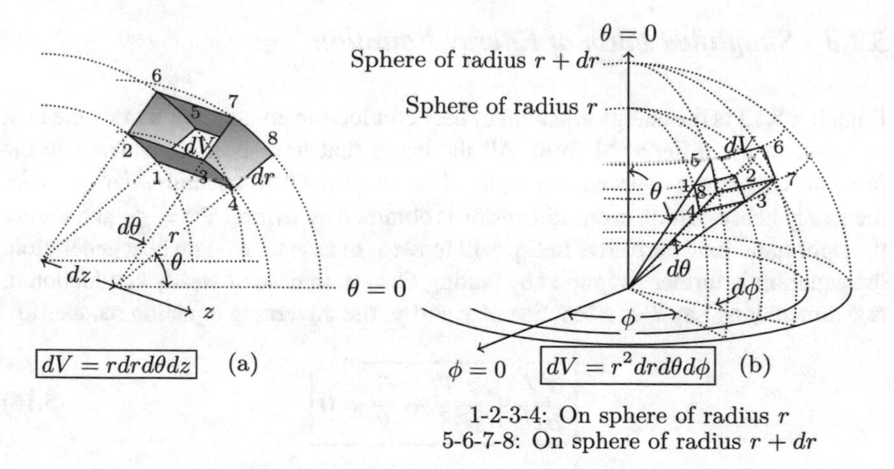

Fig. 5.3 Conduction in three dimensions—**a** differential volume in cylindrical coordinates, **b** differential volume in spherical coordinates

is *independent* of the coordinate system. It holds for cylindrical as well as spherical coordinates. Only thing that needs to be done is to use the proper expanded form of the operator ∇^2 in these coordinate systems. This may actually be done by applying the integral form of the energy equation to volume elements in cylindrical coordinates (Fig. 5.3a) and spherical coordinates (Fig. 5.3b). This is left as an exercise to the interested reader. The appropriate expressions are given below for ready reference.

Laplacian in cylindrical coordinates:

$$\nabla^2 = \frac{1}{r}\frac{\partial}{\partial r}\left(r\frac{\partial}{\partial r}\right) + \frac{1}{r^2}\frac{\partial^2}{\partial \theta^2} + \frac{\partial}{\partial z^2} \tag{5.14}$$

Laplacian in spherical coordinates:

$$\nabla^2 = \frac{1}{r^2}\frac{\partial}{\partial r}\left(r^2\frac{\partial}{\partial r}\right) + \frac{1}{r^2 \sin\theta}\frac{\partial}{\partial \theta}\left(\sin\theta\frac{\partial}{\partial \theta}\right) + \frac{1}{r^2 \sin^2\phi}\frac{\partial^2}{\partial \phi^2} \tag{5.15}$$

In both the cylindrical as well as spherical coordinates, the area changes in the direction of heat flow and this accounts for the complicated looking derivatives that appear in Eqs. 5.14 and 5.15.

5.1.3 Simplified Form of Energy Equation

Equation 5.13 is the energy equation or heat conduction equation or simply the heat equation in the differential form. All the cases that have been considered in the previous chapters are special or simplified forms of the heat equation. For example, the steady heat equation in one dimension is obtained by writing $\nabla^2 = \frac{d^2}{dx^2}$ and setting the right-hand time derivative in Eq. 5.12 to zero. In case there is no heat generation, the equation is further simplified by putting $G = 0$. In case of steady conduction in two dimensions, say T is a function of x and y, the governing equation reduces to

$$\boxed{\frac{\partial^2 T}{\partial x^2} + \frac{\partial^2 T}{\partial y^2} + \frac{G}{k} = 0} \tag{5.16}$$

This equation is known as the Poisson equation. If there is no heat generation, the heat equation reduces to

$$\boxed{\frac{\partial^2 T}{\partial x^2} + \frac{\partial^2 T}{\partial y^2} = 0} \tag{5.17}$$

This equation is known as the Laplace equation in two dimensions.

The heat equation in cylindrical and spherical coordinates also may be simplified in case of one-dimensional and two-dimensional problems, as done earlier in the case of Cartesian coordinates. As an example, steady radial heat conduction in a cylinder is governed by the equation

$$\boxed{\frac{1}{r}\frac{\mathrm{d}}{\mathrm{d}r}\left(r\frac{\mathrm{d}T}{\mathrm{d}r}\right) + \frac{G}{k} = 0}$$

In the case of steady two-dimensional heat conduction in a cylinder (T is a function of r and z), the appropriate equation is

$$\boxed{\frac{1}{r}\frac{\partial}{\partial r}\left(r\frac{\partial T}{\partial r}\right) + \frac{\partial^2 T}{\partial z^2} + \frac{G}{k} = 0}$$

However, in case T is a function of r and θ, we get

$$\boxed{\frac{1}{r}\frac{\partial}{\partial r}\left(r\frac{\partial T}{\partial r}\right) + \frac{1}{r^2}\frac{\partial^2 T}{\partial \theta^2} + \frac{G}{k} = 0}$$

Both are Poisson equations in cylindrical coordinates. If, in addition, $G = 0$ these two represent Laplace equations in two dimensions and cylindrical coordinates. Similar simplifications are also possible in the case of problems in spherical coordinates.

Table 5.1 Thermal diffusivity of materials

Material	Condition	Thermal diffusivity m/s^2
Gases	Air at 300 K	0.225×10^{-4}
	Steam at 373 K and 1 atm	0.205×10^{-4}
Liquids	Engine oil, unused at 300 K	0.859×10^{-7}
	Saturated water at 373 K and 1 atm	1.683×10^{-7}
Solids	Pure aluminum	9.71×10^{-5}
–Metals	Pure copper	1.17×10^{-4}
Solids	Aluminum oxide	0.151×10^{-4}
–Insulating	Concrete	0.519×10^{-6}
materials	Common brick	0.449×10^{-6}

For example, steady radial conduction in spherical coordinates is governed by the equation

$$\boxed{\frac{1}{r^2}\frac{\mathrm{d}}{\mathrm{d}r}\left(r^2\frac{\mathrm{d}T}{\mathrm{d}r}\right)+\frac{G}{k}=0}$$

5.1.4 *Thermal Diffusivity*

The heat equation involves a single parameter, the thermal diffusivity, which characterizes the material in which thermal conduction or heat diffusion takes place. An equation that involves the Laplacian operator represents, in general, a diffusion phenomenon and hence the name. The thermal diffusivity given by $\frac{k}{\rho c}$ has units of $\frac{\mathrm{m}^2}{\mathrm{s}}$, in SI system of units. Before we take up solution of the heat equation, it is instructive to look at typical thermal diffusivity values, which are shown in Table 5.1. It is seen that the thermal diffusivity of materials covers a range of roughly three orders of magnitude. Since thermal diffusivity is a composite parameter, this range includes some five orders of magnitude variation of thermal conductivity values, toned down by the range of density specific heat product, which itself varies over some six orders of magnitude.

5.2 One-Dimensional Transient Conduction

The simplest multidimensional conduction problems, of much practical interest, are those that involve a transient variation of temperature field with respect to one space dimension. Such problems are referred to as one-dimensional transients. The transients may be either in a semi-infinite medium or in a finite domain such as a rod or slab. Short time solution (more precise definition will come later) even in a finite

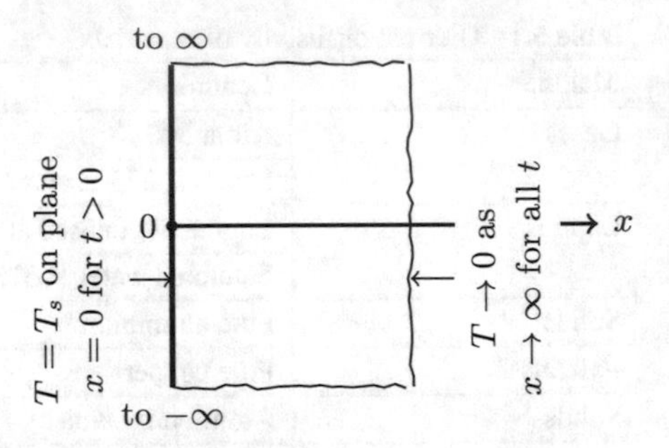

Fig. 5.4 Unsteady conduction in a semi-infinite solid subject to step change in surface temperature

domain may be treated as that in a semi-infinite domain. This has many practical applications, as, for example, in the experimental determination of thermal conductivity and thermal diffusivity of materials.

5.2.1 *Transients in a Semi-infinite Solid*

Semi-infinite solid subject to step input at surface

As a typical example, we consider a semi-infinite solid initially at zero temperature.[1] This solid is subject to a step change in surface temperature for $t > 0$, as indicated in Fig. 5.4. If we set $G = 0$ (no internal heat generation) and $\nabla^2 = \frac{\partial^2}{\partial x^2}$ (temperature variation is confined to one space dimension in Eq. 5.12), the following equation results.

$$\alpha \frac{\partial^2 T}{\partial x^2} = \frac{\partial T}{\partial t} \tag{5.18}$$

The initial and boundary conditions, as shown in Fig. 5.4 are

$$t = 0;\; T = 0 \text{ for } 0 \le x \le \infty \tag{5.19}$$

$$x = 0;\; T = T_s \text{ for } t > 0;\; T \to 0 \text{ as } x \to \infty \text{ for all } t \tag{5.20}$$

Equation 5.18 may be solved by the method of similarity. The method of similarity assumes that the time-varying temperature profile possesses a time-invariant shape, if the temperature profile is described in terms of a single composite variable that has both x and t in it. For this purpose, introduce the transformation

$$\eta = A x t^B \tag{5.21}$$

where A and B are to be determined as a part of the analysis. The requirement is that Eq. 5.18 be transformed into an ordinary differential equation in terms of the single independent variable (also known as the similarity variable) η. Transformation

[1] Since the heat equation is linear, any datum value of T may be set as zero!

(5.21) leads to the following expressions for the derivatives, based on rules of partial differentiation:

$$\left.\frac{\partial}{\partial x}\right|_t = \left.\frac{\partial \eta}{\partial x}\right|_t \frac{\mathrm{d}}{\mathrm{d}\eta} = At^B \frac{\mathrm{d}}{\mathrm{d}\eta}$$

$$\left.\frac{\partial^2}{\partial x^2}\right|_t = \left.\frac{\partial}{\partial x}\left(\frac{\partial}{\partial x}\right)\right|_t = \left.\frac{\partial}{\partial x}\left(At^B \frac{\mathrm{d}}{\mathrm{d}\eta}\right)\right|_t = At^B \left.\frac{\partial \eta}{\partial x}\right|_t \frac{\mathrm{d}^2}{\mathrm{d}\eta^2} = A^2 t^{2B} \frac{\mathrm{d}^2}{\mathrm{d}\eta^2}$$

$$\left.\frac{\partial}{\partial t}\right|_x = \left.\frac{\partial \eta}{\partial t}\right|_x \frac{\mathrm{d}}{\mathrm{d}\eta} = ABxt^{B-1} \frac{\mathrm{d}}{\mathrm{d}\eta} = B\frac{\eta}{t}\frac{\mathrm{d}}{\mathrm{d}\eta}$$

With these, the heat equation becomes

$$\alpha A^2 t^{2B} \frac{\mathrm{d}^2 T}{\mathrm{d}\eta^2} = B\frac{\eta}{t}\frac{\mathrm{d}T}{\mathrm{d}\eta} \tag{5.22}$$

Equation 5.22 is to be an ordinary differential equation, hence x and t should not appear explicitly in it, but only in the combination (5.21). This condition is satisfied if the exponent of t on both sides of equation are the same. Thus

$$2B = -1 \text{ or } B = -\frac{1}{2}$$

Under this condition, Eq. 5.22 becomes

$$\alpha A^2 \frac{\mathrm{d}^2 T}{\mathrm{d}\eta^2} = -\frac{\eta}{2}\frac{\mathrm{d}T}{\mathrm{d}\eta}$$

In addition, if we set $4\alpha A^2 = 1$ or $A = \frac{1}{2\sqrt{\alpha}}$ (this choice is arbitrary and purely for convenience) the governing equation reduces to

$$\boxed{\frac{\mathrm{d}^2 T}{\mathrm{d}\eta^2} + 2\eta\frac{\mathrm{d}T}{\mathrm{d}\eta} = 0} \tag{5.23}$$

where the similarity variable η is given by

$$\boxed{\eta = \frac{x}{2\sqrt{\alpha t}}} \tag{5.24}$$

Now we take a look at the initial and boundary conditions. These also have to be represented in terms of η if the above procedure is to make sense. In addition, since the governing equation has become a second-order ODE, the initial and two boundary conditions must become the two boundary conditions on Eq. 5.23. Thus

- Initial condition: $t = 0$ means $\eta \to \infty$ for *any* x and hence $T \to 0$
- Surface boundary condition: $x = 0$ and $t > 0$ means $\eta = 0$, and hence $T = T_s$
- Boundary condition as $x \to \infty$: and $t > 0$ means $\eta \to \infty$ and $T \to 0$

Thus the two boundary conditions to be satisfied by Eq. 5.23 are

$$\eta = 0,\ T = T_s \text{ and } \eta \to \infty,\ T \to 0 \tag{5.25}$$

Thus, the original problem has reduced to the solution of Eq. 5.23 subject to the boundary conditions (5.25). Equation 5.23 is in the variable separable form and may be written as

$$\frac{\left(\dfrac{\mathrm{d}^2T}{\mathrm{d}\eta^2}\right)}{\left(\dfrac{\mathrm{d}T}{\mathrm{d}\eta}\right)}\mathrm{d}\eta = -2\eta\mathrm{d}\eta \tag{5.26}$$

which on one integration with respect to η yields

$$\ln\frac{\mathrm{d}T}{\mathrm{d}\eta} = -\eta^2 + C_1 \text{ or } \frac{\mathrm{d}T}{\mathrm{d}\eta} = e^{(-\eta^2+C_1)}$$

A second integration with respect to η yields

$$T = \int_0^\eta e^{(-\eta^2+C_1)}\mathrm{d}\eta + C_2 \tag{5.27}$$

where C_1 and C_2 are constants of integration. Using the surface boundary condition in (5.27), we get $C_2 = T_s$. The boundary condition at $\eta \to \infty$ requires that

$$\int_0^\infty e^{(-\eta^2+C_1)}\mathrm{d}\eta + T_s = 0 \text{ or } e^{C_1} = -\frac{T_s}{\int_0^\infty e^{-\eta^2}\mathrm{d}\eta}$$

Thus, the desired solution to the problem is

$$\frac{T}{T_s} = 1 - \frac{\int_0^\eta e^{-\eta^2}\mathrm{d}\eta}{\int_0^\infty e^{-\eta^2}\mathrm{d}\eta} \tag{5.28}$$

It is easily shown that

$$\int_0^\infty e^{-\eta^2}\mathrm{d}\eta = \frac{\sqrt{\pi}}{2}$$

With this, the solution becomes

$$\frac{T}{T_s} = 1 - \frac{2}{\sqrt{\pi}} \int_0^{\eta} e^{-\eta^2} \mathrm{d}\eta = 1 - \mathrm{erf}\ \eta = \mathrm{erfc}\ \eta \tag{5.29}$$

In the above, erf η is the error function and erfc η is the complementary error function. Error function and the complementary error function are available in tabular form in handbooks of Mathematics. A short extract from such a table is given in Table 5.2. The table indicates that $\frac{T}{T_s} \rightarrow 0$ at $\eta = 3.6$ (less than a significant figure in the sixth decimal place). This means that, at any time t, the temperature has penetrated a distance δ into the material, given by the condition

$$\eta_\delta \approx 3.6 \text{ or } \delta \approx 7.2\sqrt{t}$$

Table 5.2 Error function and complementary error function

η	erf η	erfc η	η	erf η	erfc η
0	0.000000	1.000000	1.9	0.992790	0.007210
0.1	0.112463	0.887537	2	0.995322	0.004678
0.2	0.222703	0.777297	2.1	0.997021	0.002979
0.3	0.328627	0.671373	2.2	0.998137	0.001863
0.4	0.428392	0.571608	2.3	0.998857	0.001143
0.5	0.520500	0.479500	2.4	0.999311	0.000689
0.6	0.603856	0.396144	2.5	0.999593	0.000407
0.7	0.677801	0.322199	2.6	0.999764	0.000236
0.8	0.742101	0.257899	2.7	0.999866	0.000134
0.9	0.796908	0.203092	2.8	0.999925	0.000075
1	0.842701	0.157299	2.9	0.999959	0.000041
1.1	0.880205	0.119795	3	0.999978	0.000022
1.2	0.910314	0.089686	3.1	0.999988	0.000012
1.3	0.934008	0.065992	3.2	0.999994	0.000006
1.4	0.952285	0.047715	3.3	0.999997	0.000003
1.5	0.966105	0.033895	3.4	0.999998	0.000002
1.6	0.976348	0.023652	3.5	0.999999	0.000001
1.7	0.983790	0.016210	3.6	1.000000	0.000000
1.8	0.989091	0.010909			

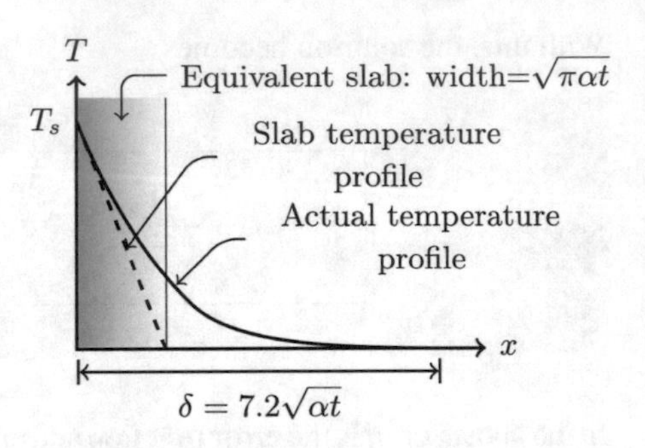

Fig. 5.5 Analogy with slab model

Let us now determine the heat flux at the front face $x = 0$ (or $\eta = 0$). We have

$$\left.\frac{\mathrm{d}T}{\mathrm{d}\eta}\right|_{\eta=0} = -\left.\frac{2}{\sqrt{\pi}}T_s e^{-\eta^2}\right|_{\eta=0} = -\frac{2}{\sqrt{\pi}}T_s$$

The surface heat flux is given by (using the definition of η)

$$q(t)|_{x=0} = -\left.k\frac{\mathrm{d}T}{\mathrm{d}x}\right|_{x=0} = -kAt^B\left.\frac{\mathrm{d}T}{\mathrm{d}\eta}\right|_{\eta=0} = -\frac{k}{2\sqrt{\alpha t}}\left[-\frac{2}{\sqrt{\pi}}T_s\right] = \frac{kT_s}{\sqrt{\pi\alpha t}} \tag{5.30}$$

It is seen that $q(t) \to \infty$ as $t \to 0$ and $q(t) \to 0$ as $t \to \infty$. Equation 5.30 may be interpreted in terms of an equivalent slab, whose thickness is increasing with time as $\sqrt{\pi\alpha t}$. This is explained in Fig. 5.5. It is easily seen from this analogy that as long as the depth of penetration δ is less than the thickness of the equivalent slab, the transient in it may be modeled as that in a semi-infinite solid.

Integrating expression (5.30) with respect to time, we get the total heat per unit area q_s that has entered the semi-infinite solid in time t.

$$q_s = \int_0^t q\Big|_{x=0} dt = \frac{kT_s}{\sqrt{\pi\alpha}}\int_0^t \frac{\mathrm{d}t}{\sqrt{t}} = 2kT_s\sqrt{\frac{t}{\pi\alpha}} \tag{5.31}$$

Thus the heat entering the body grows as square root of time. In practical terms, the penetration depth may be defined more meaningfully as indicating the depth at which the temperature is about 1% of the value at the surface, since this may be a measurable quantity in practice (see Fig. 5.5). From Table 5.2, this corresponds to $\eta \approx 1.82$. This is very close to 50% of the value quoted earlier! This value of η also indicates the time up to which a slab of given thickness may be considered as a semi-infinite slab, as will become clear from Example 5.1.

Example 5.1

A large slab of concrete of thickness equal to 200 mm is exposed to high temperature radiant environment on one side. The surface facing it attains a temperature of 125 °C almost instantaneously. The slab is initially at a temperature of 30 °C. The thermophysical properties of concrete are: density $\rho = 2150$ kg/m^3, specific heat $c = 950$ J/kg°C, thermal conductivity $k = 1.06$ W/m°C. Till what time is it reasonable to treat the concrete slab as a semi-infinite solid? How much heat would have entered the slab per square meter, in this period? Make a plot of the temperature profile in the slab, at this time.

Solution:
From the given property data, the thermal diffusivity of concrete may be ascertained.

$$\text{Thermal diffusivity of concrete: } \alpha = \frac{k}{\rho c} = \frac{1.06}{2150 \times 950} = 5.19 \times 10^{-7}\ \text{m}^2/\text{s}$$

The other pertinent data specified in the problem are listed below:

Thickness of concrete slab: $L = 200$ mm or 0.2 m
Initial temperature of slab: $T_i = 30\,°\text{C}$
Imposed surface temperature: $T_s = 125\,°\text{C}$

In order to determine the time t_{ss} up to which the semi-infinite solid assumption is valid, we equate the penetration depth (defined as the depth at which the temperature is 1% of the surface value, given by $\eta = 1.82$) to the slab thickness. Thus

$$\eta = 1.82 = \frac{L}{2\sqrt{\alpha t_{ss}}} \text{ or } t_{ss} = \frac{L^2}{4 \times \eta^2 \times \alpha}$$

$$= \frac{0.2^2}{4 \times 1.82^2 \times 5.19 \times 10^{-7}} = 5817.2s \approx 1\text{ h }37\text{ min}$$

The front surface of the slab is subject to a constant temperature given by $T_s = 125\,°\text{C}$. The initial temperature of the solid is $T_i = 30\,°\text{C}$. The latter temperature corresponds to the datum value. Hence the heat that will enter per square meter of the slab is calculated using Eq. 5.31 by replacing T_s by $T_s - T_i$.

$$q_s = 2k(T_s - T_i)\sqrt{\frac{t}{\pi\alpha}} = 2 \times 1.06 \times (125 - 30) \times \sqrt{\frac{5817.2}{\pi \times 5.19 \times 10^{-7}}}$$
$$= 12.03 \times 10^6\ \text{J/m}^2$$

Now for the temperature distribution. Using Eq. 5.29 and replacing zero temperature by the datum value

$$T(\eta) = T_i + (T_s - T_i)\text{erfc}\ \eta \text{ with } \eta = \sqrt{\frac{x}{4\alpha t_{ss}}}$$

where x varies from 0 to 0.2 m. This is plotted as shown in Fig. 5.6.

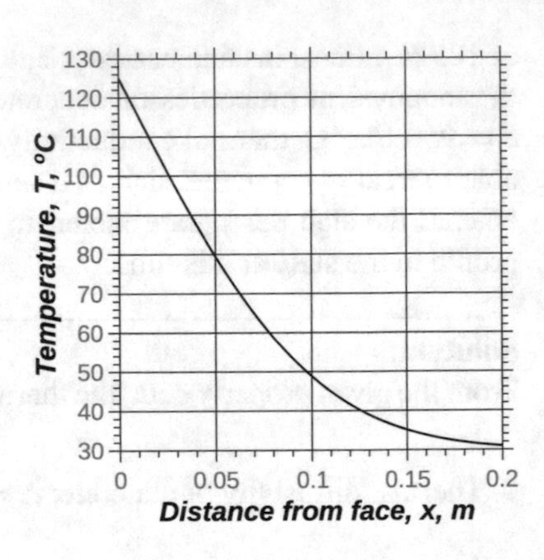

Fig. 5.6 Temperature distribution in the slab of Example 5.1 at $t = t_{ss} = 5817$ s

Example 5.2

Consider a semi-infinite solid subjected to a constant surface temperature excess of 100 °C for $t > 0$. Obtain an expression for the surface heat flux as a function of time for two materials. (a) Concrete with properties as in Example 5.1 and (b) Aluminum with the following properties: density $\rho = 2701$ kg/m^3, specific heat $c = 903$ J/kg°C, and thermal conductivity $k = 237$ W/m°C. Make a plot of surface heat flux variation with time for the two materials. Compare the energy that has entered the unit area of the two materials in 1000 s.

Solution:

The expression for heat flux is given by Eq. 5.30. The property values are substituted to get the surface heat flux in the two cases. Surface temperature excess is interpreted as $T_s - T_i$. We have

$$q_s = \frac{k(T_s - T_i)}{\sqrt{\pi \alpha t}} = (T_s - T_i)\sqrt{\frac{\rho c k}{\pi t}} = \frac{K}{\sqrt{t}}$$

where K is a constant given by $(T_s - T_i)\sqrt{\frac{\rho c k}{\pi}}$

The property values for concrete are taken from Example 5.1. The constant $K = K_c$ for concrete is obtained as

$$K_c = 100 \times \sqrt{\frac{2150 \times 950 \times 1.06}{\pi}} = 83015.5 \text{ Ws}^{\frac{1}{2}}/\text{m}^2$$

More suitable way of expressing the above will be as $K = 83.02$ kW s$^{\frac{1}{2}}$/m^2.

The property values for aluminum are given above in the Example statement. The constant $K = K_a$ for aluminum is obtained as

$$K_a = 100 \times \sqrt{\frac{2702 \times 903 \times 237}{\pi}} = 1356706 \text{ Ws}^{\frac{1}{2}}/\text{m}^2$$

More suitable way of expressing the above will be as $K = 1357$ kW $s^{\frac{1}{2}}/m^2$. Figure 5.7 shows the plot of surface heat flux history for the two materials. It is seen that the surface heat flux is much larger in the case of aluminum than concrete. Thermal properties of the material play a very important role in the transient response. Both the density and heat capacity are of comparable size for the two materials. However, thermal conductivity of aluminum is more than 200 times that of concrete and the surface heat flux mirrors this! It is interesting to compare the total energy that has entered the two materials in 1000 s for which the plot has been made. We make use of expression (5.31) for this purpose. The thermal diffusivity of concrete has been calculated in Example 5.1 as $\alpha_c = 5.19 \times 10^{-7} \text{m}^2/\text{s}$.
Heat that has entered concrete in 1000 s is calculated using Eq. 5.31 as

$$q_s(\text{concrete}) = 2 \times 1.06 \times 100 \sqrt{\frac{1000}{\pi \times 5.19 \times 10^{-7}}} = 5.25 \times 10^6 \text{ J/m}^2$$
$$= 5.25 \text{ MJ/m}^2$$

From the given data, the thermal diffusivity of aluminum is

$$\alpha_a = \frac{237}{2702 \times 903} = 9.71 \times 10^{-5} \text{ m}^2/\text{s}$$

Hence the heat that has entered aluminum in 1000 s is calculated as

$$q_s(\text{aluminum}) = 2 \times 237 \times 100 \sqrt{\frac{1000}{\pi \times 9.71 \times 10^{-5}}} = 85.82 \times 10^6$$
$$= 85.52 \text{ MJ/m}^2$$

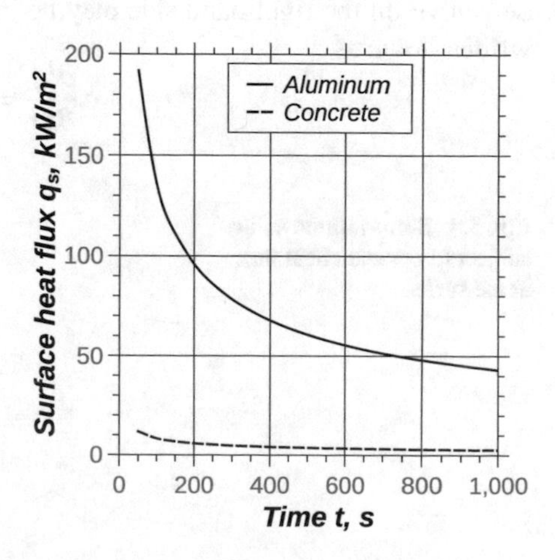

Fig. 5.7 Surface heat flux in concrete and aluminum of Example 5.2

Semi-infinite Solid Subject to Constant Heat Flux at its Surface

This case is easily realized in practice. A very large solid medium may be subject to step heating at its surface by turning on very rapidly a radiant source like a laser or a lamp. As explained earlier, the process may be assumed to be that in a semi-infinite solid if the thickness of the solid is larger than the depth of penetration. The governing equation, for this problem, is again the one-dimensional heat Eq. 5.18. The boundary conditions are specified as follows:

$$q = q_s \text{ at } x = 0; \; T \to 0 \text{ as } x \to \infty \text{ for all } t \tag{5.32}$$

The initial condition, of course, is as indicated in Fig. 5.8. The boundary condition at $x = 0$ can be rewritten in terms of T, using Fourier law, as

$$q(0, t) = -k \left. \frac{\mathrm{d}T}{\mathrm{d}x} \right|_{x=0} \tag{5.33}$$

Based on Fourier law, we also know that

$$q(x, t) = -k \left. \frac{\mathrm{d}T}{\mathrm{d}x} \right|_{x} \tag{5.34}$$

We differentiate Eq. 5.18 with respect to x to get

$$\alpha \frac{\partial}{\partial x}\left(\frac{\partial^2 T}{\partial x^2}\right) = \alpha \frac{\partial^3 T}{\partial x^3} = \frac{\partial T}{\partial x \partial t}$$

Noting from Eq. 5.34 that $\frac{\partial T}{\partial x} = -\frac{q}{k}$, and noting that order of taking the indicated derivative on the right-hand side may be interchanged, the above equation may be written down as

$$\alpha \frac{\partial^2 q}{\partial x^2} = \frac{\partial q}{\partial t} \tag{5.35}$$

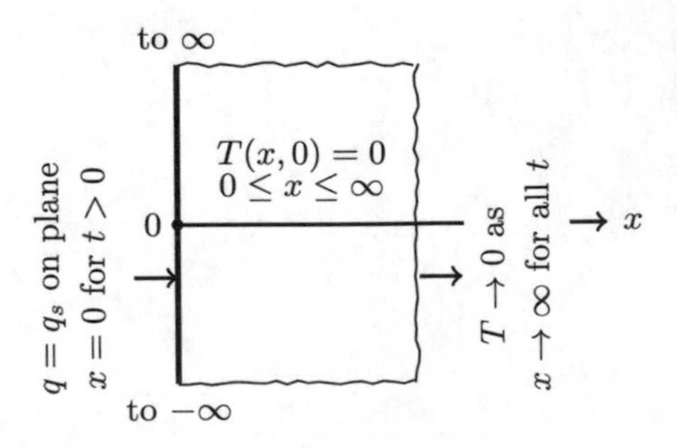

Fig. 5.8 Semi-infinite solid subject to constant heat flux at the surface

Mathematically, Eq. 5.35 along with the conditions (5.32) are the same as those encountered in the problem with step change in temperature. (The clue is, wherever q appears, replace it by T and we get the equation applicable for the previous case). Hence we conclude that the solution must be

$$q = q_s \text{erfc}\ \eta \tag{5.36}$$

where η is, the now familiar, similarity variable. We would like, however, to obtain the solution to the temperature field. This may be accomplished by substituting Eqs. 5.34 in 5.36 to get

$$-k\frac{\partial T}{\partial x} = q_s \left[1 - \frac{2}{\sqrt{\pi}} \int_0^{\eta} e^{-\eta^2} \mathrm{d}\eta\right]$$

This may be integrated once with respect to x to get

$$T = -\frac{q_s}{k} \int_0^{x} \left[1 - \frac{2}{\sqrt{\pi}} \int_0^{\eta} e^{-\eta^2} \mathrm{d}\eta\right] \mathrm{d}x + C$$

where C is a constant of integration. Noting that $T \to 0$ as $x \to \infty$, we have

$$0 = -\frac{q_s}{k} \int_0^{\infty} \left[1 - \frac{2}{\sqrt{\pi}} \int_0^{\eta} e^{-\eta^2} \mathrm{d}\eta\right] \mathrm{d}x + C$$

From the above two equations, we may eliminate C and get

$$T = \frac{q_s}{k} \int_x^{\infty} \left[1 - \frac{2}{\sqrt{\pi}} \int_0^{\eta} e^{-\eta^2} \mathrm{d}\eta\right] \mathrm{d}x$$

We note that integration with respect to x would mean that we hold t fixed. Thus, using the definition of η, we have

$$\mathrm{d}x = 2\sqrt{\alpha t} \mathrm{d}\eta \Big|_t$$

Therefore we write the above as

$$T = \left\{\frac{q_s}{k} \underbrace{\int_{\eta}^{\infty} \left[1 - \frac{2}{\sqrt{\pi}} \int_0^{\eta} e^{-\eta^2} \mathrm{d}\eta\right] \mathrm{d}\eta}_{\mathrm{I}}\right\} 2\sqrt{\alpha t}$$

Integral I within the flower braces may be obtained by repeated use of integration by parts.

$$I = \underbrace{\eta\left[1 - \frac{2}{\sqrt{\pi}}\int_0^{\eta} e^{-\eta^2}\mathrm{d}\eta\right]\Bigg|_{\eta}^{\infty}}_{\text{As } \eta\to\infty \text{ these terms go to zero}} - \int_{\eta}^{\infty} \eta\left\{-\frac{2}{\sqrt{\pi}}e^{-\eta^2}\right\}\mathrm{d}\eta$$

$$= -\eta\text{erfc }\eta + \frac{2}{\sqrt{\pi}}\int_{\eta}^{\infty} \eta e^{-\eta^2}\mathrm{d}\eta = -\eta\text{erfc }\eta + \frac{1}{\sqrt{\pi}}e^{-\eta^2} \tag{5.37}$$

Hence, we have the temperature field within the solid as

$$T = \frac{q_s}{k}\times 2\sqrt{\alpha t}\left[-\eta\text{erfc }\eta + \frac{1}{\sqrt{\pi}}e^{-\eta^2}\right] = \frac{q_s}{k}\left[\sqrt{\frac{4\alpha t}{\pi}}e^{-\eta^2} - x\text{ erfc }\eta\right] \tag{5.38}$$

It is interesting to note that the surface temperature of the medium is given by

$$\text{(a)}\quad T_s = \frac{q_s}{k}\sqrt{\frac{4\alpha t}{\pi}} \quad \text{or} \quad \text{(b)}\quad T_s = q_s\sqrt{\frac{4t}{\pi\rho c k}} \tag{5.39}$$

Example 5.3

Consider a semi-infinite solid subjected to a constant surface heat flux of 1000 W/m^2 for $t > 0$. Obtain an expression for the surface temperature as a function of time for two materials: (a) Concrete and (b) Aluminum. Use the property values given in Example 5.2. Make a plot of surface temperature versus time for the two materials.

Solution:

The surface temperature is given by expression (5.39). The property values are substituted to get the surface temperature in the two cases. Surface heat flux has been specified as $q_s = 1000$ W/m^2. Using Eq. 5.39(b), the surface temperature variation with time may be recast as $T_s = K\sqrt{t}$ where $K = 2q_s/\sqrt{\pi\rho c k}$.

In case of concrete and aluminum, the constant $K = K_c$ is and $K = K_a$ are given by

$$K_c = 2\times 1000/\sqrt{\pi\times 2150\times 950\times 1.06} = 0.767\,°\text{C/s}^{\frac{1}{2}}$$

$$K_a = 2 \times 1000/\sqrt{\pi \times 2702 \times 903 \times 237} = 0.047\,°\mathrm{C/s}^{\frac{1}{2}}$$

It is thus seen that aluminum shows much less temperature increase at its front surface as compared to concrete.

Figure 5.9 shows the plot of surface temperature history for the two materials.

Examples 5.2 and 5.3 have considered the transient response of two materials for two different boundary conditions. Concrete is classified as a poor thermal conductor with a low thermal diffusivity ($\alpha = 5.19 \times 10^{-7}$ m^2/s) while aluminum is classified as a good conductor with high thermal diffusivity ($\alpha = 9.71 \times 10^{-5}$ m^2/s). Note that the thermal diffusivity of aluminum is roughly 165 times the thermal diffusivity of concrete. In the case of step change in surface temperature, the heat entering aluminum (actually the heat that has to be supplied to maintain the surface temperature at the indicated value) is much larger than that in the case of concrete. Thermal diffusion is a process that spreads the heat within the medium. The larger the thermal diffusivity more rapid is the spreading process. In the second case where the surface flux is maintained at a constant value, because the spreading is more rapid in the case of aluminum, the surface temperature increases gradually. Concrete, on the other hand, does not transmit the heat rapidly into the bulk and hence the surface temperature increases very rapidly. The incident heat is stored right close to the front surface! People living in the tropics prefer houses with thick mud walls for the simple reason that these keep the solar heat out and keep the interior cool. The reader is encouraged to look for other examples!

Semi-infinite solid with periodic surface temperature variation

Problems with periodic heating occur in geophysical applications. The variation of the temperature below the earth's surface, due to daily, seasonal, annual, and long-term variations of temperature at its surface may be modeled this way. The governing equation is the same as Eq. 5.18. Initially the solid is at $T = 0$ throughout. For $t > 0$,

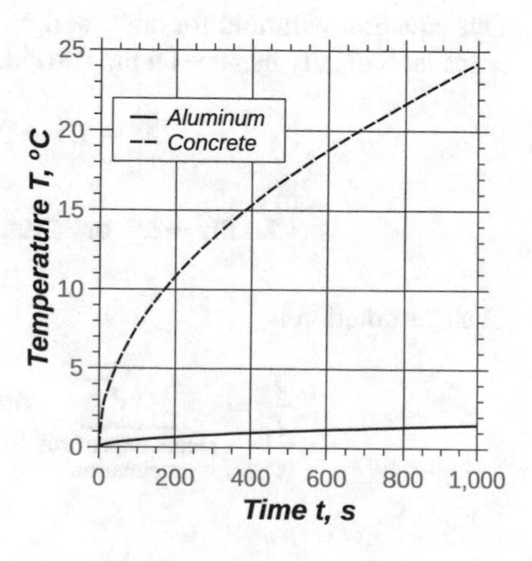

Fig. 5.9 Surface temperature history in concrete and aluminum

a sinusoidal temperature variation is imposed at the surface according to the relation.

$$T(0, t) = T_s \sin \omega t \tag{5.40}$$

Obviously, there will be an initial period in which a transient (see Chap. 3) will prevail in the solid. After a sufficiently long time, the solution should tend to a periodic variation of temperature throughout the solid. One expects the amplitude to decrease with depth x with a depth-dependent phase lag. We may assume that the solution should have the same period (this is so since the governing equation is linear) and hence we seek a solution of the form.

$$T(x, t) = T_s e^{-Ax} \sin(\omega t - Bx) \tag{5.41}$$

where A and B are positive and real constants to be determined. The form chosen for the solution automatically satisfies the boundary and the other conditions specified above. From Eq. 5.41, the following may be obtained:

$$\begin{aligned}
\frac{\partial T}{\partial x} &= -AT_s e^{-Ax} \sin(\omega t - Bx) - BT_s e^{-Ax} \cos(\omega t - Bx) \\
\frac{\partial^2 T}{\partial x^2} &= A^2 T_s e^{-Ax} \sin(\omega t - Bx) + 2ABT_s e^{-Ax} \cos(\omega t - Bx) \\
&\quad - B^2 T_s e^{-Ax} \sin(\omega t - Bx) \\
\frac{\partial T}{\partial t} &= \omega T_s e^{-Ax} \cos(\omega t - Bx)
\end{aligned}$$

Substituting these in Eq. 5.18, canceling the common factor e^{-Ax} and grouping terms, we get

$$\alpha[(A^2 - B^2) \sin(\omega t - Bx) + 2AB \cos(\omega t - Bx)] = \omega \cos(\omega t - Bx) \tag{5.42}$$

This equation will hold for any t and x only if the coefficients of "sin" and "cos" terms individually balance on the two sides. Thus

$$A^2 - B^2 = 0 \quad \text{or} \quad A = B \tag{5.43}$$

$$2AB\alpha = \omega \quad \text{or} \quad 2A^2\alpha = \omega \quad \text{or} \quad A = \sqrt{\frac{\omega}{2\alpha}} \tag{5.44}$$

Thus the solution is

$$\frac{T}{T_s} = \underbrace{e^{-x\sqrt{\frac{\omega}{2\alpha}}}}_{\text{Depth dependent attenuation}} \sin\left(\omega t - \underbrace{\sqrt{\frac{\omega}{2\alpha}}x}_{\text{Depth dependent phase lag}}\right) \tag{5.45}$$

Table 5.3 Response to periodic surface temperature variation

Material	α, m^2/s	α, m^2/h	t^*, h	Attenuation
Aluminum	8.39×10^{-5}	0.302	2.5	0.52
Steel	1.17×10^{-5}	0.042	6.7	0.17
Clay	1×10^{-6}	0.0036	23	0.0024
Wood	1.19×10^{-7}	0.00043	66	2.6×10^{-8}

The exponential term indicates attenuation with depth while the oscillatory term shows a depth-dependent phase lag. Table 5.3 shows what happens when the period of the wave is 24 h and the depth inside the medium from the surface is 1 m. The attenuation as well as the phase lag depend strongly on the thermal diffusivity of the medium. We may define a time lag t^* such that the phase lag is ωt^*. Thus t^* is given by

$$t^* = \frac{1}{\omega}\sqrt{\frac{\omega}{2\alpha}}x = \frac{x}{\sqrt{2\alpha\omega}} \tag{5.46}$$

In the case presented in Table 5.3, the time lag is $t^* = \frac{1}{\sqrt{2\alpha\omega}}$ since $x = 1$ m.

Example 5.4

The temperature below the ground is affected by the daily variations of temperature above it. The period of the daily variation may be taken as 24 h. The amplitude of the variation of the temperature at the surface is taken as the unit. The material of the top layers is known to be gravelly sand with a thermal diffusivity of 1.403×10^{-7} m^2/s. Determine the depth at which the temperature amplitude is just equal to 5% of the amplitude at the surface. What is the time lag at this depth? Make a plot of the response at a depth of 0.1 m below the surface. Assume that the problem may be treated using the semi-infinite solid model and that there is no internal heat source below the ground.

Solution:
Since the problem involves the time scale of the order of hours the problem is solved using time in hours. The thermal diffusivity needs to be converted to m^2/h. This may be done as follows.

$$\alpha = 1.403 \times 10^{-7} \times 3600 = 5.05 \times 10^{-4} \text{ m}^2/\text{h}$$

Let $x_{5\%}$ be the depth at which the amplitude is 5% of that at the ground level. The period of oscillation is given as $T = 24$ h. The circular frequency of the oscillation ω is then given by

$$\omega = \frac{2\pi}{T} = \frac{2\pi}{24} = 0.262 \text{ rad/h}$$

We then have

$$\frac{T}{T_s} = 0.05 = e^{-\sqrt{\frac{\omega}{2\alpha}}x_{5\%}}$$

Taking natural logarithms, we get

$$x_{5\%} = -\frac{\ln 0.05}{\sqrt{\frac{\omega}{2\alpha}}} = -\frac{\ln 0.05}{\sqrt{\frac{0.262}{2\times 5.05\times 10^{-4}}}} = 0.186 \text{ m}$$

The time lag at this depth is calculated using Eq. 5.46 as

$$t^*_{5\%} = \frac{x_{5\%}}{\sqrt{2\alpha\omega}} = \frac{0.186}{\sqrt{2\times 5.05\times 10^{-4}\times 0.262}} = 11.434 \text{ h}$$

Now consider the state of affairs at a depth of $x = 0.1$ m. The amplitude at this depth is given by

$$\frac{T_{0.1\text{m}}}{T_s} = e^{-0.1\times\sqrt{\frac{0.262}{2\times 5.05\times 10^{-4}}}} = 0.2$$

The time lag at this depth is calculated as

$$t^*_{0.1\text{m}} = \frac{0.1}{\sqrt{2\times 5.05\times 10^{-4}\times 0.262}} = 6.147 \text{ h}$$

The response at this depth is then given by

$$\frac{T_{0.1\text{m}}}{T_s} = 0.2\sin[0.262(t-6.147)]$$

where t is in h. This is plotted in Fig. 5.10 along with the input, i.e., the temperature variation at the surface.

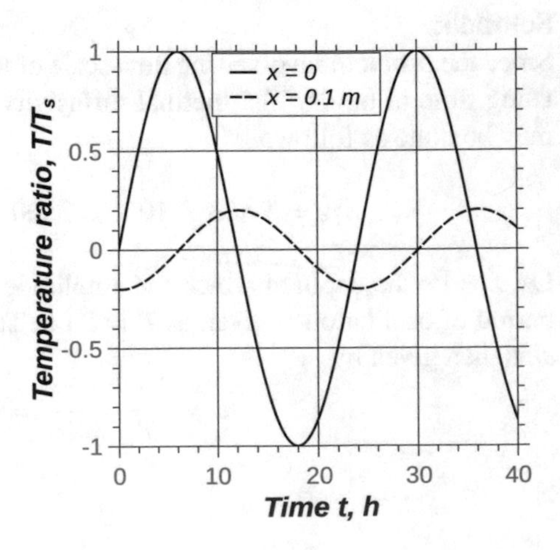

Fig. 5.10 Surface and sub-surface temperature histories in Example 5.3

Example 5.4 shows that the amplitude decays rapidly with depth for a medium with low thermal diffusivity. Cave dwellers (in Australia, China, and elsewhere) take advantage of this fact to keep themselves away from large temperature fluctuations that take place above the ground. Caves also keep away the overground noise.

5.2.2 Approximate Integral Method Due to Goodman

One-dimensional transient in a semi-infinite solid has been shown to exhibit a similarity solution in two different cases considered in Sect. 5.2.1. For example, we have alluded to an analogy with a slab problem in the case of the transient with a step change in surface temperature. An approximate method of solution based on an integral formulation is possible, and was exploited for the first time, by Goodman. Hence the method is known as Goodman's integral method. Consider the state of affairs shown in Fig. 5.11. The case corresponds to a semi-infinite solid subject to a step change in temperature at its surface, for $t > 0$. The initial temperature of the solid is zero throughout, it being the datum value, as mentioned earlier. At some positive time, t the temperature variation is as indicated by the dashed curve, with the depth of penetration being $\delta(t)$. A little later, at $t + \delta t$ the depth of penetration has increased by $\mathrm{d}\delta = \frac{\mathrm{d}\delta}{\mathrm{d}t}\mathrm{d}t$ and the temperature profile has changed as shown by the full curve. Let us look at the total energy contained in the solid.

The change in energy contained in the solid may be written in two parts. The first part is due to the change in temperature everywhere in the interval $0 \le x \le \delta(t)$ given by

$$T(x, t + \mathrm{d}t) - T(x, t) \approx \frac{\partial T}{\partial t}\mathrm{d}t$$

The first part of increase in energy is thus given by

$$\left.\frac{\mathrm{d}E}{\mathrm{d}t}\right|_I = \rho c \frac{\mathrm{d}}{\mathrm{d}t}\int_0^\delta T\,\mathrm{d}x \tag{5.47}$$

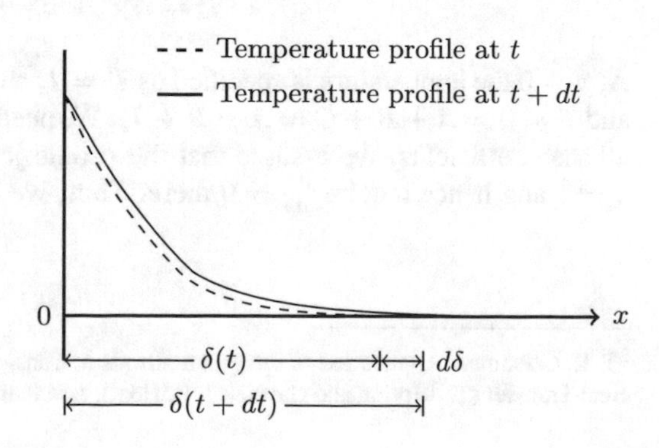

Fig. 5.11 Sketch used in arriving at the integral method

The second part is due to the fact that δ has increased by $\mathrm{d}\delta$ and hence has brought an extra thickness of material in to the heated zone. The rate of energy change due to this may be written as

$$\left.\frac{\mathrm{d}E}{\mathrm{d}t}\right|_{II} \approx \frac{\mathrm{d}\delta}{\mathrm{d}t}\Delta T \tag{5.48}$$

The integral has been replaced by a product because the change in δ is a differential quantity. The temperature change indicated, i.e., ΔT is in the region $\delta(t) \leq x \leq \delta(t+\mathrm{d}t)$ and is, in fact, zero since the temperature has the datum value beyond δ. Hence the rate of energy change for the second part is zero. The rate of energy change within the heated zone of the solid is hence given by Eq. 5.47 itself.

This rate of change of energy contained within the solid must have been brought about by q_s, the heat transfer at the surface (note that there is *no* heat transfer at δ). Using Fourier law, we thus have

$$\boxed{\rho c \frac{\mathrm{d}}{\mathrm{d}t}\int_0^{\delta} T\,\mathrm{d}x = q_s = -k\left.\frac{\partial T}{\partial x}\right|_{x=0}} \tag{5.49}$$

Equation 5.49 is the starting point for the approximate method due to Goodman.[2] The above equation itself is exact. The approximation involves the use of an assumed temperature profile in the interval 0, δ in the form $T(x,t) = T(y)$ where $y = \frac{x}{\delta}$. The assumed profile may be in the form of a suitable polynomial in y. The coefficients of the polynomial are pure numbers that will be determined partly by the boundary conditions and partly by extra or auxiliary conditions, as we shall see below. The depth of penetration is itself determined as a solution to the integral equation (5.49).

Approximate solution using a polynomial

We demonstrate the approximate solution method by using a second-degree polynomial (i.e., a quadratic) in the form

$$T(y) = Ay^2 + By + C \tag{5.50}$$

At $y = 0$ the temperature is specified as $T = T_s$ and hence $C = T_s$. At $x = \delta$, $y = 1$ and $T = 0 = A + B + C = A + B + T_s$. We need one more condition to determine all the coefficients. We assume that the profile joins smoothly the datum value at $y = 1$ and hence require $\frac{\mathrm{d}T}{\mathrm{d}y} = 0$ there. Thus, we have, $2A + B = 0$ or $B = -2A$.

[2]T. R. Goodman, Application of integral methods to transient nonlinear heat transfer, Advances in Heat Transfer (T. F. Irvine and Hartnett J. P., Eds.), Academic Press, N.Y., Vol. I, pp. 51–122, 1964.

This in the previous condition gives $A - 2A + T_s = 0$ or $A = T_s$. Hence we also have $B = -2A = -2T_s$. The polynomial eventually becomes

$$T(y) = T_s(y^2 - 2y + 1) = T_s(1 - y)^2 \tag{5.51}$$

We may now obtain the integral in Eq. 5.49 as

$$\int_0^\delta T\,\mathrm{d}x = \delta \int_0^1 T(y)\mathrm{d}y = \delta \int_0^1 T_s(1 - y)^2 \mathrm{d}y = -T_s\delta \left.\frac{(1 - y)^3}{3}\right|_0^1 = \frac{T_s\delta}{3} \tag{5.52}$$

Using Eq. 5.51, we also have

$$\left.\frac{\partial T}{\partial x}\right|_{x=0} = \left.\frac{1}{\delta}\frac{\mathrm{d}T}{\mathrm{d}y}\right|_{y=0} = -\frac{2T_s}{\delta} \tag{5.53}$$

Substituting Eqs. 5.52 and 5.53 in the integral equation we have

$$\frac{\rho c}{3}\frac{\mathrm{d}\delta}{\mathrm{d}t} = k\frac{2}{\delta} \quad \text{or} \quad \delta\frac{\mathrm{d}\delta}{\mathrm{d}t} = 6\alpha \tag{5.54}$$

where the thermal diffusivity α makes its appearance. The above is a first-order ordinary differential equation in variable separable form. It may be integrated with the initial condition $\delta(t = 0) = 0$ to get

$$\delta = \sqrt{12\alpha t} \tag{5.55}$$

With this, the variable y becomes

$$y = \frac{x}{\sqrt{12\alpha t}} \tag{5.56}$$

Using Eq. 5.53, we then have

$$q_s = k\frac{2T_s}{\delta} = \frac{kT_s}{\sqrt{3\alpha t}} \tag{5.57}$$

It is interesting to compare this with the exact value given by Eq. 5.30. The percentage error due to the approximation may be calculated as

$$\text{Error }\% = \frac{\underbrace{q_s}_{\text{(approximate)}} - \underbrace{q_s}_{\text{(exact)}}}{\underbrace{q_s}_{\text{(exact)}}} \times 100 = \frac{\left(\frac{1}{\sqrt{3}} - \frac{1}{\sqrt{\pi}}\right)}{\left(\frac{1}{\sqrt{\pi}}\right)} \times 100 = 4.72\% \tag{5.58}$$

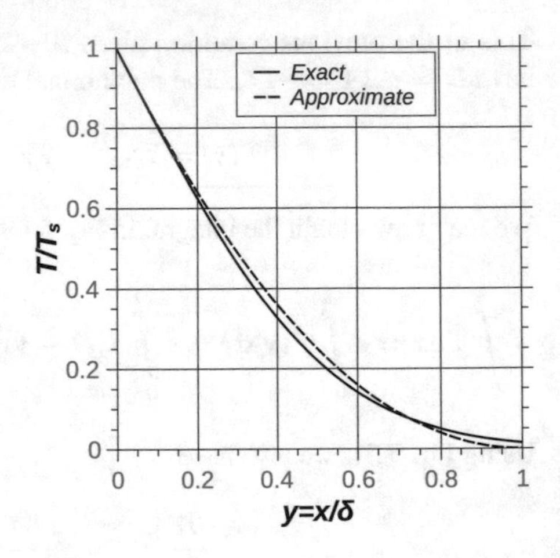

Fig. 5.12 Comparison of approximate temperature profile with the exact temperature profile

Also the depth of penetration has an error given by

$$\text{Error }\% = \frac{\underbrace{\delta}_{\text{(approximate)}} - \underbrace{\delta}_{\text{(exact)}}}{\underbrace{\delta}_{\text{(exact)}}} \times 100 = \frac{\sqrt{12} - 3.6}{3.6} \times 100 = -3.8\% \tag{5.59}$$

Even though the approximate integral solution used a very simple profile the errors in both quantities are rather small! Finally, we compare the approximate quadratic temperature profile with the exact profile in Fig. 5.12. The abscissa is x/δ where $\delta = \sqrt{12\alpha t}$. The ordinate shows the normalized nondimensional temperature $\theta = \frac{T}{T_s}$. The quadratic approximate profile appears to be a good representation of the temperature profile!

5.2.3 *One-Dimensional Transient Problem: Space Domain Finite*

One-dimensional transient conduction may occur in a slab of finite thickness or a bar of material that is insulated perfectly over the lateral surface. In both cases, appropriate boundary conditions are to be specified along with the initial temperature profile within the slab or the bar. Figure 5.13 shows the nomenclature appropriate to this case. Equation 5.18 is the governing equation for this problem also. The initial and boundary conditions that will be considered here are given below.

$$\begin{array}{ll} \underline{\text{Initial condition:}} & \underline{\text{Boundary conditions:}} \\ T(x,0) = f(x) & T(0, t > 0) = 0, \; T(L, t > 0) = 0 \end{array} \tag{5.60}$$

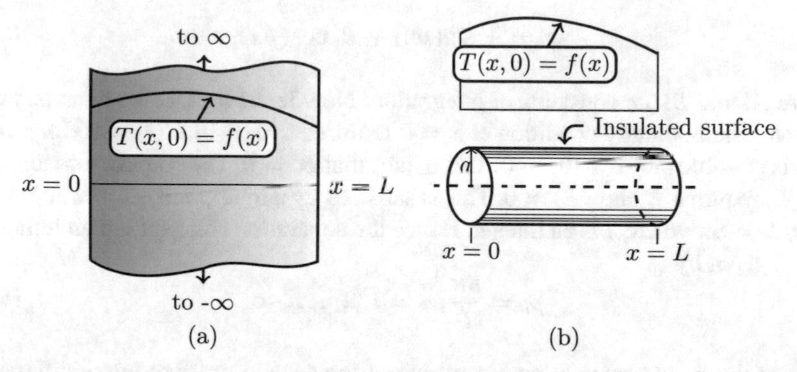

Fig. 5.13 Nomenclature for one-dimensional transient in a finite domain **a** Slab geometry **b** Bar geometry: bar is of uniform cross section

Thus both faces of the slab (or the two ends of the bar) are specified with the same temperature for $t > 0$, which is also taken as the datum value.

The technique of **separation of variables** is used to obtain the solution to the problem. For this purpose, seek a solution in the form of a product of two functions.

$$T(x, t) = F(x) \times G(t) \tag{5.61}$$

Denoting derivative with respect to t by a "dot" (i.e., $\dot{G}$ means $\frac{\mathrm{d}G}{\mathrm{d}t}$) and with respect to x by a "prime" (i.e., F' means $\frac{\mathrm{d}F}{\mathrm{d}x}$ and F'' means $\frac{\mathrm{d}^2F}{\mathrm{d}x^2}$), the governing equation becomes

$$F\dot{G} = \alpha G F'' \quad \text{or} \quad \frac{\dot{G}}{\alpha G} = \frac{F''}{F} \tag{5.62}$$

It is obvious that $\frac{\dot{G}}{\alpha G}$ is a function of "t" only and $\frac{F''}{F}$ is a function of "x" only. Hence, each of these must be equal to a constant if Eq. 5.62 should make sense. Let us call this constant a "separation constant" and let it equal $-p^2$. This choice assures that the separation constant is negative since p^2 is always positive for real p. The reason why the separation constant should be negative will become clear later on. Then Eq. 5.62 becomes equivalent to two ordinary differential equations given by

$$\text{(a)} \quad \dot{G} + \alpha p^2 G = 0 \quad \text{(b)} \quad F'' + p^2 F = 0 \tag{5.63}$$

From Eq. 5.63(a) we have, on integration with respect to t,

$$G = C' e^{-\alpha p^2 t} \tag{5.64}$$

where C' is a constant of integration. Separation constant was indeed taken as negative so that G (t) would decrease with time! The second-order equation governing F (Eq. 5.63(b)) may be integrated easily as

$$F = A' \sin(px) + B' \cos(px) \tag{5.65}$$

where A' and B' are constants of integration. Now let us look at the boundary conditions. The boundary condition at $x = 0$ requires that $F(0).G(t) = 0$ (for $t > 0$). This is possible only if $F(0) = 0$. This means that $B' = 0$. The boundary condition at $x = L$, requires $A' \sin(pL) = 0$. This is satisfied by a multiplicity of p values given by $p_n L = n\pi$ where n is an integer. Hence the separation constant has an infinity of values given by

$$p_n = \frac{n\pi}{L}, n = 1, 2, \ldots, \infty \tag{5.66}$$

Each of the p_n is known as an eigenvalue of the second-order ordinary differential equation (5.63(b)). Since every value of p_n gives a solution of form (5.65), the general solution *must* be of the form

$$F_n = A'_n \sin(p_n x) = A'_n \sin\left(\frac{n\pi x}{L}\right) \tag{5.67}$$

Since each such solution satisfies both boundary conditions, the most general solution to the second-order ODE for $F(x)$ must be given as a linear sum of the form (superposition of solutions for a linear ODE)

$$F(x) = \sum_{n=1}^{\infty} A'_n \sin\left(\frac{n\pi x}{L}\right) \tag{5.68}$$

and hence the solution to the one-dimensional heat equation should be

$$T(x,t) = \sum_{n=1}^{\infty} A_n e^{-\alpha\left(\frac{n\pi}{L}\right)^2 t} \sin\left(\frac{n\pi x}{L}\right) \tag{5.69}$$

where $A_n = A'_n C'_n$. Lastly we look at the initial condition. At $t = 0$, the exponential term becomes unity, and hence we require that

$$T(x,0) = f(x) = \sum_{n=1}^{\infty} A_n \sin\left(\frac{n\pi x}{L}\right) \tag{5.70}$$

Equation 5.70 is just the Fourier series representation of $f(x)$ in the interval $0 < x < L$. The Fourier coefficients or weights A_n may be obtained by using the orthogonal property of $\sin\left(\frac{n\pi x}{L}\right)$ in the interval $0 < x < L$.

Orthogonality property

Consider the following integral:

$$I_{mn} = \int_0^L \sin\left(\frac{n\pi x}{L}\right) \sin\left(\frac{m\pi x}{L}\right) dx \tag{5.71}$$

When $m \neq n$, we have, using trigonometric identities,

$$I_{mn} = \int_0^L \left[\cos\left\{ \frac{(m-n)\pi x}{L} \right\} - \cos\left\{ \frac{(m+n)\pi x}{L} \right\} \right] \mathrm{d}x$$

$$= \frac{\left[\frac{\sin\left\{\frac{(m-n)\pi x}{L}\right\}}{\frac{(m-n)\pi}{L}} - \frac{\sin\left\{\frac{(m+n)\pi x}{L}\right\}}{\frac{(m+n)\pi}{L}} \right]_0^L}{2} = 0 \tag{5.72}$$

However, when $m = n$, using trigonometric identities, we have

$$I_{nn} = \int_0^L \sin^2\left(\frac{n\pi x}{L}\right) \mathrm{d}x = \int_0^L \left[1 - \cos\left(\frac{2n\pi x}{L}\right) \right] \mathrm{d}x$$

$$= \frac{\left[x - \frac{\sin\left(\frac{2n\pi x}{L}\right)}{\frac{2n\pi}{L}} \right]_0^L}{2} = \frac{L}{2} \tag{5.73}$$

Equations 5.72 and 5.73 embody what is known as the orthogonal property of the function $\sin\left(\frac{n\pi x}{L}\right)$ over the interval $0 < x < L$.

Evaluation of Fourier coefficients

Evaluation of the coefficients A_n is straightforward, based on the orthogonal property of the sinusoidal function over the interval $0 < x < L$. If we multiply Eq. 5.70 by $\sin\left(\frac{m\pi x}{L}\right)$ and integrate with respect to x between $x = 0$ and $x = L$, only one term survives in the summation, viz., the term corresponding to $m = n$. All other terms vanish. Hence it is easy to see that we should have

$$A_n = \frac{2}{L} \int_0^L f(x) \sin\left(\frac{n\pi x}{L}\right) \mathrm{d}x \tag{5.74}$$

The solution to the problem will then be given by

$$\boxed{T(x, t) = \sum_1^\infty e^{-\frac{\alpha n^2 \pi^2 t}{L^2}} \left[\frac{2}{L} \int_0^L f(x) \sin\left(\frac{n\pi x}{L}\right) \mathrm{d}x \right] \sin\left(\frac{n\pi x}{L}\right)} \tag{5.75}$$

Example 5.5

A bar of length L of uniform cross section is perfectly insulated over its lateral surface. The bar is initially at a constant temperature of T_0 throughout. For $t > 0$, the two ends of the bar are maintained at zero temperature. Obtain the solution to the problem using Eqs. 5.74 and 5.75. Discuss the nature of the solution specifically with respect its dependence on time.

Solution:

Step 1 In this case $f(x) = T_0$ in the interval $0 < x < L$. The Fourier coefficients are obtained using Eq. 5.74 as

$$A_n = \frac{2}{L}\int_0^L T_0 \sin\left(\frac{n\pi x}{L}\right) \mathrm{d}x = \frac{2T_0}{L}\left\{\frac{-\cos\left(\frac{n\pi}{L}\right)}{\frac{n\pi}{L}}\right\}\Bigg|_0^L = \frac{2T_0}{n\pi}[1-\cos(n\pi)]$$

But $\cos(n\pi) = -1$ for odd values of n and $\cos(n\pi) = 1$ for even values of n. Hence, A_n will be zero whenever n is even and will be $\frac{4T_0}{n\pi}$ whenever n is odd. The solution to the temperature field in the bar then becomes, using Eq. 5.75

$$T(x,t) = \frac{4T_0}{\pi}\sum_{n=1,3,5...}^{\infty}\frac{\sin\left(\frac{n\pi x}{L}\right)}{n}e^{-\frac{\alpha n^2\pi^2 t}{L^2}} \tag{5.76}$$

Step 2 Solution (5.76) may be recast in nondimensional form by introducing the following nondimensional variables:

Non-dimensional temperature: $\theta = \frac{T(x,t)}{T_0}$
Non-dimensional x co-ordinate: $\xi = \frac{x}{L}$
Non-dimensional time or Fourier number: $Fo = \frac{\alpha t}{L^2}$

Then Eq. 5.76 takes the form

$$\theta(\xi, Fo) = \sum_{n=1,3,5...}^{\infty}\frac{\sin(n\pi\xi)}{n\pi}e^{-n^2\pi^2 Fo} \tag{5.77}$$

Step 3 In order to discuss the behavior of the solution we consider the mid-plane temperature (i.e., at $x = \frac{L}{2}$ or $\xi = 0.5$) variation with Fourier number. Since the result (5.77) is in its nondimensional form, it is applicable to any material. We compare a one-term approximation with the exact in Fig. 5.14. The one-term approximation truncates the series with just the first term and hence is given by

$$\theta(\xi, Fo) \approx \frac{\sin(\pi\xi)}{\pi}e^{-\pi^2 Fo}$$

It is seen from the plot that a one-term approximation is an excellent representation of the solution for $Fo > 0.04$ or thereabouts. The one-term approximation is also referred to as the fully developed profile since the shape of the profile in the interval $0 < \xi < 1$ remains sinusoidal (and hence invariant) with the amplitude alone varying with Fo.

The **Fourier number** may be interpreted as the ratio of actual time and a reference or a characteristic time, determined by the size and the material property, the thermal diffusivity. Thus $Fo = \frac{t}{t_{ch}}$ where $t_{ch} = \frac{L^2}{\alpha}$ is a characteristic time determined by the physical size L of the bar and thermal diffusivity α of the material. The state of affairs for $Fo < 0.04$ may be termed as the short time solution, i.e., solution for time much smaller than the characteristic time. In this region all or many terms in the Fourier summation contribute significantly to the solution, for a given ξ. However, when the time is larger than the characteristic time the first term alone contributes to the solution. This is because of the fact that exponential terms involving $\mathrm{e}^{-n^2\pi^2 Fo} \approx \mathrm{e}^{-10n^2 Fo}$ fall off rapidly.

Example 5.5, incidentally, has also demonstrated how a Fourier series can represent a periodic function. Fig. 5.15 indicates how the function $f(x) = 1$ is approximated by the summation of Fourier components. The figure shows the result of truncating the summation with a given number of terms. The function is approximated crudely when we use only a few terms. Large amplitude oscillations are present. When a large enough number of terms are included (maximum of 20 terms are shown in the figure) the oscillations die out and the function is well represented by the Fourier sum, for

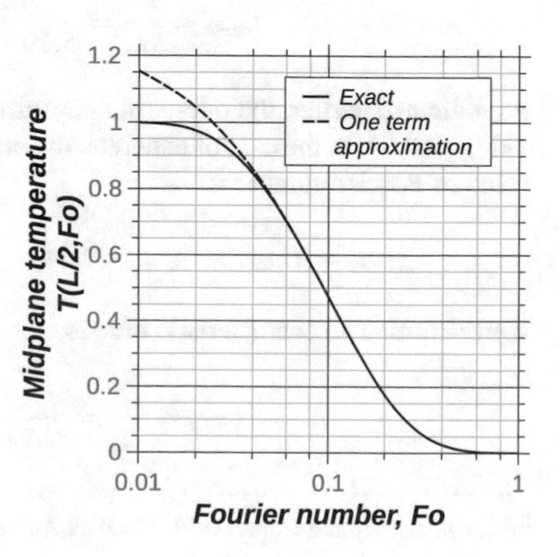

Fig. 5.14 Mid-plane temperature history in the bar

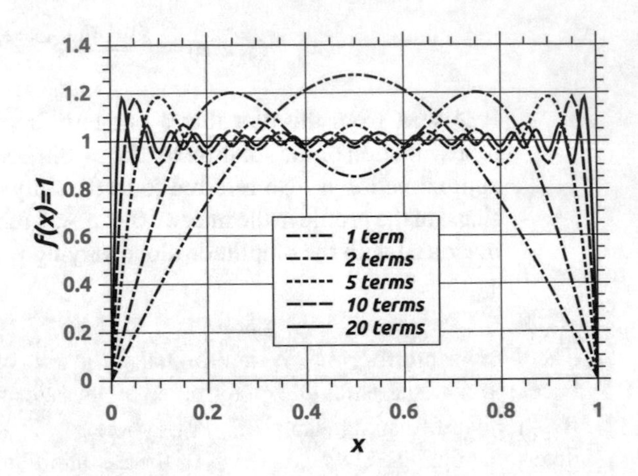

Fig. 5.15 Fourier series approximation of $f(x) = 1$

the most part of the interval. However, there will always be some oscillations (the approximate value is more than 1 or less than 1) near the two ends of the domain (near $x = 0$ and $x = L$) as seen in the figure. This is referred to as Gibb's phenomenon.[3]

Example 5.6

Consider the bar in worked Example 5.5 to be made of (a) concrete and (b) aluminum. The length of each bar is given as $L = 0.1$ m. The thermal diffusivity of concrete and aluminum are $\alpha_c = 5.19 \times 10^{-7}$ m^2/s and $\alpha_a = 9.71 \times 10^{-5}$ m^2/s, respectively. Determine the time beyond which the one-term approximation is valid, in these two materials.

Solution:

Concrete: The characteristic time is

$$t_{ch-c} = \frac{L^2}{\alpha_c} = \frac{0.1^2}{5.19 \times 10^{-7}} = 19268 \text{ s}$$

As indicated earlier, the one-term approximation is valid beyond a critical value of $Fo_{\text{crit}} = 0.04$. In the case of concrete, the actual time that corresponds to the limiting value of Fourier number is

$$t_c = Fo_{\text{crit}} \times t_{ch-c} = 0.04 \times 19268 = 770.7s \approx 13 \text{ min}$$

Aluminum: The characteristic time is

$$t_{ch-a} = \frac{L^2}{\alpha_a} = \frac{0.1^2}{9.71 \times 10^{-5}} = 103 \text{ s}$$

[3] After Josiah Willard Gibbs (1839–1903) an American scientist.

In the case of aluminum, the actual time that corresponds to the limiting value of Fourier number is

$$t_a = Fo_{\text{crit}} \times t_{ch-a} = 0.04 \times 103 = 4.12 \text{ s}$$

In order to further clarify the state of affairs, we calculate the first three terms at $t = 770.7$ s in the case of concrete or $t = 4.12$ s in the case of aluminum. The exponential terms containing Fo are given by

$$\begin{aligned} &\text{First term:} && e^{-\pi^2 \times 0.04} = 0.673825 \\ &\text{Second term:} && e^{-9\pi^2 \times 0.04} = 0.028637 \\ &\text{Third term:} && e^{-25\pi^2 \times 0.04} = 5.172319 \times 10^{-5} \end{aligned}$$

We note that the exponential factors diminish very rapidly. Thus, we have

$$\theta(\xi, 0.04) = \frac{4}{\pi}\left[0.673825 \sin(\pi\xi) + \frac{0.028637}{3}\sin(3\pi\xi) + \frac{5.172319 \times 10^{-5}}{5}\sin(5\pi\xi) + \cdots\right] \approx 0.85794 \sin(\pi\xi)$$

The error in using the one-term approximation is not more than 1.4%! Thus, we see that the solution reduces to a sinusoidal profile in as little as 4.2 s from the start, in the case of the aluminum bar.

5.3 Steady Conduction in Two Dimensions

Many practical problems of interest to the thermal engineer, under the steady state, exhibit two-dimensional temperature fields. The solution to these problems basically involves the solution of either the Laplace equation or Poisson equation in two dimensions. Typical but important problems that fall into this category are considered here.

5.3.1 *Steady Conduction in a Rectangle*

Consider a rectangular block of material as shown in Fig. 5.16. The boundary conditions are specified, for the so-called *standard* problem, as shown in this figure. The lower edge of the rectangle is maintained at unit temperature and all the other edges are maintained at zero temperature. The equation governing the temperature field is the Laplace equation in two dimensions (Cartesian form) assuming that there is no internal heat generation.

$$\frac{\partial^2 T}{\partial x^2} + \frac{\partial^2 T}{\partial y^2} = 0 \tag{5.78}$$

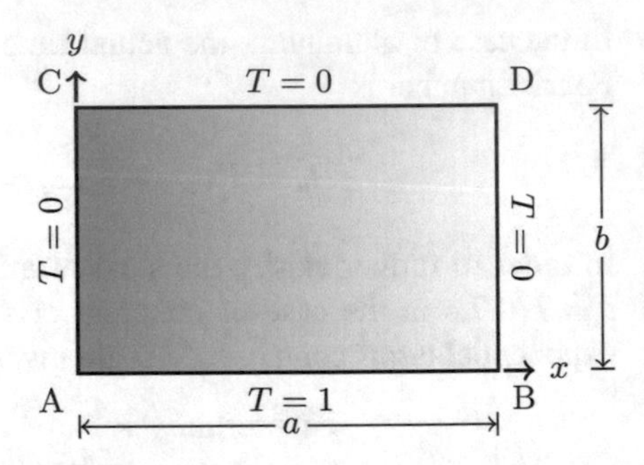

Fig. 5.16 Steady conduction in two dimensions; the *standard* problem with a rectangular domain

This equation may be solved by separation of variables technique. The procedure is similar to the one used in the case of one-dimensional transient in a finite bar. We seek the solution in the form of a product function

$$T(x, y) = F(x) \times G(y) \tag{5.79}$$

Substitute (5.79) in (5.78), divide through by $F \times G$ and rearrange to get

$$\frac{1}{F}\frac{\mathrm{d}^2 F}{\mathrm{d}x^2} = -\frac{1}{G}\frac{\mathrm{d}^2 G}{\mathrm{d}y^2} = -p^2 \tag{5.80}$$

where p^2 is again a separation constant. The negative sign used with the separation constant will be justified in due course. Equation 5.80 is equivalent to the two ordinary differential equations given by

$$\text{(a)} \quad \frac{\mathrm{d}^2 F}{\mathrm{d}x^2} + p^2 F = 0 \quad \text{(b)} \quad \frac{\mathrm{d}^2 G}{\mathrm{d}y^2} - p^2 G = 0 \tag{5.81}$$

The solution to Eqs. 5.81(a) and (b) are seen to be

$$\text{(a)} \quad F(x) = A\sin(px) + B\cos(px) \quad \text{(b)} \quad G(y) = C\sinh(py) + D\cosh(py) \tag{5.82}$$

where A, B, C, D are constants of integration.

We now look at the boundary conditions. T should vanish on line $x = 0$ as well as on line $x = a$. This means that F must vanish for $x = 0$ and $x = a$. Obviously this will require $B = 0$ and $p = \frac{n\pi}{a}$ where n is an integer. Thus Eq. 5.82(a) may be rewritten as

$$F_n(x) = A_n \sin\left(\frac{n\pi x}{a}\right) \tag{5.83}$$

Since at $y = 0$, $T = 1$ (in the interval $0 < x < a$) $F(x) \times G(0) = 1$. We can satisfy this if we take $F(x) = 1$ and $G(0) = 1$. With this, borrowing the results from the solution leading to Eq. 5.76, A_n are given as $\frac{4}{n\pi}$ for odd values of n and 0 for even values of n. With this we have

$$T(x,y)=\frac{4}{\pi}\sum_{n=1,3,5...}^{\infty}\frac{1}{n}\sin\left(\frac{n\pi x}{a}\right)\left[C_n\sinh\left(\frac{n\pi y}{a}\right)+D_n\cosh\left(\frac{n\pi y}{a}\right)\right] \quad (5.84)$$

Since $G(0)=1$, $D_n=1$ will satisfy the required condition that $T=1$ on $y=0$. However on $y=b$, we have

$$T(x,b)=\frac{4}{\pi}\sum_{n=1,3,5...}^{\infty}\frac{1}{n}\sin\left(\frac{n\pi x}{a}\right)\underbrace{\left[C_n\sinh\left(\frac{n\pi b}{a}\right)+\cosh\left(\frac{n\pi b}{a}\right)\right]}_{\text{Term=0, for each } n}=0 \quad (5.85)$$

The term shown with underbracket is set to zero to get C_n as

$$C_n=-\frac{\cosh\left(\frac{n\pi b}{a}\right)}{\sinh\left(\frac{n\pi b}{a}\right)} \quad (5.86)$$

We substitute this back in Eq. 5.84 to get

$$T(x,y)=\frac{4}{\pi}\sum_{i=1,3,5...}^{\infty}\frac{1}{n}\sin\left(\frac{n\pi x}{a}\right)\underbrace{\left[-\frac{\cosh\left(\frac{n\pi b}{a}\right)\sinh\left(\frac{n\pi y}{a}\right)}{\sinh\left(\frac{n\pi b}{a}\right)}+\cosh\left(\frac{n\pi y}{a}\right)\right]}_{\text{Simplified as shown in Eq. 5.87}}$$

The term shown with underbracket may be simplified and recast, using identities involving hyperbolic functions. The temperature distribution then takes the final form

$$\boxed{T(x,y)=\frac{4}{\pi}\sum_{n=1,3,5...}^{\infty}\frac{1}{n}\sin\left(\frac{n\pi x}{a}\right)\frac{\sinh\left\{\frac{n\pi(b-y)}{a}\right\}}{\sinh\left(\frac{n\pi b}{a}\right)}} \quad (5.87)$$

Figure 5.17 shows the isotherm pattern, wherein the bottom edge is maintained at 100 °C while all the other edges are at 0 °C. The ratio $b/a=1$ in the case shown.

5.3.2 *Steady Conduction in a Rectangle With Heat Generation*

When internal heat generation is included the governing equation becomes non-homogeneous and is given by the Poisson equation (5.16). Consider a rectangular domain of width $2L$ and height $2l$ (unit thickness in a direction perpendicular to the plane of the figure) as shown in Fig. 5.18. Uniform volumetric heat generation at rate

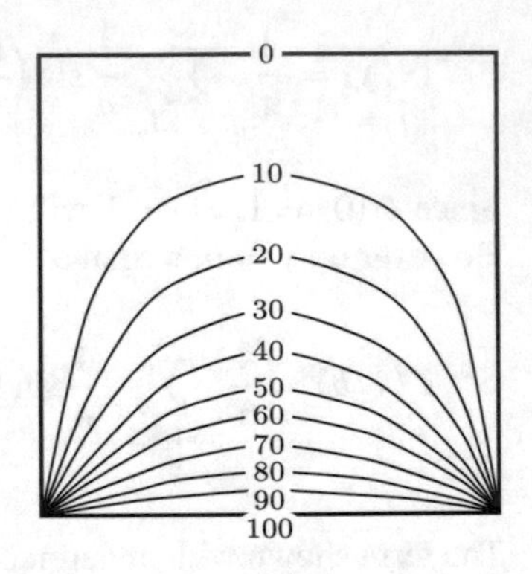

Fig. 5.17 Isotherms in the *standard* problem. Temperatures are in °C

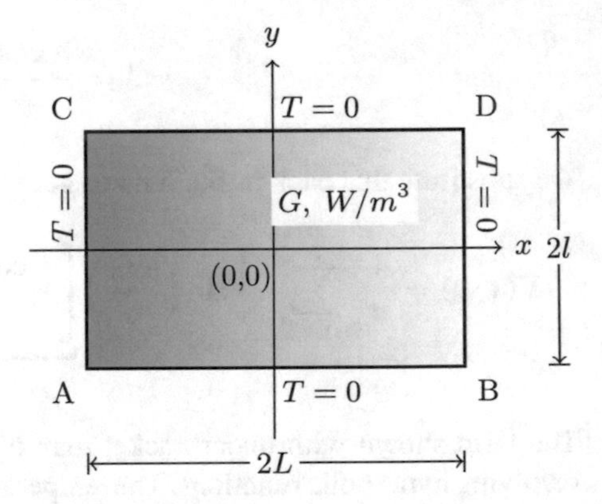

Fig. 5.18 Rectangular domain with heat generation

G takes place throughout the domain. All the boundaries are maintained at the datum value that has been taken as zero. The origin is placed at the center of the domain as shown in the figure. This case is amenable to solution by using the superposition principle and the method of separation of variables.

The equation to be solved is written down again for convenience as

$$\frac{\partial^2 T}{\partial x^2} + \frac{\partial^2 T}{\partial y^2} + \frac{G}{k} = 0 \tag{5.88}$$

The boundary conditions are specified as

$$\begin{aligned}&\text{(a) } T(x = -L, -l \le y \le l) = 0; \;\; \text{(b) } T(x = L, -l \le y \le l) = 0;\\ &\text{(c) } T(-L \le x \le L, y = -l) = 0; \;\; \text{(d) } T(-L \le x \le L, y = l) = 0\end{aligned} \tag{5.89}$$

Let us assume that the solution to Eq. 5.88 is given by[4]

$$T(x, y) = \psi(x, y) + \phi(x) \tag{5.90}$$

[4]Alternately we may take $T(x, y) = \psi(x, y) + \phi(y)$.

Substitute Eq. 5.90 in Eq. 5.88 to get

$$\frac{d^2\phi}{dx^2} + \frac{\partial^2\psi}{\partial x^2} + \frac{\partial^2\psi}{\partial y^2} + \frac{G}{k} = 0 \tag{5.91}$$

Equation 5.91 may be written down as two equations given by

$$\text{(a)}\ \frac{d^2\phi}{dx^2} + \frac{G}{k} = 0; \ \text{(b)}\ \frac{\partial^2\psi}{\partial x^2} + \frac{\partial^2\psi}{\partial y^2} = 0 \tag{5.92}$$

The boundary conditions given by Eqs. 5.89(a)–(d) may be satisfied by the following set of conditions:

$$\begin{array}{ll} \text{(a)}\ \phi(-L) = 0 & \text{(b)}\ \phi(L) = 0 \\ \text{(c)}\ \psi(-L, -l \le y \le l) = 0 & \text{(d)}\ \psi(L, -l \le y \le l) = 0 \\ \text{(e)}\ \psi(-L \le x \le L, -l) = -\phi(x) & \text{(f)}\ \psi(-L \le x \le L, l) = -\phi(x) \end{array} \tag{5.93}$$

Equation 5.92(a) may easily integrated to get $\phi(x) = Ax + B - \frac{Gx^2}{2k}$, where A and B are constants of integration. Using the boundary conditions given by Eqs. 5.93(a) and (b), the two constants may be obtained, respectively, as $A = 0$ and $B = \frac{GL^2}{2k}$. With this $\phi(x)$ is given by

$$\phi(x) = \frac{GL^2}{2k}\left[1 - \left(\frac{x}{L}\right)^2\right] \tag{5.94}$$

We may use the method of separation of variables to obtain $\psi(x, y)$. The details are not given here because the steps are similar to those used in Sect. 5.3.1. The general solution may be written down as

$$\psi(x, y) = \sum_0^\infty a_n \cos(\lambda_n x)\cosh(\lambda_n y) \tag{5.95}$$

where the eigenvalues satisfy the condition (to satisfy the boundary conditions (5.93)(c) and (d))

$$\cos(\lambda_n L) = 0 \text{ or } \lambda_n L = \frac{(2n+1)\pi}{2}, n = 0, 1, 2 \ldots \infty \tag{5.96}$$

Note that both $\phi(x)$ and $\cos(\lambda_n x)$ are even functions of x. To satisfy the boundary conditions given by Eqs. 5.93(e) and (f), we should have

$$\sum_0^\infty a_n \cos(\lambda_n x)\cosh(\lambda_n l) = -\frac{GL^2}{2k}\left[1 - \left(\frac{x}{L}\right)^2\right] \tag{5.97}$$

Again we make use of orthogonality property of the cosine function over the interval $(-L, L)$ to obtain the coefficients a_n. Thus we will be expressing the solution in terms of Fourier cosine series. Avoiding long intermediate steps, we directly give the expression for a_n as

$$a_n = -\frac{4(-1)^n}{(\lambda_n L)^3}\frac{GL^2}{2k}\frac{1}{\cosh(\lambda_n l)} \tag{5.98}$$

Example 5.7

A very long bar of a material of thermal conductivity $k = 2.5$ W/m K has a square cross section of side $L = l = 0.1$ m. Heat is generated at a uniform rate of $G = 10^4$ W/m^3 inside the bar. All the surfaces of the bar are maintained at a temperature of $T = 0$. Determine the temperature at $x = y = 0$; $x = 0.05$, $y = 0$; $x = 0$, $y = 0.05$; and $x = 0.05$, $y = 0.05$ where all lengths are in m.

Solution:
Evaluation of the desired temperatures are based on the solution presented above. With the data specified in the problem, reference temperature due to heat generation is calculated as

$$T_{\text{ref}} = \frac{GL^2}{2k} = \frac{10^4 \times 0.1^2}{2 \times 2.5} = 20 \text{ K}$$

The function $\phi(x)$ is then given by Eq. 5.94 as

$$\phi(x) = 20\left[1 - \left(\frac{x}{0.1}\right)^2\right]$$

The temperatures required to be calculated at different locations specified in the problem require the summation of Fourier series to obtain $\psi(x, y)$. Consider the location $x = y = 0$ that is at the center of the square cross section. Both the cos and cosh functions are unity at this point. Hence the function $\psi(0, 0) = \sum_0^\infty a_n$ where a_n are given by Eq. 5.98. For convergence to 5 digits after the decimal point 3 terms in the series are adequate. These are shown in the table below.

n	$\lambda_n L$	a_n
0	1.570796	−8.226191
1	4.712389	0.013734
2	7.853982	−0.000128
Sum of three terms = −8.21258		

Hence the temperature at $x = 0$, $y = 0$ is obtained as

$$T(0, 0) = \psi(0, 0) + \phi(0) = -8.21258 + 20 = 11.787417 \approx 11.79\ ^\circ\text{C}$$

Similarly temperatures at other locations specified in the problem are obtained. The cosine and hyperbolic cosine factors are to be used appropriately. As an example, the calculations are shown for $x = y = 0.05$ in the following tabulation:

n	$\lambda_n L$	a_n	Fourier terms
0	1.570796	−8.226191	−7.704980
1	4.712389	0.013734	−0.051691
2	7.853982	−0.000128	0.002301
3	10.995574	0.000002	0.000174
4	14.137167	0.000000	−0.000017
5	17.278760	0.000000	−0.000002
Sum of five terms = −7.75421			

Hence the temperature at $x = 0.05$, $y = 0.05$ is obtained as

$$T(0.05, 0.05) = \psi(0.05, 0.05) + \phi(0.05)$$
$$= -7.75421 + 20\left[1 - \left(\frac{0.05}{0.1}\right)^2\right] \approx 7.25\ ^\circ\text{C}$$

It is left as an exercise to the reader to show that the temperatures at the other two points specified in the problem are the same and equal to 9.17 °C. We show in Fig. 5.19a, b the temperature variations along two directions. First direction is along x, $y = 0$ or y, $x = 0$. The second direction is along direction $x = y$ that passes through origin. Note that the coordinates are non-dimensionalized with L as the characteristic length. r in Fig. 5.19b is defined as $r = \sqrt{x^2 + y^2} = x\sqrt{2} = y\sqrt{2}$.

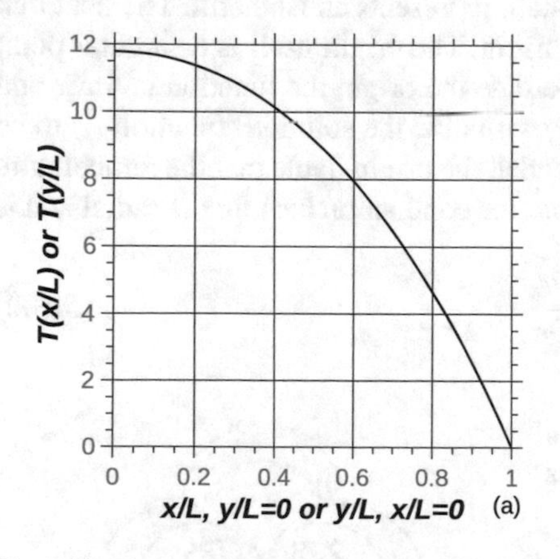

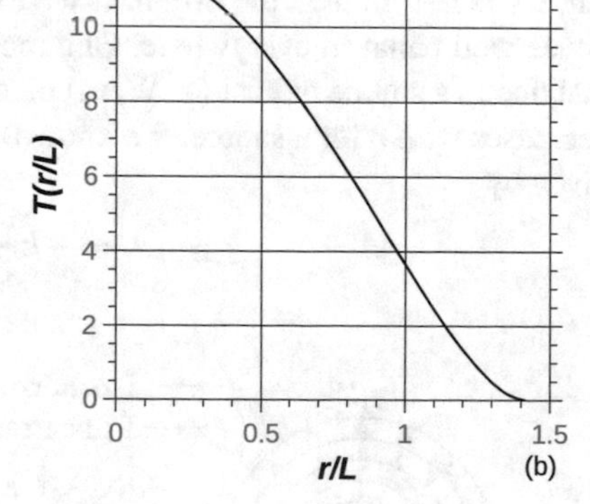

Fig. 5.19 Temperature profiles in Example 5.7

5.3.3 *Steady Two-Dimensional Conduction in Cylindrical Co-Ordinates*

Elementary solution to Laplace equation in cylindrical coordinates
Many interesting and important thermal applications involve heat transfer in cylindrical coordinates. A case that has already been considered in Chap. 2 is that of steady radial conduction in a cylinder. In that special case the problem reduced to that in one dimension. We have seen that the isotherms and heat flux lines in a cylinder are disposed as shown in Fig. 5.20a. We have seen that the temperature varies as the natural logarithm of r. The heat flux lines are obviously radial lines and are specified by lines of constant θ. The solution is, in fact, an elementary solution to Laplace equation in two dimensions and in cylindrical coordinates (Eq. 5.14 with temperature varying with r only!). This elementary solution represents the temperature field set up by a line heat source (or sink) at O, as shown in Fig. 5.20b. The solution for source or sink may be obtained as under or by using complex variable theory as shown in Appendix C. Consider the function

$$\boxed{\phi = -\frac{s}{2\pi k}\ln(r)} \tag{5.99}$$

which is the real part of the complex potential $w = -\frac{s}{2\pi k}\mathrm{Ln}(z)$. This represents a steady temperature field in cylindrical coordinates (i.e., r, θ coordinates). Of course, the temperature field happens to be a function of r alone and hence is one dimensional in nature. It is easily verified that r = constant represents an isotherm. The heat flux lines are radial lines passing through the origin. The origin itself is a singular point since ϕ is not finite at the origin. If we exclude the origin the function is finite and hence well behaved everywhere. One may visualize the solution (function ϕ) to be that due to a source of heat (s, W/m) placed at the origin. Note that the temperature decreases with r for a source. We know that the conduction heat flux is radial and is given by

$$q_r = -k\frac{\partial \phi}{\partial r} = \frac{s}{2\pi}\frac{1}{r} \tag{5.100}$$

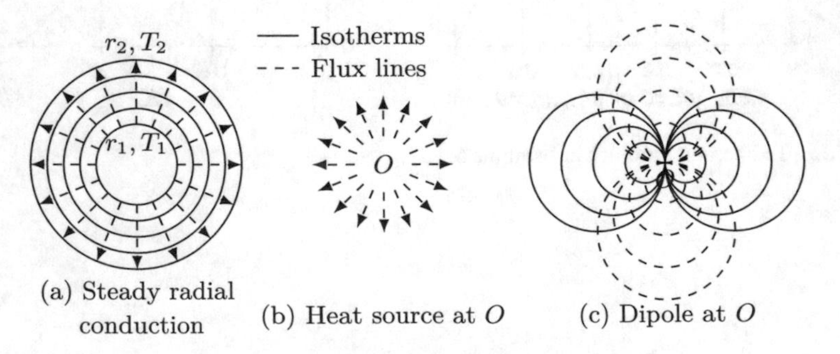

Fig. 5.20 Elementary solutions to Laplace equation in cylindrical coordinates

The total heat transfer across *any* isotherm is given by

$$q = 2\pi r q_r = s \tag{5.101}$$

This is nothing but the strength of the heat source! We may visualize the heat flux lines as radial lines given by [also visualized as being represented by the imaginary part of $w = -\frac{s}{2\pi k}\mathrm{Ln}(z)$]

$$\boxed{\psi = -\frac{s\theta}{2\pi}} \tag{5.102}$$

such that $\psi = 0$ corresponds to $\theta = 0$ and $\psi = -s$ corresponds to $\theta = 2\pi$. Thus ψ represents mathematically the heat flux lines. The isotherms and heat flux lines form an orthogonal net as indicated in Fig. 5.20a and the source is indicated as in Fig. 5.20b.

Another elementary solution

We shall look at a second elementary solution to Laplace equation in cylindrical coordinates. Consider steady heat flow, in the absence of internal heat generation, in a right-angled channel bend shown in Fig. 5.21. The bend consists of circular inner and outer boundaries with the common center at C. The bend extends to infinity in a direction perpendicular to the plane of the paper. The isotherms are now radial lines and heat flux lines are concentric circular arcs as indicated in the figure. In fact this elementary solution is represented by the same complex potential viz. $w = \mathrm{Ln}(z)$. The real and imaginary parts are interchanged, in this case, as against the previous case. Temperature is, in fact, seen to be a function of θ alone and hence the Laplace equation reduces to

$$\frac{\mathrm{d}^2 T}{\mathrm{d}\theta^2} = 0 \tag{5.103}$$

When integrated twice, we get

$$T = A\theta + B$$

where A and B are constants of integration. Use now the boundary conditions indicated in Fig. 5.21. $T = T_1$ at $\theta = 0$ requires that $B = T_1$. $T = T_2$ at $\theta = \frac{\pi}{2}$ requires

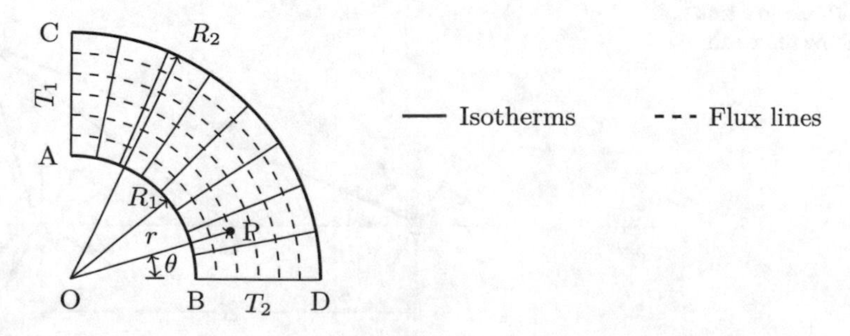

Fig. 5.21 Heat flow in a right-angled channel bend. Boundaries AB and CD are adiabatic

that $A = \frac{2(T_2-T_1)}{\pi}$. With these the solution turns out to be

$$T(\theta) = \frac{(T_2 - T_1)\theta}{\frac{\pi}{2}} + T_1 \tag{5.104}$$

We now determine the heat transfer rate by noting that the θ component of heat flux alone is nonzero and is given by

$$q_\theta = -k\frac{1}{r}\frac{\mathrm{d}T}{\mathrm{d}\theta} = -k\frac{T_2 - T_1}{\frac{r\pi}{2}} \tag{5.105}$$

The total heat transferred across the curved channel is given by integrating Eq. 5.105 with respect to r from R_1 to R_2.

$$Q = -k\frac{T_2 - T_1}{\frac{\pi}{2}}\int_{R_1}^{R_2}\frac{1}{r}\mathrm{d}r = k\frac{T_1 - T_2}{\frac{\pi}{2}}\ln\left(\frac{R_2}{R_1}\right) \tag{5.106}$$

Elementary solutions and superposition: The dipole

Elementary solutions to Laplace equation may be used to write down solutions to complex problems by superposing many elementary solutions. This is a consequence of the fact that the Laplace equation is a linear partial differential equation. When we combine elementary solutions the only thing one needs to do is to see that the resultant solution satisfies all the boundary conditions imposed in the problem. We show the use of superposition by considering a simple case shown in Fig. 5.22. A source and a sink of equal strength (magnitude s) are located at A and B as indicated in the figure, placed symmetrically with respect to the origin O. The distance between the two is $2l$. We shall assume that $l \to 0$ and $s \to \infty$ such that the product $\mu = 2ls$ is a finite quantity. The temperature at the field point $P(r, \theta)$ is obtained by adding that set up by the source and sink as (using Eq. 5.99)

$$\phi = -\frac{s}{2\pi k}\ln(r_1) + \frac{s}{2\pi k}\ln(r_2) \tag{5.107}$$

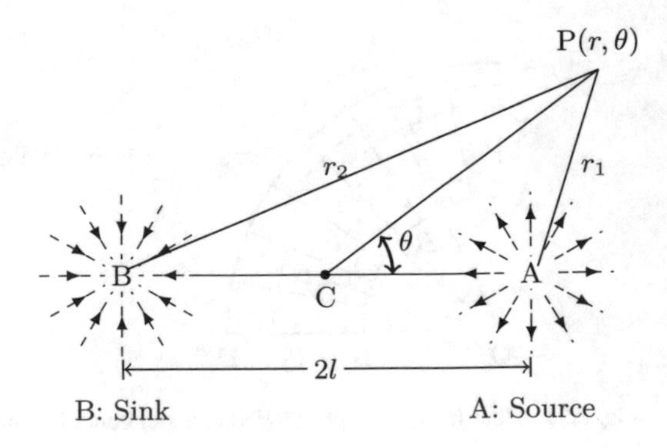

Fig. 5.22 Source sink pair lying along the x-axis

From geometry, we have

$$r_1 = \sqrt{r^2 + l^2 - 2rl\cos(\theta)}\ ;\ r_2 = \sqrt{r^2 + l^2 + 2rl\cos(\theta)}$$

These may be approximated, respectively, as

$$r_1 \approx \sqrt{r^2 - 2rl\cos(\theta)}\ ;\ r_2 \approx \sqrt{r^2 + 2rl\cos(\theta)}$$

The approximation is valid since we assume that $l \ll r$ and hence l^2 is negligible as compared to the other two terms. Introduce this in Eq. 5.107 to get[5]

$$\phi = \frac{s}{2\pi k}\ln\left[\frac{r_2}{r_1}\right] \approx \frac{s}{2\pi k}\ln\left[\sqrt{\frac{r^2 + 2rl\cos(\theta)}{r^2 - 2rl\cos(\theta)}}\right] = \frac{s}{4\pi k}\ln\left[\frac{1+\dfrac{2l\cos(\theta)}{r}}{1-\dfrac{2l\cos(\theta)}{r}}\right]$$
$$\approx \frac{s}{4\pi k}\ln\left[1 + \frac{4l\cos(\theta)}{r}\right] \approx \frac{4ls\cos(\theta)}{4\pi kr} = \frac{\mu\cos(\theta)}{2\pi kr} \tag{5.108}$$

where $\mu = 2ls$ is referred to as the dipole moment. The above expression is actually the far field temperature variation since very close to the origin the approximation that led to Eq. 5.108 will not be valid. Note that the temperature field varies inversely with r. Isotherms and flux lines shown in Fig. 5.20c are based on the solution presented in Eq. 5.108.

It is also seen that the temperature field given by Eq. 5.108 is the real part of the complex potential $w = \frac{\mu}{z}$. This is made use of in Appendix C.

Generalization to a distribution of sources

The superposition principle used above for obtaining the solution to the dipole problem may be generalized to a problem that involves a distribution of sources over a volume. Consider a volume V to contain a continuous distribution of line sources of source strength $G(r', \theta')$ per unit volume as shown in Fig. 5.23. The response is required at the field point (r, θ) represented by the vector $\vec{r}$. We make use of Eq. 5.99 to write the potential ϕ (temperature represents a potential since the conduction heat flux is defined as $\vec{q} = -k\nabla T$) at the field point P as

$$\boxed{\phi = -\frac{1}{2\pi k}\iiint\limits_V G(r', \theta')\ln(r - r')\mathrm{d}V} \tag{5.109}$$

Note that the summation (in the case of dipole we added two elementary solutions) is replaced by an integral when the sources are distributed continuously over a volume

[5] We have used $\ln\left(\frac{r_2}{r_1}\right) = \ln(r_2) - \ln(r_1)$ and $\ln(1 \pm x) \approx \pm x$ when $x << 1$ in arriving at the final expression in Eq. 5.108.

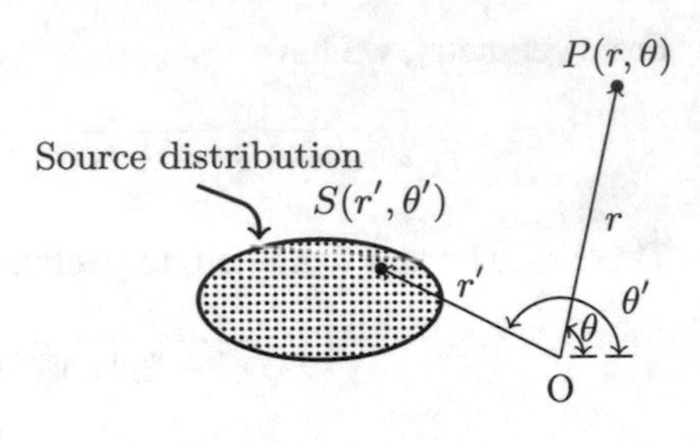

Fig. 5.23 Potential due a source distribution

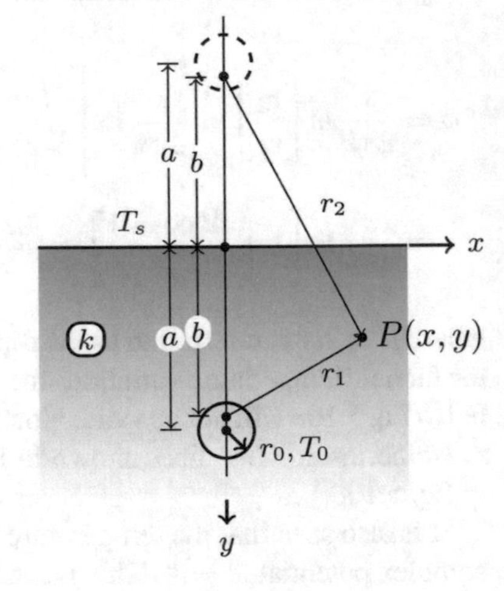

Fig. 5.24 Nomenclature for the buried cable problem showing a "ghost" sink

(note that the volume element is an elemental area in the plane of Fig. 5.23 multiplied by a unit length perpendicular to the plane of the figure).

Buried cable problem and the "conduction shape factor"

Consider a buried cable, a cylinder of radius r_0, placed at a depth a inside a firmly packed medium like soil, as shown in Fig. 5.24. The cable is generating heat at a constant rate of q per meter length. The medium surrounding the cable terminates at a level isothermal surface at $T = T_s$. The cable surface temperature is assumed to take on a steady value of T_0. These are the specified boundary conditions in the problem. The problem involves steady heat conduction in cylindrical coordinates. This means that the Laplace equation in two dimensions involving r, θ is to be solved. The procedure adopted here consists of the construction of the solution using elementary solutions to the Laplace equation and then invoking the superposition principle to obtain the complete solution.

In order to solve the buried cable problem, we consider a source of heat to be located at a distance b below the ground and a sink of equal strength to be located a distance b above the ground. We shall see later that b is only slightly different from a, the distance of the center of the cable below the ground level. The sink is entirely fictitious and is used *only* to build the solution to the problem. Consider a field point $P(x, y)$ as shown in Fig. 5.24. The radii r_1 and r_2 may be written down as

$$r_1 = \sqrt{x^2 + (b - y)^2}\ ;\ r_2 = \sqrt{x^2 + (b + y)^2}$$

We superpose the temperature induced by source sink pair by defining a temperature difference function as

$$\phi = T - T_s = -\frac{q}{2\pi k}\ln(r_1) + \frac{q}{2\pi k}\ln(r_2) = \boxed{\frac{q}{4\pi k}\ln\left[\frac{x^2 + (b + y)^2}{x^2 + (b - y)^2}\right]} \quad (5.110)$$

The difference between this solution and the solution to the dipole problem is that b does not have any condition attached to its magnitude in comparison with r. Also we have used Cartesian description of coordinates even though the problem is one in cylindrical coordinates. This is just a ploy to make the solution come out in a simple fashion. If one is able to choose the proper values for q and b such that the boundary conditions shown in Fig. 5.24 are satisfied, one has solved the buried cable problem! It is seen that $\phi = 0$, i.e., $T = T_s$ if $r_1 = r_2$! Thus one boundary condition is automatically satisfied by Eq. 5.110. In order to satisfy the second boundary condition, let $T = T_0$ in the above. Let us also define a constant C as

$$C = e^{C'} \text{ where } C' = \frac{T_0 - T_s}{\left(\frac{q}{4\pi k}\right)} \quad (5.111)$$

Then Eq. 5.110, after some algebraic manipulation, takes the form

$$C = \frac{x^2 + (b + y)^2}{x^2 + (b - y)^2} \quad (5.112)$$

We may expand the square terms and rearrange to finally get

$$x^2 + y^2 + b^2 + 2by\left[\frac{1 + C}{1 - C}\right] = 0 \quad (5.113)$$

This equation should represent the equation of the boundary on which the temperature has been specified as $T = T_0$. We are given that this should be the surface of the cable! In order to verify this, we complete squares and rewrite Eq. 5.113 as

$$x^2 + \left[y + b\left(\frac{1 + C}{1 - C}\right)\right]^2 = \frac{4b^2 C}{(1 - C)^2} \quad (5.114)$$

Equation 5.114 is the equation of a circle whose center is at

$$x = 0, y = -b\left(\frac{1 + C}{1 - C}\right) \quad (5.115)$$

and whose radius is given by

$$r_0^2 = \frac{4b^2C}{(1-C)^2} \tag{5.116}$$

In fact Eq. 5.116 determines the value of C and hence q when all other quantities in Eq. 5.111 are given or known. Equation 5.116 may be rewritten as a quadratic equation for C as

$$C^2 - 2\left[1 + \frac{2b^2}{r_0^2}\right] + 1 = 0 \tag{5.117}$$

The above equation may be solved for C to get

$$C = \left\{1 + \frac{2b^2}{r_0^2}\right\} \pm \sqrt{\left\{1 + \frac{2b^2}{r_0^2}\right\}^2 - 1} \tag{5.118}$$

Assuming that the radius of the cable r_0 is much smaller than the depth b of the source below ground level (normally satisfied in practice), we may approximate

$$\left\{1 + \frac{2b^2}{r_0^2}\right\} \text{ as } \frac{2b^2}{r_0^2}$$

Also only the bigger root makes physical sense and is given as

$$C \approx \frac{4b^2}{r_0^2} \tag{5.119}$$

Under the same condition, Eq. 5.115 shows that the center of the circle is at $y \approx -b$. Also it is seen that $b \approx a$.

Using the definition of C' and C given by Eq. 5.111, we have

$$C' = \ln(C) = \ln\left(\frac{4b^2}{r_0^2}\right) \approx \ln\left(\frac{4a^2}{r_0^2}\right) = 2\ln\left(\frac{2a}{r_0}\right) = \frac{T_0 - T_s}{\left(\frac{q}{4\pi k}\right)}$$

Solving the above for q the heat transfer per meter length of the cable, we have

$$q = k\frac{2\pi}{\ln\left(\frac{2a}{r_0}\right)}(T_0 - T_s) \tag{5.120}$$

For different values of temperature T, the corresponding value of C will yield circles centered along the y-axis and radius given by the expression derived earlier. In particular, for $T = T_s$, $C' = 0$ and hence $C = 1$ and the radius of the circle $\rightarrow \infty$, corresponding to the flat surface of the ground.

Conduction shape factor
The heat loss from the cable per meter given by Eq. 5.120 may be written in the *general* form

$$\boxed{q = k(T_0 - T_s) \times S} \tag{5.121}$$

where S is a geometric parameter called the "conduction shape factor". Since the problem has been solved from first principles, it is possible to obtain S for the buried cable problem as

$$\boxed{S_{\text{Buried cable}} = \frac{2\pi}{\ln\left(\frac{2a}{r_0}\right)}} \tag{5.122}$$

Example 5.8

A cable whose surface temperature is 100 °C has a diameter of 10 cm. Its center is at a depth of 1 m under soil whose thermal conductivity is known to be 1 W/m°C. Determine how much heat it will lose per meter length if the soil surface temperature is 30 °C. Make a plot of the temperature variation as a function of distance vertically above the center of the cable.

Solution:

Step 1 Nomenclature used in this problem follows that in Fig. 5.24. The given data is summarized as under: $a = 1$ m, $r_0 = 5$ cm or 0.05 m, $k = 1$ W/m°C, $T_0 = 100$ °C and $T_s = 30$ °C. Using this data we calculate the conduction shape factor, following Eq. 5.122 as

$$S = \frac{2 \times \pi}{\ln\left(\frac{2\times 1}{0.05}\right)} = 1.7033$$

Step 2 The heat loss per meter of cable may then be calculated, using Eq. 5.121 as

$$q = 1 \times 1.7033 \times (100 - 30) = 119.2 \text{ W/m}$$

Step 3 In order to make the plot specified in the example, the requisite data is generated using the following additional definitions (compare with Eq. 5.111). Let

$$C_T = e^{C_{T'}} \text{ where } C_{T'} = \frac{T - T_s}{\left(\frac{q}{4\pi k}\right)}$$

In terms of C_T, the coordinate of the center of the isotherm located along the y axis is given by

$$y_T = a\left[\frac{C_T + 1}{C_T - 1}\right]$$

The corresponding radius of the isotherm is given by

$$r_T = \frac{2a\sqrt{C_T}}{C_T - 1}$$

It is easily seen from Fig. 5.24 that the height of the isotherm above the center of the cable is $d_T = 1 - y_T + r_T$. The calculations are now shown in Table 5.4.

Step 4 Figure 5.25 shows the variation of the temperature of the soil vertically above the axis of the cable.

This example is typical of what happens in power stations. The results also are applicable to a buried pipe conveying hot liquids (such as hot water) and vapors (such as steam).

Table 5.4 Temperature depth data for Example 5.8

Temperature, T	100	90	80	70	60	50	40	30
Distance, y_T, m	0.05	0.081	0.134	0.217	0.341	0.517	0.742	1

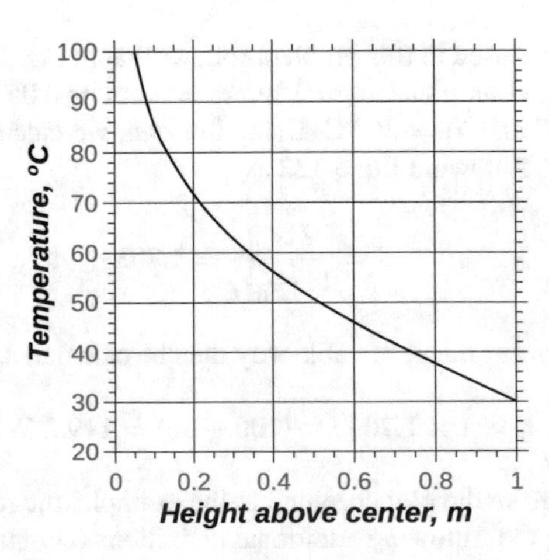

Fig. 5.25 Variation of soil temperature with height vertically above the cable in Example 5.8

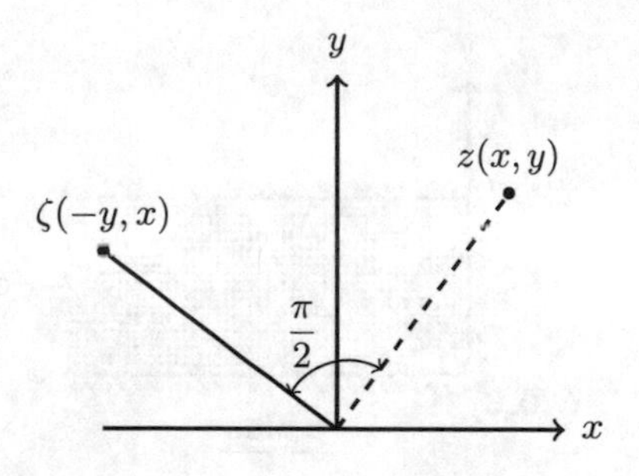

Fig. 5.26 Map of complex potential $\zeta = iz$

Illustration of use of complex variable method to an interesting problem

To round off the above, we consider a final example that uses complex potential approach to solving a problem in steady heat conduction in two dimensions. We look at an interesting transformation given by

$$w = \cosh(z) \tag{5.123}$$

This looks somewhat unfamiliar at first sight. However it is nothing but a combination of two complex potentials given by e^z and e^{-z} since $\cosh(z) = \frac{e^z+e^{-z}}{2}$. We may also look at the transformation as $w = \cosh(z) = \cos(iz)$ which is equivalent to the following two transformations:

$$\zeta = iz; \quad w_2 = \cos(\zeta) \tag{5.124}$$

The first of these is rotation of the complex number by $\frac{\pi}{2}$ while the second is mapping by the function $\cos(\zeta)$. A sketch is presented in Fig. 5.26 to demonstrate the former. Both z and ζ are shown on the same plane (plot). The dashed line indicates the point z in the z plane that is changed to the point ζ due to the transformation.

Now consider the latter transformation. We may easily write the real and imaginary parts as below:

$$\cos(\zeta) = \cos(-y + ix) = \underbrace{\cos(y)\cosh(x)}_{\text{Real part}} + i\,\underbrace{\sin(y)\sinh(x)}_{\text{Imaginary part}} \tag{5.125}$$

Let us look at the salient features of this transformation. The origin in the z plane maps on to the point

$$\phi = \cos(0)\cosh(0) = 1; \quad \psi = \sin(0)\sinh(0) = 0$$

Consider a point on the imaginary axis in the z plane given by $x = 0$, $y = \pi$. Corresponding to this we have

$$\phi = \cos(\pi)\cosh(0) = -1; \quad \psi = \sin(\pi)\sinh(0) = 0$$

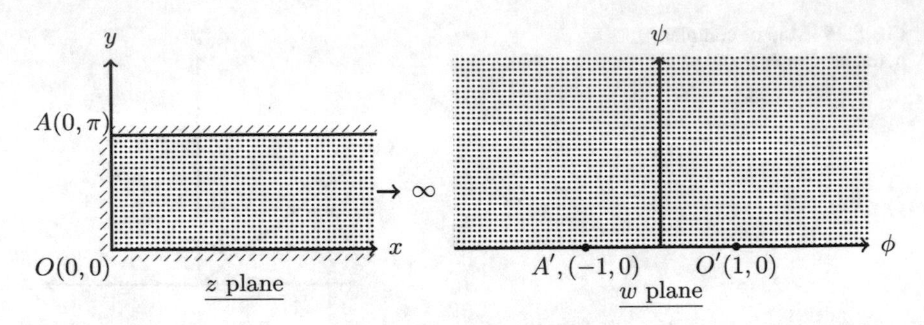

Fig. 5.27 Map of complex potential $w = \cos(\zeta)$

Thus the point has been mapped on to a point on the negative real axis. We also have the following that may be easily verified:

$$\begin{array}{lll} x \to \infty,\ y = 0 & \text{maps on to} & \phi \to \infty,\ \psi = 0 \\ x \to \infty,\ y = \pi & \text{maps on to} & \phi \to -\infty,\ \psi = 0 \\ x \geq 0, 0 \leq y \leq \pi & \text{maps on to} & \psi \geq 0 \end{array}$$

Thus the inside of the strip shown with hatched lines is mapped on to the upper half of the w plane, as shown in Fig. 5.27. Point O maps on to the point O' and point A maps on to the point A'.

Now we are ready to consider the solution of steady heat conduction in the strip. The boundary conditions are specified as

$$\begin{aligned} T = 1 \quad &\text{on} \quad x = 0 \text{ and } 0 \leq y \leq \pi \\ T = 0 \quad &\text{on} \quad y = 0 \text{ or } y = \pi \text{ and } 0 \leq x \leq \infty \end{aligned} \tag{5.126}$$

This simply means that in the w plane shown in Fig. 5.27 the potential function must take on the values given by

$$\begin{aligned} T = 1 \quad &\text{on} \quad \psi = 0 \text{ and } -1 \leq \phi \leq 1 \\ T = 0 \quad &\text{on} \quad \psi = 0 \text{ and } \phi > 1 \text{ or } \phi < -1 \end{aligned} \tag{5.127}$$

In order to satisfy the boundary conditions specified in the w plane we look at another transformation given by the function

$$w_1 = \frac{T_0}{\pi} \text{Ln} \left[\frac{w - 1}{w + 1} \right] \tag{5.128}$$

Letting $w_1 = u(\phi, \psi) + i v(\phi, \psi)$, we may write the above in the alternate form

$$w_1 = \underbrace{\frac{T_0}{2\pi} \ln \left[\frac{(\phi - 1)^2 + \psi^2}{(\phi + 1)^2 + \psi^2} \right]}_{u(\phi,\psi)} + i \underbrace{\frac{T_0}{\pi} \left[\tan^{-1} \left(\frac{\psi}{\phi - 1} \right) - \tan^{-1} \left(\frac{\psi}{\phi + 1} \right) \right]}_{v(\phi,\psi)} \tag{5.129}$$

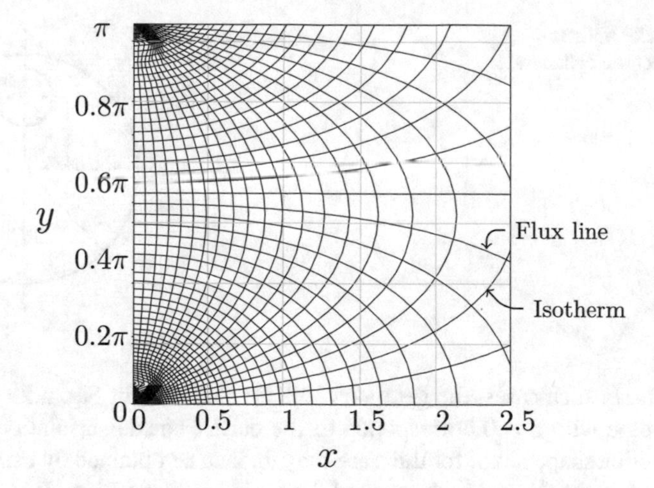

Fig. 5.28 Flux plot for the strip problem

Since w_1 is an analytic function u and v automatically satisfy the Laplace equation. In addition let us identify the complex part as the temperature function. Thus let

$$T = \frac{T_0}{\pi}\left[\tan^{-1}\left(\frac{\psi}{\phi - 1}\right) - \tan^{-1}\left(\frac{\psi}{\phi + 1}\right)\right] = \frac{T_0}{\pi}(\theta_1 - \theta_2) \qquad (5.130)$$

where $\theta_1 = \tan^{-1}\left(\frac{\psi}{\phi-1}\right)$ and $\theta_1 = \tan^{-1}\left(\frac{\psi}{\phi+1}\right)$. Consider a point on the ϕ axis that is to the right of $\phi = 1$. For such a point both angles are zero and hence $T = 0$. For a point on the ϕ axis that is in the range $-1 < \phi < 1$ $\theta_1 = \pi$ and $\theta_2 = 0$ and hence $T = 1$. For a point on the ϕ axis that is to the left of $\phi = -1$ both angles become π and hence $T = 0$. Thus the required boundary conditions are satisfied by the function T defined by Eq. 5.130. Hence this must be the solution to the temperature distribution in the strip. We may identify the real part u as the flux function to complete the solution. A flux plot made as in Fig. 5.28 shows some isotherms and iso-flux lines. The isotherms are spaced at equal intervals of 0.05. The solution shows symmetry with respect to $y = \frac{\pi}{2}$. Also the temperature is substantially the same as the top and bottom edge temperatures (i.e., zero or the datum value) for $x > 2.5$.

5.3.4 *Shape Factors for Some Useful Configurations*

Eccentric insulation on a pipe

Earlier we have studied, in Chap. 2, heat transfer in the case of a cylinder with a coaxial cylindrical insulation layer. Sometimes the insulation may have to be non-coaxial (or eccentric) because the pipe to be insulated may be in the close vicinity

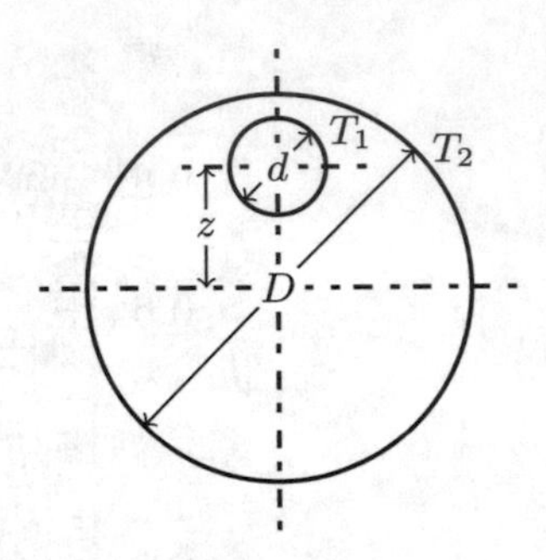

Fig. 5.29 Heat transfer between eccentric cylinders

of the walls. In such cases, the geometry will be as shown in Fig. 5.29 below. The particular case with $z = 0$ corresponds to the earlier coaxial insulation geometry. The conduction shape factor for this case may, in fact, be obtained by using the same technique as was employed in the case of the buried cable problem. It is seen that all one has to do is to identify two appropriate circular isotherms in the earlier case as corresponding to the inner and outer circles in Fig. 5.29. From such an analysis, we have

$$\boxed{S = \frac{2\pi}{\cosh^{-1}\left[\dfrac{D^2 + d^2 - 4z^2}{2Dd}\right]}} \tag{5.131}$$

When $z = 0$, the denominator of Eq. 5.131 becomes

$$\cosh^{-1}\left[\frac{D^2 + d^2 - 4z^2}{2Dd}\right] = \cosh^{-1}\left[\frac{D}{2d} + \frac{d}{2D}\right]$$

Let $x = \cosh^{-1}\left[\frac{D}{2d} + \frac{d}{2D}\right]$. Then, $\cosh x = \frac{D}{2d} + \frac{d}{2D}$ or, by definition, $\cosh x = \frac{e^x + e^{-x}}{2} = \frac{D}{2d} + \frac{d}{2D}$. Hence we have

$$e^x = \frac{D}{2d} \text{ or } x = \ln\left[\frac{D}{2d}\right] \tag{5.132}$$

Finally Eq. 5.131 reduces to

$$S = \frac{2\pi}{\ln\left[\dfrac{D}{2d}\right]} = \frac{2\pi}{\ln\left[\dfrac{R}{2r}\right]} \tag{5.133}$$

where R and r correspond to the radius of the outer and inner circle, respectively. Equation 5.133 is in agreement with the well-known relation for a cylindrical annulus given in Chap. 2.

Example 5.9

A thin-walled pipe of 38 mm OD carrying high-pressure steam at a temperature of 120 °C is insulated by a high temperature insulating material of thermal conductivity equal to 0.106 W/m°C. The insulation is eccentric with an eccentricity of 50 mm and the diameter of the outer surface of the insulation is 204 mm. Determine the heat loss per meter from steam. The outer surface of the insulation has a mean temperature of 44 °C. If the same heat loss takes place with insulation having no eccentricity, what will be the outer diameter of the insulation? Assume that there is no change in the temperature values.

Solution:

Case (a) Eccentric insulation:
Using the notation of Fig. 5.29, we have

$$D = 0.204 \text{ m}, \ d = 0.038 \text{ m}, \ z = 0.05 \text{ m}$$

The conduction shape factor is calculated, using Eq. 5.131 as

$$S = \frac{2\pi}{\cosh^{-1}\left[\dfrac{0.204^2 + 0.038^2 - 4 \times 0.05^2}{2 \times 0.204 \times 0.038}\right]} = 4.52$$

The other pertinent quantities are specified as

$$\begin{aligned} \text{Inner surface temperature: } T_1 &= 120\ °\text{C} \\ \text{Outer surface temperature: } T_2 &= 44\ °\text{C} \\ \text{Insulation thermal conductivity: } k &= 0.106\ \text{W/m°C} \end{aligned}$$

Heat loss from steam per meter length of pipe is then given by

$$q = kS(T_1 - T_2) = 0.106 \times 4.52 \times (120 - 44) = 36.4 \text{ W/m}$$

Case (b) Insulation layer without eccentricity:
If the concentric insulation loses the same amount of heat per meter of pipe length, the temperatures are the same and the thermal conductivity of the insulation material is the same, then the conduction shape factor has to be the same as in the previous case, i.e., $S = 4.52$. However, in this case the shape factor is given by Eq. 5.133. Hence the outer diameter of insulation is given by

$$D = de^{\frac{2\pi}{S}} = 0.038 \times e^{\frac{2\pi}{4.52}} = 0.153 \text{ m}$$

Thus there is much saving of insulation material if eccentricity is avoided!

Soil thermal conductivity measurement problem

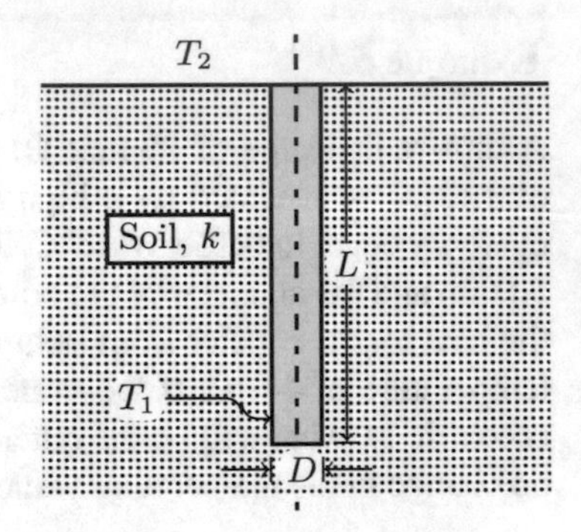

Fig. 5.30 Application: measurement of soil thermal conductivity

A vertical cylinder of diameter D and length L is inserted vertically into the ground (or any other medium) as indicated in Fig. 5.30. In the application indicated in the figure, the cylinder is heated electrically with a heater input of Q watts. When $L \gg D$ the shape factor is given by the relation

$$\boxed{S = \frac{2\pi}{\ln\left(\frac{4L}{D}\right)}} \tag{5.134}$$

The thermal conductivity is then determined by the formula

$$k = \frac{Q/L}{S(T_1 - T_2)} \tag{5.135}$$

Example 5.10

A cylindrical heating element of diameter 0.1 m and length 2 m is driven tightly into a hole drilled in the ground. Heat is produced in the heater electrically at a rate of 100 W. The heater surface temperature is measured at ten locations along the length of the heater. It is found that the average temperature is 84.9 °C. The ground surface temperature is measured some distance away from the hole and is found to be 25.5 °C. Estimate the thermal conductivity of the soil.

Solution:
Nomenclature used here is according to that in Fig. 5.30. The given data is written down as

Diameter of heating element: $D = 0.1$ m
Length of the heater: $L = 2$ m
Heat generated by the heater: $Q = 100$ W
The surface temperature of heater: $T_1 = 84.9$ °C
The surface temperature of soil: $T_2 = 25.5$ °C

We calculate the conduction shape factor using Eq. 5.134 as

$$S = \frac{2\pi}{\ln\left(\dfrac{4 \times 2}{0.1}\right)} = 1.434$$

Hence the thermal conductivity of the soil is given by Eq. 5.135 as

$$k = \frac{100/2}{1.434 \times (84.9 - 25.5)} = 0.587\ \mathrm{W/m^\circ C}$$

Heat transfer between buried pipes

Buried pipes carrying hot fluids are encountered in industrial applications such as in thermal power plants and chemical process industries. The geometry is shown in Fig. 5.31. The conduction shape factor for this configuration is given by

$$\boxed{S = \frac{2\pi}{\cosh^{-1}\left[\dfrac{4L^2 - (D_1^2 + D_2^2)}{2D_1 D_2}\right]}} \tag{5.136}$$

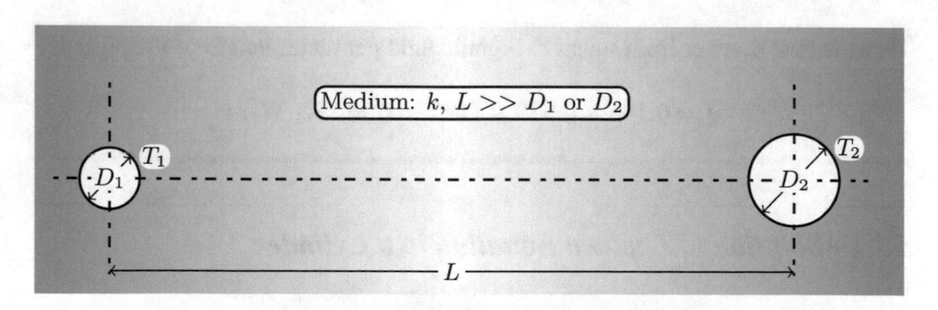

Fig. 5.31 Conduction between two long cylinders in an infinite medium

Example 5.11

In a process industry, two pipes of an equal diameter of 0.025 m run parallel to each other. The two pipes are buried under the soil of thermal conductivity equal to 0.106 W/m°C. One of the pipes carries steam at 120 °C while the other carries a cold organic fluid at −5 °C. Determine the heat transfer from steam to organic fluid per meter length of pipes if the axes of the two pipes are 0.3 m apart.

Solution:
We make use of the nomenclature in Fig. 5.31 and Eq. 5.136 for solving this problem.

Diameter of steam carrying pipe:	$D_1 = 0.025$ m
Diameter of the organic fluid carrying pipe:	$D_2 = 0.025$ m
Center to center distance between pipes:	$L = 0.3$ m
Steam temperature:	$T_1 = 120$ °C
Organic fluid temperature:	$T_2 = -5$ °C
Soil thermal conductivity:	$k = 0.106$ W/m°C

The conduction shape factor is calculated using Eq. 5.136 as

$$S = \frac{2\pi}{\cosh^{-1}\left[\dfrac{4 \times 0.3^2 - (0.025^2 + 0.025^2)}{2 \times 0.025 \times 0.025}\right]} = 0.989$$

Then the heat transfer from steam to organic fluid per meter length of the pipes is

$$q = 0.106 \times 0.989 \times (120 + 5) = 13.1 \text{ W/m}$$

5.3.5 *Solution to Laplace Equation in a Cylinder*

In Sect. 5.3.3 we have explored the possibility of solving Laplace equation in cylindrical coordinates using elementary solutions to the Laplace equation. We now consider the application of separation of variables technique for the solution of Laplace equation in a finite length cylinder. The Laplace equation needs to be solved in a cylinder of radius R and length L. The face of the cylinder at $z = L$ is maintained at a temperature distribution that varies with r alone. The left face (at $z = 0$) and the lateral surface of the cylinder ($r = R, 0 \leq z \leq L$) are maintained at zero temperature (see Fig. 5.32).

The governing differential equation is the Laplace equation given by

$$\frac{1}{r}\frac{\partial}{\partial r}\left(r\frac{\partial T}{\partial r}\right) + \frac{\partial^2 T}{\partial r^2} = 0 \tag{5.137}$$

The separation of variables technique seeks a solution to Eq. 5.137 in the product form

$$T(r,z) = \theta(r)Z(z) \tag{5.138}$$

As before, we substitute Eqs. 5.138 in 5.137, use similar manipulation to get the following two ordinary differential equations:

$$\text{(a)} \quad \frac{1}{r}\frac{\mathrm{d}}{\mathrm{d}r}\left(r\frac{\mathrm{d}\theta}{\mathrm{d}r}\right) + \lambda^2\theta = 0, \quad \text{(b)} \quad \frac{\mathrm{d}^2 Z}{\mathrm{d}z^2} - \lambda^2 Z = 0 \tag{5.139}$$

where $-\lambda^2$ is the separation constant, the sign being decided on considerations similar to the earlier ones. The function Z has the general solution given by

$$Z(z) = C_1\cosh(\lambda z) + C_2\sinh(\lambda z) \tag{5.140}$$

On $z = 0$ the temperature is zero and hence constant C_1 has to be taken as zero. The equation governing $\theta(r)$ is recognized as the Bessel equation of order zero (Appendix B). Requiring that the solution be well behaved (finite) along the axis of the cylinder, the solution is given by

$$\theta(r) = C_3 J_0(r) \tag{5.141}$$

Representing the product C_2C_3 as a, the solution to the Laplace equation is of the form

$$\boxed{T(r,z) = \sum_1^\infty a_n \sinh(\lambda_n z) J_0(\lambda_n r)} \tag{5.142}$$

where λ_n are the zeros of the Bessel function of order 0. The Fourier Bessel coefficients a_n are determined by the requirement $T(r, L) = f(r)$. Borrowing the results from Appendix B, the Fourier Bessel coefficients are determined as

$$a_n = \frac{1}{\sinh(\lambda_n L)} \times \frac{2\int_0^R r f(r) J_0(\lambda_n r)\mathrm{d}r}{R^2 J_1^2(\lambda_n R)} \tag{5.143}$$

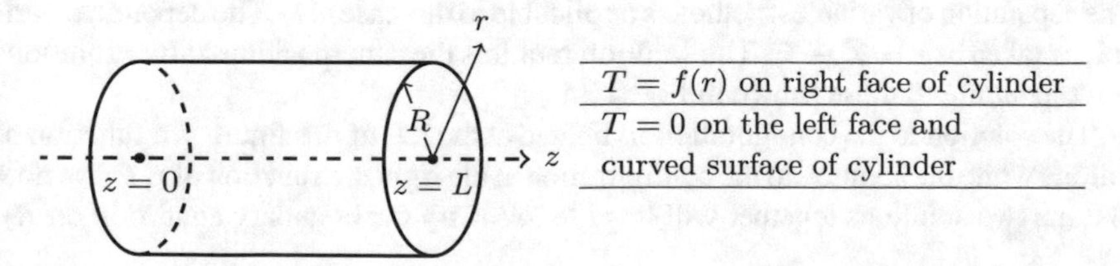

Fig. 5.32 Steady conduction in a finite cylinder

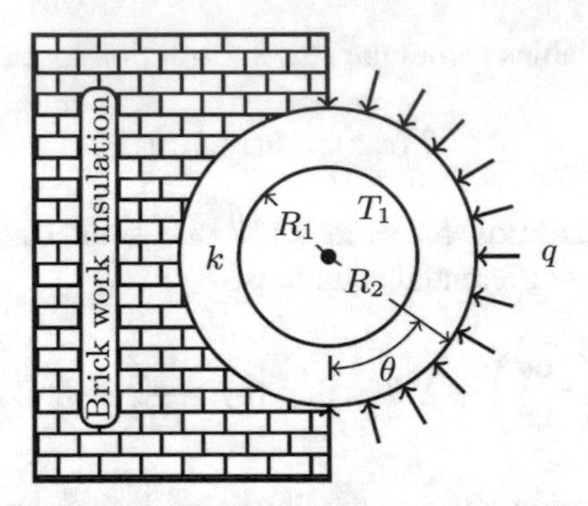

Fig. 5.33 A practical problem in cylindrical coordinates

5.3.6 Solution to a Practical Problem

An interesting application of solution to Laplace equation in cylindrical coordinates is shown in Fig. 5.33. The geometry is typical of a thick-walled boiler tube that is facing the furnace over one half ($0 < \theta < \pi$) and is insulated by the brick wall of the boiler over the other half ($\pi < \theta < 2\pi$). The temperature of the inside surface of the tube ($r = R_1, 0 < \theta < 2\pi$)is the same as the fluid temperature T_1 (probably steam) while the exposed outer surface of the tube is subject to constant heat flux q over the exposed portion. The tube material has a thermal conductivity of k. Assuming that the tube is very long in a direction perpendicular to the plane of the figure, the temperature within the tube is a function of r, θ. The governing equation is the Laplace equation (5.14) with the partial derivative with respect to z set to zero. Thus the temperature within the tube satisfies the equation

$$\frac{1}{r}\frac{\partial}{\partial r}\left(r\frac{\partial T}{\partial r}\right) + \frac{1}{r^2}\frac{\partial^2 T}{\partial \theta^2} = 0 \tag{5.144}$$

The applicable boundary conditions are

$$\begin{gathered}\text{(a) } r = R_1,\ 0 \le \theta \le 2\pi, \quad T = T_1 \\ \text{(b) } r = R_2,\ 0 \le \theta \le \pi,\ \frac{\partial T}{\partial r} = \frac{q}{k},\ \text{(c) } r = R_2,\ \pi \le \theta \le 2\pi,\ \frac{\partial T}{\partial r} = 0\end{gathered} \tag{5.145}$$

The separation of variables method is applicable to this case also. The dependent variable is taken as $z = T - T_1$. The solution requires the superposition of two solutions corresponding to those shown in Fig. 5.34.

The solution to the configuration indicated at the left of this figure is a function of r alone while the solution to the configuration at the right is a function of r, θ. We note that the two solutions together will be able to satisfy the boundary condition on R_2.

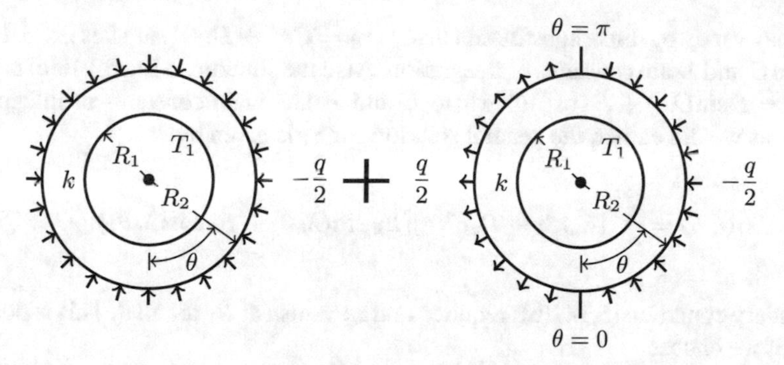

Fig. 5.34 Superposition of two configurations that is equivalent of configuration in Fig. 5.33

Let $z = h(r) + z_1(r, \theta)$ where $z_1(r, \theta) = f(r)g(\theta)$. With the already familiar procedure three ordinary differential equations are obtained, with the separation constant λ^2.

$$(a)\ \frac{\mathrm{d}}{\mathrm{d}r}\left(r\frac{\mathrm{d}h}{\mathrm{d}r}\right) = 0;\ (b)\ \frac{\mathrm{d}}{\mathrm{d}r}\left(r\frac{\mathrm{d}f}{\mathrm{d}r}\right) - \lambda^2\frac{f}{r} = 0;\ (c)\ \frac{\mathrm{d}^2 g}{\mathrm{d}\theta^2} + \lambda^2 g = 0 \quad (5.146)$$

The applicable boundary conditions are

$$(a)\ h(R_1) = 0;\ \ (b)\ \left.\frac{\mathrm{d}h}{\mathrm{d}r}\right|_{r=R_2} = \frac{q}{2k}\ \ (c)\ z_1(r = R_1, 0 \le \theta \le 2\pi) = 0$$

$$(d)\ \underbrace{\left.\frac{\partial z_1}{\partial r}\right|_{r=R_2} = \frac{q}{2k}}_{(0\le\theta\le\pi)}\quad (e)\ \underbrace{\left.\frac{\partial z_1}{\partial r}\right|_{r=R_2} = -\frac{q}{2k}}_{(\pi\le\theta\le 2\pi)|} \quad (5.147)$$

We can easily solve Eq. 5.146(a) for $h(r)$ by integrating the equation twice to get

$$h = A\ln(r) + B \quad (5.148)$$

where A and B are constants of integration. The inner boundary condition (Eq. 5.147(a)) requires that $B = -A\ln(R_1)$. The outer boundary condition (Eq. 5.147(b)) requires

$$\left.\frac{\mathrm{d}h}{\mathrm{d}r}\right|_{r=R_2} = \frac{A}{R_2} = \frac{q}{2k} \text{ or } A = \frac{qR_2}{2k}$$

With these the solution, after minor manipulation, reduces to

$$h(r) = \frac{qR_2}{2k}\ln\left(\frac{r}{R_1}\right) \quad (5.149)$$

We may verify by direct substitution that $f(r) = Cr^{\lambda} + Dr^{-\lambda}$ satisfies Eq. 5.146(b) where C and D are constants of integration. Also the solution to Eq. 5.146(c) is given by $g = E\sin(\lambda\theta) + F\cos(\lambda\theta)$ where E and F are again constants of integration. Thus, as we did earlier, the general solution for z_1 is given by

$$z_1(r,\theta) = \sum_{n=1}^{\infty}[C_n r^{\lambda_n} + D_n r^{-\lambda_n}][E_n \sin(\lambda_n\theta) + F_n \cos(\lambda_n\theta)] \tag{5.150}$$

Boundary condition (5.147(b)) requires that z_1 vanish at R_1 for all θ. This is possible only if we choose

$$C_n R_1^{\lambda_n} + D_n R_1^{-\lambda_n} = 0 \tag{5.151}$$

This requires that $C_n R_1^{\lambda_n} = -D_n R_1^{-\lambda_n} = G_n(\text{say})$. Also note that the function $f(r)$ is single valued only if λ_n is an integer, i.e., $\lambda_n = n$. Hence we may write

$$C_n = \frac{G_n}{R_1^n};\quad D_n = -G_n R_1^n \tag{5.152}$$

Letting $a_n = G_n E_n$ and $b_n = G_n F_n$, the solution may be written as

$$z_1(r,\theta) = \sum_{n=1}^{\infty}\left[\left(\frac{r}{R_1}\right)^n - \left(\frac{r}{R_1}\right)^{-n}\right][a_n \sin(n\theta) + b_n \cos(n\theta)] \tag{5.153}$$

Consider now the boundary conditions given by Eqs. 5.147(d) and (e). In order to apply these we calculate $\frac{\partial z_1}{\partial r}$ by term by term differentiation of Eq. 5.153.

$$\frac{\partial z_1}{\partial r} = \sum_{n=1}^{\infty}\frac{n}{r}\left[\left(\frac{r}{R_1}\right)^n + \left(\frac{r}{R_1}\right)^{-n}\right][a_n \sin(n\theta) + b_n \cos(n\theta)] \tag{5.154}$$

The boundary condition at $r = R_2$ requires that the infinite series

$$\sum_{n=1}^{\infty}\frac{n}{R_2}\left[\left(\frac{R_2}{R_1}\right)^n + \left(\frac{R_2}{R_1}\right)^{-n}\right][a_n \sin(n\theta) + b_n \cos(n\theta)] \tag{5.155}$$

represent the function shown in Fig. 5.35. The function is a square pulse of amplitude $\frac{q}{2k}$. it is an odd function of θ as is clear from the figure. This is represented by a Fourier sine series with $b_n = 0$ for all "n"[6] and

$$a_{(2n-1)} \cdot \frac{(2n-1)}{R_2}\left[\left(\frac{R_2}{R_1}\right)^{(2n-1)} + \left(\frac{R_2}{R_1}\right)^{-(2n-1)}\right] = \frac{q}{\pi k(2n-1)} \tag{5.156}$$

[6] The Fourier series has odd nonzero coefficients and zero even coefficients.

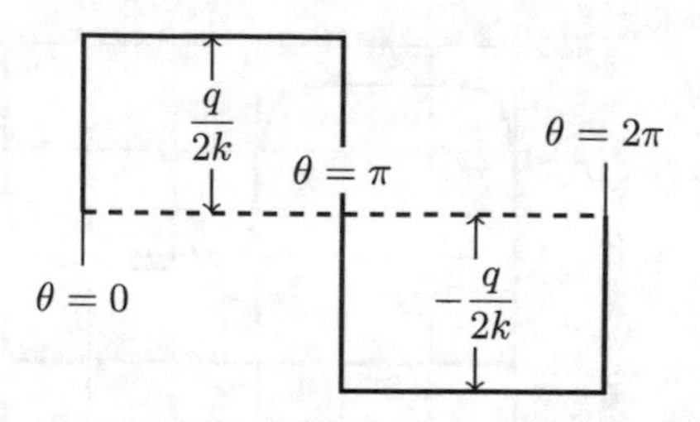

Fig. 5.35 Function to be represented by series 5.155

Finally the solution reduces to

$$z(r,\theta) = \frac{2qR_2}{k\pi}\sum_{1}^{\infty}\frac{\sin[(2n-1)\theta]}{(2n-1)^2}\times \times\left[\frac{\left(\frac{r}{R_1}\right)^{(2n-1)} - \left(\frac{r}{R_1}\right)^{-(2n-1)}}{\left(\frac{R_2}{R_1}\right)^{(2n-1)} + \left(\frac{R_2}{R_1}\right)^{-(2n-1)}}\right] + \frac{qR_2}{2k}\ln\left(\frac{r}{R_1}\right) \tag{5.157}$$

Note that $2n-1$ is $1, 3, 5, \ldots$ for $n = 0, 1, 2 \ldots$.

Example 5.12

A tube arranged on the wall of a furnace as indicated in Fig. 5.33 has an inner diameter of 25 mm and an outer diameter of 32 mm. The tube is made of a material of thermal conductivity 45 W/m°C. The heat flux incident on the exposed part is given to be 10^6 W/m^2. The inner wall temperature of the tube is 100 °C. Make a plot of the temperature variation on the outer surface of the tube as a function of angle. Also, plot the temperature variation across the thickness of the tube at $\theta = \frac{\pi}{2}$.

Solution:

Step 1 We make use of the nomenclature introduced in Fig. 5.33 and write down the given data:

Inner radius of tube: $R_1 = \frac{0.025}{2} = 0.0125$ m

Outer radius of tube: $R_2 = \frac{0.032}{2} = 0.016$ m

Thermal conductivity of tube material: $k = 45$ W/m°C

Datum temperature: $T_1 = 100$ °C

Heat flux parameter: $\frac{qR_2}{k} = \frac{10^6 \times 0.016}{45} = 355.56$ °C

Step 2 In order to make a plot of the angular variation of temperature of the tube outer surface, we make use of Eq. 5.157 with $r = R_2$. The data is calculated at 5° angular intervals from 0 to 360°. The Fourier summation needs a large

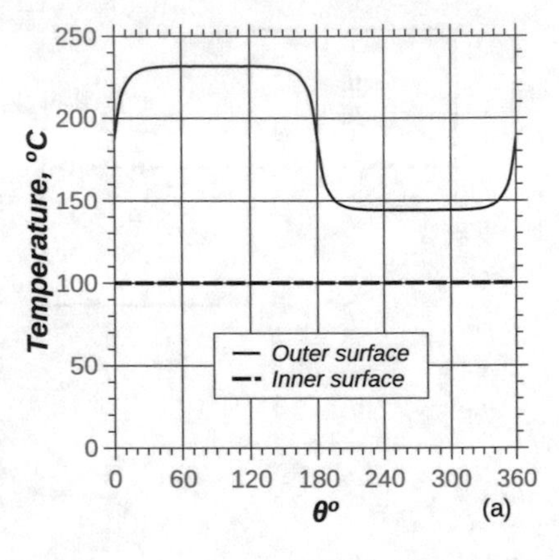

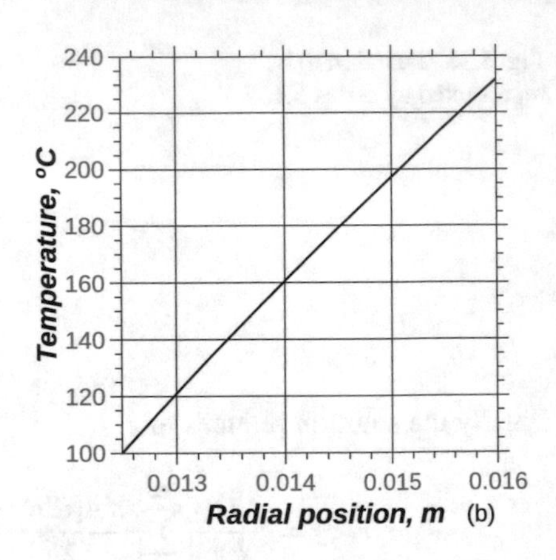

Fig. 5.36 **a** Tube outer surface temperature variation with angle **b** Radial variation of temperature in the tube at $\theta = 90°$

number of terms and the data is generated using a computer program with 1000 terms. The resulting plot is shown in Fig. 5.36a. It is observed that the maximum tube wall temperature of 231.7 °C occurs, as expected, at $\theta = 90°$.

Step 3 The second plot we make is actually for this angular location. Equation 5.157 is used to generate the data by taking the angle to be 90° and varying the radial position from R_1 to R_2. The resulting plot is shown in Fig. 5.36b). It is observed that the temperature varies very steeply within the tube wall. The radial temperature gradient is roughly 38000 °C/m!

5.3.7 *Solution to Laplace Equation in Spherical Co-ordinates*

Laplace equation in spherical coordinates is also amenable to solution by the use of the separation of variables technique. We consider the case where the temperature field is a function of r, θ only. Figure 5.37 shows the geometry that is being studied. A solid sphere has its outer surface maintained at a specified temperature variation with respect to the polar angle θ. The applicable equation is obtained by setting the derivatives with respect to ϕ to zero and $G = 0$ in the steady heat equation in spherical coordinates.

$$\frac{\partial}{\partial r}\left(r^2\frac{\partial T}{\partial r}\right) + \frac{1}{\sin\theta}\frac{\partial}{\partial\theta}\left(\sin\theta\frac{\partial T}{\partial\theta}\right) = 0 \tag{5.158}$$

The applicable boundary conditions are specified as

$$\text{(a) } T(r = 0, \theta) = \text{Finite}, \quad \text{(b) } T(r = R, \theta) = f(\theta), \quad -\frac{\pi}{2} \le \theta \le \frac{\pi}{2} \tag{5.159}$$

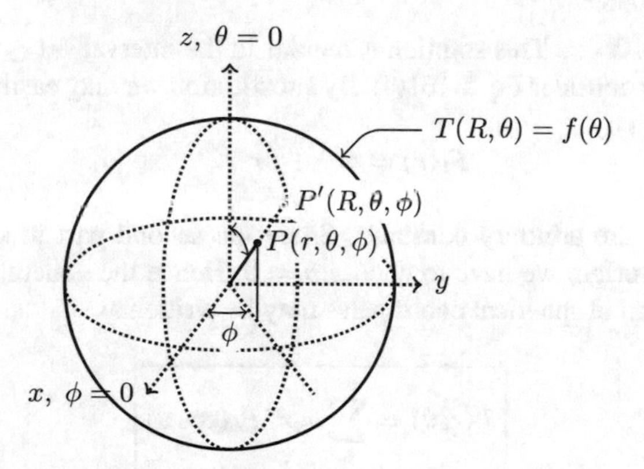

Fig. 5.37 Steady conduction in a solid sphere

We seek a solution, as in earlier cases, in the product form

$$T(r,\theta) = F_1(r)F_2(\theta) \tag{5.160}$$

Substituting this in Eq. 5.159, after the usual manipulations we get the following two ordinary differential equations that involve a separation constant λ.

$$\text{(a)}\ \frac{1}{r}\frac{\mathrm{d}}{\mathrm{d}r}\left(r^2\frac{\mathrm{d}F_1}{\mathrm{d}r}\right) - \lambda F_1 = 0, \quad \text{(b)}\ \frac{1}{\sin\theta}\frac{\mathrm{d}}{\mathrm{d}\theta}\left(\sin\theta\frac{\mathrm{d}F_2}{\mathrm{d}\theta}\right) + \lambda F_2 = 0 \tag{5.161}$$

Consider first the solution of Eq. 5.161(b). We substitute $x = \cos\theta$. We then see that $\sin\theta = \sqrt{1 - cos^2\theta} = \sqrt{1 - x^2}$. Also, we have

$$\mathrm{d}x = -\sin\theta \mathrm{d}\theta = -\sqrt{1 - x^2}\mathrm{d}\theta \text{ or } \frac{\mathrm{d}}{\mathrm{d}\theta} = -\sqrt{1 - x^2}\frac{\mathrm{d}}{\mathrm{d}x}$$

Use these in Eq. 5.161(b) to get

$$\begin{aligned} &\frac{1}{\sqrt{1-x^2}}\left[-\sqrt{1-x^2}\frac{\mathrm{d}}{\mathrm{d}x}\left(\sqrt{1-x^2}\cdot\sqrt{1-x^2}\frac{\mathrm{d}F_2}{\mathrm{d}x}\right)\right] + \lambda F_2 = 0 \\ &\text{or} \quad \frac{\mathrm{d}}{\mathrm{d}x}\left[(1-x^2)\frac{\mathrm{d}F_2}{\mathrm{d}x}\right] + \lambda F_2 = 0 \end{aligned} \tag{5.162}$$

It is normal practice to take $\lambda = n(n+1)$. With this Eq. 5.162 takes the standard form

$$(1-x^2)\frac{\mathrm{d}^2F_2}{\mathrm{d}x^2} - 2x\frac{\mathrm{d}F_2}{\mathrm{d}x} + n(n+1)F_2 = 0 \tag{5.163}$$

This equation is the same as the Legendre equation discussed in Appendix B. We know from there that the solution is given by

$$F_2(x) = F_2(\cos\theta) = P_n(x) = P_n(\cos\theta) \tag{5.164}$$

where $n = 0, 1, 2, \ldots$. This solution is regular in the interval $-1 \le x \le 1$ or $-1 \le \cos\theta \le 1$. Now consider Eq. 5.161(a). By substitution we may easily verify that the solution is given by

$$F_1(r) = ar^n + br^{-(n+1)} \tag{5.165}$$

where a and b are arbitrary constants. Since the second part in solution (5.165) diverges at the origin we have to choose $b = 0$. Hence the general solution to the Laplace equation in spherical coordinates may be written as

$$\boxed{T(r, \theta) = \sum_0^{\infty} a_n r^n P_n(\cos\theta)} \tag{5.166}$$

The boundary condition on the surface of the sphere (Eq. 5.159(b)) requires that

$$\sum_0^{\infty} a_n R^n P_n(\cos\theta) = f(\theta) \tag{5.167}$$

Coefficients a_n may be obtained by using the orthogonality property of the Legendre function in $-1, 1$ (Eq. B.16). Thus, we have

$$\boxed{a_n = \frac{(2n+1)}{2R^n} \int_{-1}^{1} f(x) P_n(x) \mathrm{d}x} \tag{5.168}$$

where $x = \cos\theta$. Note thus that the given function $f(\theta)$ needs to be represented as a function of $\cos\theta$ in the form $f(\cos\theta)$.

Concluding Remarks

We have considered two-dimensional heat conduction problems comprising of transient conduction in one dimension and also steady conduction in two dimensions. In latter case, problems in all three coordinate systems, viz., Cartesian, cylindrical, and spherical coordinates have been considered. Similarity analysis and integral method of the solution are possible in the case of one-dimensional transients. Separation of variables is otherwise the method which is applicable in all cases. Solution for temperature is represented as a Fourier series involving (a) circular functions in the Cartesian system of coordinates, (b) Bessel functions in cylindrical coordinates, and (c) Legendre polynomials in the case of spherical coordinates. Use of complex variables to solve conduction problems in two dimensions is an alternative which is also useful.

5.4 Exercises

Ex 5.1: By performing energy balance for a suitably chosen volume element derive the heat diffusion equation in three dimensions in cylindrical and spheri cal coordinate systems. Verify that the function $T(x, y) = 20\ln(x^2 + y^2)$ recast in the form $T(r) = 40\ln(r)$ is a solution to the Laplace equation written in the cylindrical coordinate system.

Ex 5.2: A semi-infinite solid of thermal diffusivity 10^{-5} m^2/s is subject to a constant surface heat flux of 10^3 W/m^2 for $t > 0$. The initial temperature of the solid is 20°C. Determine the time at which the surface temperature would be 50 °C. At this time what would be the temperature at a depth of 5 mm from the surface? What is the depth of penetration, as defined in the text, at this time? What would the answers be if the surface heat flux is increased to 10^5 W/m^2?

Ex 5.3: A semi-infinite solid initially at 30 °C has its surface temperature reduced to −10 °C for $t > 0$. Determine the surface heat flux when the temperature at a depth of 5 mm within the solid is 0 °C. Also determine the heat fluxes at the surface as well as at the interior point, at this time.

Ex 5.4: The solution to the standard problem is given by Eq. 5.87. Determine the temperature gradient at $x = \frac{a}{2}$, $y = b$ for the case with $\frac{b}{a} = 0.5$. Determine also the direction of the heat flux at (a) $x = \frac{a}{4}$, $y = b$, (b) $x = a$, $y = \frac{b}{2}$ and (c) $x = \frac{a}{4}$, $y = \frac{b}{4}$.

Ex 5.5: Consider a long rod of square cross section of side 0.05 m. The bottom side of the square is maintained at 100 °C, the left side is insulated while the other two sides are maintained at 20 °C. Show that this problem can be reduced to the standard problem considered in the text. Deduce the solution by inspection of the solution given in the text.

Ex 5.6: Consider steady two-dimensional temperature field in a rectangle of width a and height b subject to the boundary conditions indicated in Fig. 5.38. Formulate the problem and solve the governing equation using the method of separation of variables. Express the solution in terms of suitable nondimensional variables and parameters.

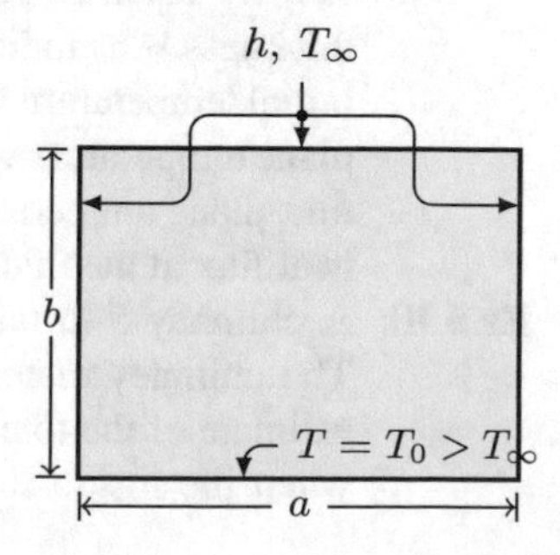

Fig. 5.38 Steady temperature field in a rectangle: Exercise 5.6

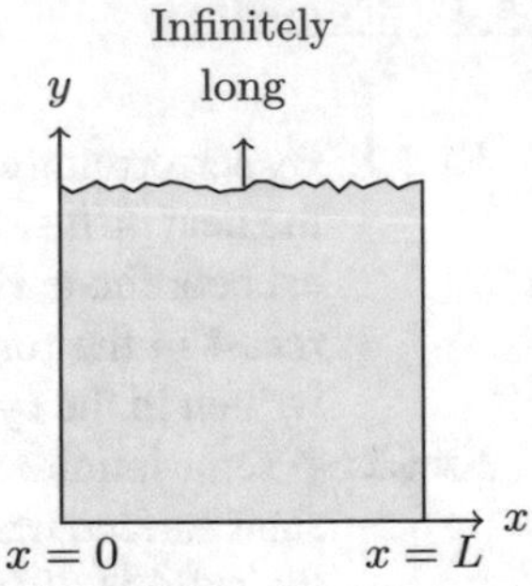

Infinitely long

y

$x = 0$ $x = L$ x

Fig. 5.39 Steady temperature in a long bar: Exercise 5.7

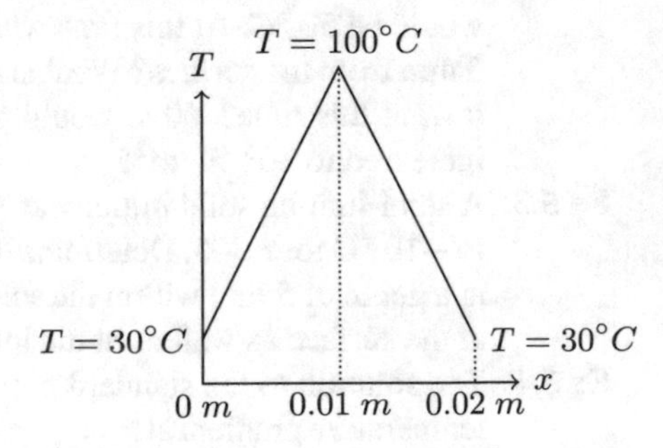

Fig. 5.40 Initial temperature profile in infinite slab of Exercise 5.9

Ex 5.7: Obtain the solution to the case shown in Fig. 5.39. The temperature goes to zero as $y \to \infty$ for all x. The temperature is specified on the edge $0 \leq x \leq L$, $y = 0$ as $T(x, 0) = T_0 \sin \frac{\pi x}{L}$ where T_0 is a constant.

Ex 5.8: A bar of uniform cross section is of length $L = 10$ cm. It is at a uniform temperature of 100 °C at $t = 0$. The lateral surface of the bar is perfectly insulated. Using the method of separation of variables determine the solution when the two ends at $x = 0$ and $x = L$ are maintained at 30 °C for $t > 0$. Plot the variation of temperature at $x = 5$ cm as a function of time. Take the thermal diffusivity of the bar material as 10^{-5} m^2/s. What is the temperature profile for a large time? Express the solution in the nondimensional form.

Ex 5.9: Consider a 0.2 m thick infinitely large wall of a material of thermal diffusivity equal to 10^{-5} m^2/s Initial wall temperature variation across its thickness is as indicated in Fig. 5.40. The two ends are maintained at the initial temperature for $t > 0$ Obtain an expression for the variation of mid-plane temperature with time. Make a plot of one-term approximation to the mid-plane temperature and indicate its validity. What is the instantaneous heat flux at the mid-plane?

Ex 5.10: A chimney 5 m tall has a uniform cross section as shown in Fig. 5.41. The chimney material has a thermal conductivity of 0.72 W/m °C. An estimate of the total rate of heat transfer through the chimney is required when the inside surface is at a uniform temperature of 100 °C and the

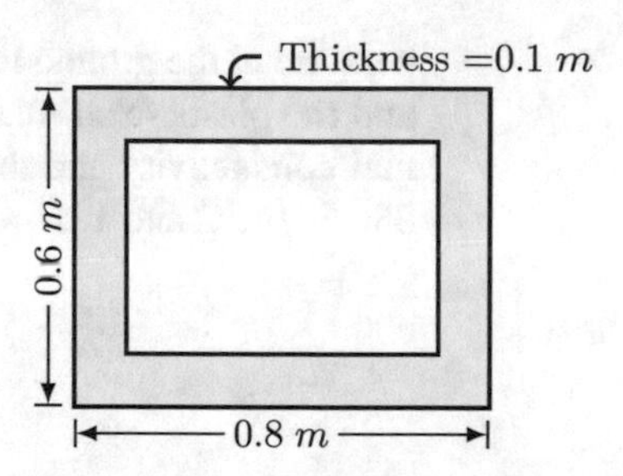

Fig. 5.41 Chimney problem of Exercise 5.10

outside surface is at a uniform temperature of 20 °C. Use the shape factor concept to solve this problem.

Ex 5.11: In a certain insulation application, a pipe is insulated by an eccentrically placed insulation of thermal conductivity 0.15 W/m°C. The pipe diameter is 50 mm while the insulation has a diameter of 150 mm. The eccentricity is 15 mm. Determine the heat loss from a meter length of pipe if the pipe is at 100 °C and the outer surface is losing heat to an ambient at 30°C via a heat transfer coefficient of 5 W/m^2 °C. Make use of thermal resistance concept in solving the problem.

Ex 5.12: A conductor of 3 mm diameter carries a current that produces 100 W/m of heat in the conductor. A coaxial layer of a material of thermal conductivity equal to 0.5 W/m°C and an outer diameter of 25 mm insulates it. What is the temperature of the conductor if the outer surface of the insulation is at 25 °C? Compare this with the case wherein the insulation is eccentrically mounted with an eccentricity of 5 mm? The outer diameter of the insulation remains the same.

Ex 5.13: A very long tube of 50 mm ID and 55 mm OD carries a hot fluid at an average temperature of 70 °C. The tube side heat transfer coefficient is 250 W/m^2 °C and the thermal conductivity of the tube wall is 45 W/m°C. The tube is buried under the soil of thermal conductivity of 0.5 W/m°C at a depth of 1 m. The soil surface is at 20 °C. Determine the heat loss per meter length of the tube. Make use of the thermal resistance concept in solving the problem.

Ex 5.14: In a power station a power cable of diameter 5 mm and a coolant carrying pipe of 50 mm diameter are buried deep in the ground. The cable and the pipe run parallel to each other at an axial distance of 0.3 m. The cable generates heat at the rate of 100 W/m. Thermal conductivity of the soil is known to be 1 W/m°C If the coolant temperature is 10 °C what is the cable temperature? Assume that all the heat generated by the cable is removed by the coolant.

Ex 5.15: A power cable that may be essentially approximated as a very long copper cylinder of 3 mm diameter surrounded by a 1 mm-thick plastic-clad material. Such a cable carries a current of 100 A and is laid 0.6 m below the ground. Determine the operating temperature of the cable if the temperature at the ground level is 30 °C Thermal conductivity of the ground and the plastic-clad material, respectively, are 0.5 and 1 W/m°C. Thermal conductivity and the specific resistance of copper are, respectively, 380 W/m°C and $1.67 \times 10^{-8}\ \Omega \cdot$ m.

Chapter 6
Multidimensional Conduction Part II

THIS chapter considers multidimensional conduction involving transients in two and three dimensions. Separation of variables technique forms the backbone of the analysis. It is shown that for large times a one-term approximation is adequate and leads to useful charts that facilitate easy calculation. Problems involving transient conduction in all three coordinate systems, viz., Cartesian, cylindrical, and spherical coordinates, are considered in this chapter.

6.1 Preliminaries

6.1.1 Introduction

In Chap. 5, attention was given to problems involving conduction in two dimensions—either steady conduction in two space dimensions or unsteady one-dimensional conduction involving one space coordinate and time. Semi-infinite, infinite, and finite spatial domains have been considered. The analysis used the similarity method, the method of separation of variables, the superposition of elementary solutions, and the method of complex variables. The problems considered were in any of the three coordinate systems, viz., Cartesian, cylindrical, or spherical. Engineering applications, specifically in material processing like heat treatment, food processing, etc., involve transient multidimensional (two and three dimensions) conduction. Such problems, that are amenable to solution by the method of separation of variables, are considered in this chapter.

S. P. Venkateshan, *Heat Transfer*,
https://doi.org/10.1007/978-3-030-58338-5_6

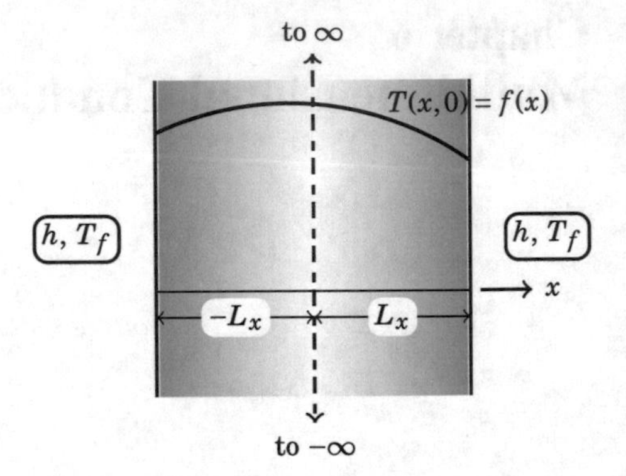

Fig. 6.1 One-dimensional transient in a slab: *the basic problem in Cartesian coordinates*

6.1.2 Basic Problem in Cartesian Coordinates

The problem that forms the basis for analyzing problems in two and three dimensions in Cartesian coordinates is the problem of transient conduction in an infinitely large slab. We consider a slab of thickness $2L_x$ initially at a temperature T_i. In general, the initial temperature profile may vary arbitrarily with respect to x as shown in Fig. 6.1. However uniform initial temperature throughout is the case that shall be of interest to us and hence T_i is treated as a constant in what follows. The two surfaces of the slab are exposed to an ambient fluid at temperature T_f different from T_i. The slab starts cooling/heating by convection at the two exposed surfaces by convection via a heat transfer coefficient h.

Placing the origin for x in the mid-plane as shown, and identifying the temperature excess $T(x,t) - T_f$ as θ_x and identifying initial temperature excess as $\theta_0 = T_i - T_f$ we define the non-dimensional temperature as $\phi_x(x,t) = \frac{\theta_x(x,t)}{\theta_0}$. The governing equation for the one-dimensional transient then is given by

$$\boxed{\frac{1}{\alpha}\frac{\partial \phi_x}{\partial t} = \frac{\partial^2 \phi_x}{\partial x^2}} \tag{6.1}$$

subject to the following initial and boundary conditions:

$$\underbrace{\phi_x(x,0) = 1}_{\text{initial condition}} \text{ and } \underbrace{\left.\frac{\partial \phi_x}{\partial x}\right|_{(\pm L_x, t)} = \mp \phi_x(\pm L_x, t)}_{\text{boundary conditions}} \tag{6.2}$$

In case symmetry exists (as will be assumed in what follows) about the slab mid-plane (i.e., $x = 0$), the problem domain may be replaced by $0 \le x \le L_x$ with the boundary conditions given by

$$\left.\frac{\partial \phi_x}{\partial x}\right|_{(0,t)} = 0; \quad \left.\frac{\partial \phi_x}{\partial x}\right|_{(L_x,t)} = \phi_x(L_x, t) \tag{6.3}$$

Solution by Separation of Variables Method

This problem may be solved in a routine way by the separation of variables method. The general solution is given by (intermediate steps may be supplied by the reader, following the details given in Chap. 5)

$$\phi_x(x, t) = \sum_{n=1}^{\infty} e^{-\lambda_n^2 \alpha t}[A_n \cos(\lambda_n x) + B_n \sin(\lambda_n x)] \tag{6.4}$$

We notice that the sine function is asymmetric with respect to the mid-plane while the cosine function is symmetric with respect to the mid-plane. Hence we put $B_n = 0$ for all n and write the solution as

$$\phi_x(x, t) = \sum_{n=1}^{\infty} e^{-\lambda_n^2 \alpha t} A_n \cos(\lambda_n x) \tag{6.5}$$

We now apply the boundary condition at $x = L_x$ to get

$$-\sum_{n=1}^{\infty} \lambda_n e^{-\lambda_n^2 \alpha t} A_n \sin(\lambda_n L_x) = -\frac{h}{k} \sum_{n=1}^{\infty} e^{-\lambda_n^2 \alpha t} A_n \cos(\lambda_n L_x) \tag{6.6}$$

This can be satisfied (for any $t \geq 0$) if and only if

$$\lambda_n \sin(\lambda_n L_x) = \frac{h}{k} \cos(\lambda_n L_x) \quad \text{or} \quad \cot(\lambda_n L_x) = \frac{\lambda_n L_x}{Bi_x} \tag{6.7}$$

In the above, Bi_x is the Biot Number defined as $Bi_x = \frac{hL_x}{k}$. The values of $X_n = \lambda_n L_x$ (λ_n are known as eigenvalues or characteristic values) that satisfy Eq. 6.7 (this is a transcendental equation) are graphically shown in Fig. 6.2a, b.

The curve $y = \cot(X_n)$ and the straight line $y = \frac{X_n}{Bi_x}$ are plotted as shown. The points of intersection of the curve and the line (three of them appear in the figure) give the first three roots of Eq. 6.7a. The value of the Biot number has been taken as 1 in the example shown in the figure. In practice, the roots are determined by a numerical procedure that will be presented in Example 6.1.

An alternate way of plotting is shown in Fig. 6.2b. Here the ordinate is the quantity $f(X) = X \sin(X) - \cos(X)$ and the abscissa is X. The advantage of this plot is that the function is well behaved without any singularities. The zero crossings (seven of

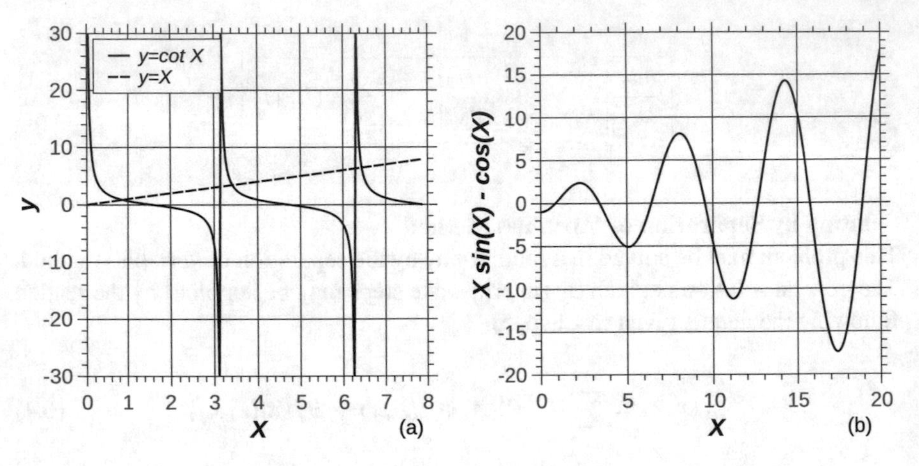

Fig. 6.2 Graphical representations of roots of Eq. 6.7 in two different ways

them occur within the range of the figure) correspond to the required roots $X_n = \lambda_n L_x$.

Using the initial condition, we have

$$1 = \sum_{n=1}^{\infty} A_n \cos(\lambda_n x) \tag{6.8}$$

A_n are given by the Fourier coefficients (this may easily be shown by the orthogonality property of cosine function over the interval of interest to us)

$$A_n = \frac{\int_0^{X_n} \cos(X) dX}{\frac{1}{2} + \frac{\sin(2X_n)}{4}} = \boxed{\frac{\sin(X_n)}{\frac{1}{2} + \frac{\sin(2X_n)}{4}}} \tag{6.9}$$

Example 6.1

Consider a large slab of thickness $2L_x$ subject to convection at its boundaries. The solution requires the roots of Eq. 6.7. Calculate the first few (say 6) roots of this equation. Determine the mid-plane temperature at $t = 50\ s$ by (a) summing six terms and (b) using only one term in the infinite series. The material has a thermal diffusivity of $\alpha = 10^{-5}\ m^2/s$ and $L_x = 0.05\ m$. Assume a value of 1 for the Biot number. The slab is initially at a uniform temperature of T_i and the fluid temperature is T_f.

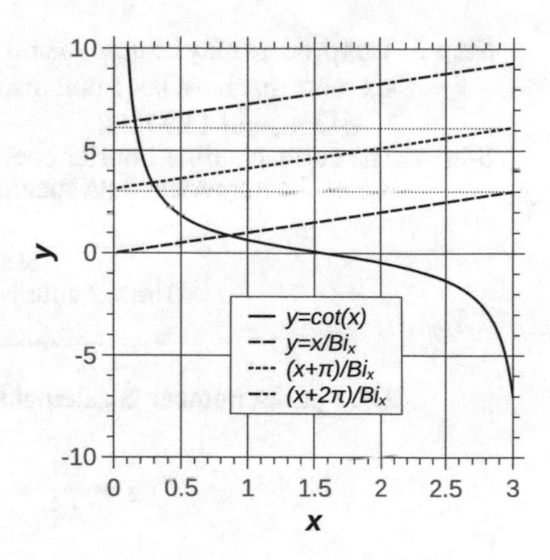

Fig. 6.3 Alternate method of visualizing the roots of Eq. 6.7

Solution:

Step 1 We use the Newton–Raphson method to determine the roots. It is seen from Fig. 6.2 that the first root lies between 0 and π and the second root between π and 2π and so on. It is also clear from Fig. 6.3 that the roots may be visualized to lie between 0 and π for all roots if the value of x is replaced by $x + n\pi$, $n = 0, 1, 2 \ldots$, on the right-hand side of the transcendental Eq. 6.7. Once the root x is found after transforming the right-hand side as shown, the actual root is given by $X_n = x + n\pi$. Let us find the first root that lies between 0 and π.

Step 2 The applicable transcendental equation is written with $Bi_x = 1$, as $f(x) = \cot(x) - x$. The derivative of the function is given by $f'(x) = -\cot^2(x) - 2$, as may be verified. We start the process of determining the root by using a guess value, say $x_g = 1$. An improved value x_b for the root is given by the rule

$$x_b = x_g - \frac{f(x_g)}{f'(x_g)} = x_g - \frac{\cot(x_g) - x_g}{-\cot^2(x_g) - 2} \tag{6.10}$$

We repeat this process by replacing the guess value by the better value. The values starting with the guess value of $x_g = 1$ converges to the correct value 0.86033 in just 4 iterations. Thus the first root of the transcendental equation is $X_1 = 0.86033$.

The root that lies between π and 2π is obtained by solving the equation $cot(x) - x - \pi = 0$ with the understanding that the x value is between 0 and π! The Newton–Raphson algorithm is used by replacing $f(x_g)$ by $cot(x_g) - x_g - \pi$. We start with a guess value of $x_g = 0.1$ (quite arbitrary) and obtain the converged value of 0.28403 in 5 iterations. The required second root is then $X_2 = 0.28403 + \pi = 3.42561$.

Step 3 This process may be used to obtain any number of roots as needed. The first six roots that have been obtained are 0.86033, 3.42561, 6.43730, 9.52934, 12.64258, and 15.77128.

Step 4 The corresponding Fourier coefficients may now be calculated. We make use of the numerical data specified in the problem.

Semi-thickness of slab:	$L_x = 0.05\ m$
Thermal diffusivity of slab material:	$\alpha = 10^{-5}\ m^2/s$
Time:	$t = 50\ s$

The Fourier number is calculated as

$$Fo_x = \frac{\alpha t}{L_x^2} = \frac{10^{-5} \times 50}{0.05^2} = 0.2$$

The Fourier coefficients are given by expression 6.9. Using the roots of the transcendental equation obtained above, the first six Fourier coefficients, corresponding to $f(x) = 1$, are calculated and given in the form of a table.

n	Root X_n	Fourier Coefficient A_n
1	0.86003	1.11905
2	3.42561	−0.15169
3	6.43730	0.04659
4	9.52934	−0.02167
5	12.64258	0.01197
6	15.77128	−0.00799

Step 5 (a) **Six-term sum for mid-plane temperature:**
The mid-plane corresponds to $x = 0$ and hence $\cos(\lambda_n x) = 1$ for all n. The six-term Fourier sum is obtained for $Fo_x = 0.2$ as

$$\phi(0, 0.2) \approx \sum_{n=1}^{6} A_n e^{-X_n^2 Fo_x} = \sum_{n=1}^{6} A_n e^{-0.2X_n^2} = 0.95064$$

Step 6 (b) **One-term approximation for mid-plane temperature:**
The one-term approximation is given by the first term alone in the Fourier series as

$$\phi(0.0.2) \approx A_1 e^{-X_1^2 Fo_x} = 1.11905 \times e^{-0.86003^2 \times 0.2} = 0.96514$$

Step 7 The percentage error in using the one-term approximation is thus given by

$$\text{Percentage error} = \frac{0.95064 - 0.96514}{0.95064} \times 100 = -1.525\%$$

In engineering terms, the one-term approximation is an acceptable estimate for the mid-plane temperature at $Fo_x = 0.2$. In general, the error will be smaller than the above for $Fo_x > 0.2$ and hence the one-term approximation is an acceptable approximation for $Fo_x > 0.2$.

Heat Loss Fraction

In applications, it is necessary to determine the total amount of energy that has entered or left the slab in time t. This may be done in any one of two ways.

1. Calculate the heat transfer at the boundary and integrate this with respect to time from $t = 0$ to t
2. Calculate the energy $E_x(t)$ contained within the slab at time t and the difference between the initial energy Q_{x0} and $E_x(t)$ gives the heat transferred in time t

The latter yields more accurate results since, at any time t, the temperature distribution within the slab is known with good accuracy, say the one-term approximation. The result is thus *not* dependent on the heat transfer history at the slab surface.

We base the calculation on the one-term approximation. Note that the non-dimensional temperature for any x within the slab is then given by

$$\boxed{\phi(x,t) \approx A_1 e^{-X_1^2 Fo_x} \cos(X_1 \xi)} \tag{6.11}$$

where $\xi = \frac{x}{L_x}$. The total energy within the solid $E_x(t)$ at any time t may be calculated by integrating the above with respect to x, after multiplying by the product of density of the slab material ρ, specific heat of the slab material c and the temperature difference $T_i - T_f$, i.e., by $\rho c L_x(T_i - T_f)$. The resulting quantity represents the energy excess or deficit within the slab with respect to that at a datum temperature of T_f. We thus have

$$\begin{aligned} E_x(t) &= 2\rho c L_x(T_i - T_f) \int_0^1 A_1 e^{-X_1^2 Fo_x} \cos(X_1 \xi) d\xi \\ &= 2\rho c L_x(T_i - T_f) A_1 e^{-X_1^2 Fo_x} \frac{\sin(X_1)}{X_1} \end{aligned} \tag{6.12}$$

Factor 2 is to account for the thickness of the slab of $2L_x$. The energy contained in the slab at $t = 0$ or $Fo_x = 0$ is simply given by

$$Q_x(0) = Q_{x0} = 2\rho c L_x(T_i - T_f)$$

The energy that has entered/left at the slab surface, by energy balance, is the difference between Q_{x0} and $E(t)$. The ratio, viz., $\frac{Q_{x0} - E_x(t)}{Q_{x0}}$ is usually of interest to us and is given by

$$\frac{Q_{x0} - E_x(t)}{Q_{x0}} = \frac{Q_x(t)}{Q_{x0}} = 1 - A_1 e^{-X_1^2 Fo_x} \frac{\sin(X_1)}{X_1} \tag{6.13}$$

where $Q_x(t)$ represents the total heat exchange (per unit area) at the slab surface. It is normalized with respect to the energy initially present within the slab material. The fraction $\frac{Q_x(t)}{Q_{x0}}$ is referred to as the heat loss fraction, basically with $T_i > T_f$ in mind.

Example 6.2

Consider again the data given in Example 6.1. Calculate the heat that would enter or leave the slab per unit area in (a) 100 s and (b) 1000 s.

Solution :
Case (a): $t = 100\ s$
Fourier number at the end of 100 s is given by

$$Fo_x(t = 100) = \frac{\alpha t}{L_x^2} = \frac{10^{-5} \times 100}{0.05^2} = 0.4$$

Then we have

$$\frac{Q_x(t)}{Q_{x0}} = 1 - 1.11913 \times e^{-0.86033^2 \times 0.4} \times \frac{\sin(0.86033)}{0.86033} = 0.267$$

About 27% of total heat transfer that would take place if we were to allow cooling to continue for $t \to \infty$ has taken place in the first 100 s.
Case (b): $t = 1000\ s$
Fourier number at the end of 1000 s is given by

$$Fo_x(t = 1000) = \frac{\alpha t}{L_x^2} = \frac{10^{-5} \times 1000}{0.05^2} = 4$$

Then we have

$$\frac{Q_x(t)}{Q_{x0}} = 1 - 1.11913 \times e^{-0.86033^2 \times 4} \times \frac{\sin(0.86033)}{0.86033} = 0.949$$

About 95% of total heat transfer that would take place if we were to allow cooling to continue for $t \to \infty$ has taken place in the first 1000 s.

6.1.3 *Basic Problem in Cylindrical Coordinates*

We consider an infinitely long cylinder of radius R that is initially at a uniform temperature T_i throughout. It is subject to a convective environment at T_f subject to a convective heat transfer coefficient of h. The governing equation is the heat equation in cylindrical coordinates with $\phi_r(r,t) = \frac{T(r,t)-T_f}{T_i-T_f}$, given by

$$\boxed{\frac{\partial \phi_r}{\partial t} = \alpha\left(\frac{\partial^2 \phi_r}{\partial r^2} + \frac{1}{r}\frac{\partial \phi_r}{\partial r}\right)} \tag{6.14}$$

The initial condition is specified as

$$\boxed{\phi_r(r,0) = 1} \tag{6.15}$$

The boundary conditions are specified as

$$\boxed{\phi_r(0,t) \text{ is finite; } \quad \left.\frac{\partial \phi_r}{\partial r}\right|_{r=R} = -\frac{h}{k}\phi_r(R,t)} \tag{6.16}$$

Solution by Separation of Variables Method

Using the familiar method of separation of variables the solution is sought in the form

$$\phi_r(r,t) = \sum_{n=1}^{\infty} f_n(r)e^{-\alpha\lambda_n^2 t} \tag{6.17}$$

On substitution in Eq. 6.14, we get

$$\frac{d^2 f_n}{dr^2} + \frac{1}{r}\frac{df_n}{dr} + \lambda_n^2 f_n = 0 \tag{6.18}$$

Solution to this equation is given in terms of Bessel function of order zero, as may be verified from Appendix A. The general solution may thus be written down as

$$\boxed{\phi_r(r,t) = \sum_{n=1}^{\infty} A_n e^{-\alpha\lambda_n^2 t} J_0(\lambda_n r)} \tag{6.19}$$

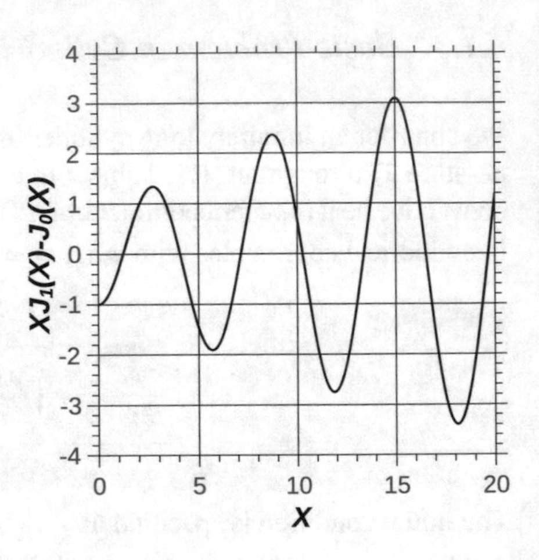

Fig. 6.4 Transcendental equation plotted for the cylinder problem

where A_n are recognized as the Fourier Bessel coefficients. In writing the above, we have set to zero the coefficients of $Y_0(\lambda_n r)$ so as to have a finite value for the temperature at $r = 0$. The eigenvalues λ_n are determined by satisfying the boundary condition at the surface of the cylinder. We obtain the following equation (compare with the slab case considered in Sect. 6.1.2) in order that the solution satisfies the surface boundary condition.

$$\left.\frac{dJ_0}{dr}\right|_{r=R} = -\frac{h}{k}J_0(\lambda_n R) \tag{6.20}$$

Using the results of Appendix A, we may rewrite this in the form of the transcendental equation

$$\boxed{XJ_1(X) - Bi_r J_0(X) = 0} \tag{6.21}$$

where $X = \lambda_n R$ and $Bi_r = \frac{hR}{k}$. A plot of this function shown in Fig. 6.4 for $Bi_r = 1$ looks similar to the plot shown in the slab case (compare with Fig. 6.2a, b).

The roots of the transcendental equation correspond to the zero crossings in Fig. 6.4b. The first six roots, obtained by the use of Newton–Raphson method are: 1.25578, 4.07948, 7.15580, 10.27099, 13.39840, 16.53116. The Fourier coefficients are then evaluated as

$$\boxed{A_n = \frac{\int\limits_0^{X_n} XJ_0(X)dX}{\int\limits_0^{X_n} XJ_0^2(X)dX}} \tag{6.22}$$

Correspondingly the first six Fourier coefficients turn out to be 1.20709, −0.29015, 0.12891, −0.07557, 0.05088, and −0.03718. Again the first term alone survives for $Bi_r \geq 0.2$ and we have the familiar one-term approximation given by

$$\boxed{\phi_r(0, Fo_r) \approx 1.20709e^{-1.25578^2 Fo_r}} \quad (6.23)$$

Heat Loss Fraction

The arguments made in the slab problem hold in this case also. The energy contained within the cylinder at any time $t > 0$ is calculated by integrating the temperature profile after multiplication by the material density ρ, heat capacity c, and the initial temperature difference $T_i - T_f$. The elemental volume, in this case is given by $2\pi r dr$. Hence we have

$$E_r(t) = \rho c(T_i - T_f) \int_0^R 2\pi r \phi(r, t) dr \quad (6.24)$$

The one-term approximation is used for $\phi_r(r, t)$ as $\phi_r(r, t) = A_1 e^{-X_1^2 Fo_r} J_0(X_1 \rho')$ where $\rho' = \frac{r}{R}$. Substituting this in Eq. 6.24 and performing the indicated integration gives

$$E_r(t) = 2A_1 \pi R^2 \rho c(T_i - T_f) e^{-X_1^2 Fo_r} \int_0^1 \rho' J_0(X_1 \rho') d\rho'$$

$$= Q_{r0} 2A_1 e^{-X_1^2 Fo_r} \frac{\int_0^{X_1} x J_0(x) dx}{X_1^2} = Q_{r0} 2A_1 e^{-X_1^2 Fo_r} \frac{J_1(X_1)}{X_1} \quad (6.25)$$

The above result follows from the following two observations:

1. Q_{r0} is the energy initially contained in the cylinder given by $\pi R^2 \rho c(T_i - T_f)$.
2. $\int_0^x x J_0(x) dx = x J_1(x)$—based on properties of the Bessel function.

As in the case of the slab problem, we may then write for the heat loss fraction

$$\boxed{\frac{Q_r(t)}{Q_{r0}} = 1 - \frac{E_r(t)}{Q_0} = 1 - 2A_1 e^{-X_1^2 Fo_r} \frac{J_1(X_1)}{X_1}} \quad (6.26)$$

6.1.4 Basic Problem in Spherical Co-Ordinates

We consider a sphere of radius R initially at a uniform temperature T_i throughout that is subject to convection at its surface to an ambient fluid at T_f via a heat transfer coefficient h for $t > 0$. Heat transfer within the sphere is by unsteady conduction in the radial direction. The governing equation for $\phi = \frac{T(r,t)-T_f}{T_i-T_f}$ is given by the one-dimensional heat equation

$$\boxed{\frac{1}{r^2}\frac{\partial}{\partial r}\left[r^2\frac{\partial \phi}{\partial r}\right] = \frac{1}{\alpha}\frac{\partial \phi}{\partial t}} \tag{6.27}$$

The initial and boundary conditions are given by

$$\boxed{\phi(r, 0) = 1} \tag{6.28}$$

and

$$\boxed{\phi(r = 0, t) \text{ is finite;} \quad \left.\frac{\partial \phi}{\partial r}\right|_{(R,t>0)} = -\frac{h}{k}\phi(R, t > 0)} \tag{6.29}$$

Solution by the Separation of Variables Method

The separation of variables method is used by seeking the solution in the form

$$\phi(r, t) = \sum_{n=1}^{\infty} A_n f_n(r) e^{-\lambda_n^2 \alpha t} \tag{6.30}$$

On substitution in Eq. 6.27, the equation that governs $f_n(r)$ is obtained as

$$\frac{1}{r^2}\frac{d}{dr}\left[r^2\frac{df_n}{dr}\right] = -\lambda_n^2 f_n(r) \tag{6.31}$$

It may be verified by actual substitution that the general solution to Eq. 6.31 is given by

$$f_n(r) = A_n\frac{\sin(\lambda_n r)}{r} + B_n\frac{\cos(\lambda_n r)}{r} \tag{6.32}$$

Since the solution that goes as $\frac{\cos(\lambda_n r)}{r}$ is singular at $r = 0$, the solution contains only the $\frac{\sin(\lambda_n r)}{r}$ part. Hence we have

$$\boxed{\phi(r, t) = \sum_{n=1}^{\infty} A_n\frac{\sin(\lambda_n r)}{r}e^{-\lambda_n^2 \alpha t}} \tag{6.33}$$

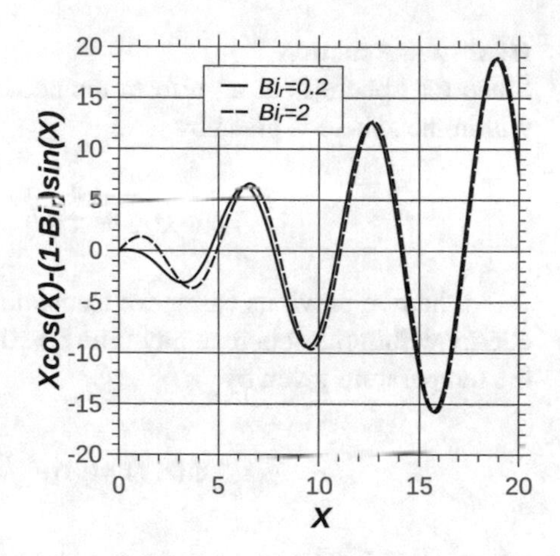

Fig. 6.5 Transcendental equation plotted for the sphere problem

where A_n are Fourier coefficients. The boundary condition requires that

$$\frac{\lambda_n \cos(\lambda_n R)}{R} - \frac{\sin(\lambda_n R)}{R^2} = -\frac{h}{k}\frac{\sin(\lambda_n R)}{R} \text{ for all n.} \tag{6.34}$$

Letting $\lambda_n R = X_n$, we may recast Eq. 6.34 as

$$\boxed{X_n \cos(X_n) - (1 - Bi_r)\sin(X_n) = 0} \tag{6.35}$$

where $Bi_r = \frac{hR}{k}$ is the Biot number. The transcendental Eq. 6.35 plots is shown in Fig. 6.5. The zero crossings are the desired roots to show the role played by the Biot number, we have plotted two cases corresponding, respectively, to $Bi_r = 2$ and $Bi_r = 0.2$. The roots are obtained easily by Newton–Raphson method. The first six roots, for $Bi_r = 2$, are given by 2.02876, 4.91318, 7.97867, 11.08554, 14.20744, and 17.33638. The root located at the origin leads to a trivial solution and hence is of no consequence. The Fourier coefficients A_n are obtained by noting that the function $r f_n(r) = \sin(\lambda_n r)$ is orthogonal in the interval $0, \lambda_n R$, i.e., $0, X_n$. The Fourier coefficients are thus obtained from the initial condition 6.28 as

$$A_n = \frac{\int_0^{X_n} X \sin(X) dX}{\int_0^{X_n} \sin^2(X) dX} = \boxed{\frac{4[\sin(X_n) - X_n \cos(X_n)]}{[2X_n - \sin(2X_n)]}} \tag{6.36}$$

The first six Fourier coefficients for $Bi = 2$ are obtained as 1.47932, −0.76726, 0.48988, −0.35650, 0.27948, and −0.22958. At a value of $Fo_r = 0.2$, the non-dimensional center temperature of the sphere $\phi(0, Fo_r)$ obtained with the six terms is 0.64334 as compared to one-term value of 0.64948. The difference is small and indicates that, in this case also, a one-term approximation is adequate for $Fo_r \geq 0.2$.

Heat Loss Fraction

Since the sphere is at a uniform temperature at $t = 0$, the total energy contained within the sphere is given by

$$Q_{r0} = \frac{4}{3}\pi R^3 \rho c(T_i - T_f) \tag{6.37}$$

As in the two previous cases, we calculate the heat loss fraction by calculating the energy within the sphere at any time $t > 0$ by using the one-term approximation to the temperature given by

$$\phi(r,t) = A_1 e^{-X_1^2 Fo_r} \frac{\sin(X_1 \rho')}{X_1 \rho'}$$

where $\rho' = \frac{r}{R}$. The energy contained in the sphere is then calculated as

$$E_r(t) = 4\pi R^3 \rho c(T_i - T_f) A_1 e^{-X_1^2 Fo_r} \int_0^1 \rho'^2 \frac{\sin(X_1\rho')}{X_1\rho'} d\rho'$$

$$= 3Q_{r0} A_1 e^{-X_1^2 Fo_r} \int_0^{X_1} x \frac{\sin(x)}{X_1^3} dx = 3Q_{r0} A_1 e^{-X_1^2 Fo_r} \left[\frac{\sin(X_1) - X_1\cos(X_1)}{X_1^3}\right] \tag{6.38}$$

The heat loss fraction is then seen to be given by

$$\boxed{\frac{Q_r(t)}{Q_{r0}} = 1 - 3A_1 e^{-X_1^2 Fo_r}\left[\frac{\sin(X_1) - X_1\cos(X_1)}{X_1^3}\right]} \tag{6.39}$$

Example 6.3

A long cylinder of an alloy of diameter 0.104 m emerges from a furnace with a uniform temperature across the cross section of 200 °C in to an ambient at 65°C. The ambient medium imposes a convection heat transfer coefficient of 30 W/m^2 °C at the cylinder surface. Determine the temperature at the axis of the cylinder 5 minutes after its exposure to the ambient. What is the surface temperature of the cylinder at this time? Properties of the alloy have been specified as follows:
Density: $\rho = 4500$ kg/m^3, Specific heat: c= 320.5 J/kg °C, Thermal conductivity: $k = 26$ W/m°C.

Solution : (a) **Calculation of temperature at axis:**

Step 1 The given data is written down using the already familiar notation.

Initial cylinder temperature: $T_i = 200°C$
Ambient temperature: $T_f = 65°C$
Heat transfer coefficient: $h = 30\ W/m^2\ °C$
Radius of cylinder: $R = \frac{D}{2} = \frac{0.104}{2} = 0.052\ m$
Time at which temperatures are needed: $t = 5$ min. $= 300\ s$

From the property data specified in the problem, the thermal diffusivity is calculated as

$$\alpha = \frac{k}{\rho c} = \frac{26}{4500 \times 320.5} = 1.803 \times 10^{-5}\ m^2/s$$

The Biot number is determined as

$$Bi = \frac{hR}{k} = \frac{30 \times 0.052}{26} = 0.06$$

The Fourier number is calculated as

$$Fo = \frac{\alpha t}{R^2} = \frac{1.803 \times 10^{-5} \times 300}{0.052^2} = 2$$

Step 2 From Fig. D.4 (Heisler chart), the non-dimensional axial temperature is read off as $\phi = 0.8$. The corresponding temperature is given by

$$T(0, 300) = T_f + (T_i - T_f)\phi = 65 + (200 - 65) \times 0.8 = 173°C$$

(b) **Surface temperature calculation:**

Step 3 The surface is characterized by $\frac{r}{R} = 1$. The reciprocal Biot number is given by $\frac{1}{Bi} = \frac{1}{0.06} = 16.67$. The correction factor is now read off Figure D.5 in Sect. 6.2 as 0.971. Thus we have

$$\begin{aligned} T(0.052, 300) &= T_f + (T_i - T_f)\phi \times \text{Correction factor} \\ &= 65 + (200 - 65) \times 0.8 \times 0.971 = 169.9°C \end{aligned}$$

Step 4 **Comment:** It is seen that there is a temperature difference of a mere 3.1°C between the axis and surface of the rod. This is not surprising since the Biot number of 0.06 is very small and hence the cylinder behaves very closely as a lumped system along the radial direction.

Example 6.4

A spherical cannon ball of diameter 0.104 m is initially at a uniform temperature of 300 °C. It is exposed to an ambient at 35 °C for $t > 0$. The ambient medium imposes a convection heat transfer coefficient of 120 W/m^2 °C at the surface of the cannon ball. Determine the center temperature of the cannon ball 10 minutes after its exposure to the ambient. What is the surface temperature of the cannon ball at this time? What percentage of heat originally in the sphere has been lost by the sphere? Properties of the material of the cannon ball have been specified as under: Density, $\rho = 7800$ kg/m^3, Specific heat, $c = 450$ J/kg °C, Thermal conductivity, $k = 75$ W/m °C.

Solution :

(a) **Axial temperature calculation**:

Step 1 The given data is written down using the already familiar notation.

Initial cannon ball temperature:	$T_i = 300°C$
Ambient temperature:	$T_f = 35°C$
Heat transfer coefficient:	$h = 120$ W/m^2 °C
Radius of cannon ball:	$R = \frac{D}{2} = \frac{0.104}{2} = 0.052$ m
Time at which temperatures are needed:	$t = 10$ min. $= 600\ s$

From the property data specified in the problem, the thermal diffusivity is calculated as

$$\alpha = \frac{k}{\rho c} = \frac{75}{7800 \times 450} = 2.14 \times 10^{-5}\ m^2/s$$

The Biot number is determined as

$$Bi = \frac{hR}{k} = \frac{120 \times 0.052}{75} = 0.083$$

The Fourier number is calculated as

$$Fo = \frac{\alpha t}{R^2} = \frac{2.14 \times 10^{-5} \times 600}{0.052^2} = 4.75$$

Step 2 From Figure D.7 in Sect. 6.2, the non-dimensional center temperature is read off as $\phi = 0.34$. The corresponding temperature is given by

$$T(0, 600) = T_f + (T_i - T_f)\phi = 35 + (300 - 35) \times 0.34 = 125.1°C$$

(b) **Surface temperature calculation:**

Step 3 The surface is characterized by $\frac{r}{R} = 1$. The reciprocal Biot number is given by $\frac{1}{Bi} = \frac{1}{0.083} = 12.05$. The correction factor is now read off in Figure D.8 as 0.96. Thus we have

$$\begin{aligned} T(0.052, 600) &= T_f + (T_i - T_f)\phi \times \text{Correction factor} \\ &= 35 + (300 - 35) \times 0.34 \times 0.96 = 121.5°C \end{aligned}$$

(c) **Percent heat loss from the shell:**
From Figure D.8 in Sect. 6.2, we read off the heat loss fraction as 59%. This means 41% of total heat loss that would eventually take place has taken place in the first 10 minutes.

Step 4 **Comment**: In this example also the sphere is behaving, more or less, as a lumped system.

6.2 One-Term Approximation and Heisler Charts

In all the three cases involving transient one-dimensional conduction that were treated in Sects. 6.1.1–6.1.4, a one-term approximation has been shown to be adequate. Hence it is customary to give the one-term solution by appropriate plots, known as Heisler charts,[1] for convenience in solving problems. These charts are also useful in dealing with transient conduction in more than one dimension. Heisler charts are available in Appendix D.

6.3 Transient Conduction in More Than One Dimension

6.3.1 Introduction

Till now we have considered transient conduction in one dimension, in a slab, a cylinder, or a sphere subject to convection at the exposed boundary. In engineering applications, transient conduction may take place in more than one space dimension. For example, in a long bar of rectangular cross section, transient temperature field is a function of both x and y. In a block of material in the form of a brick, the transient temperature field is a function of x, y, z. In the case of a cylinder of finite length,

[1]M.P. Heisler, Temperature charts for induction heating and constant temperature, Trans. ASME, Vol. 69, pp. 227–236, 1947.

the transient temperature field is a function of r and z. These problems are also amenable to solution by the use of Heisler charts. The present effort is to develop the methodology for treating such cases.

6.3.2 Transient Conduction in an Infinitely Long Rectangular Bar

Consider the case of an infinitely long bar of rectangular cross section of width $2L_x$ and height $2L_y$ as shown in Fig. 6.6. The bar is surrounded by a medium at T_f that removes heat by convection subject to a convective heat transfer coefficient h. Initially, the temperature distribution in the bar is specified as a given function of x and y (as usual, the bar is isothermal at $T(x, y, 0) = T_i$). The problem involves a temperature field which is a function of three variables −2 space variables x and y and one-time variable t. Choose the origin to lie at the intersection of the two mid-planes as shown in Fig. 6.6. The governing equation, initial and boundary conditions are Governing equation:

$$\frac{1}{\alpha}\frac{\partial \phi}{\partial t} = \frac{\partial^2 \phi}{\partial x^2} + \frac{\partial^2 \phi}{\partial y^2} \tag{6.40}$$

where $\phi(x, y, t) = \frac{T(x,y,t)-T_f}{T_i-T_f}$.

Initial condition:

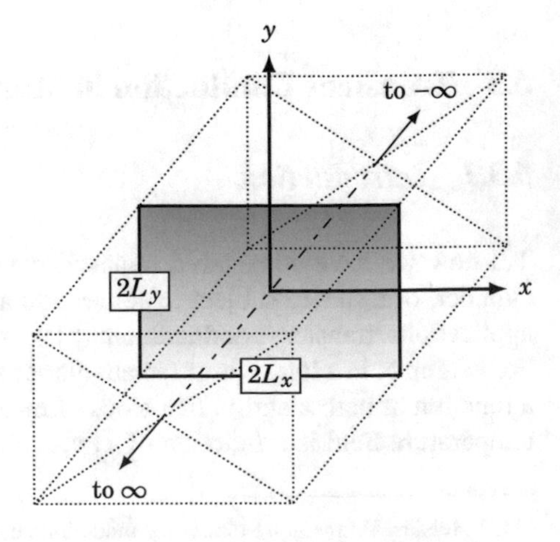

Fig. 6.6 Transient conduction in an infinitely long rectangular bar

$$\boxed{t = 0; \quad \phi(x, u, 0) = 1} \tag{6.41}$$

Boundary conditions for $t > 0$:

$$x = \mp L_x, -L_y \leq y \leq L_y; \quad \frac{\partial \phi}{\partial x} = \pm \frac{h}{k}\phi \tag{6.42}$$

$$y = \mp L_y, -L_x \leq y \leq L_x; \quad \frac{\partial \phi}{\partial y} = \pm \frac{h}{k}\phi \tag{6.43}$$

The $\pm$ sign indicates that the positive sign holds at the first boundary and the negative sign holds at the second boundary. There are situations in which the initial temperature distribution $F(x, y)$ is expressible in the form $F_1(x) \times F_2(y)$. An example is one wherein the entire cross section is initially at a constant temperature. Then both $F_1(x)$ and $F_2(y)$ are constants. We shall examine, under such a situation, whether $\phi(x, y, t)$ is expressible as $\phi_x(x, t) \times \phi_y(y, t)$. On substitution into Eq. 6.40, we should then have,

$$\frac{1}{\alpha}\left[\phi_y \frac{\partial \phi_x}{\partial t} + \phi_x \frac{\partial \phi_y}{\partial t}\right] = \left[\phi_y \frac{\partial^2 \phi_x}{\partial x^2} + \phi_x \frac{\partial^2 \phi_y}{\partial y^2}\right] \tag{6.44}$$

On re-arrangement, we should have

$$\phi_y \left[\frac{1}{\alpha}\frac{\partial \phi_x}{\partial t} - \frac{\partial^2 \phi_x}{\partial x^2}\right] = \phi_x \left[\frac{1}{\alpha}\frac{\partial \phi_y}{\partial t} - \frac{\partial^2 \phi_y}{\partial y^2}\right] \tag{6.45}$$

This equation is equivalent to two equations given by

$$\text{(a)} \quad \boxed{\frac{1}{\alpha}\frac{\partial \phi_x}{\partial t} = \frac{\partial^2 \phi_x}{\partial x^2}} \quad \text{(b)} \quad \boxed{\frac{1}{\alpha}\frac{\partial \phi_y}{\partial t} = \frac{\partial^2 \phi_y}{\partial y^2}} \tag{6.46}$$

Now look at the initial and boundary conditions. Initial condition 6.41 requires that

$$\boxed{\phi_x(x, 0) = F_1(x) = 1; \quad \phi_y(y, 0) = F_2(y) = 1} \tag{6.47}$$

The boundary conditions require the following to hold:

$$\text{(a)} \quad \boxed{x = \mp L_x, \frac{\partial \phi_x}{\partial x} = \pm \frac{h}{k}\phi_x} \quad \text{(b)} \quad \boxed{y = \mp L_y, \frac{\partial \phi_y}{\partial y} = \pm \frac{h}{k}\phi_y} \tag{6.48}$$

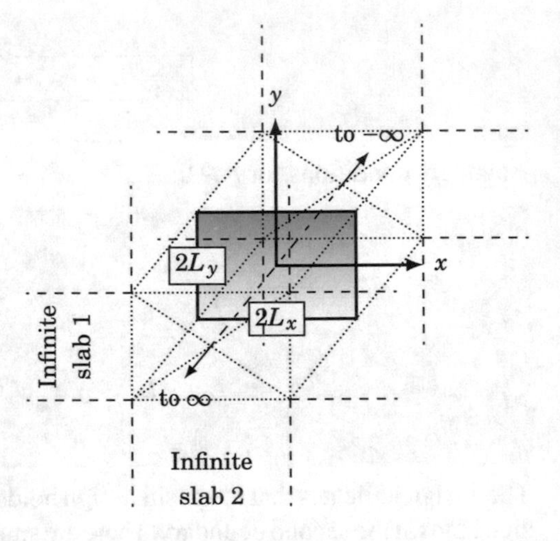

Fig. 6.7 Connection between the rectangular bar transient and the infinite slab transient

Thus the original two-dimensional transient problem has been reduced to two one-dimensional transients given by Eqs. 6.46(a) and (b).

Interpretation
Each of the one-dimensional transients given by Eqs. 6.46(a) and (b), along with the respective initial and boundary conditions (6.47, 6.48(a) and (b)) are analogous with the basic case considered in Sect. 6.1.2. Thus, the situation may be graphically shown by the equivalence, given in Fig. 6.7. A bar of infinite length and rectangular section, is obtained, by the intersection of two infinitely large perpendicular slabs, as shown in the figure.

Consider now the special case in which the temperature in the bar initially is uniform at T_i. We introduce non-dimensional temperature functions given by

$$\text{(a)}\ \phi_x = \frac{T_1(x,t) - T_f}{T_i - T_f} \quad \text{(b)}\ \phi_y = \frac{T_2(y,t) - T_f}{T_i - T_f} \tag{6.49}$$

Thus both ϕ_x and ϕ_y are unity at $t = 0$, i.e., $F_1(x) = 1$ and $F_2(y) = 1$. The two solutions may be obtained by the method of separation of variables as in Sect. 6.1.2. Each of these solutions may be written down, using the one-term approximation as

$$\text{(a)}\ \boxed{\phi_x = A_{11} e^{-X_{11}^2 Fo_x} \cos(X_{11}\xi)} \quad \text{(b)}\ \boxed{\phi_y = A_{21} e^{-X_{21}^2 Fo_y} \cos(X_{21}\eta)} \tag{6.50}$$

where $\xi = \frac{x}{L_x}$, $\eta = \frac{y}{L_y}$, $Fo_x = \frac{\alpha t}{L_x^2}$ and $Fo_y = \frac{\alpha t}{L_y^2}$. The Fourier coefficients as also the eigenvalues are double subscripted and have the following meanings:

$$X_{11} = \lambda_{11} L_x, \quad X_{21} = \lambda_{21} L_y$$

where λ_{11} is the first root of the equation $\cot(\lambda_{11} L_x) = \frac{\lambda_{11} L_x}{Bi_x}$ and λ_{21} is the first root of the equation $\cot(\lambda_{21} L_y) = \frac{\lambda_{21} L_y}{Bi_y}$. Two Biot numbers make their appearance given by $Bi_x = \frac{hL_x}{k}$ and $Bi_y = \frac{hL_y}{k}$. With these, the Fourier coefficients may be evaluated (see developments leading to Eq. 6.8 and 6.9) to yield

$$\boxed{A_{11} = \frac{\sin(X_{11})}{\frac{1}{2} + \frac{\sin(2X_{11})}{4}}, \quad A_{21} = \frac{\sin(X_{21})}{\frac{1}{2} + \frac{\sin(2X_{21})}{4}}} \tag{6.51}$$

The one-term approximation is valid as long as $Fo_x > 0.2$ and $Fo_y > 0.2$. Under these conditions, the charts presented in Figures D.1–D.3 may be used for obtaining the solution. The Fourier and Biot numbers need to be interpreted as the respective Fourier and Biot numbers, for the two directions.

Heat Loss Fraction

In the case of the rectangular bar, the initial energy within it is given by

$$Q_{xy0} = 4\rho c(T_i - T_f)L_x L_y \tag{6.52}$$

where $4L_x L_y$ is the volume per unit length of the bar. At any time $t > 0$, the energy contained within the bar is given by

$$E_{xy}(t) = \rho c(T_i - T_f) \int_{-L_x}^{L_x} \int_{-L_y}^{L_y} \phi_x \phi_y dx dy \tag{6.53}$$

We may substitute the expressions for the ϕ_x and ϕ_y from 6.49 and perform the indicated integration to get

$$E_{xy}(t) = 4\rho c(T_i - T_f)L_x L_y A_{11} A_{21} e^{(-X_{11}^2 Fo_x - X_{21}^2 Fo_y)} \cdot \frac{\sin X_{11}}{X_{11}} \cdot \frac{\sin X_{21}}{X_{21}} \tag{6.54}$$

or, alternately as

$$\frac{E_{xy}(t)}{Q_{xy0}} = \frac{E_x(t)}{Q_{x0}} \frac{E_y(t)}{Q_{y0}} \tag{6.55}$$

where $\frac{E_x(t)}{Q_{x0}}$ and $\frac{E_y(t)}{Q_{y0}}$ are given, respectively, based on Eq. 6.13 by

$$\frac{E_x(t)}{Q_{x0}} = 1 - \frac{Q_x(t)}{Q_{x0}}; \quad \frac{E_y(t)}{Q_{y0}} = 1 - \frac{Q_y(t)}{Q_{y0}} \tag{6.56}$$

Note that both $\frac{Q_x(t)}{Q_{x0}}$ and $\frac{Q_y(t)}{Q_{y0}}$ are read off Figure D.3 but by using the respective Fourier and Biot numbers, for the two directions. With these the heat loss ratio for the two-dimensional case becomes

$$\frac{Q_{xy}(t)}{Q_{xy0}} = 1 - \frac{E_x(t)}{Q_{x0}}\frac{E_y(t)}{Q_{y0}} = 1 - \left(1 - \frac{Q_x(t)}{Q_{x0}}\right)\left(1 - \frac{Q_y(t)}{Q_{y0}}\right) \quad (6.57)$$

Example 6.5

A long rectangular bar of cross section $200 \times 100\ mm$ is initially at a temperature of $30\,°C$. It is convectively heated by an ambient at $250\,°C$ subject to a constant heat transfer coefficient of $67\ W/m^{2\circ}C$. Properties of the bar material are:
Density $\rho = 3970\ kg/m^3$, specific heat $c = 765\ J/kg°C$, and thermal conductivity $k = 33.4\ W/m°C$.
(a) What is the temperature at $x = 0,\ y = 0$ after 10 minutes from the beginning of the heating process? (b) What is the amount of heat transfer to the bar in the same time? (c) What are the temperatures at the center of adjacent faces? (d) Determine the temperature along the edges of the rectangle.

Solution :

Step 1 The data is written down with the notation introduced in the text.

Initial temperature of the bar:	$T_i = 30°C$
Ambient temperature:	$T_f = 250°C$
Half width of bar:	$L_x = 0.1\ m$
Half height of bar:	$L_y = 0.05\ m$
Time at which results are needed:	$t = 10\ min = 600\ s$

Step 2 From the material property data, the thermal diffusivity is calculated as

$$\alpha = \frac{33.4}{3970 \times 765} = 1.1 \times 10^{-5}\ m^2/s$$

Biot and Fourier numbers that characterize the solution are now determined.

$$Bi_x = \frac{67 \times 0.1}{33.4} = 0.2; \quad Bi_y = \frac{67 \times 0.05}{33.4} = 0.1$$

$$Fo_x = \frac{1.1 \times 10^{-5} \times 600}{0.1^2} = 0.66; \quad Fo_y = \frac{1.1 \times 10^{-5} \times 600}{0.05^2} = 2.64$$

Step 3 From Heisler chart D.1, we read off the following:

$$\phi_x(0, 0.66) = 0.91; \quad \phi_y(0, 2.64) = 0.78$$

Step 4 (a) **Temperature at** 0,0 **at** t=600 s: The temperature at (0, 0, 600) is then obtained as

$$T(0,0,600) = T_f + (T_i - T_f)\phi_x(0,0.66)\phi_y(0,2.64)$$
$$= 250 + (30 - 250) \times 0.91 \times 0.78 = 93.8°C$$

Step 5 (b) **Heat transferred to the bar in 600 s**: From Heisler chart D.3 we read off the following:

$$\frac{Q_x(600)}{Q_{x0}} = 0.12; \quad \frac{Q_y(600)}{Q_{y0}} = 0.21$$

Using Eq. 6.57, we then have

$$\text{Heat loss ratio} = 1 - (1 - 0.12)(1 - 0.21) = 0.305$$

The total energy contained within the bar per unit length, at $t = 0$ is (Eq. 6.52)

$$Q_{xy0} = 4 \times 0.1 \times 0.05 \times 3970 \times 765(30 - 250) = -1.336 \times 10^7 \ J/m$$

The negative sign above indicates that heat is transferred to the bar. The heat transferred to the bar in 600s may now be calculated as

$$Q_{xy}(600) = -0.305 \times 1.336 \times 10^7 = -4.073 \times 10^6 \ J/m \approx -4 \ MJ/m$$

Step 6 (c) **Temperatures at the centers of adjacent faces**: The center on the short face is located at $\pm 0.1, 0$. These points correspond to $\frac{x}{L_x} = \pm 1$ and $\frac{y}{L_y} = 0$. The reciprocal Biot number is

$$\frac{1}{Bi_x} = \frac{1}{0.2} = 5$$

Correction factor is needed for ϕ_x alone and is read off Fig. D.2 as x-Correction factor $c_x = 0.91$. The temperature $T(\pm 0.1, 0, 600)$ is then given by

$$T(\pm 0.1, 0, 600) = T_f + (T_i - T_f)\phi_x(0, 0.66)\phi_y(0, 2.64) \times c_x$$
$$= 250 + (30 - 250) \times 0.91 \times 0.78 \times 0.91 = 107.9\,°C$$

The center on the long face is located at $0, \pm 0.05$. These points correspond to $\frac{x}{L_x} = 0$ and $\frac{y}{L_y} = \pm 1$. The reciprocal Biot number is

$$\frac{1}{Bi_y} = \frac{1}{0.1} = 10$$

Correction factor is needed for ϕ_y alone and is read off in Fig. D.2 as y-Correction factor $c_y = 0.99$. The temperature $T(0, \pm 0.05, 600)$ is then given by

$$\begin{aligned} T(0, \pm 0.05, 600) &= T_f + (T_i - T_f)\phi_x(0, 0.66)\phi_y(0, 2.64) \times c_y \\ &= 250 + (30 - 250) \times 0.91 \times 0.78 \times 0.99 = 95.4\,°C \end{aligned}$$

Step 7 (d) **Rectangle edge temperature**: All the four edges are at the same temperature. Correction factors are required for both directions and are the values determined above. The edge temperature $T(\pm 0.1, \pm 0.05, 600)$ is then given by

$$\begin{aligned} T(\pm 0.1, \pm 0.05, 600) &= T_f + (T_i - T_f)\phi_x(0, 0.66)\phi_y(0, 2.64) \times c_x \times c_y \\ &= 250 + (30 - 250) \times 0.91 \times 0.78 \times 0.91 \times 0.99 = 109.3°C \end{aligned}$$

6.3.3 *Transient Heat Conduction in a Rectangular Block in the form of a brick*

Consider a rectangular block of a material as shown in Fig. 6.8. In the light of the discussion above with reference to Fig. 6.7, it is clear that the block (a rectangular parallelepiped) may be visualized as being obtained by the intersection at right angles to one another of three infinitely large slabs of thickness $2L_x$, $2L_y$, and $2L_z$ parallel to the three coordinate directions. When the boundary conditions are appropriate as in the previous case, this case is also amenable to a product-type representation of the solution in terms of three one-dimensional slab solutions. When the one-term approximation is valid, the results given in Figs. D.1–D.3 may be used for this case also. Three Biot numbers (Bi_x, Bi_y, and Bi_z) and three Fourier numbers (Fo_x, Fo_y,

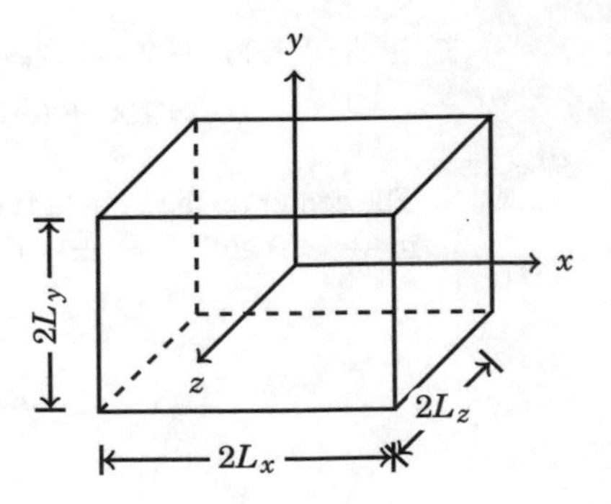

Fig. 6.8 Transient heat conduction in a rectangular block in the form of a brick

and Fo_z) will be involved in this case, since there are three characteristic length dimensions involved in describing the geometry. The analysis follows the one that was made for the rectangular bar problem. We summarize the results by indicating the additions and changes that are to be incorporated in the results given in Sect. 6.3.2. To the list of variables given in Eq. 6.49 we add the following.

$$(c) \quad \phi_z = \frac{T_3(y,t) - T_f}{T_i - T_f} \tag{6.58}$$

Eq. 6.50 is augmented by

$$\boxed{\phi_z = A_{31} e^{-X_{31}^2 Fo_z} \cos(X_{31}\zeta)} \tag{6.59}$$

where $\zeta = \frac{z}{L_z}$, $Fo_z = \frac{\alpha t}{L_z^2}$. Equation 6.51 is augmented by

$$X_{31} = \lambda_{31} L_z \tag{6.60}$$

where λ_{31} is the first root of the equation $\cot(\lambda_{31} L_z) = \frac{\lambda_{31} L_z}{Bi_z}$ with $Bi_z = \frac{hL_z}{k}$. Add to Eq. 6.51

$$(c) \quad \boxed{A_{31} = \frac{\sin(X_{31})}{\frac{1}{2} + \frac{\sin(2X_{31})}{4}}} \tag{6.61}$$

In the case of the rectangular block the initial energy within it is given by

$$Q_{xyz0} = 8\rho c(T_i - T_f) L_x L_y L_z \tag{6.62}$$

where $8L_x L_y L_z$ is the volume of the block. At any time $t > 0$, the energy contained within the block is given by

$$E_{xyz}(t) = \rho c(T_i - T_f) \int_{-L_x}^{L_x} \int_{-L_y}^{L_y} \int_{-L_z}^{L_z} \phi_x \phi_y \phi_z dx dy dz \tag{6.63}$$

We may substitute the expressions for the ϕ_x, ϕ_y from 6.49 and ϕ_z from 6.59, perform the indicated integration to get

$$E_{xyz}(t) = c(T_i - T_f) L_x L_y L_z A_{11} A_{21} A_{31} e^{(-X_{11}^2 Fo_x - X_{21}^2 Fo_y - X_{31}^2 Fo_z)} \times \frac{\sin X_{11}}{X_{11}} \cdot \frac{\sin X_{21}}{X_{21}} \cdot \frac{\sin X_{31}}{X_{31}}$$

or, alternately as

$$\frac{E_{xyz}(t)}{Q_{xyz0}} = \frac{E_x(t)}{Q_{x0}}\frac{E_y(t)}{Q_{y0}}\frac{E_z(t)}{Q_{z0}} \tag{6.64}$$

where $\frac{E_z(t)}{Q_{z0}}$ is given, based on Eq. 6.13 by

$$\frac{E_z(t)}{Q_{z0}} = 1 - \frac{Q_z(t)}{Q_{z0}} \tag{6.65}$$

Note that $\frac{Q_x(t)}{Q_{x0}}$, $\frac{Q_y(t)}{Q_{y0}}$ and $\frac{Q_z(t)}{Q_{z0}}$ are read off using Figure D.3 but by using the respective Fourier and Biot numbers, for the three directions. With these the heat loss ratio for the three-dimensional case becomes

$$\frac{Q_{xyz}(t)}{Q_{xyz0}} = 1 - \frac{E_x(t)}{Q_{x0}}\frac{E_y(t)}{Q_{y0}}\frac{E_z(t)}{Q_{z0}} = 1 - \left(1 - \frac{Q_x(t)}{Q_{x0}}\right)\left(1 - \frac{Q_y(t)}{Q_{y0}}\right)\left(1 - \frac{Q_z(t)}{Q_{z0}}\right) \tag{6.66}$$

Example 6.6

A rectangular block of steel $0.1 \times 0.075 \times 0.05m$ is initially at room temperature of $30°C$. It is hung in a preheated oven at $220°C$ and thus starts heating up subject to a moderate level of convection with a heat transfer coefficient of 45 $W/m^2\,°C$. Determine the center temperature of the block after 600 s. Also, determine the temperature at the center of all the faces of the block and at all the corners at this time. The thermal diffusivity of steel is 1.15×10^{-5} m^2/s and the thermal conductivity is 43 $W/m°C$. Determine also the amount of heat added to the block in $t = 600$ s.

Solution :

Step 1 This example requires calculations similar to those in Example 6.5. The required parameters are calculated or read off in Heisler charts and arranged in the following table:

Direction	Length	Biot number	Fourier number	ϕ	c	Heat loss fraction
x	0.05	0.052	2.64	0.92	0.96	0.12
y	0.0375	0.039	4.69	0.83	0.97	0.17
z	0.025	0.026	10.56	0.74	0.98	0.24

Step 2 The temperature at the center of the block is given by

$$T(0, 0, 0, 600) = 220 + (30 - 220) \times 0.92 \times 0.83 \times 0.74 = 112.6°C$$

Step 3 Centers of faces require the use of correction factors. The faces that are parallel to the x axis require the correction factor c_x and so on. The correction factors needed are shown in the last but first column of the table. We then have

$$T(\pm 0.05, 0, 0, 600) = 220 + (30 - 220) \times 0.92 \times 0.83 \times 0.74 \times 0.96 = 116.9\,°C$$
$$T(0, \pm 0.0375, 0, 600) = 220 + (30 - 220) \times 0.92 \times 0.83 \times 0.74 \times 0.97 = 115.9\,°C$$
$$T(0, 0, \pm 0.025, 600) = 220 + (30 - 220) \times 0.92 \times 0.83 \times 0.74 \times 0.98 = 114.8\,°C$$

Step 4 Corner temperatures are given by

$$T(\pm 0.05, \pm 0.0375, \pm 0.025, 600) = 220 + (30 - 220) \times 0.92 \times 0.83 \times 0.74 \times 0.96 \times 0.97 \times 0.98 = 122\,°C$$

Step 5 We use Eq. 6.66 and the respective directional heat loss fractions from the last column of the table to get

$$\frac{Q_{xyz}(600)}{Q_{xyz0}} = 1 - (1 - 0.12)(1 - 0.17)(1 - 0.24) = 0.445$$

Since α and k are given, the density-specific product ρc is given by

$$\rho c = \frac{k}{\alpha} = \frac{43}{1.1 \times 10^{-5}} = 3.909 \times 10^6\ J/m^3\,°C$$

With this we have

$$Q_{xyz0} = 8 \times 0.05 \times 0.0375 \times 0.025 \times 3.909 \times 10^6 \times (30 - 220) = -278520\ J$$

Thus the heat loss fraction in absolute terms is

$$Q_{xyz}(600) = 0.445 \times (-278520) = -0.124\ MJ$$

6.3.4 *Transient Heat Conduction in a Circular Cylinder of Finite Length*

An example of considerable interest is the two-dimensional transient conduction in a circular cylinder of finite length. The problem represents transient heat conduction in cylindrical coordinates. The initial temperature throughout the cylinder is assumed to be uniform at T_i. All the surfaces of the cylinder are subject to convection via a heat transfer coefficient h to an ambient at T_f. As shown in Fig. 6.9, the solution may be obtained by considering an intersection between an infinite plane of thickness equal to the length of the cylinder and an infinitely long cylinder of radius equal to the radius of the cylinder. The solution is sought in the form of a product solution

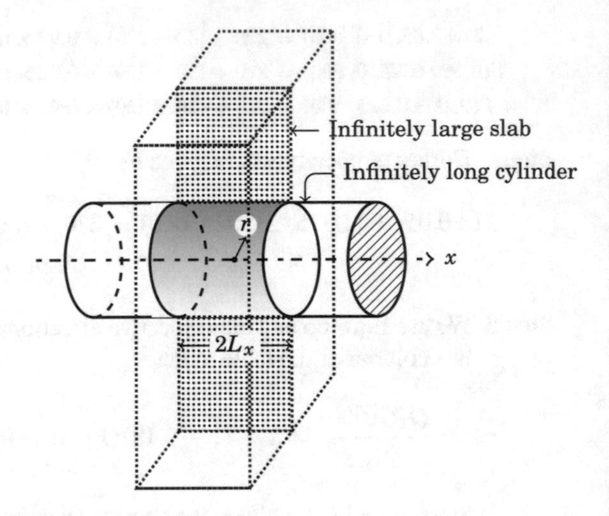

Fig. 6.9 Transient conduction in a short cylinder

involving these two cases. Placing the origin at the mid-plane of the infinite slab and at the axis of the cylinder as shown, the solution may be represented by the product solution $\phi(x, r, t) = \phi_x(x, t) \times \phi_r(r, t)$ where (i) $\phi_x(x, t)$ represents one-dimensional transient in an infinite slab and (ii) $\phi_r(r, t)$ represents one-dimensional transient in an infinite cylinder. Each of these solutions has been obtained earlier in Sects. 6.1.2 and 6.1.3 by the separation of variables method. The slab part of the solution (i) involves exponential time terms and circular functions of x. The slab solution is approximated by the one-term approximation and is available in Figures D.1–D.3. The radial part of the solution (ii) involves exponential time terms and Bessel functions of r. Infinite cylinder solution is also available in the form of charts given in Figures D.4–D.6 . The solutions involve two Biot numbers (Bi_x and Bi_r) and two Fourier numbers (Fo_x and Fo_r).

The heat loss fraction is obtained as follows. The total energy contained in the cylinder at zero time is given by

$$Q_{xr0} = 2\pi R^2 L_x \rho c(T_i - T_f) \tag{6.67}$$

where the cylinder radius and length are respectively equal to R and $2L_x$. The energy contained within the cylinder is obtained by using the slab and cylinder solutions as

$$E_{xr}(t) = \rho c(T_i - T_f) 2\pi \int_0^R \int_{-L_x}^{L_x} \phi_r(r, t)\phi_x(x, t) r\,dr\,dx \tag{6.68}$$

Performing the indicated integration, borrowing results from the infinite slab and long cylinder transients, the above may be recast in the form of heat loss ratio as

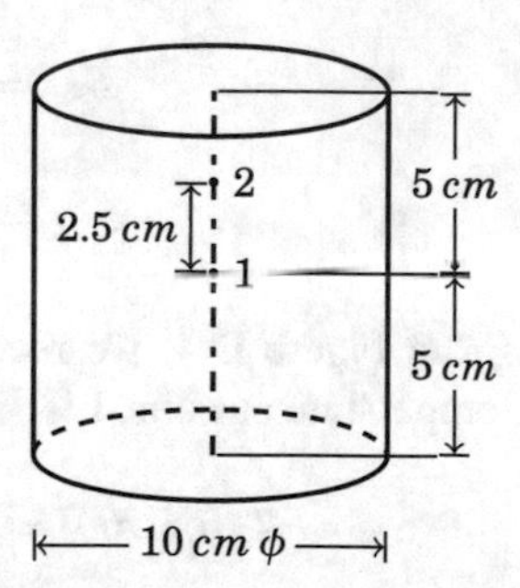

Fig. 6.10 Figure for Example 6.7

$$\frac{Q_{xr}(t)}{Q_{xr0}} = 1 - \frac{E_x(t)}{Q_{x0}}\frac{E_r(t)}{Q_{r0}} = 1 - \left(1 - \frac{Q_x(t)}{Q_{x0}}\right)\left(1 - \frac{Q_r(t)}{Q_{r0}}\right) \tag{6.69}$$

Example 6.7

A low-carbon steel cylinder of diameter 10 cm and 10 cm tall is initially at 200 °C throughout. It is plunged into a quenching oil bath at 30 °C which provides convection cooling subject to a heat transfer coefficient of $h = 250\ W/m^{2\circ}C$. Temperatures at points 1 and 2 (Fig. 6.10) are desired at $t = 60\ s$ and $t = 300\ s$ respectively. Also required is the amount of heat transfer that has taken place in 300 s. Properties of low-carbon steel are $k = 54\ W/m°C$ and $\alpha = 1.474 \times 10^{-5}\ m^2/s$.

Solution :

Case (1): Temperature at Point 1 at $t = 60\ s$

Slab part of solution:

$$L_y = 5\ cm = 0.05\ m;\ \ y = 0 \text{ or } \eta = 0$$

The Biot and Fourier numbers are calculated as

$$Bi_y = \frac{hL_y}{k} = \frac{250 \times 0.05}{54} = 0.232$$

$$Fo_y = \frac{\alpha t}{L_y^2} = \frac{1.474 \times 10^{-5} \times 60}{0.05^2} = 0.354$$

From Figure D.1, we read off the temperature as $\phi_y(0, 60) = 0.96$

Cylinder part of solution:

The radius of the cylinder is $R = 5\ cm = 0.05\ m$ and the temperature is desired on the axis with $\rho' = 0$. The Biot and Fourier numbers are calculated as

$$Bi_r = \frac{hR}{k} = \frac{250 \times 0.05}{54} = 0.232$$

$$Fo_r = \frac{\alpha t}{R^2} = \frac{1.474 \times 10^{-5} \times 60}{0.05^2} = 0.354$$

From Figure D.4, we read off the temperature as $\phi_r(0, 60) = 0.92$. The desired temperature at Point 1 is

$$T(0, 0, 60) = T_f + (T_i - T_f)\phi_y(0, 60)\phi_r(0, 60)$$
$$= 30 + (200 - 30) \times 0.96 \times 0.92 = 180.1^\circ C$$

Case (2): Temperature at Point 2 at $t = 300\ s$:
It is seen that point 2 is off mid-plane with respect to the slab solution and on axis with respect to the cylinder solution. Hence we apply a correction factor for the slab part alone using Figure D.2. Since t is different from Case 1, we redo the calculations as under.
Slab part of solution:
The Biot number $Bi_x = 0.232$ does not change. However the Fourier number is calculated as

$$Fo_y = \frac{\alpha t}{L_y^2} = \frac{1.474 \times 10^{-5} \times 300}{0.05^2} = 1.77$$

From Figure D.1, we read off the temperature as $\phi_y(0, 300) = 0.70$. Position indicated is such that $y = 2.5\ cm = 0.025\ m$. Hence the correction factor to the slab solution is required at $\frac{y}{L_y} = \frac{0.025}{0.05} = 0.5$. The reciprocal Biot number is $\frac{1}{Bi} = \frac{1}{0.232} = 7.63$. The position correction factor is read from Figure D.2 as $c_y = 0.94$. Hence

$$\phi_y(0.025, 300) = c_y\phi(0, 300) = 0.70 \times 0.94 = 0.658$$

Cylinder part of solution:
As earlier, the Biot number based on cylinder radius is $Bi_r = 0.232$. The Fourier number is given by

$$Fo_r = \frac{\alpha t}{R^2} = \frac{1.474 \times 10^{-5} \times 300}{0.05^2} = 1.77$$

From Figure D.4 we read off the temperature as $\phi_r(0, 300) = 0.48$. The desired temperature at Point 2 is

$$T(0, 0, 300) = T_f + (T_i - T_f)\phi_y(0.025, 300)\phi_r(0, 300)$$
$$= 30 + (200 - 30) \times 0.658 \times 0.48 = 83.7^\circ C$$

Heat transfer to the cylinder in 300 s:
From the given property data, the density-specific heat product for high-carbon steel

may be calculated as

$$\rho c = \frac{k}{\alpha} = \frac{54}{1.474 \times 10^{-5}} = 3663500 \; J/m^3\,°C$$

The quantity Q_{yr0} is calculated as

$$\begin{aligned} Q_{yr0} &= 2\rho c\pi R^2 L_y (T_i - T_f) \\ &= 2 \times 3663500 \times \pi \times 0.05^2 \times 0.05(200 - 30) = 489142 \; J \end{aligned}$$

From Fig. D.2 we have $\frac{Q_y(300)}{Q_{y0}} = 0.32$. From Fig. D.6 we have $\frac{Q_r(300)}{Q_{r0}} = 0.56$. using Eq. 6.69 the heat loss ratio for the cylinder is obtained as

$$\frac{Q_{yr}(300)}{Q_{yr0}} = 1 - \left(1 - 0.32\right)\left(1 - 0.56\right) = 0.7$$

We then calculate the total heat transfer in 300 s as

$$Q_{yr}(300) = 489142 \times 0.7 = 342399 \; J$$

Concluding Remarks

This chapter has considered transient conduction in one, two and three dimensions and in all three corodinate systems viz. Cartesian, cylindrical and spherical coordinates. Based on separation of variables technique a one term approximation valid for $Fo > 0.2$ has been evolved. Heisler charts, based on one term approximation, have been presented for all one dimensional transients. It is shown that these charts are also useful for transients in two and three dimensions.

6.4 Exercises

Ex 6.1 A large slab of steel 15 mm is initially at a uniform temperature of 350°C. It is exposed to a neutral gas stream at 120°C for $t > 0$. The gas stream imposes a convection heat transfer coefficient of 40 $W/m^2\,°C$ on both surfaces of the slab. Determine the time at which the mid-plane temperature is 150°C. What is the surface temperature at this time? Do you think it is reasonable to treat the slab as a lumped system?

Ex 6.2 A very large block of metal of thickness 0.2 m is initially at a uniform temperature of 400°C. It is placed on a bed of mineral insulation and is exposed to ambient air at 45°C on the other side. The thermal properties of the metal are $k = 15$ W/m°C and $\alpha = 10^{-6}\ m^2/s$. Calculate the temperature difference across its thickness at intervals of 5 min and make a plot of the same. Make use of the appropriate chart given in the text. Mineral insulation is expected to prevent heat loss from the metal surface in contact with it. Assume a suitable heat transfer coefficient for heat loss from the surface to air.

Ex 6.3 A long cylinder of radius 200 mm of a material having a thermal conductivity of $k = 170\ W/m°C$ and thermal diffusivity of $\alpha = 9.05 \times 10^{-7}\ m^2/s$ is initially at a uniform temperature of 650°C. For heat treatment purposes the cylinder is quenched in a medium at 75°C with a heat transfer coefficient of 1700 $W/m^2\,°C$. It is desired to prolong the process till the temperature at a depth of 20 mm from its surface reaches a temperature of 250°C. What is the time at which the process should terminate? What is the temperature along the axis of the cylinder at this time?

Ex 6.4 A short brass cylinder of $L = D = 0.05\ m$ is initially isothermal at a temperature of 100°C. For $t > 0$, the surface of the cylinder is subject to convection to an environment at 20°C with a heat transfer coefficient of 67 $W/m^2\,°C$. Determine the temperatures at Points 1 and 2 indicated in Fig. 6.11 after 2 and 20 minutes from the start.

Ex 6.5 A short cylinder has a diameter of 0.1 m and a length of 0.2 m. The material properties are: $k = 18\ W/m°C$ and $\alpha = 9 \times 10^{-6}\ m^2/s$. The cylinder is initially at a uniform temperature of 250°C. It is exposed to an ambient fluid at 60°C that cools the cylinder by convection with a heat transfer coefficient of 67 $W/m^2\,°C$. Compare the center temperature on any one of the flat sides with the mid-plane temperature on the surface of the curved portion of the cylinder after 5 min from the start of the cooling process. Comment based on your observation.

Ex 6.6 Temperature difference between the center of a sphere and its surface is measured using a differential thermocouple. The smallest temperature difference that may be measured in this arrangement is 0.5°C. Sphere of radius 0.1 m is made of a material of thermal diffusivity of $9.07 \times 10^{-6}\ m^2/s$ and thermal conductivity of 6 $W/m°C$. Initially, the sphere is

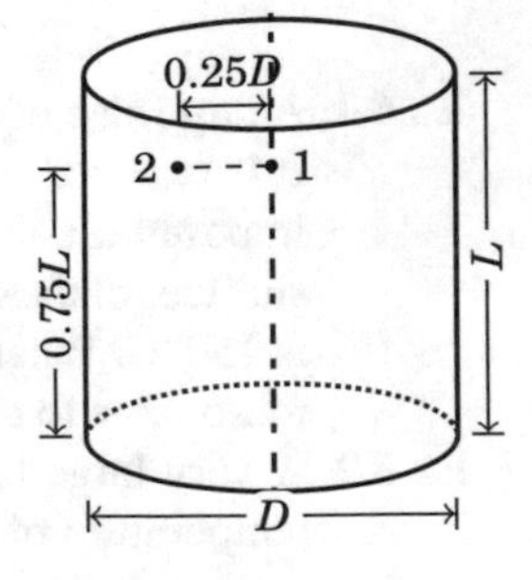

Fig. 6.11 Transient in a short cylinder: Exercise 6.4

heated to a uniform temperature of $150°C$. It is then exposed to convection cooling to an ambient air stream at $30°C$ with a heat transfer coefficient of $36\ W/m^2\,°C$. After some time the differential thermocouple reads $2.5°C$. What is the time at which you would expect this reading? What would be the temperature at the center of the sphere at this time?

Ex 6.7 A sensor used in unsteady measurements may be idealized as a sphere of radius 6 mm with the following thermal properties: $k = 45\ W/m°C$ and $\alpha = 10^{-5}\ m^2/s$. Under what condition is it possible to consider the sphere as a lumped system if lumping is justified when the temperature difference between its center and the surface is not more than $0.5\,°C$? The sensor initial temperature and the fluid temperature may be taken, respectively, as $95\,°C$ and $15\,°C$. Assume a suitable value for the heat transfer coefficient typical of natural convection to a liquid such as water.

Ex 6.8 A spherical steel ball of diameter equal to $100mm$ initially at a uniform temperature of $250\,°C$ is dropped into a vessel full of water at $30\,°C$. It is observed that water boils at the surface of the sphere as long as the surface temperature of the sphere is more than $108\,°C$ and during the boiling process the heat transfer coefficient remains constant at $5000\ W/m^2\,°C$. Determine the time at which boiling at the surface of the sphere stops.

Ex 6.9 An experiment is performed in which the temperature at the center of a sphere is measured as a function of time in order to estimate the heat transfer coefficient. The following data has been gathered:

Initial uniform temperature of the sphere:	$T_i = 250°C$
Diameter of sphere:	$D = 50$ mm
Thermal conductivity of material of sphere:	$k =$ 18.5 W/m°C
Thermal diffusivity of material of sphere:	$\alpha = 9.7 \times 10^{-7}\ ^2/s$
Fluid temperature:	$T_f = 55$°C
Time of measurement of sphere center temperature:	$t = 150$ s
Center temperature of sphere at this time:	$T_0 = 121.5$° C

Estimate the heat transfer coefficient from the above data.

Ex 6.10 A solid sphere, a solid cylinder of diameter equal to height and a solid cube all have the same volume of 25 cm^3 and are all made of brass. All these are heated to a uniform temperature of 200 °C in an oven and are then exposed to a convection environment at 35 °C with a heat transfer coefficient of 25 W/m^2 °C for $t > 0$. What is the temperature at $t = 5$ min (a) at the center of the sphere, (b) a point on the axis positioned at mid-height of the cylinder and (c) at the meeting point of all the diagonals of the cube? Comment on the result. How much heat would have crossed the boundary in each case in this time?

Chapter 7
Numerical Solution of Conduction Problems

Conduction Problems amenable to analytical solution have been considered in Chaps. 1–6. These have included both steady and transient conduction in one or more dimensions and in three different coordinate systems. In the present chapter, we discuss commonly employed numerical methods for the solution of problems not amenable to exact analysis. Emphasis is on the finite difference method. However, it is possible to use other methods such as the finite volume and finite element methods to solve conduction problems. The reader may refer to advanced texts for these methods.

7.1 Introduction

In the previous chapters, attention has been given to the analytical solution of heat conduction problems. Many simplifying assumptions need to be made if one is bent upon obtaining analytical closed-form solutions. Analytical solutions are possible for only specific types of initial and boundary conditions. Variation of properties with temperature could be taken in to account in only the simplest of cases, in one-dimensional steady heat conduction. In spite of this the solution required the introduction of special functions like the Bessel functions and Legendre polynomials.

In recent times, numerical methods have become more and more popular because of the improvements and easy access to computational facilities. Canned programs are available with user friendly features. The user does not need to be an expert in numerical methods to make use of the programs. Fairly complex problems are treatable routinely by an average practicing engineer!

Traditionally heat diffusion problems have been used to introduce one to computational methods. The present text is no exception to this. Even though several numerical schemes are available, attention is directed to the finite difference method

S. P. Venkateshan, *Heat Transfer*,
https://doi.org/10.1007/978-3-030-58338-5_7

for solution of the heat equation. The underlying mathematics is very simple and a beginner can write a program for solving heat conduction problems within a couple of hours of learning the basic ideas.

7.1.1 A Simple Example: One-Dimensional Steady Conduction

A bar of material of length L and uniform cross section area A, is insulated laterally while the two ends are maintained at specified temperatures. Heat is generated within the bar at the rate of G per unit volume. The geometry is shown in Fig. 7.1. The equation that governs the problem is well known (see Chap. 2) and is given by

$$\boxed{\frac{d^2T}{dx^2} + \frac{G}{k} = 0} \tag{7.1}$$

The finite difference method consists in approximating derivatives by differences after dividing the domain into discrete parts. The nodes are arranged as shown in Fig. 7.1. The temperatures are specified at nodal points, which are placed at the centers of each sub-domain or part (typical sub-domain bracketing the node 3 is shown by the shaded rectangle). The nodal interval as well as the size of each part is Δx. Expand $T(x)$ in a Taylor expansion around x as:

$$T(x + \Delta x) = T(x) + \left.\frac{dT}{dx}\right|_x \Delta x + \frac{1}{2!}\left.\frac{d^2T}{dx^2}\right|_x (\Delta x)^2 + O(\Delta x)^3 \tag{7.2}$$

where O indicates the order of magnitude of the term. Similarly,

$$T(x - \Delta x) = T(x) - \left.\frac{dT}{dx}\right|_x \Delta x + \frac{1}{2!}\left.\frac{d^2T}{dx^2}\right|_x (\Delta x)^2 - O(\Delta x)^3 \tag{7.3}$$

Subtracting Eq. 7.3 from Eq. 7.2, we get

$$T(x + \Delta x) - T(x - \Delta x) = 2\left.\frac{dT}{dx}\right|_x \Delta x + O(\Delta x)^3$$

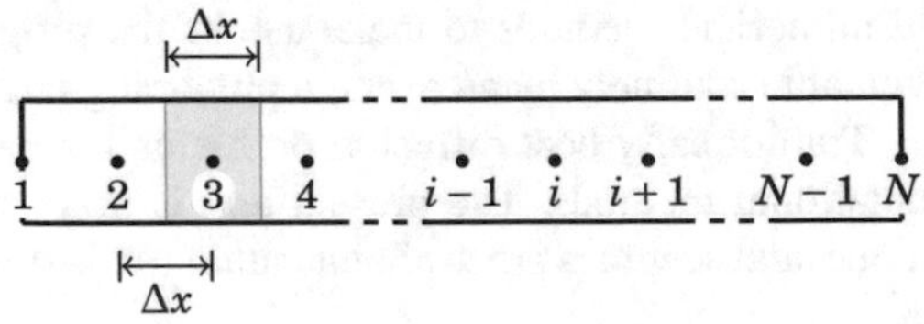

Fig. 7.1 One-dimensional conduction by finite difference approximation—Nodes are numbered from 1 to N

or on rearrangement

$$\boxed{\left.\frac{dT}{dx}\right|_x = \frac{T(x+\Delta x) - T(x-\Delta x)}{2\Delta x} + O(\Delta x)^2} \tag{7.4}$$

Adding Eqs. 7.2 and 7.3, we get

$$T(x+\Delta x) + T(x-\Delta x) = 2T(x) + \left.\frac{d^2T}{dx^2}\right|_x (\Delta x)^2 + O(\Delta x)^4$$

or on rearrangement

$$\boxed{\left.\frac{d^2T}{dx^2}\right|_x = \frac{T(x+\Delta x) - 2T(x) + T(x-\Delta x)}{(\Delta x)^2} + O(\Delta x)^2} \tag{7.5}$$

Expression 7.4–7.5, on ignoring terms of order $O(\Delta x)^2$ and above are respectively the central difference approximation to the first and second derivatives at x (or node i where $x = (i-1) \times \Delta x$) in Fig. 7.1. Both expressions are second order accurate since the error is proportional to $(\Delta x)^2$. The finite difference method consists in approximating the derivatives as given above, in the governing Eq. 7.1, and solving for the discrete set of nodal temperature values by a suitable method. Using expression 7.5 to approximate the second derivative term and assuming that G is replaced by the nodal value $G(x)$, we replace the governing equation by

$$\begin{aligned} &\frac{T(x+\Delta x) - 2T(x) + T(x-\Delta x)}{(\Delta x)^2} + \frac{G(x)}{k} = 0 \\ \text{or } &T(x+\Delta x) - 2T(x) + T(x-\Delta x) = -\frac{G(x)(\Delta x)^2}{k} \end{aligned} \tag{7.6}$$

This may be rewritten, identifying the temperatures by nodal indices, in the alternative form

$$\boxed{T_{i-1} - 2T_i + T_{i+1} = -\frac{G_i(\Delta x)^2}{k}} \tag{7.7}$$

Equation 7.7 may be written for each interior node ($i = 2$ to $N-1$). For the nodes 1 and N, we use the specified boundary conditions (these may be of any of the three types of boundary conditions we are already familiar with). Finally we get a set of simultaneous equations for obtaining all the nodal temperatures. We demonstrate this by considering a typical example.

Example 7.1

Consider a bar 0.1 m long and of uniform cross section of a material of thermal conductivity equal to 45 W/m°C. Heat is uniformly generated at the rate of 10^6 W/m^3. The two ends of the bar are held at 30 °C. Determine the steady temperature distribution in the bar by finite differences using 2 cm long elements. Compare the finite difference solution with the exact solution.

Solution: (i) **Finite difference solution**

Step 1 With $\Delta x = 2$ cm or 0.02 m each, there are 5 sub-domains and 6 nodes in this problem. The heat generation is constant and contributes the following to the nodal equations:

$$\frac{G_i(\Delta x)^2}{k} = \frac{10^6 \times (0.02)^2}{45} = 8.89°C$$

for $i = 2 - 5$. The boundary nodes use the specified boundary conditions. Hence $T_1 = T_6 = 30°C$.

Step 2 The nodal equations for $i = 2 - 5$ are given by

$$T_{i-1} - 2T_i + T_{i+1} = -8.89$$

We notice from symmetry with respect to the middle of the bar (i.e., $x = 0.5$) that $T_2 = T_5$ and $T_3 = T_4$. Hence there are only two unknown temperatures governed by the following two equations:

$$\text{(a)} \quad T_1 - 2T_2 + T_3 = -8.89$$
$$\text{(b)} \quad T_2 - 2T_3 + T_4 = -8.89 \quad \text{or} \quad T_2 - T_3 = -8.89$$

Step 3 From (a) we have $2T_2 = T_1 + T_3 + 8.89 = 30 + T_3 + 8.89 = 38.89 + T_3$. This may be recast as $T_2 - T_3 = \frac{38.89 - T_3}{2}$. Substituting this in (b) we obtain $\frac{38.89 - T_3}{2} = -8.89$ or $T_3 = 38.89 + 2 \times 8.89 = 56.67°C$. With this in (b) we have $T_2 = T_3 - 8.89 = 56.67 - 8.89 = 47.78°C$. Thus we have the nodal temperatures given by the entries in the following table.

Node No	Temperature °C	Node No	Temperature °C
1	30	6	30
2	47.78	5	47.78
3	56.67	4	56.67

(ii) **Exact solution**:

Step 4 Eq. 7.1 may be integrated twice with respect to x to get

$$T(x) = Ax + B - \frac{Gx^2}{2k}$$

Using the boundary condition at $x = 0$ we have $B = 30°C$. The second boundary condition at $x = 0.1$ m requires that $30 = 0.1A + B - \frac{10^6 \times 0.1^2}{2 \times 45}$ or $A = 1111.1°C/m$. Thus the temperature is given by

$$T(x) = 30 + 1111.1x + 11111.1x^2$$

Step 5 The temperatures at the nodal points may be calculated using the above expression. The reader may verify that the values thus calculated are the same as those shown in the table.

Step 6 **Reason**: The central difference formula is second-order accurate. The exact solution to the problem is a quadratic (2nd degree polynomial). Therefore, the solution obtained by the central difference formula is identical to the exact solution, in this particular case.

7.1.2 Numerical Solution of a Fin Problem

Conducting convecting fins or extended surfaces have been considered in Chap. 4 in some detail. The appropriate governing equations were solved by available exact analytical methods. It is instructive to consider numerical solution of the appropriate equations, as a viable alternative. In Appendix E a typical case of uniform area fin has been considered using the "shooting method" in combination with the fourth order Runge Kutta method. The same example, numerics may be different, is considered here, using the finite difference method.

Consider steady heat transfer in a straight fin of uniform cross sectional area as shown in Fig. 7.2. The fin may be in the form of a flat plate or a rod of uniform cross section. Consider the element surrounding node i. An energy balance for this element may be written as

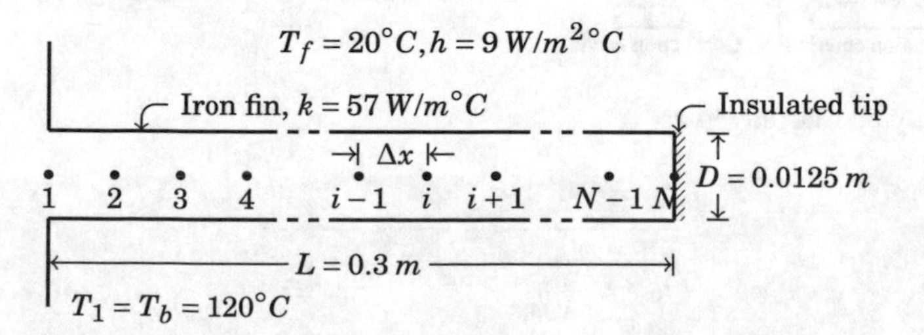

Fig. 7.2 Uniform area fin problem: the numerical values are considered in Example 7.2

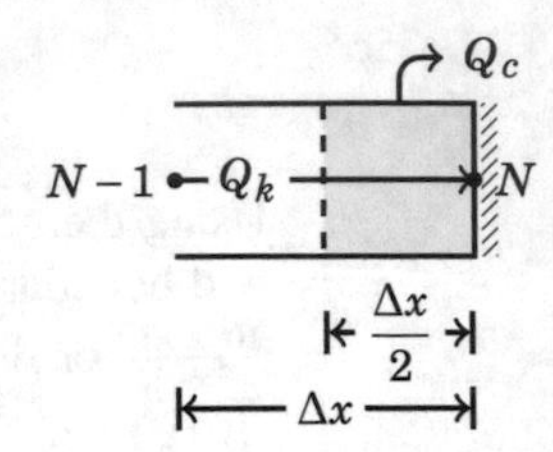

Fig. 7.3 Energy balance for half element at the tip

$$\underbrace{Q_{k,i-1\to i}}_{\text{Conduction entering}} = \underbrace{Q_{k,i\to i+1}}_{\text{Conduction leaving}} + \underbrace{Q_{c,i}}_{\text{Convection leaving}} \tag{7.8}$$

Various terms in the above equation are given by the following:

$$\text{(a) } Q_{k,i-1\to i} = -kA\frac{\theta_i - \theta_{i-1}}{\Delta x};\ \text{(b) } Q_{k,i\to i+1} = -kA\frac{\theta_{i+1} - \theta_i}{\Delta x};\ \text{(c) } Q_{c,i} = hP\Delta x\theta_i \tag{7.9}$$

where $\theta_i = T_i - T_f$ is the nodal temperature excess with respect to the ambient fluid temperature. With these, Eq. 7.8 becomes, on minor rearrangement,

$$\boxed{\theta_{i-1} - \left(2 + m^2\Delta x^2\right)\theta_i + \theta_{i+1} = 0} \tag{7.10}$$

where m is the familiar fin parameter given by $\sqrt{\frac{hP}{kA}}$. Expression 7.10 is used for the interior nodes $i = 2$ to $i = N - 1$. At the first node, the temperature is specified as $\theta_1 = T_b - T_f$. For the node N, the second kind boundary condition may be realized by performing energy balance for a half-element, as shown in Fig. 7.3. The energy balance performed on a half element ensures that the formulation is 2nd order accurate. This will ensure a consistency between the formulation for interior nodes and also for the tip (boundary) node. We have

$$\underbrace{Q_{k,N-1\to N}}_{\text{Conduction entering}} = \underbrace{Q_{c,N}}_{\text{Convection leaving}} \quad \text{or} \quad -kA\frac{\theta_N - \theta_{N-1}}{\Delta x} = hP\frac{\Delta x}{2}\theta_N \tag{7.11}$$

which may be rearranged as

$$\boxed{\theta_{N-1} - \left(1 + \frac{m^2\Delta x^2}{2}\right)\theta_N = 0} \tag{7.12}$$

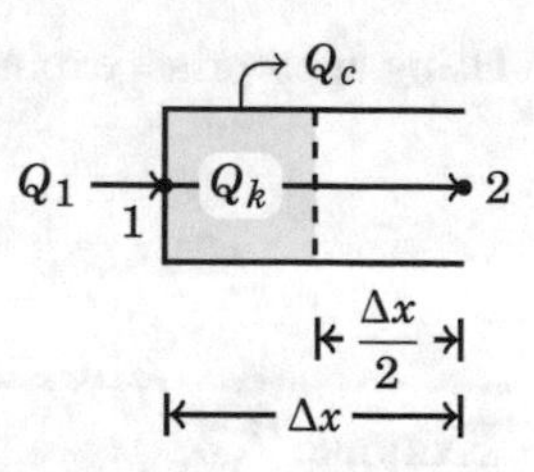

Fig. 7.4 Fin heat transfer calculation through energy balance over a half element adjacent to the base

Iterative Method of Solution

The Gauss iteration scheme presented in Appendix E will now be applied to solve the nodal equations in the fin problem. Equations 7.10 and 7.12 are arranged as follows to apply the Gauss iteration scheme:

$$\text{(a)}\quad \theta_i^{new} = \frac{\theta_{i-1}^{old} + \theta_{i+1}^{old}}{2 + m^2\Delta x^2},\ 2 \le i \le N-1; \quad \text{(b)}\quad \theta_N^{new} = \frac{\theta_{N-1}^{old}}{1 + \frac{m^2\Delta x^2}{2}} \tag{7.13}$$

However, if the intention is to apply the Gauss–Seidel iteration scheme, the equations are written as

$$\text{(a)}\quad \theta_i^{new} = \frac{\theta_{i-1}^{new} + \theta_{i+1}^{old}}{2 + m^2\Delta x^2},\ 2 \le i \le N-1; \quad \text{(b)}\quad \theta_N^{new} = \frac{\theta_{N-1}^{new}}{1 + \frac{m^2\Delta x^2}{2}} \tag{7.14}$$

The superscripts *old* and *new* represent, respectively, the nodal values before and after the iteration. The iteration starts with guess values for all the nodal temperatures. The iteration process is terminated when the following convergence criterion is met at all the nodes.

$$\left| \frac{\theta_i^{new} - \theta_i^{old}}{\theta_i^{new}} \right| \le \epsilon \tag{7.15}$$

where ϵ is a chosen tolerance (small).

Heat Transferred from the Fin

Consistent with the central difference scheme, the fin heat transfer is calculated by an energy balance on a half element adjacent to the base as shown Fig. 7.4. Note that in the figure, $T_1 - T_2$ is the same as $\theta_1 - \theta_2$ or $\theta_b - \theta_2$. Heat balance requires that

$$Q_1 = Q_{k,1\to 2} + Q_{c,1} \tag{7.16}$$

Using appropriate expressions for the various Q's, we have

$$\boxed{Q_1 = kA\frac{\theta_b - \theta_2}{\Delta x} + hP\theta_b\frac{\Delta x}{2}} \tag{7.17}$$

Example 7.2

Consider an iron fin with $k = 57$ W/m°C of circular cross section of diameter $D = 1.25$ cm and length $L = 30$ cm. The base excess temperature is 100°C. It is subject to convection over the lateral surface via a heat transfer coefficient of $h = 9$ W/m^2°C. Obtain the temperature profile by finite differences, using an iterative method of solution. Use 7 equidistant nodes along the length of the fin. The tolerance for terminating the iteration process may be taken as 0.01%. Also perform a grid sensitivity study and comment on the results.

Solution :

Step 1 The given data is specified using the notation shown in Fig. 7.2.

Diameter of pin fin:	$D = 1.25\,\text{cm} = 0.0125$ m
Length of pin fin:	$L = 30\,\text{cm} = 0.3$ m
Thermal conductivity of fin material:	$k = 57\,\text{W/m° C}$
Convection heat transfer coefficient:	$h = 9\,\text{W/m}^2\,°C$
Number of nodes:	$N = 7$
Nodal spacing:	$\Delta x = \frac{L}{N-1} = \frac{0.3}{6} = 0.05$ m

Step 2 We calculate the fin parameter as

$$m = \sqrt{\frac{hP}{kA}} = \sqrt{\frac{4h}{kD}} = \sqrt{\frac{4 \times 9}{57 \times 0.0125}} = 7.10819$$

Step 3 Since the Gauss–Seidel scheme is proposed to be used here, Eqs. 7.14(a) and (b) become:

$$\theta_i^{new} = \frac{\theta_{i-1}^{new} + \theta_{i+1}^{old}}{2 + (7.10819 \times 0.05)^2} = \frac{\theta_{i-1}^{new} + \theta_{i+1}^{old}}{2.12632}$$

and

$$\theta_N^{new} = \frac{\theta_{N-1}^{new}}{1 + \frac{(7.10819 \times 0.05)^2}{2}} = \frac{\theta_{N-1}^{new}}{1.06312}$$

The iterations start with a linear temperature profile with the temperature excess of $100°C$ at $x = 0$ or $i = 1$ to temperature excess $\theta_7 = 0$ at $x = L$. The solution which has been obtained by the above iterative procedure is shown in Table 7.1. The tolerance was set equal to 0.01%

Table 7.1 Iterative solution—fin problem of Example 7.2

Node Number	x m	θ_i Numerical	θ_i Exact
1	0	100.00	100.00
2	0.05	71.24	71.10
3	0.10	51.47	51.26
4	0.15	38.21	37.98
5	0.20	29.78	29.54
6	0.25	25.12	24.87
7	0.30	23.62	23.38

Tolerance = 0.01%
Iterations for Convergence = 34

(as prescribed) and the number of iterations needed for convergence was 34.

The exact values are calculated using the analytical expression derived in Chap. 4.

Step 4 The heat loss from the fin is calculated based on Eq. 7.17 as

$$Q_1 = 57 \times \frac{\pi \times 0.0125^2}{4} \times \frac{100 - 71.24}{0.05} + 9 \times \pi \times 0.0125 \times 100 \times \frac{0.05}{2}$$
$$= 4.907\ W$$

The exact value is calculated based on the fin efficiency concept developed in Chap. 4 and is

$$Q_1(Exact) = \pi D L \theta_b \frac{\tanh(mL)}{mL}$$
$$= \pi \times 0.0125 \times 0.3 \times 100 \frac{\tanh(7.10819 \times 0.3)}{7.10819 \times 0.3} = 4.834\ W$$

Step 5 The grid sensitivity analysis consists in changing the number of nodes and looking at the results, both the temperature profile as well as the fin heat loss. The above iterative procedure of solution is performed repeatedly with various N values. In each case the value of Δx changes and hence also the coefficient matrix. As N is increased, there is an improvement in the numerical solution as indicated by the data presented in Table 7.2. Both the fin heat loss Q_1 and the tip temperature θ_N show improvement as N is increased. Note that the number of iterations required for convergence also depends on N.

Table 7.2 Improvement of heat loss and tip temperature with number of nodes in Example 7.2. In all cases, tolerance $\epsilon = 0.01\%$

N	Q_1	θ_N	N	Q_1	θ_N
7	4.908	23.62	15	4.851	23.37
9	4.876	23.50	17	4.848	23.33
11	4.862	23.44	19	4.847	23.31
13	4.855	23.40	21	4.846	23.28

7.1.3 Solution of Nodal Equations by TDMA

An alternate method of solution of the nodal equations is to solve for all the nodal equations simultaneously, noting that the nodal equations form a set of linear equations with a banded structure for the coefficient matrix.

The base of the fin is at a fixed temperature and hence $\theta_1 = T_b - T_\infty = \theta_b$. At interior nodes, the nodal equations are given by Eq. 7.10. AT the tip node Eq. 7.12 holds. All the equations may be written down as a set of simultaneous equations as given below.

$$\left.\begin{aligned} \theta_1 &= \theta_b \\ \theta_1 - \left(2 + m^2\Delta x^2\right)\theta_2 + \theta_3 &= 0 \\ \theta_2 - \left(2 + m^2\Delta x^2\right)\theta_3 + \theta_4 &= 0 \\ &\cdots\cdots \\ \theta_{i-1} - \left(2 + m^2\Delta x^2\right)\theta_i + \theta_{i+1} &= 0 \\ &\cdots\cdots \\ \theta_{N-1} - \left(1 + \tfrac{m^2\Delta x^2}{2}\right)\theta_N &= 0 \end{aligned}\right\} \qquad (7.18)$$

The above equations may be recast in the form of a matrix equation

$$[\mathbf{A}]\{\theta\} = \{\mathbf{B}\}$$

Matrix $\{\theta\}$ is a $1 \times N$ column vector of nodal temperatures while $\{\mathbf{B}\}$ is a $1 \times N$ column vector of forcing function. These are respectively given by

$$\{\theta\}^{\mathbf{T}} = \{\theta_1 \quad \theta_2 \quad \cdots \quad \theta_i \quad \cdots \quad \theta_{N-1} \quad \theta_N\}$$

and

$$\{\mathbf{B}\}^{\mathbf{T}} = \{\theta_b \quad 0 \quad \cdots \quad 0 \quad \cdots \quad 0 \quad 0\}$$

Coefficient matrix $[\mathbf{A}]$ is an $N \times N$ square matrix given by

$$[\mathbf{A}] = \begin{bmatrix} 1 & 0 & 0 & 0 & 0 & 0 & 0 \\ -1 & (2+m^2\Delta x^2) & -1 & 0 & 0 & 0 & 0 \\ \cdots & \cdots & \cdots & \cdots & \cdots & \cdots & \cdots \\ 0 & 0 & -1 & (2+m^2\Delta x^2) & -1 & 0 & 0 \\ \cdots & \cdots & \cdots & \cdots & \cdots & \cdots & \cdots \\ 0 & 0 & 0 & 0 & -1 & (2+m^2\Delta x^2) & -1 \\ 0 & 0 & 0 & 0 & 0 & -1 & \left(1+\frac{m^2\Delta x^2}{2}\right) \end{bmatrix}$$

Matrix **A** is a banded matrix which is referred to as a tridiagonal matrix since the matrix has non-zero elements along only three diagonals. The set of equations may be solved by TDMA - TriDiagonal Matrix Algorithm - the details of which is given in Appendix E. Example 7.2 is reworked using TDMA in Example 7.3.

Example 7.3

Redo Example 7.2, for $N = 7$ using the TDMA.

Solution: We have to determine the six unknown nodal temperatures $\theta_2 - \theta_7$ by solving the nodal equations by the use of TDMA. The coefficient matrix is given, using the parameters calculated in Example 7.2, as

$$[\mathbf{A}] = \begin{bmatrix} 1 & 0 & 0 & 0 & 0 & 0 & 0 \\ -1 & 2.12632 & -1 & 0 & 0 & 0 & 0 \\ 0 & -1 & 2.12632 & -1 & 0 & 0 & 0 \\ 0 & 0 & -1 & 2.12632 & -1 & 0 & 0 \\ 0 & 0 & 0 & -1 & 2.12632 & -1 & 0 \\ 0 & 0 & 0 & 0 & -1 & 2.12632 & -1 \\ 0 & 0 & 0 & 0 & 0 & -1 & 1.06316 \end{bmatrix}$$

The nodal temperatures are represented by the column vector

$$\{\theta\}^{\mathbf{T}} = [\theta_1 \quad \theta_2 \quad \theta_3 \quad \theta_4 \quad \theta_5 \quad \theta_6 \quad \theta_7]$$

Matrix [**B**] is given by

$$\{\mathbf{B}\}^{\mathbf{T}} = [\theta_b \quad 0 \quad 0 \quad 0 \quad 0 \quad 0 \quad 0]$$

Negative of leading diagonal elements are written as a_i, the upper diagonal elements are written as b_i and the lower diagonal elements as c_i. Negative of the elements on the right hand side matrix **B** are written as d_i. The auxiliary quantities P_i and Q_i are calculated from these. The solution is then obtained by back substitution. All the parameters that have been computed in Example 7.2 are used in writing down the coefficients a, b, c and d as shown in Table 7.3. The auxiliary quantities P and Q are shown as columns 6 and 7 of the same table. The nodal temperatures obtained by back substitution are shown in the last column.

Table 7.3 Table showing application of TDMA in Example 7.3

i	a_i	b_i	c_i	d_i	P_i	Q_i	θ_i
1	1	0	0	100	0	100	100.00
2	2.12632	1	1	0	0.4703	47.0297	71.24
3	2.12632	1	1	0	0.6039	28.3993	51.48
4	2.12632	1	1	0	0.6568	18.6536	38.22
5	2.12632	1	1	0	0.6805	12.6940	29.79
6	2.12632	1	1	0	0.6917	8.7799	25.13
7	1.06316	0	1	0	0	23.6334	23.63

The comparison of the finite difference solution with the exact solution (refer to last column of Table 7.1) shows that the numerical values are remarkably close to the exact values, even with just 6 nodes in the domain! Also, the TDMA is easy to apply and yields the solution with very little effort. The iteration method involved much larger amount of computational effort, as in Example 7.2.

7.1.4 Steady Radial Conduction in a Cylinder

A second simple example we consider is one-dimensional radial conduction in a cylinder. The intent is to show that the finite difference method can be used for conduction with variable area. Consider a long cylindrical annulus of inner radius r_{in} and outer radius r_{out} with inner boundary maintained at T_{in} and outer boundary exposed to a convective environment at T_∞ subject to a heat transfer coefficient h. Heat is generated at a uniform rate of G per unit volume within the annulus. We shall derive the finite difference form of the governing equation directly by making flux balance for an element bracketing the ith node (Fig. 7.5a). The interval $r_{in} \le r \le r_{out}$ is divided by uniformly spaced nodes in to a number of elements of radial thickness Δr as shown in the figure. We make energy balance, assuming steady state to prevail, as

$$Q_k(i-1 \to i) + Q_k(i+1 \to i) + Q_g(i) = 0 \tag{7.19}$$

where subscript k refers to conduction and g refers to heat generation. Energy balance equation is written as a sum of all fluxes that enter the elemental volume. Fluxes entering will automatically turn out to be positive while those that leave will automatically become negative. The sum of all fluxes would then give zero. The various terms appearing in Eq. 7.19 are written down as below:

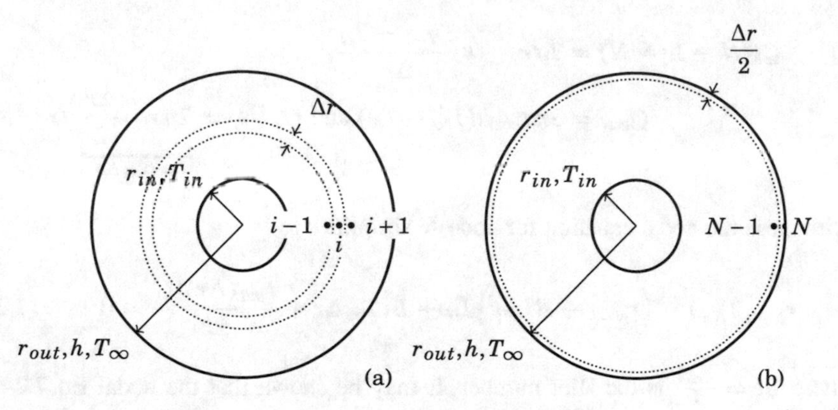

Fig. 7.5 One-dimensional steady radial conduction in an annulus. **a** Interior node **b** Surface node

$$Q_k(i-1 \to i) = 2\pi r_{i-\frac{1}{2}} k \frac{T_{i-1} - T_i}{\Delta r}$$

$$Q_k(i+1 \to i) = 2\pi r_{i+\frac{1}{2}} k \frac{T_{i+1} - T_i}{\Delta r}$$

$$Q_g(i) = 2\pi r_i \Delta r G$$

where $r_{i-\frac{1}{2}}$ and $r_{i+\frac{1}{2}}$ are the radii, respectively at the middle of nodes $i - i - 1$ and nodes $i - i + 1$. With the above, the heat balance Eq. 7.19 will take the form

$$2\pi k r_{i-\frac{1}{2}} \frac{T_{i-1} - T_i}{\Delta r} + 2\pi k r_{i+\frac{1}{2}} \frac{T_{i+1} - T_i}{\Delta r} + 2\pi r_i \Delta r G = 0 \tag{7.20}$$

This may be rearranged as

$$r_{i-\frac{1}{2}} T_{i-1} - (r_{i-\frac{1}{2}} + r_{i+\frac{1}{2}}) T_i + r_{i+\frac{1}{2}} T_{i+1} + \frac{r_i (\Delta r)^2 G}{k} = 0 \tag{7.21}$$

It is easily verified that $r_{i-\frac{1}{2}} + r_{i+\frac{1}{2}} = 2r_i$. Hence the nodal equation becomes

$$r_{i-\frac{1}{2}} T_{i-1} - 2r_i T_i + r_{i+\frac{1}{2}} T_{i+1} + \frac{r_i (\Delta r)^2 G}{k} = 0 \tag{7.22}$$

These are valid for $2 \le i \le (N-1)$. At node 1 the temperature is specified. For node N we consider energy balance for a half element (Fig. 7.5b), which is contained between the nodes $N-1$ and N. We have

$$Q_k(N-1 \to N) + Q_{c,N} + Q_g(N) = 0 \tag{7.23}$$

In the above, subscript c indicates convection at the surface node. The various quantities may be written down as

$$Q_k(N-1 \to N) = 2\pi r_{N-\frac{1}{2}} k \frac{T_{N-1} - T_N}{\Delta r},$$

$$Q_{c,N} = 2\pi r_{out} h(T_\infty - T_N) \text{ and } Q_g(N) = \underbrace{2\pi r_{out} \frac{\Delta r}{2}}_{=\pi r_{out}\Delta r} G$$

With these the nodal equation for node N simplifies to

$$r_{N-\frac{1}{2}} T_{N-1} - \left(r_{N-\frac{1}{2}} + Bi\,\Delta r\right) T_N + Bi\,T_\infty \Delta r + \frac{r_{out}(\Delta r)^2 G}{2k} = 0 \qquad (7.24)$$

where $Bi = \frac{h r_{out}}{k}$ is the Biot number. It may be shown that the nodal Eq. 7.24 is second order accurate and is consistent with the second order accurate interior nodal equations given by 7.22.

Equations 7.22 and 7.24 along with the inner boundary condition provide enough equations to determine all the nodal temperatures. Example 7.4 demonstrates this.

Example 7.4

A very long cylindrical annulus has inner and outer radii of $r_{in} = 0.025$ m and $r_{out} = 0.05$ m respectively. The inner boundary is maintained at a temperature of $T_{in} = 30$°C by passing cold water through it. The outer surfaces is perfectly insulated. Heat is generated internally in the annulus at a uniform volumetric rate of $G = 10^6$ W/m^3. Thermal conductivity of the cylinder material is $k = 15$ W/m°C. Obtain the temperature distribution within the annulus by finite differences. Use a step size of $\Delta r = 0.005$ m. Determine the amount of heat transfer per meter at the inner boundary numerically by using a half element adjacent to the inner boundary.

Solution: Since the outer surface of the cylinder is perfectly insulated we put $h = 0$ in the nodal equation for T_N. This is equivalent to taking $Bi = 0$ in the corresponding nodal equation. With $\Delta r = 0.005$ m, there are 6 nodes and 5 elements. The heat generation parameter that appears in the nodal equations is given by

$$\frac{G\Delta r^2}{k} = \frac{10^6 \times 0.005^2}{15} = 1.666667\ K$$

The nodal equations require the coefficients to be calculated. We shall normalize these by dividing each of them with r_{in}. We show the calculation for the second node as a typical example. We have

$$r_1 = 0.025\,m,\ r_2 = r_{in} + \Delta r = 0.025 + 0.005 = 0.030\,m$$
$$\frac{r_2}{r_1} = \frac{0.030}{0.025} = 1.2$$
$$r_3 = r_2 + \Delta r = 0.030 + 0.005 = 0.035\,m$$

$$\frac{r_3}{r_1} = \frac{0.035}{0.025} = 1.4$$

From these we also have

$$r_{2-\frac{1}{2}} = 0.030 - 0.0025 = 0.0275, \quad \frac{r_{2-\frac{1}{2}}}{r_{in}} = \frac{0.0275}{0.025} = 1.1$$

$$r_{2+\frac{1}{2}} = 0.030 + 0.0025 = 0.0325, \quad \frac{r_{2+\frac{1}{2}}}{r_{in}} = \frac{0.0325}{0.025} = 1.3$$

We substitute these in the nodal Eq. 7.22 for $i = 2$ to get

$$r_{2-\frac{1}{2}}T_1 - 2r_2T_2 + r_{2+\frac{1}{2}}T_3 + \frac{r_2(\Delta r)^2 G}{k} = 0$$

$$\text{or} \quad \frac{r_{2-\frac{1}{2}}}{r_{in}}T_1 - 2\frac{r_2}{r_{in}}T_2 + \frac{r_{2+\frac{1}{2}}}{r_{in}}T_3 + \frac{\frac{r_2}{r_{in}}(\Delta r)^2 G}{k} = 0$$

$$\text{or} \quad 1.1T_1 - 2.4T_2 + 1.3T_3 + 2 = 0$$

The nodal equation for node 6 follows from Eq. 7.24 as

$$r_{5-\frac{1}{2}}T_5 - \left(r_{5-\frac{1}{2}} + Bi\,\Delta r\right)T_6 + Bi\,T_\infty \Delta r + \frac{r_{out}(\Delta r)^2 G}{2k} = 0$$

$$\text{or} \quad \frac{r_{5-\frac{1}{2}}}{r_{in}}T_5 - \left(\frac{r_{5-\frac{1}{2}}}{r_{in}} + \frac{Bi\,\Delta r}{r_{in}}\right)T_6 + \frac{Bi\,T_\infty \Delta r}{r_{in}} + \frac{r_{out}(\Delta r)^2 G}{2kr_{in}} = 0$$

$$\text{or} \quad 1.9T_5 - 1.9T_6 + 1.666667 = 0$$

The other nodal equations may be derived in a similar fashion. The nodal temperatures thus follow a set of linear equations that have to solved simultaneously. The equations may be written in the form of a matrix equation $\mathbf{AT} = \mathbf{B}$ where $\mathbf{A}$ is a square (6×6 in this case) tridiagonal matrix, $\mathbf{T}$ is a column vector of unknown temperatures and $\mathbf{B}$ is a column vector involving source terms.

$$\begin{bmatrix} -1 & 0 & 0 & 0 & 0 & 0 \\ 1.1 & -2.4 & 1.3 & 0 & 0 & 0 \\ 0 & 1.3 & -2.8 & 1.5 & 0 & 0 \\ 0 & 0 & 1.5 & -3.2 & 1.7 & 0 \\ 0 & 0 & 0 & 1.7 & -3.6 & 1.9 \\ 0 & 0 & 0 & 0 & 1.9 & -1.9 \end{bmatrix} \begin{bmatrix} T_1 \\ T_2 \\ T_3 \\ T_4 \\ T_5 \\ T_6 \end{bmatrix} = \begin{bmatrix} -30 \\ -2 \\ -2.33333 \\ -2.66667 \\ -3 \\ -1.66667 \end{bmatrix}$$

Again we make use of TDMA to solve the nodal equations. The results are shown in Table 7.4. Consider a half element adjacent to the inner boundary. It is bounded by the inner boundary and a circle of radius $r_{in} + \frac{\Delta r}{2} = 0.025 + \frac{0.005}{2} = 0.0275$ m. The conduction flux crossing into the element may be written as

Table 7.4 Tabulation of results of TDMA for Example 7.4

i	a_i	b_i	c_i	d_i	P_i	Q_i	r_i	T_i
1	1	0	0	30	0	30	0.025	30.00
2	2.4	1.3	1.1	2	0.5417	14.5833	0.03	40.61
3	2.8	1.5	1.3	2.33333	0.7157	10.1590	0.035	48.04
4	3.2	1.7	1.5	2.66667	0.7995	8.4203	0.04	52.93
5	3.6	1.9	1.7	3	0.8479	7.7265	0.045	55.68
6	1.9	0	1.9	1.66667	0	56.5531	0.05	56.55

$$Q_{c,2\to1} = 2\pi r_{2-\frac{1}{2}} \frac{k(T_2 - T_1)}{\Delta r} = 2 \times \pi \times 0.0275 \frac{15(40.61 - 30)}{0.005} = 5499.83 \text{ W/m}$$

Heat generated within the volume element is

$$\begin{aligned} Q_g &= 2\pi r_{in} \frac{\Delta r}{2} G \\ &= 2 \times \pi \times 0.025 \times \frac{0.005}{2} \times 10^6 = 392.7 \text{ W/m} \end{aligned}$$

By energy balance for the half element we get the total heat transfer at the inner boundary as

$$Q_{in} = Q_{c,2\to1} + Q_g = 5499.83 + 392.7 = 5892.53 \text{ W/m}$$

Since the outer boundary is insulated, this should also represent the total heat generated within the annulus $Q_{g,t}$per meter length.

$$Q_{g,t} = \pi(r_{out}^2 - r_{in}^2)G = \pi \times (0.05^2 - 0.025^2) \times 10^6 = 5890.49 \text{ W/m}$$

The two results agree very closely and hence the finite difference solution with just 6 nodes gives satisfactory result!

The reader should note that the problem may easily be solved by analytical methods to get a closed form solution to the problem. The reader is encouraged to obtain such a solution and make comparisons with the numerical solution presented in the example here.

7.1.5 Steady Radial Conduction in a Spherical Shell

We consider a typical problem in spherical coordinates, viz. steady radial heat conduction in a spherical shell with uniform internal heat generation of $G\ W/m^3$. We may use Fig. 7.5a and b for this case also. Interpret the element with node i as a spherical shell instead of an annular element as in the case of cylindrical annulus. The heat transfer area associated with node i is given by $4\pi r_i^2$ in this case. Heat balance requires that

$$Q_k(i-1 \rightarrow i) + Q_k(i+1 \rightarrow i) + Q_g(i) = 0 \tag{7.25}$$

where subscript k refers to conduction and g refers to heat generation. The various terms appearing in Eq. 7.25 are written down as below:

$$\begin{aligned} Q_k(i-1 \rightarrow i) &= 4\pi r_{i-\frac{1}{2}}^2 k \frac{T_{i-1} - T_i}{\Delta r} \\ Q_k(i+1 \rightarrow i) &= 4\pi r_{i+\frac{1}{2}}^2 k \frac{T_{i+1} - T_i}{\Delta r} \\ Q_g(i) &= 4\pi r_i^2 \Delta r G \end{aligned}$$

where $r_{i-\frac{1}{2}}$ and $r_{i+\frac{1}{2}}$ are the radii, respectively at the middle of nodes $i\ -\ i-1$ and nodes $i\ -\ i+1$. With the above, the heat balance Eq. 7.25 will take the form

$$4\pi k r_{i-\frac{1}{2}}^2 \frac{T_{i-1} - T_i}{\Delta r} + 4\pi k r_{i+\frac{1}{2}}^2 \frac{T_{i+1} - T_i}{\Delta r} + 4\pi r_i^2 \Delta r^2 G = 0 \tag{7.26}$$

This may be rearranged as

$$r_{i-\frac{1}{2}}^2 T_{i-1} - (r_{i-\frac{1}{2}}^2 + r_{i+\frac{1}{2}}^2) T_i + r_{i+\frac{1}{2}} T_{i+1} + \frac{r_i^2 (\Delta r)^2 G}{k} = 0 \tag{7.27}$$

It is easily verified that $r_{i-\frac{1}{2}}^2 + r_{i+\frac{1}{2}}^2 \approx 2r_i^2$. Hence the nodal equation becomes

$$\boxed{r_{i-\frac{1}{2}}^2 T_{i-1} - 2r_i^2 T_i + r_{i+\frac{1}{2}}^2 T_{i+1} + \frac{r_i^2 (\Delta r)^2 G}{k} = 0} \tag{7.28}$$

These are valid for $2 \leq i \leq (N-1)$. At node 1 the boundary is specified to be an adiabatic boundary. A half spherical shell element is used adjacent to the inner boundary. Heat balance requires

$$Q_k(2 \rightarrow 1) + Q_g(1) = 0 \tag{7.29}$$

This may be rewritten as

$$Q_k(2 \rightarrow 1) = 4\pi r_{1+\frac{1}{2}}^2 k \frac{T_2 - T_1}{\Delta r}$$
$$Q_g(1) = 4\pi r_1^2 \frac{\Delta r}{2} G$$

This may be simplified to read

$$\boxed{-r_{1+\frac{1}{2}}^2 T_1 + r_{i+\frac{1}{2}}^2 T_2 + \frac{r_1^2 \Delta r^2 G}{2k} = 0} \tag{7.30}$$

For node N we consider energy balance for a half element (Fig. 7.5b), which is contained between the nodes $N - 1$ and N. We have

$$Q_k(N-1 \rightarrow N) + Q_{c,N} + Q_g(N) = 0 \tag{7.31}$$

In the above, subscript c indicates convection at the surface node. The various quantities may be written down as

$$Q_k(N-1 \rightarrow N) = 4\pi r_{N-\frac{1}{2}}^2 k \frac{T_{N-1} - T_N}{\Delta r}$$
$$Q_{c,N} = 4\pi r_{out}^2 h(T_\infty - T_N)$$
$$Q_g(N) = 4\pi r_{out}^2 \frac{\Delta r}{2} G$$

With these the nodal equation for node N simplifies to

$$\boxed{r_{N-\frac{1}{2}}^2 T_{N-1} - \left(r_{N-\frac{1}{2}}^2 + r_{out}^2 Bi_{\Delta r}\right) T_N + r_{out}^2 Bi_{\Delta r} T_\infty + \frac{r_{out}^2 (\Delta r)^2 G}{2k} = 0} \tag{7.32}$$

where $Bi_{\Delta r} = \frac{h\Delta r}{k}$ is the elemental Biot number. It may be shown that the nodal Eq. 7.32 is second order accurate and is consistent with the second order accurate interior nodal equations given by 7.28.

Equations 7.28 and 7.30 along with the inner boundary condition provide enough equations to determine all the nodal temperatures.

Example 7.5

A spherical shell has inner and outer radii of 0.025 and 0.05 m respectively. The inner boundary is perfectly insulated. The outer surface loses heat by convection to an environment at 30 °C via a heat transfer coefficient of 67 W/m^2°C. Thermal conductivity of shell material is 15 W/m°C. Heat is generated internally at a uniform volumetric heat generation rate of 10^5 W/m^3. Obtain the temperature distribution within the shell by finite differences. Use a step size of $\Delta r = 0.0025$ m.

Solution :

Step 1 The given data is summarized below:

Inner radius of shell:	$r_{in} = r_1 = 0.025$ m
Outer radius of shell:	$r_{out} = r_N = 0.05$ m
Thickness of shell elements:	$\Delta r = 0.0025$ m
Maximum node number:	$N = 1 + \frac{0.05-0.025}{0.0025} = 11$
Ambient temperature:	$T_\infty = 30°C$
Thermal conductivity:	$k = 15W/m°C$
Heat transfer coefficient:	$h = 67\ W/m^2°C$
Heat generation rate:	$G = 10^5\ W/m^3$

Step 2 Parameters that enter the problem are calculated as

$$\text{Heat generation parameter: } \frac{G\Delta r^2}{k} = \frac{10^5 \times 0.0025^2}{15} = 0.04167$$

$$\text{Elemental Biot number: } Bi_{\Delta r} = \frac{h\Delta r}{k} = \frac{67 \times 0.0025}{15} = 0.011167$$

Step 3 Nodal equations may be written down using Eqs. 7.28, 7.30 and 7.32. The coefficients are divided by r_{in}^2 so that the coefficient matrix involves numbers of order unity. Student may refer to Example 7.4 where the nodal equations were written down showing all the intermediate steps. The calculations are best done with a spread sheet program. The nodal equations are keyed in as formulae in the cells of a worksheet of spreadsheet program. The calculations may be done by copying down the formulae. In the present case we summarize the results in the form of an extract from a worksheet as shown in Table 7.5.

Step 4 The solution, as can be seen, has been obtained by TDMA. The analytical solution may easily be obtained by integrating the governing equation. It is left as an exercise to the reader. In the last column shown as "T_i, Analytical", we have given the analytically obtained nodal temperatures. These compare very well with the numerically determined values.

Table 7.5 Tabulation of TDMA results for Example 7.5 and comparison with the analytical solution

Node i	a_i	b_i	c_i	d_i	P_i	Q_i	T_i	T_i Analytical
1	1.1025	1.1025	0	0.0208	1	0.0189	53.17	53.16
2	2.425	1.3225	1.1025	0.0504	1	0.0539	53.15	53.14
3	2.885	1.5625	1.3225	0.0600	1	0.0840	53.09	53.08
4	3.385	1.8225	1.5625	0.0704	1	0.1107	53.01	53.00
5	3.925	2.1025	1.8225	0.0817	1	0.1348	52.90	52.89
6	4.505	2.4025	2.1025	0.0938	1	0.1570	52.76	52.75
7	5.125	2.7225	2.4025	0.1067	1	0.1777	52.61	52.60
8	5.785	3.0625	2.7225	0.1204	1	0.1973	52.43	52.42
9	6.485	3.4225	3.0625	0.1350	1	0.2160	52.23	52.22
10	7.225	3.8025	3.4225	0.1504	1	0.2339	52.02	52.00
11	3.847	0	3.8025	1.4233	0	51.78	51.78	51.77
	Coeffcients in nodal equations				Auxiliary quantities			

7.2 Conduction in Two Dimensions

Conduction in two dimensions may involve steady heat conduction in two space dimensions or unsteady heat conduction in one space dimension, as we have seen already in Chap. 5. The coordinate frames of reference in the former case may be any one of three coordinate systems viz. Cartesian, cylindrical or spherical. Numerical solution applicable to all these three coordinate systems will be dealt with here. Similarly unsteady one-dimensional problem may involve transient heat transfer in a slab, cylinder or a sphere. We shall look at all these cases in what follows.

7.2.1 *Steady Heat Conduction in Two Dimensions: Cartesian Coordinates*

The Standard Problem

As a typical example we consider the standard problem of steady heat conduction in a rectangle (see Chap. 5) as shown in Fig. 7.6.

The governing equation is Laplace equation in two dimensions given by

$$\frac{\partial^2 T}{\partial x^2} + \frac{\partial^2 T}{\partial y^2} = 0 \tag{7.33}$$

The domain is divided into rectangular elements by choosing a step size of Δx along the x direction and a step size of Δy along the y direction. Consider an element of

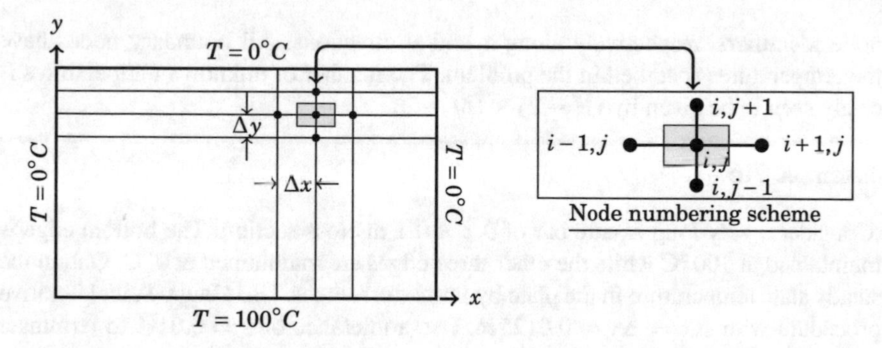

Fig. 7.6 Steady conduction in a rectangular bar of infinite length

size $\Delta x \times \Delta y$ with the node (i, j) at its center. Central difference gives the required partial derivatives as

$$
\begin{aligned}
\frac{\partial^2 T}{\partial x^2} &= \frac{T_{i-1,j} - 2T_{i,j} + T_{i+1,j}}{\Delta x^2} \\
\frac{\partial^2 T}{\partial y^2} &= \frac{T_{i,j-1} - 2T_{i,j} + T_{i,j+1}}{\Delta y^2}
\end{aligned} \tag{7.34}
$$

Denoting the ratio $\left(\frac{\Delta y}{\Delta x}\right)^2$ as r, the Laplace Eq. 7.33 will read as

$$
\boxed{r\left(T_{i-1,j} - 2T_{i,j} + T_{i+1,j}\right) + \left(T_{i,j-1} - 2T_{i,j} + T_{i,j+1}\right) = 0} \tag{7.35}
$$

This may be rearranged in a form suitable for Gauss or Gauss–Seidel iteration as

$$
\boxed{\text{Gauss: } T_{i,j}^{new} = \frac{r\left(T_{i-1,j}^{old} + T_{i+1,j}^{old}\right) + T_{i,j-1}^{old} + T_{i,j+1}^{old}}{2(1+r)}} \tag{7.36}
$$

$$
\boxed{\text{Gauss-Seidel: } T_{i,j}^{new} = \frac{r\left(T_{i-1,j}^{new} + T_{i+1,j}^{old}\right) + T_{i,j-1}^{new} + T_{i,j+1}^{old}}{2(1+r)}} \tag{7.37}
$$

Equations such as 7.35 may be written for all the interior nodes, i.e., for $2 \le i \le N-1$ and $2 \le j \le M-1$ where N and M stand for the largest values for the node identifiers, respectively along x and y directions. All boundary nodes have the temperatures specified in the problem. The number of unknown temperatures is easily seen to be given by $(N - 2) \times (M - 2)$.

Example 7.6

Consider a very long square bar of 0.1×0.1 m cross section. The bottom edge is maintained at 100 °C while the other three edges are maintained at 0 °C. Obtain the steady state temperature in the plate by finite differences. Use Gauss-Seidel iterative procedure with $\Delta x = \Delta y = 0.0125$ m. Use a tolerance of $\epsilon = 0.01\%$ to terminate the iteration process.

Solution : With $\Delta x = \Delta y = 0.0125$ ratio $r = 1$. The number of nodes along the two directions are the same and equal to 9. Unknown interior nodes are $(9 - 2) \times (9 - 2) = 49$. The interior nodal equations are obtained by putting $r = 1$ in Eq. 7.37.

$$T_{i,j}^{new} = \frac{T_{i-1,j}^{new} + T_{i+1,j}^{old} + T_{i,j-1}^{new} + T_{i,j+1}^{old}}{4}$$

Because of symmetry with respect to the plane $x = 0.05$ m the solution needs to consider the rectangle $0 < x < 0.05$ and $0 < y < 0.1$. For nodes along the symmetry line, the temperature $T_{i-1,j} = T_{i+1,j}$ and hence Eq. 7.37 takes the form

$$T_{i,j}^{new} = \frac{2T_{i-1,j}^{new} + T_{i,j-1}^{new} + T_{i,j+1}^{old}}{4}$$

The iteration process starts with zero temperature at all the interior nodes. Table 7.6 gives the nodal temperatures at the very beginning of the iteration process. Starting with the values shown in Table 7.6 the Gauss-Seidel iteration is performed once, row-wise to get the first updates for the nodal temperatures shown in Table 7.7. Note that all the interior nodal temperatures change to non-zero values after just one application of iteration step. This process is continued as many times as necessary to achieve convergence within the specified tolerance. The number of iterations needed to achieve the specified tolerance on the converged nodal temperatures was 78. The converged nodal temperatures are presented in Table 7.8. *Note that the thermal conductivity of the material has no role to play in determining the steady temperature distribution in the plate.*

Heat Transfer at a Convective Boundary

Again, we take the case of a rectangle as shown in Fig. 7.7. In order to write the finite difference equation, we consider an element shown shaded (compare this with what we did in the case of the tip element of a fin). We direct our attention to node 1. There are four fluxes that enter this node as indicated in the enlarged sketch of the element at right. We have, for conservation of energy, the following.

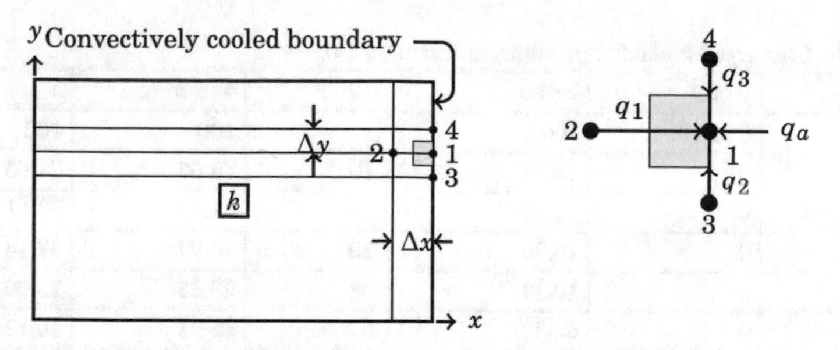

Fig. 7.7 A rectangle with convective cooling along one edge: 'half element'

$$q_1 + q_2 + q_3 + q_a = 0$$

or

$$k\frac{T_2 - T_1}{\Delta x}\Delta y + k\frac{T_3 - T_1}{\Delta y}\frac{\Delta x}{2} + k\frac{T_4 - T_1}{\Delta y}\frac{\Delta x}{2} + h(T_\infty - T_1)\Delta y = 0 \tag{7.38}$$

Divide the above through by $k\sqrt{r}$ to recast Eq. 7.38 as

$$(T_2 - T_1) + \frac{T_3 - T_1}{2r} + \frac{T_4 - T_1}{2r} + \underbrace{\frac{h\Delta y}{k}}_{Bi_{\Delta y}}\frac{1}{\sqrt{r}}(T_\infty - T_1) = 0 \tag{7.39}$$

Table 7.6 Initial temperatures for starting Gauss–Seidel iteration in Example 7.6

$j \downarrow i \rightarrow$	1 or 9	2 or 6	3 or 7	4 or 8	5
1	0 or 100	100	100	100	100
2–9	0	0.00	0.00	0.00	0.00

Table 7.7 Temperatures after one Gauss-Seidel iteration in Example 7.6

$j \downarrow i \rightarrow$	1 or 9	2 or 6	3 or 7	4 or 8	5
1	25	100	100	100	100
2	0	25.00	25.00	25.00	25.00
3	0	6.25	7.81	8.20	10.35
4	0	1.56	2.34	2.64	3.91
5	0	0.39	0.68	0.83	1.39
6	0	0.10	0.20	0.26	0.48
7	0	0.02	0.05	0.08	0.16
8	0	0.01	0.02	0.02	0.05
9	0	0	0	0	0

Table 7.8 Converged nodal temperatures in Example 7.6

$j \downarrow i \rightarrow$	1 or 9	2 or 6	3 or 7	4 or 8	5
1	0 or 100	100	100	100	100
2	0	48.26	66.10	73.05	74.93
3	0	26.93	43.11	51.18	53.61
4	0	16.36	28.20	34.97	37.14
5	0	10.29	18.38	23.35	25.00
6	0	6.44	11.69	15.03	16.17
7	0	3.77	6.89	8.93	9.63
8	0	1.74	3.20	4.15	4.48
9	0	0	0	0	0

where $Bi_{\Delta y}$ is the elemental Biot number. Grouping terms, we may write the nodal temperature T_1 as

$$\boxed{T_1 = \frac{T_2 + \dfrac{T_3 + T_4}{2r} + \dfrac{Bi_{\Delta y} T_\infty}{\sqrt{r}}}{1 + \dfrac{1}{r} + \dfrac{Bi_{\Delta y}}{\sqrt{r}}}} \tag{7.40}$$

We note in passing that these are second order accurate and hence are compatible with the central difference formulation for interior nodes. Adiabatic boundary condition is realized by putting $Bi_{\Delta y} = 0$ in Eq. 7.40. Correspondingly the nodal equation will become

$$\boxed{T_1 = \frac{T_2 + \dfrac{T_3 + T_4}{2r}}{1 + \dfrac{1}{r}}} \tag{7.41}$$

If $\Delta x = \Delta y$, the ratio $r = 1$ and Eqs. 7.40 and 7.41 take the respective simpler forms given by

$$\boxed{T_1 = \frac{T_2 + \dfrac{T_3 + T_4}{2} + Bi_{\Delta y} T_\infty}{2 + Bi_{\Delta y}}} \tag{7.42}$$

and

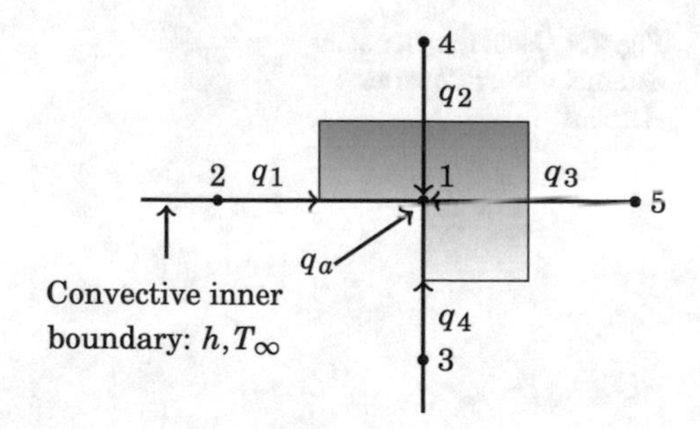

Fig. 7.8 Heat transfer at an internal corner: 'three fourth element'

$$\boxed{T_1 = \frac{T_2 + \dfrac{T_3 + T_4}{2}}{2}} \tag{7.43}$$

Equations such as 7.40–7.41 or 7.42–7.43 are written for all the boundary nodes.

Heat Transfer at Corners

Many times it is necessary to consider the finite difference analog of the governing equations at a corner, either an external or an internal corner.

(a) Internal corner:

The state of affairs at an internal corner is as shown in Fig. 7.8. Consider an element (three fourth element) shown shaded. We have, for energy balance

$$q_1 + q_2 + q_3 + q_4 + q_a = 0$$

Substituting the finite difference representation of the individual heat transfer rates we get

$$\begin{aligned} k\frac{T_2 - T_1}{\Delta x}\frac{\Delta y}{2} + k\frac{T_4 - T_1}{\Delta y}\Delta x + k\frac{T_5 - T_1}{\Delta x}\Delta y + k\frac{T_3 - T_1}{\Delta y}\frac{\Delta x}{2} \\ + h(T_\infty - T_1)\frac{(\Delta x + \Delta y)}{2} = 0 \end{aligned} \tag{7.44}$$

Divide the above through by $k\sqrt{r}$ to get

$$\frac{T_2 - T_1}{2} + \frac{T_4 - T_1}{r} + (T_5 - T_1) + \frac{T_3 - T_1}{2r} + \frac{\left(Bi_{\Delta x} + Bi_{\Delta y}\right)}{2}\frac{(T_\infty - T_1)}{\sqrt{r}} = 0 \tag{7.45}$$

where $Bi_{\Delta x} = \frac{h\Delta x}{k}$ and $Bi_{\Delta y} = \frac{h\Delta y}{k}$. The nodal equation for the corner node may then be written as

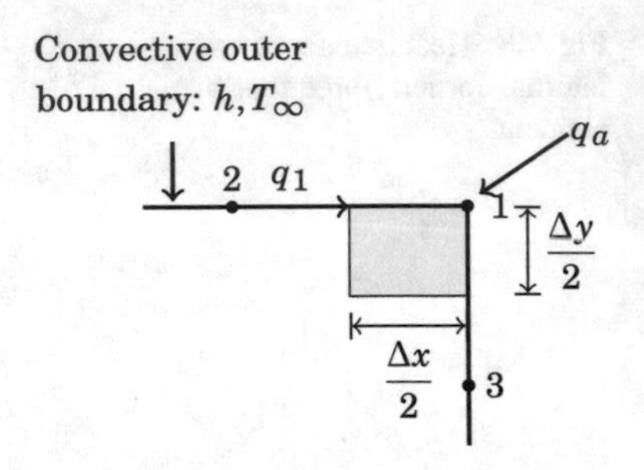

Fig. 7.9 Heat transfer at an external corner: 'quarter element'

$$T_1 = \frac{\frac{T_2}{2} + \frac{T_3}{2r} + \frac{T_4}{r} + T_5 + \frac{\left(Bi_{\Delta x} + Bi_{\Delta y}\right)}{2\sqrt{r}} T_\infty}{\frac{3}{2} + \frac{3}{2r} + \frac{\left(Bi_{\Delta x} + Bi_{\Delta y}\right)}{2\sqrt{r}}} \tag{7.46}$$

If $\Delta x = \Delta y$, the ratio $r = 1$ and $Bi_{\Delta x} = Bi_{\Delta y} = Bi_{\Delta}$ (say), Eq. 7.46 reduces to

$$T_1 = \frac{\frac{T_2}{2} + \frac{T_3}{2} + T_4 + T_5 + Bi_{\Delta} T_\infty}{3 + Bi_{\Delta}} \tag{7.47}$$

(b) External corner:

An external corner element is shown shaded in Fig. 7.9. The element is a quarter element in this case. We have, for energy balance, the following.

$$q_1 + q_2 + q_a = 0$$

Substituting the finite difference representation of the individual heat transfer rates we get

$$k\frac{T_2 - T_1}{\Delta x}\frac{\Delta y}{2} + k\frac{T_3 - T_1}{\Delta y}\frac{\Delta x}{2} + h(T_\infty - T_1)\frac{\Delta x + \Delta y}{2} = 0 \tag{7.48}$$

Divide the above through by $k\sqrt{r}$ to get

$$\frac{T_2 - T_1}{2} + \frac{T_3 - T_1}{2r} + \frac{Bi_{\Delta x} + Bi_{\Delta y}}{2\sqrt{r}}(T_\infty - T_1) = 0 \tag{7.49}$$

The nodal equation for the corner node may then be written as

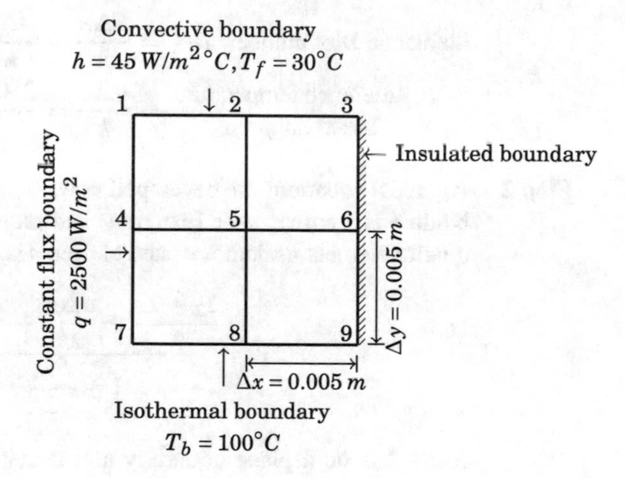

Fig. 7.10 Steady conduction in a rectangle with different boundary conditions along its four edges

$$T_1 = \frac{\dfrac{T_2}{2} + \dfrac{T_3}{2r} + \dfrac{Bi_{\Delta x} + Bi_{\Delta y}}{2\sqrt{r}} T_\infty}{\dfrac{1}{2} + \dfrac{1}{2r} + \dfrac{Bi_{\Delta x} + Bi_{\Delta y}}{2\sqrt{r}}} \tag{7.50}$$

Again, as a special case, for $\Delta x = \Delta y$, (i.e., $r = 1$), $Bi_{\Delta x} = Bi_{\Delta y} = Bi_\Delta$ (say) and Eq. 7.50 reduces to

$$T_1 = \frac{\dfrac{T_2}{2} + \dfrac{T_3}{2} + Bi_\Delta T_\infty}{1 + Bi_\Delta} \tag{7.51}$$

Example 7.7

A long bar of material of thermal conductivity $k = 1.5\,\mathrm{W/m^\circ C}$ is of square section as shown in Fig. 7.10. The four sides of the bar are subjected to the different boundary conditions indicated in the figure. Obtain all unknown temperatures. What is the heat transfer by convection from the upper surface to the ambient? Steady conditions prevail.

Solution :

Step 1 The input data along with the nomenclature is shown in the Fig. 7.10. Two parameters make their appearance in the nodal equations. These are given by

$$\text{Elemental Biot number: } Bi_\Delta = \frac{h\Delta x}{k} = \frac{h\Delta y}{k} = \frac{45 \times 0.005}{1.5} = 0.15$$

$$\text{Reference temperature based on } q: T_q = \frac{q\Delta y}{k} = \frac{2500 \times 0.005}{1.5} = 8.33°C$$

Step 2 All nodal equations are developed now.

Node 1 is a corner node (external) and requires energy balance over a quarter element as demonstrated earlier. The appropriate equation is

$$T_1 = \frac{\frac{T_2 + T_4}{2} + \frac{q\Delta y}{2k} + \frac{Bi_\Delta}{2}T_f}{1 + \frac{Bi_\Delta}{2}}$$

Node 2 is on a plane boundary and is subject to convection. Energy balance is made over a half element. The nodal equation may be written based on Eq. 7.42 as

$$T_2 = \frac{T_5 + \frac{T_1 + T_3}{2} + Bi_\Delta T_\infty}{2 + Bi_\Delta}$$

Node 3 is an external corner. Energy balance is performed over a quarter element. The right boundary is insulated and hence this node is like node 1 but with $q = 0$. The nodal equation may easily be written down as

$$T_3 = \frac{\frac{T_2 + T_6}{2} + \frac{Bi_\Delta}{2}T_f}{1 + \frac{Bi_\Delta}{2}}$$

Node 4 requires energy balance to be performed over a half element. The nodal equation may easily developed as

$$T_4 = \frac{\frac{T_1 + T_7}{2} + \frac{q\Delta y}{k} + T_5}{2}$$

Node 5 is an interior node and the nodal equation is easily seen to be

$$T_5 = \frac{T_2 + T_6 + T_8 + T_4}{4}$$

Node 6 is again treated by performing energy balance over a half element. This is similar to that for node 4, with $q = 0$. This may also be considered as an internal node with symmetry along the right boundary. Thus we have

$$T_6 = \frac{\frac{T_3 + T_9}{2} + T_5}{2}$$

Nodes 7-9 all have specified temperature values of 100 °C. Six nodal equations need to be solved simultaneously to obtain the unknown temperatures.

Step 3 Introducing the numerical values, the six nodal equations are written down, in a form suitable for Gauss–Seidel iteration.

$$T_1^{new} = \frac{\frac{T_2^{old} + T_4^{old}}{2} + 6.4166667}{1.075}; \quad T_2^{new} = \frac{T_5 + \frac{T_1^{new} + T_3^{old}}{2} + 4.5}{2.15}$$

$$T_3^{new} = \frac{\frac{T_2^{new} + T_6^{old}}{2} + 2.25}{1.075}; \quad T_4^{new} = \frac{\frac{T_1^{new} + 100}{2} + 8.33 + T_5^{old}}{2}$$

$$T_5^{new} = \frac{T_2^{new} + T_6^{old} + 100 + T_4^{new}}{4}; \quad T_6^{new} = \frac{\frac{T_3^{new} + 100}{2} + T_5^{new}}{2}$$

Step 3 The nodal equations have been solved by Gauss–Seidel iteration with a convergence criterion of 0.01%. The number of iterations for convergence was 15. The results are given in Table 7.9.

Step 4 Convection heat transfer rate from the upper surface per unit length perpendicular to the plane of figure is obtained as

$$Q_{top} = h\left[\frac{(T_1 - T_f)\Delta x}{2} + (T_2 - T_f)\Delta x + \frac{(T_3 - T_f)\Delta x}{2}\right]$$

$$= 45 \times \left[\frac{(94.45 - 30) \times 0.005}{2} + (89.30 - 30) \times 0.005 + \frac{(87.88 - 30) \times 0.005}{2}\right] = 27.1\ \text{W/m}$$

Table 7.9 Nodal temperatures in Example 7.7 during Gauss-Seidel iteration

Node Number	Initial Values	Iteration 1	Iteration 2	...	Iteration 14	Iteration 15
1	90	89.69	90.77	...	94.44	94.45
2	90	85.74	85.99	...	89.29	89.30
3	90	83.83	85.11	...	87.87	87.88
4	90	96.59	98.40	...	100.94	100.95
5	90	93.08	94.22	...	96.34	96.35
6	90	92.50	93.39	...	95.14	95.14

7.2.2 Steady Heat Conduction in Two Dimensions: Cylindrical Coordinates

We consider steady heat conduction in two dimensions and cylindrical coordinates. The temperature field is a function of r, θ. A typical example is shown in Fig. 7.11. The domain is an annulus with inside boundary I at a specified temperature T_I, part of the boundary ARB at specified temperature T_R and the part of the boundary ALB is perfectly insulated. We notice that the plane LR is a plane of symmetry and hence it is sufficient to consider half annulus as indicated in Fig. 7.12. In order to apply the finite difference method we divide the domain in to elements as shown in this figure. Step size along the r direction is taken as Δr and the step size along θ direction is $\Delta\theta$. In the figure the number of nodes along r is 5 while it is 9 along the θ direction. In general, we may choose a value such as M for the number of nodes along the r direction and N for the number of nodes along the θ direction. The nodal equations may be written down by considering the four types of elements indicated in Fig. 7.12.

Interior element

The interior element is represented by the node i, j. The element is a curvilinear rectangle with radial thickness Δr, inner boundary of length $\left(r_i - \frac{\Delta r}{2}\right)\Delta\theta$ and outer boundary of length $\left(r_i + \frac{\Delta r}{2}\right)\Delta\theta$. Heat balance for the element is given by

$$k\left(r_i - \frac{\Delta r}{2}\right)\Delta\theta\frac{T_{i-1,j} - T_{i,j}}{\Delta r} + k\left(r_i + \frac{\Delta r}{2}\right)\Delta\theta\frac{T_{i+1,j} - T_{i,j}}{\Delta r}$$
$$+k\Delta r\frac{T_{i,j-1} - Ti, j}{r_i\Delta\theta} + k\Delta r\frac{T_{i,j+1} - Ti, j}{r_i\Delta\theta} = 0 \tag{7.52}$$

The above equation may be written in the explicit form

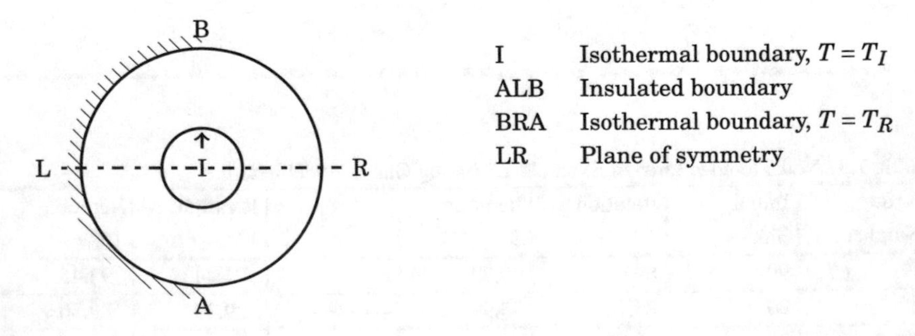

Fig. 7.11 Steady two-dimensional conduction in an annulus

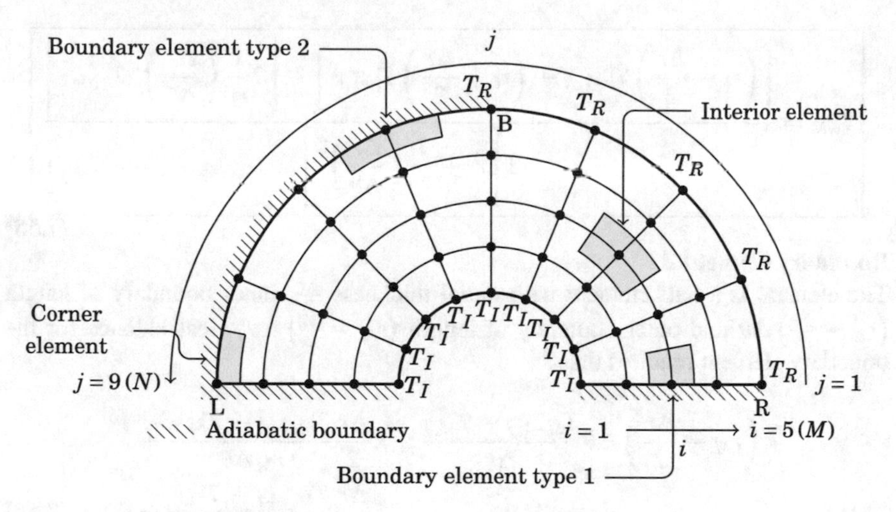

Fig. 7.12 Four types of elements occurring in the annulus problem

$$T_{i,j}=\frac{\left[\left(r_i-\frac{\Delta r}{2}\right)T_{i-1,j}+\left(r_i+\frac{\Delta r}{2}\right)T_{i+1,j}\right]+\left[\frac{1}{r_i}\left(\frac{\Delta r}{\Delta\theta}\right)^2\left(T_{i,j-1}+T_{i,j+1}\right)\right]}{2\left[r_i+\frac{1}{r_i}\left(\frac{\Delta r}{\Delta\theta}\right)^2\right]} \tag{7.53}$$

Boundary element 1

The element considered for heat balance is as shown in Fig. 7.12. The element is a half element with radial thickness equal to Δr, inner boundary of size $\frac{\left(r_i-\frac{\Delta r}{2}\right)\Delta\theta}{2}$ and outer boundary of size $\frac{\left(r_i+\frac{\Delta r}{2}\right)\Delta\theta}{2}$. Since the insulated boundary is a plane of symmetry we may write the nodal equation by putting $T_{i,j-1}=T_{i,j+1}$ in Eq. 7.53. Note also that $j=1$ along the boundary.

$$T_{i,1}=\frac{\left[\left(r_i-\frac{\Delta r}{2}\right)T_{i-1,1}+\left(r_i+\frac{\Delta r}{2}\right)T_{i+1,1}\right]+\left[2\frac{1}{r_i}\left(\frac{\Delta r}{\Delta\theta}\right)^2T_{i,2}\right]}{2\left[r_i+\frac{1}{r_i}\left(\frac{\Delta r}{\Delta\theta}\right)^2\right]} \tag{7.54}$$

It is seen that similar considerations apply to an element on the boundary with $j=N$. Equation 7.54 will have to be recast as

$$T_{i,N} = \frac{\left[\left(r_i - \frac{\Delta r}{2}\right) T_{i-1,N} + \left(r_i + \frac{\Delta r}{2}\right) T_{i+1,N}\right] + \left[2\frac{1}{r_i}\left(\frac{\Delta r}{\Delta\theta}\right)^2 T_{i,N-1}\right]}{2\left[r_i + \frac{1}{r_i}\left(\frac{\Delta r}{\Delta\theta}\right)^2\right]} \tag{7.55}$$

Boundary element 2

The element is a half element with radial thickness $\frac{\Delta r}{2}$, inner boundary of length $\left(r_M - \frac{\Delta r}{2}\right)\Delta\theta$ and outer boundary of length $\left(r_M + \frac{\Delta r}{2}\right)\Delta\theta$. Heat balance for the boundary element requires that

$$k\left(r_M - \frac{\Delta r}{2}\right)\Delta\theta \frac{T_{M-1,j} - T_{M,j}}{\Delta r} + k\frac{\Delta r}{2}\frac{T_{M,j-1} - T_{M,j}}{r_M\Delta\theta} + k\frac{\Delta r}{2}\frac{T_{M,j+1} - T_{M,j}}{r_M\Delta\theta} = 0 \tag{7.56}$$

This may be simplified to get the nodal temperature as

$$T_{M,j} = \frac{2\left(r_M - \frac{\Delta r}{2}\right) T_{M-1,j} + \left[\frac{1}{r_M}\left(\frac{\Delta r}{\Delta\theta}\right)^2 (T_{M,j-1} + T_{M,j+1})\right]}{2\left[r_M - \frac{\Delta r}{2} + \frac{1}{r_M}\left(\frac{\Delta r}{\Delta\theta}\right)^2\right]} \tag{7.57}$$

Nodal equations may thus be written for all the nodes where the temperature is unknown. The solution may be obtained by the now familiar Gauss–Seidel iteration scheme.

Corner element

The only corner element that requires a special treatment is the external corner with node identifier M, N. The element to be considered is a quarter element as shown in Fig. 7.12. Energy balance requires that

$$k\frac{T_{M-1,N} - T_{M,N}}{\Delta r}\left(r_M - \frac{\Delta r}{2}\right)\frac{\Delta\theta}{2} + k\frac{T_{M,N-1} - T_{M,N}}{r_M\Delta\theta}\frac{\Delta r}{2} = 0 \tag{7.58}$$

This equation may be rearranged to get the nodal equation

$$T_{M,N} = \frac{T_{M-1,N}\left(r_M - \frac{\Delta r}{2}\right) + T_{M,N-1}\frac{1}{r_M}\left(\frac{\Delta r}{\Delta\theta}\right)^2}{r_M - \frac{\Delta r}{2} + \frac{1}{r_M}\left(\frac{\Delta r}{\Delta\theta}\right)^2} \tag{7.59}$$

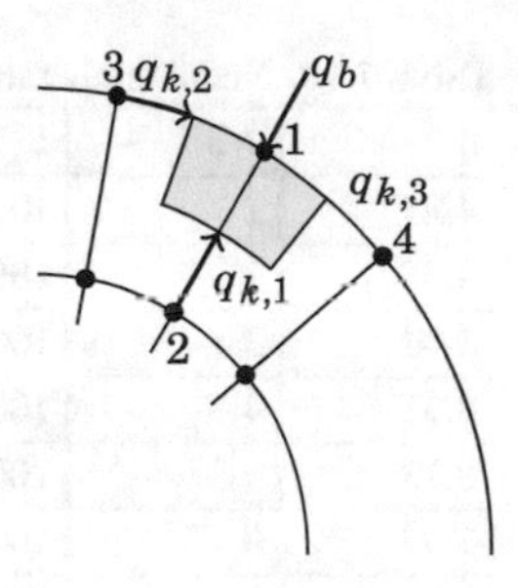

Fig. 7.13 Template for heat transfer calculation at the boundary

Heat transfer at the boundary

Heat transfer at a boundary may be estimated using a half element as shown in Fig. 7.13. We see that by energy balance heat transfer at the boundary may be written as

$$q_b = -q_{k,1} - q_{k,2} - q_{k,3} \tag{7.60}$$

Using the notation in Fig. 7.12 the nodes 1–4 correspond respectively to nodes M, j; $M-1, j$; $M, j+1$ and $M, j-1$. Introducing finite difference expressions for the conduction heat transfer rates, we have

$$\boxed{q_b = -k\left[\frac{T_{M-1,j} - T_{M,j}}{\Delta r}\left(r_M - \frac{\Delta r}{2}\right) + \frac{T_{M,j+1} - T_{M,j}}{r_M \Delta\theta}\frac{\Delta r}{2} + \frac{T_{M,j-1} - T_{M,j}}{r_M \Delta\theta}\frac{\Delta r}{2}\right]} \tag{7.61}$$

This expression will have to be modified suitably for corner nodes, an exercise left to the reader.

Example 7.8

The geometry for this problem is given in Fig. 7.12. The inner radius of the annulus is 0.04 m while the outer radius is 0.2 m. The inner boundary is maintained at 100 °C while the isothermal part of the outer boundary is maintained at 30 °C. Take $\Delta r = 0.04$ m and $\Delta\theta = 11.25° = 0.19635$ rad. Obtain all the unknown nodal temperatures by finite differences. Make use of Gauss–Seidel iteration with a stopping criterion of 0.01% as the maximum allowable change from iteration to iteration. Also obtain the heat transfer rate at the isothermal boundary maintained at 30 °C.

Solution : The given data is written down as follows:

Inner radius of annulus:	$r_i = 0.04$ m
Outer radius of annulus:	$r_o = 0.2$ m
Temperature of inner boundary:	$T_I = 100$°C
Temperature of isothermal part of outer boundary:	$T_R = 30$°C
Radial step size:	$\Delta r = 0.04$ m
Angular step size:	$\Delta\theta = 0.19635$ rad

M and N are obtained as

Table 7.10 Nodal temperatures in Example 7.8

q	$j \downarrow i \rightarrow$	1	2	3	4	5	q
4.25	1	100.00	71.17	53.28	40.23	30.00	−4.52
8.48	2	100.00	71.22	53.33	40.26	30.00	−9.07
8.42	3	100.00	71.40	53.51	40.35	30.00	−9.15
8.33	4	100.00	71.72	53.83	40.53	30.00	−9.31
8.18	5	100.00	72.22	54.37	40.84	30.00	−9.58
7.97	6	100.00	72.95	55.22	41.37	30.00	−10.05
7.67	7	100.00	73.96	56.56	42.31	30.00	−10.88
7.27	8	100.00	75.32	58.63	44.12	30.00	−12.47
6.76	9	100.00	77.04	61.75	47.89	30.00	−26.43
6.17	10	100.00	79.05	66.07	56.59	50.87	0.00
5.55	11	100.00	81.15	70.48	64.15	61.80	0.00
4.97	12	100.00	83.12	74.37	69.93	68.67	0.00
4.47	13	100.00	84.84	77.52	74.19	73.36	0.00
4.06	14	100.00	86.21	79.92	77.24	76.60	0.00
3.77	15	100.00	87.21	81.59	79.29	78.75	0.00
3.59	16	100.00	87.80	82.57	80.47	79.99	0.00
1.77	17	100.00	88.01	82.90	80.86	80.39	0.00

First and the last columns show the heat transfer rates at the boundary nodes

$$M = 1 + \frac{r_o - r_i}{\Delta r} = 1 + \frac{0.2 - 0.04}{0.04} = 5$$

$$N = 1 + \frac{\pi}{\Delta \theta} = 1 + \frac{\pi}{0.19635} = 17$$

The nodal equations are written down using the cases considered above, in a form suitable for the application of the Gauss–Seidel iteration scheme. The equations are arranged in the form of 5 columns corresponding to $i = 1$ to 5 and 17 rows corresponding to $j = 1$ to 17. The specified nodal temperatures are 17 on the inner boundary and 9 along the isothermal part of the outer boundary. Hence the nodal temperatures that are to be determined are 59 in number. All these are initialized as $50°C$ to start the iteration process. Point by point iteration starts at node 2, 1 and proceeds serially, skipping the nodes at which the temperatures are known, to the last node 5, 17.

Table 7.10 shows the results of Gauss–Seidel solution to the nodal equations. Note that the temperatures at the adiabatic boundaries vary with boundary location and are obtained as a part of the solution.

The heat transfer rate at the isothermal boundary may be obtained by the use of expression 7.61 written for all the nodes starting from $M, 1$ to $M, 9$. Of course the extreme nodes require quarter elements. The last column of table shows the heat transfer rates in W/m per unit thermal conductivity. The negative sign is indicative of the fact that the heat transfer is away from the boundary. The heat transfer rate is zero for the adiabatic nodes as required by the boundary conditions specified in the problem. The first column shows the heat transfer rates at the inner isothermal boundary. These are positive indicating that the heat transfer is in to the boundary.

The total heat transfer at the inner boundary is obtained by adding all the elements in the first column to get $q_I = 101.67$ W/m per unit thermal conductivity. The total heat transfer at the boundary R is given by the sum of all the elements in the last column to get $q_R = -101.44$ W/m per unit thermal conductivity. Since the calculations have been made for the top half of the annulus, these values must be multiplied by a factor of two for the full annulus. Energy balance for the annulus requires $q_I + q_r = 0$. It is seen that the energy residue is only 0.23 in approximately 102, which is satisfactory.

7.2.3 One-Dimensional Transient in a Bar

Transient conduction in one space dimension is also a two-dimensional problem in heat conduction. Consider transient conduction in a bar where the temperature is a function of x and t. The lateral surface of the bar is insulated and boundary conditions of third kind are specified at $x = 0$ and $x = L$. The domain is divided into $N - 1$ elements of Δx each, as shown in Fig. 7.14. Consider the element shown with the node i at its center. We know that heat diffusion is represented by second derivative with respect to x, in the heat equation. In the central difference approximation to the second derivative with respect to x, the nodal equation will involve T_{i-1}, T_i and T_{i+1}. Since the T's are also functions of t, there are several ways of writing the finite difference form of the heat equation. The heat equation appropriate to the present case is the one-dimensional heat equation in Cartesian coordinates given by

$$\frac{1}{\alpha}\frac{\partial T}{\partial t} = \frac{\partial^2 T}{\partial x^2} \tag{7.62}$$

The time derivative is invariably written as

$$\frac{\partial T}{\partial t} = \frac{T(x, t + \Delta t) - T(x, t)}{\Delta t} \quad \text{or} \quad \underbrace{\frac{\partial T}{\partial t} = \frac{T_{i,j+1} - T_{i,j}}{\Delta t}}_{\text{Nodal form}} \tag{7.63}$$

where the time step j is defined such that $t = (j - 1)\Delta t$ with $j = 1$ corresponding to $t = 0$. The second derivative may be approximated in different ways as given below.

Explicit Formulation

In the explicit formulation, the second derivative with respect to x is evaluated with all the *known* temperatures at t. Thus,

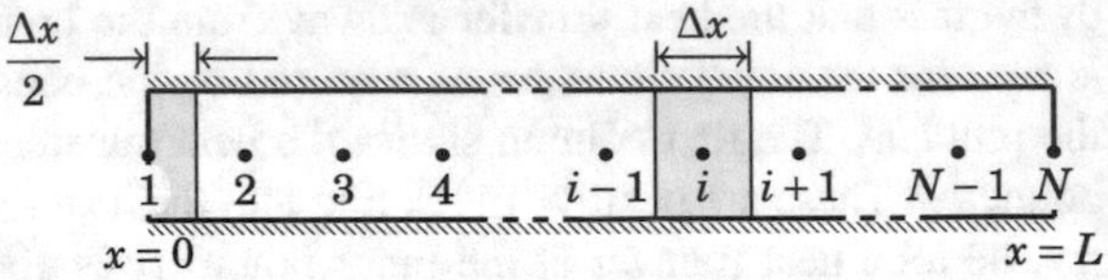

Fig. 7.14 One-dimensional transient conduction in a bar of uniform cross section

$$\left.\frac{\partial^2 T}{\partial x^2}\right|_{\text{Explicit}} = \frac{T(x-\Delta x, t) - 2T(x,t) + T(x+\Delta x, t)}{\Delta x^2}$$

or in the nodal form

$$\left.\frac{\partial^2 T}{\partial x^2}\right|_{\text{Explicit}} = \frac{T_{i-1,j} - 2T_{i,j} + T_{i+1,j}}{\Delta x^2} \tag{7.64}$$

Combining expressions 7.63 and 7.64, we directly evaluate $T(x, t+\Delta t) = T_{i,j+1}$ in terms of the known temperatures at t. We then have

$$T_{i,j+1} = \frac{\alpha \Delta t}{\Delta x^2}[Ti-1, j+Ti+1, j] + \left[1 - 2\frac{\alpha \Delta t}{\Delta x^2}\right] T_{i,j} \tag{7.65}$$

We notice that the elemental Fourier number $Fo_e = \frac{\alpha \Delta t}{\Delta x^2}$ appears in the above equation. Equation 7.65 may then be written in the form

$$\boxed{T_{i,j+1} = Fo_e[T_{i-1,j} + T_{i+1,j}] + [1 - 2Fo_e]T_{i,j}} \tag{7.66}$$

Explicit scheme is simple to apply because the solution can start from the initial profile and obtain all future temperatures by simple arithmetic operations. The third kind of boundary condition is specified at the two ends of the bar wherein the ends convectively interact with an environment at temperature T_∞ via a heat transfer coefficient h. Consider the left boundary node $i = 1$. A half element of thickness $\frac{\Delta x}{2}$ is used for writing the appropriate nodal equation. In the explicit formulation all the fluxes are calculated with known temperatures $T_{1,j}$ and $T_{2,j}$. We have the following:

$$\text{Convection to the half element} = h(T_\infty - T_{1,j}) \tag{7.67}$$

$$\text{Conduction to the half element} = k\frac{T_{2,j} - T_{1,j}}{\Delta x} \tag{7.68}$$

$$\text{Change in internal energy of the half element} = \rho c \frac{T_{1,j+1} - T_{1,j}}{\Delta t}\frac{\Delta x}{2} \tag{7.69}$$

For heat balance the sum of expressions 7.67–7.68 must equal the expression 7.69. The nodal equation may then be obtained, after some simplification as

$$\boxed{T_{1,j+1} = T_{1,j}[1 - 2Fo_e(1 + Bi_e)] + 2Fo_eBi_eT_\infty + 2Fo_eT_{2,j}} \tag{7.70}$$

where Bi_e is the elemental Biot number given by $Bi_e = \frac{h\Delta x}{k}$.

Stability Condition for the Explicit Scheme

We do not expect transient heat conduction problem to involve temperatures that are outside the range of initial temperatures excepting when internal heat generation is involved. Thus one expects temperature range to decrease with time and hence the solution shows stability. However, numerical solution may become unstable because it is an approximate solution that may show unstable behavior because of numerically induced instability.

To find out if the scheme is numerically stable we assume that the temperature field is given by any one term of a Fourier series representation given by

$$T(x, t) = A(t)\cos(\omega x) \tag{7.71}$$

where $A(t)$ is the amplitude at time t and the distribution with respect to space is given by the cosine function with wavenumber $\omega = \frac{2\pi}{\lambda}$ where λ is the wavelength. Using Eq. 7.71 we may now write down the terms that enter the explicit scheme given by Eq. 7.66 as

$$\begin{aligned}
T_{i,j+1} &= A(t + \Delta t)\cos(\omega x)\\
T_{i-1,j} &= A(t)\cos(\omega x - \omega\Delta x) = A(t)[\cos(\omega x)\cos(\omega\Delta x) - \sin(\omega x)\sin(\omega\Delta x)]\\
T_{i+1,j} &= A(t)\cos(\omega x + \omega\Delta x) = A(t)[\cos(\omega x)\cos(\omega\Delta x) + \sin(\omega x)\sin(\omega\Delta x)]\\
T_{i,j} &= A(t)\cos(\omega x)
\end{aligned}$$

We have made use of well known trigonometric identities in writing the above. These are introduced in to Eq. 7.66 to get, after some minor manipulation, the following expression for the Gain defined by the ratio $\frac{A(t+\Delta t)}{A(t)}$.

$$\boxed{G = 1 - 2Fo_x + 2Fo_x\cos(\omega\Delta x)} \tag{7.72}$$

We note that $-1 \le \cos(\omega\Delta x) \le 1$. If we assume that $\cos(\omega\Delta x) = 1$, $G = 1$ for all Fo_x. However if we assume that $\cos(\omega\Delta x) = -1$, then $G = 1 - 4Fo_x$. Thus $|G| > 1$ if $Fo_x > \frac{1}{2}$. Absolute value of gain should be less than or equal to one for the solution to be stable and not magnify any disturbances! At once we see that the explicit scheme is stable only if $Fo_x \leq \frac{1}{2}$. Since finite differences require small Δx for the approximation to be good, very small value of Δt will be required, if this condition is to be satisfied.

Implicit Formulation

In the implicit formulation, the spatial derivative is evaluated using temperatures at $t + \Delta t$ (which are as yet unknown). The second derivative with respect to x is then written as

$$\left.\frac{\partial^2 T}{\partial x^2}\right|_{\text{Implicit}} = \frac{T(x - \Delta x, t + \Delta t) - 2T(x, t + \Delta t) + T(x + \Delta x, t + \Delta t)}{\Delta x^2}$$

or, in the alternate nodal form as

$$\left.\frac{\partial^2 T}{\partial x^2}\right|_{\text{Implicit}} = \frac{T_{i-1,j+1} - 2T_{i,j+1} + T_{i+1,j+1}}{\Delta x^2} \tag{7.73}$$

From Eqs. 7.63 and 7.73, we obtain

$$\boxed{-Fo_e T_{i-1,j+1} + [1 + 2Fo_e]T_{i,j+1} - Fo_e T_{i+1,j+1} = T_{i,j}} \tag{7.74}$$

We write such equations for all the interior nodes. For node 1 (similarly for node N) expressions 7.67 and 7.68 are recast as

$$\text{Convection to the half element} = h(T_\infty - T_{1,j+1}) \tag{7.75}$$

$$\text{Conduction to the half element} = k\frac{T_{2,j+1} - T_{1,j+1}}{\Delta x} \tag{7.76}$$

Expression 7.69 however remains unchanged. The nodal equation will then be recast, after some simplification as

$$\boxed{-[1 - 2Fo_e(1 + Bi_e)]T_{1,j+1} + 2Fo_e Bi_e T_{2,j+1} = -T_{1,j} - 2Bi_e Fo_e T_\infty} \tag{7.77}$$

Thus we have a set of simultaneous equations which may be solved for the temperatures at $t + \Delta t$. The matrix of coefficients is a sparse matrix and is tridiagonal in nature. Such equations can be solved using TDMA introduced earlier. The limitation on Fo_e as in the case of explicit formulation is not there in the implicit method. The method is unconditionally stable. This may be seen by noting that Eq. 7.72 will take the form

$$G = \frac{1}{1 + 2Fo_x - 2Fo_x \cos(\omega \Delta x)} \tag{7.78}$$

It may be verified that $|G| < 1$ for all $Fo_{\Delta x}$ and hence the implicit scheme does not have any restrictive condition on the magnitude of the elemental Fourier number.

Semi-Implicit or Crank–Nicolson (CN) Scheme

This scheme evaluates the spatial derivatives as an average of the second derivative at times t and $t + \Delta t$. This scheme is referred to also as a balanced scheme. Thus, in nodal form, we have

$$\left.\frac{\partial^2 T}{\partial x^2}\right|_{CN} = \frac{T_{i-1,j+1} - 2T_{i,j+1} + T_{i+1,j+1}}{2\Delta x^2} + \frac{T_{i-1,j} - 2T_{i,j} + T_{i+1,j}}{2\Delta x^2} \tag{7.79}$$

From Eqs. 7.63 and 7.79, we then have, on rearrangement

$$-\frac{Fo_e}{2}T_{i-1,j+1} + (1 + Fo_e)T_{i,j+1} - \frac{Fo_e}{2}T_{i+1,j+1} = \frac{Fo_e}{2}(T_{i-1,j} + T_{i+1,j}) + (1 - Fo_e)T_{i,j} \tag{7.80}$$

Equation 7.80 written for all the internal nodes leads to a tridiagonal system of equations. The scheme is unconditionally stable. This may be verified by noting that Eq. 7.72 will take the form

$$G = \frac{1 + \frac{Fo_{\Delta x}}{2}[2\cos(\omega \Delta x) - 2]}{1 - \frac{Fo_{\Delta x}}{2}[2\cos(\omega \Delta x) - 2]} \tag{7.81}$$

It may be verified that $|G| < 1$ for all $Fo_{\Delta x}$ and hence the CN scheme *does not* have any restrictive condition on the magnitude of the elemental Fourier number. The CN scheme is said to be unconditionally stable.

For node 1 (similarly for node N) expressions 7.67 and 7.68 are recast as

$$\text{Convection to the half element} = h\left(T_\infty - \frac{T_{1,j} + T_{1,j+1}}{2}\right) \tag{7.82}$$

$$\text{Conduction to the half element} = k\left(\frac{\frac{T_{2,j} + T_{2,j+1}}{2} - \frac{T_{1,j} + T_{1,j+1}}{2}}{\Delta x}\right) \tag{7.83}$$

Expression 7.69 however remains unchanged. The nodal equation will then be recast, after some simplification as

$$\boxed{-[1 + Fo_e(1 + Bi_e)]T_{1,j+1} + Fo_e T_{2,j+1} = -[1 - Fo_e(1 + Bi_e)]T_{1,j} - 2Fo_e Bi_e T_\infty - Fo_e T_{2,j}} \quad (7.84)$$

Example 7.9

Consider a uniform cross section bar of stainless steel of length $L = 100\,\text{mm}$. The bar is initially at $T_i = 30°\text{C}$ and its two ends are exposed for $t > 0$ to a convective environment at $T_\infty = 100°\text{C}$ subject to a convective heat transfer coefficient of $h = 67\,\text{W/m}^2{}°\text{C}$. Obtain the solution for a few time steps by an explicit method.

Solution :

Step 1 Properties of stainless steel (usual notation) are taken as $\rho = 7900\,\text{kg/m}^3, c = 477\,\text{J/kg°C}$ and $k = 14.9\,\text{W/m°C}$. The thermal diffusivity of stainless steel is then calculated as

$$\alpha = \frac{k}{\rho c} = \frac{14.9}{7900 \times 477} = 3.95 \times 10^{-6}\,\text{m}^2/s$$

Step 2 Choose element thickness of $\Delta x = 10$ mm $= 0.01$ m. The explicit method will give convergent solution if the elemental Fourier number is less than or equal to 0.5. We choose a value of $Fo_e = 0.25$ and hence the time step will be

$$\Delta t = \frac{Fo_e \Delta x^2}{\alpha} = \frac{0.25 \times 0.01^2}{3.95 \times 10^{-6}} = 6.32\,\text{s}$$

With $\Delta x = 0.01$ m, there are 11 nodes −9 interior nodes and 2 boundary nodes. Since there is symmetry with respect to node number 6 we consider only the nodes 1–6. The elemental Biot number is calculated from the given data.

$$Bi_e = \frac{h\Delta x}{k} = \frac{67 \times 0.01}{15.9} = 0.045$$

Step 3 An extract a spreadsheet written for the present example is given in Table 7.11 on page 293. Node identifiers as well as the coordinates of nodes are given in the Table. Column 3 shows the Fourier number defined as $Fo = \frac{\alpha t}{L^2}$. The entries in the first row ($j = 1$) are the initial nodal temperature values specified in the problem. Entries starting with the second row ($j = 2$) in the columns identified by i values are the

Table 7.11 Tabulation of nodal temperatures for Example 7.9

			Spatial node, i or Location along bar, x, m					
Temporal node, j	Time s	Fo	$i = 1$ $x = 0$	2 0.01	3 0.02	4 0.03	5 0.04	6 0.05
1	0	0	30	30	30	30	30	30
2	6.32	0.0025	31.58	30.00	30.00	30.00	30.00	30.00
3	12.64	0.005	32.33	30.39	30.00	30.00	30.00	30.00
4	18.96	0.0075	32.88	30.78	30.10	30.00	30.00	30.00
5	25.28	0.01	33.34	31.13	30.24	30.02	30.00	30.00
6	31.6	0.0125	33.74	31.46	30.41	30.07	30.01	30.00
7	37.92	0.015	34.09	31.77	30.59	30.14	30.02	30.00
8	44.24	0.0175	34.41	32.05	30.77	30.22	30.05	30.01
9	50.56	0.02	34.71	32.32	30.96	30.32	30.08	30.03
10	56.88	0.0225	34.99	32.58	31.14	30.42	30.13	30.06

nodal temperatures calculated by the explicit scheme. At time $t = 44.24$ s or $Fo = 0.0175$ the temperature at the middle of the bar has started responding. The temperature of the bar at its ends have increased by about 5 °C at $t \approx 55$ s. The reader may compare the solution obtained here with the one term approximation presented in Chap. 6.

We have mentioned earlier that the explicit scheme does not work if $Fo_e > 0.5$. This is demonstrated by doing the calculations in Example 7.9 with $Fo_e = 0.3$ and $Fo_e = 0.6$. The solution at $Fo = 0.054$ obtained in these two cases are compared in Fig. 7.15. While the solution obtained with $Fo_e = 0.3$ is well behaved the solution obtained with $Fo_e = 0.6$ shows large oscillations. The solution obtained with the larger time step is thus entirely useless!

Example 7.10

Consider a bar of length $L = 0.1$ m of uniform cross section which is insulated on its lateral surfaces. The bar is initially at a uniform temperature of 100 °C and its ends are cooled to 30 °C instantaneously and are maintained at that temperature for $t > 0$. Making use of the Crank–Nicolson method, obtain the temperature in the bar for the first two time steps taken such that the elemental Fourier number is equal to one. Take the thermal diffusivity of the material of the bar to be 10^{-6} m²/s. Take $\Delta x = 0.0125$ m.

Solution: The elemental length is $\Delta x = 0.0125$ m. Thermal diffusivity of the material of the bar is $\alpha = 10^{-6}$ m^2/s. Elemental Fourier number is to be taken equal to

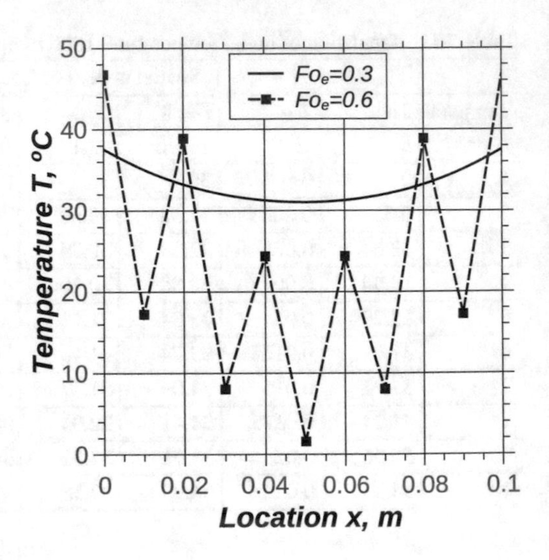

Fig. 7.15 Comparison of solutions obtained with two different time steps in Example 7.9

$Fo_e = 1$. The time step for the numerical calculation is obtained as

$$\Delta t = \frac{\Delta x^2}{\alpha} = \frac{0.0125^2}{10^{-6}} = 156.25\,s$$

Let the initial temperature in the bar be $T_0 = 100°C$ and the two end temperatures for $t > 0$ be $T_e = 30°C$. We shall define the non-dimensional temperature as

$$\phi = \frac{T - T_e}{T_0 - T_e}$$

Initially all the nodal temperatures are $\phi = 1$ and the ends attain $\phi = 0$ temperature for $t > 0$. With $\Delta x = 0.0125$ m there are 9 nodes along the bar. The two end nodes have the temperature specified. However, since these two temperatures are identical and the initial temperature in the bar is uniform, there is symmetry with respect to the node at $x = 0.05$ m (node number 5). This means that we may use insulated boundary condition for this node. The equations governing the interior nodes 2–5 are obtained by letting $Fo_e = 1$ in Eq. 7.84. Hence, for nodes 2 to 5 we have

$$\begin{aligned}
&\text{Node 2:} & 4\phi_{2,new} - \phi_{3,new} &= \phi_{3,old}\\
&\text{Node 3:} & -\phi_{2,new} + 4\phi_{3,new} - \phi_{4,new} &= \phi_{2,old} + \phi_{4,old}\\
&\text{Node 4:} & -\phi_{3,new} + 4\phi_{4,new} - \phi_{5,new} &= \phi_{3,old} + \phi_{5,old}\\
&\text{Node 5:} & -2\phi_{4,new} + 4\phi_{5,new} &= 2\phi_{4,old}
\end{aligned}$$

Subscripts *old* represents the values at the end of the previous time step and *new* represents the values after the current time step. The last of the equations for node 5 is based on either a half element as was done in the fin problem (near the tip) or

Table 7.12 Crank–Nicolson results, for example, 7.10, first two time steps

First time step								
Matrix elements				Auxiliary quantities			Nodal temperatures	
a	b	c	d	P	Q	x	$\phi(Old)$	$\phi(New)$
4	1	0	1	0.25	0.25	0.0125	1	0.4639
4	1	1	2	0.2667	0.6	0.025	1	0.8557
4	1	1	2	0.2679	0.6964	0.0375	1	0.9588
4	0	2	2	0	0.9794	0.05	1	0.9794
Second time step								
Matrix elements				Auxiliary quantities			Nodal temperatures	
a	b	c	d	P	Q	x	$\phi(Old)$	$\phi(New)$
4	1	0	0.8557	0.25	0.2139	0.0125	0.4639	0.3797
4	1	1	1.4227	0.2667	0.4364	0.025	0.8557	0.6633
4	1	1	1.8351	0.2679	0.6084	0.0375	0.9588	0.8508
4	0	2	1.9175	0	0.9048	0.05	0.9794	0.9048

by invoking symmetry condition $\phi_{4,old} = \phi_{6,old}$ and $\phi_{4,new} = \phi_{6,new}$ for all t. The appropriate interior nodal equation will then transform to the one that is shown.

The calculations start with $\phi_{i,old} = 1$ for $i = 2$ to 5 at $t = 0$ and the solution of the simultaneous equations yield the values at $t = 156.25$ s. It is noted that the equations are in the tridiagonal form and hence the TDMA may be made use of. The results of TDMA are shown in Table 7.12 for the first two time steps. The first four columns represent the augmented matrix. Columns 5 and 6 are the auxiliary quantities P and Q that are calculated from the elements of the augmented matrix. The last two columns represent the temperatures at the beginning and the end of the time interval. The Crank–Nicolson method does not show any oscillations in the solution even though we have used a value of 1 for the elemental Fourier number.

The calculations have been continued up to $Fo = 0.0625$ and the nodal temperatures at this time are compared with those obtained analytically using the one term approximation in Table 7.13. The nodal temperatures are in very good agreement even though we have used coarse grids with only 9 nodes in the computational domain.

Further comparison is made between the present numerical solution and the exact analytical solution in Fig. 7.16. The comparison is between the mid plane temperature (i.e., ϕ_5) as a function of time, represented by Fo. The comparison appears to be really good.

Table 7.13 Tabulation of results for Example 7.10 and comparison with the one term approximation

x	Fourier number					One term
	0	0.01563	0.03125	0.04688	0.0625	Approximation $Fo = 0.0625$
0	0	0	0	0	0	0
0.0125	1	0.4639	0.3797	0.3077	0.2627	0.2629
0.025	1	0.8557	0.6633	0.5675	0.4832	0.4858
0.0375	1	0.9588	0.8508	0.7317	0.6308	0.6348
0.05	1	0.9794	0.9048	0.7912	0.6812	0.6871
0.0625	1	0.9588	0.8508	0.7317	0.6308	0.6348
0.075	1	0.8557	0.6633	0.5675	0.4832	0.4858
0.0875	1	0.4639	0.3797	0.3077	0.2627	0.2629
0.1	0	0	0	0	0	0

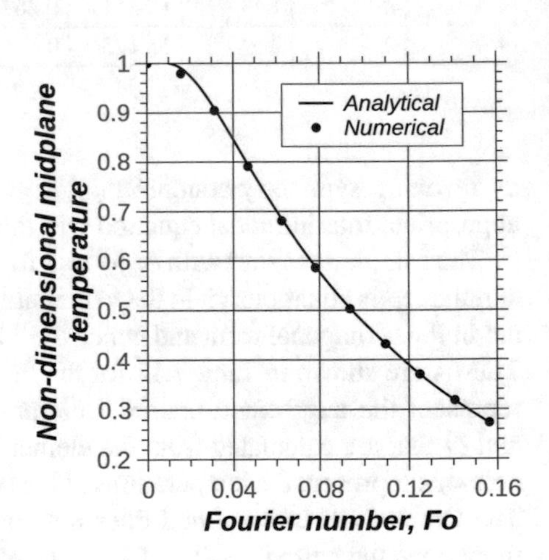

Fig. 7.16 Midplane temperature variation with time: comparison of numerical with exact solution in Example 7.10

7.2.4 *Transient Heat Transfer in a Conducting Convecting Fin*

Transient heat transfer in a fin of uniform area is considered. The fin is in the form of a flat plate of thickness 2δ, length L and initially at a uniform temperature T_b. For $t > 0$ the lateral surfaces of the fin lose heat to an ambient medium at temperature T_∞ via a heat transfer coefficient h. The fin tip is assumed to satisfy the insulated tip condition while the base remains at T_b. Numerical solution is desired using the finite difference formulation. The explicit method is intended to be made use of. Figure 7.17a is useful in deriving the nodal equation for an internal node. For energy balance we have the following:

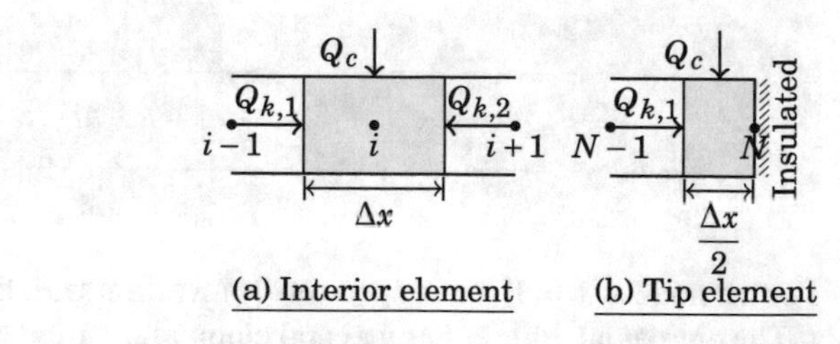

Fig. 7.17 Elements for deriving nodal equations in a uniform area fin

$$Q_{k1} + Q_{k2} + Q_c = \underbrace{\rho c 2\delta \Delta x \frac{T_{i,j+1} - T_{i,j}}{\Delta t}}_{\text{Energy stored}} \tag{7.85}$$

The conduction heat transfer rates are given by

$$Q_{k1} = 2\delta k \frac{T_{i-1,j} - T_{i,j}}{\Delta x}, \quad Q_{k2} = 2\delta k \frac{T_{i+1,j} - T_{i,j}}{\Delta x}$$

Convection heat transfer rate is given by

$$Q_c = 2\Delta x h(T_\infty - T_{i,j})$$

We substitute these in Eq. 7.85 and rearrange to get the nodal equation

$$\boxed{T_{i,j+1} = T_{i,j} + Fo_e(T_{i-1,j} - 2T_{i,j} + T_{i+1,j}) + Fo_e Bi_{\Delta x}(T_\infty - T_{i,j})\left(\frac{\Delta x}{\delta}\right)} \tag{7.86}$$

Figure 7.17b helps in deriving the equation for the tip node. For energy balance we have the following:

$$Q_{k1} + Q_c = \underbrace{\rho c 2\delta \frac{\Delta x}{2} \frac{T_{N,j+1} - T_{N,j}}{\Delta t}}_{\text{Energy stored}} \tag{7.87}$$

The conduction and convection heat transfer rates are given by

$$Q_{k1} = 2\delta k \frac{T_{N-1,j} - T_{N,j}}{\Delta x}; \quad Q_c = 2\frac{\Delta x}{2} h(T_\infty - T_{i,j})$$

We substitute these in Eq. 7.87 and rearrange to get the nodal equation

$$T_{N,j+1} = T_{N,j} + 2Fo_e(T_{N-1,j} - T_{N,j}) + Fo_e Bi_{\Delta x}(T_\infty - T_{N,j})\left(\frac{\Delta x}{\delta}\right) \tag{7.88}$$

For the node at $x = 0$ the temperature is invariant with time and hence $T_1 = T_b$ for all t. The elemental Fourier number and elemental Biot number that appear in Eqs. 7.86 and 7.88 are defined as

$$Fo_e = \frac{\alpha \Delta t}{\Delta x^2}; \quad Bi_{\Delta x} = \frac{h \Delta x}{k}$$

As always the value of the elemental Fourier number has to be less than 0.5. We leave the formulation of Crank–Nicolson scheme for the present problem as an exercise to the reader.

Example 7.11

Consider a fin in the form of a flat plate of thickness $2\delta = 3$ mm and length $L = 100$ mm. The fin is made of aluminum with $k = 237$ W/m°C, $\rho = 2702$ kg/m^3 and $c = 903$ J/kg°C. The fin loses heat to a background at $T_\infty = 30°C$ via a heat transfer coefficient of $h = 30$ W/m^2°C for $t > 0$. The fin is initially at a temperature of $T_0 = 70°C$ throughout and the fin base will remain at this temperature for all t. Assume insulated tip condition. Obtain numerically the temperature profile in the fin. Use 11 nodes along the fin length.

Solution:

Step 1 The elemental Fourier number is taken as $Fo_e = 0.25$ for this simulation. With $N = 11$, we have $\Delta x = \frac{0.1}{11-1} = 0.01$ m. The thermal diffusivity of aluminum is calculated as

$$\alpha = \frac{k}{\rho c} = \frac{237}{2702 \times 903} = 9.71 \times 10^{-5} \text{ m}^2/\text{s}$$

The time step is then given by

$$\Delta t = \frac{Fo_e \Delta x^2}{\alpha} = \frac{0.25 \times 0.01^2}{9.71 \times 10^{-5}} = 0.26 \text{ s}$$

The elemental Biot number is calculated as

$$Bi_{\Delta x} = \frac{h \Delta x}{k} = \frac{30 \times 0.01}{237} = 0.001266$$

The ratio $\frac{\Delta x}{\delta}$ is given by

$$\frac{\Delta x}{\delta} = \frac{0.01}{0.0015} = 6.666667$$

Step 2 The nodal equations are written down based on Eqs. 7.86 and 7.88. For the interior nodes we have

$$\begin{aligned} T_{i,j+1} &= T_{i,j} + 0.25\left(T_{i-1,j} - 2T_{i,j} + T_{i+1,j}\right) \\ &\quad + 0.25 \times 0.001266 \times 6.66667\left(30 - T_{i,j}\right) \\ &= T_{i,j} + 0.25\left(T_{i-1,j} - 2T_{i,j} + T_{i+1,j}\right) + 0.00211\left(30 - T_{i,j}\right) \end{aligned}$$

For the tip node we have

$$\begin{aligned} T_{11,j+1} &= T_{11,j} + 0.5\left(T_{10,j} - T_{11,j}\right) \\ &\quad + 0.25 \times 0.001266 \times 6.66667\left(30 - T_{11,j}\right) \\ &= T_{11,j} + 0.5\left(T_{10,j} - T_{11,j}\right) + 0.00211(30 - T_{11,j}) \end{aligned}$$

Step 3 The solution has been obtained by writing the formulae in to an worksheet of a spreadsheet program. The results are tabulated in Table 7.14 on page 300. The solution shows many interesting features. The steady state is reached in about 375 s or about 7 min. The numerical solution, in the steady state, is in excellent agreement with the analytical values, obtained by the well known solution given by

$$T(x) = \frac{\cosh\{m(L - x)\}}{\cosh\{mL\}}$$

where m, the fin parameter is given by $m = \sqrt{\frac{h}{k\delta}} = \sqrt{\frac{30}{237 \times 0.0015}} = 9.1863\ \mathrm{m}^{-1}$.

7.2.5 *One-Dimensional Transient in a Solid Cylinder*

One-dimensional (radial) transient heat transfer in a solid cylinder subject to convection at its boundary is considered now. The cylinder has a radius of R and is of a material with specified k and α. Initial temperature field in the cylinder is specified as a function $T(r, 0) = f(r)$. Figure 7.5a may be used for deriving the nodal equation. We assume that there is no internal heat generation in the solid cylinder.

Table 7.14 Tabulation of nodal temperatures for Example 7.11

$j \rightarrow$	1	2	3	4	5	121	386	1446	
t, s $\rightarrow$	0	0.26	0.52	0.78	1.04	31.2	100.1	375.7	$t \rightarrow \infty$
$i \downarrow$									
1	70	70	70	70	70	70	70	70	70
2	70	69.92	69.85	69.8	69.75	68.25	67.58	67.5	67.5
3	70	69.92	69.83	69.75	69.68	66.8	65.48	65.32	65.32
4	70	69.92	69.83	69.75	69.66	65.61	63.68	63.44	63.43
5	70	69.92	69.83	69.75	69.66	64.65	62.15	61.83	61.83
6	70	69.92	69.83	69.75	69.66	63.89	60.88	60.50	60.50
7	70	69.92	69.83	69.75	69.66	63.3	59.85	59.42	59.42
8	70	69.92	69.83	69.75	69.66	62.87	59.07	58.60	58.59
9	70	69.92	69.83	69.75	69.66	62.57	58.52	58.01	58.01
10	70	69.92	69.83	69.75	69.66	62.39	58.19	57.66	57.66
11	70	69.92	69.83	69.75	69.66	62.34	58.08	57.54	57.54

Explicit Formulation

Interior Node

For heat balance we have

$$\underbrace{Q_k(i-1 \rightarrow i) + Q_k(i+1 \rightarrow i)}_{\text{Conductive fluxes}} = \underbrace{\rho c 2\pi r_i \Delta r \frac{T_{i,j+1} - T_{i,j}}{\Delta t}}_{\text{Energy stored}} \tag{7.89}$$

In the explicit formulation quantities on the left hand side of Eq. 7.89 are evaluated using the known temperatures at time given by j. The terms on the left side are given by

$$Q_k(i-1 \rightarrow i) = 2k\pi r_{i-\frac{1}{2}} \frac{T_{i-1,j} - T_{i,j}}{\Delta r}; \quad Q_k(i+1 \rightarrow i) = 2k\pi r_{i+\frac{1}{2}} \frac{T_{i+1,j} - T_{i,j}}{\Delta r}$$

With these, and after minor simplification, we have

$$\boxed{T_{i,j+1} = Ti,\, j + Fo_e \left[\frac{r_{i-\frac{1}{2}}}{r_i} \left(T_{i-1,j} - T_{i,j} \right) + \frac{r_{i+\frac{1}{2}}}{r_i} \left(T_{i+1,j} - T_{i,j} \right) \right]} \tag{7.90}$$

where $Fo_e = \frac{\alpha \Delta t}{\Delta r^2}$ is the elemental Fourier number.

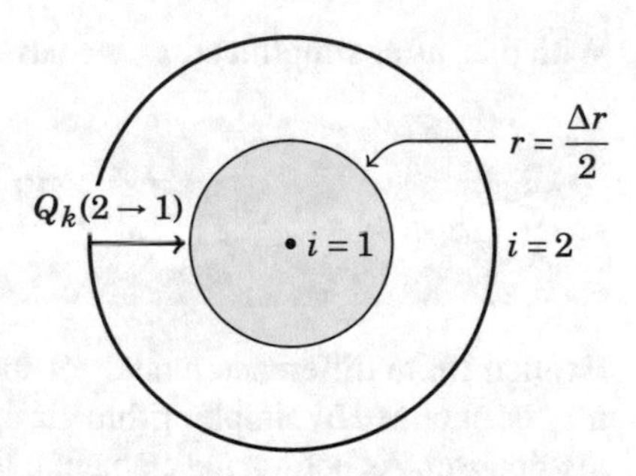

Fig. 7.18 Energy balance terminology for the center node

Surface Node

To derive the nodal equation, as usual, we make use of a half element adjacent to the boundary. We may use Fig. 7.5b to write

$$\underbrace{Q_k(N-1 \to N)}_{\text{Conductive flux}} + \underbrace{Q_c(N)}_{\text{Convective flux}} = \underbrace{\rho c 2\pi r_N \frac{\Delta r}{2} \frac{T_{N,j+1} - T_{N,j}}{\Delta t}}_{\text{Energy stored}} \tag{7.91}$$

The terms on the left hand side are given by

$$Q_k(N-1 \to N) = 2k\pi r_{N-\frac{1}{2}} \frac{T_{N-1,j} - T_{N,j}}{\Delta r}; \quad Q_c(N) = 2\pi r_N h(T_\infty - T_{N,j})$$

Introducing these in Eq. 7.91 and after minor simplification we get

$$\boxed{T_{N,j+1} = T_{N,j} + 2Fo_e(T_{N-1,j} - T_{N,j})\frac{r_{N-\frac{1}{2}}}{r_N} + 2Fo_e Bi_{\Delta r}(T_\infty - T_{N,j})} \tag{7.92}$$

where $Bi_{\Delta r}$ is the elemental Biot number.

Center Node

The node at the center requires energy balance for half element of radius $\frac{\Delta r}{2}$ surrounding the center node at $i = 1$. Figure 7.18 helps in understanding the situation. For energy balance we have

$$Q_k(2 \to 1) = \rho c \pi r^2_{1+\frac{1}{2}} \frac{T_{1,j+1} - T_{1,j}}{\Delta t} = \rho c \pi \left(\frac{\Delta r}{2}\right)^2 \frac{T_{1,j+1} - T_{1,j}}{\Delta t} \tag{7.93}$$

The heat transfer rate on the left hand side of Eq. 7.93 is written down as

$$Q_k(2 \to 1) = 2k\pi \frac{\Delta r}{2} \frac{T_{2,j} - T_{1,j}}{\Delta r}$$

With this, after simplification, we have the nodal equation

$$\boxed{T_{1,j+1} = T_{1,j} + 4Fo_e(T_{2,j} - T_{1,j})} \tag{7.94}$$

Explicit finite difference model for the problem is thus complete and the solution may be obtained by simple arithmetic operations, starting from the initial temperature distribution. As before, the elemental Fourier number has to be chosen less than 0.5.

Crank–Nicolson Scheme

The reader should be able to derive the finite difference equations for both the implicit as well as the Crank–Nicolson schemes, following the procedure given for the slab problem. For example, in the Crank–Nicolson case, Eqs. 7.90, refeq:7e91, and 7.94 are modified as given by the following equations:

Interior node:

$$-\frac{Fo_e}{2}\frac{r_{i-\frac{1}{2}}}{r_i}T_{i-1,j+1} + (1 + Fo_e)T_{i,j+1} - \frac{Fo_e}{2}\frac{r_{i+\frac{1}{2}}}{r_i}T_{i+1,j+1}$$
$$= \frac{Fo_e}{2}\left(\frac{r_{i-\frac{1}{2}}}{r_i}T_{i-1,j} + \frac{r_{i+\frac{1}{2}}}{r_i}T_{i+1,j}\right) - (1 - Fo_e)r_i T_{i,j} \tag{7.95}$$

Surface node:

$$-\frac{Fo_e}{2}\frac{r_{N-\frac{1}{2}}}{r_N}T_{N-1,j+1} + \left(1 + Fo_e\frac{r_{N-\frac{1}{2}}}{r_N} + Fo_e Bi_{\Delta r}\right)T_{N,j+1}$$
$$= T_{N,j}\left(1 - Fo_e\frac{r_{N-\frac{1}{2}}}{r_N} - Fo_e Bi_{\Delta r}\right) + 2Fo_e Bi_{\Delta r}T_\infty \tag{7.96}$$

Center node:

$$T_{1,j+1}(1 + 2Fo_e) - 2Fo_e T_{2,j+1} = T_{1,j}(1 - 2Fo_e) + 2Fo_e T_{2,j} \tag{7.97}$$

7.2.6 *One-Dimensional Transient in a Solid Sphere*

One-dimensional (radial) transient heat transfer in a solid sphere subject to convection at its boundary is considered now. The sphere has a radius of R and is of a material with specified k and α. Initial temperature field in the sphere is specified as a function $T(r, 0) = f(r)$. Figure 7.5a may be used for deriving the nodal equation. We assume that there is no internal heat generation in the solid sphere.

Explicit Formulation

Interior Node

For heat balance we have

$$\underbrace{Q_k(i-1 \to i) + Q_k(i+1 \to i)}_{\text{Conductive fluxes}} = \underbrace{\rho c 4\pi r_i^2 \Delta r \frac{T_{i,j+1} - T_{i,j}}{\Delta t}}_{\text{Energy stored}} \tag{7.98}$$

In the explicit formulation quantities on the left hand side of Eq. 7.89 are evaluated using the known temperatures at time given by j. The terms on the left side is given by

$$Q_k(i-1 \to i) = 4k\pi r_{i-\frac{1}{2}}^2 \frac{T_{i-1,j} - T_{i,j}}{\Delta r}; \quad Q_k(i+1 \to i) = 4k\pi r_{i+\frac{1}{2}}^2 \frac{T_{i+1,j} - T_{i,j}}{\Delta r}$$

With these, and after minor simplification, we have

$$\boxed{T_{i,j+1} = Ti, j + Fo_e \left[r_{i-\frac{1}{2}}^2 \left(T_{i-1,j} - T_{i,j} \right) + r_{i+\frac{1}{2}}^2 \left(T_{i+1,j} - T_{i,j} \right) \right]} \tag{7.99}$$

where $Fo_e = \frac{\alpha \Delta t}{\Delta r^2}$ is the elemental Fourier number.

Surface Node

To derive the nodal equation, as usual, we make use of a half element adjacent to the boundary. We may use Fig. 7.5b as a guide to write

$$\underbrace{Q_k(N-1 \to N) +}_{\text{Conductive flux}} \underbrace{Q_c(N)}_{\text{Convective flux}} = \underbrace{\rho c 4\pi r_N^2 \frac{\Delta r}{2} \frac{T_{N,j+1} - T_{N,j}}{\Delta t}}_{\text{Energy stored}} \tag{7.100}$$

The terms on the left hand side is given by

$$Q_k(N-1 \to N) = 4k\pi r_{N-\frac{1}{2}}^2 \frac{T_{N-1,j} - T_{N,j}}{\Delta r}; \quad Q_c(N) = 4\pi r_N^2 h (T_\infty - T_{N,j})$$

Introducing these in Eq. 7.100 and after minor simplification we get

$$T_{N,j+1} = T_{N,j} + 2Fo_e\left(T_{N-1,j} - T_{N,j}\right)\frac{r^2_{N-\frac{1}{2}}}{r^2_N} + 2Fo_e Bi_{\Delta r}\left(T_\infty - T_{N,j}\right) \tag{7.101}$$

where $Bi_{\Delta r}$ is the elemental Biot number.

Center Node

The node at the center requires energy balance for half element of radius $\frac{\Delta r}{2}$ surrounding the center node at $i = 1$. Figure 7.18 helps in understanding the situation, in this case also. or energy balance we have

$$\underbrace{Q_k(2 \to 1)}_{\text{Conductive flux}} = \underbrace{\rho c \frac{4}{3}\pi r^3_{1+\frac{1}{2}} \frac{T_{1,j+1} - T_{1,j}}{\Delta t}}_{\text{Energy stored}} = \underbrace{\rho c \frac{4}{3}\pi \left(\frac{\Delta r}{2}\right)^3 \frac{T_{1,j+1} - T_{1,j}}{\Delta t}}_{\text{Energy stored}} \tag{7.102}$$

The heat transfer rate on the left hand side of Eq. 7.93 is written down as

$$Q_k(2 \to 1) = 4k\pi\left(\frac{\Delta r}{2}\right)^2 \frac{T_{2,j} - T_{1,j}}{\Delta r}$$

With this, after simplification, we have the nodal equation

$$T_{1,j+1} = T_{1,j} + 6Fo_e\left(T_{2,j} - T_{1,j}\right) \tag{7.103}$$

Explicit finite difference model for the problem is thus complete and the solution may be obtained by simple arithmetic operations, starting from the initial temperature distribution. As before, the elemental Fourier number has to be chosen less than 0.5.

Crank–Nicolson Scheme

The reader should be able to derive the finite difference equations for both the implicit as well as the Crank–Nicolson schemes, following the procedure given for the slab problem. For example, in the Crank–Nicolson case, Eqs. 7.99, 7.101, and 7.103 are modified as given by the following equations:

Interior node:

$$-\frac{Fo_e}{2}r^2_{i-\frac{1}{2}}T_{i-1,j+1} + (1 + Fo_e)r_i^2 T_{i,j+1} - \frac{Fo_e}{2}r^2_{i+\frac{1}{2}}T_{i+1,j+1}$$
$$= \frac{Fo_e}{2}\left(r^2_{i-\frac{1}{2}}T_{i-1,j} + r^2_{i+\frac{1}{2}}T_{i+1,j}\right) - (1 - Fo_e)r_i^2 T_{i,j} \tag{7.104}$$

Surface node:

$$-\frac{Fo_e}{2}\frac{r^2_{N-\frac{1}{2}}}{r^2_N}T_{N-1,j+1} + \left(1 + Fo_e\frac{r^2_{N-\frac{1}{2}}}{r^2_N} + Fo_e Bi_{\Delta r}\right)T_{N,j+1}$$

$$= T_{N,j}\left(1 - Fo_e\frac{r^2_{N-\frac{1}{2}}}{r^2_N} - Fo_e Bi_{\Delta r}\right) + 2Fo_e Bi_{\Delta r}T_\infty \quad (7.105)$$

Center node:

$$T_{1,j+1}(1 + 3Fo_e) - 3Fo_e T_{2,j+1} = T_{1,j}(1 - 3Fo_e) + 3Fo_e T_{2,j} \quad (7.106)$$

Example 7.12

A solid sphere of radius 0.1 m is initially at a temperature of 100 °C throughout. For $t > 0$ its surface is subject to convective cooling by an ambient fluid at 30 °C via a heat transfer coefficient of 30 W/m^2 °C. Consider 9 uniformly spaced nodes along the radius of the sphere. The thermal conductivity of the material of the sphere is 15 W/m°C while the thermal diffusivity is given to be 9.76×10^{-7} m^2/s. Obtain the solution for a few time steps using an elemental Fourier number of 0.25. Use an explicit scheme.

Solution : The given data is written down using the familiar notation:

$T_0 = 100$°C	$T_\infty = 30$°C
$k = 15$ W/m°C	$h = 30$ W/m^2°C
$R = 0.1$ m; N = 9	$\Delta r = 0.0125$ m
$\alpha = 10^{-6}$m^2/s, $Fo_e = 0.25$	$\Delta t = \frac{0.25 \times 0.0125^2}{9.76 \times 10^{-7}} = 40$ s
$Bi = \frac{30 \times 0.1}{15} = 0.2$	$Bi_{\Delta r} = \frac{30 \times 0.0125}{15} = 0.025$

The nodal temperatures are calculated based on nodal Eqs. 7.99, 7.101 and 7.103. The calculations have been performed using a spreadsheet and the results are given in Table 7.15. The center temperature evaluated by the numerical method is compared with the one term approximation for three values of the Fourier number (viz. $Fo =$ 0.1, 0.3 and 0.5). The comparison can be termed as good. Fourier number is defined as $\frac{\alpha t}{R^2}$ where $t = (j - 1)\Delta t$. Improvement in the explicit solution may be demonstrated

Table 7.15 Tabulation of nodal temperatures for Example 7.12

		$r_i \rightarrow$	0	0.0125	0.025	0.0375	0.05	0.0625	0.075	0.0875	0.1
		$i \rightarrow$	1	2	3	4	5	6	7	8	9
$j \downarrow$	$t \downarrow$	$Fo \downarrow$									
1	0	0	100	100	100	100	100	100	100	100	100
2	40	0.004	100	100	100	100	100	100	100	100	99.13
3	80	0.008	100	100	100	100	100	100	100	99.75	98.65
4	120	0.012	100	100	100	100	100	100	99.93	99.49	98.27
5	160	0.016	100	100	100	100	100	99.98	99.81	99.23	97.95
6	200	0.02	100	100	100	100	99.99	99.93	99.68	98.99	97.67
7	240	0.024	100	100	100	100	99.98	99.87	99.53	98.76	97.40
8	280	0.028	100	100	100	99.99	99.95	99.79	99.37	98.54	97.16
9	320	0.032	100	100	100	99.98	99.90	99.69	99.21	98.32	96.92
10	360	0.036	100	100	99.99	99.96	99.85	99.59	99.05	98.11	96.70
11	400	0.04	100	99.99	99.98	99.93	99.79	99.48	98.89	97.91	96.49
…	…	…	…	…	…	…	…	…	…	…	…
26	1000	0.1	99.28 (99.99)	99.21	99.03	98.70	98.18	97.46	96.50	95.29	93.83
…	…	…	…	…	…	…	…	…	…	…	…
76	3000	0.3	92.64 (92.37)	92.55	92.29	91.84	91.21	90.39	89.4	88.24	86.91
…	…	…	…	…	…	…	…	…	…	…	…
126	5000	0.5	86.01 (85.57)	85.93	85.70	85.30	84.73	84.00	83.11	82.07	80.88

Figures in bracket are analytical one term approximation values

with smaller nodal spacing and correspondingly smaller time step size. In practice one would perform a grid dependence study to determine the required nodal spacing (see Sect. 7.1.2).

The reader is encouraged to use the CN scheme and redo the problem.

7.3 Transient Conduction in Two and Three Dimensions

In Chap. 6 we have already dealt with unsteady conduction in two and three dimensions by the use of analytical methods leading to the one term approximation. The discussion here will be the solution of such problems by the finite difference method, and without the limitations that go with the one term approximation.

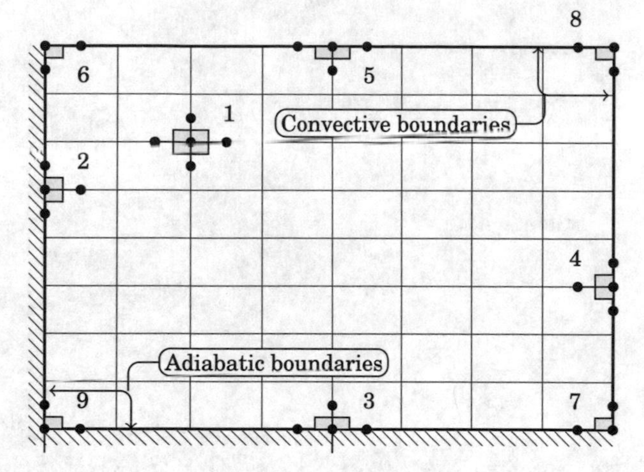

Fig. 7.19 Typical nodes for numerical solution of transient heat conduction in a rectangle

7.3.1 Transient Conduction in a Rectangle: Explicit Formulation

The problem geometry is described by Fig. 6.6. One quarter of the physical domain is treated as the computational domain based on the indicated symmetries. Figure 7.19 shows the domain along with representative elements that need to be modeled numerically. The following types of nodes are to be considered:

Interior node (1)	Face nodes (2,3,4,5)
External corners (6,7,8)	Internal corner (9)

Nodal equations may be derived, in the case of an explicit formulation, using procedure that has been followed till now. The reader is encouraged to verify that the following equations will result from such an exercise. The following parameters are defined in writing the finite difference equations:

$$\text{Number of nodes along } x : M = \frac{L_x}{\Delta x} + 1$$
$$\text{Number of nodes along } y : N = \frac{L_y}{\Delta y} + 1$$
$$\text{Elemental Fourier number based on } \delta x : Fo_{\Delta x} = \frac{\alpha \Delta t}{\Delta x^2}$$

$$\text{Elemental Fourier number based on } \delta y : Fo_{\Delta y} = \frac{\alpha \Delta t}{\Delta y^2}$$
$$\text{Elemental Biot number based on } \delta x : Bi_{\Delta x} = \frac{h \Delta x}{k}$$
$$\text{Elemental Biot number based on } \delta y : Bi_{\Delta y} = \frac{h \Delta y}{k}$$

Interior node (1):

$$T_{i,j}^{k+1} - T_{i,j}^{k} = Fo_{\Delta x}\left(T_{i-1,j}^{k} - T_{i,j}^{k} + T_{i+1,j}^{k}\right) + Fo_{\Delta y}\left(T_{i,j-1}^{k} - 2T_{i,j}^{k} + T_{i,j+1}^{k}\right) \tag{7.107}$$

where $2 \leq i \leq M-1$ and $2 \leq j \leq N-1$.

Surface node (2):

$$T_{1,j}^{k+1} - T_{1,j}^{k} = 2Fo_{\Delta x}\left(T_{2,j}^{k} - T_{i,j}^{k}\right) + Fo_{\Delta y}\left(T_{1,j-1}^{k} - 2T_{1,j}^{k} + T_{1,j+1}^{k}\right) \tag{7.108}$$

where $2 \leq j \leq N-1$.

Surface node (3):

$$T_{i,1}^{k+1} - T_{i,1}^{k} = Fo_{\Delta x}\left(T_{i-1,1}^{k} - T_{i,1}^{k} + T_{i+1,1}^{k}\right) + 2Fo_{\Delta y}\left(T_{i,2}^{k} - T_{i,j}^{k}\right) \tag{7.109}$$

where $2 \leq i \leq M-1$.

Face node (4):

$$\begin{aligned} T_{M,j}^{k+1} - T_{M,j}^{k} = {} & 2Fo_{\Delta x}\left(T_{M-1,j}^{k} - T_{M,j}^{k}\right) + Fo_{\Delta y}\left(T_{M,j-1}^{k} - 2T_{M,j}^{k} + T_{M,j+1}^{k}\right) \\ & +2\sqrt{Fo_{\Delta x}Fo_{\Delta y}}Bi_{\Delta y}\left(T_{\infty} - T_{M,j}^{k}\right) \end{aligned} \tag{7.110}$$

where $2 \leq j \leq N-1$.

Face node (5):

$$\begin{aligned} T_{i,N}^{k+1} - T_{i,N}^{k} = {} & Fo_{\Delta x}\left(T_{i-1,N}^{k} - 2T_{i,N}^{k} + T_{i+1,N}^{k}\right) + 2Fo_{\Delta y}\left(T_{i,N-1}^{k} - T_{i,N}^{k}\right) \\ & +2\sqrt{Fo_{\Delta x}Fo_{\Delta y}}Bi_{\Delta x}(T_{\infty} - T_{i,N}^{k}) \end{aligned} \tag{7.111}$$

where $2 \leq i \leq M-1$.

Corner node (6):

$$\begin{aligned} T_{1,N}^{k+1} - T_{1,N}^{k} = {} & 2Fo_{\Delta x}\left(T_{2,N}^{k} - T_{1,N}^{k}\right) + 2Fo_{\Delta y}\left(T_{1,N-1}^{k} - T_{1,N}^{k}\right) \\ & +2\sqrt{Fo_{\Delta x}Fo_{\Delta y}}Bi_{\Delta x}\left(T_{\infty} - T_{1,N}^{k}\right) \end{aligned} \tag{7.112}$$

Corner node (7):

$$\begin{aligned} T_{M,1}^{k+1} - T_{M,1}^{k} = {} & 2Fo_{\Delta x}\left(T_{M-1,1}^{k} - T_{M,1}^{k}\right) + 2Fo_{\Delta y}\left(T_{M,2}^{k} - T_{M,1}^{k}\right) \\ & +2\sqrt{Fo_{\Delta x}Fo_{\Delta y}}Bi_{\Delta y}\left(T_{\infty} - T_{M,1}^{k}\right) \end{aligned} \tag{7.113}$$

Corner node (8):

$$T_{M,N}^{k+1} - T_{M,N}^{k} = 2Fo_{\Delta x}(T_{M-1,N}^{k} - T_{M,N}^{k}) + 2Fo_{\Delta y}(T_{M,N-1}^{k} - T_{M,N}^{k}) + 2\sqrt{Fo_{\Delta x}Fo_{\Delta y}}\left(Bi_{\Delta x} + Bi_{\Delta y}\right)\left(T_{\infty} - T_{M,N}^{k}\right) \quad (7.114)$$

Corner node (9):

$$T_{1,1}^{k+1} - T_{1,1}^{k} = 2Fo_{\Delta x}\left(T_{2,1}^{k} - T_{1,1}^{k}\right) + 2Fo_{\Delta y}\left(T_{1,2}^{k} - T_{1,1}^{k}\right) \quad (7.115)$$

In Eqs. 7.107–7.115 the superscript k is the time step index defined such that $t = (k-1)\Delta t$. As before the explicit method works only if $Fo_{\Delta x} + Fo_{\Delta y} \le \frac{1}{2}$. However the explicit method is simple to implement since the temperatures may be updated in time by the application of the above nodal equations, starting with the initial temperature values.

7.3.2 ADI Method

A superior scheme for solving multi-dimensional transient conduction problems is the Alternate Direction Implicit or the ADI scheme.[1] This scheme is unconditionally stable and this is the most important *advantage* of the method.

The explicit formulation presented above has shown that the nodal temperature (for an internal node such as 1 in Fig. 7.19) at $t + \Delta t$ depends on four nearest neighbor temperatures that are prescribed at t. In case we were to use an implicit scheme the four nearest neighbors would have to be updated along with the middle nodal temperature and hence we would have a matrix equation with a banded coefficient matrix that is not tridiagonal. However it is possible to reduce the matrix to tridiagonal form if the two directions are treated alternately as implicit and explicit, as given below. This in essence is the idea behind the ADI scheme. Another way of looking at the ADI is to realize that a spatially two-dimensional problem *is reduced* to two spatially one-dimensional problems.

The time step Δt is divided in to two half steps of $\frac{\Delta t}{2}$ each. Thus we define nodal temperatures at $t + \frac{\Delta t}{2}$ or $k + \frac{1}{2}$ also. During the first half time step, i.e., from k to $k + \frac{1}{2}$ the spatial derivatives with respect to x are treated implicitly while the spatial derivative with respect to y is treated explicitly. Thus the Laplacian is written as

$$\nabla^2 T = \frac{T_{i-1,j}^{k+\frac{1}{2}} - 2T_{i,j}^{k+\frac{1}{2}} + T_{i+1,j}^{k+\frac{1}{2}}}{\Delta x^2} + \frac{T_{i,j-1}^{k} - 2T_{i,j}^{k} + T_{i,j+1}^{k}}{\Delta y^2} \quad (7.116)$$

[1] D.W. Peaceman and H.H. Rachford, The numerical solution of parabolic and elliptic differential equations, SIAM, Vol. 3, pp. 28–41, 1955; see interesting note in SIAM News, Vol. 39, No. 2, March 2006.

The heat equation is thus written for an internal node (such as Node 1 in Fig. 7.19 as

$$-\frac{Fo_x}{2}T_{i-1,j}^{k+\frac{1}{2}} + (1+Fo_x)T_{i,j}^{k+\frac{1}{2}} - \frac{Fo_x}{2}T_{i+1,j}^{k+\frac{1}{2}} = T_{i,j}^{k} + \frac{Fo_y}{2}\left(T_{i,j-1}^{k} - 2T_{i,j}^{k} + T_{i,j+1}^{k}\right) \tag{7.117}$$

During the second half time step, i.e., from $k+\frac{1}{2}$ to $k+1$ the spatial derivatives with respect to y are treated implicitly while the spatial derivative with respect to x is treated explicitly. The Laplacian takes the form

$$\nabla^2 T = \frac{T_{i-1,j}^{k+\frac{1}{2}} - 2T_{i,j}^{k+\frac{1}{2}} + T_{i+1,j}^{k+\frac{1}{2}}}{\Delta x^2} + \frac{T_{i,j-1}^{k+1} - 2T_{i,j}^{k+1} + T_{i,j+1}^{k+1}}{\Delta y^2} \tag{7.118}$$

The heat equation is thus written for an internal node as

$$-\frac{Fo_y}{2}T_{i,j-1}^{k+1} + \left(1+Fo_y\right)T_{i,j}^{k+1} - \frac{Fo_y}{2}T_{i,j+1}^{k+1} = T_{i,j}^{k+\frac{1}{2}} + \frac{Fo_x}{2}\left(T_{i-1,j}^{k+\frac{1}{2}} - 2T_{i,j}^{k+\frac{1}{2}} + T_{i+1,j}^{k+\frac{1}{2}}\right) \tag{7.119}$$

Equations 7.117 and 7.119 are valid for $2 \le i \le M-1$ and $2 \le j \le N-1$. The reader may derive the equations for all the elements shown in Fig. 7.19 by following the above procedure. The final form of nodal equations are given below.

Surface node 2, First half time step:

$$(1+Fo_x)T_{1,j}^{k+\frac{1}{2}} - Fo_x T_{2,j}^{k+\frac{1}{2}} = T_{1,j}^{k} + \frac{Fo_y}{2}\left(T_{1,j-1}^{k} - 2T_{1,j}^{k} + T_{1,j+1}^{k}\right) \tag{7.120}$$

where $2 \le j \le N-1$.

Surface node 2, Second half time step:

$$-\frac{Fo_y}{2}T_{1,j-1}^{k+1} + (1+Fo_y)T_{1,j}^{k+1} - \frac{Fo_y}{2}T_{1,j+1}^{k+1} = T_{1,j}^{k+\frac{1}{2}} + Fo_x\left(-T_{1,j}^{k+\frac{1}{2}} + T_{2,j}^{k+\frac{1}{2}}\right) \tag{7.121}$$

Surface node 3, First half time step:

$$-\frac{Fo_x}{2}T_{i-1,1}^{k+\frac{1}{2}} + (1+Fo_x)T_{i,1}^{k+\frac{1}{2}} - \frac{Fo_x}{2}T_{i+1,1}^{k+\frac{1}{2}} = T_{i,1}^{k} + Fo_y\left(-T_{i,1}^{k} + T_{i,2}^{k}\right) \tag{7.122}$$

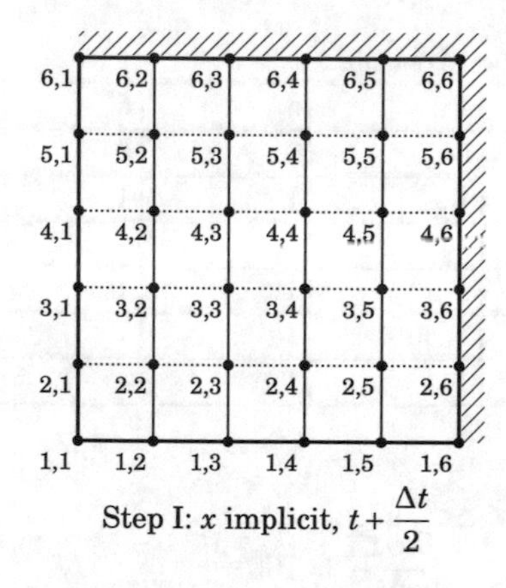

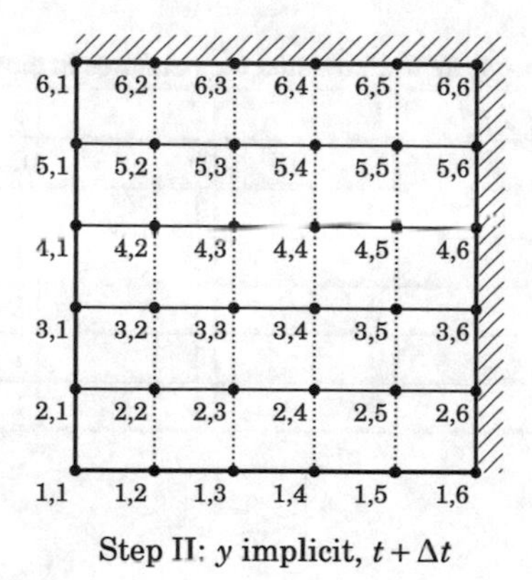

Fig. 7.20 Defining sketch for Example 7.13

where $2 \leq i \leq M - 1$.

Surface node 3, Second half time step:

$$(1 + Fo_y)T_{i,1}^{k+1} - Fo_y T_{i,2}^{k+1} = T_{i,1}^{k+\frac{1}{2}} + \frac{Fo_x}{2}\left(T_{i-1,1}^{k+\frac{1}{2}} - 2T_{i,1}^{k+\frac{1}{2}} + T_{i+1,1}^{k+\frac{1}{2}}\right) \tag{7.123}$$

Example 7.13

A very long bar of square section 0.1×0.1 m is initially at a uniform temperature of 1 throughout. For $t > 0$ all the surfaces of the bar are brought to 0 temperature and held fixed at that value. Using the ADI method obtain the temperature distribution in the bar for one ADI cycle. Choose 11 nodes along each direction and an elemental Fourier number of 1. The material of the bar has a thermal diffusivity of 10^{-6} m^2/s.

Solution:

Step 1 Referring to Fig. 7.20, we invoke symmetry and consider only quarter of the bar as shown. The number of nodes within the *computational domain* will be six in each direction. The elemental thicknesses are

$$\Delta x = \frac{0.1}{11 - 1} = 0.01 \text{ m and } \Delta y = \frac{0.1}{11 - 1} = 0.01 \text{ m}$$

With $\alpha = 10^{-6}$ m^2/s and elemental Fourier number $Fo_e = 1$ (based on either Δx or Δy) the time step is given by

Table 7.16 Initial nodal temperatures in the computational domain

$i \downarrow j \rightarrow$	1	2	3	4	5	6
1	0	0	0	0	0	0
2	0	1	1	1	1	1
3	0	1	1	1	1	1
4	0	1	1	1	1	1
5	0	1	1	1	1	1
6	0	1	1	1	1	1

$$\Delta t = \frac{Fo_e \Delta x^2}{\alpha} = \frac{1 \times 0.01^2}{10^{-6}} = 100\,\text{s}$$

Step 2 The ADI method will thus split the time step in to two half steps of 50 s each. The top and right boundaries will be imposed adiabatic boundary conditions while the left and bottom are isothermal as shown in the figure. With $Fo_e = 1$ the finite difference form of the nodal equation, during the first step is written based on Eq. 7.117 as

$$-T_{i-1,j}^{k+\frac{1}{2}} + 4T_{i,j}^{k+\frac{1}{2}} - T_{i+1,j}^{k+\frac{1}{2}} = T_{i,j-1}^{k} + T_{i,j+1}^{k}$$

The above is valid for $2 \le i \le 5$ and $2 \le j \le 5$ The nodal equation for the node on the adiabatic boundary, as for example $6, j$ is

$$-2T_{5,j}^{k+\frac{1}{2}} + 4T_{6,j}^{k+\frac{1}{2}} = T_{6,j-1}^{k} + T_{6,j+1}^{k}$$

for $2 \le j \le 5$. For node 6, 6 itself we have

$$-2T_{5,6}^{k+\frac{1}{2}} + 4T_{6,6}^{k+\frac{1}{2}} = 2T_{6,5}^{k}$$

Step 3 The set of equations along each row is in tridiagonal matrix form and may be solved by TDMA. The initial nodal temperatures are specified as in Table 7.16.

The row-wise calculation during the first step is shown for the second row, as an example. Note that the first row does not need any calculation. The nodal equations in matrix form are given, along with the solution obtained by TDMA in Table 7.17.

We complete the calculations for all the rows and get the nodal temperatures at $t = 50$ s as given in Table 7.18.

Step 4 During the second step, the calculations proceed column-wise and the procedure is similar to what has been used above. The nodal equations are

Table 7.17 Temperatures on Row 2 at $t = 50$ s

a_i	b_i	c_i	d_i	P	Q	x	$T_{2,j}^{\frac{3}{2}}$
4	1	0	1	0.25	0.25	0.01	0.3660
4	1	1	1	0.2667	0.3333	0.02	0.4641
4	1	1	1	0.2679	0.3571	0.03	0.4903
4	1	1	1	0.2679	0.3636	0.04	0.4972
4	0	2	1	0	0.4986	0.05	0.4986

Table 7.18 Nodal temperatures in the computational domain at $t = 50$ s

$i \downarrow j \rightarrow$	1	2	3	4	5	6
1	0	0	0	0	0	0
2	0	0.3660	0.4641	0.4903	0.4972	0.4986
3	0	0.7320	0.9282	0.9807	0.9945	0.9972
4	0	0.7320	0.9282	0.9807	0.9945	0.9972
5	0	0.7320	0.9282	0.9807	0.9945	0.9972
6	0	0.7320	0.9282	0.9807	0.9945	0.9972

$$-T_{i,j-1}^{k+1} + 4T_{i,j}^{k+1} - T_{i,j+1}^{k+1} = T_{i-1,j}^{k+\frac{1}{2}} + T_{i_1,j}^{k+\frac{1}{2}}$$

The above is valid for $2 \leq i \leq 5$ and $2 \leq j \leq 5$ The nodal equation for the node on the adiabatic boundary, as for example i, 6 is

$$-2T_{i,5}^{k+1} + 4T_{i,6}^{k+1} = T_{i-1,6}^{k+\frac{1}{2}} + T_{i+1,6}^{k+\frac{1}{2}}$$

for $2 \leq i \leq 5$. For node 6, 6 itself we have

$$-2T_{6,5}^{k+1} + 4T_{6,6}^{k+1} = 2T_{5,6}^{k+\frac{1}{2}}$$

Step 5 The final nodal temperatures at $t = 100$ s alone are given in Table 7.19. This completes one ADI cycle that takes the solution from $t = 0$ to $t = 100$ s. Note that the solution shows symmetry with respect to the diagonal, as it should.

Step 6 The procedure is continued for as many ADI cycles as needed to complete the solution.

Table 7.19 Nodal temperatures in the computational domain at $t = 100$ s

$i \downarrow j \rightarrow$	1	2	3	4	5	6
1	0	0	0	0	0	0
2	0	0.2154	0.3974	0.4461	0.4590	0.4615
3	0	0.3974	0.7333	0.8232	0.8469	0.8516
4	0	0.4461	0.8232	0.9241	0.9507	0.9560
5	0	0.4590	0.8469	0.9507	0.9780	0.9835
6	0	0.4615	0.8516	0.9560	0.9835	0.9890

7.3.3 Modification of the ADI Method for Three Dimensional Transient Conduction

The ADI method presented above is conditionally stable for three dimensional problems. Douglas and Gunn method[2] modifies the method to make it unconditionally stable for three dimensional conduction problems. Consider transient heat conduction in a rectangular parallelepiped. Temperature varies with x, y, z and t. Time step is split in to three parts as shown in Fig. 7.21. In the interval $t, t + \frac{\Delta t}{3}$ the second derivative with respect to x is considered semi-implicitly while the derivatives in the y and z directions are considered explicitly. Thus, in Step I, at the node i, j, k (node indices are i for the x, j for the y, k for the z directions and l for time) the heat equation is represented in finite difference form as

$$\boxed{\frac{T_{i,j,k}^{l+\frac{1}{3}} - T_{i,j,k}^{l}}{\frac{1}{3}\Delta t} = \frac{\delta_x^2 T_{i,j,k}^{l+\frac{1}{3}} + \delta_x^2 T_{i,j,k}^{l}}{2} + \delta_y^2 T_{i,j,k}^{l} + \delta_z^2 T_{i,j,k}^{l}} \tag{7.124}$$

where δ^2 stands for the central difference for the second derivative. For example, $\delta_x^2 T_{i,j,k}^{l} = \frac{T_{i-1,j,k}^{k} - 2T_{i,j,k}^{l} + T_{i+1,j,k}^{l}}{\Delta x^2}$ and similarly for the other two directions. We note that the above differs from the ADI in that the second derivative with respect to x is taken as the mean of the values at l and $l + \frac{1}{3}$.

In Step II, the time changes from $t + \frac{\Delta t}{3}$ to $t + \frac{2\Delta t}{3}$ and the heat equation is written as

$$\frac{T_{i,j,k}^{l+\frac{2}{3}} - T_{i,j,k}^{l}}{\frac{2\Delta t}{3}} = \frac{\delta_x^2 T_{i,j,k}^{l+\frac{1}{3}} + \delta_x^2 T_{i,j,k}^{l}}{2} + \frac{\delta_y^2 T_{i,j,k}^{l+\frac{2}{3}} + \delta_y^2 T_{i,j,k}^{l}}{2} + \delta_z^2 T_{i,j,k}^{l} \tag{7.125}$$

Note that y direction is now treated as the semi-implicit direction. Finally, in Step III, the z direction is treated as the semi-implicit direction and we write the heat equation

[2] J. Douglas Jr. and J.E. Gunn, A general formulation of alternate direction methods—part i. parabolic and hyperbolic problems, Numerische Mathematik, Vol. 6, pp. 428–453, 1964.

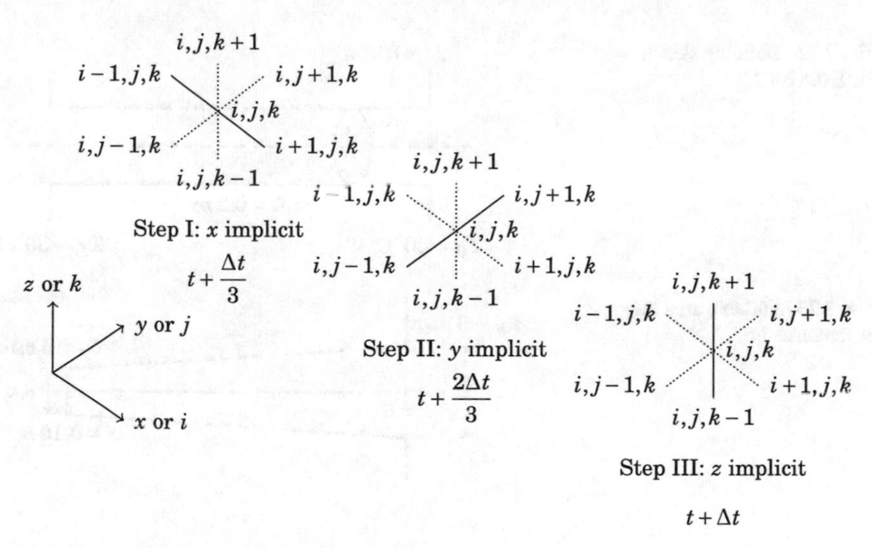

Fig. 7.21 Douglas Gunn scheme for three dimensional problem

in the form

$$\frac{T^{l+1}_{i,j,k}-T^{l}_{i,j,k}}{\Delta t}=\frac{\delta^2_x T^{l+\frac{1}{3}}_{i,j,k}+\delta^2_x T^{l}_{i,j,k}}{2}+\frac{\delta^2_y T^{l+\frac{2}{3}}_{i,j,k}+\delta^2_y T^{l}_{i,j,k}}{2}+\frac{\delta^2_z T^{l+1}_{i,j,k}+\delta^2_x T^{l}_{i,j,k}}{2} \tag{7.126}$$

Concluding Remarks

Numerical solution of heat conduction equation has been treated in adequate detail. One dimensional space/time problems are modeled by ordinary differential equations (ODEs) that are solved by a suitable ODE solver such as the Runge Kutta method. These may also be solved by finite difference methods. Problems in more than one dimension requires the solution of partial differential equations. Finite difference method is the simplest to understand and is presented in this chapter. Explicit, implicit and Crank Nicolson schemes have been discussed in detail. Question of numerical stability has been discussed. Finite Element and Finite Volume methods are other alternatives which are best learned from specialized books on these topics.

7.4 Exercises

Ex 7.1 The temperature of a first order system satisfies the equation $\frac{dT}{dt}+0.05T=0.75[1+\sin(0.05t)]$. The temperature is in °C and time t is in s. Initial system temperature is 30 °C. Obtain the response of the system at 25 s by numerically solving the differential equation by

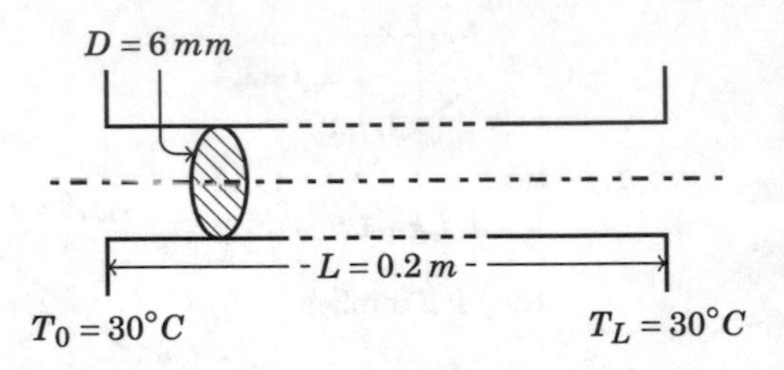

Fig. 7.22 Defining sketch for Exercise 7.2

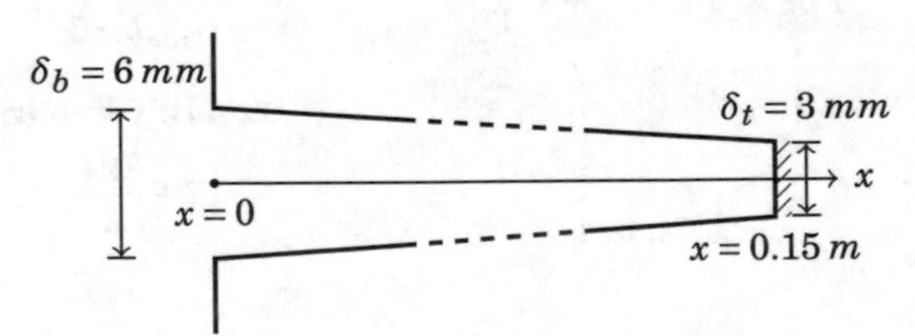

Fig. 7.23 Variable area fin of Exercise 7.5

choosing a proper time step. You may make use of the 4th order Runge Kutta method. Compare this with the exact analytical value.

Ex 7.2 A circular rod of a material (see Fig. 7.22) of thermal conductivity 110 W/m°C has a diameter of 6 mm and is 200 mm long. The two ends of the rod are maintained at 30 °*C*. The lateral surface loses heat to an ambient at 35°*C* subject to a heat transfer coefficient of 26 W/m^2 °C. A current passes through the material and generates heat at a constant rate of 10^5 W/m^3. Assume that the temperature field is one-dimensional. Determine the maximum temperature in the rod. Use the finite difference method with a suitable number of nodes along the length of the rod.

Ex 7.3 Consider a circular disk of radius $R = 0.05$ m and thickness $\delta =$ 0.003 m made of a material of thermal conductivity $k = 110$ W/m°C. It receives a constant heat flux of $q = 10^4$ W/m^2 on one side while the other side is perfectly insulated. The edge of the disk is maintained at a temperature of $T_e = 15$°C. What is the temperature at the center of the disk? You have to obtain the solution by the finite difference method. You may divide the disk in the radial direction by taking steps of $\Delta r = 0.01$ m. Use TDMA to obtain the solution to the nodal equations.

Ex 7.4 A fin made of aluminum having a thermal conductivity of 200 W/m°C is in the form of a flat plate that is 150 *mm* long and 6 *mm* thick throughout. The base is held at a temperature excess of 100 °C with respect to the ambient that is at 25 °C. Neglect heat loss from the tip. Choose 6 uniformly spaced nodes from end to end along the length of the fin, use central difference approximation and obtain the nodal temperatures. Determine the heat loss from the fin per unit width. Compare this with the exact value.

Ex 7.5 A variable area fin shown in Fig. 7.23 has a longitudinal section in the form of a trapezium as shown in the figure. The material of the fin has a thermal conductivity of 45 W/m°C. The lateral surfaces lose heat to ambient air at 25 °C with a heat transfer coefficient of 67 W/m^2 °C. The base of thickness 0.006 m is at 76 °C while the tip of thickness 0.003 m is perfectly insulated. Set up finite difference equations for nodal temperatures by taking $\Delta x = 1.25$ cm. Solve for the nodal temperatures using TDMA. What is the total heat loss from the fin? What is the fin efficiency?

Ex 7.6 Solve the equation $\frac{d^2T}{dx^2} - 2T = 0$ in the interval $0 \leq x \leq 1$ with the boundary conditions $T = 1$ at $x = 0$ and $\frac{dT}{dx} = 0$ at $x = 1$ using the finite difference method.

- Formulate the nodal equations so that an iterative solution is possible for the resulting equations. Carry out the calculations with $\Delta x = 0.2$ and stop the iteration when the function converges to within 0.1%.
- Solve the problem using the TDMA.
- Compare the temperature at $x = 1$ obtained by the above two procedures with the analytical value. Comment on the numerical solution in light of the comparison.
- Compare the base heat flux obtained by all the three methods.

The reader is encouraged to use a PC for solving this problem. A spreadsheet program may be useful for this purpose.

Ex 7.7 A very long rod of 100 mm diameter and of a material of thermal conductivity equal to 1.5 W/m°C is generating heat internally at a constant rate of 10^5 W/m^3. The surface of the rod is exposed to a convective environment at 30°C via a heat transfer coefficient of 45 W/m^2 °C. Obtain the maximum temperature in the rod by the following two methods and compare them: (a) analytical solution of the governing equation and (b) numerical solution for the temperature distribution by finite differences. In the latter case choose at least 6 nodes along the radial direction.

Ex 7.8 A certain steady two-dimensional temperature field has been analyzed using finite differences based on equal step sizes of 0.1 m in the x and y directions. Some of the temperatures are indicated in Fig. 7.24. Determine the temperatures at grid points labeled 1 and 2.

Ex 7.9 Figure 7.25 shows the grid pattern used in a finite difference analysis of steady conduction in two dimensions. The domain is subject to the boundary conditions specified on the figure. Some of the temperatures are indicated at the grid points. Determine the temperatures at grid points 1, 2 and 3 using second order accurate analysis. The grid spacing is 0.1 m along both the x and y directions.

Ex 7.10 Steady two-dimensional conduction in a rectangle subject to the indicated boundary conditions is shown in Fig. 7.26. The left side is isothermal at 100°C while the right side is isothermal at 50°C. The bottom

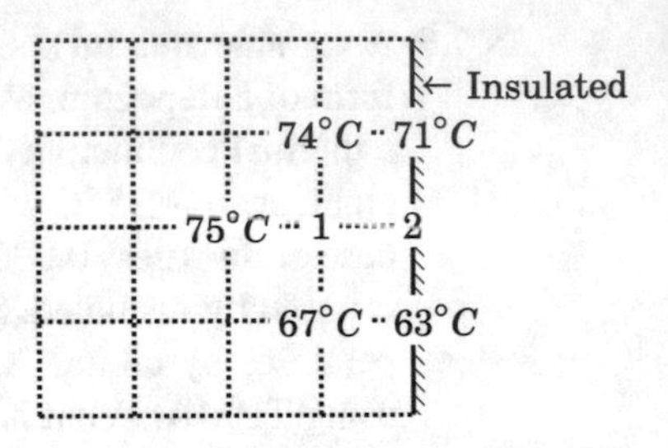

Fig. 7.24 Finite difference example of Exercise 7.8

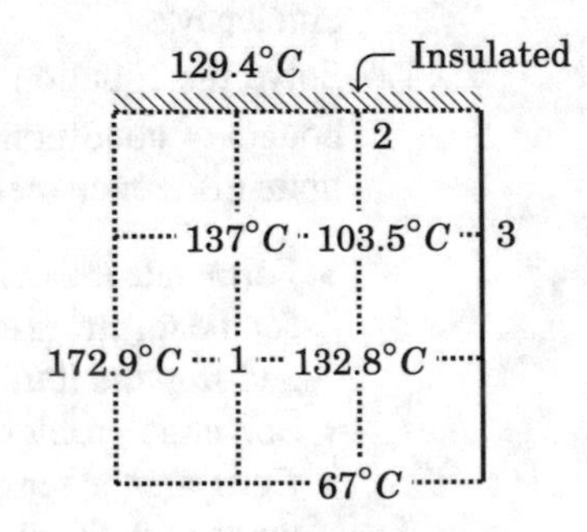

Fig. 7.25 Finite difference example of Exercise 7.9

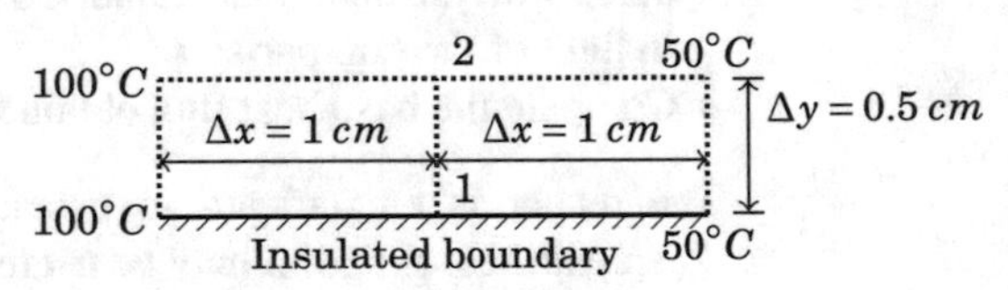

Fig. 7.26 Finite difference example of Exercise 7.10

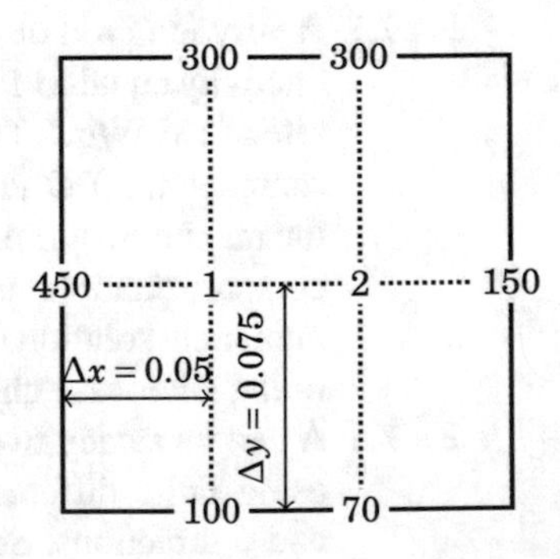

Fig. 7.27 Finite difference example of Exercise 7.11

side is perfectly insulated while the top is losing heat to an ambient at $30°C$ via a convection heat transfer coefficient of 200 W/m^2°C. Heat is being generated at a uniform rate of 0.3 MW/m^3 in the material that has a thermal conductivity of 20 W/m°C. Determine the temperatures at nodes 1 and 2. Also estimate the heat lost from the top.

What will be the temperatures at nodes 1 and 2 if there is no heat generation?

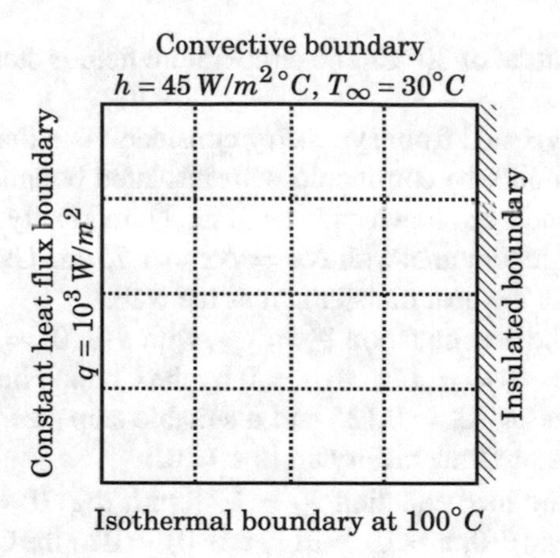

Fig. 7.28 Finite difference example of Exercise 7.12

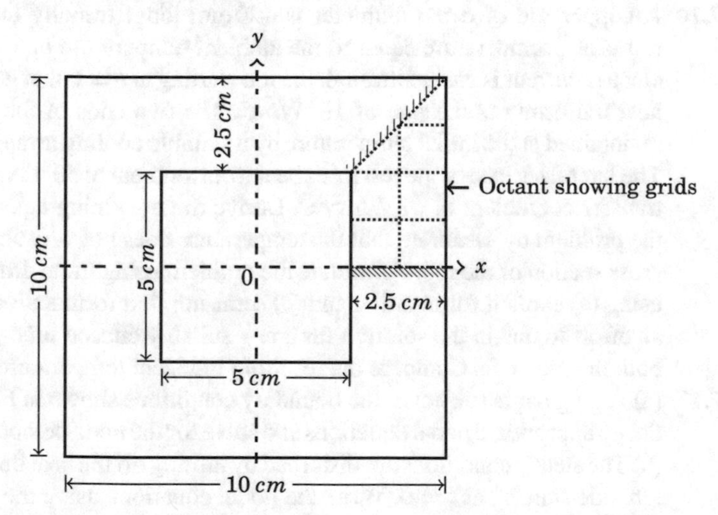

Fig. 7.29 Finite difference example of Exercise 7.13

Ex 7.11 Laplace equation is satisfied by the function $T(x, y)$ in the rectangular domain shown in Fig. 7.27. What are the values of T at the nodal points 1 and 2? All quantities are in consistent units.

Ex 7.12 A long bar of square cross section of side 0.1 m is subject to the boundary conditions shown in Fig. 7.28. Formulate the problem using finite differences. Set up nodal equations for solution by Gauss elimination method. Continue iteration till the nodal temperatures change by less than 0.1% from iteration to iteration.

Ex 7.13 Figure 7.29 shows the section of a very long duct of a material of thermal conductivity equal to 15 W/m°C. The inner boundary is maintained at a temperature of 70 °C while the outer boundary is maintained at a temperature of 30 °C. The temperature field is steady.

It is suggested from symmetry considerations that $\frac{1}{8}^{th}$ of the physical domain only be considered with insulated boundary conditions along the boundaries shown with hatching. Numerically obtain all the interior nodal temperatures with $\Delta x = \Delta y = 1.25$ cm. Use central differences. Estimate the heat transfer across the wall.

Ex 7.14 Solve the heat equation $\frac{\partial \theta}{\partial \tau} = \frac{\partial^2 \theta}{\partial \xi^2}$, with $\theta(\xi, 0) = 1$ for $0 < \xi < 1$ and $\theta(0, \tau > 0) = \theta(1, \tau > 0) = 0$ by the Crank–Nicolson method. Use a step size of $\Delta\xi = 0.125$ and a suitable step size for $\Delta\tau$ to obtain the temperature time history up to $\tau = 0.1$.

Ex 7.15 Solve the heat equation $\frac{\partial \theta}{\partial \tau} = \frac{\partial^2 \theta}{\partial \xi^2}$, with $\theta(\xi, 0) = 4\xi(1 - \xi)$ for $0 < \xi < 1$ and $\theta(0, \tau > 0) = \theta(1, \tau > 0) = 0$ by the Crank–Nicolson method. Obtain the solution up till $\tau = 1$.

Ex 7.16 A copper rod of 6 mm diameter is 400 mm long. Initially the entire rod is at a temperature equal to the ambient temperature of 30 °C. An electric current is passed through the rod starting at $t = 0$ that generates heat uniformly at the rate of 10^4 W/m^3. The two ends of the rod are maintained at the initial temperature by a suitable cooling arrangement. The lateral surface of the rod loses heat to an ambient at 30 °C via a heat transfer coefficient of 67 W/m^2 °C. Derive the governing equation for the problem by assuming that the temperature does not vary across the cross section of the rod. Formulate the problem using finite differences using (a) explicit formulation and (b) semi-implicit formulation. Make an effort to obtain the solution for a few suitably chosen time steps by both the methods. Compare the resulting transient temperature fields.

Ex 7.17 (a) A long bar is subject to the boundary conditions shown in Fig. 7.30. Set up appropriate nodal equations and solve for the nodal temperatures. (b) The steady conditions are disturbed by turning off the heat flux on the left side starting at $t = 0$. Write the nodal equations using the Crank–Nicolson method. Take the specific heat and density of the material of the bar as 300 J/kg°C and 2500 kg/m^3 respectively. Solve the nodal equations for a few time steps.

Ex 7.18 Consider a very long bar of diameter $D = 0.1$ m made of a material with the following properties:
Thermal conductivity $k = 14.9$ W/m°C, density $\rho = 7900$ kg/m^3 and specific heat $c = 0.477$ kJ/kg°C
The rod is initially at a uniform temperature of $T_i = 250$°C. For $t > 0$ the cylinder surface loses heat to an ambient at $T_\infty = 45$°C with a convective heat transfer coefficient of $h = 67$ W/m^2°C. Numerically obtain the temperature distribution inside the cylinder by using a uniform nodal spacing of $\Delta r = 0.005\,m$. Obtain the solution till such time that the one term approximation is expected to hold. Compare

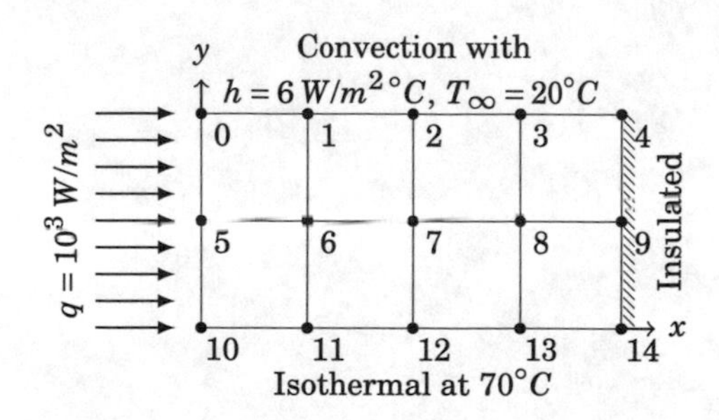

Fig. 7.30 Finite difference example of Exercise 7.17. $\Delta x = \Delta y = 0.01\,\text{m}$

the numerically obtained solution with the one term approximation. Solution will require the use of computer.

Ex 7.19 A very long bar of brass has a rectangular cross section of sides $2L_x = 0.2\,\text{m}$ and $2L_y = 0.15\,\text{m}$. Initially the bar is at a uniform temperature of $T_i = 200°\text{C}$ throughout. For $t > 0$ the surface of the bar starts losing heat to an ambient fluid at $T_\infty = 30°\text{C}$ with a heat transfer coefficient of $h = 240\,\text{W/m}^2\,°\text{C}$. Obtain numerically the temperature distribution inside the bar. Compare the temperature distribution obtained numerically with that calculated using Heisler chart. Solution will require the use of computer.

Ex 7.20 A spherical shell of inner diameter $D_i = 0.02\,\text{m}$ and outer diameter $D_o = 0.06\,\text{m}$ is made of stainless steel. The shell is initially at a uniform temperature of $T_i = 300°\text{C}$ throughout. At a certain time the outer surface of the shell starts losing heat to an ambient at $T_\infty = 25°\text{C}$ with a heat transfer coefficient equal to $25\,\text{W/m}^2\,°\text{C}$. Obtain numerically the temperature time history of the shell. Make a plot of inner and outer surface temperatures of the shell as functions of time. What happens, to the temperature difference between the two surfaces of the shell, with time?

Ex 7.21 A very long cylindrical annulus of inner diameter $D_i = 0.02\,\text{m}$ and outer diameter $D_o = 0.06\,\text{m}$ is made of stainless steel. The annulus is initially at a uniform temperature of $T_i = 300°\text{C}$ throughout. At a certain time the outer surface of the annulus starts losing heat to an ambient at $T_\infty = 25°\text{C}$ with a heat transfer coefficient equal to $25\,\text{W/m}^2\,°\text{C}$. Obtain numerically the temperature time history of the annulus. Make a plot of inner and outer surface temperatures of the annulus as functions of time. What happens, to the temperature difference between the two surfaces of the annulus, with time?

Chapter 8
Basics of Thermal Radiation

RADIATION heat transfer is an important mode of heat transfer that occurs even in the absence of other modes of heat transfer. Radiation plays an important part in areas such as space, solar energy applications, the earth's weather. This chapter introduces intensity and its moments as a means of describing thermal radiation. This chapter provides the necessary background for the three chapters that follow it.

8.1 Introduction

Radiation is one of the basic mechanisms by which energy is transferred between regions at different temperatures. The distinctive feature of radiation is that a medium is not necessary for such transport of energy even though it originates in matter. In fact, the presence of a medium tends to reduce it. Radiation energy transport is a consequence of energy carrying electromagnetic waves. Sometimes the waves behave as particles called photons, which are particles of zero rest mass, moving at the speed of light. In addition, the energy carried by a photon is proportional to the frequency of the electromagnetic waves, which represents radiation in its alternative form as an electromagnetic wave.

Photons/radiation originate in matter due to the changes in their internal energy content. The radiant energy depends solely on the nature of the material that gives rise to it and its temperature. The emission of photons is independent of the surroundings. The emission process may involve different radiant energy discharge processes. One type of discharge process that is of special interest to us in connection with heat transfer phenomena is that arising as a result of microscopic thermal activities in a material close to thermal equilibrium. This type of radiant energy is called thermal radiation and we limit our study to such radiation.

S. P. Venkateshan, *Heat Transfer*,
https://doi.org/10.1007/978-3-030-58338-5_8

Thermal radiation is involved, for example, in the description of heat transfer in space applications, high-temperature plasmas, furnaces, fires, and atmospheric phenomena that involve solar and terrestrial energy processes.

8.1.1 Fundamental Ideas

Electromagnetic Radiation and Photons

Consider electromagnetic radiation propagating in a vacuum as electromagnetic waves or alternately as photons. The speed of propagation of electromagnetic radiation in vacuum is given by the speed of light, $\boxed{c_0 = 2.998 \times 10^8 \text{ m/s}}$. If the frequency of the radiation is ν (in Hz or s^{-1}), the energy E associated with the photon is

$$\boxed{E = h\nu} \tag{8.1}$$

where h is Planck's constant which has a value of 6.625×10^{-34} J s. The associated vacuum wavelength λ_0 of the electromagnetic radiation satisfies the relation

$$\boxed{\lambda_0 = \frac{c_0}{\nu}} \tag{8.2}$$

Note that the frequency of radiation ν is independent of the medium through which the radiation propagates. The speed of propagation and wavelength vary alike so that the frequency remains invariant. If the speed of propagation of radiation is c in a medium, the refractive index n of the medium is defined as

$$\boxed{n = \frac{c_0}{c}} \tag{8.3}$$

By definition $n \geq 1$, since the speed of light in a medium cannot exceed c_0.

Electromagnetic Spectrum

Electromagnetic spectrum, in principle, spans the frequency range 0 to ∞ (or the wavelength range ∞ to 0). However, in practice, thermal radiation (radiation that originates due to thermal motions in a medium) is assumed to span the frequency range 3×10^{12}–3×10^{15} Hz (or wavelength range 0.1–100 μm). Figure 8.1 indicates the nature of the electromagnetic spectrum. The figure is drawn using the wavelength of radiation along the axis (engineers use more often the wavelength and scientists use the frequency or the wave number ω, the reciprocal of wavelength in cm^{-1}). Note that wavelength of 1 μm will correspond to

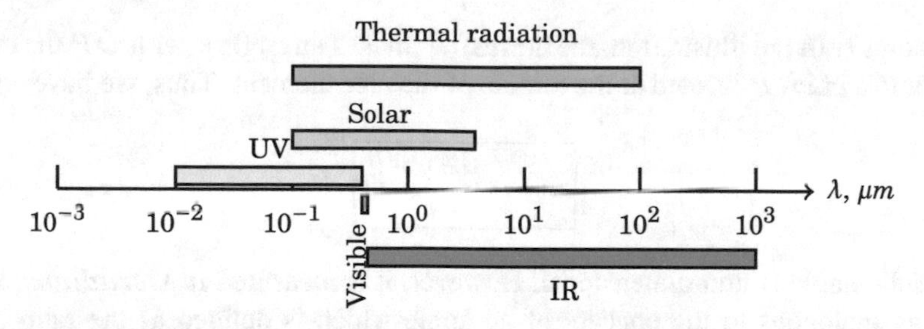

Fig. 8.1 Electromagnetic spectrum

$$1\ \mu\text{m} = 3 \times 10^{14}\ Hz = 10^4\ \text{cm}^{-1} \tag{8.4}$$

The above is based on speed of light in vacuum rounded to $c_0 \approx 3 \times 10^8$ m/s. The spectral range of interest may be further divided into the ultraviolet (UV), the visible and the infrared (IR) regions. The visible part of the spectrum (average human eye is sensitive only in this band) occupies a narrow region between 0.4 and 0.7 μm. As an example of abundant radiation, solar radiation spans the range from 0.1 to 3 μm. Also radiation emitted by earth is in the IR region and has a peak at around 10 μm. All these are indicated in Fig. 8.1.

8.1.2 Preliminaries and Definitions

Radiation propagates in three-dimensional space, and hence, radiation heat transfer takes place in three-dimensional space. From a given surface area (boundary of matter), radiation propagates in all directions. In order to describe such a propagation process, we need to define the geometric concept of the solid angle.

Figure 8.2 shows how an elemental solid angle is defined. Consider an area element dA and an arbitrary point O outside it as shown in the figure. The elemental solid angle is defined as the ratio of the area normal to the radius vector to the square of

Fig. 8.2 Definition of solid angle

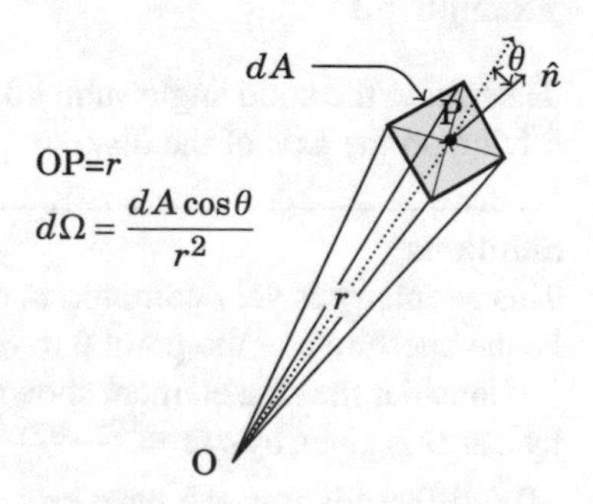

the radius r. In the illustration, the normal $\vec{n}$ makes an angle θ with OP the radius vector to a point P located at the middle of the area element. Thus, we have

$$\boxed{d\Omega = \frac{dA\cos\theta}{r^2}} \tag{8.5}$$

The solid angle is non-dimensional. However, it is measured in *steradians* or *sr*. This is analogous to the concept of an angle which is defined as the ratio of an elemental arc normal to the radius vector and the radius r. Angle is represented in *radians* or *rad*.

Solid Angle and the Sphere

Consider an area element dA on a sphere of radius r. The area element is normal to r, and hence, the elemental solid angle subtended by the area element at the center of the sphere is given by

$$d\Omega_{Sphere} = \frac{dA}{r^2} \quad \text{sr} \tag{8.6}$$

The total area of the sphere is $4\pi r^2$. Hence, the solid angle subtended by the sphere at its center 9or at *any* point within the sphere) is

$$\Omega_{Sphere} = \frac{4\pi r^2}{r^2} = 4\pi \quad sr \tag{8.7}$$

The reader should have no difficulty in showing that
1. the solid angle subtended by the sphere at *any* point inside it is 4π
2. the solid angle subtended by the hemisphere at *any* point on its base is 2π

Example 8.1

Determine the solid angle subtended by a circular disk of radius a at a point distant b lying on the axis of the disk.

Solution:

This geometry is very common in optical instruments. One may imagine the disk to be the aperture and the point 0 to be the focal point of an optical instrument.

Consider the ring element shown in Fig. 8.3. The elemental solid angle subtended by it at O is given by $d\Omega = \frac{2\pi r\cos\theta dr}{\rho^2}$. We notice that the radius ρ is equal to $\sqrt{b^2 + r^2}$. By differentiation, we have $rdr = \rho d\rho$. Also $\cos\theta = \frac{b}{\rho}$. The total solid angle is obtained then by integration as follows:

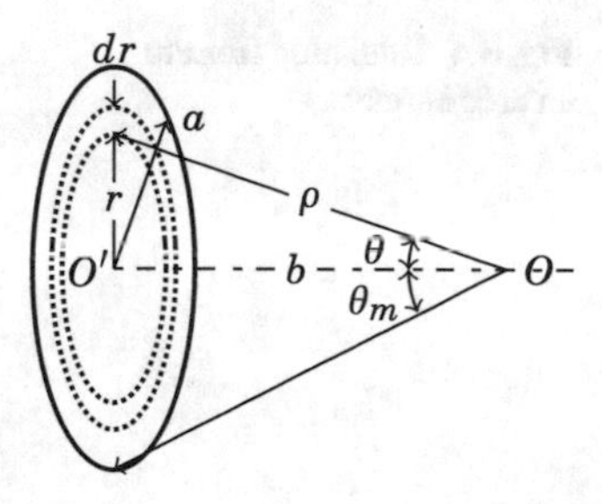

Fig. 8.3 Sketch for Example 8.1

$$\Omega = \int_0^a \frac{2\pi r \cos\theta}{\rho^2} dr = \int_b^{\sqrt{a^2+b^2}} \frac{2\pi\rho}{\rho^2}\frac{b}{\rho} d\rho$$

$$= 2\pi b \left[-\frac{1}{\rho}\right]_b^{\sqrt{a^2+b^2}} = 2\pi b \left[\frac{1}{b} - \frac{1}{\sqrt{a^2+b^2}}\right]$$

In terms of the maximum angle θ_m (see Fig. 8.3), the above expression is easily seen to be

$$\Omega = 2\pi(1 - \cos\theta_m)$$

Spectral or Monochromatic Intensity of Radiation

Intensity is a field quantity that describes radiation propagating in space. As indicated earlier, such radiation may propagate in all directions. It may also vary in different directions and from point to point in a domain of interest to us. Consider an area element dA (it may be a real area element on a surface or an imaginary area element within the domain of interest to us) as shown in Fig. 8.4. Radiant energy may be streaming through this element in all directions (represented by the angle with respect to the normal θ and the azimuthal angle ϕ). It is customary to construct a hemisphere with the center of the area element as its center, to visualize the passage of radiation across the area element.

Consider the radiant energy leaving dA in a direction (θ, ϕ), and within the solid angle $\Delta\Omega$. The intensity $I_\lambda(\theta, \phi)$ is defined as the elemental monochromatic radiant energy which leaves (or enters) the area dA normal to it due to the energy entering (or leaving) through the elemental solid angle $d\Omega$ in a time interval dt and in the spectral interval $d\lambda$. Thus, we have

$$I_\lambda(\theta, \phi) = \lim_{(\Delta t, \Delta\Omega, \Delta A, \Delta\lambda \to 0)} \frac{\Delta^4 E(\theta, \phi)}{\Delta t \Delta\Omega \Delta A \cos\theta \Delta\lambda} \tag{8.8}$$

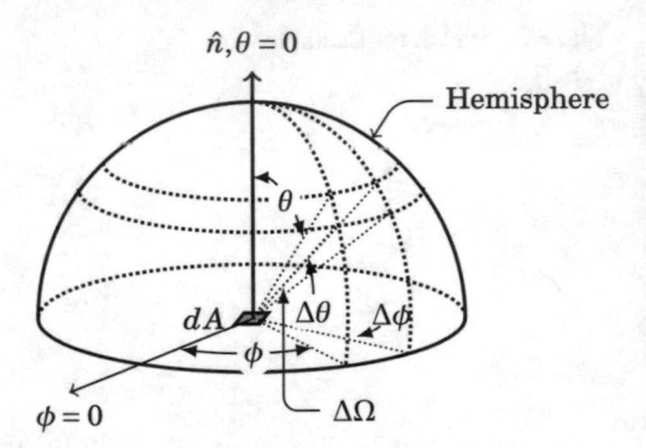

Fig. 8.4 Radiation from/to an area element

where all the indicated limits are taken simultaneously. Alternately, we may write the spectral intensity as

$$\boxed{I_\lambda(\theta, \phi) = \frac{d^4 E(\theta, \phi)}{dt d\Omega dA \cos\theta d\lambda}} \tag{8.9}$$

Superscript 4 emphasizes that the differential is dependent on four variables. Unit of intensity is $\frac{W}{m^2 \mu\text{m sr}}$ since $d\lambda$ is specified in μm even though, in SI units, this would have been $\frac{\text{W}}{\text{m}^3\ \text{sr}}$. Intensity is referred to as the *primitive* since all quantities of interest to us in radiation heat transfer are represented as *moments* of intensity, as will be shown later.

Radiant intensity depends on the direction represented by the two angles, θ the polar angle and ϕ the azimuthal angle. In words we may define the intensity as the amount of radiant power leaving (or entering) per unit area normal to it and within a unit solid angle.

We also notice from Fig. 8.4 that the elemental solid angle is given by

$$d\Omega = \frac{\text{Area element on sphere}}{r^2} = \frac{r \sin\theta d\phi r d\theta}{r^2} = \sin\theta d\theta d\phi \tag{8.10}$$

We shall introduce the notation $\mu = \cos\theta$, commonly employed in radiation heat transfer literature. We then have $d\mu = -\sin\theta d\theta$. With this, Eq. 8.10 becomes

$$\boxed{d\Omega = -d\mu d\phi} \tag{8.11}$$

The above expression for the elemental solid angle will be used extensively in what follows. The intensity may then be represented in the form $I_\lambda(\theta, \phi) = I_\lambda(\mu, \phi)$.

We now consider some derived quantities of interest in radiation heat transfer. All these quantities are moments of spectral intensity.

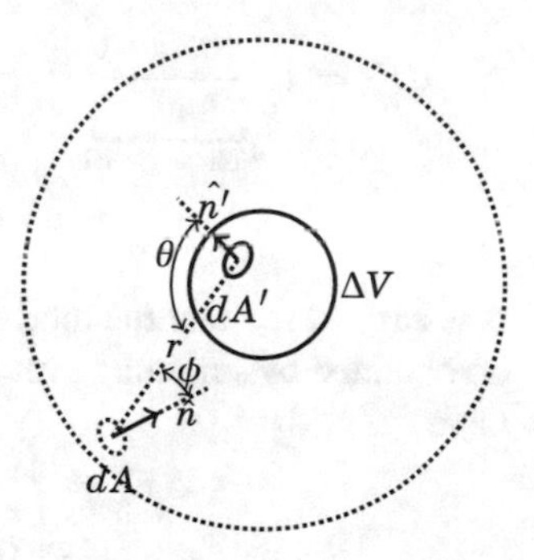

Fig. 8.5 Sketch facilitating the evaluation of energy density

In general, the nth moment of intensity is defined by the integral

$$\int_{\phi=0}^{2\pi} \int_{\mu=-1}^{1} I_\lambda(\mu, \phi)\mu^n d\mu d\phi \tag{8.12}$$

The first three moments, i.e., moments with $n = 0$, 1, and 2 have physical meaning and will be discussed below.

Spectral or Monochromatic Energy Density

Let radiation be present in a region represented by the large dotted boundary in Fig. 8.5. Even though the figure shows it as a circle of large radius, it may have any shape. Consider a small volume element ΔV within it, as indicated in the figure. Using the definition of intensity given by Eq. 8.9, we may write for the power dP_λ crossing dA, an elemental area on the big sphere and that also crosses dA', an elemental area on the small volume element as

$$dP_\lambda = [I_\lambda dA \cos\phi] \left[\frac{dA' \cos\theta}{r^2}\right] \tag{8.13}$$

Here, the first term inside the braces on the right-hand side represents the rate at which power leaves dA in a direction coincident with r. The second term inside the braces represents the solid angle subtended by dA' at the center of dA. Expression 8.13 is based on the definition on intensity introduced earlier. Let this power traverse a distance ℓ inside ΔV. Since all photons travel at a fixed speed c_0, the time for which the power is in transit within the elemental volume has to be $dt = \frac{\ell}{c_0}$. Thus, the energy that is in transit through the volume element is given by

$$dE_\lambda = I_\lambda \underbrace{\frac{\cos\phi dA}{r^2}}_{\text{Solid angle } d\Omega} \underbrace{\frac{\ell dA' \cos\theta}{c_0}}_{\substack{\text{Volume } dv \text{ intercepted in} \\ \Delta V \text{ divided by } c_0, \text{ by radiation in transit}}} = I_\lambda d\Omega dv \quad (8.14)$$

It is easy to see that the total energy in transit within the volume element ΔV is to be obtained by summing over all volume elements dv and solid angle $d\Omega$. Thus, we have

$$\Delta E_\lambda = \int_v \int_{\Omega=4\pi} \frac{I_\lambda}{c_0} d\Omega dv = \Delta V \frac{1}{c_0} \int_{\Omega=4\pi} I_\lambda d\Omega \quad (8.15)$$

From Eq. 8.15, we then have

$$\frac{\Delta E_\lambda}{\Delta V} = \frac{1}{c_0} \int_{\Omega=4\pi} I_\lambda d\Omega \quad (8.16)$$

Further, we take the limit as $\Delta V \to 0$ to define the energy density u_λ given by

$$u_\lambda = \lim_{\Delta V \to 0} \frac{\Delta E_\lambda}{\Delta V} = \frac{dE_\lambda}{dV} = \frac{1}{c_0} \int_{\Omega=4\pi} I_\lambda d\Omega \quad (8.17)$$

We may use Eq. 8.11 to write the above in the alternate form

$$\boxed{u_\lambda = \frac{1}{c_0} \int_{\phi=0}^{2\pi} \int_{-1}^{1} I_\lambda(\mu, \phi) d\mu d\phi} \quad (8.18)$$

It is seen that the integral on the right-hand side is the zeroth moment of intensity obtained by letting $n = 0$ in the general expression 8.12.

Spectral or Monochromatic Radiant Heat Flux

Radiant heat flux is a quantity of ultimate interest in most heat transfer problems involving radiation. Radiant heat flux is obtained in terms of the radiation intensity by referring to Fig. 8.4. From definition of intensity given by Eq. 8.9, the rate at which radiant energy is leaving dA in the direction (μ, ϕ), per unit area is given by

$$dq_{R\lambda} = \frac{d^3E(\mu, \phi)}{dt d\lambda dA} = I_\lambda(\mu, \phi)\mu d\Omega = -I_\lambda(\mu, \phi)\mu d\mu d\phi \quad (8.19)$$

where the last step uses Eq. 8.11 for the elemental solid angle. If we integrate expression 8.19 over all directions within the forward hemisphere,[1] we get

$$q_{R\lambda}^{+} = -\int_{\phi=0}^{2\pi}\int_{\mu=1}^{0} I_\lambda(\mu,\phi)\mu d\mu d\phi = \int_{\phi=0}^{2\pi}\int_{\mu=0}^{1} I_\lambda(\mu,\phi)\mu d\mu d\phi \tag{8.20}$$

This represents the heat flux leaving the surface and is indicated by the superscript +. The unit of radiant heat flux is $\frac{W}{m^2 \mu m}$. Since radiation may be crossing the element dA from below to above (forward stream) or from above to below (backward stream),[2] $q_{R\lambda}^{-}$ is used to describe the backward streaming radiation. This is obtained by integrating over the backward hemisphere[3] as

$$q_{R\lambda}^{-} = -\int_{\phi=0}^{2\pi}\int_{\mu=0}^{-1} I_\lambda(\mu,\phi)\mu d\mu d\phi = \int_{\phi=0}^{2\pi}\int_{\mu=-1}^{0} I_\lambda(\mu,\phi)\mu d\mu d\phi \tag{8.21}$$

It is easily seen that the net heat flux $q_{R\lambda}$ across dA may be obtained by summing expressions 8.20 and 8.21 to get

$$\begin{aligned} q_{R\lambda} &= \int_{\phi=0}^{2\pi}\int_{\mu=0}^{1} I_\lambda(\mu,\phi)\mu d\mu d\phi + \int_{\phi=0}^{2\pi}\int_{\mu=-1}^{0} I_\lambda(\mu,\phi)\mu d\mu d\phi \\ &= \int_{\phi=0}^{2\pi}\int_{\mu=-1}^{1} I_\lambda(\mu,\phi)\mu d\mu d\phi \end{aligned} \tag{8.22}$$

It is seen from the above expression for net heat flux that it is the first moment of the spectral intensity, i.e., it is obtained by taking $n = 1$ in expression 8.12.

Monochromatic Radiation Pressure

We have mentioned earlier that radiation may be considered also as particles called photons. Photon energy may be visualized as the product of momentum $\mathbf{p}_\lambda$ and

[1]Forward hemisphere is defined by $0 \le \theta \le \frac{\pi}{2}, 0 \le \phi \le 2\pi$.

[2]For a solid surface the backward stream will represent incident heat flux.

[3]Backward hemisphere is defined by $\frac{\pi}{2} \le \theta \le \pi, 0 \le \phi \le 2\pi$.

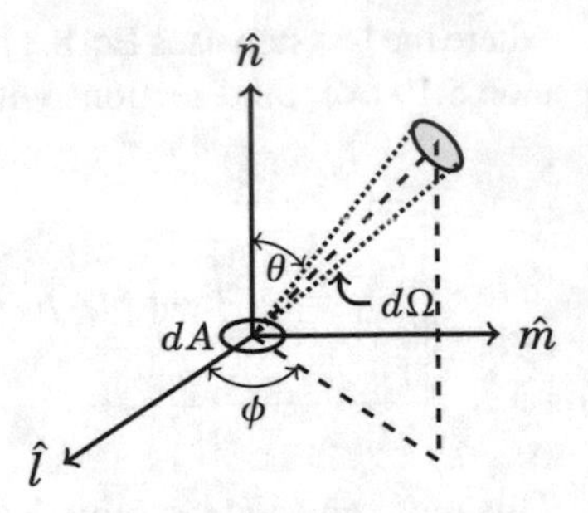

Fig. 8.6 Geometry for defining radiation stress

the speed of propagation c_0. When radiation is incident on a surface the rate at which momentum is incident normally per unit area because of radiation within the elemental solid angle may be obtained by referring to Fig. 8.6.

Rate at which energy is incident normal to area element dAdue to radiation within the solid angle $d\Omega$ is given by $\frac{dE_\lambda}{dt} \times dA \times \cos\theta = \frac{dE_\lambda}{dt} \times dA \times \mu$. If we divide this by speed of light c_0 and the area element dA, we should get the rate at which momentum is incident per unit area and normal to the area element. Hence, we have

$$\frac{d\mathbf{p}_\lambda}{dt} = \frac{\mu}{c_0}\frac{dE_\lambda}{dt} = \frac{I_\lambda(\mu, \phi)}{c_0}\mu^2 d\Omega \tag{8.23}$$

where we have linked the energy flux to the intensity introduced earlier. All we have to do is to integrate this over the forward hemisphere to get the total rate of momentum crossing normally the area element per unit area. This should be nothing but the pressure due to radiation (actually the normal stress) given by

$$(a) \quad \sigma_{n\lambda} = \int\limits_{\phi=0}^{2\pi}\int\limits_{\mu=0}^{1} \frac{I_\lambda(\mu, \phi)}{c_0}\mu^2 d\mu d\phi \tag{8.24}$$

Note that the radiation pressure involves the second moment of spectral intensity. Radiation also gives rise to tangential stresses. These are obtained by calculating the rates at which momenta are incident along $\hat{l}$ and $\hat{m}$ directions. The geometric factors for these two directions are, respectively, given by $\sin\theta\cos\phi = \sqrt{1-\mu^2}\cos\phi$ and $\sin\theta\sin\phi = \sqrt{1-\mu^2}\sin\phi$. The corresponding stresses are given by

$$(b)\quad \tau_{\hat{l}\lambda} = \int_{\phi=0}^{2\pi}\int_{\mu=0}^{1} \frac{I_\lambda(\mu,\phi)}{c_0}\sqrt{1-\mu^2}\cos\phi d\mu d\phi$$

$$(c)\quad \tau_{\hat{m}\lambda} = \int_{\phi=0}^{2\pi}\int_{\mu=0}^{1} \frac{I_\lambda(\mu,\phi)}{c_0}\sqrt{1-\mu^2}\sin\phi d\mu d\phi \tag{8.25}$$

The normal and tangential stresses are, in fact, components of the stress tensor which is symmetric.

Summary

In summary, we see that the various quantities that have been introduced above are moments of the intensity at a point. Moments involve integration over the hemisphere (or the sphere) after multiplication by factors that depend on μ and $\sin\phi$ or $\cos\phi$. There is one special case that is of much interest in the study of radiation heat transfer, the case of isotropic intensity $I_\lambda(\mu,\phi) \equiv I_\lambda$, that is independent of the direction. This case satisfies the following, as may be easily verified by using the expressions given above

$$\boxed{\begin{aligned} u_\lambda &= \frac{4\pi I_\lambda}{c_0};\ q_{R\lambda} = 0 \\ \sigma_{\hat{n}\lambda} &= \frac{4\pi I_\lambda}{3c_0};\ \tau_{\hat{l}\lambda} = \tau_{\hat{m}\lambda} = 0 \end{aligned}} \tag{8.26}$$

In this special case, the stress sensor reduces to only the normal stress which will be represented by the symbol $\sigma_{\hat{n}\lambda} = p_\lambda$. In this case, we have a relation between the energy density and the radiation pressure given by

$$\boxed{p_\lambda = \frac{u_\lambda}{3}} \tag{8.27}$$

Total Quantities

Total quantities are obtained by integrating the monochromatic quantities with respect to λ from 0 to ∞. Thus, we have

$$X = \int_{\lambda=0}^{\infty} X_\lambda d\lambda \tag{8.28}$$

where X_λ is the monochromatic quantity and X is the corresponding total quantity. In radiation heat transfer the total quantities are of interest, in the final analysis.

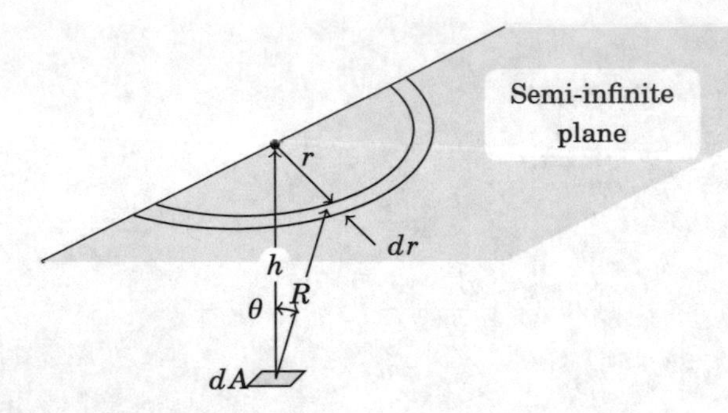

Fig. 8.7 Geometry for example 8.2

Example 8.2

Calculate the flux incident on an elemental area parallel to a semi-infinite plane slab that emits radiant energy in an isotropic manner with a direction independent total intensity I. The area element is located opposite the edge of the semi-infinite slab, as indicated in Fig. 8.7.

Solution:
Consider a ring element of radius r and width dr in the plane centered vertically above the elemental area as shown in Fig. 8.7. All points on the elemental ring are at a distance of R from the center of the elemental area. The solid angle subtended by dA at the ring is obtained as

$$d\Omega = \frac{\cos\theta dA}{R^2}$$

Radiant flux leaving ring element that is incident on dA is then given by

$$dq_R^- = I \times \underbrace{\pi r dr}_{\text{Area of ring}} \cos\theta \times \frac{\cos\theta}{R^2} = \pi I \frac{h^2}{R^4} r dr$$

since $\cos\theta = \frac{h}{R}$. But $R^2 = h^2 + r^2$ and hence $rdr = RdR$.The total flux incident on dA is obtained by integrating the above with respect to R from h to ∞.

$$q_R^- = \pi I \int_h^\infty \frac{h^2}{R^4} RdR = \frac{\pi}{2} I \int_0^\infty \frac{h^2}{R^4} dR^2$$

Letting $R^2 = X$, we have $dR^2 = dX$ and $\frac{1}{R^4} = \frac{1}{X^2}$. Performing the integration, we have

$$q_R^- = \frac{\pi}{2} I h^2 \left[-\frac{1}{X} \right] \Bigg|_{h^2}^{\infty} = \frac{\pi I}{2}$$

Alternately, the solution may be obtained by noting that the total solid angle subtended by the semi-infinite plane at dA is the same as that subtended by a quarter sphere. With the intensity independent of direction, Eq. 8.22 may be recast as

$$q_R^- = \int_{\phi=0}^{\pi}\int_{\mu=0}^{1} I(\mu,\phi)\mu d\mu d\phi = \pi I \int_{\mu=0}^{1} I\mu d\mu = \pi I \left[\frac{\mu^2}{2}\right]\Big|_0^1 = \frac{\pi I}{2}$$

8.2 Cavity or Black Body Radiation

8.2.1 Basic Ideas

Black body radiation is also referred to as equilibrium radiation. Consider an *evacuated* cavity (enclosure) whose boundary is impervious to the passage of heat and mass, as shown in Fig. 8.8. Let the wall of the cavity be maintained at a uniform temperature T_w. Since there are no temperature differences within the domain, there can be *no* net heat transfer from any elemental area on the wall of this cavity. Also the only means of heat transfer is by radiation since the cavity has been evacuated. According to the definitions given earlier, $q_{R\lambda} = 0$ for any elemental area on the boundary, (or in general any elemental area within the enclosure). This can happen only if $I_\lambda(\mu,\phi)$ is a constant, given by, say, $I_{b\lambda}$. This isotropic radiation is referred to as black body radiation.

Intensity of black body radiation depends on wavelength and temperature. The wavelength dependence will be considered later on. If we integrate the black body intensity over all wavelength, we get the total black body intensity I_b which is a function *only* of temperature. The cavity considered above is full of black body radiation in which photons are streaming in all directions and on balance the number of photons arriving (incident) at any area element and in a particular direction with

Fig. 8.8 Radiation in an evacuated enclosure

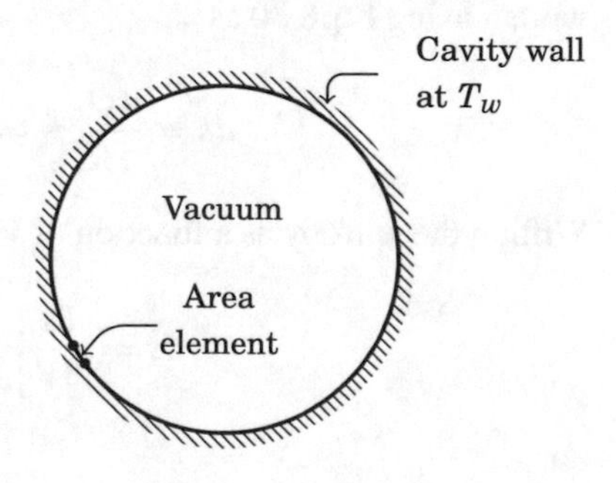

respect to the wall is balanced exactly by the number of photons leaving the area element (emission) in the same direction. Equilibrium within the enclosure is thus a dynamic equilibrium.

Since the total black body intensity is isotropic it is characterized by the relation

$$p_b = \frac{u_b}{3} \tag{8.29}$$

where the subscript *b* emphasizes that we are considering black body radiation.

If we make a very small aperture (compared to the size of the cavity) on the wall of the cavity (of course, the aperture may be a window that allows only radiation through), a small amount of radiation will escape through the aperture. The aperture *acts* as a source of black body radiation. If the escaping radiation is small, it will not disturb the equilibrium within the cavity, which can of course be made up by adding the *same* amount of heat to the cavity across its wall.

8.2.2 *Thermodynamics of Black Body Radiation*

Some interesting results may be obtained in respect of black body radiation by thermodynamic arguments. We consider black body radiation integrated over all wavelengths and governed by relation given by Eq. 8.29. It is as if the cavity is full of a substance whose total internal energy is given by u_b.

Black Body Intensity Variation with Temperature

A reversible process consisting of expansion of the cavity with heat addition may be visualized according to the relation

$$\begin{aligned} dQ_{rev} &= d(u_b V) + p_b dV \\ &= u_b dV + V du_b + p_b dV = (u_b + p_b) dV + V du_b \end{aligned} \tag{8.30}$$

where V is the volume of the cavity. The corresponding entropy change may be written using Eq. 8.30 as

$$ds = \frac{dQ_{rev}}{T} = \frac{u_b + p_b}{T} dV + \frac{V}{T} du_b \tag{8.31}$$

Writing the entropy as a function of V and u_b, we have

$$ds = \left.\frac{\partial s}{\partial V}\right|_{u_b} dV + \left.\frac{\partial s}{\partial u_b}\right|_{V} du_b \tag{8.32}$$

Comparing Eqs. 8.31 and 8.32, we have

$$(a)\ \left.\frac{\partial s}{\partial V}\right|_{u_b} = \frac{p_b + u_b}{T};\quad (b)\ \left.\frac{\partial s}{\partial u_b}\right|_{V} = \frac{V}{T} \tag{8.33}$$

Taking the derivative of Eq. 8.33(a) with respect to u_b, we get

$$\frac{\partial^2 s}{\partial V \partial u_b} = \frac{d\left[\dfrac{p_b + u_b}{T}\right]}{du_b} \tag{8.34}$$

The total differential is indicated on the right-hand side since both p_b and u_b are functions of only temperature. Taking the derivative of Eq. 8.33(b) with respect to V, we get

$$\frac{\partial^2 s}{\partial u_b \partial V} = \frac{1}{T} \tag{8.35}$$

Use Eq. 8.29 to write Eq. 8.34 as

$$\frac{\partial^2 s}{\partial V \partial u_b} = \frac{d\left[\dfrac{4u_b}{3T}\right]}{du_b} = \frac{4}{3T} - \frac{4u_b}{3T^2}\frac{dT}{du_b} \tag{8.36}$$

Since the order of differentiation should not matter expressions given by 8.35 and 8.36 should be identical and hence we have

$$\frac{4}{3T} - \frac{4u_b}{3T^2}\frac{dT}{du_b} = \frac{1}{T} \tag{8.37}$$

This may be recast as

$$\frac{du_b}{u_b} = 4\frac{dT}{T}$$

which integrates to

$$u_b = CT^4 \tag{8.38}$$

where C is a constant of integration. Thus purely thermodynamic arguments indicate a fourth power dependence for the black body energy density on temperature! The total black body intensity is then given by (using Eq. 8.26)

$$I_b = \frac{c_0}{4\pi}u_b = \frac{c_0}{4\pi}CT^4 \tag{8.39}$$

The black body heat flux q_{Rb}^+ or q_{Rb}^- which we shall represent as E_b is then given by

$$\boxed{E_b = \frac{c_0}{4}CT^4 = \sigma T^4} \tag{8.40}$$

where the quantity $\frac{c_0}{4}C$ is replaced by the constant σ, referred to as the Stefan–Boltzmann constant. This constant cannot be obtained based on thermodynamic analysis.

Isentropic Process with Black Body Radiation

Another interesting result is obtained when we consider an isentropic process involving cavity radiation. In Eq. 8.31, we set $ds = 0$ to get

$$\frac{p_b + u_b}{T}dV + \frac{V}{T}du_b = \frac{4u_b}{3}dV + Vdu_b = 0 \tag{8.41}$$

Here, Eq. 8.29 has been used to arrive at the final step. Equation 8.41 can be recast as

$$\frac{4dV}{3V} + \frac{du_b}{u_b} = 0$$

This will integrate to

$$u_b V^{\frac{4}{3}} = \text{Constant}$$

Using Eq. 8.29, this may be rewritten in the form

$$\boxed{p_b V^{\frac{4}{3}} = \text{Constant}} \tag{8.42}$$

Comparing this expression with the familiar expression $pV^\gamma = \text{Constant}$ for an isentropic process with an ideal gas, we conclude that cavity radiation behaves as an ideal gas with a gamma (ratio of specific heats) of $4/3 = 1.333$.

8.3 Wavelength Distribution of Black Body Radiation

In the previous section, we have looked at the total radiation from a black body using thermodynamic arguments. This has yielded knowledge regarding the temperature dependence of total emissive power of a black body. However, the spectral information is not available from that analysis. In order to gain this knowledge, it is necessary to look at the wave nature of radiation. A brief note on waves necessary for this is given before we turn our attention to cavity radiation.

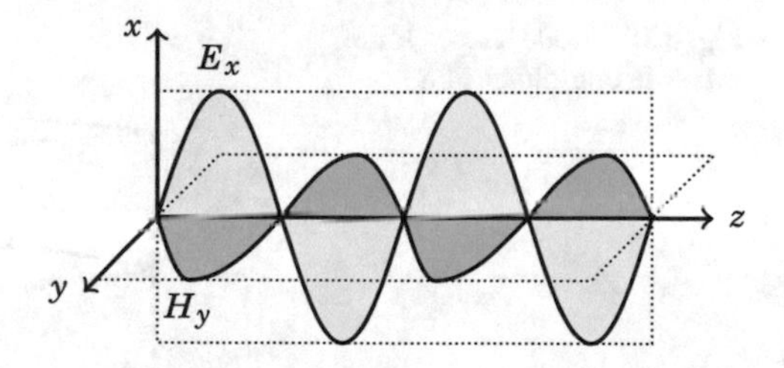

Fig. 8.9 Light—a transverse electromagnetic wave

8.3.1 About Waves

Waves in One Dimension

Waves are governed by the wave equation which may be derived starting from the Maxwell equations of electromagnetism (see Sect. 9.3 in Chap. 9). Electromagnetic waves are transverse in nature and represent periodic variation of the electric and magnetic fields in both space and time. The electric field, magnetic field, and the direction of propagation of waves are mutually perpendicular. In what follows we assume that the wave equation is known and our interest will be to look at the solution of the wave equation so that we understand enough about cavity radiation to describe the wavelength or frequency dependence of the same. The wave equation is given by

$$\nabla^2 \phi = \frac{1}{c_0^2} \frac{\partial^2 \phi}{\partial t^2} \tag{8.43}$$

where ϕ stands for the disturbance in electric or magnetic field due to wave motion and c_0 is the phase velocity in vacuum.

Consider electromagnetic waves of wavelength λ moving along the z-direction as shown in Fig. 8.9. There are two possibilities, the one shown in the figure with electric field E_x and magnetic field H_y or a second one that involves E_y and H_x. These two waves are referred to as the two polarizations. The direction of propagation is parallel to the Poynting vector given by $\overrightarrow{E} \times \overrightarrow{H}$. The electric field or the magnetic field (they are proportional to each other, according to Maxwell equations) varies with respect to z and t. The case is governed by the one-dimensional wave equation obtained by taking $\nabla^2 \equiv \frac{\partial^2}{\partial z^2}$. Consider the wave to be confined between $z = 0$ and $z = L$. Since our interest will be to describe waves of a given wavelength or frequency, the solution would be of the form[4]

$$\phi(z, t) = \phi_0 \sin\left(\frac{n\pi z}{L}\right) \sin\left(\frac{2\pi c_0 t}{\lambda}\right) \tag{8.44}$$

It may be verified by actual substitution that expression 8.44 satisfies the one-dimensional wave equation as long as

[4]Alternately one may use the method of separation of variables to obtain the solution.

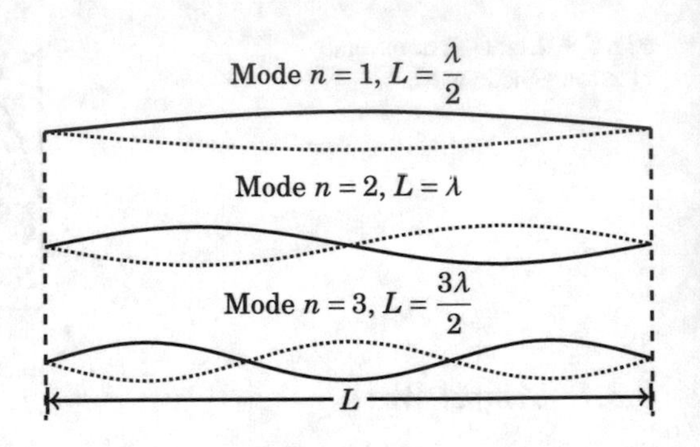

Fig. 8.10 Mode shapes for waves in one-dimension

$$n = \frac{2L}{\lambda} \quad \text{or} \quad \boxed{\frac{n\lambda}{2} = L} \tag{8.45}$$

The parameter n is referred to as the mode number and has to satisfy the condition that $\sin\left(\frac{n\pi z}{L}\right)$ vanish at the two end points, i.e., at $z = 0$ and $z = L$. This requires that n be an integer. The wave will, in fact, represent a standing wave set up over the physical domain $0 \leq z \leq L$. For a specified L and a given n there is a fixed λ. Waves of different modes with $n = 1, 2$, and 3 are shown in Fig. 8.10. Mode $n = 1$ corresponds to the fundamental and is such that there is a half wave in the interval $0, L$. Mode with $n = 3$ has three half waves in the interval $0, L$. The figure shows the values of the disturbance at the extremes corresponding to the amplitude of the waves. The shape varies continuously with time, again in a sinusoidal fashion.

Waves in More Than One Dimension

Waves in two and three dimensions have interesting characteristics that are directly relevant to the cavity radiation problem. Consider waves in two dimensions in a rectangular domain of sides L_1 and L_2. Standing waves are characterized by two mode numbers, n_1 for the x-direction, and n_2 for the y-direction. The solution may be written down as

$$\phi(x, y, t) = \phi_0 \sin\left(\frac{n_1\pi x}{L_1}\right) \sin\left(\frac{n_2\pi y}{L_2}\right) \sin\left(\frac{2\pi c_0 t}{\lambda}\right) \tag{8.46}$$

The reader may verify, by actual substitution, that Eq. 8.46 indeed satisfies the wave equation in two dimensions, viz.,

$$\frac{\partial^2\phi}{\partial x^2} + \frac{\partial^2\phi}{\partial y^2} = \frac{1}{c_0^2}\frac{\partial^2\phi}{\partial t^2} \tag{8.47}$$

The condition that is satisfied by the mode numbers, analogous to Eq. 8.45 is given by

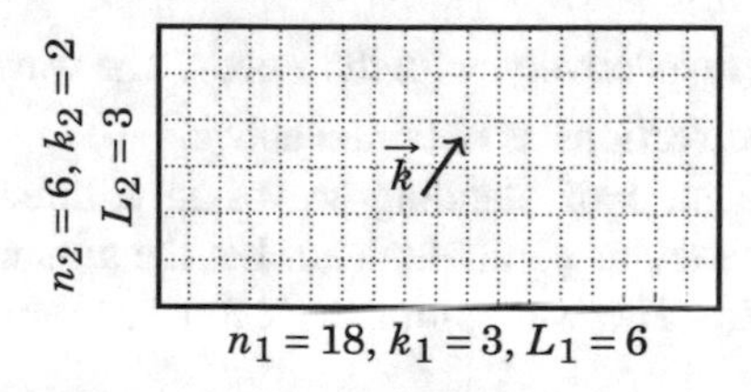

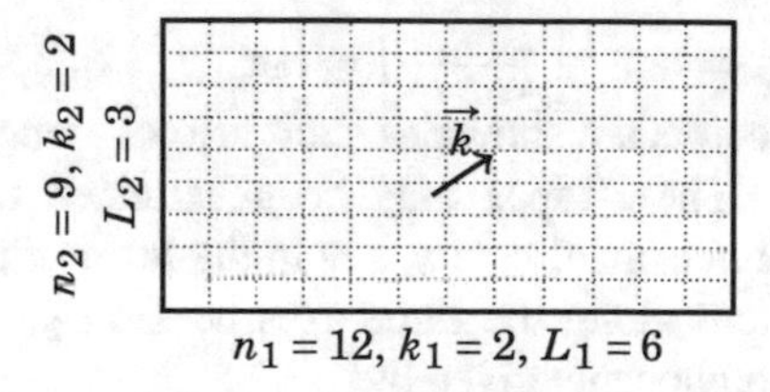

Fig. 8.11 Waves in a rectangle

$$\boxed{\left(\frac{n_1}{L_1}\right)^2 + \left(\frac{n_2}{L_2}\right)^2 = \frac{4}{\lambda^2}} \tag{8.48}$$

where n_1 and n_2 are integers. A typical case of standing waves in a rectangle of dimensions $L_1 = 6$ and $L_2 = 3$ is shown in Fig. 8.11a, with $n_1 = 18$ and $n_2 = 6$. The grid pattern is generated by drawing lines of zero disturbance parallel to the two axes. The disturbance also vanishes at all the edges of the rectangle. Using Eq. 8.48, the wavelength of the standing wave pattern with $n_1 = 18$ and $n_2 = 6$ is

$$\lambda = \sqrt{\frac{\left(\frac{18}{6}\right)^2 + \left(\frac{6}{3}\right)^2}{4}} = 0.5547$$

in the same length units as the specified lengths. What is interesting to note is that there can be more than one n_1, n_2 combination that satisfies Eq. 8.48. For example, with $L_1 = 6$ and $L_2 = 3$, the two combinations $n_1 = 18, n_2 = 6$ and $n_1 = 12, n_2 = 9$ (see Fig. 8.11b) are characterized by the same λ given by

$$\lambda = \sqrt{\frac{\left(\frac{18}{6}\right)^2 + \left(\frac{6}{3}\right)^2}{4}} = \sqrt{\frac{\left(\frac{12}{6}\right)^2 + \left(\frac{9}{3}\right)^2}{4}} = 0.5547$$

These different combinations referred to as degenerate modes set up standing waves along different directions, but having the same wavelength and hence also the frequency.

Wave Number Vector

The standing waves set up in the rectangle may be visualized by defining a wave number vector $\overrightarrow{k}$ which has components $k_1 = \frac{n_1}{L_1}$ and $k_2 = \frac{n_2}{L_2}$ along the x- and y-directions, respectively. The direction of $\overrightarrow{k}$ represents the direction along which the standing waves are set up. Thus, the directions in which standing waves are set up in these two cases correspond to $\theta_1 = \tan^{-1}\left[\frac{n_2}{L_2} \div \frac{n_1}{L_1}\right] = \tan^{-1}\left(\frac{3}{2}\right) = 56.31°$ and

$\theta_2 = \tan^{-1}\left[\frac{n_2}{L_2} \div \frac{n_1}{L_1}\right] = \tan^{-1}\left(\frac{2}{3}\right) = 33.69°$, respectively, with the x-axis. The wave vectors are shown as short arrows along the directions as obtained above.

These ideas may be generalized to the case of standing waves in a three-dimensional cavity, say in the form of a rectangular parallelepiped. Let the dimensions along the three axes be L_1, L_2, and L_3. The wave number vector has three components given by

$$\vec{k} = k_1\hat{i} + k_2\hat{j} + k_3\hat{k} = \frac{n_1}{L_1}\hat{i} + \frac{n_2}{L_2}\hat{i} + \frac{n_3}{L_3}\hat{i} \tag{8.49}$$

where $\hat{i}$, $\hat{j}$, $\hat{k}$ are unit vectors parallel to the axes, n_1, n_2, n_3 are the mode numbers along the three axes. The standing waves satisfy the condition (generalization of Eq. 8.48)

$$\boxed{\frac{4}{\lambda^2} = \left(\frac{n_1}{L_1}\right)^2 + \left(\frac{n_2}{L_2}\right)^2 + \left(\frac{n_3}{L_3}\right)^2} \tag{8.50}$$

The direction cosines of the wave number vector are

$$l = \frac{k_1}{\sqrt{k_1^2 + k_2^2 + k_3^2}},\ m = \frac{k_2}{\sqrt{k_1^2 + k_2^2 + k_3^2}},\ \text{and}\ n = \frac{k_3}{\sqrt{k_1^2 + k_2^2 + k_3^2}} \tag{8.51}$$

The standing waves show degenerate modes in this case also. Since there are three mode numbers, the possibilities are even more than in the case of two dimensions. We will look at this aspect in more detail in the ensuing section.

8.3.2 Number of Degenerates Modes in a Three-Dimensional Cavity

We consider standing waves in a parallelepiped whose dimensions are much much larger than the wavelength. This is specifically relevant to electromagnetic radiation since the wavelength range of interest of 0.1–100 μm is certainly much smaller than a cavity of reasonably big size, say of side equal to 0.1 m. The mode numbers involved are large, and hence, the number of combinations of the n's that yield the same wavelength for the standing waves is extremely large. Standing waves do not transport energy and hence describe equilibrium radiation within the cavity.

Take a specific example. Let the wavelength we are interested in be $\lambda = 1$ μm. Equation 8.50 will read as

$$k_1^2 + k_2^2 + k_3^2 = 4 \times 10^{12}$$

This represents all the points on a sphere of radius 2×10^6! Note that the k's are not necessarily integer values even though the n's are integers. Since only positive values are appropriate for the wave numbers, the points in the first octant only are relevant. If we consider radiation of wavelength between λ and $\lambda + d\lambda$, the number of modes is given by the volume of a spherical shell such that the number of modes is given by

$$\text{Number of modes} = -\frac{4\pi\left(\frac{2}{\lambda}\right)^2}{8} d\left(\frac{2}{\lambda}\right) L_1 L_2 L_3 = \frac{4\pi}{\lambda^4} d\lambda L_1 L_2 L_3$$

The $-$ sign is taken because $\frac{dn}{d\lambda}$ is negative. The factor $L_1 L_2 L_3$ occurs because of conversion from k to n. Since there are two possible polarizations (E_x, H_y or E_y, H_x), the above expression has to be multiplied by a factor of 2 to get

$$dn = \frac{8\pi}{\lambda^4} d\lambda \tag{8.52}$$

where dn is the number of modes per unit volume of the cavity, since $L_1 L_2 L_3$ is nothing but the cavity volume.

8.3.3 Planck Distribution

The black body distribution of energy over wavelength may be obtained once we know the mean energy of photons of a given wavelength λ. In order to find the mean energy, we assume that the energy of photons is quantized. We consider a number N of simple harmonic oscillators of wavelength λ having energies given by $n\frac{hc_0}{\lambda}$ where n takes on discrete integer values $0, 1, 2 \cdots$. The energy levels are populated according to the distribution given by

$$N(n) = N_0 \exp\left[-\frac{nhc_0}{\lambda kT}\right]$$

where h is Planck's constant and k is the Boltzmann constant. Letting $x = \frac{hc_0}{\lambda kT}$, the above equation may be recast as

$$N(n) = N_0 e^{-nx} \tag{8.53}$$

The total number of oscillators is then given by

$$N = \sum_{n=0}^{n=\infty} N_0 e^{-nx} \tag{8.54}$$

This is a geometric progression with common ratio e^{-x}, and the sum is given by

$$N = \frac{N_0}{1 - e^{-x}} \tag{8.55}$$

The total energy of the oscillators is given by

$$E = \sum_{n=0}^{n=\infty} N_0 n \frac{hc_0}{\lambda} e^{-nx} \tag{8.56}$$

Differentiate Eq. 8.54 with respect to x to get

$$\frac{dN}{dx} = -\sum_{n=0}^{n=\infty} N_0 n e^{-nx} \tag{8.57}$$

Comparing the above two equations, we see that

$$E = -\frac{hc_0}{\lambda}\frac{dN}{dx} \tag{8.58}$$

The required derivative may be obtained from Eq. 8.55 as

$$\frac{dN}{dx} = -\frac{N_0 e^{-x}}{(1 - e^{-x})^2}$$

Introducing this in Eq. 8.58, we finally get

$$E = \frac{hc_0}{\lambda}\frac{N_0 e^{-x}}{(1 - e^{-x})^2} \tag{8.59}$$

The mean energy of the oscillators is then obtained by taking the ratio $\frac{E}{N}$ to get, after minor simplification

$$\frac{E}{N} = \frac{\frac{hc_0}{\lambda}}{e^x - 1} = \frac{\frac{hc_0}{\lambda}}{\left(e^{\frac{hc_0}{\lambda kT}} - 1\right)} \tag{8.60}$$

The black body spectral energy density is given by using Eqs. 8.52 and 8.60 as

$$u_\lambda = \frac{8\pi}{\lambda^4} \times \frac{\left(\frac{hc_0}{\lambda}\right)}{\left(e^{\frac{hc_0}{\lambda kT}} - 1\right)} = \frac{8\pi hc_0}{\lambda^5}\frac{1}{\left(e^{\frac{hc_0}{\lambda kT}} - 1\right)} \tag{8.61}$$

We can easily show that $E_{b\lambda} = \frac{c_0}{4}u_\lambda$. Hence, the spectral emissive power of a black body is obtained using Eq. 8.61 as

$$E_{b\lambda} = \frac{2\pi h c_0^2}{\lambda^5} \frac{1}{e^{\frac{hc_0}{\lambda k T}} - 1} = \frac{C_1}{\lambda^5} \frac{1}{\left(e^{\frac{C_2}{\lambda T}} - 1\right)} \tag{8.62}$$

where

$$\begin{aligned} &(a) \quad \text{First radiation constant } C_1 = 2\pi h c_0^2 \\ &(b) \quad \text{Second radiation constant } C_2 = \frac{hc_0}{k} \end{aligned} \tag{8.63}$$

The important thing to notice is that the spectral emissive power is obtained with all the constants in terms of fundamental physical constants. Using the standard values for all the physical constants, we get

$$C_1 = 3.7413 \times 10^8 \,\frac{\text{W}\mu\text{m}^4}{\text{m}^2}; \quad C_2 = 14388\,\mu\text{m}\,\text{K}$$

These constants are obtained with the assumption that the wavelength of radiation will be specified in μm and the temperature in K. Correspondingly the spectral emissive power will be in W/m$^2\mu$m. The spectral emissive power of a black body given by Eq. 8.62 is known as the **Planck distribution function**.[5]

8.3.4 *Properties of the Planck Distribution Function*

Total Emissive Power of a Black Body

Integration of Eq. 8.62 over the wavelength range 0 to ∞ gives the total black body emissive power given by

$$E_b(T) = \int_0^\infty \frac{C_1}{\lambda^5} \frac{1}{\left(e^{\frac{C_2}{\lambda T}} - 1\right)} d\lambda \tag{8.64}$$

The integral may be recast by substituting, $\frac{C_2}{\lambda T} = x$ such that $-C_2 \frac{d\lambda}{T\lambda^2} = dx$ or $d\lambda = -\frac{T\lambda^2}{C_2} dx = -\frac{C_2 dx}{T x^2}$, as

[5] After Max Karl Ernst Ludwig Planck 1858–1947, German theoretical physicist.

$$E_b(T) = -\int_{\infty}^{0} \frac{C_1}{C_2^5} T^5 x^5 \times \frac{C_2 dx}{T x^2 (\mathrm{e}^x - 1)} = \frac{C_1}{C_2^4} T^4 \underbrace{\int_0^{\infty} \frac{x^3}{\mathrm{e}^x - 1} dx}_{P}$$

The integral labeled P in the above equation may be obtained as under. The denominator of the integrand may be rewritten as

$$\frac{1}{\mathrm{e}^x - 1} = \frac{\mathrm{e}^{-x}}{1 - e^{-x}} = e^{-x} \sum_{n=0}^{\infty} \mathrm{e}^{-nx}$$

Hence, the integral required is recast as

$$P = \sum_{n=0}^{\infty} \int_0^{\infty} x^3 \mathrm{e}^{-(n+1)x} dx = \sum_{n=1}^{\infty} \int_0^{\infty} x^3 \mathrm{e}^{-nx} dx$$

Let $u = nx$ such that $dx = \frac{du}{n}$ and $\mathrm{e}^{-nx} = e^{-u}$. The integral may then be written down as

$$P = \sum_{n=1}^{\infty} \int_0^{\infty} \frac{u^3}{n^3} \mathrm{e}^{-u} \frac{du}{n} = \sum_{n=1}^{\infty} \frac{1}{n^4} \int_0^{\infty} u^3 \mathrm{e}^{-u} du$$

The integral appearing in the above equation may be recognized as $\Gamma(4)$, on comparison with Eq. A.10. The summation may be considered separately now and is nothing but the Riemann Zeta function given by $\zeta(4) = \sum_{n=1}^{\infty} \frac{1}{n^4} = \frac{\pi^4}{90}$. Thus, we have the important result

$$P = \zeta(4)\Gamma(4) = \frac{\pi^4}{90} \times 6 = \frac{\pi^4}{15} \tag{8.65}$$

With this the total black body emissive power is given by

$$E_b(T) = \frac{C_1 \pi^4}{15 C_2^4} T^4 \tag{8.66}$$

The factor $\frac{C_1 \pi^4}{15 C_2^4}$ appearing in Eq. 8.66 may be written in terms of fundamental physical constants, using Eqs. 8.63 as $\frac{2\pi^5 k^4}{15 h^3 c_0^2}$. It is normal practice to refer to it as the Stefan–Boltzmann constant σ. Hence, the total black body emissive power is

$$\boxed{E_b(T) = \sigma T^4} \tag{8.67}$$

The numerical value of σ is

$$\boxed{\sigma = 5.67 \times 10^{-8}\ \frac{\text{W}}{\text{m}^2\text{K}^4}} \tag{8.68}$$

In Sect. 8.2.2, the fourth power dependence of emissive power on temperature was predicted using purely thermodynamic arguments. The constant was left undefined. The Planck distribution function yields the value of the constant in terms of fundamental physical constants.

Universal Form of Planck Function

Planck's distribution function may be cast in the universal form

$$\frac{E_{b\lambda}}{\sigma T^5} = \underbrace{\frac{C_1}{\sigma}\frac{1}{(\lambda T)^5}\frac{1}{\left(\mathrm{e}^{\frac{C_2}{\lambda T}} - 1\right)}}_{\text{Function of} \lambda T} \tag{8.69}$$

In this form, the Planck function plots as a single curve as shown in Fig. 8.12. In this form, the Planck distribution function is a function of λT only. The Planck function vanishes for both $\lambda T \to 0$ and $\lambda T \to \infty$, is monotonic and has a single maximum at

$$\boxed{\lambda T = 2897.8\ \mu\text{m K}} \tag{8.70}$$

The above relation is known as **Wein's displacement law**.[6] The above result is easily obtained by differentiating the universal black body function with respect to λT and setting it to zero. The equation may easily be solved using Newton Raphson method. This is left as an exercise to the reader.

Rayleigh–Jean's Approximation

In the large wavelength limit, i.e., when $\lambda T \to \infty$, $e^{\frac{C_2}{\lambda T}} \approx 1 + \frac{C_2}{\lambda T}$ and hence the Planck distribution given by Eq. 8.62 may be approximated as

$$E_{b\lambda} \approx \frac{C_1}{\lambda^5}\frac{1}{\left(1 + \frac{C_2}{\lambda T} - 1\right)} = \frac{C_1}{C_2\lambda^5}\lambda T = \boxed{\frac{2\pi c_0}{\lambda^4}kT} \tag{8.71}$$

[6]After Wilhelm Carl Werner Otto Fritz Franz Wien 1864–1928, German physicist.

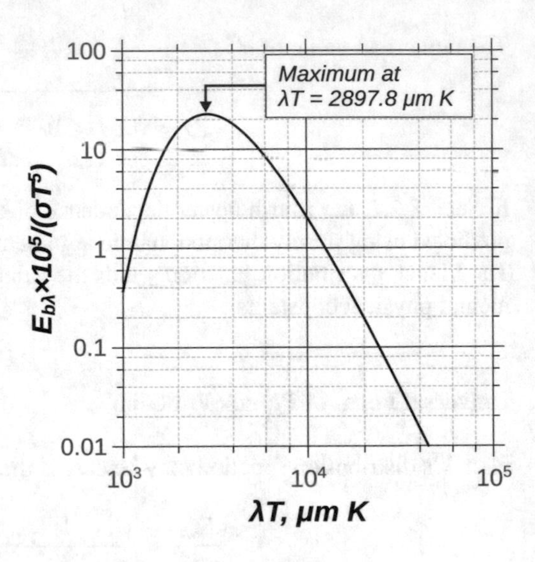

Fig. 8.12 Universal black body distribution

The last step is obtained by substituting for C_1 and C_2 in terms of the physical constants using Eq. 8.63. Equation 8.71 is known as the Rayleigh–Jeans approximation.[7]

Wein's Approximation

In the small wavelength limit, i.e., when $\lambda T \to 0$, the exponential term in the denominator of Eq. 8.62 is much larger than 1 and hence the Planck distribution may be approximated as

$$E_{b\lambda} \approx \frac{C_1}{\lambda^5}\frac{1}{e^{\frac{C_2}{\lambda T}}} = \boxed{\frac{C_1 e^{-\frac{C_2}{\lambda T}}}{\lambda^5}} \tag{8.72}$$

This is referred to as Wein's approximation. This approximation is used in optical pyrometry, the measurement of high temperatures using radiation emitted by surfaces. This approximation is good enough (to within 1%)for temperature as high as 5000 K when using visible radiation at 0.66 μm, as is common in pyrometry.

Fraction of Energy of a Black Body in $0 - \lambda T$
The universal black body function may be used to define the function $f_{0-\lambda T}$, the fraction of the radiation emitted in the range $0 - \lambda T$ as

[7] After John William Strutt, 3rd Baron Rayleigh 1842–1919, British physicist; Sir James Hopwood Jeans 1877–1946, British physicist, astronomer, and mathematician.

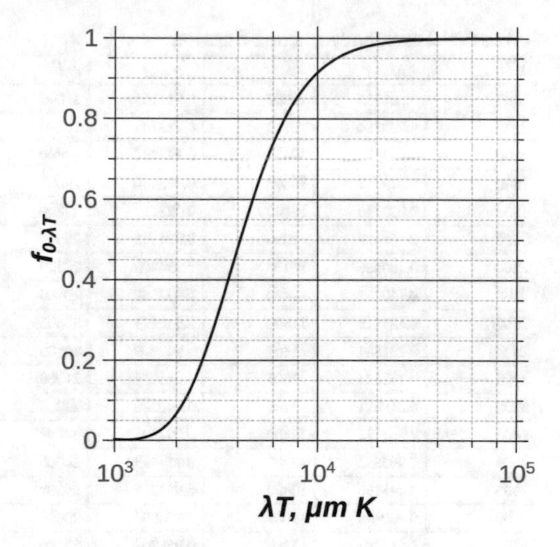

Fig. 8.13 Fraction black body function

$$f_{0-\lambda T} = \frac{\int_0^{\lambda T} E_{b\lambda} d\lambda}{\sigma T^4} = \int_0^{\lambda T} \frac{E_{b\lambda}}{\sigma T^5} d\lambda T \tag{8.73}$$

The integration indicated in Eq. 8.73 may be performed after substituting expression 8.69 in it. As before the integral may be expressed as an infinite series that converges rapidly. Term by term integration then leads to the required result. Figure 8.13 shows a plot of the fraction function as a function of λT.

For computational purposes, it is desirable to tabulate the fraction function as shown in Table 8.1.

Planck Distribution over Frequencies

The Planck distribution may be represented in terms of frequency ν instead of λ. Noting that $\lambda = \frac{c_0}{\nu}$, $I_{b\lambda}$ given by Eq. 8.62 may be written as

$$E_{b\lambda} = \frac{C_1 \nu^5}{c_0^5} \frac{1}{\left(e^{\frac{C_2 \nu}{c_0 T}} - 1\right)} \tag{8.74}$$

The power radiated by the black body in the wavelength band $d\lambda$ around λ is simply given by $E_{b\lambda} d\lambda$. Hence, we have

Table 8.1 Black body fraction function

λT	$f_{0-\lambda T}$	λT	$f_{0-\lambda T}$	λT	$f_{0-\lambda T}$	λT	$f_{0-\lambda T}$
1000	0.000321	10000	0.914155	19000	0.983405	56000	0.999210
1200	0.002134	10200	0.918135	19200	0.983867	57000	0.999250
1400	0.007790	10400	0.921877	19400	0.984312	58000	0.999287
1600	0.019718	10600	0.925400	19600	0.984741	59000	0.999321
1800	0.039340	10800	0.928717	19800	0.985155	60000	0.999354
2000	0.066727	11000	0.931845	20000	0.985553	61000	0.999384
2200	0.100886	11200	0.934795	20200	0.985938	62000	0.999413
2400	0.140253	11400	0.937580	20400	0.986309	63000	0.999439
2600	0.183115	11600	0.940211	20600	0.986668	64000	0.999464
2800	0.227884	11800	0.942699	20800	0.987014	65000	0.999488
3000	0.273223	12000	0.945052	21000	0.987348	66000	0.999510
3200	0.318091	12200	0.947280	22000	0.988858	67000	0.999531
3400	0.361722	12400	0.949391	23000	0.990137	68000	0.999551
3600	0.403592	12600	0.951392	24000	0.991229	69000	0.999570
3800	0.443366	12800	0.953291	25000	0.992166	70000	0.999587
4000	0.480858	13000	0.955092	26000	0.992974	71000	0.999604
4200	0.515993	13200	0.956804	27000	0.993675	72000	0.999620
4400	0.548774	13400	0.958431	28000	0.994286	73000	0.999635
4600	0.579256	13600	0.959977	29000	0.994821	74000	0.999649
4800	0.607534	13800	0.961449	30000	0.995291	75000	0.999663
5000	0.633720	14000	0.962850	31000	0.995706	76000	0.999676
5200	0.657942	14200	0.964185	32000	0.996074	77000	0.999688
5400	0.680330	14400	0.965458	33000	0.996401	78000	0.999699
5600	0.701016	14600	0.966671	34000	0.996693	79000	0.999710
5800	0.720126	14800	0.967828	35000	0.996954	80000	0.999721
6000	0.737785	15000	0.968933	36000	0.997188	81000	0.999731
6200	0.754106	15200	0.969989	37000	0.997399	82000	0.999740
6400	0.769199	15400	0.970998	38000	0.997589	83000	0.999749
6600	0.783163	15600	0.971962	39000	0.997762	84000	0.999758
6800	0.796092	15800	0.972884	40000	0.997918	85000	0.999766
7000	0.808071	16000	0.973767	41000	0.998060	86000	0.999774
7200	0.819179	16200	0.974612	42000	0.998189	87000	0.999781
7400	0.829488	16400	0.975421	43000	0.998308	88000	0.999789
7600	0.839063	16600	0.976196	44000	0.998416	89000	0.999795
7800	0.847965	16800	0.976940	45000	0.998515	90000	0.999802
8000	0.856248	17000	0.977653	46000	0.998606	91000	0.999808
8200	0.863962	17200	0.978337	47000	0.998690	92000	0.999814
8400	0.871153	17400	0.978994	48000	0.998767	93000	0.999820
8600	0.877863	17600	0.979624	49000	0.998838	94000	0.999826
8800	0.884130	17800	0.980230	50000	0.998904	95000	0.999831
9000	0.889987	18000	0.980812	51000	0.998965	96000	0.999836
9200	0.895467	18200	0.981371	52000	0.999021	97000	0.999841
9400	0.900599	18400	0.981910	53000	0.999074	98000	0.999845
9600	0.905408	18600	0.982427	54000	0.999123	99000	0.999850
9800	0.909920	18800	0.982926	55000	0.999168	100000	0.999854

$$E_{b\lambda}d\lambda = -\frac{C_1\nu^5}{c_0^5}\frac{1}{\left(e^{\frac{C_2\nu}{c_0T}}-1\right)}\frac{c_0}{\nu^2}d\nu \tag{8.75}$$

However, by definition the energy contained within the band $d\nu$ around ν must be given by $I_{b\nu}d\nu$. The negative sign is absorbed in the integration since $d\nu$ and $d\lambda$ are of opposite sign! Hence, it is appropriate to define the black body energy distribution over frequency by the relation

$$E_{b\nu} = \frac{C_1\nu^5}{c_0^5}\frac{1}{\left(e^{\frac{C_2\nu}{c_0T}}-1\right)}\frac{c_0}{\nu^2} = \frac{C_1\nu^3}{c_0^4}\frac{1}{\left(e^{\frac{C_2\nu}{c_0T}}-1\right)} \tag{8.76}$$

Using Eqs. 8.63, this may be recast in the universal form

$$\boxed{\frac{E_{b\nu}}{T^3} = \frac{2\pi k^3}{h^2c_0^2}\frac{\left(\frac{h\nu}{kT}\right)^3}{\left(e^{\frac{h\nu}{kT}}-1\right)}} \tag{8.77}$$

This distribution also vanishes at both the limits, i.e., $\nu = 0$ and $\nu \to \infty$. It is monotonic within this range and has a maximum at $\nu_{max} = 5.879 \times 10^{10}\,T$ where ν_{max} is in Hz. Note that this frequency *is not* the frequency that corresponds to the wavelength that occurs for the maximum of the wavelength distribution.

Black Bodies of Interest in Engineering Application

Common radiation sources of interest to us are the radiation from earth and the sun. The corresponding temperatures of interest, respectively, are 300 K and 5800 K. Important features of these sources are given in Table 8.2.

Figure 8.14 shows the black body distribution for three sources at different temperatures of 300, 1500, and 5800 K. Wein's displacement law appears as a straight line (note that the plot is made as a log-log graph) as shown therein.

Example 8.3

A surface emits thermal radiation with the intensity varying with angle θ to the normal given by $I = I_0\mu^{0.2}$ where I_0 is a constant. Determine the emitted flux in terms of I_0. If I_0 corresponds to the intensity of a black body at 2300 K, calculate the emitted flux in W/m^2. Compare this with the emitted flux from a black body at 2300 K.

Table 8.2 Black bodies of interest

	Earth	Sun
Temperature K	300	5800
E_b—W/m^2	459.2	64.16×10^6
λ_{max}, μm	9.66	0.5
Fractions		
UV	0	0.112
Visible	0	0.456
IR	1	0.432
Fraction		
Below 4 μm	0.002	0.99
Above 4 μm	0.998	0.01

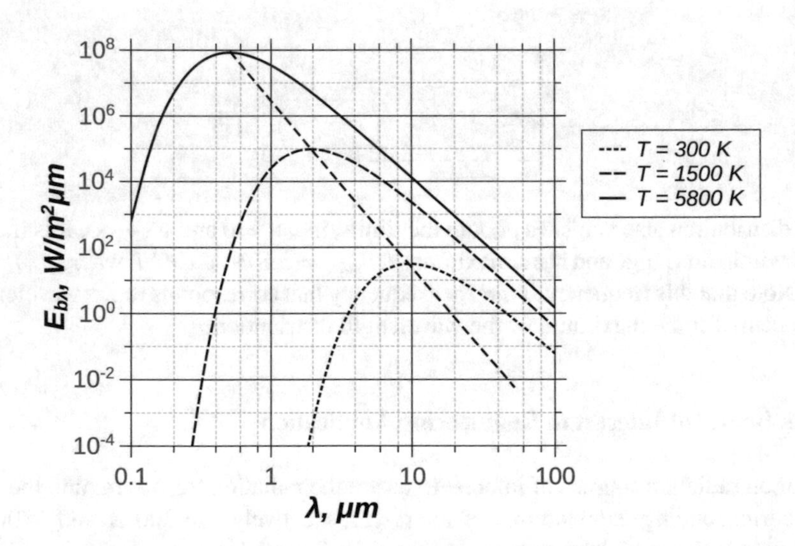

Fig. 8.14 Monochromatic emissive power of black bodies at different temperatures

---

Solution:

We make use of the definition of q^+ to get

$$q^+ = 2\pi \int_0^1 I \mu d\mu = 2\pi \int_0^1 I_0 \mu^{1.2} d\mu = 2\pi I_0 \left. \frac{\mu^{2.2}}{2.2} \right|_0^1 = \frac{\pi I_0}{1.1}$$

The intensity I_0 has been specified to be the same as that of a black body at $T = 2300$ K. Hence

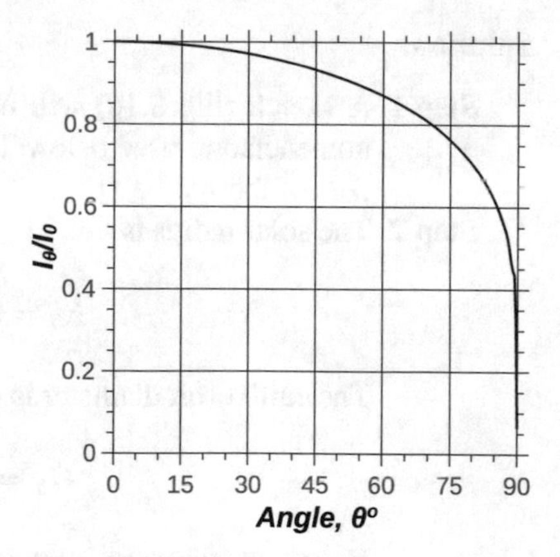

Fig. 8.15 Intensity variation with angle to the normal in Example 8.3

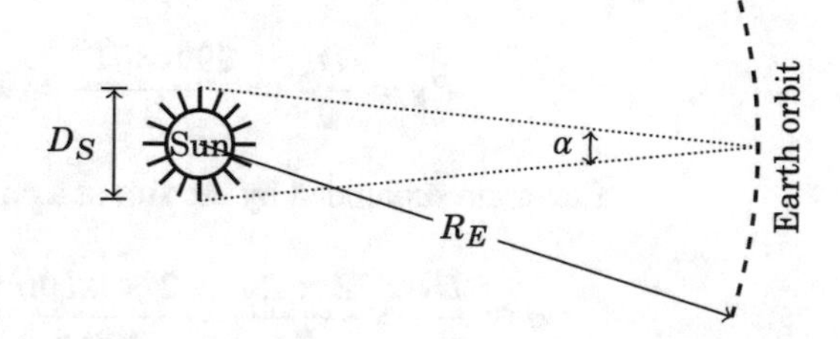

Fig. 8.16 Geometry for the sun-earth problem in Example 8.4

$$I_0 = \frac{\sigma T^4}{\pi} = \frac{5.67 \times 10^{-8} \times 2300^4}{\pi} = 505061.8 \text{ W/m}^2 \text{ sr}$$

With this, value of I_0 the radiated flux is given by

$$q^+ = \frac{\pi \times 505061.8}{1.1} = 1442453 \text{ W/m}^2 \approx 1.44 \text{ MW/m}^2$$

The emissive power of the black body at 2300 K, however, is

$$E_b = \sigma T^4 = 5.67 \times 10^{-8} \times 2300^4 = 1586698.5 \text{ W/m}^2 \approx 1.59 \text{ MW/m}^2$$

It is interesting to plot the given intensity distribution as shown in Fig. 8.15.

Example 8.4

If the solar constant, the radiant energy incident per unit area and time of a surface held normal to the sun's rays, is taken as 1354 W/m^2 after correcting for atmospheric absorption, estimate the effective sun surface temperature. The sun may be assumed to be a black body. The solar radius is 0.696×10^6 km and the diameter of the earth's orbit is 299×10^6 km.

Solution:

Step 1 A sketch (Fig. 8.16) will help in understanding the geometry and the nomenclature used below. Let the intensity of radiation leaving the surface of the sun be I_S.

Step 2 The solar radius is

$$R_S = 0.696 \times 10^6 \text{ km}$$

The earth orbit diameter is

$$D_E = 299 \times 10^6 \text{ km}$$

Hence, the distance from earth to sun is

$$R_E = \frac{D_E}{2} = \frac{299 \times 10^6}{2} = 149.5 \times 10^6 \text{ km}$$

The angle subtended by the sun at a point on the earth's orbit is

$$\alpha \approx \frac{D_S}{R_E} = \frac{2 \times R_S}{R_E} = \frac{2 \times 0.696 \times 10^6}{149.5 \times 10^6} = 0.00931 \text{ rad}$$

Step 3 This angle is quite small, and we may assume that the cosine of the half of this angle is very close to one, i.e., $\cos\alpha = 0.99999 \approx 1$. Hence, the solid angle subtended by the sun at a point on the earth's orbit may be approximated as

$$\Omega_S = \frac{\text{Solar disk area} \times \cos\alpha}{R_E^2} \approx \frac{\pi \times R_S^2}{R_E^2}$$
$$\approx \frac{\pi \times (0.696 \times 10^6)^2}{(149.5 \times 10^6)^2} = 6.77275 \times 10^{-5} \text{ sr}$$

Step 4 Hence, the solar constant should be equal to $S = I_S \Omega_S$. Hence,

$$I_S = \frac{S}{\Omega_S} = \frac{1354}{6.77275 \times 10^{-5}} = 1991877 \text{ W/m}^2 \text{ sr}$$

Since black body intensity is isotropic, the intensity is emissive power divided by π. This last quantity must equal $\frac{\sigma T_S^4}{\pi}$ where T_S is the sun surface temperature.

Step 5 Hence, the sun surface temperature is estimated as

$$T_S = \left(\frac{\pi I_S}{\sigma}\right)^{\frac{1}{4}} = \left(\frac{\pi \times 1991877}{5.67 \times 10^{-8}}\right)^{\frac{1}{4}} = 5769\,\mathrm{K}$$

This is very close to the value of 5800 K quoted in Table 8.2.

Concluding Remarks

This chapter has presented the basics of thermal radiation and sets the tone for the three chapters to follow. Description of a radiation field by radiation intensity and its moments is followed by description of black body radiation. Basic ideas about waves has been included to understand the concept of equilibrium radiation inside an evacuated enclosure.

8.4 Exercises

Ex 8.1 Calculate the solid angle subtended by a square disk of side a at a point distant H normal to the plane of the disk and lying along its axis as shown in Fig. 8.17.

Ex 8.2 Determine the solid angle subtended by a sector of a disk of radius R at a point a away from plane of the disk and positioned vertically opposite C, as shown in Fig. 8.18.

Ex 8.3 Intensity of radiation leaving a surface varies as follows:

$$I(\mu) = I_0 \mu^{0.3},\ 0 < \mu < 0.5,\ I(\mu) = 0.406 I_0,\ \mu > 0.5$$

Fig. 8.17 Geometry for Ex 8.1

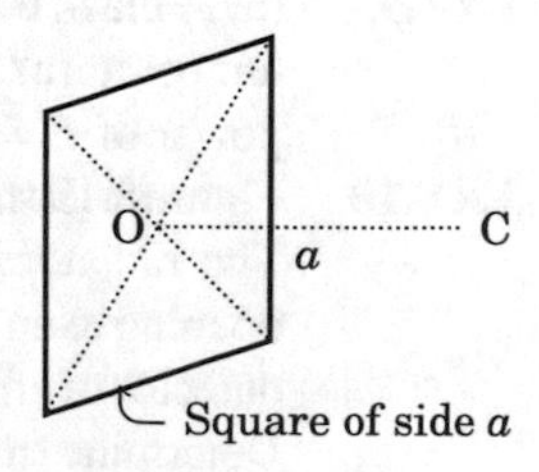

Fig. 8.18 Geometry for Ex 8.2

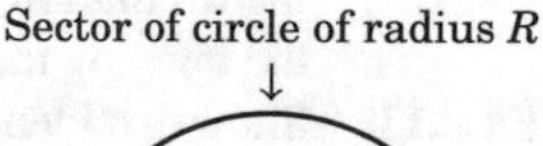

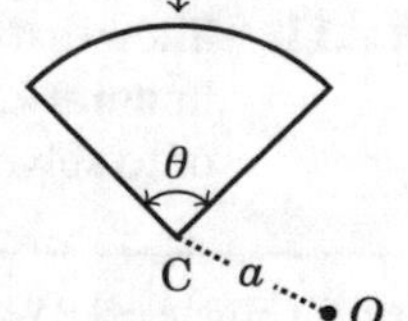

where I_0 is a constant. Determine the heat flux q^+.

Ex 8.4 A radiating surface emits radiation according to the relation $I(\theta, \phi) = 200 \cos^2 \theta$ with the units W/m^2 sr. What is the radiant heat flux in W/m^2? How does it compare with an isotropic radiator with $I = 200$ W/m$^2\cdot$ sr?

Ex 8.5 A point source of radiation emits isotropically with an intensity given by $I = 10000$ W/sr. What is the energy incident on a surface in the form of a circular plate that is held normal to the radius vector between the point source and the center of plate? The plate has a radius of $R = 2$ m and the radius vector is $L = 5$ m long.

Ex 8.6 Radiant intensity leaving a surface follows the relation $I(\mu) = 0.15 + 1.8\mu + 3\mu^2 - 2\mu^3$ W/m^{2}sr where μ has the usual meaning. What is the mean intensity?

Ex 8.7 (a) The Planck spectral distribution function is given by Eq. 8.62. When $\frac{C_2}{\lambda T} \gg 1$, the above may be approximated by dropping the -1 in the bracketed term and the resulting expression is Wein's approximation. Determine the percentage error in using Wein's approximate instead of the exact relation for $\lambda T = 2898$ K. Note that Wein's approximation is made use of in pyrometry.[8]

(b) When $\frac{C_2}{\lambda T} \ll 1$, the Planck distribution may be approximated by the Rayleigh–Jean approximation given in the text. Calculate the error in using the Rayleigh–Jean approximation instead of the exact relation for $\lambda T = 10^5$ μm K.

Ex 8.8 Filament of an electric bulb is maintained at a temperature of 2900 K and may be approximated as a gray body with an emissivity of 0.85. What is the wavelength at which the monochromatic emissive power of the filament is a maximum? Determine the fraction of the radiation emitted in the visible part of the spectrum. If the surface area of the filament has been estimated to be 10^{-5} m^2, what are the total emission and the emission in the visible part of the spectrum?

Ex 8.9 (a) A black body has total emissive power of $E_b = 500$ W/m^2. What is its temperature? (b) A gray surface of emissivity $\varepsilon = 0.8$ has total emissive power of $E = 500$ W/m^2. What is its temperature?

Ex 8.10 Figure 8.19 shows a cavity radiator whose walls are maintained at 1000 K. The radiator has an opening 20 mm in diameter. You may assume the opening to emit as a black body at the cavity wall temperature. A radiation detector of effective diameter of 20 mm is placed as indicated in the figure. Determine the following: (a) the total radiation leaving the opening, (b) the irradiation on the detector, and (c) the fraction of the power leaving the opening that is received by the detector.

Ex 8.11 Show that Planck's distribution function in terms of wavelength has a maximum at $\lambda_{max} T = 2898$ μm K. Newton Raphson method may be made use of to solve the appropriate equation.

---

[8] See, for example S.P. Venkateshan, *Mechanical Measurements*, 2nd Edition, ANE Books Pvt. Ltd. India or Athena Academic—Wiley, 2015.

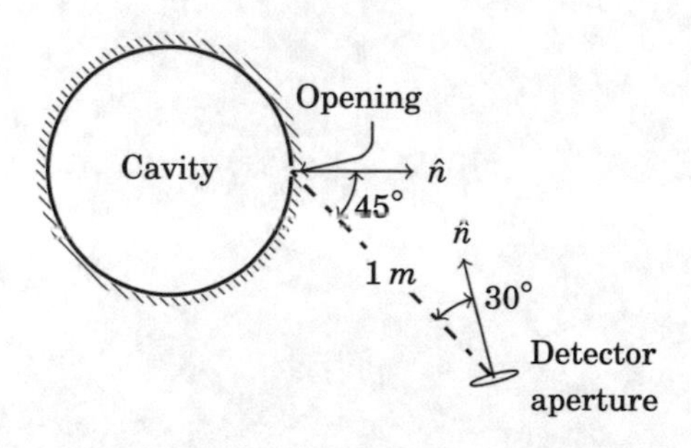

Fig. 8.19 Geometry for Exercise 8.10

Ex 8.12 Show that Planck's distribution function in terms of frequency has a maximum at $\frac{\nu_{max}}{T} = 5.879 \times 10^{10}$ Hz/K. Newton Raphson method may be made use of to solve the appropriate equation.

Ex 8.13 Consider an evacuated isothermal cavity whose walls are maintained at 3000 K.

- Calculate the wavelength at which the spectral intensity of radiation within the cavity is the maximum.
- Calculate the fraction of radiation that is below this wavelength.
- What is the wavelength below which 50% of radiation is present?
- Calculate the wavelength below which 95% of the radiation is contained.

Ex 8.14 A closed insulated cavity whose walls are maintained at 2000 K is evacuated to a residual pressure of 10 Pa. What is the radiation pressure as a fraction of this residual pressure? Also determine (a) the radiation intensity and (b) the radiation density inside the enclosure. Assume that the residual gas inside the cavity does not affect the radiation within the cavity.

Chapter 9
Surface Radiation

RADIATION from surfaces is important in most engineering applications. Walls of rooms, external surfaces of a satellite (while in outer space), and walls of furnaces and ovens are familiar examples where surface radiation plays an important role in energy transfer. Even though engineering surfaces are hard to characterize, it is possible to describe the radiative properties of optically smooth surfaces using a theoretical model based on the electromagnetic theory. Real and complex indices of refraction are used as macroscopic properties that characterize the interaction of radiation with a surface. Such a theory is described in detail in this chapter. The results are then extended to engineering surfaces.

9.1 Introduction

Engineering applications require the modeling of thermal radiation leaving surfaces such as those that occur in rooms, oven walls, furnace exterior walls, solar absorbers, and so on. The surrounding ambient usually is low temperature air, which may be assumed to be neutral to the passage of radiation, in that radiation passes through it without any attenuation. Any surface we may be interested in is essentially an interface between a medium whose boundary the surface is, and ambient air. For all practical purposes, the ambient air may be treated as vacuum, except of course that air may conduct heat or transfer heat by convection. For present purposes, we ignore both these modes of heat transfer so that the basics of surface radiation may be developed without complicating matters.

S. P. Venkateshan, *Heat Transfer*,
https://doi.org/10.1007/978-3-030-58338-5_9

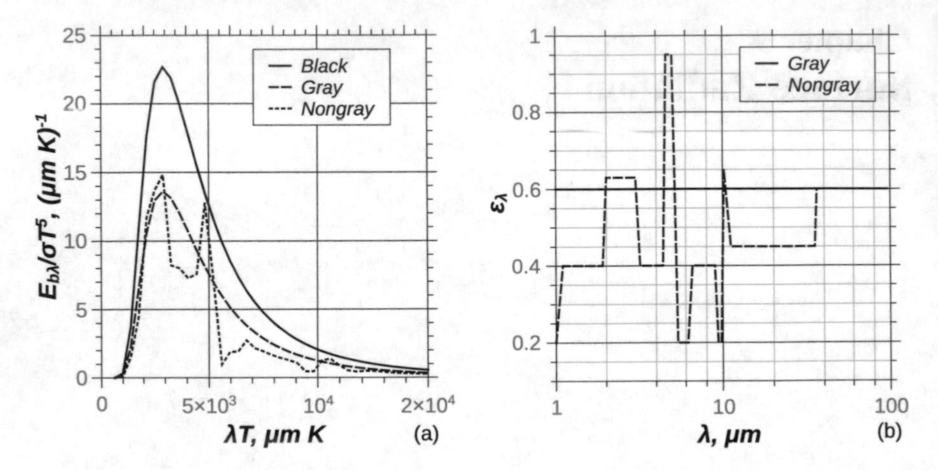

Fig. 9.1 **a** Comparison of emission from black, gray, and non-gray surfaces: all surfaces are at $T_S = 1000$ K **b** Emissivity of gray and non-gray surfaces in (**a**)

9.1.1 Surface Types

In general, a surface emits radiation (because of its temperature) with an intensity that will vary with direction and with wavelength. We look at the wavelength dependence first. A surface maintained at temperature (T_S) cannot emit more radiation per unit area than a black body at the same temperature as may be easily shown by thermodynamic arguments. Thus, the best we may expect is that the surface emits as a black body. We refer to such a surface as a **black** surface. Most surfaces emit less than a black surface at the same temperature as the surface. If the emitted intensity is a constant fraction, independent of wavelength, of that emitted by a black body at the same temperature, the surface is termed as a **gray** surface. If the fraction is dependent on a wavelength, we refer to the surface as a **non-gray** surface. This last type is the most general type and hardest to characterize.

Figure 9.1 shows the three types of surfaces, all maintained at a common temperature of $T_S = 1000$ K. The wavelength dependence of black body radiation is that given by the Planck distribution. Emission from the gray surface is a scaled version of black body radiation with a constant scaling factor (ε—total hemispherical emissivity, to be defined later) between 0 and 1. The scale factor (ε_λ—spectral hemispherical emissivity, to be defined later) varies with wavelength for the non-gray surface. The illustration shows a non-gray surface that is characterized by a band model that will be discussed later in more detail.

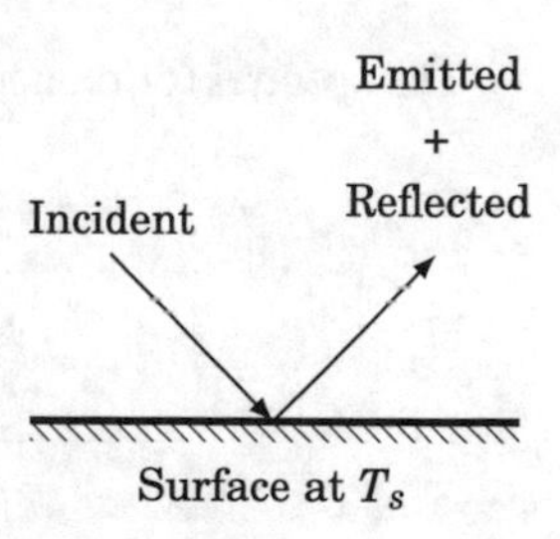

Fig. 9.2 Incoming and outgoing radiation at a surface

9.2 Spectral and Hemispherical Surface Properties

The radiation leaving a surface may be due to emission from the surface or that part of incident radiation which is reflected by the surface. Such radiation depends, in general, on two angles, viz., the polar angle θ and the azimuthal angle ϕ. Detailed knowledge of the angular dependence of the leaving intensity is seldom required in engineering radiation heat transfer. Hence, one is usually content in defining hemispherical quantities such as hemispherical emissivity and hemispherical reflectivity. Of course, these quantities will depend on the angular distribution, an aspect that will receive attention later on.

9.2.1 Spectral Hemispherical Quantities

Spectral Hemispherical Emissivity

Consider a surface whose temperature is T_S.

This surface, by virtue of its temperature, will emit radiation to the surroundings into a hemisphere (of solid angle 2π) given by, say $E_\lambda(T_S)$ per unit area (shown as "Emitted" in Fig. 9.2). In order to describe this radiation, the obvious thing to do would be to compare it with the power radiated by a black body at the same temperature. We know that the black body will emit $E_{b\lambda}(T_S)$ into the hemisphere. Then, the spectral (or monochromatic) hemispherical emissivity is defined as

$$\boxed{\varepsilon_{\lambda,h} = \frac{E_\lambda(T_S)}{E_{b\lambda}(T_S)}} \tag{9.1}$$

Spectral Hemispherical Reflectivity

The surface under consideration may be receiving radiation from the hemisphere given by $q_\lambda^- = G_\lambda$ (say) shown as "Incident" in Fig. 9.2. G_λ is the incident radiant flux referred to as spectral irradiation. A part of this irradiation is reflected and shown as "Reflected" in Fig. 9.2.

The spectral (or monochromatic) hemispherical reflectivity is defined as

$$\boxed{\rho_{\lambda,h} = \frac{\text{Spectral reflected radiant heat flux}}{\text{Spectral irradiation}} = \frac{q^{+}_{\lambda,r}}{G_{\lambda}}} \tag{9.2}$$

Spectral Hemispherical Radiosity
The radiation leaving the surface—the radiosity—into the hemisphere is the sum of the emitted radiation and reflected radiation as indicated by "Emitted + Reflected" in Fig. 9.2. Using defining expressions 9.1 and 9.2, we have the spectral radiosity J_{λ} given by

$$\boxed{J_{\lambda,h} = \varepsilon_{\lambda,h} E_{b\lambda,h}(T_S) + \rho_{\lambda,h} G_{\lambda}} \tag{9.3}$$

Spectral Hemispherical Absorptivity
For an opaque surface, the radiation, which is not reflected, must be absorbed by it. Hence, we define the spectral absorptivity $\alpha_{\lambda,h}$ as

$$\alpha_{\lambda,h} = 1 - \rho_{\lambda,h} \tag{9.4}$$

Using Eq. 9.4 in 9.3, we also have

$$\boxed{J_{\lambda,h} = \varepsilon_{\lambda,h} E_{b\lambda}(T_S) + (1 - \alpha_{\lambda,h}) G_{\lambda}} \tag{9.5}$$

Kirchhoff's Law
Imagine the surface that we have described above to be made a part of the wall of an evacuated cavity maintained at temperature T_S. The cavity is full of equilibrium radiation with q^{+}_{λ} equal to $E_{b\lambda}(T_S)$ everywhere on the cavity surface. Also the irradiation q^{-}_{λ} anywhere on the surface is equal to $E_{b\lambda}(T_S)$ so that there is no net heat transfer by radiation. Hence, we conclude that $J_{\lambda,h}$ must also be equal to $E_{b\lambda}(T_S)$. Thus, Eq. 9.5 must read

$$J_{\lambda,h} = \varepsilon_{\lambda,h} E_{b\lambda}(T_S) + (1 - \alpha_{\lambda,h}) E_{b\lambda}(T_S) = E_{b\lambda}(T_S)$$

which may be rearranged, after canceling the common factor $E_{b\lambda}(T_S)$, as

$$\varepsilon_{\lambda,h} + 1 - \alpha_{\lambda,h} = 1 \text{ or } \boxed{\varepsilon_{\lambda,h} = \alpha_{\lambda,h}} \tag{9.6}$$

Table 9.1 Monochromatic emissivities of representative surfaces

	λ, μm			
Surface	0.6	0.95	3.6	9.3
Chromium	0.49	0.43	0.26	0.08
Steel (Polished)	0.45	0.37	0.14	0.07
White Paper	0.28	0.25	0.82	0.95
Platinum Black		0.97	0.97	0.93
Graphite	0.73		0.54	0.41

Thus, we have the very important result that the spectral hemispherical emissivity of *any* opaque surface is equal to the spectral hemispherical absorptivity. This result is known as Kirchhoff's law. This does not apply automatically to the total quantities, as we shall see later. Monochromatic or spectral emissivities of several surfaces are given in Table 9.1.

9.2.2 Total Hemispherical Quantities

Total Hemispherical Emissivity

Starting with spectral hemispherical emissivity given by Eq. 9.1, integration with respect to wavelength gives

$$E(T_S) = \int_0^\infty E_\lambda(T_S)d\lambda = \int_0^\infty \varepsilon_{\lambda,h} E_{b\lambda}(T_S)d\lambda \tag{9.7}$$

It is customary to define the total hemispherical emissivity ε_h such that $E(T_S) = \varepsilon_h \sigma T_S^4$. We may hence define the total hemispherical emissivity through the relation

$$\boxed{\varepsilon_h = \frac{1}{\sigma T_S^4}\int_0^\infty \varepsilon_{\lambda,h} E_{b\lambda}(T_S)d\lambda} \tag{9.8}$$

We shall refer to ε_h *also* as the equivalent gray emissivity at temperature T_S. It is easily noticed that ε_h is a function of temperature T_s.

Total Hemispherical Reflectivity

From Eq. 9.2, we have $q^+_{\lambda,r} = \rho_{\lambda,h} G_\lambda$. Integrating this with respect to a wavelength, we have for the total quantity the relation

$$q_r^+ = \int_0^\infty q_{\lambda,r}^+ d\lambda = \int_0^\infty \rho_{\lambda,h} G_\lambda d\lambda$$

Defining the total irradiation as $G = \int_0^\infty G_\lambda d\lambda$, the above integral is written as $\rho_h G$ where ρ_h is the total hemispherical reflectivity. Thus, we have

$$\rho_h = \frac{1}{G} \int_0^\infty \rho_{\lambda,h} G_\lambda d\lambda \tag{9.9}$$

We shall refer to ρ_h also as the equivalent gray reflectivity at temperature T_B, assuming that the irradiation is due to a black body at this temperature.

Is Kirchhoff's Law Valid for Total Quantities?

The answer, in general, is a no! ρ_h depends on the nature of the incident radiation and hence it is not related to emissivity at all. When the incident radiation is black body radiation with an effective temperature of T_B, Eq. 9.9 will read as

$$\rho_h = \frac{1}{E_b(T_B)} \int_0^\infty \rho_{\lambda,h} E_{b\lambda}(T_B) d\lambda \tag{9.10}$$

This will obviously be a function of T_B and hence ρ_h will also be a function of T_B. In view of this, the total hemispherical absorptivity of the surface α_h will also be a function of T_B since $\alpha_h(T_B) = 1 - \rho_h(T_B)$ in order that energy is conserved, assuming that the surface is opaque.

However, for a gray surface Kirchhoff's law is valid for total quantities also.

9.2.3 Band Model for a Non-gray Surface

We shall introduce the band model using a typical example, the emissivity of aluminum oxide. A typical IR emission spectrum of aluminum oxide is shown in Fig. 9.3.

The advantage of the band approximation is that the equivalent gray emissivity (or absorptivity) may easily be calculated when the temperature is known. In the band approximation the emission spectrum is replaced by bands of constant emissivity but of variable width such that the collection of bands approximates the spectrum as closely as desired. The integration indicated in Eq. 9.8 is written as a sum of integrals

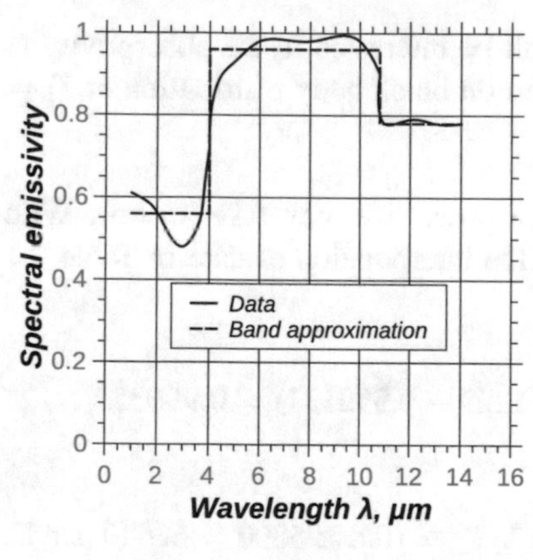

i	λ_i, μm	Band No.	Emissivity
0	0		-
1	4	1	0.56
2	10.8	2	0.96
3	∞	3	0.78

Fig. 9.3 Spectral emissivity of aluminum oxide and the band approximation

over the bands. The integrals over the bands are obtained by using the black body fraction function introduced in Sect. 8.3.4. Thus the equivalent gray emissivity of a non-gray surface is calculated by

$$\varepsilon_h(T_S) = \sum_{i=1}^{N} \varepsilon_{h,(i-1)\,-\,(i)} f_{\lambda_{(i-1)}T_S - \lambda_{(i)}T_S} \tag{9.11}$$

where there are N bands with constant band emissivities given by $\varepsilon_{i-1,i}$. Note also that we may write the fraction function appearing in Eq. 9.11 as

$$f_{\lambda_{(i-1)}T_S - \lambda_{(i)}T_S} = f_{0-\lambda_{(i)}T_S} - f_{0-\lambda_{(i-1)}T_S} \tag{9.12}$$

The f's appearing in the summation are the fractional black body function given in Table 8.1.

Example 9.1

Consider an anodized aluminum surface whose spectral emissivity is given in Fig. 9.3. The corresponding band data is shown in the inset table in the same figure. Determine the equivalent gray emissivity of such a surface at $T_1 = 5800$ K and $T_2 = 300$ K.

Solution:

Step 1 Three bands as shown in Fig. 9.3 approximate the emissivity of anodized aluminum. With the band approximation the integrals appearing in the calculation of absorptivity and emissivity may be replaced by summation over the bands using the integral black body function presented in Table 8.1.

Step 2 The emissivity (in practice, we shall be interested in the absorptivity at this temperature) is calculated based on black body distribution at $T_1 = 5800$ K.

*Band*1:

$\lambda_1 = 4$ μm and for all practical purposes, we may take $\lambda_0 = 0$. With $\lambda_1 T_1 = 4 \times 5800 = 23200$ μm K, the interpolation of data in Table 8.1 yields

$$f_{0-23200} = 0.990137 + \frac{200}{1000}(0.991229 - 0.990137) = 0.990355$$

Band 2:

$\lambda_2 = 10.8$μm and $\lambda_1 = 4$ μm. With $\lambda_2 T_1 = 10.8 \times 5800 = 62640$ μm K, the interpolation of data in Table 8.1 yields

$$f_{0-62640} = 0.999143 + \frac{640}{1000}(0.999439 - 0.999143) = 0.999332$$

Hence, we have

$$f_{23200-62640} = f_{0-62640} - f_{0-23200} = 0.999332 - 0.990355 = 0.008977$$

Band 3: $\lambda_3 = \infty$ and $\lambda_2 = 10.8$ μm. With $\lambda_3 T_1 = \infty$, we have $f_{0-\infty} = 1$. Hence, we have

$$f_{62640-\infty} = f_{0-\infty} - f_{0-62640} = 1 - 0.999332 = 0.000668$$

The emissivity of the surface at 5800 K and hence the absorptivity are then obtained by the use of Eq. 9.11 as

$$\begin{aligned}\varepsilon_h(5800K) &= 0.56 \times 0.990355 + 0.96 \times 0.008977 + 0.78 \times 0.000668 \\ &= 0.563738 \approx 0.564\end{aligned}$$

Step 3 The emissivity is now calculated based on black body distribution at $T_2 = 300$ K.

Band 1:

$\lambda_1 = 4$ μm and for all practical purposes, we may take $\lambda_0 = 0$. With $\lambda_1 T_2 = 4 \times 300 = 1200$ μm K, data in Table 8.1 yields

$$f_{0-1200} = 0.002134$$

Band 2:

$\lambda_2 = 10.8$ μm and $\lambda_1 = 4$ μm. With $\lambda_2 T_2 = 10.8 \times 300 = 3240$ μm K, the interpolation of data in Table 8.1 yields

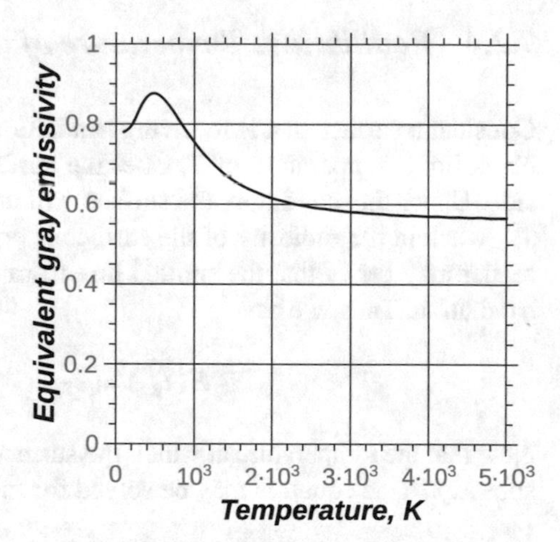

Fig. 9.4 Equivalent gray emissivity of anodized aluminum

$$f_{0-3240} = 0.318091 + \frac{40}{200}(0.361722 - 0.318091) = 0.326817$$

Hence, we have

$$f_{1200-3240} = f_{0-3240} - f_{0-1200} = 0.326817 - 0.002134 = 0.324683$$

<u>*Band*</u> 3: $\lambda_3 = \infty$ and $\lambda_2 = 10.8\,\mu$m. With $\lambda_3 T_2 = \infty$, we have $f_{0-\infty} = 1$. Hence, we have

$$f_{3240-\infty} = f_{0-\infty} - f_{0-3240} = 1 - 0.324683 = 0.675317$$

The emissivity of the surface at 300 K and hence the absorptivity are then obtained by the use of Eq. 9.11 as

$$\begin{aligned} \varepsilon_h(300K) &= 0.56 \times 0.002134 + 0.96 \times 0.324683 + 0.78 \times 0.675317 \\ &= 0.839638 \approx 0.84 \end{aligned}$$

Step 4 It is clear that the equivalent gray emissivity is a function of temperature. In order to appreciate this, Fig. 9.4 shows the variation of equivalent gray emissivity of anodized aluminum from room temperature to 5000 K. The "hump" in the variation is a consequence of the different bands having a primary effect on the gray emissivity as the temperature is varied! The reader should try to visualize this based on the nature of the spectral variation of black body emissive power with temperature.

9.2.4 Equilibrium Temperature of a Surface

Consider a surface that is receiving irradiation G from a background at an effective black body temperature of T_B. Let the surface be perfectly insulated at the back side. Under this condition, the surface will attain an equilibrium temperature $T_S = T_{eq}$ wherein the radiosity of the surface is equal to the irradiation. Alternately, the reader may verify that the emitted flux from the surface should equal the absorbed irradiation. Thus, we have[1]

$$E(T_{eq}) = \varepsilon_{T_{eq}} \sigma T_{eq}^4 = \alpha_{T_B} G$$

Note that the temperature at which the surface property is calculated is used now as subscript. This equation may be solved for the equilibrium temperature to get

$$\boxed{T_{eq} = \left(\frac{\alpha_{T_B} G}{\varepsilon_{T_{eq}} \sigma} \right)^{\frac{1}{4}}} \tag{9.13}$$

It is interesting to note that for a gray surface $\alpha_{T_B} = \varepsilon_{T_{eq}}$, independent of temperature, and hence the equilibrium temperature for a gray surface is given by

$$\boxed{T_{eq} = \left(\frac{G}{\sigma} \right)^{\frac{1}{4}}} \tag{9.14}$$

Example 9.2

An anodized aluminum surface is receiving solar flux of 1400 W/m^2 on one side while the other side is perfectly insulated. Determine its equilibrium temperature. Use the anodized aluminum data given in the inset table in Fig. 9.3.

Solution:
The irradiation is from the sun that may be characterized by an effective black body temperature of $T_B = 5800$ K. Hence the absorptivity of the surface is equal to the equivalent gray emissivity $\varepsilon_{5800} = 0.564$ at 5800 K that has been calculated in Example 9.1.

Since the surface temperature is unknown, the surface emissivity is also unknown. Hence Eq. 9.13 cannot be used directly. We may use this equation with an assumed value for T_{eq} to calculate $\varepsilon_{T_{eq}}$ to get an estimate for T_{eq}. This may then be used to correct the assumed value of emissivity and the procedure continued till a converged value of T_{eq} is obtained.

[1] From this point onwards, subscript h for hemispherical will be dropped.

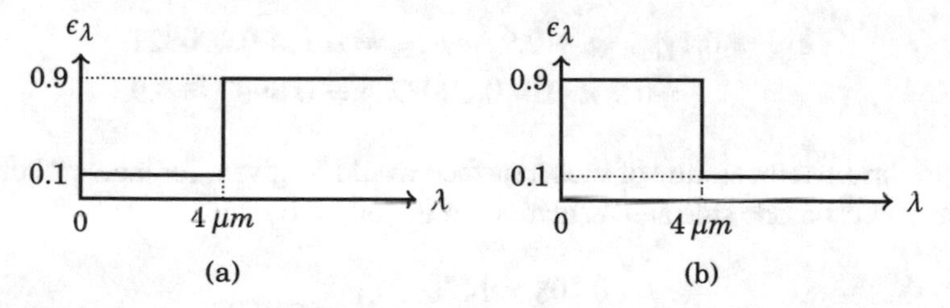

Fig. 9.5 Ideal selective surfaces

Let us assume $T_S = T_{eq}^0 = 300$ K as the first guess. The equivalent gray emissivity at this temperature is also available from Example 9.1 and is $\alpha_{300} = 0.84$. We use Eq. 9.13 to get a better value for T_{eq}^1 as

$$T_{eq}^1 = \left(\frac{0.564 \times 1400}{5.67 \times 10^{-8} \times 0.84} \right)^{\frac{1}{4}} = 358.83 \text{ K}$$

We may use this temperature to calculate a fresh value for the solar absorptivity and continue the iteration. The reader may continue the iteration process and show that the converged value of equilibrium temperature is 350.5 K.

9.2.5 *Selective Surfaces*

Non-gray surface properties play an important role in the thermal control of spacecraft systems and in solar energy devices. These surfaces are known as selective surfaces. The reader should ponder over why they are referred to as selective surfaces! In the former, the aim is to keep solar heat away while in the latter the aim is to gather as much of it as possible. Let us look first at a surface that is an ideal element for keeping the solar heat away. Consider an ideal surface having low emissivity up to 4 μm and high emissivity beyond 4 μm as shown in Fig. 9.5a. The non-gray behavior is modeled by a two-band approximation. Since solar radiation is negligibly small beyond 4 μm (see Table 8.2), the solar absorptivity of the surface is close to the first band value. We may calculate the equivalent gray absorptivity for solar radiation using Eq. 9.11 as

$$\alpha_S = 0.1 f_{0-23200} + 0.9 f_{23200-\infty} = 0.1 \times 0.9899$$
$$+0.9 \times (1 - 0.0.9899) = 0.108$$

where the subscript S stands for "Solar". Similarly, the emissivity for low frequency radiation from the surface at its temperature that will be close to room temperature (say 300 K), referred to as ε_{IR}, is

$$\varepsilon_{IR} = 0.1 f_{0-1200} + 0.9 f_{1200-\infty} = 0.1 \times 0.000321 \\ +0.9 \times (1 - 0.000321) = 0.8997 \approx 0.9$$

The equilibrium temperature of such a surface would be given, for incident solar flux of 1400 W/m^2 on one side and insulation on the other, by

$$T_{eq} = \left(\frac{0.108 \times 1400}{0.9 \times 5.67 \times 10^{-8}} \right)^{\frac{1}{4}} = 233.3\,\text{K} \tag{9.15}$$

This means that the surface will remain cool while dissipating some heat it may be allowed to gain from the side that is assumed to be adiabatic. This is referred to as a *passive cooler* since the cooling effect is obtained by the manipulation of the radiative property of the surface.

Now we look at a solar collecting surface. Consider the ideal non-gray surface with two bands as shown in Fig. 9.5b. In this case, the emissivity is high up to 4 μm and low beyond 4 μm. We may run through a calculation similar to the one made for calculating the solar absorptivity. It is left as an exercise to the reader! The appropriate numbers are (the iterative solution is called for; see Example 9.2)

$$\alpha_S = 0.892; \;\; \varepsilon_{IR} = 0.197 \text{ and} T_{eq} = 574\,\text{K} \tag{9.16}$$

Of course, the temperature will be much less than what is indicated because there will always be heat losses by conduction and convection in terrestrial applications (note that the solar collector usually is on the roof top). Ideal surfaces with two bands as indicated in Fig. 9.5 are not naturally available. Surfaces that come close to these are made by the use of surface coatings (by painting or by vacuum deposition) and by chemical treatment of the surfaces.

Example 9.3

Optical solar reflector (OSR), used in spacecraft temperature control, typically has maximum absorptivity of 0.085 between 0.25–2.5 μm. The emissivity of the OSR in the range 5–50 μm is 0.87.[2] Determine the amount of heat it can dissipate if it runs at 300 K and receives 1400 W/m^2 of solar radiation on the sun-facing side.

Solution:

Step 1 We calculate the fraction of solar radiation between 0.25–2.5 μm first based on $T_B = 5800$ K, $\lambda_{min} T_B = 0.25 \times 5800 = 1450$ μm K, and $\lambda_{max} T_B = 2.5 \times 5800 = 14500$ μm K. Interpolation using Table 8.1 yields

$$f_{0-1450} = 0.00779 + \frac{50}{200}(0.019718 - 0.00779) = 0.010772$$

[2]Plain OSR(CMO)—PS 614 supplied by Qioptiq Space Technology, U.K.,www.qiotiq.co.uk.

$$f_{0-14500} = 0.965458 + \frac{100}{200}(0.966671 - 0.965458) = 0.966065$$

Only 3.4% of irradiation is outside the band considered. Irradiation absorbed by unit area of the surface is taken as

$$Q_a = \varepsilon_{T_B} G = 0.085 \times 1400 = 119 \text{ W/m}^2$$

Step 2 Black body energy in the 5–50 μm band is calculated using $T_S = 300$ K with $\lambda_{min} T_S = 5 \times 300 = 1500$ μm K and $\lambda_{max} T_S = 50 \times 300 = 15000$ μm K. The fraction functions are obtained using Table 8.1.

$$f_{0-1500} = 0.00779 + \frac{100}{200}(0.019718 - 0.00779) = 0.013754$$

$$f_{0-15000} = 0.968933$$

Emission from the surface per unit area is then given by

$$Q_e = \varepsilon_{T_S}(f_{0-15000} - f_{0-1500})\sigma T_S^4$$
$$= 0.85 \times (0.968933 - 0.013754) \times 5.67 \times 10^{-8} \times 300^4 = 372.88 \text{ W/m}^2$$

The fraction emitted is not rounded to 1 so that the calculation is conservative.

Step 3 It is thus clear that the surface will run at a temperature below 300 K if the surface is insulated at the back. The surface will be able to support a heat gain of $372.9 - 119 = 253.9$ W/m^2 on the back side. If sun-facing side of a satellite is covered completely with the OSR, it will be able to dissipate about 250 W/m^2 of heat gained from within the satellite.

Example 9.4

Titanium-oxy-nitride[3] is a coating used in solar absorber applications. It has solar absorptivity of $\alpha_S = 0.9$ and an infrared emissivity of $\varepsilon_{IR} = 0.05$. This surface is exposed to solar irradiation of 800 W/m^2, on a partly cloudy day. Determine the amount of heat that may be rejected by a square meter of surface to a coolant if the absorber plate is at a mean temperature of $T_S = 370$ K.

---

[3] TiNOX GmbH, Germany.

Solution:
Heat absorbed per unit area of the absorber is given by

$$Q_a = \alpha_S G = 0.9 \times 800 = 720 \text{ W/m}^2$$

Heat emitted by the absorber is given by

$$Q_e = \varepsilon_{IR} \sigma T_S^4 = 0.05 \times 5.67 \times 10^{-8} \times 370^4 = 53.1 \text{ W/m}^2$$

The heat rejected to the coolant is obtained by energy balance as

$$Q_{\text{rejected}} = Q_a - Q_e = 720 - 53.1 = 666.9 \text{ W/m}^2$$

In practice, the heat rejected to the coolant may be less than this because of heat losses that have not been accounted for.

Note on selective surfaces: Selective surfaces are used extensively in solar energy applications and for passive thermal control of satellites. Selective surfaces are also used in temperature control of buildings and in cryogenic systems. Many selective surfaces have been developed, specifically with these applications in mind, over the past 60 years. The surfaces are typically characterized by the $\alpha_S - \varepsilon_{IR}$ ratio. The available information gathered from various sources is shown in Table 9.2. The surfaces are arranged in increasing order of $\alpha_S - \varepsilon_{IR}$ ratio. The reader will note that the values in the last column will guide in the selection of the surface for a specific application.

Selective properties may also be obtained by a suitable choice of thickness of a film of a material deposited on a substrate. Desirable optical properties are obtained by a combination of transmittance and reflectance based on absorption as well as the interference of light. Table 9.3 shows the properties of a typical sun film (supplied by 3M company under the description "Night Vision 15") used for controlling heat and light transmission through windows in automobiles and buildings.

As indicated in the table, a solar film also reduces the glare. Camera and spectacle lenses are coated with thin anti-reflection coatings that are based on the optical properties of thin films. Refractive index and thickness play a role in reducing reflection at a particular wavelength. It is usual to use multiple layers of different thicknesses and refractive indices to achieve low reflectance over a desirable part of the spectrum such as from 0.4–0.7 μm (visible radiation). The reader may refer to books on Optics for more information on this topic.

Table 9.2 Table of surface properties of selective surfaces

Surface	α_S	ε_{IR}	$\frac{\alpha_S}{\varepsilon_{IR}}$
Barium Sulfate with Polyvinyl Alcohol	0.06	0.88	0.07
Aluminum anodized	0.14	0.84	0.17
Titanium Oxide White Paint with Potassium Silicate	0.17	0.92	0.18
OSR (Optical Solar Reflector)	0.16	0.81	0.19
Concrete	0.6	0.88	0.68
Anodize Black	0.88	0.88	1
3M Black Velvet Paint	0.97	0.91	1.07
Dull brass, copper, galvanized steel, aluminum	0.40–0.65	0.20–0.30	2.1
Silver, Highly polished	0.06–0.09	0.02–0.03	3
Aluminum polished	0.09	0.03	3
Aluminum foil	0.15	0.05	3
Metal, plated Nickel oxide	0.92	0.08	11

Table 9.3 Typical sun control film properties: Performance Results on 0.25 inch (6 mm) clear glass

Visible Light Transmitted	15%
Total Solar Energy Rejected	72%
UV Light Rejected	99.9%
Glare Reduction	83%
Visible Light Reflected: Interior	11%
Visible Light Reflected: Exterior	38%
Solar Heat Reduction	66 %

Example 9.5

A thin aluminum wall 2 × 2 m in size is covered partly with OSR and partly left as polished aluminum. Solar flux is illuminating one side with a uniform flux of 250 W/m^2 while internal heat is dumped at a uniform rate of 100 W/m^2. The aluminum wall may be assumed to be isothermal since heat spreads over it. What should be the area covered with OSR if the wall temperature is to be 85 °C?

Solution:

Step 1 We make use of surface properties given in Table 9.2 to solve the problem. The required data is written down below:

Surface	α_S	ε_{IR}
OSR	0.16	0.81
Polished Al	0.09	0.03

Area of the wall is obtained as $A = 2 \times 2 = 4\ \text{m}^2$. Surface temperature is given to be $T_w = 85\,°\text{C} = 85 + 273 = 358$ K. Emissive power of a black body at this temperature is given by $E_w = \sigma T_w^4 = 5.67 \times 10^{-8} \times 358^4 = 931.35\ \text{W/m}^2$.

Step 2 Solar flux on the wall is specified to be $S = 250\ \text{W/m}^2$. Let the fraction of the surface covered with OSR be f. Then the heat absorbed Q_a by the surface exposed to solar radiation is given by

$$Q_a = \left(\underbrace{f \times \alpha_S}_{\text{OSR}} + \underbrace{(1-f) \times \alpha_S}_{\text{Polished Al}}\right) AS = [0.16f + 0.09(1-f)] \times 4 \times 250$$
$$= (70f + 90)\ \text{W}$$

Heat gain by the wall due to that dumped into it Q_d is

$$Q_d = 100 \times 4 = 400\ \text{W}$$

Step 3 We assume that heat is radiated only from the exposed wall. Hence, the heat loss due to emission Q_e is

$$Q_e = E_w A \left(\underbrace{f\varepsilon_{IR}}_{\text{OSR}} + \underbrace{(1-f)\varepsilon_{IR}}_{\text{Polished Al}}\right) = 931.35 \times 4(f \times 0.81 + (1-f) \times 0.03)$$
$$= 2905.8f + 111.8$$

Step 4 Under thermal equilibrium we should have $Q_a + Q_d = Q_e$. Hence, we get

$$70f + 90 + 400 = 2905.8f + 111.8 \text{or} \boxed{f = \frac{378.2}{2835.9} = 0.133}$$

Thus, the OSR covered fraction is 13.3% of the surface area exposed to sun.

9.3 Angle-Dependent Surface Properties

In this section, we pay attention to angle-dependent surface properties. There are two extremes with reference to angle dependence: (a) Specular surface and (b) Diffuse surface. In the case of the former, the angle dependence is easy to understand and characterize using the laws of geometrical optics along with fundamental ideas from the electromagnetic theory.[4] In the case of the latter, the surfaces reflect without any direction bias and the irradiation is scattered in to all the directions of the forward hemisphere. The emerging or reflected radiation is isotropic. The more general surface is one that is neither specular nor diffuse but has an angular dependence that is somewhere in between. This case is the most difficult to characterize. We now consider the above three cases in more detail.

9.3.1 Some Results from Electromagnetic Theory

We start with Maxwell's equations written for charge-free space given by

$$\begin{aligned}&\text{(a) Faraday's Law:} \nabla \times \vec{E} = -\frac{\partial \vec{B}}{\partial t} \quad \text{(c) } \nabla \cdot \vec{D} = 0\\&\text{(b) Ampere's law: } \nabla \times \vec{H} = \mathbf{i}_v + \frac{\partial \vec{D}}{\partial t} \quad \text{(d) } \nabla \cdot \vec{B} = 0\end{aligned} \tag{9.17}$$

Here, $\vec{E}$ is the electric field, $\vec{B}$ is the magnetic induction, $\vec{H}$ is the magnetic intensity, $\vec{i}_v$ is the current flux, and $\vec{D}$ is the electrical displacement vector. In a charge-free or neutral medium, both $\vec{D}$ and $\vec{B}$ are solenoidal, i.e., vectors with zero divergence. In addition to the above, the material properties yield the following relations[5]:

$$\boxed{(a)\ \vec{D} = \epsilon \vec{E},\ (b)\ \vec{B} = \mu \vec{H},\ (c)\ \vec{i}_v = \sigma \vec{E}} \tag{9.18}$$

where ϵ is the permittivity or the dielectric constant, μ is the permeability and σ is the conductivity of the medium. Equation 9.18(c) assumes that the material is electrically conducting and follows Ohm's law. Additionally, the force experienced by a charge moving with a velocity $\vec{v}$ is given by

$$\boxed{\vec{F} = q[\vec{E} + \mathbf{v} \times \vec{B}]} \tag{9.19}$$

---

[4] Refer to J. D. Jackson, "Classical Electrodynamics", 3rd Edition, Academic Press, NY.

[5] Symbol σ should not be confused with the Stefan–Boltzmann constant.

Table 9.4 Units of quantities appearing in Eqs. 9.17–9.19

Quantity	Symbol	Dimension	Unit*
Electric field	E	Force/Charge	N/C
Electric displacement	D	Charge/Area	C/m^2
Magnetic induction	B	Force/Velocity	Wb/m^2 or (V· s)/m^2
Magnetic intensity	H	Current/Length	A/m
Dielectric constant	ϵ		F/m or C/(V · m)
Permeability	μ		H/m or (V · s)/(A · m)
Electrical conductivity	σ		S/m or 1/(Ω · m)

C is Coulomb, *Wb* is weber, *F* is farad, *H* is Henry, *S* is Siemens

where $\overrightarrow{v}$ is the velocity vector. Units of various quantities appearing in Eqs. 9.17 and 9.18 are given in Table 9.4.

In case of vacuum the dielectric constant is $\epsilon_0 = \frac{1}{36\pi \times 10^9} \frac{F}{m}$ and permeability is $\mu_0 = 4\pi \times 10^{-7} \frac{H}{m}$.

Substitute Eq. 9.18(b) in Eq. 9.17(a) and take the curl of the resulting expression to get

$$\nabla \times \nabla \times \overrightarrow{E} = -\mu \frac{\partial \nabla \times \overrightarrow{H}}{\partial t} \tag{9.20}$$

Introduce Eq. 9.18(a) in to Eq. 9.17(b) to get

$$\nabla \times \overrightarrow{H} = \sigma \overrightarrow{E} + \epsilon \frac{\partial \overrightarrow{E}}{\partial t} \tag{9.21}$$

This in Eq. 9.20 will yield

$$\nabla \times \nabla \times \overrightarrow{E} = -\mu\sigma \frac{\partial \overrightarrow{E}}{\partial t} - \epsilon\mu \frac{\partial^2 \overrightarrow{E}}{\partial t^2} \tag{9.22}$$

We make use of the vector identity $\nabla \times \nabla \times \overrightarrow{E} = \nabla(\nabla \cdot \overrightarrow{E}) - \nabla^2 \overrightarrow{E}$, and the solenoidal property of the electric field to get

$$\boxed{\nabla^2 \overrightarrow{E} = \mu\sigma \frac{\partial \overrightarrow{E}}{\partial t} + \epsilon\mu \frac{\partial^2 \overrightarrow{E}}{\partial t^2}} \tag{9.23}$$

It may easily be shown, using similar manipulations of the Maxwell equations, that the magnetic intensity $\overrightarrow{H}$ also satisfies Eq. 9.23. Thus

$$\nabla^2 \vec{H} = \mu\sigma \frac{\partial \vec{H}}{\partial t} + \epsilon\mu \frac{\partial^2 \vec{H}}{\partial t^2} \tag{9.24}$$

We shall now look at plane electromagnetic waves that propagate parallel to, for example, the $z-$axis. Accordingly, $\vec{E}$ and $\vec{H}$ are functions of z and t only. The condition $\nabla \cdot \vec{E} = 0$ then translates to

$$\frac{\partial E_x}{\partial x} + \frac{\partial E_y}{\partial y} + \frac{\partial E_z}{\partial z} = 0 \tag{9.25}$$

The first two terms on the left-hand side are individually zero since $\vec{E}$ is a function of z and t only. Hence E_x, the x component of $\vec{E}$, and E_y, the y component of $\vec{E}$, are functions of z and t only and E_z, the z component of $\vec{E}$, is at best a constant independent of z. We may take $E_z = 0$ without loss of generality. The vector Eq. 9.23 may then be written as the following set of two equations:

$$\begin{aligned}
&\text{(a)}\ \frac{\partial^2 E_x}{\partial z^2} = \mu\sigma \frac{\partial E_x}{\partial t} + \mu\epsilon \frac{\partial^2 E_x}{\partial t^2} \\
&\text{(b)}\ \frac{\partial^2 E_y}{\partial z^2} = \mu\sigma \frac{\partial E_y}{\partial t} + \mu\epsilon \frac{\partial^2 E_y}{\partial t^2}
\end{aligned} \tag{9.26}$$

Using similar arguments, we may also write for the magnetic intensity the following two equations:

$$\begin{aligned}
&\text{(a)}\ \frac{\partial^2 H_x}{\partial z^2} = \mu\sigma \frac{\partial H_x}{\partial t} + \mu\epsilon \frac{\partial^2 H_x}{\partial t^2} \\
&\text{(b)}\ \frac{\partial^2 H_y}{\partial z^2} = \mu\sigma \frac{\partial H_y}{\partial t} + \mu\epsilon \frac{\partial^2 H_y}{\partial t^2}
\end{aligned} \tag{9.27}$$

Equations 9.17(a) and 9.17(b) indicate that vectors $\vec{E}$ and $\vec{H}$ are related to each other and also are perpendicular to each other. Hence, we associate E_x with H_y and E_y and H_x. In fact, we may easily show from Eq. 9.17(a) that

$$\text{(a)}\ \frac{\partial E_x}{\partial z} = -\mu \frac{\partial H_y}{\partial t};\ \text{(b)}\ \frac{\partial E_y}{\partial z} = \mu \frac{\partial H_x}{\partial t} \tag{9.28}$$

As indicated in Chap. 8, the solution to Eqs. 9.27 and 9.28 are transverse electromagnetic waves, indicated schematically in Fig. 8.9. We shall look at these in more detail below.

Fig. 9.6 Reflection and transmission of radiation at an interface between two media

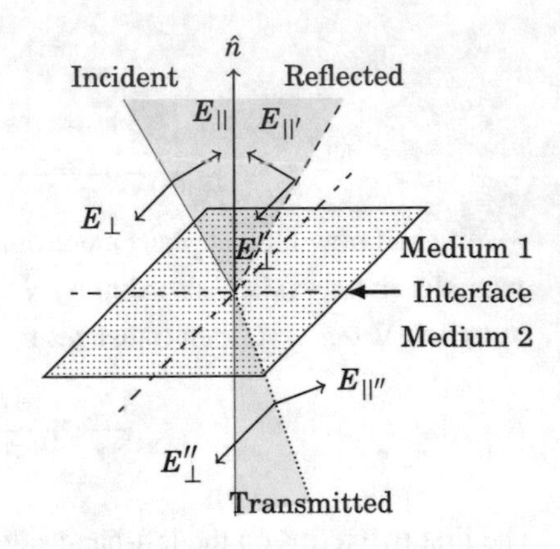

9.3.2 *Specular Surface*

Consider a perfectly smooth interface (shown with filled dot pattern) between two media as shown in Fig. 9.6.

At such a surface, the reflection is specular with the angle of incidence θ equal to the angle of reflection θ'. Two waves propagate along the direction of incidence (think of this direction as the z-direction, in our terminology; see Fig. 8.9) as shown. The first one is the parallel-polarized wave, characterized by the electric field $E_{\parallel}$—the electrical field vector of the incident wave lies in a plane consisting of the incident direction and the normal to the plane (plane is shown with gray fill). This is also referred to as the transverse magnetic mode or the TM mode. In our earlier terminology, $E_{\parallel}$ is E_x. The second is the perpendicularly polarized wave, characterized by the electric field $E_{\perp}$—the electrical field vector of the incident wave is normal to the plane containing the incident direction and the normal to the surface. This is also referred to as the transverse electric mode or the TE mode.In our earlier terminology, $E_{\perp}$ is E_y. These are accompanied by the corresponding magnetic fields $H_{\parallel}$ (H_y—in earlier terminology) and $H_{\perp}$ (H_x—in earlier terminology). The parallel and perpendicularly polarized waves behave differently, as far as reflection and transmission are considered, and hence need to be considered separately.

More on Polarization

(a) Plane-polarized radiation:

Consider the parallel and perpendicular waves expressed as

$$(a)E_x = A\sin(z - ct);\ (b)E_y = B\sin(z - ct) \tag{9.29}$$

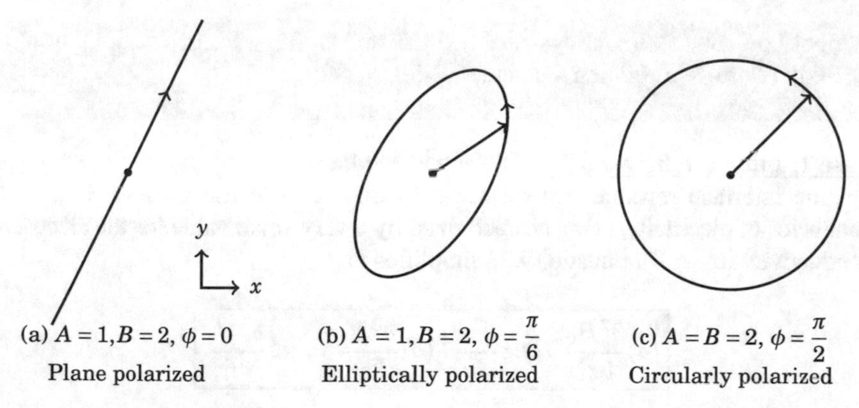

Fig. 9.7 Polarized radiation

The two rays traveling along the z−axis at speed c are in phase, with different amplitudes. At a fixed z, the amplitude of each wave varies sinusoidally with respective amplitudes A and B. The resultant vector always points along the direction given by $\theta_p = \tan^{-1}\left(\frac{A}{B}\right)$ with respect to E_y, with an amplitude of $\sqrt{A^2 + B^2}$ (see Fig. 9.7a. The radiation is said to be plane polarized.

(b) Elliptically polarized radiation:

Consider the parallel and perpendicular waves expressed as

$$(a) E_x = A\sin(z - ct)(b) E_y = B\sin(z - ct - \phi) \tag{9.30}$$

The two waves may be visualized as phasors with the resultant phasor as indicated in Fig. 9.7b. The phase difference ϕ between E_x and E_y may have any value in the interval 0, $\frac{\pi}{2}$. The resultant vector traces out an ellipse as indicated. The radiation is said to be elliptically polarized.

(c) Circularly polarized radiation:

In the special case when $A = B$ and $\phi = \frac{\pi}{2}$, the resultant phasor traces out a circle, and the radiation is said to be circularly polarized and is as shown in Fig. 9.7c.

(d) Unpolarized or natural radiation:

In case there is no definite relation between the amplitudes and phases, we have unpolarized or natural radiation. In practice, radiation sources are not continuous and the phase relationship between the two components, the parallel and the perpendicular, may vary with time. In such a case, the polarization state is determined by the use of time-averaged quantities. This aspect is covered in books on optics (see, for example, M.Born and E. Wolf, Principles of Optics, Sixth (Corrected) Edition, Pergamon Press, 1989).

Case 1: Interface Between Two Dielectric Media

Let the interface separate two dielectric media. A dielectric medium is a non-conductor of electricity, being characterized by a very small value for the electrical conductivity ($\sigma \rightarrow 0$). Equation 9.27 simplifies to

$$\boxed{(a)\frac{\partial^2 H_x}{\partial z^2} = \mu\epsilon\frac{\partial^2 H_x}{\partial t^2}(b)\frac{\partial^2 H_y}{\partial z^2} = \mu\epsilon\frac{\partial^2 H_y}{\partial t^2}} \tag{9.31}$$

Comparing these with Eq. 8.43, it is clear that the phase velocity, call it c in the present case, is given by

$$c = \frac{1}{\sqrt{\mu\epsilon}} \tag{9.32}$$

for any medium and by

$$c_0 = \frac{1}{\sqrt{\mu_0\epsilon_0}} \tag{9.33}$$

in vacuum. The refractive index may then be defined as

$$n = \frac{c_0}{c} = \sqrt{\frac{\mu\epsilon}{\mu_0\epsilon_0}} \tag{9.34}$$

Consider now the propagation of radiation from vacuum or air in to dielectric such as glass. The interface is assumed to be perfectly flat.

The Following Hold in this Case:

- Incident ray, reflected ray, and the transmitted ray (refracted ray) all lie in the same plane;
- Specular reflection means that

Snell's law is valid with the incident and reflected rays lying on opposite sides of the normal to the surface and

$$\boxed{\theta = \theta'}$$

Refracted and incident ray are related by the relation

$$\boxed{n_1 \sin\theta = n_2 \sin\theta''}$$

Here, n_1 and n_2 are the refractive indices of the two media that are separated by the interface.

- In case of an air–glass interface $n_1 \approx 1$ and $\frac{n_2}{n_1} \approx n_2 = n$ (say) is the relative index of refraction of glass, typically a value close to 1.5. The reason the parallel and perpendicular waves behave differently is due to the fact that while the perpendicular wave $E_\perp$ lies on the interface at the point of incidence, the parallel wave $E_\parallel$ makes an angle of $(90 - \theta)$ with respect to the interface, at the point of incidence.
- At the interface, we require that the electric and magnetic fields be continuous, as may be shown by the laws of electromagnetic theory.

Continuity of electric field:

$$\text{(a) } E_\perp + E'_\perp = E''_\perp \text{(b) } (E_\parallel - E'_\parallel)\cos\theta = E''_\parallel \cos\theta'' \tag{9.35}$$

Continuity of magnetic field:

$$\text{(a) } H_\parallel + H'_\parallel = H''_\parallel \text{(b) } (H_\perp - H'_\perp)\cos\theta = H''_\perp \cos\theta'' \tag{9.36}$$

From the electromagnetic theory, we can show that E and H are related as below:

$$\begin{aligned} &\text{(a) } H_\parallel = n_1 E_\parallel;\ H'_\parallel = n_1 E'_\parallel;\ H''_\parallel = n_2 E''_\parallel \\ &\text{(b) } H_\perp = n_1 E_\perp;\ H'_\perp = n_1 E'_\perp;\ H''_\perp = n_2 E''_\perp \end{aligned} \tag{9.37}$$

Introducing these in Eq. 9.36(a) and (b), we have

$$\text{(a) } E_\parallel + E'_\parallel = nE''_\parallel \text{(b) } (E_\perp - E'_\perp)\cos\theta = nE''_\perp \cos\theta'' \tag{9.38}$$

where $n = \frac{n_2}{n_1}$. Also, Snell's law relates angles θ and θ'' as $\sin\theta = n\sin\theta''$. Hence $\cos\theta''$ may be written as

$$\cos\theta'' = \sqrt{1 - \sin^2\theta''} = \sqrt{1 - \frac{\sin^2\theta}{n^2}}$$

From Eq. 9.35(b), we then have

$$E_\parallel - E'_\parallel = E''_\parallel \frac{\cos\theta''}{\cos\theta} = E''_\parallel \frac{\sqrt{1 - \frac{\sin^2\theta}{n^2}}}{\cos\theta} \tag{9.39}$$

Adding Eqs. 9.38(a) and 9.39, and after simplification we get

$$\frac{E''_{\|}}{E_{\|}} = \frac{2n\cos\theta}{n^2\cos\theta + \sqrt{n^2 - \sin^2\theta}} \tag{9.40}$$

Subtracting Eq. 9.39 from Eq. 9.38(a), and after simplification we also get

$$\frac{E'_{\|}}{E_{\|}} = \frac{n^2\cos\theta - \sqrt{n^2 - \sin^2\theta}}{n^2\cos\theta + \sqrt{n^2 - \sin^2\theta}} \tag{9.41}$$

Similarly we may derive the following two relations.

$$\frac{E''_{\perp}}{E_{\perp}} = \frac{2\cos\theta}{\cos\theta + \sqrt{n^2 - \sin^2\theta}} \tag{9.42}$$

$$\frac{E'_{\perp}}{E_{\perp}} = \frac{\cos\theta - \sqrt{n^2 - \sin^2\theta}}{\cos\theta + \sqrt{n^2 - \sin^2\theta}} \tag{9.43}$$

The four relations given by Eqs. 9.40–9.43 are known as the Fresnel relations.[6] These relate the strengths of the transmitted and the reflected fields to the incident electric fields for the two polarizations. Since the intensities are proportional to the squares of the corresponding electric fields, the reflectances are given by

$$\text{(a) } \rho_{\|} = \left[\frac{n^2\cos\theta - \sqrt{n^2 - \sin^2\theta}}{n^2\cos\theta + \sqrt{n^2 - \sin^2\theta}}\right]^2 \quad \text{(b) } \rho_{\perp} = \left[\frac{\cos\theta - \sqrt{n^2 - \sin^2\theta}}{\cos\theta + \sqrt{n^2 - \sin^2\theta}}\right]^2 \tag{9.44}$$

Since energy is conserved, the corresponding transmittances are given by

$$\text{(c)} t_{\|} = 1 - \rho_{\|} \quad \text{(d)} t_{\perp} = 1 - \rho_{\perp} \tag{9.45}$$

---

[6]Named after Augustin-Jean Fresnel 1788–1827, French engineer and physicist who made important contributions to wave optics.

Special Cases:

Normal Incidence With $\theta = \theta'' = 0$

In this case, both polarizations behave the same way since all the electric fields lie on the interface, at the point of incidence. From Eq. 9.44(a) and (b), we have

$$\boxed{\rho_{\parallel} = \rho_{\perp} = \left[\frac{n-1}{n+1}\right]^2} \tag{9.46}$$

Brewster Angle

$\rho_{\perp}$ is a monotonic function of θ with the reflectance becoming unity at $\theta = \frac{\pi}{2}$, i.e., for grazing incidence. However, $\rho_{\parallel}$ is a non-monotonic function that has a zero at an angle between $0 \leq \theta \leq 90^o$. For this purpose we set the numerator of Eq. 9.44(a) to zero to get

$$n^4 \cos^2\theta = n^2 - \sin^2\theta = n^2 - (1 - \cos^2\theta)$$

This may be solved for $\cos\theta$ to get $\cos^2\theta = \frac{n^2-1}{n^4-1} = \frac{1}{1+n^2}$. Hence, $\cos\theta = \frac{1}{\sqrt{1+n^2}}$. It is then easily seen that $\sin\theta = \sqrt{1-\cos^2\theta} = \sqrt{1 - \frac{1}{1+n^2}} = \sqrt{\frac{n^2}{1+n^2}} = \frac{n}{\sqrt{1+n^2}}$. Thus, we have the important result

$$\boxed{\tan\theta_B = \frac{\sin\theta}{\cos\theta} = n} \tag{9.47}$$

where θ_B is known as the Brewster angle.

As an example, we consider an interface between air and glass with a relative index of refraction of $n = 1.5$. Figure 9.8 shows a plot of $\rho_{\parallel}$, $\rho_{\perp}$, and ρ, the mean of the two reflectances as a function of the incidence angle θ in degrees. Note that both components have a common reflectance of $\rho_{\parallel} = \rho_{\perp} = \left[\frac{1.5-1}{1.5+1}\right]^2 = 0.04$ for normal incidence. The parallel-polarized component has a reflectance of zero at $\theta_B = \tan^{-1} 1.5 = 56.3°$, the Brewster angle.

Consider incident radiation to be plane polarized with both components having the same magnitude. Polarization angle is $\frac{\pi}{4}$ rad or 45°. Using the Fresnel relations, it is easy to see that the reflected radiation remains plane polarized but with the polarization angle, given by $\theta_p = \tan^{-1}\sqrt{\frac{\rho_{\parallel}}{\rho_{\perp}}}$, varying with incident angle θ as shown in Fig. 9.9. For normal and grazing incidences, the plane of polarization remains at 45°. At incident angle equal to the Brewster angle, the polarization angle is zero. The polarization angle varies between 0 and 45° otherwise.

Total Internal Reflection and the Critical Angle

An interesting phenomena known as the total internal reflection takes place when $n = \frac{n_2}{n_1} < 1$, i.e., when $n_2 < n_1$. Light is propagating from a high refractive index material to a low refractive index material. An example will be light propagation from glass to air or vacuum, from water to air, etc. Using Snell's law, it is easily seen that $\sin\theta'' = \frac{n_1}{n_2}\sin\theta = \frac{\sin\theta}{n}$ will take on the maximum value of unity ($\theta'' =$

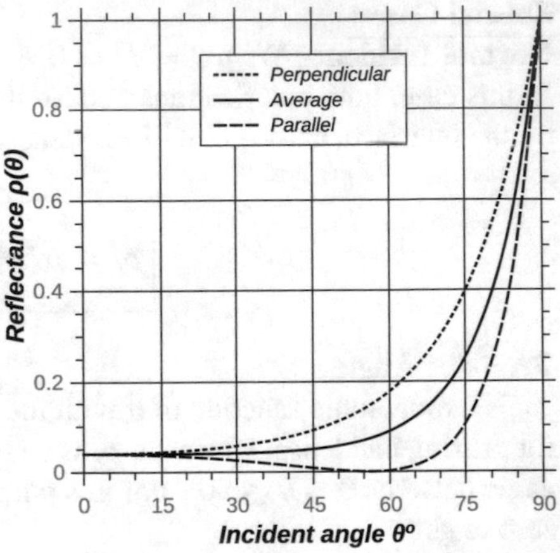

Fig. 9.8 Reflectance of glass with $n = 1.5$

Fig. 9.9 Variation of polarization angle of reflected radiation

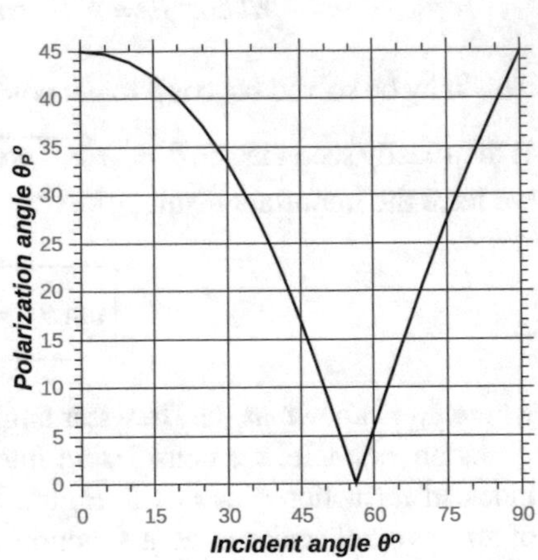

$\pi/2$) when $\sin\theta = n$ or $\theta = \sin^{-1} n$. This angle is referred to as the critical angle θ_c. Transmitted (refracted) radiation is absent for $\theta \geq \theta_c$. For $\theta \geq \theta_c$, the incident radiation is completely reflected and hence the *total internal reflection* takes place.

Consider as example reflection at an interface between glass and air/vacuum. In this case, $n = \frac{1}{1.5} = \frac{2}{3}$. We make use of Fresnel relations to compute the reflectances as a function of incident angle θ and make a plot as shown in Fig. 9.10. The figure clearly shows that all reflectances are unity at and beyond the critical angle of $\theta_c = \sin^{-1}\left(\frac{2}{3}\right) = 41.81°$ or $\theta_c = 0.73$ rad. The total internal reflectance is used for the measurement of the refractive index of dielectric materials as will be discussed below as one of the applications.

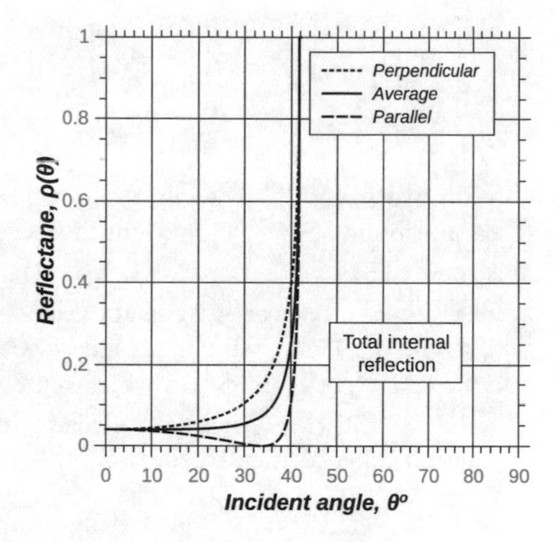

Fig. 9.10 Reflectance at an interface between glass and air, $n = \frac{2}{3}$

Some Applications

In what follows, we consider two applications that are based on the theory presented above for interface phenomena between two dielectric materials. The first application discusses anti-reflection coatings used in optical elements such as lenses.

Anti-reflection Coating:
Let the wavelength of incident radiation be λ. At the front surface, a part of the incident radiation is reflected and emerges with a phase change of π as shown in Fig. 9.11. This figure is drawn for normal incidence of incoming light. The radiation that is reflected from the interface has a change in the phase of π due to reflection as well as a change in phase due to the travel through double the anti-reflection coating thickness, viz., 2δ where δ is the thickness of the coating.The change of phase is due to change in the speed of light in the anti-reflection coating as compared to air. The additional change of phase is due to path length $2n_{ar}\delta$. If the additional phase change is equal to π and hence the additional path length is $\frac{\lambda}{2}$, the light reflected from the interface and that from the front surface will destructively interfere and the intensity of reflected light will be reduced. We then have

$$2n_{ar}\delta = \frac{\lambda}{2} \text{or } \boxed{\delta = \frac{\lambda}{4n_{ar}}} \tag{9.48}$$

If the destructive interference is to be perfect, the front and interface reflectivities must be equal. For normal incidence, we should then have

$$\left[\frac{n_{ar}-1}{n_{ar}+1}\right]^2 = \left[\frac{n_g - n_{ar}}{n_g - n_{ar}}\right]^2 \tag{9.49}$$

It is easily shown that $\boxed{n_{ar} = \sqrt{n_g}}$ for this to happen. This is based on the refractive index of air being taken as 1. In general, if the refractive index of a medium above the anti-reflection coating is n_a, it may be shown that $\boxed{n_{ar} = \sqrt{n_a n_g}}$. We may choose n_{ar} such that destructive interference takes place for a particular wavelength. Since visible radiation spans the region $0.4 \leq \lambda \leq 0.7\ \mu$m, we choose the wavelength at the middle of this band, viz., $\lambda = \frac{0.4+0.7}{2} = 0.55\ \mu$m which represents green color. Choosing $n_{ar} = \sqrt{1.5} = 1.225$ (MgF_2 is a typical coating material) as shown in Fig. 9.11, the anti-reflection coating thickness is

$$\delta = \frac{0.55}{4 \times 1.225} = 0.112\ \mu\text{m} \tag{9.50}$$

A change of phase will also take place for other wavelengths. However, complete destructive interference will not take place and hence reduction in reflection will be less for these wavelengths. In order to improve the purpose of the anti-reflection coating to serve over a range of wavelengths, multiple layers of anti-reflection coatings of different refractive index values are used.

The above argument holds for normal incidence. However, when the incident radiation is at an angle the path length will be enhanced by $\frac{1}{\cos\theta}$, and hence interference will become less effective in reducing reflected radiation. In most optical instruments where anti-reflection coatings are used, the maximum incident angle may be between 20–30°. If n_{ar} is λ independent, destructive interference will take place for higher wavelengths (toward red color) when the incident is off normal.

Anti-reflection coatings are also used in the infrared part of the spectrum. Recently, anti-reflection coatings have also been used over silicon-based solar cells. Since silicon has a very high reflectivity of ~ 0.3, anti-reflection coatings of Silicon dioxide (SiO_2) and Titanium dioxide (TiO_2) are deposited over the solar cell to cut down reflectivity to a very small value.

The second application is in the measurement of the refractive index of dielectric materials.

Measurement of Refractive Index by Total Internal Reflection
The total internal reflection has been seen to be dependent on the refractive index of the dielectric material. If the critical angle is accurately measured,

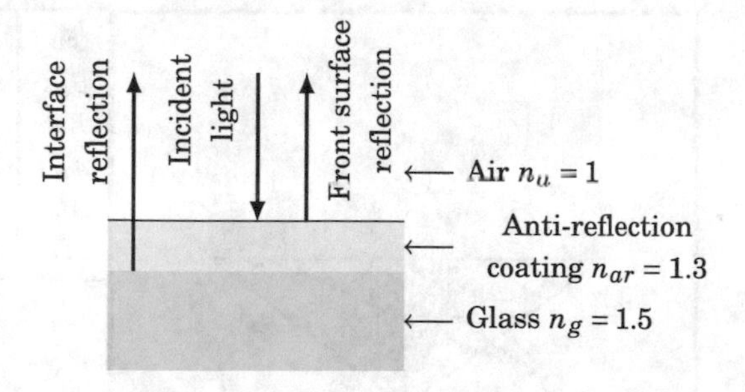

Fig. 9.11 Normal reflection in the presence of anti-reflection coating on a glass substrate

one would be able to get an accurate estimate of the index of refraction of the material.

An arrangement that is suitable for the measurement of the refractive index of solid/liquid is shown schematically in Fig. 9.13. The dielectric sample whose index of refraction n_s needs to be determined is placed adjacent to the base of a prism of the index of refraction $n_p > n_s$ as shown in the figure. A point source of light illuminates one of the faces of the prism. Light passes through this face and is incident on the base of the prism. Light will be partly reflected and partly transmitted as long as the angle of incidence is less than the critical angle. For example, rays 1 and 2 are incident at angles less than the critical angle, and hence both transmitted T rays (T_1 and T_2) and reflected R rays (R_1 and R_2) are present. After reflection, the light falls on the other face of the prism as shown in the figure. For an incident angle equal to or greater than the critical angle, the total internal reflection takes place and falls on the other face of the prism (e.g., R_3 and R_4).

Thus, the face of the prism that receives reflected rays has a low-level illumination till just before the critical angle and high-level illumination thereafter. By placing a high-resolution CCD camera close to the prism face, we may determine the location where the reflection just reaches a maximum by digital image analysis. This corresponds to the critical angle. Hence, the critical angle may be ascertained, and correspondingly the refractive index of the sample (actually $n_p - n_s$) can be determined. With a known value of n_p, the index n_s may be determined.

Case 2: Interface Between a Dielectric and a Metal

A conductor has free electrons that interact strongly with the incident radiation. The movement of these free electrons, under the influence of the incident radiation, leads to Joule heating thus dissipating the radiation as heat. Because of this, the radiation is strongly attenuated as it passes from a dielectric (vacuum or air) into the conducting material (metal).

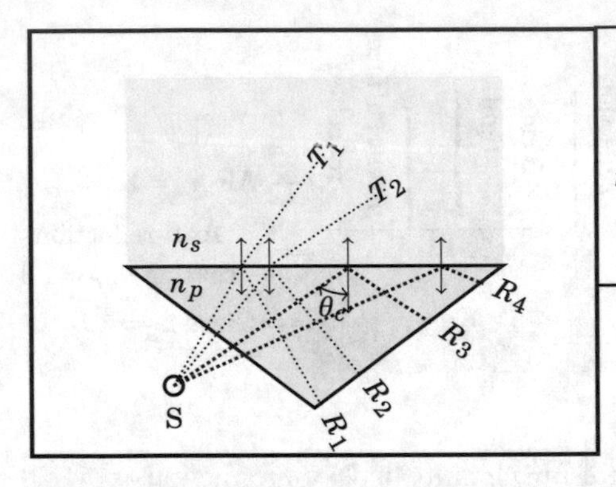

S:	Diffuse point source of light
θ_c:	Critical angle
n_s:	Refractive index of sample
n_p:	Refractive index of prism
R:	Reflected light
T:	Transmitted light

Fig. 9.12 Angular reflectance of Platinum for 10 μm radiation

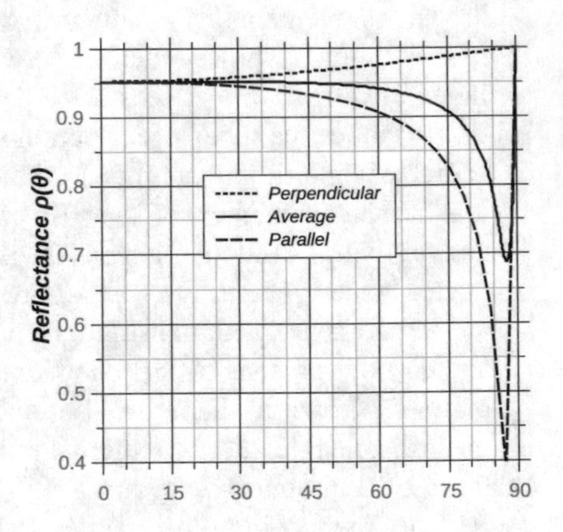

Fig. 9.13 Total internal reflection at an interface between two dielectrics for $n_s < n_p$

The index of refraction is complex with the imaginary part accounting for the attenuation of radiation as it passes through into the conducting medium. The attenuation is very strong and takes place in a very small thickness in the vicinity of the interface. This thickness is a few microns at the most and hence the interaction of radiation with the conducting medium may be assumed to be a surface phenomenon. Also, the magnitude of the index of refraction is much larger than one (typical value is 50) so that $\theta'' \approx 0$.

We start with Eq. 9.26(either one of them would do, say 9.26(a)) in which the first derivative of the electric field with respect to time plays an important role. This term accounts for attenuation of radiation as we shall see below. A solution that is periodic in time is sought in the form $E_x(z, t) = V(z)\mathrm{e}^{i\omega t}$ where the factor $V(z)$ is expected to be a function of z. We then have

$$\frac{\partial^2 E_x}{\partial z^2} = \mathrm{e}^{i\omega t}\frac{d^2 V}{dz^2}, \frac{\partial E_x}{\partial t} = i\omega V \mathrm{e}^{i\omega t}, \frac{\partial^2 E_x}{\partial t^2} = -\omega^2\, V \mathrm{e}^{i\omega t}$$

These in Eq. 9.26(a) yield

$$\frac{d^2V}{dz^2} - (i\mu\sigma\omega - \omega^2\mu\epsilon)V \text{or} \boxed{\frac{d^2V}{dz^2} - \alpha^2 V = 0} \tag{9.51}$$

where

$$\boxed{\alpha^2 = i\mu\sigma\omega - \omega^2\mu\epsilon} \tag{9.52}$$

Equation 9.51 has a solution that decreases with z given by $V(z) = V_0 e^{-\alpha z}$ where α is a complex number with a positive real part. The solution for the function $V(z)$ is then given by

$$V(z) = e^{-\alpha z} \tag{9.53}$$

The Electric field $E_x(z, t)$ is then given by

$$E_x(z, t) = e^{-\alpha z + i\omega t} = e^{-\alpha(z - i\frac{\omega}{\alpha}t)} \tag{9.54}$$

This represents a wave propagating in the $+z$ direction with amplitude decreasing with z. We may identify the wave speed as

$$c = \frac{i\omega}{\alpha} \tag{9.55}$$

By squaring the above, we get

$$c^2 = -\frac{\omega^2}{\alpha^2} \tag{9.56}$$

The index of refraction may then be defined as

$$n_c^2 = \frac{c_0^2}{c^2} = -\frac{c_0^2\alpha^2}{\omega^2} \tag{9.57}$$

The index of refraction n_c is complex and is usually represented in the form[7]

$$n_c = n - ik \tag{9.58}$$

By squaring n_c, using Eq. 9.52, we have

$$n_c^2 = (n^2 - k^2 - 2ink) = -\frac{c_0^2[i\mu\sigma\omega - \omega^2\mu\epsilon]}{\omega^2} \tag{9.59}$$

With $c_0^2 = \frac{1}{\mu_0\epsilon_0}$, the real and imaginary parts of the refractive index satisfy the following relations:

$$\text{(a)}\, n^2 - k^2 = \frac{\mu\epsilon}{\mu_0\epsilon_0}, \text{(b)}\, 2nk = \frac{\mu\sigma}{\omega\mu_0\epsilon_0} \tag{9.60}$$

[7] The imaginary part is sometimes represented as $k = n\kappa$.

Eliminating k from these two equations, we get, after some minor simplification, the following quadratic equation.

$$n^4 - \frac{\mu\epsilon}{\mu_0\epsilon_0}n^2 - \frac{\mu^2\sigma^2}{4\omega^2\mu_0^2\epsilon_0^2} = 0 \tag{9.61}$$

This may be solved for n^2 to get a physically meaningful result (note that n^2 should be positive)

$$n^2 = \frac{\mu\epsilon}{2\mu_0\epsilon_0}\left[\sqrt{1 + \frac{\sigma^2}{\omega^2\epsilon^2}} + 1\right] \tag{9.62}$$

Using this in Eq. 9.60(a), we also get

$$k^2 = \frac{\mu\epsilon}{2\mu_0\epsilon_0}\left[\sqrt{1 + \frac{\sigma^2}{\omega^2\epsilon^2}} - 1\right] \tag{9.63}$$

Noting that α is a complex quantity that may be represented as $\alpha = a + ib$, we use Eqs. 9.56,9.62, and 9.63 to arrive at the following

$$\text{(a)}\, a = \frac{\omega k}{c_0},\ \text{(b)}\, b = \frac{\omega n}{c_0} \tag{9.64}$$

where both n and k are positive. Using this in Eq. 9.54, we see that the wave is attenuated by the exponential factor $\mathrm{e}^{-az} = \mathrm{e}^{-\frac{\omega k}{c_0}z}$. Thus, the imaginary part of the refractive index accounts for the attenuation of the wave as it progresses into the metal. It is customary to introduce an absorption coefficient κ as

$$\kappa = \frac{2\omega k}{c_0} = \frac{4\pi k}{\lambda} \tag{9.65}$$

where the last part is obtained by expressing the radiation in terms of its vacuum wavelength and also noting that radiant power is proportional to the square of the electric field intensity. For highly conducting metals—very large σ—the attenuation is very well approximated by the relation

$$\mathrm{e}^{-az} \approx \mathrm{e}^{-\sqrt{\frac{\omega\sigma\mu}{2}}z} \tag{9.66}$$

noting that $k \approx \sqrt{\frac{\mu\sigma}{2\mu_0\epsilon_0\omega}}$. Attenuation takes place with a length scale given by $\delta = \sqrt{\frac{2}{\omega\sigma\mu}}$. The electric field is reduced by a factor of $\frac{1}{\mathrm{e}}$ over this length, referred to as the "skin depth". In the case of copper, for example, we have

$$\mu \approx \mu_0 = 4\pi \times 10^{-7}\ \text{H/m};\ \sigma = 5.8 \times 10^7\ \text{S/m}$$

The skin depth for radiation at 10 GHz is 6.6×10^{-7} m! The radiation is completely absorbed within such a short distance from the interface that it may be considered to be at the interface itself. Thus, metals are opaque and highly reflecting as we shall see from the discussion below.

Consider an interface between a dielectric of refractive index n_1 and a metal of refractive index $n_c = n_2 - ik_2$. We then define the relative index m as

$$m = \frac{n_2}{n_1} - i\frac{k_2}{n_1} = n - in' \tag{9.67}$$

in case the dielectric is vacuum or possibly air, $n_1 \approx 1$ and hence $n \approx n_2$ and $n' \approx k_2$. Since the index of refraction is complex, the refraction angle θ'' is also complex. Since hardly any radiation is transmitted across the boundary, we shall look at the reflected radiation only. Consider the perpendicularly polarized component. The reflectivity given in Eq. 9.44(d) is valid with the relative index of refraction n replaced by the complex index of refraction m and taking only the magnitude. Thus we should have

$$\text{Reflectance: } \boxed{\rho_\perp = \left[\frac{\cos\theta - \sqrt{m^2 - \sin^2\theta}}{\cos\theta + \sqrt{m^2 - \sin^2\theta}}\right]^2} \tag{9.68}$$

Since m is complex, let $a - ib = \sqrt{m^2 - \sin^2\theta}$. The magnitude of the numerator and denominator are then seen, respectively, to be $\sqrt{(a - \cos\theta)^2 + b^2}$ and $\sqrt{(a + \cos\theta)^2 + b^2}$. The reflectance then becomes

$$\rho_\perp = \frac{(a - \cos\theta)^2 + b^2}{(a + \cos\theta)^2 + b^2} \tag{9.69}$$

By squaring, we have

$$\begin{aligned}(a - ib)^2 &= (a^2 - b^2) - i2ab = m^2 - \sin^2\theta\\ &= (n - in')^2 - \sin^2\theta = n^2 - n'^2 - \sin^2\theta - i2nn'\end{aligned}$$

Equating the real and imaginary parts, we get the following two equations:

$$\text{(a) } a^2 - b^2 = n^2 - n'^2 - \sin^2\theta, \text{ (b) } ab = nn' \tag{9.70}$$

We may solve these two equations for a and b to obtain the following relations:

$$\begin{aligned}(a)\, a^2 &= \frac{1}{2}\left[\sqrt{(n^2 - n'^2 - \sin^2\theta)^2 + 4n^2n'^2} + (n^2 - n'^2 - \sin^2\theta)\right] \\ (b)\, b^2 &= \frac{1}{2}\left[\sqrt{(n^2 - n'^2 - \sin^2\theta)^2 + 4n^2n'^2} - (n^2 - n'^2 - \sin^2\theta)\right]\end{aligned} \tag{9.71}$$

It may also be shown that the reflectance for the parallel-polarized wave is given by

$$\text{Reflectance: } \boxed{\rho_{\parallel} = \rho_{\perp}\frac{(a - \sin\theta\tan\theta)^2 + b^2}{(a + \sin\theta\tan\theta)^2 + b^2}} \tag{9.72}$$

In the case of metals, both the real and imaginary parts of the refractive index are very large. For example, in the case of Platinum, $m = 37 - i41$ for the radiation of wavelength equal to 10μm. Since $\sin\theta$ and $\cos\theta$ are bounded between 0 and 1, Eq. 9.71 indicates that $a \approx n$ and $b \approx n'$. The reflectances are approximated then as

$$\boxed{(a)\rho_{\perp} \approx \frac{(n - \cos\theta)^2 + n'^2}{(n + \cos\theta)^2 + n'^2};\ (b)\rho_{\parallel} = \rho_{\perp}\frac{(n - \sin\theta\tan\theta)^2 + n'^2}{(n + \sin\theta\tan\theta)^2 + n'^2}} \tag{9.73}$$

Special Case:

Normal Incidence with $\theta = 0$

In this case, both polarizations behave the same way since all the electric fields lie on the interface, at the point of incidence. Equations 9.72 and 9.73 reduce to

$$\boxed{\rho_{\parallel} = \rho_{\perp} = \frac{(n - 1)^2 + n'^2}{(n + 1)^2 + n'^2}} \tag{9.74}$$

The angular variations given by the expressions 9.72 and 9.73 are quite unlike the angular dependence for the dielectric case considered earlier and presented in Fig. 9.8. The angular reflectivity plot shown in Fig. 9.12 brings this out. The case considered is for the incidence of 10 μm (mid-IR) radiation from air on to an optically smooth Platinum surface. The n and n' values are 37 and 41, respectively. It is seen that the metal reflects significantly at all angles and particularly strongly as $\theta \to 0$ as well as when $\theta \to 90°$. The perpendicularly polarized component shows a monotonic variation with incident angle while the parallel-polarized radiation shows a non-monotonic behavior.

As another example we show, in Fig. 9.14, the angular reflectance from a gold surface with $n = 0.181$ and $n' = 3.068$ for incident radiation from a Helium Neon laser at a wavelength of 0.6328 μm (visible). In this case, also the reflectance is very high for all incident angles.

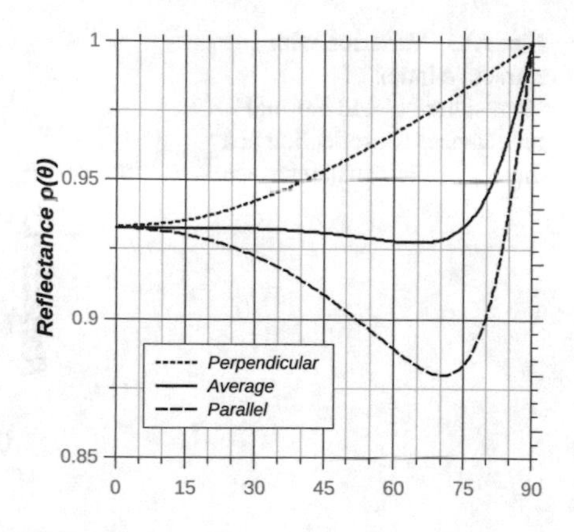

Fig. 9.14 Angular reflectance of Gold for 0.6328 μm radiation from a Helium Neon laser

Interface Between a Dielectric and a Semi-transparent Medium:
The case of the dielectric–dielectric interface considered previously is typical of a transparent material like glass. The dielectric–metal case represents an opaque material. When the transparent material such as glass contains some impurities, radiation is weakly absorbed by the material and the material is translucent or semi-transparent. Very thin metal layers also transmit radiation such as when the thickness is less than the skin depth that was alluded to earlier. Radiation transfer across thin films requires the application of wave optics accounting for the interference of light. This is usually presented in books on optics.

Semi-transparent media require the consideration of the attenuation of transmitted radiation by invoking the concept of the absorption coefficient. The treatment is similar to that used in gas radiation and will hence be considered later.

9.3.3 *Hemispherical Reflectance*

In engineering applications, it is unlikely that the surface will be illuminated at a fixed angle. The incident radiation is likely to be from all directions within a hemisphere. In case the incident intensity variation with direction is known, it is possible to calculate the total reflected radiation in to the hemisphere. The simplest of the incident intensity distributions is the case of isotropic intensity, as for example, unpolarized radiation from a black body background at a specified temperature. We may use the reflectances for dielectric–dielectric and dielectric–metal interfaces given earlier to calculate the hemispherical reflectance. Let $\rho_\lambda(\theta)$ be the monochromatic reflectance of the surface for radiation incident at angle θ with respect to the normal. This is given by

$$\rho_\lambda(\theta) = \frac{\rho_{\lambda,\parallel}(\theta) + \rho_{\lambda,\perp}(\theta)}{2} \tag{9.75}$$

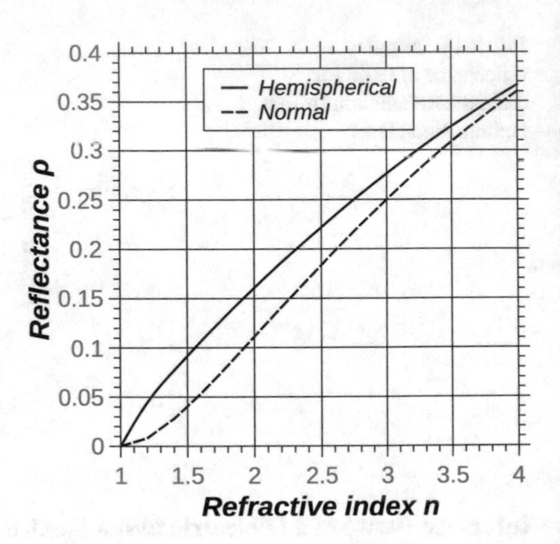

Fig. 9.15 Variation with refractive index of Hemispherical and Normal reflectances for reflection at a dielectric–dielectric interface

Let the incident intensity be constant independent of the incident direction. The monochromatic irradiation is given by $G_\lambda = \pi I_\lambda$. The reflected monochromatic radiant flux is given by

$$q_\lambda^- = \int_{2\pi} \rho_\lambda(\theta) G_\lambda \cos\theta d\Omega = 2\pi I_\lambda \int_0^1 \rho_\lambda(\mu)\mu d\mu \tag{9.76}$$

where we have used the relation $\mu = \cos\theta$. The spectral hemispherical reflectance of the surface is then given by

$$\rho_{\lambda,h} = \frac{q_\lambda^-}{G_\lambda} = 2\int_0^1 \rho_\lambda(\mu)\mu d\mu \tag{9.77}$$

The indicated integration may be done numerically for both dielectric–dielectric and dielectric–metal cases. Closed form expressions are also available, for sufficiently large values of n and n'.[8] In the case of Platinum, the hemispherical reflectance is calculated for 10 μm radiation as 0.94. In the case of gold, it is computed by numerical integration as 0.932, for 0.6328 μm radiation. The corresponding normal reflectances are calculated, based on Eq. 9.74, respectively, as 0.953 and 0.933.

It is interesting to look at the variations of hemispherical and normal reflectances of a transparent material with the relative index of refraction as shown in Fig. 9.15.

The highest refractive index shown in the figure corresponds to that of silicon. Both reflectances increase with the index of refraction. The hemispherical and normal reflectances are of comparable magnitude. For example, a material like glass with $n = 1.5$ has hemispherical reflectance of 0.092 while the normal reflectance is 0.04.

[8]R. V. Dunkle, pp. 39–44, NASA SP-55, 1965.

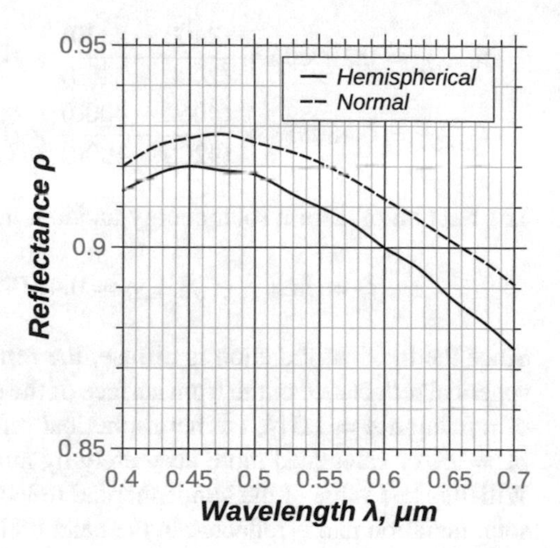

Fig. 9.16 Variation with the wavelength of Hemispherical and Normal reflectances of polished aluminum

Diamond has a refractive index of 2.42 for visible radiation corresponding to a wavelength of 0.59 μm. Correspondingly, the hemispherical and normal reflectances are 0.213 and 0.172, respectively. Such a high reflectance makes diamonds very precious as gemstones.

As mentioned earlier, metals are highly reflecting. Literature data on the index of refraction of aluminum [9] is used to determine the hemispherical and normal reflectances of aluminum as functions of wavelength in the visible part of the electromagnetic spectrum. Figure 9.16 presents the result. The normal reflectance appears to be higher than the hemispherical reflectance throughout the wavelength range as shown in the figure.

Example 9.6

A typical glass sheet has a mean refractive index of 1.514 in the visible part of the spectrum between 0.4 and 0.7 μm. Assume that diffuse solar radiation is incident on it, with spectral characteristics that of a black body at 5800 K. Determine the fraction of incident diffuse solar radiation that is reflected from the front face of the glass sheet.

Solution:
The fraction of incident flux that is contained within the wavelength range $\lambda_1 = 0.4\mu\text{m}$ and $\lambda_2 = 0.6\mu\text{m}$ is calculated first. Equivalent black body temperature characterizing solar radiation is $T_s = 5800K$. We then have

$$\lambda_1 T_s = 0.4 \times 5800 = 2320\ \mu\text{m K};\ \lambda_2 T_s = 0.7 \times 5800 = 4060\ \mu\text{m K}$$

The linear interpolation of entries in Table 8.1 give the following:

[9] Shiles, E., Sasaki, T., Inokuti, M., and Smith, D. Y., Phys. Rev. Sect. B, 22, p-1612, 1980.

$$f_{0-\lambda_1 T_s} = 0.100886 + \frac{(2400 - 2320)}{(2400 - 2200)} \times (0.140253 - 0.100886) = 0.11663$$

$$f_{0-\lambda_2 T_s} = 0.480858 + \frac{(4060 - 4000)}{(4200 - 4000)} \times (0.515993 - 0.480858) = 0.49140$$

The fraction of diffuse solar energy incident in the band under consideration is

$$f_s = f_{0-\lambda_2 T_s} - f_{0-\lambda_1 T_s} = 0.49140 - 0.11663 = 0.37477$$

Since the incident radiation is diffuse, the reflected part is just given by the hemispherical reflectance of the front surface of the glass sheet. For glass of relative index of refraction $n = 1.514$, the hemispherical reflectance is read off from Fig. 9.15 as $\rho_h = 0.1$ or calculated more accurately by numerical integration as $\rho_h = 0.09383$. With this last value of the hemispherical reflectance, the fraction of incident diffuse solar radiation that is reflected in the band under consideration is

$$f_r = f_s \rho_h = 0.37477 \times 0.09383 = 0.03516$$

Example 9.7

Diffuse solar radiation is incident on a windowpane made of transparent glass of mean refractive index of 1.5 in the visible wavelength range 0.4–0.7 μm. The incident flux is known to be 200 W/m^2. Determine the amount of diffuse radiation that is reflected by the window pane per square meter of its surface. Assume that the glass sheet is thin and of a very large area. We need to consider the reflection and transmission of radiation across the pane by taking in to account the reflection from the front and the back surfaces of the glass pane. Figure 9.17 shows the state of affairs.

Solution:

Multiple reflections are involved from both surfaces of the pane. Because the incident radiation is diffuse, we use the hemispherical quantities in the calculations. Since the glass sheet is transparent, there is no reduction in intensity due to absorption within the glass sheet. The fraction transmitted is obtained by summing all the transmitted fractions[10] given by the series

$$t_{h,\text{eff}} = (1 - \rho_h)^2 + \rho_h^2(1 - \rho_h)^2 + \rho_h^4(1 - \rho_h)^2 + \cdots\cdots = (1 - \rho_h)^2 \sum_0^{\infty} \rho_h^{2n}$$

[10] If the glass sheet is not large, the series will truncate since the incident radiation walks along the surface and will terminate at an edge.

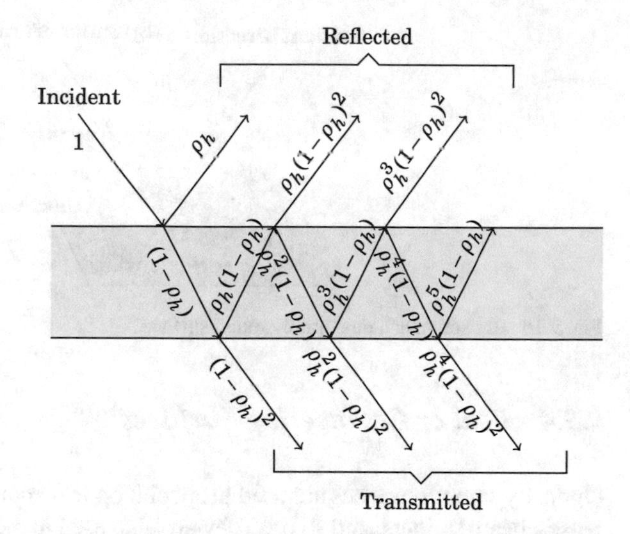

Fig. 9.17 Reflection and transmission of radiation across a thin window pane

This series is a geometric series with common ratio ρ_h^2 and hence

$$t_{h,\text{eff}} = (1-\rho_h)^2 \cdot \frac{1}{1-\rho_h^2} = \frac{1-\rho_h}{1+\rho_h}$$

For the windowpane, we have $n = 1.5$ and hence the hemispherical reflectance is read off from Fig. 9.15 as $\rho_h = 0.09$ or calculated more accurately by integration as $\rho_h = 0.092$. With this, the transmittance of the windowpane is

$$t_{h,\text{eff}} = \frac{1-0.092}{1+0.092} = 0.832$$

The effective reflectance of the window pane is then obtained as

$$\rho_{h,\text{eff}} = 1 - t_{h,\text{eff}} = 1 - 0.832 = 0.168$$

The incident diffuse solar flux in the 0.4–0.7 μm band is given to be $G_S = 200\ \text{W/m}^2$. Hence, the effective reflected flux in this band is given by

$$R_S = \rho_{h,\text{eff}} G_S = 0.168 \times 200 = 33.7\ \text{W/m}^2$$

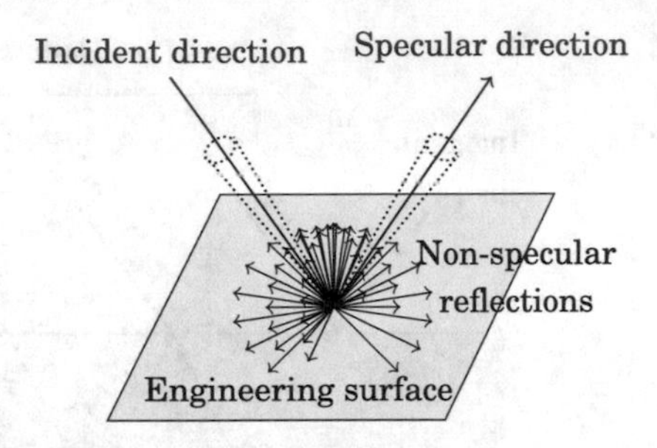

Fig. 9.18 Reflection at a moderately rough surface

9.3.4 Real or Engineering Surfaces

Optically smooth surfaces are used in special optical applications involving mirrors, lenses, beam splitters, and so on. They are also used in metrology. Most engineering applications either do not need perfectly smooth surfaces or the cost precludes the use of such surfaces. Engineering surfaces are basically not optically smooth and hence the angular dependence of surface properties is different from the specular cases we have considered in the previous sections.

A moderately rough surface tends to reflect as shown in Fig. 9.18. Even though the reflection is random, there is a preferential reflection close to the specular direction. If the surface roughness elements have regularity, such as when parallel grooves are etched on the surface, the reflection may show regular features. If the roughness elements are small and comparable to the wavelength of the incident radiation, it is likely that interference effects may be observable. Gratings used in spectrometry, in fact, use fine parallel grooves that disperse radiation of different wavelengths along different directions. Most surfaces used in engineering applications, however, have three-dimensional features and hence reflect radiation in a fairly random fashion along all the directions of a hemisphere.

For a planar real or engineering surface, the mean normal is oriented as shown in Fig. 9.19a. However, the local normal is oriented along different directions because of surface roughness. For a given direction of incidence, even though the reflection is specular over a small area element, the change in the direction of the local normal makes the reflected rays travel in different directions. The reflected light is thus thrown into all directions in a random fashion.

If the incident radiation is reflected perfectly randomly, the reflected intensity is independent of direction and is referred to as diffuse reflection. The rough or engineering surface is then referred to as a diffuse surface. Such surfaces are met with in radiation heat transfer very often and hence will be pursued in more detail.

Mildly rough surfaces are fairly common in many applications such as in space applications. The roughness is measured by the standard deviation of the surface with respect to the mean plane of the surface as indicated by σ in Fig. 9.19b. The surface roughness is also characterized by the mean slope $m = \tan\phi$ defined as indicated in

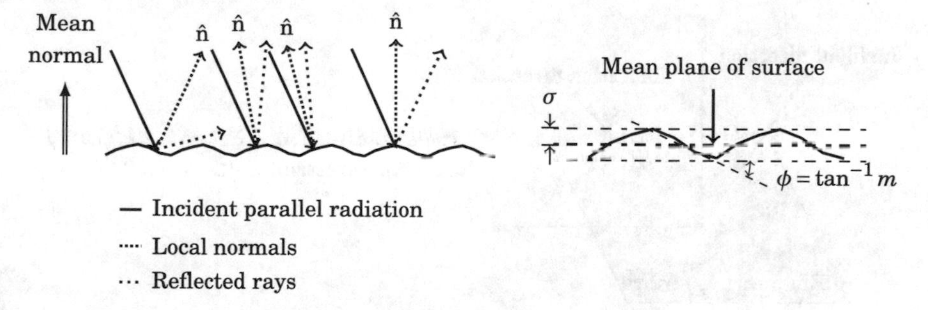

Fig. 9.19 Reflection at a moderately rough surface

the figure. The slope is a measure of the ratio of the deviation of the surface with respect to the mean plane and the autocorrelation distance α, the distance over which the feature extends. Thus, we have

$$\frac{m}{\sqrt{2}} = \frac{\sigma}{\alpha} \tag{9.78}$$

Larger the m, the higher the deviation out of a plane or shorter the autocorrelation distance. An optically rough surface is one for which $\frac{\sigma}{\lambda} \gg 1$. An optically smooth surface is one for which $\frac{\sigma}{\lambda} \ll 1$. A slightly rough surface is one for which $\frac{\sigma}{\lambda} \leq 1$.

The reflectance distribution from a surface uses the nomenclature shown in Fig. 9.20. The incident direction is represented by θ, ϕ while the reflected direction is represented by θ', ϕ'. The elemental solid angles are represented by $d\Omega$ and $d\Omega'$ as indicated in the figure. The reflectance of a rough surface is modeled in terms of a bidirectional reflectance distribution function (BRDF) as proposed by Beckman and Spizzichino.[11]

$$f(\theta, \phi, \theta', \phi') = f_{sp}(\theta)U(\delta) + f_{ic}(\theta, \phi, \theta', \phi') \tag{9.79}$$

where the subscript sp stands for the specular part and subscript ic stands for the incoherent part. While the specular part depends only on the incident angle θ, the incoherent part is a function of the incident and reflected directions. In Eq. 9.79, the unit delta function $U(\delta)$ is given by

$$U(\delta) = \delta(\theta - \theta')\frac{\delta(\phi - (\phi' + \pi))}{\cos\theta d\Omega} \tag{9.80}$$

[11] P. Beckman and A. Spizzichino, *The Scattering of Electromagnetic Waves from Rough Surfaces*, Pergamon Press, 1963.

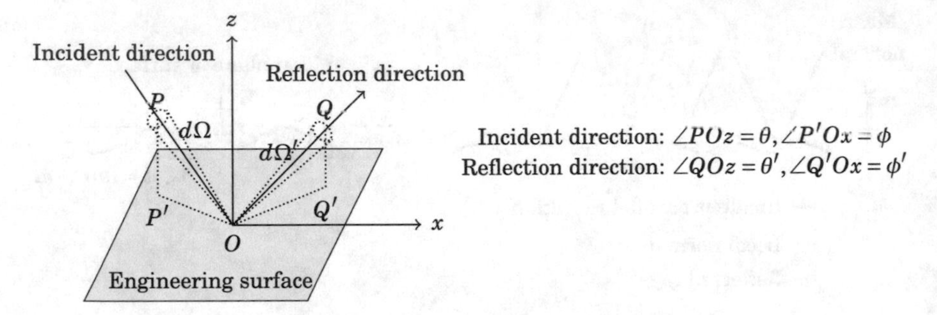

Fig. 9.20 Nomenclature for reflectance from a rough surface

Since the BRDF is difficult to use in radiation heat transfer calculations, a simplified model suggested by Schornhurst and Viskanta,[12] Toor and Viskanta,[13], and modified with suggestion made by Torrance and Sparrow[14] is given below. The reflectance is specified as

$$f(\theta) = [1 - \varepsilon(\theta, m)]\left[\frac{1 - g(\theta)}{\pi} + g(\theta)U(\delta)\right] \tag{9.81}$$

where

$$g(\theta) = \exp\left[-\left\{\frac{4\pi\sigma\cos\theta}{\lambda}\right\}^2\right] \tag{9.82}$$

and $\varepsilon(\theta, m)$, known as the directional emissivity, is the surface emissivity that depends on angle θ and the RMS slope m. Birkebak and Abdulkadir[15] suggest the way the directional emissivity is calculated relating it to the blocking effect due to surface undulations. The directional emissivity of a metal surface (or a surface coated with a layer of metal) such as aluminum with $n = 6.1$ and $k = 4.984$ at $\lambda = 4\,\mu$m varies with θ as shown in Fig. 9.21a. It is seen that the directional emissivity approaches a diffuse behavior as m increases. A non-metallic coating (typically like aluminum oxide) exhibits a different behavior as seen from Fig. 9.21b. The literature is available on how to use the above model in radiation heat transfer calculations.[16]

In summary, (a) specular reflection takes place when the surface is optically smooth, i.e., when the surface roughness is much smaller than the wavelength of the incident radiation and (b) perfectly random or diffuse reflection occurs when the surface is extremely rough, i.e., the surface roughness is much larger

[12] J. R. Schornhurst and R. Viskanta, AIAA Journal, 6,pp. 1450–1455, 1968.

[13] J. S. Toor and R. Viskanta, Int. J. Heat and Mass transfer, 11, pp. 883–897, 1968.

[14] K. E. Torrance and E. M. Sparrow, J. Heat Transfer, 87, pp. 283–292, 1965.

[15] R. C. Birkebak and A. Abdulkadir, Int. J. Heat and Mass transfer, 19, pp. 1039–1043, 1976.

[16] E. M. Sparrow and S. H. Lin, Int. J. Heat and Mass transfer, 8, pp. 769–779, 1965.

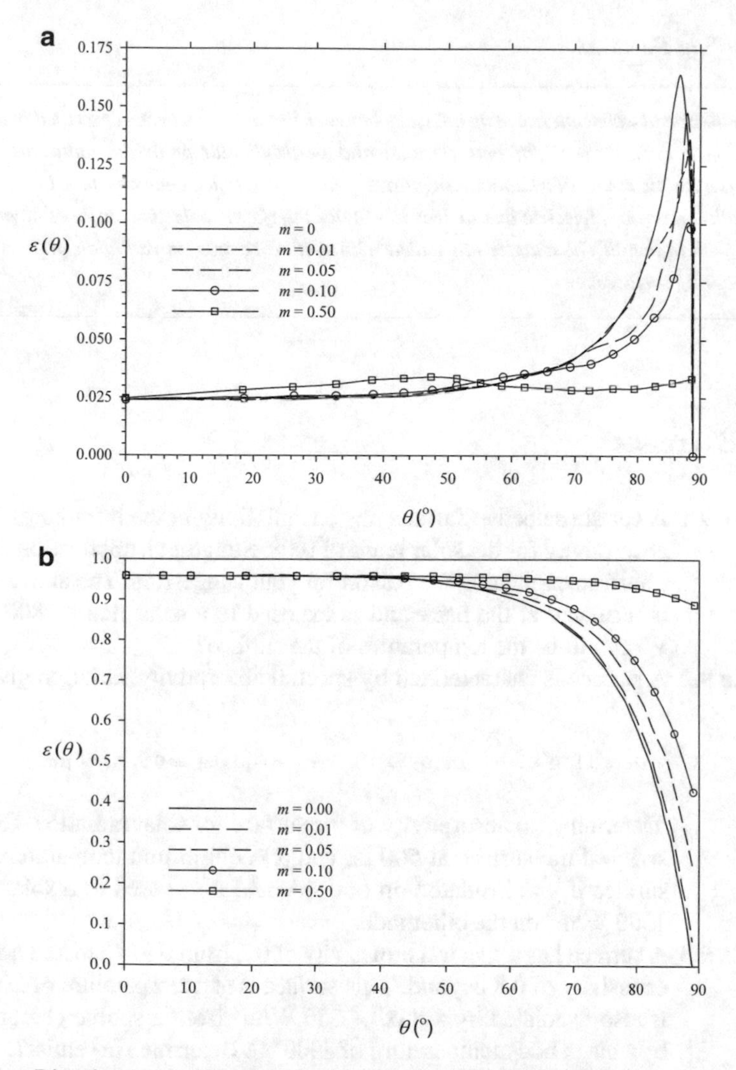

Fig. 9.21 **a** Directional emissivity of an aluminum surface **b** Directional emissivity of an aluminum oxide surface

than the wavelength of incident radiation. These are the two extreme limits that one encounters in practice. Optical elements like prisms, lenses, and mirrors used in optical instruments are polished to a high degree of smoothness and are optically smooth. Surfaces encountered in heat transfer applications like furnace walls, walls of buildings, etc., are rough and may be considered as diffuse reflectors. These two cases are simple to model. However, some surfaces may not belong to either category. Reflection from such surfaces are difficult to model.

Concluding Remarks

Reflectance and transmittance at an interface between two media has been described in detail. These have been derived, for both parallel and perpendicular polarized radiation, using electromagnetic theory of radiation. Angular dependence of reflectance has been brought out for both dielectric -dielectric and dielectric - metal interfaces. Selective surfaces have been dealt with in detail. The chapter has ended with a short section on radiation properties of engineering surfaces.

9.4 Exercises

Ex 9.1 A certain selective surface has an emissivity in the infrared of 0.6 and absorptivity for the solar band of 0.08. Suggest an application for such a surface explaining the reason for your suggestion. The above surface is insulated at the back and is exposed to a solar flux of 800 W/m^2. What will be the temperature of the surface?

Ex 9.2 A surface is characterized by spectral absorptivity variation given by

$$\alpha_\lambda = 0.1,\ 0 < \lambda < 2\ \mu\text{m};\ \alpha_\lambda = 0.8,\ 2 < \lambda < 4\ \mu\text{m};\ \alpha_\lambda = 0.5,\ \lambda > 4\ \mu\text{m}$$

Determine (a) absorptivity of the surface for solar radiation, (b) emissivity of the surface at 500 K, and (c) equilibrium temperature of the surface if it is insulated on one side and is exposed to a solar flux of 1300 W/m^2 on the other side.

Ex 9.3 A surface has a spectral emissivity of 0.2 from 0 to 4μm and a spectral emissivity of 0.8 beyond. This surface is at a temperature of 200 °C. It is also irradiated by a flux of 350 W/m^2 from a source characterized by a black body temperature of 2000 °C. Determine (a) emissive power of the surface, (b) absorptivity of the surface for the incident radiation, and (c) radiosity of the surface.
What will be a suitable application for this selective surface?

Ex 9.4 A real surface has spectral properties specified by three bands as shown below.

$$\alpha_\lambda = 0.1,\ 0 < \lambda < 2\ \mu\text{m};\ \alpha_\lambda = 0.9,\ 2 < \lambda < 4\ \mu\text{m};\ \alpha_\lambda = 0.4,\ \lambda > 4\ \mu\text{m}$$

This surface receives a total radiant flux of 500 W/m^2 from a background characterized by a black body temperature equal to 1500 K. The surface is insulated at the back and is hence at its equilibrium temperature that needs to be estimated. Assume a starting guess value of

350 K and perform iterations till the value converges. You may use a tolerance of 0.1 K to stop the iterations.

Ex 9.5 The hemispherical spectral emissivity of a certain opaque surface may be approximated by the distribution given below:

$$\varepsilon_\lambda = 0.2,\ 0 < \lambda < 2.5\ \mu\text{m};\ \varepsilon_\lambda = 0.8,\ 2.5 < \lambda < 6\ \mu\text{m};\ \varepsilon_\lambda = 0.4,\ \lambda > 6\ \mu\text{m}$$

The surface temperature is varied between 600 K and 1800 K in steps of 100 K. Evaluate the total emissive power of the surface at these temperatures. Make a suitable plot.

Ex 9.6 A sheet of translucent material is 3 mm thick. The material has a monochromatic complex refractive index for radiation at a wavelength of 1 μm of $n_\lambda = 1.6 - i0.05$. The monochromatic radiation of the wavelength of 1 μm is incident normal to the first surface. Determine the amount of radiation transmitted by the material.

Ex 9.7 Repeat Exercise 9.4 for an incident angle of 30° to the normal. All other data remains unchanged.

Ex 9.8 Make a literature search for optical data of aluminum, gold, nickel, and copper. Using the data, grade these materials in the order of their reflectances for visible light of 0.6 μm wavelength. Assume, in each case, that the material is in the form of a perfectly smooth foil.

Ex 9.9 Thin mylar films with aluminum deposited over them are used very often in space applications. Why is this so? Give credible arguments. Discuss how a blanket made of a large number of such films may be used as an insulation.

Ex 9.10 Passive temperature control is a possible option in many satellite thermal management applications. Consult the relevant literature and prepare a note on such an application.

Ex 9.11 Solar energy applications make use of selective surfaces, as has been indicated before. Study the relevant literature and prepare a note on useful selective surfaces for solar energy applications. Grade them according to their performance. Why are glass cover plates used in solar collectors? What should be the optical properties of such glass cover plates? Discuss how a solar film (used on windows of rooms and automobiles) works. What are the desirable optical properties of such films?

Chapter 10
Radiation in Enclosures

RADIATION heat transfer models all problems as enclosure problems. All the surfaces that take part in radiant heat exchange are assumed to form an enclosure. In case all the surfaces are diffuse and gray, it is possible to calculate radiant interchange by separating geometric and thermal aspects. Shape factors/view factors/angle factors take care of the geometric part. Governing equations are written in terms of the radiosity of each surface in the enclosure. The band model is useful in modeling diffuse non-gray surfaces. Specular surfaces are treated using the concept of the exchange factor. When uniform radiosity assumption is invalid, it is necessary to subdivide surfaces into a large number of divisions to perform the analysis.

10.1 Introduction

The background regarding radiation interaction at different types of surfaces has been developed in Chap. 9. Engineering applications deal mainly with radiation heat transfer among such surfaces arranged to form an *enclosure*. In many of these applications, the enclosure may contain atmospheric air which may be assumed, without loss of accuracy, to be transparent to the passage of radiation. Ignoring conduction and convection heat transfer in the enclosed medium as being not of significance, the enclosure analysis may proceed by considering *only* surface radiation, assuming that the enclosure is evacuated!

In most cases, the surfaces that constitute an enclosure are opaque. Also, the surfaces are rough and hence diffusely reflecting except in specific applications that may use specular surfaces. The first part of this chapter will hence deal with enclosures having gray and diffusely reflecting walls. Later these restrictions will be relaxed and we shall also consider enclosures with non-gray, semi-transparent as well as specularly reflecting surfaces.

S. P. Venkateshan, *Heat Transfer*,
https://doi.org/10.1007/978-3-030-58338-5_10

The case of an enclosure with radiatively participating (absorbing and emitting) medium will be taken up in Chap. 11.

10.2 Evacuated Enclosure with Gray Diffuse Walls

10.2.1 Assumptions

The following assumptions are generally made in enclosure analysis.

- Since engineering surfaces are generally rough, the diffuse assumption seems to be well justified, as indicated toward the end of Chap. 9. This means that the intensity of radiation leaving any surface is isotropic or angle independent.
- Surfaces that are *more or less isothermal* also have a *uniform* radiosity everywhere on the surface.
- Surfaces are assumed to be gray. This means that we can treat total quantities in the analysis without worrying about the distribution of radiation over wavelength.

10.2.2 Diffuse Radiation Interchange Between Two Surfaces

Diffuse Radiation Interchange Between Two Area Elements
Consider the radiation leaving an area element dA_j and incident on an area element dA_i as shown in Fig. 10.1. Because of diffuse assumption, the intensity of radiation leaving dA_j, (i.e., I_j) is $\frac{J_j}{\pi}$ where J_j is the radiosity. In addition, J_j is assumed to be uniform over A_j, with dA_j being an elemental area situated anywhere on A_j. The total power leaving dA_j is

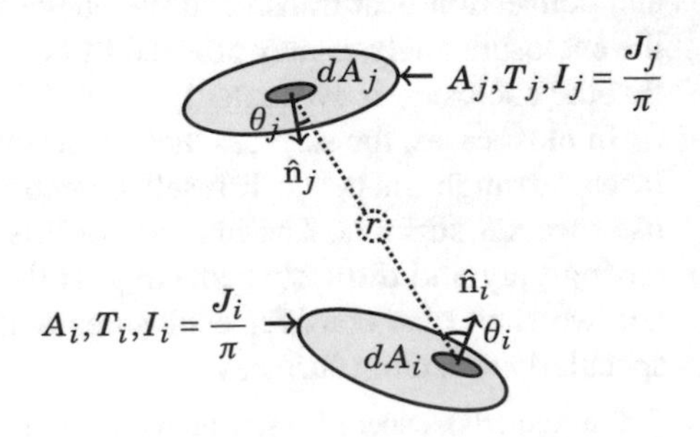

Fig. 10.1 Diffuse radiant interchange between two area elements

$$dP_j = J_j dA_j = \pi I_j dA_j \tag{10.1}$$

The power incident on dA_i from the total power leaving dA_j is

$$dP_i = I_j dA_j \cos\theta_j d\Omega \tag{10.2}$$

where $d\Omega$ is the solid angle subtended by dA_i at dA_j and is seen to be given by $d\Omega = \frac{\cos\theta_i dA_i}{r^2}$, where r is the length of the line joining the mid-points of dA_j and dA_i. Thus, the power incident on dA_i is given as

$$dP_i = I_j \frac{dA_i dA_j \cos\theta_i \cos\theta_j}{r^2} \tag{10.3}$$

The ratio $\frac{dP_i}{dP_j}$ obtained from Eqs. 10.1 and 10.3 as

$$\boxed{\frac{dP_i}{dP_j} = \frac{I_j\left(\frac{dA_i dA_j \cos\theta_i \cos\theta_j}{r^2}\right)}{\pi I_j dA_j} = \underbrace{\frac{\cos\theta_i \cos\theta_j dA_i}{\pi r^2}}_{F_{dA_j - dA_i}}} \tag{10.4}$$

is a pure geometric parameter. This is referred to as the differential (or elemental) angle factor (or view factor or shape factor) for diffuse radiant interchange between two elemental areas and is symbolically represented as $F_{dA_j - dA_i}$. This parameter represents the fraction of radiation that is leaving dA_j that is intercepted by dA_i. If we run through the above derivation, considering radiation leaving area element dA_i and incident on dA_j, it is clear that we should get

$$\boxed{F_{dA_i - dA_j} = \frac{\cos\theta_j \cos\theta_i dA_j}{\pi r^2}} \tag{10.5}$$

Equation 10.5 gives the differential angle factor for radiant interchange between dA_i and dA_j. It is clear from Eqs. 10.4 and 10.5 that elemental area–elemental shape factor product satisfies the relation

$$\boxed{\underbrace{\overbrace{dA_i \times F_{dA_i - dA_j}}^{E_{dA_i - dA_j}} = \overbrace{dA_j \times F_{dA_j - dA_i}}^{E_{dA_i - dA_j}}}_{\text{Reciprosity relation}}} = \frac{\cos\theta_i \cos\theta_j dA_i dA_j}{\pi r^2} \tag{10.6}$$

This relation is referred to as the reciprocity relation. The area angle factor product is referred to as the exchange area for diffuse radiant interchange, denoted by $E_{dA_i-dA_j}$ or $E_{dA_j-dA_i}$.

Diffuse Radiation Interchange Between an Area Element and a Finite Area
An integration of Eq. 10.3 over the area A_i gives the power leaving dA_j which is incident on A_i. Thus, power leaving dA_j that is incident on A_i is

$$P_i = I_j \int\limits_{A_i} \frac{dA_i dA_j \cos\theta_i \cos\theta_j}{r^2} = J_j \int\limits_{A_i} \frac{dA_i dA_j \cos\theta_i \cos\theta_j}{\pi r^2} \tag{10.7}$$

But the total power leaving dA_j is $P_{dA_j} = J_j dA_j$. Hence, the fraction of the power leaving dA_j that is incident on A_i is

$$F_{dA_j-A_i} = \int\limits_{A_i} \frac{\cos\theta_i \cos\theta_j}{\pi r^2} dA_i \tag{10.8}$$

It is clear (how?) that the power leaving A_i that will be incident on dA_j is

$$F_{A_i-dA_j} = \frac{dA_j}{A_i} \int\limits_{A_i} \frac{\cos\theta_i \cos\theta_j}{\pi r^2} dA_i \tag{10.9}$$

Diffuse Radiation Interchange Between Two Finite Areas
In this case, the integration is to be performed over both the areas A_i and A_j. We may follow a similar procedure to show that the fraction of the total power leaving A_i that is incident on A_j (this may easily be obtained by integrating Eq. 10.9 over A_j) is

$$\boxed{F_{A_i-A_j} = \frac{1}{A_i} \int\limits_{A_i} \int\limits_{A_j} \frac{\cos\theta_i \cos\theta_j}{\pi r^2} dA_i dA_j} \tag{10.10}$$

Expression 10.10 is purely a geometric factor for radiant interchange between finite areas and is called the angle factor (or view factor or shape factor). Similarly, we can show that

$$F_{A_j-A_i} = \frac{1}{A_j}\int_{A_i}\int_{A_j}\frac{\cos\theta_i\cos\theta_j}{\pi r^2}dA_i dA_j \tag{10.11}$$

From Eqs. 10.10 and 10.11, we see that

$$A_i \cdot F_{A_i-A_j} = A_j \cdot F_{A_j-A_i} = E_{A_i-A_j} = E_{A_j-A_i} \tag{10.12}$$

where $E_{A_i-A_j}$ or $E_{A_j-A_i}$ is the exchange area for diffuse radiation interchange. The above equality is referred to as the *reciprocity relation*. It shows that the exchange area for diffuse interchange between two finite areas is the same irrespective of the direction of radiant interchange.

Sum Rule

Consider a total of *N* surfaces in an enclosure. Conservation of radiant energy requires that the sum of all the fractions of energy leaving the surface *i* and reaching surface *j* should add up to unity. In this summation, we should include the *self*-angle factor $F_{A_i-A_i}$ in case the ith surface is concave.[1] Thus, we have the so-called sum rule that states that

$$\sum_{j=1}^{N} F_{A_i-A_j} = 1 \tag{10.13}$$

Reciprocity relations for surfaces taken in pairs, sum rule for each surface in an enclosure, and the use of additional angle factor rules (presented in the next section) constitute *angle factor algebra*.

10.2.3 Angle Factor Algebra and Its Applications

In most enclosure problems, we have radiant interchange among a set of surfaces that comprise the enclosure. Assume that there are *N* diffuse surfaces that make up the enclosure. The number of angle factors that are required for analysis is seen to be N^2, i.e., all $F_{A_i-A_j}$'s for $i = 1, 2, \cdots N$ and $j = 1, 2, \cdots N$. However, for each surface the sum rule may be written such that there are *N* such relations available.

[1]Self-angle factor represents the fraction of energy leaving surface A_i that is incident on itself. The reader should note that $F_{A_i-A_i}$=0 for a flat or a convex surface.

Taken two at a time, there are ${}^{N}C_2$ reciprocity relations among the angle factors. The number of angle factors that are to be determined independently N_{in} (say, by the application of Eq. 10.10) is given by

$$N_{\text{in}} = N^2 - N -{}^{N} C_2 = N(N-1) - \frac{N!}{(N-2)!2!} \\ = N(N-1) - \frac{N(N-1)}{2} = \boxed{\frac{N(N-1)}{2}} \tag{10.14}$$

In addition, if all the N surfaces are convex or flat, all the self-angle factors are zero, i.e., $F_{A_i-A_i} = 0$ for $i = 1, 2, \cdots N$. Hence, in this case the number of independent angle factors that need to be determined is further reduced to

$$N_{\text{in}} = \frac{N(N-1)}{2} - N = \boxed{\frac{N(N-3)}{2}} \tag{10.15}$$

It is thus clear that all angle factors may be determined by the use of angle factor algebra when the number of surfaces is 3 or less (the lowest N is of course 2).

Another useful rule, the decomposition rule, is discussed below.

Decomposition Rule

Consider the three surfaces shown in Fig. 10.2a, b.

While Fig. 10.2a is the most general case, Fig. 10.2b is a typical application of interest. We shall represent angle factor between A_1 and A_2 as F_{12}, between A_1 and A_3 as F_{13}. Noting that the angle factors represent the fraction of energy leaving a surface that is incident on another surface, we see that the fraction of energy leaving A_1 that reaches both A_2 and A_3 must be

$$F_{1(2+3)} = F_{12} + F_{13}$$

This may be recast in more useful form

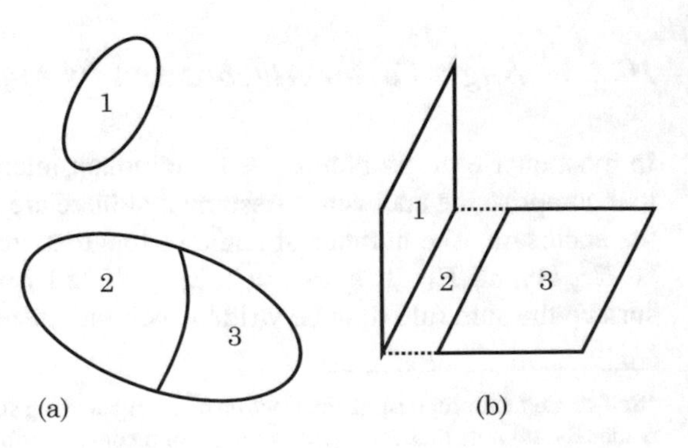

Fig. 10.2 Illustration sketch for decomposition rule

$$\boxed{F_{13} = F_{1(2+3)} - F_{12}} \tag{10.16}$$

where it is presumed that both the angle factors on the right-hand side of Eq. 10.16 are available or easily evaluated.

Two-Surface Enclosures

The simplest enclosure one can think of is a two-surface enclosure shown in its two possible variants in Fig. 10.3a, b. In the first case, the cavity wall is concave and is covered by a flat lid. In the second case, a convex surface is completely enclosed within a concave surface. In both cases, angle factors may be obtained by angle factor algebra. Since all the radiation leaving surface 1 reaches surface 2, the angle factor $F_{12} = 1$. This is because surface 1 is convex (or flat) and hence $F_{11} = 0$. Now we may use the reciprocity rule to write

$$A_1 F_{12} = A_2 F_{21} \text{ or } F_{21} = \frac{A_1}{A_2}$$

The sum rule is used now to get the self-angle factor of surface 2 as

$$F_{22} = 1 - F_{21} = 1 - \frac{A_1}{A_2}$$

We may represent the angle factors in a matrix form as shown below:

$$\boxed{\begin{bmatrix} F_{11} & F_{12} \\ F_{21} & F_{22} \end{bmatrix} = \begin{bmatrix} 0 & 1 \\ \frac{A_1}{A_2} & 1 - \frac{A_1}{A_2} \end{bmatrix}}$$

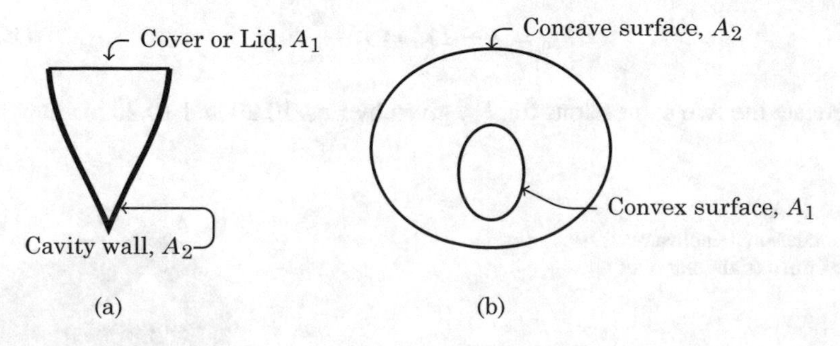

Fig. 10.3 Typical two-surface enclosures

Two-Dimensional Enclosures in the Form of Long Ducts

Triangular Duct

Angle factors are relatively easily evaluated for two-dimensional enclosures. These occur very often, in engineering practice, in applications like ducts carrying conditioned air in buildings, ducts transporting high-temperature gases in power plants, and so on. If the fluid flowing in the duct is assumed to be transparent to radiation, radiation transfer between the walls of the duct takes place purely due to surface radiation. The simplest case is a very long duct of the triangular cross section as shown in Fig. 10.4. Since $N = 3$, all the angle factors may be determined by the application of angle factor algebra.

The sum rule is applied to each side of the triangular duct:

$$\text{(a)}\;\; F_{ab} + F_{ac} = 1, \quad \text{(b)}\;\; F_{ba} + F_{bc} = 1, \quad \text{(c)}\;\; F_{ca} + F_{cb} = 1 \qquad (10.17)$$

Three reciprocity relations are written down:

$$\text{(a)}\;\; aF_{ab} = bF_{ba}, \quad \text{(b)}\;\; bF_{bc} = cF_{cb}, \quad \text{(c)}\;\; aF_{ac} = cF_{ca} \qquad (10.18)$$

From Eq. 10.17a, we have $F_{ac} = 1 - F_{ab}$. Using Eq. 10.18(c), we then get

$$F_{ca} = \frac{a}{c}F_{ac} = \frac{a}{c}(1 - F_{ab}) \qquad (10.19)$$

From Eq. 10.18(b), using Eq. 10.17(c) we have $F_{cb} = \frac{b}{c}F_{bc} = 1 - F_{ca}$. With Eq. 10.19, this becomes

$$\frac{b}{c}F_{bc} = 1 - \frac{a}{c}(1 - F_{ab})$$

This may be recast as

$$F_{bc} = \frac{c}{b}\left[1 - \frac{a}{c}(1 - F_{ab})\right] \qquad (10.20)$$

Using Eqs. 10.17(a) and (b), we also have

$$F_{bc} = 1 - F_{ba} = 1 - \frac{a}{b}F_{ab} \qquad (10.21)$$

We equate the two expressions for F_{bc} given by Eqs. 10.20 and 10.20 to get

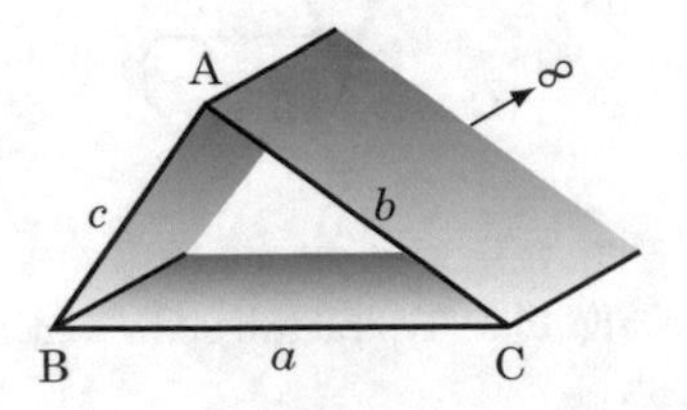

Fig. 10.4 A two-dimensional enclosure in the form a triangular duct

Fig. 10.5 A two-dimensional enclosure in the form a Trapezoidal duct

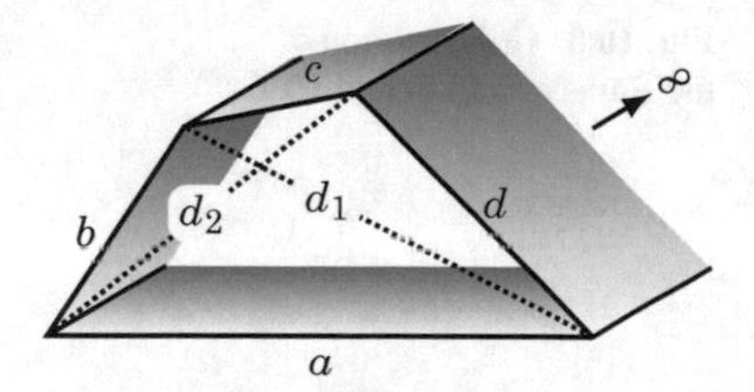

$$1-\frac{a}{b}F_{ab}=\frac{c}{b}-\frac{a}{b}(1-F_{ab})\text{ or }2\frac{a}{b}F_{ab}=1+\frac{a}{b}-\frac{c}{b}$$

$$\text{or (a)}\quad \boxed{F_{ab}=\frac{a+b-c}{2a}} \tag{10.22}$$

In words, F_{ab} is the ratio of the sum of lengths of sides a (radiation leaving surface) and b (radiation receiving surface) minus the third side c divided by twice the length of side a. Using this, it is easily seen that the following should hold:

$$\boxed{\text{(b)}\quad F_{ac}=\frac{a+c-b}{2a}}\quad\text{and}\quad\boxed{\text{(c)}\quad F_{bc}=\frac{b+c-a}{2b}} \tag{10.23}$$

Reciprocity relations then give the other angle factors.

Polygonal Duct

The angle factors obtained above may be used as the basis for evaluating the angle factors in a polygonal duct containing more than 3 surfaces. As an example, we consider a duct of quadrilateral section as shown in Fig. 10.5. We demonstrate the procedure by calculating the angle factor between opposite sides of the quadrilateral, as for example, F_{ac}. By drawing the diagonals, we apply the triangle formula 10.22 to triangle with sides a, b, and d_1 to get F_{ab} as

$$F_{ab}=\frac{a+b-d_1}{2a} \tag{10.24}$$

We apply the triangle formula 10.22 to a triangle with sides a, d, and d_2 to get F_{ad} as

$$F_{ad}=\frac{a+d-d_2}{2a} \tag{10.25}$$

Applying now the sum rule to side a, we have

$$\begin{aligned}F_{ac}&=1-F_{ab}-F_{ad}\\&=1-\frac{a+b-d_1}{2a}-\frac{a+d-d_2}{2a}=\boxed{\frac{(d_1+d_2)-(b+d)}{2a}}\end{aligned} \tag{10.26}$$

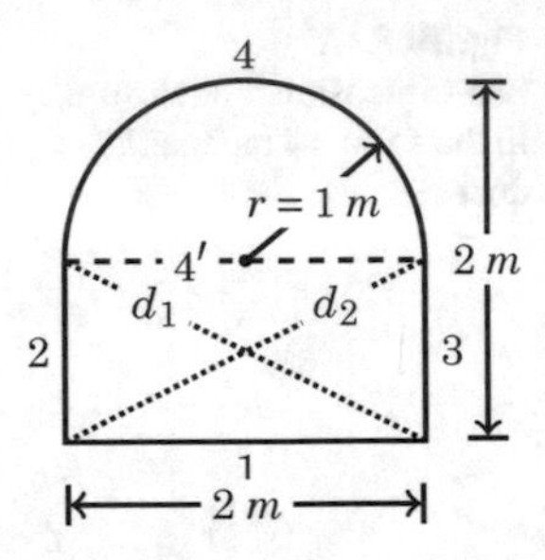

Fig. 10.6 Cross section of the furnace in Example 10.1

The numerator is the difference between the sum of the crossed lengths (imagine a string being held taut between the left corner of a and the right corner of c or the right corner of a and the left corner of c) and the straight lengths (imagine a string being held taut between the left corner of a and the left corner of c or the right corner of a and the right corner of c). Expression 10.26 embodies what is called the **Hottel's crossed string method.**[2]

Example 10.1

A very long furnace has the cross section shown in Fig. 10.6. The cross section has a rectangular part with a circular domed top. The surfaces are numbered as shown in the figure. Determine all the angle factors for diffuse radiant interchange.

Solution:

Step 1 Surfaces 1,2, and 3 are flat and hence $F_{11} = F_{22} = F_{33} = 0$. By symmetry, it is clear that $F_{12} = F_{13}$. Each of these may be determined by the triangle formula. We denote the length of side 1 as $l_1 = 2$ m, sides 2 and 3 as $l_2 = l_3 = 1$ m, and the circular dome as $l_4 = \pi r = \pi$ m. The required diagonals are obtained as

$$d_1 = d_2 = \sqrt{l_1^2 + l_2^2} = \sqrt{2^2 + 1^2} = \sqrt{5}\text{ m}$$

Step 2 Applying the triangle rule, we then get

$$F_{12} = \frac{l_1 + l_2 - d_1}{2l_1} = \frac{2 + 1 - \sqrt{5}}{2 \times 2} = 0.191$$

Hence, F_{13} is also equal to 0.191. The reciprocity rule gives the following:

$$F_{21} = F_{31} = \frac{l_1}{l_2} F_{12} = \frac{2}{1} \times 0.191 = 0.382$$

[2]H. C. Hottel and A. F. Sarofim, *Radiative Transfer*, McGraw Hill, 1967.

Step 3 We may determine the angle factor between 1 and 4′ by the application of the crossed string method. We then have

$$F_{14'} = \frac{(d_1 + d_2) - (l_2 + l_3)}{2l_1} = \frac{2 \times \sqrt{5} - 2 \times 1}{2 \times 2} = 0.618$$

Note that the last result could also have been obtained by applying the sum rule to side l_1.

Step 4 What is desired is the angle factor F_{14}. We note that any radiation that passes across the line 4′ from the bottom has to arrive on 4. Hence, the angle factor between 1 and 4 should be the same as the angle factor between 1 and 4′. Hence we have $F_{14} = 0.618$.

Step 5 The angle factor between 4 and 1 may be obtained by reciprocity as

$$F_{41} = \frac{l_1}{l_4} F_{14} = \frac{2}{\pi} \times 0.618 = 0.393$$

Step 6 The crossed string rule is applied to get either F_{23} or F_{32} which are equal to each other because of symmetry. We have

$$F_{23} = F_{32} = \frac{(d_1 + d_2) - (l_1 + l_4')}{2l_2} = \frac{2\sqrt{5} - 2 \times 2}{2 \times 1} = 0.236$$

Step 7 By symmetry, it is clear that $F_{24'} = F_{34'} = F_{21} = F_{31} = 0.382$.

Step 8 Again we see that the angle factor between 2 and 4′ or 3 and 4′ is the same as that between 2 and 4 or 3 and 4. Hence, we have $F_{24} = F_{24'} = F_{34} = F_{34'} = 0.382$.

Step 9 Reciprocity will now yield the remaining angle factors.

$$F_{42} = F_{43} = \frac{l_2}{l_4} F_{24} = \frac{1}{\pi} \times 0.382 = 0.122$$

Step 10 The application of the sum rule to surface 4 will now yield its self-angle factor as

$$F_{44} = 1 - F_{41} - F_{42} - F_{43} = 1 - 0.393 - 2 \times 0.122 = 0.363$$

The angle factors may be presented in the matrix form.

$$\begin{bmatrix} F_{11} & F_{12} & F_{13} & F_{14} \\ F_{21} & F_{22} & F_{23} & F_{24} \\ F_{31} & F_{32} & F_{33} & F_{34} \\ F_{41} & F_{42} & F_{43} & F_{44} \end{bmatrix} = \begin{bmatrix} 0 & 0.191 & 0.191 & 0.618 \\ 0.382 & 0 & 0.236 & 0.382 \\ 0.382 & 0.236 & 0 & 0.382 \\ 0.393 & 0.122 & 0.122 & 0.363 \end{bmatrix}$$

Example 10.2

A very long tunnel has the shape of an isosceles triangle as shown in Fig. 10.7. The tunnel has a thin partition wall down the middle as shown. Determine all the relevant angle factors for diffuse radiant interchange among the surfaces that make up the tunnel. Also, bring out the effect of shadowing by comparing F_{AB-DC} in the presence and absence of the partition.

Solution:

In all, there are 8 surfaces in this two-dimensional enclosure, considering the two sides of the partition (represented by DE_l facing ABD and DE_r facing ACD) and the gap (represented by AE_l facing ABD and AE_r facing ACD) as individual surfaces.

Step 1 The lengths of all the sides are first calculated. Since the triangle ABC is an isosceles triangle, $AB = AC$. Also $BD = DC = \frac{BC}{2} = \frac{2}{2} = 1$ m since the partition runs down the middle of the tunnel. Then $AB = AC = \sqrt{AD^2 + BD^2} = \sqrt{2^2 + 1^2} = 2.236$ m. In the calculations, we also require BE and CE that are equal to each other and given by $BE = CE = \sqrt{DE^2 + BD^2} = \sqrt{1^2 + 1^2} = 1.414$ m. Now we are ready to calculate the angle factors.

Step 2 Since all surfaces are flat, all self-angle factors are zero. Because of symmetry, we note the following:

$$F_{BD-AB} = F_{DC-AC} \quad F_{BD-DE_l} = F_{DC-DE_r}$$
$$F_{BD-AC} = F_{DC-AB} \quad F_{AB-DE_l} = F_{AC-DE_r}$$
$$F_{AB-BD} = F_{AC-DC} \quad F_{AB-AC} = F_{AC-AB}$$
$$F_{AB-DC} = F_{AC-BD}$$

Reciprocity will indicate other similar equalities. The reader may draw up the complete list based on these.

We look at individual angle factors now.

Step 3 Angle factor F_{AB-BD} :

This may be obtained by applying the triangle formula to triangle ABD as

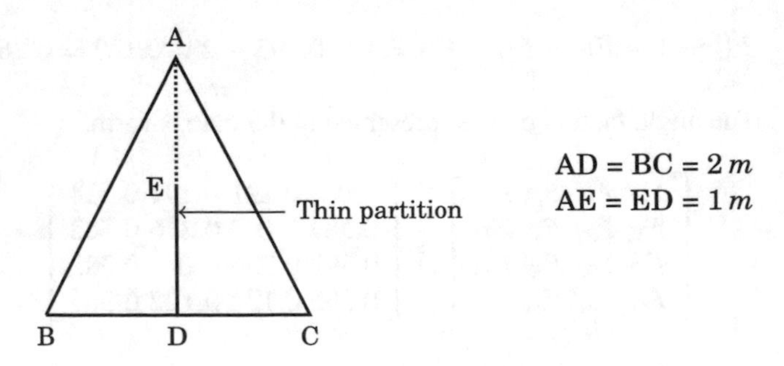

Fig. 10.7 Cross section of the tunnel in Example 10.2

$$F_{AB-BD} = \frac{AB + BD - AD}{2AB} = \frac{2.236 + 1 - 2}{2 \times 2.236} = 0.276$$

Step 4 Angle factor F_{AB-DE_l}:

This requires the application of the decomposition rule. We determine first the angle factors F_{AB-AD_l} and F_{AB-AE_l}. The former is obtained by the use of the sum rule applied to surface AB.

$$F_{AB-AD_l} = 1 - F_{AB-BD} = 1 - 0.276 = 0.724$$

The latter is obtained by the use of the triangle formula applied to triangle ABE.

$$F_{AB-AE_l} = \frac{AB + AE_l - BE}{2AB} = \frac{2.236 + 1 - 1.414}{2 \times 2.236} = 0.407$$

Using the decomposition rule, we get

$$F_{AB-DE_l} = F_{AB-AD_l} - F_{AB-AE_l} = 0.724 - 0.407 = 0.317$$

Step 5 Angle factor $F_{BD-AC} = F_{DC-AB}$:

From the figure, it is clear that this angle factor is the same as $F_{BD-AE_l} = F_{DC-AE_r}$ which may be obtained again by the use of the decomposition rule as done below. Applying the triangle formula to triangle ABD, we have

$$F_{BD-AD_l} = \frac{BD + AD_l - AB}{2 \times BD} = \frac{1 + 2 - 2.236}{2 \times 1} = 0.382$$

Applying the triangle formula to triangle BED, we have

$$F_{BD-ED_l} = \frac{BD + ED_l - BE}{2 \times BD} = \frac{1 + 1 - 1.414}{2 \times 1} = 0.293$$

By the decomposition rule, we then have

$$F_{BD-AE_l} = F_{BD-AD_l} - F_{BD-ED_l} = 0.382 - 0.293 = 0.089$$

We thus have $F_{BD-AC} = F_{DC-AB} = 0.089$. Using the reciprocity rule, we then have

$$F_{AB-DC} = F_{AC-BD} = \frac{DC}{AB} F_{DC-AB} = \frac{1}{2.236} \times 0.089 = 0.040$$

Step 6 Angle factor $F_{AB-AC} = F_{AC-AB}$:

By inspection of the figure, it is clear that $F_{AB-AE_l} = F_{AB-(AC+DC)} = 0.407$ as obtained earlier. By decomposition rule, we then have

$$F_{AB-AC} = F_{AB-(AC+DC)} - F_{AB-DC} = 0.407 - 0.040 = 0.367$$

Alternately, this angle factor may be determined by using the crossed string method. Assume that a string is tightly stretched between B and C. Obviously, the string will be represented by BEC. $AB - BEC - CA$ may be treated as a triangle that has three convex sides. The triangle rule may be applied to this to get

$$F_{AB-AC} = \frac{AB + AC - (BE + EC)}{2AB}$$
$$= \frac{2AB - 2BE}{2AB} = \frac{2.236 - 1.414}{2.236} = 0.367$$

Table 10.1 summarizes the results for this example.

Step 7 Angle factor F_{AB-DC} intheabsenceofthepartition :

We need to use the decomposition rule to obtain the desired angle factor. By applying the triangle formula to ABC, we get

$$F_{AB-BC} = \frac{AB + BC - AC}{2AB} = \frac{2.236 + 2 - 2.236}{2 \times 2.236} = 0.447$$

Applying the triangle formula to triangle ABD, we get

$$F_{AB-BD} = \frac{AB + BD - AD}{2AB} = \frac{2.236 + 1 - 2}{2 \times 2.236} = 0.276$$

By the decomposition rule we then get, in the absence of the partition,

$$F_{AB-DC} = F_{AB-BC} - F_{AB-BD} = 0.447 - 0.276 = 0.171$$

Step 8 Observation: On comparison, it is seen that the the effect of the partition (shadowing body) is to reduce the angle factor between AB and DC from 0.171 to a mere 0.04!

Table 10.1 Angle factors in Example 10.2

F_{AB-BD}	0.276	F_{BD-AB}	0.618	F_{DE_l-AB}	0.707
F_{AC-CD}		F_{DC-AC}		F_{DE_r-AC}	
F_{AB-DE_l}	0.317	F_{BD-DE_l}	0.293	F_{DE_l-BD}	0.293
F_{AC-DE_r}		F_{DC-DE_r}		F_{DE_r-DC}	
F_{AB-AC}	0.367	F_{BD-AC}	0.089		
F_{AC-AB}		F_{DC-AB}			
F_{AB-DC}	0.04				
F_{AC-BD}					

Example 10.3

Cross sections of two hexagonal ducts are shown in Fig. 10.8. Lengths of some of the sides are specified as under:

- Hexagon at left: $AB = CD = 0.75$; $AC = BD = 1$; $AE = EC = BF = DF = 0.625$
- Hexagon at right: $AB = CD = 0.75$; $AE = CE = BF = DF = 0.52$

Determine the diffuse angle factor between AB and CD in each case.

Solution :

Hexagon at left; Fig. 10.8a:

We shall use the crossed string method for solving the problem. We consider the rectangle $ABDC$ and note that the appropriate crossed lengths are:

$$BC = \sqrt{AB^2 + AC^2} = \sqrt{0.75^2 + 1^2} = 1.25$$
$$AD = \sqrt{AB^2 + BD^2} = \sqrt{0.75^2 + 1^2} = 1.25$$

Then the crossed string method gives

$$F_{AB-CD} = \frac{(AD + BC) - (AC + BD)}{2AB} = \frac{(2 \times 1.25) - (2 \times 1)}{2 \times 0.75} = 0.333$$

Hexagon at right; Fig. 10.8b:

The crossed lengths remain the same. The straight lengths are replaced by $AE + EC$ and $BF + DF$. We apply the crossed string method to the polygon $AE - EC - CD - DF - FB - BA$ to get

$$F_{AB-CD} = \frac{(AD + BC) - (AE + EC + BF + DF)}{2AB}$$
$$= \frac{(2 \times 1.25) - (2 \times 1.04)}{2 \times 0.75} = 0.28$$

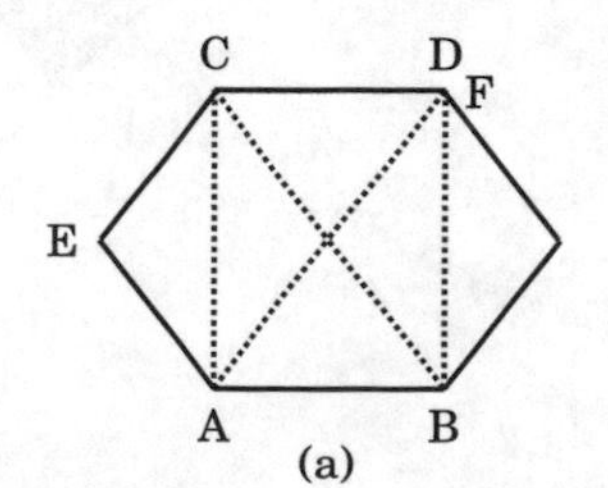

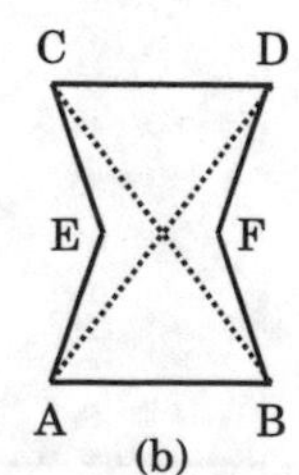

Fig. 10.8 Cross sections of ducts in Example 10.3

Comment: The reduction in the angle factor is due solely to the shadowing effect in the second case!

10.2.4 Three-Dimensional Enclosures

Enclosures encountered in engineering applications generally involve surfaces oriented in three-dimensional space such that the angle factor determination is more complex than the simple analysis that involved only lengths of sides in the case of two-dimensional enclosures. A catalog of angle factors and the vast literature available on these have been made by Howell.[3] In Appendix G, we present angle factors for commonly encountered geometries.

Rule of Corresponding Corners:
Apart from the data presented in Appendix G, angle factor algebra and angle factor rules provide additional resources for evaluating angle factors. One of these, the decomposition rule, has already been presented. Another useful rule dealing with rectangles is the rule of corresponding corners that is illustrated using Fig. 10.9.

Consider the small rectangles shown numbered from 1 through 4. These smaller rectangles are obtained by drawing lines parallel to the edges of the larger rectangles as indicated. By writing down the angle factor relations explicitly in terms of double integrals over each of the areas that make up the corresponding corner configurations, it is possible to show that the following hold:

$$\boxed{A_1 F_{12} = A_3 F_{34}} \quad \text{and} \quad \boxed{A_2 F_{21} = A_4 F_{43}} \tag{10.27}$$

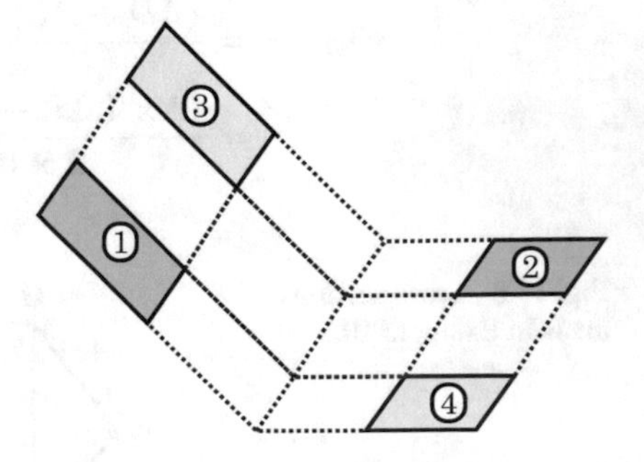

Fig. 10.9 Rule of corresponding corners

[3]Catalog is accessible at http://www.me.utexas.edu/~howell/index.html.

Example 10.4

Consider an enclosure consisting of diffuse walls in the form of a cube. Determine all the required angle factors.

Solution:
Refer to Fig. G.1. For a cubical enclosure, $a = b = c$ and hence $X = \frac{a}{b} = 1$ and $Y = \frac{b}{c} = 1$. All angle factors between adjacent sides are equal and those between opposite sides are equal. If we determine any one of them, all the angle factors may be determined by angle factor algebra.

Consider the angle factor between opposite sides. The angle factor is read from Fig. G.2 with $X = Y = 1$ to get $F_{\text{Opposite sides}} \approx 0.2$.[4] The sum rule now gives

$$F_{\text{Adjacent sides}} = \frac{(1 - F_{\text{Opposite sides}})}{4} = \frac{1 - 0.2}{4} = 0.2$$

A more accurate value would be 0.20004. The values read off the graph are sufficiently accurate for our purpose here. However, the use of accurate values are important in analysis as pointed out in the appropriate literature.

Evaluation of View Factors in Complex Arrangements

Configuration involving rectangles

We shall look at some configurations for which the graphical data of angle factors between rectangles may be made use of. Angle factors between rectangles that are shown in Fig. 10.10a, b are considered. The first configuration consists of two perpendicular rectangles touching at one corner. The angle between the planes of the two rectangles is 90°. The second configuration consists of two parallel rectangles with an offset. Both these configurations are easily handled using graphical data given in

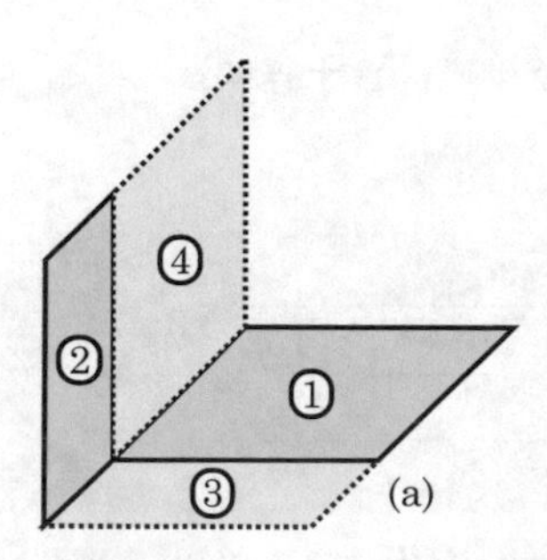

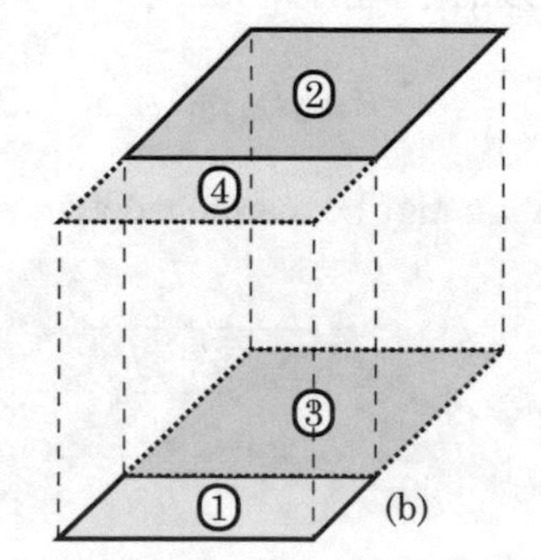

Fig. 10.10 Configurations involving rectangles

[4] A more accurate value is obtained by calculation as 0.19982.

Figs. G.2 and G.3 and angle factor algebra. The following development is common to both configurations.

With reference to Fig. 10.10a, it is clear that F_{14}, F_{32}, and $F_{(1+3)-(2+4)}$ are evaluated using the results for two perpendicular rectangles sharing a common edge. With reference to Fig. 10.10b, it is clear that F_{14}, F_{32}, and $F_{(1+3)-(2+4)}$ are evaluated using the results for two parallel rectangles of equal size. From the law of corresponding corners, it is also clear that the following should hold:

$$A_1 F_{12} = A_3 F_{34} \tag{10.28}$$

Using the decomposition rule, we have

$$A_{1+3} F_{(1+3)-(2+4)} = A_{1+3} F_{(1+3)-2} + A_{1+3} F_{(1+3)-4} \tag{10.29}$$

Using reciprocity, followed by the decomposition rule, the first term on the right-hand side of Eq. 10.29 is written as

$$A_{1+3} F_{(1+3)-2} = A_2 F_{2-(1+3)} = A_2 (F_{21} + F_{23}) \tag{10.30}$$

Using the reciprocity rule, the above may also be rewritten as

$$A_{1+3} F_{(1+3)-2} = A_1 F_{12} + A_3 F_{32} \tag{10.31}$$

A similar procedure is used to write the last term in Eq. 10.29 as

$$A_{1+3} F_{(1+3)-4} = A_4 (F_{41} + F_{43}) = A_1 F_{14} + A_3 F_{34} \tag{10.32}$$

Introducing 10.31, 10.32 in 10.29, we then have

$$A_{1+3} F_{(1+3)-(2+4)} = A_1 F_{12} + A_3 F_{32} + A_1 F_{14} + A_3 F_{34}$$

which becomes, with Eq. 10.28,

$$A_{1+3} F_{(1+3)-(2+4)} = 2A_1 F_{12} + A_1 F_{14} + A_3 F_{32}$$

This equation may be rearranged as

$$\boxed{F_{12} = \frac{A_{1+3}}{A_1} F_{(1+3)-(2+4)} - \frac{F_{14}}{2} - \frac{A_3}{2A_1} F_{32}} \tag{10.33}$$

All the angle factors on the right-hand side of Eq. 10.33 are read off either from Figs. G.2 or G.3.

Fig. 10.11 Configuration for Example 10.5

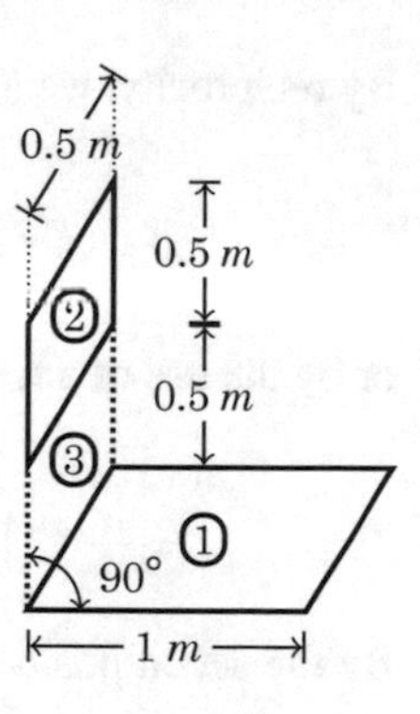

Example 10.5

Determine the angle factor F_{12} for the configuration shown in Fig. 10.11. Use the data presented in Fig. G.3 and angle factor algebra.

Solution:

Angle factor $F_{(2+3)-1}$:

These are two perpendicular rectangles that share a common edge. Hence, we may use Fig. G.3 to determine $F_{(2+3)-1}$. We have $a = 1$ m, $b = 0.5$ m, and $c = 1$ m. Thus $Z = \frac{a}{b} = \frac{1}{0.5} = 2$ and $Y = \frac{b}{c} = \frac{0.5}{1} = 0.5$ or $Y' = \frac{1}{Y} = 2$. From Fig. G.3, we read off the angle factor as $F_{(2+3)-1} = 0.15$.[5] A more accurate value from Table G.4 is 0.14930.

Angle factor F_{31} :

Again, these are two perpendicular rectangles that share a common edge. We have for this configuration $a = 1$ m, $b = 0.5$ m, and $c = 0.5$ m. Thus, we have $Z = \frac{a}{b} = \frac{1}{0.5} = 2$ and $Y = \frac{b}{c} = \frac{0.5}{0.5} = 1$ or $Y' = \frac{1}{Y} = 1$. Again, we read off the angle factor as $F_{31} = 0.24$ from Fig. G.3. A more accurate value from Table G.3 is 0.23285 (comment in the previous footnote is valid here also).

View factor F_{12} :

By reciprocity, we then get

$$F_{1-(2+3)} = \frac{A_{2+3}}{A_1} F_{(2+3)-1} = \frac{1 \times 0.5}{1 \times 0.5} \times 0.15 = 0.15$$

or by the use of a more accurate value

$$F_{1-(2+3)} = \frac{A_{2+3}}{A_1} F_{(2+3)-1} = \frac{1 \times 0.5}{1 \times 0.5} \times 0.1493 = 0.1493$$

[5] Note that $F_{1-(2+3)}$ may be obtained directly by interchanging Z and Y' values.

By reciprocity, we also get

$$F_{13} = \frac{A_3}{A_1} F_{31} = \frac{0.5 \times 0.5}{1 \times 0.5} \times 0.24 = 0.12$$

or by the use of a more accurate value

$$F_{13} = \frac{A_3}{A_1} F_{31} = \frac{0.5 \times 0.5}{1 \times 0.5} \times 0.23285 = 0.11643$$

By the decomposition rule, we then get

$$F_{12} = F_{1-(2+3)} - F_{13} = 0.15 - 0.12 = 0.03$$

However, if we use the more accurate tabulated data, the above result is modified as

$$F_{12} = F_{1-(2+3)} - F_{13} = 0.14930 - 0.11643 = 0.03287 \approx 0.033$$

Configuration involving circular disks

An example of a configuration involving parallel disk geometry is shown in Fig. 10.12. It is desired to determine F_{13}, the angle factor between the top circle 1 and the annulus 3. This is quite simply done by the use of the decomposition rule. We have

$$F_{13} = F_{1-(2+3)} - F_{12} \tag{10.34}$$

Both angle factors on the right-hand side of Eq. 10.34 may be obtained by the use of Fig. G.4. Using reciprocity, we may easily obtain also the angle factor between the annulus and the top disk as

$$F_{31} = \frac{A_1}{A_3} (F_{1-(2+3)} - F_{12}) \tag{10.35}$$

Now, we look at how to obtain the angle factor between the annulus and the two curved surfaces labeled 4 and 5. We note that all the radiation that leaves 3 and

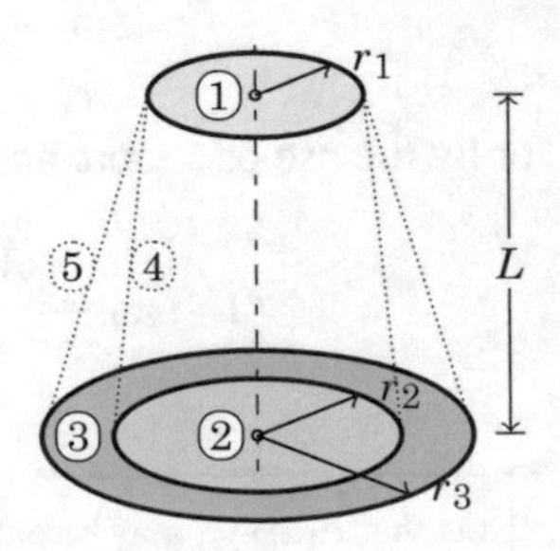

Fig. 10.12 Configuration involving circular disks

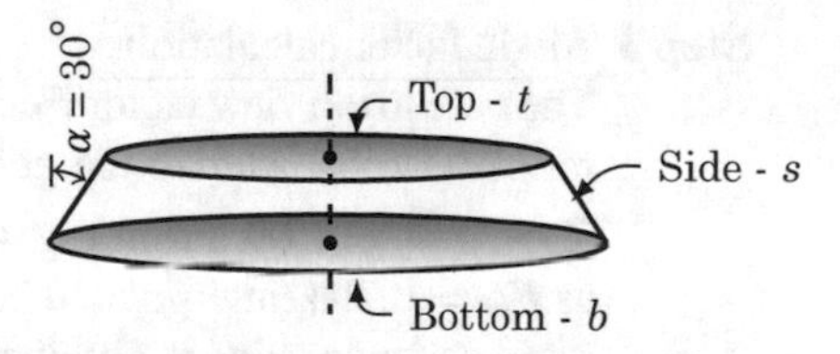

Fig. 10.13 Configuration for Example 10.6

reaches 1 should necessarily pass through 4. Hence, the angle factor between 3 and 1 is also the angle factor between 3 and 4, i.e., $F_{31} = F_{34}$. The sum rule applied to surface 3 will then give the important result

$$F_{35} = 1 - F_{31} = 1 - \frac{A_1}{A_3}(F_{1-(2+3)} - F_{12}) \tag{10.36}$$

The application again of the sum rule to surface 1 yields the angle factor F_{15} as

$$F_{15} = 1 - F_{1-(2+3)} \tag{10.37}$$

It is left as an exercise to the reader to obtain all the other angle factors that are required in this case.

Example 10.6

A frustum of a cone (see Fig. 10.13) has a base of diameter $D_b = 1$ m, height $H = 0.3$ m, and semi-cone angle $\alpha = 30°$. Determine all the diffuse angle factors for this configuration.

Solution:

Step 1 The base diameter is specified as $D_b = 1\ m$. The height of the frustum of cone is given as $H = 0.3$ m. The semi-cone angle is specified as $\alpha = 30°$. Let D_t be the diameter of the top of the frustum. From the geometry, we have

$$\tan\alpha = \frac{D_b - D_t}{2H} \quad \text{or} \quad D_t = D_b - 2H\tan\alpha$$

$$D_t = 1 - 2 \times 0.3 \times \tan 30° = 0.654\ m$$

Step 2 Area calculations:

$$\text{Base: } A_b = \frac{\pi D_b^2}{4} = \frac{\pi \times 1^2}{4} = 0.785\ m^2$$

$$\text{Top: } A_t = \frac{\pi D_t^2}{4} = \frac{\pi \times 0.654^2}{4} = 0.336\ m^2$$

$$\text{Curved side: } A_s = \frac{\pi(D_b + D_t)}{2}\frac{H}{\cos\alpha} = \frac{\pi(1 + 0.654)}{2}\frac{0.3}{\cos 30°} = 0.9\ m^2$$

Step 3 Angle factor calculations:

The base to top view factor F_{bt} is first calculated. The two non-dimensional ratios that characterize the geometry are $\frac{1}{R_1} = \frac{H}{r_b} = \frac{0.3}{0.5} = 0.6$ and $R_2 = \frac{r_t}{h} = \frac{0.654}{2\times0.3} = 1.09$. From Fig. G.4, we read off the base to top view factor as $F_{bt} = 0.284$ (interpolated between curves for $R_2 = 1$ and $R_2 = 1.5$). A more accurate value is obtained by the use of Table G.5 as

$$F_{bt} = 0.24388 + \frac{(0.47491 - 0.24388) \times 0.09}{0.5} = 0.28622 \approx 0.286$$

The value interpolated from the figure is good enough. Using reciprocity, the top to base view factor is obtained as

$$F_{tb} = \frac{A_b}{A_t} F_{bt} = \frac{0.785}{0.336} \times 0.284 = 0.664$$

Applying the sum rule to the bottom surface, we then have

$$F_{bs} = 1 - F_{bt} = 1 - 0.284 = 0.716$$

Applying the sum rule to the top surface, we have

$$F_{ts} = 1 - F_{tb} = 1 - 0.664 = 0.336$$

The view factors from the curved side to the other two sides may be calculated by the application of reciprocity rule. Thus

$$F_{sb} = \frac{A_b}{A_s} F_{bs} = \frac{0.785}{0.9} \times 0.716 = 0.624$$

$$F_{st} = \frac{A_t}{A_s} F_{ts} = \frac{0.336}{0.9} \times 0.336 = 0.125$$

Since the curved surface s is concave, it is able to see itself and hence the self-view factor is F_{ss} is non-zero. The self-view factor for the curved side is obtained by the application of the sum rule to it.

$$F_{ss} = 1 - F_{sb} - F_{st} = 1 - 0.624 - 0.125 = 0.251$$

Step 4 Thus all the angle factors are available. These are presented in the form of a matrix:

$$\begin{bmatrix} F_{bb} & F_{bt} & F_{bs} \\ F_{tb} & F_{tt} & F_{ts} \\ F_{sb} & F_{st} & F_{ss} \end{bmatrix} = \begin{bmatrix} 0 & 0.284 & 0.716 \\ 0.664 & 0 & 0.336 \\ 0.624 & 0.125 & 0.251 \end{bmatrix}$$

Note that only one angle factor was calculated using the graph presented in Fig. G.4. All other angle factors have been deduced using angle factor algebra.

10.3 Radiation Heat Transfer in Enclosures with Gray Diffuse Walls

Radiation heat transfer calculations require the development of the basic methodology, when the number of surfaces in the enclosures is not limited to a small number like two or three. In this section, we treat simple cases first, in order to understand the basic ideas before presenting general methods applicable to an enclosure with any number of surfaces. The method of detailed balancing is used in a typical simple application to begin our study.

10.3.1 Method of Detailed Balancing

Consider two isothermal infinite parallel planes as shown in Fig. 10.14. Consider a test area element on surface 2 as shown.

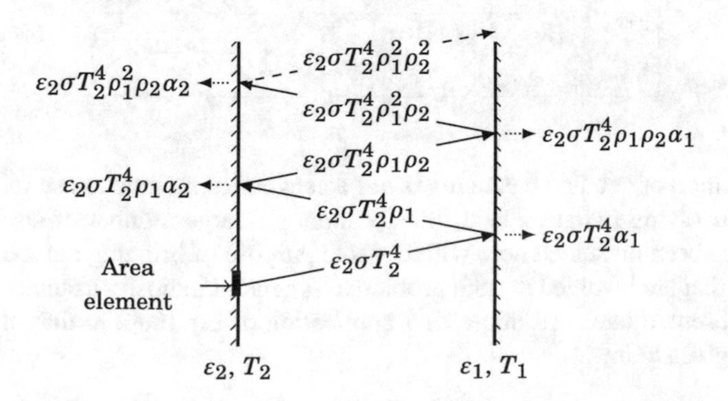

Fig. 10.14 Two infinite parallel planes exchanging radiation

Emitted power from this area element per unit area is given by $\varepsilon_2\sigma T_2^4$. This is incident totally on surface 1 because the angle factor $F_{12} = F_{21} = 1$. An amount of power equal to $\varepsilon_2\sigma T_2^4 \times \rho_1$ is reflected by surface 1 while the balance equal to $\varepsilon_2\sigma T_2^4 \times \alpha_1$ is absorbed by surface 2. The reflected part, viz., $\varepsilon_2\sigma T_2^4 \times \rho_1$ undergoes a second reflection at surface 2 and an amount equal to $\varepsilon_2\sigma T_2^4 \times \rho_1 \times \rho_2$ is incident on surface 1. Surface 1 absorbs $(\varepsilon_2\sigma T_2^4 \times \rho_1 \times \rho_2)\alpha_1$ out of this. This process continues endlessly so that the amount of power emitted by surface 2 that is absorbed by surface 1 is

$$P_{21} = \varepsilon_2\sigma T_2^4\alpha_1[1 + \rho_1\rho_2 + (\rho_1\rho_2)^2 + \cdots] \tag{10.38}$$

The quantity in square brackets is a geometrical progression with common ratio $\rho_1\rho_2$ that has the sum (to infinite terms) given by $\frac{1}{1-\rho_1\rho_2}$. Thus, power emitted by surface 2 that is absorbed by surface 1, per unit area, is

$$P_{21} = \varepsilon_2\sigma T_2^4\alpha_1 \times \frac{1}{1-\rho_1\rho_2} \tag{10.39}$$

Similarly, the power emitted by surface 1 that is absorbed by surface 2 is given by

$$P_{12} = \varepsilon_1\sigma T_1^4\alpha_2 \times \frac{1}{1-\rho_1\rho_2} \tag{10.40}$$

Hence, the net Radiant Flux transferred from 1 to 2 is obtained as

$$q_{1-2} = P_{12} - P_{21} = \frac{\varepsilon_1\sigma T_1^4\alpha_2 - \varepsilon_2\sigma T_2^4\alpha_1}{1-\rho_1\rho_2} \tag{10.41}$$

Since the two surfaces are gray, we have $\alpha_1 = \varepsilon_1$ and $\alpha_2 = \varepsilon_2$. Also $\rho_1 = 1 - \varepsilon_1$ and $\rho_2 = 1 - \varepsilon_2$. With these, Eq. 10.41 may be simplified to read

$$\boxed{q_{1-2} = \frac{\varepsilon_1\varepsilon_2\sigma(T_1^4 - T_2^4)}{1-(1-\varepsilon_1)(1-\varepsilon_2)} = \frac{\sigma(T_1^4 - T_2^4)}{\left(\frac{1}{\varepsilon_1} + \frac{1}{\varepsilon_2} - 1\right)}} \tag{10.42}$$

The method of detailed balancing is *not* a convenient method to use for a general problem involving radiation heat transfer among a large number of surfaces. The method has been presented here with the sole purpose of bringing out the details of processes that are involved in such problems. A general radiosity irradiation method will be presented later. An interesting application of Eq. 10.42 to that of radiation shields is given below.

10.3.2 Radiation Shields

A radiation shield is useful in reducing the heat transfer between two surfaces that have specified surface properties and temperatures. Assume, for simplicity, $\varepsilon_1 = \varepsilon_2 = \varepsilon$. Introduce a third infinite plane (thin foil) in between surfaces 1 and 2 as shown in Fig. 10.15. Let the foil have a common emissivity of ε_3 on both its sides.

Under steady conditions, it is obvious that we should have $q_{1-3} = q_{3-2} = q_{1-2}$. Each of the q's may be evaluated using a formula of type given by Eq. 10.42. Thus, we have

$$q_{1-3} = \frac{\sigma(T_1^4 - T_3^4)}{\left(\frac{1}{\varepsilon} + \frac{1}{\varepsilon_3} - 1\right)}, \quad \text{and} \quad q_{3-2} = \frac{\sigma(T_3^4 - T_2^4)}{\left(\frac{1}{\varepsilon_3} + \frac{1}{\varepsilon} - 1\right)}$$

In both of the above equations, the unknown temperature T_3 will adjust to a value such that these fluxes are equal. Since the denominators in these two relations are identical, the numerators must be equal and hence

$$T_1^4 - T_3^4 = T_3^4 - T_2^4 \quad \text{or} \quad \boxed{T_3^4 = \frac{T_1^4 + T_2^4}{2}} \tag{10.43}$$

Substituting this in expressions for either q_{1-3} or q_{3-2}, we get the important result

$$q_{1-2} = q_{1-3} = q_{3-2} = \frac{\sigma\left(T_1^4 - \frac{T_1^4+T_2^4}{2}\right)}{\left(\frac{1}{\varepsilon} + \frac{1}{\varepsilon_3} - 1\right)} = \frac{\sigma(T_1^4 - T_2^4)}{2\left(\frac{1}{\varepsilon} + \frac{1}{\varepsilon_3} - 1\right)} \tag{10.44}$$

In case all the ε's are the same, the presence of a radiation shield reduces the heat flux by a factor of 2.

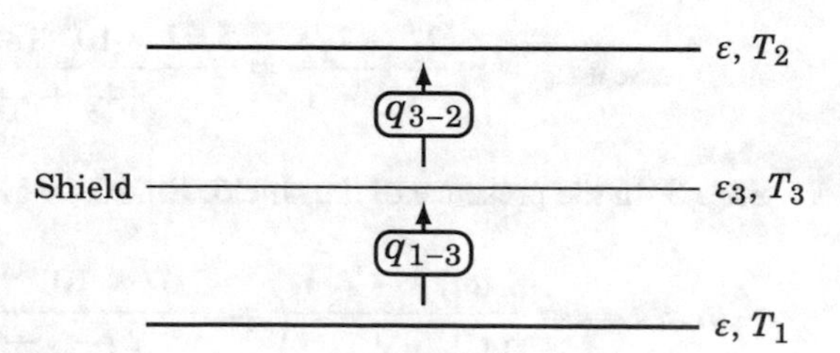

Fig. 10.15 Sketch for radiation shield analysis

Multi-layer Insulation or MLI

In practice, it is common to use many radiation shields in the form of a Multi-Layer Insulation (MLI) blanket consisting of polyimide or polyester layers (5 to 30 layers) separated by very small diameter thread made of polyester or nylon (in the form of a net). Each layer has vapor-deposited aluminum on both sides thus providing very low emissivity. In principle, the analysis given above may be extended, assuming that the layers of the MLI do not contact each other except through the spacer threads and provide conduction paths, to get a reduction of radiation transfer by the factor $\frac{1}{n+1}$ where n is the number of layers in the blanket. MLI blankets are commonly used in satellite applications.

Example 10.7

Two very large surfaces are maintained at 300 K and 400 K, respectively. Both surfaces have an emissivity of 0.85. A third very thin plate of very large dimensions is placed midway between the two surfaces. This surface has an emissivity of 0.05 on both sides. Compare the heat transfer per unit area in the two cases, with and without the third surface. Determine the temperature of the third surface.

Solution:

Step 1 The nomenclature used in this example is as given in Fig. 10.15. Given data:

$$T_1 = 300\,\text{K},\ T_2 = 400\,\text{K},\ \varepsilon_1 = \varepsilon_2 = 0.85,\ \varepsilon_3 = 0.05$$

Step 2 In the absence of the third surface, the heat transfer from 1 to 2 per unit area (using Eq. 10.42) is

$$q_{1-2,\text{No shield}} = \frac{\sigma(T_1^4 - T_2^4)}{\frac{1}{\varepsilon_1} + \frac{1}{\varepsilon_2} - 1} = \frac{5.67 \times 10^{-8}(300^4 - 400^4)}{\frac{1}{0.85} + \frac{1}{0.85} - 1} = -733.4\ \text{W/m}^2$$

Step 3 In the presence of the shield, Eq. 10.44 gives

$$q_{1-2,\text{Shield}} = \frac{\sigma(T_1^4 - T_2^4)}{2\left(\frac{1}{\varepsilon_1} + \frac{1}{\varepsilon_3} - 1\right)} = \frac{5.67 \times 10^{-8}(300^4 - 400^4)}{2\left(\frac{1}{0.85} + \frac{1}{0.05} - 1\right)} = -24.6\ \text{W/m}^2$$

The negative sign indicates that radiation transfer is from surface 2 to surface 1 in both cases.

Step 4 The temperature of the third surface from Eq. 10.43 is

$$T_3 = \left(\frac{T_1^4 + T_2^4}{2}\right)^{\frac{1}{4}} = \left(\frac{300^4 + 400^4}{2}\right)^{\frac{1}{4}} = 360.3 \text{ K}$$

Step 5 The heat transfer, in the presence of the shield, is roughly $\frac{1}{30}$ of the heat transfer in the absence of the shield!

10.3.3 Radiosity Irradiation Method of Enclosure Analysis

The radiosity irradiation formulation is a general method applicable to any number of isothermal diffusely radiating, gray surfaces that make up an enclosure.

Assumptions made in the formulation are

- Each surface of an enclosure is at a uniform temperature.
- Radiosity of such a surface is uniform over its extent.
- Irradiation over such a surface is also uniform over its extent.

The second and third assumptions are somewhat difficult to justify. However, these assumptions are made to make the formulation tractable. When these assumptions are relaxed, the formulation requires more advanced treatment that will be considered later on.

Radiosity formulation

Consider an evacuated enclosure consisting of N diffusely radiating, gray and isothermal surfaces. The radiosity of the ith surface J_i, is defined as the sum of its emissive power E_i and the reflected irradiation G_i. Thus, all the quantities are defined per unit area of the ith surface. We may thus write the expression for radiosity as

$$J_i = E_i + \rho_i G_i = \varepsilon_i \sigma T_i^4 + (1 - \varepsilon_i) G_i \tag{10.45}$$

where the gray assumption has been made to write the reflectance in terms of emissivity of the surface. We know that the total radiation leaving a surface j in the enclosure is $A_j J_j$. The fraction of this radiation that reaches the ith surface is $F_{ji} A_j J_j$. Thus, the radiation that leaves the jth surface and reaches the ith surface per unit area of surface i is $\frac{F_{ji} A_j J_j}{A_i}$. Using the angle factor identity $A_j F_{ji} = A_i F_{ij}$, this expression simplifies to $F_{ij} J_j$. Since the surface receives radiation from all surfaces in the enclosure (possibly including itself), the radiation received by surface i per unit area, i.e., the irradiation G_i is given by

$$G_i = \sum_{j=1}^{N} F_{ij} J_j \tag{10.46}$$

Using Eqs. 10.46 in 10.45, we get the radiosity of surface i in terms of radiosities of all the surfaces as

$$J_i = \varepsilon_i \sigma T_i^4 + (1 - \varepsilon_i) \sum_{j=1}^{N} F_{ij} J_j \tag{10.47}$$

In enclosure analysis, it is possible to have two types of surfaces. In the first case, we may specify the temperature in which case equation such as 10.47 is written for it. In the second case, the surface may have a specified heat flux. For example, a surface that is heated by an external means, such as in a furnace, has the heat flux specified, but the temperature is unknown. A special case of the latter is a surface that is adiabatic, in which case the heat flux is zero and hence the temperature is unknown.

To take care of the second case, we shall look at the heat flux at the ith surface. The heat flux at surface i is obviously given by the difference between the radiosity and the irradiation. Thus,

$$q_i = J_i - G_i = J_i - \sum_{j=1}^{N} F_{ij} J_j \tag{10.48}$$

where the irradiation has been written in terms of radiosities using Eq. 10.46. Alternately, we may also solve for G_i using Eq. 10.45 to get the heat flux as

$$q_i = J_i - \frac{J_i - \varepsilon_i \sigma T_i^4}{1 - \varepsilon_i} = \frac{\varepsilon_i}{1 - \varepsilon_i} \cdot [\sigma T_i^4 - J_i] \tag{10.49}$$

when $\varepsilon_i \neq 1$, i.e., if the surface is not black. However, for a black surface with $\varepsilon_i = 1$, from Eq. 10.45 we have the trivial result $J_i = \sigma T_i^4$. Thus, equation such as Eq. 10.48 is written down for a surface with specified heat flux such that the unknown temperature of the surface does not appear in the formulation. The heat flux for a surface with specified temperature is obtained directly by the use of Eq. 10.49. The temperature of the surface with specified heat flux, however, is determined by solving Eq. 10.49

for temperature in terms of the radiosity. Thus, we have the temperature of the heat flux specified surface as

$$\boxed{T_i = \left[\frac{J_i}{\sigma} + \frac{1-\varepsilon_i}{\varepsilon_i} \cdot \frac{q_i}{\sigma}\right]^{\frac{1}{4}}} \tag{10.50}$$

In the special case of an adiabatic surface (the surface is said to be a reradiating surface), $q_i = 0$ and the temperature is given by the expression

$$\boxed{T_i = \left[\frac{J_i}{\sigma}\right]^{\frac{1}{4}}} \tag{10.51}$$

which is independent of ε_i!

Solution of radiosity equations:

In summary, we write equations based on 10.47 for all the surfaces in the enclosure with specified temperatures. For those surfaces for which the heat flux is specified, equations such as 10.48 may be written down. We will then have N simultaneous linear equations for the N unknown radiosities. These equations are solved to get the radiosities of all the surfaces in the enclosure. Then the heat flux for T specified surfaces are obtained by using expression 10.49. For q specified surfaces, the corresponding temperatures are obtained by using 10.50.

Application of Radiosity Method to Two-Surface Enclosures

We consider a simple two-surface enclosure consisting of a smaller convex surface placed inside a larger concave surface as shown in Fig. 10.16a. This applies to a two-dimensional enclosure in which the figure represents the cross section of a duct with internal and external boundaries represented by the outer and inner curves. It also applies to the case where the inner surface is that of a smaller convex shell and the outer surface is the inner closed surface of an outer concave shell. As shown in Fig. 10.16b, the former case is typical of a very long cylinder of radius R_1 inside an

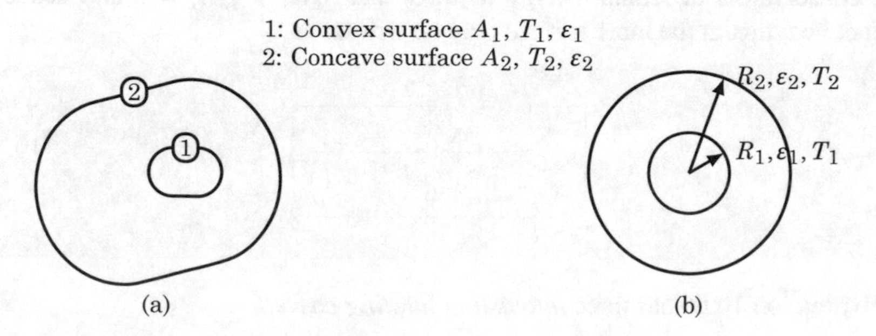

Fig. 10.16 Two-surface enclosure configurations

outer very long cylinder of radius R_2. An example of the latter is the case of a smaller sphere of radius R_1 placed inside a hollow sphere of larger radius R_2.

Required angle factors are easily evaluated. Surface 1 is convex and hence $F_{11} = 0$ and $F_{12} = 1$. By reciprocity, we get $F_{21} = \frac{A_1}{A_2}$. Hence, by the sum rule, $F_{22} = 1 - F_{21} = 1 - \frac{A_1}{A_2}$. Note that the placement of the inner surface anywhere within the outer surface is all that is required for these to hold. Note that in the case of cylinders $\frac{A_1}{A_2} = \frac{R_1}{R_2}$ and in the case of spheres $\frac{A_1}{A_2} = \left(\frac{R_1}{R_2}\right)^2$.

Since the temperatures are specified for both surfaces, radiosity equations are written down using Eq. 10.47 as

$$J_1 = \varepsilon_1 \sigma T_1^4 + (1 - \varepsilon_1) J_2 \tag{10.52}$$

$$J_2 = \varepsilon_2 \sigma T_2^4 + (1 - \varepsilon_2) \frac{A_1}{A_2} J_1 + \left(1 - \frac{A_1}{A_2}\right) J_2 \tag{10.53}$$

Using Eq. 10.52 for J_1 in 10.53, we get

$$J_2 = \varepsilon_2 \sigma T_2^4 + (1 - \varepsilon_2) \frac{A_1}{A_2} [\varepsilon_1 \sigma T_1^4 + (1 - \varepsilon_1) J_2] + \left(1 - \frac{A_1}{A_2}\right) J_2$$

This may be solved for J_2 as

$$J_2 = \frac{\varepsilon_2 \sigma T_2^4 + \varepsilon_1 (1 - \varepsilon_2) \cdot \frac{A_1}{A_2} \sigma T_1^4}{\varepsilon_2 + \varepsilon_1 (1 - \varepsilon_2) \cdot \frac{A_1}{A_2}} \tag{10.54}$$

The heat flux at surface 2 may at once be evaluated using 10.49 as

$$q_2 = \frac{\varepsilon_2}{1 - \varepsilon_2} \left[\sigma T_2^4 - \frac{\varepsilon_2 \sigma T_2^4 + \varepsilon_1 (1 - \varepsilon_2) \cdot \frac{A_1}{A_2} \sigma T_1^4}{\varepsilon_2 + \varepsilon_1 (1 - \varepsilon_2) \cdot \frac{A_1}{A_2}} \right] = \frac{\sigma (T_2^4 - T_1^4) \cdot \frac{A_1}{A_2}}{\frac{1}{\varepsilon_1} + \frac{A_1}{A_2} \left(\frac{1}{\varepsilon_2} - 1\right)} \tag{10.55}$$

The conservation of radiant energy requires that $q_1 A_1 + q_2 A_2 = 0$ and hence the radiant heat flux at the inner surface may be written as

$$\boxed{q_1 = \frac{\sigma (T_1^4 - T_2^4)}{\frac{1}{\varepsilon_1} + \frac{A_1}{A_2} \left(\frac{1}{\varepsilon_2} - 1\right)}} \tag{10.56}$$

Expression 10.56 has three *interesting limiting cases*.

1. When $\frac{A_1}{A_2} \rightarrow 0$, i.e., when a small object is placed inside a large room or enclosure, $q_1 \approx \varepsilon_1 \sigma (T_1^4 - T_2^4)$. The emissivity of the second surface has no role to play!
2. When $\frac{A_1}{A_2} \rightarrow 1$, the case corresponds to two very large surfaces with a very narrow gap and Eq. 10.56 reduces to Eq. 10.42 obtained earlier by the method of detailed balancing for the case of two large parallel planes.
3. In case both surfaces are black, i.e., $\varepsilon_1 = \varepsilon_2 = 1$, Eq. 10.56 reduces to $q_1 = \sigma(T_1^4 - T_2^4)$.

Example 10.8

A large thermos flask consists of two thin-walled coaxial cylinders with a narrow gap such that the radius ratio $\frac{R_1}{R_2} = 0.95$. The surfaces that face each other are silvered and have an emissivity of 0.05 each. The annulus is evacuated. The length of the flask is $L = 0.3$ m and the inner radius is $R_1 = 0.05$ m. Calculate the heat loss when the thermos flask is filled with hot water such that $T_1 = 373$K while the outer cylinder is at a temperature of $T_2 = 300$ K. Also calculate the instantaneous rate of change of the temperature of hot water, assuming it to be a lumped system.

Solution:

Step 1 The radius of the inner wall is specified as $R_1 = 0.05$ m. The radius ratio is 0.95 and hence the radius of the outer wall is $R_2 = \frac{R_1}{0.95} = \frac{0.05}{0.95} = 0.0526$ m. The annular gap is thus obtained as $s = R_2 - R_1 = 0.0526 - 0.05 = 0.0026$ m. Since the length of the flask is some 6 times the radius and 100 times the gap, it is admissible to ignore end effects in the calculation. Thus, the two surfaces may be assumed to form a two-dimensional duct-like enclosure. The area ratio may hence be assumed as $\frac{A_1}{A_2} = \frac{R_1}{R_2} = 0.95$. The given data is written down as $\varepsilon_1 = \varepsilon_2 = 0.05$; $T_1 = 373K$; $T_2 = 300$K.

Step 2 Expression 10.56 may now be used to calculate the heat loss.

$$q_1 = \frac{5.67 \times 10^{-8}(373^4 - 300^4)}{\frac{1}{0.05} + 0.95 \times \left(\frac{1}{0.05} - 1\right)} = 16.77 \text{ W/m}^2$$

The total heat loss is a product of the heat flux q_1 and the area A_1. The area is given by

$$A_1 = 2\pi R_1 L = 2 \times \pi \times 0.05 \times 0.3 = 0.0942 \text{ m}^2$$

Hence, the heat loss from surface 1 (and thus the heat loss from hot water within the flask) is

$$Q_1 = q_1 A_1 = 16.77 \times 0.0942 = 1.58 \text{ W}$$

Step 3 The initial rate of cooling of hot water (temperature =373 K or 100° C) contained inside the flask may be calculated now. The properties of water at 373 K are taken as
Density of water:$\rho = 958.4$ kg/m^3, Specific heat of water:$C_p = 4211$ J/kg K. Volume of water in the tank is calculated as

$$V = \pi R_1^2 L = \pi \times 0.05^2 \times 0.3 = 0.002356 \text{ m}^3$$

Hence, the mass of water in the flask is

$$M = \rho V = 958.4 \times 0.002356 = 2.258 \text{ kg}$$

Using the lumped system approximation, we then have the initial cooling rate given by

$$\left.\frac{dT}{dt}\right|_{t=0} = -\frac{Q_1}{MC_p} = -\frac{1.58}{2.258 \times 4211} = -1.662 \times 10^{-4}\,{}^{\circ}C/s \text{ or } = 0.6^{\circ}C/h$$

Cavity Radiator—Sources of Black Body Radiation:

The analysis of a two-surface enclosure, by the radiosity irradiation method, is useful in modeling a cavity radiator that is used as a source of black body radiation. A typical laboratory source of black body radiation is shown schematically in Fig. 10.17. The source consists of a cavity (in this case, conical) that is electrically heated. The cavity has an insert of a material like graphite. The cavity is encased in insulation as shown. A small aperture as shown provides an outlet for radiation. A thermocouple is used to measure the source temperature as well as for heater control to keep the temperature at the desired level. As long as $\frac{A_a}{A_s} << 1$, as will be shown by the analysis below, the radiation emanating from the aperture approximates closely the characteristics of black body radiation. It is easily seen that the cavity radiator is essentially a two-surface enclosure with the aperture acting as a convex surface (imagine that the surface is a tightly stretched diaphragm across the opening) that is transparent to radiation and is essentially a black surface. This is so since any radiation passing out through the aperture does not return and hence is effectively absorbed by it! We may identify the aperture A_a as surface 1 in our earlier discussion on the two-surface enclosure problem. Similarly, we may identify A_s, the cavity surface, as surface 2. If the surroundings are at a temperature much lower than the cavity wall temperature, we may assume that the radiosity of the aperture is effectively zero, i.e., $J_a = 0$. Hence, the radiosity of the cavity surface may be written down as

$$J_s = \varepsilon_s \sigma T_s^4 + (1 - \varepsilon_s)\left(1 - \frac{A_a}{A_s}\right) J_s$$

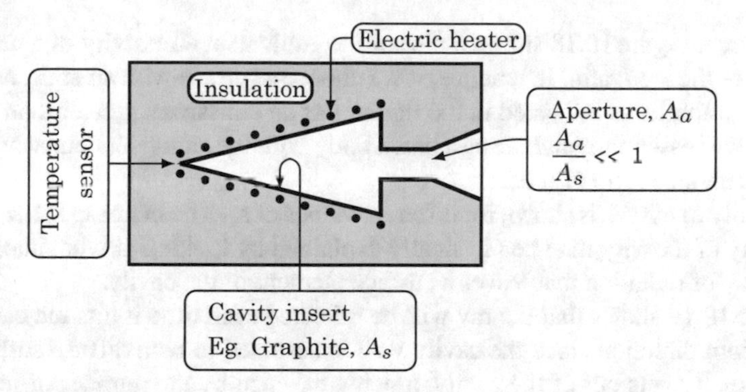

Fig. 10.17 Black body radiation cavity

where the self-angle factor of the cavity surface has been written in terms of the area ratio. We may solve the above for J_s to get

$$J_s = \frac{\varepsilon_s \sigma T_s^4}{1 - (1 - \varepsilon_s)\left(1 - \frac{A_a}{A_s}\right)} \tag{10.57}$$

It is easily seen that the irradiation for the aperture is $F_{as} J_s = J_s$. Since its radiosity is zero, the heat transfer (or loss) through the opening is given by

$$Q_a = q_a A_a = -A_a J_s = \boxed{-A_a \sigma T_s^4 \underbrace{\left[\frac{\varepsilon_s}{1 - (1 - \varepsilon_s)\left(1 - \frac{A_a}{A_s}\right)}\right]}_{\text{Equivalent emissivity of cavity}}} \tag{10.58}$$

The term within the large brackets on the right-hand side of Eq. 10.58 is referred to as ε_{eq}, the equivalent emissivity of the cavity. Two limiting cases that are relevant are

1. When the aperture area and the cavity surface area are equal, the equivalent emissivity is the same as the emissivity of the cavity surface.
2. When the aperture area is very small compared to the area of the cavity surface, the equivalent emissivity approaches unity, for any non-zero ε_s. Thus, the aperture appears as a black body!

It is clear that ε_s close to 1 and $\frac{A_a}{A_s} \to 0$ are conducive to getting ε_{eq} very close to 1. Thus, a deep cavity made of a material with high emissivity is desirable as a black body source. Figure 10.18 shows a plot of the equivalent emissivity of a cavity as a function of the area ratio. In practice, if we choose a surface with an emissivity close to unity (graphite, as indicated in the figure, has an emissivity greater than 0.9), the aperture will essentially radiate as a black body with the equivalent emissivity being 0.995 for an area ratio of 0.05.

A family of curves is shown for different values of ε_s. The increase in the effective emissivity of a cavity may be physically explained by looking at what happens to a typical ray of radiation that leaves a surface element of the cavity.

Figure 10.19 shows that the ray will be reflected each time it hits the cavity wall in a random direction since the cavity wall is assumed to be a diffuse surface. The ray eventually gets out of the cavity, much weakened by the number of reflections that take place. This argument may be justified by looking at what happens if the arrows indicating the direction of the rays are reversed. A ray entering the cavity will eventually be absorbed after a large number of reflections. Of course, there is a small chance that a part of it may go back through the aperture. The multiple reflections effectively make the cavity a black body!

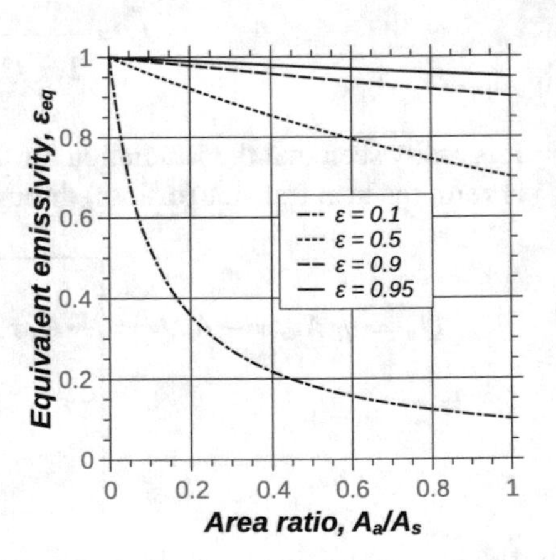

Fig. 10.18 Characteristics of a cavity radiator

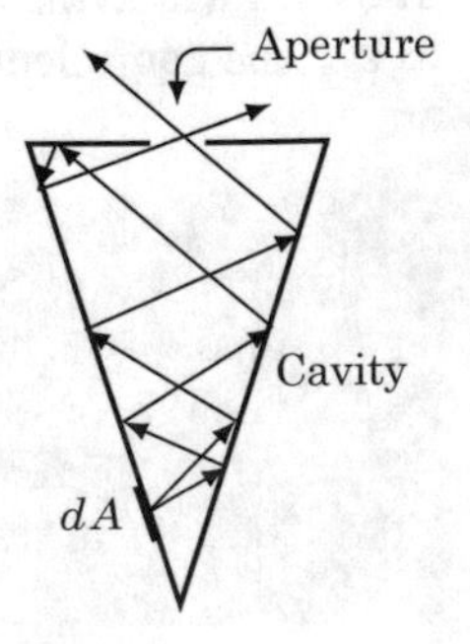

Fig. 10.19 Cavity effect due to multiple reflections

Rough surfaces may be visualized as having a large number of small cavities adjacent to the surface. These minute cavities increase the effective emissivity of the surface by the cavity effect and hence rough surfaces have higher emissivity as opposed to smooth surfaces.

Example 10.9

A typical example of an enclosure with gray and diffuse walls is shown in Fig. 10.20. The enclosure is a rectangular parallelepiped of length 1 m, width 0.5 m, and height 0.5 m. The top face identified as surface 1 is supplied with a heat input of 50 kW. The bottom face identified as surface 2 is maintained at a temperature of 500 K. The right face identified as surface 3 is maintained at a temperature of 600 K. The other three sides (the front, the left, and the back faces) are specified as reradiating surfaces and hence are clubbed together as a single surface 4 (it is concave). All surfaces are specified to have a common emissivity of 0.8. Determine the temperatures of those surfaces for which the heat fluxes are specified and the heat fluxes at the surfaces for which the temperatures are specified.

Solution:

The solution is in two parts. In the first geometric part, the required angle factors are evaluated. The second part uses the radiosity irradiation formulation to obtain surface radiosities. The third part uses these radiosities to get the required answers.

1. Geometric part:

All the surfaces are either rectangles or consist of rectangles and hence the areas are easily calculated and tabulated as shown in Table 10.2. There are 4 surfaces and hence the total number of angle factors is $4^2 = 16$. However, there are ${}^4C_2 = \frac{4!}{2!2!} = 6$ reciprocity relations. Four expressions due to the sum rule are available. Out of 4 surfaces, 3 are flat and hence self-angle factors for these are zero. That is, there are 3 angle factors, which are known. Because of symmetry $F_{13} = F_{23}$. Thus, the number of angle factors that have to be independently determined is $16 - 6 - 4 - 3 - 1 = 2$. F_{12} and F_{31} are identified as the two angle factors that need to be independently

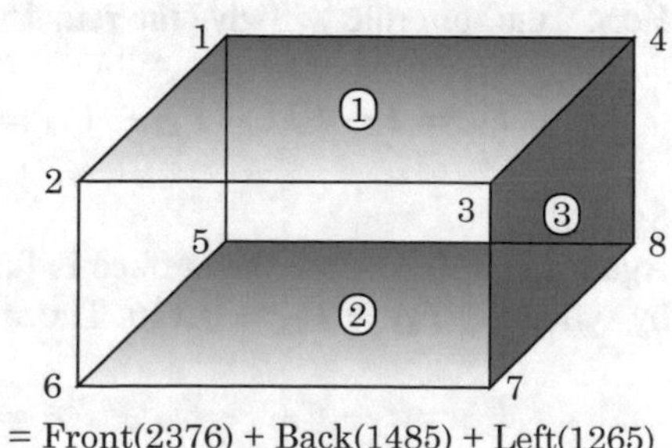

Fig. 10.20 An enclosure with four surfaces for Example 10.9

Table 10.2 Areas in Example 10.8

Surface	Area m^2
1	$1 \times 0.5 = 0.5$
2	$1 \times 0.5 = 0.5$
3	$0.5 \times 0.5 = 0.25$
4	$1 \times 0.5 + 1 \times 0.5 + 0.5 \times 0.5 = 1.25$

determined. Surfaces 1 and 2 are two parallel and equal rectangles. Using the notation of Fig. G.1, we have $a = 1$ m, $b = 0.5$ m, and $c = 0.5$ m. Hence, we have

$$X = \frac{a}{c} = \frac{1}{0.5} = 2; \; Y = \frac{b}{c} = \frac{0.5}{0.5} = 1$$

From Fig. G.2, we read off the required angle factor as $F_{12} = 0.28$. However, if we use the results of Table G.1, the value is $F_{12} = 0.28588 \approx 0.286$.

Surfaces 3 and 1 are two perpendicular rectangles sharing a common edge. Using the notation of Fig. G.1, we have $a = 1$ m, $b = 0.5$ m, and $c = 0.5$ m. Hence, we have

$$Z = \frac{a}{b} = \frac{1}{0.5} = 2; \; Y = \frac{c}{b} = \frac{0.5}{0.5} = 1 \text{ or } Y' = \frac{1}{Y} = 1$$

From Table G.4, we read off the required angle factor as $F_{31} = 0.23285 \approx 0.233$. All other angle factors are obtained by any one of the following:

1. Symmetry condition
2. Reciprocity relation
3. Sum rule

Surface 1:

Since it is flat, the self-angle factor $F_{11} = 0$. The angle factor F_{12} has already been obtained. By the use of the reciprocity rule, the angle factor between 1 and 3 is obtained as

$$F_{13} = \frac{A_3}{A_1} F_{31} = \frac{0.5 \times 0.5}{1 \times 0.5} \times 0.233 = 0.116$$

Hence, the sum rule will give the remaining view factor as

$$F_{14} = 1 - F11 - F_{12} - F_{13} = 1 - 0 - 0.286 - 0.116 = 0.598$$

Surface 2:

Again $F_{22} = 0$ because the surface is flat. By symmetry $F_{21} = F_{12} = 0.286$. Also, by symmetry $F_{23} = F_{13} = 0.116$. The sum rule then gives

$$F_{24} = 1 - F22 - F_{21} - F_{23} = 1 - 0 - 0.286 - 0.116 = 0.598$$

Surface 3:
$F_{33} = 0$ since it is flat. By symmetry $F_{32} = F_{31} = 0.233$. Lastly, by the sum rule we get

$$F_{34} = 1 - F33 - F_{32} - F_{31} = 1 - 0 - 0.233 - 0.233 = 0.534$$

Surface 4:
By reciprocity, we have

$$F_{41} = \frac{A_1}{A_4} F_{14} = \frac{0.5}{1.25} \times 0.598 = 0.239$$

By symmetry $F_{42} = F_{41} = 0.239$. By reciprocity

$$F_{43} = \frac{A_3}{A_4} F_{34} = \frac{0.25}{1.25} \times 0.534 = 0.107$$

Finally, by the sum rule we get

$$F_{44} = 1 - F41 - F_{42} - F_{43} = 1 - 0.239 - 0.239 - 0.107 = 0.415$$

The angle factor data is conveniently given in the form of a matrix. The two entries shown in bold face are the only values obtained independently, i.e., without the use of angle factor algebra.

$$\begin{bmatrix} F_{11} & F_{12} & F_{13} & F_{14} \\ F_{21} & F_{22} & F_{23} & F_{24} \\ F_{31} & F_{32} & F_{33} & F_{34} \\ F_{41} & F_{42} & F_{43} & F_{44} \end{bmatrix} = \begin{bmatrix} 0 & \mathbf{0.286} & 0.116 & 0.598 \\ 0.286 & 0 & 0.116 & 0.598 \\ \mathbf{0.233} & 0.233 & 0 & 0.534 \\ 0.239 & 0.239 & 0.107 & 0.415 \end{bmatrix}$$

2. Radiosity formulation and solution:
Surface 1: Heat flux specified
With $Q_1 = 50$ kW and $A_1 = 0.5$ m^2, the heat flux at surface 1 is

$$q_1 = \frac{Q_1}{A_1} = \frac{50000}{0.5} = 100000 \text{ W/m}^2$$

But $q_1 = J_1 - G_1$ and hence

$$\text{(a)} \quad q_1 = J_1 - G_1 = J_1 - 0.286 J_2 - 0.116 J_3 - 0.598 J_4 = 100000$$

Surface 2: Temperature specified
Temperature has been specified on this surface as $T_2 = 500$ K whose emissivity is $\varepsilon_2 = 0.8$. Hence, the emissive power of this surface is

$$E_2 = \varepsilon_2 \sigma T_2^4 = 0.8 \times 5.67 \times 10^{-8} \times 500^4 = 2835 \text{ W/m}^2$$

The irradiation on this surface is

$$G_2 = F_{21} J_1 + F_{23} J_3 + F_{24} J_4 = 0.286 J_1 + 0.116 J_3 + 0.598 J_4$$

The radiosity equation for surface 2 is then written down as

$$J_2 = E_2 + (1 - \varepsilon_2) G_2 = 2835 + (1 - 0.8)(0.286 J_1 + 0.116 J_3 + 0.598 J_4)$$

This simplifies to

$$\text{(b)} \quad -0.057 J_1 + J_2 - 0.023 J_3 - 0.120 J_4 = 2835$$

Surface 3: Temperature specified

Temperature has been specified on this surface as $T_3 = 600$ K whose emissivity is $\varepsilon_3 = 0.8$. Hence, the emissive power of this surface is

$$E_3 = \varepsilon_3 \sigma T_2^4 = 0.8 \times 5.67 \times 10^{-8} \times 600^4 = 5878.7 \text{ W/m}^2$$

The irradiation on this surface is

$$G_3 = F_{31} J_1 + F_{32} J_2 + F_{34} J_4 = 0.233 J_1 + 0.233 J_2 + 0.534 J_4$$

The radiosity equation for surface 3 is then written down as

$$J_3 = E_3 + (1 - \varepsilon_2) G_3 = 5878.7 + (1 - 0.8)(0.233 J_1 + 0.233 J_2 + 0.534 J_4)$$

This simplifies to

$$\text{(c)} \quad -0.047 J_1 - 0.047 J_2 + J_3 - 0.107 J_4 = 5878.7$$

Surface 4: Reradiating surface

Since this surface is reradiating, the heat flux $q_4 = 0$ and hence $J_4 = G_4$. This may be written as

$$\begin{aligned} J_4 = G_4 &= F_{41} J_1 + F_{42} J_2 + F_{43} J_3 + F_{44} J_4 \\ &= 0.239 J_1 + 0.239 J_2 + 0.107 J_3 + 0.415 J_4 \end{aligned}$$

This may be rearranged as

$$\text{(d)} \quad -0.239 J_1 - 0.239 J_2 - 0.107 J_3 + 0.585 J_4 = 0$$

The four equations (*a*) – (*d*) are solved simultaneously to obtain the four radiosities. One possible way of obtaining the solution is to eliminate J_4 and reduce the number of equations to three for the radiosities of the other three surfaces. The three equations may be solved by Kramer's rule. Another method of solution would be to use the Gauss method of iteration. This has been discussed earlier while dealing with numerical methods for solving heat conduction problems. The iteration would start with assumed radiosities for surfaces 1 to 3. For surfaces 2 and 3, one starts with $J_2 = E_2$ and $J_3 = E_3$. For surface 1 the starting value would be $J_1 = q_1$. Equation (d) would then be used to get the first guess for J_4. The iteration would stop when the radiosities have converged to the desired level. This procedure was indeed used to arrive at the following radiosity values:

$$J_1 = 153765.6;\ J_2 = 21169.4;\ J_3 = 22096 \text{ and } J_4 = 75526 \text{ all in W/m}^2$$

3. Required answers:

Now, we are in a position to evaluate the desired answers set out in the problem. Applying Eq. 10.50 to surface 1, the unknown temperature T_1 is obtained as

$$T_1 = \left[\frac{\frac{(1 - 0.8) \times 100000}{0.8} + 153765.6}{5.67 \times 10^{-8}} \right]^{\frac{1}{4}} = 1332.5 \text{ K}$$

We apply Eq. 10.51 to surface 4 to obtain the unknown temperature T_4 as

$$T_4 = \left[\frac{75526}{5.67 \times 10^{-8}} \right]^{\frac{1}{4}} = 1074.3 \text{ K}$$

Applying Eq. 10.49 to surface 2, we get the unknown heat flux q_2 as

$$q_2 = \frac{0.8}{1 - 0.8} \times (5.67 \times 10^{-8} \times 500^4 - 21169.4) = -70502.7 \text{ W/m}^2$$

Hence, the heat transfer from surface 2 is given by

$$Q_2 = q_2 A_2 = -70502.7 \times (1 \times 0.5) = -35251.4 \text{ W}$$

Similarly, we get q_3 as

$$q_3 = \frac{0.8}{1 - 0.8} \times (5.67 \times 10^{-8} \times 600^4 - 22096) = -58990.9 \text{ W/m}^2$$

Hence, the heat transfer from surface 3 is given by

$$Q_3 = q_3 A_3 = -58990.9 \times (0.5 \times 0.5) = -14747.7 \text{ W}$$

We may check the calculations by performing power balance for the enclosure. Thus

$$\sum_{i=1}^{4} q_i A_i = 50000 - 35251.4 - 14747.7 = 0.9 \text{ W}$$

Indeed the calculations have been performed with sufficient precision!

10.3.4 Electrical Analogy

The radiosity irradiation formulation may be approached alternately by the use of electrical analogy. Starting from Eq. 10.48, the heat flux for the ith surface in an N surface enclosure may be written as

$$\begin{aligned} q_i = J_i - G_i &= J_i - \sum_{j=1}^{N} F_{ij} J_j \\ &= J_i \times \underbrace{\sum_{j=1}^{N} F_{ij}}_{\text{equal to 1 by sum rule applied to } i} - \sum_{j=1}^{N} F_{ij} J_j = \sum_{j=1}^{N} F_{ij}(J_i - J_j) \end{aligned} \tag{10.59}$$

In writing the above, the sum rule has been applied to surface i. Equation 10.59 may be recast in the form analogous to Ohm's law as

$$Q_i = q_i A_i = \sum_{j=1}^{N} \frac{J_i - J_j}{\left(\dfrac{1}{A_i F_{ij}}\right)} = \sum_{j=1}^{N} Q_{ij} \tag{10.60}$$

With $J_i - J_j$ representing a potential difference, Q_{ij} representing current, we may identify $R_{ij} = \frac{1}{A_i F_{ij}}$ as a resistance that is usually referred to as the "space resistance".

Equation 10.49 may also be interpreted in analogy with Ohm's law by writing it as

$$Q_i = \frac{E_{bi} - J_i}{\left(\dfrac{1 - \varepsilon_i}{A_i \varepsilon_i}\right)} \tag{10.61}$$

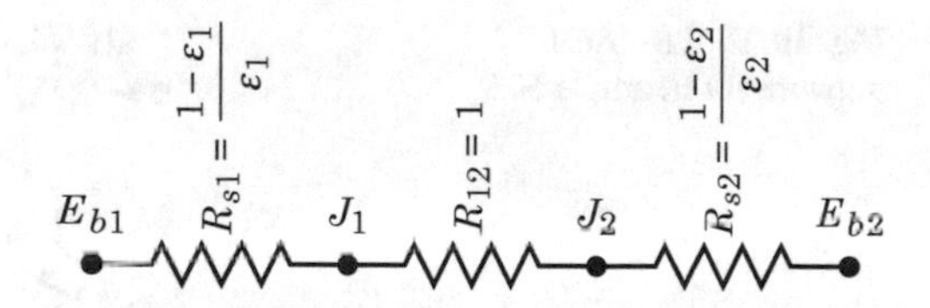

Fig. 10.21 Electrical network for a two-surface enclosure

Here, E_{bi} is the emissive power of a black body at the temperature of the surface T_i. The potential difference is identified as the difference between E_{bi} and J_i, the current with Q_i, and hence the resistance known as "surface resistance" is identified with $R_{si} = \frac{1-\varepsilon_i}{A_i \varepsilon_i}$. Note that the surface resistance vanishes for a black surface and the radiosity of such a surface is the same as black body emissive power at the specified surface temperature.

Application of Electrical Analogy to Two-Surface Enclosure

We shall consider the problem of two infinite parallel planes considered earlier by the method of detailed balancing. We consider all quantities on a per unit area basis, taking unit area for both surfaces, i.e., $A_1 = A_2 = 1$. An electrical network may be constructed as shown in Fig. 10.21. In this simple case of an enclosure with two surfaces, the appropriate resistances are written down as

$$\text{(a)}\ R_{s1} = \frac{1-\varepsilon_1}{\varepsilon_1} \quad \text{(b)}\ R_{12} = 1 \quad \text{(c)}\ R_{s2} = \frac{1-\varepsilon_2}{\varepsilon_2} \tag{10.62}$$

The three resistances are in series and hence the current (i.e., the heat flux) between the two nodes 1 and 2 is given by

$$\boxed{q_{1-2} = \frac{E_{b1} - E_{b2}}{R_{s1} + R_{12} + R_{s2}} = \frac{E_{b1} - E_{b2}}{\frac{1-\varepsilon_1}{\varepsilon_1} + 1 + \frac{1-\varepsilon_2}{\varepsilon_2}} = \frac{E_{b1} - E_{b2}}{\frac{1}{\varepsilon_1} + \frac{1}{\varepsilon_2} - 1}} \tag{10.63}$$

which agrees with the earlier result.

As is clear from this simple example, the electrical analogy is a good alternative in case the network is easily analyzed. In the case of a complex enclosure, say with more number of surfaces, the network may not be a simple alternative. In such a case, the radiosity irradiation formulation is made use of.

Electrical Analogy for a Four-Surface Enclosure

We consider the four-surface enclosure of Example 10.9. The electrical network for this example is given in Fig. 10.22.

The number of nodes where more than one electrical resistance comes together in the electrical network is equal to the number of surfaces as shown in the figure. These nodes have the respective radiosities as the potentials. Since there is no current in the surface resistance R_{s4}, the corresponding node 4 "floats" as shown. All the nodes are interconnected by surface resistances as indicated. Surfaces 2 and 3 have specified

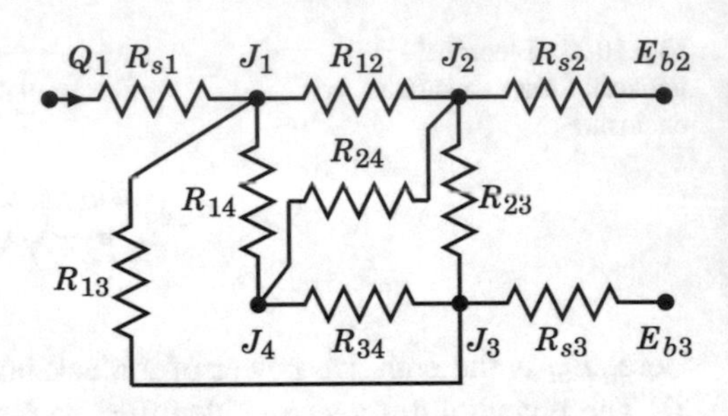

Fig. 10.22 Electrical network for Example 10.9

temperatures and hence the corresponding surface potentials are E_{b2} and E_{b3} as indicated. Surface 1 has specified heat flux and hence the current is as indicated. The potential at surface 1 is to be determined.

Equations for the unknown radiosities are obtained by applying Kirchhoff's Current Law to the nodes which states that the sum of all the currents entering a node is zero. For node 1, we have

$$Q_1 + \frac{J_2 - J_1}{R_{12}} + \frac{J_3 - J_1}{R_{13}} + \frac{J_4 - J_1}{R_{14}} = 0$$

This may be rearranged as

$$J_1\left(\frac{1}{R_{12}} + \frac{1}{R_{13}} + \frac{1}{R_{14}}\right) - \frac{J_2}{R_{12}} - \frac{J_3}{R_{13}} - \frac{J_4}{R_{14}} = Q_1 \qquad (10.64)$$

Similarly, for node 4 we have

$$\frac{J_4 - J_1}{R_{14}} + \frac{J_4 - J_2}{R_{24}} + \frac{J_3 - J_4}{R_{34}} = 0$$

This may be rearranged as

$$-\frac{J_1}{R_{14}} - \frac{J_2}{R_{24}} - \frac{J_3}{R_{34}} + J_4\left(\frac{1}{R_{14}} + \frac{1}{R_{24}} + \frac{1}{R_{34}}\right) = 0 \qquad (10.65)$$

For node 2, we have

$$\frac{E_{b2} - J_2}{R_{s2}} + \frac{J_1 - J_2}{R_{12}} + \frac{J_3 - J_2}{R_{23}} + \frac{J_4 - J_2}{R_{24}} = 0$$

This may be rearranged as

$$-\frac{J_1}{R_{12}} + J_2\left(\frac{1}{R_{s2}} + \frac{1}{R_{12}} + \frac{1}{R_{23}} + \frac{1}{R_{24}}\right) - \frac{J_3}{R_{23}} - \frac{J_4}{R_{24}} = \frac{E_{b2}}{R_{s2}} \qquad (10.66)$$

Similarly, we may write the equation for node 3 as

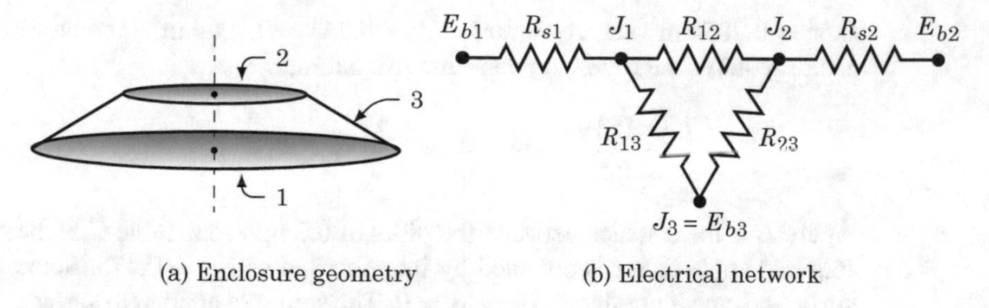

Fig. 10.23 Three enclosure geometry of Example 10.10

$$-\frac{J_1}{R_{13}} - \frac{J_2}{R_{23}} + J_3\left(\frac{1}{R_{s3}} + \frac{1}{R_{13}} + \frac{1}{R_{23}} + \frac{1}{R_{34}}\right) - \frac{J_4}{R_{34}} = \frac{E_{b3}}{R_{s3}} \tag{10.67}$$

Equations 10.64–10.67 have to be solved simultaneously to obtain the surface radiosities. The solution of these equations is similar to that in the radiosity irradiation formulation, and hence the procedure is more or less the same!

Example 10.10

A circular disk of diameter 1 m and at 600 K is kept directly opposite to a disk of diameter 0.5 m and 400 K. The distance between the two disks is 0.3 m. Both the disks have an emissivity of 0.6. The room in which the two disks are placed is very large and is at a temperature of 300 K. Determine the heat transfer from each surface. Use electrical analogy to solve the problem.

Solution:

Step 1 The nomenclature used in the problem is introduced through Fig. 10.23a. The electrical analog is shown in Fig. 10.23b. The given data is written down as

$$r_1 = \frac{1}{2} = 0.5 \text{ m};\ T_1 = 600 \text{ K};\ \varepsilon_1 = 0.6;\ r_2 = \frac{0.5}{2} = 0.25 \text{ m};\ T_2 = 400 \text{ K};\ \varepsilon_2 = 0.6$$

Node 3 may be visualized as a black surface at the temperature of the room, with zero surface resistance.

$$\varepsilon_3 = 1;\ T_3 = 300 \text{ K and } J_3 = E_{b3} = \sigma T_3^4 = 5.67 \times 10^{-8} \times 300^4 = 459.3 \text{ W/m}^2$$

Step 2 Calculation of geometric parameters:

Areas of surfaces 1 and 2 are, respectively, given by $A_1 = \pi r_1^2 = \pi \times$

$0.5^2 = 0.7854\ \text{m}^2$ and $A_2 = \pi r_2^2 = \pi \times 0.25^2 = 0.1964\ \text{m}^2$. The angle factor requires the following non-dimensional ratios:

$$\frac{L}{r_1} = \frac{0.3}{0.5} = 0.6; \quad \frac{r_2}{L} = \frac{0.25}{0.3} = 0.8333$$

where L is the distance between the disks of 0.3 m. From Table G.5, the required angle factor is obtained by interpolation as $F_{12} = 0.175$. Since surfaces 1 and 2 are flat, $F_{11} = F_{22} = 0$. The sum rule applied to surface 1 gives $F_{13} = 1 - F_{12} = 1 - 0.175 = 0.825$. Reciprocity is used to get

$$F_{21} = \frac{A_1}{A_2} F_{12} = \frac{0.7854}{0.1964} \times 0.175 = 0.6998 \approx 0.7$$

Applying the sum rule to surface 2, we get $F_{23} = 1 - F_{21} = 1 - 0.7 = 0.3$.

Step 3 The space resistances are calculated now.

$$R_{12} = \frac{1}{A_1 F_{12}} = \frac{1}{0.7854 \times 0.175} = 7.276\ \text{m}^{-2}$$
$$R_{13} = \frac{1}{A_1 F_{13}} = \frac{1}{0.7854 \times 0.825} = 1.543\ \text{m}^{-2}$$
$$R_{23} = \frac{1}{A_2 F_{23}} = \frac{1}{0.1964 \times 0.3} = 16.977\ \text{m}^{-2}$$

Surface resistances are calculated next.

$$R_{s1} = \frac{1 - \varepsilon_1}{A_1 \varepsilon_1} = \frac{1 - 0.6}{0.7854 \times 0.6} = 0.849\ \text{m}^{-2}$$
$$R_{s2} = \frac{1 - \varepsilon_2}{A_2 \varepsilon_2} = \frac{1 - 0.6}{0.0.1963 \times 0.6} = 3.395\ \text{m}^{-2}$$

Step 4 Radiation network and radiosity equations:
All the resistances are now available for defining the radiation network shown in Fig. 10.23b. The black body emissive powers at temperatures of surface 1 and 2 are

$$E_{b1} = \sigma T_1^4 = 5.67 \times 10^{-8} \times 600^4 = 7348.3\ \text{W/m}^2$$
$$E_{b2} = \sigma T_2^4 = 5.67 \times 10^{-8} \times 400^4 = 1451.5\ \text{W/m}^2$$

Equations in the radiosity formulation are obtained from the electrical network by the application of Kirchhoff's law to nodes 1 and 2. For node 1, we have

$$\frac{E_{b1} - J_1}{R_{s1}} + \frac{J_2 - J_1}{R_{12}} + \frac{J_3 - J_1}{R_{13}} = 0$$

This may be rearranged as

$$J_1\left(\frac{1}{R_{s1}}+\frac{1}{R_{12}}+\frac{1}{R_{13}}\right)-\frac{J_2}{R_{12}}=\frac{E_{b1}}{R_{s1}}+\frac{J_3}{R_{13}}$$

J_3, being known, has been taken to the right-hand side. Analogously, we may write the radiosity equation for surface 2 as

$$-\frac{J_1}{R_{12}}+J_2\left(\frac{1}{R_{s2}}+\frac{1}{R_{12}}+\frac{1}{R_{23}}\right)=\frac{E_{b2}}{R_{s2}}+\frac{J_3}{R_{23}}$$

Introducing the numerical values for the various quantities appearing in the above, we get the following two simultaneous equations for the radiosities of surfaces 1 and 2.

$$1.9635J_1 - 0.1374J_2 = 8954.6$$
$$-0.1374J_1 + 0.4909J_2 = 454.6$$

These equations are easily solved (use Kramer's rule) to get

$$J_1 = 4717.8\ \text{W/m}^2,\quad J_2 = 2247.0\ \text{W/m}^2$$

Step 5 Surface heat transfer rates:

The heat transferred from the three surfaces, respectively, are

$$Q_1 = \frac{E_{b1}-J_1}{R_{s1}} = \frac{7348.3-4717.8}{0.849} = 3099.0\ \text{W}$$

$$Q_2 = \frac{E_{b2}-J_2}{R_{s2}} = \frac{1451.5-2247.0}{3.395} = -234.3\ \text{W}$$

$$Q_3 = \frac{J_3-J_1}{R_{13}}+\frac{J_3-J_2}{R_{23}} = \frac{459.3-7348.3}{1.553}+\frac{459.3-2247.0}{16.977} = -2864.7\ \text{W}$$

Overall energy balance for the enclosure gives

$$\sum_{i=1}^{3} Q_i = 3099 - 234.3 - 2864.7 = 0\ \text{W}$$

The solution is indeed very satisfactory.

10.4 Enclosure Analysis Under Special Circumstances

Enclosure analysis that has been presented thus far has been based on several assumptions that may not always be valid. Gray assumption is alright as long as the surface temperatures are not too widely different. When, for example, solar radiation is involved, the gray assumption is inadequate and a non-gray analysis is warranted. The diffuse assumption has been made under the premise that the surfaces in the enclosure were engineering surfaces that have been obtained by normal machining and fabrication techniques where surface finish is what may be termed as "rough". However, some of the surfaces may have been subject to refined methods of fabrication where the surfaces may indeed turn out to be "smooth". In such a case, the surface may have to be treated as specular. The uniform radiosity assumption that has been made thus far may not be valid when surfaces receive radiation from surfaces that are at widely different temperatures and are located at widely differing distances.

This section is devoted to treating these exceptional cases.

10.4.1 Enclosure Containing Diffuse Non-gray Surfaces

In very simple cases with a broad-banded structure in emissivity, it is possible to use the band approximation introduced in Chap. 9 to analyze enclosure problems consisting of non-gray surfaces. If an enclosure contains surfaces with different emissivity patterns, it is necessary to use a set of common bands for the analysis. Let us take a simple case in which two surfaces are involved with different emissivity patterns as shown in Fig. 10.24.

It is clear from the figure that the common bands would be defined by the series of λ's indicated on the λ axis. The radiosity irradiation formulation will have to be made separately for each of the bands using the emissivity values shown in Table 10.3. We may generalize the above to any number of non-gray surfaces in an enclosure; say m. Plotting the band emissivities as in Fig. 10.24 and looking for all λ's that are needed

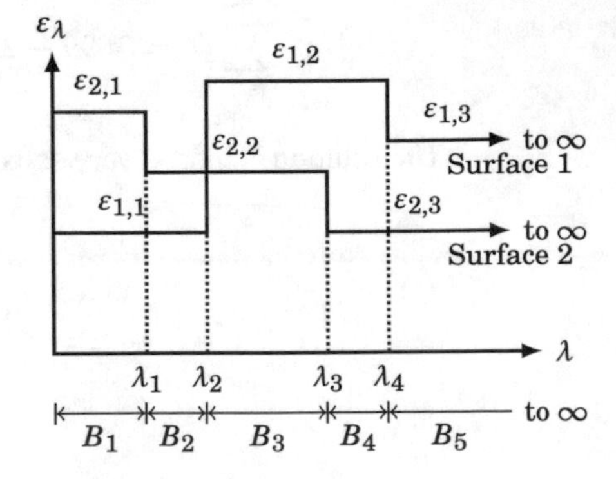

Fig. 10.24 Two non-gray surfaces showing common bands. Note that B stands for "Band"

Table 10.3 Common bands and corresponding emissivities for enclosure with two non-gray surfaces

Band No.	Wavelength Range	Emissivity value	
		Surface 1	Surface 2
B_1	$0 - \lambda_1$	$\varepsilon_{1,1}$	$\varepsilon_{2,1}$
B_2	$\lambda_1 - \lambda_2$	$\varepsilon_{1,1}$	$\varepsilon_{2,2}$
B_3	$\lambda_2 - \lambda_3$	$\varepsilon_{1,2}$	$\varepsilon_{2,2}$
B_4	$\lambda_3 - \lambda_4$	$\varepsilon_{1,2}$	$\varepsilon_{2,3}$
B_5	$\lambda_4 - \infty$	$\varepsilon_{1,3}$	$\varepsilon_{2,3}$

to describe the emissivities of all the surfaces, we identify a set of n common bands. Within each band, the surfaces behave as gray surfaces. If we replace the total black body emissive power by the fraction of the black body power in the ith band, the radiosity for the jth surface in the ith band may be defined as

$$\boxed{J_{j,i} = \varepsilon_{j,i} f_{(\lambda_i - \lambda_{i-1})T_j} \sigma T_j^4 + (1 - \varepsilon_{j,i}) \sum_k F_{i,k} J_{k,i}} \tag{10.68}$$

In the example shown in Fig. 10.24, the radiosities of surfaces 1 and 2 in the 2nd band are given by

$$\begin{aligned} J_{1,2} &= \varepsilon_{1,1} f_{(\lambda_3 - \lambda_2)T_1} \sigma T_1^4 + (1 - \varepsilon_{1,1}) \left[F_{11} J_{1,2} + F_{12} J_{2,2} \right] \\ J_{2,2} &= \varepsilon_{2,2} f_{(\lambda_3 - \lambda_2)T_2} \sigma T_2^4 + (1 - \varepsilon_{2,2}) \left[F_{21} J_{1,2} + F_{22} J_{2,2} \right] \end{aligned} \tag{10.69}$$

It is immediately clear that a set of radiosity equations may be written down for all the surfaces and for each of the bands. The set of equations—for a given band and all the surfaces in the enclosure—represents a gray enclosure problem. Thus the original non-gray problem reduces to n gray problems. These may be solved for the n band radiosities for each surface. These band radiosities are used to determine the heat fluxes at each of the surfaces. It may easily be shown that the heat flux for the jth surface is

$$\boxed{q_j = \sum_{i=1}^{n} \frac{\varepsilon_{j,i}}{(1 - \varepsilon_{j,i})} \left[f_{(\lambda_i - \lambda_{i-1})T_j} \sigma T_j^4 - J_{j,i} \right]} \tag{10.70}$$

Thus we see that, in principle, the non-gray enclosure problem is analyzed using the same method as that for the gray enclosure problem, but with more effort! The electrical analogy may also be invoked for each of the common bands and the analysis performed by making use of the electrical network. A simple two-surface enclosure problem is solved below.

Example 10.11

Two very large parallel planar surfaces are maintained at 400 K and 800 K, respectively. The surfaces are non-gray and have the emissivity variation given below:

Surface 1: $\varepsilon_{1,1} = 0.6$ for $0 < \lambda < 3$ μm; $\varepsilon_{1,2} = 0.2$ for $\lambda > 3$ μm

Surface 2: $\varepsilon_{2,1} = 0.2$ for $0 < \lambda < 3$ μm; $\varepsilon_{2,2} = 0.6$ for $\lambda > 3$ μm

Determine the heat transfer between the two planes per unit area using a non-gray analysis. Compare this with a gray analysis based on the equivalent gray emissivities of the two surfaces at their respective temperatures.

Solution:

Step 1 There are two common bands in this problem. The first band spans the wavelength region 0 to 3 μm and the second band spans the wavelength region from 3 μm to ∞. Let $f_{j,i}$ represent the fraction of black body power for the jth surface and the ith band.

Step 2 Surface 1:
With $T_1 = 800$ K and $\lambda_1 = 3$ μm, we have $\lambda_1 T_1 = 2400$ μm K. From Table 8.1, we then have $f_{0-2400} = 0.150256$ and hence $f_{1,1} = 0.150256$. With $\lambda_2 = \infty$ we then have $f_{1,2} = 1 - f_{1,1} = 1 - 0.150256 = 0.859744$.

Surface 2:
With $T_2 = 400$ K and $\lambda_1 = 3$ μm, we have $\lambda_1 T_2 = 1200$ μm K. From Table 8.1, we then have $f_{0-1200} = 0.002134$ and hence $f_{2,1} = 0.002134$. With $\lambda_2 = \infty$, we then have $f_{2,2} = 1 - f_{2,1} = 1 - 0.002134 = 0.997866$.

Step 3 Heat transfer from surface 1 to surface 2 per unit area:
It is calculated using expression 10.42 for each band. For band 1, we have

$$q_1 = \frac{\sigma(f_{1,1}T_1^4 - f_{2,1}T_2^4)}{\dfrac{1}{\varepsilon_{1,1}} + \dfrac{1}{\varepsilon_{2,1}} - 1}$$

$$= \frac{5.67 \times 10^{-8}(0.150256 \times 800^4 - 0.002134 \times 400^4)}{\dfrac{1}{0.6} + \dfrac{1}{0.2} - 1} = 581.7 \text{ W/m}^2$$

Similarly, for band 2 we have

$$q_2 = \frac{\sigma(f_{1,2}T_1^4 - f_{2,2}T_2^4)}{\dfrac{1}{\varepsilon_{1,2}} + \dfrac{1}{\varepsilon_{2,2}} - 1}$$

$$= \frac{5.67 \times 10^{-8}(0.859744 \times 800^4 - 0.997866 \times 400^4)}{\dfrac{1}{0.2} + \dfrac{1}{0.6} - 1} = 3260.6 \text{ W/m}^2$$

Hence, the net heat transfer from surface 1 to surface 2 is

$$q = 581.7 + 3260.6 = 3842.3 \text{ W/m}^2$$

Heat transfer under the gray model:

The gray model calculation is based on the equivalent gray body emissivities of the two surfaces at the respective surface temperatures.

$$\varepsilon_1 = f_{1,1}\varepsilon_{1,1} + f_{1,2}\varepsilon_{1,2} = 0.150256 \times 0.6 + 0.002134 \times 0.2 = 0.257$$
$$\varepsilon_2 = f_{2,1}\varepsilon_{2,1} + f_{2,2}\varepsilon_{2,2} = 0.859744 \times 0.2 + 0.997866 \times 0.6 = 0.599$$

The gray calculation thus yields

$$q = \frac{\sigma(T_1^4 - T_2^4)}{\frac{1}{\varepsilon_1} + \frac{1}{\varepsilon_2} - 1} = \frac{5.67 \times 10^{-8}(800^4 - 400^4)}{\frac{1}{0.257} + \frac{1}{0.599} - 1} = 4771.8 \text{ W/m}^2$$

The example shows that a gray analysis is grossly in error.

10.4.2 Gray Enclosures Containing Diffuse and Specular Surfaces

When an enclosure contains a combination of diffuse and specular surfaces, the analysis is complicated by the nature of reflection at a specular surface. In order to simplify the analysis, we assume that the specular reflection is angle independent. This is strictly not valid, as we have seen from the results of the electromagnetic theory. However, the analysis becomes extremely complex if this assumption is *not made*. Also, the interest here is to discuss the method to be used when specular surfaces are involved and this purpose is satisfied even with this assumption. The method is developed with a two-dimensional enclosure since view factor calculations are simple in this case. However, the method is applicable without any change to the general case of a three-dimensional enclosure.

Consider a two-dimensional enclosure (the length of the surfaces in a direction perpendicular to the plane of the figure is infinite) consisting of a diffuse surface (1), a specular surface (2), and an opening (3) as shown in Fig. 10.25. The ambient temperature is assumed to be very small compared to the temperature of surface (1) or (2), and hence the opening is essentially a black surface at 0 K. The diffuse surface (1) is maintained at a temperature of T_1 and has an emissivity of ε_1. We assume that the specularly reflecting surface (2) is a reradiating surface with $q_2 = 0$ and a specular reflectivity of ρ_2.

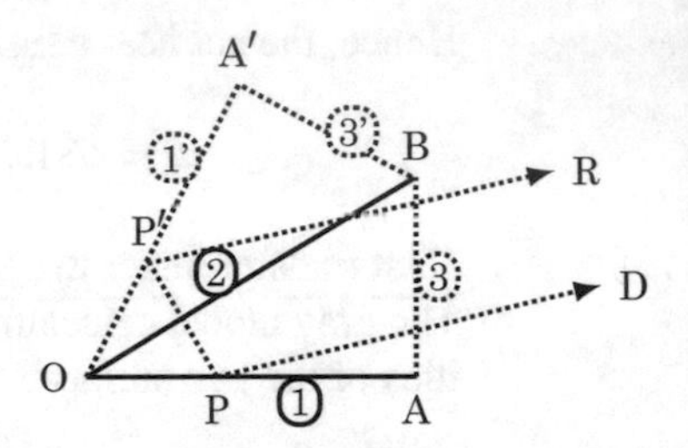

Fig. 10.25 Three-surface enclosure with one specular surface

Consider a small area element located on surface (1) at P as shown in Fig. 10.25. The element radiates diffusely, and we can see that a ray such as PD goes out of the opening directly. A ray such as $PP'R$ also goes out of the opening, but after a reflection at the specular surface. This ray appears to come from an image point P' as indicated in the figure. Thus, the analysis has to take into account the direct as well as reflected radiation in writing the appropriate radiosity equations.

Consider energy leaving (1) and incident on (3). The fraction that is directly incident is given by the diffuse view factor F_{13}. As explained above, the reflected radiation appears to come from the image (1'), and hence the fraction of the reflected radiation that is incident on (3) is given by $F_{1'3}$. It is clear from the figure that this is also equal to $F_{13'}$. We may thus write for the fraction of energy that leaves surface (1) per unit area and incident on (3) as $J_1[F_{13} + \rho_2 F_{13'}]$. Similarly, we see that the fraction of the radiation leaving (1) per unit area and incident on itself is $J_1[\rho_2 F_{11'}]$ because the surface is able to see its image. The bracketed terms in these expressions are referred to as exchange factors[6] and represented as

$$E_{11} = \rho_2 F_{11'};\ \ E_{13} = F_{13} + \rho_2 F_{13'} \tag{10.71}$$

In the present case, there is only one specular reflection since only one surface in the enclosure is specularly reflecting. In general, when several surfaces in an enclosure reflect specularly, many images are formed and the exchange factor should include the contribution of each one of the reflections. Ray tracing or the detailed balancing method is required in taking care of all the specular reflections. The radiosity irradiation formulation is similar to that given earlier with the view factors replaced by the exchange factors. A simple numerical example is presented below to demonstrate the procedure.

Example 10.12

Consider the enclosure shown in Fig. 10.25. Use the following data and calculate the heat loss from surface (1) per unit length. What is the temperature of surface (2)? Compare the above with an enclosure in which the surface (2) is also a diffuse surface.

[6]Not to be confused with exchange area which is also represented by the same symbol.

(a) Surface 1: Diffuse surface with $T_1 = 600$ K, $\varepsilon_1 = 0.6$ and $OA = 0.4$ m; (b) Surface 2: Reradiating, specularly reflecting surface with $\rho_2 = 0.75,\ OB = 0.5$ m; (c) Surface 3: Opening effectively at $T_3 = 0$ K and $\varepsilon_3 = 1$

Solution:

In the first part of the solution, we use the exchange factor concept introduced in the text. The opening is taken as a black surface with zero radiosity, i.e., $J_3 = 0$. The view factors needed are all calculated using the triangle rule or the crossed string method and view factor algebra. OAB is a right-angled triangle and hence $AB = \sqrt{OB^2 - OA^2} = \sqrt{0.5^2 - 0.4^2} = 0.3$ m

Step 1 Calculation of view factors :

Using the triangle rule, we get

$$F_{12} = \frac{OA + OB - AB}{2OA} = \frac{0.4 + 0.5 - 0.3}{2 \times 0.4} = 0.75$$

$$F_{13} = \frac{OA + AB - OB}{2OA} = \frac{0.4 + 0.3 - 0.5}{2 \times 0.4} = 0.25$$

$$F_{23} = \frac{OB + AB - OA}{2OB} = \frac{0.5 + 0.3 - 0.4}{2 \times 0.5} = 0.4$$

We now use reciprocity relations to get

$$F_{21} = \frac{OA}{OB} F_{12} = \frac{0.4}{0.5} \times 0.75 = 0.6$$

$$F_{31} = \frac{OA}{AB} F_{13} = \frac{0.4}{0.3} \times 0.25 = 0.333$$

$$F_{32} = \frac{OB}{AB} F_{23} = \frac{0.5}{0.3} \times 0.4 = 0.667$$

Also, the self-view factors for all the surfaces are zero.

The crossed string method is used for determining the angle factor $F_{13'}$. The crossed string method requires the calculation of the diagonal AA'. The angle AOB is obtained as

$$\theta = \cos^{-1}\left(\frac{OA}{OB}\right) = \cos^{-1}\left(\frac{0.4}{0.5}\right) = 0.6435 \text{ rad}$$

The diagonal AA' is then given by (note that line AA' cuts OB at right angle)

$$AA' = 2OA \sin\theta = 2 \times 0.4 \times \sin(0.7435) = 0.48 \text{ m}$$

Using the crossed string method, we then have

$$F_{13'} = \frac{OB + AA' - OA - AB}{2OA} = \frac{0.5 + 0.48 - 0.4 - 0.3}{2 \times 0.4} = 0.35$$

Using the triangle rule, we also have

$$F_{11'} = \frac{OA + OA' - AA'}{2OA} = \frac{0.4 + 0.4 - 0.48}{2 \times 0.4} = 0.4$$

Step 2 Calculation of Exchange factors :
The required exchange factors are then given by

$$E_{11} = \rho_2 F_{11'} = 0.75 \times 0.4 = 0.3$$
$$E_{13} = F_{13} + \rho_2 F_{13'} = 0.25 + 0.75 \times 0.35 = 0.5125$$

Step 3 Radiosity irradiation formulation: The radiosity equations are written down now.

$$J_1 = \varepsilon_1 \sigma T_1^4 + (1 - \varepsilon_1)(E_{11} J_1 + F_{12} J_2)$$
$$J_2 = F_{21} J_1$$

Substituting the latter in the former and simplifying, we get

$$J_1 = \frac{\varepsilon_1 \sigma T_1^4}{(1 - \varepsilon_1)(E_{11} + F_{12} F_{21})}$$

Step 4 Results for specular case:
Using the data, we then have

$$J_1 = \frac{0.6 \times 5.67 \times 10^{-8} \times 600^4}{1 - (1 - 0.6)(0.3 + 0.75 \times 0.6)} = 6298.6 \text{ W/m}^2$$

The heat loss from surface (1) is then given by

$$Q_1 = OA \cdot \frac{\varepsilon_1}{1 - \varepsilon_1}(E_{b1} - J_1)$$
$$= 0.4 \times \frac{0.6}{1 - 0.6} \times (5.67 \times 10^{-8} \times 600^4 - 6298.6) = 629.8 \text{ W/m}$$

The radiosity of surface (2) is

$$J_2 = F_{21} J_1 = 0.6 \times 6298.6 = 3779.2 \text{ W/m}^2$$

The temperature of surface (2) is then given by

$$T_2 = \left(\frac{J_2}{\sigma}\right)^{\frac{1}{4}} = \left(\frac{3779.2}{5.67 \times 10^{-8}}\right)^{\frac{1}{4}} = 508.1 \text{ K}$$

Step 5 Results for diffuse case:

In this case, the exchange factor $E_{11} = 0$ and hence the results are

$$J_1 = \frac{0.6 \times 5.67 \times 10^{-8} \times 600^4}{1 - (1 - 0.6)(0.75 \times 0.6)} = 5376.8 \text{ W/m}^2$$

Hence, the heat loss from surface (1) is given by

$$Q_1 = OA \cdot \frac{\varepsilon_1}{1 - \varepsilon_1}(E_{b1} - J_1) \quad = 0.4 \times \frac{0.6}{1 - 0.6} \times (5.67 \times 10^{-8} \times 600^4 - 5376.8) = 1182.9 \text{ W/m}$$

$$J_2 = F_{21} J_1 = 0.6 \times 5376.8 = 3226.1 \text{ W/m}^2$$

The temperature of surface (2) is then given by

$$T_2 = \left(\frac{J_2}{\sigma}\right)^{\frac{1}{4}} = \left(\frac{3226.1}{5.67 \times 10^{-8}}\right)^{\frac{1}{4}} = 488.4 \text{ K}$$

10.4.3 *Enclosure Analysis with Surfaces of Non-uniform Radiosity*

In all the cases that were considered thus far, an enclosure is assumed to contain surfaces that are isothermal and of uniform radiosity over the extent of each surface. This assumption cannot be justified when the irradiation of a surface is from different surfaces that are unevenly distributed spatially and have radiosities that vary widely. In such a case, one has to subdivide each surface such that the uniform radiosity assumption may be justified over each of the subdivisions. Of course, in the limit the subdivisions are infinitesimally small, the radiosity is represented by a continuous function and the summations in the radiosity equations are replaced by integrals. We shall consider a simple example of a two-dimensional cavity first and later discuss the general case.

Consider the two-dimensional cavity shown in Fig. 10.26. The analysis will make use of two surfaces labeled as 1 and 2 in the simplest possible analysis, as indicated in Fig. 10.26(a). Surface 1, the surface of the cavity, is at a uniform temperature of T_c and is assumed to have a uniform radiosity J_1. The opening is also assumed to be described by uniform radiosity given by the emissive power of a black body at the background temperature T_b.

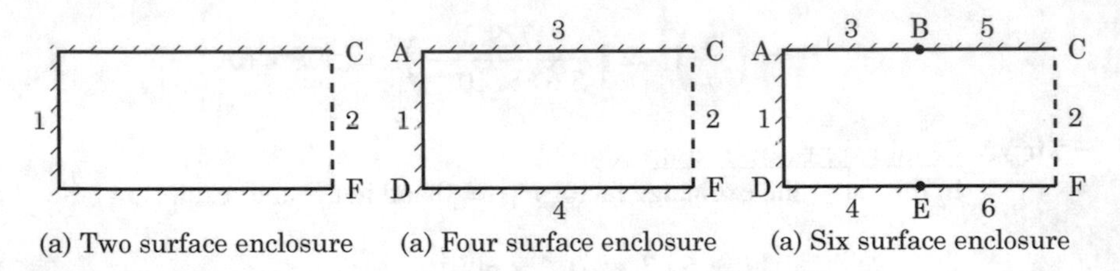

Fig. 10.26 Three different ways of analyzing a two-dimensional enclosure

A better analysis would be the one in which the bottom surface of cavity 1 is assumed to have different radiosity as compared to the radiosities of the two sides labeled as 3 and 4 as indicated in Fig. 10.26(b). Of course, because of symmetry we have $J_3 = J_4$. The enclosure analysis thus considers four surfaces in the analysis.

The analysis may further be improved by assuming the sides of the cavity to be divided into parts as shown in Fig. 10.26c. We note that there is essentially only one more radiosity than in the previous case because $J_3 = J_4$ and $J_5 = J_6$.

If it is desired that the bottom also is a surface with non-uniform radiosity, it must be divided in to at least three parts—say three equal parts. Dividing in to two equal parts will not be enough since both the halves will have the same radiosity.

Since the cavity is two-dimensional, all the diffuse angle factors may be calculated using the triangle formula or the crossed string method.

Example 10.13

Consider a two-dimensional cavity as shown in Fig. 10.27. The cavity surface is maintained at a uniform temperature of T_c while the opening is looking at a background at T_b. The cavity surface may be assumed to be gray with an emissivity of $\varepsilon_c = 0.6$. Calculate the heat transfer per unit length of cavity to the background. Model the enclosure in both the variants shown in the figure.

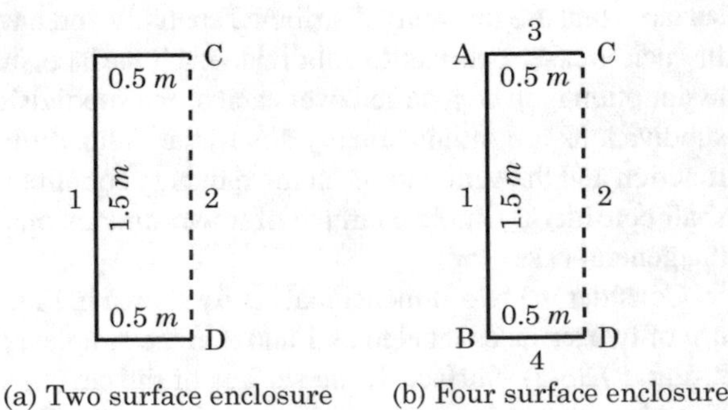

Fig. 10.27 Two different ways of analyzing a two-dimensional enclosure

Table 10.4 Angle factors for configurations in Fig. 10.27

Surface	Case (a)			
Number	1	2		
1	0.4	0.6		
2	1	0		
Surface	Case (b)			
Number	1	2	3	4
1	0	0.72076	0.13962	0.13962
2	0.72076	0	0.13962	0.13962
3	0.41886	0.41886	0	0.16228
4	0.41886	0.41886	0.16228	0

Note The reader may verify the entries

Solution:

The analysis is presented for the two cases below. The results will be compared so that the need or otherwise of the non-uniform radiosity analysis becomes clear.

Case (a):

The analysis is very simple in this case. The radiosity of surface 1 is written down as

$$J_1 = \varepsilon_c E_b(T_c) + (1 - \varepsilon_c)(F_{11} J_1 + F_{12} J_2)$$
$$= \varepsilon_c E_b(T_c) + (1 - \varepsilon_c)(F_{11} J_1 + [1 - F_{11}] J_2)$$

where E_b represents the emissive power of a black body. This may be rearranged in the form

$$J_1 - J_2 = \varepsilon_c [E_b(T_c) - J_2] + (1 - \varepsilon_c) F_{11} [J_1 - J_2]$$

Representing the radiosity difference $J_1 - J_2$ as $J_{d,1}$, noting that $J_2 = E_b(T_b)$ we have

$$J_{d,1} = \varepsilon_c [E_b(T_c) - E_b(T_b)] + (1 - \varepsilon_c) F_{11} J_{d,1}$$

The angle factors shown in Table 10.4 will be made use of to get

$$J_{d,1} = 0.6[E_b(T_c) - E_b(T_b)] + 0.16 J_{d,1} \quad \text{or} \quad J_{d,1} = 0.714[E_b(T_c) - E_b(T_b)]$$

Heat loss from the cavity is easily seen to be given by

$$q_2 = J_2 - G_2 = J_2 - J_1 = -J_{d,1} = -0.714[E_b(T_c) - E_b(T_b)]$$

This agrees with the value predicted by Eq. 10.58 when we note that A_s there corresponds to A_1 and A_a there corresponds to A_2. The black body emissive power there is to be replaced by the difference of black body emissive powers in the present case.

Case (b): Because of symmetry, $J_3 = J_4$ and hence only three radiosity equations are to be written down. We may again define radiosity differences as $J_{d,i} = J_i - J_2$, where $i = 1, 3$, or 4. Making use of the sum rule for each surface, the radiosity equations may be written down as

$$J_{d,1} = \varepsilon_c[E_b(T_c) - E_b(T_b)] + (1 - \varepsilon_c) \cdot 2F_{13}J_{d,3}$$
$$J_{d,3} = \varepsilon_c[E_b(T_c) - E_b(T_b)] + (1 - \varepsilon_c)(F_{31}J_{d,1} + F_{34}J_{d,3})$$

Introducing the numerical values for the angle factors and the surface emissivity, the above equations become

$$J_{d,1} - 0.1117J_{d,3} = 0.6[E_b(T_c) - E_b(T_b)]$$
$$-0.1675J_{d,1} + 0.9351J_{d,3} = 0.6[E_b(T_c) - E_b(T_b)]$$

The two equations may be solved to get

$$J_{d,1} = 0.6854[E_b(T_c) - E_b(T_b)]; \quad J_{d,3} = 0.7645[E_b(T_c) - E_b(T_b)]$$

The radiosities of surfaces 1 and 3 are certainly not the same and hence clubbing these, as in Case (a), is not correct.

The heat loss from the cavity may be calculated as

$$q_2 = J_2 - G_2 = -(F_{21}J_{d,1} + 2F_{23}J_{d,3}) = -0.708[E_b(T_c) - E_b(T_b)]$$

Since the results of Case (a) and Case (b) are very close to each other, there is no need to extend the analysis any further. Since the heat loss from the cavity is our concern, the simple two-surface enclosure analysis may itself be adequate in the present case.

Wedge-Type Cavity with Surfaces of Non-uniform Radiosity

An interesting application of the above ideas is to the case of a two-dimensional wedge cavity with specified temperature variations along its surfaces. Figure 10.28 shows the geometry. The wedge cavity consists of two surfaces 2 and 3 forming a wedge with included angle θ as shown. The opening may be treated as a black surface at the background temperature T_b. Distances are measured as x along surface 3 while they are measured by y along surface 2. The surfaces of the wedge are divided in to small elements as shown. For an element i situated on surface 3, the radiosity difference defined as $J_{d,i} = J_i - E_b(T_b)$ is written down as

$$J_{d,i} = \varepsilon_3[E_b(T_i) - E_b(T_b)] + (1 - \varepsilon_3) \sum_{j=1}^{m} F_{ij}J_{d,j} \tag{10.72}$$

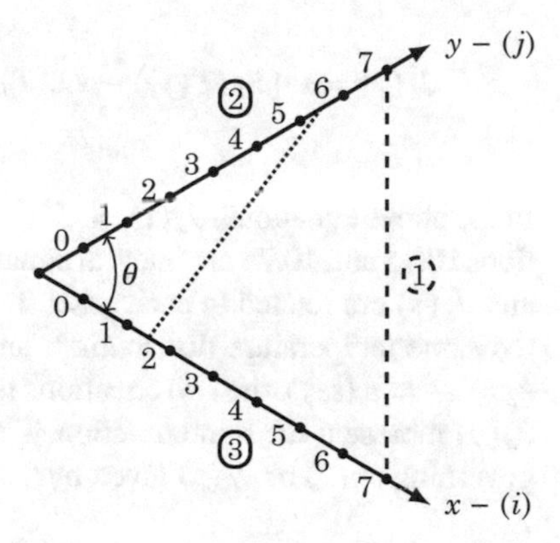

Fig. 10.28 Wedge cavity nomenclature

where T_i is the mean temperature of element i, and F_{ij} is the diffuse angle factor between element i on surface 3 and element j on surface 2. Equations such as 10.72 may be written down for all the elements $1 \le i \le n$. Similarly, we may write down the radiosity of element j on surface 2 as

$$J_{d,j} = \varepsilon_2[E_b(T_j) - E_b(T_b)] + (1 - \varepsilon_2)\sum_{i=1}^{n} F_{ji} J_{d,i} \tag{10.73}$$

All the symbols used may be interpreted analogously with Eq. 10.72. Equations such as 10.72 may be written down for all the elements $1 \le j \le m$. All the angle factors may be obtained by the application of either the triangle rule or the crossed string method. $m + n$ unknown radiosities are obtained by the simultaneous solution of all the equations given above. The heat loss from the cavity is seen to be given by

$$q_1 = J_1 - G_1 = -\left[\sum_{i=1}^{n} F_{1i} J_{d,i} + \sum_{j=1}^{m} F_{1j} J_{d,j}\right] \tag{10.74}$$

Now consider what would happen if the size of the elements is shrunk such that the size of each element is given by the differential amount dx or dy. We shall assume that each surface of the wedge is L long. In this case, the elemental angle factors become differential angle factors given by dF_{dx-dy}, dF_{dy-dx}. The summations are replaced by integrals and Eqs. 10.72 and 10.73 are replaced by the following:

$$J_d(x) = \varepsilon_3[E_b(T(x)) - E_b(T_b)] + (1 - \varepsilon_3)\int_{x=0}^{L} J_d(y) dF_{dx-dy} \tag{10.75}$$

$$J_d(y) = \varepsilon_2[E_b(T(y)) - E_b(T_b)] + (1 - \varepsilon_2) \int_{y=0}^{L} J_d(x) dF_{dy-dx} \tag{10.76}$$

In the above equations, $J_d(x) = J(x) - E_b(T_b)$ and $J_d(y) = J(y) - E_b(T_b)$. Equations 10.75 and 10.76 are integral equations that are coupled to each other since $J_d(x)$ and $J_d(y)$ are related to each other. In the special case where the two surfaces have the same temperature distribution, and the two surfaces have the same emissivity $\varepsilon_2 = \varepsilon_3 = \varepsilon$ (say), the two equations reduce to one integral equation since $J_d(x)$ and $J_d(y)$ represent the same function. Thus, we have, in this special case, the equation governing $J_d(x)$ or $J_d(y)$ given by

$$J_d(x) = \varepsilon[E_b(T(x)) - E_b(T_b)] + (1 - \varepsilon) \int_{y=0}^{L} J_d(y) dF_{dx-dy} \tag{10.77}$$

The formulation of the integral equation will be complete if the required elemental angle factor is obtained. The required elemental angle factor is available from the angle factor catalog compiled by Howell and is given by

$$dF_{dx-dy} = \frac{xy \sin^2\theta}{(x^2 + y^2 - 2xy\cos\theta)^{\frac{3}{2}}} dy \tag{10.78}$$

Introducing Eq. 10.78 in Eq. 10.77, the integral equation is finally of the form

$$J_d(x) = \varepsilon[E_b(T(x)) - E_b(T_b)] + (1 - \varepsilon) \int_{y=0}^{L} \frac{J_d(y) xy \sin^2\theta}{(x^2 + y^2 - 2xy\cos\theta)^{\frac{3}{2}}} dy \tag{10.79}$$

The above equation is of the general form

$$f(x) = u(x) + C \int_{y=a}^{b} K(x, y) f(y) dy \tag{10.80}$$

Here, $u(x)$ is a specified function of x, $K(x, y)$ is a function of x and y, and is referred to as the Kernel, C is a constant, a and b are specified limits, and the integral equation is known as the Fredholm integral equation of second kind.[7] The reference cited also deals with several methods of solving such equations. Here, we shall discuss only one method, an iterative method for the solution of the integral equation.

---

[7]For more details, see F.B. Hildebrand, *Methods of Applied Mathematics*, Prentice Hall, 1965.

Iterative Method of Solution

Let us assume that $f(x) = f^{(0)}(x)$ is a first guess for the solution. Introduce this in the right-hand side of Eq. 10.80 to get a superior approximation to the solution $f^{(1)}(x)$ as

$$f^{(1)}(x) = u(x) + C \int_{y=a}^{h} K(x, y) f^{(0)}(y) dy \tag{10.81}$$

It would be ideal if the integral on the right could be obtained in a closed analytical form. The process is continued by substituting $f^{(1)}(y)$ within the integral sign to improve the solution to $f^{(2)}(x)$ and so on.

Example 10.14

Consider a wedge cavity as shown in Fig. 10.28 with a wedge angle of 60° such that the enclosure has the shape of an *equilateral* triangle. Consider two cases:

1. Each surface of the wedge is at a uniform temperature of 500 K
2. The temperature of each surface of the wedge varies from 500 K at the vertex to a temperature of 400 K at the tip.

The opening of the wedge may be treated as a black body at 300 K. The wedge surfaces are gray having equal emissivities of $\varepsilon = 0.6$. Obtain the heat loss from the wedge per unit wedge length with uniform and non-uniform radiosity assumptions. Draw conclusions from the results. Solve the integral equation by a a numerical method such as the Gauss–Seidel method (refer to Chap. 7) after writing the integral as a sum over strips. Obtain all the required angle factors by the use of the triangle rule.

Solution:

(1) Uniform wedge temperature case: Comparisons may be based on the improvement made with reference to the two-surface enclosure model familiar to us already, applied to the case where the wedge surfaces are at a uniform temperature of 500 K. The opening is represented as surface 0 while the wedge surfaces are represented by a single surface 1. In this model, the angle factors are obtained by angle factor algebra and are given by

$$F_{11} = 0.5; \quad F_{10} = 0.5; \quad F_{01} = 1; \quad \text{and} \quad F_{00} = 0$$

The radiosity of the wedge surface is given by

$$J_1 = \frac{0.6 \times 5.67 \times 10^{-8} \times 500^4 + (1 - 0.6)5.67 \times 10^{-8} \times 300^4}{1 - (1 - 0.6)0.5} = 2428.2 \text{ W/m}^2$$

Based on Eq. 10.58, the heat flux leaving the cavity is given by

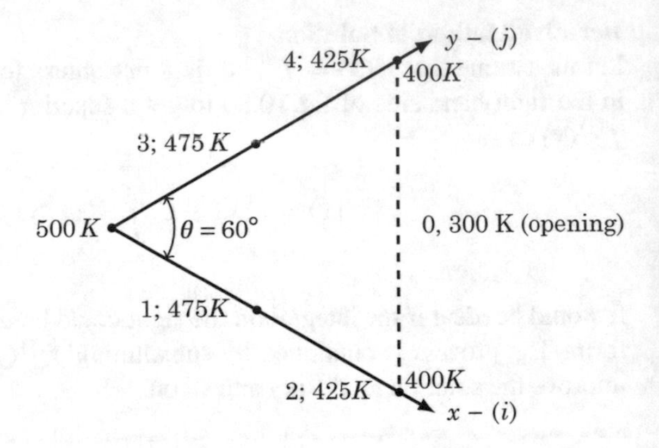

Fig. 10.29 Wedge cavity with four strips of non-uniform temperature

$$q = \frac{0.6 \times 5.67 \times 10^{-8} \times (500^4 - 300^4)}{1 - (1 - 0.6) \times 0.5} = 2313.4\text{W/m}^2$$

(2) Variable wedge temperature case: If we apply the two-surface model to this case, the temperature of each surface of the wedge is taken as the mean between the apex and the tip of the wedge and is given by $T_1 = 450\text{K}$. The radiosity of the wedge surface is given by

$$J_1 = \frac{0.6 \times 5.67 \times 10^{-8} \times 450^4 + (1 - 0.6)5.67 \times 10^{-8} \times 300^4}{1 - (1 - 0.6)0.5} = 1514.2\ \text{W/m}^2$$

The heat loss from the wedge is then given by

$$q = \frac{0.6 \times 5.67 \times 10^{-8} \times (450^4 - 300^4)}{1 - (1 - 0.6) \times 0.5} = 1399.3\ \text{W/m}^2$$

The case where the wedge temperature varies significantly along its length cannot be solved with the two-surface enclosure model since it is a gross approximation. The solution may be improved by considering the radiosity variation by dividing the wedge to a number of strips each of which is considered isothermal. For example, each wedge is divided in to two strips of equal width, as an improvement over the two-surface analysis. This case is shown in Fig. 10.29 and the appropriate calculations are shown below.

All the angle factors may be calculated by the use of the triangle formula or the crossed string method. The reader is encouraged to verify the angle factors given in Table 10.5. Since $J_1 = J_3$ and $J_2 = J_4$, we have only two radiosity equations that may be written down as

$$J_1 = \varepsilon E_{b1} + (1 - \varepsilon)(F_{10} E_{b0} + F_{13} J_1 + F_{14} J_2)$$
$$J_2 = \varepsilon E_{b2} + (1 - \varepsilon)(F_{20} E_{b0} + F_{23} J_1 + F_{24} J_2)$$

Table 10.5 Angle factors for configurations in Fig. 10.29

Surface No.	0	1	2	3	4
0	0	0.1830	0.3170	0.1830	0.3170
1	0.3660	0	0	0.5	0.1340
2	0.6340	0	0	0.1340	0.2321
3	0.3660	0.5	0.1340	0	0
4	0.6340	0.1340	0.2321	0	0

Introducing the numerical values and rearranging, we get

$$0.8J_1 - 0.05359J_2 = 2193.5$$
$$-0.05359J_1 + 0.90718J_2 = 2242.7$$

These two simultaneous equations are solved to get $J_1 = 2348.7$ and $J_2 = 1490.6$. The radiosity indeed varies significantly along the wedge length! The heat loss from the wedge cavity may then be obtained as

$$\begin{aligned} q &= 2(F_{01}J_1 + F_{02}J_2) - E_{b0} \\ &= 2(0.183 \times 2348.7 + 0.317 \times 1490.6) - 459.27 \\ &= 1345.4 \text{ W/m}^2 \end{aligned}$$

The heat loss from the cavity has also changed significantly between the two- and the four-surface models!

The procedure given at the end in Example 10.14 may be carried on with a larger number of strips along each wedge surface. As an example, we consider the case where each surface of the wedge is divided in to ten strips of equal width. The strips are numbered from 1 to 10 along the bottom surface of the wedge and from 1′ to 10′ along the top surface of the wedge. Each strip is isothermal at a mean temperature obtained by using the specified temperature variation between the apex and the tip of the wedge. The radiosity variation along the two wedge surfaces are identical and hence there are ten radiosity equations that need to be solved. The angle factor matrix is symmetrical and is given in Table 10.6 on page 466. The radiosity equations are written following Eq. 10.73. The resulting equations are solved by the Gauss iteration method familiar to us from Chap. 7. This method is essentially the same as the method suggested earlier for the solution of the Fredholm integral equation, except that the integral is replaced by a sum. The iterations start with the radiosities of the strip taken as $J_j = \varepsilon\sigma T_j^4$ where $1 \le j \le 10$. The iteration scheme converges rapidly and the desired solution is obtained after 9 iterations as indicated in Table 10.7 on page 466. We observe that the radiosity values continuously decrease from the apex of the

Table 10.6 Angle factors for configurations for wedge geometry with ten strips

Strip No.	0	1	2	3	4	5	6	7	8	9	10
0	0	0.0270	0.0313	0.0362	0.0415	0.0471	0.0529	0.0585	0.0639	0.0687	0.0730
1′	0.2697	0.5	0.1340	0.0432	0.0201	0.0115	0.0074	0.0052	0.0038	0.0029	0.0023
2′	0.3129	0.1340	0.2321	0.1340	0.0707	0.0411	0.0263	0.0181	0.0132	0.0100	0.0078
3′	0.3615	0.0432	0.1340	0.1458	0.1064	0.0707	0.0477	0.0334	0.0244	0.0185	0.0144
4′	0.4148	0.0201	0.0707	0.1064	0.1056	0.0854	0.0642	0.0477	0.0359	0.0276	0.0217
5′	0.4712	0.0115	0.0411	0.0707	0.0854	0.0826	0.0706	0.0570	0.0452	0.0359	0.0288
6′	0.5288	0.0074	0.0263	0.0477	0.0642	0.0706	0.0678	0.0599	0.0507	0.0421	0.0347
7′	0.5852	0.0052	0.0181	0.0334	0.0477	0.0570	0.0599	0.0574	0.0519	0.0453	0.0389
8′	0.6385	0.0038	0.0132	0.0244	0.0359	0.0452	0.0507	0.0519	0.0498	0.0458	0.0409
9′	0.6871	0.0029	0.0100	0.0185	0.0276	0.0359	0.0421	0.0453	0.0458	0.0440	0.0409
10′	0.7303	0.0023	0.0078	0.0144	0.0217	0.0288	0.0347	0.0389	0.0409	0.0409	0.0394

Table 10.7 Gauss iteration results for radiosities (W/m^2) of strips

Strip No.→ Iteration No. ↓	0	1	2	3	4	5	6	7	8	9	10
1	459.3	2042.5	1882.4	1731.8	1590.5	1458.1	1334.1	1218.1	1109.9	1009.1	915.3
2	459.3	2660.9	2421.3	2209.6	2019.6	1847.0	1689.6	1545.7	1414.2	1294.1	1184.3
3	459.3	2829.8	2558.3	2324.1	2116.7	1929.9	1760.4	1606.0	1465.6	1337.7	1221.4
4	459.3	2874.5	2592.1	2350.7	2138.3	1947.6	1775.0	1618.2	1475.7	1346.2	1228.5
5	459.3	2886.1	2600.3	2356.9	2143.2	1951.5	1778.2	1620.8	1477.8	1347.9	1230.0
6	459.3	2889.0	2602.3	2358.3	2144.3	1952.4	1778.9	1621.3	1478.3	1348.3	1230.3
7	459.3	2889.8	2602.8	2358.7	2144.5	1952.6	1779.0	1621.5	1478.4	1348.4	1230.4
8	459.3	2889.9	2602.9	2358.8	2144.6	1952.7	1779.1	1621.5	1478.4	1348.5	1230.4
9	459.3	2890.0	2602.9	2358.8	2144.6	1952.7	1779.1	1621.5	1478.4	1348.5	1230.4

wedge to the tip of the wedge. This variation is significant and hence there is a case made out for the non-uniform radiosity analysis carried out here.

It is interesting to see the convergence of the heat loss from the cavity with the number of strips. The following table shows the results for both cases with the number of strips.

Number of strips	Uniform T case	Variable T case
1	2313.4	1399.3
2	2285.8	1345.4
4	2278.5	1329.0
6	2277.2	1325.7
8	2276.6	1324.4
10	2276.3	1323.8

Concluding Remarks

General enclosure analysis for the calculation of radiant heat exchange amongst surfaces has been presented. Diffuse gray, diffuse nongray and specular surfaces have been considered in the analysis. Both uniform and nonuniform radiosity cases have been dealt with. Solution methods presented include the method of detailed balancing, radiosity-irradiation formulation and electrical analogy based on resistance network.

10.5 Exercises

Ex 10.1 An enclosure has an inside area of 100 cm^2 and its surface is polished aluminum with an emissivity of 0.05. A small opening on the wall of the enclosure has an area of 0.5 cm^2. The radiant power leaving the enclosure through the opening is 0.07 W. What is the temperature of the surface of the enclosure? The gray calculation is expected to be adequate.

Ex 10.2 In a rectangular box-type enclosure, there are 36 individual view factors between the 6 surfaces that constitute the enclosure. How many independent view factors need to be determined?

Ex 10.3 In a cubical enclosure, there are 36 individual view factors between the 6 surfaces that constitute the enclosure. How many independent view factors need to be determined?

Ex 10.4 A long duct is in the form of a regular pentagon. How many shape factors need to be determined independently? Explain your answer. Determine the shape factor between any one side and all the other sides.

Ex 10.5 A very long duct has a rectangular cross section with the sides in the ratio of 1.5 : 1. Determine all the view factors.

Ex 10.6 A long duct has a regular hexagonal cross section. Determine the view factor between opposite sides by view factor algebra.

Ex 10.7 In the duct of Exercise 10.6, a partition is placed as shown in Fig. 10.30. Determine all the view factors.

Ex 10.8 A cube with 1 m edges is located inside a second cube with 2 m edges. If the two cubes do not touch each other, determine the diffuse shape factor between the inner surface of the outer cube and the outer surface of the inner cube.

Ex 10.9 Two rectangular areas are arranged as shown in Fig. 10.31. Determine the view factor F_{12}. Use suitable tables and the decomposition rule presented in the text.

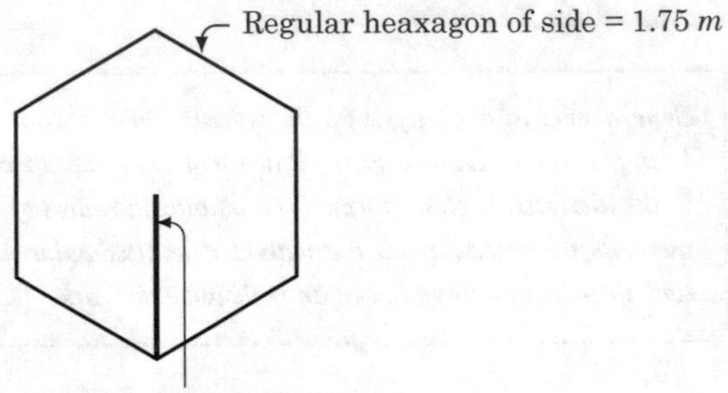

Fig. 10.30 Duct with partition in Exercise 10.7

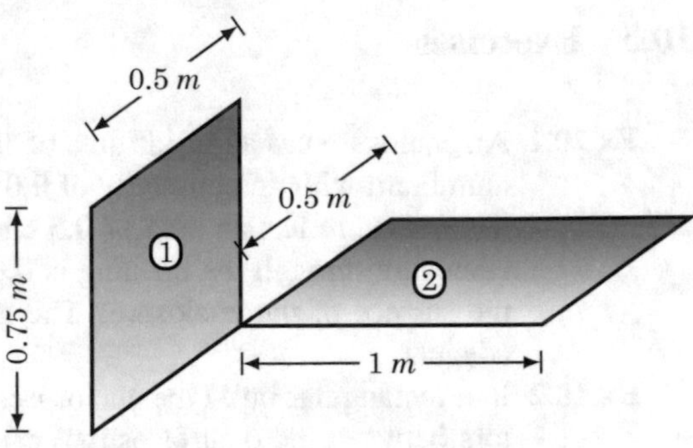

Fig. 10.31 Geometry for Exercise 10.9. Planes 1 and 2 are perpendicular to each other

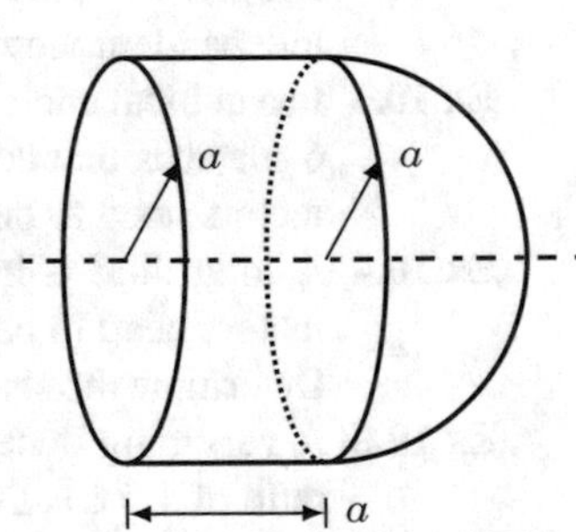

Fig. 10.32 Geometry of space capsule in Exercise 10.10

Ex 10.10 A space capsule has the shape shown in Fig. 10.32. Determine
(a) Solid angle subtended by the hemispherical cap at 0.
(b) Solid angle subtended by the cylindrical part of the space capsule at 0.
(c) Diffuse view factor between the hemispherical cap and the base of the cylinder.
(d) Diffuse view factor between the hemispherical cap and its own base.
(e) Diffuse view factor between the hemispherical cap and the curved surface of the cylindrical part of the space capsule.
Hint: Parts c to e requires intelligent use of view factor algebra.

Ex 10.11 A short solid cylinder of diameter D and height H is placed on the ground with its axis vertical. What is the view factor between the exposed surface of the cylinder and (a) the ground and (b) the sky?

Ex 10.12 Two infinite parallel planes, one at 400° C and the other at 150° C with common gray emissivity of $\varepsilon = 0.3$, are to be shielded by placing a third plane (with $\varepsilon = 0.05$ on each side) between them, and allowing it to come to thermal equilibrium. Find the radiant flux between the planes before and after the insertion of the third plane, and find the equilibrium temperature of the shield.

Ex 10.13 Two very large parallel gray planes are maintained at 100° C and 25° C, respectively. The two planes have an equal emissivity of 0.15. Determine the heat transfer per unit area between them. Now a third surface of equal area and very small thickness is placed in between to form a shield. Both the sides of this shield have the same emissivity equal to 0.04. What is the heat transfer per unit area now?

Ex 10.14 Liquid oxygen (boiling point $= -183$° C) is to be stored in a spherical container of outer diameter equal to 0.3 m. The system is insulated by an evacuated space between the inner sphere and a surrounding 45 cm ID concentric sphere. Both spheres are of polished aluminum with emissivity equal to 0.05. The temperature of the outer sphere is -1° C. Estimate the rate of heat flow by radiation to oxygen.

Ex 10.15 A thermos flask may be idealized as a long double-walled cylinder, ignoring the end effects due to the stopper. The space between the two walls is evacuated and the surfaces that face each other are silvered to obtain an effective emissivity of 0.035. The OD of the inner cylinder is 10.5 cm while the ID of the outer cylinder is 11 cm. The two walls of the double-walled cylinder are 0.5 mm thick and the thermal conductivity of each is 0.2 W/m° C. The cylinder is filled with boiling water at 100° C. What is the heat loss per *cm* length of the double-walled cylinder? Assume that the outer wall of the flask is exposed to an ambient at 20° C via a heat transfer coefficient of 6 $W/m^2\,^\circ C$.

Ex 10.16 A plane surface of a material of gray emissivity 0.6 has been machined to obtain parallel grooves 0.1 mm wide and 0.2 mm deep. The grooves have planar sides. The center-to-center distance between the grooves is also 0.1 mm. Determine the effective emissivity of the grooved surface.

Ex 10.17 An evacuated enclosure is in the shape of a right circular cylinder of diameter 0.3 m and height 0.45 m. The curved side is gray, has an emissivity of 0.6, and is maintained at 450 K. At both the flat sides (top and bottom having an emissivity of 0.8), a uniform cooling rate of 500 W/m^2 is maintained. Determine the temperature of the two flat sides.

Ex 10.18 A long cavity radiator has the cross section shown in Fig. 10.33. Surface 1 is the active surface, which is maintained at 800 K and has a gray emissivity of 0.85. Surface 2 is a reradiating surface. Opening 3 may be assumed to be black and at a very low temperature. Determine

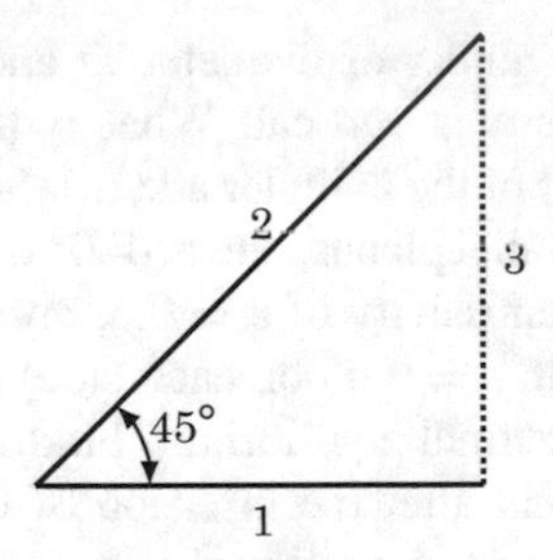

Fig. 10.33 Geometry of cavity radiator in Exercise 10.18

(a) Heat flux leaving surface 1, (b) Temperature of surface 2, and (c) Equivalent emissivity of the cavity radiator.

Ex 10.19 A cavity is constructed in the form of a cylinder cone combination as shown by its cross section as in Fig. 10.34. The cone has an included angle of 120° and is integrated with the cylinder part which has a diameter $D = 25$ mm and length $L = 100$ mm. The front of the cylinder has an aperture of diameter $d = 3$ mm. All the internal surfaces of the cavity have a common gray diffuse emissivity of $\varepsilon = 0.5$. Determine the effective emissivity of the aperture? Consider the conical part as surface 1, the cylindrical part and the front cover as surface 2, and the aperture as surface 3.

Ex 10.20 A cavity radiator is in the form of a cylinder with one end open as shown in Fig. 10.35. The walls of the cavity are gray and diffuse with an emissivity of 0.85. The geometry of the cavity may be changed

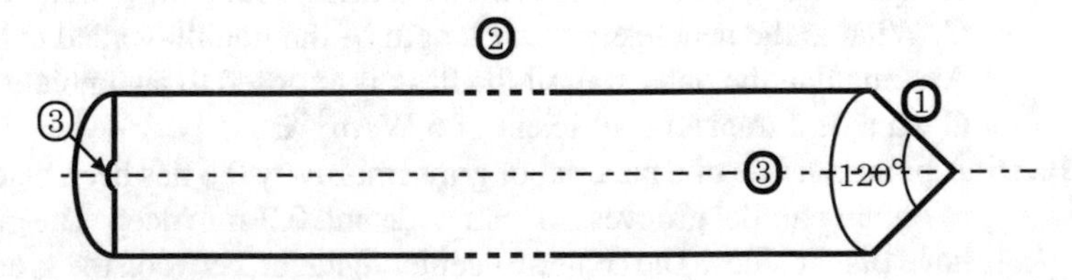

Fig. 10.34 Cavity configuration in Exercise 10.19

Fig. 10.35 Cylindrical cavity in Exercise 10.20

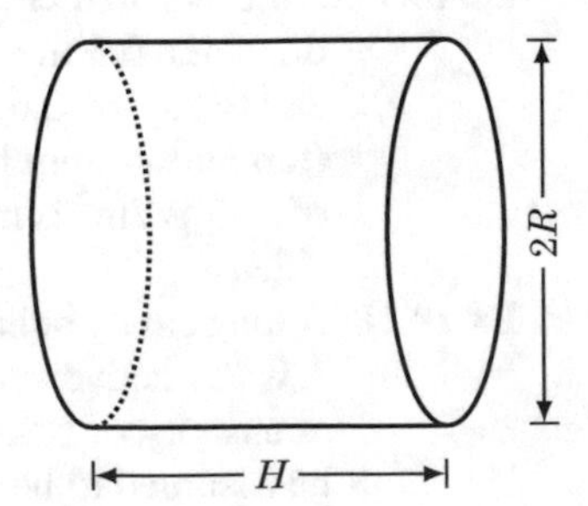

by varying the ratio of its depth to the radius of the opening, i.e., by changing the ratio $\frac{H}{R}$. Obtain an expression for the equivalent emissivity of the cavity radiator in terms of this ratio in the range 0.25–5. Make a plot of the equivalent emissivity as a function of $\frac{H}{R}$.
Hint: Consider the cylinder inside the surface as a single zone.

Ex 10.21 A cavity is in the form of a cylinder of diameter 0.05 m and height 0.05 m. The bottom of the cavity is heated to a temperature of 1500 K while the curved side is perfectly insulated. The top of the cylinder is open to a low temperature ambient that may be taken effectively to be at 0 K. What is the power leaving the cavity and what is the temperature of the curved side of the cylinder? Assume that all actual surfaces are gray, diffuse, and have an emissivity of 0.85.

Ex 10.22 A long furnace has a 3 m square cross section. The roof is maintained at 2000 K by the circulation of hot combustion gases, while the floor is at 800 K. The side walls are well insulated with refractory bricks. Calculate the radiant heat flux into the floor if the roof and side walls have an emissivity of 0.7 and the floor has an emissivity of 0.4.

Ex 10.23 The cross section of a very long tunnel is shown in Fig. 10.36. The required geometric data is also provided in the figure. (a) It is desired to determine the shape factors F_{AB-GH}, F_{AC-GH}, and F_{CE-GH}. Make use of the triangle rule, Hottel's crossed string method, and view factor algebra. (b) If all the surfaces internal to the tunnel are maintained at a uniform temperature and if all the surfaces have the same emissivity of $\varepsilon = 0.6$, what will be the effective emissivity of the opening?

Ex 10.24 A cavity is in the form of a vertical cylinder of 8 cm diameter and 16 cm length and is open at the top to black surroundings at 300 K. The bottom end is heated electrically and is maintained at 1900 K. If the side walls are at a uniform temperature of 1500 K, calculate the power input to the heater and the heat loss to the surroundings through the open end. Assume that all the inner surfaces have a gray emissivity of $\varepsilon = 0.85$.

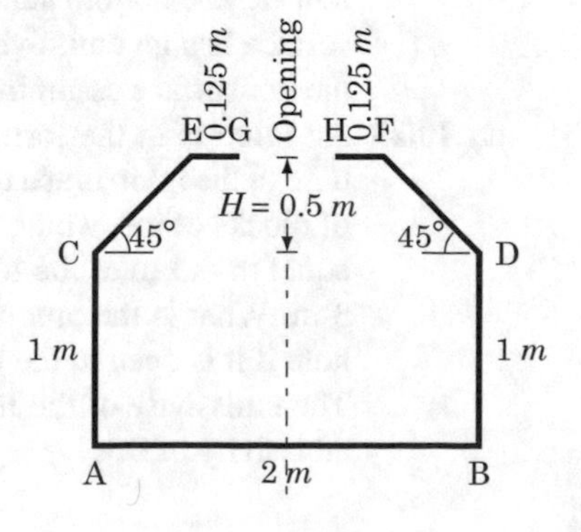

Fig. 10.36 Cross section of the very long tunnel in Exercise 10.23

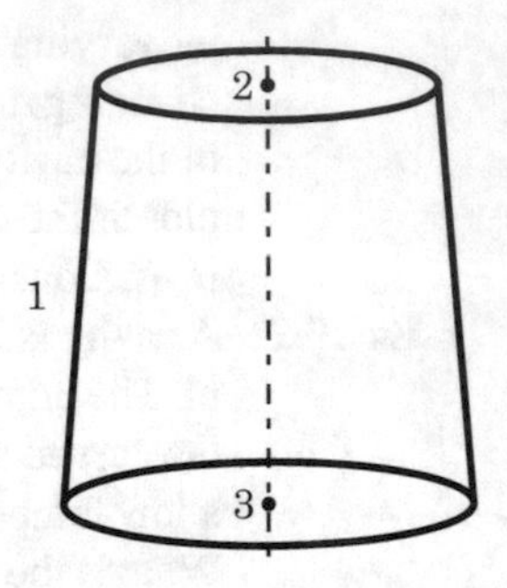

Fig. 10.37 Evacuated enclosure in Exercise 10.27

Ex 10.25 The floor of a large furnace acts as a plane at $800°C$, emissivity 0.6, and the ceiling is a plane at 250° C, emissivity 0.8. The furnace is a rectangular box 3 m wide, 4 m long, and 5 m tall. All the other surfaces act as a single reradiating surface. What is the heat transfer from the hot floor and the equilibrium temperature of the reradiating surfaces?

Ex 10.26 An evacuated enclosure is in the shape of a right circular cylinder of diameter 0.3 m and height 0.45 m. The curved side is gray, has an emissivity of 0.6, and is maintained at 450 K. At both the flat sides (top and bottom having an emissivity of 0.8), a uniform cooling rate of 500 W/m^2 is maintained. Determine the temperature of the two flat sides.

Ex 10.27 An evacuated enclosure is in the shape of a frustum of a cone of base diameter $D_3 = 0.3$ m, top diameter $D_2 = 0.25$ m, and height $L = 0.3$ m. The curved surface 1 is gray, has an emissivity of $\varepsilon_1 = 0.6$, and is maintained at $T_1 = 450$ K. At both the flat sides (top and bottom having common emissivity of $\varepsilon_2 = \varepsilon_3 = 0.8$), a uniform cooling rate of $q_2 = q_3 = -500$ W/m^2 is maintained. Determine the temperatures of the two flat sides. Use electrical analogy for solving this problem.

Ex 10.28 An enclosure is in the form of a short cylinder of diameter 1 m and height 0.5 m. The top surface is reradiating while the bottom surface and the lateral curved surface are maintained, respectively, at 500 K and 350 K. The bottom surface has an emissivity of 0.45 while the curved surface has an emissivity of 0.8. Determine the heat transferred from the hot surface assuming that all surfaces are diffuse.

Ex 10.29 An igloo is in the form of a hemisphere as shown in Fig. 10.38. The base of the igloo is at a uniform temperature of 30° C. The igloo is made of blocks of ice, which are at −5° C. There is a small hole of diameter equal to 0.3 m at the top for ventilation. The diameter of the igloo is 3 m. What is the amount of radiation leaving through the ventilation hole if it is open to the night sky at an equivalent temperature of 5° C? The emissivity of the floor of the igloo is 0.3 while the walls have an emissivity of 0.1.

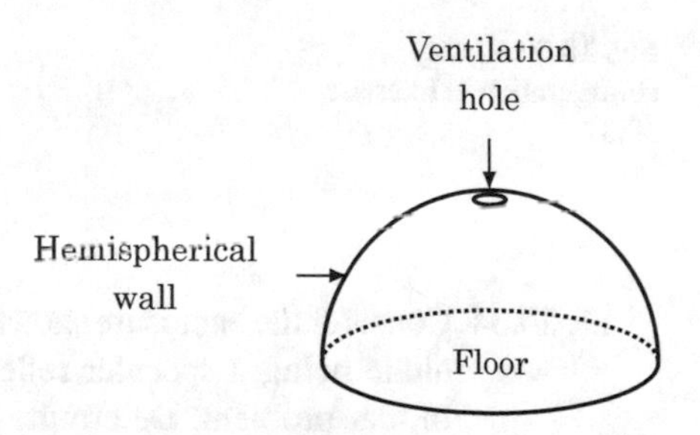

Fig. 10.38 Igloo in Exercise 10.29

Ex 10.30 A rectangular box has dimensions of $1 \times 0.25 \times 0.5$ m. One of the $1\times$ m sides is maintained at a temperature of 450 K while the other $1\times$ m side is maintained at 350 K. Both these surfaces have an emissivity of 0.8. All the other sides are in radiant balance, i.e., the radiant heat flux is zero on all these surfaces. You will notice that the box may be considered as an enclosure made of three surfaces. Calculate the heat fluxes at the two surfaces where the temperatures are specified and the temperature of the reradiating surface. Assume that all surfaces are diffuse and gray.

Ex 10.31 An enclosure is in the form of a frustum of a cone. The base of the frustum has a diameter of 2 m while the top of the frustum has a diameter of 1 m. The height of the enclosure is 0.5 m. Determine all the view factors for this enclosure.
In the above enclosure, the bottom is maintained at 400 K while the top is maintained at 300 K. The emissivities of these are, respectively, 0.5 and 0.8. The curved side is assumed to be reradiating. Calculate the net heat transfer from the bottom to the top. What is the temperature of the curved side?

Ex 10.32 A thermocouple is used to measure the temperature of a hot gas in a combustion chamber. If the thermocouple temperature is 1033 K and the walls of the combustion chamber are at 700 K, what is the temperature of the hot gas? Assume that all surfaces are black and the convection heat transfer coefficient between the hot gas and the thermocouple is 568 $\text{W/m}^2\,^\circ C$. Ignore conduction through the thermocouple lead wires. Hint: Note that areas for convection and radiation heat transfer are the same.

Ex 10.33 Consider the enclosure shown in Exercise 10.28 with the curved surface being a selective surface with the emissivity given by

$$\varepsilon_\lambda = 0.8,\ 0 < \lambda < 4\ \mu\text{m}$$
$$\varepsilon_\lambda = 0.4,\ \lambda > 4\ \mu\text{m}$$

All other data remains the same as in that problem. Determine the heat transferred from the hot surface assuming that all surfaces are diffuse.

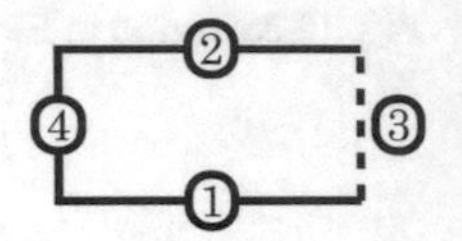

Fig. 10.39 Cavity configuration in Exercise 10.37

Ex 10.34 Consider the enclosure shown in Exercise 10.28 with the bottom surface alone being a specular reflector. All other data remains the same as in that problem. Determine the heat transferred from the hot surface assuming that other surfaces are diffuse.

Ex 10.35 A long open channel is 0.2 m deep and 0.1 m wide. The bottom of the channel is maintained at a temperature of 500° C. The vertical faces of the channel have a linear temperature along them with the bottom end at 500° C and the top end being at 300° C. Determine the heat loss through the opening per unit length of the channel. The channel opening views a background at 30° C. Divide the non-isothermal vertical sides of the channel suitably into several uniform radiosity parts and solve the resulting equations numerically.

Ex 10.36 A deep open cylindrical cavity is 0.5 m deep and 0.2 m in diameter. The bottom of the cavity is maintained at a temperature of 600° C. The curved surface of the cavity has a linear temperature variation along its height with the bottom end at 600° C and the top end being at 100° C. Determine the heat loss through the opening. The channel opening views a background at 10° C. Divide the non-isothermal curved surface of the cavity suitably into several uniform radiosity parts and solve the resulting equations numerically.

Ex 10.37 A very long open channel has a rectangular section as shown in Fig. 10.39. Surfaces identified as 1 and 2 are both 0.1 m wide, gray and diffuse with temperature and emissivities of $T_1 = 600$ K, $\varepsilon_1 = 0.8$ and $T_2 = 500$ K, $\varepsilon_2 = 0.6$, respectively. The bottom surface shown as 4 is 0.05 m wide, specular, gray, and reradiating with a reflectivity of $\rho_4 = 0.8$. The opening may be assumed to be black at a very low temperature. Determine the heat loss from the cavity per unit length. Make use of the exchange area concept.

Chapter 11
Radiation in Participating Media

RADIATION in participating media is important in applications such as in furnaces, nozzles used in space applications, in the study of atmosphere and weather, oceans, in glassmaking, etc. Since gases emit and absorb radiation in specific bands, gas radiation is by force non-gray in nature. Modeling such radiation is made complex because of the large path lengths over which such a process can take place. Both geometric and thermal aspects are involved and the present chapter intends to present the most important aspects of radiation in participating media.

11.1 Introduction

A medium that absorbs, emits, or scatters radiation is referred to as a participating medium. The medium may be in any of the three phases—solid, liquid, or gas. Otherwise, transparent solid may absorb and emit radiation when impurities are present in it. In the case of optical instruments, it is imperative that we account for the absorption of radiation by optical elements such as lenses and optical windows in interpreting data. Liquids may absorb radiation as it passes through them. Radiation transfer to the depths of the ocean or lakes is important in geophysical studies. Gases absorb and emit radiation in definite bands depending on the nature of the gas molecules. A gas laden with dust particles will also scatter radiation. The study of radiation transfer through participating media is important because of its engineering applications such as in combustion chambers. It is also important in atmospheric science where the nature of absorption of solar and terrestrial radiation by the atmospheric constituents affects overall energy balance and also global weather. In recent times, there is a lot of concern about global warming that is primarily due to the way gases interact with radiation.

S. P. Venkateshan, *Heat Transfer*,
https://doi.org/10.1007/978-3-030-58338-5_11

The reason why we club all the above in this chapter is due to the fact that the underlying physics and the analysis methods are the same in all the cases.

11.2 Preliminaries

11.2.1 Definitions

Absorption and emission of radiation by a non-scattering participating medium are described by the following quantities.

Definitions

- κ_λ—Monochromatic or spectral absorption coefficient (unit m^{-1}). If incident monochromatic radiation is characterized by the intensity I_λ, the absorption by the medium per unit volume and solid angle is given by $\kappa_\lambda I_\lambda$.
- ε_λ—Monochromatic or spectral emission coefficient (unit m^{-1}). The emission by the medium per unit volume and solid angle is given by $\varepsilon_\lambda I_{b\lambda}(T_m)$ where $I_{b\lambda}(T_m)$ is the black body intensity at temperature T_m of the medium.

Note that in the case of a surface, the symbol ε_λ represents the spectral emissivity which has no dimensions.

11.2.2 Equation of Transfer

The equation of radiation transfer (or simply the equation of transfer) governs the passage of radiation through a volume of participating medium. This equation describes the variation of radiation intensity with position, taking into account both the emission and absorption of radiation by the medium. In order to derive the equation, consider an elemental volume of the participating medium in the form of a "pillbox" as shown in Fig. 11.1.

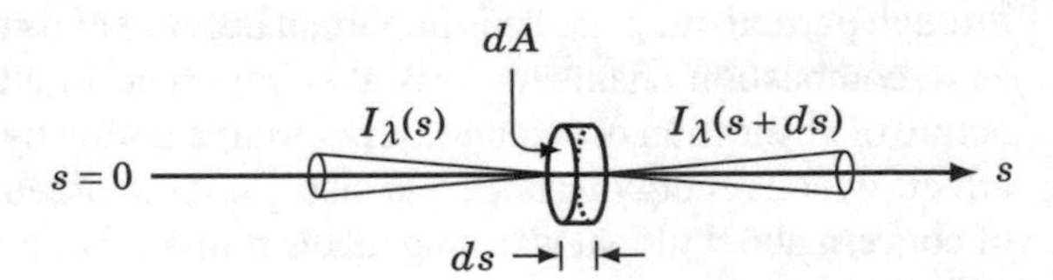

Fig. 11.1 Variation of radiation intensity as it traverses a volume element of a participating medium

Consider radiation traveling along the s direction. Orient the area element dA such that it is normal to the s direction. Then we have the following:

1. Power absorbed by the volume element, according to the definition of κ_λ given earlier, is $\kappa_\lambda I_\lambda dAds$ where $dAds$ is the volume of the pillbox. The absorption process tends to decrease the intensity.
2. Power emitted by the volume element, according to the definition of ε_λ given earlier, is $\varepsilon_\lambda I_{b\lambda}(T_m)dAds$. The emission of radiation by the volume element tends to increase the intensity.
3. If the intensity changes as indicated in Fig. 11.1, the change in monochromatic radiant power per unit solid angle is given by

$$I_\lambda(s+ds)dA - I_\lambda(s)dA = \frac{dI_\lambda}{ds}dAds$$

   The last result is obtained by retaining the first-order term in a Taylor expansion of intensity I_λ around s.

From (1)–(3) it is clear that, after canceling the common factor $dAds$, we have

$$\boxed{\frac{dI_\lambda}{ds} = \varepsilon_\lambda I_{b\lambda}(T_m) - \kappa_\lambda I_\lambda} \tag{11.1}$$

We may now imagine the participating medium to be placed within an isothermal cavity whose walls are at temperature $T_c = T_m$. Then the intensity I_λ is isotropic and invariant within the cavity and is equal to $I_{b\lambda}(T_m)$. The derivative on the left-hand side of Eq. 11.1 is zero and hence

$$0 = \varepsilon_\lambda I_{b\lambda}(T_m) - \kappa_\lambda I_{b\lambda}(T_m) \tag{11.2}$$

Thus, $\varepsilon_\lambda = \kappa_\lambda$ which is the familiar Kirchhoff's law. With this, Eq. 11.1 takes the form

$$\boxed{\frac{dI_\lambda}{ds} + \kappa_\lambda I_\lambda = \kappa_\lambda I_{b\lambda}(T_m)} \tag{11.3}$$

11.3 Absorption of Radiation in Different Media

11.3.1 Transmittance of a Solid Slab

Consider now a simple case of a slab of a solid material of thickness L which has an absorption coefficient κ_λ and refractive index n_λ. The absorption coefficient is assumed to be small so that the imaginary part of the index of refraction is small. This assumption is usually good in the case of glasses with low levels of impurities. For example, soda lime glass (used in window panes) may have a very low level of iron as contamination. The refractive index at a wavelength of $\lambda = 0.6328\,\mu$ m (Helium Neon laser wavelength) is 1.507 and the absorption coefficient is typically $\kappa_\lambda = 24\,\text{m}^{-1}$. The imaginary part of the complex index of refraction is very small and is given by

$$n' = \frac{\kappa_\lambda \times \lambda}{4\pi} = \frac{24 \times 0.6328 \times 10^{-6}}{4\pi} = 1.19 \times 10^{-6}$$

based on Eq. 9.65. If the temperature of the medium is small such that the emission by the medium is insignificant, the transmission of a laser beam through the medium follows the equation

$$\frac{dI_\lambda}{ds} + \kappa_\lambda I_\lambda = 0 \tag{11.4}$$

With intensity specified as $I_\lambda = I_{\lambda,0}$ at $s = 0$, the top surface of the glass slab, the intensity at any distance inside the medium along the s direction is obtained as

$$I_\lambda(s) = I_{\lambda,0}\mathrm{e}^{-\kappa_\lambda s} \tag{11.5}$$

The transmittance is then defined as

$$\tau_\lambda(s) = \frac{I_\lambda(s)}{I_{\lambda,0}} = \mathrm{e}^{-\kappa_\lambda s} \tag{11.6}$$

If we choose the s direction to be along a normal to the face of the glass sheet, the transmittance of the sheet in this direction is obtained as

$$\tau_\lambda(L) = \frac{I_\lambda(L)}{I_{\lambda,0}} = \mathrm{e}^{-\kappa_\lambda L} \tag{11.7}$$

Expression 11.7 is referred to as Beer's law.[1]

Consider now the case shown in Fig. 11.2. Radiation is incident at an angle onto the top surface of a glass sheet and the interest is in determining the amount that is transmitted across the sheet. We assume that the incident radiation is unpolarized

[1] After August Beer, 1825–1863, a German physicist.

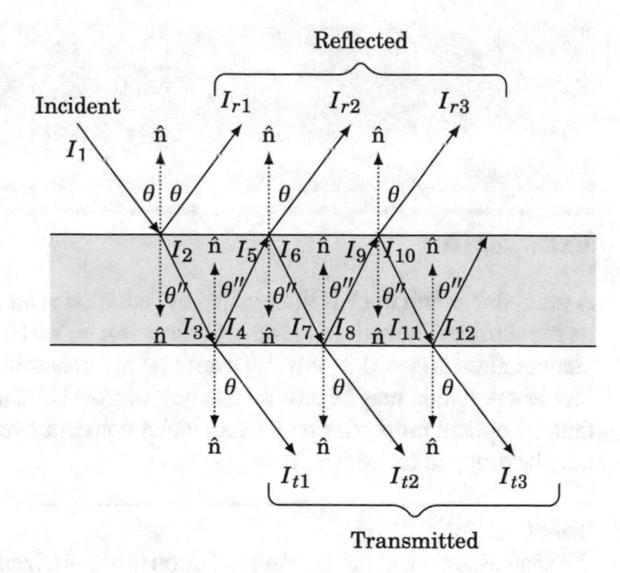

Fig. 11.2 Transmission of radiation across a semi-transparent slab

and hence all the optical quantities will be calculated as the mean for the parallel and perpendicular components. Let ρ be the reflectance at each interface (we drop the subscript λ for convenience).

The processes that place at each surface of the glass sheet are indicated in detail in the figure. The intensities indicated in the figure are given by the following:

$I_2 = I_1(1-\rho)$	$I_3 = I_1\tau(1-\rho)$	$I_4 = I_1\rho\tau(1-\rho)$
$I_5 = I_1\rho\tau^2(1-\rho)$	$I_6 = I_1\rho^2\tau^2(1-\rho)$	$I_7 = I_1\rho^2\tau^3(1-\rho)$
$I_8 = I_1\rho^3\tau^3(1-\rho)$	$I_9 = I_1\rho^3\tau^4(1-\rho)$	$I_{10} = I_1\rho^4\tau^4(1-\rho)$
$I_{11} = I_1\rho^4\tau^5(1-\rho)$	$I_{12} = I_1\rho^5\tau^5(1-\rho)$	.... and so on

We also note from the figure that the transmitted intensities are given by

$$I_{t1} = I_1\tau(1-\rho)^2;\ I_{t2} = I_1\rho^2\tau^3(1-\rho)^2;\ I_{t3} = I_1\rho^4\tau^5(1-\rho)^2;\ \ldots\ldots \tag{11.8}$$

The total transmitted intensity is given by the following summation.

$$I_t = I_1\tau(1-\rho)^2[1+\rho^2\tau^2+\rho^4\tau^4+\cdots] = I_1\frac{\tau(1-\rho)^2}{(1-\rho^2\tau^2)} \tag{11.9}$$

The last result is seen to arise out of the fact that the summation within the brackets represents a geometric series with 1 as the first term and the common ratio $\rho^2\tau^2$. The transmittance of the glass sheet t is then given by

$$t = \frac{I_t}{I_1} = \frac{\tau(1-\rho)^2}{(1-\rho^2\tau^2)} \tag{11.10}$$

Example 11.1

A glass slab 3 mm thick is illuminated by radiation at an incident angle of $\theta = 30°$ at its top surface. The refractive index of glass is $n = 1.514$ while the absorption coefficient of glass is $\kappa = 0.24\ \text{cm}^{-1}$. Determine the transmittance of the glass sheet. The incident radiation may be assumed to be a narrow band around $\lambda = 0.632\ \mu$ m such that the optical properties may be assumed constant over the band. The subscript λ may be dropped for convenience.

Solution:

We shall assume that the incident radiation is unpolarized and all properties are taken as the mean of those for the two polarizations. The transmittance of the glass sheet is determined by the combined effects of multiple reflections at the two surfaces and by the transmittance of the glass sheet for radiation traversing endlessly across it due to multiple reflections. Figure 11.2 shows the details of the process that takes place. For $n = 1.514$ and $\theta = 30°$ we have, by Snell's law,

$$\theta'' = \sin^{-1}\left(\frac{\sin\theta}{n}\right) = \sin^{-1}\left(\frac{\sin 30°}{1.514}\right) = 19.28°$$

The reflectance at each surface is calculated based on Eq. 9.44(b) and (d) for reflection at an interface between two dielectrics (air and glass).

$$\rho_\perp = \left[\frac{1.514^2\cos 30° - \sqrt{1.514^2 - \sin^2 30°}}{1.514^2\cos 30° + \sqrt{1.514^2 - \sin^2 30°}}\right]^2 = 0.0265$$

$$\rho_\parallel = \left[\frac{\cos 30° - \sqrt{1.514^2 - \sin^2 30°}}{\cos 30° + \sqrt{1.514^2 - \sin^2 30°}}\right]^2 = 0.0602$$

$$\rho = \frac{0.0265 + 0.0602}{2} = 0.0434$$

Radiation has a slant path of length L' given by

$$L' = \frac{L}{\cos\theta''} = \frac{0.003}{\cos 19.28°} = 0.00346\ m$$

With the given value of the absorption coefficient, the transmittance τ is obtained as

$$\tau = e^{-\kappa L'} = e^{-0.24\times 100\times 0.00346} = 0.9203$$

The transmittance of the glass sheet for incident radiation may then be calculated using Eq. 11.10 as

$$t = \frac{0.9203(1 - 0.0434)^2}{(1 - 0.0434^2 \times 0.9203^2)} = 0.844$$

This may be compared with the transmittance in the absence of absorption by the glass sheet (i.e., in case the sheet is perfectly transparent) given by (refer to Example 9.7 or by letting $\tau = 1$ in Eq. 11.10)

$$t = \frac{1-\rho}{1+\rho} = \frac{1 - 0.0434}{1 + 0.0434} = 0.958$$

The absorption of the glass sheet cuts down the transmittance by an additional 12% as compared to the case where the glass is transparent and transmission loss is due only to reflection at the two interfaces.

11.3.2 Absorption of Radiation by Liquids

Liquids absorb radiation by the interaction of radiation at the molecular level and also with a cluster of molecules that are characteristic of liquids in bulk form. The absorption coefficient shows large variations with the wavelength. Typical of such a behavior is that of water for which the absorption data is shown in Fig. 11.3.[2] The existence of peaks in the spectrum is indicative of absorption bands. These are

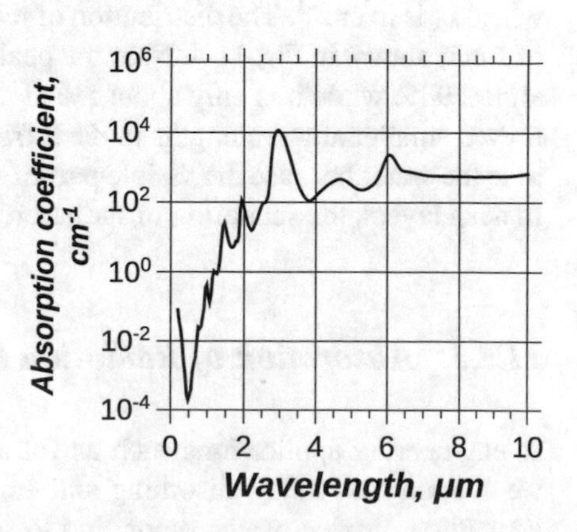

Fig. 11.3 Monochromatic absorption coefficient of distilled water

[2] S. A. Sullivan, J. Opt. Soc. America, Vol. 53, No. 8, pp. 962–968, 1963; W. M. Irvine, J. B. Pollack, Icarus, 8, pp. 324–360, 1968.

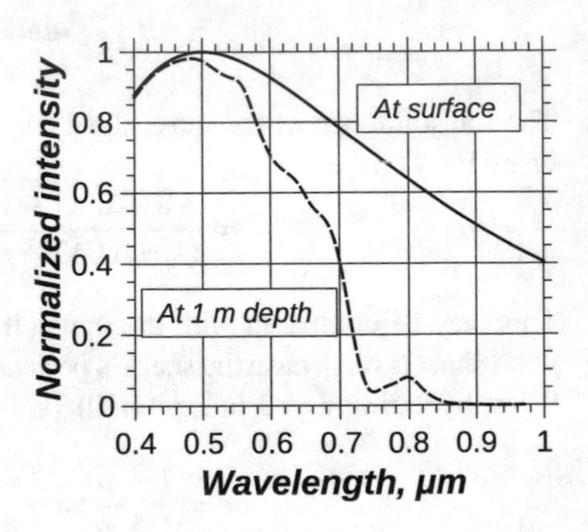

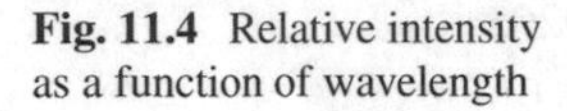
Fig. 11.4 Relative intensity as a function of wavelength

ascribed to harmonics of the fundamental frequencies and combinations of these. We shall have more to say on this while discussing absorption by water vapor, which is water in the vapor form. However, the bands are wide and do not show any fine structure since it is expected that the rotational motions of molecules are inhibited in the liquid state. In the case of water, hydrogen bonding between water molecules to form large clusters also accounts for the broadening of the spectral features. Water absorbs strongly in the ultraviolet region due to electronic transitions. It is weakly absorbing in the visible part of the spectrum. The absorption is again strong in the infrared region.

It is interesting to look at what happens when solar radiation (with emphasis on the visible and near-infrared part of the spectrum between 0.4 and 1 μ m) is incident at a free surface of a layer of water. The relative intensity (defined as the ratio $I_\lambda/I_{\lambda_{max}}$) of the incident radiation follows the black body distribution corresponding to a black body at 5800 K. The relative intensity at a depth of 1 m is obtained by multiplying the relative intensity at the surface by the monochromatic transmittance $\tau_\lambda = \mathrm{e}^{-100\kappa_\lambda}$ where κ_λ is in cm^{-1}. The distribution of relative intensity at the surface and at a depth of 1 *m* is shown in Fig. 11.4. Near the peak wavelength ($\approx$ 0.5 μ m), transmission is almost 98% while it is only about 5% at a wavelength of 0.8 μ m. The transmission is even smaller at wavelengths in the infrared part of the spectrum. We can "see" far in to the water because the visible part of the spectrum is weakly absorbed by water. In deep layers, the scattering of radiation also needs attention.

11.3.3 Absorption of Radiation by Gases

In engineering applications such as furnaces, rocket plumes, and the atmosphere, we come across many absorbing and emitting gases. The main gaseous absorbers are carbon dioxide, water vapor, and to some extent carbon monoxide and unburnt hydrocarbons. The last two constituents are present in combustion devices whenever combustion is incomplete. At elevated temperatures, nitrogen in the air may also form oxides. These oxides are also radiatively active, i.e., they absorb and emit radiation. The absorption by these gases is complicated by their inherent non-gray nature. Absorption and emission take place over bands that have a characteristic fine structure. Also, the absorption bands of different species overlap and hence when a gas mixture is involved, the absorption pattern gets further complicated. A clear understanding of gas radiation needs a background in quantum mechanics and some understanding of the fundamentals of molecular spectroscopy. A very brief description is given below so that the basic principles may be appreciated, at least to a limited but desired extent.

Gas molecules that are of interest to us may be classified as

- Diatomic molecules like oxygen, nitrogen, carbon monoxide, and nitric oxide;
- Polyatomic molecules like carbon dioxide, water vapor, and nitrous oxide.

Oxygen and nitrogen are further classified as homo-nuclear molecules, and they do not display any absorption properties and hence are said to be infrared inactive. Only heteronuclear molecules like carbon monoxide and nitric oxide (the two atoms in the diatomic molecule are different) are infrared active. The absorption and emission spectra of diatomic molecules are very simple since they exhibit a single fundamental band. However, polyatomic molecules show complex structures since they have several fundamental and combination bands. We consider examples to bring out the salient features of the absorption and emission of diatomic and polyatomic molecules.

Example of Diatomic Molecule, Carbon Monoxide

A molecule of carbon monoxide consists of a carbon atom and an oxygen atom held together. The equilibrium distance between the two atoms is $r_e = 1.18$Å with 1Å $= 10^{-8}\ m = 0.01\ \mu$ m. If this distance is made to change, a restoring force comes to play and hence the two atoms act as a spring–mass system. The potential energy versus the displacement is a parabola for small amplitudes and hence the oscillations are harmonic. If the displacement from the equilibrium distance between the atoms is $r - r_e$, the potential V is given by $V = \frac{k}{2}(r - r_e)^2$ where k is referred to as the spring constant. The potential is shown schematically in Fig. 11.5. The spring–mass analogy immediately tells us that the molecule can vibrate, and it does so with discrete energy levels (follows from the quantum mechanical description) given by

$$\boxed{E_v = \left(v + \frac{1}{2}\right) h\nu_0} \tag{11.11}$$

where v is the vibrational quantum number which is either zero or an integer, h is the Planck constant, and ν_0 is the fundamental frequency of the harmonic oscillator.

When $v = 0$, the energy of the oscillator is given by $\frac{1}{2}h\nu_0$. This is known as the zero point energy. If energy equal to D, the dissociation energy, is supplied to the molecule, it will dissociate into an atom of carbon and an atom of oxygen. At any given temperature, the molecules are distributed over the various vibrational energy levels according to the Boltzmann distribution. Apart from the vibrational

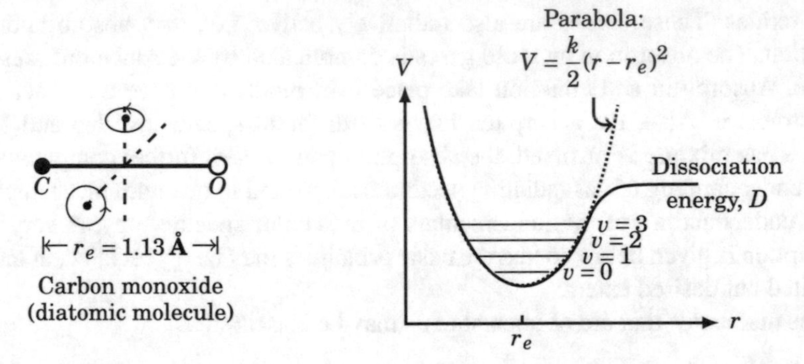

Fig. 11.5 Potential energy diagram for a diatomic molecule: carbon monoxide

motion, a carbon monoxide molecule can also perform rotational motion along two perpendicular axes as shown in Fig. 11.5. The allowed rotational energy values are again specified by angular momentum quantum number J (it is is either zero or an integer) such that the rotational energy of the molecule in the state J is given by

$$\boxed{E_r = BhJ(J+1)} \tag{11.12}$$

where B is known as the rotational constant of the molecule. The rotational constant is given by

$$B = \frac{h}{8\pi^2 I} \tag{11.13}$$

In Eq. 11.13, the moment of inertia I is given by

$$I = Mr_e^2 \tag{11.14}$$

where M is the reduced mass given by

$$M = \frac{m_C m_O}{m_C + m_O} \tag{11.15}$$

with m_C and m_O representing, respectively, the mass of carbon atom and oxygen atom. Imagine a volume of isothermal gas at a temperature of T_g. At this temperature, a certain proportion of molecules will occupy, say the excited state $v = 1$. Further, these excited molecules are distributed over the accessible rotational states according to the Boltzmann distribution given by

$$\boxed{N_J = N_0(2J+1)\exp\left[-\frac{\theta_R J(J+1)}{T_g}\right]} \tag{11.16}$$

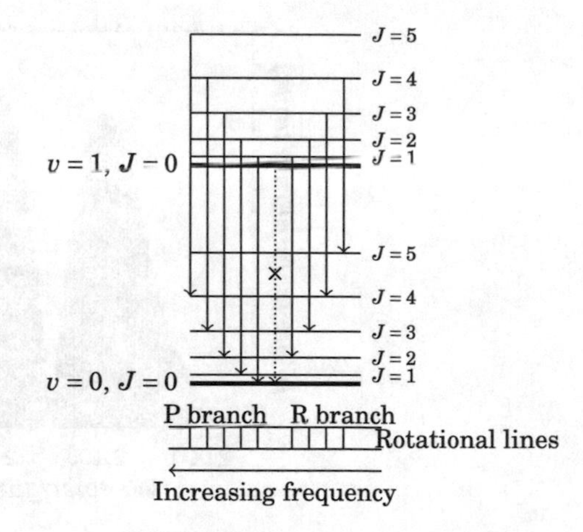

Fig. 11.6 Allowed transitions and the vibrational rotational lines of carbon monoxide

where N_J is the number of molecules in the vibrationally excited state and the factor $(2J + 1)$ is the degeneracy of the state with rotational quantum number J. The quantity $\theta_R = \frac{Bhc_0}{k}$ is the characteristic rotational temperature and has a unit of K. Here k is the Boltzmann constant. Some of these molecules will undergo a transition to the ground state $v = 0$, subject to the condition $\Delta J = \pm 1$. This is shown schematically in Fig. 11.6. During such a transition, both the vibrational as well as the rotational state of the molecule change and hence a photon is emitted with the energy given by

$$\begin{aligned} E &= h\nu_0 - 2BhJ \quad \text{for} \quad \Delta J = -1 \quad - \quad \text{P Branch} \\ E &= h\nu_0 + 2BhJ \quad \text{for} \quad \Delta J = 1 \quad - \quad \text{R Branch} \end{aligned} \tag{11.17}$$

These transitions are shown schematically at the bottom part of Fig. 11.6. It is noticed that a transition with $\Delta J = 0$ is not allowed and hence there is no emission corresponding to frequency ν_0 at all. Only the rotational lines governed by Eq. 11.17 appear in the emission spectrum. It is easily seen that the rotation lines are regularly spaced with a spacing of $2B$ with a missing line corresponding to ν_0. This last observation is valid as long as there is no interaction between the vibrational and rotational states.

A synthetic emission spectrum of carbon monoxide gas at T_g=1000 K is shown in Fig. 11.7. The parameters appropriate to carbon monoxide are given by $\nu_0 = 2143\,\text{cm}^{-1}, B = 1.897\,\text{cm}^{-1}$ and $\theta_R = 2.73$ K.[3] At low resolution, the spectrum will appear as an emission band indicated by the continuous curve while a high-resolution spectrum will show the fine structure in the form of rotational lines. The maximum in the spectrum occurs at a J value given by the nearest integer of the quantity $0.59\sqrt{\frac{T_g}{B}}$. For the spectrum shown in Fig. 11.7, the maximum indeed occurs at $J = 13$ as predicted by this formula. The point worth *noting* is that the width of the spectrum increases as the square root of the gas temperature.

[3]Note that energy is proportional to the photon wavenumber and is also expressible in temperature units.

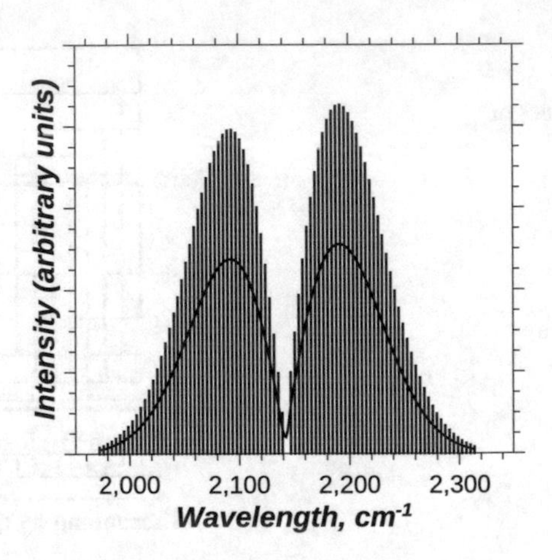

Fig. 11.7 Synthetic emission spectrum of carbon monoxide at 1000 K

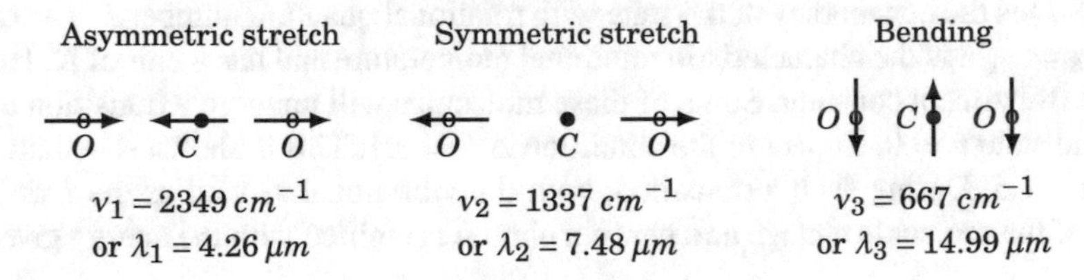

Fig. 11.8 Fundamental modes of a water vapor molecule

Note that the contributions from $v = 2$ and higher have not been included while constructing the synthetic emission spectrum. At 1000 K, these will also be populated and hence will contribute to the emission. To an extent, anharmonicity will then complicate the matters. The figure shows that the band occupies the region from 1950 to 2320 cm^{-1}. The width of the band is a function of temperature and that is one reason why the emission from carbon monoxide shows temperature dependence. The intensities of the lines have been indicated as being in terms of arbitrary units. This will depend on the number of molecules of carbon monoxide in a given volume and hence will also affect the total emission. From the point of view of radiation heat transfer, this should also depend on the geometry of the containing vessel. Thus, the gas emissivity is expected to depend on the gas pressure (and hence the density), the temperature of the gas, and the geometry of the enclosure containing the gas.

Example of Polyatomic Molecule, Carbon Dioxide

Carbon dioxide is a *linear* triatomic molecule as shown in Fig. 11.9. The molecule does not possess an intrinsic dipole moment, and hence the symmetric stretch mode is infrared inactive. It has three fundamental vibrational modes as indicated. In the asymmetric stretch mode, one of the bonds is increasing in length while the other one is decreasing. In the symmetric stretch vibration, both the bond lengths increase or decrease together. In the bending mode, the angle between the two bonds changes. The molecule also displays combination modes at 1.9 μ m and 2.7 μ m.

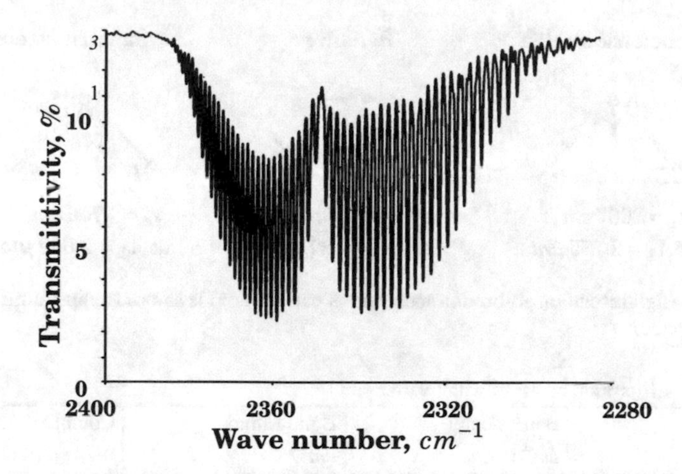

Fig. 11.9 Vibrational modes of a carbon dioxide molecule

A typical emission spectrum of carbon dioxide will span a range 500–6000 cm^{-1}. The bands that appear in the infrared spectrum are the asymmetric stretch centered at 2349 cm^{-1} (4.26 μ m) and the bending mode at 667 cm^{-1} (14.99 μ m). The symmetric stretch mode is infrared inactive. In the case of the bending mode, transitions with $\Delta J = 0$ are allowed and hence the spectrum shows three branches, the P branch corresponding to $\Delta J = -1$, the Q branch corresponding to $\Delta J = 0$, and the R branch corresponding to $\Delta J = 1$. Absorption corresponding to the band center is now strong since many of the rotational lines merge in the Q branch. Each band contains a rotational fine structure. Again, as we saw in the case of the carbon monoxide molecule, the width of each band varies with temperature.

At elevated temperatures, several vibrational states may be populated to a significant extent and hence combination bands also appear in the carbon dioxide spectrum. These bands are centered at 3690 cm^{-1} (2.7 μ m) and 5260 cm^{-1} (1.9 μ m). Thus unlike a diatomic gas for which only one band was seen, a polyatomic gas shows a number of bands in its spectrum. A typical high-resolution absorption spectrum of carbon dioxide is shown in Fig. 11.10 corresponding to the asymmetric stretch vibration of the molecule.

Example of Polyatomic Molecule, Water Vapor

Water molecule is a polar molecule that has a bent shape as shown in Fig. 11.8. Since it has an intrinsic dipole moment, it is infrared active in all the three fundamental bands shown in the figure. Also, absorption is observed due to rotational transitions. Because of this, the absorption of water vapor spans over a wide spectrum. Water vapor exhibits a large number of absorption bands as indicated in Table 11.1. Before proceeding with the modeling of radiation from gases, we shall consider a simple example involving radiation transfer across a slab of gray gas to bring out the role of geometric factors in gas radiation. Subsequently, it will be possible to deal with non-gray radiation from gases in a comprehensive fashion.

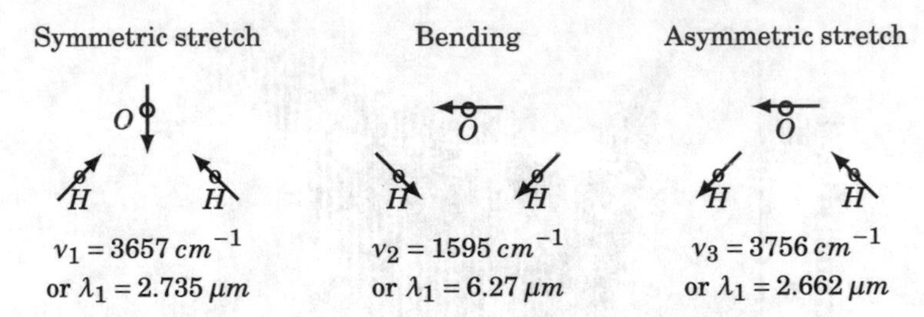

Fig. 11.10 High-resolution absorption spectrum of carbon dioxide at room temperature (asymmetric stretch mode)

Table 11.1 Absorption bands of water vapor

Band μm	Band Center cm^{-1}	Band Limits cm^{-1}	Combination (n_1, n_2, n_3)*
		$< 1000\,\text{cm}^{-1}$	Rotational
6.3	1587	1150–2050	(0, 1, 0)
4.7	2130		(1,−1,0)† (0, −1, 1)
2.7	3760	2800–4400	(0, 2, 0) (1, 0, 0) (0, 0, 1)
1.87	5351	4800–5900	(0, 1, 1)
1.38	7246	6500–8000	(1, 0, 1)
1.1	9091	8300–9300	(1, 1, 1)
0.94	10638	10100–11500	(2, 0, 1) (0, 0, 3)

*The n's are zero or integers such that $\nu = n_1\nu_1 + n_2\nu_2 + n_3\nu_3$
† Negative n requires upper vibrational states to be populated
The band is referred to as a difference band

11.3.4 Radiation in an Isothermal Gray Gas Slab and the Concept of Mean Beam Length

A plane isothermal gas layer problem is introduced through Fig. 11.11. An infinitely large black wall at temperature T_w oriented normal to the x-axis is located at $x = 0$. The plane layer of gas at temperature T_g is of thickness L as shown.

Intensity Variation with Angle
Consider an area element located at $x = L$ as indicated in the figure. The radiation incident on this element consists of two parts.

- Radiation leaving the wall at $x = 0$ which passes through the layer and then is incident on the area element;
- Radiation emitted by the intervening gas that is incident on the area element.

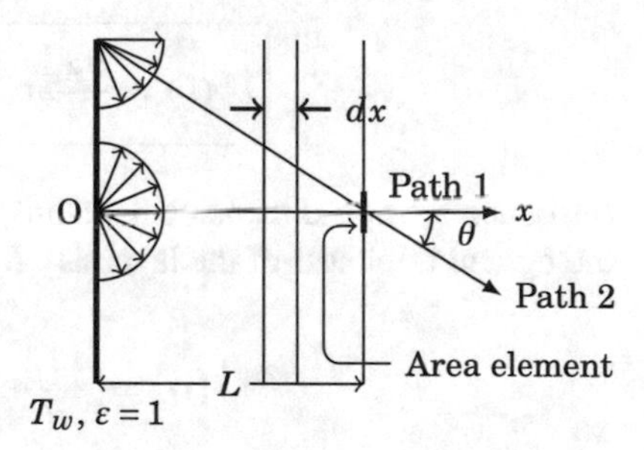

Fig. 11.11 Gas Radiation in an isothermal slab of gas

In order to calculate these, let us consider two paths.

1. Straight path 1 coinciding with the x-axis;
2. Slant path 2, a representative one is as shown in the figure. Note that an infinite number of slant paths are possible!

1. *Straight path*

The equation of transfer for path 1 is simply given by Eq. 11.3 where we replace T_m by T_g, I_λ as I^+, $I_{b\lambda}(T_g)$ by $\frac{\sigma T_g^4}{\pi}$, and κ_λ by κ the gray absorption coefficient. The equation now reads as

$$\frac{dI^+}{dx} + \kappa I^+ = \frac{\kappa\sigma T_g^4}{\pi} \tag{11.18}$$

The solution to the equation consists of two parts. The first part, the complementary function, is obtained by solving the homogeneous equation, i.e., by setting the right-hand side term in Eq. 11.18 to zero. Thus, we have

$$I^+(x) = A\mathrm{e}^{-\kappa x} \tag{11.19}$$

Here A is a constant of integration. The second part, a particular integral, is verified by actual substitution to be given by $I^+ = \frac{\sigma T_g^4}{\pi}$. The general solution is obtained by adding the above two solutions. Thus

$$I^+(x) = A\mathrm{e}^{-\kappa x} + \frac{\sigma T_g^4}{\pi} \tag{11.20}$$

At $x = 0$, the intensity is nothing but the black body intensity at the wall temperature, i.e., $I^+(0) = \frac{\sigma T_w^4}{\pi}$. Substituting this in Eq. 11.20, we have $A = \frac{\sigma}{\pi}(T_w^4 - T_g^4)$. Substituting this expression for A in Eq. 11.20, and on rearrangement, we get

$$I^+(x) = \frac{\sigma T_w^4}{\pi}\mathrm{e}^{-\kappa x} + \frac{\sigma T_g^4}{\pi}(1 - \mathrm{e}^{-\kappa x}) \tag{11.21}$$

The intensity at $x = L$ is then given by

$$I^+(L) = \frac{\sigma T_w^4}{\pi} e^{-\kappa L} + \frac{\sigma T_g^4}{\pi} \left(1 - e^{-\kappa L}\right) \quad (11.22)$$

Introduce the non-dimensional quantity $\tau = \kappa x$, the optical thickness, and represent the optical thickness of the layer as $\kappa L = \tau_L$, to recast Eq. 11.22 as

$$I^+(\tau_L) = \frac{\sigma T_w^4}{\pi} e^{-\tau_L} + \frac{\sigma T_g^4}{\pi} (1 - e^{-\tau_L}) \quad (11.23)$$

From the figure it is clear that along path 2, the slant length $\frac{x}{\cos\theta} = \frac{x}{\mu}$ replaces x along the straight path. Hence, for the slant path we have the optical thickness equal to $\frac{\tau_L}{\mu}$. In the solution given by Eq. 11.23, we simply replace τ_L by $\frac{\tau_L}{\mu}$ to get the intensity incident on the area element at angle θ as

$$I^+\left(\frac{\tau_L}{\mu}\right) = \frac{\sigma T_w^4}{\pi} e^{-\frac{\tau_L}{\mu}} + \frac{\sigma T_g^4}{\pi} \left(1 - e^{-\frac{\tau_L}{\mu}}\right) \quad (11.24)$$

Equations 11.23 and 11.24 may be recast in the form

$$I_{\tau_L}^+(\mu) = \frac{\sigma T_w^4}{\pi} e^{-\frac{\tau_L}{\mu}} + \frac{\sigma T_g^4}{\pi} \left(1 - e^{-\frac{\tau_L}{\mu}}\right) \quad (11.25)$$

Expression 11.25 represents the angular variation of intensity at $x = L$. In spite of the fact that both the emission from the wall and the gas are isotropic, the intensity at $x = L$ is not isotropic! This is due to the variation of path length or the optical thickness with angle. Thus, the geometry plays an important role in determining the angular variation of intensity.

Radiant Heat Flux

Based on the above solution, we calculate the radiant fluxes of interest to us. At $x = 0$, the radiant flux leaving the black wall is $q^+(x = 0) = q_0^+ = \sigma T_w^4$. At $x = L$, the heat flux is calculated, using the definition of heat flux and Eq. 11.25, as

$$q_L^+ = 2\pi \int_0^1 I_{\tau_L}^+(\mu)\mu d\mu = 2\pi \left[\frac{\sigma T_w^4}{\pi} \int_0^1 e^{-\frac{\tau_L}{\mu}} \mu d\mu + \frac{\sigma T_g^4}{\pi} \int_0^1 (1 - e^{-\frac{\tau_L}{\mu}})\mu d\mu \right] \quad (11.26)$$

We note that $\int_0^1 \mu d\mu = \frac{1}{2}$ and the integral $\int_0^1 e^{-\frac{\tau_L}{\mu}} \mu d\mu = E_3(\tau_L)$, the exponential integral of order 3, using the results from Appendix F. We may thus write Eq. 11.26 as

$$\boxed{q_L^+ = 2\sigma T_w^4 E_3(\tau_L) + \sigma T_g^4[1 - 2E_3(\tau_L)]} \tag{11.27}$$

The heat flux at $x = L$ thus consists of two terms. The first term involves the wall radiation multiplied by the factor $2E_3(\tau_L)$. The second term contains the contribution due to emission from the gas that is a product of black body emissive power and the factor $1 - 2E_3(\tau_L)$. These two factors are referred to, respectively, as the transmittivity t_g and the emissivity ε_g of the gas. Thus, we may rewrite Eq. 11.27 as

$$\boxed{q_L^+ = t_g \sigma T_w^4 + \varepsilon_g \sigma T_g^4} \tag{11.28}$$

where

$$\boxed{t_g = 2E_3(\tau_L),\ \varepsilon_g = 1 - 2E_3(\tau_L)} \tag{11.29}$$

Thus, we see that the emissivity and transmittivity of the gas depend on the absorption coefficient as well as the path length through the gas. Consider the transmittivity of the gas. It is given by twice the Exponential function of order 3 and having argument τ_L or κL. The exponential integral function itself integrates an exponential function over paths that are at different angles across the gas slab and hence different path lengths which range from L when $\theta = 0$ to ∞ when $\theta = 90°$. It is therefore physically meaningful to visualize it as an exponential with argument κL_m where L_m is a mean of all the possible paths—or simply the *mean beam length*. Thus, we introduce the concept of mean beam length such that

$$\boxed{2E_3(\kappa L) = \mathrm{e}^{-\kappa L_m}} \tag{11.30}$$

Further, this makes sense in that when $\kappa L \ll 1$, then $\mathrm{e}^{-\kappa L_m} = 1 - \kappa L_m$ and $2E_3(\kappa L) = 1 - 2\kappa L_m$ and hence the mean beam length, in this case of an *optically thin medium*, i.e., $\kappa L \ll 1$, is just twice the slab thickness. We also see from Appendix F that the Exponential integral function of order 3 is very closely approximated by an exponential function given by $\frac{1}{2}\mathrm{e}^{-1.8t}$. Hence, we may identify $1.8L$ as the mean beam length in a more general sense. The main conclusion to be drawn from the above treatment is that the gas transmittivity may be calculated based on a suitably defined mean beam length L_m such that

$$\boxed{t_g = \mathrm{e}^{-\kappa L_m}} \quad \text{and} \quad \boxed{\varepsilon_g = 1 - \mathrm{e}^{-\kappa L_m}} \tag{11.31}$$

Example 11.2

Consider a gray gas with an absorption coefficient of $\kappa = 0.2\,\text{m}^{-1}$. It is maintained at a temperature of $350\,K$ and is 0.5 m thick. A black wall at $400\,K$ is located at $x = 0$. Determine the intensity at $x = 0.5$ m for a straight path as well as for a slant path at 45°. Make a plot of intensity as a function of angle at $x = 0.5$ m

Solution :

Step 1 The given data is written down as

$$T_w = 400\,\text{K},\ T_g = 350\,\text{K},\ L = 0.5\,\text{m and}\ \kappa = 0.2\,\text{m}^{-1}\ \tau_L = 0.5 \times 0.2 = 0.1$$

Step 2 The intensity of radiation leaving the wall is

$$I_0^+ = \frac{\sigma T_w^4}{\pi} = \frac{5.67 \times 10^{-8} \times 400^4}{\pi} = 462.03\ \text{W/m}^2\ sr$$

Step 3 *Straight path*: We make use of Eq. 11.23 to evaluate the intensity at $x = 0.5\ m$ or $\tau_L = 0.1$.

$$I_L(\mu = 1) = \frac{5.67 \times 10^{-8} \times 400^4}{\pi} e^{-0.1} + \frac{5.67 \times 10^{-8} \times 350^4}{\pi}(1 - e^{-0.1}) = 443.84\ \text{W/m}^2 \cdot sr$$

Step 4 *Slant path at $\theta = 45°$ or $\mu = 0.707$*: The intensity at $x = L$ for the slant path is obtained by setting $\mu = 0.707$ and hence the effective optical thickness as $\tau = \frac{0.1}{0.707} = 0.1414$ in Eq. 11.25. Thus

$$I_L(\mu = 1) = \frac{5.67 \times 10^{-8} \times 400^4}{\pi} e^{-0.1414} + \frac{5.67 \times 10^{-8} \times 350^4}{\pi}(1 - e^{-0.1414}) = 436.82\ \text{W/m}^2 \cdot sr$$

Step 5 The intensities in other directions may be calculated to get the plot of intensity as a function of the slant path angle shown in Fig. 11.12. When $\theta = \frac{\pi}{2}$, the intensity at $x = L$ is due only to the radiation from the gas as the wall radiation is totally attenuated. The intensity tends to

$$I_L(\mu = 0) = \frac{\sigma T_g^4}{\pi} = \frac{5.67 \times 10^{-8} \times 350^4}{\pi} = 270.84\ \text{W/m}^2 \cdot sr$$

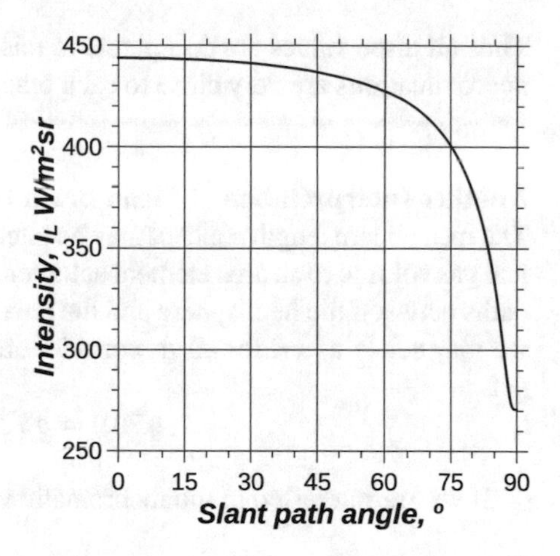

Fig. 11.12 Variation of radiation intensity with angle at $x = 0.5$ m in Example 11.2

Example 11.3

Consider the data given in Example 11.2. Use the exact, small t approximation as well as the exponential approximation for evaluating the heat flux at $x = 0.5$ m and comment on the results.

Solution :

From the data, the optical thickness of the gas layer is $\tau_L = 0.2 \times 0.5 = 0.1$. The exact value of the exponential integral of order 3 is read off Table F.1 as 0.416291. Using this value in Eq. 11.27, the exact heat flux at $x = L = 0.5$ m is

$$q_L^+ = 5.67 \times 10^{-8}[2 \times 400^4 \times 0.416291 + 350^4(1 - 2 \times 0.416291)]$$
$$= 1350.96 \text{ W/m}^2$$

The small t approximation yields $E_3(0.1) \approx \frac{1}{2} - 0.1 = 0.4$. Using this value in Eq. 11.27, the heat flux at $x = L = 0.5$ m is

$$q_L^+ = 5.67 \times 10^{-8}[2 \times 400^4 \times 0.4 + 350^4(1 - 2 \times 0.4)] = 1331.39 \text{ W/m}^2$$

The exponential approximation yields $E_3(0.1) \approx \frac{1}{2}e^{-1.8\times0.1} = 0.417635$. Using this value in Eq. 11.27, the heat flux at $x = L = 0.5$ m is

$$q_L^+ = 5.67 \times 10^{-8}[2 \times 400^4 \times 0.417635 + 350^4(1 - 2 \times 0.417635)]$$
$$= 1352.57 \text{ W/m}^2$$

Thus all three values are acceptable in this problem. The exact and the exponential approximations are very close to each other.

Another Interpretation of Mean Beam Length

The mean beam length concept may be interpreted using radiation from a hemispherical gas volume to an area element at its center as shown in Fig. 11.13. In this case, all paths between the hemisphere and the area element have the same length and hence we may set $\frac{\tau_L}{\mu}$ as κR for all μ, consider only emission from the gas in Eq. 11.27 to get

$$q^-(0) = \sigma T_g^4[1 - \mathrm{e}^{-\kappa R}] \tag{11.32}$$

If we are interested in monochromatic values, Eq. 11.32 would read as

$$q_\lambda^-(0) = \pi I_{b\lambda}(T_g)[1 - \mathrm{e}^{-\kappa_\lambda R}] \tag{11.33}$$

The center of the hemisphere is indicated by the argument 0 and the flux is shown with a superscript—to indicate that it is incident on the area element. Since only the gas radiation is assumed to be important, the non-grayness of the gas does not pose any problem. In the optically thin case, Eq. 11.33 may be replaced by

$$q_\lambda^-(0) \approx \pi I_{b\lambda}(T_g)\kappa_\lambda R \tag{11.34}$$

Thus for a hemispherical gas volume, the monochromatic gas emissivity, under the optically thin approximation, is given by

$$\varepsilon_\lambda = \kappa_\lambda R \tag{11.35}$$

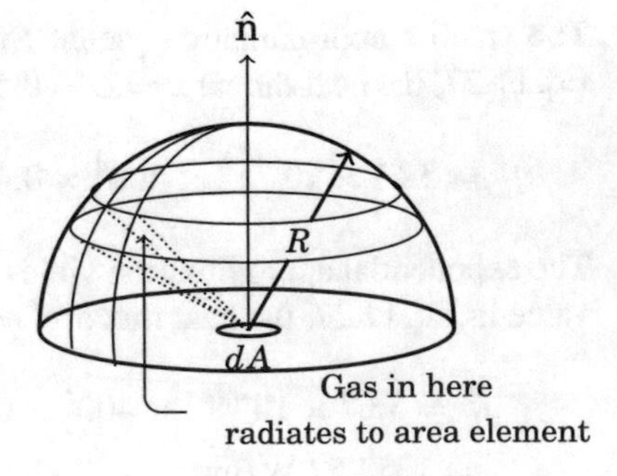

Fig. 11.13 Radiation from a hemispherical gas volume at temperature T_g to an area element at its center

Thus the mean beam length is R itself. We may thus *interpret* the mean beam length for an arbitrarily shaped gas volume as the radius of a hemispherical gas volume radiating to an area element at its center.

It is possible to associate a mean beam length with commonly encountered gas shapes as given in Table 11.2.

Table 11.2 Mean beam length for radiation from the entire gas volume in common geometries

No.	Geometry	L_m
1	Sphere of diameter D radiating to inside surface	0.65D
2	Hemisphere of diameter D radiating to its center	0.5D
3	Infinitely long circular cylinder of diameter D radiating to inside cylindrical surface	0.95D
4	Semi-infinite circular cylinder of diameter D radiating to	
(a)	Element at the center of its base	0.9D
(b)	Entire base	0.65D
5	Circular cylinder of diameter D and height $H = 2D$ radiating to	
(a)	Plane end	0.6D
(b)	Cylindrical surface	0.76D
(c)	Entire surface	0.72D
6	Circular cylinder of diameter D and height $H = 0.5D$ radiating to	
(a)	Plane end	0.43D
(b)	Cylindrical surface	0.46D
(c)	Entire surface	0.45D
7	Volume between two infinite parallel planes spaced L apart radiating to an element on one face	1.8L
8	Cube of edge L radiating to any of its six faces	0.6L
9	Gas volume outside an infinite bank of tubes with tube diameter D and center to center distance S radiating to a single tube	
(a)	Equilateral triangle array $S = 2D$	3 D
(b)	Equilateral triangle array S = 3D	7.6 D
(c)	Square array $S = 2D$	3.5 D
10	Arbitrary volume V of surface area A	3.6V/A

11.4 Modeling of Gas Radiation

The modeling of gas radiation is essential because of its manifold applications in engineering and in many natural processes that occur in the earth's atmosphere. In most engineering applications such as furnaces, the path length is finite and may at most be several meters. However, in atmosphere-related applications, the path lengths may be several kilometers. Because of this, even very weak absorption lines will play an important role in modifying radiation passing through the atmosphere. As mentioned earlier, solar energy passing in through the atmosphere and long-wavelength terrestrial radiation passing out through the atmosphere are two cases of great importance in determining global weather as well as the long-term happenings within the earth's atmosphere. One may also use such radiation to interrogate the atmosphere to determine its state by remote sensing using satellites. While heat transfer applications may be satisfactorily carried out by average absorption over absorption bands of atmospheric gases, the study of the behavior of the atmosphere may require line-by-line calculations. The latter is performed using databases such as the LOWTRAN [4] or the HITRAN[5] accessible on the web. We shall consider only heat transfer applications in what follows. The interested reader may consult the appropriate literature for atmospheric radiation.

11.4.1 Basics of Gas Radiation Modeling

The absorption of radiation by gases is characterized by the spectral absorption coefficient over the spectral region in which a particular gas may absorb (or emit). The amount absorbed depends, as mentioned already, on the amount of gas present, the path length as well as the temperature. As seen previously, gas absorption takes place over discrete bands with a fine structure due to rotational lines. No absorption or emission would take place with lines of zero "width" since the probability of a photon having the correct transition frequency becomes vanishingly small. However, these lines have a finite "width" because of the broadening of the lines due to three reasons.

(1) *Natural broadening*:
The absorption or emission of radiation takes place by the transition of a molecule from one energy level to another energy level. According to the uncertainty principle, the transition cannot all take place with precisely the same energy change and hence the absorbed or emitted photons will have a range of values around that calculated using the energies of the two states. The line thus gets smeared or broadened.

(2) *Collision broadening*:
The gas molecules are continually in thermal motion and have frequent collisions with other gas molecules. Collisions disrupt the process of transition between energy

[4] http://www1.ncdc.noaa.gov/pub/software/lowtran.

[5] http://www.cfa.harvard.edu/hitran.

states involved and hence lead to the broadening of the absorption or emission line. During each collision, the excess or deficit energy with respect to the line center is taken care of by the collision process.

Both of the above mechanisms lead to a line shape given by the Lorentz profile. While contribution due to natural broadening is very small, broadening due to collisions is important. If the line central frequency is ν_0, γ_c is the line half width, and S is the integrated absorption coefficient or the integrated line strength (instead of κ_λ, we use κ_ν—these are interconvertible), the Lorentz line shape is given by

$$\kappa_\nu = \frac{S}{\pi\gamma_c}\frac{1}{\left[\frac{\nu-\nu_0}{\gamma_c}\right]^2+1} \tag{11.36}$$

where

$$S = \int\limits_{\Delta\nu} \kappa_\nu d\nu \tag{11.37}$$

where $\Delta\nu$ represents the interval of integration. Even though the integral is essentially from $-\infty$ to ∞, the interval is finite and is over several half widths only since the integrand becomes very small beyond this range. The number of collisions depends on the concentration (number density and hence on the pressure) as well as the temperature. The number density is essentially dependent on $\frac{p}{T}$ where p is the pressure. The mean molecular speed varies as the square root of temperature T and hence the number of collisions varies as $\sqrt{T}$. Hence, the line half width dependence on p and T is of the form

$$\gamma_c = \gamma_{c0}\frac{p}{p_0}\sqrt{\frac{T_0}{T}} \tag{11.38}$$

where the subscript 0 refers to a suitable reference state of the gas.

(3) *Doppler broadening*:
The absorption or emission of photons by moving molecules is shifted due to the Doppler effect. If the central frequency is ν_0, and if the molecule undergoing transition is moving with a speed v in a direction parallel to the incident radiation (may be positive or negative), the Doppler shift is given by

$$\nu = \nu_0\left(1+\frac{v}{c}\right) \text{ or } \nu-\nu_0 = \nu_0\frac{v}{c} \tag{11.39}$$

where c is the speed of incident light. The line consequently gets broadened. Since the molecular speed distribution follows the Boltzmann distribution, the line shifts are also distributed according to the same distribution. With the Doppler half width represented by γ_D, the Doppler broadened line shape is given by

$$\kappa_\nu = \sqrt{\frac{\ln 2}{\pi}}\frac{S}{\gamma_D}\exp\left\{-\ln 2\left(\frac{\nu-\nu_0}{\gamma_D}\right)^2\right\} \tag{11.40}$$

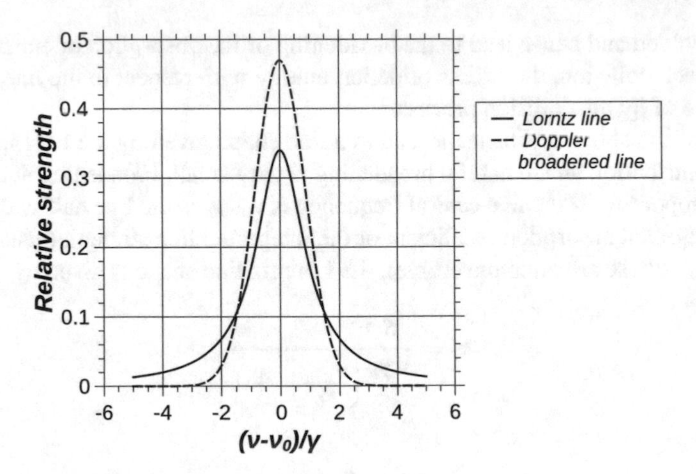

Fig. 11.14 Lorentz and Doppler broadened lines

where the Doppler half width is given by

$$\gamma_D = \frac{\nu_0}{c}\sqrt{\frac{2kT}{m}\ln 2} \tag{11.41}$$

In Eq. 11.41, m is the mass of the molecule and k is the Boltzmann constant. In Fig. 11.14, we compare Lorentz and Doppler line shapes, assuming that the half width of the two are the same.
It is seen that the Collision broadened line is broader than the Doppler broadened line.

Consider the equation of transfer 11.3 again. The equation has been written for a particular wavelength λ or the corresponding ν. In application, this equation has to be integrated over a line or a band if we are interested in the change in the integrated intensity over a line or a band as the case may be. The integration may be carried out using the strategy given below.

The line width associated with the broadening mechanisms described above is usually of the order of a fraction of a wave number. For example, the rotational line of carbon monoxide has a typical width of $\gamma = 0.01$ cm^{-1}. The broadened line has typically a width of $\pm 5\gamma$=0.05 cm^{-1}. The Planck function (or the intensity) may be assumed to be a constant over the line. Hence, the term representing absorption over a line in Eq. 11.3 may be written as the product of the intensity at the line center wavelength and the integral over the line as

$$\int\limits_{\Delta\nu} I_\nu \kappa_\nu d\nu = I_\nu \int\limits_{\Delta\nu} \kappa_\nu d\nu \tag{11.42}$$

where $\Delta\nu$ represents the width of the line which may be taken as some $\pm 5\gamma$ around the center of the line. Now consider the absorption of radiation over a vibrational rotational band. Again, for example, the width of the band for carbon monoxide is from $\nu_1 = 1950 - \nu_2 = 2350$ cm^{-1}. The Planck function is very closely a linear function of ν over the entire band. In case one would like to find the term representing gas emission over the carbon monoxide band in Eq. 11.3, the following may be done:

$$\int_{\nu_1}^{\nu_2} I_{b\nu}\kappa_\nu d\nu = I_{b\nu_c}\int_{\nu_1}^{\nu_2} \kappa_\nu d\nu \tag{11.43}$$

where ν_c corresponds to the band center. The indicated integration may be made if we know the positions of rotational lines and their shapes. The point is that the integration over the frequency may be made independent of the Planck function. To the extent that the shape of the line is dependent on the gas temperature and pressure, the integrated absorption over the band will depend on these parameters.

The integration over the band is itself based on a suitable model for the position of each line within the band and the strength of each line. One may use different models for these and thus there is some variation in the description of absorptivity or emissivity of a gas. The geometric part may be accounted for by using the mean beam length coupled with an exponential dependence of the absorptivity or the emissivity on the mean beam length absorption coefficient (integrated over a line or an absorption band) product.

11.4.2 Band Models

Various band models have been used to describe the properties of a participating medium. The vibration rotation band is modeled as being made up of a number of lines using one of the following:

Narrow band models: The narrow band models account for the rapid variation of spectral absorptivity due to rotational lines and the more gentle variation due to the variation of line intensity due to various broadening mechanisms dealt with earlier.
Wide band models: The wide band models describe the variation of mean absorption coefficient over the band.

However, at a simpler level of description suitable for radiative transfer calculations, it is usual to describe the gas properties in terms of total quantities. This description leads to tables, charts, or correlations as given by Hottel[6] and later by Leckner,[7] or the weighted sum of the gray gas model (WSGG).[8]

Leckner Model

This model uses detailed spectral data for common combustion gases, carbon dioxide and water vapor, to calculate the total emissivity by a statistical model with spectral intervals of 5–25 cm^{-1}. The lines within such intervals are treated as being randomly positioned and line strengths are exponentially distributed. Literature data is used in the calculations. The emissivity of the gas is represented as

$$\varepsilon_i = 1 - \exp\left[-\frac{\kappa_i X}{\sqrt{1 + \frac{\kappa_i X}{4a_i}}}\right] \tag{11.44}$$

where κ_i is the mean value of the absorption coefficient in the interval in units of $(bar\ cm)^{-1}$, reduced to a reference temperature of $T_0 = 273$ K and a reference pressure of $P_0 = 1\ bar$. The quantity X is given by

$$X = \frac{pL}{P_0}\frac{T_0}{T} \tag{11.45}$$

which is the optical path length in *bar cm* reduced to reference temperature and pressure, p is the partial pressure of the gas specie in the gas mixture in *bar*, L is the geometrical path length in *cm* and T is the gas temperature in K. a_i, the fine structure parameter for the interval, is defined as

$$a_i = \frac{\gamma}{d_i} \tag{11.46}$$

where d_i is the mean line spacing in the interval in cm^{-1}, γ is the mean line half width of the gas at T, p and total pressure P_T. The line half widths, as we have seen earlier, depend on the broadening mechanisms, and are functions of pressure and temperature. The gas mixture is assumed to consist of carbon dioxide, water vapor, and nitrogen. The line widths for carbon dioxide and water vapor are given by

$$\gamma_{CO_2} = 0.07 P_T \sqrt{T_0 T}\left(1 + 0.28\frac{p_c}{P_T}\right) \tag{11.47}$$

$$\gamma_{H_2O} = 0.09 P_T \sqrt{T_0 T} + 0.44 p_w \frac{T_0}{T} \tag{11.48}$$

where subscripts c and w stand, respectively, for CO_2 and H_2O. The total emissivity of the gas mixture is then given by

---

[6] H. C. Hottel and A. F. Sarofim, *Radiative Transfer*, McGraw Hill, NY, 1967.

[7] B. Leckner, Combustion and Flame, Vol. 19, pp. 33–48, 1972.

[8] T. F. Smith et al., ASME Journal of Heat Transfer, Vol. 104, pp. 602–608, 1982.

$$\varepsilon_g = \frac{\sum_i f_{\Delta\nu}(T_g)\varepsilon_i \Delta\nu}{\sigma T_g^4} \tag{11.49}$$

where $f_{\Delta\nu}$ is the integrated Planck function over the interval $\Delta\nu$. The emissivity obtained as above is referred to as the Planck Mean since Eq. 11.49 is nothing but an expression of the form

$$\varepsilon_g = \frac{\int_{\nu=0}^{\infty} E_b(\nu)\varepsilon_\nu d\nu}{\int_{\nu=0}^{\infty} E_b(\nu) d\nu} \tag{11.50}$$

Leckner also takes into account the overlapping of bands of CO_2 and H_2O and uses correction factors. Finally, the total emissivity data is represented in terms of polynomials of second degree as given in the previously cited reference. The calculations also have been presented in the form of charts there. However, the emissivity calculations may be performed using the spreadsheet program presented by Ronney.[9] The spreadsheet is appropriately named "PlanckMeanAndLeckner".

Input data entered in the designated cells consist of the following:

Total pressure	P_T in *atm*
Path length	L in m
Gas temperature	T_g in K
Background (wall) temperature	T_w in K
Mole fraction of water vapor	$X_w = \frac{p_w}{P_T}$
Mole fraction of carbon dioxide	$X_c = \frac{p_c}{P_T}$

The program output appears in the appropriate cells as

Emissivity	ε_w of H_2O
Emissivity	ε_c of CO_2
Emissivity correction	$\Delta\varepsilon$
Gas mixture emissivity	ε_g
Gas mixture absorptivity	α_g for T_g, T_w combination
Total absorption coefficient	κ

WSGG Model

According to the WSGG model, the emissivity or absorptivity of a gas is considered to be a sum of gray gas emissivities (independent of temperature) weighted with temperature-dependent factors. The total emissivity is hence written in the form

$$\varepsilon_g = \sum_{i=1}^{I} a_i(T_g)[1 - e^{-\kappa_i p_i L}] \tag{11.51}$$

[9] http://carambola.usc.edu/spreadsheets.

where i identifies the ith gray gas in the gas mixture with κ_i the absorption coefficient and $p_i L$ the partial pressure mean beam length product. The coefficient a_i may be interpreted as the fractional amount of radiation that exists in the absorption band of the ith gray gas. To account for "windows" in the absorption spectrum of the gas mixture, the gray gas $i = 0$ (clear gas) is assigned a value of $a_0 = 0$. Since all fractions must add to unity, we additionally have

$$\sum_{i=1}^{I} a_i = 1 \tag{11.52}$$

The WSGG model is derived by the use of literature data on the absorption of various gases that comprise the gas mixture. For example, if the gas mixture contains three components (two gray gases and one clear gas), the three values of κ_i are, respectively, 0, κ_1 and κ_2 and the corresponding weights are 0, a_1 and a_2. With Eq. 11.52 providing a condition, we require to essentially obtain three quantities. The weights are themselves functions of temperature and are represented typically in the polynomial form

$$a_i = \sum_{j=1}^{J} b_{i,j} T^{j-1} \tag{11.53}$$

where the polynomial coefficients are additional parameters that need to be determined. In case our interest is to determine the absorptivity of the gas, the weights will be functions of the temperature of the surface from which radiation originates and hence we should have polynomials of the form

$$a_i = \sum_{j=1}^{J} \left(\sum_{k=1}^{K} c_{i,j,k} T^{k-1} \right) T^{j-1} \tag{11.54}$$

All the parameters that are required in the WSGG model are determined by using appropriate data and using curve fitting procedures. More details may be obtained from the reference cited earlier.

Example 11.4

Consider a gas mixture at a total pressure of 1 *atm* containing equal proportions of water vapor and carbon dioxide with the total partial pressure of 0.2 *atm*. The rest may be assumed to be a non-absorbing constituent. The gas mixture is at a temperature of 1000 K and background radiation is from a black wall at an effective temperature of 600 K. The mean beam length may be taken as 1 m. Determine the emissivity and absorptivity of the gas using Leckner and WSGG models.

Solution :

Results based on Leckner model: The given data may be arranged as

$$T_g = 1000 \text{ K},\ P = 1 \text{ atm},\ p_w = p_c = 0.1 \text{ atm},\ T_w = 600 \text{ K},\ L = 1 \text{ m}$$

Hence, the mole fractions of water vapor and carbon dioxide are given by

$$X_w = X_c = \frac{0.1}{1} = 0.1$$

We make use of the spreadsheet prepared by Ronney to obtain the desired output as

$$\varepsilon_w = 0.14980,\quad \varepsilon_c = 0.10460,\quad \Delta\varepsilon = 0.01688$$
$$\varepsilon_g = 0.23752,\quad \alpha_g = 0.29554$$

Results based on WSGG: The values are based on the graph presented by Smith et al. The parameters on the graph are

$$\text{Gas temperature} = 1000 \text{ K}$$
$$\frac{p_w + p_c}{P} = \frac{0.05 + 0.05}{1} = 0.1$$
$$\frac{(p_w + p_c)L}{P} = \frac{(0.05 + 0.05) \times 1}{1} = 0.1$$

The desired emissivity of the gas mixture is read from the graph as

$$\varepsilon_g = 0.22$$

Two significant digits is all that one can expect from the graph. Similarly, the gas absorptivity is obtained as

$$\alpha_g = 0.28$$

The two ways of calculation are in good agreement.

Example 11.5

Two very large gray planes are placed parallel to each other. The top plane 1 is at 1200 K and has an emissivity of 0.85. The bottom plane 2 is at a temperature of 800 K and has an emissivity of 0.65. The gap between the planes is 1.5 m and is filled with a gray gas at 1050 K with an absorption coefficient of 0.074 m^{-1}. Determine the heat transfer at each of the boundaries. Also, determine the net heat transfer to the gas. What would be the heat transfer across the gap in the absence of the gas?

Solution: This problem consists of two surfaces (number of surface zones is (2) with an enclosed isothermal gas volume (number of gas zones is (1). Since all the radiation leaving surface zone 1 reaches surface zone 2, the angle factor area product is the area itself. We may base the calculations on a per unit area basis.

Step 1 Given data is written down as below.

Surface 1:	$T_1 = 1200K$	$\varepsilon_1 = 0.85$
Surface 2:	$T_1 = 800K$	$\varepsilon_1 = 0.65$
Gas:	$T_g = 1050$ K	$\kappa = 0.074$ m^{-1}
		$L = 1.5$ m

From Table 11.2, the mean beam length for this geometry is

$$L_m = 1.8L = 1.8 \times 1.5 = 2.7 \text{ m}$$

The optical thickness of the gas is calculated as

$$\tau_L = \kappa L_m = 0.074 \times 2.7 = 0.1998$$

Step 2 The optically thin approximation is not valid. Using the data given in Table F.1, we have $E_3(0.1998) = 0.352$. The gas emissivity and transmittivity are given by

$$\varepsilon_g = 1 - 2E_3(0.1998) = 1 - 2 \times 0.352 = 0.296$$
$$t_g = 2E_3(0.1998) = 2 \times 0.352 = 0.704$$

Step 3 Emissive powers:
The black body emissive powers are calculated as

$$E_{b1} = \sigma T_1^4 = 5.67 \times 10^{-8} \times 1200^4 = 117573.1 \text{ W/m}^2$$
$$E_{b2} = \sigma T_2^4 = 5.67 \times 10^{-8} \times 800^4 = 23224.3 \text{ W/m}^2$$
$$E_{bg} = \sigma T_g^4 = 5.67 \times 10^{-8} \times 1050^4 = 68919.2 \text{ W/m}^2$$

Step 4 Radiosity formulation:
For surface 1, we have

$$\begin{aligned} J_1 &= \varepsilon_1 E_{b1} + (1-\varepsilon_1)(\varepsilon_g E_{bg} + t_g J_2) \\ &= 0.85 \times 117573.1 + (1-0.85)(0.296 \times 68919.2 + 0.704 J_2) \\ &= 102997.1 + 0.1056 J_2 \end{aligned}$$

For surface 2, we have

$$\begin{aligned} J_2 &= \varepsilon_2 E_{b2} + (1-\varepsilon_2)(\varepsilon_g E_{bg} + t_g J_1) \\ &= 0.65 \times 23224.3 + (1-0.65)(0.296 \times 68919.2 + 0.704 J_1) \\ &= 22235.8 + 0.2464 J_2 \end{aligned}$$

The two radiosity equations may be written in the form of a matrix equation

$$\begin{bmatrix} 1 & -0.1056 \\ -0.2464 & 1 \end{bmatrix} \begin{Bmatrix} J_1 \\ J_2 \end{Bmatrix} = \begin{Bmatrix} 102997.1 \\ 22235.8 \end{Bmatrix}$$

Using Kramer's rule, the solution to the above equations is

$$J_1 = 108159.5 \text{ W/m}^2, \quad J_2 = 48886.3 \text{ W/m}^2$$

Step 5 Heat fluxes are calculated now.
q_1, the heat flux at the hot surface, is

$$q_1 = \frac{\varepsilon_1}{1-\varepsilon_1}(E_{b1} - J_1) = \frac{0.85}{1-0.85}(117573.1 - 108159.5) = 53343.7 \text{ W/m}^2$$

q_2, the heat flux at the cold surface, is

$$q_2 = \frac{\varepsilon_2}{1-\varepsilon_2}(E_{b2} - J_2) = \frac{0.65}{1-0.65}(23224.3 - 48886.3) = -47658 \text{ W/m}^2$$

The heat transfer to the gas q_g may be determined by energy balance.

$$q_g = -q_1 - q_2 = -53343.7 + 47658 = 5685.7 \text{ W/m}^2$$

It is seen that heat leaving the hot wall is transferred partly to the gas and partly to the cold wall.

Step 6 Results in the absence of gas:
If the gap between the slabs is evacuated, the heat transfer may be calculated using a simple two-surface enclosure analysis presented earlier. The heat transfer across the enclosure is given by Eq. 10.42 and hence

$$q_{1-2} = \frac{\sigma\left(1200^4 - 800^4\right)}{\left(\frac{1}{0.85} + \frac{1}{0.65} - 1\right)} = 55016.1\ \mathrm{W/m^2}$$

The heat transfer across the gap is more in the case of an evacuated enclosure as compared to the gas-filled enclosure.

The calculations may easily be extended to the case of a gas mixture whose properties may be evaluated by using the Leckner or the WSGG model. We consider a typical case to show how the calculation is performed.

Example 11.6

Two very large black plane surfaces are 0.3 m apart and the space between them is filled with a gas mixture containing 25% carbon dioxide, 25% water vapor, and rest nitrogen, by volume. The total pressure of the gas mixture is 1 atmosphere. One of the surfaces is at 1200 K while the other is at 600 K. Calculate the following:

1. The effective emissivity of the gas mixture at its temperature of 900 K
2. The effective absorptivity of the gas mixture to radiation from the two walls at their respective temperatures
3. The net rate of heat transfer from the hot wall to the cold wall.

Solution :

The given data is written down as below.

Black Wall 1:	$T_1 = 1200$ K	Black Wall 2:	$T_2 = 600$ K
Gas:	$T_g = 900$ K	Thickness:	$L = 0.3\ m$
	$X_c = 0.25$	$X_w = 0.25$	$P = 1\ atm$

The mean beam length may be calculated as

$$L_m = 1.8L = 1.8 \times 0.3 = 0.54\ \mathrm{m}$$

Radiation properties of the gas mixture are calculated using the spreadsheet "PlanckMeanAndLeckner". For the first wall gas combination, the appropriate properties are obtained as

$$\varepsilon_g = 0.290, \ \alpha_{g1} = 0.242$$

where α_{g1} is the gas absorptivity for radiation emanating from Wall 1. For the second wall gas combination, we have

$$\varepsilon_g = 0.290, \ \alpha_{g2} = 0.342$$

where α_{g2} is the gas absorptivity for radiation emanating from Wall 2. The radiosities of the two walls are simply the corresponding black body emissive powers given by

$$J_1 = E_{b1} = 5.67 \times 10^{-8} \times 1200^4 = 117573.1 \text{ W/m}^2$$
$$J_2 = E_{b2} = 5.67 \times 10^{-8} \times 600^4 = 7348.3 \text{ W/m}^2$$

Black body emissive power at gas temperature is

$$E_{bg} = 5.67 \times 10^{-8} \times 900^4 = 37200.9 \text{ W/m}^2$$

The irradiations may be calculated as

$$\begin{aligned} G_1 &= J_2 t_{g2} + \varepsilon_g E_{bg} \\ &= 7348.3 \times (1 - 0.342) + 0.290 \times 37200.9 = 15612.5 \text{ W/m}^2 \\ G_2 &= J_1 t_{g1} + \varepsilon_g E_{bg} \\ &= 117573.1 \times (1 - 0.242) + 0.290 \times 37200.9 = 99882.5 \text{ W/m}^2 \end{aligned}$$

Heat fluxes at the two walls are now calculated.

$$q_1 = J_1 - G_1 = 117573.1 - 15612.5 = 101960.6 \text{ W/m}^2$$
$$q_2 = J_2 - G_2 = 7348.3 - 99882.5 = -92534.1 \text{ W/m}^2$$

The gas absorbs an amount equal to $101960.6 - 92534.1 = 9426.5 \text{ W/m}^2$. The net heat transfer between the two walls is 92534.1 W/m^2.

We present yet another example, this time an enclosure with a single wall enclosing a gas mixture. The shape of the enclosure is taken as a sphere; its surface is gray and diffuse.

Example 11.7

A spherical vessel of diameter 0.4 m encloses a gas mixture at a total pressure of $P = 2$ atm. The gas mixture contains nitrogen at a partial pressure of 1 atm, water vapor at a partial pressure of 0.4 *atm*, and carbon dioxide at a partial pressure of 0.6 atm. The gas is at a temperature of $T_g = 800$ K, while the sphere surface is at a temperature of $T_s = 400$ K. The sphere is gray with an emissivity of $\varepsilon_s = 0.5$. Determine the radiant heat transfer to the sphere. Figure 11.15 gives a graphical description of the problem.

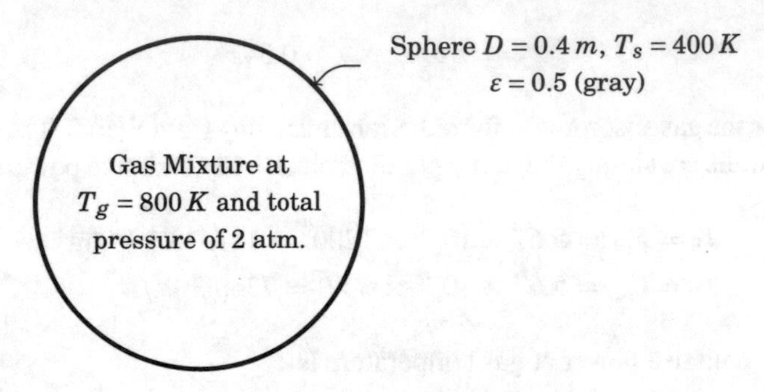

Fig. 11.15 Gas radiation with a single surface—Example 11.7

Solution :

In order to evaluate the gas emissivity and absorptivity, we need the mean beam length for this configuration. From Table 11.2, the mean beam length for a spherical gas volume is $L_m = \frac{2}{3}D$. Thus

$$L_m = \frac{2}{3} \times 0.4 = 0.2667\ m$$

Gas properties are now calculated using the Leckner model. The given data is arranged as

Total pressure	$P_T = 2$ atm
Path length	$L_m = 0.2667$ m
Gas temperature	$T_g = 800$ K
Sphere temperature	$T_s = 400$ K
Mole fraction of water vapor	$X_w = \frac{p_w}{P_T} = \frac{0.4}{2} = 0.2$
Mole fraction of carbon dioxide	$X_c = \frac{p_c}{P_T} = \frac{0.6}{2} = 0.3$

The output from spreadsheet "PlanckMeanAndLeckner" is given below:

Emissivity of water vapor:	$\varepsilon_w = 0.2264$
Emissivity of carbon dioxide:	$\varepsilon_c = 0.1152$
Emissivity correction:	$\Delta\varepsilon = 0.0208$
Gas mixture emissivity:	$\varepsilon_g = 0.2264 + 0.1152$ $-0.0208 = 0.3207$
Gas mixture absorptivity:	α_g=0.3925

The radiosity formulation may now be made noting that all the radiation that leaves the surface is incident on itself. However, it is attenuated by gas absorption and augmented by gas emission. Thus we have

$$J_s = \varepsilon_s E_{bs} + (1 - \varepsilon_s)(\varepsilon_g E_{bg} + t_g J_s)$$

We may solve this equation for J_s to get

$$J_s = \frac{\varepsilon_s E_{bs} + (1-\varepsilon_s)\varepsilon_g E_{bg}}{1-(1-\varepsilon_s)t_g}$$

We should note that $t_g = 1 - \alpha_g = 1 - 0.3925 = 0.6075$ in this case. The two emissive powers are given by

$$E_{bs} = 5.67 \times 10^{-8} \times 400^4 = 1451.5 \text{ W/m}^2$$
$$E_{bg} = 5.67 \times 10^{-8} \times 800^4 = 23224.3 \text{ W/m}^2$$

With these, thc radiosity of the surface is obtained as

$$J_s = \frac{0.5 \times 1451.5 + (1-0.5)0.3207 \times 23224.3}{1-(1-0.5) \times 0.6075} = 6391 \text{ W/m}^2$$

The irradiation on the surface is obtained as

$$G_s = \varepsilon_g E_{bg} + t_g J_s = 0.3207 \times 23224.3 + 0.6075 \times 6391 = 11330.5 \text{ W/m}^2$$

The wall heat flux may then be calculated as

$$q_s = J_s - G_s = 6391 - 11330.5 = -4940.5 \text{ W/m}^2$$

The negative sign indicates that the surface gains heat.

11.5 Radiation in a Non-isothermal Participating Medium

We are now ready to relax the condition that the gas volume is isothermal. We consider a simple but important problem of radiative transfer in a slab of gas wherein the temperature is a function of only one coordinate, the distance measured perpendicular to one of the boundaries. Most atmospheric models are based on the plane parallel layer model. The analysis is presented for a gray gas to make it simple. However, it is possible to extend the analysis to a non-gray gas whose properties may be determined by any of the methods presented earlier.

11.5.1 Radiation Transfer in a Gray Slab:

Consider a slab of gray gas of absorption coefficient κ of thickness L bounded by gray diffuse walls at $x = 0$ and $x = L$. Let ε_0, T_0 and ε_L, T_L be the specified emissivity temperature combinations at the two walls. Let us represent the radiosities of the two walls, respectively, as J_0 and J_L. Let the gas temperature at any location x be $T(x)$. The equation of transfer is written for the forward intensity along the $\mu(=\cos\theta)$ direction as

$$\mu\frac{dI^+}{dx}+\kappa I^+=\frac{\kappa}{\pi}E_b(x) \tag{11.55}$$

where $E_b(x)$ is the black body emissive power at the local gas temperature $T_g(x)$. We may integrate this equation by making use of the integrating factor given by $\mathrm{e}^{\frac{\kappa x}{\mu}}$ to get

$$I^+(x)=A\mathrm{e}^{-\frac{\kappa x}{\mu}}+\frac{\kappa}{\pi}\int_0^x \mathrm{e}^{\frac{\kappa(x'-x)}{\mu}}E_b(x')\frac{dx'}{\mu} \tag{11.56}$$

where A is an integration constant that is determined by equating $I^+(x=0)$ to $\frac{J_0}{\pi}$ at the left boundary. Hence, the solution is

$$I^+(x)=\frac{J_0}{\pi}\mathrm{e}^{-\frac{\kappa x}{\mu}}+\frac{\kappa}{\pi}\int_0^x \mathrm{e}^{\frac{\kappa(x'-x)}{\mu}}E_b(x')\frac{dx'}{\mu} \tag{11.57}$$

Introducing the optical thickness $\tau=\kappa x$, the above may be recast as

$$I^+(\tau)=\frac{J_0}{\pi}\mathrm{e}^{-\frac{\tau}{\mu}}+\frac{1}{\pi}\int_0^\tau \mathrm{e}^{\frac{\tau'-\tau}{\mu}}E_b(\tau')\frac{d\tau'}{\mu} \tag{11.58}$$

Using now the definition of the forward flux as $q^+=2\pi\int_0^1 I^+\mu d\mu$, Eq. 11.58 may be recast as

$$q^+(\tau)=2\int_0^1\left(J_0e^{-\frac{\tau}{\mu}}+\int_0^\tau \mathrm{e}^{\frac{\tau'-\tau}{\mu}}E_b(\tau')\frac{d\tau'}{\mu}\right)\mu d\mu \tag{11.59}$$

A similar procedure is used for obtaining the backward flux ($q^-=2\pi\int_0^{-1} I^-\mu d\mu$) to get

$$q^-(\tau)=2\int_0^1\left(J_Le^{-\frac{\tau_L-\tau}{\mu}}+\int_\tau^{\tau_L} \mathrm{e}^{\frac{\tau-\tau'}{\mu}}E_b(\tau')\frac{d\tau'}{\mu}\right)\mu d\mu \tag{11.60}$$

Using Exponential integrals defined by Eq. F.2, Eqs. 11.59 and 11.60 may be recast as

$$q^+(\tau) = 2\left(J_0 E_3(\tau) + \int_0^{\tau} E_2(\tau-\tau')E_b(\tau')d\tau'\right)$$
$$q^-(\tau) = 2\left(J_L E_3(\tau_L-\tau) + \int_{\tau}^{\tau_L} E_2(\tau'-\tau)E_b(\tau')d\tau'\right) \tag{11.61}$$

The net radiant flux is then obtained as

$$q = q^+ - q^- = 2J_0E_3(\tau) + 2\int_0^{\tau} E_2(\tau-\tau')E_b(\tau')d\tau'$$
$$-2J_LE_3(\tau_L-\tau) - 2\int_{\tau}^{\tau_L} E_2(\tau'-\tau)E_b(\tau')d\tau' \tag{11.62}$$

11.5.2 Radiation Equilibrium

If the gas is in radiative equilibrium with the walls the net heat flux, which in general is a function of τ, should be a constant. Hence $\frac{dq}{d\tau} = 0$. Taking the derivative of Eq. 11.62 with respect to τ and setting it to zero, we get

$$0 = -2J_0E_2(\tau) - 2\int_0^{\tau} E_1(\tau-\tau')E_b(\tau')d\tau' + 2E_2(0)E_b(\tau) -$$
$$2J_LE_2(\tau_L-\tau) - 2\int_{\tau}^{\tau_L} E_1(\tau'-\tau)E_b(\tau')d\tau' + 2E_2(0)E_b(\tau) \tag{11.63}$$

In writing the above, we have made use of Leibnitz' rule according to which

$$\frac{d}{dx}\int_{a(x)}^{b(x)} F(x-x')dx' = \int_{a(x)}^{b(x)} \frac{dF(x-x')}{dx}dx'$$
$$+ F[x-b(x)]\frac{db}{dx} - F[x-a(x)]\frac{da}{dx}$$

and Eq. F.2 for the derivative of the Exponential integral function. We introduce the non-dimensional emissive power given by

$$e_b = \frac{E_b - J_L}{J_0 - J_L} \tag{11.64}$$

Consider the integrals appearing in Eq. 11.63. We have, in terms of the non-dimensional emissive power,

$$\int_0^{\tau} E_1(\tau-\tau')E_b(\tau')d\tau' = \int_0^{\tau} E_1(\tau-\tau')[(J_0-J_L)e_b(\tau')+J_L]d\tau'$$

$$=(J_0-J_L)\int_0^{\tau} E_1(\tau-\tau')e_b(\tau')d\tau' + J_L[E_2(0)-E_2(\tau)]$$

where the last part has used Eq. F.4 for integrating the Exponential integral over the interval 0, τ. Similarly, we have

$$\int_{\tau}^{\tau_L} E_1(\tau'-\tau)E_b(\tau')d\tau' = \int_{\tau}^{\tau_L} E_1(\tau'-\tau)[(J_0-J_L)e_b(\tau')+J_L]d\tau'$$

$$=(J_0-J_L)\int_{\tau}^{\tau_L} E_1(\tau'-\tau)e_b(\tau')d\tau' + J_L[E_2(0)-E_2(\tau_L-\tau)]$$

Introducing these in Eq. 11.63, noting that $E_2(0)=1$, combining the two integrals and after simplification, we get

$$\boxed{2e_b(\tau) = E_2(\tau) + \int_0^{\tau_L} E_1(|\tau-\tau'|)e_b(\tau')d\tau'} \tag{11.65}$$

Define non-dimensional heat flux as

$$q^* = \frac{q}{J_0-J_L} \tag{11.66}$$

Noting that q and hence q^* are constant under radiative equilibrium, it may be evaluated at any τ and hence say at $\tau=0$, in Eq. 11.62. The reader may show that the non-dimensional heat flux is

$$\boxed{q^* = 1-2\int_0^{\tau_L} e_b(\tau')E_2(\tau')d\tau'} \tag{11.67}$$

11.5.3 Solution of Integral Equation

Equation 11.65 is easily recognized as the Fredholm integral equation of the second kind. The integral equation is linear and the solution depends only on one parameter, 0., $\tau_L = \kappa L$. The Kernel in the present case is $E_1(|\tau - \tau'|)$ which is singular at $\tau = \tau'$, i.e., $E_1(|\tau - \tau'|) \to \infty$ as $\tau \to \tau'$. However, this does not create any problem as far as the solution of the integral equation is concerned. Heaslet and Warming[10] have solved the integral equation using tabulated functions and hence their solution may be termed "exact". Usiskin and Sparrow[11] presented a numerical solution by an iterative scheme.

In order to appreciate the procedure, we shall solve the integral equation by assuming the emissive power to be given by a piecewise constant representation of the non-dimensional emissive power. Consider the case where $\tau_L = 0.1$. The domain is divided in to eight parts and the mid-points of these elements are assigned the unknown emissive powers. The integral appearing in Eq. 11.65 is replaced by a sum. The mid-points of the elements are located at $\tau_1 = 0.00625$, $\tau_2 = 0.01875$, $\tau_3 = 0.03125$, $\tau_4 = 0.04375$, $\tau_5 = 0.05625$, $\tau_6 = 0.06875$, $\tau_7 = 0.08125$, and $\tau_8 = 0.09375$. The integral equation is written for these 8 values of τ to get 8 simultaneous equations for the nodal emissive powers e_{bi} where $1 \le i \le 8$. As an example, consider the first node at $\tau_1 = 0.00625$. The integral equation applied to this point will read as

$$
\begin{aligned}
2e_{b1} = E_2(0.00625) &+ \int\limits_0^{0.00625} E_1(0.00625 - \tau')e_{b1}(\tau')d\tau' \\
&+ \int\limits_{0.00625}^{0.0125} E_1(\tau' - 0.00625)e_{b1}(\tau')d\tau' + \sum_{i=2}^{8} e_{bi}I_i
\end{aligned}
\tag{11.68}
$$

where I_i are integrals given by

$$
I_i = \int\limits_{\tau_i - 0.00625}^{\tau_i + 0.00625} E_1(\tau' - 0.00625)d\tau' \tag{11.69}
$$

These integrals may be obtained easily by using Eq. F.4 as

$$
I_i = \int\limits_{\tau_i - 0.00625}^{\tau_i + 0.00625} E_1(\tau' - 0.00625)d\tau' = E_2(u) - E_2(l)
$$

[10] M. A. Heaslet and R. F. Warming, Int. J. Heat and Mass Transfer, Vol. 8, pp. 979–994, 1965.
[11] C. M. Usiskin and E. M. Sparrow, Int. J. Heat and Mass Transfer, Vol. 1, pp. 28–36, 1960.

where $u = \tau_i + 0.0125$ and $l = \tau_i$ are the upper and lower limits in the integral. We may use the same procedure for all the other values of τ and write the integral equation as a set of simultaneous equations given by

$$[A]\{e_b\} = \{B\} \tag{11.70}$$

where $[A]$ is an 8×8 square matrix, $\{e_b\}$ is a column matrix of unknown emissive powers, and $\{B\}$ is a column matrix of E_2's. The matrix $[A]$ is a symmetric matrix as can be easily seen by looking at the I_i's. All the diagonal elements are the same, as may be easily verified. We thus have the following:

$$[A] = \begin{bmatrix} 1.9312 & 0.0483 & 0.0393 & 0.0344 & 0.0309 & 0.0282 & 0.0261 & 0.0243 \\ 0.0483 & 1.9312 & 0.0483 & 0.0393 & 0.0344 & 0.0309 & 0.0282 & 0.0261 \\ 0.0393 & 0.0483 & 1.9312 & 0.0483 & 0.0393 & 0.0344 & 0.0309 & 0.0282 \\ 0.0344 & 0.0393 & 0.0483 & 1.9312 & 0.0483 & 0.0393 & 0.0344 & 0.0309 \\ 0.0309 & 0.0344 & 0.0393 & 0.0483 & 1.9312 & 0.0483 & 0.0393 & 0.0344 \\ 0.0282 & 0.0309 & 0.0344 & 0.0393 & 0.0483 & 1.9312 & 0.0483 & 0.0393 \\ 0.0427 & 0.0282 & 0.0309 & 0.0344 & 0.0393 & 0.0483 & 1.9312 & 0.0483 \\ 0.0243 & 0.0427 & 0.0282 & 0.0309 & 0.0344 & 0.0393 & 0.0483 & 1.9312 \end{bmatrix}$$

$$\{e_b\} = \begin{bmatrix} e_{b1} \\ e_{b2} \\ e_{b3} \\ e_{b4} \\ e_{b5} \\ e_{b6} \\ e_{b7} \\ e_{b8} \end{bmatrix} \quad \text{and} \quad \{B\} = \begin{bmatrix} 0.9656 \\ 0.9173 \\ 0.8780 \\ 0.8436 \\ 0.8128 \\ 0.7845 \\ 0.7584 \\ 0.7341 \end{bmatrix}$$

All entries have been rounded to four significant digits after decimals. However, machine calculation proceeds with the available number of digits. The solution of the set of linear equations is easily accomplished by the Gauss iteration. The coefficient matrix exhibits diagonal dominance and hence the iterative scheme will succeed in yielding the desired solution. The Gauss iteration requires "old" and "new" values—a total of 16 entries to be stored. Table 11.3 shows the results of the Gauss iteration with 0.5 as the starting value for each of the unknown emissive powers.

It is seen that the Gauss iteration converges in just 3 iterations. To demonstrate that the procedure is good over a large range of optical thicknesses, the calculations were performed for $\tau_L = 1$ also. The results are given in Table 11.4.

The dimensionless heat flux may be obtained by the use of Eq. 11.67 and performing the indicated integration by using the discrete values of the dimensionless emissive power. The results in the two cases considered above are

$$q^* = 0.9157 \text{ for } \tau_L = 0.1; \quad q^* = 0.5550 \text{ for } \tau_L = 1$$

Table 11.3 Gauss iteration results

Iteration Number			
1	2	3	4
0.5	0.5537	0.5535	0.5535
0.5	0.5411	0.5414	0.5414
0.5	0.5242	0.5245	0.5245
0.5	0.5080	0.5082	0.5082
0.5	0.4920	0.4919	0.4920
0.5	0.4758	0.4756	0.4756
0.5	0.4631	0.4633	0.4633
0.5	0.4444	0.4449	0.4449

Table 11.4 Results for two optical thicknesses

$\frac{\tau}{\tau_L}$	$e_b,\ \tau_L = 0.1$	$e_b,\ \tau_L = 1$
0	**0.5630**	**0.7493**
0.0625	0.5535	0.7145
0.1875	0.5414	0.6486
0.3125	0.5245	0.5879
0.4375	0.5082	0.5291
0.5625	0.4920	0.4709
0.6875	0.4756	0.4121
0.8125	0.4633	0.3514
0.9375	0.4449	0.2855
1	**0.4378**	**0.2507**

Bold entries are extrapolated values

All the results obtained by the above numerical method are in very good agreement with the results reported in the two references cited above. While the solution corresponding to $\tau_L = 0.1$ is almost linear, the solution for $\tau_L = 1$ is represented excellently by a third-degree polynomial. The mean value of e_b in both the cases is 0.5. Figure 11.16 brings out these observations.

The extrapolated values shown in Table 11.4 are also in excellent agreement with the values given in the two references. It is observed that the e_b values at $\tau = 0$ and $\tau = \tau_L$ are not equal to the radiosities of the walls. Thus, the gas emissive power is discontinuous at the two boundaries and there is a "slip" at the two walls. Of course, the slip decreases with an increases in τ_L and vanishes for $\tau_L \to \infty$. The presence of slip is because of the fact that the gas has been assumed to be non-conducting!

In the optically thin case, the dimensionless emissive power is very nearly constant and is close to the mean value of 0.5. Thus, the dimensional emissive power is just the mean of the radiosities of the two walls. In case the two walls are black, the emissive

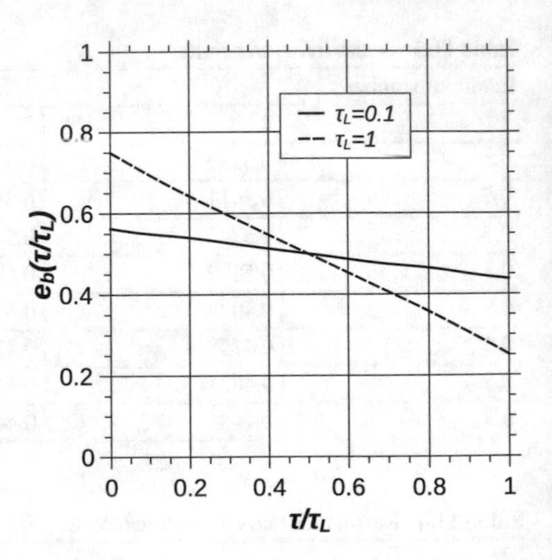

Fig. 11.16 Variation of emissive power across the slab

power of the gas will be the mean of the black body emissive powers corresponding to the temperatures of the two bounding walls.

Example 11.8

Consider two large parallel black walls enclosing a gray gas. The temperatures of the two walls are, respectively, 1000 K and 700 K. The spacing between the two walls is 0.5567 m and the gas has an absorption coefficient of 0.998 m^{-1}. Make a plot of the temperature of the gas as a function of the position. What is the heat transfer between the two walls?

Solution :

From the given data, we have the following.

Hot wall temperature: $T_1 = 1000\ K$;	Cold wall temperature: $T_2 = 700$ K
Radiosity of hot wall: $J_0 = E_{b1} = \sigma T_1^4 = 5.67 \times 10^{-8} \times 1000^4 = 56700\ \text{W/m}^2$	
Radiosity of cold wall: $J_L = E_{b2} = \sigma T_2^4 = 5.67 \times 10^{-8} \times 700^4 = 13613.7\ \text{W/m}^2$	
$\kappa = 0.998\ m^{-1}$, $L = 0.5567\ m$,	$L_m = 1.8 \times 0.5567 = 1.00006 \approx 1\ m$

Since the optical thickness $\tau_{L_m} = 0.998 \times 1.00006 \approx 1$, the non-dimensional emissive power is basically given by the values shown in the last column of Table 11.4. The local black body emissive power E_b is obtained by the use of Eq. 11.64 as

$$E_b = e_b(J_0 - J_L) + J_L$$
$$= e_b(56700 - 13613.7) + 13613.7 = 43086.3e_b + 13613.7$$

The local gas temperature is then given by

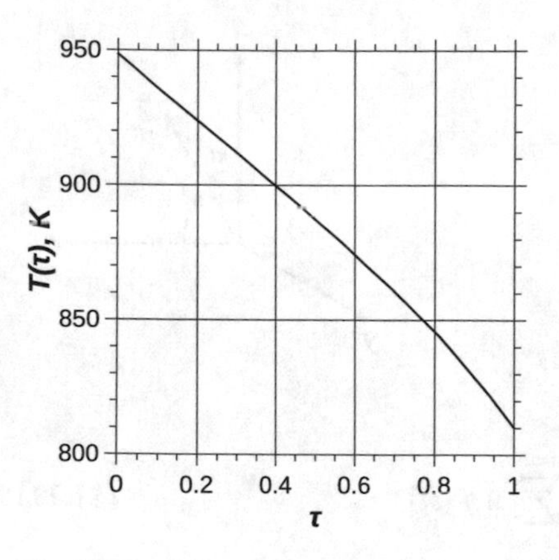

τ	$E_b(\tau)$, W/m^2	$T(\tau)$, K
0	45898.3	948.5
0.0625	44400.1	940.7
0.1875	41557.3	925.3
0.3125	38943.7	910.4
0.4375	36411.7	895.2
0.5625	33902.0	879.3
0.6875	31370.0	862.4
0.8125	28756.3	843.9
0.9375	25913.5	822.2
1	24415.4	810.1

Fig. 11.17 Variation of temperature across gas slab in Example 11.8

$$T = \left(\frac{E_b}{\sigma}\right)^{\frac{1}{4}}$$

The local temperature data is presented in a tabular form in the table shown at the right in Fig. 11.17. The temperature "profile" is also shown as a plot in the same figure. Noting that the non-dimensional heat transfer is $q* = 0.5550$, the heat transfer between the two walls is obtained as

$$q = q^*(J_0 - J_L) = 0.5550(56700 - 13613.7) = 23912.8 \text{ W/m}^2$$

11.5.4 *Discrete Ordinate Method*

Background

We have seen in Example 11.2 that the analytical solution required calculation for *all* (infinite in number) slant paths from the left wall to the right gas boundary and a subsequent integration over all directions. An approximate method of calculation that reduces the number of slant paths to a small number is the idea behind the discrete order method or DOM, for short. Integration over angle is replaced by a weighted sum in the DOM based on a suitable quadrature rule.

Central Idea

The quadrature rule replaces integration of a function with respect to a solid angle by a summation given by

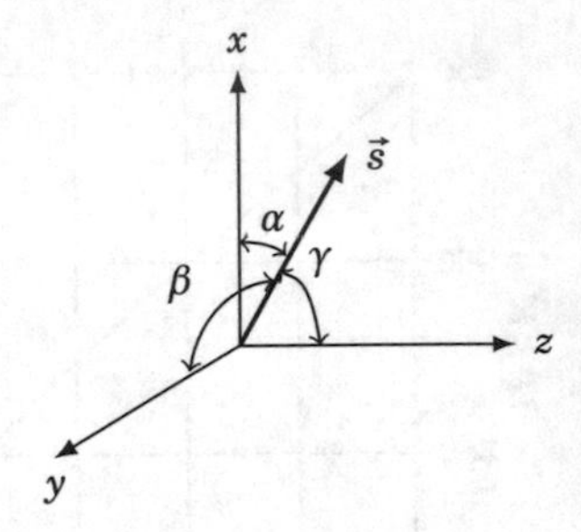

Fig. 11.18 Description of intensity of radiation propagating along direction $\vec{s}$

$$\int_0^{4\pi} f(\vec{s})d\Omega = \sum_{i=-n}^{i=n} w_i f(\vec{s}_i) \tag{11.71}$$

where w_i are the weights (weight may be looked upon as an effective solid angle associated with the direction $\vec{s}_i$).

Consider the radiation intensity propagating along $\vec{s}$ with direction cosines μ, η and ζ (μ represents the cosine of polar angle α measured with respect to the x-axis, η represents the cosine of polar angle β measured with respect to the y-axis, and ζ represents the cosine of polar angle γ measured with respect to the z-axis) as shown in Fig. 11.18.

Simplest Case

The simplest case corresponds to one in which we have a single direction in each octant of the sphere (Fig. 11.19). We require that on rotation by $\frac{\pi}{2}$, the direction remains the same. This will satisfy the no flux condition when the intensity is isotropic. It is then easy to see that the direction cosines are to be chosen such that the direction has equal polar angles $\alpha = \beta = \gamma$ and hence

$$\mu = \eta = \beta = \frac{1}{\sqrt{3}} \tag{11.72}$$

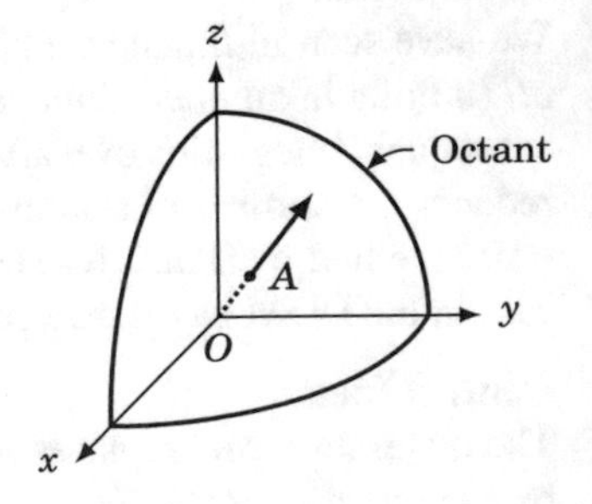

Fig. 11.19 Discrete ordinates in the S_2 method. There is one discrete ordinate direction in the octant indicated by direction OA

The only weight *must* be the solid angle subtended by one octant, i.e., $w = \frac{\pi}{2}$. This is referred to as the S_2 (symmetric) case. The squares of the direction cosines add up to one. When we consider the whole sphere, the eight directions are symmetrically located, one direction per octant. The zeroth moment of isentropic intensity I will yield $4\pi I$ as it should. Due to symmetry, the first moment of intensity will vanish, as it should, satisfying the zero flux condition.

In the literature, we also come across the S_2 (asymmetry) case. In this case, the three directions are chosen as

$$\mu = \frac{1}{\sqrt{2}}; \quad \eta = \beta = 0.5 \tag{11.73}$$

The sums of squares of direction cosines add up to unity. The weight remains unchanged at $w = \frac{\pi}{2}$.

More Useful Case: S_4

Consider now the S_4 method. In this method, we have $4 \times (4 + 2) = 24$ [in general, S_n has $n(n + 2)$] discrete ordinates or directions in the entire sphere. These directions are chosen such that they remain invariant with respect to rotation by $\frac{\pi}{2}$ and hence yield only 3 ordinates in an octant of the sphere. It is enough to consider these three directions for describing the S_4 method.

Because of symmetry, it is easy to see that the weights associated with the three directions are the same and equal to $w_1 = w_2 = w_3 = \frac{\pi}{6}$ so that the sum of weights is equal to $\frac{\pi}{2}$, the solid angle subtended by the octant at its center. The three directions are such that they correspond to the following:

$$\begin{bmatrix} \mu_1 \ \eta_1 \ \zeta_1 \\ \mu_2 \ \eta_2 \ \zeta_2 \\ \mu_3 \ \eta_3 \ \zeta_3 \end{bmatrix} = \begin{bmatrix} c_2 \ c_2 \ c_1 \\ c_2 \ c_1 \ c_2 \\ c_1 \ c_2 \ c_2 \end{bmatrix} \tag{11.74}$$

Thus, there are only two parameters c_1 and c_2 to be determined. We impose the condition that the sum of squares of the direction cosines yield unity, i.e.,

$$c_1^2 + c_2^2 + c_2^2 = c_1^2 + 2c_2^2 = 1 \tag{11.75}$$

In view of these, the three directions in an octant are as shown in Fig. 11.20. Now consider the case of isotropic intensity. In this case, azimuthal variation is absent and hence we can consider the directions in terms of any one of the direction cosines, say μ. We require that the quadrature formula that uses these three directions also satisfies the following half moment:

$$\int_0^1 \mu d\mu = \frac{\pi}{2} \quad \text{or} \quad \frac{\pi}{6}c_1 + 2\frac{\pi}{6}c_2 = \frac{\pi}{2} \quad \text{or} \quad c_1 + 2c_2 = \frac{3}{2} \tag{11.76}$$

From Eq. 11.76 we have, solving for c_1, the relation

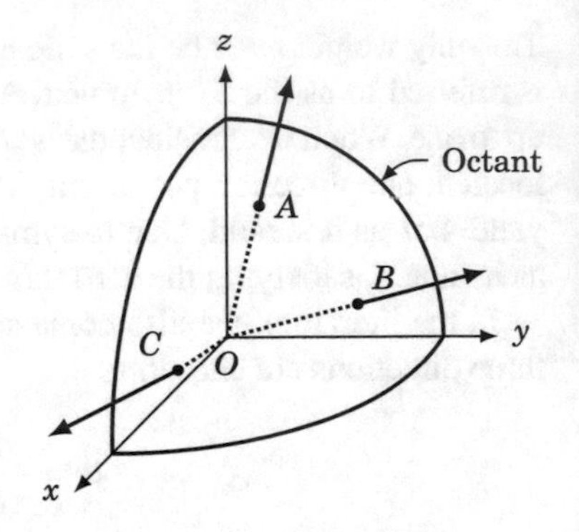

Fig. 11.20 Discrete ordinates in the S_4 method. There are three discrete ordinate directions in the octant indicated by direction OA, OB, OC

$$c_1 = \frac{3}{2} - 2c_2$$

Introduce this in Eq. 11.75 to get

$$\left(\frac{3}{2} - 2c_2\right)^2 + 2c_2^2 = \frac{9}{4} + 4c_2^2 - 6c_2 + 2c_2^2 = \frac{9}{4} + 6c_2^2 - 6c_2 = 1$$

The above reduces to the quadratic equation

$$c_2^2 - c_2 + \frac{5}{24} = 0$$

which has two solutions given by

$$c_2 = \frac{1 \pm \sqrt{1 - \frac{5}{6}}}{2} = \frac{1}{2} \pm \frac{1}{\sqrt{6}} \tag{11.77}$$

We choose the larger of the two roots obtained by taking the + sign in the above. Using this, we then obtain c_1 as

$$c_1 = \frac{3}{2} - 2 \times \left(\frac{1}{2} + \frac{1}{\sqrt{6}}\right) = \frac{1}{2} - \frac{1}{\sqrt{24}} \tag{11.78}$$

It is interesting to note that the above choice guarantees the satisfaction of another moment, viz., $\int_0^1 \mu^2 d\mu = \frac{2\pi}{3}$. We may obtain DOM with a larger number of directions, using similar arguments.[12] We choose even n (this will avoid singularity associated with zero direction cosine) such that the DOM is defined by a method with n

[12]Discrete Ordinates Quadrature Generator, PSR-110 DOQDP, Oak Ridge National Laboratory, RSICC Peripheral Shielding Routine Collection, September 1977.

Table 11.5 Directions and weights in the S_n method

DO #	μ	η	ζ	Weight
S_2—*Symmetric*				
1	0.577350269	0.577350269	0.577350269	6.283185307
S_2—*Asymmetric*				
1	0.500000000	0.500000000	0.707106781	6.283185307
S_4				
1	0.295875855	0.295875855	0.908248291	2.094395102
2	0.295875855	0.908248291	0.295875855	2.094395102
3	0.908248291	0.295875855	0.295875855	2.094395102
S_6				
1	0.183867109	0.183867109	0.965601249	0.643807273
2	0.183867109	0.965601249	0.183867109	0.643807273
3	0.965601249	0.183867109	0.183867109	0.643807273
4	0.183867109	0.695051396	0.695051396	1.450587830
5	0.695051396	0.183867109	0.695051396	1.450587830
6	0.183867109	0.695051396	0.695051396	1.450587830
S_8				
1	0.142255532	0.142255532	0.979554351	0.684943622
2	0.142255532	0.979554351	0.142255532	0.684943622
3	0.979554351	0.142255532	0.142255532	0.684943622
4	0.142255532	0.577350269	0.804008725	0.396913784
5	0.577350269	0.804008725	0.142255532	0.396913784
6	0.804008725	0.142255532	0.577350269	0.396913784
7	0.142255532	0.804008725	0.577350269	0.396913784
8	0.804008725	0.577350269	0.142255532	0.396913784
9	0.577350269	0.142255532	0.804008725	0.396913784
10	0.577350269	0.577350269	0.577350269	1.846871738

segmented angles such as $n = 6$, $n = 8$..... Directions and weights are tabulated in Table 11.5 for S_2 through S_8.

Special Case of Radiation in One Dimension

We have considered an earlier one-dimensional problem in the form of radiation transfer in a slab of participating medium. In such a case, azimuthal symmetry exists with reference to the independent variable x. Discrete ordinate directions are represented with reference to only one of the three polar angles and hence one of the direction cosines. Assuming that the direction cosine of interest is represented as μ, the discrete ordinates now are represented in terms of only the μ values. For example, in the case of S_2-symmetric, we will have only one direction corresponding to $\mu = 0.577350269$ with the corresponding weight of $w = 2\pi = 6.283185307$.

In the case of S_4, we have only two directions corresponding to $\mu = 0.2958758548$ and $\mu = 0.9082482905$. Directions 1 and 2 in Table 11.5 merge, and the total weight

Table 11.6 Directions and weights in the S_n method—One-dimensional case

DO #	μ	Weight	DO #	μ	Weight
S_2			S_4		
1	0.577350269	6.283185307	1	0.295875855	4.188790205
			2	0.908248290	2.094395102
S_6			S_8		
1	0.183867109	2.738202375	1	0.142255532	2.163714812
2	0.695051396	2.901175660	2	0.577350269	2.640699306
3	0.965601249	0.643807273	3	0.804008725	0.793827568
			4	0.979554351	0.684943622

is the sum of these weights and hence given by $w = 4.188790205$. With this background, it is possible to arrange the directions and weights as shown in Table 11.6.

Application of DOM to a Slab of Participating Medium

Having considered the basis for the DOM, we demonstrate its use in the case of a simple slab problem where radiation transfer is along the thickness of the slab. For simplicity, we assume the slab to be gray with an absorption coefficient of κ. The slab is L thick and the temperature variation across the slab is specified. At $x = 0$, there is a black wall with a temperature of T_w.

Consider an elemental absorbing and emitting gas slab of thickness Δx as shown in Fig. 11.21. Radiation of intensity $I_x^+(\mu_i)$ is incident on the left (west) face of the gas slab at one of the angles chosen in the discrete ordinate method, viz., μ_i. The intensity at $x + \Delta x$ (east face) is related to the intensity at x as given by the following.

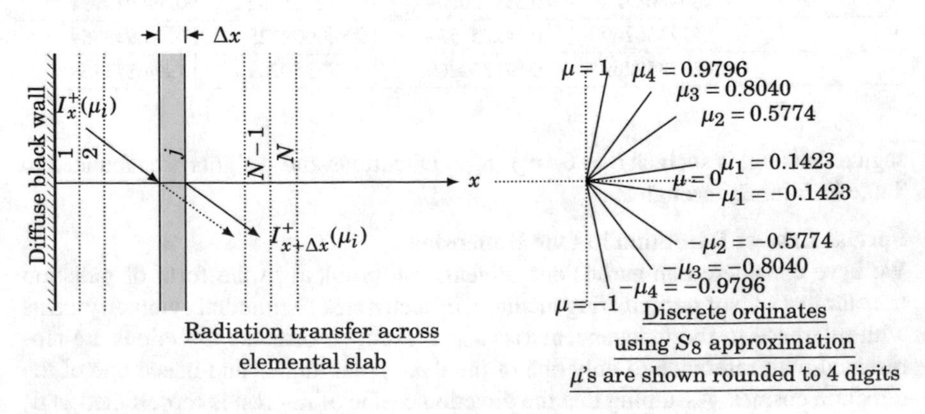

Fig. 11.21 Discrete ordinate method applied to an elemental slab

$$I^+_{x+\Delta x}(\mu_i) = I^+_x(\mu_i)\mathrm{e}^{-\frac{\kappa\Delta x}{\mu_i}} + \frac{\sigma T^4_{g,j}(x)}{\pi}\left(1 - \mathrm{e}^{-\frac{\kappa\Delta x}{\mu_i}}\right) \tag{11.79}$$

Here, $T_{g,j}$ is the temperature of the elemental gas layer j. The radiant flux is then given by the following summation, where S_8 approximation is used.

$$q^+(x + \Delta x) = \sum_{i=1}^{4} \mu_i I^+_{x+\Delta x}(\mu_i) w_i \tag{11.80}$$

The above process starts at the wall situated at $x = 0$. For the first elemental gas slab, we have

$$I^+_{\Delta x}(\mu_i) = \frac{\sigma T^4_w}{\pi}\mathrm{e}^{-\frac{\kappa\Delta x}{\mu_i}} + \frac{\sigma T^4_{g,1}}{\pi}\left(1 - \mathrm{e}^{-\frac{\kappa\Delta x}{\mu_i}}\right) \tag{11.81}$$

For subsequent elemental slabs, we make use of Eq. 11.79 for obtaining the intensities along the discrete ordinates and Eq. 11.80 for obtaining the forward-going flux. Eventually, we will be able to determine the flux leaving the gas from the east face of the gas layer identified as N.

Example 11.9

Consider Example 11.3 again. Compare the exact result obtained there with those obtained using various S_n's. Since the temperature of the gas is uniform across the slab, the calculations may be made by treating the entire slab as a single elemental slab. We shall calculate the flux leaving at the right using the S_4 method. Similar calculations may be performed for other S_n's also.

Solution :

We make use of the directions and weights given in Table 11.6. The intensity of radiation leaving the black wall at the left is

$$I(\tau = 0) = \frac{5.67 \times 10^{-8} \times 400^4}{\pi} = 462.03 \text{ W/m}^2\text{sr}$$

The black body intensity corresponding to gas temperature is given by

$$I_g = \frac{5.67 \times 10^{-8} \times 350^4}{\pi} = 270.84 \text{ W/m}^2\text{sr}$$

Along the slant path $\mu_1 = 0.295876$, the intensity leaving at the right is given by

$$I_{\mu_1}(\tau = 0.1) = 462.03\mathrm{e}^{-\frac{0.1}{0.295876}} + 270.84\left(1 - \mathrm{e}^{-\frac{0.1}{0.295876}}\right) = 407.20 \text{ W/m}^2\text{sr}$$

Along the slant path $\mu_2 = 0.908248$, the intensity leaving at the right is given by

$$I_{\mu_2}(\tau = 0.1) = 462.03e^{-\frac{0.1}{0.908248}} + 270.84\left(1 - e^{-\frac{0.1}{0.908248}}\right) = 442.10 \text{ W/m}^2\text{sr}$$

With these two values of leaving intensities, the forward flux leaving the gas slab is given by

$$q^+(\tau = 0.1) = 4.188790 \times 0.295876 \times 407.20 + 2.094395 \times \\ \times 0.908248 \times 442.10 = 1345.64 \text{ W/m}^2$$

This compares pretty well with the exact value obtained in Example 11.3 of $q^+(\tau = 0.1, \text{exact}) = 1352.57$ W/m^2. The following table compares the exact and the various S_n results .

DOM	$q^+(\tau = 0.1)$	Error %
S_2-Symmetric	1565.77	15.76
S_2-Asymmetric	1342.64	−0.73
S_4	1345.64	−0.51
S_6	1347.78	−0.35
S_8	1348.73	−0.28
Exact	1352.57	0

It appears that S_4 may be adequate in this problem.

Example 11.10

Consider a gray gas with an absorption coefficient of $\kappa = 0.5$ m^{-1}. The temperature of the gas varies linearly from $x = 0$ to $x = L = 1$ m according to the relation $T = 400 - 100x$ where x is in *m*. Radiation emanates from a black wall at $x = 0$ maintained at a temperature of $T_w = 400$ K. Determine the radiant flux leaving the gas slab at $x = 1$ m. Make use of the discrete ordinate method with S_8 for solving this problem.

Solution:

Step 1 With the given data, we may calculate the intensity of radiation leaving the black wall at $x = 0$ as $I_w^+ = \frac{\sigma T_w^4}{\pi} = \frac{5.67 \times 10^{-8} \times 400^4}{\pi} = 462.03$ W/m^{2}sr.

Step 2 We divide the slab into 10 elemental slabs of thickness $\Delta_x = \frac{L}{10} = \frac{1}{10} = 0.1$ *m*. The optical thickness of the slab is $\tau_L = \kappa L = 0.5 \times 1 = 0.5$ and that of an elemental slab is $\Delta\tau = \kappa L = 0.5 \times 0.1 = 0.05$.

Step 3 We associate the gas temperature with the mid-point of each elemental slab. The locations and temperature are calculated based on the linear temperature variation specified in the problem and are tabulated as given below.

j	1	2	3	4	5	6	7	8	9	10
x, m	0.05	0.15	0.25	0.35	0.45	0.55	0.65	0.75	0.85	0.95
T_j, K	395	385	375	365	355	345	335	325	315	305

Note that the temperature profile is approximated by a staircase-type profile that is familiar to us from the finite difference method presented in conduction heat transfer.

Step 4 Consider the very first element next to the black wall. We calculate the intensities at $\Delta x = 0.1$ m along the four discrete ordinate directions using Eq. 11.81. For example, along $\mu_1 = 0.14226$ we have

$$I^+_{\Delta x}(\mu_1) = \frac{\sigma T_w^4}{\pi} e^{-\frac{\kappa \Delta x}{\mu_1}} + \frac{\sigma T_{g,1}^4}{\pi}\left(1 - e^{-\frac{\kappa \Delta x}{\mu_1}}\right) = 462.03 \times e^{-\frac{0.5\times 0.1}{0.14226}}$$

$$+\frac{5.67 \times 10^{-8} \times 395^4}{\pi}\left(1 - e^{-\frac{0.5\times 0.1}{0.14226}}\right) = 455.31 \text{ W/m}^2\text{sr}$$

Similarly, we get the following values along the other discrete ordinates.

$\mu_2 = 0.57735$	$I^+_{\Delta x}(\mu_2) = 460.15$ W/m^{2}sr
$\mu_3 = 0.80401$	$I^+_{\Delta x}(\mu_3) = 460.67$ W/m^{2}sr
$\mu_4 = 0.97955$	$I^+_{\Delta x}(\mu_4) = 460.90$ W/m^{2}sr

Step 5 We make use of these and Eq. 11.79 to calculate the corresponding values at $x = 2\Delta x$. For example, along $\mu_1 = 0.14226$ we have

$$I^+_{2\Delta x}(\mu_1) = I^+_{\Delta x}(\mu_1) e^{-\frac{\kappa \Delta x}{\mu_1}} + \frac{\sigma T_{g,2}^4}{\pi}\left(1 - e^{-\frac{\kappa \Delta x}{\mu_1}}\right) = 455.31 \times e^{-\frac{0.5\times 0.1}{0.14226}} +$$

$$\frac{5.67 \times 10^{-8} \times 385^4}{\pi}\left(1 - e^{-\frac{0.5\times 0.1}{0.14226}}\right) = 450.59 \text{ W/m}^2\text{sr}$$

Similarly, we calculate for the other ordinate directions also.

Step 6 The above calculations are continued till we reach $j = 10$. The results are tabulated below as computed by a spreadsheet program.

j	x, m	T_j	$I^+(\mu_i)$ $i=1$	$i=2$	$i=3$	$i=4$
1	0.05	395	455.31	460.15	460.67	460.90
2	0.15	385	450.59	458.43	459.38	459.83
3	0.25	375	434.57	453.29	455.59	456.68
4	0.35	365	411.55	445.30	449.64	451.72
5	0.45	355	384.52	434.93	441.85	445.18
6	0.55	345	355.51	422.63	432.49	437.29
7	0.65	335	325.93	408.78	421.83	428.25
8	0.75	325	296.70	393.72	410.10	418.25
9	0.85	315	268.45	377.77	397.51	407.46
10	0.95	305	241.55	361.17	384.26	396.03

Step 7 The radiant flux leaving the slab at $x = 1\ m$ is obtained as a weighted sum of the intensities shown in the last row of the table.

$$\boxed{q_L^+ = \sum_{i=1}^{4} I^+(\mu_i)\mu_i w_i} = 241.55 \times 0.14226 \times 2.16371 +$$

$$361.17 \times 0.57735 \times 2.64070 + 384.26 \times 0.80401 \times 0.79382 +$$

$$396.03 \times 0.97955 \times 0.68494 = \boxed{1135.95\ \mathrm{W/m^2}}$$

11.6 Enclosure Analysis in the Presence of an Absorbing and Emitting Gas

We develop the analysis method for a gray gas bounded by gray enclosing walls. The gas contained in the enclosure may in general be non-isothermal. In that case, we divide the gas volume in to small volume elements within which the temperature is assumed to be uniform. The gray diffuse walls are also considered as being isothermal, if necessary, by dividing a surface in to small elements. The method that will be detailed is the so-called "zone" method.[13] The development here follows closely that given by Goyhénèche and Sacadura.[14]

[13]H. C. Hottel and H. S. Cohen, AIChE Journal, Vol. 4, pp. 3–14, 1958.

[14]J.M.Goyhénèche and J.F.Sacadura, ASME Journal of Heat Transfer, Vol. 124, pp. 696–703, 2002.

11.6.1 *Zone Method*

Direct Exchange Areas (DEA)

We make use of Fig. 11.22 in deriving the basic geometric factors that govern radiation transfer in a gas-filled enclosure.

Consider two bounding surfaces i and j as shown in Fig. 11.22a. Each surface is referred to as a zone. Consider the radiation that leaves zone i of area A_i and is incident on zone j of area A_j by direct transfer. Similar to the procedure adapted in treating surface radiation in an evacuated enclosure, we associate direct exchange area (DEA—we referred to it as exchange area there) between these two zones represented by $\overline{s_i s_j}$ defined through the following relation:

$$\boxed{\overline{s_i s_j} = \int_{A_i}\int_{A_j} \frac{\mu_i \mu_j e^{-\kappa r_{ij}}}{\pi r_{ij}^2} dA_i dA_j = \overline{s_j s_i}} \quad (11.82)$$

We have used the notation $\mu = \cos\theta$ in writing the above equation. Note that the above will be nothing but the area angle factor product (i.e., exchange area) in the absence of the gas. The exponential factor accounts for the attenuation of radiation as it transits between the zones through the gas. The symmetry of the integrand is an indication that the DEA satisfies the reciprocity relation. Now consider radiant interchange between gas zone k and surface zone j—refer to Fig. 11.22(b). We introduce a DEA for this case represented as $\overline{g_k s_j}$ through the relation

$$\boxed{\overline{g_k s_j} = \int_{V_k}\int_{A_j} \frac{\mu_j \kappa e^{-\kappa r_{ij}}}{\pi r_{kj}^2} dV_k dA_j = \overline{s_j g_k}} \quad (11.83)$$

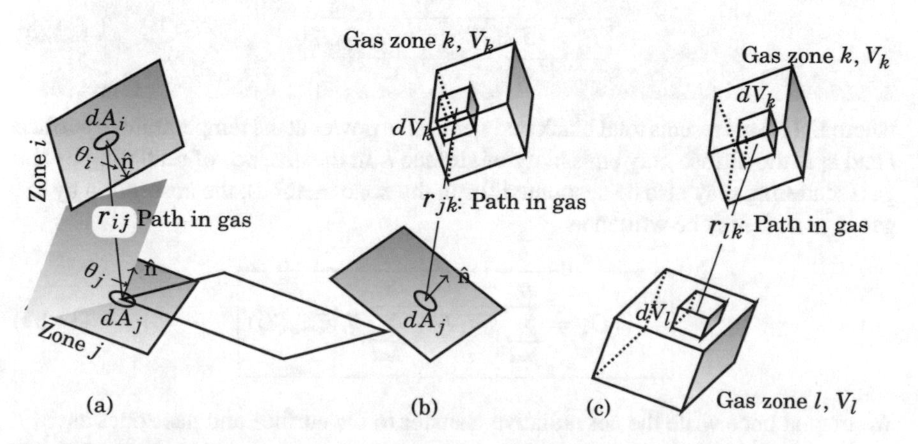

Fig. 11.22 Radiation transfer in an enclosure with a confined absorbing and emitting gas

Again the exponential factor accounts for the transmittance of the intervening gas path. The reciprocity rule is satisfied because of the symmetry of the integrand.

Consider now the radiation from gas zone k of volume V_k that reaches the gas zone l of volume V_l as shown in Fig. 11.22c . The emission by elemental volume ΔV_k is given by $\kappa \Delta V_k$. The amount of this absorbed by elemental volume ΔV_l is given by $\kappa \Delta V_l$. Hence, we define the direct exchange area $\overline{g_k g_l}$ between gas zones k and l as

$$\overline{g_k g_l} = \int_{V_k} \int_{V_l} \frac{\kappa^2 e^{-\kappa r_{kl}}}{\pi r_{kl}^2} dV_k dV_l = \overline{g_l g_k} \tag{11.84}$$

Again the exponential factor accounts for the transmittance of the intervening gas path. No μ factors are involved while dealing with gas emission since emission is isotropic. The reciprocity rule is satisfied because of the symmetry of the integrand. Note that κ has a unit of reciprocal length.

Radiosity Formulation

The radiosity formulation uses the DEA's obtained above. For surface zone i, the irradiation G_i may be written down as

$$A_i G_i = \sum_{j=1}^{N_s} \overline{s_j s_i} J_j + \sum_{k=1}^{N_g} \overline{g_k s_i} E_b(T_k) \tag{11.85}$$

where J_j represents the radiosity of a surface j and $E_b(T_k)$ represents the total black body emissive power at the temperature of the gas in zone k. N_s and N_g represent, respectively, the number of surface and gas zones used in the enclosure analysis. The radiosity of surface zone i may then be written down as

$$J_i = \varepsilon_i E_b(T_i) + (1 - \varepsilon_i) G_i \tag{11.86}$$

where $E_b(T_i)$ represents total black body emissive power at the temperature of surface i and ε_i is the diffuse gray emissivity of surface i. In the absence of scattering by the gas (scattering may also be accounted for in the zone method), the absorption by the gas in zone k may be written as

$$4\kappa V_k G_k = \sum_{j=1}^{N_s} \overline{s_j g_k} J_j + \sum_{l=1}^{N_g} \overline{g_l g_k} E_b(T_l) \tag{11.87}$$

We may at once write the net radiative transfer to the surface and gas zones as

$$\text{(a)}\ Q_i = A_i(J_i - G_i), \quad \text{(b)}\ Q_k = 4\kappa V_k[E_b(T_k) - G_k] \tag{11.88}$$

The factor $4\kappa V_i$ accounts for radiation in to the sphere of solid angle 4π. If we consider an isothermal enclosure, it is clear that $Q_k = 0$ and $J_i = E_b(T_k)$ for all i and k. Then we have from Eqs. 11.87 and 11.88 the following important result:

$$\boxed{\sum_{j=1}^{N_s} \overline{s_j g_l} + \sum_{k=1}^{N_g} \overline{g_k g_l} = 4\kappa V_l} \tag{11.89}$$

for $l = 1, 2 \ldots N_g$.

We thus see that the formulation closely follows that of an evacuated enclosure. The radiosity formulation for an evacuated enclosure will be obtained by setting $\kappa = 0$ in the equations presented above. Introducing the irradiation from Eq. 11.85 in to Eq. 11.86 will yield, as before, a set of simultaneous equations for the radiosities of the surface zones. As before, we may have surface zones with specified temperatures or specified fluxes (for example, reradiating surfaces). These may be treated in a fashion identical to that used in the case of the evacuated enclosure. Additional factors due to gas radiation need to be incorporated. These need to be solved before the surface and gas fluxes are calculated.

Simple Case with One Gas Zone and Several Surface Zones

In this case, we assume that the entire confined gas is at a uniform temperature of T_g, i.e., $N_g = 1$. We shall use subscript g to indicate the gas volume. Only surface-to-surface and surface-to-gas DEA's are involved in this problem. While the expression for surface-to-surface DEA remains unchanged from the definition given in Eq. 11.82, the gas-to-surface DEA is written down as

$$\overline{g_g s_j} = \int\limits_{V_g}\int\limits_{A_j} \frac{\mu_j \kappa e^{-\kappa r_{gj}}}{\pi r_{gj}^2} dV_g dA_j = \overline{s_j g_g} \tag{11.90}$$

Now consider a further special case where the gas is optically thin.

Special Case: Single Zone of an Optically Thin Gas

In the special case of an optically thin gas, it is possible to replace the exponential factor $e^{-\kappa r_{ij}}$ by $1 - \kappa r_{ij}$. The DEA between surfaces becomes

$$\overline{s_i s_j} \approx \int_{A_i}\int_{A_j} \frac{\mu_i \mu_j (1 - \kappa r_{ij})}{\pi r_{ij}^2} dA_i dA_j = \overline{s_j s_i} \tag{11.91}$$

The integral separates to two integrals given by

$$\text{(a)}\ A_i F_{ij} = \int\limits_{A_i}\int\limits_{A_j} \frac{\mu_i \mu_j}{\pi r_{ij}^2} dA_i dA_j \quad \text{(b)}\ A_i L_{ij} = \int\limits_{A_i}\int\limits_{A_j} \frac{\mu_i \mu_j}{\pi r_{ij}} dA_i dA_j \tag{11.92}$$

The former will be recognized as the area angle factor product. The second integral has a unit of volume and is the mean area beam length product for radiation transfer between zones i and j through the intervening gas. We may then rewrite Eq. 11.90 as

$$\overline{s_i s_j} \approx A_i F_{ij}\left(1 - \frac{\kappa L_{ij}}{F_{ij}}\right) = A_i F_{ij} t_{ij} \tag{11.93}$$

where t_{ij} is the transmittance of the gas for radiant interchange between surfaces i and j, given by the term within the brackets. Consider now the gas-to-surface DEA. We may rewrite it using the optically thin approximation as

$$\overline{g_g s_j} \approx \int\limits_{V_{g,1}} \int\limits_{A_{s,j}} \frac{\mu_j \kappa\left(1 - \kappa r_{1j}\right)}{\pi r_{1j}^2} dV_{g,1} dA_{s,j} \tag{11.94}$$

This integral is interpreted as the product of gas emissivity $\varepsilon_{g,j}$ and the area of the surface A_j. This is surface zone-specific and hence the subscript g, j on the gas emissivity. Hence, Eq. 11.94 may be rewritten as

$$\overline{g_g s_j} = \varepsilon_{g,j} A_j \tag{11.95}$$

11.6.2 Example of Zone Analysis

We present zone analysis for a single gas zone enclosed within a cubical enclosure. The angle factors between various surfaces of the enclosure may easily be obtained by using Figs. G.2 and G.3. The mean beam lengths are calculated based on formulae and charts presented in Appendix G. The gas emissivity is based on the choice of a suitable mean beam length using Table 11.2.

Example 11.11

Combustion gases entering a cubical enclosure at $T_{g,i} = 1370$ K may be treated as having a gray absorption coefficient of $\kappa = 0.125$ m^{-1}. The base of the cube labeled 1 (see Fig. 11.23) is a gray diffuse heat sink with an emissivity of $\varepsilon_1 = 0.8$. All the other sides labeled $2 - 6$ are reradiating. The cube is of $0.8 \times 0.8 \times 0.8$ m size. The gas flow rate is $\dot{m} = 0.024$ kg/s and the gas mixture has a specific heat capacity of $c = 1200$ J/kg · K. Calculate the exit temperature of the gas? Assume that the entire gas volume is a single zone at a temperature that is the mean of the entering and leaving temperature.

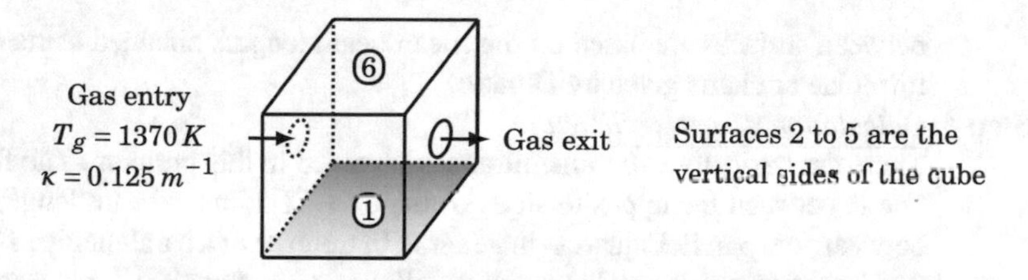

Fig. 11.23 Geometry for Example 11.11

Solution :

Step 1 We shall assume that the holes through which the gas enters and leaves the enclosure are very small compared to the faces of the cube. Since the gas exit temperature is not known, we start the computations with an assumed value for the mean gas temperature of $T_{g,m} = 1150$ K. We aim to get the correct value by an iterative process.

Step 2 Given data is written down now:

Length of sides of the cubical enclosure:	$L = 0.8$ m
Gray absorption coefficient of the combustion gas:	$\kappa = 0.125$ m^{-1}
Gray emissivity of floor 1:	$\varepsilon_1 = 0.8$
Gas inlet temperature:	$T_{g,i} = 1370$ K
Mass flow rate of combustion gases:	$\dot{m} = 0.02$ kg/s
Specific heat capacity of combustion gases:	$c = 1200$ J/kg K

Step 3 *Calculation of gas emissivity*:
Mean beam length for emission from the gas volume to any one face of the cube may be calculated based on the recipe given in Table 11.2 as $L_m = 0.6L = 0.6 \times 0.8 = 0.48$ m. The corresponding optical thickness is

$$\tau = \kappa L_m = 0.125 \times 0.48 = 0.06$$

Since $\tau < 0.15$, the optically thin approximation may be assumed to be valid. The gas emissivity may hence be calculated as

$$\varepsilon_g = \tau = 0.06$$

Surface 1 is assigned a radiosity of J_1. We note from the geometry that the radiosities of surfaces 2–5 are all the same and hence we denote each of them as J_2. Surface 6 is assigned a radiosity of J_3. The transmittances between surfaces are based on the mean beam lengths obtained using the formulae or charts given by Dunkle.

Step 4 *Calculation of transmittances*:

There are basically two transmittances involved in this problem. The first one is between the opposite sides of the cube. The mean beam length is between two parallel squares (in general, in the case of a parallelepiped the configuration corresponds to two parallel equal rectangles). The second one is between adjacent sides of the cube. The mean beam length is between two perpendicular squares that share a common edge (in general, in the case of a parallelepiped the configuration corresponds to two perpendicular rectangles sharing a common edge).

For the cube case, the non-dimensional ratios that go into the calculation of mean beam lengths as well as the angle factors are both equal to 1 (Appendix G). For the parallel squares configuration, we obtain

$$Z_{par} = \frac{F_{par} A r_{par}}{V} = 0.2218$$

where the notation follows the abovementioned Appendix.

$$Z_{per} = \frac{F_{per} A r_{per}}{V} = 0.1112$$

where the notation follows the abovementioned Appendix. In the case of a cube, both angle factors are ≈ 0.2. With this, the mean beam lengths between opposite sides of the cube are

$$r_{par} = \frac{Z_{par} V}{A F_{par}} = \frac{0.2218 \times 0.8^3}{0.8^2 \times 0.2} = 0.8872 \text{ m}$$

The mean beam lengths between adjacent sides of the cube are

$$r_{per} = \frac{Z_{per} V}{A F_{per}} = \frac{0.1112 \times 0.8^3}{0.8^2 \times 0.2} = 0.4448 \text{ m}$$

The corresponding optical thicknesses are obtained as

$$\tau_{par} = \kappa r_{par} = 0.125 \times 0.8872 = 0.1110$$
$$\tau_{per} = \kappa r_{per} = 0.125 \times 0.4448 = 0.0556$$

The corresponding transmittances t_{par} and t_{per} are obtained, using the optically thin approximation as

$$t_{par} = 1 - \tau_{par} = 1 - 0.1110 = 0.8890$$
$$t_{per} = 1 - \tau_{per} = 1 - 0.0556 = 0.9444$$

Step 5 *Radiosity equations*:

Radiosity equation for the floor:

$$J_1 = \varepsilon_1 E_{b1} + (1 - \varepsilon_1)(\varepsilon_g E_{bg} + 4F_{per}t_{per}J_2 + F_{par}t_{par}J_3)$$

Radiosity equation for any of the faces 2–5:
Radiation balance requires that $J_2 = G_2$, and hence we have

$$J_2 = \varepsilon_g E_{bg} + F_{per}t_{per}J_1 + (2F_{per}t_{per} + F_{par}t_{par})J_2 + F_{per}t_{per}J_3$$

Radiosity equation for face 6:
Radiation balance requires that $J_3 = G_3$, and hence we have

$$J_3 = \varepsilon_g E_{bg} + F_{par}t_{par}J_1 + 4F_{per}t_{per}J_2$$

Using the numerical values given in the data and the calculated angle factors, gas emissivity, and the transmittances, we may write the three equations in the matrix form.

$$\begin{bmatrix} 1 & -0.1511 & -0.03556 \\ -0.1888 & 0.4444 & -0.1888 \\ -0.1778 & -0.7555 & 1 \end{bmatrix} \times \begin{Bmatrix} J_1 \\ J_2 \\ J_3 \end{Bmatrix} = \begin{Bmatrix} 19800 \\ 5950 \\ 5950 \end{Bmatrix}$$

Step 6 These three equations may be solved for the radiosities to get

$$J_1 = 28049, \; J_2 = 44166 \; J_3 = 44306$$

where the J's are in W/m^2.

Step 7 *Desired answers*:

Heat transfer to the floor may be calculated as

$$Q_1 = A_1 \frac{\varepsilon_1}{1 - \varepsilon_1}(E_{b1} - J_1)$$
$$= 0.8 \times 0.8 \times \frac{0.8}{1 - 0.8}(5.67 \times 10^{-8} \times 800^4 - 28049) = -12351 \text{ W}$$

Since all other surfaces are in radiant balance, this heat must come from the gas as it cools in passing through the enclosure. We calculate the gas exit temperature $T_{g,e}$ by energy balance as

$$-Q_1 = \dot{m}c(T_{g,i} - T_{g,e})$$

$$or$$

$$T_{g,e} = T_{g,i} + \frac{Q_1}{\dot{m}c}$$

$$= 1370 + \frac{-12351}{0.024 \times 1200} = 941 \text{ K}$$

The mean gas temperature may then be calculated as

$$T_{g,m} = \frac{T_{g,i} + T_{g,e}}{2} = \frac{1370 + 941}{2} = 1155.5 \text{ K}$$

This is close to the assumed value of 1150 K and hence it is not necessary to perform any iterations.
The solution indicates that J_2 and J_3 are very close to each other. This means that the temperatures of all reradiating surfaces are more or less the same. We may calculate the temperatures as

$$T_{2-5} = \left(\frac{J_2}{\sigma}\right)^{0.25} = \left(\frac{44166}{5.67 \times 10^{-8}}\right)^{0.25} = 939.5 \text{ K}$$

$$T_6 = \left(\frac{J_3}{\sigma}\right)^{0.25} = \left(\frac{44306}{5.67 \times 10^{-8}}\right)^{0.25} = 940.2 \text{ K}$$

11.6.3 Application of DOM to Two-Surface Enclosure with a Non-isothermal Participating Medium

Consider a participating medium between two large parallel surfaces as shown in Fig. 11.24. Both surfaces are gray and diffuse with temperatures and emissivities as indicated in the figure.

DOM requires the solution of the RTE along discrete ordinates such as the ones shown in the figure. Since the RTE is a first-order differential equation, it can be solved *only* as an initial value problem. We may start from the left boundary with specified radiosity $J_1^{(0)}$ (or assumed radiosity, to start the process) there and integrate the equation along the "forward pass" with a chosen μ_i corresponding to one of those in the DOM. Because the wall is assumed to be diffuse, the assumed radiosity will yield a uniform intensity independent of μ given by $I_{\mu_i}^{+(0)}(x = 0) = \frac{J_1^{(0)}}{\pi}$. Since the intervening participating medium is non-isothermal, it is necessary to perform the integration by dividing the medium into a number of thin isothermal elemental layers

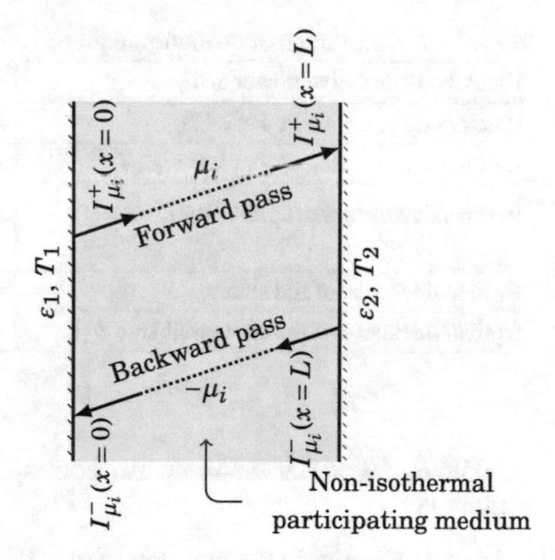

Fig. 11.24 Radiation in a two-surface enclosure with a participating medium

as before. At the end of this pass, we will get the intensity $I^{+(0)}_{\mu_i}(x = L)$. Performing similar integrations along all paths specified by DOM, we will be able to determine the irradiation on the right wall given by $G_2^{(0)} = \sum_{i=1}^{n} w_i \mu_i I^{+(0)}_{\mu_i}(x = L)$.

With the irradiation calculated at the end of the forward pass, it is possible to compute the radiosity $J_2^{(0)}$ of the second surface. The intensity leaving the surface is then given by $I^{-(0)}_{\mu_i}(x = L) = \frac{J_2^{(0)}}{\pi}$. We may use this as the starting value and integrate the RTE along the "backward pass" as indicated in the figure. At the end of this pass, we will be able to compute the irradiation on the left surface $G_1^{(0)}$. With this, it is possible to get a better value for the radiosity of the first surface as $J_1^{(1)}$. If the radiosity change is small, we may stop the iterative process or we may continue with another forward–backward pass combination to get a better value. An example is worked out below to demonstrate the above methodology.

Example 11.12

Consider a gray gas with an absorption coefficient of $\kappa = 0.5\ \text{m}^{-1}$. The temperature of the gas varies linearly from $x = 0$ to $x = L = 1$ m according to the relation $T_g(x) = 400 - 100x$ where x is in m. A diffuse gray wall of emissivity $\varepsilon = 0.8$ is located at $x = 0$ and is maintained at a temperature of $T_1 = 400$ K. A second diffuse gray wall of $\varepsilon = 0.6$ is located at $x = 1$ m and is maintained at a temperature of $T_2 = 300$ K. What is the heat transfer across the gas slab? Use the discrete ordinate method with S_4 to solve the problem.

Solution :

The solution follows the procedure described above. The S_4 method involves two discrete directions along the forward and backward paths. They are given by $\mu_1 =$

Table 11.7 Calculation of quantities required before starting iteration

Black body emissive power at T_1	$E_{b1} = 5.67 \times 10^{-8} \times 400^4 = 1451.52$ W/m^2
Black body emissive power at T_2	$E_{b2} = 5.67 \times 10^{-8} \times 300^4 = 459.27$ W/m^2
Initial guess for radiosity of left wall $J_1^{(0)}$	$J_1^{(0)} = \varepsilon_1 E_{b1} = 0.8 \times 1161.22$ W/m^2
Initial value of forward intensity $I_\mu^+(x=0)$	$I_\mu^+(x=0) = \frac{J_1^{(0)}}{\pi} = \frac{1451.52}{\pi} =$ 459.27 W/m^2 *sr*
Optical thickness of the slab τ_L	$\tau_L = 0.5 \times 1 = 0.5$
Optical thickness of the elemental slab $\tau_{\Delta L}$	$\tau_{\Delta L} = 0.5 \times 0.1 = 0.05$

0.295876, $\mu_2 = 0.908248$ with corresponding weights $w_1 = 4.188790$, $w_2 =$ 2.094395.

Step 1 We divide the gas slab in to 10 slabs of elemental thickness each of $\Delta L =$ 0.1 m such that each slab may be assumed to be isothermal. Temperatures of elemental slabs are as given below:

x_j m	0.05	0.15	0.25	0.35	0.45
T_j, K	395	385	375	365	355
$E_b(T_j)$|; W/m^2	1380.29	1245.74	1121.26	1006.36	900.53

x_j m	0.55	0.65	0.75	0.85	0.95
$T_{g,j}$, K	345	335	325	315	305
$E_b(T_{g,j})$|; W/m^2	803.27	714.11	632.58	558.25	490.66

Using the given data, the quantities required before starting the iteration process for obtaining the wall radiosities are calculated and are shown in Table 11.7. Note that the forward going intensity is isotropic, and hence it is the same along both directions in the S_4.

Step 2 We make use of Eqs. 11.79 and 11.81 to calculate the variation of the forward intensity along μ_1 and μ_2. The calculations terminate when we reach the right wall. In the present case, Table 11.8 shows what happens during the forward pass.

Step 3 From the values in the last two columns, we calculate the irradiation on the right wall as

$$G_2^{(0)} = w_1\mu_1 I_{\mu_1}^+(x=L) + w_2\mu_2 I_{\mu_2}^+(x=L) = 4.188790 \times 0.295876 \times 264.76$$
$$+2.094395 \times 0.908248 \times 326.50 = 949.21 \text{ W/m}^2$$

The radiosity of surface 2 may then be evaluated as

$$J_2^{(0)} = \varepsilon_2 E_{b2} + (1-\varepsilon_2)G_2^{(0)} = 0.6 \times 459.27 + (1-0.6) \times 949.21 = 655.25 \text{ W/m}^2$$

Since surface 2 is diffuse, the intensity of radiation leaving this wall is independent of μ and is given by $I_\mu^-(x=L) = \frac{655.25}{\pi} = 208.57$ W/m^2 *sr*.

Table 11.8 Forward Pass

Element No., j	x, m	$T_{g,j}$ K	$E_b(T_{g,j})$ W/m^2	$I_b(T_{g,j})$	$I^+_{\mu 1}$	$I^+_{\mu 2}$
	0	Guess value $I^+_\mu(x = 0)$:			369.63	369.63
1		395	1380.29	439.36		
	0.1				380.47	373.36
2		385	1245.74	396.53		
	0.2				382.97	374.60
3		375	1121.26	356.91		
	0.3				378.91	373.66
4		365	1006.36	320.34		
	0.4				369.81	370.80
5		355	900.53	286.65		
	0.5				356.88	366.29
6		345	803.27	255.69		
	0.6				341.14	360.37
7		335	714.11	227.31		
	0.7				323.44	353.24
8		325	632.58	201.36		
	0.8				304.46	345.10
9		315	558.25	177.70		
	0.9				284.75	336.14
10		305	490.66	156.18		
	1	Final value $I^+_\mu(x = L)$:			264.76	326.50

Note: Intensities in the last three columns are in $W/m^2\ sr$

Using this value, we perform the backward pass and tabulate the results as in Table 11.9.

Step 4 From the values in the last two columns, we calculate the irradiation on the left wall as

$$G_1^{(0)} = w_1\mu_1 I^-_{\mu_1}(x = L) + w_2\mu_2 I^-_{\mu_2}(x = L) = 4.188790 \times 0.295876 \times 276.84$$
$$+2.094395 \times 0.908248 \times 364.24 = 1013.40\ \text{W/m}^2$$

The radiosity of surface 1 may then be updated as

$$J_1^{(1)} = \varepsilon_1 E_{b1} + (1 - \varepsilon_1)G_1^{(0)} = 0.8 \times 1451.52 + (1 - 0.8) \times 1013.40$$
$$= 1363.90\ \text{W/m}^2$$

Since surface 1 is diffuse, the intensity of radiation leaving this wall is independent of μ and is given by $I^+_\mu(x = 0) = \frac{1363.90}{\pi} = 434.14\ \text{W/m}^2\ \text{sr}$.

Table 11.9 Backward Pass

Element No., j	$x,\ m$	$T_{g,j}\ K$	$E_b(T_{g,j})\ W/m^2$	$I_b(T_{g,j})$	$I^-_{\mu 1}$	$I^-_{\mu 2}$
	1	Guess value $I^-_{\mu}(x = L)$:			208.57	208.57
10		305	490.66	156.18		
	0.9				200.43	352.95
9		315	558.25	177.70		
	0.8				197.59	335.94
8		325	632.58	201.36		
	0.7				202.21	330.12
7		335	714.11	227.31		
	0.6				210.52	326.14
6		345	803.27	255.69		
	0.5				222.36	324.02
5		355	900.53	286.65		
	0.4				237.59	323.82
4		365	1006.36	320.34		
	0.3				256.14	325.60
3		375	1121.26	356.91		
	0.2				277.97	329.40
2		385	1245.74	396.53		
	0.1				303.06	335.29
1		395	1380.29	439.36		
	0	Final value $I^+_{\mu}(x = 0)$:			276.84	364.24

Note: Intensities in the last three columns are in $W/m^2\ sr$

Step 5 After one forward and one backward pass the change in radiosity of surface 1 is

$$\Delta J_1 = J_1^{(1)} - J_1^{(0)} = 1363.90 - 1161.22 = 202.68\ \text{W/m}^2$$

The change is substantial and hence we continue the iteration process. Iteration converges fast as seen in the following table.

Iter. No	J_1	J_2	G_1	G_2
1	1161.22	655.25	1013.40	949.21
2	1363.90	689.46	1027.83	1034.73
3	1366.78	689.94	1028.04	1035.95
4	1366.82	689.95	1028.04	1035.97

Step 6 We calculate the radiative fluxes at the two walls now.

$$q_1 = J_1 - G_1 = 1366.82 - 1028.04 = 338.78 \text{ W/m}^2$$
$$q_2 = J_2 - G_2 = 689.95 - 1035.97 = -346.02 \text{ W/m}^2$$

Concluding Remarks

Radiation transfer through a participating medium has been considered in some detail in this chapter. The process of absorption or emission has been discussed with reference to solid, liquid, and gaseous media. The presence of discrete absorption bands in gaseous media makes them non-gray and complex to model. The equation of transfer is to be solved and several methods have been presented. The most useful, from an application point of view, is the discrete order method that has been discussed in sufficient detail. A one-dimensional problem has been considered for demonstrating its use. It may be extended to two- and three-dimensional problems also.

11.7 Exercises

Ex 11.1 Compare the transmittance of two glass sheets (labeled A and B) for normal incidence of monochromatic radiation at a vacuum wavelength of $\lambda = 0.5$ μ m. Each glass sheet is 3 mm thick. Take into account multiple reflections from both surfaces of the glass sheets. Assume the incident radiation to be unpolarized. What would happen for incidence at 40° to the normal, all other things remaining the same? The real and imaginary indices of refraction are specified as $n = 1.6$, $n' = 10^{-5}$ for sheet A while they are $n = 1.8$, $n' = 10^{-5}$ for sheet B.

Ex 11.2 A black surface at a temperature T_B is observed through a plane layer containing a gray absorbing and emitting gas at temperature T_g. Indicate what would be the nature of the response of a detector kept at the observation point that responds to the radiant heat flux, if the temperature of the black surface is varied continuously from a value less than T_g to a value greater than T_g. Take the reading of the detector in the absence of the gas as the datum.

Ex 11.3 Redo Exercise 11.7 for the case of a non-gray gas that absorbs and emits only over a narrow band $\Delta\lambda$ centered around λ. The detector may be assumed to respond only to the radiation within the narrow band.

Ex 11.4 A gas of volume $V = 1$ m^3 has an effective surface area of $A = 5$ m^2. What is the mean beam length for the gas? If the gas is carbon dioxide at

$T_g = 2000$ K and $p_g = 1$ *atm*. what is the emissivity of the gas? What will be your answer if the gas is a mixture of carbon dioxide and water vapor with partial pressures given by $p_c = 0.7$ *atm*. and $p_w = 0.3$ atm., the temperature remaining the same?

Ex 11.5 A spherical vessel of 0.3 m diameter is filled with carbon dioxide at a pressure of 1.5 atm. What is the absorptivity of the gas to radiation characterized by a black body temperature of (a) 300 K and b) 1000 K?

Ex 11.6 A cubical enclosure of side 1 *m* contains a gas mixture containing 50% by volume of carbon dioxide and 50% by volume of nitrogen. The total pressure of the gas mixture is 2 atm. The walls of the enclosure are at 400 K while the gas mixture is at 600 K. Determine the emissivity of the gas at its temperature. Also, determine its absorptivity to radiation from the walls of the enclosure.

Ex 11.7 Two very large black plane surfaces are 0.3 m apart and the space between them is filled with a gas mixture containing 24% carbon dioxide, 25% water vapor, and rest nitrogen, by volume. The total pressure of the gas mixture is 1 atm.. One of the surfaces is at 1300 K while the other is at 300 K. Calculate the following:

(a) The effective emissivity of the gas mixture at its temperature.
(b) The effective absorptivity of the gas mixture to radiation from the two walls at their respective temperatures.
(c) The net rate of heat transfer from the hot wall to the cold wall.

Ex 11.8 Two very large black walls are 0.3 m apart and are maintained at temperatures of 800 K and 400 K, respectively. The intervening space between the walls is filled with a gray gas having an absorption coefficient of $\kappa = 0.3\ \text{m}^{-1}$. What is the net heat transfer between the hot wall and the cold wall if the gas temperature at a certain instant is 500 K? What is the net heat transfer to the gas at the same instant?

Ex 11.9 Two very large black walls are 0.3 *m* apart and are maintained at temperatures 800 K and 400 K, respectively. The intervening space between the walls is filled with a gray gas having an absorption coefficient of $\kappa = 0.3\ \text{m}^{-1}$. The gas is in radiation equilibrium. What is the net heat transfer between the walls?

Ex 11.10 A cube of side $L = 0.4$ m contains a mixture of 25% carbon dioxide, 25% water vapor, and 50% nitrogen (all percentages by volume) at a total pressure of 2 atm. The gas mixture is at a temperature of $T_g = 1000$ K while the surface of the cube, which is gray, having an emissivity of $\varepsilon = 0.6$ is maintained at $T_s = 400$ K.

- What is the emissivity of the gas?
- What is the absorptivity of the gas?
- Determine the net heat transfer to the surface of the cube
- Is it reasonable to assume that the gas volume is optically thin? Explain.

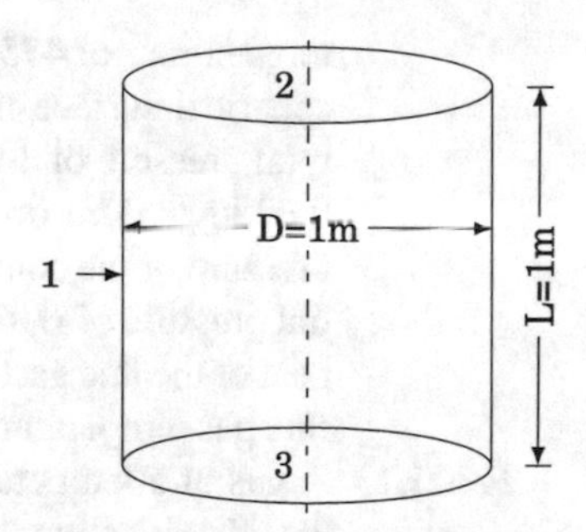

Fig. 11.25 Cylindrical enclosure in Exercise 11.11

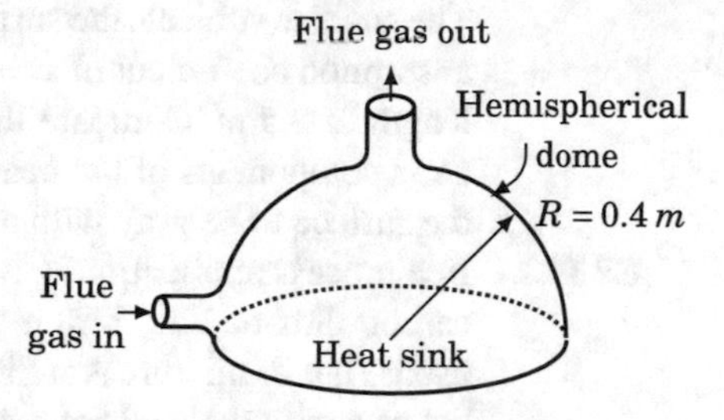

Fig. 11.26 Hemispherical furnace in Exercise 11.12

Take mean beam length as $L_m = 0.6L$ for calculating the emissivity as well as the absorptivity of the gas mixture.

Ex 11.11 An enclosure is in the form of a right circular cylinder (see Fig. 11.25) of diameter $D = 1$ m and height $L = 1$ m. Top surface labeled 1 is maintained at temperature $T_1 = 800$ K and has a gray emissivity of $\varepsilon_1 = 0.6$. The curved surface labeled 2 is maintained at temperature $T_2 = 600$ K and has a gray emissivity of $\varepsilon_2 = 0.8$. The bottom surface labeled 3 is a reradiating surface. The cylinder contains a gray gas at temperature $T_g = 400$ K and gray absorption coefficient $\kappa = 0.5$ m^{-1}. Take the mean beam length for calculating gas emissivity as 0.6 D. For calculating the transmittivity between surfaces 1 and 3, the mean beam length may be taken as 1.15 D. For calculating the transmittivity between surfaces 1 and 2 or 3 and 2, the mean beam length may be taken as 0.8 D.

- Calculate the matrix of diffuse shape factor values.
- Calculate the gas emissivity and transmittivity.
- Write a suitable network to represent the problem and obtain all the resistances.
- Formulate the radiosity equations for all the surfaces.
- Solve the resulting equations for all the radiosities.
- Obtain the heat transfer from/to each surface in the enclosure.
- What is the temperature of the reradiating surface?

Ex 11.12 A furnace is in the form of a hemispherical enclosure as shown in Fig. 11.26. The base of the hemisphere is a gray diffuse heat sink at a temperature of 475 K and has a surface emissivity of 0.85. The hemispherical surface is under radiant balance. Flue gas, at 750 K and a total pressure of 1 atm., passes through the furnace at a steady rate of 0.02 kg/s. The flue gas composition is given in terms of the partial pressure of the constituents as partial pressure of $CO_2 = 0.1$ atm., partial pressure of $H_2O = 0.2$ atm. and the rest is N_2. The average specific heat of the flue gases may be taken as 1024 $J/kg \cdot K$. Calculate the exit flue gas temperature and the useful heat output of the furnace.

Ex 11.13 A gas at a temperature of 1000 K flows past a surface maintained at 300 K subjecting it to convection heat transfer with $h = 45$ W/m$^2\,°C$. The gas also subjects the surface to a radiant heat flux since it has a gray absorption coefficient of $\kappa = 0.2$ m^{-1} and has an effective mean beam length of 0.5 m. Compare the magnitudes of the convection and radiation components of the heat flux and comment on the result. Assume the surface to be gray with an emissivity of $\varepsilon_s = 0.6$.

Ex 11.14 In a process application, a mixture of volumetric composition of 50% carbon dioxide and 50% nitrogen and at a total pressure of 1 atm. is used. The gas mixture is at a temperature of 1000 K. The gas is in radiant balance with the walls of a gray container of emissivity $\varepsilon_c = 0.65$ that is at a temperature of $T_c = 760$ K. Determine the mean beam length for the gas.

Ex 11.15 Combustion generated gases at 1800 K pass through a long circular duct of 0.3 m diameter. The volumetric gas composition is $CO_2 = 75\%$ and $H_2O = 25\%$. The total pressure of the gas is 1 atm. The walls of the duct that are black are maintained at a temperature of 500 K. What is the net radiant flux on the duct wall? It has also been determined independently that the heat transfer by convection is governed by a heat transfer coefficient of 26 W/m^2 °C. What is the convection heat flux on the duct wall? Comment on the results.

Ex 11.16 Consider a gray gas slab of optical thickness 0.3. The boundary surfaces are gray and have a common emissivity value of 0.5. The temperatures of the boundary surfaces are $T_{\tau=0} = 600$ K and $T_{\tau=0.3} = 400$ K. Obtain the temperature variation in the gas assuming that it is under equilibrium. Make use of the discrete ordinate method.

Ex 11.17 Figure 11.27 shows a gas-filled enclosure. Bottom surface labeled 1 has an area of 0.5×0.5 m, temperature $T_1 = 400$ K, and $\varepsilon_1 = 0.6$. Top surface labeled 2 has an area of 0.5×0.5 m, temperature $T_2 = 800$ K, and $\varepsilon_2 = 0.8$. The back of the enclosure is surface 3 that has an area of 0.5×0.5 m and is reradiating. The other three sides are openings that communicate to a very low temperature background via transparent windows.

Fig. 11.27 Gas-filled enclosure in Exercise 11.17

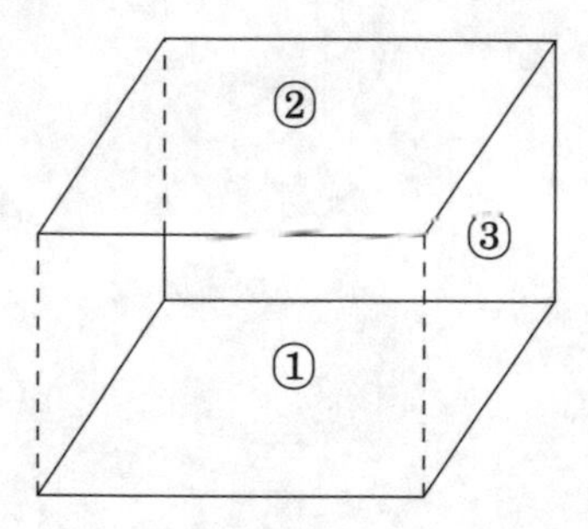

The enclosure is filled with a gray gas at $T_g = 600$ K and $\kappa = 0.15$ m^{-1}. Assume that the gas may be treated as an optically thin medium. Determine the total radiation transfer to surfaces 1 and 2. Also determine the heat loss to the surroundings via the transparent windows. (Hint: Treat the gas volume as a single zone and use Dunkle's charts to formulate the radiosity equations.)

Chapter 12
Laminar Convection In Internal Flow

We commence the study of convection heat transfer in this chapter. After looking at fluid properties of interest in convection heat transfer, we present notion of similarity to understand scaling principles that play a crucial role in convection heat transfer. Laminar fully developed flow and heat transfer in internal flows are covered in great detail. Useful relations are presented for heat transfer in the developing region.

12.1 Introduction

In Chap. 1, we have introduced the concept of convection through a phenomenological description by introducing h, the convection heat transfer coefficient. The heat transfer coefficient was introduced through the so-called "Newton's law of cooling". In many problems encountered in conduction heat transfer, we have made use of a suitable "h" value to describe what happens at a boundary between a solid and the ambient fluid. However, no effort was made to describe the basis for choosing a particular value of h. In what follows we would like to calculate h by using fundamental heat transfer principles that are involved in the case of a flowing fluid.

12.1.1 Classification of Flows

The main goal of the study of convection heat transfer is to understand the dependence of the convection heat transfer coefficient on (1) The nature of the fluid, (2) The nature of flow, and (3) The type of flow.

S. P. Venkateshan, *Heat Transfer*,
https://doi.org/10.1007/978-3-030-58338-5_12

(1) The nature of fluid

Basically, the nature of the fluid is mirrored by its physical and transport properties. Also, the variation of these properties within the flow domain decides the method of analysis. The fluid may be described by any of the following models, depending on the circumstance.

Incompressible fluid: Fluid density remains fixed irrespective of variations in pressure and temperature.

Compressible fluid: Fluid density varies with position and time due to changes in pressure or temperature. In high speed flows (flow speed comparable to the speed of sound in the fluid), the compressibility effects may become significant. However, the same fluid may be treated as incompressible if the fluid speed is small compared to the speed of sound in the medium.

Inviscid or non-viscous and non-heat conducting fluid: This is also referred to as an ideal fluid. A flow, far away from boundaries, even when the fluid has a non-zero viscosity, may sometimes be treated this way.

Viscous and heat conducting fluid: The fluid is referred to as a real fluid. As a subset of this, the fluid may be Newtonian or non-Newtonian. Newtonian fluid has a linear relationship between shear stress and velocity gradient while the non-Newtonian fluid has a more complex relationship. We consider *only* a Newtonian fluid in this text.

Fluid with constant thermo-physical properties: For such a fluid the properties like viscosity and thermal conductivity have very insignificant variation with temperature and pressure. In flows with small variation of temperature constant, property assumption may be justified.

Fluid with variable thermo-physical properties: The fluid properties such as viscosity and thermal conductivity vary significantly in the flow domain. Most important variation that needs to be considered is with respect to fluid temperature. Variation with pressure is seldom significant. Constant property assumption is not necessarily connected with the variation or otherwise of the fluid density.

(2) Nature of flow and attendant heat transfer

The nature of the flow is important since it affects heat transfer to a great extent. In practical applications, it is usual to look for flow conditions that enhance heat transfer significantly in comparison with conduction heat transfer that will take place in a stationary fluid.

Compressible high-speed flow: High speed means M, the Mach number (the ratio of fluid velocity to the speed of sound in the fluid) is large. Incompressible, low speed flow approximation is valid, in gases, for $M \leq 0.3$.

Laminar flow. Laminar flow is orderly or "streamline" flow. Laminar flow is also characterized by weak mixing except in regions of flow close to boundaries.

Turbulent flow: Turbulent flow exhibits temporal variations in velocity and temperature fields even when the flow is steady. Rapid mixing normal to the flow direction is a characteristic of turbulent flow.

Forced Convection: Flow is created or forced to take place by an external agency like a pump. The pump creates a pressure gradient that promotes and maintains the flow.

Free or natural convection: Flow is generated by temperature differences and the consequent density differences within the flowing medium. The flow may be assumed to be incompressible except for the buoyancy effect.

Mixed convection: Forced and free convection occur simultaneously and are of comparable importance. The buoyancy effects may either aid or oppose the forced flow.

(3) Type of Flow

The flow may also be classified according to the following types.

Internal Flow: Flow inside tubes and ducts. These occur in applications such as air handling systems, heat exchangers, energy conversion devices like turbines, engines, etc.

External Flow: Flow over extended surfaces, flow past a tube bundle in a heat exchanger, flow past vehicles, etc.

Steady flow: Velocity and temperature fields do not change with time.

Unsteady flow: Velocity and temperature fields change with time.

12.1.2 Fluid Properties and Their Variation

Thermo-physical properties of the fluid influence flow and the consequent heat transfer. Details of flow and temperature fields are affected by the properties as well as their variations with temperature and pressure of the flowing fluid. Hence, we shall look at some of the important thermo-physical properties and their variations with temperature and pressure in this section.

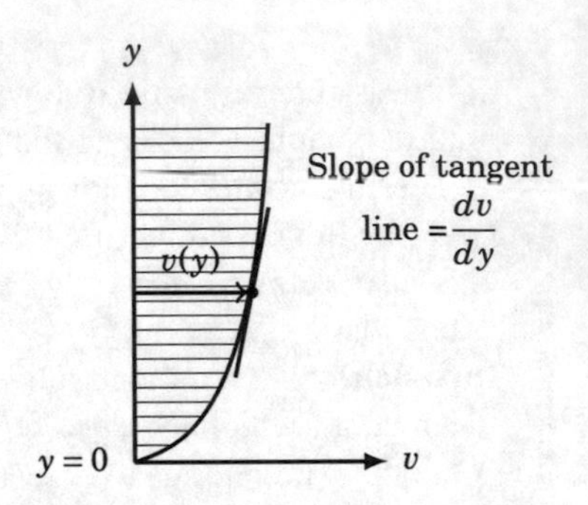

Fig. 12.1 Viscosity of a Newtonian fluid

(1) Fluid Viscosity

For a Newtonian fluid, the dynamic viscosity μ is defined through a linear relation between the shear stress and the velocity gradient.

$$\boxed{\tau = \mu \frac{dv}{dy}} \tag{12.1}$$

Here, τ = shear stress, μ = dynamic viscosity, v = velocity, y = coordinate normal to v. The velocity field varies with y as shown in Fig. 12.1 when a viscous fluid flows past a boundary. The fluid at lower velocity tends to decelerate the flow with a higher velocity. The unit of dynamic viscosity μ is given by

$$\boxed{[\mu] = \frac{[\tau]}{\frac{dv}{dy}} = \frac{\frac{\mathrm{N}}{\mathrm{m}^2}}{\frac{\mathrm{m/s}}{\mathrm{m}}} = \frac{\mathrm{kg}}{\mathrm{m\ s}} = \left[\frac{M}{LT}\right]} \tag{12.2}$$

In Eq. 12.2, the brackets indicate that the unit of the quantity within the brackets is being considered and not the magnitude. The last entry indicates the dimensions as will be explained later on.

Newton's law of viscosity resembles Hooke's law in solid mechanics and Fourier law of thermal conduction. For gases, μ increases with temperature. At 300 K, air has a dynamic viscosity of 18.46×10^{-6} kg/m s which increases to 42.4×10^{-5} kg/m s when the temperature changes to 1000 K. Dynamic viscosity of liquids decrease with temperature. For saturated liquid water, the dynamic viscosity decreases from 8.67×10^{-4} kg/m s at 300 K to 9.01×10^{-5} kg/m s at 573 K.

(2) Kinematic Viscosity

This is defined as the ratio of dynamic viscosity of the fluid and its density ρ.

Table 12.1 Variation of thermal conductivity with temperature

	Air		Water		
T, K	300	2000	300	400	580
k, W/m °C	0.0267	0.1149	0.611	0.685	0.516

$$\boxed{\nu = \frac{\mu}{\rho}} \tag{12.3}$$

It may be verified that the unit of kinematic viscosity is m^2/s. The reader may note that the same unit also characterizes the thermal diffusivity encountered in conduction heat transfer. Generally, the kinematic viscosity of gases increases with temperature. However, for liquids, the kinematic viscosity decreases with temperature. For example, the kinematic viscosity of air increases from 15.89×10^{-6} m^2/s at 300 K to 12.9×10^{-5} m^2/s at 1000 K. The kinematic viscosity of saturated water decreases from 8.004×10^{-7} m^2/s at 303 K to 1.265×10^{-7} m^2/s at 573 K.

(3) Fluid Thermal Conductivity

Fourier law (already familiar to us from conduction heat transfer study) introduces the conductivity of the fluid. The unit of thermal conductivity is either W/m °C or W/m K. Thermal conductivity of gases increases with temperature while it may show increasing as well as decreasing trends in the case of liquids. Examples are given in Table 12.1.

(4) Thermal Diffusivity of a Fluid

This is defined in the usual way as $\alpha = \frac{k}{\rho c}$ where c is the specific heat capacity of the fluid in J/kg °C or J/kg K. Thermal diffusivity has the units of m^2/s.

(5) Prandtl Number

The ratio of kinematic viscosity to thermal diffusivity occurs very often in heat and fluid flow problems and hence is given a specific name, the Prandtl number, Pr.[1] Thus,

$$\boxed{Pr = \frac{\nu}{\alpha} = \frac{\mu/\rho}{k/\rho c} = \frac{\mu c}{k}} \tag{12.4}$$

It has no dimensions. The ranges of Pr values are given in Table 12.2.

[1]Named in honor of Ludwig Prandtl, 1875–1953, a German engineer. He proposed the boundary layer theory which is successful in explaining pressure drop and heat transfer in the flow of a viscous heat conducting fluid and gave impetus for much development in Fluid Mechanics.

Table 12.2 Ranges of Prandtl number of various fluids

Fluid	*Pr* range	Remarks
Liquid Metals-Hg, Na, K, etc.	0.001–0.05	Decrease with temperature
Gases-H_2, He, Air, CO_2, etc.	0.5–1	More or less independent of temperature
Liquids—Water, Organic liquids, etc.	5–30,000	Decrease with temperature

Table 12.3 Prandtl number of two common liquids

	Saturated Water			Unused engine oil		
T, K	300	400	580	280	300	400
Pr	5.9	1.4	0.94	27,000	6600	154

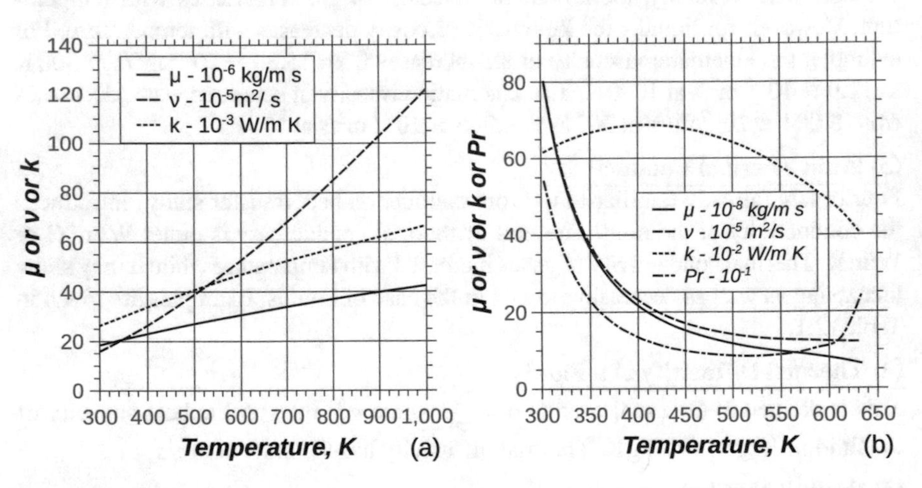

Fig. 12.2 **a** Variation of properties of air with temperature **b** Variation of properties of saturated liquid water with temperature

Table 12.3 shows Prandtl number variation for two common liquids.

Since air and water are commonly used in heat transfer applications, their property variations with temperature are shown in Fig. 12.2a, b on p. 550. While the properties are for air at 1 atm, the properties of water are for saturated water at the indicated temperatures. Prandtl number of air varies very little with temperature and hence is not included in Fig. 12.2a. However, Prandtl number of water varies significantly with temperature and hence has been included in Fig. 12.2b.

12.2 Dimensional Analysis and Similarity

Non-dimensional parameters are useful in discussing the behavior of thermal systems. They naturally evolve while solving the governing differential equations, as we have already seen in the case of conduction problems. We have already seen how parameters such as the Biot and Fourier numbers evolve while solving conduction problems in one and two dimensions. We have introduced a non-dimensional parameter, the Prandtl number in Sect. 12.1. Many more non-dimensional parameters become appropriate in fluid flow and heat transfer problems. These are discussed with the concept of similarity in mind. Similarity may be of two types:

1. Geometric similarity
2. Dynamic similarity
   - involves motion, forces, temperatures, heat fluxes, etc.

These two concepts are elucidated below using examples from fluid flow and heat transfer.

12.2.1 Dimensional Analysis of a Flow Problem

The first example we consider is a flow problem in which a viscous fluid flows steadily through a straight tube of circular cross section. Two fluid flow situations are shown in Fig. 12.3. At the left is a circular tube of diameter D_1 carrying a fluid 1. At the right is a circular tube of diameter D_2 carrying a fluid 2. Geometric similarity would require that both be circular tubes. If one tube is straight, the other also should be straight. However, dynamic similarity requires that suitable non-dimensional parameters remain the same for the two cases.

The quantity of interest to us is the pressure drop between stations 1 and 2 or stations $1'$ and $2'$. We first identify all the variables that enter the problem and also write out the units of these variables, using the SI system of units and also the length, mass, time system (refer to Table 12.4). In this last method, $[M]$ will represent mass dimension, $[L]$ will represent length dimension, and $[T]$ will represent time dimension. Buckingham π theorem (π theorem because each non-dimensional parameter was represented by the symbol π) states that the number of non-dimensional parameters that characterize the problem are $(n - r)$ where "n" is the number of variables (= 6 in

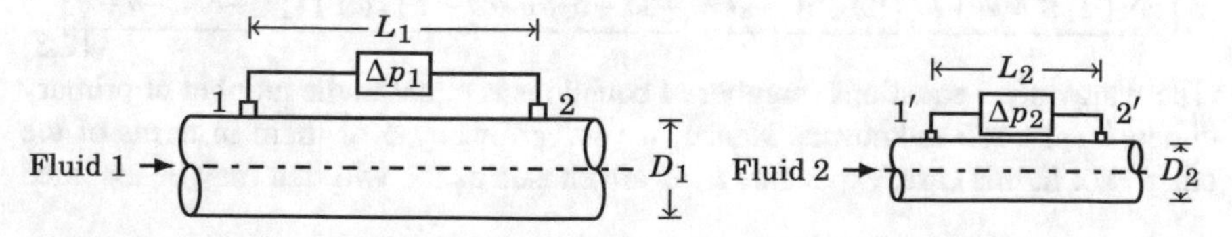

Fig. 12.3 Pressure drop in a fluid flowing in a straight tube

Table 12.4 Physical quantities and their dimensions

Physical quantity	Unit/Dimension
Δp: Pressure drop across L_1 or L_2	$\frac{\text{N}}{\text{m}^2} = Pa = \frac{\text{kg m/s}^2}{\text{m}^2} = \frac{\text{kg}}{\text{m s}^2} = \left[\frac{M}{LT^2}\right]$
ρ: Fluid density	$\frac{\text{kg}}{\text{m}^3} = \left[\frac{M}{L^3}\right]$
μ: Fluid viscosity	$\frac{\text{kg}}{\text{m s}} = \left[\frac{M}{LT}\right]$
V: Fluid velocity	$\frac{\text{m}}{\text{s}} = \left[\frac{L}{T}\right]$
L: Tube length	$m = \left[L\right]$
D: Tube diameter	$m = \left[L\right]$

the present case) and "r" is the number of primary dimensions involved (= 3; Mass, Length, Time or $[M]$, $[L]$ and $[T]$).[2] Thus, we expect three non-dimensional parameters to characterize the problem. In order to obtain these parameters, we represent Δp as a function of all the other variables that occur in the problem. Thus,

$$\Delta p = f(\rho, \mu, V, L, D) \tag{12.5}$$

It is possible as it happen many times that the functional relation is of the type

$$f(\rho, \mu, V, L, D) = K\left\{\rho^a \mu^b v^c L^d D^e\right\} \tag{12.6}$$

where K is a numerical constant and a, b, c, d, e are numerical exponents. If indeed this is valid, then the unit of Δp must be the same as the unit of the quantity inside the flower brackets in Eq. 12.6. This may be written using units of various quantities as

$$\begin{aligned} &\text{Units}: \left(\frac{\text{kg}}{\text{m s}^2}\right) = \left(\frac{\text{kg}}{\text{m}^3}\right)^a \left(\frac{\text{kg}}{\text{m s}}\right)^b \left(\frac{\text{m}}{\text{s}}\right)^c (\text{m})^d (\text{m})^e \\ &\text{Dimensions}: \left[\frac{M}{LT^2}\right]^1 = \left[\frac{M}{L^3}\right]^a \left[\frac{M}{LT}\right]^b \left[\frac{L}{T}\right]^c [L]^d [L]^e \end{aligned} \tag{12.7}$$

Dimensional homogeneity requires that the left-hand side and right-hand side of Expression 12.7 have identical dimensional units. This will require the right choice of all the exponents $a - e$. This may be done by equating the exponent of each primary dimension mass, length, and time on the two sides. These lead to the following three equations:

$$\boxed{\text{(a) [M]}: 1 = a + b} \quad \boxed{\text{(b) [L]}: \; -1 = -3a - b + c + d + e} \quad \boxed{\text{(c) [T]}: \; -2 = -b - c} \tag{12.8}$$

There are only 3 equations (number of equations is equal to the number of primary dimensions) but 5 unknowns. Hence, we *solve* for any 3 of them in terms of the other two. In this case, exponents b, d are chosen as the two that may be assigned

[2] Edgar Buckingham, 1867–1940, American physicist.

arbitrary/suitable numerical values and exponents a, c, e are solved in terms of them. From Eq. 12.8(a) we have $a = 1 - b$. From Eq. 12.8(c), we have $c = 2 - b$. Substitute these in Eq. 12.8(b) to get

$$e = 3a + b - c - d - 1 = 3(1 - b) + b - (2 - b) - d - 1 = -b - d$$

With these Eq. 12.5 may be rewritten, using Eq. 12.6 as

$$\Delta p = K\rho^{1-b}\mu^{b}V^{2-b}L^{d}D^{-b-d} = K\left(\frac{\mu}{\rho v D}\right)^{b}\left(\frac{L}{D}\right)^{d}\rho V^{2} \tag{12.9}$$

or, on rearrangement,

$$\boxed{\frac{\Delta p}{\rho V^2} = K\left(\frac{\mu}{\rho v D}\right)^{b}\left(\frac{L}{D}\right)^{d} \quad \text{or} \quad Eu = K\,Re_D^{-b}\left(\frac{L}{D}\right)^{d}} \tag{12.10}$$

Dimensional analysis cannot give the values of K, b, and d. They have to be determined from solution of appropriate equations that govern the fluid flow problem or from experiments. Both these alternates are used in practice. These will be presented later on.

We notice that Eq. 12.10 contains three non-dimensional parameters. They are

- Euler number Eu:

$$Eu = \frac{\Delta p}{\rho V^2} \tag{12.11}$$

  Euler number is nothing but a non-dimensional pressure drop that uses the "dynamic head" $\frac{\rho V^2}{2}$ as the reference pressure drop. The factor $\frac{1}{2}$ may appropriately be absorbed in the coefficient K.
- Reynolds number Re_D:

$$Re_D = \frac{\rho V D}{\mu} \tag{12.12}$$

  The subscript D is used to indicate that the Reynolds number is based on diameter of tube as the characteristic "length scale" in the problem.
- Length to diameter ratio or the non-dimensional length L':

$$L' = \frac{L}{D} \tag{12.13}$$

12.2.2 *Notion of "Similarity"*

Equation 12.9 may be interpreted as follows using the concept of similarity. Apart from the geometric similarity that was alluded to earlier, dynamic similarity requires additional conditions to be satisfied. For example, if we compare the two cases shown in Fig. 12.3 with L_1, D_1 fluid 1 and L_2, D_2 fluid 2, the non-dimensional pressure drop

$$\frac{\Delta p_1}{\rho_1 V_1^2} = \frac{\Delta p_2}{\rho_2 V_2^2} \text{ or } Eu_1 = Eu_2 \tag{12.14}$$

if and only if

$$\text{(a) } \frac{\rho_1 V_1 D_1}{\mu_1} = \frac{\rho_2 V_2 D_2}{\mu_2} \text{ or } Re_{D_1} = Re_{D_2}; \text{ (b) } \frac{L_1}{D_1} = \frac{L_2}{D_2} \tag{12.15}$$

Alternately, we may state that dynamic similarity exists if and only if the Reynolds numbers and length to diameter ratios are the same in the two cases. A typical example shows the utility of this concept.

Example 12.1

Air at atmospheric pressure and at a temperature of 300 K flows in a 2 m long smooth circular tube of 25 mm inner diameter. The velocity is adjusted such that the Reynolds number is 15,000. What is the velocity? What is the mass flow rate? The pressure drop is measured to be 100 Pa. If the fluid flowing in the tube is replaced by water at 300 K what will be the mass flow rate and the corresponding pressure drop?

Solution:

Step 1 Since the concept of similarity applies to the cases, the following parameters are common to both cases.

Diameter of tube: $D_1 = D_2 = 0.025$ m
Length of tube: $L_1 = L_2 = 2$ m
Reynolds number: $Re_{D_1} = Re_{D_2} = 15{,}000$

Case (a) Fluid is air

Step 2 The air properties are taken from table of properties at $T = 300$ K. All quantities are shown with a subscript 1 to indicate that the fluid is air.

$$\rho_1 = 1.1614 \text{ kg/m}^3; \ \nu_1 = 15.89 \times 10^{-6} \text{ m}^2/\text{s}$$

Step 3 Using the given value of Reynolds number, air velocity in the tube is

$$V_1 = \frac{Re_{D_1}\nu_1}{D_1} = \frac{15000 \times 15.89 \times 10^{-6}}{0.025} = 9.534 \text{ m/s}$$

Step 4 The mass flow rate of air is then given by

$$\dot{m} = \rho_1 \cdot \frac{\pi D_1^2}{4} \cdot V_1 = 1.1614 \times \frac{\pi \times 0.025^2}{4} \times 9.534 = 0.00545 \text{ kg/s}$$

Step 5 It is given that the pressure drop has been measured with air as $\Delta p_1 = 100$ Pa. Hence, the Euler number (the non-dimensional pressure drop) may be calculated as

$$Eu_1 = \frac{\Delta p_1}{\rho_1 V_1^2} = \frac{100}{1.1614 \times 9.534^2} = 0.9473$$

Case (b) Fluid is water

Step 6 The properties of water are taken from tables of properties at 300 K. All quantities are shown with a subscript 2 to indicate that the fluid is water.

$$\rho_2 = 995.7 \text{ kg/m}^3; \ \nu_2 = 8.004 \times 10^{-7} \text{ m}^2\text{/s}$$

Step 7 Using the given value of Reynolds number, water velocity in the tube is

$$V_2 = \frac{Re_{D_2}\nu_2}{D_2} = \frac{15000 \times 8.004 \times 10^{-7}}{0.025} = 0.48 \text{ m/s}$$

Step 8 The mass flow rate of water is then given by

$$\dot{m} = \rho_2 \cdot \frac{\pi D_2^2}{4} \cdot V_2 = 995.7 \times \frac{\pi \times 0.025^2}{4} \times 0.48 = 0.235 \text{ kg/s}$$

Step 9 The two cases satisfy dynamic similarity since the length to diameter ratio and the Reynolds number are unchanged. Hence, the Euler number is the same for the two cases. With this, we can calculate the pressure drop with water as

$$\Delta p_2 = \rho_2 V_2^2 Eu_2 = \rho_2 V_2^2 Eu_1 = 995.7 \times 0.48^2 \times 0.9473 = 217.3 \text{ Pa}$$

12.2.3 Dimensional Analysis of Heat Transfer Problem

Consider fluid flow in a tube with heat addition to the fluid as shown in Fig. 12.4.

We shall think of some average temperature difference ΔT_{ref} as a representative temperature difference applicable to this problem. Then, we can define a suitable mean heat transfer coefficient h based on a representative area S_{ref} as $h = \frac{Q}{S_{ref}\Delta T_{ref}}$. Variables entering the problem along with their dimensions are given in Table 12.5.

The tube length drops out of consideration since our interest is on the mean heat transfer coefficient defined for the entire length of the tube. There are thus $r = 7$ parameters that govern the problem. We use $n = 4$ in the M, L, T, θ—mass, length, time, temperature—system. By Buckingham π theorem, there are $n - r = 7 - 4 = 3$ non-dimensional parameters that describe the problem. Let us assume that the functional relation we seek is of form

$$h = K\rho^a \mu^b V^c c^d k^e D^f \tag{12.16}$$

Hence, the dimensional equation may be written in the form

$$\left[\frac{M}{T^3\theta}\right]^1 = \left[\frac{M}{L^3}\right]^a \left[\frac{M}{LT}\right]^b \left[\frac{L}{T}\right]^c \left[\frac{L^2}{T^2\theta}\right]^d \left[\frac{ML}{T^3\theta}\right]^e [L]^f \tag{12.17}$$

Dimensional homogeneity requires the following balances.

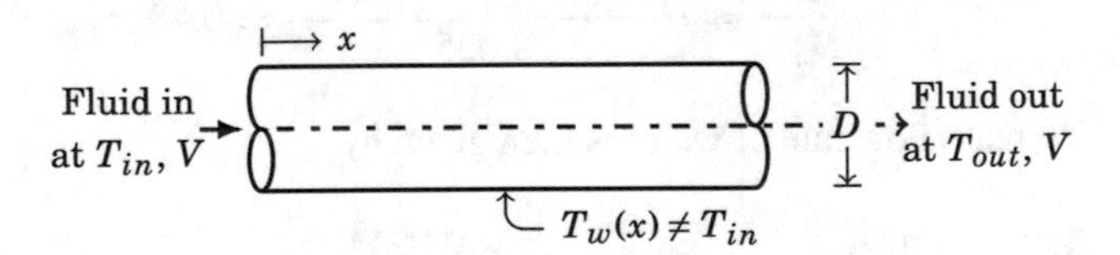

Fig. 12.4 Tube flow with heat addition

Table 12.5 Physical quantities and dimensions

Physical quantity	Unit/Dimension
h: Heat transfer coefficient	$\frac{\text{W}}{\text{m}^2\text{ K}} = \frac{\text{N m}}{\text{s m}^2\text{ K}} = \frac{\text{kg m}^2}{\text{s}^3\text{ m}^2\text{ K}} = \frac{\text{k}}{\text{s}^3\text{ K}} = \left[\frac{M}{T^3\theta}\right]$ where $[\theta]$ is the dimension of the fourth primary quantity, temperature
ρ: Fluid density	$\frac{\text{kg}}{\text{m}^3} = \left[\frac{M}{L^3}\right]$
μ: Fluid viscosity	$\frac{\text{kg}}{\text{m s}} = \left[\frac{M}{LT}\right]$
V: Fluid velocity	$\frac{\text{m}}{\text{s}} = \left[\frac{L}{T}\right]$
c: Fluid specific heat	$\frac{\text{J}}{\text{kg K}} = \frac{\text{N m}}{\text{kg K}} = \frac{\text{kg m m}}{\text{s}^2\text{ kg·K}} = \frac{\text{m}^2}{\text{s}^2\text{ K}} = \left[\frac{L^2}{T^2\theta}\right]$
k: Fluid thermal conductivity	$\frac{\text{W}}{\text{m K}} = \frac{\text{N m}}{\text{s m K}} = \frac{\text{kg m}^2}{\text{s}^3\text{ m K}} = \frac{\text{kg m}}{\text{s}^3\text{ K}} = \left[\frac{ML}{T^3\theta}\right]$
D: Tube diameter	$m = [L]$

$$[M] \text{ balance: } 1 = a + b + e \text{ or } b = (1 - a - e) \tag{12.18}$$

$$[L] \text{ balance: } 0 = -3a - b + c + 2d + e + f \text{ or } f = +3a + b - c - 2d - e \tag{12.19}$$

$$[T] \text{ balance: } -3 = -b - c - 2d - 3e \tag{12.20}$$

$$[\theta] \text{ balance: } -1 = -d - e \text{ or } d = 1 - e \tag{12.21}$$

There are 6 unknowns and 4 equations (equal to the number of fundamental units). We solve for four of the unknowns, b, c, d, f in terms of a and e. Using Eqs. 12.18 and 12.21 in Eq. 12.19 gives

$$f = 3a + (1 - a - e) - c - 2(1 - e) - e = 2a - c - 1 \tag{12.22}$$

From Eq. 12.20, using Eq. 12.21, we have

$$b = 3 - c - 2(1 - e) - 3e = 1 - c - e$$

Comparing this with Eq. 12.18, we conclude that $a = c$. Using this in Eq. 12.22, we finally get

$$f = 3a + (1 - a - e) - a - 2(1 - e) - e = a - 1 \tag{12.23}$$

Substituting all these back in Eq. 12.16, we have

$$h = K\rho^a \mu^{1-a-e} V^a c^{1-e} k^e D^{a-1} \tag{12.24}$$

Grouping terms with the same exponent, Eq. 12.24 takes the form

$$h = K\left(\frac{\rho V D}{\mu}\right)^a \left(\frac{k}{\mu c}\right)^{1-e} \left(\frac{k}{D}\right) \text{ or } \left(\frac{hD}{k}\right) = K\left(\frac{\rho V D}{\mu}\right)^a \left(\frac{k}{\mu c}\right)^{1-e}$$

This may be recast in terms of non-dimensional groups as

$$\boxed{Nu_D = K Re_D^a Pr^{e-1}} \tag{12.25}$$

The above relation links the three non-dimensional parameters that are important in the problem. Two of these, the Reynolds number $Re_D = \frac{VD}{\nu}$ and the Prandtl number given by $Pr = \frac{\mu c}{k}$ are already familiar to us. The third non-dimensional parameter that appears here is the Nusselt number given by $Nu_D = \frac{hD}{k}$ which is based again on the tube diameter as the characteristic length.[3] Note that the Nusselt number is

[3]Named after Ernst Kraft Wilhelm Nusselt, 1882–1957, a German engineer.

similar to the Biot number that was defined in problems involving conduction with convection at a boundary. However, the Biot number is based on the solid thermal conductivity while the Nusselt number is based on the fluid thermal conductivity. Similarity, in this case means that the Nusselt number is invariant if and only if $f(Re_{D_1}, Pr_1) = f(Re_{D_2}, Pr_2)$. Note that K, a, and e are not obtainable by dimensional analysis alone. Either experiments or analysis will have to give these.

The **Nusselt number** may be given a physical interpretation. It is the ratio of two heat fluxes, the convective heat flux in the moving medium to the conductive heat transfer in the stationary fluid. We may easily verify this by writing the Nusselt number as

$$Nu_D = \frac{hD}{k} = \frac{(h\Delta T_{\mathrm{ref}})}{\left(\frac{k\Delta T_{\mathrm{ref}}}{D}\right)} = \frac{q_c}{q_k} \tag{12.26}$$

The numerator q_c is a representative convective heat flux, and the denominator q_k is a representative conductive heat flux. Since Nu_D is invariably greater than unity, convection *enhances* heat transfer to a value bigger than the representative conductive flux.

Example 12.2

Consider the situation described in Example 12.1. It is estimated that the heat transfer coefficient with air is 46 W/m^2 K. The Prandtl number of the fluid is expected to affect the Nusselt number by a factor proportional to $Pr^{0.36}$. What will be the heat transfer coefficient when the fluid flowing in the tube is changed to water?

Solution:
The data specified in Example 12.1 is reproduced below for ready reference. These are fluid independent.

$$Re_{D_1} = Re_{D_2} = 15{,}000,\ \ D_1 = D_2 = 0.025\ \text{m},\ L_1 = L_2 = 2\ \text{m}$$

Nusselt number with air as the fluid :
The heat transfer coefficient with air as the fluid is given as $h_1 = 46$ W/m^2 K. From table of properties of air, we have, at 300 K,

$$k_1 = 0.0267\ \text{W/m K}, \quad Pr_1 = 0.71$$

The Nusselt number with air as the flowing medium is then calculated as

$$Nu_1 = \frac{h_1 D_1}{k_1} = \frac{46 \times 0.025}{0.0267} = 43.07$$

Nusselt number with water as the fluid :

Since the Reynolds number and the length to diameter ratio are held fixed, the Nusselt number is affected only by the change in the Prandtl number when the fluid is changed from air to water. We have the following property values for water at 300 K.

$$k_2 = 0.611\ \text{W/m K}, \quad Pr_2 = 5.9$$

Using similarity law given by Eq. 12.25, we may identify the exponent $e - 1$ as 0.36. Hence, the Nusselt number, Nu_2 with water as the fluid follows the relation

$$\frac{Nu_2}{Nu_1} = \left(\frac{Pr_2}{Pr_1}\right)^{0.36} = \left(\frac{5.9}{0.71}\right)^{0.36} = 2.143$$

Hence, the Nusselt number with water is

$$Nu_2 = 43.07 \times 2.143 = 92.31$$

Heat transfer coefficient with water as the fluid is then obtained as

$$h_2 = \frac{Nu_2 k_2}{D_2} = \frac{92.31 \times 0.611}{0.025} = 2256\ \text{W/m}^2\ \text{K}$$

There is thus a dramatic increase in the heat transfer coefficient when the fluid is changed from air to water keeping all other things the same!

12.3 Internal Flow Fundamentals

Convection heat transfer involves an interaction between flow (velocity) and temperature fields. Hence, it is not possible to discuss convection heat transfer without a clear understanding of fundamentals of fluid flow. As mentioned earlier in Sect. 12.1 there are several ways of classifying a flow. Here our interest will be the steady flow of a real (viscous and heat conducting) incompressible fluid. We attempt to understand laminar flow. Subsequently, in a later chapter, the attention will be directed toward internal as well as external turbulent flow. Special cases like compressible flows will also be taken up later on in Chap. 17.

12.3.1 Fundamentals of Steady Laminar Tube Flow

Consider steady laminar fluid flow in a straight tube of circular cross section. Experiments indicate that Laminar flow prevails in the tube for $Re_D < 2300$ based on the mean velocity U and the tube diameter D. Assume that the fluid enters the tube at $z = 0$ with a uniform velocity profile, i.e., the velocity is uniform across the tube cross section. Thus, the velocity u_z in the axial direction is equal to a constant given $u_z(r, 0) = U =$ constant.

Figure 12.5 shows the details of how the velocity profile changes from entry down the length of the tube. Because of viscosity, the fluid velocity becomes zero at the tube wall and the flow field varies with r and z as indicated. Boundary layer—nonuniform velocity region near the boundary is referred to as boundary layer—develops from the periphery of the tube such that the velocity profile is non-uniform in the boundary layer and uniform in the core. Since the velocity is $<U$ near the tube wall, the velocity in the core region is $>U$, to guarantee that the volume flow rate (the flow is incompressible) across the tube cross section is the same for all z. The boundary layer occupies the entire tube cross section for $z \geq L_{\text{dev}}$, where L_{dev} is referred to as the entry length. Beyond $z = L_{\text{dev}}$, the velocity profile remains invariant with respect to z. Thus, the velocity profile is a function of "r" only for $z > L_{\text{dev}}$. Experiments and analysis indicate that the entry length depends on Re_D and is given by

$$\boxed{\frac{L_{\text{dev}}}{D} = 0.05\ Re_D} \tag{12.27}$$

The flow beyond $z = L_{\text{dev}}$ is referred to as fully developed flow. Analysis of the flow in this region is fairly simple and will be done below by two methods. First method derives the appropriate equation governing fully developed tube flow starting from the first principles. The second method starts with the Navier Stokes (NS) equations presented in Appendix H and obtains the governing equation by simplifying them.

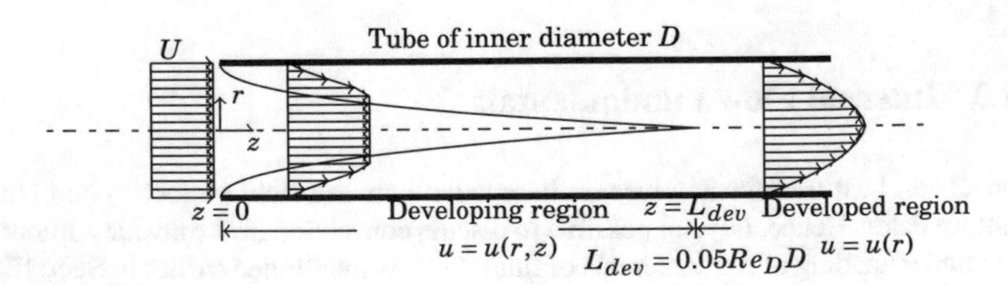

Fig. 12.5 Fluid flow in a straight tube

12.3.2 Governing Equation Starting from First Principles

The fluid element in the form of a cylinder

Consider force balance on a cylindrical fluid element as shown in Fig. 12.6a. The fluid element is located in the fully developed region, is of radius r and is of length Δz as shown in the figure. Under the fully developed condition, there is *no change* in the velocity u_z with z. Hence, the rate at which momentum enters the cylinder through the left face of the cylindrical fluid element is the same as that leaving through the right face. Hence, there is no net momentum change for the fluid across the element length dz. Thus, the forces that are acting on the fluid element are as shown in Fig. 12.6a. The forces are the pressure forces at the two end faces and the shear stress on the curved cylindrical portion. All forces involved are along the z-direction. Force balance requires the following:

$$\pi r^2 \; p(z) + 2\pi r \Delta z \; \tau = \pi r^2 \; p(z + \Delta z) \tag{12.28}$$

Note that the shear stress is shown pointing toward $+z$. The *convention* is that the axial velocity u_z is an increasing function with r. Using Taylor expansion, we have

$$p(z + \Delta z = p(z) + \left.\frac{dp}{dz}\right|_z \Delta z + O(\Delta z^2)$$

Inserting these in Eq. 12.28, we get

$$\pi r^2 \; p(z) + 2\pi r \Delta z \tau = \pi r^2 \; p(z) + \left.\frac{dp}{dz}\right|_z \Delta z$$

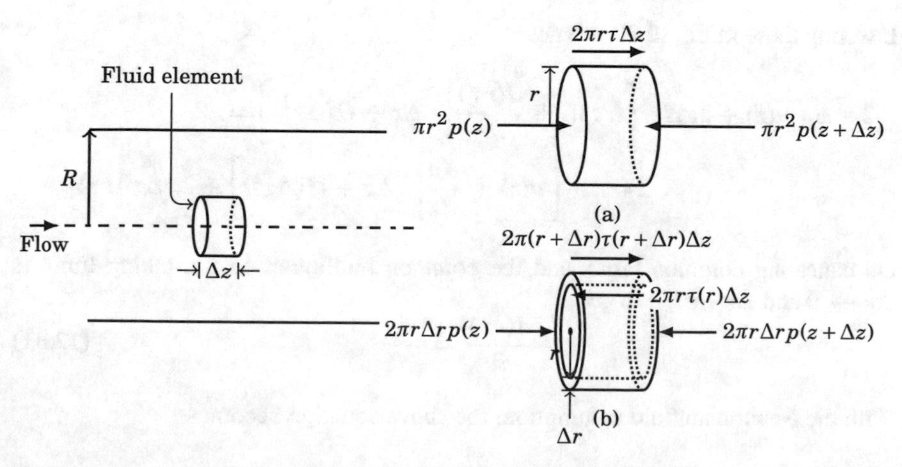

Fig. 12.6 Sketches to help in force balance from first principles **a** cylindrical fluid element **b** cylindrical annular fluid element

Assuming the fluid to be a Newtonian fluid, the shear stress is related to the derivative of velocity with respect to r as $\tau = \mu \frac{du_z}{dr}$. On substituting this in the previous equation and on simplification, taking limit as $\Delta z \to 0$, we get

$$\boxed{\frac{du_z}{dr} = \frac{r}{2\mu}\frac{dp}{dz}} \tag{12.29}$$

We note that the governing equation is a *first-order differential equation*. This equation is also obtained if we integrate Eq. 12.32, once with respect to r!

Fluid element in the form of a thin cylindrical shell

Consider force balance on a cylindrical shell element of length Δz and thickness Δr. The comments made while describing the cylindrical fluid element also apply in the present case. Thus, the forces are the pressure forces at the two ends and the shear stresses on the cylindrical portions. All forces involved are along the z-direction. Force balance requires the following:

$$2\pi r \Delta r \; p(z) + 2\pi \Delta z \; (r\tau)|_{r+\Delta r} = 2\pi r \Delta r \; p(z+\Delta z) + 2\pi \Delta r \Delta z \; (r\tau)|_r \tag{12.30}$$

Using Taylor expansion, we have

$$p(z+\Delta z) = p(z) + \left.\frac{dp}{dz}\right|_z \Delta z + O(\Delta z^2)$$

$$(r\tau)|_{r+\Delta r} = (r\tau)|_r + \left.\frac{d(r\tau)}{dr}\right|_r \Delta r + O(\Delta r^2)$$

Inserting these in Eq. 12.30, we get

$$2\pi r \Delta r \; p(z) + 2\pi \Delta z \left[(r\tau)|_r + \left.\frac{d(r\tau)}{dr}\right|_r \Delta r + O(\Delta r^2)\right] =$$
$$2\pi r \Delta r \left[p(z) + \left.\frac{dp}{dz}\right|_z \Delta z + O(\Delta z^2)\right] + 2\pi \Delta z \; (r\tau)|_r$$

On canceling common terms and the common multiplier $\Delta r \Delta z$, taking limit as $\Delta r \to 0$ and $\Delta z \to 0$, we get

$$\frac{d(r\tau)}{dr} = r\frac{dp}{dz} \tag{12.31}$$

With the Newtonian fluid assumption, the above equation becomes

$$\boxed{\frac{d}{dr}\left(r\frac{du_z}{dr}\right) = \frac{r}{\mu}\frac{dp}{dz}} \quad (12.32)$$

We note that the governing equation is a *second-order ordinary differential equation*.

12.3.3 Governing Equation Starting with the NS Equations

Equations of motion of an incompressible fluid in steady $\left(\frac{\partial}{\partial t} \equiv 0\right)$ laminar flow are given by the Navier Stokes Equations. The present case involves axisymmetric flow for which the appropriate equations are given by Eqs. H.31 and H.32 since we are considering only the flow problem here. In the fully developed region the velocity component $u_r \equiv 0$, the velocity u_z is a function of only r. With these, the equation of continuity is identically satisfied. The r momentum equation (Eq. H.31) reduces on taking $u_r \equiv 0$ and $\frac{\partial u_z}{\partial z} = 0$ to

$$-\frac{\partial p}{\partial r} = 0 \quad (12.33)$$

thus showing that the pressure is a function of z alone. The z momentum equation (Eq. H.32) then simplifies to

$$0 = -\frac{1}{\rho}\frac{dp}{dz} + \nu\frac{1}{r}\frac{d}{dr}\left(r\frac{du_z}{dr}\right)$$

or, on rearrangement to

$$\frac{d}{dr}\left(r\frac{du_z}{dr}\right) = \frac{r}{\mu}\frac{dp}{dz} \quad (12.34)$$

Note that, for obvious reasons, all partial derivatives are now changed to total derivatives. Equation 12.34 is identical to Eq. 12.32 derived from first principles.

12.3.4 Solution

The governing equation for fully developed flow requires two boundary conditions or one boundary condition depending on whether we use the second-order equation or the first-order equation. In the first case, the two boundary conditions are specified as

$$\text{Tube wall: } u_z = 0 \text{ at } r = R; \quad \text{Tube axis: } u_z \text{ is finite at } r = 0 \quad (12.35)$$

The first of these boundary condition corresponds to "no slip" at the tube wall. In the second case, only the tube wall boundary condition needs to be imposed. The other conditions are automatically satisfied.

We may integrate Eq. 12.29 with respect to r, noting that $\frac{dp}{dz}$ is independent of r, to get

$$u_z(r) = \frac{1}{2\mu}\frac{dp}{dz}\frac{r^2}{2} + A \tag{12.36}$$

where A is a constant of integration. In general, A could have been a function of z. However, it as well as $\frac{dp}{dz}$ cannot be functions of z since the velocity profile is invariant with respect to z. We apply the boundary condition at the tube wall. We then have

$$0 = \frac{1}{2\mu}\frac{dp}{dz}\frac{R^2}{2} + A \tag{12.37}$$

Subtracting Eq. 12.37 from 12.36 the constant of integration gets eliminated and hence

$$u_z(r) = \frac{r^2}{4\mu}\frac{dp}{dz} - \frac{R^2}{4\mu}\frac{dp}{dz} = -\frac{R^2}{4\mu}\frac{dp}{dz}\left[1 - \left(\frac{r}{R}\right)^2\right] \tag{12.38}$$

We notice that at $r = 0$, i.e., at the axis of the tube, u_z has the maximum value given by, say $u_z(r = 0) = u_{\text{max}}$. The maximum value is obtained by putting $r = 0$ in Eq. 12.38 as

$$u_{\text{max}} = -\frac{R^2}{4\mu}\frac{dp}{dz} \tag{12.39}$$

This will be a positive quantity if the pressure decreases in the direction of flow! Equation 12.38 may be recast in the non-dimensional form

$$\boxed{\frac{u_z}{u_{\text{max}}} = 1 - \left(\frac{r}{R}\right)^2} \tag{12.40}$$

The relationship between velocity and radius is a parabolic relation and is referred to as the Hagen–Poiseuille solution.[4] The average velocity U is defined such that the volume flow rate through the tube is $\dot{V} = \pi R^2 U$. Note that U is also the uniform velocity at entry to the tube. To conserve mass flow across the tube this must also be equal to the volume flow rate at any z. The volume flow rate in the fully developed region may be obtained the fully developed velocity profile given by Eq. 12.40.

[4] Gotthilf Heinrich Ludwig Hagen, 1797–1884, German physicist and hydraulic engineer and Jean Léonard Marie Poiseuille, 1797–1869, French physicist and physiologist.

$$\dot{V} = \int_0^R \underbrace{u_z(r)}_{\text{Local velocity}} \underbrace{2\pi r dr}_{\text{Elemntal area}} \tag{12.41}$$

Using the parabolic velocity profile, taking $\frac{r}{R}$ as ζ, the above expression becomes

$$\dot{V} = 2\pi R^2 u_{\max} \int_0^1 \left(1 - \zeta^2\right) \zeta d\zeta = 2\pi R^2 u_{\max} \left[\frac{\zeta^2}{2} - \frac{\zeta^4}{4}\right]\Bigg|_0^1 = \frac{\pi R^2 u_{\max}}{2} \tag{12.42}$$

Equating the volume flow rate obtained above with $\dot{V} = \pi R^2 U$, we see that the mean velocity is just half the maximum velocity, i.e.,

$$\boxed{U = \frac{u_{\max}}{2}} \tag{12.43}$$

The pressure gradient may now be obtained in terms of the mean velocity, using Eq. 12.39.

$$\frac{dp}{dz} = -\frac{4\mu u_{\max}}{R^2} = -\frac{8\mu U}{R^2} \tag{12.44}$$

The pressure gradient is a constant as already indicated. Hence, we may write it as the ratio of pressure drop Δp over a length L in the fully developed region. Thus, we also have

$$\frac{dp}{dz} = \frac{\Delta p}{L} = -\frac{8\mu U}{R^2} \tag{12.45}$$

It is customary to define a Darcy friction factor f such that the pressure drop is given by

$$\Delta p = -f \times \underbrace{\frac{L}{D}}_{\substack{\text{Length to}\\ \text{diameter ratio}}} \times \underbrace{\frac{\rho U^2}{2}}_{\text{Dynamic head}} \tag{12.46}$$

We notice then that $-f\frac{L}{2D}$ is the Euler number that was obtained by the use of Buckingham π theorem in Sect. 12.2. We also note that the present analysis provides the undetermined exponents in the expression obtained by dimensional arguments. The friction factor may be expressed as

$$f = -\frac{\frac{\Delta p}{\rho U^2}}{\frac{L}{2D}} = \frac{\frac{8\mu UL}{R^2\rho U^2}}{\frac{L}{2D}} = \frac{\frac{32\mu U}{D^2\rho U^2}}{\frac{1}{2D}} = \frac{64\mu}{\rho UD} = \frac{64}{Re_D} \text{ or } \boxed{f\,Re_D = 64} \tag{12.47}$$

using Eq. 12.45 and by noting that $R = \frac{D}{2}$. With these, we may write for the Euler number the relation

$$\boxed{Eu = \frac{32}{Re_D}\frac{L}{D}} \tag{12.48}$$

Comparing this with Eq. 12.10, we identify the constant K as 32, exponent b as 1, and exponent d as 1.

Example 12.3

Engine oil at 20 °C is made to flow in a tube of 12 mm diameter. What is the maximum mass flow rate if the Reynolds number is not to exceed 10? What is the pressure drop in a length of 10 m under this flow condition?

Solution:

Step 1 The density and kinematic viscosity of engine oil are taken from table of properties.

$$\rho = 885.23 \text{ kg/m}^3,\ \nu = 0.0009 \text{ m}^2/\text{s}$$

The tube diameter and length are given as $L = 10$ m, $D = 12$ mm $=$ 0.012 m.The Reynolds number based on the diameter is taken as the limiting value of $Re_D = 10$ given in the problem.

Step 2 Velocity calculation:
The mean velocity corresponding to this Reynolds number is obtained as

$$U = \frac{Re_D\,\nu}{D} = \frac{10 \times 0.0009}{0.012} = 0.75 \text{ m/s}$$

Step 3 The mass flow corresponding to this flow velocity is obtained as

$$\dot{m} = \rho\frac{\pi D^2}{4}U = 885.23 \times \frac{\pi \times 0.012^2}{4} \times 0.75 = 0.075 \text{ kg/s}$$

Step 4 Pressure drop calculation:
It is seen that the flow is laminar. The friction factor is calculated, using Eq. 12.47 as

$$f = \frac{64}{Re_D} = \frac{64}{10} = 6.4$$

The flow development length is calculated based on Eq. 12.27 as

$$L_{\text{dev}} = 0.058 Re_D D = 0.058 \times 10 \times 0.012 = 0.00696 \text{ m}$$

The tube length of 10 m is much much larger than the development length, and hence, we make the assumption that the pressure drop is based on the fully developed assumption throughout the length of the tube.

Step The pressure drop is calculated using Eq. 12.46 as

$$\Delta p = 6.4 \times \frac{10}{0.012} \times \frac{885.23 \times 0.75^2}{2} = 1.328 \times 10^6 \text{ Pa} \approx 13 \text{ atm}$$

12.3.5 Fully Developed Flow in a Parallel Plate Channel

Governing equation

Consider steady laminar flow of a viscous incompressible fluid between two parallel plates with a spacing of $2b$, as an example of flow in Cartesian coordinates. The coordinate axes are chosen such that the origin is at the center of the entry plane and the x-axis is parallel to the two plates. The governing equation for fully developed flow may be derived starting from first principles. Consider a fluid element of thickness Δy and length Δx as shown (enlarged for clarity) in Fig. 12.7. Let the thickness of the element in a direction perpendicular to the plane of the figure be one unit.

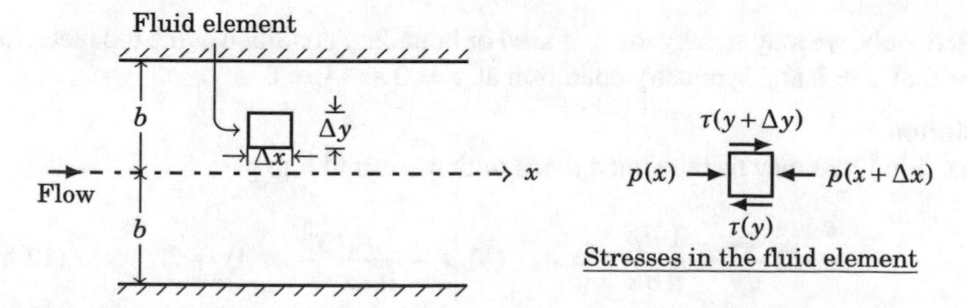

Fig. 12.7 Laminar fluid flow between two parallel plates

The velocity u along the x-direction varies only with y while the pressure p varies only with x. A force balance may be made on the element as follows.

$$\tau(y+\Delta y)\Delta x + p(x)\Delta y = \tau(y)\Delta x + p(x+\Delta x)\Delta y \tag{12.49}$$

Using Taylor expansion, we have the following.

$$\tau(y+\Delta y) = \tau(y) + \frac{d\tau}{dy}\Delta y + O(\Delta y^2);\;\; p(x+\Delta x) = p(x) + \frac{dp}{dx}\Delta x + O(\Delta x^2) \tag{12.50}$$

Substitute these in Eq. 12.49 to get

$$\frac{d\tau}{dy}\Delta y\Delta x + O(\Delta y^2\Delta x) = \frac{dp}{dx}\Delta x\Delta y + O(\Delta x^2\Delta y)$$

The common factor $\Delta y\Delta x$ (this is nothing but the volume of the element) is removed and in the limit $\Delta x \to 0$, $\Delta y \to 0$ we obtain

$$\frac{d\tau}{dy} - \frac{dp}{dx} = 0 \tag{12.51}$$

Using Newton's law of viscosity, we then get

$$\boxed{\mu\frac{d^2u}{dy^2} - \frac{dp}{dx} = 0} \tag{12.52}$$

The same equation may be obtained by starting with the NS equations in cartesian coordinates and by suitable simplification. This is left as an exercise to the reader.

Boundary conditions

Since the governing equation is a second-order equation, we need to specify two boundary conditions. These are specified by the no slip conditions at the two boundaries, i.e.,

$$u = 0 \text{ at } y = -b \text{ and } y = b \tag{12.53}$$

Alternately, we may specify the first kind of boundary condition at the top plate, i.e., $u = 0$ at $y = b$ and symmetry condition at $y = 0$ as $\frac{du}{dy} = 0$.

Solution

Equation 12.52 may be integrated twice with respect to y to get

$$(a)\; \frac{du}{dy} - \frac{1}{\mu}\frac{dp}{dx}y = A;\;\; (b)\; u - \frac{1}{\mu}\frac{dp}{dx}\frac{y^2}{2} = Ay + B \tag{12.54}$$

where A and B are constants of integration to be determined by the use of the boundary conditions. The symmetry condition at $y = 0$ requires that the constant A

be set to zero. The constant B is then obtained by using the no slip at top (or bottom) wall as

$$0 - \frac{1}{\mu}\frac{dp}{dx}\frac{b^2}{2} = B \tag{12.55}$$

Substituting Eqs. 12.55 in 12.54(b), we get

$$\boxed{u = -\frac{1}{\mu}\frac{dp}{dx}\left(\frac{b^2}{2} - \frac{y^2}{2}\right)} \tag{12.56}$$

The maximum velocity $u_{\max}$ obviously occurs at $y = 0$ and is given by

$$u_{\max} = -\frac{1}{\mu}\frac{dp}{dx}\frac{b^2}{2} \tag{12.57}$$

Mean velocity
Let us denote the mean velocity as U. It is defined such that the volume flow rate $\dot{V} = 2bU$ (per m length in a direction perpendicular to the plane of the figure) is equal to that obtained with the actual velocity profile given by Eq. 12.56. Thus, we have

$$\begin{aligned}\dot{V} = 2bU = -\frac{1}{\mu}\frac{dp}{dx}\int_{-b}^{b}\left(\frac{b^2}{2} - \frac{y^2}{2}\right)dy = -\frac{1}{\mu}\frac{dp}{dx}\left(\frac{b^2 y}{2} - \frac{y^3}{6}\right)\Bigg|_{-b}^{b} \\ = -\frac{1}{\mu}\frac{dp}{dx} \times 2\left(\frac{b^3}{2} - \frac{b^3}{6}\right) = -\frac{1}{\mu}\frac{dp}{dx}\frac{2b^3}{3}\end{aligned} \tag{12.58}$$

The mean velocity is thus given by

$$U = -\frac{1}{\mu}\frac{dp}{dx}\frac{b^2}{3} \tag{12.59}$$

Using Eqs. 12.57 and 12.59, we have the important relation

$$\boxed{U = \frac{2}{3}u_{\max}} \tag{12.60}$$

Friction factor
The Darcy friction factor f is defined through the relation

$$f = -\frac{\frac{\Delta p}{\rho U^2}}{\frac{L}{2D_H}} \tag{12.61}$$

where $\frac{\Delta p}{L} = \frac{dp}{dx}$, D_H is the hydraulic diameter given by $\frac{A_c}{P_w}$ where A_c is the flow area and P_w is the wetted perimeter, i.e., the wall in contact with the fluid. In the case of the channel, the area is given by $2b$ and the wetted perimeter is 2. Hence, the hydraulic diameter is $D_H = \frac{4\times 2\times b}{2} = 4b$. Hence, the friction factor may be written using Eq. 12.59 as

$$f = \frac{(3U)(2D_H)}{\rho U^2 \mu b^2} = \frac{96UD_H}{\rho U^2 \mu D_H^2} = \frac{96}{Re_{D_H}} \text{ or } \boxed{f \times Re_{D_H} = 96} \tag{12.62}$$

12.3.6 *Concept of Fluid Resistance*

Fluid resistance R_f is introduced by treating the mass flow rate $\dot{m}$ through the tube/channel as a current and the pressure drop Δp across the length L of the tube/channel as the potential difference.

Resistance in tube flow
Based on Eq. 12.45, the pressure drop is given by $-\Delta p = \frac{8\mu UL}{R^2}$. The mass flow rate is obtained by using the definition of mean velocity as $\dot{m} = \rho\pi R^2 U$. Fluid resistance R_f is then defined as

$$R_f = \frac{-\Delta p}{\dot{m}} = \frac{\left(\frac{8\mu UL}{R^2}\right)}{\rho\pi R^2 U} = \frac{8\mu L}{\pi\rho R^4} \tag{12.63}$$

This expression may also be written based on the tube diameter D as the characteristic length as

$$\boxed{R_f = \frac{128\mu L}{\pi\rho D^4}} \tag{12.64}$$

We see that the fluid resistance is directly proportional to tube length and inversely proportional to the fourth power of diameter of the tube.

Resistance in channel flow
Using Eq. 12.59, the pressure drop is given by $-\Delta p = \frac{3\mu UL}{b^2}$. The mass flow rate is obtained by using the definition of mean velocity as $\dot{m} = 2\rho bU$. Thus, we have by definition

$$R_f = \frac{-\Delta p}{\dot{m}} = \frac{\frac{3\mu U L}{b^2}}{2\rho b U} = \frac{3\mu L}{\rho b^4} \quad (12.65)$$

The above expression may be recast, using the characteristic length $D_H = 4b$, as

$$\boxed{R_f = \frac{768\mu L}{\rho D_H^4}} \quad (12.66)$$

We see that the fluid resistance is directly proportional to channel length and inversely proportional to the fourth power of the hydraulic diameter.

Example 12.4 demonstrate the use of resistance concept in fluid flow distribution in two tubes in parallel.

Example 12.4

A highly viscous oil flows under a head of 0.5 m of water through two tubes that are arranged in parallel. The first tube has a diameter of 3 mm and the second has a diameter of 4 mm. Both tubes are 1 m long. Determine the volume flow rates in the two tubes. The viscosity of oil may be taken as 3 times the viscosity of water and the relative density of oil is 0.8. Take water properties at 30 °C.

Solution:

Step 1 Water properties at 30 °C are taken from table of properties of water. They are

$$\rho_w = 995.7\ \text{kg/m}^3,\ \mu_w = 7.97 \times 10^{-4}\ \text{kg/m s}$$

The flowing fluid is oil with the following properties:

Viscosity: $\mu_{oil} = 3 \times \mu_w = 3 \times 7.97 \times 10^{-4} = 2.39 \times 10^{-3}$ kg/m s
Density: $\rho_{oil} = 0.8 \times \rho_w = 0.8 \times 995.7 = 797$ kg/m^3

Step 2 The given data is written down as below

Tube 1: Diameter: $D_1 = 0.003$ m Length: $L_1 = 1$ m
Tube 2: Diameter: $D_2 = 0.004$ m Length: $L_2 = 1$ m

Step 3 Available pressure drop is given to be equal to a head of water of $h = 0.5$ m. The corresponding pressure drop is given by

$$\Delta p = \rho_w g h = 995.7 \times 9.81 \times 0.5 = 4884\ \text{Pa}$$

where we have used the standard value for the acceleration due to gravity of $g = 9.81$ m/s^2. We shall assume that the flow through both tubes is laminar. Of course we shall verify it later on.

Step 4 The flow resistance of the tubes may be obtained using Eq. 12.64.

$$\underline{\text{Tube 1:}}\ R_{f1} = \frac{128 \times 2.39 \times 10^{-3} \times 1}{\pi \times 797 \times 0.003^4} = 1.510 \times 10^6 \text{ Pa s/kg}$$

$$\underline{\text{Tube 2:}}\ R_{f2} = \frac{128 \times 2.39 \times 10^{-3} \times 1}{\pi \times 797 \times 0.004^4} = 4.777 \times 10^5 \text{ Pa s/kg}$$

Step 5 Using the definition of flow resistance, the mass flow rates in the two tubes may be calculated now.

$$\underline{\text{Tube 1:}}\ \dot{m}_1 = \frac{\Delta p}{R_{f1}} = \frac{4884}{1.510 \times 10^6} = 3.235 \times 10^{-3} \text{ kg/s}$$

$$\underline{\text{Tube 2:}}\ \dot{m}_2 = \frac{\Delta p}{R_{f2}} = \frac{4884}{4.777 \times 10^5} = 1.022 \times 10^{-2} \text{ kg/s}$$

The corresponding oil velocities in the two cases are given by

$$\underline{\text{Tube 1:}}\ U_1 = \frac{\dot{m}_1}{\rho_{\text{oil}} A_1} = \frac{4\dot{m}_1}{\rho_{\text{oil}} \pi D_1^2} = \frac{4 \times 3.235 \times 10^{-3}}{797 \times \pi \times 0.003^2} = 0.575 \text{ m/s}$$

$$\underline{\text{Tube 2:}}\ U_2 = \frac{\dot{m}_2}{\rho_{\text{oil}} A_2} = \frac{4\dot{m}_2}{\rho_{\text{oil}} \pi D_2^2} = \frac{4 \times 1.022 \times 10^{-2}}{797 \times \pi \times 0.004^2} = 1.021 \text{ m/s}$$

Step 6 We now verify whether the flow is laminar in the two cases. This is done by making sure that the larger of the two Reynolds numbers is less than 2300. The Reynolds number in the case of 4 mm tube is the larger of the two and is

$$Re_{D_2} = \frac{\rho_{\text{oil}} U_2 D_2}{\mu_{\text{oil}}} = \frac{797 \times 1.021 \times 0.004}{2.39 \times 10^{-3}} = 1361$$

The flow is indeed laminar and the use of laminar flow resistance formula is justified.

Step 7 The volume flow rates are obtained now.

$$\underline{\text{Tube 1:}}\ \dot{V}_1 = \frac{\dot{m}_1}{\rho_{\text{oil}}} = \frac{3.235 \times 10^{-3}}{797} = 4.06 \times 10^{-6} \text{ m}^3/\text{s}$$

$$\underline{\text{Tube 2:}}\ \dot{V}_2 = \frac{\dot{m}_2}{\rho_{\text{oil}}} = \frac{1.022 \times 10^{-2}}{797} = 12.83 \times 10^{-6} \text{ m}^3/\text{s}$$

12.4 Laminar Heat Transfer in Tube Flow

Heat transfer to or from a fluid flowing in a tube is of great importance since this configuration is very common in heat transfer devices such as heat exchangers. Even though laminar flow is not very common, the analysis of laminar flow provides an opportunity to learn about convection in internal flow using simple mathematics. Two boundary conditions that are easily achieved in practice are the constant heat flux and the constant wall temperature conditions. The former is obtained by electrical heating of a highly conducting tube and the latter by having condensing or evaporating fluid in contact with the outside of the tube wall.

12.4.1 Bulk Mean Temperature

Recall from the discussion in Sect. 12.3.4 where the mean velocity for flow in a tube was defined. The fluid flowing at the mean velocity transports a constant amount of fluid per unit time along the tube. In a heat transfer application, we would be interested in determining the rate at which enthalpy is transported across any cross section of the tube. This is easily done by introducing the so called bulk mean temperature (also known as the mixing cup temperature). The rate at which enthalpy $\dot{H}(z)$ is transported across any section of the tube is obtained by the following integral:

$$\dot{H}(z) = \int_0^R C_p T(r,z) d\dot{m}$$

where $d\dot{m}$ is the mass flow rate through an elemental area given by $2\pi r dr$ and $C_p T(r,z)$ is the magnitude of the enthalpy of the fluid entering the elemental area. The elemental mass flow rate itself is obtained as the product of density, area, and the velocity as

$$d\dot{m} = \rho \times 2\pi r dr \times u_z(r,z)$$

Combining these we get

$$\dot{H}(z) = \int_0^R 2\rho\pi r u_z(r,z) C_p T(r,z) dr \tag{12.67}$$

We shall equate the rate of enthalpy crossing the tube section by introducing the mean velocity introduced earlier and the bulk mean temperature $T_B(z)$ such that $\dot{H}(z) = \dot{m} C_p T_B(z) = (\pi R^2 \rho U) C_p T_B(z)$. Note that this is the product of the mass flow rate across the section and the mean value of enthalpy of the entering fluid. Thus, we get for a constant property fluid

$$\left(\pi R^2 \rho U\right)\left(C_p T_B(z)\right) = \int_0^R 2\rho\pi r u_z(r,z) C_p T(r,z) dr$$

$$\text{or} \quad \boxed{T_B(z) = \frac{2}{\pi R^2 U} \int_0^R u_z(r,z) T(r,z) dr} \tag{12.68}$$

Note that the bulk mean temperature as defined above is valid at any z along the flow and may, in fact, vary with z. However, U is independent of z because of mass conservation, even though u_z may be a function of r and z. In what follows we shall be interested in applying the above to the fully developed region where u_z will be a function of r alone.

12.4.2 Variation of the Bulk Mean Temperature

The bulk mean temperature varies with z, and this variation depends on the condition applicable at the tube wall. In most practical applications the tube wall is thin, and hence it is customary to neglect axial heat conduction in the tube wall, i.e., heat conduction along the z-direction. Hence, heat transfer across the tube wall is assumed to be radial. This heat transfer may be subject to a very small temperature variation across the tube wall if it is thin and made of a material with a high thermal conductivity. Hence, it is possible to make a simple analysis assuming that heat transfer to the fluid or away from the fluid takes place radially and is specified at the fluid–solid interface.

The analysis may be made using the control volume shown in Fig. 12.8. The control volume is taken in the form of a short cylinder of length Δz and of radius R, equal to the inner radius of the tube.

Heat balance may be made for the control volume as follows:

$$\begin{bmatrix}\text{Heat convected}\\ \text{across left boundary}\end{bmatrix}+\begin{bmatrix}\text{Heat transfer entering}\\ \text{at tube wall}\end{bmatrix}=\begin{bmatrix}\text{Heat convected}\\ \text{across right boundary}\end{bmatrix}$$

The heat transfer by convection entering through the left boundary is obtained by the use of the bulk mean temperature as $\rho\pi R^2UC_pT_B(z)$. The heat transfer by convection leaving through the right boundary may be written as $\rho\pi R^2UC_pT_B(z+\Delta z)=\rho\pi R^2UC_p\left[T_B(z)+\frac{dT_B}{dz}\Delta z\right]$. We have made use of the Taylor expansion and retained only the first-order term. The heat transfer entering at the tube wall is given by $2\pi Rq_w\Delta z$. Introducing these in the heat balance equation and simplifying, we get

$$\boxed{\frac{dT_B}{dz}=\frac{2q_w}{\rho UC_pR}} \tag{12.69}$$

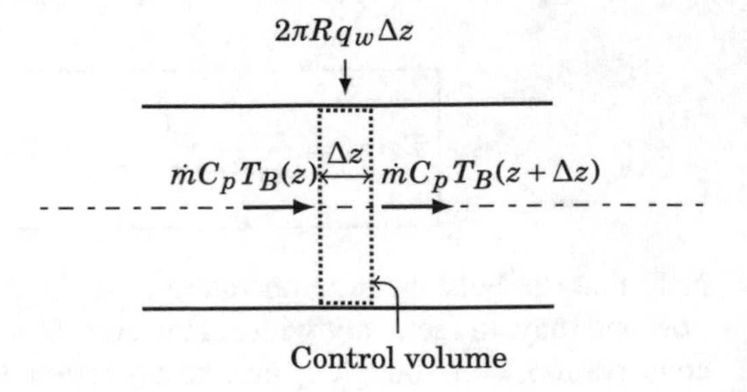

Fig. 12.8 Control volume for heat transfer analysis

The above equation is general in that it applies to *any variation* of q_w with z. In the special case in which q_w is constant, the bulk mean temperature increases (or decreases if q_w is negative, i.e., heat is lost from the fluid element to the tube wall) linearly with z.

12.4.3 Tube Flow with Uniform Wall Heat Flux

Consider tube flow with heat transfer as indicated in Fig. 12.9. The fluid enters with a uniform temperature T_0 as indicated. The wall is subjected to a constant heat flux q_w. There is a thermal entry length L'_{dev} over which the temperature distribution develops just as the flow development would take place over an entry length L_{dev} discussed earlier. For laminar flow, the entry length is given by $L'_{\text{dev}}/D = 0.05 Re_D Pr = 0.05 Pe$ where the Reynolds number Prandtl number product has been represented as Pe, the Peclet number.[5] For $z > L'_{\text{dev}}$ the temperature is fully developed, and for $q_w =$ constant, both T_w and T_B increase linearly at the same rate, keeping a constant difference between the two. Here, T_B is the bulk mean temperature of the fluid, as defined earlier through Eq. 12.68.

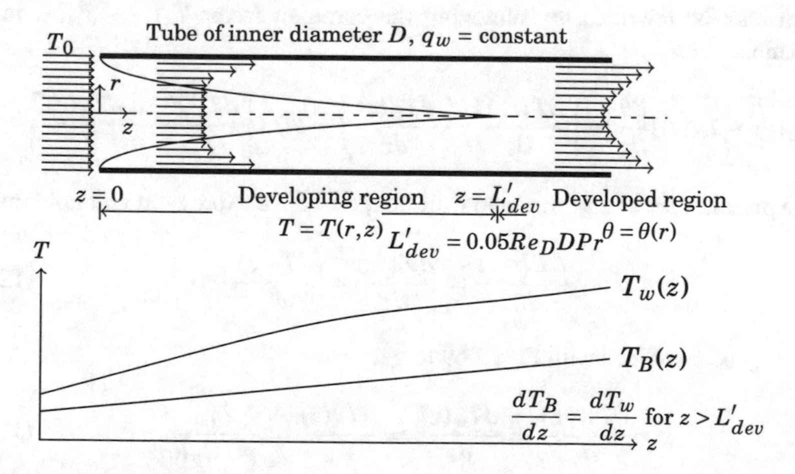

Fig. 12.9 Tube flow with constant heat flux at its surface

[5]Named after Jean Claude Eugène Péclet 1793–1857, a French physicist

12.4.4 *Fully Developed Temperature with Uniform Wall Heat Flux*

The idea of fully developed temperature profile is analogous to the fully developed velocity profile considered earlier. We look for a suitably defined non-dimensional temperature profile that is a function of r only, being thus independent of z. This is in spite of the fact that the temperature of the fluid varies with both r and z. Consider the non-dimensional temperature ratio given by

$$\theta(r) = \frac{T(r,z) - T_w(z)}{T_B(z) - T_w(z)} \tag{12.70}$$

where $T_w(z)$ stands for the wall temperature and $T_B(z)$ is the bulk mean temperature of the fluid. As indicated in Eq. 12.70, θ is a function of only r and hence $\frac{\partial \theta}{\partial z} \equiv 0$. This requires that

$$\frac{\partial \theta}{\partial z} = \frac{\frac{\partial T(r,z)}{\partial z} - \frac{dT_w(z)}{dz}}{T_B(z) - T_w(z)} - \frac{T(r,z) - T_w(z)}{[T_B(z) - T_w(z)]^2}\left(\frac{dT_B(z)}{dz} - \frac{dT_w(z)}{dz}\right) = 0 \tag{12.71}$$

which may be rewritten, by removing the common factor $T_B(z) - T_w(z)$ in the denominator, as

$$[T_B(z) - T_w(z)]\frac{\partial \theta}{\partial z} = \left[\frac{\partial T(r,z)}{\partial z} - \frac{dT_w(z)}{dz}\right] - \theta(r)\left[\frac{dT_B(z)}{dz} - \frac{dT_w(z)}{dz}\right] = 0 \tag{12.72}$$

In the present case of uniform tube wall flux, the above expression will hold only if

$$\frac{\partial T(r,z)}{\partial z} = \frac{dT_w(z)}{dz} = \frac{dT_B(z)}{dz} \tag{12.73}$$

This may be combined with Eq. 12.69 to get

$$\frac{\partial T(r,z)}{\partial z} = \frac{dT_w(z)}{dz} = \frac{dT_B(z)}{dz} = \frac{2q_w}{\rho U C_p R} \tag{12.74}$$

where the wall heat flux q_w is a constant independent of z. Hence, the axial temperature gradient $\frac{\partial T(r,z)}{\partial z}$ is a constant, and hence the second derivative of $T(r,z)$ with respect to z is zero. This means that the axial heat conduction does not change with z and hence the axial diffusion term drops off.

Governing equation

The governing equation may be developed either from the energy equation in cylindrical coordinates (see Appendix H) or from first principles as is done here. Consider energy balance over an annular element as shown in Fig. 12.10.

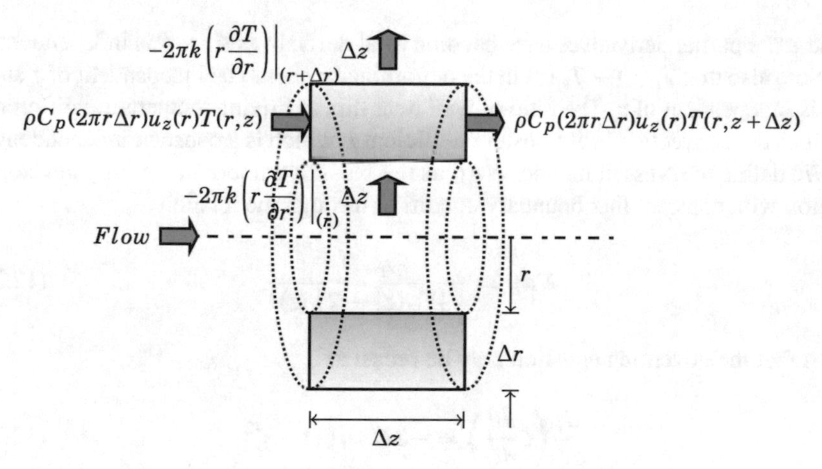

Fig. 12.10 Annular control volume for developing the governing equation

Since conduction flux along the axis does not change with z, net convection crossing the control volume in the axial direction is balanced by net conduction in the radial direction. With this in mind, the fluxes crossing the control volume are as shown in the figure. Energy balance may be spelt out in words as follows:

$$\begin{bmatrix}\text{Conduction}\\ \text{leaving at}\\ \text{outer boundary}\end{bmatrix} - \begin{bmatrix}\text{Conduction}\\ \text{entering at inner}\\ \text{boundary}\end{bmatrix} = \begin{bmatrix}\text{Convection}\\ \text{leaving across}\\ \text{right boundary}\end{bmatrix} - \begin{bmatrix}\text{Convection}\\ \text{entering across}\\ \text{left boundary}\end{bmatrix}$$

As usual we use Taylor expansion retaining first-order terms to write, after simplification, the following governing equation.

$$k\frac{\partial}{\partial r}\left(r\frac{\partial T}{\partial r}\right) = \rho u_z C_p r\frac{\partial T}{\partial z} \tag{12.75}$$

We shall assume now that the velocity profile is given by the fully developed profile (see Eq. 12.40). We also use the variation of temperature along z given by Eqs. 12.69 and 12.74 to write the governing equation as

$$\frac{\partial}{\partial r}\left(r\frac{\partial T}{\partial r}\right) = \frac{4q_w r}{kR}\left[1 - \left(\frac{r}{R}\right)^2\right] \tag{12.76}$$

We may recast this equation in terms of the non-dimensional temperature $\theta(r)$ introduced through Eq. 12.70 as

$$\boxed{\frac{d}{dr}\left(r\frac{d\theta}{dr}\right) = \frac{4q_w r}{kR\{T_B(z) - T_w(z)\}}\left[1 - \left\{\frac{r}{R}\right\}^2\right]} \tag{12.77}$$

where the partial derivatives have become total derivatives since θ is independent of z. Note also that $T_B(z) - T_w(z)$ in the denominator should be independent of z since q_w is independent of z. The ratio of wall heat flux to driving temperature difference defines the convection heat transfer coefficient h which is a constant independent of z. We define the Nusselt number Nu_H as the Nusselt number in the fully developed region with constant flux boundary condition through the relation

$$Nu_H = \frac{q_w D}{k\{T_w(z) - T_B(z)\}} \tag{12.78}$$

such that the governing equation may be recast as

$$\frac{d}{d\zeta}\left(\zeta\frac{d\theta}{d\zeta}\right) = -2Nu_H\zeta(1-\zeta^2) \tag{12.79}$$

The accompanying boundary conditions are specified as

$$\theta \text{ is finite at } \zeta = 0; \text{ and } \frac{q_w}{T_B(z) - T_w(z)} - k\frac{d\theta}{dr} = 0 \text{ at } r = R \tag{12.80}$$

in dimensional form. The boundary condition at tube wall is a statement of the fact that the heat flux is continuous across the boundary. This may be rewritten in non-dimensional form, using the Nusselt number defined above as

$$\frac{d\theta}{d\zeta} + \frac{Nu_H}{2} = 0 \text{ at } \zeta = 1 \tag{12.81}$$

Solution

Equation 12.76 is integrated once with respect to ζ to get

$$\frac{d\theta}{d\zeta} = -2Nu_H\left(\frac{\zeta}{2} - \frac{\zeta^3}{4}\right) + \frac{C_1}{\zeta}$$

where C_1 is a constant of integration. The boundary condition at $\zeta = 0$ requires that we choose C_1 as 0. The resulting equation is integrated once more with respect to ζ to get

$$\theta = -2Nu_H\left(\frac{\zeta^2}{4} - \frac{\zeta^4}{16}\right) + C_2 \tag{12.82}$$

where C_2 is a second constant of integration. It is seen that the constant of integration, in fact, represents the non-dimensional temperature θ_0 at the axis of the tube, that is not known as of now. Thus, we write Eq. 12.82 as

$$\theta - \theta_0 = \underbrace{\phi(\zeta)}_{\text{Define}} = -2Nu_H\left(\frac{\zeta^2}{4} - \frac{\zeta^4}{16}\right) \tag{12.83}$$

The boundary condition at the tube wall is not available to us since it has been implicitly used in deriving Eq. 12.69 by overall energy balance. Consider the following integral:

$$I_n = \int_0^R u_z(r)\phi(r)rdr = R^2 \int_0^1 2U(1-\zeta^2)\phi(\zeta)\zeta d\zeta$$

Using the non-dimensional temperature profile given by Eq. 12.83, the above integral is written as

$$\begin{aligned} I_n &= -4UR^2 Nu_H \int_0^1 (1-\zeta^2)\left(\frac{\zeta^2}{4} - \frac{\zeta^4}{16}\right)\zeta d\zeta = -UR^2 Nu_H \int_0^1 \left[\zeta^3 - \frac{5\zeta^5}{4} + \frac{\zeta^7}{4}\right] d\zeta \\ &= -UR^2 Nu_H \left[\frac{\zeta^4}{4} - \frac{5\zeta^6}{24} + \frac{\zeta^8}{32}\right]\Bigg|_{\zeta=0}^1 = -UR^2 Nu_H \left(\frac{7}{96}\right) \end{aligned} \tag{12.84}$$

Consider also the integral $I_d = \int_0^R u_z(r)rdr$. We may easily obtain this integral as

$$I_d = 2UR^2 \int_0^1 (\zeta - \zeta^3)d\zeta = \frac{UR^2}{2} \tag{12.85}$$

Finally, by division, we get

$$\frac{I_n}{I_d} = -\frac{7}{48} Nu_H \tag{12.86}$$

We recognize this to represent $\theta_B - \theta_0$. We may obtain from this the difference $\theta_B - \theta_w$ as

$$\theta_B - \theta_w = 1 = [\theta_B - \theta_0] - [\theta_w - \theta_0]$$

where the relation $\theta_B - \theta_w = 1$ follows from the definition of the non-dimensional temperatures. The second term on the right-hand side is obtained by evaluating Eq. 12.83 at $\zeta = 1$ as

$$\theta - \theta_0 = -2Nu_H \left(\frac{1}{4} - \frac{1}{16}\right) = -\frac{3}{8} Nu_H \tag{12.87}$$

With these, we get

$$\theta_B - \theta_w = 1 = -\frac{7}{48} Nu_H + \frac{3}{8} Nu_H$$

or

$$\boxed{Nu_H = \frac{48}{11} = 4.364} \tag{12.88}$$

Hence, the Nusselt number is a constant equal to 4.364 (the heat transfer coefficient is also a constant) in fully developed tube flow with constant wall flux. Obviously, the Nusselt number is not what is important when the wall heat flux is specified or known. The above equation is useful in determining the difference between the wall and bulk fluid temperature as

$$T_w(z) - T_B(z) = \frac{11}{48}\frac{q_w D}{k} \tag{12.89}$$

The non-dimensional temperature variation across the tube may now be represented using Eqs. 12.83 and 12.87 as

$$\theta_w - \theta = (\theta_w - \theta_0) - (\theta - \theta_0) = -2Nu_H\left(\frac{\zeta^2}{4} - \frac{\zeta^4}{16}\right) + \frac{3}{8}Nu_H$$

Using the known value of Nu_H, the above becomes

$$\boxed{\theta_w - \theta = \frac{24}{11}\left\{\frac{3}{4} - \zeta^2 + \frac{\zeta^4}{4}\right\}} \tag{12.90}$$

12.4.5 Tube Flow with Constant Wall Temperature

As mentioned earlier the constant wall temperature case is typical of what happens when the outer wall of the tube is in contact with a fluid undergoing phase change, such as in a condenser of a steam power plant. The tube side fluid (i.e., the fluid that flows inside the tube) is usually water. The flow velocity and the tube diameter are such that the flow in the tube is invariably turbulent. However, it is instructive to look at the laminar flow case since fundamental ideas involved in heat transfer are the same in the laminar case also. Schematic of tube flow with constant wall temperature is as shown in Fig. 12.11. The temperature field undergoes a development over an entry length L'_{dev}. The temperature in the core remains constant at T_0 till $z = L'_{\text{dev}}$. Thereafter the thermal boundary layer fills the entire tube. The bulk temperature varies as indicated graphically at the bottom of Fig. 12.11. Assuming that the fluid in the tube is getting heated, T_B will continually increase but the rate of heat transfer continuously reduces since the driving temperature difference continuously decreases with z. We shall see later that the temperature difference reduces exponentially with z when the heat transfer coefficient is constant.

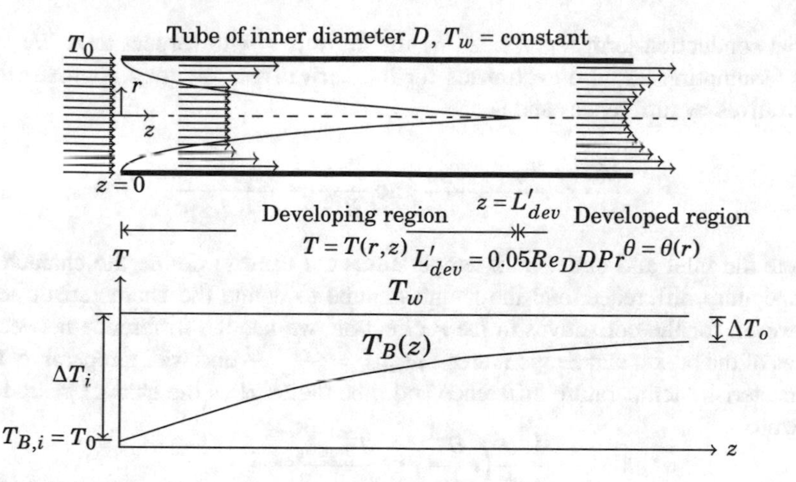

Fig. 12.11 Tube flow with constant wall temperature

12.4.6 Fully Developed Tube Flow with Constant Wall Temperature

Let us see what happens in the fully developed temperature region. We go back to Eq. 12.72 and notice that the fully developed condition holds only if

$$\theta(r)\frac{dT_B}{dz} = \frac{\partial T(r,z)}{\partial z} \tag{12.91}$$

since $\frac{dT_w}{dz} = 0$.

We shall look at this condition after deriving the appropriate equation that governs the temperature field.

Governing equation

We derive the governing equation starting with the energy Eq. H.38 in cylindrical coordinates given in Appendix H. Since the flow is steady $\frac{\partial}{\partial t} \equiv 0$. The flow velocity component along the axis of the tube alone is non-zero, and hence, the convective term consists of only the term $u_z \frac{\partial T(r,z)}{\partial z}$. The diffusion terms (terms appearing in the energy equation that account for conduction in the fluid) will involve both derivatives with respect to r and z and the governing equation becomes

$$u_z \frac{\partial T}{\partial z} = \alpha\left[\frac{1}{r}\frac{\partial}{\partial r}\left(r\frac{\partial T}{\partial r}\right) + \frac{\partial^2 T}{\partial z^2}\right] \tag{12.92}$$

On the right-hand side of Eq. 12.92, we have the axial diffusion represented by the second derivative of T with respect to z. In the case of tube flow with constant wall heat flux this term *dropped off* since $\frac{\partial T}{\partial z}$ was a constant. In the present case, we shall assume that this axial conduction term is negligibly small when compared to the radial conduction term represented by the derivative with respect to r. We justify this assumption based on estimates for the derivatives. We may approximate the derivatives by differences and hence

$$\frac{\partial T}{\partial z} \sim \frac{T_{B,o} - T_{B,i}}{L} \text{ and } \frac{\partial^2 T}{\partial z^2} \sim \frac{T_{B,o} - T_{B,i}}{L^2}$$

where the inlet and outlet bulk temperatures are used to define the characteristic temperature difference, and the length of tube to define the characteristic length. However, for the derivatives in the r direction, we use the difference between the mean of the bulk mean temperatures $T_{B,\,\text{mean}} = \frac{T_{B,o}+T_{B,i}}{2}$ and wall temperature as the characteristic temperature difference and tube radius R as the characteristic length to write

$$\frac{1}{r}\frac{\partial}{\partial r}\left(r\frac{\partial T}{\partial r}\right) \sim \frac{T_{B,\,\text{mean}} - T_w}{R^2}$$

In applications, invariably the temperature difference of the fluid between the entry and exit is smaller than that between the fluid and the wall. For example, the bulk temperature difference may be 15 °C while the temperature difference between the fluid and the wall may be 50 °C. Also the length of the tube L is normally much larger than the radius R of the tube. For example, with a tube Reynolds number of 1000 fully developed conditions are obtained with $\frac{L}{D} > \frac{L'_{\text{dev}}}{D} = 0.05 \times 1000 \times 5 = 250$ or $\frac{L}{R} > 500$ where the Prandtl number has been assumed to have a value of 5, typical of water. With $R = 0.005$ m, the corresponding L is about 2.5 m. The axial and radial diffusion terms are typically given by

$$\text{Axial diffusion term: } \frac{T_{B,o} - T_{B,i}}{L^2} \approx \frac{15}{2.5^2} = 2.4\ °\text{C/m}^2$$

$$\text{Radial diffusion term: } \frac{50}{0.005^2} = 2 \times 10^6\ °\text{C/m}^2$$

It is thus clear that the axial diffusion term is much smaller than the radial diffusion term, thus justifying the assumption suggested above. Hence, we approximate the governing equation, neglecting axial conduction, as

$$u_z\frac{\partial T}{\partial z} \approx \alpha\frac{1}{r}\frac{\partial}{\partial r}\left(r\frac{\partial T}{\partial r}\right) \tag{12.93}$$

Further, we shall assume that u_z is given by the fully developed velocity profile specified by Eq. 12.40. Additionally, making use of the fully developed temperature condition Eq. 12.91 and θ defined by Eq. 12.70, we simplify the governing equation to

$$\alpha\frac{1}{r}\frac{\partial}{\partial r}\left(r\frac{\partial T}{\partial r}\right) = 2U\left[1 - \left(\frac{r}{R}\right)^2\right]\theta\frac{dT_B}{dz} \tag{12.94}$$

By definition, the fully developed temperature profile is a function of r alone and hence

$$\frac{\partial T}{\partial r} = \underbrace{(T_B - T_w)}_{\substack{\text{Depends only}\\ \text{on } z}} \frac{d\theta}{dr}$$

where Eq. 12.70 has been made use of. Also the radial diffusion term takes the form

$$\alpha \frac{1}{r}\frac{\partial}{\partial r}\left(r\frac{\partial T}{\partial r}\right) = (T_B - T_w)\alpha \frac{1}{r}\frac{d}{dr}\left(r\frac{d\theta}{dr}\right)$$

Thus, the governing equation takes the form of an *ordinary* differential equation given by

$$\frac{1}{r}\frac{d}{dr}\left(r\frac{d\theta}{dr}\right) = 2\frac{U}{\alpha}\left[1-\left(\frac{r}{R}\right)^2\right]\frac{\frac{dT_B}{dz}}{(T_B - T_w)}\theta \quad (12.95)$$

We immediately see that $\frac{\frac{dT_B}{dz}}{(T_B-T_w)}$ should be independent of z. This is, in fact, the real import of the fully developed temperature profile. Using the relationship between wall heat flux and the driving temperature difference given by Eq. 12.69, we have

$$\frac{U}{\alpha}\frac{\frac{dT_B}{dz}}{(T_B - T_w)} = -\frac{2q_w(z)}{kR(T_w - T_B)} = -\frac{Nu_T}{R^2} \quad (12.96)$$

where Nu_T is the constant Nusselt number in the fully developed region in the constant wall temperature case. Using the non-dimensional variable $\zeta = \frac{r}{R}$, the governing equation takes the form

$$\boxed{\frac{1}{\zeta}\frac{d}{d\zeta}\left(\zeta\frac{d\theta}{d\zeta}\right) = \frac{d^2\theta}{d\zeta^2} + \frac{1}{\zeta}\frac{d\theta}{d\zeta} = -2Nu_T(1-\zeta^2)\theta} \quad (12.97)$$

This equation is to be solved with the boundary conditions given by

$$\theta \text{ is finite at } \zeta = 0, \text{ and } \theta = 0 \text{ at } \zeta = 1 \quad (12.98)$$

Solution

Since the governing equation is an ordinary differential equation with variable coefficients, the solution may be obtained by using an infinite series to represent the temperature field.[6] The solution that is finite at the origin will have only positive powers of ζ in the series. Since the parameter Nu_H is not known, the solution will involve this as a parameter. The boundary condition at the tube wall will determine Nu_H as we shall soon see. Since the solution is axisymmetric, only even powers of ζ will occur in the series solution. Hence, let the solution be represented by the series given by

$$\theta = \sum_{n=0}^{\infty} C_{2n}\zeta^{2n} \tag{12.99}$$

On substitution in Eq. 12.97, using term by term differentiation, collecting terms containing same powers of ζ, we get the following:

$$\begin{aligned}
\zeta^{-2}: &\quad C_0 \times 0 \text{ Hence } C_0 \neq 0\\
\zeta^{0}: &\quad 4C_2 + \lambda^2 C_0\\
\text{Hence} &\quad C_2 = -\frac{\lambda^2}{4}C_0\\
&\dots\\
\zeta^{2n}: &\quad (2n)^2 C_{2n} - \lambda^2\left(C_{2n-4} - C_{2n-2}\right)\\
\text{Hence} &\quad C_{2n} = -\frac{\lambda^2}{(2n)^2}\left(C_{2n-4} - C_{2n-2}\right)
\end{aligned} \tag{12.100}$$

where, for convenience, λ^2 stands for $2Nu_T$. Hence, the solution may be written as

$$\theta = C_0\left[1 - \frac{\lambda^2}{4}\zeta^2 + \frac{\lambda^2}{16}\left(1 + \frac{\lambda^2}{4}\right)\zeta^4 - \frac{\lambda^2}{36}\left\{\frac{\lambda^2}{4} + \frac{\lambda^2}{16}\left(1 + \frac{\lambda^2}{4}\right)\right\}\zeta^6 - + \cdots\right] \tag{12.101}$$

Note that both C_0 and λ are unknown as of now. The non-dimensional temperature has to vanish at the tube wall, and hence, the series given by Eq. 12.101 should vanish at $\zeta = 1$. Luckily for us the series converges rapidly, and it is necessary to take only 10 terms. Since C_0 is non-zero, the sum of terms within the braces have to vanish. By trial, it may be verified that the sum vanishes for $\lambda^2 = 7.313588$, and hence the value of the Nusselt number is given by

$$\boxed{Nu_T = \frac{\lambda^2}{2} = \frac{7.313588}{2} = 3.656794 \approx 3.657} \tag{12.102}$$

The value of the unknown constant C_0 may be determined by using the heat flux continuity condition at $\zeta = 1$. This requires that (Eq. 12.81 with Nu_H replaced by Nu_T)

[6] M.S. Bhatti,"Fully developed temperature distribution in a circular tube with uniform wall temperature", Unpublished paper, Owens-Corning Fiberglass Corporation, Ohio, 1985 as cited by S. Kakac and R.K. Shah, *Handbook of Single Phase Convective Heat Transfer*, John Wiley, NY, 1987.

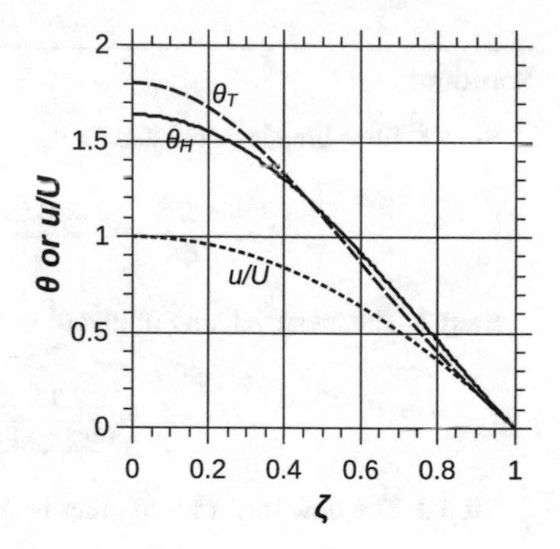

Fig. 12.12 Fully developed velocity and temperature profiles

$$\left.\frac{d\theta}{d\zeta}\right|_{\zeta=1} = -\frac{Nu_T}{2} = -\frac{3.656794}{2} = -1.828397 \tag{12.103}$$

The derivative required may be calculated by term by term differentiation of series given by Eq. 12.101 and inserting $\zeta = 1$ to get

$$\frac{d\theta}{d\zeta} = C_0 \times (-1.01428) = -1.828397 \text{ or } C_0 = 1.802652 \tag{12.104}$$

The fully developed temperature (constant wall heat flux and constant wall temperature cases) and velocity profiles are shown in Fig. 12.12. While the velocity profile is quadratic in $\zeta = \frac{r}{R}$ the temperature profile is a quartic in ζ, in the case of constant heat flux case (identified as θ_H) while it is given by an infinite series in the case of the constant wall temperature case (identified as θ_T). The maximum non-dimensional temperature difference occurs between the wall and the fluid at the tube axis, in both cases. The maximum velocity occurs along the tube axis.

Example 12.5

Ethylene glycol is flowing in a $D = 6$ mm diameter thin-walled copper tube heated electrically such that the wall heat flux is $q_w = 1000\,\text{W/m}^2$. At a certain section, glycol has a bulk mean temperature of 70 °C. The volume flow rate of glycol has been measured to be $\dot{V} = 15$ ml/s. Determine the wall temperature at this location. Also determine rate of change of the bulk temperature of glycol with axial distance. Glycol properties may be taken as constant and are specified as below Density $\rho = 1109\,\text{kg/m}^3$, Dynamic viscosity $\mu = 0.0144$ kg/m s, Thermal conductivity $k = 0.2814$ W/m °C, and Prandtl number $Pr = 124.4$.

Solution:

Step 1 Flow area is calculated as

$$A = \frac{\pi D^2}{4} = \frac{\pi \times 0.006^2}{4} = 2.82743 \times 10^{-5}\ \text{m}^2$$

Step 2 The mean velocity of glycol in the tube is then obtained as

$$U = \frac{\dot{V}}{A} = \frac{15 \times 10^{-6}}{2.82743 \times 10^{-5}} = 0.531\ \text{m/s}$$

Step 3 The flow Reynolds number is determined as

$$Re_D = \frac{\rho U D}{\mu} = \frac{1109 \times 0.531 \times 0.006}{0.0144} = 245$$

Since the Reynolds number is less than 2300, the flow is laminar. The results of preceding analysis of fully developed tube flow with constant wall heat flux are used to get the desired results.

Step 4 The Nusselt number has the fully developed value of $Nu_H = \frac{48}{11} = 4.364$. Using the definition of Nusselt number, the corresponding heat transfer coefficient may be obtained as

$$h = \frac{Nu_H k}{D} = \frac{4.364 \times 0.2814}{0.006} = 204.65\ \text{W/m}^2\,°\text{C}$$

Step 5 The driving temperature difference at any z in the fully developed region is

$$T_w - T_B = \frac{q_w}{h} = \frac{1000}{204.65} = 4.89\,°\text{C}$$

Step 6 It is given that the bulk temperature at a certain location along the tube is $T_B = 70$ °C. Hence, the corresponding wall temperature is

$$T_w = 70 + 4.89 = 74.89\,°\text{C}$$

Step 7 The specific heat of glycol may be obtained by making use of the thermo-physical properties specified in the problem as

$$C_p = \frac{Pr \cdot k}{\mu} = \frac{124.4 \times 0.2814}{0.0144} = 2431\ \text{J/kg}\,°\text{C}$$

Step 8 To determine the axial temperature gradient, we make use of Eq. 12.74 to get

$$\frac{dT_B}{dz} = \frac{2q_w}{\rho U C_p R} = \frac{2 \times 1000}{1109 \times 0.531 \times 2431 \times 0.003} = 0.47\,°\text{C/m}$$

Example 12.6

Air is heated by passing it through a copper tube of 2.5 *mm* ID that is steam jacketed with steam at 100 °C. The properties of air may be taken at a mean temperature of 40 °C. The steam side heat transfer coefficient is extremely large, and hence, the wall of the tube may be assumed to be essentially at the steam temperature. At a certain location along the tube, both flow and temperature are fully developed. Determine the axial gradient of the bulk mean temperature at this location if the bulk mean temperature is 60 °C when the mass flow of rate of air is 0.05 g/s.

Solution:
Air properties at 40 °C are

Density: $\rho = 1.1169$ kg/m^3
Specific heat: $C_p = 1005$ J/kg°C
Dynamic viscosity: $\mu = 1.91 \times 10^{-5}$ kg/m s
Thermal conductivity: $k = 0.0274$ W/m°C
Prandtl number: $Pr = 0.699$

Other data specified in the problem are

Tube diameter: $D = 2.5$ mm $= 0.0025$ m
Wall temperature: $T_w = 100$°C
Bulk mean temperature: $T_B = 60$°C
Mass flow rate of air: $\dot{m} = 0.05$ g/s $= 5 \times 10^{-5}$ kg/s

Air velocity in the tube may be calculated as

$$U = \frac{\dot{m}}{\rho A} = \frac{4\dot{m}}{\rho \pi D^2} = \frac{4 \times 5 \times 10^{-5}}{1.1169 \times \pi \times 0.0025^2} = 9.12 \text{ m/s}$$

Tube Reynolds number is then given by

$$Re_D = \frac{\rho U D}{\mu} = \frac{1.1169 \times 9.12 \times 0.0025}{1.91 \times 10^{-5}} = 1340$$

Since the Reynolds number is less than 2300, the flow is laminar. Hence, we may use the results of analysis presented previously to obtain the axial temperature gradient. In particular, we make use of Eq. 12.96 to get

$$\frac{dT_B}{dz} = -\frac{Nu_T \alpha (T_B - T_z)}{U R^2}$$

The thermal diffusivity α appearing in the above is obtained as

$$\alpha = \frac{k}{\rho C_p} = \frac{0.0274}{1.1169 \times 1005} = 2.441 \times 10^{-5}\ \mathrm{m^2/s}$$

Under the fully developed condition the Nusselt number Nu_T is equal to 3.657. Hence, the axial gradient of the bulk mean temperature may be obtained as

$$\frac{dT_B}{dz} = -\frac{3.657 \times 2.441 \times 10^{-5}(60 - 100)}{9.12 \times 0.00125^2} = 250.58\ ^\circ\mathrm{C/m}$$

12.5 Laminar Fully Developed Flow and Heat Transfer in Non-circular Tubes and Ducts

12.5.1 Introduction

Tubes and ducts of non-circular cross section are used in many heat transfer applications. The concept of flow and temperature development applies equally to these cases. The Reynolds and Nusselt numbers are based on suitably defined characteristic lengths. The characteristic length is also known as the hydraulic diameter in the case of the flow problem and the energy diameter in the case of the heat transfer problem. These two may or may not be the same, for a given duct or tube of non-circular cross section.

We have earlier seen how the friction factor for a parallel plate channel is expressed using the hydraulic diameter as the characteristic length scale. Figure 12.13 shows how the hydraulic diameter D_H is defined, for the case of a duct or tube of *any* cross section. For the flow problem, the hydraulic diameter uses the so-called wetted perimeter P_w—the perimeter over which there is contact between the flowing

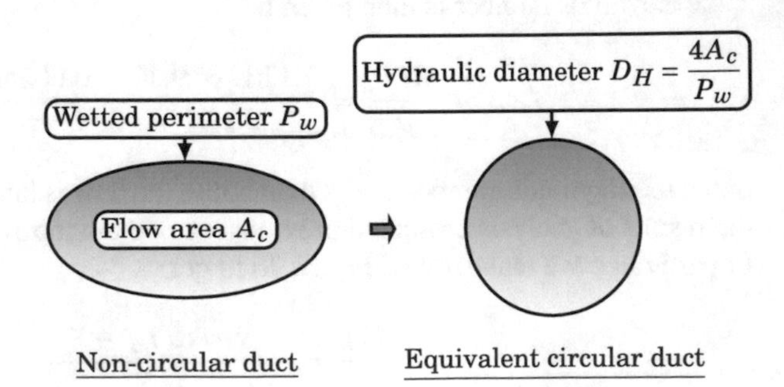

Fig. 12.13 Non-circular duct nomenclature—the hydraulic diameter

fluid and the solid wall—where viscous shear is manifest. In the case of a tube of circular diameter, the wetted perimeter is obviously the circumference of the circle representing the cross section of the tube. The flow area is the cross-sectional area A_c of the tube. In case of an annulus—the flow takes place in the region between an inner and outer tube—the wetted perimeter is the sum of the circumferences of the outer surface of the inner tube and the inner surface of the outer tube. The flow area is the area of the annulus.

The hydraulic diameter D_H is defined by the following relation:

$$D_H = \frac{4A_c}{P_w} \tag{12.105}$$

In the case of a circular cross-sectional tube, the hydraulic and actual diameter are the same. In the case of an annulus with inner diameter D_i and outer diameter D_o, we have

$$P_w = \pi\,(D_i + D_o); \quad A_c = \frac{\pi\left(D_o^2 - D_i^2\right)}{4}; \quad D_H = \frac{4\dfrac{\pi\left(D_o^2 - D_i^2\right)}{4}}{\pi\,(D_i + D_o)} = (D_o - D_i)$$

12.5.2 Parallel Plate Channel with Asymmetric Heating

The fully developed flow in this geometry has been considered in Sect. 12.3.5. We shall now consider the case of fully developed temperature problem. Detailed solution is worked out for the case where the top wall is subject to uniform heat flux q_w while the bottom wall is adiabatic (refer Fig. 12.7). The energy equation may be written for the present case starting from the cartesian form of equation given in Appendix H. This is left as an exercise to the reader. The appropriate equation in non-dimensional form is

$$\frac{d^2\theta}{d\zeta^2} = -\frac{3}{16}Nu_H(1 - \zeta^2) \tag{12.106}$$

where the velocity field has been replaced using the fully developed profile given by

$$\frac{u}{U} = 1 - \zeta^2$$

Nu_H in this case is defined as $\frac{4bq_w}{k(T_w - T_B)}$ where T_w is the top (heated) wall temperature and $4b$ is the hydraulic diameter. The boundary conditions are specified as

$$\text{Top wall: } \theta|_{\zeta=1} = 0$$
$$\text{Bottom wall: } \left.\frac{d\theta}{d\zeta}\right|_{\zeta=-1} = 0 \tag{12.107}$$

Integrating the governing Eq. 12.106 and applying the boundary conditions Eq. 12.107, we can easily show that the solution is

$$\theta(\zeta) = Nu_H\left[\frac{13}{64} - \frac{\zeta}{8} - \frac{3\zeta^2}{32} + \frac{\zeta^4}{64}\right] \tag{12.108}$$

To determine the unknown Nusselt number, we use a procedure similar to that in the case of fully developed temperature problem in the case of a circular tube with constant wall heat flux considered in Sect. 12.4.4. We utilize the velocity and temperature profiles to obtain the bulk-wall temperature difference and hence show that $Nu_H = 5.38459$.

12.5.3 Parallel Plate Channel with Symmetric Heating

The case where both walls are subject to uniform heat flux is easily considered by a few modifications to the above analysis. The governing equation is written by modifying Eq. 12.106 as

$$\frac{d^2\theta}{d\zeta^2} = -\frac{3}{8}Nu_H(1-\zeta^2) \tag{12.109}$$

The boundary conditions are recast as

$$\begin{aligned} \text{Top wall: } \theta|_{\zeta=1} &= 0 \\ \text{Bottom wall: } \theta|_{\zeta=-1} &= 0 \end{aligned} \tag{12.110}$$

Again the solution is obtained easily as

$$\theta(\zeta) = \frac{3}{16}Nu_H\left[\frac{5}{6} - \zeta^2 + \frac{\zeta^4}{6}\right] \tag{12.111}$$

By a similar procedure as in Sect. 12.4.4, the Nusselt number may be shown to be $Nu_H = 8.23529$.

To highlight the differences in the asymmetric and symmetric heating cases, the temperature profiles have been plotted in Fig. 12.14. The fully developed velocity profile is also shown in the figure.

12.5.4 Fully Developed Flow in a Rectangular Duct

As an example of a non-circular section, we consider fully developed flow in a duct of rectangular section of sides $2a$ and $2b$ parallel, respectively, to the x- and y-axes.

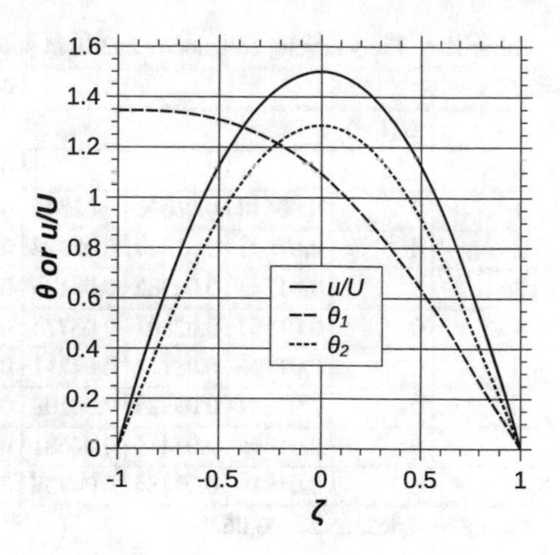

Fig. 12.14 Temperature profiles with asymmetric and symmetric heating in the case of parallel plate channel **a** θ_1—Asymmetric heating **b** θ_2—Symmetric heating

The origin is placed at the bottom left hand corner of the rectangle. Fluid velocity, under the fully developed condition, is now a function of x and y, being independent of z. The governing equation may be written down as,

$$\frac{\partial^2 u_z}{\partial x^2} + \frac{\partial^2 u_z}{\partial y^2} = \frac{1}{\mu}\frac{dp}{dz}$$

representing balance between viscous and pressure forces. The velocity vanishes along the four sides of the rectangle. We note that the right-hand side is a constant being related to the pressure drop per unit length of the duct. Introduce the following non-dimensional coordinates:

$$X = \frac{x}{2a},\ Y = \frac{y}{2a} \tag{12.112}$$

Introduce also a non-dimensional velocity given by

$$U = \frac{u_z(x, y)}{-\frac{4a^2}{\mu}\frac{dp}{dz}} \tag{12.113}$$

The governing equation takes the form

$$\frac{\partial^2 U}{\partial X^2} + \frac{\partial^2 U}{\partial Y^2} = -1 \tag{12.114}$$

The boundary conditions are now specified as

Table 12.6 Fully developed velocity matrix in $\frac{1}{4}$ section of a square duct

$j \rightarrow$	0	1	2	3	4	5	6	7	8
$i \downarrow$	$U(i, j)$								
0	0	0	0	0	0	0	0	0	0
1	0	0.00618	0.01041	0.01345	0.01567	0.01726	0.01834	0.01896	0.01916
2	0	0.01041	0.01809	0.02382	0.02807	0.03114	0.03323	0.03444	0.03484
3	0	0.01345	0.02382	0.03176	0.03774	0.04211	0.04508	0.04681	0.04738
4	0	0.01567	0.02807	0.03774	0.04512	0.05055	0.05427	0.05644	0.05716
5	0	0.01726	0.03114	0.04211	0.05055	0.05681	0.06111	0.06363	0.06446
6	0	0.01834	0.03323	0.04508	0.05427	0.06111	0.06584	0.06860	0.06951
7	0	0.01896	0.03444	0.04681	0.05644	0.06363	0.06860	0.07151	0.07247
8	0	0.01916	0.03484	0.04738	0.05716	0.06446	0.06951	0.07247	0.07345

Note $X = 0.0625i,\ Y = 0.0625j$

$$U(X, 0) = 0, \quad 0 \leq X \leq 1; \; U\left(X, \frac{b}{a}\right) = 0, \quad 0 \leq X \leq 1;$$

$$U(0, Y) = 0, \quad 0 \leq Y \leq \frac{b}{a}; \; U(1, Y) = 0, \quad 0 \leq Y \leq \frac{b}{a} \tag{12.115}$$

The governing equation thus is the Poisson equation in two dimensions. This equation is easily solved by finite differences using the methods discussed earlier, using equi-spaced nodes along the two directions, with $\Delta X = \Delta Y = 0.0625$. As a particular example, we consider the fully developed flow in a square duct for which $\frac{b}{a} = 1$. The hydraulic diameter for this section is $D_H = 2a$ as may be easily verified. The Poisson equation was solved by finite differences and the resulting $U(X, Y)$ is given in Table 12.6 as a matrix. Since the flow is symmetrical with respect to $X = 0.5$ and $Y = 0.5$, only the velocities in $\frac{1}{4}$ section of the square are presented in the table. By numerical integration using Simpson rule (second-order accurate—as is the finite difference method used in the solution of the Poisson equation), the mean velocity may be obtained as

$$\overline{U} = \int\limits_{X=0}^{1} \int\limits_{Y=0}^{1} U(X, Y) dX dY = 0.03502$$

The maximum velocity occurs at the center of the section, i.e., $X = Y = 0.5$ and is given by $U_{\max} = 0.07345$. Thus, the ratio of mean to maximum velocity is given by

$$\frac{\overline{U}}{U_{\max}} = \frac{0.03502}{0.07345} \approx 0.477$$

The actual mean velocity is then given by

$$\overline{u_z} = -0.03502\frac{4a^2}{\mu}\frac{dp}{dz} \tag{12.116}$$

As before, we replace $\frac{dp}{dz}$ by $\frac{-\Delta p}{L}$ and introduce friction factor f such that $\frac{-\Delta p}{L} = \frac{f\rho\overline{u}^2}{2D_H}$. Introduce this in Eq. 12.116 to get

$$\overline{u}_z = 0.03502\frac{4a^2}{\mu}\frac{f\rho\overline{u}_z^2}{2D_H}$$

With $4a^2 = D_H^2$, the above equation may be recast as

$$f = \frac{2}{0.03502}\frac{\mu}{\rho\overline{u}_z D_H} \approx \frac{57}{Re_{D_H}} \tag{12.117}$$

12.5.5 Fully Developed Heat Transfer in a Rectangular Duct: Uniform Wall Heat Flux Case

The corresponding heat transfer problem, with uniform wall heat flux, may be worked out in a manner analogous to the flow problem. The governing equation may be shown, following a method similar to that in the case of a circular tube, to be

$$\frac{\partial^2\theta}{\partial X^2} + \frac{\partial^2\theta}{\partial Y^2} = \frac{u_z(x,y)}{\overline{u_z}} = \frac{U(X,Y)}{\overline{U}} \tag{12.118}$$

where

$$\theta = \frac{\phi - \phi_w}{\phi_B - \phi_w} \text{ with } \phi = \frac{T}{\left(\frac{8aq_w}{k}\right)} \tag{12.119}$$

where q_w is the constant heat flux at the duct boundary. The subscript w represents the wall, and subscript B refers to the bulk mean value. We see that the temperature problem is also governed by Poisson equation but with the source term varying with X, Y. Since θ vanishes along the four sides of the duct cross section, we have

$$\theta(X,0) = 0, \quad 0 \le X \le 1;\ \theta\left(X, \frac{b}{a}\right) = 0, \quad 0 \le X \le 1;$$
$$\theta(0,Y) = 0, \quad 0 \le Y \le \frac{b}{a};\ \theta(1,Y) = 0, \quad 0 \le Y \le \frac{b}{a} \tag{12.120}$$

The solution, in the specific case of a square duct, has been numerically obtained and is given in matrix form in Table 12.7.

Table 12.7 Fully developed temperature matrix in $\frac{1}{4}$ section of a square duct

$j \rightarrow$	0	1	2	3	4	5	6	7	8
$i \downarrow$	$\phi(i, j)$								
0	0	0	0	0	0	0	0	0	0
1	0	0.00089	0.00289	0.00534	0.00779	0.00996	0.01164	0.01269	0.01305
2	0	0.00289	0.00973	0.01832	0.02708	0.03489	0.04098	0.04482	0.04614
3	0	0.00534	0.01832	0.03499	0.05221	0.06771	0.07982	0.08749	0.09013
4	0	0.00779	0.02708	0.05221	0.07846	0.10223	0.12095	0.13284	0.13692
5	0	0.00996	0.03489	0.06771	0.10223	0.13373	0.15858	0.17442	0.17986
6	0	0.01164	0.04098	0.07982	0.12095	0.15858	0.18841	0.20740	0.21392
7	0	0.01269	0.04482	0.08749	0.13284	0.17442	0.20740	0.22844	0.23567
8	0	0.01305	0.04614	0.09013	0.13692	0.17986	0.21392	0.23567	0.24316

Note $X = 0.0625i,\ Y = 0.0625j$

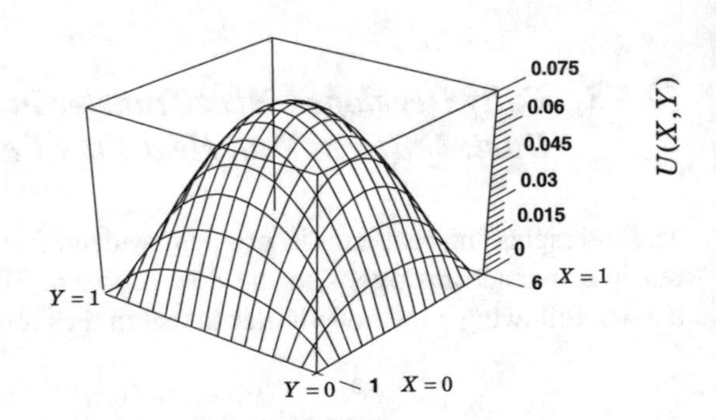

Fig. 12.15 3D plot of velocity in the square duct

It may easily be shown that the Nusselt number is related to the integral of the product of $\frac{U}{\overline{U}}$ and ϕ over the cross section of the duct represented in the form

$$\phi_B = \int\limits_{X=0}^{1} \int\limits_{Y=0}^{1} \left(\phi \times \frac{U}{\overline{U}} \right) dXdY \tag{12.121}$$

The above integral is evaluated numerically and is equal to 0.069559. The Nusselt number is then given by $Nu_H = \frac{1}{4 \times 0.069559} \approx 3.6$. Note that the characteristic length used in the Nusselt number definition is the hydraulic diameter $D_H = 2a$.

To complete this discussion, we present 3D plots of $U(X.Y)$ and $\phi(X, Y)$ in Figs. 12.15 and 12.16. Both figures indicate symmetry that was referred to earlier. The temperature variations with respect to X for a given Y or with respect to Y for a given X are close to being quadratic. The maximum velocity as well as the temperature occurs at the center of the square duct.

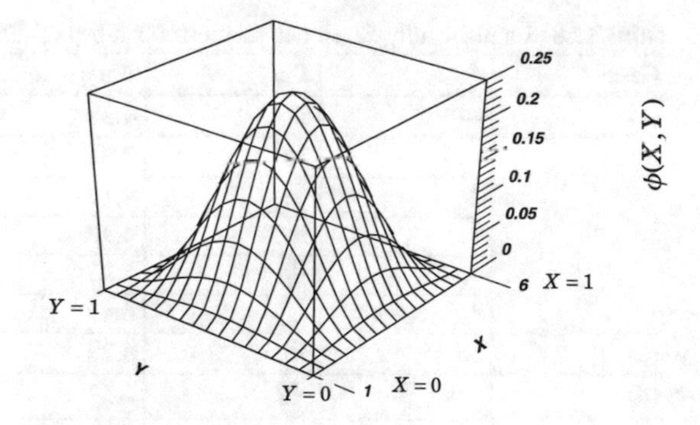

Fig. 12.16 3D plot of temperature in the square duct

12.5.6 Fully Developed Flow and Heat Transfer Results in Several Important Geometries

Non-circular sections are many times used in applications like air handling systems, power plants, etc. A non-circular duct may be treated in terms of an equivalent duct of circular cross section with the diameter given by the hydraulic diameter D_H. Figure 12.17 shows several cases that are important. Laminar friction coefficient—Nusselt number results for all these cases, in the fully developed region, are shown in Table 12.8. The table also gives expressions for the appropriate hydraulic diameters. The reader will recognize that a few of the results in the table have been worked out in detail in previous sections.

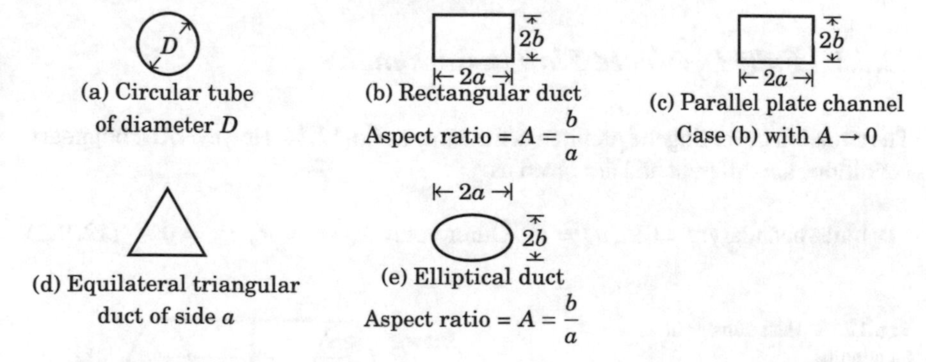

Fig. 12.17 Ducts of different useful cross sections

Table 12.8 Laminar fully developed relations for tubes of different cross sections

Case	$\frac{b}{a}$	D_H	Nu_H	Nu_T	$f \cdot Re_{D_H}$
(a)	—	D	4.36	3.66	64
(b)	1	$2a$	3.61	2.98	57
	2	$\frac{8a}{3}$	4.12	3.39	62.4
	3	$3a$	4.79	3.96	68.8
	4	$\frac{16a}{5}$	5.33	4.44	73.2
	8	$\frac{32a}{9}$	6.49	5.6	82.8
(c)	∞	$4b$	8.23	7.54	96
(d)	–	$\frac{a}{\sqrt{3}}$	3.11	2.47	53
(e)	0.9	$1.893a$	5.1	3.66	74.8

Case identifiers as in Fig. 12.17

12.6 Laminar Fully Developed Heat Transfer to Fluid Flowing in an Annulus

Flow in an annulus is quite common in heat exchanger applications, such as in the case of "tube in tube" heat exchanger.

In this case, the hot fluid may flow inside the inner tube of outer radius R_i while the coolant flows in the annular region between the inner tube and an outer tube of inner radius R_o as shown in Fig. 12.18. The outer tube is normally insulated on the outside so that heat transfer takes place only across the inner tube wall.

12.6.1 Fully Developed Flow in an Annulus

The equation governing the problem is the same as Eq. 12.34. However, the boundary conditions are different and are given as

$$\text{Inner boundary: } r = R_i, u_z = 0 \quad \text{Outer boundary: } r = R_o, u_z = 0 \tag{12.122}$$

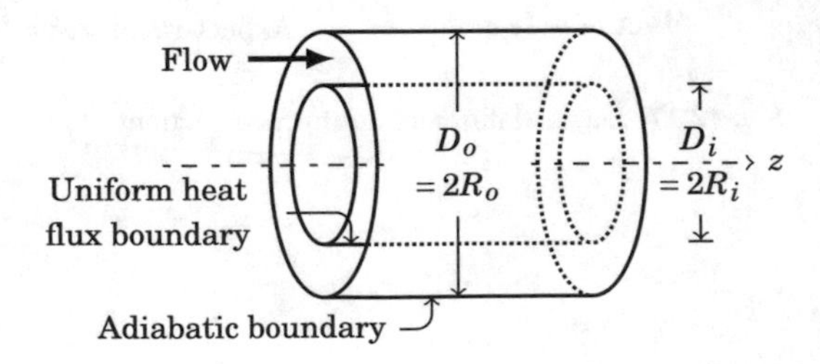

Fig. 12.18 Heat transfer in an annulus

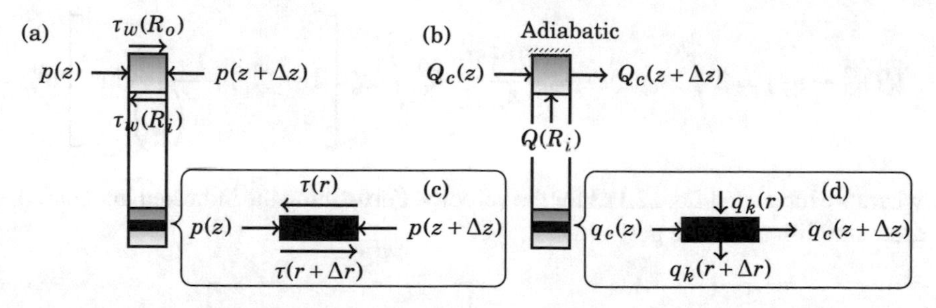

Fig. 12.19 Fluid elements in an annulus for performing force and energy balances

The reader may derive the governing equation by making a force balance on a differential element as shown in Fig. 12.19c. We may integrate the governing equation and apply the boundary conditions to get

$$u_z(\zeta) = -\frac{R_0^2}{4\mu}\frac{dp}{dz}\left[1-\zeta^2+\frac{1-a^2}{\ln\left(\frac{1}{a}\right)}\ln\zeta\right] \tag{12.123}$$

where $\zeta = \frac{r}{R_0}$ and $a = \frac{R_i}{R_0}$. The appearance of the logarithmic term is the main difference between the flow in a circular tube and an annulus. The maximum velocity occurs at a location given by

$$\frac{du_z}{d\zeta} = 0 \quad \text{or} \quad \underbrace{-2\zeta + \frac{1-a^2}{\ln\left(\frac{1}{a}\right)}\frac{1}{\zeta}}_{\text{Differentiating terms in square bracket in Eq. 12.123 with respect to } \zeta} = 0 \quad \text{or} \quad \zeta = \sqrt{\frac{1-a^2}{2\ln\left(\frac{1}{a}\right)}}$$

For example, when the radius ratio $a = 0.5$, the maximum velocity occurs at $\zeta = 0.73552 \approx 0.736$. Note that the inner boundary corresponds to $\zeta = 0.5$ in this case. Correspondingly the maximum velocity is given by

$$u_{\max} = -\frac{R_0^2}{4\mu}\frac{dp}{dz}\left[1-0.73552^2+\frac{1-0.5^2}{\ln\left(\frac{1}{0.5}\right)}\ln 0.73552\right] = -0.12664\frac{R_0^2}{4\mu}\frac{dp}{dz}$$

The mean velocity may be obtained by using the usual definition by equating the volume flow rates as

$$U(R_0^2 - R_i^2) = 2\int_{R_i}^{R_0} r u_z(r) dr = -\frac{R_0^2}{4\mu}\frac{dp}{dz} R_0^2 \int_a^1 2\zeta \left[1 - \zeta^2 + \frac{1 - a^2}{\ln\left(\frac{1}{a}\right)} \ln\zeta\right] d\zeta$$

where we have used Eq. 12.123 for the velocity. Performing the indicated integration, after simplification, we get

$$U(1 - a^2) = -\frac{R_0^2}{4\mu}\frac{dp}{dz}\left[\frac{1}{2} - \frac{a^4}{2} - \frac{(1 - a^2)^2}{2\ln\left(\frac{1}{a}\right)}\right] \tag{12.124}$$

For the case with $a = 0.5$, we have $U = -0.08399\frac{R_0^2}{4\mu}\frac{dp}{dz}$, and hence the ratio of mean velocity to the maximum velocity is equal to $\frac{U}{u_{max}} = \frac{0.08399}{0.12664} = 0.66322$. The important thing to note is that the velocity profile may be represented in the non-dimensional form as

$$\frac{u_z}{U} = \frac{1 - \zeta^2 + \frac{1 - a^2}{\ln\left(\frac{1}{a}\right)} \ln\zeta}{\frac{1}{2} + \frac{a^2}{2} - \frac{(1 - a^2)}{2\ln\left(\frac{1}{a}\right)}} \tag{12.125}$$

An overall force balance may be made for the fluid contained in an element of length Δz of the annulus as shown in Fig. 12.19a. The net pressure force acting on the element may be seen to be

$$\text{Net Pressure Force} = \frac{\Delta p}{\Delta z}\pi(R_0^2 - R_i^2)\Delta z \tag{12.126}$$

This is in the negative z-direction. The net force due to viscous shear at the two boundaries may be deduced as

$$\begin{bmatrix}\text{Net viscous}\\ \text{force}\end{bmatrix} = \begin{bmatrix}\text{Force at}\\ \text{outer boundary}\end{bmatrix} - \begin{bmatrix}\text{Force at}\\ \text{inner boundary}\end{bmatrix}$$

or in terms of the derivatives of velocity as

$$\text{Net Viscous Force} = (2\pi R_i \Delta z)\mu \left.\frac{du_z}{dr}\right|_{r=R_i} - (2\pi R_0 \Delta z)\mu \left.\frac{du_z}{dr}\right|_{r=R_0}$$

In terms of the non-dimensional velocity and radial coordinates, the above equation may be recast as

$$\text{Net Viscous Force} = (2\pi R_i \Delta z)\mu \frac{U}{R_0} \frac{d\left(\frac{u_z}{U}\right)}{d\zeta}\Bigg|_{\zeta=a} - (2\pi R_0 \Delta z)\mu \frac{U}{R_0} \frac{d\left(\frac{u_z}{U}\right)}{d\zeta}\Bigg|_{\zeta=1}$$

Note that the net viscous force is in the negative z-direction. The pressure gradient term is negative as it should be. We may now use Eq. 12.125 to obtain the derivatives in the above equation as

$$\frac{d\left(\frac{u_z}{U}\right)}{d\zeta}\Bigg|_{\zeta=a} = \frac{\left\{-2a + \dfrac{1-a^2}{a\ln\left(\frac{1}{a}\right)}\right\}}{\dfrac{1}{2}\left\{1+a^2 - \dfrac{1-a^2}{\ln\left(\frac{1}{a}\right)}\right\}};\quad \frac{d\left(\frac{u_z}{U}\right)}{d\zeta}\Bigg|_{\zeta=1} = \frac{\left\{-2 + \dfrac{1-a^2}{\ln\left(\frac{1}{a}\right)}\right\}}{\dfrac{1}{2}\left\{1+a^2 - \dfrac{1-a^2}{\ln\left(\frac{1}{a}\right)}\right\}} \tag{12.127}$$

Combining Eqs. 12.126 and 12.127, we may derive an expression for $\frac{\Delta p}{\Delta z}$ that is the same as $\frac{\Delta p}{L}$ where L is the length of the annulus in the fully developed region of the flow. Again we introduce the familiar friction factor to represent the pressure drop in terms of the dynamic pressure. The Reynolds number is represented in terms of the hydraulic diameter $D_H = 2(R_0 - R_i) = 2R_0(1-a)$. The reader may supply the intermediate steps to get the following expression:

$$\boxed{f Re_{D_H} = \frac{64(1-a)^2}{\left\{1+a^2 - \dfrac{(1-a^2)}{\ln\left(\frac{1}{a}\right)}\right\}}} \tag{12.128}$$

This is shown as a plot in Fig. 12.20 for various values of a. Note that the friction factor Reynolds number product tends to 96 as $a \to 1$. In this limit, the annulus behaves as a parallel plate channel. As $a \to 0$ the value tends to 64 that for a circular tube.

12.6.2 Fully Developed Temperature in an Annulus

We consider now fully developed region with constant heat flux q_w specified at the inner boundary. Energy balance over a short length element of the annulus (Fig. 12.19b) will indicate that the z derivative of the bulk fluid temperature follows the relation

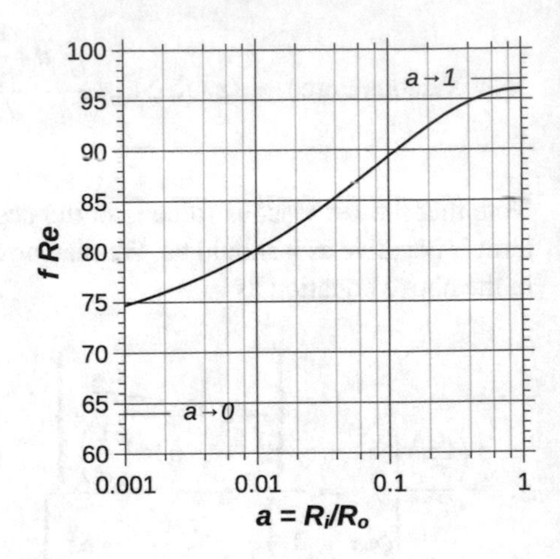

Fig. 12.20 Variation of $f \cdot Re$ with a for an annulus

$$\frac{\frac{dT_B}{dz}}{T_w - T_B} = \frac{\alpha Nu_H}{U R_0^2} \frac{a}{(1-a)^2(1+a)} \tag{12.129}$$

where the Nusselt number is based on the hydraulic diameter and T_w is the inner wall temperature. By performing energy balance over an elemental volume element shown in Fig. 12.19d, it is possible to show that the non-dimensional temperature θ is governed by the following equation.

$$\frac{1}{\zeta}\frac{d}{d\zeta}\left(\zeta\frac{d\theta}{d\zeta}\right) = -Nu_H \frac{a}{(1-a)^2(1+a)}\frac{u_z}{U} \tag{12.130}$$

The boundary conditions are given by

$$\theta = 0 \text{ at } \zeta = a; \quad \frac{d\theta}{d\zeta} = 0 \text{ at } \zeta = 1 \tag{12.131}$$

The velocity ratio is the fully developed value given by Eq. 12.125. The governing equation along with the boundary conditions may be integrated twice with respect to ζ to get the following solution.

$$\frac{\theta}{K} = \frac{1}{4}\left[1 - \frac{(1-a^2)}{\ln\left(\frac{1}{a}\right)}\right]\ln\left(\frac{\zeta}{a}\right) - \left[\frac{(\zeta^2 - a^2)}{4} - \frac{(\zeta^4 - a^4)}{16} + \frac{(1-a^2)}{4\ln\left(\frac{1}{a}\right)}\left\{\zeta^2(\ln\zeta - 1) - a^2(\ln a - 1)\right\}\right] \tag{12.132}$$

where

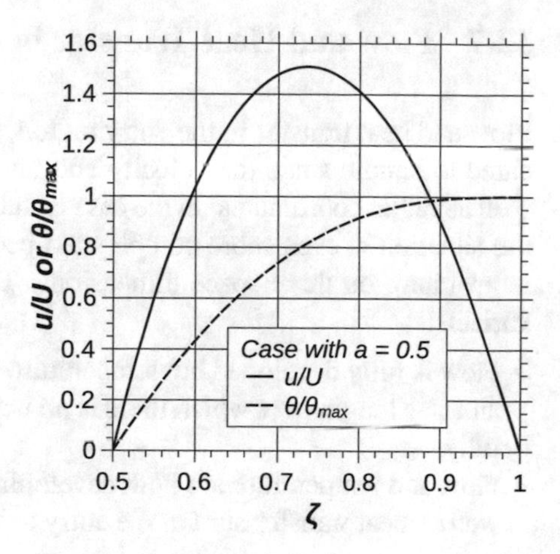

Fig. 12.21 Fully developed velocity and temperature profiles in an annulus

$$K = 2Nu_H \times \underbrace{\frac{a}{(1-a)^2(1+a)\left\{1 + a^2 - \frac{(1-a^2)}{\ln\left(\frac{1}{a}\right)}\right\}}}_{=g(a)} \tag{12.133}$$

The Nusselt number is determined by requiring that the weighted mean value of $\frac{\theta}{K}$ is $\frac{1}{K}$, i.e.,

$$\overline{\left(\frac{\theta}{K}\right)} = \frac{1}{K} = \frac{\int_a^1 \theta \frac{u_z}{U} \zeta d\zeta}{\int_a^1 \frac{u_z}{U} \zeta d\zeta} = f(a) \tag{12.134}$$

The ratio of the integrals is written as $f(a)$ to stress the point that it depends on a. Note that K is also a function of a and contains the Nusselt number as a factor. Hence the Nusselt number is obtained as

$$Nu_H = \frac{1}{f(a) \cdot g(a)} \tag{12.135}$$

As an example, we consider the specific case of an annulus with $a = 0.5$. The velocity and temperature profiles, normalized suitably are shown in Fig. 12.21. The friction factor is given by $f = \frac{95.25}{Re_{D_H}}$, and the Nusselt number turns out to be $Nu_H = 6.18$.

12.7 Flow and Heat Transfer in Laminar Entry Region

Flow and heat transfer in the entry region, i.e., $z < L_{dev}$; $z < L'_{dev}$ is more complicated to handle since the velocity and temperature fields are functions of axial as well as radial coordinates, in the case of tube flow. In the case of non-circular ducts, the situation is even more complicated because of the dependence of velocity and temperature on three space dimensions. The problem may occur in the following variants:

- Flow is fully developed but temperature is developing—the tube is provided with an entry length over which there is no heat transfer, the flow is allowed to develop fully.
- Flow and temperature are both developing simultaneously—flow development as well as heat transfer start at the entry to the tube.

The former case is handled more easily than the latter. The entry region heat transfer problem is referred to as the Graetz problem.[7] Contrary to the constant Nusselt number observed in the fully developed region, the Nusselt number varies with z in the developing region. The governing equations are solved, under suitable assumptions, by separation of variables, the solution being expressed in terms of eigenfunctions and eigenvalues.

12.7.1 *Heat Transfer in Entry Region of Fully Developed Tube Flow*

As an example, we consider the case where the flow is fully developed but the temperature starts developing from $z = 0$. The governing equation is written down using the energy equation given in Appendix H as Eq. H.33. The flow and temperature fields are steady, and hence, the time derivative does not occur. The velocity component u_z alone is non-zero, and hence, the convective term is restricted to that involving the axial derivative of temperature. Thus, we have

$$u_z \frac{\partial T}{\partial z} = \alpha \left[\frac{1}{r} \frac{\partial}{\partial r} \left(r \frac{\partial T}{\partial r} \right) + \frac{\partial^2 T}{\partial z^2} \right] \tag{12.136}$$

Introduce now the following non-dimensional variables:

Non-dimensional axial co-ordinate: $Z = \frac{z}{R \cdot Re_D Pr}$

Non-dimensional radial co-ordinate: $\zeta = \frac{r}{R}$

Non-dimensional velocity: $u^+ = \frac{u_z}{U} = 2(1 - \zeta^2)$

Non-dimensional temperature: $\theta = \frac{T_w - T}{T_w - T_0}$

[7] Named after Leo Graetz, 1856–1941, German physicist

where T_0 is the uniform temperature of the fluid at $z = 0$. It is customary to refer to $\frac{1}{Z}$ as the Graetz number Gz. Equation 12.136 is then recast as

$$(1-\zeta^2)\frac{\partial\theta}{\partial Z} = \frac{1}{\zeta}\frac{\partial}{\partial\zeta}\left(\zeta\frac{\partial\theta}{\partial\zeta}\right) + \frac{1}{(Re_D Pr)^2}\frac{\partial^2\theta}{\partial Z^2} \tag{12.137}$$

The second term on the right-hand side is small even for moderate values of $Re_D Pr$ and hence may be neglected in comparison with the axial derivatives. Hence, the governing equation is simplified as

$$\boxed{\frac{1}{\zeta}\frac{\partial}{\partial\zeta}\left(\zeta\frac{\partial\theta}{\partial\zeta}\right) = (1-\zeta^2)\frac{\partial\theta}{\partial Z}} \tag{12.138}$$

The following initial and boundary conditions may be specified:

$$\begin{aligned} \text{Entry:} \quad & \theta(\zeta, 0) = 1 \text{ for } 0 \le \zeta \le 1 \\ \text{Boundary condition:} \quad & \theta(1, Z) = 0 \\ \text{Boundary condition:} \quad & \theta(0, Z) \text{ is finite} \end{aligned} \tag{12.139}$$

Equation 12.138 subject to conditions Eq. 12.139 may be solved by using the separation of variables technique. The solution is sought in the form $\theta(\zeta, Z) = f(\zeta)\cdot g(Z)$. The governing equation then may be written as two equations given by

$$\frac{d^2 f}{d\zeta^2} + \frac{1}{\zeta}\frac{df}{d\zeta} + \lambda^2(1-\zeta^2)f = 0; \quad \frac{dg}{dZ} + \lambda^2 g = 0 \tag{12.140}$$

where $-\lambda^2$ is the separation constant. It is clear that the solution shows an exponentially decreasing dependence on Z. The dependence on ζ is through a set of orthogonal functions over the interval 0, 1. Details of the solution including the eigenvalues λ are available from the literature.[8] We present here graphically the variation of Nusselt number in Fig. 12.22, for both the constant wall temperature and constant wall heat flux cases. Asymptotically these tend, respectively, to 3.66 and 4.36.

We notice that the Nusselt number is theoretically infinite at $z = 0$ and decreases rapidly as z increases. It is also seen from the figure that the fully developed values are obtained for $\frac{1}{Gz} = Z \approx 0.1$ or for $\frac{z}{R} \approx 0.1 Re_D Pr$. It is seen that the result for the constant wall temperature case is always below that for the constant wall heat flux case.

[8] J.R. Sellars, M. Tribus and J.S. Klein, Trans. ASME, Vol.78, pp. 441–448,1956.

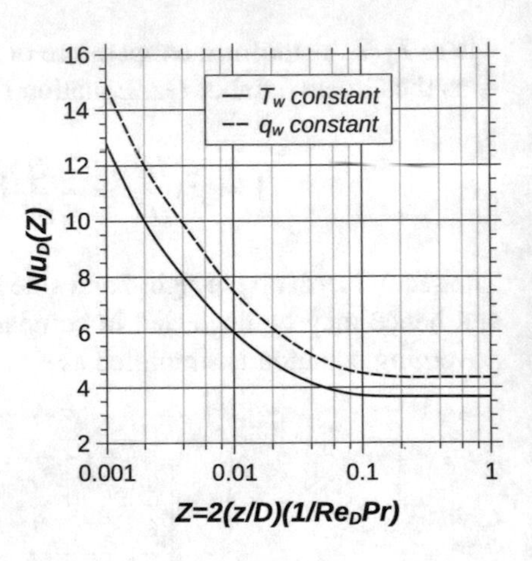

Fig. 12.22 Nusselt number variation in entry region of a tube

12.7.2 Mean Nusselt Number and Useful Correlations

By definition, the heat transfer coefficient is given by $h(z) = \frac{q_w}{T_w - T_B}$. In general, q_w, T_B, and T_w are all functions of z, and hence h is a function of z. The local Nusselt number at any z is defined as $Nu_D(z) = \frac{h(z)z}{k}$. Hence, the Nusselt number is simply a scaled local heat transfer coefficient. Consider, as an example, the case of T_w = constant. In this case, in the developing region, the variation of q_w with z is *different* from the variation of $T_w - T_B$ with z. Hence, h and Nu_D vary with z. As $z \to \infty$, q_w and $T_w - T_B$ vary alike with z and hence the Nusselt number tends to a constant value (3.66 in this case). If the tube is of length L, we may define an average Nusselt number as

$$\overline{Nu_D}(L) = \frac{1}{L} \int_0^L Nu_D(z)dz$$

Using the non-dimensional z coordinate, the above may be recast as

$$\overline{Nu_D}(L^*) = \frac{1}{L^*} \int_0^{L^*} Nu_D(Z)dZ \tag{12.141}$$

where $L^* = \frac{L}{R}\frac{1}{Re_D Pr} = \frac{2L}{D \cdot Re_D Pr}$. The mean Nusselt number variation with tube length is shown in Fig. 12.23 for both the constant wall temperature and the constant wall heat flux cases.

For the case of constant wall temperature, Hausen[9] has given a formula for $\overline{Nu_D}$ as a function of L^*.

[9]H. Hausen, Z. VDI Beih. Verfahrenstech., Vol. 4, pp. 91–98, 1943.

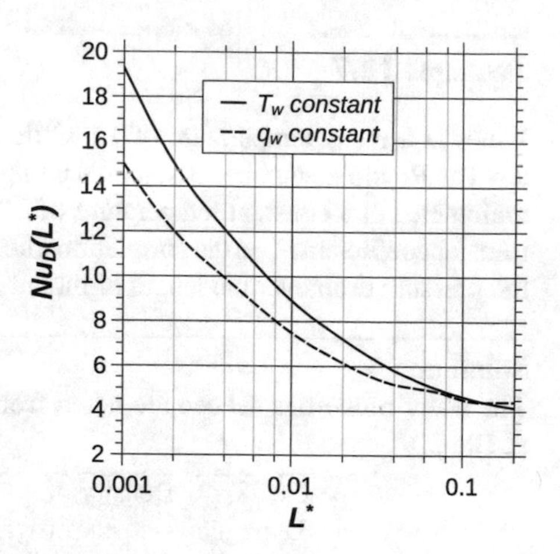

Fig. 12.23 Mean Nusselt number for short tubes

$$\overline{Nu}_D = 3.66 + \frac{\frac{0.1336}{L^*}}{1 + \frac{0.0635}{(L^*)^{\frac{2}{3}}}} \tag{12.142}$$

This formula is for a fluid whose properties remain constant and hence is applicable to problems in which the variation of fluid temperature is not large. The above is valid for the case where the velocity profile has already developed and only the temperature profile is developing.

For the combined entry length case, taking into account the variation of properties with temperature, Sieder and Tate[10] give the following relation.

$$\overline{Nu}_D = 2.34(L^*)^{-\frac{1}{3}} \left(\frac{\mu}{\mu_w}\right)^{\frac{1}{4}} \tag{12.143}$$

This is valid under the following conditions:

- All properties are evaluated at the mean bulk fluid temperature except μ_w which is evaluated at the wall temperature.
- $0.48 \le Pr \le 16,700; \quad 0.0044 < \frac{\mu}{\mu_w} < 9.75$.

[10]E.N. Seider and G.E. Tate, Ind. Eng. Chem., Vol.28, pp. 1429–1435, 1936.

Example 12.7

Water at a mean temperature of 10 °C flows in a 3 m long tube of 12 mm diameter. The Reynolds number based on the tube diameter is 500. The wall of the tube is maintained at a constant temperature of 30 °C. What is the mean value of the heat transfer coefficient? Use the correlation due to Hausen, if appropriate. Also, calculate the pressure drop over the length of the tube.

Solution:
The water properties needed are taken from tables of properties of saturated water at 10 °C.

Density of water: $\rho = 999.2$ kg/m^3
Dynamic viscosity of water: $\mu = 0.00131$ kg/m s
Thermal conductivity of water: $k = 0.585$ W/m°C
Prandtl number of water: $Pr = 9.4$

The geometrical parameters specified in the problem are

$$\text{Tube length: } L = 3 \text{ m}; \quad \text{Tube diameter: } D = 12 \text{ mm or } 0.012 \text{ m}$$

The Reynolds number for the flow is given to be $Re_D = 500$. The development length for temperature (assuming that the flow is fully developed at $z = 0$) is estimated as

$$L'_{dev} = (0.05 Re_D Pr)D = (0.05 \times 500 \times 9.4) \times 0.012 = 2.82 \text{ m}$$

The development length is comparable to the tube length. Hence, we make use of the correlation due to Hausen. The tube length parameter L^* is calculated as

$$L^* = 2\frac{L}{D}\frac{1}{Re_D Pr} = 2 \times \frac{3}{0.012} \times \frac{1}{500 \times 9.4} = 0.1064$$

Hausen correlation Eq. 12.142 gives

$$\overline{Nu_D} = 3.66 + \frac{\frac{0.1336}{0.1064}}{1 + \frac{0.0635}{0.1064^{\frac{2}{3}}}} = 4.64$$

The mean value of the heat transfer coefficient is then given by

$$\bar{h} = \frac{\overline{Nu_D} k}{D} = \frac{4.64 \times 0.585}{0.012} = 226.2 \text{ W/m}^2\text{°C}$$

Thus, it would be unwise to use the fully developed value for the heat transfer coefficient!

The pressure drop may be calculated assuming that the flow is fully developed across tube length. The friction factor is given by

$$f = \frac{64}{Re_D} = \frac{64}{500} = 0.128$$

The mean water velocity across the tube section is calculated from the Reynolds number as

$$U = \frac{\mu Re_D}{\rho D} = \frac{0.00131 \times 500}{999.2 \times 0.012} = 0.055 \text{ m/s}$$

The pressure drop over the length of the tube is then given by

$$\Delta p = \frac{f L \rho U^2}{2D} = \frac{0.128 \times 3 \times 999.2 \times 0.055^2}{2 \times 0.012} = 48.4 \text{ Pa}$$

Concluding Remarks

Study of convection heat transfer has been initiated in this chapter. Fundamental ideas regarding laminar internal flow and heat transfer are covered here. Useful results are presented for fully developed and developing flow and heat transfer.

12.8 Exercises

Ex 12.1: In the case of tube flow, the following 10 parameters have a role to play:

Fluid density ρ, fluid specific heat C_p, fluid viscosity μ, fluid thermal conductivity k, mean velocity of the fluid U, the mean temperature of the fluid T_m, the wall temperature T_w, the total heat transfer to the fluid over the tube length Q_w, the tube diameter D, and the tube length L.

By defining a suitable mean heat transfer coefficient show that the number of parameters may be reduced to 8. Indicate what these parameters are. Make use of mass $[M]$, length $[L]$, time $[T]$, and temperature $[\theta]$ as the four primary dimensions and perform a dimensional analysis of the problem. Obtain the relevant non-dimensional parameters that govern the problem.

Ex 12.2: The velocity distribution in laminar flow between two parallel planes is expressed as $u(y) = ay(s - y)$ where a is a constant, s is the distance between the planes, and y is the coordinate measured normal to the plane with $y = 0$ representing the bottom plane. Determine the ratio of average velocity to the maximum velocity. Based on the above velocity profile determine an expression for the friction factor.

Ex 12.3: (a) In a laminar pipe flow that is fully developed the axial velocity distribution is parabolic. What is the rate at which momentum is transferred across the tube at any section? Compare this with the momentum carried across the tube by the fluid moving at the mean velocity.

(b) The temperature profile in the above case varies linearly from the tube wall to a maximum value at $r = 0.5R$ (where R is the tube radius) and then remains constant. What is the energy flux across the tube, in each case?

Ex 12.4: A certain oil has a specific gravity of 0.862. It flows at a mass flow rate of 0.2 kg/s in a tube 1.2 cm inner diameter. At a temperature of 370 K the pressure drop in a length of 3 m is 31 kPa. Calculate: (a) the dynamic viscosity and (b) the kinematic viscosity of the oil. Justify your answer.

Ex 12.5: Consider laminar fully developed flow between two infinite parallel planes with a gap of $2b$. Obtain the velocity profile starting from first principles. What is the friction factor? Does it agree with the value indicated in Table 12.8?

Obtain the Nusselt number for fully developed conditions, in the same case, assuming that the walls are subject to a uniform heat flux. Does it agree with the value indicated in Table 12.8?

> Hint: Energy balance over an elemental length of the fluid will indicate that the wall temperature, local fluid temperature, and the bulk mean temperature of the fluid all vary at a constant rate. Use this information to arrive at the governing equation.

Ex 12.6: (a) Air at 30 °C is flowing in a circular tube of inner diameter 25 mm. It is known that the flow may be considered laminar if $Re_D < 2000$. What is the largest mass flow that the tube can support in laminar flow?

(b) If the air temperature at entry to the tube is 30 °C and the wall of the tube is maintained at a constant temperature of 90 °C what is the average heat transfer coefficient, assuming the tube to be very long?

(c) What is the outlet air temperature if the tube is 15 m long?
(d) What is the pressure drop between the entry and the exit?

Ex 12.7: A liquid metal (has a very low Prandtl number and hence $\nu \ll \alpha$) may be assumed to flow with a uniform velocity across a tube of radius R since the velocity field undergoes very little change. This model for liquid metal flow is referred to as a plug flow model. For such a flow, with a specified constant wall heat flux, determine the Nusselt number in the thermally fully developed condition. Compare this with the value that is obtained for the case of a fluid using the parabolic velocity distribution.

Ex 12.8: Table 12.8 presents the laminar pressure drop and heat transfer results for fully developed conditions in terms of the hydraulic diameter as the appropriate characteristic length scale, for several cases. In each case verify the expression for the hydraulic diameter given in the table.

Ex 12.9: A duct is of rectangular cross section of height $2b = 0.04$ m and width $2a = 0.02$ m. Air flows through this duct with a mean velocity of 1.5 m/s. The air enters at a bulk temperature of 30 °C and leaves the duct at 70 °C. Determine the length of the duct required for this. The wall of the duct is maintained at a constant temperature of 90 °C.

Ex 12.10: Table 12.8 indicates that $Nu_H > Nu_T$. Justify this from physical considerations.

Ex 12.11: A fluid flows with an average velocity of 1 m/s in a circular tube of 0.05 m diameter. If the same fluid flows in a square duct of side 0.05 m and has the same Reynolds number, what is the average velocity of the fluid in the square duct? Compare the volume flow rates in the two cases? Which of the two cases will involve a bigger pressure drop per unit length, assuming that the flow is laminar and fully developed, in both cases?

Ex 12.12: Air at atmospheric pressure and 30 °C flows at 3 m/s through a 1 cm ID pipe. An electrical resistance heater surrounds 20 cm length of tube toward its discharge end and supplies a constant heat flux to raise the temperature of air to 90 °C. What is the power input? What is the mean value of the heat transfer coefficient? Based on the above determine the mean temperature difference between the tube wall and the fluid.

Ex 12.13: Consider the fully developed temperature problem with wall temperature held fixed at a temperature different from the initial uniform temperature of the fluid. Assume that the velocity profile is given by the fully developed parabolic distribution. The resulting equation governing the non-dimensional temperature variation with r has been derived in the text. Solve this equation numerically using the finite difference method. Use ten uniformly spaced nodes between the center of the tube and the periphery of the tube. Derive the Nusselt number from the solution.

Ex 12.14: A copper tube of inner diameter 50 mm and outer diameter 55 mm is 10 m long. Hot water enters it at an average velocity of 0.9 m/s and a uniform temperature of 60 °C and loses heat to an ambient surrounding the pipe at a temperature of 15 °C. The heat transfer coefficient between the tube outer surface and the ambient may be taken as constant equal to 7.5 W/m^2 °C. What is the mean temperature of the water as it exits the tube?
A 60 mm thick layer of insulation of thermal conductivity 0.6 W/m °C is installed on the outside of the tube. The heat transfer coefficient to the ambient may be assumed to remain the same. What is the water exit temperature in this case?

Ex 12.15: A fluid flows at constant temperature through an annulus of inner diameter D_i and outer diameter D_o. Assume that the flow is laminar and fully developed. Formulate the governing differential equation for the problem. Specify appropriate boundary conditions. Obtain the velocity distribution in the annulus. Where does the maximum velocity occur? What is the magnitude of the maximum velocity in terms of the mean velocity?
From the solution obtain an expression for the pressure gradient in the annulus. Express the result in terms of a suitably defined friction factor.

Ex 12.16: Consider the fully developed flow in the annulus as in Exercise 11.17. Assume that the flow is fully developed, the inner surface of the annulus is maintained at a constant temperature different from the fluid entry temperature and the outer surface of the annulus is adiabatic. Formulate the governing energy equation for the problem in the non-dimensional form and specify the appropriate boundary conditions. Solve the equation using the finite difference method.

Ex 12.17: A fluid forced through its interstices cools a porous medium. The differential equations governing the temperature of the porous medium, T_m, and the temperature of the coolant, T_c are

$$k_m \frac{d^2 T_m}{dx^2} = h_i (T_m - T_c); \quad G_c C_{pc} \frac{dT_c}{dx} = h_i (T_m - T_c)$$

In the above h_i is an internal volumetric heat transfer coefficient, G_c is the mass flow of coolant per unit area, C_{pc} is the specific heat of the coolant, and k_m is the thermal conductivity of the porous medium. The coolant travels through the porous medium of thickness L. At $x = 0$ the porous medium is at T_{m0} and the coolant enters at T_{c0}. Measurement shows that the coolant leaves at a temperature equal to T_{cL}. Obtain an expression for the temperature of the coolant at any location x inside the porous medium.

Chapter 13
Laminar Convection in External Flow

HEAT transfer from an object to a fluid flowing past it—external flow and heat transfer—forms the topic of this chapter. The flow and heat transfer phenomena are described by the use of boundary layer theory. Flows with and without externally imposed axial pressure gradient are dealt with. Both exact and approximate methods are described in detail.

13.1 Introduction

Flow over bodies and heat transfer therefrom are important in many engineering applications such as heat exchangers, walls of buildings, external surfaces of ducts and so on. Heat exchange may take place across a flat surface or across curved surfaces such as tubes—either a single tube or a bundle of tubes such as in a heat exchanger. For example, a tube in cross flow is idealized as a cylinder, the flow taking place normal to the axis. Heat exchange may also be between a fluid flowing past a bank of tubes that carry a second fluid inside the tubes. Figure 13.1a–c depict these situations.

In Fig. 13.1a, a fluid stream at a uniform velocity of U in a direction parallel to the x-axis (and hence parallel to the surface) and a uniform temperature of T_∞ flows past a flat plate held at a fixed temperature T_w that is different from T_∞. Heat transfer will take place between the wall and the fluid. As the fluid flows past the plate, it picks up heat from the plate and carries it along in the x-direction. In contrast to the case of internal flow where we could expect fully developed flow and temperature fields, in the case of external flow the flow and temperature fields *continue to develop* in the x- direction. In view of this, the heat transfer coefficient or the Nusselt number continues to change with x. Also, as we shall see later the flow may become turbulent beyond some x value.

S. P. Venkateshan, *Heat Transfer*,
https://doi.org/10.1007/978-3-030-58338-5_13

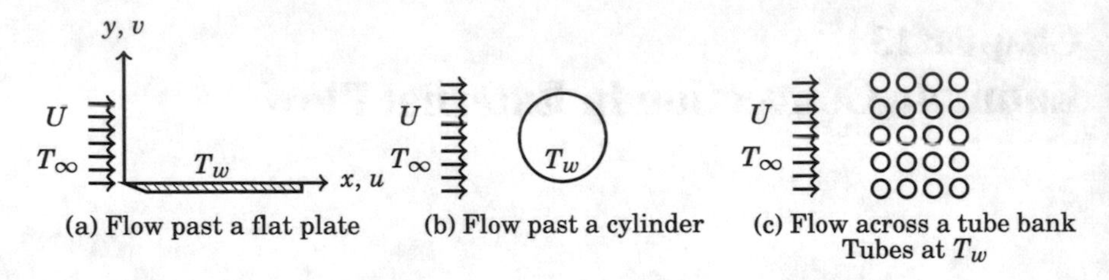

Fig. 13.1 Schematic representation of typical external flows

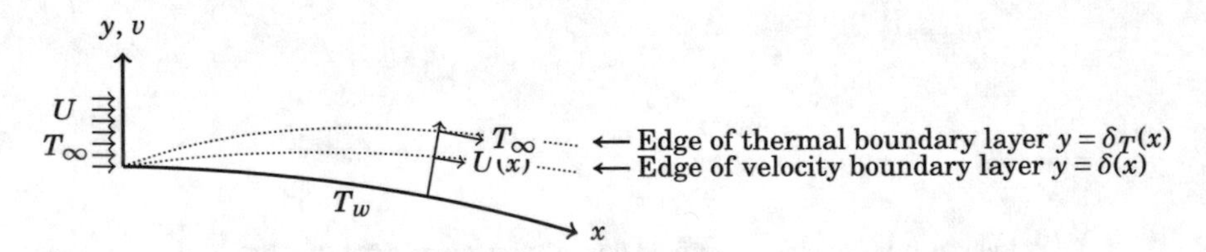

Fig. 13.2 Boundary layer flow past a surface

Flow past a cylinder shown schematically in Fig. 13.1b is even more complex. The flow may separate from the boundary, inducing a recirculating zone downstream, near the backward portion of the cylinder. The heat transfer coefficient and hence the Nusselt number varies with location measured by the angle downstream of the stagnation point that occurs at the forward part of the cylinder.

Flow over a tube bank shown schematically in Fig. 13.1c shows more interesting features. The flow past each of the cylinder that makes a row in the tube bundle has a wake region where there is considerable recirculation. If the tube spacing in a row is comparable to the recirculation zone behind a tube in the row there will be considerable interaction between the flows past adjacent tubes. The heat transfer coefficient will therefore depend on the tube pitch in a row. It will also depend on the distance between successive rows, because of interaction of the wake flows in a row and the tubes in the succeeding row. Further complications are due to the change of flow regime from laminar to turbulent, the transition being dependent on the Reynolds number. Flow past a flat plate is analyzed more easily than the other two cases. Hence, we shall take recourse to theoretical and experimental studies to complement each other.

13.2 Laminar Boundary Layer Flow Past a Surface

We consider flow of a viscous heat-conducting fluid adjacent to a surface shown schematically in Fig. 13.2.

The surface may be at a uniform temperature T_w different from T_∞. The flow shows non-uniform velocity and temperature fields in a small layer near the surface as shown in the figure. This region of non-uniform velocity and temperature fields is known as the boundary layer.[1] The changes in the velocity and temperature of the fluid are brought out, respectively, by fluid viscosity and fluid thermal conductivity. The flow upstream of the surface is characterized by a velocity U_∞ and uniform temperature T_∞. The velocity far away from the surface (for any x and $y >> \delta(x)$) may be a function of x and specified as $U(x)$. Correspondingly the temperature is specified by the free stream value T_∞. Since the boundary layer is very thin, the curvature of the surface does not play a significant role and the boundary layer flow is described by writing the governing equations using x defined as the coordinate along the surface and y defined as the coordinate normal to the surface.

In the presence of viscosity and thermal conductivity, fluid velocity and temperature at $y = 0$ are the same for the fluid as well as the wall. Hence a boundary layer, or a region of nonuniform velocity and temperature, forms adjacent to the plate surface, as indicated schematically in Fig. 13.2. Since viscosity and thermal conductivity of fluids encountered in engineering applications are *small* (see the values for air and water, given in Chap. 12), the effect of these is limited to regions of large gradients (normal to the surface) close to the wall—and hence large shear stress and conduction flux—at $y = 0$ and all x.

The boundary layer approximation assumes that the thickness of the boundary layer $\delta(x)$ is very small compared to x (of course, this will be acceptable some distance away from $x = 0$). The boundary layer is laminar for $x < x_c$ such that $Re_{x_c} < 5 \times 10^5$. Beyond $x = x_c$, the boundary layer undergoes a transition to turbulent flow. The turbulent part of the flow will be considered later on. Justification for the boundary layer model in the laminar region of the flow will be given below, based on an order of magnitude analysis.

13.2.1 Order of Magnitude Analysis and the Boundary Layer Approximation

NS Equations in Two Dimensions

We consider the laminar flow of an incompressible, viscous, heat conducting fluid in two dimensions, with constant thermo-physical properties. The velocity components along the x- and y- directions are taken, respectively, as u and v. The governing equations are the Navier–Stokes equations given in Appendix H represented in Cartesian coordinates by Eqs. H.1, H.6 and H.7. The equations, however, reduce to the following equations, by putting $w = 0$ and by setting $\frac{\partial}{\partial t} \equiv 0$.

Continuity equation:

$$\boxed{\frac{\partial u}{\partial x} + \frac{\partial v}{\partial y} = 0} \tag{13.1}$$

[1] Boundary Layer Theory was proposed by L.Prandtl (1904). English translation of German original available as NACA-TM-452.

x momentum equation:

$$u\frac{\partial u}{\partial x}+v\frac{\partial u}{\partial y}=-\frac{1}{\rho}\frac{\partial p}{\partial x}+\nu\left[\frac{\partial^2 u}{\partial x^2}+\frac{\partial^2 u}{\partial y^2}\right] \tag{13.2}$$

y momemntum equation:

$$u\frac{\partial v}{\partial x}+v\frac{\partial v}{\partial y}=-\frac{1}{\rho}\frac{\partial p}{\partial y}+\nu\left[\frac{\partial^2 v}{\partial x^2}+\frac{\partial^2 v}{\partial y^2}\right] \tag{13.3}$$

Energy equation:

$$u\frac{\partial T}{\partial x}+v\frac{\partial T}{\partial y}=\alpha\left[\frac{\partial^2 T}{\partial x^2}+\frac{\partial^2 T}{\partial y^2}\right] \tag{13.4}$$

Identification of the Scales in the Problem

An **order of magnitude analysis** is possible once we identify the scales for various quantities that appear in the governing equations given above.

- In case the surface is of finite length L, the surface length becomes a natural length scale in the problem, for representing x.
- Since our interest is to understand the flow and temperature variations within the boundary layer, the natural length scale along the y-direction is $\delta(L)$, the thickness of the boundary layer at L. In Fig. 13.2, the dashed curve represents the extent of the boundary layer normal to the surface at any x.
- The appropriate scale for u, the velocity along x is the free stream velocity U_∞.
- The temperature of the incoming fluid, T_∞ may be used as suitable reference temperature. The temperature difference $T_w - T_\infty$ may be used as a representative temperature difference.
- Since constant property assumption has been made, all the thermo-physical properties may be taken as those in the free stream, i.e., the incoming fluid or fluid far away from the surface. Thus, the required properties are $\nu = \nu(T_\infty) = \nu_\infty$, $\alpha = \alpha(T_\infty) = \alpha_\infty$.
- The free stream pressure p_∞ may be taken as a measure of the fluid pressure.

The only scale that has not been identified is that with reference to the normal (in a direction normal to the plate surface) velocity v. The appropriate scale is obtained by a bit of analysis.

For this purpose, we shall look at the continuity equation. The order of magnitude of the two terms in Eq. 13.1 are given by

$$\frac{\partial u}{\partial x} \sim \left(\frac{U_\infty}{L}\right), \quad \frac{\partial u}{\partial x} \sim \left(\frac{V}{\delta(L)}\right) \tag{13.5}$$

where V is as yet unknown scale for v. The continuity equation requires that the orders of magnitude of the two terms be the same. Otherwise the flow will degenerate to a simple flow such as ideal parallel flow. Thus, we require that

$$\left(\frac{U_\infty}{L}\right) = \left(\frac{V}{\delta(L)}\right) \quad \text{or} \quad V \sim \frac{U_\infty \delta(L)}{L} \tag{13.6}$$

Order of Magnitude Analysis

The order of magnitude analysis consists in determining the relative magnitudes of all the terms in all the equations that govern the problem and retaining terms of equal importance while neglecting terms that are of lower importance. The procedure will become clear when we look at, for example, the terms in the x momentum equation.

1. Inertia terms: $u\frac{\partial u}{\partial x} \sim U_\infty \frac{U_\infty}{L} = \frac{U_\infty^2}{L}$; $v\frac{\partial u}{\partial y} \sim \frac{U_\infty \delta(L)}{L}\frac{U_\infty}{\delta(L)} = \frac{U_\infty^2}{L}$. Thus both inertia terms are of equal order.
2. Pressure gradient term: $\frac{1}{\rho}\frac{\partial p}{\partial x} \sim \frac{1}{\rho_\infty}\frac{p_\infty}{L}$.
3. Viscous terms: $\nu_\infty \frac{\partial^2 u}{\partial x^2} \sim \nu_\infty \frac{U_\infty}{L^2}$; $\nu_\infty \frac{\partial^2 u}{\partial y^2} \sim \nu_\infty \frac{U_\infty}{\delta(L)^2}$. The second viscous term due to velocity variation normal to the surface is of larger order of magnitude compared to the viscous term due to variation of velocity parallel to the plate surface.

We use the magnitude of the inertia terms to normalize the other terms to write the order of magnitude of the terms in the x momentum equation as

$$\frac{U_\infty^2}{L} : \frac{U_\infty^2}{L} : \frac{1}{\rho_\infty}\frac{p_\infty}{L} : \nu_\infty \frac{U_\infty}{L^2} : \nu_\infty \frac{U_\infty}{\delta(L)^2}$$

$$1 : 1 : \frac{p_\infty}{\rho_\infty U_\infty^2} : \frac{\nu_\infty}{U_\infty L} : \frac{\nu_\infty}{U_\infty L}\left[\frac{L}{\delta(L)}\right]^2$$

or

$$\boxed{1 : 1 : Eu : \frac{1}{Re_L} : \frac{1}{Re_L}\left[\frac{L}{\delta(L)}\right]^2} \tag{13.7}$$

where we have replaced the quantity $\frac{p_\infty}{\rho_\infty U_\infty^2}$ by the familiar non-dimensional parameter, the Euler number and $\frac{U_\infty L}{\nu_\infty}$ by the Reynolds number Re_L based on plate length. We recognize that the order of magnitude analysis is to be made assuming that the Reynolds number is very large, but smaller than the critical value alluded to earlier. Thus, when $Re_L >> 1$, the viscous term due to the velocity change along x is small and may be neglected. The other viscous term—due to velocity variation with y—cannot be neglected since the viscous effects would then be completely absent, and it will not be possible to satisfy the no slip condition at the surface. Hence, we retain this term as being of order unity by requiring that

$$\frac{1}{Re_L}\left[\frac{L}{\delta(L)}\right]^2 \sim 1 \quad \text{or} \quad \frac{\delta(L)}{L} \sim Re_L^{-\frac{1}{2}} \tag{13.8}$$

The x momentum equation may hence be approximated by retaining the dominant terms as

$$\boxed{u\frac{\partial u}{\partial x} + v\frac{\partial u}{\partial y} = -\frac{1}{\rho_\infty}\frac{\partial p}{\partial x} + \nu_\infty\frac{\partial^2 u}{\partial y^2}} \tag{13.9}$$

An order of magnitude analysis of the y momentum equation may be made following similar steps as in the case of the x momentum equation. The resulting order of magnitudes are

$$\begin{aligned} &1 : 1 : \frac{p_\infty}{\rho_\infty U^2}\left(\frac{L}{\delta(L)}\right)^2 : \frac{\nu_\infty}{U_\infty L} : \frac{\nu_\infty}{U_\infty L}\left[\frac{L}{\delta(L)}\right]^2 \\ \text{or} \quad &1 : 1 : Eu \cdot Re_L : \frac{1}{Re_L} : 1 \end{aligned} \tag{13.10}$$

where we have used the order of magnitude of $\delta(L)$ determined earlier by order of magnitude analysis of the x momentum equation. We see that the pressure gradient term is the dominant term in this equation, and hence, the y momentum equation simplifies to

$$\boxed{-\frac{1}{\rho_\infty}\frac{\partial p}{\partial y} = 0} \tag{13.11}$$

Thus, the fluid pressure is a function of x alone, and hence the pressure gradient term in the x momentum Eq. 13.9 may be replaced by $\frac{dp}{dx}$. More importantly, the pressure does *not* vary within the boundary layer and hence the pressure at any x is *determined* by the flow outside the boundary layer—in other words that determined by the shape of the body.

A similar order of magnitude analysis is made for the energy equation also. Such an analysis (the reader may do this) indicates that the diffusion term along x may be neglected as compared to the other terms in the energy equation. Hence, the energy equation may be written down as

$$\boxed{u\frac{\partial T}{\partial x} + v\frac{\partial T}{\partial y} = \alpha_\infty \frac{\partial^2 T}{\partial y^2}} \tag{13.12}$$

Note that the order of magnitude analysis for the energy equation may use a thermal boundary layer thickness $\delta_T(L)$ instead of the hydrodynamic boundary layer thickness $\delta(L)$. The thermal boundary layer thickness is also assumed to be small compared to the plate length, noting that it depends on thermal diffusivity instead of kinematic viscosity. Later this aspect will receive more attention.

As derived above the boundary layer equations (Eqs. 13.1, 13.9 and 13.12) are valid within the boundary layer that is of thickness of order of $\delta(L)$. For $y > \delta(L)$, the variations of velocity and temperature are governed by inviscid, non-heat conducting flow obtained by taking $\nu = 0$ and $\alpha = 0$ in the NS equations. Thus, the flow outside the boundary layer corresponds to that of an ideal fluid. Momentum equations reduce to Euler equations. Outside the boundary layer we may ignore the viscous terms altogether, and Eq. 13.9 is written as

$$u\frac{\partial u}{\partial x} + v\frac{\partial u}{\partial y} = -\frac{1}{\rho_\infty}\frac{dp}{dx} \tag{13.13}$$

The flow outside the boundary layer is governed by $u \approx U$ and $v \approx 0$. Hence, the above equation shows that

$$U\frac{dU}{dx} = -\frac{1}{\rho_\infty}\frac{dp}{dx} \tag{13.14}$$

The boundary layer momentum equation may thus be recast as

$$\boxed{u\frac{\partial u}{\partial x} + v\frac{\partial u}{\partial y} = U\frac{dU}{dx} + \nu_\infty \frac{\partial^2 u}{\partial y^2}} \tag{13.15}$$

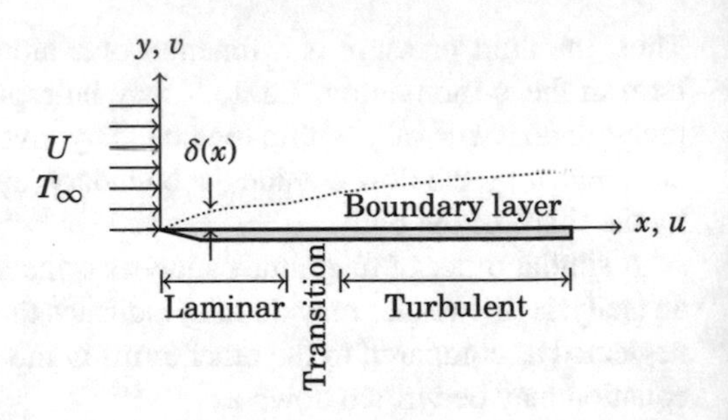

Fig. 13.3 Boundary layer flow past a flat plate

13.2.2 *Laminar Boundary Layer over a Flat Plate: Velocity Boundary Layer*

Flow parallel to a flat plate is shown schematically in Fig. 13.3. The flow velocity outside the boundary layer remains the same and equal to U_∞ for all x. Hence, the pressure remains the same throughout the flow field. Equation 13.15 further simplifies as

$$\boxed{u\frac{\partial u}{\partial x}+v\frac{\partial u}{\partial y}=\nu_\infty\frac{\partial^2 u}{\partial y^2}} \tag{13.16}$$

Boundary Conditions

As far as the velocity boundary conditions are concerned, we should have

$$\boxed{u=v=0,\ y=0,\ \text{for } 0<x<L\ ;\ u=U_\infty \text{ as } y\rightarrow\infty \text{ for } 0<x<L} \tag{13.17}$$

Boundary conditions at the wall assure that the velocity components are continuous at the interface between the fluid and the solid. Far away outside the boundary layer the velocity becomes uniform and equal to the free stream velocity. As far as the temperature boundary conditions are concerned, we should have

$$\boxed{T=T_w\ y=0,\ \text{for } 0<x<L\ ;\ T=T_\infty \text{ as } y\rightarrow\infty \text{ for } 0<x<L} \tag{13.18}$$

Boundary condition at the wall assures that temperature is continuous at the interface between the fluid and the solid. Far away outside the boundary layer, the temperature becomes uniform and equal to the free stream temperature.

Wall Shear and Surface Heat Flux

The velocity components vanish at the plate surface. The shear stress at the wall may be obtained by

$$\tau_w(x) = \underbrace{\tau_{w,x}}_{\text{Alternately}} = \mu_\infty \left.\frac{\partial u}{\partial y}\right|_{y=0} \tag{13.19}$$

The heat flux at the wall is given by the conduction flux at the wall since the fluid is at rest there. Thus, we have

$$q_w(x) = \underbrace{q_{w,x}}_{\text{Alternately}} = -k_\infty \left.\frac{\partial T}{\partial y}\right|_{y=0} \tag{13.20}$$

We may use the wall-free stream temperature difference as the driving potential to define the heat transfer coefficient as

$$h(x) = \underbrace{h_x}_{\text{Alternately}} = \frac{q_{w,x}}{T_w - T_\infty} = -\frac{k_\infty \left.\frac{\partial T}{\partial y}\right|_{y=0}}{[T_w - T_\infty]} \tag{13.21}$$

The local Nusselt number $Nu(x)$ is based on the current location x as the characteristic length. It is hence given by

$$Nu(x) = \underbrace{Nu_x}_{\text{Alternately}} = -\frac{\left.\frac{\partial T}{\partial y}\right|_{y=0}}{[T_w - T_\infty]}\, x \tag{13.22}$$

Solution to the Flow Problem-Similarity Analysis

Boundary layer flow past a flat plate may easily be dealt with by the similarity method that is familiar to us (see Chap. 5, Sect. 5.2.1). Since the flow is two-dimensional, it is possible to introduce a stream function $\psi(x, y)$ such that

$$u(x, y) = \frac{\partial \psi(x, y)}{\partial y}\ ;\ v(x, y) = -\frac{\partial \psi(x, y)}{\partial x} \tag{13.23}$$

With this, the equation of continuity is identically satisfied. The advantage of introducing the stream function is that we will be dealing with a scalar function instead of the two components of the velocity vector. Also, we need to consider only the momentum Eq. 13.16 now. We define a suitable composite variable such that the stream function is governed by an ordinary differential equation. We shall consider the following similarity variable to continue with the analysis:

$$\eta = \frac{y}{\sqrt{\frac{2\nu_\infty x}{U_\infty}}} \tag{13.24}$$

The stream function is introduced as a product function given by

$$\psi(x, y) = \psi(x, \eta) = \sqrt{2\nu_\infty x U_\infty}\; f(\eta) \tag{13.25}$$

where $f(\eta)$ is a function of the single composite variable η. The terms appearing in the momentum boundary layer equation are then recast using the following:

$$\frac{\partial}{\partial x} = \left.\frac{\partial}{\partial x}\right|_y = \left.\frac{\partial \eta}{\partial x}\right|_y \frac{d}{d\eta} = \frac{y}{\sqrt{\frac{2\nu_\infty}{U_\infty}}}\left(-\frac{x^{-\frac{3}{2}}}{2}\right)\frac{d}{d\eta} = -\frac{\eta}{2x}\frac{d}{d\eta} \tag{13.26}$$

$$\frac{\partial}{\partial y} = \left.\frac{\partial}{\partial y}\right|_x = \left.\frac{\partial \eta}{\partial y}\right|_x \frac{d}{d\eta} = \sqrt{\frac{U_\infty}{2\nu_\infty x}}\frac{d}{d\eta} \tag{13.27}$$

The above follow from the definitions of partial differentials and the chain rule of partial differentiation. The reader may refer back to Chap. 5 for any clarification. The reader will also recognize that the present similarity variable is the same as that used in the conduction problem with the variable t being replaced by the time like variable $\frac{x}{U_\infty}$. We use Expressions 13.25–13.27 to rewrite the terms in the momentum equation as follows.

Velocity component u:

$$u = \frac{\partial \psi}{\partial y} = \frac{\partial}{\partial y}\left[\sqrt{2\nu_\infty x U_\infty} f(\eta)\right] = \sqrt{2\nu_\infty x U_\infty}\sqrt{\frac{U_\infty}{2\nu_\infty x}}\frac{df}{d\eta} = U_\infty \frac{df}{d\eta} \tag{13.28}$$

Velocity component v:

$$\begin{aligned} v = -\frac{\partial \psi}{\partial x} &= -\frac{\partial}{\partial x}\left[\sqrt{2\nu_\infty x U_\infty} f(\eta)\right] = -f(\eta)\sqrt{2\nu_\infty U_\infty}\frac{x^{-\frac{1}{2}}}{2} \\ &-\sqrt{2\nu_\infty x U_\infty}\left(\frac{-\eta}{2x}\right)\frac{df}{d\eta} = \sqrt{\frac{\nu_\infty U_\infty}{2x}}\left[\eta\frac{df}{d\eta} - f\right] \end{aligned} \tag{13.29}$$

Various derivatives of u:

$$(a)\ \frac{\partial u}{\partial x} = -\frac{\eta U_\infty}{2x}\frac{d^2 f}{d\eta^2};\ (b)\ \frac{\partial u}{\partial y} = U_\infty\sqrt{\frac{U_\infty}{2\nu_\infty x}}\frac{d^2 f}{d\eta^2};$$
$$(c)\ \frac{\partial^2 u}{\partial y^2} = U_\infty\sqrt{\frac{U_\infty}{2\nu_\infty x}}\sqrt{\frac{U_\infty}{2\nu_\infty x}}\frac{d^3 f}{d\eta^3} = \frac{U_\infty^2}{2\nu_\infty x}\frac{d^3 f}{d\eta^3} \tag{13.30}$$

We now introduce all these in Eq. 13.16 to get

$$U_\infty\frac{df}{d\eta}\left[-\frac{\eta U_\infty}{2x}\frac{d^2 f}{d\eta^2}\right] + \sqrt{\frac{\nu_\infty U_\infty}{2x}}\left[\eta\frac{df}{d\eta} - f\right]U_\infty\sqrt{\frac{U_\infty}{2\nu_\infty x}}\frac{d^2 f}{d\eta^2} = \frac{\nu_\infty U_\infty^2}{2\nu_\infty x}\frac{d^3 f}{d\eta^3}$$

On simplification, this equation reduces to

$$\boxed{\frac{d^3 f}{d\eta^3} + f\frac{d^2 f}{d\eta^2} = 0} \tag{13.31}$$

This is a third-order non-linear ordinary differential equation known as the Blasius equation.[2] We shall look now at the boundary conditions.

1. $y = 0$ for a fixed x corresponds to $\eta = 0$. The wall boundary conditions may then be written, using Eqs. 13.28 and 13.29, respectively, as

$$\underbrace{\frac{df}{d\eta} = 0}_{\text{i.e., } u=0};\quad \text{and}\quad \underbrace{f = 0}_{\text{Hence } v=0}\ \text{at } \eta = 0 \tag{13.32}$$

2. $y \to \infty$ for a fixed x corresponds to $\eta \to \infty$. The condition $u \to U_\infty$ translates to the condition

$$\frac{df}{d\eta} \to 1 \text{ as } \eta \to \infty \tag{13.33}$$

We have thus succeeded in replacing the original partial differential equation along with its boundary conditions by ordinary differential Eq. 13.31 along with appropriate boundary conditions given by Eqs. 13.32 and 13.33. The Blasius equation cannot be solved by elementary methods. Numerical solution is obtained by using the Runge–Kutta method to arrive at the solution. Since the Runge–Kutta method is an initial value solver, the two point boundary value problem needs to be converted to an initial value problem.

[2]H. Blasius (1908), English translation of original German available as NACA-TM-1256.

One way of doing this is as follows:

- The domain of η is semi-infinite. Computationally $\eta \to \infty$ may be interpreted as some large value of $\eta = \eta_\infty$. The boundary condition at $\eta \to \infty$ is, in fact, applied at different assumed values of η_∞, and the final choice is that value that does not affect the results.
- Since the Runge–Kutta method is an initial value problem solver, we need to solve the Blasius equation with all three initial values, i.e., f, $\frac{df}{d\eta}$ and $\frac{d^2f}{d\eta^2}$ specified at $\eta = 0$.
- Since $\frac{d^2f}{d\eta^2}$ at $\eta = 0$ is not known, we have to use a method such as the Secant method to arrive at the correct initial value of the second derivative, by an iterative procedure.

An alternate method is possible using the following property of the Blasius equation. Consider the following transformation

$$f(\eta) = cF(Y);\; Y = c\eta \tag{13.34}$$

where c is a constant to be determined. We note that, under this transformation

$$\frac{d}{d\eta} = c\frac{d}{dY}$$

Hence, we have the following:

$$\frac{d^3f}{d\eta^3} = c^4\frac{d^3F}{dY^3} \quad \text{and} \quad \frac{d^2f}{d\eta^2} = c^3\frac{d^2F}{dY^2}$$

The Blasius equation is hence transformed to

$$\boxed{\frac{d^3F}{dY^3} + F\frac{d^2F}{dY^2} = 0} \tag{13.35}$$

It is of the same form as the original Blasius equation. The boundary condition at the surface of the plate becomes

$$F = \frac{dF}{dY} = 0 \text{ at } Y = 0$$

The boundary condition far away from the plate surface becomes

$$\frac{dF}{dY} = \frac{1}{c^2} \text{ as } Y \to \infty$$

The second derivative of the Blasius function at $\eta = 0$ becomes

$$\left.\frac{d^2 f}{d\eta^2}\right|_{\eta=0} = c^3 \left.\frac{d^2 F}{dY^2}\right|_{Y=0}$$

- **Details regarding solution to the Blasius equation**:
- The important point to note is that we may choose arbitrarily a value for $\frac{d^2F}{dY^2}$ at $Y = 0$, say $\frac{d^2F}{dY^2} = 1$.
- Equation 13.35 is solved with the initial conditions $F = \frac{dF}{dY} = 0$ and $\frac{d^2F}{dY^2} = 1$ at $Y = 0$ using fourth-order Runge–Kutta method. A fairly large value such as $Y = 6$ is chosen to represent $Y \to \infty$. A couple of trials will give the proper value.
- We choose a fairly small step size such as $\Delta Y = 0.05$. Calculation indicates that the first derivative of F tends to 1.65519 at $Y = 6$. By the above argument, this value represents $\frac{1}{c^2}$. We require that the the first derivative $\frac{df}{d\eta}$ be unity. Thus, the value of c is obtained as $c = \frac{1}{\sqrt{1.65519}} = 0.777277$. At once we see that the initial value for the second derivative that is required is $\left.\frac{d^2f}{d\eta^2}\right|_{\eta=0} = c^3 \left.\frac{d^2F}{dY^2}\right|_{Y=0} = c^3 = 0.777277^3 = 0.4696$.
- The results shown in Table 13.1 on p. 624 were generated by treating the Blasius problem as an initial value problem, taking the second derivative at $\eta = 0$ to be known and given as $\frac{d^2f}{d\eta^2} = 0.4696$.

Noting that $\frac{df}{d\eta} = \frac{u}{U_\infty}$, the results from the table may also be plotted to obtain the velocity profile within the boundary layer as shown in Fig. 13.4. Measurements of velocity profiles indicate very good agreement with the Blasius solution. The boundary layer thickness $\delta(x)$ is defined such that $\frac{u}{U_\infty} = 0.99$ at $y = \delta(x)$. The table indicates that this happens at about $\eta = 3.5$. Using the definition of the similarity variable, this may be written as

$$\eta = 3.5 = \delta(x)\sqrt{\frac{U_\infty}{2\nu_\infty x}} = \frac{\delta(x)}{x}\sqrt{\frac{U_\infty x}{2\nu_\infty}} \quad \text{or} \quad \frac{\delta(x)}{x} = 3.5\sqrt{\frac{2}{Re_x}} \approx \frac{5}{\sqrt{Re_x}} \tag{13.36}$$

We use Eq. 13.19 and 13.30 to calculate the shear stress at the wall as

$$\tau_{w,x} = \mu_\infty \frac{\partial u}{\partial y} = \mu_\infty U_\infty \sqrt{\frac{U_\infty}{2\nu_\infty x}} \left.\frac{d^2 f}{d\eta^2}\right|_{\eta=0} = 0.469604\mu_\infty U_\infty \sqrt{\frac{U_\infty}{2\nu_\infty x}} \tag{13.37}$$

Table 13.1 Solution to the Blasius equation

η	$\frac{df}{d\eta}$	$\frac{d^2f}{d\eta^2}$	η	$\frac{df}{d\eta}$	$\frac{d^2f}{d\eta^2}$	η	$\frac{df}{d\eta}$	$\frac{d^2f}{d\eta^2}$
0	0	0.46960	1.7	0.72994	0.32195	3.4	0.98798	0.03054
0.1	0.04696	0.46957	1.8	0.76106	0.30045	3.5	0.99071	0.02442
0.2	0.09391	0.46931	1.9	0.79	0.27825	3.6	0.99289	0.01933
0.3	0.14081	0.46861	2	0.8167	0.25567	3.7	0.99461	0.01515
0.4	0.18761	0.46726	2.1	0.84113	0.23301	3.8	0.99595	0.01176
0.5	0.23423	0.46503	2.2	0.86331	0.21058	3.9	0.99698	9.04E-3
0.6	0.28058	0.46174	2.3	0.88327	0.18867	4	0.99778	6.87E-3
0.7	0.32654	0.45718	2.4	0.90107	0.16756	4.1	0.99837	5.18E-3
0.8	0.37197	0.45119	2.5	0.91681	0.14748	4.2	0.99882	3.86E-3
0.9	0.41672	0.44363	2.6	0.93061	0.12861	4.3	0.99916	2.85E-3
1	0.46064	0.43438	2.7	0.94258	0.11112	4.4	0.99940	2.08E-3
1.1	0.50354	0.42337	2.8	0.95288	0.09511	4.5	0.99958	1.51E-3
1.2	0.54525	0.41057	2.9	0.96166	0.08064	4.6	0.99971	1.08E-3
1.3	0.58559	0.39599	3	0.96906	0.06771	4.7	0.99980	7.67E-4
1.4	0.62439	0.37969	3.1	0.97525	0.05631	4.8	0.99986	5.39E-4
1.5	0.66148	0.36181	3.2	0.98037	0.04637	4.9	0.99991	3.75E-4
1.6	0.6967	0.34249	3.3	0.98457	0.03781	5	0.99996	4.57E-4

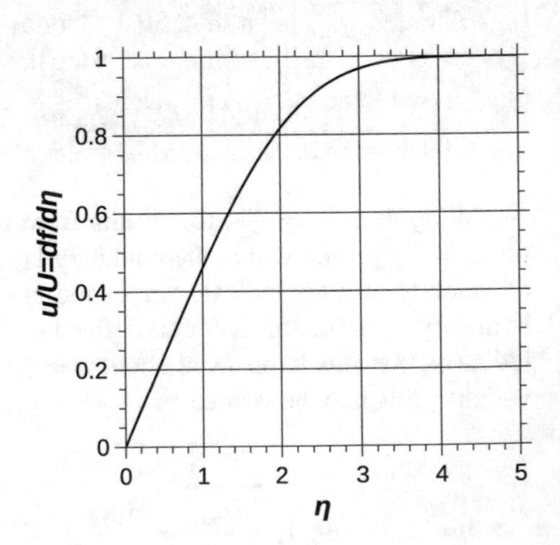

Fig. 13.4 Velocity profile in the laminar boundary layer over a flat plate

We define a friction coefficient $C_{f,x}$ by non-dimensionalizing $\tau_w(x)$ using the dynamic pressure $\frac{\rho_\infty U_\infty^2}{2}$ to get

$$C_{f,x} = \frac{0.46960\mu_\infty U_\infty \sqrt{\frac{U_\infty}{2\nu_\infty x}}}{\frac{\rho_\infty U_\infty^2}{2}} = \frac{0.66412}{\sqrt{Re_x}} \tag{13.38}$$

Example 13.1

Consider the flow of air at $T_\infty = 300\,\text{K}$ past a flat plate that is $L = 0.3\,\text{m}$ long. The free stream velocity of air is known to be $U_\infty = 4.5\,\text{m/s}$. What is the velocity boundary layer thickness at the trailing edge of the flat plate?

Solution:

Step 1 The kinematic viscosity of air at 300 K is taken from air tables as $\nu_\infty = 15.69 \times 10^{-6}\,\text{m}^2/\text{s}$.

Step 2 The Reynolds number based on plate length is calculated as

$$Re_L = \frac{U_\infty L}{\nu_\infty} = \frac{4.5 \times 0.3}{15.69 \times 10^{-6}} = 86{,}042$$

Since this value is less than $Re_c = 5 \times 10^5$, the critical Reynolds number, the flow is laminar throughout.

Step 3 The boundary layer thickness at $x = L$ is obtained, using Eq. 13.36, as

$$\delta_L = \frac{5L}{\sqrt{Re_L}} = \frac{5 \times 0.3}{\sqrt{86042}} = 0.0051\,\text{m or just around 5 mm!}$$

Remark: This example shows that all the viscous effects are confined to a very narrow region adjacent to the plate surface. The boundary layer approximation is very satisfactory.

Example 13.2

Consider again the data given in Example 13.1. What is the value of friction coefficient at $x = L$? Determine also the shear stress at the wall at $x = L$. What is the total shear force (drag) experienced by the plate, per unit width of plate in a direction perpendicular to the flow direction?

Solution:

Step 1 Apart from the data of Example 13.1, the air density is needed and is taken from the table of properties as $\rho_\infty = 1.1769\,\text{kg/m}^3$.

Step 2 We make use of the results of the Blasius solution to solve the problem. The friction coefficient at $x = L$ is obtained by using Eq. 13.38 as

$$C_{f,L} = \frac{0.66412}{\sqrt{Re_L}} = \frac{0.66412}{\sqrt{86042}} = 0.002264$$

Step 3 The shear stress at the wall at $x = L$ is then given by

$$\tau_{w,L} = C_f(L)\left(\frac{\rho_\infty U_\infty^2}{2}\right) = 0.002264 \times \left(\frac{1.1769 \times 4.5^2}{2}\right) = 0.027\ \text{Pa}$$

Step 4 The total shear force (drag force) experienced by the plate is obtained by integrating the shear stress distribution at the wall with respect to x from $x = 0$ to $x = 0.3\,\text{m}$. Elemental shear force is given by $dF = \tau_w(x)dx = C_{f,x}\left(\frac{\rho_\infty U_\infty^2}{2}\right)dx = \frac{0.66412}{\sqrt{Re_x}}\left(\frac{\rho_\infty U_\infty^2}{2}\right)dx$. But $x = \frac{\nu_\infty Re_x}{U_\infty}$ and hence $dx = \frac{\nu_\infty dRe_x}{U_\infty}$. Thus,

$$dF = \frac{0.66412}{\sqrt{Re_x}}\left(\frac{\rho_\infty U_\infty^2}{2}\right)\frac{\nu_\infty dRe_x}{U_\infty} = 0.33206\mu_\infty U_\infty \frac{dRe_x}{\sqrt{Re_x}}$$

This has to be integrated between $Re_x = 0$ and $Re_x = Re_L$ to obtain the total drag force experienced by the plate. Thus, we have

$$F = 0.33206\mu_\infty U_\infty \int_0^{Re_L} \frac{dRe_x}{\sqrt{Re_x}}$$

$$= 0.33206\mu_\infty U_\infty \left.\frac{Re_x^{-\frac{1}{2}+1}}{-\frac{1}{2}+1}\right|_0^{Re_L} = 0.66412\mu_\infty U_\infty \sqrt{Re_L}$$

Thus, the total drag force experienced by the plate is

$$F = 0.66412 \times 1.1769 \times 15.69 \times 10^{-6} \times 4.5 \times \sqrt{86042} = 0.0162\ \text{N/m}$$

Example 13.3

Consider water at 293 K flowing parallel to a flat plate 0.3 m long. What is the maximum allowable velocity if the flow just turns critical at $x = 0.3$ m? What is the thickness of the velocity boundary layer at this location? What is the drag force experienced by the plate per unit width in a direction perpendicular to the flow direction?

Solution:
Water properties needed for the calculation are taken from table of properties of water.

$$\rho_\infty = 998.2 \text{ kg/m}^3, \ \mu_\infty = 9.93 \times 10^{-4} \text{ kg/m s}$$

The Reynolds number at $x = L = 0.3$ m must be just critical, i.e., $Re_L = Re_c = 5 \times 10^5$. The maximum allowable velocity is hence given by

$$U_\infty = \frac{Re_L \mu_\infty}{\rho_\infty L} = \frac{5 \times 10^5 \times 9.93 \times 10^{-4}}{998.2 \times 0.3} = 1.658 \text{ m/s}$$

From the Blasius solution, the boundary layer thickness at $x = L = 0.3$ m is calculated as

$$\delta(L) = \frac{5L}{\sqrt{Re_L}} = \frac{5 \times 0.3}{\sqrt{5 \times 10^5}} = 0.00212 \text{ m} \approx 2 \text{ mm}$$

Friction coefficient at $x = L = 0.3$ m is given by

$$C_{f,L} = \frac{0.66412}{\sqrt{Re_L}} = \frac{0.66412}{\sqrt{5 \times 10^5}} = 0.000939$$

Example 13.2 has shown that the drag force over the entire length of the plate may be calculated based on a mean friction coefficient that is just twice the friction coefficient at L. Thus, as in Example 13.2, we have

$$F = 2C_{f,L}\frac{\rho_\infty U_\infty^2}{2}L = 2 \times 0.000939 \times \frac{998.2 \times 1.658^2}{2} \times 0.3 = 0.773 \text{ N/m}$$

13.2.3 *Laminar Thermal Boundary Layer over a Flat Plate*

Since the constant property assumption has been made in formulating the problem, the flow and temperature fields are not coupled. Hence the flow problem, as it has been done above, is calculated first and used as an input in the boundary layer energy

Eq. 13.12. Let us define a non-dimensional temperature function as

$$\theta = \frac{T - T_w}{T_\infty - T_w} \tag{13.39}$$

and assume that this is a function of η only. With this assumption, we have

$$\frac{\partial T}{\partial x} = (T_\infty - T_w)\left(-\frac{\eta}{2x}\right)\frac{d\theta}{d\eta} \tag{13.40}$$

where we have made use of Eq. 13.26.

$$\frac{\partial T}{\partial y} = (T_\infty - T_w)\sqrt{\frac{U_\infty}{2\nu_\infty x}}\frac{d\theta}{d\eta} \tag{13.41}$$

where we have made use of Eq. 13.27.

$$\frac{\partial^2 T}{\partial y^2} = (T_\infty - T_w)\frac{U_\infty}{2\nu_\infty x}\frac{d^2\theta}{d\eta^2} \tag{13.42}$$

where again we have made use of Eq. 13.27. Substitute these along with expressions for u and v given, respectively, by Eqs. 13.28 and 13.29 into Eq. 13.12 to get

$$U_\infty \frac{df}{d\eta}(T_\infty - T_w)\left(-\frac{\eta}{2x}\right)\frac{d\theta}{d\eta} + \sqrt{\frac{\nu_\infty U_\infty}{2x}}\left[\eta\frac{df}{d\eta} - f\right](T_\infty - T_w)\sqrt{\frac{U_\infty}{2\nu_\infty x}}\frac{d\theta}{d\eta}$$
$$= \alpha_\infty (T_\infty - T_w)\frac{U_\infty}{2\nu_\infty x}\frac{d^2\theta}{d\eta^2}$$

This equation simplifies to

$$\boxed{\frac{d^2\theta}{d\eta^2} + Pr\, f\frac{d\theta}{d\eta} = 0} \tag{13.43}$$

where Pr is the Prandtl number given by $Pr = \frac{\alpha_\infty}{\nu_\infty}$. The difference between the flow problem and the temperature problem is that the former gives a universal solution while the temperature field is dependent on a parameter, the Prandtl number. The solution to the temperature field is thus fluid specific. Since the energy equation has reduced to an ordinary differential equation, we infer that our assumption that the non-dimensional temperature field is a function of only the similarity variable η is valid. The boundary conditions are written down easily as

$$\boxed{\theta = 0 \text{ at } \eta = 0, \quad \theta \to 1 \text{ as } \eta = \infty} \tag{13.44}$$

Solution to the Temperature Problem

Solution to Eq. 13.43 is quite straightforward. Let $\frac{d\theta}{d\eta} = Z(\eta)$. Then Eq. 13.43 becomes

$$\frac{dZ}{d\eta} + Pr\ fZ = 0 \tag{13.45}$$

This equation is in variable separable form, and hence, we have

$$\frac{dZ}{Z} = -Pr\ f d\eta$$

Integrate this equation to get

$$\ln Z = -\int_0^\eta Pr\ f d\eta + A' \text{ or } Z = Ae^{-\int_0^\eta Pr\ f d\eta} = \frac{d\theta}{d\eta} \tag{13.46}$$

where A' is a constant of integration and $A = \ln A'$. Equation 13.46 may be integrated with respect to η once to get

$$\theta = A \int_0^\eta e^{-\int_0^{\eta_1} Pr\ f d\eta_2} d\eta_1 + B \tag{13.47}$$

where B is a second integration constant. The constants A and B are determined by using the boundary conditions. In order to satisfy the boundary condition at $\eta = 0$ given in Eq. 13.44, it is necessary to have $B = 0$. The second boundary condition in Eq. 13.44 requires that

$$1 = A \int_0^\infty e^{-\int_0^{\eta_1} Pr\ f d\eta_2} d\eta_1$$

Using this in Eq. 13.47, we have the solution as

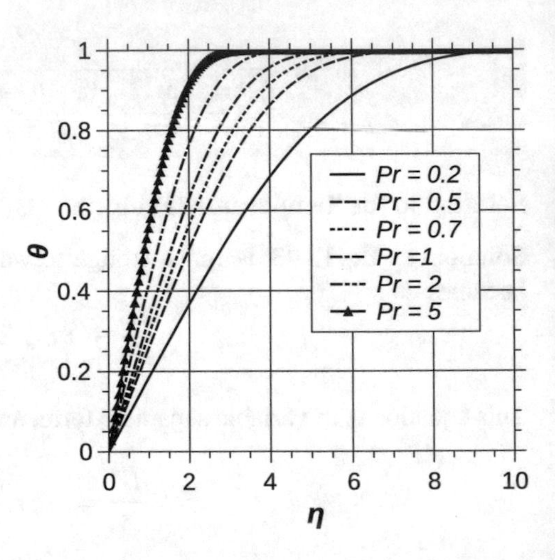

Fig. 13.5 Boundary layer temperature profiles for various Prandtl numbers

$$\theta(\eta) = \frac{\int_0^{\eta} e^{-\int_0^{\eta_1} Pr\, f d\eta_2} d\eta_1}{\int_0^{\infty} e^{-\int_0^{\eta_1} Pr\, f d\eta_2} d\eta_1} \tag{13.48}$$

The indicated integration may be performed numerically using the Blasius function that is already available, for a fixed Prandtl number. Boundary layer temperature profiles thus obtained are shown for various Prandtl numbers in Fig. 13.5. The thermal boundary layer thickness is seen to be a function of the Prandtl number. When $Pr < 1$, thermal boundary layer is thicker than the viscous boundary layer. When $Pr = 1$, the two boundary layer thicknesses are equal. When $Pr > 1$, the thermal boundary layer is thinner than the viscous boundary layer thickness.

Wall Heat Flux

From Eq. 13.20, using the definition of non-dimensional temperature and the similarity variable, the wall heat flux may be written down as

$$q_w = -\, k_\infty \frac{\partial T}{\partial y}\bigg|_{y=0} = -k_\infty (T_\infty - T_w)\sqrt{\frac{U_\infty}{2\nu_\infty x}}\, \frac{d\theta}{d\eta}\bigg|_{\eta=0}$$

From this, the local heat transfer coefficient h_x may be obtained as

$$h_x = \frac{q_w}{T_w - T_\infty} = k_\infty \sqrt{\frac{U_\infty}{2\nu_\infty x}} \left.\frac{d\theta}{d\eta}\right|_{\eta=0}$$

Hence, the local Nusselt number Nu_x is

$$Nu_x = \frac{h_x x}{k_\infty} = \frac{h_x x}{k_\infty} = k_\infty \sqrt{\frac{U_\infty}{2\nu_\infty x}} \left.\frac{d\theta}{d\eta}\right|_{\eta=0} \frac{x}{k_\infty} = \sqrt{\frac{Re_x}{2}} \left.\frac{d\theta}{d\eta}\right|_{\eta=0} \tag{13.49}$$

Nusselt Number Dependence on Prandtl Number

We have seen above that the Prandtl number plays an important role in that it determines the relative thicknesses of the velocity and temperature boundary layers. Since the wall heat transfer is directly related to the temperature gradient at the wall, the heat flux would mirror the effect of Prandtl number on the thermal boundary layer thickness.

For large Prandtl number, i.e., $Pr_\infty \gg 1$, the thermal boundary layer is thinner than the velocity boundary layer. In the limit $Pr \to \infty$, thermal boundary layer is very thin, and hence we may assume that $f' = C\eta$ or $f(\eta) = C\frac{\eta^2}{2}$ where C is a constant, inside the thermal boundary layer. Solution 13.48 then indicates that the ratio $\frac{Nu_x}{\sqrt{Re_x}}$ varies as $Pr_\infty^{\frac{1}{3}}$.

For vanishingly small Prandtl number, i.e., $Pr_\infty \ll 1$, the thermal boundary layer is thicker than the velocity boundary layer. In the limit, we may assume that the velocity is equal to free stream velocity throughout the thermal boundary layer. The solution given by Eq. 13.48 indicates that the ratio $\frac{Nu_x}{\sqrt{Re_x}}$ varies as $Pr_\infty^{\frac{1}{2}}$. The solution is the same as the solution we encountered for a semi-infinite solid subject to step change in surface temperature, with the proviso that the time variable there is interpreted as the ratio $\frac{x}{U_\infty}$.

Figure 13.6 shows the results of variation of $\frac{Nu_x}{\sqrt{Re_x}}$ with Prandtl number. The points are calculated by the use of solution obtained above. The asymptotes shown correspond to the following:

$$Pr_\infty \ll 1; \quad \frac{Nu_x}{\sqrt{Re_x}} \approx 0.5\sqrt{Pr_\infty} \tag{13.50}$$

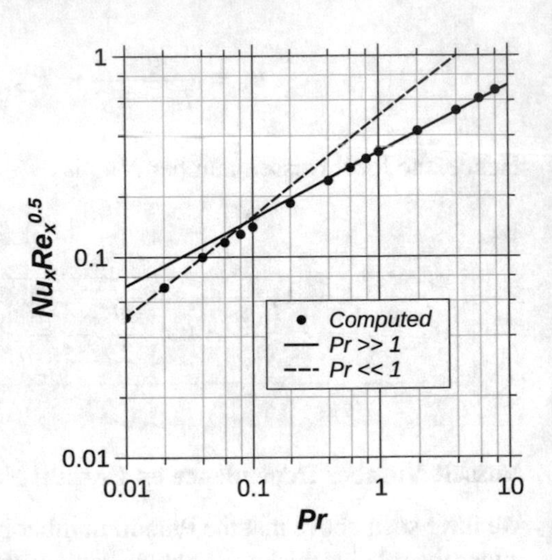

Fig. 13.6 Influence of Prandtl number on heat transfer from a flat plate

$$Pr_\infty \gg 1; \quad \frac{Nu_x}{\sqrt{Re_x}} \approx 0.332\ Pr_\infty^{\frac{1}{3}} \tag{13.51}$$

For $0.6 \leq Pr \leq 50$, the derivative of the non-dimensional temperature at $\eta = 0$ is very well approximated by the formula

$$\left.\frac{d\theta}{d\eta}\right|_{\eta=0} \approx 0.469604 Pr^{\frac{1}{3}} \tag{13.52}$$

Hence, the local Nusselt number is given by

$$Nu_x = 0.469604\sqrt{\frac{Re_x}{2}}Pr^{\frac{1}{3}} = 0.33206\ Re_x^{\frac{1}{2}}\ Pr^{\frac{1}{3}} \tag{13.53}$$

which is the asymptote shown in Fig. 13.6 as full line (valid for $Pr > 0.6$).

Average Wall Heat Flux

Equation 13.53 gives the local Nusselt number, and the local heat transfer coefficient may be obtained therefrom as

$$h_x = \frac{Nu_x k_\infty}{x} = \frac{0.33206 Re_x^{\frac{1}{2}} Pr^{\frac{1}{3}} k_\infty}{x} = 0.33206\sqrt{\frac{U_\infty}{\nu_\infty}} Pr^{\frac{1}{3}} k_\infty x^{-\frac{1}{2}} \tag{13.54}$$

We may calculate an average heat transfer coefficient $\overline{h}_L$ over a length L of the plate by defining it as

$$\overline{h}_L = \frac{\int_0^L h_x dx}{L} = \frac{0.33206}{L}\sqrt{\frac{U_\infty}{\nu_\infty}}Pr^{\frac{1}{3}}k_\infty \int_0^L x^{-\frac{1}{2}}dx \tag{13.55}$$

This may be recast, after performing the indicated integration, in the form

$$\boxed{\overline{Nu_L} = \frac{\overline{h}_L L}{k_\infty} \approx 0.664\ Re_L^{\frac{1}{2}}\ Pr^{\frac{1}{3}}} \tag{13.56}$$

Thus, the average Nusselt number over a length L of the plate is twice the local Nusselt number at $x = L$.

Example 13.4

Consider flow of air at $U_\infty = 3$ m/s parallel to a flat plate that is of length $L = 0.3$ m. The plate is maintained at $T_w = 400$ K while the incoming air is at $T_\infty = 300$ K and at 1 atmosphere pressure. Determine the heat flux at the trailing edge of the plate. Also determine the total heat transferred to air. Consider unit width of plate in a direction normal to the direction of incoming stream.

Solution:

Step 1 Since the wall–free stream temperature difference is 100 K the constant property assumption may not be very good. In order to account for the variation of properties, the suggested practice is to take the properties at the so-called film temperature T_f that is the arithmetic mean of the wall and free stream temperatures, i.e., $T_f = \frac{T_w+T_\infty}{2} = \frac{400+300}{2} = 350$ K. We take the air properties at a pressure of 1 atmosphere and at a temperature of 350 K from table of properties.

Density: $\rho_f = 0.995$ kg/m^3
Kinematic viscosity: $\nu_f = 20.82 \times 10^{-6}$ m^2/s
Thermal conductivity: $k_f = 0.030$ W/m°C
Prandtl number: $Pr_f = 0.7$*

*Pr of air varies very little with temperature

Step 2 The Reynolds number at $x = L = 0.3$ m is then calculated as

$$Re_L = \frac{U_\infty L}{\nu_f} = \frac{3 \times 0.3}{20.82 \times 10^{-6}} = 43227.7$$

Since the Reynolds number Re_L is less than 5×10^5, the flow remains laminar throughout.

Step 3 The Nusselt number at the trailing edge (i.e., at $x = L$) of the plate is then calculated using Eq. 13.53 as

$$Nu_L = 0.332 \times 43227.7^{\frac{1}{2}} \times 0.7^{\frac{1}{3}} = 61.29$$

The heat transfer coefficient at the trailing edge then is

$$h_L = \frac{Nu_L k_f}{L} = \frac{61.29 \times 0.030}{0.3} = 6.129 \text{ W/m}^2 {}^\circ\text{C}$$

Step 4 The heat flux at the trailing edge is then obtained as

$$q_{w,L} = h_L(T_w - T_\infty) = 6.129 \times (400 - 300) = 612.9 \text{ W/m}^2$$

Step 5 The average Nusselt number is twice the value of Nusselt number at the trailing edge. Hence, the average heat transfer coefficient over the length of the plate is twice the heat transfer coefficient determined above. Hence,

$$\overline{h}_L = 2h_L = 2 \times 6.129 = 12.26 \text{ W/m}^2 {}^\circ\text{C}$$

The total heat transferred from the plate per unit width is then given by

$$Q_w = \overline{h}_L(T_w - T_\infty)L = 12.26 \times (400 - 300) \times 0.3 = 367.8 \text{ W/m}$$

Example 13.5

Water at $T_\infty = 15$ °C flows parallel to a flat plate of length $L = 0.3$ m and $W = 0.6$ m wide with a velocity of $U_\infty = 0.4$ m/s. The plate surface is maintained at $T_w = 95$ °C. Determine the heat loss from both sides of the plate to water.

Solution:

Step 1 Water properties are evaluated at the film temperature of

$$T_f = \frac{T_w + T_\infty}{2} = \frac{95 + 15}{2} = 55 \text{ °C}$$

Step 2 Water properties at T_f are taken from table of properties.

Dynamic viscosity: $\mu_f = 4.89 \times 10^{-4}$ kg/m s
Density: $\rho_f = 984.3$ kg/m^3
Thermal conductivity: $k_f = 0.65$ W/m°C
Prandtl number: $Pr_f = 3.15$

Hence, the kinematic viscosity is obtained as

$$\nu_f = \frac{\mu_f}{\rho_f} = \frac{4.89 \times 10^{-4}}{984.3} = 4.968 \times 19^{-7} \text{ m}^2/\text{s}$$

Step 3 The Reynolds number based on the plate length is obtained as

$$Re_L = \frac{U_\infty L}{\nu_f} = \frac{0.4 \times 0.3}{4.968 \times 19^{-7}} = 214546$$

Since $Re_L < 5 \times 10^5$ the flow is laminar throughout.

Step 4 The mean Nusselt number over the plate length is given by (using Eq. 13.56)

$$\overline{Nu_L} = 0.664 \times 214546^{\frac{1}{2}} \times 3.15^{\frac{1}{3}} = 478.4$$

The average heat transfer coefficient over the plate length is then given by

$$\overline{h}_L = \frac{\overline{Nu_L} k_f}{L} = \frac{478.4 \times 0.65}{0.3} = 1036.5 \text{ W/m}^{2\circ}\text{C}$$

Step 5 Since the plate loses heat from both sides, the heat transfer area is $A = 2LW = 2 \times 0.3 \times 0.6 = 0.36\,\text{m}^2$. The total heat transferred from both sides of the plate then is

$$Q_w = A\overline{h}_L(T_w - T_\infty) = 0.36 \times 1036.5 \times (95 - 15) = 29851 \text{ W} \approx 30 \text{ kW}$$

Special Case of Laminar Boundary Layer Flow with $Pr = 1$– Reynolds Analogy

Rearranging the Blasius Eq. 13.31, we have

$$f = -\frac{d^3 f}{d\eta^3} / \frac{d^2 f}{d\eta^2}$$

The exponential term appearing in the solution to the boundary layer energy equation (Eq. 13.48) may be written as

$$e^{-\int f\,Pr\,d\eta} = \exp\left[-\int\left\{Pr\frac{\frac{d^3 f}{d\eta^3}}{\frac{d^2 f}{d\eta^2}}\right\}d\eta\right] = \exp\left[-Pr\ \ln\left(\frac{d^2 f}{d\eta^2}\right)\right] = -\left[\frac{d^2 f}{d\eta^2}\right]^{Pr}$$

Introducing this in Eq. 13.48, we get an alternate expression for the boundary layer temperature profile as

$$\theta(\eta) = \frac{\int\limits_0^\eta \left[\frac{d^2 f}{d\eta^2}\right]^{Pr} d\eta}{\int\limits_0^\infty \left[\frac{d^2 f}{d\eta^2}\right]^{Pr} d\eta} \tag{13.57}$$

The above is valid for any Prandtl number. In the special case when $Pr = 1$, we have

$$\int\limits_0^\eta \left[\frac{d^2 f}{d\eta^2}\right]^{Pr} d\eta = \int\limits_0^\eta \frac{d^2 f}{d\eta^2} d\eta = \left.\frac{df}{d\eta}\right|_0^\eta = \left.\frac{df}{d\eta}\right|_\eta = \frac{u}{U_\infty}$$

and

$$\int\limits_0^\infty \left[\frac{d^2 f}{d\eta^2}\right]^{Pr} d\eta = \int\limits_0^\infty \frac{d^2 f}{d\eta^2} d\eta = \left.\frac{df}{d\eta}\right|_0^\infty = \left.\frac{df}{d\eta}\right|_{\eta=\infty} = 1$$

The temperature profile given by Eq. 13.57 will then be recast as

$$\theta(\eta) = \frac{T - T_w}{T_\infty - T_w} = \frac{\int\limits_0^\eta \left[\frac{d^2 f}{d\eta^2}\right] d\eta}{\int\limits_0^\infty \left[\frac{d^2 f}{d\eta^2}\right] d\eta} = \frac{u}{U_\infty} \tag{13.58}$$

Thus, the temperature and velocity profiles (both of them suitably normalized) in the boundary layer are identical. This is referred to as Reynolds analogy. From Eq. 13.58, we have by differentiation

$$\frac{\left.\frac{\partial T}{\partial y}\right|_{y=0}}{T_\infty - T_w} = \frac{\left.\frac{\partial u}{\partial y}\right|_{y=0}}{U_\infty} \tag{13.59}$$

The partial derivatives are recast in terms of wall quantities of interest to us as follows:

$$\left.\frac{\partial T}{\partial y}\right|_{y=0} = -\frac{q_w}{k_\infty} \text{ and } \left.\frac{\partial u}{\partial y}\right|_{y=0} = \frac{\tau_{w,x}}{\mu_\infty}$$

Further, the first of these may be rewritten in terms of the local heat transfer coefficient as

$$\frac{q_w}{k_\infty} = \frac{h_x(T_w - T_\infty)}{k_\infty}$$

Introduce these in Eq. 13.59 to get

$$\frac{h_x}{k_\infty} = \frac{\tau_{w,x}}{\mu_\infty U_\infty}$$

Since the Prandtl number is equal to one, we may write thermal conductivity as $k_\infty = \mu_\infty C_p$. Also the wall shear stress may be written in terms of the friction factor as $\tau_{w,x} = C_{f.x}\frac{\rho_\infty U_\infty^2}{2}$. With these, we have

$$\frac{h_x}{k_\infty} = \frac{h_x}{\mu_\infty C_p} \text{ and } \frac{\tau_{w,x}}{\mu_\infty U_\infty} = \frac{C_{f,x}\rho_\infty U_\infty}{2\mu_\infty}$$

Hence, we obtain the important result

$$\frac{h_x}{\rho_\infty U_\infty C_p} = \frac{C_{f,x}}{2} \text{ or } \boxed{St_x = \frac{C_{f,x}}{2}} \tag{13.60}$$

We have introduced the local Stanton number $St_x = \frac{h_x}{\rho_\infty U_\infty C_p}$ in the above.[3] The reader may note that the above equation would also hold for the mean values over the plate length.

13.3 Boundary Layer Flow in the Presence of Stream-Wise Pressure Variation

As an example of boundary flow with stream-wise pressure gradient, we consider viscous flow past a wedge as shown schematically in Fig. 13.7.

The flow far away from the wedge is parallel flow with uniform velocity U_∞, as shown. The flow outside the boundary layer is characterized by inviscid flow with typical streamline pattern shown in the figure. We place the origin 0 at the apex of the wedge and lay the x-axis parallel to the wedge surface and the y-axis is normal

[3]It is to be noted that St_x may be represented as $St_x = \frac{Nu_x}{Re_x Pr}$ in terms of non-dimensional parameters already familiar to the reader.

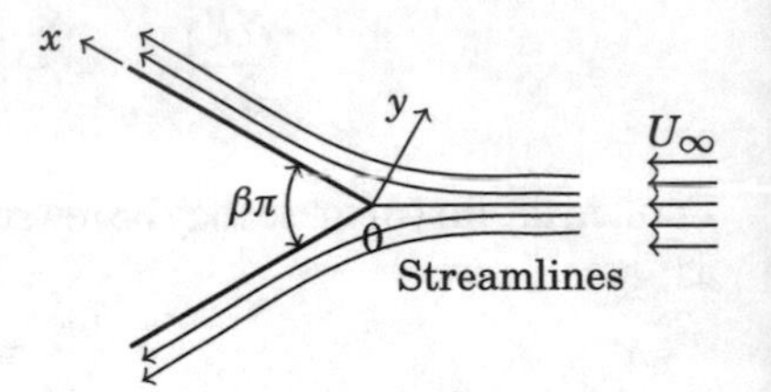

Fig. 13.7 Flow past a wedge

to it. The included angle of the wedge is $\beta\pi$ where β is a constant for a particular wedge.

13.3.1 Inviscid Flow Past the Wedge

The inviscid flow is governed by the Euler equation, and, in the present two-dimensional case, the stream function satisfies the Laplace equation. We use the complex potential to solve the problem without much effort. For this purpose consider the flow past the top surface of the wedge, essentially half the wedge. The external angle of the half wedge α is seen to be $\alpha = \pi - \frac{\beta\pi}{2} = \frac{\pi(2-\beta)}{2}$. Consider the complex potential given by

$$w = cz^{\frac{\pi}{\alpha}} \tag{13.61}$$

where c is a constant. Represent the complex variable z in its polar form given by $z = re^{i\theta}$. Substituting this in Eq. 13.61, we get

$$w = c(re^{i\theta})^{\frac{\pi}{\alpha}} = cr^{\frac{\pi}{\alpha}}e^{i\frac{\theta\pi}{\alpha}} \tag{13.62}$$

Explicitly, we can write the complex potential as

$$w = cr^{\frac{\pi}{\alpha}}\left\{\cos\left(\frac{\theta\pi}{\alpha}\right) + i\sin\left(\frac{\theta\pi}{\alpha}\right)\right\} \tag{13.63}$$

Consider the stream function $\psi(r, \theta)$ given by the imaginary part of w, i.e.,

$$\psi(r, \theta) = \sin\left(\frac{\theta\pi}{\alpha}\right) \tag{13.64}$$

This vanishes for $\theta = 0$ as well as for $\theta = \alpha$, i.e., along incoming direction passing through the apex as well as on the surface of the wedge. Thus, this stream function is the desired solution to the inviscid flow past the wedge. In order to obtain the velocity, we note that

$$\frac{dw}{dz} = \frac{\partial\phi}{\partial x} + i\frac{\partial\psi}{\partial x} = u - iv$$

where u and v are the velocity components. This result is a consequence of the Cauchy Reimann (CR) conditions. We thus notice that the magnitude of the velocity is given by the magnitude of the derivative of w with respect to z. Thus, in the present case, we have

$$\left|\frac{dw}{dz}\right| = \frac{c\pi}{\alpha} z^{\left(\frac{\pi}{\alpha}-1\right)} = \frac{c\pi}{\alpha} r^{\left(\frac{\pi}{\alpha}-1\right)} \underbrace{\left|e^{i\theta\left(\frac{\pi}{\alpha}-1\right)}\right|}_{\text{Magnitude is 1}} \tag{13.65}$$

Any point on the surface of the wedge corresponds to $\theta = \alpha$ and $r = x$. Hence, on the wedge surface the velocity (which perforce is parallel to the wedge surface) is given by

$$u = \frac{c\pi}{\alpha} x^{\frac{\pi}{\alpha}-1} \tag{13.66}$$

Using the definition of α, we have

$$\frac{\pi}{\alpha} - 1 = \frac{2}{2-\beta} - 1 = \frac{\beta}{2-\beta} = m(\text{say}) \tag{13.67}$$

With this the desired velocity *outside*, the boundary layer on the wedge is

$$\boxed{U(x) = Kx^m} \tag{13.68}$$

where K is a constant equal to $\frac{c\pi}{\alpha}$. Note that this velocity goes to zero at $x = 0$, the stagnation point.

General wedge flow solution represents typically the following special cases:

1. When $\beta = 1$, the flow turns by angle $\frac{\beta\pi}{2} = \frac{\pi}{2}$ and the corresponding wedge parameter is $m = \frac{1}{2-1} = 1$. The flow is known as stagnation point flow in two dimensions.
2. When $\beta = 0$ the flow proceeds with no direction change and the flow corresponds to flat plate flow considered earlier.
3. The flow is referred to as wedge flow for $0 \leq \beta \leq 2$ or $0 \leq m \leq \infty$. The turning angle is between 0 and π.
4. For positive β, the derivative of pressure with respect to x is negative since $\frac{dp}{dx} = -\rho_\infty U \frac{dU}{dx} < 0$. The flow is said to encounter a *favorable* pressure gradient.
5. For negative β, the derivative of pressure with respect to x is positive since $\frac{dp}{dx} = -\rho_\infty U \frac{dU}{dx} > 0$. The flow is said to encounter an *adverse* pressure gradient.

The flow in the boundary layer is amenable to similarity analysis as shown below.

13.3.2 Flow Within the Boundary Layer

Assume that the velocity within the boundary layer is given by an expression of form

$$u(x, y) = U(x)\frac{df}{d\eta} \tag{13.69}$$

where $\eta = \eta(x, y)$ is as of now undetermined. It is required that $\frac{df}{d\eta} \to 1$ as $\eta \to \infty$. We also require that the as of now unknown function $f(\eta)$ satisfies the conditions $f(\eta = 0) = \left.\frac{df}{d\eta}\right|_{\eta=0} = 0$. From Eq. 13.69, we have

$$(a)\ \frac{\partial u}{\partial x} = \frac{dU}{dx}\frac{df}{d\eta} + U(x)\frac{d^2 f}{d\eta^2}\frac{\partial \eta}{\partial x};\quad (b)\ \frac{\partial u}{\partial y} = U(x)\frac{d^2 f}{d\eta^2}\frac{\partial \eta}{\partial y} \tag{13.70}$$

In case of wedge flow, we know that $U(x) = Kx^m$. Let us consider the composite variable η to be given by

$$\eta(x, y) = Cyx^a \tag{13.71}$$

where C and a are constants to be determined as a part of the analysis. Indicating, for sake of convenience, differentiation with respect to η by $'$, we have the following:

$$\frac{\partial u}{\partial x} = mKx^{m-1}f' + Kx^m f'' \cdot aCyx^{a-1} = Kx^{m-1}\left[mf' + a\eta f''\right]$$

or

$$u\frac{\partial u}{\partial x} = K^2 f' x^{2m-1}\left[mf' + a\eta f''\right] \tag{13.72}$$

Using the equation of continuity, we have

$$\frac{\partial v}{\partial y} = -\frac{\partial u}{\partial x} = -Kx^{m-1}\left[mf' + a\eta f''\right]$$

Integrate this with respect to y to get

$$v = -Kx^{m-1}\int_0^y \left[mf' + a\eta f''\right]dy = -Kx^{m-1}\int_0^\eta \left[mf' + a\eta f''\right]d\eta\frac{1}{Cx^a}$$

$$= -\frac{K}{C}x^{m-a-1}\left[mf + a\eta f' - \int_0^\eta af' d\eta\right] = -\frac{K}{C}x^{m-a-1}\left[mf + a\eta f' - af\right]$$

$$= -\frac{K}{C}x^{m-a-1}\left[(m-a)f + a\eta f'\right]$$

We also have

$$\frac{\partial u}{\partial y} = Kx^m f'' C x^a = KCx^{m+a} f''$$

With these, we have

$$\begin{aligned} v\frac{\partial u}{\partial y} &= -\frac{K}{C}x^{m-a-1}\left[(m-a)f + a\eta f'\right] \times KCx^{m+a} f'' \\ &= -K^2 x^{2m-1} f''\left[(m-a)f + a\eta f'\right] \end{aligned} \tag{13.73}$$

Expressions 13.72 and 13.73 contribute to the LHS of boundary layer momentum Eq. 13.15.

We have, with the given functional form of the inviscid velocity $U(x)$,

$$U\frac{dU}{dx} = Kx^m \times Kmx^{m-1} = K^2 m x^{2m-1} \tag{13.74}$$

Lastly, we have

$$\nu_\infty \frac{\partial^2 u}{\partial y^2} = \nu_\infty KCx^{m+a} f''' C x^a = \nu_\infty KC^2 x^{m+2a} f''' \tag{13.75}$$

Expressions 13.74 and 13.75 contribute to the RHS of the boundary layer momentum Eq. 13.15. We notice that these four terms will have no *explicit* x dependence if

$$2m - 1 = m + 2a \quad \text{or} \quad a = \frac{m-1}{2} \tag{13.76}$$

With the above, the factors containing x and its powers drop off in the boundary layer momentum equation, which may be written as

$$\begin{aligned} K^2 f'[mf' + \frac{m-1}{2} f''\eta] - K^2 f''\left[\left(m - \frac{m-1}{2}\right) + \eta f' \frac{m-1}{2}\right] \\ = K^2 m + \nu_\infty C^2 \frac{f'''}{K} \end{aligned}$$

Further we shall choose $\nu_\infty \frac{C^2}{K} = \frac{m+1}{2}$ (this is an arbitrary, but convenient choice), to simplify the above equation to

$$f''' + ff'' - \frac{2m}{m+1}\left[(f')^2 - 1\right] = 0$$

The constant appearing in the above equation may be recast using Eq. 13.67 as

$$\frac{2m}{m+1} = \frac{2\dfrac{\beta}{2-\beta}}{\dfrac{\beta}{2-\beta}+1} = \beta$$

Thus, the boundary layer momentum equation takes the final form of a non-linear third-order ordinary differential equation

$$\boxed{f''' + ff'' - \beta\left[(f')^2 - 1\right] = 0} \tag{13.77}$$

The similarity variable η is given by

$$\eta = Cyx^a = \sqrt{\frac{(m+1)}{2}\frac{K}{\nu_\infty}}yx^{\frac{(m-1)}{2}} = y\sqrt{\frac{(m+1)}{2}\frac{Kx^m}{\nu_\infty x}} = \boxed{y\sqrt{\frac{(m+1)}{2}\frac{U(x)}{\nu_\infty x}}} \tag{13.78}$$

Equation 13.77 is known as the Falkner–Skan equation.[4] This equation is subject to the boundary conditions given by

$$\boxed{f(0) = f'(0) = 0; \;\; f' \to 1 \text{ as } \eta \to \infty} \tag{13.79}$$

The Falkner–Skan equation is hence a boundary value problem that can be solved only numerically, for example, by the use of the "shooting" method. Alternately the solution may be obtained numerically by the finite difference method, after a transformation of the coordinates, as discussed by Asaithambi.[5] What is required in practice, in calculating the wall shear, is the second derivative of the function f at $\eta = 0$. Table of values shown as inset in Fig. 13.8 gives the required data for typical values of β. η_∞ is the location of the "edge" of the boundary layer. Figure 13.8 shows the velocity profiles within the boundary layer for all cases in the table.

It is noticed from the figure and the data in the table that the velocity gradient is very nearly zero for a wedge with $\beta = -0.1988$. The shear stress at the wall vanishes and the condition represents incipient separation. This corresponds to flow over a diverging corner with wedge angle of 11.4°. It has been observed that multiple solutions are observed for $\beta < -0.1988$. For negative values of β the flow decelerates with x, pressure increases with x and the flow is subject to adverse pressure gradient. The case $\beta = 0$ corresponds to flow past a flat plate. The solution shown is the same

[4] V. M. Falkner and S. W. Skan, "Some Approximate Solutions of the Boundary Layer Equations" British ARC, R.& M. No. 1314, 1930.

[5] A. Asaithambi, "A finite difference method for the Falkner Skan Equation", Applied Mathematics and Computation, Vol. 92, pp. 135–141, 1998.

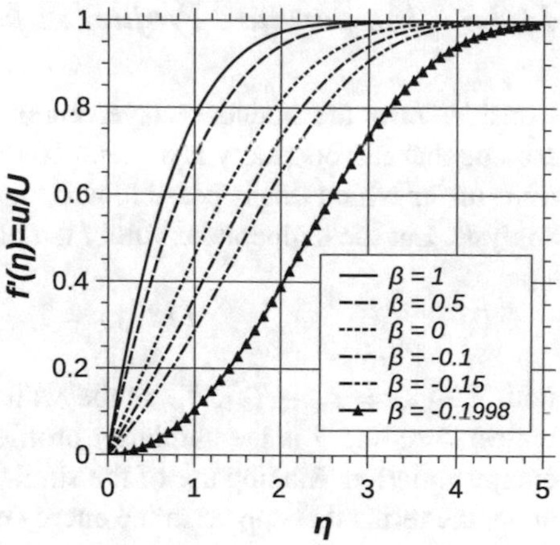

Fig. 13.8 Boundary layer velocity profiles for wedge parameters shown in the table at right

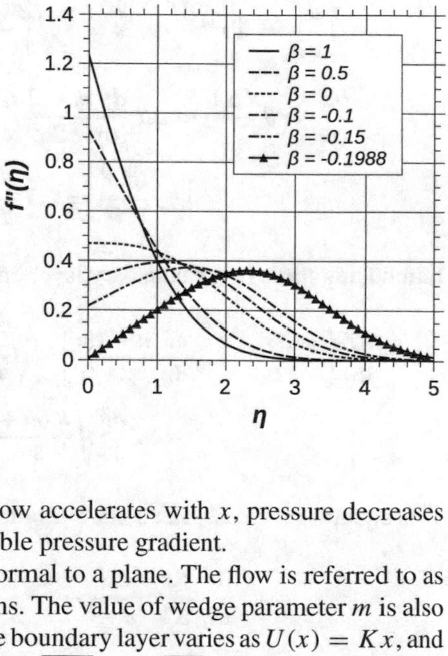

Fig. 13.9 Shear stress distributions in the boundary layer for wedge angles given in Fig. 13.8

as the Blasius solution. For $\beta > 0$, the flow accelerates with x, pressure decreases with x, and the flow is subject to a favorable pressure gradient.

The case $\beta = 1$ corresponds to flow normal to a plane. The flow is referred to as the stagnation point flow in two dimensions. The value of wedge parameter m is also equal to one. The velocity at the edge of the boundary layer varies as $U(x) = Kx$, and the similarity variable is $\eta = y\sqrt{\frac{U(x)}{2\nu_\infty x}} = y\sqrt{\frac{Kx}{2\nu_\infty x}} = y\sqrt{\frac{K}{2\nu_\infty}}$ which is just a scaled y! The boundary layer is hence of constant thickness.

The viscous shear stress is proportional to the second derivative of the function f with respect to η. The variation of the shear stress across the boundary layer is indicated by making a plot of f'' versus η as shown in Fig. 13.9. Figure shows the shear stress variation for all the wedge angles given in Fig. 13.8.

13.3.3 Temperature Profiles in Falkner–Skan Flows

Consider now the boundary layer energy equation given by Eq. 13.12. We shall assume that the boundary layer temperature profile also exhibits similarity. Conditions under which this is possible will, of course, be explored as we go through the analysis. Let the temperature within the boundary layer be given by

$$T(x, y) = T_{ref} + \Delta T(x)\theta(\eta) \tag{13.80}$$

with $\Delta T(x) = T_w - T_{ref}$. T_w is the wall temperature, T_{ref} is a suitable reference temperature and θ is the similarity profile that we are seeking as a solution to the energy equation. Making use of the similarity variable given by Eq. 13.78, we write down the terms that appear in the energy equation as under

$$u = U\frac{df}{d\eta} = Kx^m\frac{df}{d\eta}; \quad v = -\sqrt{\frac{2\nu_\infty K}{m+1}}x^{\frac{m-1}{2}}\left[\frac{m+1}{2}f + \frac{m-1}{2}\eta\frac{df}{d\eta}\right];$$

$$\frac{\partial T}{\partial x} = \theta\frac{d\Delta T}{dx} + \Delta T\frac{d\theta}{d\eta}\frac{m-1}{2}\frac{\eta}{x}; \quad \frac{\partial T}{\partial y} = \Delta T\frac{d\theta}{d\eta}\sqrt{\frac{K(m+1)}{2\nu_\infty}}x^{\frac{m-1}{2}};$$

$$\frac{\partial^2 T}{\partial y^2} = \Delta T\frac{d^2\theta}{d\eta^2}\frac{K(m+1)}{2\nu_\infty}x^{m-1}$$

Introducing these in the boundary layer energy equation, we have

$$Kx^m\frac{df}{d\eta}\left[\theta\frac{d\Delta T}{dx} + \Delta T\frac{d\theta}{d\eta}\frac{m-1}{2}\frac{\eta}{x}\right] - \sqrt{\frac{2\nu_\infty K}{m+1}}x^{\frac{m-1}{2}}\left[\frac{m+1}{2}f + \frac{m-1}{2}\eta\frac{df}{d\eta}\right]\times$$
$$\times\Delta T\frac{d\theta}{d\eta}\sqrt{\frac{K(m+1)}{2\nu_\infty}}x^{\frac{m-1}{2}} = \alpha_\infty\Delta T\frac{d^2\theta}{d\eta^2}\frac{K(m+1)}{2\nu_\infty}x^{m-1}$$

This equation is recast after some simplification as

$$\frac{df}{d\eta}\frac{\theta x}{\Delta T}\frac{d\Delta T}{dx} - \frac{m+1}{2}f\frac{d\theta}{d\eta} = \frac{m+1}{2Pr_\infty}\frac{d^2\theta}{d\eta^2} \tag{13.81}$$

Similarity solution is possible only if x does not appear explicitly in Eq. 13.81. Thus, the condition that has to be satisfied is

$$\frac{x}{\Delta T}\frac{d\Delta T}{dx} = n, \text{ a constant} \tag{13.82}$$

The above may be integrated with respect to x to yield

$$\Delta T = K_1 x^n \tag{13.83}$$

where K_1 is a constant. With this, Eq. 13.81 may be recast as

$$\boxed{\frac{d^2\theta}{d\eta^2} + Pr_\infty f \frac{d\theta}{d\eta} = \frac{2n Pr_\infty}{m+1}\theta \frac{df}{d\eta}} \tag{13.84}$$

It is evident that the θ depends on three parameters $m,\ n,$ and Pr_∞. The case $n = 0$ corresponds to constant ΔT which is of particular interest. In this case, the above equation further simplifies to

$$\boxed{\frac{d^2\theta}{d\eta^2} + Pr_\infty f \frac{d\theta}{d\eta} = 0} \tag{13.85}$$

From defining Eq. 13.80, we have

$$\theta = \frac{T - T_{\text{ref}}}{T_w - T_{\text{ref}}} \tag{13.86}$$

At $\eta = 0$, this has a value of 1. As $\eta \to \infty$ the temperature may be taken as the free stream temperature T_∞ and therefore $\theta = \frac{T_\infty - T_{\text{ref}}}{T_w - T_{\text{ref}}}$. Hence, we may set $T_{\text{ref}} = T_\infty$ to have $\theta = 0$ as $\eta \to \infty$. Equation 13.85 may be integrated by the method used in the case of boundary layer over a flat plate (see developments leading to Eq. 13.48) to get

$$\boxed{\theta = 1 - \frac{\int\limits_0^\eta e^{-\int\limits_0^{\eta_1} Pr\, f\, d\eta_2} d\eta_1}{\int\limits_0^\infty e^{-\int\limits_0^{\eta_1} Pr\, f\, d\eta_2} d\eta_1}} \tag{13.87}$$

The solution depends only on the two parameters m and Pr_∞. Dependence on m (or β) is through the dependence on f on m.

The special case of $Pr_\infty = 1$ has been considered, with three values of β, viz., 0.5, 0, and -0.10, to work out the temperature profile within the boundary layer using the solution given by Eq. 13.87. The results are shown plotted in Fig. 13.10. It is observed that the thermal boundary layer thickness increases with decrease in β. The Prandtl number dependence has already been discussed in the case of boundary layer over a flat plate. Similar observations hold for wedge flows. As an example, we show the temperature variation within the boundary layer for $\beta = 1$, stagnation point flow in two dimensions. Results are shown in Fig. 13.11 for three Prandtl numbers. The boundary layer thickness is larger for a smaller Prandtl number.

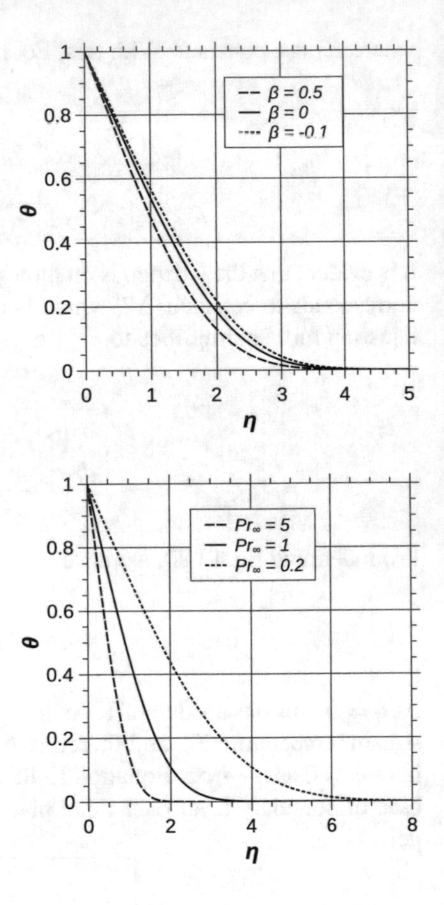

Fig. 13.10 Temperature distribution within the boundary layer over a wedge for $Pr_\infty = 1$ and three different values of β

Fig. 13.11 Temperature distribution within the boundary layer over a wedge for $\beta = 1$ and three different Prandtl numbers

Wall Heat Flux

The wall heat flux may be calculated by applying Fourier law at $y = 0$. Using the temperature profile given by Eq. 13.87, it is easily shown that the wall heat flux is

$$q_{w,x} = -k\frac{\partial T}{\partial y}\bigg|_{y=0} = k_\infty\sqrt{\frac{m+1}{2}\frac{U}{\nu_\infty x}}\frac{T_w - T_\infty}{\int\limits_0^\infty e^{-\int\limits_0^{\eta_1} Prf\,d\eta_2}\,d\eta_1}$$

Local Nusselt number may then be defined as

$$Nu_x = \frac{q_w}{T_w - T_\infty} \frac{x}{k_\infty} = \frac{\sqrt{\frac{m+1}{2} \frac{U}{\nu_\infty x}}}{\int\limits_0^\infty e^{-\int\limits_0^{\eta_1} Prf d\eta_2} d\eta_1} \tag{13.88}$$

Introducing the local Reynolds number $Re_x = \frac{Ux}{\nu_\infty}$, we may rewrite Eq. 13.88 as

$$\frac{Nu_x}{\sqrt{Re_x}} = \frac{\sqrt{\frac{m+1}{2}}}{\int\limits_0^\infty e^{-\int\limits_0^{\eta_1} Prf d\eta_2} d\eta_1} \tag{13.89}$$

If we substitute $m = 0$, i.e., $\beta = 0$ in the above, the local Nusselt number in the case of flat plate flow is obtained. Representing f with $\beta = 0$ as f_0 and f for any β as f_β, the ratio of Nusselt numbers turns out to be

$$\frac{Nu_{x,\beta}}{Nu_{x,0}} = \sqrt{m+1} \cdot \frac{\int\limits_0^\infty e^{-\int\limits_0^{\eta_1} Pr\, f_0 d\eta_2} d\eta_1}{\int\limits_0^\infty e^{-\int\limits_0^{\eta_1} Pr\, f_\beta d\eta_2} d\eta_1} \tag{13.90}$$

Here, $Nu_{x,\beta}$ represents the local Nusselt number for flow over the wedge while $Nu_{x,0}$ represents the local Nusselt number for flow over the flat plate, both for the same local Reynolds number. It is observed that the ratio is greater than 1 for $\beta > 0$ and less than 1 for $\beta < 0$, for a given Prandtl number. Figure 13.12 shows the variation of Nusselt number ratio with β for three representative Prandtl numbers. The trends are opposite for positive and negative values of β when Prandtl number increases.

13.4 Integral Form of Boundary Layer Equations

The boundary layer equations may be represented in an alternate integral form by integrating the boundary layer equations across the boundary layer, i.e., from the wall to the edge of the boundary layer. The resulting integral equations are useful in gaining physical understanding as well as in obtaining approximate solutions for the boundary layer velocity and temperature fields. Such solutions are surprisingly close to the exact solutions and hence this approach is useful, even from a practical point of view.

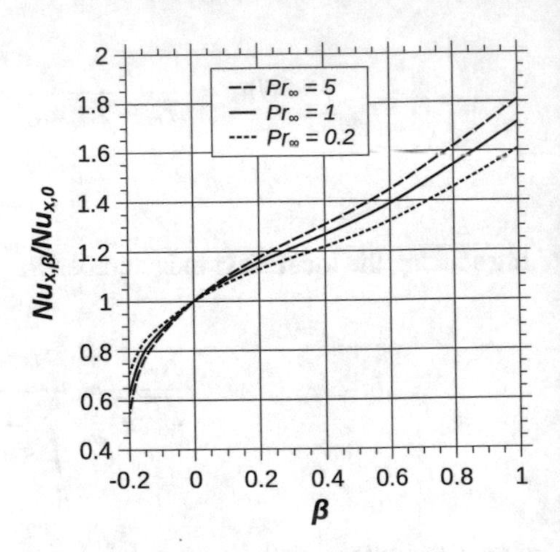

Fig. 13.12 Variation of Nusselt number ratio with β for three Prandtl numbers

13.4.1 Momentum and Energy Integral Equations

Integral Form of Boundary Layer Momentum Equation

Consider the boundary layer momentum equation given by Eq. 13.15. We integrate each term with respect to y from $y = 0$ to $y = \delta(x)$ to get

$$\int_0^\delta u\frac{\partial u}{\partial x}dy + \int_0^\delta v\frac{\partial u}{\partial y}dy - \int_0^\delta U\frac{dU}{dx}dy = \nu_\infty \int_0^\delta \frac{\partial^2 u}{\partial y^2}dy \tag{13.91}$$

In the second term on the left-hand side of Eq. 13.91, we may write $v\frac{\partial u}{\partial y}$ as $\frac{\partial uv}{\partial y} - u\frac{\partial v}{\partial y}$. Using the equation of continuity (Eq. 13.1), the latter term may be recast as $u\frac{\partial v}{\partial y} = -u\frac{\partial u}{\partial x}$. Hence, we have

$$v\frac{\partial u}{\partial y} = \frac{\partial uv}{\partial y} + u\frac{\partial u}{\partial x}$$

With this, the left-hand side of the integral of momentum equation becomes

$$\text{LHS of Eq. 13.19} = \int_0^\delta u\frac{\partial u}{\partial x}dy + \int_0^\delta \left[\frac{\partial uv}{\partial y} + u\frac{\partial u}{\partial x}\right]dy - \int_0^\delta U\frac{dU}{dx}dy$$
$$= 2\int_0^\delta u\,\frac{\partial u}{\partial x}dy + (uv)\Big|_0^{y=\delta} - \frac{dU}{dx}\int_0^\delta U dy = \int_0^\delta \frac{\partial u^2}{\partial x}dy + Uv_\delta - \frac{dU}{dx}\int_0^\delta U dy \tag{13.92}$$

where v_δ stands for the y component of velocity at the edge of the boundary layer. This velocity itself may be obtained by integrating the continuity equation across the boundary layer to get

$$\int_0^\delta \frac{\partial u}{\partial x} dy + \int_0^\delta \frac{\partial v}{\partial y} dy = \int_0^\delta \frac{\partial u}{\partial x} dy + v_\delta = 0 \quad \text{or} \quad v_\delta = -\int_0^\delta \frac{\partial u}{\partial x} dy$$

With this, we then have

$$U v_\delta = -U \int_0^\delta \frac{\partial u}{\partial x} dy = -\int_0^\delta \frac{\partial uU}{\partial x} dy + \frac{dU}{dx} \int_0^\delta u dy$$

Introducing these in Eq. 13.92 and clubbing terms appropriately, we get

$$\int_0^\delta \frac{\partial u^2}{\partial x} dy + U v_\delta - \frac{dU}{dx} \int_0^\delta U dy = \int_0^\delta \frac{\partial u^2}{\partial x} dy - \int_0^\delta \frac{\partial uU}{\partial x} dy$$
$$+ \frac{dU}{dx} \int_0^\delta u dy - \frac{dU}{dx} \int_0^\delta U dy = \underline{\int_0^\delta \frac{\partial}{\partial x} \{u(u - U)\} dy} + \frac{dU}{dx} \int_0^\delta (u - U) dy \tag{13.93}$$

We notice that the integrand in the first term (shown underlined) vanishes for $y = 0$ as well as for $y \geq \delta$. Hence, invoking Leibnitz rule[6] the first integral may be replaced by $\frac{d}{dx} \int_0^\delta u(u - U) dy$. Now, we may simplify the RHS of Eq. 13.91 as

$$\text{RHSofEq. 13.91} = \nu_\infty \int_0^\delta \frac{\partial^2 u}{\partial y^2} dy = \nu_\infty \left. \frac{\partial u}{\partial y} \right|_0^\delta = -\nu_\infty \left. \frac{\partial u}{\partial y} \right|_0 = -\frac{\tau_{w,x}}{\rho_\infty} \tag{13.94}$$

where it is assumed that the velocity gradient vanishes at the edge of the boundary layer. The boundary layer momentum equation in its integral form is finally obtained by combining all these as

[6] Leibnitz rule states that $\frac{d}{dx} \int_{a(x)}^{b(x)} f(x, y) dy = \int_{a(x)}^{b(x)} \frac{\partial f}{\partial x} dy + f(x, b(x)) \frac{db}{dx} - f(x, a(x)) \frac{da}{dx}$.

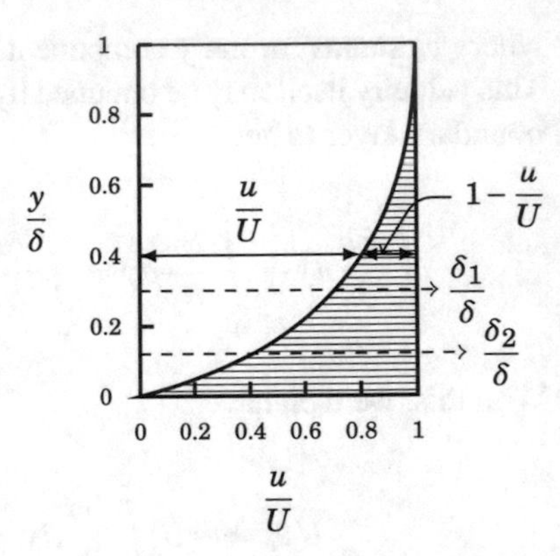

Fig. 13.13 Boundary layer velocity profile

$$\frac{d}{dx}\int_0^\delta u(u-U)dy + \frac{dU}{dx}\int_0^\delta (u-U)dy = -\frac{\tau_{w,x}}{\rho_\infty} \tag{13.95}$$

The integral equation given by Eq. 13.95 may be interpreted using two appropriate length scales that will be discussed below.

Displacement Thickness

The first length scale is the displacement thickness. Consider the velocity distribution within the boundary layer shown in Fig. 13.13, at some location x along the surface.

The actual mass flow rate across the station x is given by

$$\dot{m}_{\text{actual}}(x) = \int_{y=0}^{\infty} \rho_\infty u dy = \int_{y=0}^{\delta} \rho_\infty u dy + \int_{y=\delta}^{\infty} \rho_\infty U dy$$

In the absence of viscous effects, the mass flow rate across station x would have been

$$\dot{m}_{\text{ideal}}(x) = \int_{y=0}^{\infty} \rho_\infty U dy = \int_{y=0}^{\delta} \rho_\infty U dy + \int_{y=\delta}^{\infty} \rho_\infty U dy$$

The change in mass flow rate due to viscous effects is thus given by

$$\Delta \dot{m} = \dot{m}_{\text{actual}}(x) - \dot{m}_{\text{ideal}}(x) = \int_{y=0}^{\delta} \rho_\infty (u - U) dy \tag{13.96}$$

This mass rate reduction may be visualized as a certain height of fluid layer δ_1—the displacement thickness—traveling with velocity U such that the area below the velocity profile, indicated by horizontal lines, is equivalent to the area under the rectangle lying below the line $y = \delta_1$ in the figure. Thus, we have

$$\int_{y=0}^{\delta} \rho_\infty (u - U) dy = -\rho_\infty U \delta_1 \quad \text{or} \quad \delta_1 = vy = 0^\delta \left(1 - \frac{u}{U}\right) dy$$

Further, the integral on the right may be written down as $\delta \int_{\eta=0}^{1} \left(1 - \frac{u}{U}\right) d\eta$ where $\eta = \frac{y}{\delta}$ and the velocity ratio $\frac{u}{U}$ is represented as a function of η. Then the ratio of displacement thickness to the boundary layer thickness at the same station x is obtained as

$$\boxed{\frac{\delta_1}{\delta} = \int_0^1 \left(1 - \frac{u}{U}\right) d\eta} \tag{13.97}$$

Momentum Thickness

Now consider the rate at which momentum $\dot{p}$ crosses the station at x. This is obtained by integrating product of an elemental mass flow rate at y and the fluid velocity u as

$$\dot{p}_{\text{actual}} = \int_{y=0}^{\infty} \rho_\infty u^2 dy = \int_{y=0}^{\delta} \rho_\infty u^2 dy + \int_{y=\delta}^{\infty} \rho_\infty U^2 dy$$

Had there been no viscous effects, the rate at which momentum would cross section x would be given by the actual mass flow rate U product, i.e.,

$$\dot{p}_{\text{ideal}} = \int_{y=0}^{\infty} \rho_\infty u U dy = \int_{y=0}^{\delta} \rho_\infty u U dy + \int_{y=\delta}^{\infty} \rho_\infty U^2 dy$$

Again the deficit in the rate of momentum transfer across station x is obtained as

$$\Delta \dot{p} = \dot{p}_{\text{actual}}(x) - \dot{p}_{\text{ideal}}(x) = \int_{y=0}^{\delta} \rho_\infty u(u-U)dy \tag{13.98}$$

This momentum rate reduction may be visualized as a certain height of fluid layer δ_2—the momentum thickness[7] such that

$$\int_{y=0}^{\delta} \rho_\infty u(u-U)dy = -\rho_\infty U^2 \delta_2 \quad \text{or} \quad \delta_2 = \int_{y=0}^{\delta} \frac{u}{U}\left(1 - \frac{u}{U}\right) dy$$

Further, the integral on the right may be written down as $\delta \int_{\eta=0}^{1} \frac{u}{U}\left(1-\frac{u}{U}\right) d\eta$. Then, the ratio of momentum thickness to the boundary layer thickness at the same station x is obtained as

$$\boxed{\frac{\delta_2}{\delta} = \int_0^1 \frac{u}{U}\left(1 - \frac{u}{U}\right) d\eta} \tag{13.99}$$

With these length scales, the momentum integral Eq. 13.95 may be recast in the form

$$\boxed{\frac{d(U^2\delta_2)}{dx} + \delta_1 U \frac{dU}{dx} = \frac{\tau_{w,x}}{\rho_\infty}} \tag{13.100}$$

Special Case of Flat Plate Flow

In the case of flat plate flow, the free stream velocity is independent of x and the second term on the left-hand side of Eq. 13.100 vanishes. Also the velocity U is replaced by U_∞ to get the momentum integral for flat plate flow as

$$\frac{d}{dx}\int_0^{\delta} u(u - U_\infty)dy = -\frac{\tau_{w,x}}{\rho_\infty} \tag{13.101}$$

This may be rewritten as

[7] In fluid mechanics literature, the momentum thickness is represented by the symbol θ. This symbol is avoided here since θ is reserved the temperature.

$$\boxed{\frac{d\delta_2}{dx} = \frac{\tau_{w,x}}{\rho_\infty U_\infty^2} = \frac{C_{f,x}}{2}} \tag{13.102}$$

Integral Form of Boundary Layer Energy Equation

Consider now the boundary layer energy equation given by Eq. 13.12. We integrate this equation between the wall $y = 0$ and $y = \delta_T(x)$, the thermal boundary layer thickness at a location x along the surface.

$$\int_0^{\delta_T} u\frac{\partial T}{\partial x}dy + \int_0^{\delta_T} v\frac{\partial T}{\partial y}dy = \int_0^{\delta_T} \alpha_\infty \frac{\partial^2 T}{\partial y^2}dy \tag{13.103}$$

The second term on the LHS of this equation may be rewritten as

$$\int_0^{\delta_T} v\frac{\partial T}{\partial y}dy = \int_0^{\delta_T} \frac{\partial vT}{\partial y}dy - \int_0^{\delta_T} T\frac{\partial v}{\partial y}dy = \int_0^{\delta_T} \frac{\partial vT}{\partial y}dy + \int_0^{\delta_T} T\frac{\partial u}{\partial x}dy$$

where the last operation has made use of the equation of continuity. Further, we perform the indicated integration to get

$$\int_0^{\delta_T} \frac{\partial vT}{\partial y}dy = (vT)|_0^{\delta_T} = v_{\delta_T}T_\infty = -T_\infty \int_0^{\delta_T} \frac{\partial u}{\partial x}dy$$

where v_{δ_T} has been obtained by integrating the continuity equation across the thermal boundary layer.

Introducing the above, the LHS of Eq. 13.103 may be written as

$$\begin{aligned} \text{LHS} &= \int_0^{\delta_T} u\frac{\partial T}{\partial x}dy + \int_0^{\delta_T} v\frac{\partial T}{\partial y}dy = \int_0^{\delta_T} u\frac{\partial T}{\partial x}dy + \int_0^{\delta_T} T\frac{\partial u}{\partial x}dy - T_\infty \int_0^{\delta_T} \frac{\partial u}{\partial x}dy \\ &= \int_0^{\delta_T} \frac{\partial uT}{\partial x}dy - T_\infty \int_0^{\delta_T} \frac{\partial u}{\partial x}dy = \int_0^{\delta_T} \frac{\partial u(T - T_\infty)}{\partial x}dy = \frac{d}{dx}\int_0^{\delta_T} u(T - T_\infty)dy \end{aligned} \tag{13.104}$$

where the very last step has made use of the Leibnitz rule. The indicated integration is easily performed on the RHS of Eq. 13.103 to get

$$\alpha_\infty \int_0^{\delta_T} \frac{\partial^2 T}{\partial y^2} dy = \alpha_\infty \frac{\partial T}{\partial y}\bigg|_0^{\delta_T} = -\alpha_\infty \frac{\partial T}{\partial y}\bigg|_0 = \frac{q_{w,x}}{\rho_\infty C_{p\infty}} \tag{13.105}$$

Combining the above two equations, we finally obtain the boundary layer energy integral equation as

$$\boxed{\frac{d}{dx}\int_0^{\delta_T} u(T - T_\infty)dy = \frac{q_{w,x}}{\rho_\infty C_{p\infty}}} \tag{13.106}$$

13.4.2 *Approximate Solution for Boundary Layer Flow Past a Flat Plate Using a Polynomial Profile for Velocity*

We have already used the integral method to obtain approximate solution in the case of transient conduction in one dimension (see Sect. 5.2.2 of Chap. 5). The procedure employed here is similar to the one used there. The method of solution using the integral method in the case of viscous boundary layer is known as the von Karman Pohlhausen method.[8]

We shall assume that the velocity profile is given in the form of a polynomial of form

$$\frac{u}{U_\infty} = \overline{u} = f\left(\frac{y}{\delta}\right) = \sum_{i=0}^{n} a_i \left(\frac{y}{\delta}\right)^i = f(\eta) = \sum_{i=0}^{n} a_i \eta^i \tag{13.107}$$

where $\overline{u} = \frac{u}{U_\infty}$ and $\eta = \frac{y}{\delta}$. There are $(n+1)$ constants a_i that characterize the velocity profile in the boundary layer that need to be determined, apart from the thickness of the boundary layer $\delta(x)$. As in the Goodman's method, the last mentioned quantity has to be obtained by using the momentum integral Eq. 13.101. The constants are determined by using the available boundary conditions on velocity, and other auxiliary conditions that are to be specified as will be done below.

Let us take, as an example, a quartic profile (fourth degree polynomial) for the velocity with $n = 4$ in Eq. 13.107. Since the velocity has to vanish at $y = 0$, or correspondingly at $\eta = 0$, $a_0 = 0$. At $y = \delta$ or $\eta = 1$, the velocity is equal to the free stream velocity U_∞ or $\overline{u} = 1$. Hence, we get

---

[8]Th. von Karman and K. Pohlhausen published two independent papers in the same issue of the Journal Z. Angew. Math. Mech. which form the basis for the method which is named after them. They are 1. Th. von Karman, Über laminare und turbulente reiburg, Z. Angew. Math. Mech., Vol. 1, p. 233, 1921 and 2. K. Pohlhausen, Zur integration der differential gleichung der laminare grenzschichten, Vol. 1, p. 252, 1921.

$$a_1 + a_2 + a_3 + a_4 = 1 \tag{13.108}$$

The available boundary conditions have been exhausted. Hence, it is necessary to impose three more *auxiliary* conditions so that all the coefficients are determined. The first auxiliary condition is obtained by assuming that the boundary layer momentum Eq. 13.16 is satisfied at $y = 0$. This requires that $\frac{\partial^2 u}{\partial y^2} = 0$ at $y = 0$ since $u = v = 0$ there. It also means that $\frac{d^2\overline{u}}{d\eta^2} = 0$ at $\eta = 0$. From Eq. 13.107 with $n = 4$, we have

$$\frac{d^2\overline{u}}{d\eta^2} = 2a_2 + 6a_3\eta + 12a_4\eta^2$$

Putting $\eta = 0$ this simply means that $a_2 = 0$. Hence, Eq. 13.108 takes the from

$$a_1 + a_3 + a_4 = 1 \tag{13.109}$$

The other two auxiliary conditions are imposed at $y = \delta$ or $\eta = 1$ by requiring that the velocity profile merges smoothly with the free stream velocity there. Thus, we require that the first and second derivative of u with respect to y (i.e., first and second derivatives of $\overline{u}$ with respect to η) be zero at the edge of the boundary layer, i.e., at $\eta = 1$. These conditions require that

$$(a)\ a_1 + 3a_3 + 4a_4 = 0; \quad (b)\ 6a_3 + 12a_4 = 0 \tag{13.110}$$

From Eq. 13.110(b), we have $a_3 = -2a_4$. Eliminate a_1 from Eqs. 13.109 and 13.110 to get $-2a_3 - 3a_4 = 1$. This may be rewritten as $-2a_3 - 3a_4 = -2(-2a_4) - 3a_4 = a_4 = 1$. Thus, we have

$$\begin{aligned} a_0 = 0; \quad a_1 &= -3a_3 - 4a_4 = -3(-2a_4) - 4a_4 = 2a_4 = 2; \\ a_2 = 0; \quad a_3 &= -2a_4 = -2 \quad \text{and} \quad a_4 = 1 \end{aligned} \tag{13.111}$$

Introduce all these in Eq. 13.107 to get

$$\boxed{\overline{u} = 2\eta - 2\eta^3 + \eta^4} \tag{13.112}$$

Having obtained all the coefficients defining the polynomial, it is now necessary to obtain $\delta(x)$. We do this by using the momentum integral equation. The momentum thickness is written down using Eq. 13.99 as

$$\delta_2 = \delta \int_0^1 \left[2\eta - 2\eta^3 + \eta^4\right]\left[1 - \left(2\eta - 2\eta^3 + \eta^4\right)\right] d\eta$$

$$= \delta \int_0^1 \left[2\eta - 2\eta^3 + \eta^4 - 4\eta^2 - 4\eta^6 - \eta^8 + 8\eta^4 + 4\eta^7 - 4\eta^5\right] d\eta$$

$$= \delta \left[\eta^2 - 2\frac{\eta^4}{4} + \frac{\eta^5}{5} - 4\frac{\eta^3}{3} - 4\frac{\eta^7}{7} - \frac{\eta^9}{9} + 8\frac{\eta^5}{5} + 4\frac{\eta^8}{8} - 4\frac{\eta^6}{6}\right]\Bigg|_0^1$$

$$= \delta \left[1 - \frac{2}{4} + \frac{1}{5} - \frac{4}{3} - \frac{4}{7} - \frac{1}{9} + \frac{8}{5} + \frac{4}{8} - \frac{4}{6}\right] = \frac{37}{315}\delta$$

The shear stress at the wall is given by

$$\tau_{w,x} = \mu_\infty \frac{\partial u}{\partial y}\bigg|_{y=0} = \frac{\mu_\infty U_\infty}{\delta} \frac{d\left(\frac{u}{U_\infty}\right)}{d\eta}\Bigg|_{\eta=0}$$

Using the non-dimensional velocity profile given by Eq. 13.112, we have

$$\frac{d\left(\frac{u}{U_\infty}\right)}{d\eta}\Bigg|_{\eta=0} = \left[2 - 6\eta^2 + 4\eta^3\right]\Big|_{\eta=0} = 2$$

Hence, we have

$$\frac{\tau_{w,x}}{\rho_\infty U_\infty^2} = \frac{2\mu_\infty}{\rho_\infty U_\infty \delta}$$

Momentum integral Eq. 13.102 will then take the form

$$\boxed{\frac{37}{315}\frac{d\delta}{dx} = \frac{2\nu_\infty}{U_\infty \delta}} \tag{13.113}$$

This is a first-order ordinary differential equation for the boundary layer thickness $\delta(x)$. We supply an initial condition by *assuming* that the boundary layer thickness is zero at $x = 0$.

Equation 13.113 is integrated with respect to x to get

$$\int_0^\delta \delta d\delta = \frac{2 \times 315\nu_\infty}{37U_\infty} \int_0^x dx \quad \text{or} \quad \frac{\delta^2}{2} = \frac{2 \times 315\nu_\infty x}{37U_\infty}$$
$$\text{or} \quad \boxed{\delta = 2 \times \sqrt{\frac{315}{37}}\sqrt{\frac{\nu_\infty x}{U_\infty}} = 5.836\sqrt{\frac{\nu_\infty x}{U_\infty}}} \tag{13.114}$$

The functional form of variation of boundary layer thickness, thus, is in agreement with the similarity solution obtained earlier. The numerical coefficient is not the same as that given by the Blasius profile, it being a consequence of the assumed velocity profile within the boundary layer. We may calculate the friction coefficient as

$$\frac{2\tau_{w,x}}{\rho_\infty U_\infty^2} = \frac{4\mu_\infty}{\rho_\infty U_\infty \delta} = \frac{4\mu_\infty}{\rho_\infty U_\infty \times 5.836\sqrt{\frac{\nu_\infty x}{U_\infty}}} = \boxed{\frac{0.686}{\sqrt{Re_x}}} \tag{13.115}$$

This is in excess of the exact Blasius value (Eq. 13.38) by about 3%!

Example 13.6

Consider laminar flow of a viscous fluid parallel to a flat plate. Obtain the friction coefficient using the integral method and a cubic velocity profile inside the boundary layer.

Solution:

Step 1 We make use of the notation in the text and seek the velocity profile as $\frac{u}{U_\infty} = A + B\eta + C\eta^2 + D\eta^3$. For satisfying the no slip at the wall A is set to zero. The requirement that the profile satisfy the governing equation at the wall requires C to be taken as zero. Hence, the profile is chosen as $\frac{u}{U} = B\eta + D\eta^3$. Use smoothness condition at $\eta = 1$ to obtain $B + 3D = 0$ or $B = -3D$. Using the velocity condition at $\eta = 1$ we have $B + D = -3D + D = -2D = 1$ or $D = -\frac{1}{2}$. We then have $B = -3D = \frac{3}{2}$. The cubic profile for velocity is thus given by

$$\frac{u}{U_\infty} = \frac{3}{2}\eta - \frac{1}{2}\eta^3$$

Step 2 In order to obtain the boundary layer thickness based on the integral method, we obtain $\frac{\delta_2}{\delta}$ as the integral below (reader is encouraged to verify this).

$$\frac{\delta_2}{\delta} = \int_0^1 \left(\frac{3}{2}\eta - \frac{1}{2}\eta^3\right)\left(1 - \frac{3}{2}\eta + \frac{1}{2}\eta^3\right) d\eta = \frac{39}{280}$$

Step 3 From the cubic velocity profile, the shear stress at the wall is obtained as $\tau_w(x) = \frac{3\mu_\infty}{2\delta}$.

Step 4 Using the results in the above two steps, we make use of the momentum integral, integrate with respect to x and simplify to obtain the boundary layer thickness as

$$\delta = \sqrt{\frac{280}{13}}\frac{x}{\sqrt{Re_x}} = 4.641\frac{x}{\sqrt{Re_x}}$$

Step 5 The friction coefficient is then obtained as

$$C_{f,x} = \frac{\tau_w(x)}{\rho_\infty U_\infty^2} = \frac{3\mu_\infty}{2\delta} = \frac{3\mu_\infty}{24.641\frac{x}{\sqrt{Re_x}}} = \frac{0.323}{\sqrt{Re_x}}$$

Step 6 Comparing with the exact value due to Blasius, the above is in error by

$$\text{Error} = \frac{C_{f,x}(\text{Approximate}) - C_{f,x}(\text{Exact})}{C_{f,x}(\text{Exact})} \times 100 = \frac{0.323 - 0.332}{0.332} \times 100 = -2.64\%$$

13.4.3 Approximate Solution for Boundary Layer Temperature Profile for Flow Past a Flat Plate Using a Polynomial Profile for Temperature

We have seen earlier that the velocity and temperature profiles, expressed in suitable non-dimensional form, are the same for a fluid with Prandtl number equal to unity. Hence, we expect to represent the temperature in the thermal boundary using a polynomial similar to the one used in the case of the velocity boundary layer. However, there are two possibilities. If the Prandtl number is greater than one, the velocity boundary layer thickness is larger than the thermal boundary layer thickness. If the Prandtl number is less than one, the velocity boundary layer thickness is smaller than the thermal boundary layer thickness. As the more important case, we look at the $Pr > 1$ case. We represent the ratio of these as

$$\Delta = \frac{\delta_T}{\delta} \tag{13.116}$$

where δ_T is the thermal boundary layer thickness. Note that $\Delta < 1$ in what follows. We define the non-dimensional temperature θ as

$$\theta = \frac{T - T_w}{T_\infty - T_w} \tag{13.117}$$

Hence, the boundary conditions satisfied by the non-dimensional temperature are

$$T = T_w \text{ or } \theta = 0 \text{ at } y = 0; \quad T = T_\infty \text{ or } \theta = 1 \text{ at } y = \delta_T \tag{13.118}$$

Thus, the boundary conditions are identical to those satisfied by $\frac{u}{U_\infty}$.

We now consider approximating the temperature profile within the thermal boundary layer by a polynomial of fourth degree. As in the case of the velocity boundary layer, this requires, apart from the two boundary conditions given by Eq. 13.118, auxiliary conditions.

The auxiliary conditions that may be used are two smoothness conditions at $y = \delta_T$ and the condition that the boundary layer energy equation (Eq. 13.12) is exactly satisfied at $y = 0$. Since both velocity components are zero at the wall, the energy equation is satisfied with $\left.\frac{\partial^2\theta}{\partial y^2}\right|_{y=0} = 0$. Further if we approximate the temperature profile by a polynomial of fourth degree in terms of $\zeta = \frac{y}{\delta_T}$, the profile is identical to the velocity profile Eq. 13.112 except that η there is replaced by ζ. Thus, the appropriate temperature profile is

$$\boxed{\theta = 2\zeta - 2\zeta^3 + \zeta^4} \tag{13.119}$$

The energy integral Eq. 13.106 requires the evaluation of the integral given by

$$E = \int_0^{\delta_T} u(T - T_\infty)dy$$

Introducing θ defined through Eq. 13.117, we have

$$T - T_\infty = (T - T_w) - (T_\infty - T_w) = (T_\infty - T_w)\left[\frac{T - T_w}{T_\infty - T_w} - 1\right] = (T_\infty - T_w)(\theta - 1)$$

Then, we have

$$E = U_\infty(T_\infty - T_w)\delta_T \int_0^1 \frac{u}{U_\infty}(\theta - 1)d\zeta \tag{13.120}$$

The upper limit of integration is limited to the thermal boundary layer thickness since the integrand vanishes beyond it ($\theta = 1$ for $y > \delta_T$). Note also that the velocity profile should be expressed as a function of ζ in order to evaluate the integral. Since the thermal boundary layer thickness is smaller than the velocity boundary layer thickness, the velocity profile is written as

$$\frac{u}{U_\infty} = 2\Delta\zeta - 2\Delta^3\zeta^3 + \Delta^4\zeta^4 \tag{13.121}$$

which is obtained by substituting $\eta = \frac{y}{\delta} = \frac{y}{\delta_T}\frac{\delta_T}{\delta} = \zeta\Delta$ in Eq. 13.112. Introducing Eqs. 13.119 and 13.121 into the energy Eq. 13.120, we have

$$\frac{E}{U_\infty(T_\infty - T_w)\delta_T} = \int_0^1 (2\Delta\zeta - 2\Delta^3\zeta^3 + \Delta^4\zeta^4)(2\zeta - 2\zeta^3 + \zeta^4 - 1)d\zeta$$

The above integration may easily be performed to get

$$\frac{E}{U_\infty(T_\infty - T_w)\delta_T} = -\left[\frac{2}{15}\Delta - \frac{3}{140}\Delta^3 + \frac{1}{180}\Delta^4\right] \tag{13.122}$$

The heat flux at the wall may be calculated by using Fourier law as

$$q_{w,x} = -k_\infty \frac{\partial T}{\partial y}\bigg|_{y=0} = -k_\infty \frac{T_\infty - T_w}{\delta_T}\frac{d\theta}{d\zeta}\bigg|_{\zeta=0} = -2k_\infty \frac{T_\infty - T_w}{\delta_T} \tag{13.123}$$

Noting that $\Delta < 1$ we retain only the linear term in Eq. 13.122 and write the energy integral as

$$\frac{2}{15}\Delta\delta_T \frac{d\delta_T}{dx} = 2\frac{k_\infty}{\rho_\infty C_{p\infty}}$$

This equation may be rewritten using the already known velocity boundary layer thickness δ as

$$\frac{2}{15}\Delta^3\delta\frac{d\delta}{dx} = 2\frac{k_\infty}{\rho_\infty C_{p\infty}}$$

From Eq. 13.114, we have $\delta\frac{d\delta}{dx} = \frac{630}{37}\frac{\nu_\infty}{U_\infty}$. We may thus obtain Δ as

$$\boxed{\Delta = \left[\frac{15 \times 37}{630 \times Pr}\right]^{\frac{1}{3}} = \frac{0.959}{Pr^{\frac{1}{3}}}} \tag{13.124}$$

It is left as an exercise to the reader, using Eqs. 13.114,13.123, and 13.124, to show that the Nusselt number is given by

$$\boxed{Nu_x = 0.357 Re_x^{\frac{1}{2}} Pr^{\frac{1}{3}}} \tag{13.125}$$

The integral solution thus shows that the Prandtl number dependence is through $\frac{1}{3}$rd power.[9] This is in agreement with the behavior shown by the exact solution also. However, on comparison with the exact value given by Eq. 13.53, the coefficient is in error by around +8%.

[9]Note that the $\frac{1}{3}$rd power dependence is an approximate representation of the solution to the energy boundary layer equation. In fact, if we take into account all the terms in the energy integral Eq. 13.122, we may solve for Δ for a given Pr by the Newton Raphson method. The resulting relationship between Δ and Pr is well represented by the relation $\Delta \approx \frac{0.998}{Pr^{0.35}}$.

13.4.4 *Integral Method Applied to Boundary Layer Flow with Axial Pressure Gradient*

Flow Problem

We consider now wedge flow by the application of integral method, using a suitable velocity profile within the boundary layer. We have seen in Sect. 13.4.2 that the velocity profile for flow past a flat plate is represented by a quartic equation given by Eq. 13.112. For flow over a wedge, we shall assume that the appropriate velocity profile is a sum of two polynomials, the first one given by Eq. 13.112 corresponding to flow with zero axial pressure gradient and the second one represented as $\overline{v}(\eta)$ such that

$$\boxed{\overline{u} = \left[2\eta - 2\eta^3 + \eta^4\right] + A\Lambda\overline{v}(\eta)} \tag{13.126}$$

where A is a numerical constant and Λ is a factor that will be chosen as the analysis proceeds, by relating it to the axial pressure gradient. This velocity profile has to satisfy the boundary and auxiliary conditions that are considered now. At $\eta = 0$ the velocity vanishes, and assuming that $A \neq 0$ and $\Lambda \neq 0$, $\overline{v}$ also should vanish at $\eta = 0$. At the edge of the boundary layer, $\overline{u} = 1$ and hence $\overline{v} = 0$. We assume that the boundary layer equation is satisfied by the velocity profile at the wedge surface. Thus, we should satisfy

$$0 = U\frac{dU}{dx} + \nu_\infty \frac{U}{\delta^2}\frac{d^2\overline{u}}{d\eta^2}\bigg|_{\eta=0} = U\frac{dU}{dx} + \nu_\infty \frac{U}{\delta^2}A\Lambda\frac{d^2\overline{v}}{d\eta^2}\bigg|_{\eta=0} \tag{13.127}$$

The last step results from the observation that the flat plate profile gives zero second derivative at $\eta = 0$. Consider now a possible candidate polynomial $\overline{v} = \eta(1-\eta)^3$. This vanishes at both $\eta = 0$ and $\eta = 1$. The second derivative of $\overline{v}$ with respect to η is obtained as $\frac{d^2\overline{v}}{d\eta^2} = -6(1-\eta)^2 + 6\eta(1-\eta)$, and has a value of -6 at $\eta = 0$. Hence, Eq. 13.127 becomes

$$-\frac{\delta^2}{\nu_\infty}\frac{dU}{dx} = -6A\Lambda \tag{13.128}$$

We choose conveniently a value of $A = \frac{1}{6}$ and $\Lambda = \frac{\delta^2}{\nu_\infty}\frac{dU}{dx}$ to satisfy all the conditions listed above. The required velocity profile is thus given by

$$\boxed{\overline{u} = \left[2\eta - 2\eta^3 + \eta^4\right] + \frac{\Lambda}{6}\left[\eta(1-\eta)^3\right]} \tag{13.129}$$

The parameter Λ is referred to as the shape parameter since it brings in the axial pressure gradient that is dependent on the shape of the body.

Specific Cases

Flat plate profile:
We know that $\Lambda = 0$ corresponds to flow past a flat plate. This case has been considered in detail earlier.

Separating profile:
A separating profile is obtained when $\frac{d\bar{u}}{d\eta} = 0$ at $\eta = 0$. Using the profile given above, we see that

$$\left.\frac{d\bar{u}}{d\eta}\right|_{\eta=0} = \left.\left[2 - 6\eta^2 + 4\eta^3 + \frac{\Lambda}{6}\left\{(1-\eta)^3 + 3\eta(1-\eta)^2\right\}\right]\right|_{\eta=0} = 2 + \frac{\Lambda}{6} = 0$$

Hence, we obtain a value of $\Lambda = -12$ for a separating profile.

Stagnation point flow:
We know that the velocity outside the boundary layer follows the linear law $U(x) = Kx$. The Falkner–Skan solution has indicated that the similarity variable is simply a scaled y. Also, we have seen that the boundary layer thickness is constant. Consequently the momentum and displacement thicknesses are also constant. Let $\frac{\delta_1}{\delta} = r_1$ and $\frac{\delta_2}{\delta} = r_2$, each being a constant dependent on the parameter Λ. We also see that the parameter $\Lambda = \frac{\delta^2}{\nu_\infty}\frac{dU}{dx} = \frac{K\delta^2}{\nu_\infty}$ must be a constant. Let this be Λ_0. By the use of expression 13.129 for the boundary layer velocity, we may easily show that

$$(a)\ r_1(\Lambda) = \frac{3}{10} + \frac{\Lambda}{120}; \quad (b)\ r_2(\Lambda) = \frac{37}{315} - \frac{\Lambda}{945} - \frac{\Lambda^2}{9072} \tag{13.130}$$

Now we look at the terms appearing in the momentum integral (Eq. 13.100). With $U = Kx$, we have

$$U\frac{dU}{dx} = Kx\frac{d(Kx)}{dx} = K^2x$$

Then, we have

$$\delta_1 U\frac{dU}{dx} = K^2 x r_1 \delta$$

Also

$$\frac{d(U^2\delta_2)}{dx} = \frac{d(Kx^2\delta_2)}{dx} = 2Kx\delta_2 = 2Kxr_2\delta$$

since δ_2 is independent of x. The wall shear stress term is

$$\frac{\tau_{w,x}}{\rho_\infty} = \left.\frac{\nu_\infty U}{\delta}\frac{d\bar{u}}{d\eta}\right|_{\eta=0} = \frac{Kx\nu_\infty}{\delta}\left(2 + \frac{\Lambda}{6}\right)$$

Substituting these in the momentum integral and simplifying we get the algebraic equation

$$2\Lambda r_2(\Lambda) + \Lambda r_1(\Lambda) = 2 + \frac{\Lambda}{6} \tag{13.131}$$

It is easily verified that $\Lambda_0 = 7.052$ satisfies this algebraic equation. The boundary layer thickness then works out to

$$\delta_0 = \sqrt{\frac{\Lambda_0 \nu_\infty}{K}} = \sqrt{\frac{7.052\nu_\infty}{K}} = 2.656\sqrt{\frac{\nu_\infty}{K}} \tag{13.132}$$

The shear stress at the wall divided by density is given by

$$\frac{\tau_{w,x}}{\rho_\infty} = \frac{Kx\nu_\infty}{\delta}\left(2 + \frac{\Lambda_0}{6}\right) = \frac{Kx\nu_\infty}{2.656\sqrt{\frac{\nu_\infty}{K}}}\left(2 + \frac{7.052}{6}\right) = 1.1955\sqrt{\nu_\infty K}Kx$$

The local friction coefficient may then be obtained as

$$\boxed{\frac{\tau_w}{\frac{1}{2}\rho U^2} = \frac{1.1955\sqrt{\nu_\infty K}Kx}{\frac{1}{2}K^2x^2} = \frac{2.3911}{x}\sqrt{\frac{\nu_\infty}{K}}} \tag{13.133}$$

Based on the Falkner–Skan solution with $\beta = 1$, picking the value of $\frac{d^2 f}{df^2} = 1.232589$ from table in Fig. 13.8, the corresponding exact expression for the friction coefficient has the numerical constant 2.3911 replaced by 2.4652. The approximate value is in error with respect to the exact value by about -3%.

13.4.5 *Thwaites's Method*

Flow Problem

It is possible to recast the integral method of solution in terms of profile independent solution using Thwaite's method.[10] The method is based on available exact solutions to the wedge problem for various wedge angles and deriving the exact values for the displacement and momentum thicknesses based on the exact solutions. From these, it is observed that the solutions can be represented in terms of universal functions involving the parameter λ, (introduced in the next paragraph) which plays the role of a shape parameter. Thwaites's method is considered in detail now.

Introduce the quantity $\lambda = \frac{\delta_2^2}{\nu_\infty}\frac{dU}{dx}$. We see that this is related to Λ through the relation $\lambda = r_2^2\Lambda$. The momentum integral Eq. 13.100 is written in the form

[10] N.Curle, *The Laminar Boundary Layer Equations*, Oxford at the Clarendon Press, 1962.

$$\frac{\tau_{w,x}\delta_2}{\mu_\infty U} = \frac{U\delta_2}{\nu_\infty}\frac{d\delta_2}{dx} + \lambda\left(2 + \frac{\delta_1}{\delta_2}\right)$$

or

$$l(\lambda) = \frac{U\delta_2}{\nu_\infty}\frac{d\delta_2}{dx} + \lambda(H + 2)$$

where $l(\lambda) = \frac{\tau_{w,x}\delta_2}{\mu_\infty U}$ and $H(\lambda) = \frac{\delta_1}{\delta_2}$. The last equation may be rearranged as

$$\frac{U}{\nu_\infty}\frac{d\delta_2^2}{dx} = 2[l - \lambda(H + 2)] = L(\lambda)$$

Thwaites observed that the right-hand side in the above equation is very nearly linear and is well approximated by $L(\lambda) = 0.45 - 6\lambda$. This observation was based on all the available exact solutions. Thus, we have the equation

$$\frac{U}{\nu_\infty}\frac{d\delta_2^2}{dx} = 0.45 - 6\frac{\delta_2^2}{\nu_\infty}\frac{dU}{dx}$$

This equation may be rearranged as

$$\frac{U}{\nu_\infty}\frac{d\delta_2^2}{dx} + 6\frac{\delta_2^2}{\nu_\infty}\frac{dU}{dx} = 0.45$$

On multiplication by U^5 the left-hand side becomes $\frac{1}{\nu_\infty}\frac{d(U^6\delta_2^2)}{dx}$. Hence, this equation is analytically integrated to get

$$U^6\delta_2^2 = 0.45\nu_\infty\int_0^x U^5 dx + A$$

where A is a constant of integration. The constant of integration may be taken as zero since U is zero at $x = 0$ if the body is bluff and $\delta_2 = 0$ if the body has a sharp edge at $x = 0$. Thus, the approximate solution according to Thwaites's method reduces the problem to a simple quadrature given by

$$\boxed{\delta_2^2 = \frac{0.45\nu_\infty}{U^6}\int_0^x U^5 dx} \tag{13.134}$$

Once the momentum thickness is obtained from the above expression, the other pertinent quantities are obtained as below.

Table 13.2 Approximate expressions for $H(\lambda)$ and $l(\lambda)$

(a) $0 \le \lambda \le 0.25$	
$H(\lambda) = 2.61 - 3.75\lambda + 5.24\lambda^2$	$l(\lambda) = 0.22 + 1.57\lambda - 1.8\lambda^2$
(b) $-0.1 \le \lambda \le 0$	
$H(\lambda) = 2.088 + \frac{0.0731}{\lambda+0.14}$	$l(\lambda) = 0.22 + 1.402\lambda + \frac{0.018\lambda}{\lambda+0.107}$

The functions $l(\lambda)$ and $H(\lambda)$ are universal functions and are tabulated in the reference cited earlier. These are very well approximated by the following expressions: From the definition of $l(\lambda)$, we also obtain

$$C_f = \frac{\tau_w}{\frac{1}{2}\rho_\infty U^2} = \frac{\mu_\infty l(\lambda)}{\frac{1}{2}\rho_\infty U\delta_2} = \frac{2\nu_\infty l(\lambda)}{U\delta_2} \tag{13.135}$$

Table 13.3 presents the universal functions and compares them with the approximate values obtained using expressions in Table 13.2.

Example 13.7

Obtain the solution to the stagnation flow case using Thwaites's method. Compare the results with those obtained using the Pohlhausen method as well as the exact solution.

Solution:

Step 1 The velocity at the edge of the boundary layer is given by $U = Kx$. Introducing this in Eq. 13.134, we have

$$\delta_2^2 = \frac{0.45\nu_\infty}{K^6x^6}\int_0^x K^5x^5dx = \frac{0.45\nu_\infty}{6K} \text{ or } \delta_2 = \sqrt{\frac{0.45}{6}}\sqrt{\frac{\nu_\infty}{K}} = 0.274\sqrt{\frac{\nu_\infty}{K}}$$

Step 2 Using this and by the definition of λ, we have

$$\lambda = \frac{\delta_2^2}{\nu_\infty}\frac{dU}{dx} = \frac{0.45\nu_\infty}{6K}\frac{K}{\nu_\infty} = 0.075$$

Step 3 Corresponding to this λ, the function $l(\lambda)$ is calculated as

$$l(\lambda) = 2.61 + 1.57 \times 0.075 - 1.8 \times 0.075^2 = 0.328$$

Step 4 The skin friction coefficient is then given by

Table 13.3 Universal functions for use with Thwaites's method

λ	$l(\lambda)$	$l(\lambda)^{\dagger}$	$H(\lambda)$	$H(\lambda)^{\dagger}$	$L(\lambda)$	$L(\lambda)^{\ddagger}$
0.250	0.500	0.500	2.00	2.00	−1.000	−1.050
0.200	0.463	0.462	2.07	2.07	−0.702	3.750
0.140	0.404	0.405	2.18	2.19	−0.362	−0.390
0.120	0.382	0.382	2.23	2.24	−0.251	−0.270
0.100	0.359	0.359	2.28	2.29	−0.138	−0.150
0.080	0.333	0.334	2.34	2.34	−0.028	−0.030
0.064	0.313	0.313	2.39	2.39	0.064	0.066
0.048	0.291	0.291	2.44	2.44	0.156	0.162
0.032	0.268	0.268	2.49	2.50	0.249	0.258
0.016	0.244	0.245	2.55	2.55	0.342	0.354
0.000	0.22	0.22	2.61	2.61	0.440	0.450
−0.016	0.195	0.194	2.67	2.68	0.539	0.546
−0.032	0.168	0.167	2.75	2.76	0.64	0.642
−0.04	0.153	0.153	2.81	2.82	0.691	0.690
−0.048	0.138	0.138	2.87	2.88	0.744	0.738
−0.056	0.122	0.122	2.94	2.96	0.797	0.786
−0.060	0.113	0.113	2.99	3.00	0.825	0.81
−0.064	0.104	0.103	3.04	3.05	0.853	0.834
−0.068	0.095	0.093	3.09	3.10	0.882	0.858
−0.072	0.085	0.082	3.15	3.16	0.912	0.882
−0.076	0.072	0.069	3.22	3.23	0.937	0.906
−0.08	0.056	0.055	3.3	3.31	0.960	0.930
−0.084	0.038	0.036	3.39	3.39	0.982	0.954
−0.086	0.027	0.026	3.44	3.44	0.99	0.966
−0.088	0.015	0.013	3.49	3.49	0.996	0.978
−0.090	0	−0.001	3.55	3.55	0.999	0.990

†—Based on expressions in Table 13.2, ‡—Thwaites

$$C_{f,x} = \frac{2\nu_\infty l}{Kx\delta_2} = \frac{2\nu_\infty \times 0.328}{Kx0.274\sqrt{\frac{\nu_\infty}{K}}} = 2.3942\sqrt{\frac{\nu_\infty}{K}} \cdot \frac{1}{x}$$

Step 5 This may be compared with the exact value based on Falkner–Skan solution of $C_{f,x} = 2.4652\sqrt{\frac{\nu_\infty}{K}} \cdot \frac{1}{x}$ and the Pohlhausen value of $C_{f,x} = 2.3911\sqrt{\frac{\nu_\infty}{K}} \cdot \frac{1}{x}$. The approximate value obtained by Thwaite's method is very satisfactory.

Temperature Problem

We now turn our attention to the thermal boundary layer problem in the presence of an axial pressure gradient. Analogous to Thwaites's method we develop a solution method, starting with the energy equation in integral form given by Eq. 13.106. We shall introduce a thermal energy thickness δ_{1T} such that the energy crossing station x is given by

$$\int_0^{\delta_T} u(T - T_\infty)dy = U(T_w - T_\infty)\delta_{1T} \tag{13.136}$$

which may be rewritten in the form

$$\delta_{1T} = \int_0^{\delta_T} \left(\frac{u}{U}\right)\left(\frac{T - T_\infty}{T_w - T_\infty}\right) dy \tag{13.137}$$

With this definition, the thermal boundary layer energy integral Eq. 13.106 is written as

$$\frac{d}{dx}[(T_w - T_\infty)U\delta_{1T}] = \frac{q_{w,x}}{\rho_\infty C_{p\infty}} \tag{13.138}$$

We further introduce the following quantities:

$$\text{(a)}\quad \zeta_T(x) = \frac{\delta_{1T}^2}{\nu_\infty}U \quad \text{(b)}\quad \Gamma_T(x) = \frac{\delta_{1T}^2}{\nu_\infty}\frac{dU}{dx} \tag{13.139}$$

To be general, we shall assume that both the wall heat flux and wall temperature vary with x. Differentiate Eq. 13.139(a) with respect to x to get

$$\frac{d\zeta_T}{dx} = \frac{\delta_{1T}^2}{\nu_\infty}\frac{dU}{dx} + \frac{U}{\nu_\infty}\frac{d\delta_{1T}^2}{dx}$$

Using Eq. 13.139(b), this becomes

$$\frac{d\zeta_T}{dx} = \Gamma_T(x) + 2\frac{U\delta_{1T}}{\nu_\infty}\frac{d\delta_{1T}}{dx} \tag{13.140}$$

Expanding the left-hand side of Eq. 13.138, we have

$$\frac{d}{dx}[(T_w - T_\infty)U\delta_{1T}] = U(T_w - T_\infty)\frac{d\delta_{1T}}{dx} + \delta_{1T}(T_w - T_\infty)\frac{dU}{dx} + U\delta_{1T}\frac{dT_w}{dx}$$

Multiply this by $2\delta_{1T}$ and use Eqs. 13.138 and 13.139 to get

$$2U(T_w - T_\infty)\delta_{1T}\frac{d\delta_{1T}}{dx} + 2\nu_\infty(T_w - T_\infty)\Gamma_T + 2\nu_\infty\zeta_T\frac{dT_w}{dx} = \frac{2q_{w,x}\delta_{1T}}{\rho_\infty C_{p\infty}}$$

Using Eq. 13.140, the above equation is recast as

$$\frac{d\zeta_T}{dx} = \frac{2\delta_{1T}q_{w,x}}{\mu_\infty C_{p\infty}(T_w - T_\infty)} - \left[\Gamma_T + \frac{2\zeta_T\frac{dT_w}{dx}}{T_w - T_\infty}\right]$$

or

$$\frac{d\zeta_T}{dx} = \frac{2\delta_{1T}q_{w,x}}{\mu_\infty C_{p\infty}(T_w - T_\infty)} - \left[1 + \frac{2U}{\frac{dU}{dx}}\frac{\frac{dT_w}{dx}}{T_w - T_\infty}\right]\Gamma_T \tag{13.141}$$

Approximating the right-hand side of Eq. 13.141 by a linear relation of the form

$$\frac{d\zeta_T}{dx} = F_T(\Gamma_T) = a_T - b_T\Gamma_T \tag{13.142}$$

and using $\zeta_T = 0$ at $x = 0$, we get ζ_T (following the procedure that was used in Thwaites's method) as

$$\boxed{\zeta_T(x) = \frac{a_T}{U^{b_T}}\int_0^x U^{b_T}\,dx} \tag{13.143}$$

The two constants a_T and b_T depend on the Prandtl number of the fluid and on the nature of the boundary condition specified at the surface. These constants are determined by requiring that the solution be exact in the case of flow parallel to a flat plate and the stagnation point flow in two dimensions.

Consider, as an example, the constant wall temperature case. Equation 13.141 may be rearranged, after substituting $\frac{dT_w}{dx} = 0$ as

$$\frac{2\delta_{1T}q_{w,x}}{\mu_\infty C_{p\infty}(T_w - T_\infty)} = \frac{d\zeta_T}{dx} + \gamma_T \tag{13.144}$$

By definition of Prandtl number, we have

$$\mu_\infty C_{p\infty} = k_\infty Pr_\infty$$

With this, the left hand term of Eq. 13.144 may be written as

$$\frac{2\delta_{1T}q_{w,x}}{\mu_\infty C_{p\infty}(T_w - T_\infty)} = 2\delta_{1T}\frac{q_{w,x}}{Pr_\infty k_\infty(T_w - T_\infty)} = \frac{2\delta_{1T}}{Pr_\infty}\frac{h_x}{k_\infty} = \frac{2\delta_{1T}}{Pr_\infty}\frac{Nu_x}{x}$$

We may also write γ_T as $\zeta_T\frac{\frac{dU}{dx}}{U}$. With these, Eq. 13.144 may be recast as

$$Nu_x = \frac{x}{2\delta_{1T}}\left[\frac{d\zeta_t}{dx} + \zeta_T\frac{\frac{dU}{dx}}{U}\right]$$

This may further be simplified using Eq. 13.142 and by substituting $\sqrt{\frac{\zeta_T\nu_\infty}{U}}$ for δ_{1T} as

$$Nu_x = \frac{Pr_\infty x}{2}\sqrt{\frac{U}{\zeta_T\nu_\infty}}\left[a_T + \zeta_T(1-b_T)\frac{\frac{dU}{dx}}{U}\right] \tag{13.145}$$

Application to Wedge Problem

Consider the general wedge problem now. We know that the velocity U is given by $U = Kx^m$ where $m = \frac{\beta}{2-\beta}$. Hence, we also have $\frac{\frac{dU}{dx}}{U} = \frac{Kmx^{m-1}}{Kx^m} = \frac{m}{x}$. Using Eq. 13.143, we get

$$\zeta_T(x) = \frac{a_T}{(Kx^m)^{b_T}}\int_0^x (Kx^m)^{b_T}dx = \frac{a_T x}{(mb_T+1)}$$

Substitute this in Eq. 13.145 to get

$$\begin{aligned} Nu_{x,\beta} &= \frac{Pr_\infty x}{2}\sqrt{\frac{U(mb_T+1)}{a_T x\nu_\infty}}\left[a_T + \frac{a_T x}{(mb_T+1)}(1-b_T)\frac{m}{x}\right] \\ &= \frac{Pr_\infty}{2}\sqrt{Re_x a_T(mb_T+1)}\left[1+\frac{m(1-b_T)}{(mb_T+1)}\right] \end{aligned} \tag{13.146}$$

Flow Parallel to a Flat Plate

Consider now the case of flow parallel to a plate maintained at a uniform temperature with $U(x) = U_\infty$, a constant. This case corresponds to $m = 0$, and Eq. 13.146 simplifies to

$$Nu_{x,0} = \frac{Pr_\infty}{2}\sqrt{Re_x a_T}$$

In order to determine a_T, we require that the above agree with the heat transfer results of the Blasius solution. For example, for $Pr_\infty = 1$, we know that this should equal $0.332\sqrt{Re_x}$. Hence, we have

$$\frac{\sqrt{a_T}}{2} = 0.332 \text{ or } a_T = (2\times 0.332)^2 = 0.441 \tag{13.147}$$

Stagnation Point Flow

In order to determine b_T, we consider stagnation point flow. In this case, we have $\beta = 1$, $m = 1$ and hence Eq. 13.146 becomes

$$Nu_{x,1} = \frac{Pr_\infty}{2}\sqrt{Re_x a_T (b_T + 1)}\left[1 + \frac{(1 - b_T)}{(b_T + 1)}\right] = Pr_\infty \sqrt{\frac{a_T Re_x}{(1 + b_T)}}$$

This may be recast in the form

$$\frac{Nu_{x,\beta=1}}{Nu_{x,\beta=0}} = \frac{2}{\sqrt{1 + b_T}} \tag{13.148}$$

Using the Falkner–Skan solution, the above ratio has already been obtained and shown in Fig. 13.12. We thus have for $Pr_\infty = 1$

$$\frac{2}{\sqrt{1 + b_T}} = 1.71805 \text{ or } b_T = \left[\frac{2}{1.718025}\right]^2 - 1 \text{ or } b_T = 0.355 \tag{13.149}$$

With these, it is possible to recast Eq. 13.146 in the more useful form

$$\boxed{\frac{Nu_{x,\beta}}{Nu_{x,0}} = \sqrt{mb_T + 1}\left[1 + \frac{m(1 - b_T)}{(mb_T + 1)}\right]} \tag{13.150}$$

Schlichting [11] has tabulated the a_T, b_T values for various Prandtl numbers and for both the constant wall temperature and constant wall heat flux cases. The values derived above for specific cases agree with the tabulated values.

13.5 Cylinder in Cross Flow

13.5.1 Introduction

Heat transfer to a fluid flowing normal to the axis of a cylinder occurs in many interesting applications such as heat exchangers, pin fins in a heat sink, thermometer well in process temperature measurement, and so on. Flow normal to a cylinder is mathematically more complex than the flow across a flat plate. The fluid flow schematic is as shown in Fig. 13.1(b). The free stream approaches the cylinder with the condition specified by p_∞, ρ_∞, T_∞, and U_∞. In the case of incompressible flow, which we are considering, the density remains constant at ρ_∞. For a fluid with constant properties, these also represent the property values throughout the flow. In

[11] Table 9.2, p. 224, H. Schlichting and K. Gersten, *Boundary Layer Theory*, 8th Ed., Springer, 2000.

practice, however, it is usual to account for fluid property variations by taking the thermo-physical properties at the film temperature T_f taken equal to $\frac{T_w+T_\infty}{2}$ where T_w is the constant temperature specified on the surface of the cylinder. Hence, we shall use the subscript f to represent all the thermo-physical properties of the fluid.

Inviscid Flow Normal to a Cylinder

If the cylinder temperature is the same as the fluid free stream temperature, we need to only look at the flow problem. Because of the body shape the pressure and velocity both change with position. Based on inviscid flow theory, we may write the stream function as (see Appendix C, Eq. C.22)

$$\psi(r,\theta) = c\sin\theta\left[r - \frac{a^2}{r}\right] \tag{13.151}$$

where we have replaced v there with ψ to represent the stream function. Note that a represents the radius of the cylinder or $\frac{D}{2}$ where D is the cylinder diameter. On the surface of the cylinder, the velocity is tangential to the cylinder and hence is equal to the component u_θ. This is obtained as

$$u_\theta(a,\theta) = \left.\frac{\partial\psi}{\partial r}\right|_{r=a} = c\sin\theta\left.\left[1 + \frac{a^2}{r^2}\right]\right|_{r=a} = 2c\sin\theta \tag{13.152}$$

This velocity vanishes at both $\theta = 0$ and $\theta = \pi$ corresponding, respectively, to the forward and backward stagnation points. The velocity becomes a maximum equal to $2c$ at $\theta = \frac{\pi}{2}$. Consider the radial velocity now. It is given by

$$u_r(r,\theta) = -\frac{1}{r}\frac{\partial\psi}{\partial\theta} = -c\cos\theta\left[1 - \frac{a^2}{r^2}\right] \rightarrow -c \text{ as } r \rightarrow \infty \text{ for } \theta = 0$$

Hence, we may set the constant c as U_∞. With this, the velocity variation along the surface of the cylinder becomes

$$u_\theta(a,\theta) = 2U_\infty\sin\theta \tag{13.153}$$

Let us now look at the pressure distribution on the surface of the cylinder. For an inviscid fluid, Bernoulli equation is valid which states that

$$p + \frac{1}{2}\rho V^2 = p_0 = \text{constant}$$

The constant p_0 is referred to as the stagnation pressure since this is the pressure at the stagnation point where the velocity is zero. V represents the magnitude of the velocity vector. In the present case, at any θ and on the surface of the cylinder $V = u_\theta(a,\theta) = 2U_\infty\sin\theta$. Hence, we have

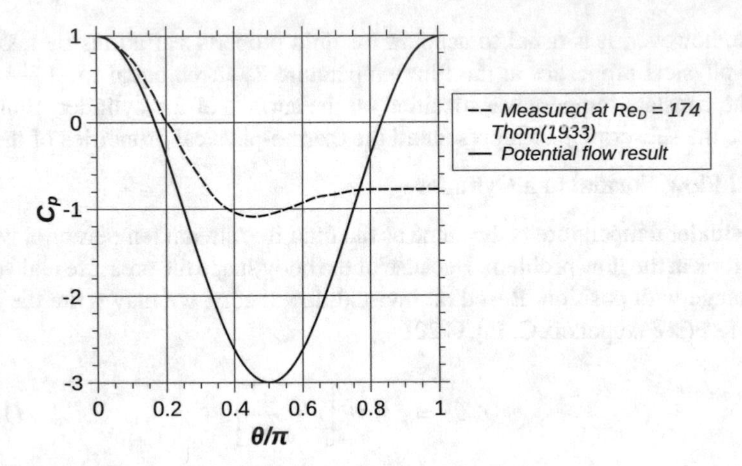

Fig. 13.14 Pressure distribution around a cylinder: potential and laminar flow cases

$$p + \frac{1}{2}\rho_\infty[4U_\infty^2 \sin^2\theta] = p + 2\rho_\infty U_\infty^2 \sin^2\theta = p_0$$

Hence, a pressure coefficient (it is the same as the Euler number) may be defined as $C_p = \frac{p - p_\infty}{\frac{1}{2}\rho_\infty U_\infty^2}$ to get

$$C_p(\theta) = \frac{(p_0 - p_\infty) - 2\rho_\infty U_\infty^2 \sin^2\theta}{\frac{1}{2}\rho_\infty U_\infty^2}$$

Note that $p_0 - p_\infty$ is equal to $\frac{1}{2}\rho_\infty U_\infty^2$. Hence, we have

$$C_p(\theta) = 1 - 4\sin^2\theta \tag{13.154}$$

The inviscid flow pressure distribution around the cylinder is shown in Fig. 13.14 as solid line. It is symmetric with respect to $\theta = \pm\frac{\pi}{2}$. The pressure coefficient is 1 at the forward and backward stagnation points. The pressure coefficient has a minimum of -3 at $\theta = \pm\frac{\pi}{2}$. The pressure distribution as measured is indicated by the dashed line.[12]

13.5.2 Laminar Flow Normal to a Cylinder

Within the framework of the boundary layer theory, the inviscid solution presented above represents the flow outside the thin boundary layer that forms adjacent to the surface of the cylinder. The flow within the boundary layer may be represented by

[12] A. Thom, Proc. Roy. Soc. London, Series A, Vol. 141, pp. 651–659, No. 845, 1933.

the boundary layer equations in two dimensions presented earlier. The coordinate x will correspond to distance measured along the surface of the cylinder with $x = 0$ corresponding to the forward stagnation point. Thus, we have $x = a\theta = \frac{D\theta}{2}$. y is measured along any radius, away from the surface of the cylinder that corresponds to $y = 0$. Thus, $y = r - a = r - \frac{D}{2}$. The boundary layer flow is subject to a pressure gradient that is determined by the inviscid flow.

It is seen that C_p decreases away from the forward stagnation point up until $\theta = \frac{\pi}{2}$. The pressure gradient is thus negative in this region. Beyond $\theta = \frac{\pi}{2}$ the pressure increases with θ and hence the pressure gradient is positive. Thus boundary layer flow is subject to favorable and unfavorable pressure gradients as the flow proceeds around the cylinder.

13.5.3 Laminar Boundary Layer Flow Past a Cylinder

We apply Thwaites's method to analyze the laminar boundary layer in the case of a cylinder in cross flow. Let the incoming flow far away from the cylinder be parallel flow with a uniform velocity U_∞. From the potential flow solution, it is known that the velocity outside the boundary layer is given by $U(x) = 2U_\infty \sin\theta$ where $\theta = \frac{x}{a}$ is the angle measured from the stagnation point. Using Eq. 13.134 from Thwaites's method, we have

$$\frac{\delta_2^2}{\nu_\infty} = \frac{0.45a}{(2U_\infty \sin\theta)^6}\int_0^\theta (2U_\infty \sin\theta)^5 d\theta = \frac{0.45a}{2U_\infty \sin^6\theta}\int_0^\theta \sin^5\theta\, d\theta \tag{13.155}$$

The integration indicated in Eq. 13.155 may be analytically accomplished to write

$$I(\theta) = \int_0^\theta \sin^5\theta\, d\theta = (1-\cos\theta) + \frac{(1-\cos^5\theta)}{5} - \frac{2(1-\cos^3\theta)}{3} \tag{13.156}$$

Noting that $\frac{dU}{dx} = \frac{dU}{d\theta}\frac{d\theta}{dx} = \frac{d(2U_\infty \sin\theta)}{d\theta}\frac{d\left(\frac{x}{a}\right)}{dx} = \frac{2U_\infty \cos\theta}{a}$, the shape parameter λ may be obtained as

$$\lambda = \frac{\delta_2^2}{\nu_\infty}\frac{dU}{dx} = \frac{0.45\cos\theta}{\sin^6\theta}I(\theta) \tag{13.157}$$

The friction coefficient may now be obtained using Eq. 13.135 as

$$C_{f,\theta} = \frac{2\nu_\infty l(\lambda)}{U\delta_2} = \frac{2\nu_\infty l(\lambda)}{2U_\infty \sin\theta\sqrt{\frac{0.45a\nu_\infty I(\theta)}{2U_\infty \sin^6\theta}}} = \frac{2}{\sqrt{Re_D}}\sqrt{\frac{1}{0.45I(\theta)}}l(\lambda)\sin^2\theta \tag{13.158}$$

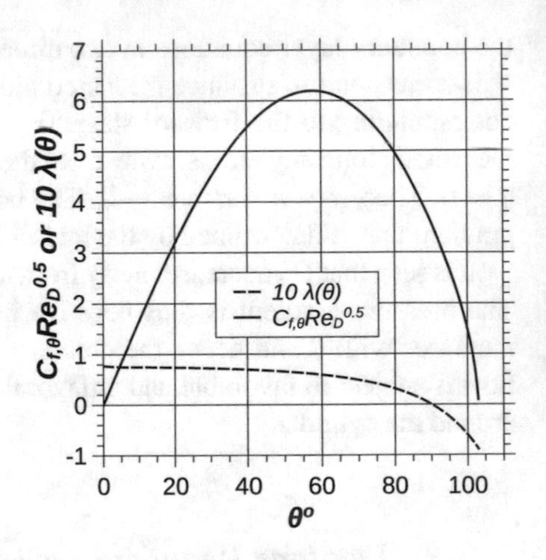

Fig. 13.15 Variation of shape parameter and friction coefficient for boundary layer flow past a cylinder

where $Re_D = \frac{U_\infty D}{\nu_\infty} = \frac{2U_\infty a}{\nu_\infty}$ is the Reynolds number based on the free stream velocity U_∞ and the cylinder diameter $D = 2a$. It is convenient to base the friction coefficient on velocity head $\frac{1}{2}\rho_\infty U_\infty^2$ instead of $\frac{1}{2}\rho_\infty U^2$. This modification requires that the expression given by Eq. 13.158 be multiplied by $(2\sin\theta)^2$. Note that the shape parameter λ as well as the friction coefficient $C_{f,\theta}$ vary with the angle θ. The variation of these with θ is shown in Fig. 13.15. Close to $\theta = 104°$ the boundary layer separates and the integral formulation is incapable of calculation beyond this point.

It is interesting to look at the variation of velocity within and outside the boundary layer. The velocity tangential to the surface of the cylinder is designated as $u_\theta(a, \theta)$ as before, and, in general the component of the velocity along θ is indicated as $u_\theta(r, \theta)$. From the potential flow, this velocity is given by

$$u_\theta(r, \theta) = U_\infty \sin\theta \left[1 + \frac{a^2}{r^2}\right] \tag{13.159}$$

This represents the velocity perpendicular to a radial line emanating from the center of the cylinder. This velocity has a magnitude of $2U_\infty \sin\theta$ at the surface of the cylinder and decreases to $U_\infty \sin\theta$ as $r \to \infty$. Now consider the state of affairs at an angular position of 45° from the stagnation point. For this θ the θ velocity reduces from a value of $\sqrt{2}U_\infty$ at the cylinder surface to $\frac{U_\infty}{\sqrt{2}}$ as $r \to \infty$. Within the boundary layer, the velocity has to vary between 0 on the cylinder surface to a value equal to $\sqrt{2}U_\infty$ at the edge of the boundary layer. Based on the integral method (Thwaites's method), we may easily show that $\lambda = 0.068$ for $\theta = 45°$. The corresponding shape parameter Λ is obtained as 5.9 and the boundary layer thickness $\frac{\delta}{a} = \frac{\Lambda}{\sqrt{Re_D}}$. For example, with $Re_D = 200$ we have $\frac{\delta}{a} = \frac{5.9}{\sqrt{200}} = 0.172$. Note that correspondingly the boundary layer edge is located at $\frac{r_\delta}{a} = \frac{a+\delta}{a} = 1.172$. The velocity profile within the thin boundary layer follows the polynomial relation given by Eq. 13.129 with $\Lambda = 5.9$. This profile will start with $\frac{u}{U_\infty} = 0$ at $r = a$ or $\frac{r}{a} = 1$ and remains constant

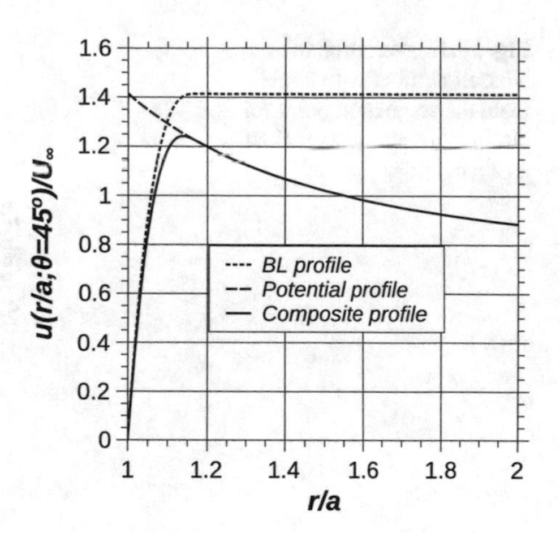

Fig. 13.16 Velocity variation with radius at an angular location of $\theta = 45°$ from the forward stagnation point

at $\frac{u}{U_\infty} = \sqrt{2}$ as $\frac{r}{a} \to \infty$. The velocity variation within the boundary layer is shown by the short dashed curve in Fig. 13.16. The potential velocity is indicated by the long dashed line in the same figure. The composite profile indicated by the full line is the actual velocity profile, obtained by summing the boundary layer and potential velocities and subtracting the common part given by $\frac{u}{U_\infty} = \sqrt{2}$. It is seen that the composite velocity profile smoothly transitions from the boundary layer velocity profile to the potential velocity variation outside the boundary layer.

Now we turn our attention to the thermal boundary layer. Consider, as an example, flow past cylinder of a fluid with $Pr_\infty = 1$. The constants a_T and b_T are, respectively, 0.441 and 0.355. We use Eq. 13.143 to get

$$\zeta_T(\theta) = \frac{a_T}{(2U_\infty \sin\theta)^{b_T}} \int_0^\theta (2U_\infty \sin\theta)^{b_T} a \, d\theta \quad \text{or} \quad \frac{\zeta_T(\theta)}{a} = \frac{a_T I(\theta)}{\sin^{b_T}\theta} \qquad (13.160)$$

where$I(\theta) = \int_0^\theta \sin^{b_T}\theta \, d\theta$. This integral needs to be obtained numerically. The derivative of the velocity $\frac{dU}{dx}$ may be written as $\frac{1}{a}\frac{dU}{d\theta} = \frac{1}{a}2U_\infty \cos\theta$. With this, we have

$$\frac{1}{U}\frac{dU}{dx} = \frac{1}{a}\frac{\cos\theta}{\sin\theta} = \frac{\cot\theta}{a}$$

The local Nusselt number Nu_x given by Eq. 13.145 may be rephrased based on the cylinder diameter as the appropriate length scale to define $Nu_{D,\theta}$, the local Nusselt number based on D. With $x = a\theta$, $D = 2a$, it is easily seen that Eq. 13.145 will become (intermediate steps are left to the reader)

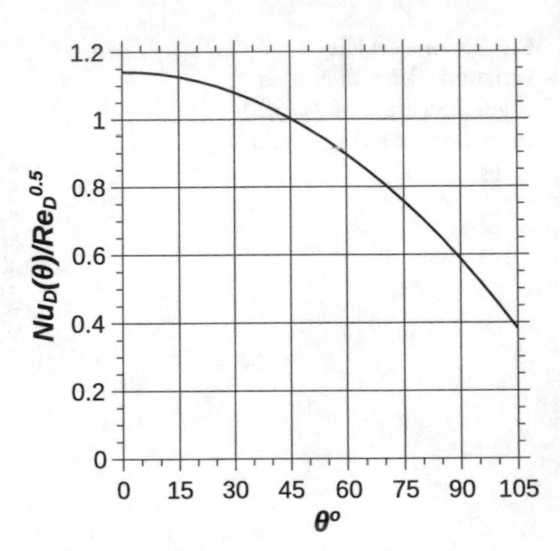

Fig. 13.17 Variation of Nusselt number with angle from the stagnation point for laminar boundary layer flow past a cylinder

$$\boxed{Nu_{D,\theta} = \sqrt{\frac{\sin\theta}{\frac{\zeta_T}{a}}}\left[a_T + \frac{\zeta_T}{a}(1 - b_t)\cot\theta\right]\sqrt{Re_D}} \tag{13.161}$$

A plot of the variation of local Nusselt number with angle is shown in Fig. 13.17. The plot stops at the separation point. Beyond this point the integral method can not be used.

13.5.4 Effect of Pressure Gradient on Boundary Layer Flow

We have seen based on the integral solution that the potential flow pressure variation affects the flow within the boundary layer. When the pressure decreases with x, such as when the potential flow is accelerating, the potential flow is said to impose a favorable pressure gradient on the boundary layer. Shear force at the wall due to friction leads to an increase in the boundary layer thickness as the flow develops along x. We also see that this has an effect of reducing the wall shear with x and hence the derivative of x velocity with y. At a certain location, the velocity gradient actually becomes zero and flow separation takes place, thereafter. Beyond the separation point the fluid separates with a reversed flow region next to the boundary. The pressure variation departs from the ideal or potential flow pressure variation. Pressure varies very little beyond the separation point.

As an example, we look at the flow past a cylinder. The pressure variation is symmetric with respect to $\theta = 90°$, in the case of potential flow, as shown by the full line in Fig. 13.14. In case of laminar viscous flow past the cylinder,

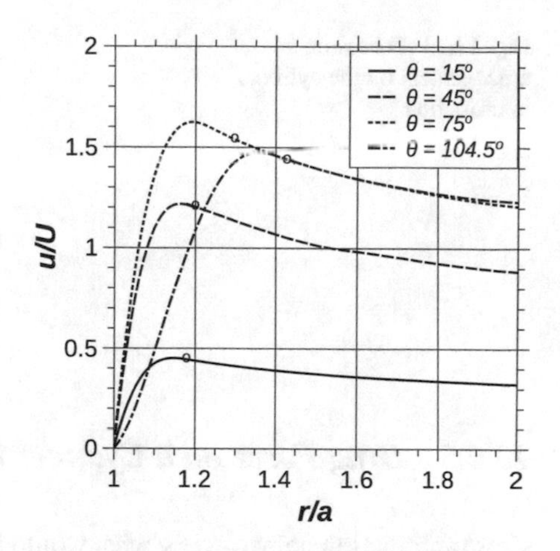

Fig. 13.18 Variation of laminar boundary layer velocity profiles with θ according to the integral method

the pressure variation departs from the potential flow value as shown by the dotted line. The pressure coefficient decreases with θ and the flow separates at around $\theta = 82°$ even before the flow reaches the minimum pressure point at $\theta = 90°$. The pressure head is used up in supporting the acceleration of the flow as well as in making up for the *momentum loss* due to wall shear. Beyond the separation point the pressure does not recover and stays more or less constant. The flow separates from the body since the pressure variation is incapable of supporting the flow were it to follow the contour of the body. The zero streamline separates from the body such that the pressure variation is supportable by the potential flow. Close to the separation point the boundary layer assumptions are actually violated and hence the flow cannot be computed using the boundary layer equations.

In order to explain the above observations, we show the velocity profiles computed by the integral method for flow past a cylinder in Fig. 13.18. The flow past the cylinder is from left to right. Velocity profiles are shown at four angular locations as indicated. The boundary layer thickness at each station is indicated by the—o—indicated in each of the profiles. We see that the boundary layer thickness increases with θ. The case shown corresponds to a flow with $Re_D = 200$. The flow separates at around $\theta = 105°$. This is an over-prediction by about 30% as compared to the value of $\theta \approx 82°$, observed experimentally.

The important point to notice is that the potential flow pressure distribution that is symmetric is changed by the presence of viscosity to a distribution that is not symmetric. This has the effect of imposing a drag on the cylinder which is referred to as the form drag. The boundary layer also imposes a viscous drag on the cylinder via the wall shear stress.

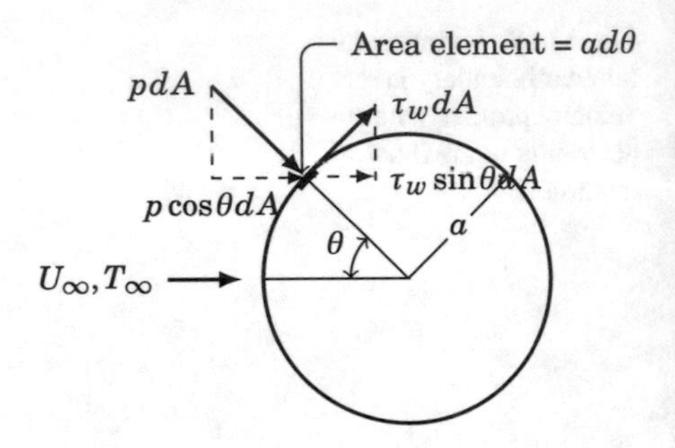

Fig. 13.19 Forces on an area element on the cylinder in cross flow

13.5.5 Drag Force on a Cylinder in Cross Flow

From application point of view what would be of interest is the total drag experienced by a cylinder in cross flow. The drag force is made of two parts:

- Force experienced by the cylinder due to the variation of the fluid pressure along its surface
- Drag force due to fluid friction

Figure 13.19 shows an area element on the surface of the cylinder along with the forces acting on it. The fluid pressure acts radially inwards and has a force component $p\cos\theta\, dA$ parallel to the direction of U_∞. Integration with respect to θ gives the total force acting on the cylinder due to fluid pressure.

$$F_p = \int_0^{2\pi} p\cos\theta\, ad\theta = 2a\int_0^{\pi} p\cos\theta\, d\theta \tag{13.162}$$

The second step follows from the fact that the flow is symmetric with respect to the horizontal axis. It is customary to represent the drag force via a Drag coefficient defined as

$$F_D = C_D \frac{\rho_\infty U_\infty^2}{2} A \tag{13.163}$$

where A is the projected area facing the flow given by $A = D \times 1 = 2a$ assuming unit length of cylinder. The form drag may then be characterized by a form drag coefficient given by

$$C_{Dp} = \frac{F_p}{\frac{\rho_\infty U_\infty^2}{2} A} = \frac{\int_0^{\pi} p\cos\theta d\theta}{\frac{\rho_\infty U_\infty^2}{2}}$$

The pressure may be written in terms of the pressure coefficient as $p = p_\infty + C_p \frac{\rho_\infty U_\infty^2}{2}$. The above equation may then be recast as

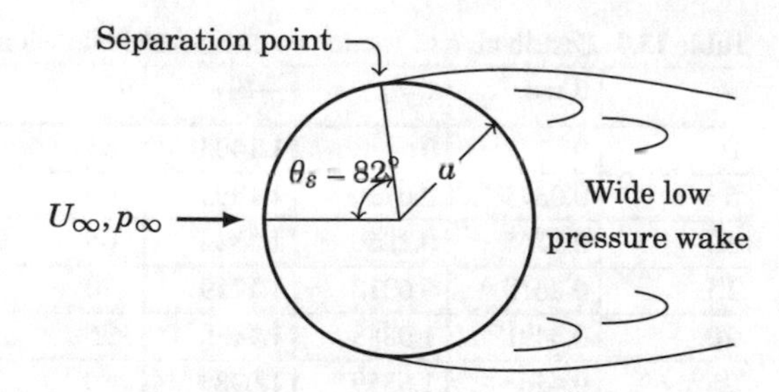

Fig. 13.20 Schematic of separating laminar flow past a cylinder

$$C_{Dp} = \int_o^\pi \frac{\left(p_\infty + C_p \frac{\rho_\infty U_\infty^2}{2}\right)}{\frac{\rho_\infty U_\infty^2}{2}} \cos\theta d\theta = \int_0^\pi C_p \cos\theta d\theta \tag{13.164}$$

since the p_∞ part integrates to zero. The pressure distribution over the cylinder may be integrated to get the form drag.

If the flow is also symmetrical with respect to the vertical axis, as in the case of potential flow the net pressure force is zero and hence there is no pressure or form drag. In the limit of Reynolds number approaching zero, inertia forces are negligible and the flow tends to be symmetric with respect to the vertical axis and the form drag vanishes. The drag is then dominated by viscous drag or friction drag. In the case of laminar boundary layer flow (as the Reynolds number increases), we have seen that the flow separates at around $\theta = 82^\circ$, and there is a recirculating flow at the back of the cylinder where the pressure is low. The wake region is relatively wide as indicated in Fig. 13.20. Hence, there is a net force acting against the cylinder giving rise to form drag. The form drag dominates over the viscous drag.

The shear stress acts tangentially as shown and has a force component $\tau_w \sin\theta$ parallel to the direction of U_∞. The total viscous drag force is obtained by integration as

$$F_f = 2\int_0^\pi \tau_w \sin\theta \, a d\theta$$

Again this may be recast in the form of a Drag coefficient C_{Df} given by

$$C_{Df} = \frac{2\int_0^\pi \tau_w \sin\theta \, a d\theta}{\frac{\rho_\infty U_\infty^2}{2} A} = \frac{\int_0^\pi \tau_w \sin\theta \, d\theta}{\frac{\rho_\infty U_\infty^2}{2}}$$

Introducing now the friction coefficient $C_f = \frac{\tau_w}{\frac{\rho_\infty U_\infty^2}{2}}$ the above equation may be recast as

Table 13.4 Distribution of friction coefficient and Nusselt number on a cylinder in laminar flow

$\theta°$	θ rad	$C_{f,x}$	$\frac{Nu_D}{Re_D^{0.5}}$	$\theta°$	θ rad	$C_{f,x}$	$\frac{Nu_D}{Re_D^{0.5}}$
0	0	0	1.1403	55	0.9599	5.0593	0.9345
5	0.0873	0.0726	1.1393	60	1.0472	5.3536	0.8953
10	0.1745	0.2866	1.1344	65	1.1345	5.4977	0.8527
15	0.2618	0.6312	1.1259	70	1.2217	5.4755	0.8067
20	0.3491	1.0888	1.1139	75	1.3090	5.2766	0.7572
25	0.4363	1.6359	1.0983	80	1.3963	4.8963	0.7041
30	0.5236	2.2444	1.0795	85	1.4835	4.3343	0.6474
35	0.6109	2.8824	1.0573	90	1.5708	3.5926	0.5870
40	0.6981	3.5164	1.0317	95	1.6581	2.6696	0.5228
45	0.7854	4.1121	1.0027	100	1.7453	1.4791	0.4545
50	0.8727	4.6366	0.9703	103	1.7977	0.0915	0.4040

$$C_{Df} = \int_0^{\pi} C_f \sin\theta \, d\theta \tag{13.165}$$

As an example consider the case considered earlier by the application of Thwaites's method. The x-component (this is the usual x-direction—not along the surface of the cylinder) of the friction coefficient is shown tabulated in Table 13.4 for various angles measured from the forward stagnation point.

The point of separation is identified at $\theta = 103°$. Beyond the point of separation experiments indicate that the shear stress is negligibly small and hence the integration indicated in Eq. 13.165 may be terminated at θ_s, the angle at separation point. The indicated integration needs to be performed numerically and the resulting expression for the frictional drag coefficient is

$$\boxed{C_{Df} = \frac{5.626}{\sqrt{Re_D}}} \tag{13.166}$$

Example 13.8

Air flows normal to the axis of an infinitely long cylinder of diameter $D = 1$ cm with a velocity of $U_\infty = 0.5$ m/s. The pressure and temperature of air stream may be taken, respectively, as $p_\infty = 1$ atm and $T_\infty = 20$ °C. Determine the drag force per unit length of the cylinder. If the cylinder is maintained at a temperature of $T_w = 40$ °C how much heat per unit length will be lost from the cylinder?

Solution:

Step 1 From the given data, air properties are taken at a temperature of $T_f = \frac{T_w+T_\infty}{2} = \frac{40+20}{2} = 30\,°\text{C}$.

$\rho_f = 1.1552\ \text{kg/m}^3$; $\nu_f = 16.115 \times 10^{-6}\ \text{m}^2/\text{s}$; $k_f = 0.0266\ \text{W/m°C}$; $Pr_f = 0.71$

Step 2 The Reynolds number based on the cylinder diameter is then given by

$$Re_D = \frac{U_\infty D}{\nu_\infty} = \frac{0.5 \times 0.01}{16.115 \times 10^{-6}} = 310.3$$

Step 3 The Drag coefficient at this Reynolds number is $C_D = 1.3$ as read from Fig. 14.14. Hence, the drag force experienced by the cylinder per unit length is

$$F_D = (C_D)\left(\frac{\rho_f U_\infty^2}{2}\right) D = (1.3)\left(\frac{1.1552 \times 0.5^2}{2}\right) \times 0.01 = 0.002\ \text{N/m}$$

Step 4 The heat transfer calculation is based on the observation that the Prandtl number dependence of the Nusselt number appears through a factor $Pr^{\frac{1}{3}}$ such that the calculation presented earlier leads to an overall Nusselt number given by

$$Nu_D = \left(0.525 Re_D^{\frac{1}{2}} + 0.001 Re_D\right) Pr_f^{\frac{1}{3}}$$

where the first term in the braces is the contribution up to the point of separation and the second term due to heat transfer from the separated region. The latter is based on experimental observations. Hence, we have

$$Nu_D = \left(0.525 \times 310.3^{\frac{1}{2}} + 0.001 \times 310.3\right) 0.71^{\frac{1}{3}} = 9.525$$

Step 5 The average heat transfer coefficient is then calculated as

$$\overline{h} = \frac{Nu_D k_f}{D} = \frac{9.525 \times 0.0266}{0.01} = 25.34\ \text{W/m}^{2\circ}\text{C}$$

Step 6 Heat transfer from unit length of the cylinder then is

$$Q = \overline{h} D (T_w - T_\infty) = 25.34 \times 0.01 \times (40 - 20) = 5.07\ \text{W/m}$$

Concluding Remarks

Flow past a body and heat transfer from the body to the flowing fluid have been dealt with in detail. Detailed description of boundary layer hypothesis and the solutions obtained therefrom are discussed. Approximate integral method of solution is also described in great detail. Flows with and without axial pressure gradient are considered with flow past a cylinder being the most important case where pressure changes over the surface.

13.6 Exercises

Ex 13.1: An electric air heater consists of a horizontal array of thin metal strips each 10 mm long in the direction of an air stream that is in parallel flow over the top of the strips. Each strip is 0.2 m wide and 25 strips are arranged side by side forming a continuous and smooth surface over which the air flows at 2 m/s. During operation each strip is maintained at 150 °C and the air is at 25 °C.

- What is the convection heat transfer rate from the first strip?
- What is the convection heat transfer rate from all of the strips?

Ex 13.2: (a) The velocity profile in a laminar boundary layer flow over a flat plate is represented by a quadratic function given by $\frac{u}{U} = A\eta^2 + B\eta + C$, where U is the free stream velocity, A, B, and C are constants, and $\eta = \frac{y}{\delta}$ where δ is the momentum boundary layer thickness. Obtain an expression for the friction coefficient C_f using the above profile and the momentum integral. Compare this with the value obtained from the Blasius solution.
(b) Make a plot showing the approximate velocity profile as well as the Blasius profile and comment on it.

Ex 13.3: (a) The velocity profile in a laminar boundary layer flow over a flat plate is represented by the relation $\frac{u}{U} = A \sin\left(\frac{\pi\eta}{2}\right)$, where U is the free stream velocity, A is a constant, and $\eta = \frac{y}{\delta}$ where δ is the momentum boundary layer thickness. Obtain an expression for the friction coefficient C_f using the above profile and the momentum integral. Compare this with the value obtained from the Blasius solution.
(b) Make a plot showing the approximate velocity profile as well as the Blasius profile and comment on it.

Ex 13.4: A thin metal plate of size 0.2×0.4 m is held in water with $\nu = 10^{-6}$ m^2/s and $\rho = 1000$ kg/m^3. The water velocity is 1 m/s. What is the drag force acting on the plate? Which edge, the longer or the shorter, should be aligned with the flow to get a lower drag force?

Ex 13.5: A thin flat plate of size $0.3 \times 0.6\,\text{m}$ is maintained at a constant temperature of 50 °C by removing the heat gained by it from an air stream at 100 °C flowing parallel to its surface. In the first arrangement, the longer side is aligned parallel to the air stream moving with such a speed that Reynolds number based on its length is 1.5×10^5. What is the rate at which heat is to be removed from the plate?
In a second arrangement, the same plate is placed in an air stream with the shorter side aligned parallel to the air stream. The temperatures of the stream and the plate are as given in the previous arrangement. What is the air velocity in this case if the rate at which heat is removed remains the same?

Ex 13.6: Evaluate the velocity and momentum boundary layer thicknesses at the trailing edge in the following two cases:

- Air at 30 °C flows parallel to a flat plate with a free stream speed of 10 m/s and the plate length is such that the flow is just critical at its trailing edge.
- Water at 30 °C flows parallel to the same flat plate such that the flow is just critical at its trailing edge.

The plate temperature is the same and equal to 10 °C in both cases.

Ex 13.7: Air at a free stream temperature of 30 °C flows parallel to a thin flat plate with a velocity of 2 m/s. The plate that is 2.5 m long is maintained at a constant temperature of 80 °C. Determine the drag offered by the plate and the total heat transfer from the plate to the air stream.

Ex 13.8: Solve the Blasius equation numerically using the 4th Runge–Kutta method using the procedure suggested in the text.

Ex 13.9: Using the Blasius solution obtained in Exercise 13.6 obtain the temperature profile within the flat plate thermal boundary layer for one Prandtl number less than 1 and a second Prandtl number more than 1.

Ex 13.10: An air stream flowing with a velocity of 2 m/s encounters a wedge at an angle of 30° to it. Obtain the boundary layer parameters at a distance of 0.05 m from the apex of the wedge using (a) Falkner–Skan solution and (b) Thwaite's method. Compare the two results.

Ex 13.11: Consider flow past a wedge of included angle of 75°. What is the wedge parameter m in this case? Obtain solution to the Falkne–Skan equation for this value of m numerically using the fourth-order Runge–Kutta method. Based on the numerical solution obtain all the boundary layer parameters at a distance of 0.05 m from the apex of the wedge. The fluid is air at atmospheric pressure and 30 °C and the free stream velocity is 5 m/s. The wedge surface is maintained at a uniform temperature of 60 °C.

Ex 13.12: Air at a free stream temperature of 30 °C flows normal to a cylinder of 0.2 m diameter maintained at a temperature of 70 °C. The velocity in the free stream is 10 m/s. Using Thwaite's method, determine the wall shear stress and heat flux at $\theta = 45°$ from the forward stagnation point.

Chapter 14
Convection in Turbulent Flow

STUDY of heat transfer in turbulent flow is important because of the prevalence of turbulent flow in most engineering applications. We deal with both internal and external flow situations. Since rigorous analysis is of limited value in an elementary treatment of the subject, we treat the topic using mostly empirical data from experimental work done over the last 100 years or so. Useful correlations are presented so that many important applications may be analyzed.

14.1 Introduction

Turbulent flow and heat transfer is more common in engineering applications than laminar flow and heat transfer that has been the subject of the previous two chapters. Analysis of turbulent flow is substantially more complex, as will become clear later on, and hence our understanding of turbulent flows is largely through experimental observations. In this chapter, we make an attempt to give a first level treatment of a rather complex subject just so that the reader may be able to do some simple but application-oriented problems in heat transfer. More advanced texts may be consulted by the interested reader.[1] In view of this both internal and external turbulent flows are considered in this chapter.

[1]H. Schlichting & K. Gersten, *Boundary Layer Theory*, 8th Edition, Springer 2000.

S. P. Venkateshan, *Heat Transfer*,
https://doi.org/10.1007/978-3-030-58338-5_14

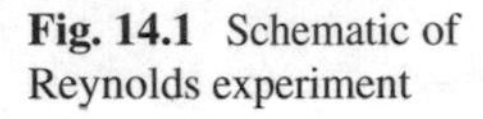

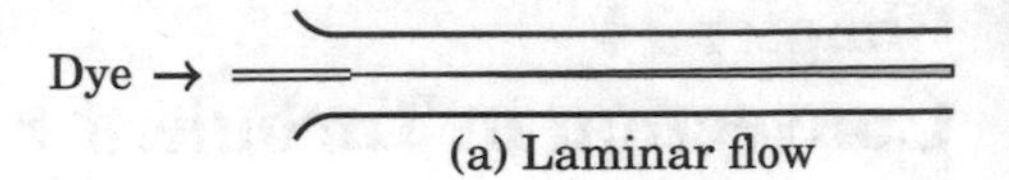

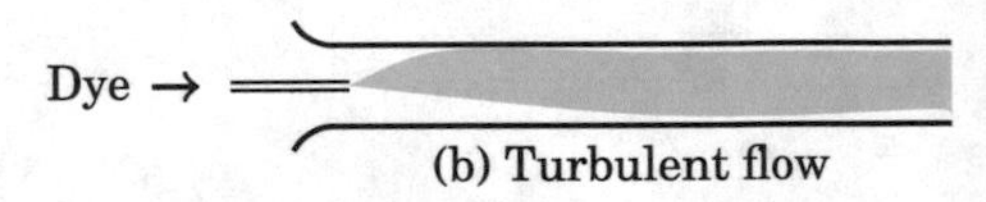

Fig. 14.1 Schematic of Reynolds experiment

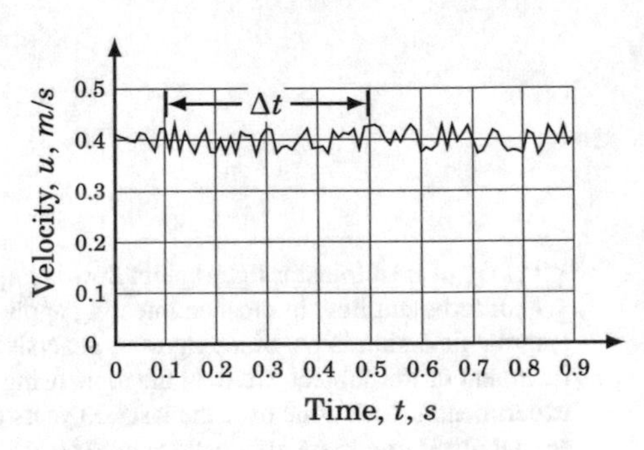

Fig. 14.2 Velocity fluctuations with respect to time in *steady* turbulent flow

Experiments performed by Reynolds[2] indicate that Reynolds number (after him) affects the structure of flow. The experiment essentially consists in observing the path of a dye introduced into a tube flow. When the Reynolds number based on the pipe diameter is less than about 2300, the dye flows along in an orderly fashion with very little lateral mixing. The state of affairs is as shown in Fig. 14.1a. As the fluid velocity is increased and the Reynolds number is in the range 2300–10,000, the dye starts showing some irregularity in its structure and the lateral mixing shows signs of increase. If the flow velocity is increased further and the Reynolds number is greater than 10,000 the dye injected into the flow rapidly diffuses (or, to put it more properly, mixes) in the radial direction and the entire flow field becomes colored as indicated in Fig. 14.1b. The flow is said to be fully turbulent.

In the case of turbulent flow, as opposed to laminar flow, there are unsteady velocity and temperature fluctuations even when the flow is steady. The velocity, for example, will have a steady component that may depend on the location, accompanied by a fluctuating component. A typical velocity trace at a given location will appear as shown in Fig. 14.2.

The velocity at a point in the flow, say u the x component, at any time t, may be written as a sum of a mean value $\overline{u}$ and a fluctuating component u'. Similar is the case with other velocity components, and also pressure and temperature at any point in the flow field. If we average the signal over a large enough time interval, say Δt, the fluctuations will average out to zero indicating that they are randomly distributed in time. Thus, we have

[2]Osborne Reynolds, 1842–1912, British Fluid Dynamicist. The original experimental setup by Reynolds is still at the University of Manchester where he spent his entire career.

$$\overline{u} = \frac{1}{\Delta t}\int_{t}^{t+\Delta t} u dt = \frac{1}{\Delta t}\int_{t}^{t+\Delta t} (\overline{u} + u')dt; \text{ with } \frac{1}{\Delta t}\int_{t}^{t+\Delta t} u' dt = \overline{u'} = 0$$

Thus, we may generalize and say that all fluctuating components have zero means. However, quantities like $\overline{u'^2}$, $\overline{u'v'}$, $\overline{u'T'}$, etc., have non-zero values. Such quantities are expected to differentiate turbulent flows from laminar flows, since such terms make their appearance when the governing momentum and energy equations are time averaged. The process of averaging defined above is referred to as Reynolds averaging.[3] It is easy to visualize that $\frac{\rho(\overline{u'^2}+\overline{v'^2}+\overline{w'^2})}{2}$ is nothing but the mean kinetic energy per unit volume, due to fluctuations in the velocity components with respect to time. The magnitude of the turbulent energy is related to the intensity of turbulence. The origin or production of turbulent kinetic energy and the dissipation of this provides the framework for describing turbulent flows. We shall see later on that the quantity $\rho\overline{u'v'}$ relates to turbulent shear stress and $\overline{v'T'}$ relates to turbulent heat flux.

14.2 Time-Averaged Equations

In order to understand the intricacies involved in turbulent flows, and at the same time to keep the discussion simple, we consider the turbulent boundary layer flow in greater detail here. The starting point for the study is the boundary layer equations given earlier as Eqs. 13.1, 13.15 and 13.12.[4] In the case of turbulent flow, each velocity component is replaced by the sum of the mean value along with the random fluctuating component. For example, the time averaging of the inertia term $u\frac{\partial u}{\partial x}$ will be written as

$$\overline{u\frac{\partial u}{\partial x}} = \overline{(\bar{u} + u')\frac{\partial(\bar{u} + u')}{\partial x}} = \bar{u}\frac{\partial \bar{u}}{\partial x} + \overline{u'\frac{\partial u'}{\partial x}} + \bar{u}\frac{\partial \overline{u'}}{\partial x} + \overline{u'}\frac{\partial \bar{u}}{\partial x}$$

The last two terms drop off because the time average of the fluctuating components are zero. Hence, the above may be rewritten as

$$\overline{u\frac{\partial u}{\partial x}} = \bar{u}\frac{\partial \bar{u}}{\partial x} + \overline{u'\frac{\partial u'}{\partial x}} = \bar{u}\frac{\partial \bar{u}}{\partial x} + \frac{\partial \overline{\frac{u'^2}{2}}}{\partial x} \tag{14.1}$$

The second inertia term may also be time averaged, in an analogous manner to get

[3] Time-averaged governing equations are referred to as Reynolds Averaged Navier Stokes Equations or RANS Equations.

[4] Even though we are considering the Cartesian form here, it is possible to do a similar analysis using equations in the other two coordinate systems, viz., cylindrical and spherical coordinates.

$$\overline{v\frac{\partial u}{\partial y}} = \bar{v}\frac{\partial \bar{u}}{\partial y} + \overline{v'\frac{\partial u'}{\partial y}}$$

We may write the second term as

$$\overline{v'\frac{\partial u'}{\partial y}} = \frac{\partial \overline{u'v'}}{\partial y} - \overline{u'\frac{\partial v'}{\partial y}}$$

The equation of continuity requires that

$$\frac{\partial \bar{u}}{\partial x} + \frac{\partial \bar{v}}{\partial y} + \frac{\partial \bar{u'}}{\partial x} + \frac{\partial \bar{v'}}{\partial y} = 0$$

Of course the mean velocity components should satisfy the continuity equation. Hence, the first two terms add up to zero. Hence, we have

$$\frac{\partial \bar{u'}}{\partial x} + \frac{\partial \bar{v'}}{\partial y} = 0 \text{ or } \frac{\partial \bar{u'}}{\partial x} = -\frac{\partial \bar{v'}}{\partial y}$$

Hence, we have

$$\overline{v\frac{\partial u}{\partial y}} = \bar{v}\frac{\partial \bar{u}}{\partial y} + \frac{\partial \overline{u'v'}}{\partial y} + \overline{u'\frac{\partial u'}{\partial x}} = \bar{v}\frac{\partial \bar{u}}{\partial y} + \frac{\partial \overline{u'v'}}{\partial y} + \frac{\partial \overline{\frac{u'^2}{2}}}{\partial x} \tag{14.2}$$

Similarly, we may write for the other terms in the momentum equation as

$$\nu_\infty \overline{\frac{\partial^2 u}{\partial y^2}} = \nu_\infty \frac{\partial^2 \bar{u}}{\partial y^2}; \quad -\frac{1}{\rho_\infty}\overline{\frac{dp}{dx}} = -\frac{1}{\rho_\infty}\frac{d\bar{p}}{dx} \tag{14.3}$$

Combining expressions 14.1–14.3, we obtain the time-averaged momentum equation in the boundary layer as

$$\bar{u}\frac{\partial \bar{u}}{\partial x} + \bar{v}\frac{\partial \bar{u}}{\partial y} = -\frac{1}{\rho_\infty}\frac{d\bar{p}}{dx} + \nu_\infty \frac{\partial^2 \bar{u}}{\partial y^2} - \frac{\partial \overline{u'^2}}{\partial x} - \frac{\partial \overline{u'v'}}{\partial y} \tag{14.4}$$

Time averaging has thus yielded two extra terms (last two terms on the RHS of Eq. 14.4) in the momentum equation. The first of these two terms is axial derivative of the time average of u'^2 while the latter is the normal derivative of the time average of $u'v'$. The time-averaged u'^2 and $u'v'$ are expected to be of the same order of magnitude. However, if we invoke the boundary layer assumption the x derivative should be much smaller than the y derivative, and hence we may ignore the term containing the time-averaged u'^2 in Eq. 14.4. With this proviso, we approximate the time-averaged momentum equation as

$$\bar{u}\frac{\partial \bar{u}}{\partial x} + \bar{v}\frac{\partial \bar{u}}{\partial y} = -\frac{1}{\rho_\infty}\frac{d\bar{p}}{dx} + \nu_\infty \frac{\partial^2 \bar{u}}{\partial y^2} - \frac{\partial \overline{u'v'}}{\partial y} \tag{14.5}$$

Similarly the energy Eq. 13.12 may be written in the time-averaged form as

$$\bar{u}\frac{\partial \bar{T}}{\partial x} + \bar{v}\frac{\partial \bar{T}}{\partial y} = \alpha_\infty \frac{\partial^2 \bar{T}}{\partial y^2} - \frac{\partial \overline{v'T'}}{\partial y} \tag{14.6}$$

14.2.1 Turbulent Shear Stress and Turbulent Heat Flux

Comparison of Eqs. 14.5 and 14.6 with the laminar boundary layer equations shows that additional terms make their appearance in the turbulent time-averaged equations. Note that the viscous term $\nu_\infty \frac{\partial^2 \bar{u}}{\partial y^2}$ may be interpreted as $\frac{1}{\rho_\infty}\frac{\partial \tau_{lam}}{\partial x}$ where τ_{lam} is the viscous shear stress . The last result follows from Newton's law of viscosity. We may similarly interpret the term $-\frac{\partial \overline{u'v'}}{\partial y}$ as $\frac{1}{\rho_\infty}\frac{\partial \tau_{turb}}{\partial y}$, where $\tau_{turb} = -\rho_\infty \overline{u'v'}$ is the turbulent shear stress (also known as the Reynolds stress).[5] In view of this, we may write the time-averaged momentum equation as

$$\bar{u}\frac{\partial \bar{u}}{\partial x} + \bar{v}\frac{\partial \bar{u}}{\partial y} = -\frac{1}{\rho_\infty}\frac{d\bar{p}}{dx} + \frac{1}{\rho_\infty}\frac{\partial(\tau_{lam} + \tau_{turb})}{\partial y} \tag{14.7}$$

In a similar fashion, we can interpret $\overline{v'T'}$ as being related to a turbulent heat flux q_{turb} arising from fluctuating velocity and fluctuating temperature. In fact, we would like to write the right-hand side of the time-averaged energy equation as

$$\alpha_\infty \frac{\partial^2 \bar{T}}{\partial y^2} - \frac{\partial \overline{v'T'}}{\partial y} = \frac{1}{\rho_\infty C_{p\infty}}\frac{\partial(q_{lam} + q_{turb})}{\partial y} \tag{14.8}$$

In order to make the above possible and also make the turbulent quantities appear *like* the laminar quantities, two quantities, ε—the eddy viscosity and ε_H—the eddy diffusivity are introduced. These are defined through the relations

---

[5] If we were to Reynolds average the Navier Stokes equations, we would obtain a turbulent stress tensor represented as a 3×3 matrix.

$$\varepsilon = \frac{-\overline{u'v'}}{\dfrac{\partial \bar{u}}{\partial y}}, \ \varepsilon_H = \frac{\overline{v'T'}}{\dfrac{\partial \bar{T}}{\partial y}} \tag{14.9}$$

With these definitions, the turbulent shear stress and turbulent heat flux are given by

$$\boxed{\tau_{\text{turb}} = \rho_\infty \varepsilon \frac{\partial \bar{u}}{\partial y}; \quad q_{\text{turb}} = -\rho_\infty C_{p\infty} \varepsilon_H \frac{\partial \bar{T}}{\partial y}} \tag{14.10}$$

14.2.2 *Turbulent Boundary Layer Equations*

The turbulent boundary layer equations are finally written in the following form:

$$\begin{gathered}
\text{(a)} \frac{\partial \bar{u}}{\partial x} + \frac{\partial \bar{v}}{\partial y} = 0 \text{(b)} \quad \bar{u}\frac{\partial \bar{u}}{\partial x} + \bar{v}\frac{\partial \bar{u}}{\partial y} = -\frac{1}{\rho_\infty}\frac{d\bar{p}}{dx} + (\nu_\infty + \varepsilon)\frac{\partial^2 \bar{u}}{\partial y^2} \\
\text{(c)} \bar{u}\frac{\partial \bar{T}}{\partial x} + \bar{v}\frac{\partial \bar{T}}{\partial y} = (\alpha_\infty + \varepsilon_H)\frac{\partial^2 \bar{T}}{\partial y^2}
\end{gathered} \tag{14.11}$$

It is to be noted that while ν_∞ and α_∞ are properties of the fluid while the corresponding turbulent quantities ε and ε_H are not. These depend actually on the flow and also on the location within the flow. The Reynolds averaged equations have thrown up new quantities ε and ε_H which have to be determined or specified in order to complete the formulation. This action constitutes what amounts to providing "closure" to the formulation.

Turbulence far away from bounding walls is not inhibited by high viscous forces and is of isotropic nature. Near wall turbulence is affected by proximity to walls since viscous effects provide damping for the turbulent fluctuations. At once we see that turbulent flow is affected differently in different regions of flow. Several models are available for describing these as described below.

14.3 Turbulence Models

The goal of all turbulence models is to provide a means of describing the variation of turbulent stresses and fluxes within the flow field. Models may be very simple using algebraic expressions or very complex involving the solution of partial differential equations. Generally, these models require numerical solution of the governing equations and hence cannot be covered in the present book. All these models are based on experimental observations made by various investigators over the past 100 years or so.

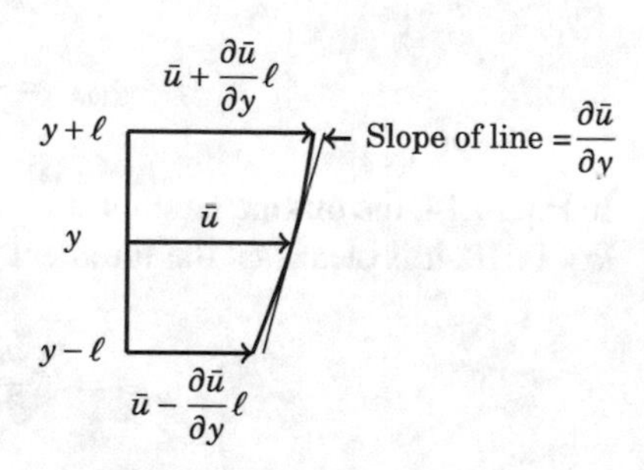

Fig. 14.3 Mixing in turbulent flow characterized by the mixing length

Within the boundary layer, three regions may be distinguished according to the relative magnitude of turbulent viscosity and turbulent diffusivity, in comparison, respectively, with the fluid viscosity and the thermal diffusivity. To carry out the analysis, we shall look at the simplest of models of turbulence, viz., Prandtl's mixing length theory.

14.3.1 Prandtl's Mixing Length Theory

Any analysis of turbulent flow should be able to formulate the basic equations in terms of the properties of the time-varying fluctuations in velocity and temperature. In fact, ε and ε_H are only shortcut symbols for these and just do not, as yet, reflect any physics. A method or model for obtaining these is necessary if Eqs. 14.11(a)–(c) are to be of any use. Prandtl argued that the turbulent viscosity and diffusivity arise out of large-scale motions due to turbulence just as molecular level fluctuations explain the transport properties through the concept of mean free path.

We assume that the fluctuating components u',v', and T' retain their identity over a length l called the mixing length. This is shown schematically in Fig. 14.3.

The fluctuation of the velocity components at y is assumed to be due to the difference in the mean velocities at $y + l$ and $y - l$ from that prevailing at y. Thus, the fluctuation in velocity u at y is given by

$$u' \propto l \frac{\partial \bar{u}}{\partial y} \tag{14.12}$$

The fluctuations in v' is also assumed to be of the same order of magnitude, and hence the turbulent shear stress may be written as

$$\tau_{\text{turb}} = \rho_\infty l^2 \left[\frac{\partial \bar{u}}{\partial y}\right]^2 \tag{14.13}$$

In an analogous manner, we may write an expression for turbulent heat flux as

$$q_{\text{turb}} = -\rho_\infty l_H^2 \left[\frac{\partial \bar{T}}{\partial y}\right]^2 \tag{14.14}$$

In Eq. 14.14, the mixing length for temperature is taken as l_H. Comparing these with Eq. 14.10, it is clear that the turbulent viscosity and diffusivity are given by

$$\varepsilon = l^2 \frac{\partial \bar{u}}{\partial y},\ \varepsilon_H = l_H^2 \frac{\partial \bar{T}}{\partial y} \tag{14.15}$$

Analogous to $Pr = \frac{\nu}{\alpha}$, we may also define a turbulent Prandtl number given by

$$Pr_{\text{turb}} = \frac{\varepsilon}{\varepsilon_H} = \frac{l^2 \frac{\partial \bar{u}}{\partial y}}{l_H^2 \frac{\partial \bar{T}}{\partial y}} \tag{14.16}$$

Experimental data indicates that the turbulent Prandtl number is close to one.

14.3.2 Universal Velocity Distribution

Let us consider the mixing length concept in more detail. It is clear that very close to a solid wall the mixing length has to be small and should actually be zero at the wall. The simplest variation of the mixing length variation with y that satisfies this requirement is a linear variation in the form $l = Cy$ where C is a constant. Also if it is assumed that the shear stress is constant at its value τ_w at the wall, we have, using Eq. 14.13

$$\tau_w = \rho_\infty C^2 y^2 \left[\frac{\partial \bar{u}}{\partial y}\right]^2 \tag{14.17}$$

Introducing the friction velocity $v^* = \sqrt{\frac{\tau_w}{\rho_\infty}}$, this may be recast in the form

$$\frac{\partial \bar{u}}{\partial y} = \frac{v^*}{Cy} \tag{14.18}$$

Equation 14.18 may be integrated once to get

$$\bar{u} = \frac{v^*}{C} \ln y + C_2' = C_1' \ln y + C_2' \tag{14.19}$$

For convenience, we have set $\frac{v^*}{C} = C_1'$. C_2' is a constant of integration. We introduce a non-dimensional velocity $u^+ = \frac{\bar{u}}{v^*}$ and non-dimensional y coordinate given by

$y^+ = \frac{v^* y}{\nu_\infty}$ (y^+ - Reynolds number based on friction velocity) and adjust the constants suitably to get

$$\boxed{u^+ = C_1 \ln y^+ + C_2} \tag{14.20}$$

The velocity profile thus shows the characteristic logarithmic form. Constants appearing in Eq. 14.20 need to be obtained from measurements.

We identify three flow regions adjacent to the surface as follows:

1. Laminar (or viscous) sub-layer in the range $0 \le y^+ \le 5$ where $\varepsilon \ll \nu_\infty$ and $\varepsilon_H \ll \alpha_\infty$
2. Buffer layer spanning $5 \le y^+ \le 30$ in which $\varepsilon \approx \nu_\infty$ and $\varepsilon_H \approx \alpha_\infty$
3. Fully turbulent region for $y^+ > 30$ where $\varepsilon \gg \nu_\infty$ and $\varepsilon_H \gg \alpha_\infty$

Within the laminar sub-layer, velocity varies linearly with distance. Inside the buffer layer, the constants in the logarithmic profile are chosen so as to show continuity at the interfaces with the laminar sub-layer and the fully turbulent regions. In the turbulent region, the constants in the logarithmic velocity profile are adjusted to be in agreement with experiments. Thus, we get the following logarithmic velocity profile which is referred to as the "universal velocity profile".

$$\underbrace{u^+ = y^+}_{\text{(a) Laminar sub layer}} \quad \underbrace{u^+ = 5\ln(y^+) - 3.05}_{\text{(b) Buffer layer}} \quad \underbrace{u^+ = 2.5\ln(y^+) + 5.5}_{\text{(c) Fully turbulent region}} \tag{14.21}$$

A plot of the "universal velocity profile" is shown in Fig. 14.4.

14.3.3 Velocity Profiles in Pipe Flow

The three-layer model presented above also describes the state of affairs in pipe flow. Consider a pipe of diameter with the pipe Reynolds number greater than 10,000. The flow is turbulent, and the region $0 \le y \le R$ from the pipe wall is divided in to three zones as in the case of the turbulent boundary layer. The fully turbulent region is referred to as the turbulent core. Since the Reynolds number is very large, the flow inside the pipe has all the characteristics of boundary layer flow with significant gradients near the wall. The flow development is very rapid, and becomes fully developed within a few pipe diameters. The velocity profile given in Fig. 14.4 then applies in this case also.

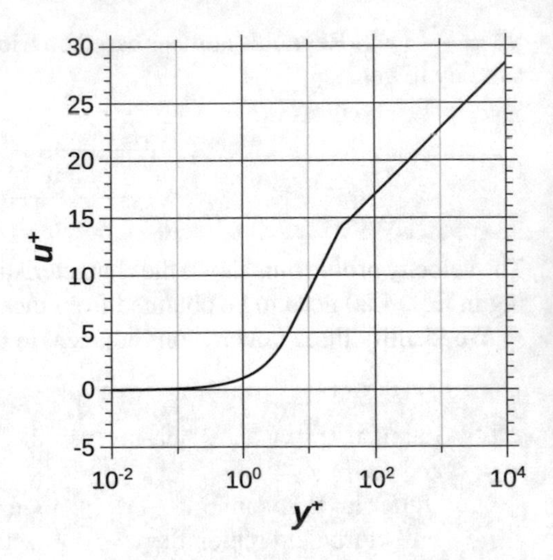

Fig. 14.4 Universal velocity profile for turbulent flow

Power law velocity profiles

Power law velocity profiles have also been used to represent fully developed velocity profiles in pipe flow. Noting that $y = R - r$, the velocity profile is represented in the form

$$\frac{\bar{u}}{U_{\max}} = \left(1 - \frac{r}{R}\right)^{\frac{1}{n}} \tag{14.22}$$

where $U_{\max}$ is the centerline velocity. The exponent n varies with the pipe Reynolds number, between about 6 at $Re_D = 2 \times 10^4$ and 10 at $Re_D = 3 \times 10^6$. Denoting the mean velocity across the pipe cross section as U, the following expression may easily be obtained.

$$\boxed{\frac{U}{U_{\max}} = \frac{1}{R^2}\int_0^R 2\frac{\bar{u}}{U_{\max}} r dr = \frac{2n^2}{(n+1)(2n+1)}} \tag{14.23}$$

We compare the turbulent velocity profiles for $n = 6$ and $n = 10$ with the laminar fully developed profile in Fig. 14.5. It is observed that the turbulent profiles are more "full" with very steep gradients near the pipe wall and largely uniform velocity in the core. In comparison, the laminar profile varies significantly across the full section of the pipe. The mean velocities for the turbulent profiles, according to Eq. 14.23 are $\frac{U}{U_{\max}} = 0.791$ for $n = 6$, $\frac{U}{U_{\max}} = 0.866$ for $n = 10$ while it is $\frac{U}{U_{\max}} = 0.5$ for laminar flow. Another interesting comparison is that between the universal logarithmic profile and the power law profile. We consider, as an example, pipe flow with $Re_D = 10^4$. The shear stress at the wall of the pipe can be related to the axial pressure drop by making a momentum balance over an elemental length of the pipe. Representing the pressure drop using the friction factor concept, we can easily show that

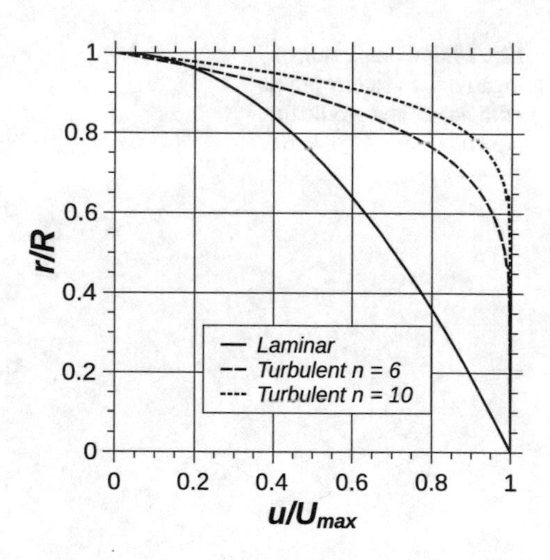

Fig. 14.5 Comparison of power law turbulent profiles with laminar velocity profile

$$\frac{\tau_w}{\rho} = \frac{fU^2}{8} = v^{*2}$$

This may be recast as

$$\frac{v^*}{U} = \sqrt{\frac{f}{8}}$$

If we assume that the distance measured away from the pipe surface toward the centerline is y, we may write y^+ as

$$y^+ = \frac{yv^*}{\nu_\infty} = \frac{\sqrt{\frac{f}{8}}Uy}{\nu_\infty} = \sqrt{\frac{f}{8}}Re_y$$

For pipe Reynolds number of 10^4, the measured value of friction factor is 0.0309 (more about turbulent friction factor in the next section). With $Re_D = 10{,}000$, the relation between y^+ and Re_y may be written as

$$y+ = \sqrt{\frac{f}{8}}Re_y = \sqrt{\frac{0.0309}{8}}Re_y = 0.0621Re_y$$

The maximum value of y^+ corresponds to the pipe axis with $y = R$ and hence

$$y^+_{\max} = 0.0621Re_R = 0.0621\frac{Re_D}{2} = 0.0621 \times 5000 = 310.7$$

The data for the logarithmic profile may be generated for $0 \le y \le R$ using Eqs. 14.21(a)–(c) and made in to a plot as shown in Fig. 14.6. For comparison, the

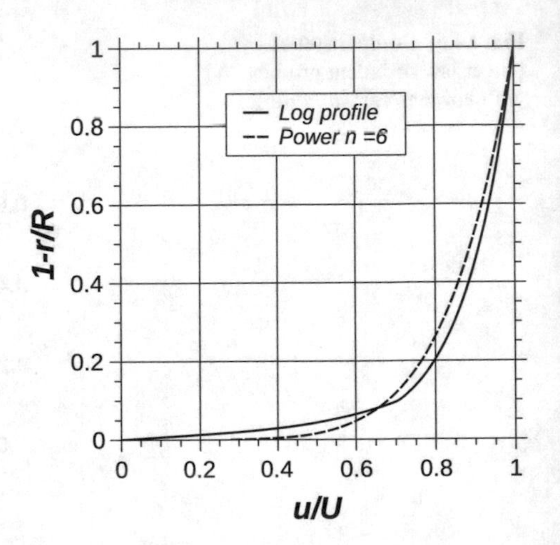

Fig. 14.6 Comparison of logarithmic velocity profile with power law profile for pipe flow with $Re_D = 10^4$

power law profile with $n = 6$ is also calculated and shown in the same plot. It is seen that the power law profile does not resolve the laminar sub-layer properly. However, it agrees very closely with the logarithmic profile away from the pipe wall.

14.4 Pressure Drop and Heat Transfer in Turbulent Pipe Flow

14.4.1 Pressure Drop in Turbulent Pipe Flow

Pipes/tubes/ducts are commonly employed in engineering applications. Depending on the manufacturing process and the material used, pipes may be either "smooth" or "rough". This classification is based on the relative height of surface imperfections or surface roughness elements in comparison with the laminar sub-layer thickness. If the roughness ϵ is much smaller than the laminar sub-layer thickness, i.e., $\epsilon^+ << 5$ the pipe is classified as smooth. If the roughness is comparable to the laminar sub-layer thickness the pipe is classified as rough. The surface is considered to be fully rough if $\epsilon^+ > 70$. In the case of a fully rough surface, the friction at the wall becomes independent of Reynolds number. The shear stress at the wall is attributed to pressure drag because of the flow past roughness elements. A full discussion of these as also the appropriate modifications to the logarithmic velocity profile are given in Kays and Crawford.[6] Roughness plays an important role in determining the frictional pressure drop in turbulent flow and hence plays the role of a second parameter, apart

[6] W.M.Kays and M.E.Crawford, *Convective Heat and Mass Transfer*, McGraw Hill International Edition, 1993.

from Reynolds number. Roughness plays a minor role as far as laminar pipe flow is concerned.

The wall shear stress and hence the axial pressure drop may be determined if the Reynolds number dependence of the velocity gradient in the laminar sub-layer is known. This information comes basically from experimental measurements. For smooth pipes, the turbulent friction factor is well represented by the formulae due to Blasius.

$$\text{(a)}\ f = \frac{0.316}{Re_D^{0.25}} \text{ for } Re_D < 2 \times 10^4 : \quad \text{(b)}\ f = \frac{0.184}{Re_D^{0.2}} \text{ for } Re_D > 2 \times 10^4 \tag{14.24}$$

Rough pipe data is obtained from experiments on artificially roughened pipes with inner walls coated with sand particles of the desired size. The friction factor is evaluated using Colebrook–White equation.[7]

$$\frac{1}{\sqrt{f}} = 1.14 - 2\log\left(\frac{\epsilon}{D} + \frac{9.35}{Re_D\sqrt{f}}\right) \tag{14.25}$$

This correlation requires iterative solution for f since it is in the form of a transcendental equation. However, it may be solved with just one iteration if the initial guess f_0 is taken as

$$f_0 = \frac{1}{4}\left[\log\left(\frac{\epsilon}{D} + \frac{21.238}{Re_D^{0.9}}\right) - 0.5682\right]^{-2} \tag{14.26}$$

The friction factor data is also available in the form of a graph known as the Moody chart[8] or Moody diagram. The friction factor is plotted for smooth as well as rough tubes as shown in Fig. 14.7.

Example 14.1

Air at a temperature of 20 °C flows through a 19 mm ID tube of copper which may be assumed to be smooth. The mass flow rate through the tube is 0.03 kg/s. Is the flow laminar or turbulent? Calculate the pressure loss across a meter length of tube if the mean pressure is 2 bar.

[7] Colebrook, C. F. and White, C. M., "Experiments with Fluid Friction in Roughened Pipes", Proceedings of the Royal Society of London. Series A, Mathematical and Physical Sciences 161 (906): 367–381, 1937.

[8] L.F. Moody, Trans. ASME, Vol. 66, p. 671, 1944.

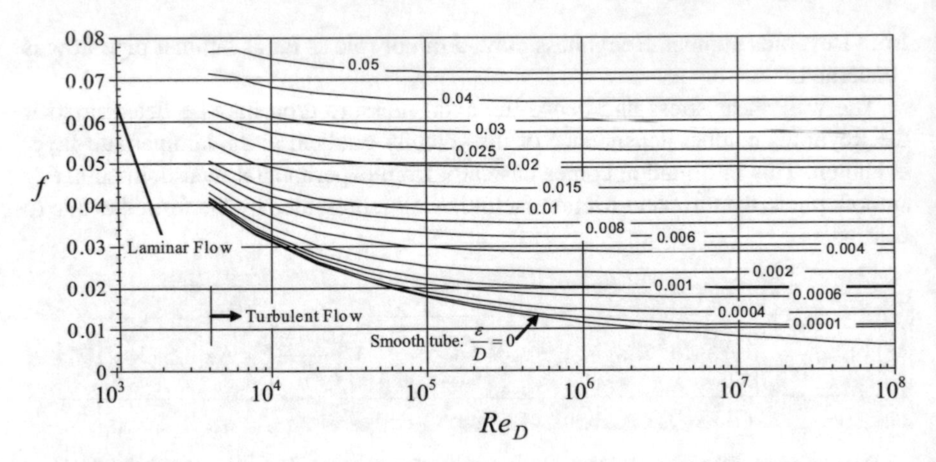

Fig. 14.7 Moody chart for friction factor in smooth and rough tubes

---

Solution:
The properties of air are taken at $p = 2\ bar = 2 \times 10^5\ Pa$, $T = 20\,°\mathrm{C} = 290\,\mathrm{K}$. Assuming air to behave as an ideal gas with $R = 287$ J/kg K, we have

$$\rho = \frac{p}{RT} = \frac{2 \times 10^5}{287 \times 290} = 2.791\ \mathrm{kg/m^3}$$

From air tables (pressure equal to one atmosphere and temperature of 20 °C), we have the dynamic viscosity of air as $\mu = 18.09 \times 10^{-6}$ kg/m s. The kinematic viscosity of air is then obtained as

$$\nu = \frac{\mu}{\rho} = \frac{18.09 \times 10^{-6}}{2.791} = 6.482 \times 10^{-6}\ \mathrm{m^2/s}$$

The diameter of the tube is given as $D = 19$ mm $= 0.019$ m. The mass flow of air is specified as $\dot{m} = 0.03$ kg/s. Hence, the air velocity may be calculated as

$$U = \frac{\dot{m}}{\rho\pi\frac{D^2}{4}} = \frac{0.03}{2.791 \times \pi \times \frac{0.019^2}{4}} = 37.91\ \mathrm{m/s}$$

The tube Reynolds number is then calculated as

$$Re_D = \frac{UD}{\nu} = \frac{37.91 \times 0.019}{6.482 \times 10^{-6}} = 111124$$

The flow is turbulent. Hence, the friction factor may be calculated using Eq. 14.24(b) as

$$f = \frac{0.184}{111124^{0.2}} = 0.018$$

The reader may verify that this value of friction factor agrees with that read off Moody chart. The pressure drop over a meter length of tube is then calculated as

$$-\frac{\Delta p}{L} = \frac{f\rho U^2}{2D} = \frac{0.018 \times 2.791 \times 37.91^2}{2 \times 0.019} = 1900 \text{ Pa/m}$$

Example 14.2

Consider Example 14.1 again but assuming that the tube has become rough with build up of scale and is characterized by a roughness of $\epsilon = 0.05$ mm. Calculate the pressure drop per meter length in this case if all other data remains the same.

Solution:
The friction factor alone changes in this case as compared to the previous example. With $\epsilon = 0.05$ mm, we have $\frac{\epsilon}{D} = \frac{0.05}{19} = 0.0026$. The friction factor calculation is based on the Colebrook–White equation. The initial value is calculated using Eq. 14.26 as

$$f_0 = \frac{1}{4}\left[\log\left(\frac{0.05}{19} + \frac{21.238}{111124^{0.9}}\right) - 0.5682\right]^{-2} = 0.0267$$

Using this in Eq. 14.25, we get a better value of friction factor given by

$$\frac{1}{\sqrt{f}} = 1.14 - 2\log\left(\frac{0.05}{19} + \frac{9.35}{111124 \times \sqrt{0.0267}}\right) = 6.1396$$

or

$$f = \frac{1}{6.1396^2} = 0.0265$$

The reader may note that this also agrees with Moody chart. The pressure drop per meter of tube may hence be calculated as

$$-\frac{\Delta p}{L} = \frac{f\rho U^2}{2D} = \frac{0.0265 \times 2.791 \times 37.91^2}{2 \times 0.019} = 2797 \text{ Pa/m}$$

14.4.2 *Heat Transfer in Turbulent Pipe Flow*

Tubes or pipes are used in most heat exchangers to convey fluids as heat transfer takes place either from or to the fluid flowing inside them. Since turbulent flow and temperature development takes place over a length of a few diameters close to the entrance invariably, the fully developed heat transfer rates are required in applications. An analysis of fully developed temperature problem may be made by starting from the Reynolds averages equations in cylindrical coordinates as applicable to the fully developed region. The velocity profile in the fully developed region follows the universal logarithmic profile as discussed earlier. Analogously one would expect the temperature profile also to show a similar behavior with the turbulent diffusivity showing three regions within the flow field. For details, the reader should refer to advanced books on heat transfer. Here, we shall present some useful correlations.

A characteristic of turbulent pipe flow is that the developmental length is limited to a few diameters. Dimensional arguments have shown that the Nusselt number should be a function of Reynolds and Prandtl numbers. Fully developed turbulent heat transfer correlation is given by the Dittus Boelter equation. [9]

$$Nu_D = 0.023 Re_D^{0.8} Pr^n \tag{14.27}$$

where $n = 0.4$ for heating (i.e., $T_w > T_B$) and $n = 0.3$ for cooling (i.e., $T_w < T_B$). The above correlation is valid within the following range of parameters:

$$Re_D \geq 10^4;\ 0.7 \leq Pr \leq 160;\ \text{and}\ \frac{L}{D} > 10$$

The Dittus Boelter equation is valid only in case property variations of the fluid is not significant. In case property variations are important, the following correlation due to Sieder and Tate[10] is recommended:

$$Nu_D = 0.027 Re_D^{0.8} Pr^{\frac{1}{3}} \left(\frac{\mu}{\mu_w}\right)^{0.14} \tag{14.28}$$

In the above all properties except μ_w are evaluated at the mean fluid temperature. μ_w alone is evaluated at the wall temperature. The range of parameters for which Eq. 14.28 is valid is given below

$$Re_D \geq 10^4;\ 0.7 \leq Pr \leq 16700;\ \text{and}\ \frac{L}{D} > 10$$

[9]F.W. Dittus and L.M.K. Boelter, Univ. of Calif, Berkeley, Publications on Eng., Vol. 2, pp. 443–461, 1930.

[10]Sieder, E.N. and Tate, G.E., Heat transfer and pressure drop of liquids in tubes, Ind. Eng. Chem., Vol. 28, p. 1429, 1936.

14.4.3 *Application of Average Heat Transfer Coefficient Concept to a Practical Application*

Consider a flow in a tube whose wall is maintained at a constant temperature T_w. Energy balance for an elemental control volume shown in Fig. 14.8 gives

$$dQ = \dot{m}C_p dT_B = h(x) \cdot (2\pi R dx)(T_w - T_B)$$

Rearranging this, we have

$$\frac{dT_B}{T_w - T_B} = \frac{2\pi R}{\dot{m}C_p} h(x)dx \tag{14.29}$$

Integrate this between $x = 0$ (entry) and $x = L$ (exit) to get

$$\ln\left[\frac{T_w - T_{B,0}}{T_w - T_{B,L}}\right] = \frac{2\pi R}{\dot{m}C_p}\int_0^L h(x)dx = \frac{2\pi R\bar{h}L}{\dot{m}C_p} \tag{14.30}$$

where the mean heat transfer coefficient has been defined as $\bar{h} = \frac{1}{L}\int_0^L h(x)dx$. Equation 14.30 may be written in the alternate form

$$\frac{T_w - T_{B,L}}{T_w - T_{B,0}} = \mathrm{e}^{-\frac{\bar{h}S_L}{\dot{m}C_p}} \tag{14.31}$$

where $S_L = 2\pi RL = \pi DL$ is the total heat transfer area from pipe entry to exit. Using the relation $\dot{m} = \rho U \frac{\pi D^2}{4}$, we have

$$\frac{\bar{h}S_L}{\dot{m}C_p} = \frac{\bar{h}\pi DL}{\rho U \times \dfrac{\pi D^2}{4} \times C_p} = 4 \times \frac{\bar{h}}{\rho U C_p} \times \frac{L}{D}$$

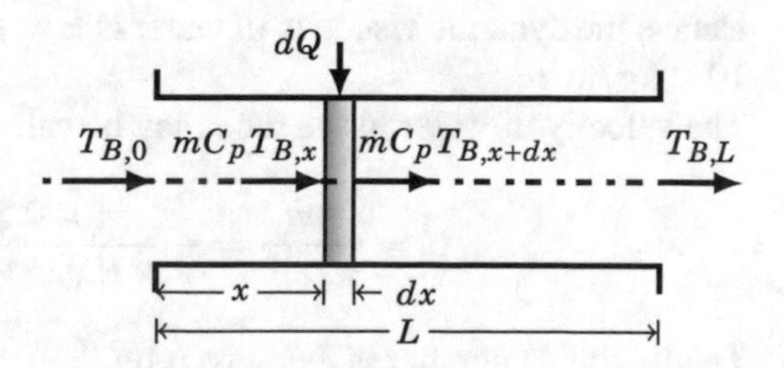

Fig. 14.8 Energy balance for a fluid element

Defining the non-dimensional quantity $\frac{\bar{h}}{\rho U C_p}$ as Stanton number[11] St we get

$$\frac{T_w - T_{B,L}}{T_w - T_{B,0}} = \mathrm{e}^{-\left(4 \times St \times \frac{L}{D}\right)} \tag{14.32}$$

Stanton number is related to the non-dimensional parameters that we are already familiar with as shown under

$$St = \frac{\bar{h}}{\rho U C_p} = \frac{\bar{h} D}{k} \times \frac{k}{\mu C_p} \times \frac{\mu}{\rho U D} = \frac{\overline{Nu_D}}{Re_D Pr} \tag{14.33}$$

Example 14.3

Water at 40 °C at a mass flow rate of 0.5 kg/s enters a 2.5 cm ID tube whose wall is maintained at a uniform temperature of 90 °C. Calculate the tube length required for heating the water to 60 °C. Also determine the pressure drop across the length of pipe.

Solution:
Given data is written down

$$\dot{m} = 0.5 \text{ kg/s}, \; D = 25 \text{ mm} = 0.025 \text{ m}, \; T_{B,0} = 40\,°\text{C}, \; T_{B,L} = 60\,°\text{C}, \; T_w = 90\,°\text{C}$$

Water properties are evaluated at $T_m = \frac{T_{B,0}+T_{B,L}}{2} = \frac{40+60}{2} = 50\,°\text{C}$. From table of properties of water, we have

$$\nu = 0.568 \times 10^{-6} \text{ m}^2/\text{s}, \; k = 0.064 \text{ W/m}\,°\text{C}, \; Pr = 3.68 \text{ and } \rho = 990 \text{ kg/m}^3$$

Hence, the dynamic viscosity of water is $\mu = \rho\nu = 990 \times 0.568 \times 10^{-6} = 5.623 \times 10^{-4}$ kg/m s.
The velocity of water in the tube may be calculated as

$$U = \frac{4\dot{m}}{\pi D^2 \rho} = \frac{4 \times 0.5}{\pi \times 0.025 \times 990} = 1.029 \text{ m/s}$$

The Reynolds number is then given by

[11]Named after Thomas Edward Stanton, 1865–1931, British engineer who studied under Osborne Reynolds.

$$Re_D = \frac{UD}{\nu} = \frac{1.029 \times 0.025}{0.568 \times 10^{-6}} = 45285$$

The flow is turbulent and hence we may use the Sieder and Tate correlation, taking into account the variation of properties, to calculate the Nusselt number. From tables, we read off the kinematic viscosity and density at wall temperature of 90 °C as

$$\nu_w = 0.329 \times 10^{-6}\ \text{m}^2/\text{s}'\ \rho_w = 967.4\ \text{kg/m}^3$$

Hence, the dynamic viscosity of water at the wall temperature is

$$\mu_w = \rho_w \nu_w = 967.4 \times 0.329 \times 10^{-6} = 3.183 \times 10^{-4}\ \text{kg/m s}$$

Hence, we have

$$\frac{\mu}{\mu_w} = \frac{5.623 \times 10^{-4}}{3.183 \times 10^{-4}} = 1.767$$

Clearly the constant property assumption would not be valid. The Sieder and Tate correlation is appropriate. Hence, using Eq. 14.28, we have

$$Nu_D = 0.027 \times 45285^{0.8} \times 3.68^{\frac{1}{3}} \times 1.767^{0.14} = 239.5$$

The mean value of the heat transfer coefficient is then given by

$$\bar{h} = \frac{Nu_D k}{D} = \frac{239.5 \times 0.640}{0.025} = 6131.2\ \text{W/m}^2\,°\text{C}$$

With the various temperatures that are specified, we may calculate the temperature differences ratio as

$$\frac{T_w - T_{B,L}}{T_w - T_{B,0}} = \frac{90 - 60}{90 - 40} = \frac{30}{50} = 0.6$$

The Stanton number may be calculated as

$$St = \frac{Nu_D}{Re_D Pr} = \frac{239.5}{45285 \times 3.68} = 0.001437$$

Hence, using Eq. 14.32,

$$e^{-4 \cdot St \cdot \frac{L}{D}} = 0.6$$

or solving for pipe length

$$L = -\frac{0.025 \times \ln(0.6)}{4 \times 0.001437} = 2.22\ \text{m}$$

Pressure drop is calculated assuming that the tube is smooth. Equation 14.24(b) is appropriate and hence

$$f = \frac{0.184}{45285^{0.2}} = 0.0216$$

The pressure drop over the length of the tube is then given by

$$-\Delta p = f \cdot \frac{L}{D} \cdot \frac{\rho U^2}{2} = 0.0216 \times \frac{2.2}{0.025} \times \frac{990 \times 1.029^2}{2} = 1005.3 \text{ Pa}$$

14.5 Turbulent Boundary Layer over a Flat Plate

14.5.1 Approximate Analysis of Turbulent Flow Parallel to a Flat Plate

Turbulent velocity profile

Extensive measurements indicate that the local friction coefficient in turbulent flow is well correlated by the relation

$$\boxed{C_{fx} = 0.045 \left[\frac{\nu_\infty}{U\delta}\right]^{\frac{1}{4}}} \tag{14.34}$$

valid in the Reynolds number range $5 \times 10^5 \leq Re_x \leq 10^7$. In Eq. 14.34, δ is the local turbulent boundary layer thickness. The velocity profile within the turbulent boundary layer is very well represented by the $\left(\frac{1}{7}\right)$ th power law

$$\boxed{\frac{u}{U_\infty} = \left[\frac{y}{\delta}\right]^{\frac{1}{7}}} \tag{14.35}$$

This relation is obviously not adequate immediately adjacent to the wall since the flow must be laminar there. In fact, the derivative of expression 14.35 at $y = 0$ is indeterminate. In order to circumvent this problem what is done is to evaluate the wall shear stress based on expression 14.34. We make use of the momentum integral to solve for the local boundary layer thickness. The term on the right-hand side of the momentum integral is

$$-\frac{\tau_{w,x}}{\rho_\infty} = -C_{f,x} \times U_\infty^2 = -\frac{0.045}{2} U_\infty^2 \left[\frac{\nu_\infty}{U_\infty \delta}\right]^{\frac{1}{4}} \tag{14.36}$$

The left-side of the momentum integral requires the evaluation of δ_2, the momentum thickness. This may be obtained using the velocity profile given by Eq. 14.35.

$$\begin{aligned}\frac{\delta_2}{U_\infty^2\delta} &= \int_0^1 \frac{u}{U_\infty}\left(\frac{u}{U_\infty}-1\right)d\left(\frac{y}{\delta}\right)\\ &= \int_0^1 \eta^{\frac{1}{7}}(\eta^{\frac{1}{7}}-1)d\eta = \left[\frac{\eta^{\frac{9}{7}}}{\frac{9}{7}} - \frac{\eta^{\frac{8}{7}}}{\frac{8}{7}}\right]\Bigg|_0^1 = -\frac{7}{72}\end{aligned} \tag{14.37}$$

The momentum integral equation (Equation 13.100) may now be simplified to read

$$\frac{7}{72}\frac{d\delta}{dx} = \frac{0.045}{2}\left[\frac{\nu_\infty}{U_\infty\delta}\right] \tag{14.38}$$

Integration of this equation yields

$$\frac{7}{72}\cdot\frac{\delta^{\frac{5}{4}}}{\frac{5}{4}} = \frac{0.045}{2}\left[\frac{\nu_\infty}{U_\infty}\right]^{\frac{1}{4}} x + A \tag{14.39}$$

where A is a constant of integration. We *assume* that the turbulent boundary layer thickness is zero at $x = 0$ and hence choose $A = 0$. The local boundary layer thickness is thus given by

$$\boxed{\frac{\delta}{x} = \frac{0.371}{Re_x^{\frac{1}{5}}}} \tag{14.40}$$

We thus see that the boundary layer thickness varies as $x^{\frac{4}{5}}$ or as $x^{0.8}$. This variation is more rapid than the $x^{0.5}$ growth in the case of the laminar boundary layer. Substitute expression 14.40 back in expression 14.34 to get

$$\boxed{C_{f,x} = 0.045\left[\frac{\nu_\infty Re_x^{\frac{1}{5}}}{0.371 U_\infty x}\right] = 0.0583 Re_x^{-\frac{1}{5}}} \tag{14.41}$$

In order to appreciate the changes that take place in the velocity profile when flow changes from laminar to turbulent flow, we compare laminar and turbulent velocity profiles assuming that the two have the same boundary layer thickness. Figure 14.9 shows that the turbulent velocity profile is "more full" as compared to the laminar velocity profile.

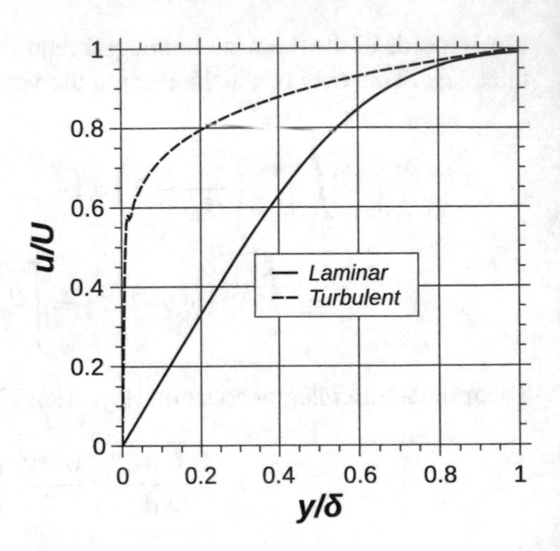

Fig. 14.9 Comparison of laminar and turbulent velocity profiles

14.5.2 Heat Transfer in the Turbulent Boundary Layer over a Flat Plate

The approximate analysis of the velocity boundary layer problem is usually followed by an analysis based on "analogy". Different analogies are discussed below. The common refrain in these is that we link the velocity and temperature fields.

Modified Reynolds or Colburn analogy

The simplest of the analogies is the modified Reynolds or the Colburn analogy. This is a modification of the Reynolds analogy that is strictly valid for a fluid with $Pr_\infty = 1$. Colburn analogy is based on experimental observations and consists in introducing a Prandtl number term in the Reynolds analogy (Eq. 13.60—laminar flow case) as

$$St \cdot Pr_\infty^{\frac{2}{3}} = \frac{C_{f,x}}{2} \tag{14.42}$$

The friction factor in the case of turbulent boundary layer flow given by Eq. 14.41 is modified to $C_{f,x} = 0.0592 Re_x^{-\frac{1}{5}}$ to bring it closer to experimental observations and substituted in the above equation to get

$$St \cdot Pr_\infty^{\frac{2}{3}} = \frac{Nu_x}{Re_x Pr_\infty} \cdot Pr_\infty^{\frac{2}{3}} = \frac{Nu_x}{Re_x Pr_\infty^{\frac{1}{3}}} = \frac{0.0592}{2} Re_x^{-\frac{1}{5}}$$

This may be recast in the form

$$\boxed{Nu_x = 0.0296 Re_x^{0.8} Pr_\infty^{\frac{1}{3}}} \tag{14.43}$$

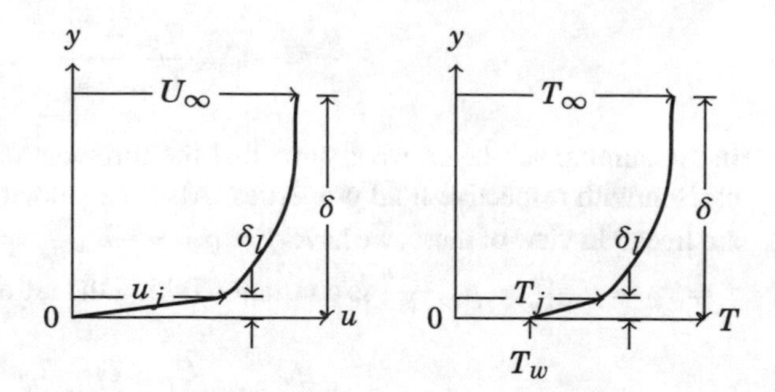

Fig. 14.10 Schematic of the velocity and temperature variations in the turbulent boundary layer

This relation is valid in the Reynolds number range $5 \times 10^5 < Re_x < 10^7$ and within 15% for Reynolds number up to 10^8. The Prandtl number may vary between 0.6 and 60. Note that this expression is very similar to the Dittus Boelter equation that was presented while discussing turbulent tube flow.

We shall discuss two more analogies in the context of heat transfer from a flat plate to a fluid in turbulent flow. The first one, the Prandtl analogy recognizes the existence of a laminar sub-layer and an outer turbulent layer. The second one, the von Karman analogy recognizes the existence of three layers that includes also a buffer layer in between the laminar sub-layer and the outer turbulent layer.

Prandtl analogy

We have earlier presented the universal velocity profile in Fig. 14.4. It is clear that very close to the wall the velocity varies linearly and in a markedly non-linear fashion outside it, in what we may consider as the fully turbulent layer. We assume that the thermal boundary layer will also *show* a conduction layer close to the wall and a fully turbulent region outside it. The velocity and temperature profiles are expected to look like those indicated in Fig. 14.10. In the turbulent boundary layer, the shear stress and heat flux are given by

$$\tau = \rho_\infty(\nu_\infty + \varepsilon)\frac{\partial u}{\partial y}; \quad q = -\rho_\infty C_{p\infty}(\alpha_\infty + \varepsilon_H)\frac{\partial T}{\partial y} \tag{14.44}$$

where we have indicated the time averaged velocity and temperature without the “bar”. In the turbulent part of the boundary layer, we assume that the eddy viscosity and eddy diffusivity are much larger than, respectively, the fluid viscosity and thermal diffusivity. Since ε and ε_H are basically due to similar phenomena, it is reasonable to assume that the turbulent Prandtl number is unity. With these, we have

$$\frac{q}{\tau} = -\frac{C_{p\infty}}{Pr_{\text{turb}}}\frac{\partial T}{\partial u} \approx -C_{p\infty}\frac{\partial T}{\partial u} \tag{14.45}$$

Assuming that the temperature and velocity vary “gently” in the turbulent region, the derivative may be approximated as $\frac{\partial T}{\partial u} = \frac{T_\infty - T_j}{U_\infty - u_j}$. Hence, we have in the turbulent region

$$\frac{q}{\tau} \approx -C_{p\infty}\frac{T_\infty - T_j}{U_\infty - u_j} \tag{14.46}$$

In the laminar sub-layer, we assume that the turbulent quantities are small in comparison with respective fluid properties. Also the velocity and temperature profiles are linear. In view of these, we have $q = q_w = -k_\infty \frac{\partial T}{\partial y} = -k_\infty \frac{T_j - T_w}{\delta_l}$ is constant and $\tau = \tau_w = \mu_\infty \frac{\partial u}{\partial y} = \mu_\infty \frac{u_j - 0}{\delta_l}$ is constant . Taking the ratio of these, we get

$$\frac{q}{\tau} = \frac{q_w}{\tau_w} = -\frac{C_{p\infty}}{Pr_\infty}\frac{T_j - T_w}{u_j} \tag{14.47}$$

We *assume* that the ratio of heat flux to shear stress in the laminar sub-layer is equal to the value in the turbulent layer. From Eq. 14.46, we have

$$T_j = \frac{q_w}{\tau_w}\frac{U_\infty - u_j}{C_{p\infty}} + T_\infty$$

From Eq. 14.47, we have

$$T_j = \frac{q_w}{\tau_w}\frac{u_j Pr_\infty}{C_{p\infty}} + T_w$$

Equating the above two expressions we get

$$\frac{q_w}{\tau_w}\left[\frac{U_\infty - u_j}{C_{p\infty}} - \frac{u_j Pr_\infty}{C_{p\infty}}\right] = T_w - T_\infty$$

Defining the heat transfer coefficient as $h = \frac{q_w}{T_w - T_\infty}$, and rearranging, the above equation becomes

$$h = \frac{\frac{C_{p\infty}\tau_w}{U_\infty}}{1 + \frac{u_j}{U_\infty}(Pr_\infty - 1)} \tag{14.48}$$

Within the laminar sub-layer, we know from our earlier treatment in Chap. 12 that $u^+ = y^+$ and hence $u_j^+ = y_j^+ = 5$. Thus, $u_j = 5u^* = 5\sqrt{\frac{\tau_w}{\rho_\infty}}$. Also the shear stress may be written in terms of the friction coefficient as $\tau_w = C_{fx}\frac{\rho_\infty U_\infty^2}{2}$. Thus, we have

$$\frac{u_j}{U_\infty} = 5\sqrt{\frac{\tau_w}{\rho_\infty U_\infty^2}} = 5\sqrt{\frac{C_{fx}}{2}}$$

Equation 14.48 may then be simplified to read

Prandtl analogy	$: St_x = \dfrac{\frac{C_{fx}}{2}}{1 + 5\sqrt{\frac{C_{fx}}{2}}(Pr_\infty - 1)}$

$$(14.49)$$

Equation 14.49 is a mathematical representation of Prandtl analogy.

von Karman analogy

In this case, the three-layer model is made use of. Skipping the details, the final result is given as

von Karman analogy	$: St_x = \dfrac{\frac{C_{fx}}{2}}{1 + 5\sqrt{\frac{C_{fx}}{2}}\left[(Pr_\infty - 1) + \ln\left(1 + \frac{5}{6}(Pr_\infty - 1)\right)\right]}$

$$(14.50)$$

Example 14.4

Consider flow of air at atmospheric pressure and 300 K parallel to a flat plate 2 m long. The velocity of air far away from the plate is 10 m/s. The plate surface is held at a constant temperature of 400 K. Determine the heat transfer coefficient at the trailing edge of the plate using the formulae based on the various analogies presented in the text. Comment on the results.

Solution:

Step 1 Given data is written down first.

Plate length:	$L = 2\,\text{m}$
Plate temperature:	$T_w = 400\,\text{K}$
Free stream velocity:	$U_\infty = 10\,\text{m/s}$
Free stream temperature:	$T_\infty = 300\,\text{K}$

Step 2 Air properties are taken from tables of properties at the film temperature of $T_f = \frac{400+300}{2} = 350\,\text{K}$.

Density:	$\rho_f = 0.995\,\text{kg/m}^3$
Kinematic viscosity:	$\nu_f = 20.92 \times 10^{-6}\,\text{m}^2\text{/s}$
Thermal conductivity:	$k_f = 0.030\,\text{W/m K}$
Prandtl number:	$Pr_f = 0.7$

Step 3 Then, the Reynolds number at the trailing edge of the plate is

$$Re_L = \frac{U_\infty L}{\nu_f} = \frac{10 \times 2}{20.92 \times 10^{-6}} = 9.56 \times 10^5$$

The flow is turbulent since the Reynolds number is greater than the critical Reynolds number $Re_c = 5 \times 10^5$. The heat transfer coefficient at the trailing edge of the plate is now estimated using the three analogies given in the text.

Step 4 **(a) Colburn analogy:** The Nusselt number is obtained using Eq. 14.43 as

$$Nu_x = 0.0296 \times (9.56 \times 10^5)^{0.8} \times 0.7^{\frac{1}{3}} = 1599.7$$

The heat transfer coefficient is then given by

$$h_L = \frac{Nu_L k_f}{L} = \frac{1599.7 \times 0.030}{2} = 24\ \text{W/m}^2\text{K}$$

Step 5 **(b) Prandtl analogy:** The friction factor at $x = L$ is calculated using Eq. 14.41 with the constant modified to 0.0592, as mentioned in the text, as

$$C_{f,L} = 0.0592 \times (9.56 \times 10^5)^{-\frac{1}{5}} = 0.003769$$

The Stanton number is then obtained using Eq. 14.49 as

$$St_L = \frac{\frac{0.003769}{2}}{1 + 5\sqrt{\frac{0.003769}{2}}(0.7 - 1)} = 0.002016$$

Noting that $St_L = \frac{Nu_L}{Re_L Pr_f}$, we have

$$Nu_L = St_L Re_L Pr_f = 0.002016 \times 9.56 \times 10^5 \times 0.7 = 1349$$

The heat transfer coefficient is then given by

$$h_L = \frac{Nu_L k_f}{L} = \frac{1349 \times 0.030}{2} = 20.2\ \text{W/m}^2\text{K}$$

Step 6 **(c) von Karman analogy:** The friction factor calculated above may now be used in Eq. 14.50 to get the Stanton number as

$$St_L = \frac{\frac{0.003769}{2}}{1 + 5\sqrt{\frac{0.003769}{2}}\left[(0.7 - 1) + \ln\left(1 + \frac{5}{6}(0.7 - 1)\right)\right]} = 0.00216$$

Nu_L is then calculated as

$$Nu_L = St_L Re_L Pr_f = 0.00216 \times 9.56 \times 10^5 \times 0.7 = 1445.5$$

The heat transfer coefficient is then given by

$$h_L = \frac{Nu_L k_f}{L} = \frac{1445.5 \times 0.030}{2} = 21.7 \text{ W/m}^2\text{K}$$

All three analogies give heat transfer coefficient values within a 9% band around a mean value of 22 W/m^{2}K.

The boundary layer in Example 14.4 is visualized as shown in Fig. 14.11. The boundary layer is laminar up to the critical point indicated by the vertical line marked x_c. The boundary layer thickness follows the lower curve till this point. Beyond this point the boundary layer thickness follows the upper curve that has been obtained by assuming that it passes through the origin. Even though the figure makes it appear as though the boundary layer thickness is discontinuous at the critical distance, in reality, there is a small region around the critical point where the boundary layer changes from being laminar to turbulent.

Similar statements may be made with reference to the local friction coefficient as well as the local heat transfer coefficients (h alone is in SI units). Scale factors have been used to fit all the curves within the graph. It is generally observed that there is an increase in both friction and heat transfer after transition from laminar to turbulent flow.

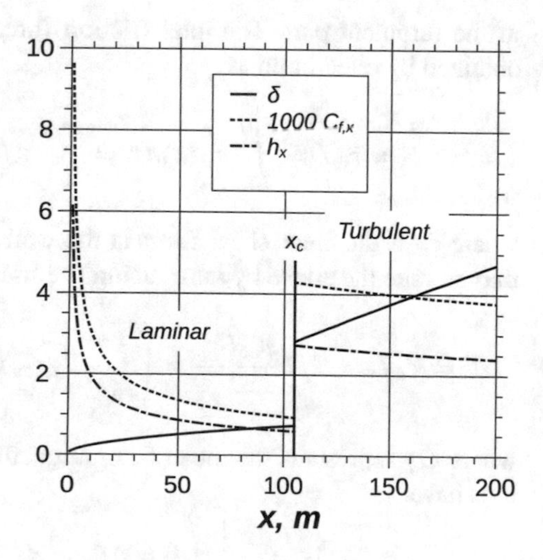

Fig. 14.11 δ, $C_{f,x}$, and h_x variations with x in Example 14.4

14.5.3 Calculation of Drag with Flow Being Partly Laminar and Partly Turbulent

When flow takes place over a flat plate, it may be partly laminar and partly turbulent as seen in Example 14.4. In case there is a transition to turbulent flow, it takes place at a location $x = x_c$ such that $Re_{x_c} = 5 \times 10^5$. For our purpose, it is acceptable to assume that the transition takes place abruptly at $x = x_c$. We may use the laminar friction coefficient from the leading edge up to x_c and the turbulent friction coefficient from $x = x_c$ to the trailing edge at x = L, not withstanding the fact that the friction coefficient is discontinuous at $x = x_c$. The total drag force is obtained by integrating the shear stress at the wall with respect to x, by writing the integral as a sum of two integrals, to allow for the laminar and turbulent parts.

By definition, we have

$$\tau_{w,\mathrm{Lam}}(x) = C_{f,\mathrm{Lam}}(x)\frac{\rho_\infty U_\infty^2}{2} = 0.664 Re_x^{-\frac{1}{2}} \frac{\rho_\infty U_\infty^2}{2}$$

in the laminar part and

$$\tau_{w,\mathrm{Turb}}(x) = 0.0592 Re_x^{-\frac{1}{5}} \frac{\rho_\infty U_\infty^2}{2}$$

in the turbulent part. The total friction force per unit width of the plate may be obtained by integration as

$$F = \overline{\tau}_w L = \int_0^L \tau_w(x)dx = \int_0^{x_c} \tau_{w,\mathrm{Lam}}(x)dx + \int_{x_c}^L \tau_{w,\mathrm{Turb}}(x)dx$$

where $\overline{\tau}_w$ is the mean shear stress at the wall over the entire length of the plate. We may rewrite the above by introducing the friction coefficient as

$$F = \overline{\tau}_w L = \overline{C}_f L \frac{\rho_\infty U_\infty^2}{2} = \left[\int_0^{x_c} C_{f,\mathrm{Lam}}(x)dx + \int_{x_c}^L C_{f,\mathrm{Turb}}(x)dx\right]\frac{\rho_\infty U_\infty^2}{2}$$

where $\overline{C}_f$ represents the mean friction coefficient over the length of the plate. We thus have

$$\overline{C}_f = \frac{1}{L}\left[\int_0^{x_c} 0.664 Re_x^{-\frac{1}{2}} dx + \int_{x_c}^{L} 0.0592 Re_x^{-\frac{1}{5}} dx\right]$$

We may rewrite the above, after multiplying both the numerator and the denominator by $\frac{U_\infty}{\nu_\infty}$ as

$$\overline{C}_f = \frac{1}{Re_L}\left[\int_0^{Re_{x_c}} 0.664 Re_x^{-\frac{1}{2}} dRe_x + \int_{Re_{x_c}}^{Re_L} 0.0592 Re_x^{-\frac{1}{5}} dRe_x\right]$$

The indicated integration may be performed easily and the resulting expression for the mean friction coefficient is

$$\overline{C}_f = \frac{1}{Re_L}\left[0.664 \left.\frac{Re_x^{\frac{1}{2}}}{\frac{1}{2}}\right|_0^{Re_{x_c}} + 0.0592 \left.\frac{Re_x^{\frac{4}{5}}}{\frac{4}{5}}\right|_{Re_{x_c}}^{Re_L}\right]$$

$$= \frac{1}{Re_L}\left[1.328 Re_{x_c}^{\frac{1}{2}} + 0.074 Re_L^{\frac{4}{5}} - 0.074 Re_{x_c}^{\frac{4}{5}}\right]$$

Substituting $Re_{x_c} = 5 \times 10^5$, we get

$$\boxed{\overline{C}_f = \frac{0.074}{Re_L^{0.2}} - \frac{1743}{Re_L}} \qquad (14.51)$$

Calculation of heat transfer with flow being partly laminar and partly turbulent

We look at the calculation of the average heat transfer coefficient for flow past a flat plate wherein the flow is partly laminar and partly turbulent. Assuming that the transition takes place abruptly at $x = x_c$, as in the above, the average heat transfer coefficient may be defined as

$$\overline{h} = \frac{1}{L}\left[\int_0^{x_c} h_{\text{Lam}}(x)dx + \int_{x_c}^{L} h_{\text{Turb}}(x)dx\right]$$

To keep the algebra simple, we shall make use of Colburn analogy to treat heat transfer in turbulent flow. The heat transfer coefficients appearing on the right-hand side of the above equation are given by Eq. 13.53 (based on the solution to the Blasius equation) and Eq. 14.43 (based on Colburn analogy). Using these expressions, the heat transfer coefficients in the laminar and turbulent regions of flow are written down as

$$h_{\rm Lam}(x) = \frac{Nu_x k}{x} = \frac{U_\infty k}{\nu_\infty}\frac{Nu_x}{Re_x}$$
$$= \frac{U_\infty k}{\nu_\infty} \times \frac{0.332 Re_x^{\frac{1}{2}} Pr_\infty^{\frac{1}{3}}}{Re_x} = \frac{U_\infty k}{\nu_\infty} \times 0.332 Re_x^{-\frac{1}{2}} Pr_\infty^{\frac{1}{3}}$$
$$h_{\rm Turb}(x) = \frac{U_\infty k}{\nu_\infty} \times \frac{0.0296 Re_x^{\frac{4}{5}} Pr_\infty^{\frac{1}{3}}}{Re_x} = \frac{U_\infty k}{\nu_\infty} \times 0.0296 Re_x^{-\frac{1}{5}} Pr_\infty^{\frac{1}{3}}$$

Substitute these in the equation defining the mean heat transfer coefficient to represent the mean Nusselt number as

$$\overline{Nu} = \frac{\overline{h}L}{k_\infty} = Pr_\infty^{\frac{1}{3}} \left[0.332 \int_0^{Re_{x_c}} \frac{dRe_x}{Re_x^{\frac{1}{2}}} + 0.0296 \int_{Re_{x_c}}^{Re_L} \frac{dRe_x}{Re_x^{\frac{1}{5}}} \right] \quad (14.52)$$

Performing the indicated integration, using $Re_{x_c} = 5 \times 10^5$, we arrive at the following expression for the average Nusselt number:

$$\boxed{\overline{Nu} = \left[0.037 Re_L^{0.8} - 871.3\right] Pr_\infty^{\frac{1}{3}}} \quad (14.53)$$

Example 14.5

Consider air at a free stream temperature of 10 °C and free stream velocity of 15 m/s flowing parallel to a flat plate 1.5 m long and held at a temperature of 90 °C. Calculate the heat transfer from one side to the air stream. Also what is the drag force experienced by the plate? Consider unit width of the plate.

Solution:

Step 1 Given data is written down as

$$U_\infty = 15 \text{ m/s},\ T_\infty = 10\,°\text{C},\ T_w = 90\,°\text{C},\ L = 1.5 \text{ m}$$

Step 2 Air properties are based on the film temperature of $T_f = \frac{10+90}{2} = 50\,°\text{C}$. They are

Density:	$\rho_f = 1.088\,\text{kg/m}^3$
Kinematic viscosity:	$\nu_f = 18.65 \times 10^{-6}\,\text{m}^2/\text{s}$
Thermal conductivity:	$k_f = 0.0281\,\text{W/m K}$
Prandtl number:	$Pr_f = 0.703$

Step 3 Reynolds number based on plate length L is

$$Re_L = \frac{U_\infty L}{\nu_f} = \frac{15 \times 1.5}{18.65 \times 10^{-6}} = 1.2 \times 10^6$$

Since the Reynolds number based on plate length is greater than the critical Reynolds number, the flow is partly laminar and partly turbulent.

Step 4 We use Eq. 14.53 to calculate the mean Nusselt number.

$$\overline{Nu} = \left[0.037 \times (1.2 \times 10^6)^{0.8} - 871.3\right] 0.703^{\frac{1}{3}} = 1627$$

The average heat transfer coefficient is then obtained as

$$\overline{h} = \frac{\overline{Nu} k_f}{L} = \frac{1627 \times 0.0281}{1.5} = 30.5 \text{ W/m}^2\text{K}$$

Step 5 Heat transferred from one side of the plate is then given by

$$Q = \overline{h} L (T_w - T_\infty) = 30.5 \times 1.5(90 - 10) = 3658 \text{ W/m}$$

Step 6 The drag force experienced by the plate is calculated based on the mean friction coefficient given by expression 14.51.

$$\overline{C}_f = \frac{0.074}{(1.2 \times 10^6)^{0.2}} - \frac{1743}{1.2 \times 10^6} = 0.00305$$

The mean shear stress at the wall is given by

$$\overline{\tau}_w = \overline{C}_f \frac{\rho_f U_\infty^2}{2} = 0.00305 \times \frac{1.088 \times 15^2}{2} = 0.373 \text{ Pa}$$

The drag force experienced by the plate is then given by

$$F = \overline{\tau}_w L = 0.373 \times 1.5 = 0.560 \text{ N/m}$$

14.6 Cylinder in Cross Flow

In Chap. 13, we have considered in detail laminar flow past a cylinder. Here our concern will be with turbulent flow past a cylinder. Since the analysis of turbulent flow is fairly complex, we shall present some physical features of turbulent flow and present correlations that will be of use in solving problems.

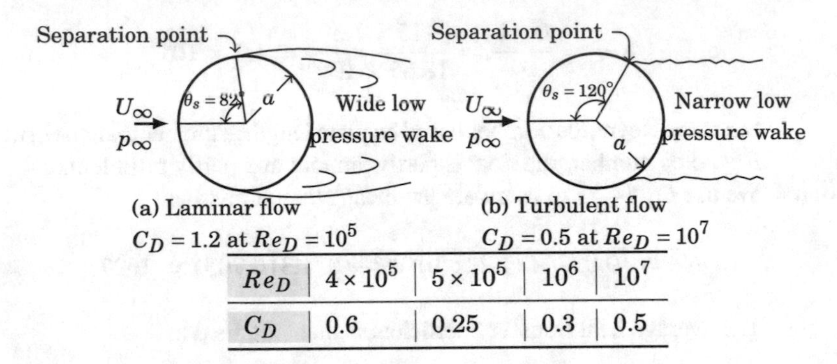

Re_D	4×10^5	5×10^5	10^6	10^7
C_D	0.6	0.25	0.3	0.5

Fig. 14.12 Change of flow pattern close to transition to turbulent flow of a fluid flowing normal to a cylinder **a** laminar flow **b** turbulent flow

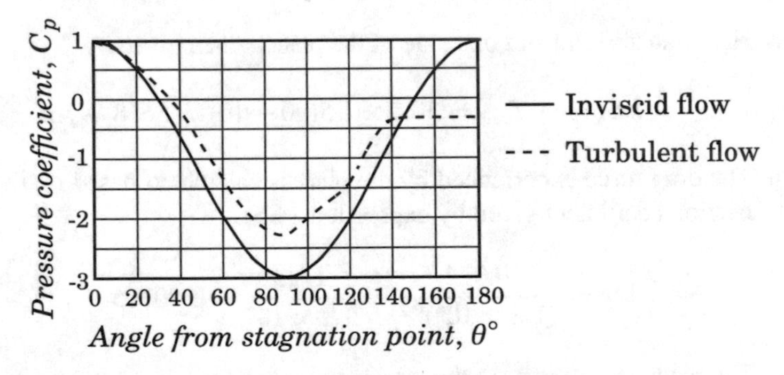

Fig. 14.13 Pressure distribution around a cylinder in cross flow: comparison of potential and turbulent flow pressure distributions

Consider the differences in flow pattern when laminar flow across a cylinder changes to turbulent flow at about $Re_D = 4 \times 10^5$ as shown in Fig. 14.12. The separation point that occurs around $\theta = 82°$ in laminar flow moves downstream to $\theta = 120°$ when the flow becomes turbulent. There is thus a smaller wake with low pressure and hence the form drag reduces as the flow becomes turbulent. The pressure distribution under turbulent conditions based on measured data is shown in Fig. 14.13. The C_D values shown in Fig. 14.12 clearly show this effect. There is an abrupt reduction in the C_D value at $Re_D = 4 \times 10^5$ from 0.6 to 0.25. These facts are also clear from the C_D versus Re_D curve shown in Fig. 14.14.

We have already seen that the flow and temperature fields show analogous variations. Thus, we expect the heat transfer coefficient to vary with angle θ in a way that mirrors the variation in the flow pattern. In applications it is seldom that we are interested in the local Nusselt number. It is the mean Nusselt number that is important to us. The Reynolds number and the Nusselt number are usually based on the cylinder diameter. The mean Nusselt number is correlated by an equation of form

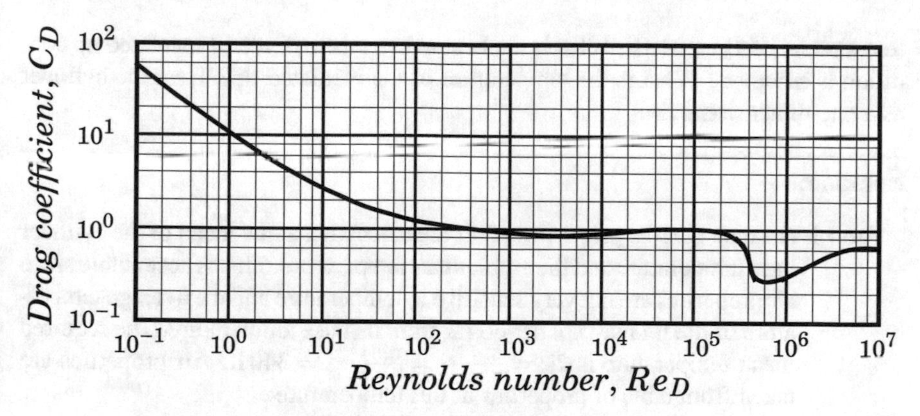

Fig. 14.14 Variation of drag coefficient with Reynolds number for flow across a smooth cylinder

Table 14.1 Constants in the Zhukauskas correlation

Re_D range	C	m
1–40	0.75	0.4
$40-10^3$	0.51	0.5
$10^3-2\times10^5$	0.26	0.6
$2\times10^5-10^6$	0.076	0.7

$$Nu_D = C Re_D^m Pr^n \left(\frac{Pr_\infty}{Pr_w}\right)^{\frac{1}{4}} \tag{14.54}$$

Recently, based on experimental data, Zhukauskas[12] has given the set of C, m, n values that are to be used in Eq. 14.54 and the range of validity of these. Table 14.1 gives the values of the constants along with the ranges of applicability.

The above correlation is valid for $0.7 \leq Pr \leq 500$, $1 \leq Re_D \leq 10^6$. n is specified as 0.37 for $Pr \leq 10$ and as 0.36 for $Pr > 10$. All properties are calculated at the mean temperature, $\frac{T_w+T_\infty}{2}$. Pr_∞ and Pr_w are evaluated at free stream temperature T_∞ and wall temperature T_w, respectively.

Example 14.6

Air at a temperature of $T_\infty = 370\,\text{K}$ is flowing normal to a cylinder at an average velocity of $U = 10\,\text{m/s}$. The cylinder made of aluminum has a diameter of $D = 5\,\text{mm}$ and is $L = 7.5\,\text{cm}$ long. The base of the cylinder is maintained at a

[12] A. Zhukauskas, in *Advances in Heat Transfer*, J.P. Hartnett and T.F. Irvine Jr (Eds.), Vol. 8, pp. 93–160, Academic Press N.Y.

temperature of $T_b = 310\,\text{K}$. What is the heat gain by the cylinder if insulated tip condition is assumed? What is the temperature of the insulated tip? Treat the cylinder as a one-dimensional fin.

Solution:

Step 1 Assume that the properties of air are calculated at the mean of the cylinder base temperature and the ambient air temperature. This is acceptable since air properties are not very sensitive to temperature and the average temperature of the fin may not be too far from its base temperature. The required mean temperature is $T_m = \frac{T_b+T_\infty}{2} = \frac{310+370}{2} = 340\,\text{K}$. Air properties are taken from table of properties at this temperature:

Density:	$\rho_m = 1.0382\,\text{kg/m}^3$
Kinematic viscosity:	$\nu_m = 19.55 \times 10^{-6}\,\text{m}^2/\text{s}$
Thermal conductivity:	$k_m = 0.0293\,\text{W/m}^\circ\text{C}$
Prandtl number:	$Pr_m = 0.7$

Thermal conductivity of aluminum, the cylinder material is taken as $k_{Al} = 207\,\text{W/m}^\circ\text{C}$.

Step 2 Reynolds number based on the cylinder diameter is calculated as

$$Re_D = \frac{UD}{\nu_m} = \frac{10 \times 0.005}{19.55 \times 10^{-6}} = 2558$$

Step 3 Zhukauskas correlation is used now to calculate the convective heat transfer coefficient. For the Reynolds number range that brackets the above value, Table 14.1 gives $C = 0.26$, $m = 0.6$, and $n = 0.37$. Even though the cylinder length is finite, we assume that the flow is largely two-dimensional since $\frac{L}{D} = \frac{0.075}{0.005} = 15$ is large. Hence, we have

$$Nu_D = 0.26 \times 2558^{0.6} \times 0.7^{0.37} = 25.3$$

The average heat transfer coefficient then is

$$\overline{h} = \frac{Nu_D k_m}{D} = \frac{25.3 \times 0.0293}{0.005} = 148\ \text{W/m}^2\,{}^\circ\text{C}$$

Step 4 The non-dimensional fin parameter is calculated now as

$$\mu = L\sqrt{\frac{4\overline{h}}{k_{Al} D}} = 0.075\sqrt{\frac{4 \times 148}{207 \times 0.005}} = 1.793$$

Step 5 The fin efficiency is calculated as

$$\eta = \frac{\tanh \mu}{\mu} = \frac{\tanh 1.793}{1.793} = 0.528$$

Step 6 The heat gain by the cylinder is then calculated as

$$Q = \pi D L \bar{h}(T_\infty - T_b)\eta = \pi \times 0.005 \times 0.075 \times 148(370 - 310) \times 0.528$$
$$= 5.52 \text{ W}$$

Step 7 The non-dimensional tip temperature is given by

$$\theta_L = \frac{1}{\cosh \mu} = \frac{1}{\cosh 1.793} = 0.324$$

But the non-dimensional tip temperature is $\theta_L = \frac{T_L - T_\infty}{T_b - T_\infty}$. Hence, we have

$$T_L = T_\infty + \theta_L(T_b - T_\infty) = 370 + (310 - -370) \times 0.324 = 350.6 \text{ K}$$

14.6.1 *Heat Transfer for Flow Normal to a Tube Bank*

Most heat exchangers have more than one cylinder arranged such that the flow of one fluid takes place past them while a second fluid flows through them. In cross flow heat exchangers (see Chap. 15), the flow is normal to the tube bank and hence practical formulae for calculation of heat transfer on the outside surfaces of a tube bank are required. These are based on experimental data and are of the general form

$$\overline{Nu_D} = C Re_D^m Pr^{0.36} \left(\frac{Pr_\infty}{Pr_w}\right)^{\frac{1}{4}} \tag{14.55}$$

where tube diameter is used as the characteristic length scale. The other symbols have their usual meanings. The factor containing the ratio of Prandtl numbers accounts for variation of thermo-physical properties of the fluid. The relation given above depends on the range of parameters as well as the arrangement of tubes in the bank. Two possible arrangements of tubes in a bank are (a) Aligned or in line arrangement and (b) Staggered arrangement, as shown in Fig. 14.15.

The bank of tubes are described by the transverse pitch S_T, the longitudinal pitch S_L, and the diagonal pitch S_D, as appropriate. The heat transfer is affected by the flow through the free space available between the tubes. A downstream row of tubes

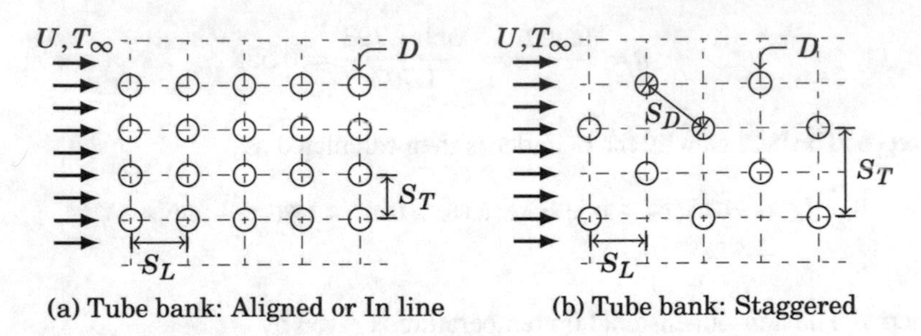

(a) Tube bank: Aligned or In line (b) Tube bank: Staggered

Fig. 14.15 Two different arrangements of tubes in a tube bank

Table 14.2 Constants for use with correlating Eq. 14.55

	Aligned		Staggered	
Re_D range	C	m	C	m
$10 - 10^2$	0.8	0.4	0.9	0.4
$10^2 - 10^3$	Use single tube formula			
$10^3 - 2 \times 10^5$	$\frac{S_T}{S_L} < 0.7$		$\frac{S_T}{S_L} < 2$	
	Do not use		0.35	0.6
	$\frac{S_T}{S_L} > 0.7$		$\frac{S_T}{S_L} > 2$	
	0.27	0.63	0.4	0.6
$2 \times 10^5 - 10^6$	0.021	0.84	0.022	0.84

will be affected by the flow in the wake of the tubes in the upstream row and these effects depend on the various geometric parameters mentioned above. The constants in the correlation Eq. 14.55 are chosen from the values indicated in Table 14.2. The correlation given above is valid under the following conditions:

$$\boxed{10 \leq Re_D \leq 10^6, \ 0.7 \leq Pr \leq 500} \tag{14.56}$$

Example 14.7

Air at free stream temperature of $T_\infty = 90\,°\text{C}$ flows past a bank of tubes at a free stream velocity of $U = 3\,\text{m/s}$. The tubes in the bank are of diameter $D = 0.018\,\text{m}$ each and arranged in a staggered arrangement with $S_T = 2D$, $S_L = 2D$. Determine the mean heat transfer coefficient if the tubes are maintained at a mean temperature of $T_w = 30\,°\text{C}$. What happens if the tubes are aligned with the same transverse and longitudinal pitches?

Solution:

Step 1 Since properties of air are not very sensitive to temperature, the factor involving the ratio of Prandtl numbers may be taken as unity. The fluid properties, as usual, are evaluated at the mean temperature given by $T_m = \frac{T_\infty + T_w}{2} = \frac{90+30}{2} = 60\,°\text{C}$.

Density:	$\rho_m = 1.0496\,\text{kg/m}^3$
Kinematic viscosity:	$\nu_m = 19.08 \times 10^{-6}\,\text{m}^2/\text{s}$
Thermal conductivity:	$k_m = 0.029\,\text{W/m°C}$
Prandtl number:	$Pr_m = 0.7$

Step 2 The Reynolds number may be determined as

$$Re_D = \frac{UD}{\nu_m} = \frac{3 \times 0.018}{19.08 \times 10^{-6}} = 2830$$

Staggered tube arrangement

Step 3 The ratio of transverse to longitudinal pitch is given by

$$\frac{S_T}{S_L} = \frac{2D}{2D} = 1$$

The constants in the Nusselt number correlation are chosen from Table 14.2 as $C = 0.35,\ m = 0.6$. The Nusselt number is obtained using Eq. 14.55 as

$$\overline{Nu}_D = 0.35 \times 2830^{0.6} \times 0.7^{0.36} = 36.3$$

Step 4 The mean heat transfer coefficient may then be obtained as

$$\bar{h} = \frac{\overline{Nu}_D k_m}{D} = \frac{36.3 \times 0.029}{0.018} = 58.4\,\text{W/m}^2\,°\text{C}$$

Aligned tube arrangement

Step 5 The constants in the Nusselt number correlation are chosen from Table 14.2 as $C = 0.27,\ m = 0.63$. The Nusselt number is obtained using Eq. 14.55 as

$$\overline{Nu}_D = 0.27 \times 2830^{0.63} \times 0.7^{0.36} = 35.5$$

Step 6 The mean heat transfer coefficient may then be obtained as

$$\bar{h} = \frac{\overline{Nu}_D k_m}{D} = \frac{35.5 \times 0.029}{0.018} = 57.2\,\text{W/m}^2\,°\text{C}$$

Example 14.8

Water at free stream temperature of $T_\infty = 60\,°\text{C}$ flows past a bank of tubes at a free stream velocity of $U = 0.5\,\text{m/s}$. The tubes in the bank are of diameter $D = 0.018\,\text{m}$ each and arranged in a staggered arrangement with $S_T = 2D,\ S_L = 2D$. Determine the mean heat transfer coefficient if the tubes are maintained at a mean temperature of $T_w = 20\,°\text{C}$.

Solution:

Step 1 The fluid properties, as usual, are evaluated at the mean temperature given by $T_m = \frac{T_\infty + T_w}{2} = \frac{60+20}{2} = 40\,°\text{C}$.

Density:	$\rho_m = 992.3\,\text{kg/m}^3$
Kinematic viscosity:	$\nu_m = 6.564 \times 10^{-7}\,\text{m}^2/\text{s}$
Thermal conductivity:	$k_m = 0.630\,\text{W/m°C}$
Prandtl number:	$Pr_m = 4.32$

The Prandtl numbers at the wall and free stream temperatures are

$$Pr_\infty = 2.967,\quad Pr_w = 6.957$$

Step 2 The Reynolds number may be determined as

$$Re_D = \frac{UD}{\nu_m} = \frac{0.5 \times 0.018}{6.564 \times 10^{-7}} = 13711$$

The ratio of transverse to longitudinal pitch is given by

$$\frac{S_T}{S_L} = \frac{2D}{2D} = 1$$

Step 3 The constants in the Nusselt number correlation are chosen from Table 14.2 as $C = 0.35,\ m = 0.6$. The Nusselt number is obtained using Eq. 14.55 as

$$\overline{Nu}_D = 0.35 \times 13711^{0.6} \times 4.32^{0.36} \left(\frac{2.967}{6.957}\right)^{0.25} = 145.4$$

Step 4 The mean heat transfer coefficient may then be obtained as

$$\bar{h} = \frac{\overline{Nu}_D k_m}{D} = \frac{145.4 \times 0.630}{0.018} = 5089\ \text{W/m}^2\,°\text{C}$$

Concluding Remarks

Even though this chapter has been short because turbulent flow and heat transfer analysis is complex and beyond the scope of the present book, a working knowledge is made available to the reader. Most of the results have been based on empirical experimental knowledge. The material in this chapter should be treated as a preparation for the reader to explore more advanced treatment in books that deal specifically with turbulent flow and heat transfer.

14.7 Exercises

Ex 14.1: A fluid of kinematic viscosity equal to $1.5 \times 10^{-6}\,\text{m}^2/\text{s}$ flows with an average velocity of 10 m/s in a square duct of 0.08×0.08 m cross section. What is the Reynolds number based on the hydraulic diameter? Is the flow laminar or turbulent? What is the Nusselt if the flow is fully developed and the Prandtl number is 0.7?

Ex 14.2: A 2 m long pipe carries high pressure water at an inlet pressure of 2 bars. The diameter of the pipe is 18 mm and the mass flow rate is 1 kg/s. What is the pressure drop due to friction between the inlet and the outlet? If there are 1000 such tubes in parallel in the facility, what is the pumping power required?

Ex 14.3: Water flows with a mean velocity of 1.7 m/s in a boiler tube of inner diameter of 0.018 m. The tube is 2.5 m long and the wall of the tube is at a mean temperature of 90 °C. What is the pressure drop over the length of the pipe? How much heat will be gained by water as it flows through the tube length? Take into account the variation of water properties with temperature, in the usual way.

Ex 14.4: A tube carries a flow of a certain fluid at a temperature of 20 °C. Due to malfunction in the cooling system, the fluid temperature increases abruptly to 43 °C. The density ρ and viscosity μ of the fluid are temperature dependent and are given by

$$\mu = \mu_{20}\{1 - 0.0005(T - 20)\} \quad \text{and} \quad \rho = \rho_{20}\{1 + 0.015(T - 20)\}$$

where the quantities are all based on SI units, T is in °C, and the subscript 20 stands for the value of the property at 20 °C. The control system that controls the pump, however, maintains the flow velocity constant. Determine the fractional change in the pumping power. It is known that the flow remains turbulent throughout.

Ex 14.5: Saturated steam at 2 atm condenses over a tube carrying cold water. The heat transfer coefficient for condensing steam is so large that the outside wall of the tube may be assumed to be at the saturation temperature of the condensing steam. The cold water enters at 35 °C and a mean velocity of 1.7 m/s. The tube is made of an alloy of thermal conductivity 125 W/m K, has an ID of 18 mm and an OD of 21 mm. The tube is 2.08 m long. What is the exit temperature of water?

Ex 14.6: Hot air at atmospheric pressure and 350 K enters a bare duct of square cross section 15 × 15 cm that passes through a 2.5 m long room. The volume flow rate of air is 0.1 m^3/s. The duct wall is observed to be more or less at a constant temperature of 330 K. Determine the exit temperature of the air and also the pressure drop between entry and exit.

Ex 14.7: Water at a temperature of 30 °C flows with free stream velocity of 2.5 m/s parallel to a flat plate maintained at a uniform temperature of 70 °C. The plate is 2 m long and 1 m wide. Determine the total drag force on one side of the plate. Also determine the total heat loss from the plate. Determine also the thickness of the boundary layer as the flow leaves the plate.

Ex 14.8: A flat fin of 6061 Aluminum of thermal conductivity 180 W/m K and density 2760 kg/m^3 is attached to a base structure at 124 °C. It is exposed on both sides to cool air stream at 30 °C flowing with a speed of 5 m/s. The width of the fin (parallel to the direction of flow) is 0.25 m. What is the maximum heat that may be dissipated if the mass of the fin is to be limited to 0.1 kg? What is the heat loss per unit mass? Calculate a mean heat transfer coefficient modeling the flow of air as flow past a flat plate.

Ex 14.9: A gas of kinematic viscosity 10^{-6} m^2/s flows parallel to a flat plate with a velocity of 10 m/s. The plate is 0.75 m long. Determine the location where the boundary layer becomes turbulent. Also determine the thickness of the boundary layer at (a) a location 0.2 m from the leading edge and (b) at the trailing edge of the plate.

Ex 14.10: Air flows with a velocity of 5 m/s through a duct of 0.8 m diameter, in an air handling system. The duct is effectively 15 m long. Determine the pressure drop across the duct length. Assume that air enters the duct at a temperature of 70 °C and leaves the duct at a mean temperature of 50 °C.

Ex 14.11: Turbulent boundary layers may be analyzed using an approximate method proposed by Head.[13] Heat transfer in turbulent boundary layers may be analyzed using an approximate method proposed by Jander.[14] Study these two papers and apply these to turbulent flow across a cylinder.

Ex 14.12: A cylindrical pin of 12 mm diameter and 100 mm length is maintained at a constant temperature of 70 °C. The pin is placed with its axis normal to a stream of 15 °C air moving with a velocity of 5 m/s. What is the heat loss from the pin to air. The two ends of the pin are perfectly insulated.

Ex 14.13: It is desired to protect a long cylindrical bolt of steel of 25 mm diameter by a 15 mm thick layer of mineral insulation that has a thermal conductivity of 1.5 W/m°C. The outer surface of the insulation is exposed to 300 °C air flowing at a velocity of 2.5 m/s normal to the axis of the bolt. How much

---

[13] M.R. Head, A.R.C. Technical Report R. & M. No. 3152, 1960.

[14] B. Schulz-Jander, Acta Mechanica, Vol. 21, pp. 301–312, 1975.

heat is to be removed from a unit length of the bolt if the temperature along its axis is not to exceed 80 °C? What is the temperature of the surface of insulation that is in contact with air? What is the temperature at the interface between the steel bolt and the insulation layer? Assume that there is no contact resistance at the interface between the bolt and the insulation layer.

Ex 14.14: A cylindrical 150 mm long pin fin is made of an alloy with a thermal conductivity of 126 W/m°C. The diameter of the pin is 8 mm. The base of the pin is maintained at a constant temperature of 80 °C while the pin is exposed to 30 °C air stream moving with a velocity of 5 m/s normal to its axis. Determine the temperature of the pin fin tip as well as the heat loss from its surface. The air properties may be taken at the average of the base temperature and the air temperature.

Ex 14.15: A bank of tubes is arranged in an in line arrangement. There are ten rows in the bank. The tubes have an OD of 25 mm. The transverse pitch is twice the tube diameter. The longitudinal pitch is 1.25 times the transverse pitch. Air at a free stream temperature of 300 °C flows across the tube bank with a velocity of 2 m/s. The tubes may be assumed to be at an average temperature of 150 °C. What is the mean heat transfer coefficient for heat transfer between air and the tubes?

Ex 14.16: Rework Exercise 14.15 if the tube bank is arranged in the staggered form with longitudinal and transverse pitches remaining the same. All other conditions also remain the same.

Chapter 15
Heat Exchangers

THE heat transfer principles and methods learnt in the previous chapters find application here in the analysis of heat exchangers. The LMTD and $\epsilon - NTU$ approaches are described in detail in this chapter. Different types of heat exchangers used in engineering applications are described and analyzed.

15.1 Introduction

The study of "Heat Transfer" is basically undertaken with the idea that an engineer will be able to analyze or design heat transfer equipments. Also he will be able to apply the basic ideas learnt in this subject to design thermal protection systems in space applications, design insulating systems for ovens, boilers, engines, turbines and so on. At least during the preliminary stages he may idealize components by studying or analyzing them in isolation, using the methods that have been presented in this book, before putting them together. In order to carry this activity forward, he may have to equip himself with the study of design of thermal systems, a field that is covered in specialized books.

Heat transfer between two fluids separated across a boundary is very common in thermal engineering. Typical applications include thermal power plants, chemical process equipments, food processing equipments, air conditioning equipments, and so on. We may classify heat exchangers in many different ways. Firstly, they may be classified according to the nature of fluids involved. The fluids may be in a single phase, liquid, or gas (vapor). Examples are

S. P. Venkateshan, *Heat Transfer*,
https://doi.org/10.1007/978-3-030-58338-5_15

- **Liquid to liquid**: Oil cooler with oil as the hot liquid being cooled by liquid water
- **Gas to liquid**: Economizer in a steam power plant where heat exchange is between hot flue gases and boiler feed water
- **Gas to gas**: Air pre-heater in a steam power plant where flue gases heat up combustion air

One of the fluids undergoing heat exchange may be going through a change of phase:

- **Boiler**: water evaporates on being heated by gases generated by a combustion process or by electric heaters
- **Condenser**: Steam or a vapor of an organic fluid (in refrigeration applications) condenses over a surface cooled by coolant such as water
- **Distillation process**: Potable water is obtained from brackish water by distillation that involves evaporation followed by condensation

Heat exchangers may also be classified according to the direction of flow of the two fluids, within the heat transfer equipment. In the counter-current heat exchanger, the two fluids flow in opposite directions. In the co-current or parallel flow heat exchanger, the two fluids flow parallel to each other entering together at one end and exiting together at the other end. Figure 15.1a, b show the schematic of these two types of heat exchangers.

The nomenclature for heat exchangers is also indicated in the figures. Subscript h and c stand for the hot fluid and cold fluid, respectively. The subscript i and o indicate the inlet and outlet, respectively. The mass flow rate ($\dot{m}$) specific heat (C_p) product for each of the fluid streams occurs in the analysis and is indicated by $C_c = \dot{m}_c C_{pc}$ and $C_h = \dot{m}_h C_{ph}$. Temperature variation in the two types of heat exchangers are schematically shown in Fig. 15.2a, b.

We assume that all the heat that leaves the hot fluid reaches the cold fluid. This is possible if the heat exchanger is perfectly insulated from the ambient. In the co-

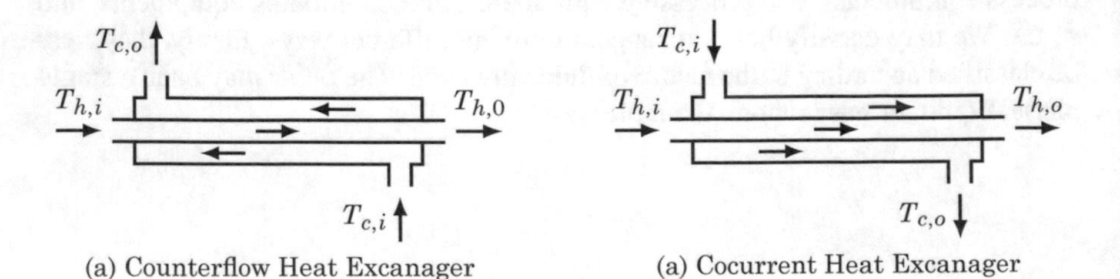

Fig. 15.1 Schematic of basic heat exchanger types

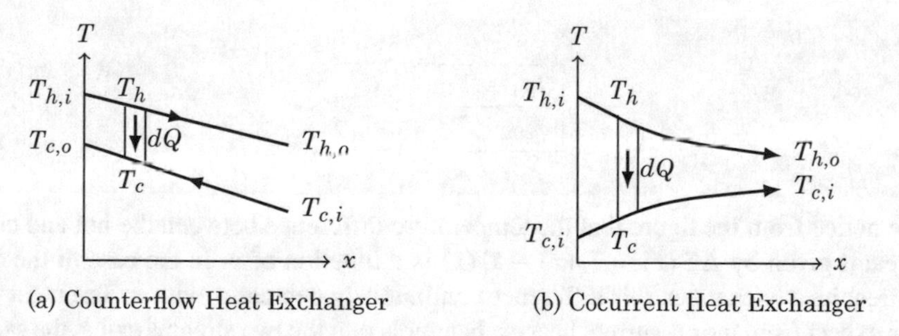

Fig. 15.2 Hot and cold fluid temperature variations in the two types of heat exchangers shown in Fig. 15.1

current heat exchanger, the driving temperature difference continuously decreases from end to end. In the case of counter-current heat exchanger, it varies much less from end to end. Later, we shall see that this has an important bearing on the performance of these heat exchangers.

15.2 Analysis of Heat Exchangers

15.2.1 Thermodynamic Analysis of a Co-Current Heat Exchanger

Heat exchanger basics are introduced here by analyzing a co-current heat exchanger shown in Fig. 15.1b. The two fluids enter at the left, proceed parallel to each other, and exit at the right. In the illustration, the hot fluid flows through the tube and the cold fluid through the annulus. With the mass flow rate specified in kg/s and specific heat specified in J/kg K, the mass flow rate specific heat product will have the unit of W/K or W/°C. The smaller of C_h and C_c will be represented as C_{min}, and the larger of the two will be represented as C_{max}. The ratio $\frac{C_{min}}{C_{max}}$ is represented by the symbol R. A heat exchanger is said to be balanced if $C_h = C_c$ and hence $R = 1$. Figure 15.2b shows the variation of temperature of the two fluids as they move in and out of the heat exchanger. Assume that there is no heat loss and that heat exchange takes place *only* between the hot and cold fluids. The following should then hold:

$$\begin{bmatrix}\text{Rate of heat gained}\\ \text{by the cold stream}\end{bmatrix} = \begin{bmatrix}\text{Rate of heat lost}\\ \text{by the hot stream}\end{bmatrix}$$

or

$$\boxed{C_c(T_{c,o} - T_{c,i}) = C_h(T_{h,i} - T_{h,o})}$$

We may rearrange this as

$$\boxed{\frac{T_{h,i} - T_{h,o}}{T_{c,o} - T_{c,i}} = \frac{C_c}{C_h}} \tag{15.1}$$

We notice from the figure that the temperature difference between the hot and cold streams given by $\Delta T(x) = T_h(x) - T_c(x)$ is a function of x. In the case of the co-current heat exchanger, this difference continuously decreases with x. The most we can expect from the co-current heat exchanger is that the two streams exit at the same temperature, i.e., $T_{h,o} = T_{c,o} = T_o$ say. Inserting this in Eq. 15.1, we may write it as

$$\frac{T_{h,i} - T_o}{T_o - T_{c,i}} = \frac{C_c}{C_h}$$

which may be solved for T_o to get

$$\boxed{T_o = \frac{C_h T_{h,i} + C_c T_{c,i}}{C_h + C_c}} \tag{15.2}$$

Thus, the common exit temperature for the two fluids is a weighted mean, the weights being the C's for the respective fluid streams. The maximum possible heat exchange between the two fluids then is given by either one of the following two expressions:

$$Q_{max} = C_c(T_o - T_{c,i}) \text{ or } Q_{max} = C_h(T_{h,i} - T_o)$$

Using Eq. 15.2 in the first of the above, we get

$$\begin{aligned} Q_{max} &= C_c \left[\frac{C_h T_{h,i} + C_c T_{c,i}}{C_h + C_c} - T_{c,i} \right] = \frac{C_c(C_h T_{h,i} + C_c T_{c,i} - C_h T_{c,i} - C_c T_{c,i})}{C_h + C_c} \\ &= \frac{C_c C_h (T_{h,i} - T_{c,i})}{C_h + C_c} = \frac{T_{h,i} - T_{c,i}}{\left[\dfrac{1}{C_h} + \dfrac{1}{C_c} \right]} \end{aligned} \tag{15.3}$$

The biggest heat exchange between the two fluids (how or whether it is possible is unimportant) is termed the absolute maximum heat transfer $Q_{max,abs}$. This takes place if the fluid having the lower C heats up through the largest temperature difference, i.e., $T_{h,i} - T_{c,i}$. Thus,

$$Q_{max,abs} = C_{min}(T_{h,i} - T_{c,i})$$

Effectiveness ϵ of the heat exchanger is defined as the ratio of actual heat exchange to the absolute maximum heat exchange. Thus,

$$\epsilon = \frac{C_c(T_{c,o} - T_{c,i})}{C_{min}(T_{h,i} - T_{c,i})} \quad \text{or} \quad \epsilon = \frac{C_h(T_{h,i} - T_{h,o})}{C_{min}(T_{h,i} - T_{c,i})} \tag{15.4}$$

The actual maximum heat transfer for a heat exchanger with co-current flow is limited to the value given by Eq. 15.3. Hence, the maximum effectiveness possible for a co-current heat exchanger is

$$\epsilon_{max} = \frac{1}{C_{min}(T_{h,i} - T_{c,i})} \cdot \frac{T_{h,i} - T_{c,i}}{\left[\frac{1}{C_h} + \frac{1}{C_c}\right]} = \frac{1}{C_{min}} \cdot \frac{1}{\left[\frac{1}{C_h} + \frac{1}{C_c}\right]} \tag{15.5}$$

If $C_{min} = C_h$ then $C_{max} = C_c$ and

$$\frac{1}{C_{min}} \cdot \frac{1}{\frac{1}{C_h} + \frac{1}{C_c}} = \frac{1}{C_{min}} \cdot \frac{1}{\frac{1}{C_{min}} + \frac{1}{C_{max}}} = \frac{1}{1 + \frac{C_{min}}{C_{max}}} = \frac{1}{1 + R}$$

If $C_{max} = C_h$ then $C_{min} = C_c$ and

$$\frac{1}{C_{min}} \cdot \frac{1}{\frac{1}{C_h} + \frac{1}{C_c}} = \frac{1}{C_{min}} \cdot \frac{1}{\frac{1}{C_{max}} + \frac{1}{C_{min}}} = \frac{1}{\frac{C_{min}}{C_{max}} + 1} = \frac{1}{1 + R}$$

Thus, irrespective of which fluid has the smaller heat capacity mass flow rate product, the maximum effectiveness of a co-current heat exchanger is

$$\epsilon = \frac{1}{1 + R} \tag{15.6}$$

15.2.2 Thermal Analysis of a Co-Current Heat Exchanger

The thermal analysis of a co-current heat exchanger requires a knowledge of the heat transfer rate between the two streams and hence the variation of the temperature of the two streams with x. Figure 15.3 helps in the analysis. Let $U(x)$ be the local overall heat transfer coefficient between the two fluid streams. For the control volume shown in the figure, the following relations may be written. Heat transfer across the boundary between the fluids is

$$dQ = U(T_h - T_c)dS \tag{15.7}$$

Heat given up by the hot and cold fluids in crossing the control volume are, respectively, given by

Fig. 15.3 Analysis of a co-current heat exchanger

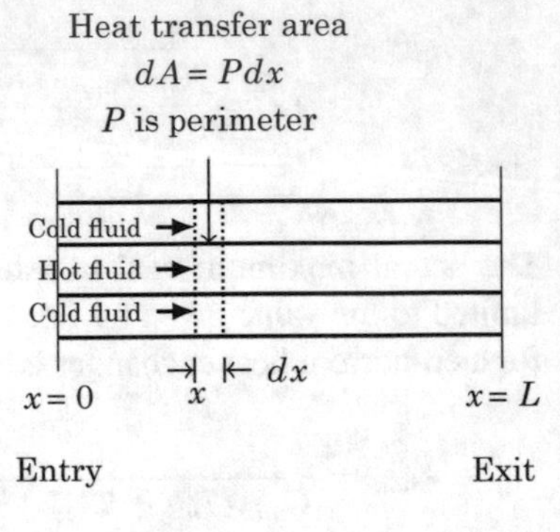

$$dQ = -C_h dT_h \quad \text{and} \quad dQ = C_c dT_c$$

The negative sign in the case of hot fluid is to make dQ positive. From the two expressions, we have

$$dT_h = -\frac{dQ}{C_h}, \quad dT_c = \frac{dQ}{C_c}$$

Hence, we get

$$dT_h - dT_c = -dQ\left[\frac{1}{C_h} + \frac{1}{C_c}\right] \tag{15.8}$$

Note that $dT_h - dT_c$ may be written as $d(T_h - T_c) = d(\Delta T)$ where $\Delta T = T_h - T_c$. Thus, Eq. 15.8 may be rewritten as

$$dQ = -\frac{d(\Delta T)}{\left[\frac{1}{C_h} + \frac{1}{C_c}\right]} \tag{15.9}$$

Comparing Eqs. 15.9 with 15.7, we conclude that

$$-\frac{d(\Delta T)}{\left[\frac{1}{C_h} + \frac{1}{C_c}\right]} = U(T_h - T_c)dS = U\Delta T dS \text{ or } -\frac{d(\Delta T)}{\Delta T} = U\left[\frac{1}{C_h} + \frac{1}{C_c}\right]dS = UCdS \tag{15.10}$$

where $C = \left[\frac{1}{C_h} + \frac{1}{C_c}\right]$. Integrate this between $x = 0$ and $x = L$ to get

$$\int_{\Delta T_0}^{\Delta T_L} \frac{d(\Delta T)}{\Delta T} = -C \int_0^{S_L} UdS \tag{15.11}$$

$\int_0^{S_L} UdS$ may be written as $\overline{U}S_L$ where $\overline{U}$ is the mean overall heat transfer coefficient based on the total heat transfer area S_L. Then, the integration indicated in Eq. 15.11 is performed to get

$$\ln\left[\frac{\Delta T_L}{\Delta T_0}\right] = -C\overline{U}S_L$$

which may be rearranged as

$$C = \frac{\ln\left[\frac{\Delta T_0}{\Delta T_L}\right]}{\overline{U}S_L} \tag{15.12}$$

Integration between $x = 0$ and $x = L$ of Eq. 15.9 gives

$$Q = \frac{\Delta T_0 - \Delta T_L}{C} \tag{15.13}$$

Eliminating C between Eqs. 15.12 and 15.13, we finally have

$$\boxed{Q = \bar{U}S_L \frac{\Delta T_0 - \Delta T_L}{\ln\left[\frac{\Delta T_0}{\Delta T_L}\right]} = \bar{U}S_L \frac{\Delta T_0 - \Delta T_L}{\overline{\Delta T}}} \tag{15.14}$$

The mean temperature difference $\overline{\Delta T}$ is referred to as the log mean temperature difference or the $LMTD$. In fact, the above is valid for both the co-current as well as a counter-current heat exchanger. This may be verified by running through the above analysis, using the temperature variation for the two streams shown in Fig. 15.2a. For these two cases, we note the following:

- **Co-current heat exchanger:**

$$\boxed{\Delta T_0 = T_{h,i} - T_{c,i},\ \Delta T_L = T_{h,o} - T_{c,o}}$$

- **Counter-current heat exchanger:**

$$\boxed{\Delta T_0 = T_{h,i} - T_{c,o},\ \Delta T_L = T_{h,o} - T_{c,i}}$$

The $LMTD$ is itself obtained as indicated in Eq. 15.14 in both cases.

15.2.3 Overall Heat Transfer Coefficient

In the thermal analysis presented in the previous section, we have introduced the overall heat transfer coefficient between the two fluids flowing on two sides of a common heat transfer surface. In a typical tube in tube heat exchanger (either the co-current or the counter-current), the overall heat transfer coefficient may be defined

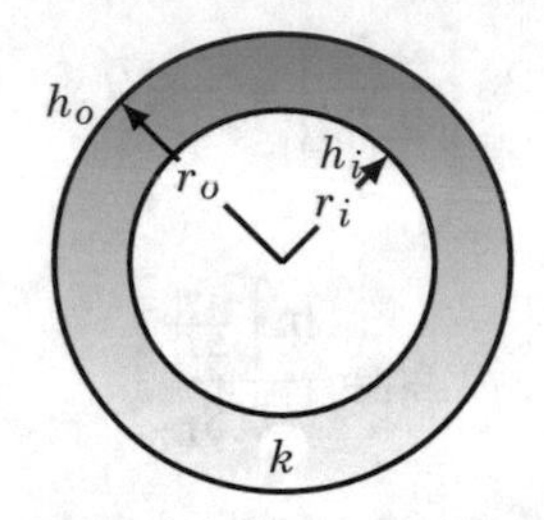

Fig. 15.4 Heat transfer across a tube wall

Table 15.1 A simple longtable example

Situation	U in $W/m^2\,°C$	Fluid	R_f in $m^2\,°C/W$
Steam condenser:	1000–5000	Sea water (>50 °C):	0.0002
Feed water heater:	1000–8000	Sea water (<50 °C):	0.0001
Water to water heat exchanger:	800–2000	Boiler feed (>50 °C):	0.0002
Water to oil heat exchanger	100–350	Fuel oil:	0.001
Finned tube heat exchanger: Water in tube, air across tubes:	30–55	Steam:	0.0001
		Industrial air:	0.0004

using Fig. 15.4. In this figure, h_i and h_o are the inside and outside heat transfer coefficients. The overall resistance to heat transfer includes the conduction resistance of the tube as well as any scale (due to corrosion after usage) that may have been formed over the heat transfer surface. The overall heat transfer coefficient without fouling is then defined through the relation

$$\boxed{\frac{1}{U_i r_i} = \frac{1}{U_o r_o} = \frac{1}{h_i r_i} + \frac{1}{k}\ln\left(\frac{r_o}{r_i}\right) + \frac{1}{h_o r_o}} \tag{15.15}$$

where U_i and U_o are the overall heat transfer coefficients based, respectively, on the inner and the outer areas. In case the fouling resistance is to be included the following relation is used:

$$\boxed{\frac{1}{U_i r_i} = \frac{1}{U_o r_o} = \frac{1}{h_i r_i} + \frac{1}{k}\ln\left(\frac{r_o}{r_i}\right) + \frac{1}{h_o r_o} + R_f} \tag{15.16}$$

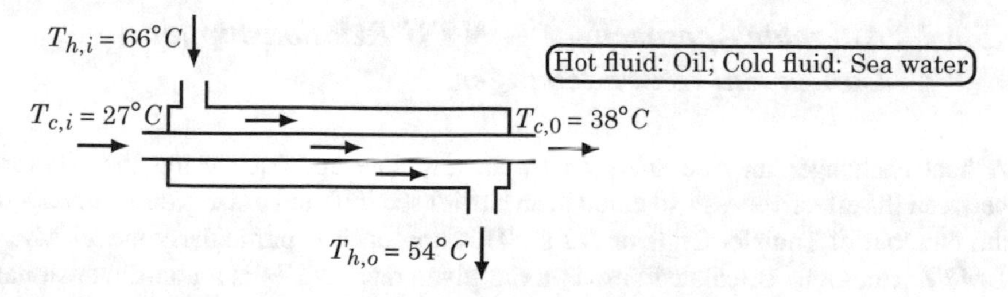

Fig. 15.5 Oil cooler for marine application of Example 15.1

where R_f is the fouling resistance. Typical overall heat transfer coefficient values and fouling resistance values are given in Table 15.1.

Example 15.1

Seawater is used to cool oil in a marine application. The various temperatures are specified in Fig. 15.5. It has been ascertained that the total heat transferred in this oil cooler is 190 kW. The overall heat transfer coefficient has also been determined to be 740 W/m^2 °C. Determine the heat transfer area in this heat exchanger.

Solution:

Step 1 Given data is written down using nomenclature shown in Fig. 15.5:

Hot fluid inlet temperature:	$T_{h,i} = 66\,°C$
Hot fluid outlet temperature:	$T_{h,o} = 54\,°C$
Cold fluid inlet temperature:	$T_{c,i} = 2\,°C$
Cold fluid outlet temperature:	$T_{c,o} = 38\,°C$
Total heat exchange:	$Q = 190$ kW $= 190,000$ W
Overall heat transfer coefficient:	$\overline{U} = 740$ W/m^2 °C

Step 2 Since all the temperatures are specified, it is possible to calculate the $LMTD$. We have $\Delta T_0 = T_{h,i} - T_{c,i} = 66 - 27 = 39\,°C$ and $\Delta T_L = T_{h,o} - T_{c,o}$
$= 54 - 38 = 16\,°C$. The $LMTD$ is calculated as

$$LMTD = \frac{\Delta T_0 - \Delta T_L}{\ln\left[\frac{\Delta T_0}{\Delta T_L}\right]} = \frac{39 - 16}{\ln\left(\frac{39}{16}\right)} = 25.8\,°C$$

Step 3 The heat exchanger area is then readily calculated as

$$S_L = \frac{Q}{U \times LMTD} = \frac{190,000}{740 \times 25.8} = 9.946\,\text{m}^2$$

15.2.4 Alternate Approach—$\epsilon - NTU$ Relationship for a Co-Current Heat Exchanger

A heat exchanger may be analyzed by an alternate approach using the relation between the effectiveness ϵ (defined in an earlier section) and a size parameter called the Number of Transfer Units or NTU. This approach is particularly useful when $LMTD$ cannot be calculated based on the given data. NTU is a non-dimensional parameter defined as

$$\boxed{NTU = \frac{\overline{U}S_L}{C_{min}}} \tag{15.17}$$

The heat exchanger effectiveness, in general, is a function of NTU, the heat capacity ratio R, and the configuration of the heat exchanger.

Consider again a co-current heat exchanger. Equation 15.12 may be written in the alternate form

$$\frac{\Delta T_L}{\Delta T_0} = \mathrm{e}^{-C\overline{U}S_L} \tag{15.18}$$

C may be written as

$$C = \frac{1}{C_c} + \frac{1}{C_h} = \frac{1+R}{C_{min}}$$

as can be verified very easily. Then, Eq. 15.18 becomes

$$\frac{\Delta T_L}{\Delta T_0} = \mathrm{e}^{\left[-\overline{U}S_L\cdot\frac{1+R}{C_{min}}\right]} = \mathrm{e}^{-NTU(1+R)}$$

With $\Delta T_0 = T_{h,i} - T_{c,i}$ and $\Delta T_l = T_{h,o} - T_{c,o}$, the above may be solved for the exit temperature of the cold stream as

$$T_{c,o} = T_{h,o} - (T_{h,i} - T_{c,i})\mathrm{e}^{-NTU(1+R)} \tag{15.19}$$

Assume any one of the streams to have the smaller $\dot{m}C_p$ product, say $C_{min} = C_h$ and hence $R = \frac{C_h}{C_c}$. The actual heat transfer may be written as

$$Q_{act} = C_h(T_{h,i} - T_{h,o})$$

The maximum possible heat exchange is of course given by

$$Q_{max} = C_h(T_{h,i} - T_{c,i})$$

Hence, the effectiveness of this heat exchanger may be written as

$$\epsilon = \frac{Q_{act}}{Q_{max}} = \frac{T_{h,i} - T_{h,o}}{T_{h,i} - T_{c,i}}$$

For the entire heat exchanger, we also have, from Eq. 15.1, for the present case

$$\frac{T_{h,i} - T_{h,o}}{T_{c,o} - T_{c,i}} = \frac{C_c}{C_h} = \frac{1}{R}$$

From this, we obtain the outlet temperature of the cold stream as

$$T_{c,o} = T_{c,i} + R(T_{h,i} - T_{h,o}) \tag{15.20}$$

Equating the two expressions for $T_{c,o}$ given by Eqs. 15.19 and 15.20, we get

$$T_{c,i} + R(T_{h,i} - T_{h,o}) = T_{h,o} - (T_{h,i} - T_{c,i})\mathrm{e}^{-NTU(1+R)}$$

$$\text{or } T_{h,o}(1 + R) = T_{c,i} + RT_{h,i} + (T_{h,i} - T_{c,i})e^{-NTU(1+R)}$$

$$\text{or } \boxed{(T_{h,i} - T_{h,o})(1 + R) = (T_{h,i} - T_{c,i})(1 - e^{-NTU(1+R)})}$$

Rearrangement of this expression is possible in the form

$$\boxed{\epsilon = \frac{T_{h,i} - T_{h,o}}{T_{h,i} - T_{c,i}} = \frac{1 - \mathrm{e}^{-NTU(1+R)}}{1 + R}} \tag{15.21}$$

Thus, the effectiveness of a co-current heat exchanger depends on the two non-dimensional parameters NTU and R. For a fixed value of R, the effectiveness asymptotically tends to $\frac{1}{1+R}$ as $NTU \rightarrow \infty$. It may be recalled that this was the thermodynamic limit derived in Sect. 15.2.1 and given by Eq. 15.6.

$\epsilon - NTU$ Plot

Equation 15.21 can be used to prepare plot of ϵ as a function of NTU with R treated as a parameter. Such a plot is shown in Fig. 15.6. The unique relationship that exists between the effectiveness and the number of transfer units may be utilized as a method for analyzing heat exchangers. This is demonstrated by solving the problem solved earlier by the $LMTD$ approach in Example 15.1 by the use of the $\epsilon - NTU$ method.

Example 15.2

Solve the oil cooler example in Example 15.1 by the $\epsilon - NTU$ method. Use the same data as given there.

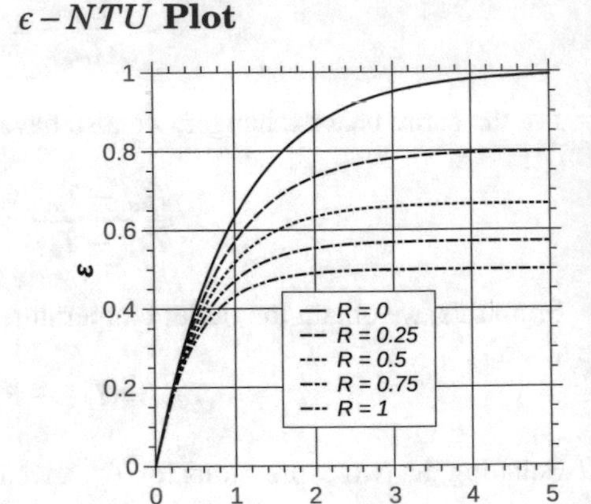

Fig. 15.6 $\epsilon - NTU$ Plot for a co-current heat exchanger

---

Solution:
Since the heat duty Q is specified along with the temperatures of both the fluids at entry and exit, it is possible to evaluate R for this heat exchanger. Since the hot oil temperature drop is larger than the temperature increase of seawater $C_{min} = C_h$. Using the condition for overall energy balance, we have

$$R = \frac{C_h}{C_c} = \frac{\Delta T_c}{\Delta T_h} = \frac{38-27}{66-54} = \frac{11}{12} = 0.917$$

We may also calculate C_h as

$$C_h = C_{min} = \frac{Q}{\Delta T_h} = \frac{190,000}{66-54} = 15833.3\,\text{W/°C}$$

By definition, the effectiveness is given by

$$\epsilon = \frac{T_{h,i} - T_{h,o}}{T_{h,i} - T_{c,i}} = \frac{66-54}{66-27} = \frac{12}{39} = 0.308$$

We use expression 15.21 to calculate the NTU for this heat exchanger.

$$NTU = -\frac{\ln\left[1 - \epsilon(1+R)\right]}{1+R} = -\frac{\ln\left[1 - 0.308(1+0.917)\right]}{1+0.917} = 0.465$$

Alternately Fig. 15.6 may be used to read off the required NTU with the known values of ϵ and R. Interpolation is called for and the NTU is obtained to a worse approximation as compared to the use of the analytical expression used above. But, by definition, $NTU = \frac{\overline{U}S_L}{C_h}$ and hence we have

$$S_L = \frac{NTU \cdot C_h}{\overline{U}} = \frac{0.465 \times 15833.3}{740} = 9.946\,\text{m}^2$$

This is in perfect agreement with the result of Example 15.1.

15.2.5 Counter-Current Heat Exchanger

The derivations of the $\epsilon - NTU$ relationship given for a co-current heat exchanger above may be easily repeated for a counter-current heat exchanger. The result of such an analysis would yield the following relationship:

$$\boxed{\epsilon = \frac{1 - \text{e}^{-NTU(1-R)}}{1 - R\,\text{e}^{-NTU(1-R)}}} \tag{15.22}$$

In this case also the effectiveness is a function of NTU and R. A plot of effectiveness as a function of NTU, for various values of R, is shown in Fig. 15.7. In the case of a counter-current heat exchanger, for any value of R, the effectiveness asymptotically tends to unity as $NTU \to \infty$.

Example 15.3

Rework Example 15.1 assuming that the heat exchanger is a counter-current heat exchanger. Use *(a)* $LMTD$ approach and *(b)* $\epsilon - NTU$ approach to solve the problem.

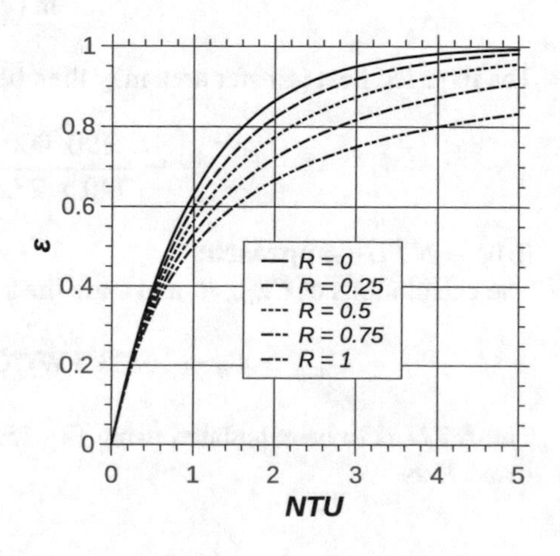

Fig. 15.7 $\epsilon - NTU$ Plot for a counter-current heat exchanger

$$T_{h,i} = 66°C,\ T_{h,o} = 54°C,\ T_{c,i} = 38°C,\ T_{c,o} = 27°C$$

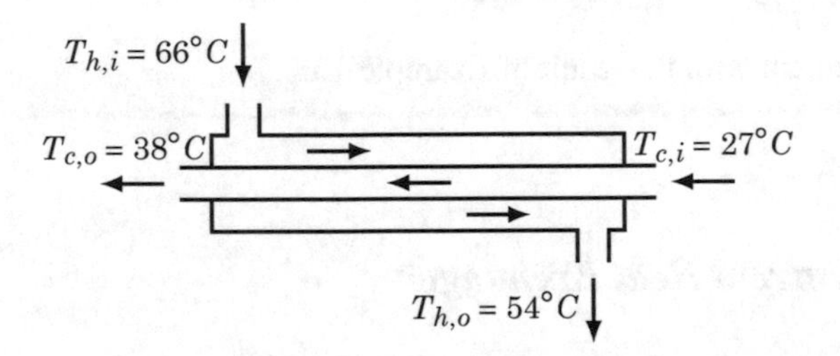

Fig. 15.8 Heat exchanger of Example 15.1 rearranged in the Counter-current arrangement

---

Solution:
A sketch of oil cooler of Example 15.1, modified for counter-current arrangement, is shown in Fig. 15.8.

(a) ***LMTD*****—approach:**
Referring to Fig. 15.8, we have

$$T_{h,i} = 66\,°\text{C},\ T_{h,o} = 54\,°\text{C},\ T_{c,i} = 38\,°\text{C},\ T_{c,o} = 27\,°\text{C}$$

Hence, the *terminal* temperature differences are

$$\Delta T_0 = T_{h,i} - T_{c,o} = 66 - 38 = 28\,°\text{C} \quad \text{and} \quad \Delta T_L = T_{h,o} - T_{c,i} = 54 - 27 = 27\,°\text{C}$$

The $LMTD$ is then calculated as

$$LMTD = \frac{28 - 27}{\ln\left(\frac{28}{27}\right)} = 27.5\,°\text{C}$$

The required heat transfer area may then be calculated as

$$S_L = \frac{190,000}{740 \times 27.5} = 9.338\,\text{m}^2$$

(b) $\epsilon - NTU$**—approach:**
The calculations of C_{min}, R, and ϵ are the same as in Example 15.2. Hence, we have

$$C_{min} = C_h = 15833.3\,\text{W/°C},\ R = 0.917,\ \epsilon = 0.308$$

The NTU is to be calculated using Eq. 15.22. We may solve for NTU in terms of ϵ and R as

Fig. 15.9 Nomenclature for heat exchanger of Example 15.4

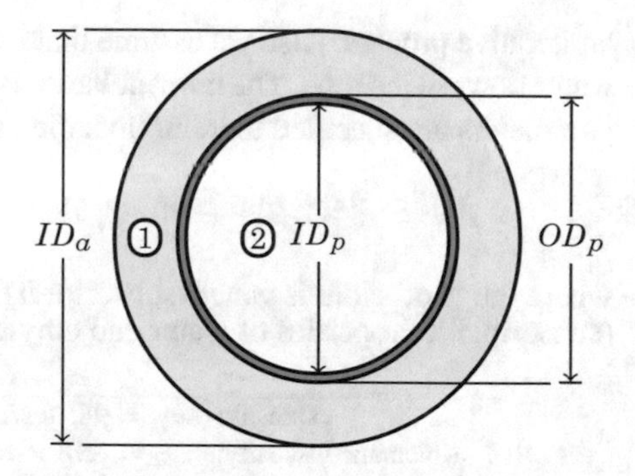

$$NTU = -\frac{\ln\left(\frac{1-\epsilon}{1-R\epsilon}\right)}{1-R} = -\frac{\ln\left(\frac{1-0.308}{1-0.917\times 0.308}\right)}{1-0.917} = 0.437$$

With this value of NTU the area of the heat exchanger is given by

$$S_L = \frac{NTU\, C_h}{\overline{U}} = \frac{0.437 \times 15833.3}{740} = 9.352\,\text{m}^2$$

The two ways of calculating the area required agree closely with each other. The counter-current heat exchanger requires smaller amount of heat transfer area as compared to the co-current arrangement.

The examples given above considered cases where the overall heat transfer cocfficient was specified. In practice, however, the flow parameters are usually specified and the heat exchanger analysis has to proceed from there. The next example demonstrates how such an analysis is made.

Example 15.4

Water at 90 °C at a mass flow rate of 2250 kg/hr is to be used to heat ethylene glycol. Ethylene glycol is available at 30 °C with a mass flow rate of 5500 kg/hr. A double pipe heat exchanger consisting of a 34.93 mm OD - 32.79 mm ID standard type M copper tubing inside of a 53.98 mm OD - 50.42 mm ID type M copper tubing is to be used. The heat exchanger is 10 m long. It is constructed by using"hair pin" type construction with four 2.5 m sections. Determine the outlet temperature of both fluids using a parallel flow arrangement. What is the total heat transferred between water and ethylene glycol? Ignore the effect of the bends.

Solution:

Since outlet temperatures of the two fluids are not known, and since the pipe temperature also is not known, we calculate the fluid properties at the average temperature of $\frac{T_{h,i}+T_{c,i}}{2} = \frac{90+30}{2} = 60\,°\text{C}$. If necessary the calculations may be revised later via an iterative process. Also we assume that ethylene glycol flows in the annulus while water flows in the tube. The nomenclature is explained in Fig. 15.9 and the geometric parameters are specified therein. Specified temperatures and flow rates are

$$T_{h,i} = 90\,°\text{C},\ T_{c,i} = 30\,°\text{C},\ \dot{m}_h = 2250\,\text{kg/hr},\ \dot{m}_c = 5500\,\text{kg/hr}$$

where the "hot" fluid is water (subscript h) while the "cold" fluid is ethylene glycol (subscript c). Properties of water and ethylene glycol are

Density:	$\rho_h = 985\ \text{kg/m}^3$	$\rho_c = 1087\ \text{kg/m}^3$
Kinematic viscosity:	$\nu_h = 4.78 \times 10^{-7}\ \text{m}^2/\text{s}$	$\nu_c = 4.75 \times 10^{-6}\ \text{m}^2/\text{s}$
Thermal conductivity:	$k_h = 0.651\ \text{W/m°C}$	$k_c = 0.260\ \text{W/m°C}$
Specific heat:	$C_{ph} = 4185\ \text{J/kg°C}$	$C_{pc} = 2562\ \text{J/kg°C}$
Prandtl number:	$Pr_h = 3.02$	$Pr_c = 51$

Annulus side calculations:
Flow area on the annuls side is given by

$$A_a = \frac{\pi}{4}(ID_a^2 - OD_p^2) = \frac{\pi}{4}(50.42^2 - 34.93^2) = 1038.4\,\text{mm}^2$$

Velocity of ethylene glycol is obtained as

$$V_c = \frac{\dot{m}_c}{\rho_c A_a} = \frac{\frac{5500}{3600}}{1087 \times 1038.4 \times 10^{-6}} = 1.354\,\text{m/s}$$

Heat transfer to the fluid flowing in the annulus takes place only on the pipe side. The *energy* diameter is different from the hydraulic diameter and is given by

$$D_{a,E} = \frac{4 \times \text{Flow Area}}{\text{Heated Perimeter}} = \frac{4A_a}{\pi OD_p} = \frac{4 \times 1038.4}{\pi \times 34.93} = 37.85\,\text{mm}$$

In order to calculate the Nusselt number from the outside surface of pipe, the flow is like the flow in a shell of a shell and tube heat exchanger without any baffles. Shell side heat transfer data is correlated by the relation

$$Nu_D = 0.0184 D_{a,E}^{0.6} Re_D^{0.6} Pr^{0.33} \tag{15.23}$$

as proposed by Donohue.[1] Note that the Reynolds and Nusselt numbers are based on the pipe diameter D or OD_p in the current context and $D_{a,E}$ is in mm. In the paper, the constant is different from that shown in Eq. 15.23 because $D_{a,E}$ was to be specified in inches.
Reynolds number for the ethylene glycol is

[1]D. A. Donohue, "Heat transfer and pressure drop in heat exchangers", Industrial and Engineering Chemistry, Vol. 41, No. 11, pp. 2499–2511, 1949.

$$Re_D = \frac{V_c\, OD_P}{\nu_c} = \frac{1.354 \times 34.93 \times 10^{-3}}{4.75 \times 10^{-6}} = 9956.9$$

The Nusselt number is obtained by using Eq. 15.23 as

$$Nu_D = 0.0184 \times 37.85^{0.6} 9956.9^{0.6} 51^{0.33} = 149.3$$

Hence, the heat transfer coefficient on the annulus side is

$$h_c = \frac{Nu_D k_c}{OD_p} = \frac{149.3 \times 0.260}{34.93 \times 10^{-3}} = 1111.3\,\mathrm{W/m^2{}^\circ C}$$

Pipe side calculations:
The flow area is

$$A_p = \frac{\pi}{4} ID_p^2 = \frac{\pi}{4} \times 32.79^2 = 844.45\,\mathrm{mm^2}$$

Velocity of water is then given by

$$V_h = \frac{\dot{m}_h}{\rho_h A_p} = \frac{\frac{2250}{3600}}{985 \times 844.45 \times 10^{-6}} = 0.751\,\mathrm{m/s}$$

Energy diameter is the same as the hydraulic diameter that is also the same as the inner diameter of the pipe. Hence, the water side Reynolds number is

$$Re_h = \frac{V_H \cdot ID_p}{\nu_h} = \frac{0.751 \times 32.79 \times 10^{-3}}{4.78 \times 10^{-7}} = 51517.3$$

The flow is certainly turbulent. The fluid is getting cooled and hence $n = 0.3$ in the Dittus–Boelter equation. The Nusselt number is given by

$$Nu_h = 0.023 Re_h^{0.8} Pr_h^{0.3} = 0.023 \times 51517.3^{0.8} \times 3.02^{0.3} = 188.49$$

Hence, the water side heat transfer coefficient is

$$h_h = \frac{Nu_h \cdot k_h}{ID_p} = \frac{188.49 \times 0.651}{32.79 \times 10^{-3}} = 3742.2\,\mathrm{W/m^2{}^\circ C}$$

Tube material is copper with a thermal conductivity of

$$k_{Cu} = 401\,\mathrm{W/m^\circ C}$$

Pipe wall conduction resistance R_p may be calculated as

$$R_p = \frac{1}{k_{Cu}} \ln \frac{OD_p}{ID_p} = \frac{1}{401} \ln \frac{34.93}{32.79} = 1.577 \times 10^{-4}\,\mathrm{m^\circ C/W}$$

Hence, the overall heat transfer coefficient is obtained using Eq. 15.15 as

$$\frac{2}{U_h \cdot ID_p} = \frac{2}{U_c \cdot OD_p} = \frac{2}{h_h \cdot ID_p} + R_p + \frac{2}{h_c \cdot OD_p}$$
$$= \frac{2}{3742.2 \times 32.79 \times 10^{-3}} + 1.577 \times 10^{-4} + \frac{2}{1111.3 \times 34.93 \times 10^{-3}}$$
$$= 0.0679\,\mathrm{m^\circ C/W}$$

We shall base the thermal analysis that follows on the outside area of the inner tube. Hence, the overall heat transfer coefficient is given by

$$U_c = \frac{2}{0.0679 \times 34.93 \times 10^{-3}} = 842.3\,\mathrm{W/m^{2\circ}C}$$

Analysis of the heat exchanger:
Mass flow rate specific heat products for the two fluids are now calculated.

$$C_h = \dot{m}_h C_{ph} = \frac{2250}{3600} \times 4184 = 2615\,\mathrm{W/^\circ C}$$

and

$$C_c = \dot{m}_c C_{pc} = \frac{5500}{3600} \times 2562 = 3914.2\,\mathrm{W/^\circ C}$$

Hence, $C_{min} = C_h = 2615\,\mathrm{W/^\circ C}$ and $C_{max} = C_c = 3914.2\,\mathrm{W/^\circ C}$. The heat capacity ratio is then given by

$$R = \frac{C_{min}}{C_{max}} = \frac{C_h}{C_c} = \frac{2615}{3914.2} = 0.668$$

Heat exchange area is calculated as

$$S_L = \pi \cdot OD_p \cdot L = \pi \times 34.93 \times 10^{-3} \times 10 = 1.097\,\mathrm{m^2}$$

The NTU is then calculated as

$$NTU = \frac{U_c S_L}{C_{min}} = \frac{842.3 \times 1.097}{2615} = 0.353$$

For the co-current heat exchanger, the effectiveness may then be calculated using Eq. 15.21 as

$$\epsilon = \frac{1 - e^{-NTU(1+R)}}{1 + R} = \frac{1 - e^{-0.353(1+0.668)}}{1 + 0.668} = 0.229$$

Thus, the total heat exchange between water and ethylene glycol is

$$Q_{act} = \epsilon C_{min}(T_{h,i} - T_{c,i}) = 0.229 \times 2615(90 - 30) = 35877\,\text{W}$$

Outlet temperatures of the two fluids, then are

$$T_{h,o} = T_{h,i} - \frac{Q_{act}}{C_h} = 90 - \frac{35877}{2615} = 76.3\,°\text{C}$$

and

$$T_{c,o} = T_{c,i} + \frac{Q_{act}}{C_c} = 30 + \frac{35877}{3914.2} = 39.2\,°\text{C}$$

15.3 Other Types of Heat Exchangers

In practice, it is not always convenient to use either the co-current or the counter-current heat exchanger. In order that standard length tubes are made use of the length of the heat exchanger may be limited to certain lengths only. The designer has no option but to use other types of heat exchangers that will be considered below. There are three types, which are useful in catering to different needs:

1. Multi-pass heat exchanger
2. Shell and tube heat exchanger
3. Cross flow heat exchanger

(1) Multi-pass heat exchanger

The schematic of a multi-pass heat exchanger is shown in Fig. 15.10. The cold fluid flows through the shell (the tubes are held within the outer enclosure which is called the shell) while the hot fluid flows through the tubes.

Fig. 15.10 Example of a multi-pass heat exchanger

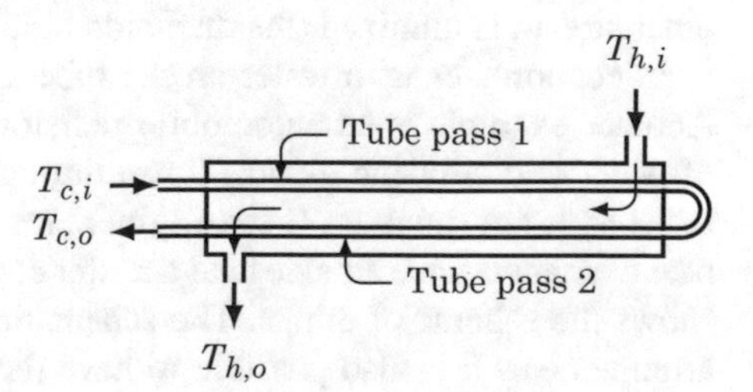

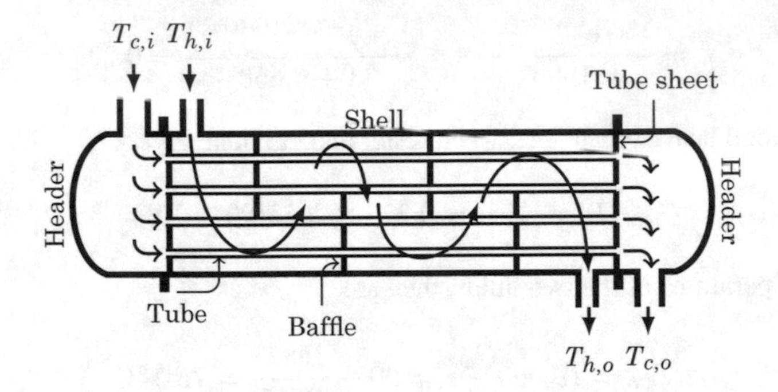

Fig. 15.11 Schematic of a shell and tube heat exchanger with 1—shell and 1—tube Pass

In this example, there are two tube passes. In fact, several *parallel* tubes may be arranged in each pass. The cold stream will divide through these parallel tubes and join again at exit and mix into a single stream. In the above illustration, tube-pass 1 experiences a counter-current heat exchange and tube-pass 2 experiences a co-current heat exchange. Thus, the performance of this heat exchanger will be between that of a counter-current and a co-current heat exchanger.

(2) Shell and tube heat exchanger
A schematic of a shell and tube heat exchanger is shown in Fig. 15.11. On the tube side, the fluid traverses through parallel tubes arranged between two mixing chambers (usually referred to as headers or boxes). The shell side fluid is made to follow a tortuous path by the placement of suitable baffles along the length of the heat exchanger, as shown in the figure. The flow cannot be considered either co-current flow or cross flow, but is a combination of the two. Since mixing takes place outside, the hot stream is unmixed while the cold stream may be considered to be mixed. Variants of the multi-pass heat exchanger are obtained by combining several two-tube pass heat exchangers. An example is shown in Fig. 15.12 where two two-tube pass heat exchangers are connected in series. In the illustration, in all there are four tube passes and two shells. Baffles are employed to guide the second fluid through a tortuous path for proper contact between the fluid and the outside surfaces of the tubes.

(3) Cross flow heat exchanger
This is an example wherein one of the streams is mixed (the tube side fluid) and the other stream is unmixed (the duct side fluid). An example of such a heat exchanger is an economizer with water on the tube side and the flue gases on the duct side. Another example is an automobile radiator where the coolant (water or a mixture of water and ethylene glycol) flows through the tubes and air is blown across the tubes by a fan running off the engine. Fins are provided on the air side to improve heat exchange since air side heat transfer coefficients are usually small. Figure 15.13 shows the scheme of things. The schematic shows the tubes arranged in an in-line arrangement. It is also possible to have the tubes arranged in a staggered manner.

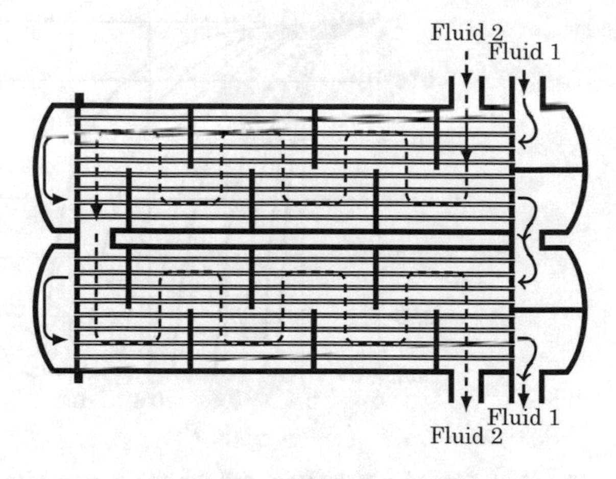

Fig. 15.12 Schematic of a heat exchanger with 4—tube passes and 2—shell passes

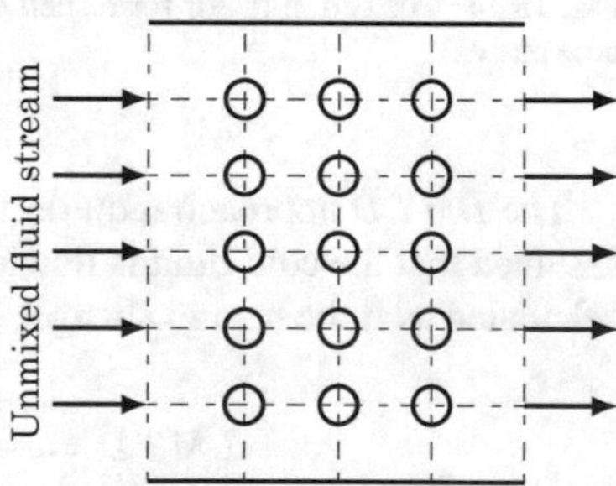

Fig. 15.13 Schematic of a cross flow heat exchanger

In the cross flow heat exchanger, the two fluid paths are normal to each other. The heat transfer on the air side may be obtained by using the heat transfer correlations appropriate to flow past a tube bank.

Cross flow heat exchangers may also involve heat exchange between two unmixed streams such as in plate heat exchangers. Plate heat exchangers involve the flow of two fluid streams in the passages between a stack of corrugated plates and the two streams may flow in co-current, counter-current, or cross flow modes. Plate heat exchangers are classified as compact heat exchangers.

15.3.1 Analysis of Shell and Tube Heat Exchanger

Analysis of shell and tube heat exchangers makes use of the $LMTD$ approach followed by the use of correction factors to account for the fact that these do not conform either to co-current or counter-current type. In fact, the flow on the shell side is very complex and the suggested method of analysis at best approximate.

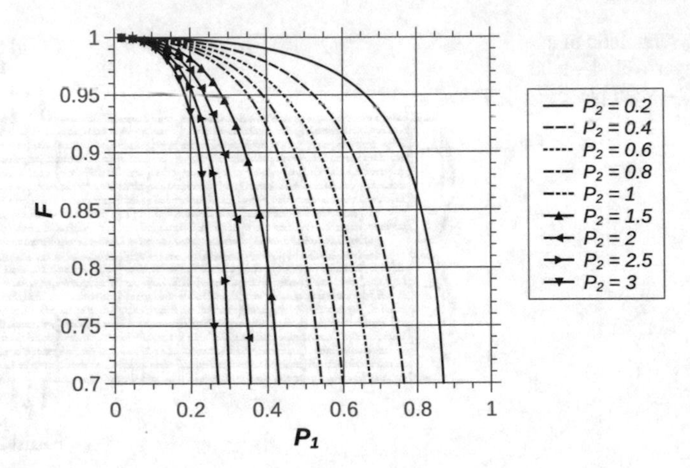

Fig. 15.14 Correction factor for a shell and tube heat exchanger with 1 shell pass and 2 or more tube passes

The $LMTD$ approach requires that all the terminal temperatures be known. It is assumed that the cold fluid is flowing through the tubes (see Fig. 15.10). $LMTD$ is calculated as if the heat exchanger is of the counter-current type. Thus, we have

$$LMTD = \frac{(T_{h,i} - T_{c,o}) - (T_{h,o} - T_{c,i})}{\ln\left(\dfrac{T_{h,i} - T_{c,o}}{T_{h,o} - T_{c,i}}\right)} \tag{15.24}$$

All temperatures are as given in Fig. 15.10. The total heat transferred is calculated as

$$Q = \overline{U} \times S \times LMTD \times F \tag{15.25}$$

where F is a correction factor.[2] The other symbols are familiar to us. The subscript on S has been dropped for convenience. The correction factor depends on two ratios defined as below

$$\text{(a) } P_1 = \frac{T_{c,o} - T_{c,i}}{T_{h,o} - T_{c,i}}, \text{ (b) } P_2 = \frac{T_{h,i} - T_{h,o}}{T_{c,o} - T_{c,i}} = \frac{C_c}{C_h} \tag{15.26}$$

The correction factor is shown plotted as a family of F vs P_1 curves for different P_2 values in Fig. 15.14.

[2]More details may be obtained from R.A. Bowman, A.C. Mueller and W.M. Nagle, Trans. ASME, Vol., pp. 283–294, May 1940.

Example 15.5

A shell and tube oil to water heat exchanger has tubes of internal diameter 12 mm and length 2 m in a single shell. Cold water ($C_{pc} = 4180\,\text{J/kg°C}$) enters the tubes at 33 °C with a flow rate of 5 kg/s and leaves at 55 °C. Oil ($C_{ph} = 2150\,\text{J/kg°C}$) flows through the shell and is cooled from 120 °C to 75 °C. The overall heat transfer coefficient is $U_i = 885\,\text{W/m}^2\,°\text{C}$ based on the internal area of the tubes. Determine the number of tubes required in this heat exchanger.

Solution:
All temperatures are specified in this problem. Hence, the $LMTD$-correction factor approach is used to solve the problem. The given data is listed first:

Hot fluid inlet temperature:	$T_{h,i} = 125\,°\text{C}$
Hot fluid outlet temperature:	$T_{h,o} = 75\,°\text{C}$
Specific heat of the hot fluid:	$C_{pc} = 2150\ \text{J/kg°C}$
Cold fluid inlet temperature:	$T_{c,i} = 33\,°\text{C}$
Cold fluid outlet temperature:	$T_{c,o} = 55\,°\text{C}$
Specific heat of the cold fluid:	$C_{pc} = 4180\ \text{J/kg°C}$
Mass flow rate of cold fluid:	$\dot{m}_c = 5\ \text{kg/s}$
Total heat exchange:	$Q = 190\ \text{kW} = 190,000\ \text{W}$
Overall heat transfer coefficient:	$\overline{U}_i = 885\ \text{W/m}^2\,°\text{C}$

Geometric data:

$$D_{pi} = 12\,\text{mm} = 0.012\,\text{m},\ L = 2\,\text{m}$$

The total heat transferred in the heat exchanger is calculated as

$$Q = \dot{m}_c C_{pc}(T_{c,o} - T_{c,i}) = 5 \times 4180(55 - 33) = 4.598 \times 10^5\ \text{W}$$

The terminal temperature differences assuming counter-current operation are

$$\Delta T_0 = T_{h,i} - T_{c,o} = 120 - 55 = 65\,°\text{C}$$
$$\Delta T_L = T_{h,o} - T_{c,i} = 75 - 33 = 42\,°\text{C}$$

The LMTD based on counter-current assumption is

$$LMTD = \frac{65 - 42}{\ln\left(\dfrac{65}{42}\right)} = 52.67\,°\text{C}$$

The parameters P_1 and P_2 are now calculated:

$$P_1 = \frac{55 - 33}{120 - 33} = 0.253,\ \ P_2 = \frac{120 - 75}{55 - 33} = 2.045$$

From Fig. 15.14, the correction factor is read off as $F = 0.94$. Based on the total heat transferred, the area required may be calculated as

$$S = \frac{Q}{U_i \times LMTD \times F} = \frac{4.598 \times 10^5}{885 \times 52.67 \times 0.94} = 10.494\,\text{m}^2$$

Heat transfer area per tube is obtained as (based on tube ID)

$$A_{pi} = \pi D_{pi} L = \pi \times 0.12 \times 2 = 0.0754\,\text{m}^2$$

The number of tubes N required is then calculated as

$$N = \frac{\text{Total area required}}{\text{Area per tube}} = \frac{10.494}{0.0754} = 139.2$$

The number may be rounded to nearest whole number $N = 140$.

It is also possible to use the NTU—ϵ method for the analysis of shell and tube heat exchangers. For details the reader should consult specialized books on heat exchangers.

15.3.2 Analysis of a Cross Flow Heat Exchanger by the $\epsilon - NTU$ Approach

Consider a cross flow heat exchanger in which one stream is unmixed and the other stream is mixed, as shown by the schematic sketch in Fig. 15.15. Typically an automobile radiator belongs to this type of a heat exchanger. The tubes carry the automobile coolant, usually a mixture of water and ethylene glycol and air is blown across the outside of the tubes, normal to the tubes.

The flow inside the tubes is characterized by a bulk temperature that varies along the tube length, the flow being said to be *mixed*. The stream enters the heat exchanger ($x = 0$) at temperature $T_{m,i}$ and exits the heat exchanger ($x = L$) at $T_{m,o}$. Air that is blown across the tubes is at a uniform temperature $T_{u,i}$ before contacting the tubes. The air passes across the tubes and emanates as a stream with outlet temperature $T_{u,o}(x)$ which is a function of x. The air stream is said to be *unmixed*. The analysis of such a heat exchanger is accomplished by looking at the simple case of a single tube shown by a square duct across which air flows, as shown in Fig. 15.16. Assume that the heat exchanger duct is of uniform section, and hence, the heat transfer area varies linearly with x such that the heat transfer area of the element of length dx shown in the figure is

$$dS = \frac{S_L}{L} dx$$

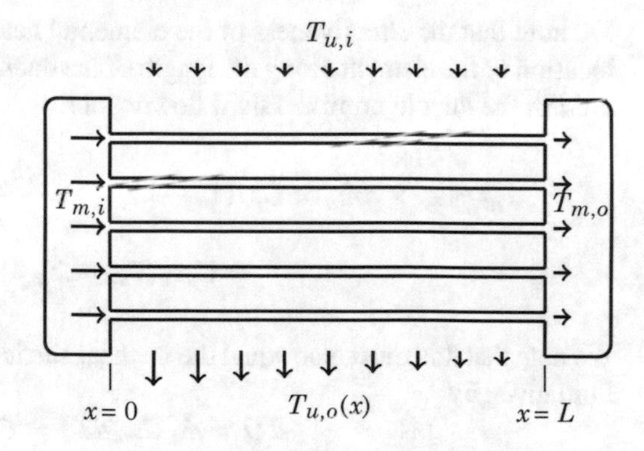

Fig. 15.15 Cross flow heat exchanger with hot stream mixed (subscript m) and the cold stream unmixed (subscript u)

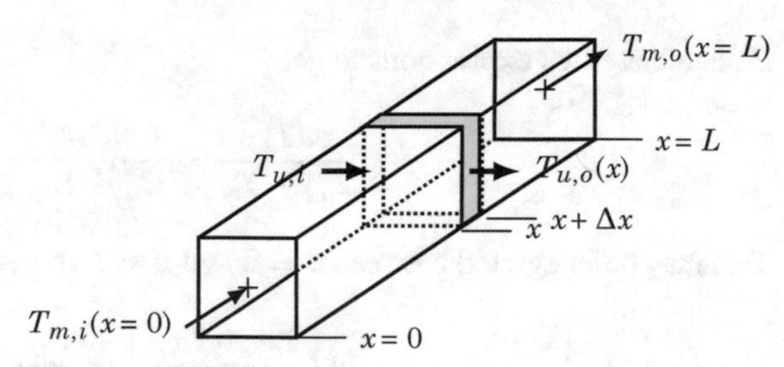

Fig. 15.16 A single duct in cross flow: simple cross flow heat exchanger. Subscript m—mixed fluid, Subscript u—unmixed fluid

As far as this element is concerned the mixed stream is at a *uniform temperature* of $T_{m,x}$ and the heat transfer takes place from an unmixed stream of mass flow

$$d\dot{m}_u = \frac{\dot{m}_u}{L}dx$$

Again we have assumed that the mass flux of the unmixed fluid varies linearly with x. The effectiveness of the elemental heat exchanger ϵ_e is given by taking $R = 0$ (for the elemental heat exchanger flow outside is elemental flow of unmixed fluid while it is the full flow for the mixed fluid) for a co-current heat exchanger (Eq. 15.21) and hence

$$\epsilon_e = 1 - \mathrm{e}^{-NTU_e}$$

where NTU_e is the NTU for this element. Obviously, the NTU_e is given by

$$NTU_e = \frac{\overline{U}dS}{d\dot{m}_u C_{pu}} = \frac{\overline{U}\frac{S_L}{L}}{\frac{\dot{m}_u}{L}C_{pu}} = \frac{\overline{U}S_L}{\dot{m}_u C_{pu}} = \frac{\overline{U}S_L}{C_u}$$

where $\bar{U}$ is the overall heat transfer coefficient. Hence, the effectiveness of the elemental heat exchanger is

$$\epsilon_e = 1 - \mathrm{e}^{-\frac{\bar{U}S_L}{C_u}} \tag{15.27}$$

We note that the effectiveness of the elemental heat exchanger is independent of the location of the element along the length of the duct. The heat transfer from the mixed fluid in the duct to unmixed fluid flowing across it may hence be written down as

$$dQ = \epsilon_e \times d\dot{m}_u \times C_{pu}(T_{u,i} - T_{m,x}) = \frac{\dot{m}_u dx}{L} \times C_{pu}\epsilon_e(T_{u,i} - T_{m,x})$$
$$= C_u\epsilon_e(T_{u,i} - T_{m,x})\frac{dx}{L} \tag{15.28}$$

We note that this must also equal the net heat carried across the element by the mixed fluid given by

$$dQ = \dot{m}_u C_{pm} dT_m = C_m dT_m \tag{15.29}$$

Equate these two expressions to get

$$\frac{dT_m}{T_{u,i} - T_m} = \frac{C_u}{C_m}\epsilon_e\frac{dx}{L}$$

This may be integrated between $x = 0$ and $x = L$ to get

$$-\ln\left[\frac{T_{u,i} - T_{m,o}}{T_{u,i} - T_{m,i}}\right] = \frac{C_u}{C_m}\epsilon_e \tag{15.30}$$

The temperature ratio occurring on the left-hand side of Eq. 15.30 may be written as

$$\frac{T_{u,i} - T_{m,o}}{T_{u,i} - T_{m,i}} = \frac{(T_{u,i} - T_{m,i}) + (T_{m,i} - T_{m,o})}{T_{u,i} - T_{m,i}} = 1 + \frac{T_{m,i} - T_{m,o}}{T_{u,i} - T_{m,i}} \tag{15.31}$$

Case (a): $C_m < C_u$
In this case, the maximum possible heat transfer is $C_m(T_{u,i} - T_{m,i})$ and the actual heat transfer is $C_m(T_{m,i} - T_{m,o})$. Hence, the temperature difference ratio in Eq. 15.31 is nothing but the effectiveness of the cross flow heat exchanger ϵ. Also $\frac{C_u}{C_m} = \frac{1}{R}$ and $NTU = \frac{\overline{U}S_L}{C_m}$. Hence, we get from Eqs. 15.30 and 15.31 the following:

$$\boxed{\epsilon = 1 - e^{-\frac{1}{R}(1-e^{-R \cdot NTU})}} \tag{15.32}$$

Case (b): $C_m > C_u$
In this case, we define the mean temperature of the unmixed fluid at exit as $T_{u,o}$ such that the actual heat transfer may be written by energy balance as

$$C_u(T_{u,o} - T_{u,i}) = C_m(T_{m,i} - T_{m,o})$$

or

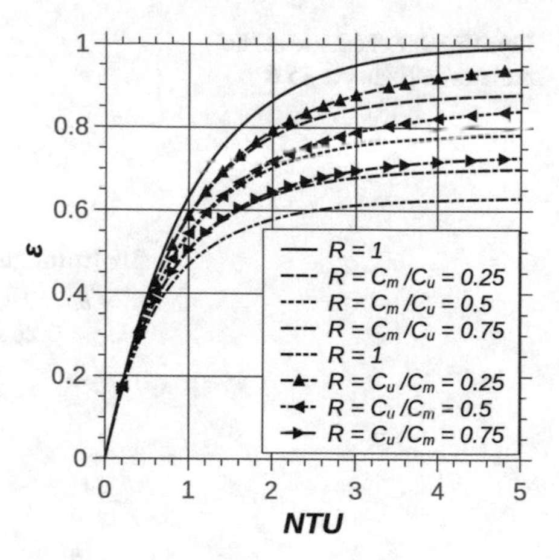

Fig. 15.17 $\epsilon - NTU$ plots for cross flow heat exchangers

$$\frac{T_{m,i} - T_{m,o}}{T_{u,i} - T_{m,i}} = \frac{C_u}{C_m}\frac{T_{u,o} - T_{u,i}}{T_{m,i} - T_{m,o}} = R\epsilon$$

Hence, we get from Eqs. 15.30 and 15.31 the following relation:

$$\boxed{\epsilon = \frac{1}{R}\left[1 - \mathrm{e}^{-R(1-\mathrm{e}^{-NTU})}\right]} \tag{15.33}$$

In the $\epsilon - NTU$ approach, a plot of the heat exchanger effectiveness as a function of NTU with R as a parameter, shown in Fig. 15.17 for both cases (a) and (b) considered above is made use of. The plain curves correspond to the case (a) wherein the mixed fluid has the smaller mass flow rate specific product. The curves with symbols correspond to case (b) wherein the unmixed fluid has the smaller mass flow rate specific heat product. In the extreme cases of $R = 0$ and $R = 1$, the two curves merge and are shown by curves without symbols.

Example 15.6

The data pertaining to an air-cooled radiator is shown in Fig. 15.18. Determine the required heat exchange area. Use the $\epsilon - NTU$ approach. Overall heat transfer coefficient is specified to be $80\,\mathrm{W/m^2\,°C}$.

Solution:

Important data for the problem is specified in the figure. We shall assume a specific heat value of $C_{pc} = 1006\,\mathrm{J/kg°C}$ for air, the cold fluid. The hot fluid (water) has a mean temperature of $T_m = \frac{T_{h,i}+T_{h,o}}{2} = \frac{95+55}{2} = 75\,°\mathrm{C}$. Specific heat of water is taken as $C_{ph} = 4193\,\mathrm{J/kg°C}$ at this temperature. The air outlet temperature may be obtained by energy balance for the radiator as

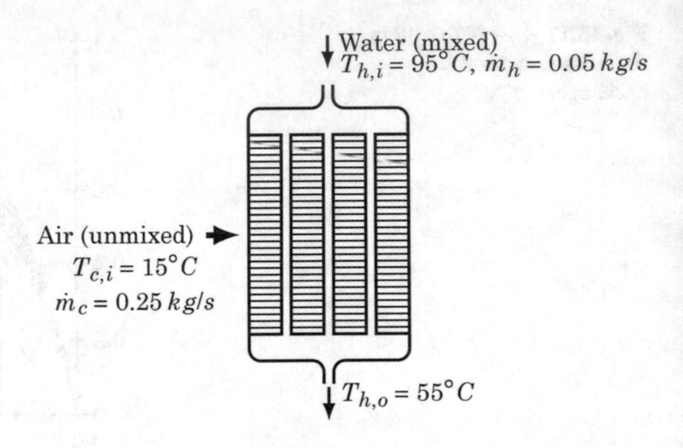

Fig. 15.18 Schematic of the radiator in Example 15.6

$$\dot{m}_c C_{pc}(T_{c,o} - T_{c,i}) = \dot{m}_h C_{ph}(T_{h,i} - T_{h,o})$$

We may solve for the *average* air exit temperature as

$$T_{c,o} = T_{c,i} + \frac{\dot{m}_h C_{ph}(T_{h,i} - T_{h,o})}{\dot{m}_c C_{pc}} = 15 + \frac{0.05 \times 4193(95 - 55)}{0.25 \times 1006} = 48.35\,°\text{C}$$

We also have $C_h = C_m = 0.05 \times 4193 = 209.65\,\text{W}/°\text{C}$ and $C_c = C_u = 0.25 \times 1006 = 251.5\,\text{W}/°\text{C}$. Hence, $C_{min} = C_m = 209.65\,\text{W}/°\text{C}$ and $C_{max} = C_u = 251.5\,\text{W}/°\text{C}$. The heat capacity ratio may then be calculated as

$$R = \frac{C_{min}}{C_{max}} = \frac{C_m}{C_u} = \frac{209.65}{251.5} = 0.834$$

With all the terminal temperatures being known, the effectiveness of the radiator may be calculated as

$$\epsilon = \frac{T_{h,i} - T_{h,o}}{T_{h,i} - T_{c,i}} = \frac{95 - 55}{95 - 15} = 0.5$$

Referring to Fig. 15.17, we get the value of $NTU = 1$. Note that this case is represented by case (a) and hence is represented by plain curves in the figure. Alternately, we may solve analytical expression 15.33 for NTU in terms of R and ϵ and get

$$NTU = -\frac{1}{R}\ln\left[1 + R\ln(1 - \epsilon)\right] = -\frac{1}{0.834}\ln(1 + 0.834\ln 0.5) = 1.035$$

Using the definition of NTU, the heat exchange area may be calculated as

$$S = \frac{C_{min} \times NTU}{U} = \frac{209.65 \times 1.035}{80} = 2.71\,\text{m}^2$$

15.3.3 Analysis of a Cross Flow Heat Exchanger by LMTD Correction Factor Approach

The heat transferred in a cross flow heat exchanger may be obtained by treating it as a counter-current heat exchanger and applying a correction factor F such that

$$Q = US_L \times LMTD \times F \tag{15.34}$$

where the $LMTD$ is given by the counter-flow heat exchanger value

$$LMTD = \frac{(T_{m,i} - T_{u,o}) - (T_{m,o} - T_{u,i})}{\ln\left[\dfrac{T_{m,i} - T_{u,o}}{T_{m,o} - T_{u,i}}\right]} \tag{15.35}$$

Energy balance for the entire heat exchanger yields the following useful relation

$$Q = C_u(T_{u,i} - T_{u,o}) = C_m(T_{m,o} - T_{m,i}) \tag{15.36}$$

where Q is the rate of heat transfer between the two fluids. The numerator Nr of Eq. 15.35 may be rewritten as

$$\begin{aligned} Nr &= (T_{m,i} - T_{u,o}) - (T_{m,o} - T_{u,i}) = (T_{m,i} - T_{m,o}) - (T_{u,o} - T_{u,i}) \\ &= (T_{m,i} - T_{m,i}) - \frac{C_m}{C_u}(T_{m,i} - T_{m,o}) = \left(1 - \frac{C_m}{C_u}\right)\frac{Q}{C_m} \end{aligned}$$

The denominator Dr of Eq. 15.35 may be rewritten as

$$Dr = \frac{T_{m,i} - T_{u,o}}{T_{m,o} - T_{u,i}} = \frac{(T_{m,i} - T_{u,i}) - \dfrac{C_m}{C_u}(T_{m,i} - T_{m,o})}{(T_{m,i} - T_{u,i}) - (T_{m,i} - T_{m,o})} = \frac{\left(1 - \dfrac{C_m}{C_u}\dfrac{\Delta T_m}{\Delta T_{max}}\right)}{\left(1 - \dfrac{\Delta T_m}{\Delta T_{max}}\right)}$$

where ΔT_m is the change in mixed fluid temperature as it passes through the heat exchanger and ΔT_{max} is the maximum temperature difference in the heat exchanger. With this the expression for $LMTD$ may be rewritten as

$$LMTD = \frac{\left(1 - \dfrac{C_m}{C_u}\right)\dfrac{Q}{C_m}}{\ln\left[\dfrac{1 - \dfrac{C_m}{C_u}\dfrac{\Delta T_m}{\Delta T_{max}}}{1 - \dfrac{\Delta T_m}{\Delta T_{max}}}\right]} \tag{15.37}$$

h We substitute $LMTD$ given by Eq. 15.37 in Eq. 15.34 to get

$$Q = US_L \times \frac{\left(1 - \frac{C_m}{C_u}\right)\frac{Q}{C_m}}{\ln\left[\frac{1 - \frac{C_m}{C_u}\frac{\Delta T_m}{\Delta T_{max}}}{1 - \frac{\Delta T_m}{\Delta T_{max}}}\right]} \cdot F$$

Canceling Q on the two sides and noting that $\frac{US_L}{C_m}$ is NTU_m based on the heat capacity of the mixed fluid, the above equation may be rearranged to get the correction factor as

$$F_{cf} = \frac{\ln\left[\frac{1 - \frac{C_m}{C_u}\frac{\Delta T_m}{\Delta T_{max}}}{1 - \frac{\Delta T_m}{\Delta T_{max}}}\right]}{NTU_m\left(1 - \frac{C_m}{C_u}\right)} \tag{15.38}$$

Equation 15.38 may be written in the alternate form

$$F = \frac{\ln\left[\frac{1 - \frac{\Delta T_u}{\Delta T_{max}}}{1 - \frac{C_u}{C_m}\frac{\Delta T_u}{\Delta T_{max}}}\right]}{NTU_m\left(1 - \frac{C_m}{C_u}\right)} \tag{15.39}$$

where the temperature ratio has been written in terms of the temperature change of the unmixed fluid. Equation 15.27 may now be manipulated to get NTU_m as follows.

$$NTU_m = NTU_u\frac{C_u}{C_m} = -\frac{C_u}{C_m}\ln(1 - \epsilon_e)$$

However, Eq. 15.30 may rearranged as

$$\epsilon_e = -\frac{C_m}{C_u}\ln\left[1 - \frac{T_{m,o} - T_{m,i}}{T_{u,i} - T_{m,i}}\right] = -\frac{C_m}{C_u}\ln\left[1 - \frac{\Delta T_m}{\Delta T_{max}}\right]$$

With this, the mixed fluid NTU is rewritten as

$$NTU_m = -\frac{C_u}{C_m}\ln\left(1 + \frac{C_m}{C_u}\ln\left[1 - \frac{\Delta T_m}{\Delta T_{max}}\right]\right) \tag{15.40}$$

Again Eq. 15.40 may also be written in the alternate form

$$NTU_m = -\frac{C_u}{C_m}\ln\left(1 + \frac{C_m}{C_u}\ln\left[1 - \frac{C_u}{C_m}\frac{\Delta T_u}{\Delta T_{max}}\right]\right) \tag{15.41}$$

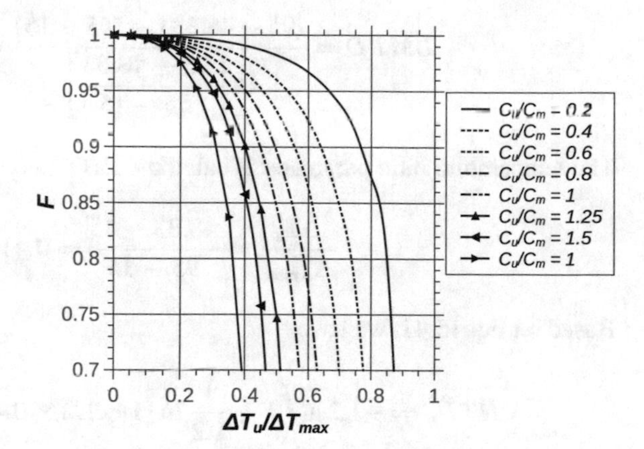

Fig. 15.19 *LMTD* correction factor for a cross flow heat exchanger

Equation 15.38 or 15.39 along with Eq. 15.40 or15.41 may be used to calculate the *LMTD* correction factor for a cross flow heat exchanger. Heat capacity ratio $\frac{C_m}{C_u}$ and the temperature difference ratio $\frac{\Delta T_m}{\Delta T_{max}}$ or $\frac{\Delta T_u}{\Delta T_{max}}$ have parametric influence on the correction factor F. Figure 15.19 shows the variation of correction factor for a useful range of parameter values.

Example 15.7

Redo Example 15.6 using the *LMTD*-correction factor method.

Solution:

Given data is written down using the notation followed in Sect. 15.3.3.

$$\dot{m}_u = 0.25\,\text{kg/s} \quad C_{pu} = 1006\,\text{J/kg°C}$$
$$\dot{m}_m = 0.05\,\text{kg/s} \quad C_{pm} = 4193\,\text{J/kg°C}$$
$$T_{u,i} = 15\,°\text{C}$$
$$T_{m,i} = 95\,°\text{C} \quad T_{m,o} = 55\,°\text{C}$$

The heat capacity ratio is now calculated as

$$\frac{C_u}{C_m} = \frac{\dot{m}_u C_{pu}}{\dot{m}_m C_{pm}} = \frac{0.25 \times 1006}{0.05 \times 4193} = 1.2$$

By energy balance, we calculate the exit temperature of the unmixed fluid (air) as

$$T_{u,o} = T_{u,i} + \frac{C_m}{C_u}(T_{m,i} - T_{m,o}) = 15 + \frac{95 - 55}{1.2} = 48.33\,°\text{C}$$

The *LMTD* on the basis of counter-current flow is obtained as

$$LMTD = \frac{(95 - 48.33) - (55 - 15)}{\ln\left[\dfrac{95 - 48.33}{55 - 15}\right]} = 43.25\,^\circ\mathrm{C}$$

The temperature ratio parameter is calculated as

$$\frac{\Delta T_u}{\Delta T_{max}} = \frac{48.33 - 15}{95 - 15} = 0.4166$$

Based on Eq. 15.41, we have

$$NTU_m = -1.2\ln\left(1 + \frac{1}{1.2}\ln\left[1 - 1.2 \times 0.4166\right]\right) = 1.0338$$

The $LMTD$ correction factor F may now be calculated using Eq. 15.39 as

$$F = \frac{\ln\left[\dfrac{1 - 0.4166}{1 - 1.2 \times 0.4166}\right]}{1.0388(1 - 1.2)} = 0.8944$$

The heat transfer area required may now be obtained using Eq. 15.34 as

$$S_L = \frac{Q}{U \cdot LMTD \cdot F_{fc}} = \frac{0.05 \times 4193(95 - 55)}{80 \times 43.25 \times 0.8944} = 2.71\,\mathrm{m}^2$$

This answer agrees with that obtained in Example 15.6.

15.3.4 General Remarks on Heat Exchangers

Before we conclude the chapter, we give some useful information regarding the heat exchanger types that have been considered in the earlier sections. We have looked at the method of analysis, starting from first principles, in the case of co-current and counter-current heat exchangers. However, in the case of shell and tube exchangers *recipes* for analyzing them have been given. The analysis is complicated by the fact that the shell side fluid in a shell and tube heat exchanger follows a tortuous path and the heat transfer coefficient may be difficult to determine. The problems we considered were with a *prescribed* heat transfer coefficient.

Two ways of obtaining heat transfer coefficient would be by the numerical solution of governing equations using commercial numerical codes or by experiments. Many research papers have appeared in recent times that use full scale numerical simulations of heat exchangers. However, the designer depends, to a large extent on experimentally obtained correlations, at least for preliminary design of heat exchangers. For example, we have referred earlier to the shell side heat transfer correlation given by Donohue. He has also given correlations applicable to shells with baffles of different types. The original paper may be referred to for useful correlations. Information regarding the dependence of ϵ on R and NTU have been given previously in the form of suitable plots. Analytical expressions have also been given for some of the cases. A useful analytical expression is provided below for a shell and tube heat exchanger:

Shell and tube heat exchanger: 1 shell pass and even number of tube passes

$$\epsilon = \frac{2}{1 + R + X\sqrt{1 + R^2}} \quad \text{with} \quad X = \frac{1 + e^{-NTU\sqrt{1+R^2}}}{1 - e^{-NTU\sqrt{1+R^2}}}$$

There are several other types of heat exchange equipment that have not been covered in this elementary treatment of the subject. For example, fixed beds of granular materials such as pebble beds are used as energy storage devices. They get heated or "charged" by passing hot gases through them for a certain length of time. Subsequently, cold gases may be passed through the bed to "discharge" the heat to the cold gases. The pebble bed heat exchanger is essentially a device that operates in the transient mode. The operation is cyclic in nature with alternate charge and discharge cycles. In order to improve the energy density, it is also possible to use phase change materials for storage of energy. These have important applications in the utilization of solar energy. See, for example, the book by S.P. Sukhatme and J.K. Nayak, *Solar Energy: Principles of Thermal Collection and Storage*, 3rd Edition, Tata McGraw Hill, 2008. We have also not covered compact heat exchangers that find application in areas such as aviation. The reader may refer to advanced books for these topics.

Concluding Remarks

This chapter has dealt with the analysis of heat exchangers that are encountered in many engineering applications. Principles involved in their operation and the specific methods of analysis have been discussed in detail. Both the $LMTD$ as well as $\epsilon - NTU$ methods have been presented. Useful formulae and charts have also been provide in this chapter.

15.4 Exercises

Ex 15.1 For a counter-current heat exchanger, the effectiveness is given by Eq. 15.22. When $R \to 1$, ϵ must be calculated by using L'Hospital rule since the above is of form $\frac{0}{0}$. Obtain, by such a procedure, the appropriate value of ϵ as a function of NTU when $R \to 1$.

Ex 15.2 A double pipe heat exchanger is to be used to cool water from 22 °C to $6°C$, using brine entering at −2 °C and leaving at 3 °C. The overall heat transfer coefficient has been estimated at 500 W/m^2 °C. Calculate the heat transfer area for a design heat load of 10 kW for both co-current and counter flow arrangements. Also determine the effectiveness of the heat exchanger in these cases, by two different ways.

Ex 15.3 Steam at 100 °C condenses over a horizontal tube subject to a heat transfer coefficient of 3500 W/m^2 °C. Cooling water enters the tube at 50 °C and leaves at 70 °C. The inner diameter of the tube is 18 mm, the tube wall thickness is 2 mm, and the tube material has a thermal conductivity of 45 W/m°C. Determine the tube length required if the water flow rate is given as 0.2 kg/s.

Ex 15.4 Air at atmospheric pressure and 30 °C flows at 3 m/s through a 10 mm ID pipe. An electrical resistance heater surrounds 30 cm length of tube toward its discharge end and supplies a constant heat flux to raise the temperature of air to 90 °C. What is the power input? What is the mean value of the heat transfer coefficient? Based on the above determine the mean temperature difference between the tube wall and the fluid.

Ex 15.5 Hot chemical products with specific heat of 2500 J/kg°C, inlet temperature 600 °C at a flow rate of 30 kg/s are used to heat cold chemical products with specific heat of 4200 J/kg°C at 100 °C and a flow rate of 30 kg/s in a parallel flow arrangement. The total heat transfer area is 50 m^2 and the overall heat transfer coefficient is 1500 W/m^2 °C. Calculate the outlet temperatures of the hot and cold products.

Ex 15.6 A counter-flow heat exchanger is to heat air to 500 °C with the exhaust gas from a turbine. Air enters the exchanger at 300 °C and at a mass flow rate of 4 kg/s while the exhaust gas enters at a temperature of 650 °C with a mass flow rate of 4 kg/s. The overall heat transfer coefficient is 1500 W/m^2 °C. The specific heat for both fluids can be taken as equal and as 1100 J/kg°C. Calculate the heat transfer area and the outlet temperature of the hot gas.

Ex 15.7 A refrigerator coil has a tube of 3 mm ID and 4 mm OD with radial fins of outer radius 10 mm and thickness 0.2 mm attached to the outside with a spacing of 2 mm. Other pertinent data are:

- Temperature of fluid flowing inside the tube: 60 °C
- Heat transfer coefficient on the tube side: 270 W/m^2 °C
- Temperature of the fluid on the outside of the tube: 30 °C

- Heat transfer coefficient on the outside: 24 W/m^2°C
- Tube and fin material thermal conductivity: 380 W/m°C

Determine the heat transfer per meter length of the coil.

Ex 15.8 A double pipe heat exchanger is used to condense steam at 6894 Pa. Water at an average temperature of 10 °C flows at 3.05 m/s through the inner pipe of copper (2.54 cm ID, 3.05 cm OD). Steam at its saturation temperature flows in the annulus formed between the outer surface of the copper tube and an outer pipe of 5.08 cm ID. The average heat transfer coefficient on the steam side is 5680 W/m^2°C. The outer surface of the copper tube has a scale formed on it whose thermal resistance is 1.76×10^{-4} m^2°C/W.

- Determine the overall heat transfer coefficient between steam and water based on the outer area of the inner copper tube
- Evaluate the temperature of the inner surface of the pipe
- Estimate the tube length required for condensing 0.45 kg/min of steam

Ex 15.9 A cross flow air to water heat exchanger with an effectiveness of 0.65 is used to heat water ($C_{pc} = 4180$ J/kg°C) with hot air ($C_{ph} =$ 1010 J/kg°C). Water enters the heat exchanger at 20 °C at 4 kg/s while air enters at 100 °C at 9 kg/s. If the overall heat transfer coefficient based on the waterside is 260 W/m^2°C, determine the heat transfer area on the waterside. Assume that both fluids are unmixed. Use suitable chart (refer to a book on heat exchangers) to solve this problem.

Ex 15.10 In a solar assisted air conditioning system 0.5 kg/s of ambient air at 270 K is to be preheated by the same amount of air leaving the system at 295 K. If a countercurrent heat exchanger has a heat transfer area of 30 m^2, and the overall heat transfer coefficient is estimated to be 25 W/m^2°C, determine the outlet temperature of the preheated air. Assume that the specific heat of air remains constant at $C_{ph} = 1000$ J/kg°C.

Ex 15.11 A long tube of wall thickness 3 mm is made of a material of thermal conductivity equal to 15 W/m°C. The tube has an ID of 23 mm. It carries a fluid flowing at a mean velocity of 3 m/s having the following properties: Density = 1000 kg/m^3, specific heat = 4180 J/kg°C, dynamic viscosity = 0.00108 kg/m · s, Prandtl number = 7.5, and thermal conductivity = 0.598 W/m°C. The outside surface of the tube is subjected to heat transfer to a condensing fluid with a known heat transfer coefficient of 14000 W/m^2°C. Determine the overall heat transfer coefficient based on the inside area of the tube. If the tube side fluid enters at 30 °C and the temperature of the condensing fluid remains constant at 86 °C, determine the length of the tube required to heat the tube side fluid to 42 °C.

Ex 15.12 A co-current heat exchanger involves heat transfer from water to water. The cold water flows on the tube side at 0.8 kg/s while the hot water flows through the annulus at 1 kg/s. Hot water enters at 70 °C and leaves at 54 °C. The cold water enters the tube at 25 °C. The overall heat transfer coefficient between the two fluids has been ascertained to be 1800 W/m^2 °C. Determine the required heat transfer area.

Ex 15.13 Water flows with a velocity of 5 m/s in a tube of 19 mm ID. The water enters the tube at 30 °C and leaves at 45 °C. The tube wall is maintained at a constant temperature of 85 °C. Using the *LMTD* concept, the estimated value of the heat transfer coefficient based on a suitable correlation, obtain the length of the tube.

Ex 15.14 Hot air at atmospheric pressure and 350 K enters a bare duct of square cross section 15 × 15 cm that passes through a room 8 m long at a volume flow rate of 0.1 m^3/s. The duct wall is observed to be more or less at a constant temperature of 330 K. Determine the exit temperature of air and also the pressure drop between duct entry and exit.

Ex 15.15 Water at 90 °C and a mass flow rate of 2250 kg/h is to be used to heat ethylene glycol. Ethylene glycol is available at 30 °C with a mass flow rate of 550 kg/h. A double pipe heat exchanger consisting of a 34.93 mm OD (32.79 mm ID) standard copper tubing inside of a 53.98 mm OD (50.42 mm ID) copper tubing is to be used. The exchanger is 10 m. It is constructed by using "hair pin" type construction with four 2.5 m sections. Determine the outlet temperature of both fluids using a counter flow arrangement. Ignore the effect of tube bends.

Ex 15.16 In a certain heat exchanger, condensing steam at 100 °C is used to heat oil from 30 °C to 45 °C. The mass flow of oil is given to be 0.2 kg/s and the specific heat of oil is constant at 1200 J/kg K. Design the heat exchanger as a shell and tube type heat exchanger. Assume that tubes are to be restricted to a length of 1 m. Note that many design variants are possible.

Ex 15.17 A test is conducted to determine the overall heat transfer coefficient in a shell and tube oil to water heat exchanger that has 96 tubes of internal diameter 12 mm and length 2 m in a single shell. Cold water ($C_{pc} = 4180$ J/kg°C) enters the tubes at 20 °C with a mass flow rate of 5 kg/s and leaves at 55 °C. Oil ($C_{ph} = 2150$ J/kg°C) flows through the shell and is cooled from 120 °C to 75 °C. Determine the overall heat transfer coefficient U_i of this heat exchanger based on the inner surface area of the tubes.

Chapter 16
Natural Convection

NATURAL or free convection heat transfer occurs due to flow generated because of density differences in the medium, in the presence of an external force field, such as gravity. It may also be due to a force field generated by rotation that gives rise to the centrifugal force field. Because natural convection takes place in the absence of an external agency, such as a pump or blower, that generates the flow, heat transfer is free in that there is no expenditure of mechanical energy. Examples of natural convection are many and are important in applications such as cooling of electronic equipment for terrestrial use. Natural convection also is important in buildings, equipments such as chimney and cooling towers that make use of natural draft due to temperature differences.

16.1 Introduction

Free or natural convection is the mode of heat transfer that takes place from a surface to a fluid, in the presence of a temperature difference between the surface and the fluid, but in the absence of any externally imposed flow. Even when the fluid can be considered as incompressible, density differences in the presence of temperature differences create a flow due to the effect of buoyancy. Of course, it is essential that an external force field like gravity be there for this to happen. It is customary to assume that the acceleration due to gravity is directed downwards. A typical example of buoyancy induced flow is the natural draft induced by a chimney as shown in Fig. 16.1. The draft induced because of the temperature difference between warm indoor and cold outdoor air operates over the height H of the chimney. If it is assumed that the warm air does not lose heat as it moves in the chimney the pressure difference induced due to buoyancy is given by

S. P. Venkateshan, *Heat Transfer*,
https://doi.org/10.1007/978-3-030-58338-5_16

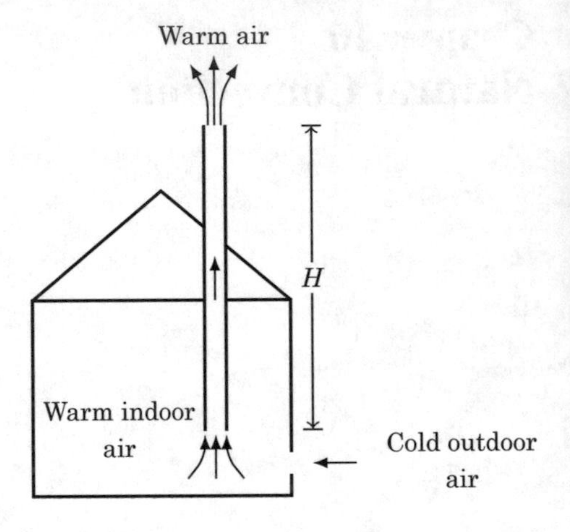

Fig. 16.1 Example of buoyancy induced flow: natural draft in a chimney

$$\Delta p = (\rho_o - \rho_i) g H \tag{16.1}$$

where subscripts o and i stand for outdoor and indoor, respectively. If we assume that the pressure drop due to entrance, friction, and exit of the chimney is $K \frac{\rho_i V^2}{2}$ where K is a suitable constant, the velocity V in the chimney is given by

$$V = \sqrt{\frac{2(\rho_o - \rho_i) g H}{K \rho_i}} \tag{16.2}$$

If the inner diameter of the chimney is D the draft Q is given by

$$Q = \frac{\pi D^2}{4} \sqrt{\frac{2(\rho_o - \rho_i) g H}{K \rho_i}} \tag{16.3}$$

A heated tube closed at the bottom, but open at the top, will support flow in the form of a movement of cold ambient air down the middle of the tube and upward movement of warm air close to the tube wall, as shown in Fig. 16.2. Free convection is due to buoyant forces that are normally neglected or negligible in the case of forced convection problems. Many examples of free convection flow are encountered in practice. A few of them are schematically indicated in Fig. 16.3a–d.

In the case of the vertical isothermal plate ($T_w > T_\infty$) shown in Fig. 16.3a the fluid close to the plate is hotter and hence lighter than the fluid far away which is lighter and heavier. Hence the fluid near the plate experiences an upward force due to buoyancy. The flow near the plate is then upwards. In case ($T_w < T_\infty$) as shown in Fig. 16.3b the flow direction is reversed as indicated. The flow near the plate is then downwards.

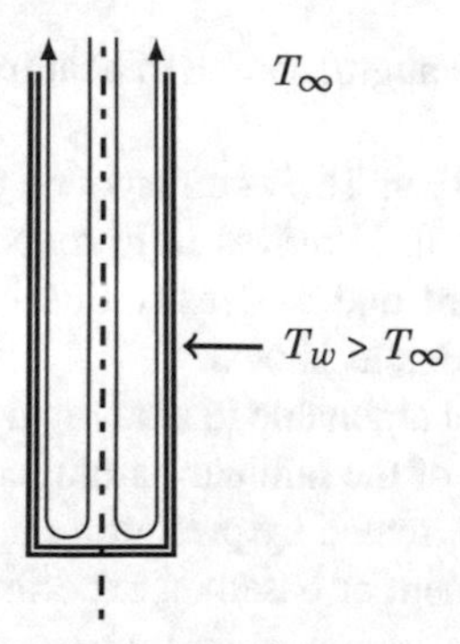

Fig. 16.2 Buoyancy induced flow in a heated tube with closed bottom

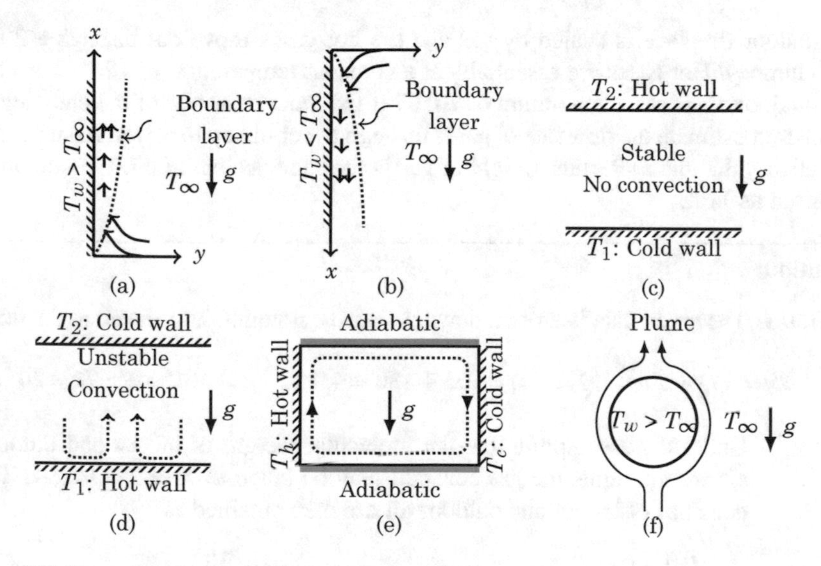

Fig. 16.3 Examples of natural convection flows

The case of a horizontal fluid layer between two large walls is shown in Fig. 16.3c. When the lower wall is cooler than the upper wall, the layer is stable and no natural convection flow is observed. The fluid layer indicates temperature stratification and the heat transfer is by conduction alone. However, when the lower wall is at a temperature higher than that of the upper wall, as shown in Fig. 16.3d, a complex natural convection flow pattern is set up. There is heat transfer augmentation due to natural convection as compared to pure conduction in the stable configuration.

A two-dimensional cavity formed by two vertical isothermal walls and two horizontal adiabatic walls is shown in Fig. 16.3e. The heated fluid moves upward near the hot wall, turns around at the adiabatic wall, and subsequently moves downward near the cold wall. A circulatory flow is thus set up. Heat is removed by the upward moving fluid near the hot wall and is transferred to the cold wall as the fluid moves down near it. Again there is an augmentation of heat transfer in comparison with the pure conduction case.

The last example shown in Fig. 16.3f corresponds to natural convection from an isothermal cylinder. The cold fluid moves in to displace the heated fluid near the surface of the cylinder upward due to density differences. A well-defined *plume* rises from the top of the cylinder, as shown.

Since all these cases are not amenable to analytical treatment, a detailed analysis is presented only for the case of the laminar natural convection from an isothermal vertical plate. Other cases are treated by presenting appropriate correlations, based on either full numerical treatment or based on experiments.

Example 16.1

An indoor fireplace is heated by gas and the hot gases move out through a 20 m tall chimney. Hot gases are essentially at a constant temperature of 180 °C. Ambient outdoor air is at a temperature of 20 °C. If the inner diameter of the chimney is 203.2 mm estimate the flow rate of gases through the chimney in m^3/h. Assume, for simplicity, that the molecular weight of gas is the same as that of air. Pressure may be taken as 1 *atm*

Solution:

Step 1 The given data is written down, using the notation introduced in the text.

$$T_o = 273 + 20 = 293\,\text{K},\quad T_i = 273 + 180 = 453\,\text{K},\quad p = 101300\,\text{Pa}\quad H = 20\,\text{m}$$

Under the assumption that the molecular weight of indoor and outdoor air are the same, the gas constant may be taken as $R = 287$ J/kg K. The densities of indoor and outdoor air are then obtained as

$$\rho_i = \frac{101300}{287 \times 453} = 0.779\,\text{kg/m}^3 \quad \text{and} \quad \rho_o = \frac{101300}{287 \times 293} = 1.205\,\text{kg/m}^3$$

Pressure developed due to buoyancy is then given by

$$\Delta p = (\rho_o - \rho_i)gH = (1.205 - 0.779) \times 9.81 \times 20 = 83.5\,\text{Pa} \qquad (16.4)$$

Step 2 In order to calculate the velocity of air in the chimney, we need to know the value of loss coefficient K. This comprises of 3 parts. First one is due to entry loss where the hot air enters the chimney. Assuming that the entry is flush or square edged contraction (no data specified and we take the worst case) the loss coefficient is $K_i = 0.5$. The second one is the exit loss that is characterized by $K_e = 1$. The third one is due to friction in the chimney. This depends on the Reynolds number and is based on the friction factor calculated by the use of an appropriate relation for tube flow. Since the

Reynolds number is based on as yet unknown velocity we need to use an iterative scheme for the solution.

We start the solution by altogether ignoring the pressure drop due to fluid friction and use an initial K value of $K^{(1)} = K_i + K_e = 1 + 0.5 = 1.5$ to obtain the first guess for the fluid velocity $V^{(1)}$ as

$$V^{(1)} = \sqrt{\frac{2\Delta p}{K^{(1)}\rho_i}} = \sqrt{\frac{2 \times 83.5}{1.5 \times 0.779}} = 11.95\ \text{m/s}$$

Step 3 With the first guess for velocity available we estimate the contribution of fluid friction to the pressure drop now. The Reynolds number is calculated, based on inner diameter of chimney of $ID = 203.2$ mm $= 0.2032$ m and the kinematic viscosity of air at temperature of $T_i = 180\,°\text{C}$ given by $\nu_i = 3.28 \times 10^{-5}\ \text{m}^2/s$. Thus we have

$$Re_{ID}^{(1)} = \frac{V^{(1)} ID}{\nu_i} = \frac{11.95 \times 0.2032}{3.28 \times 10^{-5}} = 74135.3$$

Friction factor is then calculated as (using Equation 14.24(b))

$$f^{(1)} = \frac{0.184}{\left[Re_{ID}^{(1)}\right]^{0.25}} = \frac{0.184}{74135.3^{0.25}} = 0.0112$$

Then $K_f^{(1)}$ is calculated as

$$K_f^{(1)} = \frac{4 f^{(1)} H}{ID} = \frac{4 \times 0.0112 \times 20}{0.2032} = 4.390$$

Step 4 We may now update the overall K as $K^{(2)} = K_i + K_f^{(1)} + K_e = 0.5 + 4.39 + 1 = 5.89$. We may revise the velocity calculation and continue the iteration till convergence. The results are tabulated below.

Iteration number	K	V m/s	Change ΔV, m/s
1	1.5	11.952	
2	5.89	6.032	−5.921
3	6.71	5.652	−0.380
4	6.79	5.616	−0.036
5	6.80	5.612	−0.003

Step 5 The converged value for the average velocity of flow through the chimney is taken as $V = 5.612$ m/s. With this velocity, the draft may be calculated as

$$Q = 3600 \times \frac{\pi ID^2 V}{4} = 3600 \times \frac{\pi \times 0.2032^2 \times 5.612}{4} = 655.23\ \text{m}^3/\text{h}$$

In Example 16.1, we have assumed that the gas flowing through the chimney is isothermal. This may be justified by looking at the heat loss from the gas to the chimney as it flows upwards. Since the flow velocity is substantial we may use a suitable correlation—*forced convection correlation*—such as the Dittus–Boelter equation to calculate the heat transfer coefficient. Temperature variation of the gas as it moves up may be obtained by treating the chimney as a heat exchanger. We demonstrate this by the following example.

Example 16.2

Consider the chimney in Example 16.1. The chimney is made of a double wall construction with a $t = 50.8$ mm thick Super Wool 607 Plus in the annular space. Thermal conductivity of the insulation is known to be $k_{Insul} = 0.05$ W/m°C. Assume that the chimney experiences a mild wind of 2 m/s normal to its axis.

Solution:
This example will show how we make use of the material already covered in previous chapters to model a fairly complex problem. We make use of the overall resistance concept in treating heat transfer radially across the chimney, assuming very little heat transfer to take place along the axis. Heat transfer on the outside surface of the chimney is modeled as that due to flow normal to the axis of a cylinder, using the Zhukaskas correlation. Resistance due to the insulation is considered as conduction through an annulus. The stainless sheet metal used in making the double wall chimney is very thin and may be assumed to offer zero conduction resistance. Air properties required in the analysis are tabulated below.

T, K	ρ, kg/m^3	ν, m^2/s	k, W/m°C	Pr
293	1.205	1.52×10^{-5}	0.0258	0.704
453	0.779	3.28×10^{-5}	0.0376	0.686

Step 1 We calculate all the resistances now.
(a) Conduction resistance due to the insulation is given by

$$R_{Insul} = \frac{1}{2\pi k_{Insul}} \ln\left(\frac{ID + 2t}{ID}\right)$$
$$= \frac{1}{2 \times \pi \times 0.05} \ln\left(\frac{203.2 + 2 \times 50.8}{203.2}\right) = 1.2906\ \text{m°C/W}$$

(b) Reynolds number for flow in the chimney is given, based on air properties at 453 K, by

$$Re_{ID} = \frac{5.612 \times 0.2032}{3.28 \times 10^{-5}} = 34812.5$$

Based on Dittus–Boelter equation the corresponding Nusselt number is given by

$$Nu_{ID} = 0.023 Re_{ID}^{0.8} Pr^{0.37} = 0.023 \times 34812.5^{0.8} \times 0.686^{0.37} = 88.3$$

Corresponding to the above the inside heat transfer coefficient h_i is given by

$$h_i = \frac{Nu_{ID} k_i}{ID} = \frac{88.8 \times 0.0376}{0.2032} = 16.34\ \text{W/m}^2\,{}^\circ\text{C}$$

Film resistance on the inside is then given by

$$R_i = \frac{1}{\pi I D h_i} = \frac{1}{\pi \times 0.2032 \times 16.34} = 0.0959\ \text{m}^\circ\text{C/W}$$

(c) Consider the cross flow across the chimney on the outside due to the prevailing wind. The wind velocity is specified as $U = 2$ m/s. Air properties are taken at 293 K. Reynolds number based on OD is then given by

$$Re_{OD} = \frac{2 \times 0.3048}{1.52 \times 10^{-5}} = 39994.3$$

The appropriate constants in the Zhukaskas correlation are $C = 0.26,\ \text{m} = 0.6$ and $n = 0.37$. The Nusselt number is then given by

$$Nu_{OD} = 0.26 \times 39994.3^{0.6} \times 0.704^{0.37} = 131.8$$

Then the external heat transfer coefficient is

$$h_o = \frac{Nu_{OD} k_o}{OD} = \frac{131.8 \times 0.0258}{0.3048} = 11.15\ \text{W/m}^2\,{}^\circ\text{C}$$

Hence the outside film resistance is given by

$$R_o = \frac{1}{\pi O D h_o} = \frac{1}{\pi \times 0.3048 \times 11.15} = 0.0936\ \text{m}^\circ\text{C/W}$$

Hence the overall resistance is given by

$$R_{Overall} = R_i + R_{Insul} + R_o = 0.0959 + 1.2906 + 0.0936 = 1.4801\ \text{m}^\circ\text{C/W}$$

Step 2 We model the cooling of the gas as it moves up the chimney as a heat exchanger tube with constant wall temperature that corresponds to the outdoor air temperature. The analysis is akin to that presented in section

14.4.3. $2\pi R\bar{h}$ in that analysis will correspond to $\frac{1}{R_{Overall}}$ of the present case. Taking ID as the basis, we may calculate the mean heat transfer coefficient as

$$\bar{h} = \frac{1}{\pi I D R_{Overall}} = \frac{1}{\pi \times 0.2032 \times 1.4801} = 1.0583\,\text{W/m}^2\,°\text{C}$$

The mass flow rate of the flue gas is $\dot{m} = \rho_i Q = 0.7942 \times 0.182 = 0.1445$ kg/s. We assume its specific heat to be $C_p = 1005$ J/kg K. With the heat transfer area being given by $S_H = \pi I D\, H = \pi \times 0.2032 \times 20 = 12.767$ m^2, we then have

$$\frac{\bar{h} S_H}{\dot{m} C_p} = \frac{1.4801 \times 12.767}{0.1445 \times 1005} = 0.0930$$

The temperature difference between the gas leaving the chimney and the ambient is then obtained as

$$\Delta T_H = (180 - 20)\text{e}^{-0.0930} = 145.8\,°\text{C}$$

Hence the exit temperature of the gases from the chimney is

$$T_H = 20 + 145.8 = 165.8\,°\text{C}$$

Thus the temperature drops by about 14 °C from entry to exit. Example 16.1 may be reworked by taking the temperature of the gases as the mean at entry and exit and rework all the numbers.

16.2 Laminar Natural Convection from a Vertical Isothermal Plate

A new non-dimensional parameter called the Grashof number[1] makes its appearance in natural convection flows. Consider, for example, natural convection from an object of characteristic length L with a surface to the ambient temperature difference

[1]Named after Franz Grashof 1826–1893, a German engineer.

of $T_w - T_\infty = \Delta T$. Let the gravitational acceleration be g, and oriented vertically downwards. Let the fluid be characterized by an isobaric coefficient of volume expansion of β, kinematic viscosity ν. The Grashof number Gr, as it will become clear later on, turns out to be given by

$$\boxed{Gr = \frac{g\beta \Delta T L^3}{\nu^2}} \tag{16.5}$$

The flow is basically by conduction for low values of Grashof number. For $Gr >$ 1000 or so, natural convection sets in because the buoyant forces are able to overcome viscous forces. For large values of Gr the flow exhibits boundary layer behavior, in that, significant velocities occur in a thin layer near the heated boundary. The flow remains laminar till the Grashof number crosses a critical value that is geometry specific. For example, in the case of natural convection from a vertical isothermal plate, the critical value is $Gr_c \approx 10^9$.

Laminar natural convection from a vertical isothermal plate is amenable to analytical solution—either exact as given by Ostrach[2]—or approximate by the integral method. The latter solution is presented here in detail and reference is made to Ostrach solution, for comparison purposes, later on.

16.2.1 Isothermal Vertical Plate—Integral Solution

Governing Equations

Refer again to Fig. 16.3a. Assume that the Grashof number is large enough for the boundary layer type of flow to exist. The velocities are then significant only within the boundary layer close to the surface. The density differences are usually small under the assumption that temperature differences are small. Based on this, we may assume the fluid to be basically incompressible excepting for the buoyancy term (body force term wherein the density variation is taken into account). This approximation is called the Boussinesq approximation.[3] The boundary layer equations are written down in the usual way (two-dimensional flow, x and y are measured as shown in Fig. 16.3a.

[2]S. Ostrach, NACA Report 1111, 1953.

[3]Joseph Valentin Boussinesq, 1842–1929, a French mathematician and physicist suggested this approximation.

- **Continuity:**

$$\frac{\partial u}{\partial x} + \frac{\partial v}{\partial y} = 0 \tag{16.6}$$

- **x - momentum:**

$$\rho \left[u\frac{\partial u}{\partial x} + v\frac{\partial u}{\partial y} \right] = -\rho g - \frac{dp}{dx} + \mu \frac{\partial^2 u}{\partial y^2} \tag{16.7}$$

The extra term '$-\rho g$' is because of body force.

- **Energy:**

$$u\frac{\partial T}{\partial x} + v\frac{\partial T}{\partial y} = \alpha \frac{\partial^2 T}{\partial y^2} \tag{16.8}$$

All properties including density are assumed constant except ρ in the body force term. As we approach the ambient, $\rho \rightarrow \rho_\infty$ and $u \rightarrow 0$. Hence, Eq. 16.7, as $y \rightarrow \infty$ becomes

$$0 = -\rho_\infty g - \frac{dp}{dx} \quad \text{or} \quad -\frac{dp}{dx} = \rho_\infty g$$

Introduce this in Eq. 16.7 to get

$$\rho \left[u\frac{\partial u}{\partial x} + v\frac{\partial u}{\partial y} \right] = (\rho_\infty - \rho) g + \mu \frac{\partial^2 u}{\partial y^2} \tag{16.9}$$

The pressure gradient outside the boundary layer has thus been written in terms of the body force there. The pressure variation is very mild in the case of natural convection flows since the velocities generated by density differences are small. Hence, it is reasonable to assume that density variations are due only to temperature variations. Hence, we obtain density variations using the isobaric coefficient of volumetric expansion of the fluid given by

$$\beta = -\frac{1}{\rho} \left(\frac{\partial \rho}{\partial T} \right)_p$$

With the small temperature difference approximation i.e. $T_w - T_\infty \ll \frac{T_w + T_\infty}{2}$, to a good approximation, we have

$$\frac{\partial \rho}{\partial T} \approx \frac{\rho - \rho_\infty}{T - T\infty}$$

Hence, we have

$$\beta\rho \approx \beta\rho_\infty = \frac{\rho_\infty - \rho}{T - T_\infty} \quad \text{or} \quad \rho_\infty - \rho = -\beta\rho_\infty(T_\infty - T) \tag{16.10}$$

Introduce this in Eq. 16.7, *assume* that $\rho = \rho_\infty$ excepting as far as the body force term is concerned, to get

$$u\frac{\partial u}{\partial x} + v\frac{\partial u}{\partial y} = g\beta(T - T_\infty) + \nu\frac{\partial^2 u}{\partial y^2} \tag{16.11}$$

For an ideal gas, (air may be considered to be an ideal gas, for engineering purposes)

$$p = \rho RT$$

Since p is constant, we have

$$dp = 0 = RTd\rho + R\rho dT \text{ or} -\frac{1}{\rho}\left(\frac{\partial \rho}{\partial T}\right)_p = -\frac{1}{T} \approx \frac{1}{T_\infty}$$

where the last step is possible because of the small temperature difference approximation. Note also that the temperature should be specified in K because the above is based on thermodynamic relations. With this, Eq. 16.11 finally becomes

$$\boxed{u\frac{\partial u}{\partial x} + v\frac{\partial u}{\partial y} = g\beta\left(\frac{T}{T_\infty} - 1\right) + \nu\frac{\partial^2 u}{\partial y^2}} \tag{16.12}$$

Boundary Conditions
At the surface of the plate, no slip conditions apply.

$$\boxed{y = 0 \text{ and } 0 \le x \le L,\ u = v = 0,\ T = T_w} \tag{16.13}$$

Far away from the plate, we have the ambient fluid at rest.

$$\boxed{y \to \infty \text{ and } 0 \le x \le L,\ u = v = 0,\ T = T_\infty} \tag{16.14}$$

Integral Equation
An approximate solution to the equations presented above are possible by the integral method. The method is essentially similar to the one used in the case of forced convection parallel to an isothermal flat plate. The appearance of buoyancy term

and hence the coupling between the flow and energy equations is a feature that is peculiar to natural convection flows. This simply means that momentum and energy equations have to be solved simultaneously even though fluid properties have been assumed to be constant. In view of this, we derive the momentum integral equation and the energy integral equation, starting from the boundary layer Eqs. 16.6, 16.8, and 16.12 subject to the boundary conditions 16.13 and 16.14. In this method, we integrate the boundary layer equations between $y = 0$ and $y = \delta$, the boundary layer thickness. From momentum equation we than have

$$\int_0^\delta u\frac{\partial u}{\partial x}dy + \int_0^\delta v\frac{\partial u}{\partial y}dy = g\beta\int_0^\delta\left(\frac{T}{T_\infty} - 1\right)dy + \nu\int_0^\delta\frac{\partial^2 u}{\partial y^2}dy \tag{16.15}$$

Integrate the second term on the left-hand side of Eq. 16.15 by parts to get the following:

$$\int_0^\delta v\frac{\partial u}{\partial y}dy = \underbrace{vu\Big|_0^\delta}_{\left\{\begin{array}{c}\text{Vanishes at}\\ \text{both limits}\end{array}\right\}} - \int_0^\delta u\underbrace{\frac{\partial v}{\partial y}}_{\left\{\begin{array}{c}= -\frac{\partial u}{\partial x}\\ \text{by continuity}\end{array}\right\}}dy = \int_0^\delta u\frac{\partial u}{\partial x}dy$$

Thus, the left-hand side of Eq. 16.15 becomes

$$\int_0^\delta u\frac{\partial u}{\partial x}dy + \int_0^\delta v\frac{\partial u}{\partial y}dy = 2\int_0^\delta u\frac{\partial u}{\partial x}dy = \int_0^\delta\frac{\partial u^2}{\partial x}dy = \frac{d}{dx}\int_0^\delta u^2 dy$$

The second term on right-hand side of Eq. 16.15 can be written as

$$\nu\int_0^\delta\frac{\partial^2 u}{\partial y^2}dy = \nu\left(\frac{\partial u}{\partial y}\right)\Bigg|_0^\delta = -\nu\frac{\partial u}{\partial y}\Bigg|_0 = -\frac{\tau_{w,x}}{\rho}$$

where $\tau_{w,x}$ is the shear stress at the wall. Thus, the momentum integral equation is written down as

$$\boxed{\frac{d}{dx}\int_0^\delta u^2 dy - g\beta\int_0^\delta\left(\frac{T}{T_\infty} - 1\right)dy = -\frac{\tau_{w,x}}{\rho}} \tag{16.16}$$

Integrate Eq. 16.8 with respect to y between $y = 0$ and $y = \delta$ to get

$$\int_0^\delta u\frac{\partial T}{\partial x}dy + \int_0^\delta v\frac{\partial T}{\partial y}dy = \alpha\int_0^\delta \frac{\partial^2 T}{\partial y^2}dy \tag{16.17}$$

Integrate by parts the second term on the left-hand side of Eq. 16.17 to get

$$\int_0^\delta v\frac{\partial T}{\partial y}dy = (vT)\Big|_0^\delta - \int_0^\delta T\frac{\partial v}{\partial y}dy = v_\delta T_\infty + \int_0^\delta T\frac{\partial u}{\partial x}dy$$

where v_δ is the normal velocity of the fluid at the edge of the boundary layer. We have made use of the continuity equation in the above. The two terms on the left-hand side of Eq. 16.17 may then be combined to get

$$\int_0^\delta u\frac{\partial T}{\partial x}dy + \int_0^\delta v\frac{\partial T}{\partial y}dy = \int_0^\delta u\frac{\partial T}{\partial x}dy + \int_0^\delta T\frac{\partial u}{\partial x}dy + v_\delta T_\infty = \frac{d}{dx}\int_0^\delta uT\,dy + v_\delta T_\infty$$

In order to get the normal velocity at the edge of the boundary layer, we integrate the continuity Eq. 16.6 to get

$$\int_0^\delta \frac{\partial u}{\partial x}dy + \int_0^\delta \frac{\partial v}{\partial y}dy = \int_0^\delta \frac{\partial u}{\partial x}dy - v_\delta = 0 \quad \text{or } v_\delta = \int_0^\delta \frac{\partial u}{\partial x}dy$$

Thus, we have

$$\int_0^\delta u\frac{\partial T}{\partial x}dy + \int_0^\delta v\frac{\partial T}{\partial y}dy = \int_0^\delta \frac{\partial u}{\partial x}dy - T_\infty\int_0^\delta \frac{\partial u}{\partial x}dy = \frac{d}{dx}\int_0^\delta u(T - T_\infty)dy$$

The right-hand side term in Eq. 16.17 may be written as

$$\alpha\int_0^\delta \frac{\partial^2 T}{\partial y^2} = \alpha\left(\frac{\partial T}{\partial y}\right)\Big|_0^\delta = \frac{q_{w,x}}{\rho_\infty C}$$

where $q_{w,x} = -k\left.\frac{\partial T}{\partial y}\right|_{y=0}$ is the wall heat flux. With all of these, the energy integral Eq. 16.17 becomes

$$\frac{d}{dx}\int_0^\delta u(T-T_\infty)dy = \frac{q_{w,x}}{\rho_\infty C} \quad (16.18)$$

Solution

In the integral equation approach, we assume that the velocity profile inside the boundary layer is as indicated in Fig. 16.4 and is a function of $\eta = \frac{y}{\delta}$, where δ, the boundary layer thickness is a function of x.

As shown in the figure, u vanishes at $x = 0$, as well as at $x = \delta$. Instead of using the original boundary layer equations, (partial differential Eqs. 16.6,16.7, and 16.8) we make use of the integral Eqs. 16.16 and 16.18 to obtain an approximate solution. Assume a velocity scale U (to be determined as part of the analysis) such that $\frac{u}{U}$ is a third degree polynomial in η. U is as yet an unknown function of x.

$$\frac{u}{U} = a + b\eta + c\eta^2 + d\eta^3$$

Since u vanishes at $y = 0$, we choose $a = 0$. Since u also vanishes at $y = \delta$ or $\eta = 1$, we have

$$b + c + d = 0$$

We shall use the smoothness condition $\frac{\partial u}{\partial y} = 0$ at $\eta = 1$ to get

$$b + 2c + 3d = 0$$

The above two conditions may be satisfied by $b = d$ and $c = -2d$. Constant d may be arbitrarily chosen as 1 since the scale factor U may be adjusted suitably. Thus, the velocity profile is taken as

$$\frac{u}{U} = \eta - 2\eta^2 + \eta^3 = \eta(1-\eta)^2 \quad (16.19)$$

The above profile has a zero slope at $\eta = \frac{1}{3}$. Thus, the velocity profile has the appearance of the profile indicated schematically in Fig. 16.4.

We assume that the non-dimensional temperature $\theta = \frac{T-T_\infty}{T_w-T_\infty}$ is a second degree polynomial in η satisfying the condition $T = T_w$ at $y = 0$, i.e., $\theta = 1$ at $\eta = 0$, $T = T_\infty$ at $y = \delta$, i.e., $\theta = 0$ at $\eta = 1$. In addition it satisfies the smoothness condition $\frac{d\theta}{d\eta} = 0$ at $\eta = 1$. The reader may verify that the required temperature profile is given by

$$\theta = (1-\eta)^2 \quad (16.20)$$

In the above, we have assumed that δ is the same for the momentum boundary layer, as well as the thermal boundary layer. This is a consequence of the fact that the

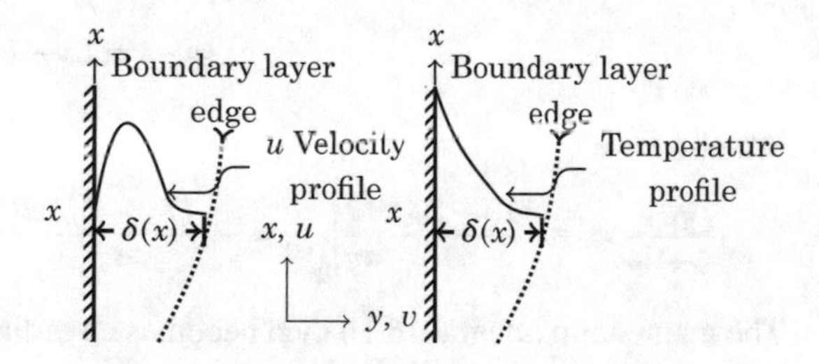

Fig. 16.4 Boundary layer velocity and temperature profiles

momentum and energy equations are coupled. Temperature appears in Eq. 16.12 and u and v in Eq. 16.8, and hence the two boundary layers are related, *irrespective* of the value of the Prandtl number. This is quite different from the forced convection case where momentum and thermal boundary layer thicknesses are distinct and depend on the Prandtl number.

Using the assumed velocity and temperature profile, we calculate the individual terms in the momentum and thermal integral equations.

$$\int_0^\delta u^2 dy = U^2\delta \int_0^1 \left(\frac{u}{U}\right)^2 d\eta = U^2\delta \int_0^1 [\eta(1-\eta)^2]^2 d\eta$$

$$= U^2\delta \int_0^1 [\eta^2(1-4\eta+6\eta^2-4\eta^3+\eta^4)]d\eta = U^2\delta \left[\frac{\eta^3}{3} - \frac{4\eta^4}{4} + \frac{6\eta^5}{5} - \frac{4\eta^6}{6} + \frac{\eta^7}{7}\right]\Big|_0^1$$

$$= U^2\delta\left[\frac{1}{3} - 1 + \frac{6}{5} - \frac{2}{3} + \frac{1}{7}\right] = \frac{U^2\delta}{105}$$

We also have

$$\int_0^\delta (T - T_\infty)dy = (T_w - T_\infty)\delta \int_0^1 \theta d\eta = (T_w - T_\infty)\delta \int_0^1 (1-\eta)^2 d\eta$$

$$= -(T_w - T_\infty)\delta \frac{(1-\eta)^3}{3}\Big|_0^1 = \frac{(T_w - T_\infty)\delta}{3}$$

The wall shear stress may be calculated using the velocity profile as

$$\frac{\tau_{w,x}}{\rho_\infty} = \nu \frac{\partial u}{\partial y}\Big|_{y=0} = \frac{\nu U}{\delta}\frac{d}{d\eta}\left(\frac{u}{U}\right)\Big|_{\eta=0} = \frac{\nu U}{\delta}[(1-\eta)^2 - 2\eta(1-\eta)]|_{\eta=0} = \frac{\nu U}{\delta}$$

Turning our attention to the terms occurring in the energy integral, we have the following:

$$\int_0^\delta u(T - T_\infty)dy = U\delta(T_w - T_\infty)\int_0^1 \frac{u}{U}\theta d\eta = U\delta(T_w - T_\infty)\int_0^1 \eta(1-\eta)^2(1-\eta)^2 d\eta$$

$$= U\delta(T_w - T_\infty)\int_0^1 \eta(1-\eta)^4 d\eta = U\delta(T_w - T_\infty)\left[-\eta \cdot \frac{(1-\eta)^5}{5}\Big|_0^1 + \int_0^1 \frac{(1-\eta)^5}{5} d\eta\right]$$

$$= -U\delta(T_w - T_\infty)\frac{(1-\eta)^6}{30}\bigg|_0^1 = \frac{U\delta(T_w - T_\infty)}{30}$$

We also have

$$\frac{q_{w,x}}{\rho_\infty C_p} = -\alpha\frac{(T_w - T_\infty)}{\delta}\frac{d\theta}{d\eta}\bigg|_{\eta=0} = -\alpha\frac{(T_w - T_\infty)}{\delta}\frac{d(1-\eta)^2}{d\eta}\bigg|_{\eta=0} = \frac{2\alpha(T_w - T_\infty)}{\delta}$$

The momentum integral 16.16 then becomes an ordinary differential equation given by

$$\boxed{\frac{d}{dx}\left(\frac{U^2\delta}{105}\right) - \frac{g\beta(T_w - T_\infty)}{3}\delta = -\frac{\nu U}{\delta}} \tag{16.21}$$

Also, the energy integral 16.18 becomes an ordinary differential equation given by

$$\boxed{\frac{d}{dx}\left(\frac{U\delta}{30}\right) = \frac{2\alpha}{\delta}} \tag{16.22}$$

The two coupled ordinary differential equations have to be solved to get the velocity scale U, as well as the boundary layer thickness δ. We expect these two to be of the form

$$\text{(a)} \quad U = Ax^m \quad \text{and} \quad \text{(b)} \quad \delta = Bx^n \tag{16.23}$$

where $A,\ B,\ n$, and m are to be so chosen that the integral equations are identically satisfied. These in Eq. 16.21 lead to

$$\frac{d}{dx}\left(\frac{A^2Bx^{2m+n}}{105}\right) - \frac{g\beta(T_w - T_\infty)}{3}Bx^n = -\frac{\nu A}{B}x^{m-n}$$

or

$$\frac{A^2B(2m+n)}{105}x^{2m+n-1} - \frac{g\beta(T_w - T_\infty)}{3}Bx^n = -\frac{\nu A}{B}x^{m-n} \tag{16.24}$$

Eq. 16.22 takes the form

$$\frac{d}{dx}\frac{ABx^{m+n}}{30} = \frac{2\alpha}{B}x^{-n} \text{ or } \left(\frac{AB(m+n)}{30}\right)x^{m+n-1} = \frac{2\alpha}{B}x^{-n} \tag{16.25}$$

We require that the x dependence on both sides of the above two equations match. Thus, from Eq. 16.24, we have

$$2m + n - 1 = n = m - n$$

From Eq. 16.25, we also should have

$$m + n - 1 = -n$$

We at once see that these conditions are mutually compatible, and hence we have

$$m + 2n = 1 \quad \text{and} \quad 2m = 1$$

The required exponents m and n are

$$m = \frac{1}{2} \quad \text{and} \quad n = \frac{1}{4} \tag{16.26}$$

With these we are ready to look at the coefficients A and B. From Eq. 16.25, we have, with the above values of m and n

$$\frac{AB}{30}\left(\frac{1}{2} + \frac{1}{4}\right) = \frac{AB}{40} = \frac{2\alpha}{B}$$

or solving for A in terms of B, $A = 80\frac{\alpha}{B^2}$. From Eq. 16.24, we have

$$\frac{A^2B(2 \times \frac{1}{2} + \frac{1}{4})}{105} - \frac{g\beta(T_w - T_\infty)}{3}B = -\frac{\nu A}{B}$$

or

$$\frac{A^2B}{84} - \frac{g\beta(T_w - T_\infty)}{3}B = -\frac{\nu A}{B}$$

This may be rewritten using A in terms of B obtained above to get

$$\frac{80^2\alpha^2}{84B^3} - \frac{g\beta(T_w - T_\infty)}{3}B = -\frac{80\nu\alpha}{B^3}$$

Noting that $Pr = \frac{\nu}{\alpha}$, we may write the above as

$$\frac{1}{B^4}\left(\frac{80^2}{84}\alpha^2 + 80\nu\alpha\right) = \frac{\alpha^2}{B^4}\left(\frac{80^2}{84} + 80Pr\right) = \frac{g\beta(T_w - T_\infty)}{3}$$

This may be solved for A to get

$$B = 3.93\left[\frac{Pr + 0.952}{g\beta(T_w - T_\infty)}\right]^{\frac{1}{4}}\sqrt{\alpha} \tag{16.27}$$

Coefficient A is then obtained as

$$A = 80\frac{\alpha}{B^2} = 5.17\left[\frac{g\beta(T_w - T_\infty)}{Pr + 0.952}\right]^{\frac{1}{2}} \tag{16.28}$$

We define a local Grashof number Gr_x with x as the characteristic length, i.e., replace L in Eq. 16.5 by x. Then we may write for δ as

$$\begin{aligned}\delta &= 3.93\left[\frac{Pr + 0.952}{g\beta(T_w - T_\infty)}\right]^{\frac{1}{4}}\sqrt{\alpha}x^{\frac{1}{4}} = 3.93\left[\frac{\alpha^2(Pr + 0.952)}{g\beta(T_w - T_\infty)}\right]^{\frac{1}{4}}\frac{x}{x^{\frac{3}{4}}}\\ &= 3.93\left[\frac{\nu^2(Pr + 0.952)}{g\beta Pr^2(T_w - T_\infty)}\right]^{\frac{1}{4}}\cdot\frac{x}{x^{\frac{3}{4}}} = 3.93\left[\frac{(Pr + 0.952)}{Pr^2}\right]^{\frac{1}{4}}\left[\frac{\nu^2}{g\beta(T_w - T_\infty)x^3}\right]^{\frac{1}{4}}\cdot x\\ &= 3.93\left[\frac{Pr + 0.952}{Pr^2}\right]^{\frac{1}{4}}\cdot\frac{x}{Gr_x^{\frac{1}{4}}}\end{aligned} \tag{16.29}$$

The velocity scale within the boundary layer is given by

$$U = Ax^m = 5.17\left[\frac{g\beta(T_w - T_\infty)}{Pr + 0.952}\right]^{\frac{1}{2}}\cdot x^{\frac{1}{2}} \tag{16.30}$$

The maximum velocity within the boundary layer occurs at $\eta = \frac{1}{3}$ as may easily be verified. Hence, the maximum velocity is given by

$$u_{max} = U\cdot\frac{1}{3}\left(1 - \frac{1}{3}\right)^2 = \frac{4}{27}U = 0.766\sqrt{\frac{g\beta(T_w - T_\infty)}{Pr + 0.952}x} \tag{16.31}$$

Wall Heat Flux and the Nusselt Number

The quantity of interest to us is the heat transfer from the plate. The local wall heat flux is given by

$$q_{w,x} = \frac{2k(T_w - T_\infty)}{\delta} = \frac{2k(T_w - T_\infty)}{3.93\left[\frac{Pr + 0.952}{Pr^2}\right]^{\frac{1}{4}}\cdot\frac{x}{Gr_x^{\frac{1}{4}}}}$$

This may be expressed in terms of local Nusselt number as

$$\boxed{Nu_x = \frac{q_w}{k(T_w - T_\infty)}\cdot x = 0.509\left[\frac{Pr^2}{Pr + 0.952}\right]^{\frac{1}{4}}Gr_x^{\frac{1}{4}}} \tag{16.32}$$

Sometimes another non-dimensional parameter, Rayleigh number Ra,[4] the product of Grashof and Prandtl numbers is used in the heat transfer literature. The above

[4]After John William Strutt, 3rd Baron Rayleigh 1842–1919, British Physicist.

equation may hence be written in the alternate form

$$\boxed{Nu_x = 0.509\left[\frac{Pr}{Pr + 0.952}\right]^{\frac{1}{4}} Ra_x^{\frac{1}{4}}} \tag{16.33}$$

Average Nusselt Number

In applications, we would like to define an average heat transfer coefficient based on height L of the plate as

$$\overline{Nu_L} = \frac{\overline{Q}_{wL}L}{k(T_w - T_\infty)}$$

where $\overline{Q}_{wL}$ is the total heat transfer from the plate. We may write $\overline{Q}_{wL}$ as

$$\overline{Q}_{wL} = \frac{1}{L}\int_0^L q_w dx == \frac{1}{L}\int_0^L \frac{2k(T_w - T_\infty)}{3.93\left[\frac{Pr + 0.952}{Pr^2}\right]^{\frac{1}{4}} \cdot \frac{x}{Gr_x^{\frac{1}{4}}}} dx$$

Hence, the average Nusselt number may be written as

$$\overline{Nu_L} = \frac{\overline{Q}_{wL}L}{k(T_w - T_\infty)} = \int_0^L \frac{2}{3.93\left[\frac{Pr + 0.952}{Pr^2}\right]^{\frac{1}{4}} \cdot \frac{x}{Gr_x^{\frac{1}{4}}}} dx = K\int_0^L x^{-\frac{1}{4}} dx$$

where K stands for all the other terms excepting that containing x. The indicated integration may be performed to get

$$\int_0^L x^{-\frac{1}{4}} dx = \left[\frac{x^{-\frac{1}{4}+1}}{-\frac{1}{4}+1}\right]\Bigg|_0^L = \frac{4}{3}L^{\frac{3}{4}}$$

When we introduce this back in the expression for the average Nusselt number, we get

$$\boxed{\overline{Nu_L} = \frac{4}{3}Nu_L = \frac{4}{3}\cdot 0.509\left[\frac{Pr}{Pr + 0.952}\right]^{\frac{1}{4}} Ra_L^{\frac{1}{4}} = 0.677\left[\frac{Pr}{Pr + 0.952}\right]^{\frac{1}{4}} Ra_L^{\frac{1}{4}}} \tag{16.34}$$

Example 16.3

Heating element of a water heater is idealized as a 0.1×0.1 m square vertical plate. The plate temperature is 27 °C while the water temperature is 15 °C. Determine the amount of heat transferred using the results of the integral method. Properties are to be taken at the mean of plate and ambient temperatures.

Solution:
The given data, in the notation used in the above section is written down first.

Heater plate temperature:	$T_w = 27\,°C$
Ambient water temperature:	$T_\infty = 15\,°C$
Plate height:	$L = 0.1$ m
Plate width:	$W = 0.1$ m

Mean temperature for calculating water propertis is

$$T_a = \frac{T_w + T_\infty}{2} = \frac{27 + 15}{2} = 21\,°C$$

Temperature difference for calculating the Grashof/Rayleigh number is

$$\Delta T = T_w - T_\infty = 27 - 15 = 12\,°C$$

Properties of water at 21 °C are taken from from table of properties

$$k_a = 0.597\,\text{W/m°C}, \;\; Pr_a = 7.0, \;\; \frac{g\beta_a}{\nu_a^2} = 2.035 \times 10^9 \text{ (SI units)}$$

The Rayleigh number based on plate height is calculated as

$$Ra_L = \frac{g\beta_a}{\nu_a^2} \times L^3 \times \Delta T \times Pr_a = 2.035 \times 10^9 \times 0.1^3 \times 12 \times 7 = 1.71 \times 10^8$$

Flow is laminar since the Rayleigh number is less than 10^9. Eq. 16.34 based on the integral solution gives

$$\overline{Nu_L} = 0.677 \left[\frac{7}{7 + 0.952}\right]^{\frac{1}{4}} (1.71 \times 10^8)^{\frac{1}{4}} = 75$$

The average heat transfer coefficient may then be calculated as

$$\overline{h}_L = \frac{\overline{Nu_L} k_a}{L} = \frac{75 \times 0.597}{0.1} = 447.8\,\text{W/m}^2\text{°C}$$

Assuming that heat transfer takes place from both sides of the plate, the total heat transfer is given by

$$Q_L = 2LW\overline{h}_L \Delta T = 2 \times 0.1 \times 0.1 \times 447.8 \times 12 = 107.5\,\text{W}$$

16.2.2 Exact Solution of Ostrach

We have seen above how an approximate solution is obtained by the integral method. The fact that the boundary layer velocity and temperature profiles could be chosen as given polynomial functions of η indicates that the solution should be possible by the method of similarity. The reader may compare the situation that prevailed in the case of transient conduction in one dimension, as well as the velocity and temperature boundary layers in laminar forced convection. Since the method of similarity, is by now, familiar to the reader, we give a short summary of the solution, leaving out the details to the original research report of Ostrach cited earlier.

Taking a hint from the integral solution, it is clear that the similarity variable for this problem should go as $\frac{y}{x^{\frac{1}{4}}}$. In fact, Ostrach has shown that the appropriate similarity variable is given by

$$\eta = \frac{y}{x}\frac{Gr_x^{\frac{1}{4}}}{4} \tag{16.35}$$

Correspondingly the velocity scales with $\sqrt{x}$ (see Eq. 16.30). Ostrach found the velocity scale to be

$$U = \frac{2\nu\sqrt{\alpha}}{x} Gr_x^{\frac{1}{2}} \tag{16.36}$$

Ostrach introduced a functions $F(\eta)$ such that the velocity is given by

$$u = \frac{2\nu\sqrt{\alpha}}{x} Gr_x^{\frac{1}{2}} \frac{dF}{d\eta} \tag{16.37}$$

The non-dimensional temperature was defined as usual as

$$\theta(\eta) = \frac{T - T_\infty}{T_w - T_\infty} \tag{16.38}$$

The similarity analysis then reduces the problem to the solution of the following two non-linear coupled ordinary differential equations.

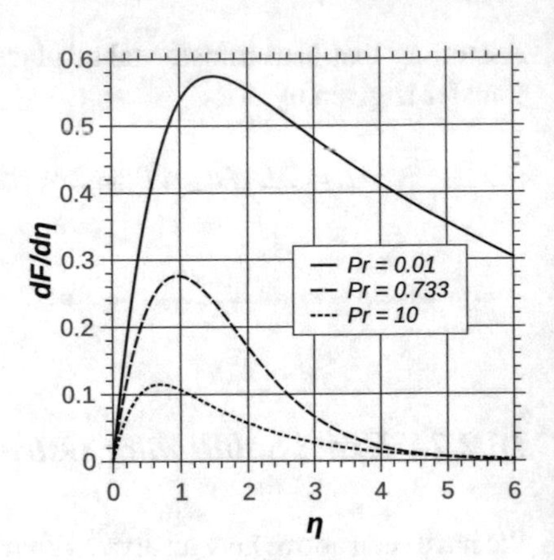

Fig. 16.5 Boundary layer velocity profiles: Ostrach solution

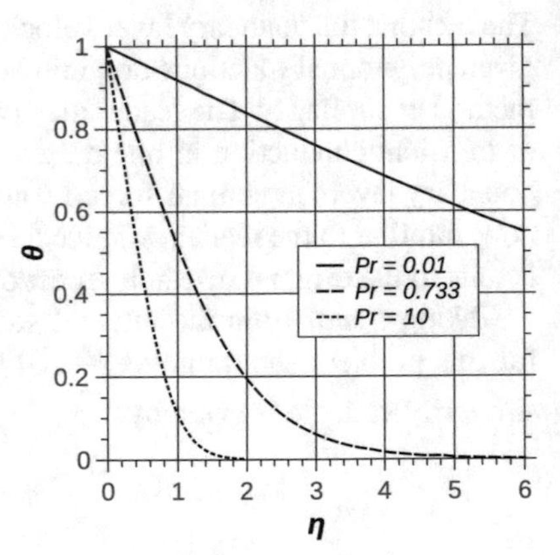

Fig. 16.6 Boundary layer temperature profiles: Ostrach solution

$$\boxed{\frac{d^3F}{d\eta^3} + 3F\frac{d^2F}{d\eta^2} - 2\left(\frac{dF}{d\eta}\right)^2 + \theta = 0} \tag{16.39}$$

$$\boxed{\frac{d^2\theta}{d\eta^2} + 3PrF\frac{d\theta}{d\eta} = 0} \tag{16.40}$$

The boundary conditions are specified as below

$$\text{(a)} \quad \eta = 0,\ F = \frac{dF}{d\eta} = 0,\ \text{and} \theta = 1, \quad \text{(b)} \quad \eta \to \infty,\ \frac{dF}{d\eta} \to 0 \quad \text{and} \quad \theta \to 1 \tag{16.41}$$

Table 16.1 Table of $C(Pr)$, Exact, Curve fit and Approximate

Pr	C	C	C
	Exact*	Curve fit	Approximate
0.01	0.2421	0.2393	0.2156
0.72	0.5165	0.5137	0.5476
0.733	0.5176	0.5147	0.5490
1	0.5347	0.5318	0.5720
2	0.5680	0.5654	0.6137
10	0.6200	0.6186	0.6616
100	0.6532	0.6523	0.6754
1000	0.6649	0.6635	0.6768

*From S. Ostrach, NACA 1111, 1953

These equations may be solved numerically by the use of the Runge–Kutta method. Since the two equations are coupled, the numerical scheme, in essence, is equivalent to solving *five* first order equations. The natural convection boundary velocity profile is shown in Fig. 16.5 and temperature profiles in Fig. 16.6, for three representative Pr values. The case with $Pr = 0.1$ applies to the case of a liquid metal. The case with $Pr = 0.733$ is applicable to gases like air. The case $Pr = 10$ is applicable to a liquid such as water. We observe that the thickness of both velocity and temperature boundary layers increase with decrease in Pr. This is quite unlike the forced convection case, where Pr does not influence the velocity profile. Average Nusselt number over length L of the plate is obtained from Ostrach solution as

$$\overline{Nu_L} = C(Pr)Ra_L^{\frac{1}{4}} \tag{16.42}$$

The coefficient $C(Pr)$ depends on the Prandtl number. Recall that the integral method also yields an equation which resembles the above, with the coefficient explicitly appearing as a closed form function of the Prandtl number (Eq. 16.34). In Table 16.1 on page 785, we compare the values of coefficient $C(Pr)$ for a range of Prandtl numbers obtained by the exact solution of Ostrach, obtained by a curve fit to the exact values and obtained by the approximate integral solution. The curve fit follows the relation

$$C(Pr) = \left[\frac{0.4Pr}{1 + 2\sqrt{Pr} + 2Pr}\right]^{\frac{1}{4}} \tag{16.43}$$

Even though the integral approximate method follows the trend of the exact solution, the largest error with respect to the exact is as much as 11% at $Pr = 0.01$.

Example 16.4

A flat surface of height 0.2 m at a uniform temperature of 65 °C is placed in still air at 27 °C. Calculate the maximum velocity within the natural convection boundary layer based on the Ostrach solution. Compare this with that estimated using the integral solution.

Solution:
Maximum velocity based on Ostrach solution:
The Ostrach solution labeled $Pr = 0.733$ is used for the purpose of evaluating the desired maximum velocity at $L = 0.2m$.

Step 1 The air properties are evaluated at the ambient temperature of $T_\infty = 27\,°\text{C} = 300\,\text{K}$.

$$\beta = \frac{1}{T_\infty} = \frac{1}{300} = 0.00333\,\text{K}^{-1}, \quad \nu = 15.89 \times 10^{-6}\,\text{m}^2/\text{s}$$

Step 2 The wall temperature is $T_w = 65\,°\text{C}$ and hence the temperature difference is

$$\Delta T = T_w - T_\infty = 65 - 27 = 38\,°\text{C}$$

The maximum velocity in the boundary layer will occur as the flow leaves at the top of the plate. Hence, we calculate the Grashof number based on plate length as

$$Gr_L = \frac{g\beta \Delta T L^3}{\nu^2} = \frac{9.81 \times 0.000333 \times 38 \times 0.2^3}{(15.89 \times 10^{-6})^2} = 3.937 \times 10^7$$

Step 3 We make use of Fig. 16.5 to evaluate the maximum velocity. We note that for $Pr = 0.733$ (typical value for air), the maximum velocity occurs at $\eta = 0.95$. The maximum value itself is read off the graph as

$$\frac{u_{max,L}}{\frac{2\nu}{L}\sqrt{Gr_L}} = 0.28$$

$$\text{or } u_{max,L} = 0.28 \times \frac{2 \times 15.89 \times 10^{-6}}{0.2}\sqrt{3.937 \times 10^7} = 0.279\,\text{m/s}$$

Maximum velocity based on integral solution:

Step 4 The cubic velocity profile is given by Eq. 16.19 as $\frac{u}{U} = \eta(1-\eta)^2$, where $\eta = \frac{y}{\delta}$. The maximum velocity occurs when

$$\frac{d}{d\eta}\left(\frac{u}{U}\right) = (1-\eta)^2 - 2\eta(1-\eta) = 0$$

This happens when $\eta = \frac{1}{3}$. Thus the maximum velocity within the boundary layer is

$$u_{max,L} = U_L \frac{1}{3}\left(1 - \frac{1}{3}\right)^2 = \frac{4}{27} U_L$$

where U_L is the velocity scale evaluated at $x = L$.

Step 5 Using Eq. 16.30, we have

$$U_L = 5.17 \left[\frac{9.81 \times 0.000333 \times 38}{0.733 + 0.952}\right]^{\frac{1}{2}} \times 0.2^{\frac{1}{2}} = 1.986 \text{ m/s}$$

Step 6 Thus, the maximum velocity according to the integral solution is

$$u_{max,L} = \frac{4}{27} \times 1.986 = 0.294 \text{ m/s}$$

The exact and the integral solution estimates are remarkably close to each other.

16.2.3 Comparison with Experimental Results

Local heat transfer coefficient, and hence the local Nusselt number in Natural convection from a heated isothermal vertical plate (in this case an aluminum plate losing heat to ambient air) may be measured easily by optical methods. A differential interferometer responds to the refractive index gradient set up by the density variations within the natural convection flow field. Typically, an interferogram appears as the fringe pattern shown in the left half of Fig. 16.7a.[5] The "bending" of the fringes near the plate is indicative of the temperature gradients present there. In the infinite fringe spacing arrangement of the differential interferometer, we get a contrast pattern that shows constant temperature gradient lines. This is shown in the right half of Fig. 16.7a. Clearly, one sees the developing boundary layer along the height of the plate.

The interferogram has been analyzed quantitatively to obtain the local Nusselt number variation along the plate height. The experimental data is compared with the distributions given by both the exact results due to Ostrach and that given by the integral method. All three are in excellent agreement with each other as shown by Fig. 16.7b.

[5]The interferogram was recorded using a digital camera by S. Prasanna in our laboratory.

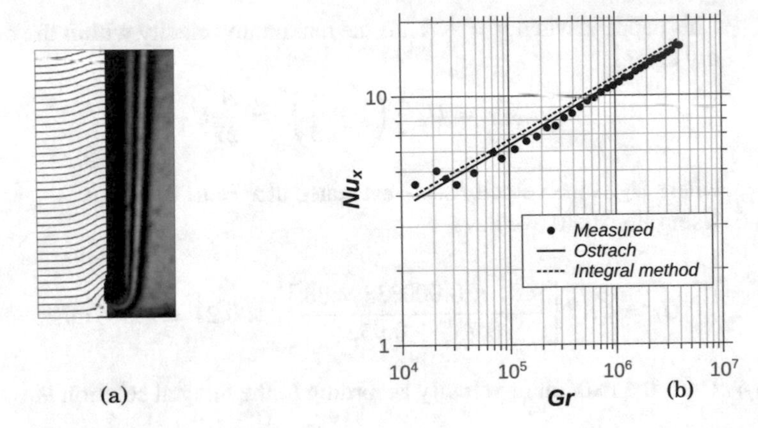

Fig. 16.7 **a** Interferograms with natural convection from a heated isothermal vertical plate—left half fringe pattern, right half—contrast pattern; **b** Local Nusselt number variation along a heated isothermal vertical plate

16.3 Turbulent Natural Convection from a Vertical Isothermal Plate

As mentioned earlier, the natural convection flow becomes turbulent when the Rayleigh number is greater than 10^9. From an engineering point of view, experimental data presented in the form of correlations, provide all that is needed by the designer. However, from a point of view of understanding the physical aspects, numerical solution of governing equations incorporating suitable turbulence models may be necessary. Sometimes approximate methods such as the integral method may also be useful. Even though the integral equations derived in the case of laminar natural convection and that for turbulent natural convection are the *same*, shear stress and heat transfer at the wall require attention in the case of turbulent natural convection. These may not be obtained by the usual process of taking derivatives of velocity and temperature as in the case of laminar flow.

16.3.1 Approximate Integral Analysis

Analysis of turbulent free convection boundary layer over a flat plate has been presented by Eckert and Jackson[6] based on the integral approach. We give a brief description of the arguments used by them to analyze the problem. We have seen that the main difference between laminar and turbulent flow is the variation of velocity and temperature in the near wall region. While in laminar flow, the variation is over the entire laminar boundary layer thickness, in the case of turbulent flow it shows

[6]E. R. G. Eckert and T. W. Jackson, NACA Report 1015, 1951.

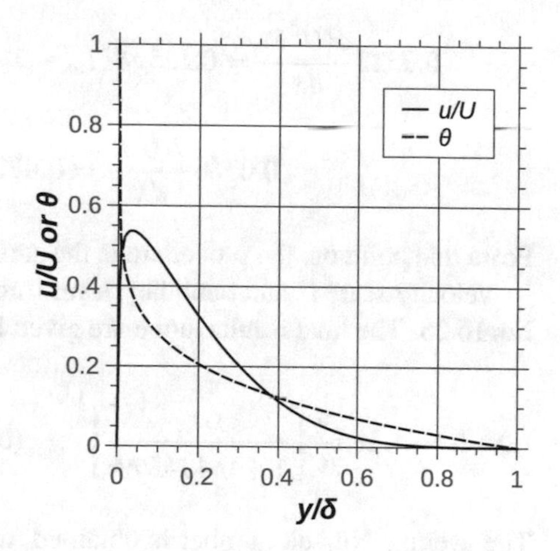

Fig. 16.8 Velocity and temperature profiles in the natural convection boundary layer

regions within the turbulent boundary layer where the variations are non-uniform. It was argued that right next to the wall, the behavior of the velocity and temperatures must be steep, showing the characteristic $\frac{1}{7}$thpower law. The velocity and temperature profiles are hence taken as

$$\text{(a)} \quad \underbrace{\frac{u}{U} = \eta^{\frac{1}{7}}(1-\eta)^4}_{\text{Velocity}}; \quad \text{(b)} \quad \underbrace{\theta = 1 - \eta^{\frac{1}{7}}}_{\text{Temperature}} \tag{16.44}$$

where η is the ratio of distance normal to the plate to the turbulent natural convection boundary layer thickness. The velocity and temperature profiles given above are shown plotted in Fig. 16.8. Both the velocity and temperature profiles indeed show a very steep near wall behavior followed by a gentler variation outside. Beyond the boundary layer thickness, both functions vanish. These profiles match with experimentally measured profiles. The shear stress at the wall is based on the relation suggested by von Karman.[7]

$$\tau_{w,x} = 0.0225\rho_\infty U^2 \left(\frac{\nu}{U\delta}\right)^{\frac{1}{4}} \tag{16.45}$$

Using arguments based on the modified Reynolds analogy, the wall heat flux is taken as

$$q_{w,x} = 0.0225 g \rho_\infty C_p (T_w - T_\infty) \left(\frac{\nu}{U\delta}\right)^{\frac{1}{4}} Pr^{-\frac{2}{3}} \tag{16.46}$$

The various integrals required in the integral Eqs. 16.16 and 16.18 are obtained by the use of the assumed velocity and temperature profiles, to finally get the following two coupled ordinary differential equations.

[7] von Karman, NACA TM 1092, 1946.

$$0.0523\frac{d(U^2\delta)}{dx} = 0.125g\beta(T_w - T_\infty)\delta - 0.0225U^2\left(\frac{\nu}{U\delta}\right)^{\frac{1}{4}} \quad (16.47)$$

$$0.0366\frac{d(U\delta)}{dx} = 0.0225U\left(\frac{\nu}{U\delta}\right)^{\frac{1}{4}} Pr^{-\frac{2}{3}} \quad (16.48)$$

From this point on, the procedure is the same as that pursued in the laminar case.

Velocity scale U and boundary layer thickness δ are sought in the form given by Eq. 16.23. The final results alone are given here

$$\text{(a)}\quad U = 1.185\frac{\nu}{x}\left[\frac{Gr_x}{1+0.494Pr^{\frac{2}{3}}}\right]^{\frac{1}{2}}, \quad \text{(b)}\quad \delta = 0.565\frac{x}{Pr^{\frac{8}{15}}}\left[\frac{1+0.494Pr^{\frac{2}{3}}}{Gr_x}\right]^{\frac{1}{10}} \quad (16.49)$$

The average Nusselt number is obtained, using a procedure similar to that used in the laminar case, as

$$\boxed{\overline{Nu_L} = 0.0246\left[\frac{Gr_L}{1+0.494Pr^{\frac{2}{3}}}\right]^{\frac{2}{5}} Pr^{\frac{7}{15}}} \quad (16.50)$$

Eckert and Jackson have compared the above with available experimental results which indicate a close agreement between them.

16.3.2 Useful Nusselt Number Correlations

For engineering calculations, it is convenient to use correlations for the average Nusselt number based on the height of the plate. In the case of laminar flow, the Ostrach solution indeed provides an appropriate relation for determining the average Nusselt number. The $\frac{1}{4}$thpower dependence of average Nusselt number on the Rayleigh number is indicated by that solution. McAdams[8] has analyzed data for free convection from isothermal vertical plates over a wide range of Rayleigh numbers from $1 - 10^{11}$. The data is based on experiments conducted by various researchers over a period of several decades. For the present purpose, we shall present a useful correlation covering a range that is encountered often. The laminar region extends up to a Rayleigh number of 3.5×10^7. The available data is correlated as

$$\boxed{\overline{Nu_L} = 0.55Ra_L^{\frac{1}{4}}} \quad (16.51)$$

[8] W. H. McAdams, *Heat Transmission*, 3rd Edition, McGraw Hill, NY 1954.

valid in the range $10^4 < Ra_L < 3.5 \times 10^7$. For $Ra_L > 3.5 \times 10^7$, the free convective boundary layer becomes turbulent. The available data is correlated as

$$\boxed{\overline{Nu_L} = 0.13 Ra_L^{\frac{1}{3}}} \tag{16.52}$$

valid in the range $3.5 \times 10^7 < Ra_L < 10^{12}$. More recently, Churchill and Chu[9] have proposed two correlations, the first one specific to laminar natural convection, while the second is valid for both laminar and turbulent cases. The first correlation valid for laminar case ($Ra_L < 10^9$) is

$$\boxed{\overline{Nu_L} = 0.68 + \frac{0.670 Ra_L^{\frac{1}{4}}}{\left[1 + \left(\frac{0.492}{Pr}\right)^{\frac{9}{16}}\right]^{\frac{4}{9}}}} \tag{16.53}$$

The second correlation is valid for $Ra_L < 10^{12}$ and is given by

$$\boxed{\overline{Nu_L} = \left\{0.825 + \frac{0.387 Ra_L^{\frac{1}{6}}}{\left[1 + \left(\frac{0.492}{Pr}\right)^{\frac{9}{16}}\right]^{\frac{8}{27}}}\right\}^2} \tag{16.54}$$

Example 16.5

A metal wall 3 mm thick made of steel with a thermal conductivity of 45 W/m°C is covered with a 10 cm layer of an insulating material of thermal conductivity 0.1 W/m°C. The temperature of the inner surface of steel wall is measured to be 60 °C. The ambient is atmospheric air at 27 °C. The wall is 0.4 m wide and 1.5 m tall. Determine the heat transfer across the composite wall. Refer to Fig. 16.9 for pictorial representation of the problem.

[9] S. W. Churchill and H. H. S. Chu, Int. J Heat Mass Transfer Vol. 18, pp. 13231329, 1975.

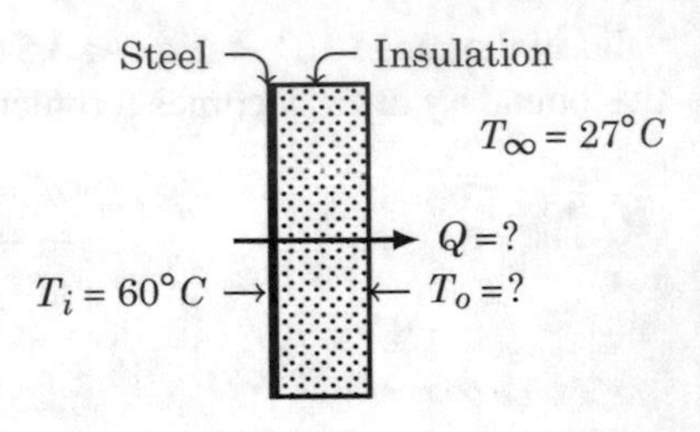

Fig. 16.9 Heat transfer through a composite wall: Example 16.5

Solution:

Figure 16.9 shows that there are two things that are not known viz. the temperature T_0 of the outer surface of the insulation and the heat transfer Q through the composite wall. An *iterative* solution is required. We expect a small temperature difference $T_0 - T_\infty$ since the insulating layer is fairly thick and it has a very low thermal conductivity. Hence, the air properties are taken at the ambient temperature of $T_\infty = 27\,°\text{C}$.

Step 1 Air properties at 27 °C are

$$\nu = 15.89 \times 10^{-6}\,\text{m}^2/\text{s}, \; k = 0.0263;\ \text{W/m°C}, \; Pr = 0.707$$

Other data specified in the problem are listed below.

Temperature of inner surface of steel:	$T_i = 60\,°\text{C}$
Thermal conductivity of steel:	$k_s = 45\,\text{W/m°C}$
Thickness of steel:	$t_s = 3\,\text{mm} = 0.003\,\text{m}$
Thermal conductivity of insulation:	$k_i = 0.1\,\text{W/m°C}$
Thickness of insulation:	$t_i = 10\,\text{cm} = 0.1\,\text{m}$

Step 2 The length scale for free convection from the outer surface of the insulation layer is the same as the height of the wall, i.e., $H = 1.5\,\text{m}$. The width of the wall is $W = 0.4\,\text{m}$, and hence the heat transfer area is $A = HW = 1.5 \times 0.4 = 0.6\,\text{m}^2$.
The isobaric expansion coefficient is

$$\beta = \frac{1}{T_\infty} = \frac{1}{273 + 27} = 0.0033\,\text{K}^{-1}$$

Temperature T_0 is unknown. We start by assuming a suitable value for this and iterate to home in on the correct value.

Step 3 Let us start with $T_0 = T_\infty + 10 = 27 + 10 = 37\,°\text{C}$. The temperature on the absolute scale is $T_0 = 273 + 37 = 310\,\text{K}$. The Rayleigh number is calculated as

$$Ra_H = \frac{g\beta(T_0 - T_\infty)H^3}{\nu^2} \times Pr$$
$$= \frac{9.81 \times 0.00033(310 - 300) \times 1.5^3}{(15.89 \times 10^{-6})^2} \times 0.707 = 3.09 \times 10^9$$

The regime is hence turbulent. According to McAdams recommended formula 16.52

$$\overline{Nu_H} = 0.13(3.091 \times 10^9)^{\frac{1}{3}} = 189.3$$

The average heat transfer coefficient over the height of the plate is then obtained as

$$\bar{h} = \frac{\overline{Nu_H}k}{H} = \frac{189.3 \times 0.0263}{1.5} = 3.32\ \mathrm{W/m^2\,^\circ C}$$

Step 4 Convective heat transfer from the outer surface of the insulation is calculated as

$$Q_c = \bar{h}A(T_0 - T_\infty) = 3.32 \times 0.6 \times 10 = 19.92\ \mathrm{W}$$

Conductive heat transfer through the composite wall is

$$Q_k = A\frac{T_i - T_0}{\frac{t_s}{k_s} + \frac{t_i}{k_i}}$$
$$= 0.6\frac{60 - 37}{\frac{0.003}{45} + \frac{0.1}{0.1}} = 13.79\ \mathrm{W}$$

Since $Q_c > Q_k$, it is clear that T_0 should be smaller than the value assumed above. We may reduce it to a lower value and repeat the calculation.

Step 5 Instead we have systematically varied T_0 between 30 and 37 °C and plotted the conductive heat transfer across the composite wall against the convective heat transfer from the exposed surface of the insulation (Fig. 16.10). It is seen that the two curves cross when $T_0 = 35.1$ °C and $Q_c = Q_k = 15$ W. These are the desired answers.

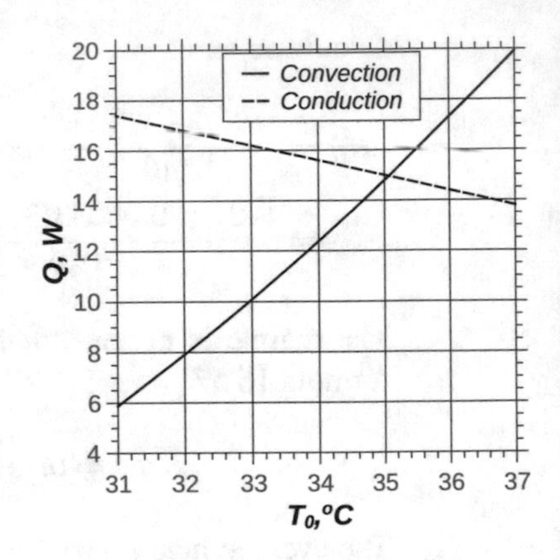

Fig. 16.10 Variation of conductive and convective heat transfers with T_0 in Example 16.5

16.4 Natural Convection from Other Geometries

In this section, we consider other commonly encountered geometric configurations. Most of the information available are based on experimental investigations carried out by researchers over the past 100 years or so. Analytical or numerical investigations have been made in recent times with the advent of the digital computer. These have led to a better understanding of the natural convection phenomena. However, useful correlations based on experiments are still the best bet for analysis.

16.4.1 Correlation for Horizontal Plates

Plates of interest in applications may have different shapes. The usual one is a rectangle. Sometimes we may be interested in a circular plate. If the plate is hot and faces upwards the layer above the plate is unstable and free convective motion is set up. For very large plates, end effects are unimportant and a cellular flow pattern is set up while for small plates, the flow is reminiscent of a chimney type flow. The cold fluid moves in from the sides to replace the hot fluid moving up. In the case of a circular plate the chimney flow is axi-symmetric while it is three-dimensional in the case of a rectangular plate. Similar situation prevails in the case of a cold plate facing downward. The cold fluid adjacent to the plate moves downwards and is replaced by the hot fluid moving in from the sides.

Fig. 16.11 shows schematically some of the situations. The Rayleigh number and the Nusselt number are based on a characteristic length L_{ch}, defined specifically for each configuration.

For a hot plate facing down or for a cold plate facing up, McAdams recommends the following correlations:

$$\overline{Nu}_{L_{ch}} = 0.27 Ra_{L_{ch}}^{\frac{1}{4}} \tag{16.55}$$

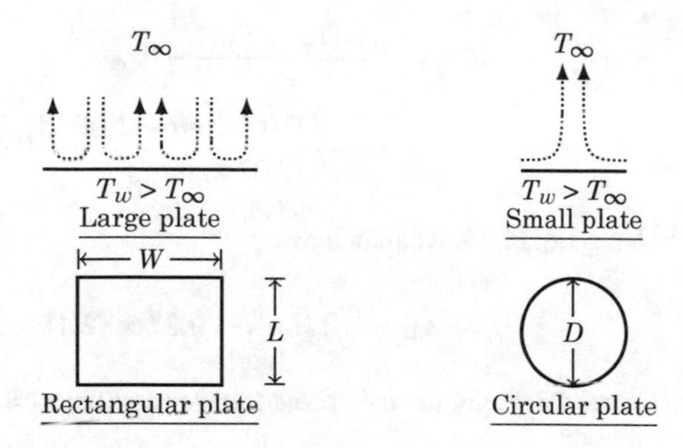

Fig. 16.11 Schematic of natural convection from horizontal plates

valid in the range $3 \times 10^5 < Ra_{L_{ch}} < 10^{10}$. In the above, $L_{ch} = \frac{W+L}{2}$ for a rectangle and $L_{ch} = D$ for a circle. However Al-Arabi and El-Riedy[10] recommend the following correlations:

$$\overline{Nu}_{L_{ch}} = 0.7 Ra_{L_{ch}}^{\frac{1}{4}} \tag{16.56}$$

in the range $2 \times 10^5 < Ra_{L_{ch}} < 4 \times 10^7$ and

$$\overline{Nu}_{L_{ch}} = 0.155 Ra_{L_{ch}}^{\frac{1}{3}} \tag{16.57}$$

when $Ra_{L_{ch}} > 4 \times 10^7$. These authors define the smaller of L or W as the characteristic length for a rectangle. However D remains the characteristic length for a circular plate.

Example 16.6

Consider a horizontal circular plate of diameter $D = 0.2$ m losing heat by natural convection to surrounding ambient air at $T_\infty = 20\,°C$. The plate is maintained at a temperature of $T_w = 60\,°C$. Compare the heat loss from this plate with a square plate of the same area, all other data remaining the same.

Solution: Case (a): Circular plate

We use the thermo-physical properties of air at the mean temperature of $T_m = \frac{T_w + T_\infty}{2} = \frac{60+20}{2} = 40\,°C$.

Kinematic viscosity:	$\nu_m = 17.07 \times 10^{-6}\ m^2/s$
Thermal conductivity:	$k_m = 0.0274\ W/m°C$
Prandtl number:	$Pr_m = 0.699$
Isobaric expansion coefficient:	$\beta_m = \frac{1}{T_m} = \frac{1}{273+40} = 0.0032\ K^{-1}$

The Rayleigh number may now be calculated with $L_{ch} = D = 0.2$ m.

[10]M. AL-Arabi and El-Riedy, Int. J Heat and Mass Transfer, Vol. 19, pp. 1399–1404, 1976.

$$Ra_D = \frac{g\beta_m(T_w - T_\infty)D^3}{\nu_m^2}Pr_m$$
$$= \frac{9.81 \times 0.0032(60 - 20)0.2^3}{(17.07 \times 10^{-6})^2}0.699 = 2.41 \times 10^7$$

Using Eq. 16.55, we than have

$$\overline{Nu}_D = 0.27Ra_D^{\frac{1}{4}} = 0.27 \times (2.41 \times 10^7)^{\frac{1}{4}} = 18.92$$

Correspondingly the mean heat transfer coefficient is

$$\bar{h} = \frac{\overline{Nu}_D k_m}{D} = \frac{18.92 \times 0.0274}{0.2} = 2.59\,\text{W/m}^2\,^\circ\text{C}$$

Heat loss from the plate to the ambient air is

$$Q = \bar{h}\frac{\pi D^2}{4}(T_w - T_\infty) = 2.59 \times \frac{\pi \times 0.2^2}{4}(60 - 20) = 3.26\,\text{W}$$

Case (b): Square plate of equal area

The only change in this case is that the characteristic dimension changes from D to $\frac{W+L}{2} = a$ where a is the side of the square plate. Since the square plate has the same area as that of the circular plate, we have

$$a = \sqrt{\frac{\pi D^2}{4}} = \sqrt{\frac{\pi \times 0.2^2}{4}} = 0.177\,\text{m}$$

The Rayleigh number in this case is

$$Ra_a = \frac{g\beta_m(T_w - T_\infty)a^3}{\nu^2}Pr_m$$
$$= \frac{9.81 \times 0.0032(60 - 20)0.177^3}{(17.07 \times 10^{-6})^2}0.699 = 1.67 \times 10^7$$

Using Eq. 16.55, we than have

$$\overline{Nu}_a = 0.27Ra_a^{\frac{1}{4}} = 0.27 \times (1.67 \times 10^7)^{\frac{1}{4}} = 17.26$$

Correspondingly the mean heat transfer coefficient is

$$\bar{h} = \frac{\overline{Nu}_a k_m}{a} = \frac{17.26 \times 0.0274}{0.177} = 2.672\,\text{W/m}^2\,^\circ\text{C}$$

Heat loss from the plate to the ambient air is

$$Q = \overline{h}a^2(T_w - T_\infty) = 2.672 \times 0.177^2(60 - 20) = 3.35\,\text{W}$$

The square plate loses marginally more heat than the circular plate.

16.4.2 Correlation for Vertical Cylinders

If the Grashof number is large enough such that the boundary layer assumptions are valid, the boundary layer thickness is much smaller than the diameter of the cylinder. Curvature of the parent surface then does not affect heat transfer and the correlations given earlier for a vertical plate are valid for vertical cylinders also. Eqs. 16.51 and 16.52 hold in the case of vertical cylinders as long as

$$\frac{D}{L} \geq \frac{35}{Gr_L^{\frac{1}{4}}} \tag{16.58}$$

where D is the cylinder diameter and L is the cylinder height.

16.4.3 Correlation for Horizontal Cylinders

Free convection from a heated horizontal cylinder is important in many applications like cartridge heaters used for heating water. The isotherm pattern and the velocity profile are as indicated schematically in Fig. 16.12. Ambient fluid moves around the cylinder, gets heated as it moves, with the boundary layer thickening with θ, the angle measured as indicated in the figure. Near the top of the cylinder, there is hardly any heat transfer from the cylinder to the fluid that goes up in the form of a hot plume.

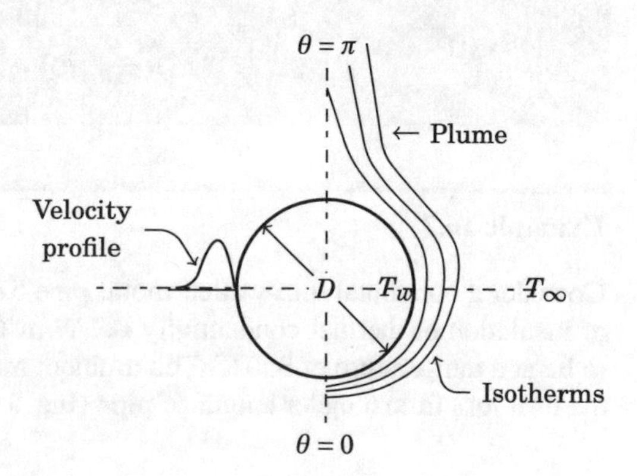

Fig. 16.12 Natural convection around a heated vertical cylinder

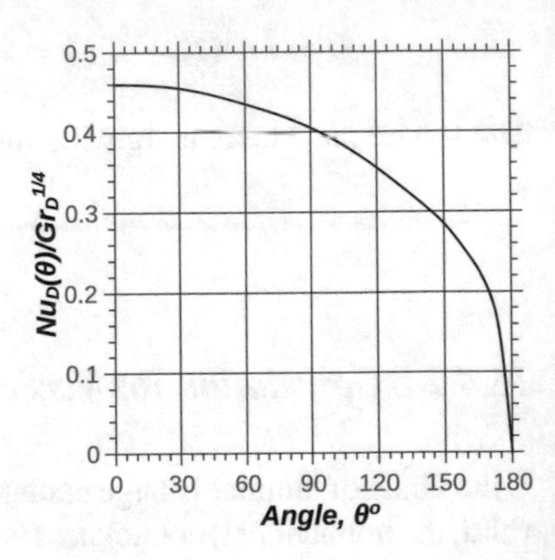

Fig. 16.13 Local Nusselt number for natural convection from an isothermal horizontal cylinder in air

We base the Nusselt number on the cylinder diameter D as the characteristic length. Nusselt number depends on angle θ, being largest at the bottom of the cylinder corresponding to $\theta = 0$ and decreasing monotonically with an increase in θ. The typical variation of local Nusselt number $Nu_D(\theta)$ is as indicated in Fig. 16.13 for laminar natural convection in air. The Rayleigh number can range from a very low value of order 1 and a high value of 10^{12} depending on the application. The low range is encountered in heated wires and the high range in practical applications like cartridge heaters used in industrial heat transfer equipment, domestic hot water system, and so on. McAdams has given a summary of available data and has given recommended formulae to be used in calculations of the average Nusselt number based on the cylinder diameter as the characteristic dimension. The appropriate formulae for laminar and turbulent ranges are given below.

Laminar range

$$\overline{Nu}_D = 0.53 Ra_D^{\frac{1}{4}},\ 10^4 < Ra_D < 10^9 \tag{16.59}$$

Turbulent range

$$\overline{Nu}_D = 0.13 Ra_D^{\frac{1}{3}},\ 10^9 < Ra_D < 10^{12} \tag{16.60}$$

Example 16.7

Consider a horizontal thin-walled metal pipe 5 cm OD surrounded by a 5 cm layer of insulation of thermal conductivity 0.1 W/m°C. The pipe outer surface is known to be at a temperature of 350 K. The ambient medium is still air at 300 K. Estimate the heat loss from a meter length of pipe (Fig. 16.14).

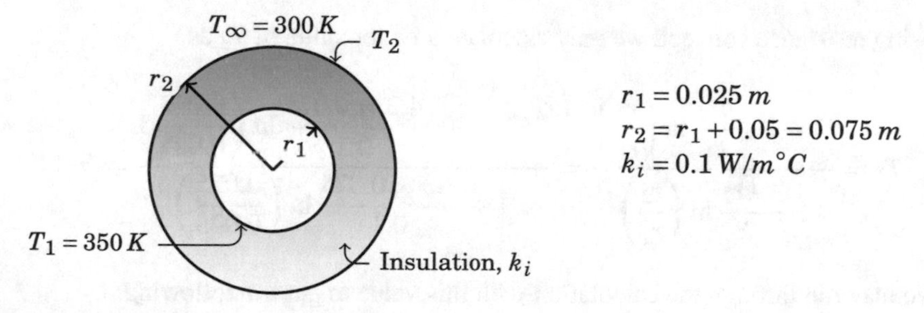

Fig. 16.14 Nomenclature used in Example 16.7

Solution:

As in Example 16.5, both the outer surface temperature of the insulation and the heat transferred across the insulation layer are unknown. An iterative method is required to solve the problem. To start the iteration, we assume a value of $T_2 = 315$ K. We use the correlation recommended by McAdams for obtaining the convective heat transfer coefficient.

Most of the data is shown in the figure itself. Air properties are taken at the ambient temperature of 300 K.

Kinematic viscosity:	$\nu = 15.89 \times 10^{-6}$ m²/s
Thermal conductivity:	$k = 0.0263$ W/m°C
Prandtl number:	$Pr = 0.7$
Isobaric expansion coefficient:	$\beta = \frac{1}{300} = 0.0033\,\text{K}^{-1}$

Both the Rayleigh number and the Nusselt number are based on $D_2 = 2r_2 = 2 \times 0.075 = 0.15$ m. The Rayleigh number is then calculated as

$$Ra_{D_2} = \frac{9.81 \times 0.0033(315 - 300)0.15^3}{(15.89 \times 10^{-6})^2} 0.7 = 4.59 \times 10^6$$

Flow is laminar and we use Eq. 16.59 to get the average Nusselt number as

$$\overline{Nu}_{D_2} = 0.53 \times (4.59 \times 10^6)^{\frac{1}{4}} = 24.53$$

The convective heat transfer coefficient is then given by

$$\overline{h} = \frac{\overline{Nu}_{D_2} k}{D_2} = \frac{24.53 \times 0.0263}{0.15} = 4.3\,\text{W/m}^2\text{°C}$$

Using resistance concept, we may calculate a better value of T_2 as

$$T_2 = \frac{T_1 + T_\infty \frac{\bar{h} r_2}{k_i} \ln\left(\frac{r_2}{r_1}\right)}{1 + \frac{\bar{h} r_2}{k_i} \ln\left(\frac{r_2}{r_1}\right)} = \frac{350 + 300 \frac{4.3 \times 0.075}{0.1} \ln\left(\frac{0.075}{0.025}\right)}{1 + \frac{4.3 \times 0.075}{0.1} \ln\left(\frac{0.075}{0.025}\right)} = 311\,\text{K}$$

We may run through the calculation with this value to get the following:

$$Ra_{D_2} = 3.366 \times 10^6,\ \overline{Nu}_{D_2} = 22.7,\ \bar{h} = 3.98\,\text{W.m}^2\,{}^\circ\text{C}$$

A better value for T_2 is then obtained as $T_2 = 311.7\,\text{K}$, at which point we stop the iteration. The corresponding heat loss per meter of pipe is given by

$$Q = \pi D_2 \bar{h}(T_2 - T_\infty) = \pi \times 0.15 \times 3.98(311.7 - 300) = 21.9\,\text{W}$$

Example 16.8

A cylinder of diameter $D = 0.1\,\text{m}$, length $L = 0.15\,\text{m}$ is maintained at a uniform temperature of $T_w = 60\,^\circ\text{C}$ in still air at $T_\infty = 20\,^\circ\text{C}$. The two flat ends of the cylinder are perfectly insulated. Compare the heat loss from the cylinder if its axis is (a) horizontal and (b) vertical.

Solution:
We base the thermo-physical properties of air at $T_m = \frac{T_w + T_\infty}{2} = \frac{60+20}{2} = 40\,^\circ\text{C} = 273 + 40 = 313\,\text{K}$.

Kinematic viscosity:	$\nu_m = 17.07 \times 10^{-6}\,\text{m}^2/\text{s}$
Thermal conductivity:	$k_m = 0.0274\,\text{W/m}^\circ\text{C}$
Prandtl number:	$Pr_m = 0.699$
Isobaric expansion coefficient:	$\beta_m = \frac{1}{313} = 0.0032\,\text{K}^{-1}$

Case (a) Cylinder axis vertical

The characteristic dimension in this case is $L_{ch} = L = 0.15\,\text{m}$. We have for the cylinder $\frac{D}{L} = \frac{0.1}{0.15} = 0.667$. The Rayleigh number is

$$Ra_L = \frac{9.81 \times 0.0032(60 - 20)0.15^3}{(17.07 \times 10^{-6})^2} 0.699 = 1.017 \times 10^7$$

The Grashof number is given by $Gr_L = \frac{Ra_L}{Pr_m} = \frac{1.017 \times 10^7}{0.699} = 1.454 \times 10^7$. We then have

$$\frac{35}{Gr_L^{\frac{1}{4}}} = \frac{35}{(1.454 \times 10^7)^{\frac{1}{4}}} = 0.567 < 0.667$$

Hence, the correlation for a vertical plate may be made use of. The flow is laminar and we make use of Eq. 16.51 to get

$$\overline{Nu_L} = 0.55 \times (1.017 \times 10^7)^{\frac{1}{4}} = 31.06$$

The average heat transfer coefficient is given by

$$\bar{h} = \frac{31.06 \times 0.0274}{0.15} = 5.67\ \text{W/m}^2\,°\text{C}$$

The heat loss from the cylinder is then given by

$$Q_{vert} = \bar{h}\pi DL(T_w - T_\infty) = 5.67 \times \pi \times 0.1 \times 0.15 \times 40 = 10.7\ \text{W}$$

Case (b) Cylinder axis horizontal

The characteristic dimension in this case is $L_{ch} = D = 0.1$ m. The Rayleigh number is

$$Ra_D = \frac{9.81 \times 0.0032(60 - 20)0.1^3}{(17.07 \times 10^{-6})^2} 0.699 = 3.013 \times 10^6$$

The flow is laminar and we make use of Eq. 16.59 to get

$$\overline{Nu_L} = 0.53 \times (3.013 \times 10^6)^{\frac{1}{4}} = 22.08$$

The average heat transfer coefficient is given by

$$\bar{h} = \frac{22.08 \times 0.0274}{0.1} = 6.05\ \text{W/m}^2\,°\text{C}$$

The heat loss from the cylinder is then given by

$$Q_{hor} = \bar{h}\pi DL(T_w - T_\infty) = 6.05 \times \pi \times 0.1 \times 0.15 \times 40 = 11.4\ \text{W}$$

The cylinder with its axis horizontal loses *marginally* more heat than when the cylinder axis is vertical.

Fig. 16.15 Benard cells in horizontal fluid layers

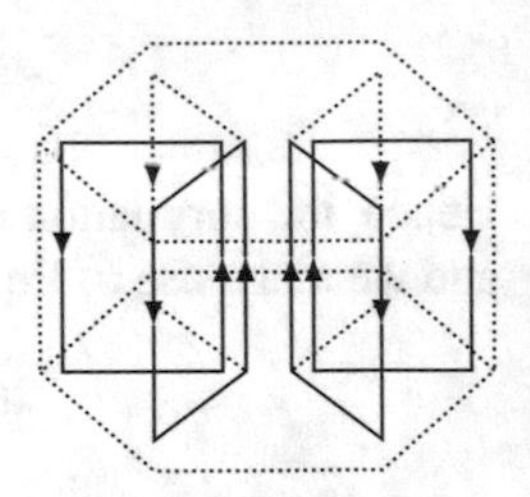

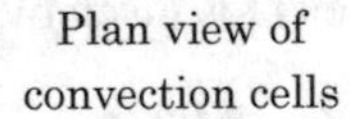

Plan view of convection cells

Typical flow lines within a cell

16.5 Heat Transfer Across Fluid Layers

A fluid layer is described by its thickness, orientation with respect to a chosen plane—usually the horizontal plane and the temperature difference across its thickness. Horizontal layers are involved while heating a fluid layer from below. All natural bodies of water belong to this category. A vertical layer is typical of fluid layer trapped within layered walls or double glazed windows. The air layer is used as a means of preventing/reducing heat transfer through the walls or windows. These are employed in energy efficient buildings. In cold countries, the heating cost comes down when such a construction is used. In the tropics, the air conditioning cost is reduced by their use. An inclined layer is typically encountered in solar collector applications. The intention here is to collect solar energy efficiently and prevent heat loss from the collector plate back to the background. More about this will be presented later on.

16.5.1 Horizontal Fluid Layers

The layer thickness s and the temperature difference $(T_1 - T_2)$ across it characterize a horizontal layer. In the absence of convection, heat transfer is by pure conduction and the Nusselt number defined as $Nu_s = \frac{hs}{k}$ is equal to unity. This holds as long as $Gr_s < 1700$. For $Gr_s > 1700$, free convection develops in the form of ordered cellular pattern, first observed by Benard[11], and hence referred to as Benard cells. In each hexagonal cell, the fluid rises in the middle and flows down at the periphery as shown in Fig. 16.15. The ordered cellular pattern changes into a disordered cell structure when the flow is turbulent. Transition to turbulence takes place at $Gr_s = 4 \times 10^5$. Jakob [12] has given the following correlations for a horizontal air layer:

[11]Henri Benard, 1874–1939, French physicist.

[12]M. Jakob, *Heat Transfer*, Vol. 1, 1st Edition, Wiley 1949, Vol. 2, 1st Edition, Wiley 1957.

$$Nu_s = 0.195 Gr_s^{\frac{1}{4}} \text{ valid for } 10^4 < Gr_s < 4 \times 10^5 \tag{16.61}$$

$$Nu_s = 0.068 Gr_s^{\frac{1}{3}} \quad \text{valid for} \quad Gr_s > 4 \times 10^5 \tag{16.62}$$

Recently, Graaf and Held[13] have proposed the following correlations.

$$\begin{aligned}
&\text{(a)} \quad Nu_s = 1 \text{ for } Gr_s < 2000 \\
&\text{(b)} \quad Nu_s = 0507 Gr_s^{0.4} \text{ for } 2000 < Gr_s < 5 \times 10^4 \\
&\text{(c)} \quad Nu_s = 3.8 \text{ for } 5 \times 10^4 < Gr_s < 2 \times 10^5 \\
&\text{(d)} \quad Nu_s = 0.426 Gr_s^{0.37} \text{ for } Gr_s > 2 \times 10^5
\end{aligned} \tag{16.63}$$

It is interesting to consider the variation of heat transfer across a horizontal air layer with the layer thickness. For this purpose, we take the case of an air layer that is at a mean temperature of $T_m = 13.5\,°\mathrm{C} = 286.5\,\mathrm{K}$ with a temperature difference across the layer of $T_1 - T_2 = 17\,°\mathrm{C}$. We systematically vary the layer thickness and calculate the heat transfer coefficient and thence the layer thermal resistance using the correlations given by Graaf and Held. Depending on the Grashof number, appropriate relation among those given by Eqs. 16.63a–d is made use of. The result is shown as a plot in Fig. 16.16. We observe a "global" maximum in the resistance of $R_{max} = 0.424\,\mathrm{m^2\,°C}$ at $s = 0.045\,\mathrm{m}$. The "minimum" heat transfer corresponds to this maximum thermal resistance, and hence the heat transfer across air layer per unit area is

$$Q = \frac{T_1 - T_2}{R_{max}} = \frac{17}{0.424} = 40.1\,\mathrm{W/m^2}$$

16.5.2 Vertical Fluid Layers

A vertical fluid layer enclosed between a vertical hot and a vertical cold wall with the top and bottom walls adiabatic is referred to as side heated cavity. Figure 16.17 shows what happens in the various regimes of flow. In the conduction regime, the flow is very weak and does not affect heat transfer significantly. The heat transfer is

[13] J. G. A. de Graaf and E. F. M. Held, Appl. Sci. Res(A), Vol. 3, p. 393–409, 1953.

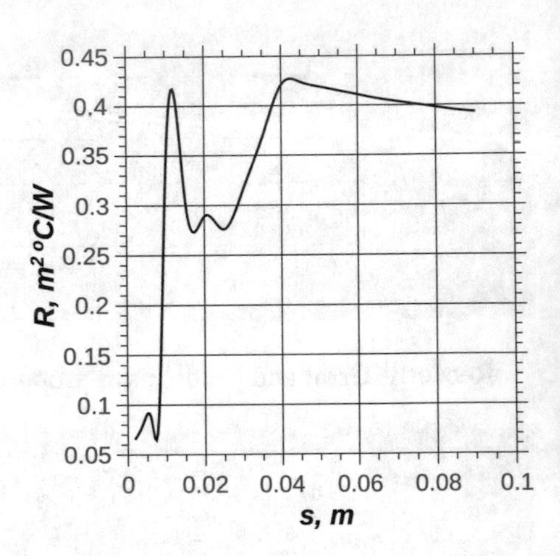

Fig. 16.16 Variation of thermal resistance of air layer with its thickness

due to fluid thermal conduction alone. When the Rayleigh number based on width of the cavity and the temperature difference $T_h - T_c$ is increased, the flow goes through a phase called asymptotic flow. The velocity and the thermal boundary layers (if they may be called so) occupy the entire width s of the layer. In the boundary layer regime, there are clearly discernible thin flow and thermal boundary layers close to the vertical walls. The heated fluid near the left hot wall rises up, turns to the right at the top, and then looses heat to the right cold wall and descends. It then takes a turn to the left and reaches the bottom of the hot wall to start the process again. The flow develops into a well-defined circulatory flow pattern in the form of a giant loop. As indicated, the flow becomes turbulent for $Ra_s > 10^7$. Heat transfer between the hot and cold wall is a function of the Grashof number based on $T_h - T_c$, the fluid Prandtl number Pr and the aspect ratio $\frac{H}{s}$, i.e.

$$\overline{Nu_s} = f\left[Gr_s, Pr, \frac{H}{s}\right]$$

Several useful correlations are available in the literature and are given below. Numerical study of Berkovsky and Polevikov [14] leads to the equation

$$\overline{Nu_s} = 0.22\left(\frac{H}{s}\right)^{-\frac{1}{4}}\left[\frac{Ra_s Pr}{Pr + 0.2}\right]^{0.28} \tag{16.64}$$

which is valid for the following range of parameters:

$$Ra_s < 10^{10}, \quad Pr < 10 \quad \text{and} \quad 2 < \frac{H}{s} < 10$$

---

[14] B. M. Berkovsky and V. K. Polevikov, In: D. B. Spalding and N. Afghan, (Eds.) *Turbulent Buoyant Flow and Convection*, Volume 2, Hemisphere, Washington, pp. 443–445, 1977.

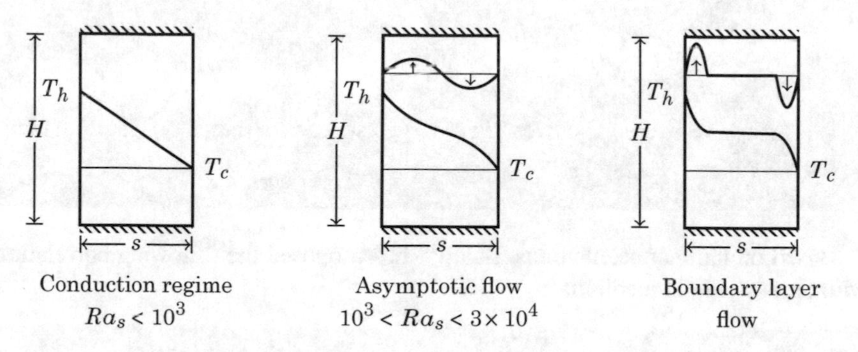

Fig. 16.17 Natural convection in vertical fluid layers.* Laminar: $3 \times 10^4 < Ra_s < 10^6$, Transition: $10^6 < Ra_s < 10^7$, Turbulent: $Ra_s > 10^7$

They propose a separate correlation for aspect ratio between 1 and 2 as

$$\overline{Nu_s} = 0.18\left[\frac{Ra_s Pr}{Pr + 0.2}\right]^{0.29} \tag{16.65}$$

which is valid for the following range of parameters:

$$\frac{Ra_s Pr}{Pr + 0.2} > 10^3, \quad 10^{-3} < Pr < 10^5 \quad \text{and} \quad 1 < \frac{H}{s} < 2$$

For tall cavities, ElShirbiny et al. [15] propose that we choose the largest of the following:

$$\overline{Nu_s} = 0.06Ra_s^{\frac{1}{3}} \ \text{or} \ \overline{Nu_s} = \left[1 + \left\{\frac{0.104Ra_s^{0.293}}{1 + \left(\frac{6310}{Ra_s}\right)^{1.36}}\right\}^3\right]^{\frac{1}{3}} \ \text{or} \ \overline{Nu_s} = 0.242\left[\frac{Ra_s}{\left(\frac{H}{s}\right)}\right]^{0.272} \tag{16.66}$$

[15] S. M. ElShirbiny, G. D. Raithby and K. G. T. Hollands, ASME J. Heat Transfer, Vol. 104, pp. 96–102, 1982.

These equations are valid for the following range of parameters:

$$10^2 < Ra_s < 10^7,\ 5 < \frac{H}{s} < 110$$

Thus, Eqs. 16.66 are valid over a very large aspect ratio range, typical of many building heat transfer applications.

Based on laminar calculations, Balaji[16] has proposed the following correlations with air as the fluid medium:

Square cavity: Aspect ratio = 1, i.e., $H = s$

$$\overline{Nu_s} = 0.13 Gr_s^{0.305} \quad \text{for} \quad 750 < Gr_s < 5 \times 10^5 \qquad (16.67)$$

Tall cavity: Aspect ratio > 1, i.e., $H > s$

$$\overline{Nu_s} = 0.215 Ra_s^{0.265} \left(\frac{H}{s}\right)^{-0.215} \quad \text{for} \quad 750 < Gr_s < 5 \times 10^5,\ 1 < \frac{H}{s} < 45 \qquad (16.68)$$

Example 16.9

A vertical double wall 2.5 m high 1 m wide has an air gap 2.5 cm thick. The internal wall faces spanning the air gap are at 305 K and 295 K, respectively. Determine the heat loss across the air gap.

Solution:

We make use of ElShirbiny et al. correlation to solve this problem. The given data, in the usual notation is

$$s = 2.5\,\text{cm} = 0.025\,\text{m},\ H = 2.5\,\text{m},\ T_h = 305\,\text{K},\ T_c = 295\,\text{K}$$

The aspect ratio of the vertical enclosure is

$$\frac{H}{s} = \frac{2.5}{0.025} = 100$$

Mean temperature of trapped air is $T_m = \frac{T_h + T_c}{2} = \frac{305+205}{2} = 300\,\text{K}$. Air properties at 300 K are

[16] C. Balaji, Ph.D. Thesis, IIT Madras 1994.

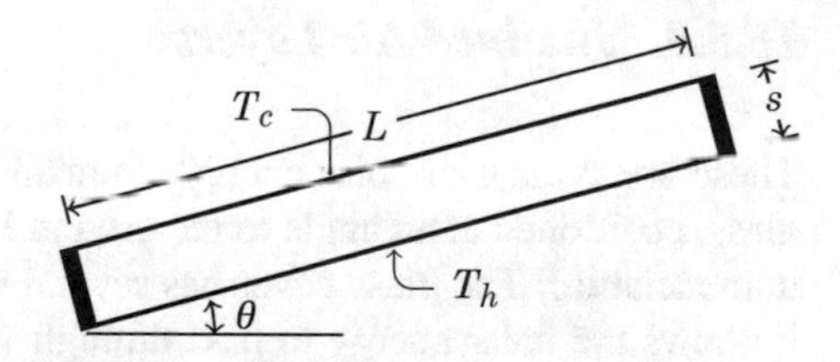

Fig. 16.18 Natural convection in inclined air layer

$$\nu_m = 15.89\times10^{-6}\,\text{m}^2/\text{s},\ Pr_m = 0.7,\ k_m = 0.0263\,\text{W/m°C},$$

$$\beta_m = \frac{1}{300} = 0.0033\,\text{K}^{-1}$$

The Rayleigh number is calculated as

$$Ra_s = \frac{9.81 \times 0.0033(305 - 295)0.025^3}{(15.89 \times 10^{-6})^2} \times 0.7 = 14165$$

We have to find the biggest of the three alternate expressions given by Eq. 16.66

$$\text{(a) } \overline{Nu_s} = 0.0605 \times 14165^{\frac{1}{3}} = 1.464$$

$$\text{(b) } \overline{Nu_s} == \left[1 + \left\{\frac{0.104 \times 14165^{0.293}}{1 + \left(\frac{6310}{14165}\right)^1 .36}\right\}^3\right]^{\frac{1}{3}} = 1.461$$

$$\text{(c) } \overline{Nu_s} = 0.242\left[\frac{14165}{100}\right]^{0.272} = 0.931$$

Choosing the biggest of these, we have $\overline{Nu_s} = 1.464$. Hence, the heat transfer coefficient is

$$\bar{h} = \frac{1.464 \times 0.0263}{0.025} = 1.54$$

The heat transfer across the air layer is

$$Q = \bar{h}HW(T_h - T_c) = 1.54 \times 2.5 \times 1(305 - 295) = 38.5\,\text{W}$$

16.5.3 Inclined Air Layers

These are typical of solar energy applications. A flat plate collector with a cover glass, positioned at an angle to the ground is an example. Figure 16.18 explains the nomenclature. The glass cover has several roles to play in this application. Firstly, it allows the solar energy to pass through with very little attenuation. Secondly, it blocks the infrared energy emitted by the collector plate from escaping back to the ambient. Thirdly, it encloses a shallow layer of air, and hence reduces the convective heat transfer across the gap between the collector plate and the glass cover. A very useful correlation for evaluating the heat transfer across the inclined air layer has been proposed by Hollands et al.[17] Introducing the notation

$$\psi = \frac{1708}{Ra_s \cos\theta}$$

the correlation is given by

$$\overline{Nu_s} = 1 + \underbrace{1.44(1-\psi)}_{\text{0 if negative}}[1 - \psi \sin^{1.6}(1.8\theta)] + \underbrace{\left[\frac{0.664}{\psi^{\frac{1}{3}}} - 1\right]}_{\text{0 if negative}} \tag{16.69}$$

The correlation is valid under the following conditions:

$$\frac{L}{s} > 10, \quad 15^\circ < \theta < 75^\circ, \quad \text{and} \quad 0 < Ra_s < 10^5$$

Example 16.10

A flat plate solar collector $L = 2$ m long and $W = 1$ m wide is inclined at $\theta = 13°$ to the horizontal. The cover plate is separated from the absorber plate by an air gap of $s = 2.5$ cm. The average temperature of the cover plate is $T_c = 305$ K, while the collector is at $T_h = 330$ K. Estimate the convective heat loss assuming air to be at 1 atm. Compare this with the solar energy input to the system at $S = 800$ W/m^2.

Solution:

Air properties are taken at the mean air temperature of $T_m = \frac{T_h+T_c}{2} = \frac{330+305}{2} =$ 317.5 K and pressure of 1 atm.

Kinematic viscosity:	$\nu_m = 17.46 \times 10^{-6}$ m^2/s
Thermal conductivity:	$k_m = 0.0276$ W/m°C
Prandtl number:	$Pr_m = 0.704$
Isobaric expansion coefficient:	$\beta_m = \frac{1}{317.5} = 0.0032$ K^{-1}

The Rayleigh number based on air layer thickness is calculated.

[17] K. G. T. Hollands, T.E. Unny, G.D. Raithby and L. Konicek, ASME J. Heat Transfer, Vol. 98, pp. 189–193, 1976.

$$Ra_s = \frac{9.81 \times 0.0032(330 - 305)0.025^3}{(17.46 \times 10^{-6})^2} \times 0.704 = 27872$$

The factor ψ used in Hollands et al. correlation is then given by

$$\psi = \frac{1708}{17872 \times \cos 13} = 0.0629$$

Using the correlating equation n16.69, we have

$$\overline{Nu}_s = 1 + 1.44(1 - 0.0629)[1 - \psi \sin^{1.6}(1.8\theta)] + \left[\frac{0.664}{0.0629^{\frac{1}{3}}} - 1\right] = 3.00$$

The convective heat transfer coefficient is then calculated as

$$\bar{h} = \frac{3 \times 0.0276}{0.025} = 3.312 \text{ W/m}^2\,°\text{C}$$

The heat loss through the air layer is then given by

$$Q_{loss} = \bar{h}LW(T_h - T_c) = 3.312 \times 2 \times 1(330 - 305) = 165.6 \text{ W}$$

Incident solar energy intercepted by the collector area is given by

$$Q_{Solar} = LWS = 2 \times 1 \times 800 = 1600 \text{ W}$$

The loss as a percentage of the incident solar flux is then given by

$$\frac{Q_{loss} \times 100}{Q_{Solar}} = \frac{165.6 \times 100}{1600} = 10.35\%$$

Concluding Remarks

We have rounded off the treatment of convection heat transfer by discussing natural or free convection in this chapter. Fundamental ideas behind natural convection have been discussed and appropriate parameters that describe natural convection have been introduced. After considering flow over an isothermal vertical plate that is amenable to analytical treatment, many interesting cases of engineering interest have been considered. Appropriate heat transfer correlations have been presented so as to facilitate simple calculations in interesting applications such as in solar collectors, hollow walls, double glazed windows.

16.6 Exercises

Ex 16.1 Perform a dimensional analysis of a problem involving natural convection from an object of characteristic dimension L. Write down a list of all the important quantities that govern the problem. How many non-dimensional parameters do you expect? Obtain all of them and discuss the physical significance of each.

Ex 16.2 A thin flat plate of height L is oriented vertically and is losing heat to an ambient by natural convection. The surface of the plate is subject to a constant heat flux. Where along the height would you expect the temperature to peak? Using dimensional arguments obtain an expression for the characteristic temperature governing the problem.

Ex 16.3 An isothermal 0.2 m long flat plate is placed vertically in an ambient medium. Its surface is maintained at a uniform temperature of 80 °C when the ambient fluid is at 30 °C. How much heat is dissipated from the plate if (a) the fluid is dry air, (b) the fluid is water? In each case what is the thickness of the boundary layer at the top edge of the plate? What is the maximum velocity in each case?

Ex 16.4 An isothermal flat plate is placed vertically in different media having Prandtl number of 1, 5, 10, and 100. In each case, the temperature levels are adjusted such that the Grashof number remains fixed at a value of 10^6. Compare the Nusselt number, boundary layer thickness, and the maximum velocity at $x = 0.1$ m in these cases.

Ex 16.5 An isothermal vertical flat plate is 0.3 m tall and is placed in air. The temperature of the surface is 120 °C, while the ambient air temperature is 20 °C. The plate surface has an emissivity of 0.85. Determine the heat loss by natural convection and radiation from a unit width of the plate. What, as a percentage of the total, is the contribution of radiation to the total heat loss? Is it reasonable to neglect radiation?

Ex 16.6 An aluminum plate is 0.005 m thick and 3.5 m tall. It is initially at a uniform temperature of 120 °C and is allowed to cool in ambient air at 30 °C. A researcher claims that the cooldown process may be modeled as that of a first order system by lumping the plate as a system at uniform temperature at any instant of time. What would be your assessment of the situation? Give proper reasons for your assessment.

Ex 16.7 The interior wall of a building is 5 m wide and 3.5 m tall. The wall temperature is uniform at 47 °C, when the room air is at a temperature of 21 °C. Heat transfer from the wall to room air may be assumed to be by natural convection and radiation. The emissivity of the wall surface has been estimated to be 0.22. Determine the convection heat loss, radiation heat loss, and the total heat loss from the wall. Is it acceptable to linearize radiation in this application? Make a comparison between the average convection heat transfer coefficient with the radiation heat transfer coefficient.

Ex 16.8 Two large plates are placed vertically forming a channel. Both the plates are isothermal and at a common temperature equal to 70 °C. The height of each plate is 2 m and the ambient fluid is room air at 30 °C. Determine the spacing between the walls that will make the two natural convection boundary layers just meet half way between the walls.

Ex 16.9 An isothermal vertical plate is 0.2 m tall and 0.15 m wide. It is maintained at a constant temperature of 85 °C in ambient air at 25 °C. Experimental data shows that the total heat loss from the plate is 32% more than that due to natural convection alone. What is the emissivity of the plate surface assuming it to be gray and diffuse?

Ex 16.10 A metal wall 3 mm thick made of steel with a thermal conductivity of 45 W/m K is covered with a 10 cm layer of an insulating material of thermal conductivity 0.1 W/m K. The inner surface of steel wall is known to be at 60 °C. The ambient air is at 27 °C. The wall is 0.4 m wide and 1.5 m tall. Determine the heat transfer across the composite wall. Model heat transfer from the exposed insulation surface as that from a vertical isothermal wall losing heat by natural convection.

Ex 16.11 Water is maintained at a temperature of 45 °C by electric heaters placed horizontally inside a rectangular tank which is 1.5 m wide, 1.5 m deep and 2.5 m high. The outside of the tank is insulated with a 10 cm thick layer of insulation having a thermal conductivity of 0.05 W/m°C. Outer surface of the insulation is coated with a special paint that has an emissivity of 0.2. The ambient temperature is 20 °C. What is the average power that needs to be dissipated by the heaters?
Make any assumption you feel are reasonable. Mention them with justification.

Ex 16.12 An electrical conductor of copper has a diameter of 0.003 m and is very long. Its surface is covered with a 0.001 m thick layer of an electrical insulator of thermal conductivity 2 W/m°C. How much heat a unit length of the wire may dissipate if the copper temperature should not exceed 85 °C? It is proposed by a consultant that the current carrying capacity may be improved by replacing the insulator by a 0.0018 m thick layer of a material of thermal conductivity equal to 3.2 W/m°C. Is the claim reasonable? The ambient fluid is air at 35 °C.

Ex 16.13 In a process application a copper tube of 18 mm inner diameter and of wall thickness 1.2 mm is made use of. The tube is insulated on the outside by a 25 mm thick insulation layer of thermal conductivity equal to 0.1 W/m°C. The process fluid is high-pressure steam at a temperature of 160 °C. Determine the heat loss from a meter length of the pipe if the ambient air is at 40 °C.

Ex 16.14 A cartridge heater 100 mm long and 25 mm diameter is immersed in water at 30 °C. What is the maximum heat that may be dissipated if the water at atmospheric pressure should not boil at the surface of the heater? What is the corresponding average heat flux based on the total surface area of the heater? How would the answers depend on the orientation of the heater?

Ex 16.15 A side heated vertical square cavity 150 × 150 mm has the two vertical walls maintained at $T_h = 75\,°C$ and $T_c = 25\,°C$. The fluid contained in the cavity is dry air. (a). Determine the heat transfer across the cavity per unit length of the cavity. (b). What will be heat transfer across the cavity if the fluid is changed to water and all other things are maintained as in part (a).

Ex 16.16 A central thin vertical partition is introduced between the vertical walls of a square cavity 200 × 200 mm. It is found that the partition may be treated as an isothermal surface with the temperature at some value in between $T_h = 90\,°C$ and $T_c = 30\,°C$. The fluid within the cavity is dry air at atmospheric pressure. Compare the heat transfer across the cavity with and without the central partition. What is the temperature of the partition?
(Hint: Heat transfer remains the same across the two cavities formed by placing the central partition)

Ex 16.17 The vertical air layer formed between two walls may be treated as a side heated cavity with a height of 0.6 m and width of 0.05 m. The two surfaces are at 20 °C and −5 °C, respectively. What is the heat transfer per unit surface area of the cavity wall? A consultant suggests that the spacing be increased to 0.07 m such that the heat transfer across the cavity is reduced. Would you agree with the consultant? Explain.

Ex 16.18 In Exercise 16.6 what would be the radiation heat transfer across the cavity if each wall may be treated as a diffuse gray surface with an emissivity of 0.65? Assume that the other two adiabatic walls of the cavity are effectively at 50 °C. Is the radiation heat transfer significant in comparison with the convection heat transfer?

Ex 16.19 A horizontal air layer is formed between two large plates placed 5 mm apart. The fluid trapped between the plates is air at atmospheric pressure. The bottom plate is at 90 °C while the top plate is losing heat to an ambient at 25 °C by convection with a heat transfer coefficient of 11.5 $W/m^2\,°C$. Determine the heat transfer across the horizontal layer per unit area and also the temperature of the top plate.

Ex 16.20 A horizontal air layer formed between two large plates spaced 15 mm apart is divided into two layers of equal thickness by placing a very thin plate in between the two plates. The bottom plate is at 90 °C , while the top plate is at 30 °C. Determine the heat transfer across the air layer in the presence of the partition and compare this with that without the partition.

Ex 16.21 A 5 mm layer of air at atmospheric pressure is formed between two large plates maintained at 90 °C and 30 °C, respectively. It is desired to determine the heat transfer per unit area across the air layer in three different arrangements: (a) air layer is horizontal, (b) air layer is vertical, and (c) the air layer is inclined at an angle of 45°. In case (a) and case (c) the hotter surface is below the colder surface.

Chapter 17
Special Topics in Heat Transfer

BASIC heat transfer processes viz. conduction, radiation, and convection have been dealt with in the previous chapters. We have dealt with the topics in sufficient detail so as to build a strong base that may now be used to discuss several special topics that are more advanced and labeled as "Special topics in heat transfer". These consist of representative problems that involve interaction between different modes of heat transfer, heat transfer in space applications, phase change—melting, boiling and condensation, mixed convection, heat transfer in particle beds and porous media, and heat transfer in high speed flows.

17.1 Introduction

Till now we have considered topics that may be termed elementary in that each heat transfer process was considered in isolation. It is seldom that heat transfer problems occur that way. In practice, there is an interplay of different heat transfer processes and we refer to them as conjugate problems. The present chapter considers the typical problems of this genre.

We have considered heat transfer involving a single phase—solid, liquid or gas (vapor). In some problems, there may be a change of phase such as in solidification and melting or condensation and boiling. These are more complex to model and we make an attempt to give a brief introductory treatment of such problems.

In all the previous chapters, we have avoided considering problems that involve variation of properties of the fluid that takes part in the heat transfer process. Such a treatment is adequate when the temperature range is small or the fluid involved has very mild variation of properties with temperature. However, such an assumption breaks down when we consider high speed flows where the fluid velocity is com-

S. P. Venkateshan, *Heat Transfer*,
https://doi.org/10.1007/978-3-030-58338-5_17

parable to the speed of sound in that medium. A typical example involving laminar high speed flow parallel to a flat plate is considered.

The reader should treat this chapter as an introduction to special topics in heat transfer. Each topic considered here has received individual attention and the relevant literature is the best way of going beyond the discussions given in the present book.

17.2 Multi-mode Problem Involving Radiation

We have come across a multi-mode problem involving conduction and convection earlier while studying heat transfer in extended surfaces and transients in a lumped system. In many applications, we encounter heat transfer by surface radiation in the presence of other modes of heat transfer such as conduction and convection. We discuss some of these in this section.

17.2.1 Transient Cooling of a Lumped System

Consider a thin shell of a high thermal conductivity material that is losing heat by radiation to an ambient at a specified temperature. Let the characteristic length be L_{ch}. Let the surface have a gray emissivity of ε. Conductive flux within the shell is given by

$$q_k = \frac{k\Delta T}{L_{\text{ch}}}$$

where ΔT is a temperature difference within the shell. Radiant flux at the surface is given by

$$q_r = \varepsilon\sigma(T^4 - T_\infty^4)$$

where T is the temperature of the shell at its surface and T_∞ is the ambient temperature. These two fluxes are essentially the same, and hence the ΔT set up within the shell is given by

$$\Delta T = \frac{\varepsilon\sigma(T^4 - T_\infty^4)L_{\text{ch}}}{k}$$

We may divide the above by $T - T_\infty$ to get

$$\frac{\Delta T}{T - T_\infty} = \frac{\varepsilon\sigma(T + T_\infty)(T^2 + T_\infty^2)L_{\text{ch}}}{k}$$

Lumping is appropriate if the above temperature ratio is very small. Consider a typical case of an aluminum shell $k = 200$ W/m°C with the surface coated with a thin film/layer of high emissivity ($\varepsilon \approx 0.85$) paint. Let the characteristic length be 0.003 m and $T = 350$ K and $T_\infty = 300$ K. The temperature ratio will then be

$$\frac{\Delta T}{T - T_\infty} = \frac{0.85 \times 5.67 \times 10^{-8}(350 + 300)(350^2 + 300^2)0.003}{200} \approx 10^{-4}$$

In fact, we may refer to this ratio as the radiation Biot number Bi_r. Hence, it is appropriate to treat the transient cooling by lumping the solid if $Bi_r \ll 1$. We consider the transient behavior of a lumped system in the form of a thin shell losing heat by radiation from its surface. The governing differential equation is easily derived as

$$\rho V C \frac{dT}{dt} = -\varepsilon\sigma S(T^4 - T_\infty^4)$$

where the symbols have the usual meanings. If we set $\theta = \frac{T}{T_\infty}$, the above equation is transformed to

$$\boxed{\frac{d\theta}{dt} = -\frac{\varepsilon\sigma S T_\infty^3}{\rho V C}(\theta^4 - 1) = -\frac{\theta^4 - 1}{\tau}} \quad (17.1)$$

where $\tau = \frac{\rho V C}{\varepsilon \sigma S T_\infty^3}$ is a characteristic time, analogous to the first order time constant we encountered in the case of a lumped system losing heat by convection. Even though Eq. 17.1 is non-linear, it is in variable separable form and hence

$$\frac{d\theta}{\theta^4 - 1} = -\frac{dt}{\tau}$$

The above is integrated to get

$$t = \tau \int_{\theta_0}^{\theta} \frac{d\theta}{\theta^4 - 1}$$

where $\theta = \theta_0$ at $t = 0$. The integral appearing in the above equation may be obtained in closed form. The integrand may be written as

$$\frac{1}{\theta^4 - 1} = \frac{1}{4}\left(\frac{1}{\theta - 1} - \frac{1}{\theta + 1}\right) - \frac{1}{2(1 + \theta^2)}$$

Term by term integration gives the solution as

$$t = \tau\left[\frac{1}{4}\ln\left\{\frac{(\theta_0 - 1)(\theta + 1)}{(\theta_0 + 1)(\theta - 1)}\right\} - \frac{1}{2}\left(\tan^{-1}\theta_0 - \tan^{-1}\theta\right)\right] \tag{17.2}$$

Note that the solution is not a simple exponential as in the earlier linear case. The solution is worked out by specifying θ and calculating the corresponding t. While plotting we consider θ as the output and t as the input!

Example 17.1

Consider a copper shell of thickness $\delta = 0.5$ mm initially at $T_0 = 333$ K. It is exposed to an ambient at $T_\infty = 300$ K. The shell has a gray surface emissivity of $\varepsilon = 0.85$. Plot the temperature as a function of time for the non-linear radiation cooling case and compare it with the linear cooling case.

Solution:
Copper shell thermo-physical properties are taken as

$$\text{Density: } \rho = 8700\ \text{kg/m}^3,\ \text{Specific heat: } C = 394\ \text{J/kg}^\circ\text{C}$$

The volume to surface area ratio for the shell is just the thickness of the shell. The characteristic time may now be calculated as

$$\tau = \frac{\rho C}{\sigma \varepsilon T_\infty^3}\frac{V}{S} = \frac{8700 \times 394}{5.67 \times 10^{-8} \times 0.85 \times 300^3} \times 0.5 \times 10^{-3} = 1318\ \text{s}$$

We make use of Eq. 17.2 to derive the time-temperature data that is shown plotted in Fig. 17.1. It is seen that the system cools down to 303 K in approximately 700 s. The graph also shows what will be the response of a first order system with linear cooling, with a time constant of $\tau = 1318$ s. It is seen that the system is at approximately 320 K at $t = 700$ s in the linear cooling case.

17.2.2 *Radiation Error in Thermometry*

Measurement of temperature of a flowing gas by a sensor may be idealized as a spherical sensor whose temperature is indicated by the electrical output of the embedded thermocouple as indicated in Fig. 17.2.

If the thermocouple lead wires are very thin we may ignore the heat conducted away from the sensor through the lead wires. If the sensor is exposed to a background,

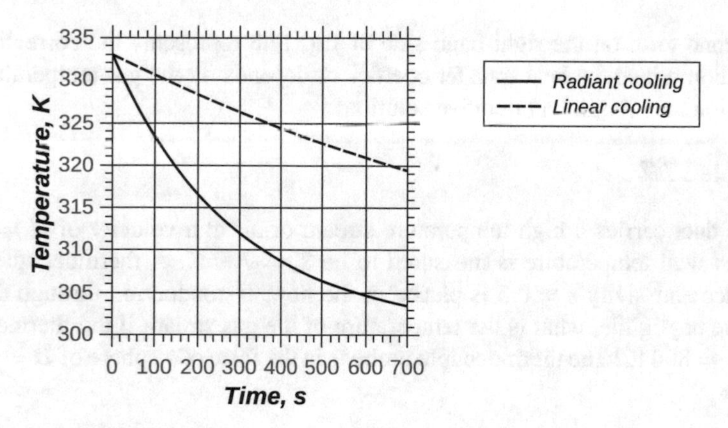

Fig. 17.1 Radiative cooling of a first order system compared with the linear cooling case

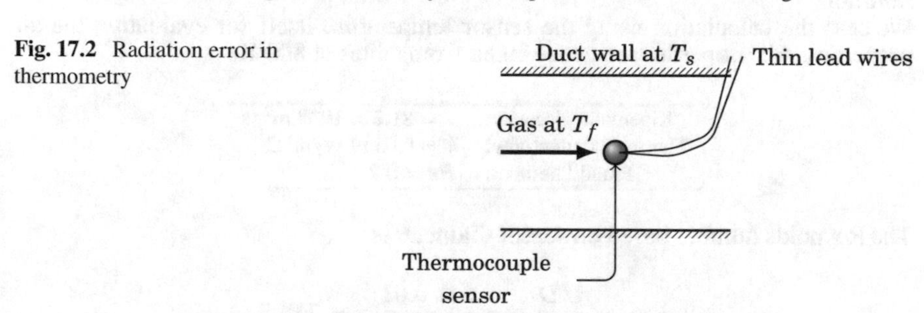

Fig. 17.2 Radiation error in thermometry

there will be radiation heat transfer from the sensor to the background. The temperature indicated by the thermocouple is then due to a balance between heat transferred to the sensor by gas convection and that lost by radiation to the background. Thus, this problem is a conjugate problem involving convection and radiation.

Let the temperature indicated by the thermocouple be T_t. If we assume that the sensor is very small compared to the size of the duct, the radiation heat loss from the sensor per unit area is given by $\varepsilon\sigma\left(T_t^4 - T_s^4\right)$, where ε is the emissivity of the sensor surface and T_s is the temperature of surfaces visible to the thermocouple. The heat gained by the sensor per unit area is $h\left(T_f - T_t\right)$, where h is the convective heat transfer coefficient between the gas and the sensor. Under thermal equilibrium, we should have

$$h\left(T_f - T_t\right) = \varepsilon\sigma\left(T_t^4 - T_s^4\right)$$

Since the temperature of the thermocouple and the background are measured, and hence known, the gas temperature may be estimated as

$$T_f = T_t + \frac{\varepsilon\sigma}{h}\left(T_t^4 - T_s^4\right) \tag{17.3}$$

The second term on the right hand side of Eq. 17.3 represents the correction due to radiation. Since the heat transfer coefficient depends on the gas temperature, the calculation may require an iterative solution!

Example 17.2

A large duct carries a high temperature stream of air at a velocity of $U = 6$ m/s. The duct wall temperature is measured to be $T_s = 700$ K. A thermocouple probe of surface emissivity $\varepsilon = 0.5$ is placed in the flow. If conduction through the lead wires are negligible, what is the temperature of the gas stream if the thermocouple reads $T_t = 800$ K? The thermocouple probe is in the form of a sphere of $D = 10$ mm diameter.

Solution:
We start the calculation using the sensor temperature itself for evaluating the air properties. Air properties are hence taken from tables at 800 K:

Kinematic viscosity:	$\nu = 81.2 \times 10^{-6}$ m^2/s
Thermal conductivity:	$k = 0.0559$ W/m°C
Prandtl number:	$Pr = 0.7$

The Reynolds number based on sensor diameter is

$$Re_D = \frac{UD}{\nu} = \frac{6 \times 0.01}{81.2 \times 10^{-6}} = 738.9$$

We make use of the Whitaker correlation[1] for evaluating the mean convective heat transfer coefficient. Accordingly, we have

$$\begin{aligned} Nu_D &= 2 + \left(0.4Re_D^{\frac{1}{2}} + 0.06Re_D^{\frac{2}{3}}\right) Pr^{0.4} \\ &= 2 + \left(0.4 \times 738.9^{\frac{1}{2}} + 0.06 \times 738.9^{\frac{2}{3}}\right) 0.7^{0.4} = 15.68 \end{aligned}$$

The heat transfer coefficient is then calculated as

$$h = \frac{Nu_D k}{D} = \frac{15.68 \times 0.0559}{0.01} = 87.65 \text{ W/m}^{2\circ}\text{C}$$

The radiation correction is now calculated using Eq. 17.3 as

$$\Delta T_{rad} = \frac{\varepsilon\sigma}{h}(T_t^4 - T_s^4) = \frac{0.5 \times 5.67 \times 10^{-8}}{87.65}(800^4 - 700s^4) = 54.82 \text{ K}$$

The air temperature is thus estimated as

[1] S. Whitaker, AIChE J., Vol. 18, pp. 361–371,1972.

$$T_f = T_t + \Delta T_{\text{rad}} = 800 + 54.82 = 854.82 \text{ K}$$

A better value may be estimated by using the film temperature for evaluating the air properties. We may take the film temperature as

$$T_m = \frac{T_f + T_t}{2} = \frac{854.82 + 800}{2} = 827.6 \text{ K}$$

The revised air properties are

Kinematic viscosity:	$\nu = 86.3 \times 10^{-6}$ m^2/s
Thermal conductivity:	$k = 0.063$ W/m°C
Prandtl number:	$Pr = 0.7$

The revised values of all the other quantities are as follows:

$$Re_D = 695.7 \text{ and } h = 96.2 \text{ W/m}^{2\circ}\text{C}$$
$$\Delta T_{\text{rad}} = 50 \; K \text{ and } T_f = 850 \text{ K}$$

The new value is approximately 5 K below the previous value. We may stop the calculation here! The estimated air temperature is thus given by $T_f = 850$ K.

17.2.3 Duct Type Space Radiator

As a second typical example of the problem that involves interaction between convection and radiation, we present the analysis of a typical duct type space radiator, shown schematically in Fig. 17.3. The analysis method closely follows that used in Chap. 14 on heat exchangers. The hot fluid enters the duct at high temperature T_{f1} and exits it at a lower temperature T_{f2} after losing heat to space by radiation via the

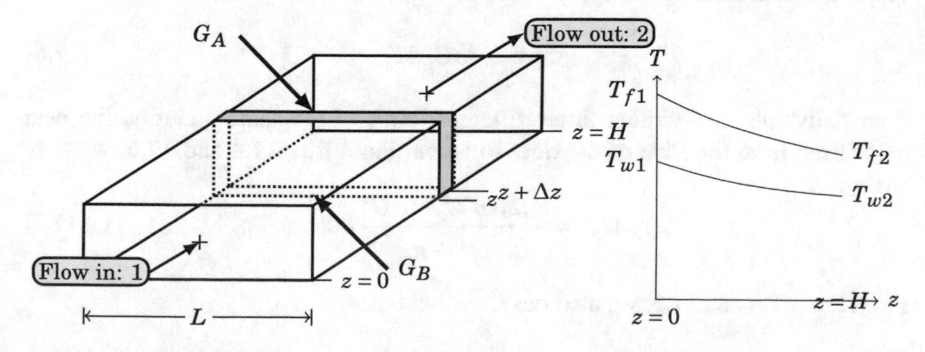

Fig. 17.3 A duct type radiator showing the fluid and wall temperature variations along its length

walls of the duct. In most applications, we may ignore the conductive resistance of the wall. The outside surfaces of the duct lose heat to space by radiation. The duct may receive radiation on both the exposed faces. We assume that the duct surfaces are selective surfaces. Hence, let α_A, ε_A and α_B, ε_B be, respectively, the absorptivity-emissivity combinations of the two exposed surfaces of the duct. Let the irradiation on the two sides of the duct be G_A and G_B, as shown in the figure. The relation below represents the radiation absorbed by the duct per unit area

$$\begin{bmatrix}\text{Radiation absorbed}\\ \text{by duct per unit area}\end{bmatrix} = \alpha_A G_A + \alpha_B G_B = 2\alpha G \text{ (say)}$$

Similarly, total emission from the two sides of the duct per unit area may be written as

$$\begin{bmatrix}\text{Radiation emitted}\\ \text{by duct per unit area}\end{bmatrix} = \varepsilon_A \sigma T_w^4 + \varepsilon_B \sigma T_w^4 = 2\varepsilon\sigma T_w^4 \text{ (say)}$$

where T_w is the duct wall temperature assumed to be the same on both sides of the duct. Hence, we may write for the net heat loss from elemental area of the duct of length $\mathrm{d}z$ as

$$\mathrm{d}q = 2\left[\varepsilon\sigma T_w^4 - \alpha G\right] L\mathrm{d}z \tag{17.4}$$

Let the mass flow of the fluid through the duct be $\dot{m}$, the specific heat be constant at C_p. Let the duct side convection heat transfer coefficient be h. The fluid to wall temperature difference is because of the film resistance on the fluid side. Consider a slice of the duct of length $\mathrm{d}z$ as shown in the figure. On the fluid side, the heat transfer to the duct walls is

$$\mathrm{d}q = (h)\ \underbrace{(2L\mathrm{d}z)}_{\substack{\text{Area}\\ \text{element}}}\ \left[T_f(z) - T_w(z)\right] \tag{17.5}$$

The heat loss from the fluid as it crosses the duct element is

$$\mathrm{d}q = -\dot{m}C_p \mathrm{d}T_f \tag{17.6}$$

Essentially, we have written three different forms of the same quantity, the heat transferred from the slice of the duct to space. From Eqs. 17.4 and 17.6, we may write

$$\mathrm{d}T_f = -\frac{2[\varepsilon\sigma T_w^4 - \alpha G]}{\dot{m}C_p} L\mathrm{d}z \tag{17.7}$$

From Eqs. 17.4 and 17.5, we also have

$$T_f - T_w = \frac{\varepsilon\sigma T_w^4 - \alpha G}{h}$$

which may be differentiated to get

$$\mathrm{d}T_f - \mathrm{d}T_w = \left[\frac{4\varepsilon\sigma T_w^3}{h}\right]\mathrm{d}T_w$$

This my be rearranged as

$$\mathrm{d}T_f = \left[\frac{4\varepsilon\sigma T_w^3}{h} + 1\right]\mathrm{d}T_w \tag{17.8}$$

Equating the two expressions (17.7) and (17.8), we obtain a relation between the wall temperature and the position along the duct as

$$\mathrm{d}z = \frac{\dot{m}C_p}{L}\frac{\left[\frac{4\varepsilon\sigma T_w^3}{h} + 1\right]}{2\left[\varepsilon\sigma T_w^4 - \alpha G\right]}\mathrm{d}T_w \tag{17.9}$$

Let the wall temperatures at entry and exit be T_{w1} and T_{w2}. These temperatures are determined as follows. Equating 17.4 and 17.5, applying the resulting equation at the entry and exit, we have

$$\text{(a)}\quad \varepsilon\sigma T_{w1}^4 - \alpha G = h\left(T_{f1} - T_{w1}\right) \quad \text{(b)}\quad \varepsilon\sigma T_{w2}^4 - \alpha G = h\left(T_{f2} - T_{w2}\right) \tag{17.10}$$

These non-linear algebraic equations provide relations between the respective wall and fluid temperatures at entry and exit. Introduce now the following non-dimensional variables and parameters:

$$\theta_w = \frac{T_w}{T_{w1}},\ \theta_{w2} = \frac{T_{w2}}{T_{w1}},\ N = \frac{\alpha G}{\varepsilon\sigma T_{w1}^4}$$

The parameter N is known as the environmental parameter. More about this parameter later. Equation 17.9 may then be rewritten as

$$\mathrm{d}z = \frac{\dot{m}C_p}{2Lh}\cdot\frac{4\theta_w^3\mathrm{d}\theta_w}{\left(N - \theta_w^4\right)} + \frac{\dot{m}C_p}{2\varepsilon\sigma T_{w1}^3 L}\cdot\frac{\mathrm{d}\theta_w}{\left(N - \theta_w^4\right)} \tag{17.11}$$

This equation may be integrated from $z = 0$ to any $z = H$ by using the inlet condition $\theta_w = 1$ at $z = 0$ and $\theta_w = \theta_{w2}$ at z = H to get

$$H = \frac{\dot{m}C_p}{2Lh}\underbrace{\int_1^{\theta_{w2}}\frac{4\theta_w^3\mathrm{d}\theta_w}{\left(N - \theta_w^4\right)}}_{\psi_f} + \frac{\dot{m}C_p}{2\varepsilon\sigma T_{w1}^3 L}\underbrace{\int_1^{\theta_{w2}}\frac{\mathrm{d}\theta_w}{\left(N - \theta_w^4\right)}}_{\psi_r} \tag{17.12}$$

We define the film resistance number ψ_f to be given by the first integral appearing on the right-hand side of Eq. 17.12. Also, we define the radiation number ψ_r to be given by the second integral on the right-hand side of Eq. 17.12. Using this notation, we may rewrite expression for the length of the heat exchanger as

$$H = \frac{\dot{m}C_p}{2Lh}\psi_f + \frac{\dot{m}C_p}{2\varepsilon\sigma T_{w1}^3 L}\psi_r \tag{17.13}$$

Both the integrals may be integrated to obtain closed form relations presented below

$$\boxed{\psi_f = \ln\left[\frac{1-N}{\theta_{w2}^4 - N}\right]} \tag{17.14}$$

$$\boxed{\psi_r = \frac{1}{4N^{\frac{3}{4}}}\ln\left[\frac{\left(N^{\frac{1}{4}}-1\right)\left(N^{\frac{1}{4}}+\theta_{w2}\right)}{\left(N^{\frac{1}{4}}+1\right)\left(N^{\frac{1}{4}}-\theta_{w2}\right)}\right] + \frac{1}{2N^{\frac{3}{4}}}\left[\tan^{-1}\left(\frac{\theta_{w2}}{N^{\frac{1}{4}}}\right) - \tan^{-1}\left(\frac{1}{N^{\frac{1}{4}}}\right)\right]} \tag{17.15}$$

The reader will recall that integral similar to the latter that has occurred earlier in the problem of a lumped system transient cooling by radiation.

The environmental parameter N plays an important role in the amount of heat transfer that can take place between the radiator and the background or the environment. If $N > 1$, there is a net radiation gain by the duct and for $N < 1$ there is a net radiation loss from the duct. Hence, for a cooling application, it is seen that $0 \leq N \leq 1$ and also $\theta_{w2} > N^{\frac{1}{4}}$. When there is no irradiation, the environmental parameter is zero. For a useful range of N, keeping cooling application in mind, and a meaningful range of θ_{w2} the radiation number and the film resistance number vary as shown in Figs. 17.4 and 17.5.

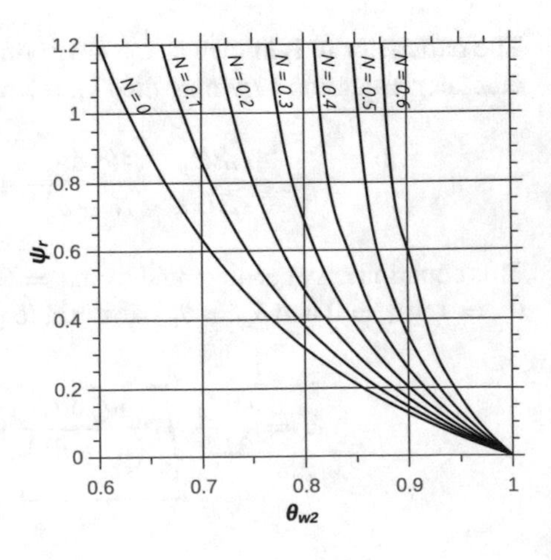

Fig. 17.4 Variation of radiation number for a cooling duct

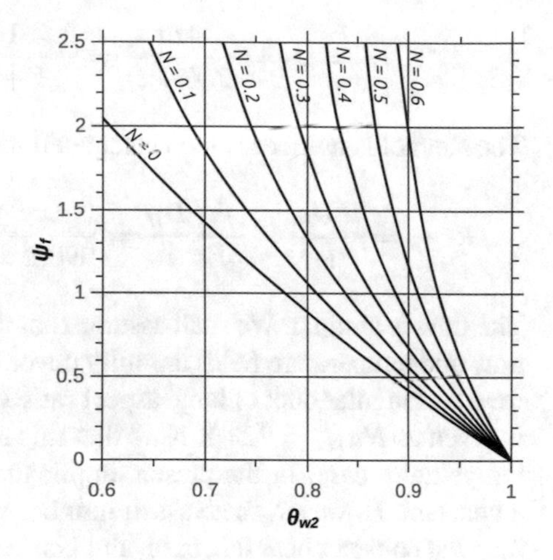

Fig. 17.5 Variation of film resistance number for a cooling duct

Example 17.3

A space radiator is in the form of a duct of 1 × 0.005 m cross section. Water flowing in the duct at the rate of 0.1 kg/s cools from an inlet temperature of 55 °C to an outlet temperature of 35 °C. The average power absorbed by the duct from the background is 200 W/m^2 from both the sides. The average emissivity of the two sides of the duct is 0.85. What is the length of the radiator?

Solution:
We make use of the notation used in the discussion just preceding this example. We have

$$L = 1\ m,\ t = 0.005\ \text{m},\ T_{f1} = 55°\text{C} = 328\ \text{K},\ T_{f2} = 35°\text{C} = 308\ \text{K},$$
$$2\alpha G = 200\ \text{W/m}^2, \varepsilon = 0.85, \dot{m} = 0.1\ \text{kg/s}$$

We shall evaluate the water properties at its mean temperature given by

$$T_m = \frac{T_{f1} + T_{f2}}{2} = \frac{55 + 35}{2} = 45\ °\text{C}$$

From table of properties of water, we then have:

Density:	$\rho_m = 990.2$ kg/m^3
Specific heat:	$C_{pm} = 4176$ J/kg°C
Kinematic viscosity:	$\nu_m = 6.11 \times 10^{-7}$ m^2/s
Thermal conductivity:	$k_m = 0.640$ W/m°C
Prandtl number:	$Pr_m = 3.9$

The hydraulic and energy diameters for the duct are identical and given by

$$D_H = D_E = \frac{4Lt}{2(L+t)} = \frac{2 \times 1 \times 0.005}{1 + 0.005} = 0.00995 \approx 0.01 \text{ m}$$

The Reynolds number is then calculated as

$$Re_{D_H} = \frac{UD_H}{\nu} = \frac{\dot{m}}{\rho L t}\frac{D_H}{\nu} = \frac{0.1}{990.2 \times 1 \times 0.005} \times \frac{0.01}{6.11 \times 10^{-7}} = 330.6$$

The flow is laminar. We shall assume that the heat exchanger is long enough that the flow and temperature fields are fully developed. From Chap. 12, the Nusselt number for a rectangular duct of large aspect ratio (it is $\frac{L}{t} = \frac{1}{0.005} = 200$ in the present case) is given as $Nu_{D_E} = 7.541$. Note that this Nusselt number is based on constant wall temperature case. In the present application, the wall temperature is certainly not a constant. However, the Nusselt number with constant wall temperature is smaller than the constant heat flux case, and hence the choice to be on the conservative side. The mean convective heat transfer coefficient is then given by

$$\overline{h} = \frac{Nu_{D_E} k}{D_E} = \frac{7.541 \times 0.640}{0.01} = 482.6 \text{ W/m}^2\,^\circ\text{C}$$

The wall temperatures at entry and exit are now calculated based on Eq. 17.10. We have

$$\frac{\alpha G}{h} = \frac{200}{482.6} = 0.4144 \quad \text{and} \quad \frac{\varepsilon\sigma}{h} = \frac{0.85 \times 5.67 \times 10^{-8}}{482.6} = 9.987 \times 10^{-11}$$

The relation between T_{f1} and T_{w1} (using Eq. 17.10) is then given by

$$T_{f1} + 0.4144 = 328.4144 = 9.987 \times 10^{-11} T_{w1}^4 + T_{w1}$$

The duct wall temperature at entry is hence given by $T_{w1} = 327.3$ K (solving the above non-linear equation). Similarly, we may obtain the duct wall temperature at exit as $T_{w2} = 307.6$ K (the reader may verify this). The parameters needed to evaluate the heat exchanger length are calculated now.
We have

$$\theta_{w2} = \frac{T_{w2}}{T_{w1}} = \frac{307.6}{327.3} = 0.9398$$

$$N = \frac{\alpha G}{\varepsilon \sigma T_w^4} = \frac{200}{0.85 \times 5.67 \times 10^{-8} \times 327.3^4} = 0.3616$$

The film resistance number is then calculated using expression (17.14) as

$$\psi_f = \ln\left[\frac{1 - 0.3616}{0.9398^4 - 0.3616}\right] = 0.4228$$

The radiation number is calculated using expression (17.14) as

$$\psi_r = \frac{1}{4 \times 0.3616^{\frac{3}{4}}} \ln\left[\frac{(0.3616^{\frac{1}{4}} - 1)(0.3616^{\frac{1}{4}} + 0.9398)}{(0.3616^{\frac{1}{4}} + 1)(0.3616^{\frac{1}{4}} - 0.9398)}\right] + \frac{1}{2 \times 0.3616^{\frac{3}{4}}}\left[\tan^{-1}\left(\frac{0.9398}{0.3616^{\frac{1}{4}}}\right) - \tan^{-1}\left(\frac{1}{0.3616^{\frac{1}{4}}}\right)\right] = 0.1158$$

Film resistance number and radiation number may also be read off the graphs shown in Figs. 17.4 and 17.5. The heat exchanger length is then given by expression (17.13) as

$$H = \frac{0.1 \times 4176}{2 \times 1 \times 482.6} \times 0.4228 + \frac{0.1 \times 4176}{2 \times 0.85 \times 5.67 \times 10^{-8} \times 327.3^3 \times 1} \times 0.1158$$
$$= 0.18 + 14.31 = 14.49 \text{ m}$$

Since the heat exchanger is very long compared to the hydraulic diameter, the fully developed assumption is valid. Also, since the temperature change along the length is very gradual, axial conduction through the walls of the duct are negligible. The second assumption has been tacitly assumed in the formulation. The justification has come now.

17.2.4 Uniform Area Fin Losing Heat by Convection and Radiation

Consider an extended surface in the form of a flat plate of uniform cross section as shown in Fig. 17.6.

Let us consider a unit width in a direction perpendicular to the plane of the paper. Consider a slice $2t \times \Delta x \times 1$ of the fin as shown. We account for all energy fluxes entering or leaving the control volume as under.

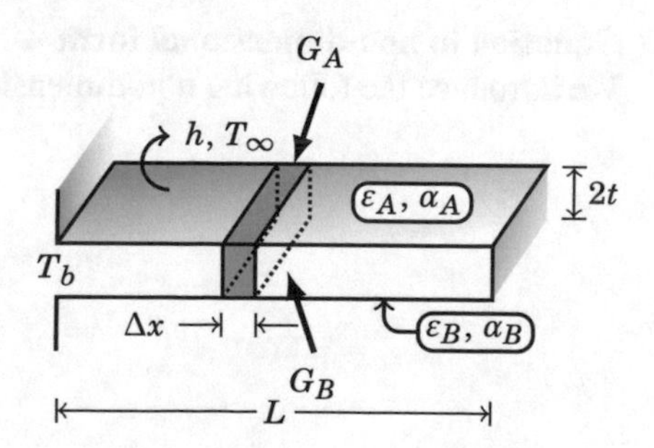

Fig. 17.6 Uniform area fin losing heat by convection and radiation

Cond. In:	$Q_{k,\text{in}} = -2tk \left.\frac{dT}{dx}\right\|_x$	Cond. Out:	$Q_{k,\text{out}} = -2tk \left.\frac{dT}{dx}\right\|_{x+\Delta x}$
Rad. In:	$Q_{r,in} = (\alpha_A G_A + \alpha_B G_B)\Delta x$	Rad. Out:	$Q_{r,\text{out}} = (\varepsilon_A + \varepsilon_B)\sigma T^4 \Delta x$
Conv. Out:	$Q_{c,\text{out}} = 2h(T - T_\infty)\Delta x$		

Under steady conditions that are assumed to prevail, energy balance requires that

$$Q_{k,\text{in}} + Q_{r,\text{in}} = Q_{k,\text{out}} + Q_{r,\text{out}} + Q_{c,\text{out}}$$

$$\text{or} \quad \left(Q_{k,\text{out}} - Q_{k,\text{in}}\right) + \left(Q_{r,\text{out}} - Q_{r,\text{in}}\right) + Q_{c,\text{out}} = 0$$

We use Fourier law for conduction flux, use Taylor expansion and retain only first order terms to get

$$Q_{k,\text{out}} - Q_{k,\text{in}} = -2tk \frac{d^2T}{dx^2} \Delta x$$

Define $\alpha_A G_A + \alpha_B G_B = 2\alpha G$ and $\varepsilon_A + \varepsilon_B = 2\varepsilon$ as done earlier in the case of the duct type radiator. Then we have

$$Q_{r,\text{out}} - Q_{r,\text{in}} = 2\left[\varepsilon\sigma T^4 - \alpha G\right]\Delta x$$

With these the energy balance equation becomes, after minor simplification,

$$\boxed{\frac{d^2T}{dx^2} - \frac{h}{kt}(T - T_\infty) - \frac{\left(\varepsilon\sigma T^4 - \alpha G\right)}{kt} = 0} \tag{17.16}$$

The accompanying boundary conditions are given by

$$\boxed{T = T_b \text{ at } x = 0, \quad \underbrace{\frac{dT}{dx} = 0 \text{ at } x = L}_{\text{Insulated tip}}} \tag{17.17}$$

Equation in non-dimensional form

We introduce the following non-dimensional variables and parameters:

Non-dimensional temperature: $\theta = \frac{T}{T_b}$

Non-dimensional distance: $\xi = \frac{x}{L}$

Absorbed irradiation-emissive power ratio: $\frac{\alpha G}{\varepsilon \sigma T_b^4} = \left(\frac{T_{ref}}{T_b}\right)^4$

Non-dimensional ambient temperature: $\theta_\infty = \frac{T_\infty}{T_b}$

In the above, the reference temperature is defined as $T_{\text{ref}} = \left(\frac{\alpha G}{\varepsilon \sigma}\right)^{\frac{1}{4}}$. Then

$$\frac{d^2\theta}{d\xi^2} - \left(\frac{hL^2}{kt}\right)(\theta - 1) - \left(\frac{\varepsilon \sigma T_b^3 L^2}{kt}\right)\left(\theta^4 - \theta_{\text{ref}}^4\right) = 0$$

Further, let $p^2 = \frac{hL^2}{kt}$ and $N_{RC} = \frac{\varepsilon \sigma T_b^3 L^2}{kt}$. Then, the governing equation becomes

$$\boxed{\frac{d^2\theta}{d\xi^2} - p^2(\theta - 1) - N_{RC}\left(\theta^4 - \theta_{\text{ref}}^4\right) = 0} \tag{17.18}$$

with the boundary conditions

$$\boxed{\theta = \theta_b = 1 \quad \text{at} \quad \xi = 0, \quad \text{and} \quad \frac{d\theta}{d\xi} = 0 \text{ at } \xi = 1} \tag{17.19}$$

The boundary condition at the tip is based on the assumption of negligible heat loss, both by convection and radiation, from the narrow strip that represents the tip.

Since the present problem involves all three modes of heat transfer, three non-dimensional parameters appear in the formulation. The first one is $p^2 = \frac{hL^2}{kt} = m^2L^2$, where $m = \sqrt{\frac{h}{kt}}$ is the familiar fin parameter appearing in conducting-convecting fins. The second one is the radiation-conduction interaction parameter N_{RC}, which may be written as

$$N_{RC} = \frac{\varepsilon \sigma T_b^3 L^2}{kt} = \underbrace{\frac{\varepsilon \sigma T_b^4}{k\frac{T_b}{L}}}_{\text{Term } (i)} \cdot \underbrace{\frac{L}{t}}_{\text{Term } (ii)} \tag{17.20}$$

We recognize Term (i) as the ratio of a representative radiant heat flux to a representative conductive heat flux. We recognize Term (ii) to be a geometric parameter, the *slenderness ratio*. In fact p^2 may be interpreted in a like manner as

$$p^2 = \frac{hL^2}{kt} = \underbrace{\frac{h(T_b - T_\infty)}{k\frac{(T_b - T_\infty)}{L}}}_{\text{Term } (i)} \cdot \underbrace{\frac{L}{t}}_{\text{Term } (ii)} \tag{17.21}$$

We recognize Term (i) as the ratio of a representative convective heat flux to a representative conductive heat flux. We recognize Term (ii) to be a geometric parameter, the *slenderness ratio*.

A third parameter that makes its appearance is $\theta_{\text{ref}} = \left(\frac{\alpha G}{\varepsilon\sigma T_b^4}\right)^{\frac{1}{4}}$, in the pressure of incident radiation. This parameter is hence referred to as the environmental parameter. Note that radiation will heat the fin if $\theta_{\text{ref}} > 1$ and cool the fin if $\theta_{\text{ref}} < 1$. In most applications, we shall be interested only in the latter case.

Special Cases

Special Case 1: Radiating-conducting fin in the absence of convection

The governing Eq. 17.18 reduces on putting $p^2 = 0$ to

$$\frac{\mathrm{d}^2\theta}{\mathrm{d}\xi^2} - N_{RC}\left(\theta^4 - \theta_{\text{ref}}^4\right) = 0 \tag{17.22}$$

with the boundary conditions remaining the same as Eq. 17.19. Equation 17.22 may be integrated once with respect to ξ after multiplying through by $2\frac{\mathrm{d}\theta}{\mathrm{d}\xi}$ to get

$$\left(\frac{\mathrm{d}\theta}{\mathrm{d}\xi}\right)^2 - 2N_{RC}\left(\frac{\theta^5}{5} - \theta_{\text{ref}}^4\theta\right) = C$$

where C is a constant of integration. Let $\theta = \theta_t$ (as yet unknown) at $\xi = 1$. The vanishing of the derivative at $\xi = 1$ then requires that

$$-2N_{RC}\left(\frac{\theta_t^5}{5} - \theta_{\text{ref}}^4\theta_t\right) = C$$

We eliminate C from these two expressions to get

$$\left(\frac{\mathrm{d}\theta}{\mathrm{d}\xi}\right)^2 = 2N_{RC}\left[\frac{\theta^5 - \theta_t^5}{5} - \theta_{\text{ref}}^4\left(\theta - \theta_t\right)\right]$$

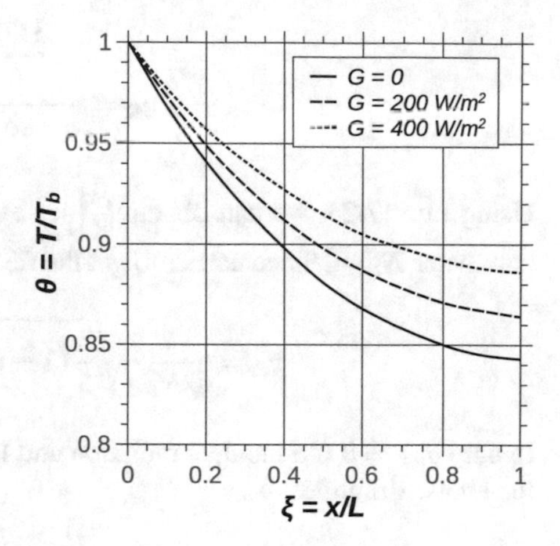

Fig. 17.7 Temperature profiles in a conducting and radiating fin

Taking square root, we then have

$$\frac{d\theta}{d\xi} = -\sqrt{\frac{2N_{RC}}{5}\left(\theta^5 - \theta_t^5\right) - 2N_{RC}\theta_{\text{ref}}^4\left(\theta - \theta_t\right)} \tag{17.23}$$

For a cooling fin, which is being considered here, it is appropriate to choose the negative sign on the right-hand side and $\theta_b > \theta_t > \theta_{\text{ref}}$. θ_t is not known as of now. Equation 17.23 will have to be solved numerically. We assume a value for θ_t and solve the first order equation starting with $\theta = 1$ at $\xi = 0$, using, for example, the fourth order Runge–Kutta method. A new value of θ_t is obtained from the numerical solution and is used to solve the equation afresh. This procedure is stopped when θ_t converges to a specific value.

Figure 17.7 shows the temperature profiles for typical cases with $T_b = 373\ K$, $k = 207$ W/m°C, $\varepsilon = 0.8$, $\alpha = 0.6$, $t = 0.0005$ m and $L = 0.15$ m. The value of radiation-conduction parameter turns out to be $N_{RC} = 0.512$. Three irradiation values as indicated in the figure legend are considered. These correspond, respectively, to environmental parameter values of 0, 0.608, and 0.723.

Radiating fin efficiency

The maximum possible heat transfer $q_{\max}$ takes place when the entire fin is at the base temperature and the environmental parameter is zero.

$$q_{max} = 2\varepsilon\sigma T_b^4 L$$

However, the actual heat transferred is given by

$$q_{\text{act}} = -\ 2tk\left.\frac{dT}{dx}\right|_{x=0} = -\ \frac{2ktT_b}{L}\left.\frac{d\theta}{d\xi}\right|_{\xi=0}$$

The radiating fin efficiency is defined as $\eta_R = \frac{q_{\text{act}}}{q_{\max}}$. Thus, we have

$$\eta_R = \frac{-\left.\dfrac{ktT_b}{L}\dfrac{\mathrm{d}\theta}{\mathrm{d}\xi}\right|_{\xi=0}}{\varepsilon\sigma T_b^4 L}$$

Using Eq. 17.23, we can obtain $\left.\frac{\mathrm{d}\theta}{\mathrm{d}\xi}\right|_{\xi=0}$ by taking $\theta = 1$. Also, we introduce the parameter N_{RC} defined earlier to get the fin efficiency as

$$\eta_R = \frac{1}{\sqrt{N_{RC}}}\sqrt{\frac{2}{5}\left(1-\theta_t^5\right) - 2\theta_{\mathrm{ref}}^4\left(1-\theta_t\right)} \quad (17.24)$$

In case $\theta_{\mathrm{ref}} = 0$ (no incident radiation and hence environmental parameter is zero), the above simplifies to

$$\eta_R = \frac{1}{\sqrt{N_{RC}}}\sqrt{\frac{2}{5}\left(1-\theta_t^5\right)} \quad (17.25)$$

In case the fin is very long, i.e., $L \to \infty$, $\theta_t \to \theta_{\mathrm{ref}}$ and expression (17.24) becomes

$$\eta_R = \frac{1}{\sqrt{N_{RC}}}\sqrt{\frac{2}{5}\left(1-\theta_{\mathrm{ref}}^5\right) - 2\theta_{\mathrm{ref}}^4\left(1-\theta_{\mathrm{ref}}\right)} \quad (17.26)$$

Further, in case there is no irradiation, i.e., $\theta_{\mathrm{ref}} = 0$, this simplifies further to

$$\eta_R = \sqrt{\frac{2}{5N_{RC}}} \quad (17.27)$$

For $L \to 0$ or $N_{RC} \to 0$, the actual maximum heat loss is limited to $2(\varepsilon\sigma T_b^4 - \alpha G)L$, and hence

$$\eta_R \to \frac{2(\varepsilon\sigma T_b^4 - \alpha G)L}{2L\varepsilon\sigma T_B^4} = 1 - \theta_{ref}^4 \quad (17.28)$$

Thus, in general, η_R is a function of radiation-conduction parameter or the *profile number* N_{RC} and the environmental parameter θ_{ref}. η_R is normally plotted as a function of N_{RC} for different θ_{ref}. Figure 17.8 shows the performance of a radiating fin in the absence of external radiation, i.e., for $\theta_{\mathrm{ref}} = 0$.

Optimum Radiating Fin

It may also be shown that a fin of optimum configuration (fin of minimum mass) satisfies the condition[2]

$$\frac{\partial \eta_R}{\partial N_{RC}} = -\frac{1}{3} \times \frac{\eta_R}{N_{RC}} \quad (17.29)$$

[2] D. B. Mackay, C. P. Bacha, *Space Radiator Analysis and Design*, ASD TR 61-30 Pt. I, Space and Information Systems Division, North American Aviation, INC, October 1961.

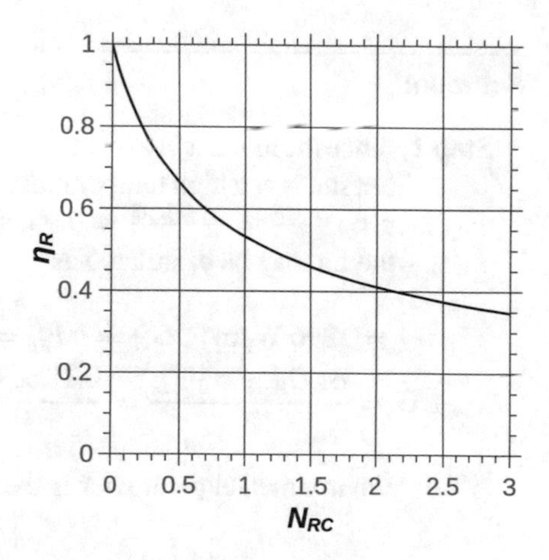

Fig. 17.8 Efficiency of a radiating fin in the absence of irradiation

Table 17.1 Radiating fin of 'theoretical profile'

θ^4_{ref}	N_{RC}	η_R	θ^4_{ref}	N_{RC}	η_R
0.900	0.382	0.069	0.400	0.630	0.363
0.800	0.446	0.131	0.300	0.670	0.412
0.700	0.496	0.191	0.200	0.712	0.459
0.600	0.542	0.250	0.100	0.753	0.511
0.500	0.583	0.305	0.000	0.801	0.554

The partial derivative is indicated to point out that the environmental parameter is to be held fixed. A fin that satisfies condition (17.29) is said to be a fin of '*theoretical profile*'. The η_R, N_{RC} combination for a fin of theoretical profile depends on the environmental parameter θ_{ref}. Table 17.1 gives the data in the form useful for design of such fins.

Example 17.4

An electrical system in space has to dissipate 1 kW to its environment from a white painted aluminum fin. The appropriate data is

Base structure temperature:	$T_b = 550$ K
Width of fin structure:	$W = 1.5$ m
Emissivity of fin surfaces:	$\varepsilon_A = \varepsilon_B = 0.95$
Absorptivity of fin surfaces:	$\alpha_A = \alpha_B = 0.18$

Heat is lost from both surfaces. One face views the moon at a surface temperature of $T_M = 374$ K while the other surface faces the sun. The solar constant may be taken as 1396 W/m^2. Determine the proportions of a minimum mass fin for this application. Compare this with the case where the fin does not receive any irradiation on either side.

Solution:

Step 1 Since the moon is at a low temperature, we assume $\varepsilon_B = \alpha_B = 0.95$. Since the sun is at a high temperature, we assume $\varepsilon_A = 0.95,\ \alpha_A = 0.18$. Thus, $\varepsilon = \frac{\varepsilon_A+\varepsilon_B}{2} = \frac{0.95+0.95}{2} = 0.95$. The radiation absorbed on the two sides of the fin may be calculated as[3]

$$G_A = 1396\ \text{W/m}^2 \quad G_B = \sigma T_M^4 = 5.67 \times 10^{-8} \times 374^4 = 1109.4\ \text{W/m}^2$$

$$\alpha G = \frac{\alpha_A G_A + \alpha_B G_B}{2} = \frac{0.18 \times 1396 + 0.95 \times 1109.4}{2} = 652.6\ \text{W/m}^2$$

Environmental parameter is then obtained as

$$\theta_{\text{ref}}^4 = \frac{\alpha G}{\varepsilon \sigma T_b^4} = \frac{652.6}{0.95 \times 5.67 \times 10^{-8} \times 550^4} = 0.132$$

Step 2 Profile number and fin efficiency for the minimum mass profile are interpolated from Table 17.1 as $N_{RC} = 0.739,\ \eta_R = 0.494$ for an environmental parameter of 0.132.

Step 3 The heat to be transferred by the fin is $Q_{\text{net}} = 1\ \text{kW} = 1000\ rmW = Q_{\text{act}}$. Using the definition of η_R, we have,

$$\eta_R = \frac{Q_{\text{act}}}{Q_{\text{max}}} = \frac{Q_{\text{act}}}{2LW\varepsilon\sigma T_b^4}$$

or

$$L = \frac{Q_{\text{act}}}{2\eta_R W \varepsilon \sigma T_b^4} = \frac{1000}{2 \times 0.494 \times 1.5 \times 0.95 \times 5.67 \times 10^{-8} \times 550^4} = 0.137\ \text{m}$$

By the definition of N_{RC} as given in Eq. 17.20, we then have

$$t = \frac{\varepsilon \sigma T_b^3 L^2}{k N_{RC}} = \frac{0.95 \times 5.67 \times 10^{-8} \times 550^3 \times 0.137^2}{230 \times 0.739} = 9.89 \times 10^{-4}\ \text{m}$$

The fin thickness is double this value and is $2t = 2 \times 9.89 \times 10^{-4} =$ 1.98 mm.

Step 4 With no irradiation on either side, $G = 0$. In this case the optimum proportioned fin is given by $N_{RC} = 0.801$ and $\eta_R = 0.554$ as given in Table 17.1 with the environmental parameter equal to zero. L and t are then given by

[3] In the analysis we have ignored the contribution of solar radiation reflected from the moon surface to the irradiation on the moon facing side

$$L = \frac{1000}{2 \times 0.554 \times 1.5 \times 0.95 \times 5.67 \times 10^{-8} \times 550^4} = 0.122 \text{ m}$$

and

$$t = \frac{0.95 \times 5.67 \times 10^{-8} \times 550^3 \times 0.122^2}{230 \times 0.801} = 7.24 \times 10^{-4} \text{ m}$$

The fin thickness is thus equal to $2t = 2 \times 7.24 \times 10^{-4}$ m $= 1.45$ mm.

Step 5 The fin masses are in the ratio of the volumes, and hence we have

$$\frac{\text{Fin mass with irradiation}}{\text{Fin mass without irradiation}} = \frac{2 \times 0.137 \times 1.98}{2 \times 0.122 \times 1.45} = 1.53$$

Thus, the fin in the former case is some 53% heavier than that in the latter case.

17.2.5 *Radiating-Conducting-Convecting Fin With Linearized Radiation*

In some applications where temperature differences are small, it is possible to linearize radiation as indicated in Chap. 1. An example of this is found in the estimation of thermometric error in a thermometer well used primarily to measure the temperature of a fluid flowing in a duct. A thermometer well is normally used to avoid direct contact between a temperature sensor and the flowing fluid. Typically, an installation using a thermometer well is as shown schematically in Fig. 17.9. The thermometer well consists of an annular cylinder introduced perpendicular to the axis of a pipe

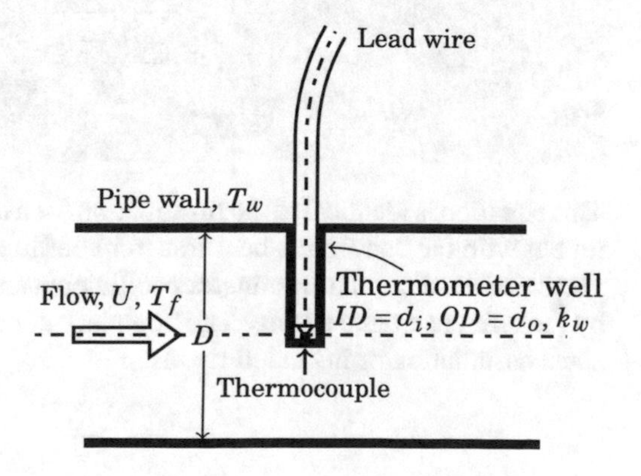

Fig. 17.9 Schematic of a thermometer well

carrying the fluid as indicated. The well material has a thermal conductivity k and has an emissivity of ε on its outside surface. A sensor is attached to the closed bottom of the well. Lead wires of small diameter take the electrical signal from the sensor to a voltmeter. The depth of the well L is referred to as the depth of immersion. The pipe wall temperature is assumed to be less than the fluid temperature, the temperature difference being such as to justify linear radiation assumption. It is seen that the sensor actually indicates the temperature T_t at the bottom of the well.

Let us assume that $T_f > T_t > T_w$. If the well is made of thin walled cylinder such that $L \gg d_0$ we may treat it as a uniform area fin that receives heat from the flowing fluid via convection, as in a cylinder in cross flow, and loses heat by radiation to the pipe wall. We assume that the flowing fluid is radiatively non-participating. The heat loss by radiation from an element of fin along its length is given by

$$q_R = \varepsilon\sigma(T^4 - T_w^4) \approx h_R(T - T_w)$$

where $h_R = 4\varepsilon\sigma T_w^3$ is the radiation heat transfer coefficient. Let the convective heat transfer coefficient for heat transfer from the fluid to the well be h. Either Eq. 17.18 may be manipulated to write the governing equation or it may derived from first principles by making energy balance for an elemental length of the well. The derivation is left as an exercise to the reader. The following equation should result

$$\frac{d^2T}{dx^2} + \frac{4h d_0}{k\left(d_0^2 - d_i^2\right)}(T_f - T) - \frac{4h_r d_0}{k\left(d_0^2 - d_i^2\right)}(T - T_w) = 0 \qquad (17.30)$$

We define a reference temperature given by

$$T_{\text{ref}} = \frac{hT_f + h_R T_w}{T_f + T_w} \qquad (17.31)$$

Note that this is the equilibrium temperature attained by the surface in the absence of fin conduction. Letting $\theta = T - T_{\text{ref}}$ Eq. 17.30 may be recast as

$$\boxed{\frac{d^2\theta}{dx^2} - \frac{4\left(h + h_R\right) d_0}{k\left(d_0^2 - d_i^2\right)}\theta = 0} \qquad (17.32)$$

This equation is identical to the fin equation for a uniform area conducting-convecting fin but with the convective heat transfer coefficient replaced by the sum of the convective and radiation heat transfer coefficients and the ambient temperature replaced by the reference temperature. The boundary conditions may be written down, by the usual assumption of insulated tip, as

$$x = 0, \ \theta = (T_w - T_{ref}); \quad x = L, \ \frac{d\theta}{dx} = 0 \tag{17.33}$$

The solution to Eq. 17.33 is straightforward and thus we have the sensor temperature given by

$$\boxed{\frac{T_t - T_{\text{ref}}}{T_w - T_{\text{ref}}} = \frac{1}{\cosh \mu}} \tag{17.34}$$

where $\mu = \sqrt{\frac{4(h+h_R)\text{d}_0}{k(\text{d}_0^2 - \text{d}_i^2)}}L$ is the non-dimensional fin parameter.

Example 17.5

Air at a temperature of $T_f = 373$ K is flowing in a tube of diameter $D = 10$ cm at an average velocity of $U = 0.5$ m/s. The tube walls are at a temperature of $T_w = 353$ K. A thermometer well of outer diameter $d_0 = 4$ mm and wall thickness $t = 1$ mm made of iron is immersed to a depth of $L = 5$ cm, perpendicular to the axis. The iron tube is dirty because of usage and has a surface emissivity of $\varepsilon = 0.85$. What will be the temperature indicated by a thermocouple that is attached to the bottom of the thermometer well? What is the consequence of ignoring radiation?

Solution:

Step 1 The inner diameter of well is calculated, using the data specified in the problem as, $\text{d}_i = \text{d}_0 - 2t = 4 - 2 \times 1 = 2$ mm.

Step 2 The air properties are taken at the air temperature without loss in accuracy. Thus

Kinematic viscosity of air:	$\nu = 23.02 \times 10^{-6}$ m^2/s
Thermal conductivity of air:	$k_f = 0.0313$ W/m°C
Prandtl number of air:	$Pr = 0.7$

The thermal conductivity of the well material is $k = 45$ W/m°C.

Step 3 Calculation of convective heat transfer coefficient:

The Reynolds number based on the outside diameter of the thermometer well is

$$Re_{\text{d}_0} = \frac{U\text{d}_0}{\nu} = \frac{0.5 \times 0.004}{23.02 \times 10^{-6}} = 86.9$$

Zhukaskas correlation is used now. For the range that brackets the above Reynolds number the constants in the Zhukaskas correlation are

$$C = 0.51,\ m = 0.5 \text{ and } n = 0.37$$

The Nusselt number based on the outside diameter of the thermometer well is then given by

$$Nu_{d_0} = C Re_{d_0}^m Pr^n = 0.51 \times 86.9^{0.5} \times 0.7^{0.37} = 4.17$$

Hence, the convective heat transfer coefficient between the air and the well is

$$h = \frac{Nu_{d_0} k_f}{d_0} = \frac{4.17 \times 0.0313}{0.004} = 32.6\ \text{W/m}^2\,°\text{C}$$

Step 4 Radiation heat transfer coefficient:
Linearized radiation is used since the fluid and wall temperatures are close to each other. The radiation heat transfer coefficient is given by

$$h_R = 4\varepsilon\sigma T_w^3 = 4 \times 0.85 \times 5.67 \times 10^{-8} \times 353^3 = 8.48\ \text{W/m}^2\,°\text{C}$$

Step 5 Reference temperature:
The reference temperature is given by (after Eq. 17.31)

$$T_{\text{ref}} = \frac{32.6 \times 373 + 8.48 \times 353}{32.6 + 8.48} = 368.9\ \text{K}$$

Step 6 Well treated as a fin:
The non-dimensional fin parameter is calculated as

$$\mu = \sqrt{\frac{4(32.6 + 8.48)0.004}{45\left(0.004^2 - 0.002^2\right)}} \times 0.05 = 1.744$$

The non-dimensional well bottom temperature is given by Eq. 17.34 as

$$\theta_t = \frac{1}{\cosh \mu} = \frac{1}{\cosh(1.744)} = 0.339$$

Hence, the well bottom temperature is

$$T_t = T_{\text{ref}} + \theta_t(T_w - T_{\text{ref}}) = 368.9 + 0.339(353 - 368.9) = 363.5\ \text{K}$$

This is the temperature indicated by the thermocouple. Hence, the thermometer error is $\delta T = 363.5 - 373 = -9.5$ K.

Step 7 If radiation is ignored, the reference temperature is the air temperature equal to $T_{\text{ref}} = 373$ K. The non-dimensional fin parameter is given by

$$\mu = \sqrt{\frac{4 \times 32.6 \times 0.004}{45(0.004^2 - 0.002^2)}} \times 0.05 = 1.553$$

Non-dimensional thermocouple indicated temperature is

$$\theta_t = \frac{1}{\cosh \mu} = \frac{1}{\cosh(1.553)} = 0.405$$

Hence, the thermocouple indicates a temperature of $T_t = T_{\text{ref}} + \theta_t(T_w - T_{\text{ref}}) = 373 + 0.405(353 - 373) = 364.9$ K. The thermometer error is $\delta T = 364.9 - 373 = -8.1$ K.

Step 8 In view of the above, radiation introduces an additional -1.4 K error in the measured temperature.

It is to be noted, in either case, that the thermometer well design needs to be improved to reduce the error.

17.3 Heat Transfer During Melting or Solidification

Transient conduction in a medium that undergoes phase change has important applications in material processing such as casting of metals, frost propagation in wet soils, in ice making, and so on. In recent times it has also been used in heat sinks for electronic cooling applications. Heat is absorbed or released during the phase change process in a region close to the phase change boundary. The rate of melting or freezing is, in fact, controlled by the rate at which the heat is brought in or removed from this region. These problems are inherently non-linear and pose a challenge since the position of the phase change front has to be determined as a part of the solution. The phase change front is also referred to as a *moving boundary*. We shall consider two simple cases referred to as Stefan and Neumann problems that involve melting or solidification in a semi-infinite medium. Subsequently, we shall look into phase change in a finite domain.

17.3.1 Stefan Problem

Consider the melting of a semi-infinite medium made of a pure substance initially at the phase change temperature T_m that is subject to a constant temperature $T_0 > T_m$ applied at its surface for $t > 0$. The state of affairs at time t is as indicated in Fig. 17.10. The governing differential equation for the region that has undergone melting is the familiar heat equation in one dimension given by

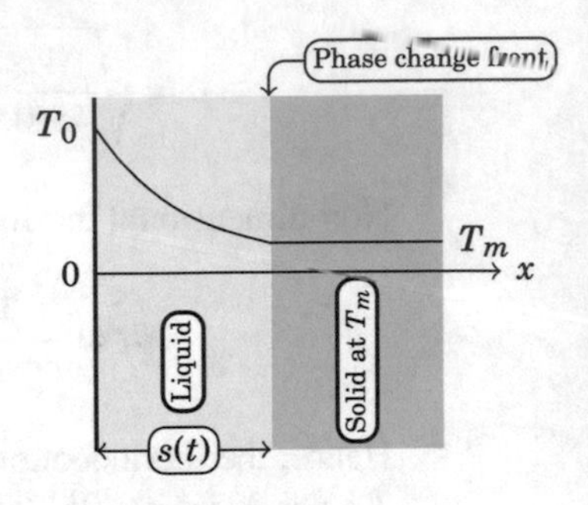

Fig. 17.10 Stefan problem for melting in semi-infinite space

$$\frac{\partial T}{\partial t} = \alpha_f \frac{\partial^2 T}{\partial x^2} \tag{17.35}$$

applicable for $0 \le x \le s(t)$ and $t > 0$. The thermal diffusivity α_f pertains to the liquid phase since heat propagation for change of phase at the phase change front is through the liquid phase. The temperature is uniform at the phase change temperature in the solid phase. The initial condition (IC) and boundary conditions (BC) that accompany Eq. 17.35 are

$$\text{IC: } T(x > 0, 0) = T_m \quad \text{BCs: } T(x = 0, t > 0) = T_0, \quad T(s, t > 0) = T_m$$

Another condition that is specified pertains to the rate at which the phase change front advances in to the solid. The rate at which the phase change front moves is such that the heat brought in through the liquid phase is able to supply the latent heat that is required to melt the solid. Thus, we have

$$\begin{bmatrix} \text{Rate heat conducted} \\ \text{in to the} \\ \text{phase change front} \end{bmatrix} = \begin{bmatrix} \text{Rate at which energy} \\ \text{is used in} \\ \text{melting the solid} \end{bmatrix}$$

This in equation form is

$$-k_f \left. \frac{\partial T}{\partial x} \right|_{x=s(t)} = \rho_s h_{sf} \frac{\mathrm{d}s}{\mathrm{d}t} \tag{17.36}$$

In the above, the thermal conductivity k_f is that of the liquid, the density ρ_s is that of the solid, and h_{sf} is the latent heat of melting in J/kg.

The Stefan problem can be solved analytically since a general solution to the heat equation is known to be given by

$$T(x, t) = A + B \operatorname{erf}\left[\frac{x}{2\sqrt{\alpha_f t}}\right] \tag{17.37}$$

Obviously, $A = T_0$ such that the boundary condition at the front surface is satisfied. In order for the boundary condition at the phase change front to be satisfied, B must be chosen as

$$B = \frac{T_m - T_0}{\text{erf}\left[\frac{s}{2\sqrt{\alpha_f t}}\right]} \tag{17.38}$$

We still have to determine $s(t)$ using the condition at the phase change front. The temperature gradient at the phase change front is obtained by differentiating expression (17.38) with respect to x and then putting $x = s(t)$. Recalling that the error function is given by $\text{erf}\,\eta = \frac{2}{\sqrt{\pi}} \int_0^{\eta} e^{-\eta^2} d\eta$, the indicated differentiation may be performed to get

$$\left.\frac{\partial T}{\partial x}\right|_{x=s} = -B \times \frac{2}{\sqrt{\pi}} \times e^{-\frac{s^2}{4\alpha_f t}} \times \frac{1}{2\sqrt{\alpha_f t}} \tag{17.39}$$

Using expressions (17.38) and (17.39), the condition at the moving boundary (17.36) may be simplified to read

$$\rho_s h_{sf} \frac{ds}{dt} = \frac{k_f}{2\sqrt{\alpha_f t}} e^{-\frac{s^2}{4\alpha_f t}} \frac{(T_m - T_0)}{\text{erf}\left[\frac{s}{2\sqrt{\alpha_f t}}\right]} \tag{17.40}$$

Fortunately, the above is satisfied for $s = m\sqrt{t}$ where m is a constant that depends only on the properties of the medium that is undergoing change of phase. If we substitute this in Eq. 17.40, time drops off! Thus, m is a solution of the algebraic equation given by

$$\rho_s h_{sf} \frac{m}{2} = \frac{k_f}{2\sqrt{\alpha_f}} e^{-\frac{m^2}{4\alpha_f}} \frac{(T_m - T_0)}{\text{erf}\left[\frac{m}{2\sqrt{\alpha_f}}\right]} \tag{17.41}$$

The position of the moving boundary is thus obtained. Further, we introduce the following non-dimensional quantities:

(1) $\theta = \frac{T - T_m}{T_0 - T_m}$ — Non-dimensional temperature

(2) $\eta = \frac{x}{2\sqrt{\alpha_f t}}$ — Similarity variable

(3) $\mu = \frac{m}{2\sqrt{\alpha_f}}$ — Non-dimensional phase change front

(4) $Ste = \frac{C_{pf}(T_0 - T_m)}{h_{sf}}$ — Stefan number (17.42)

Table 17.2 Variation with Stefan number of the phase change front

μ	Ste	μ	Ste	μ	Ste
0.01	0.0002	0.14	0.0397	0.6	0.9205
0.02	0.0008	0.16	0.0521	0.7	1.3727
0.03	0.0018	0.18	0.0662	0.8	1.9956
0.04	0.0032	0.2	0.0822	0.9	2.8576
0.05	0.00501	0.22	0.1	1	4.0602
0.06	0.00722	0.24	0.1197	1.2	8.1721
0.07	0.00983	0.26	0.1415	1.4	16.776
0.08	0.0129	0.28	0.1653	1.6	35.8174
0.09	0.0163	0.3	0.1912	1.8	80.5745
0.1	0.0201	0.4	0.3564	2	192.64
0.12	0.0291	0.5	0.5923		

Stefan number introduced above is a non-dimensional parameter that appears in phase change problems. Expression (17.41) may then be recast as

$$\boxed{Ste = \sqrt{\pi}\mu e^{\mu^2} \,\mathrm{erf}\, \mu} \tag{17.43}$$

The temperature profile within the liquid region then satisfies the equation

$$\theta = 1 - \frac{\mathrm{erf}\, \eta}{\mathrm{erf}\, \mu} \tag{17.44}$$

The easiest way of solving Eq. 17.43 is to obtain the value of Stefan number for a chosen value of μ. Table 17.2 shows the data obtained by the above method for a typical range of Stefan numbers. Stefan number is the ratio of sensible heat to latent heat. It is easily seen, on physical grounds, that the smaller the Stefan number, larger the latent heat in comparison with sensible heat and the slower is the process of melting.

Energy Utilization Rate in Melting

The rate at which energy is used in melting (per unit frontal area) is simply given by

$$q_m = \rho_s h_{sf} \frac{ds}{dt}$$

With $s = \mu\sqrt{2\alpha_f t}$, we get

$$q_m = \rho_s h_{sf} \mu \sqrt{2\alpha_f} \frac{1}{2\sqrt{t}} = \rho_s \frac{C_{pf}(T_0 - T_m)}{Ste} \mu \sqrt{\alpha_f} \frac{1}{\sqrt{t}}$$

the last part resulting from the definition of Stefan number. From the definition of the thermal diffusivity, we also have $\rho_f C_{pf} = \frac{k_f}{\alpha_f}$. Hence, we may rewrite the above expression as

$$\boxed{q_m = \left(\frac{\rho_s}{\rho_f}\right)\left(\frac{\mu}{Ste}\right)\left(\frac{k_f(T_0 - T_m)}{\sqrt{2\alpha_f t}}\right)} \tag{17.45}$$

Thus, the rate at which energy is consumed towards melting reduces inversely as the square root of time.

Problem of *solidification* in a semi-infinite region may be treated by a similar method. The property values will have to be appropriately changed to make the analysis valid. The reader is encouraged to figure out what the changes are!

17.3.2 Neumann Problem

The Neumann problem deals with the melting of a pure substance occupying the half space and initially at a temperature below the melting temperature. The surface of the solid is brought instantaneously to a temperature above the melting point and is held fixed at that value thereafter. The temperature variation will now be as shown in Fig. 17.11. The main difference between the Stefan problem and the Neumann problem is the prevalence of temperature variations in both liquid and solid phases. Also, at the phase change front, the heat conducted from the liquid side is shared between the heat consumed in melting and the heat conducted in to the solid phase. The properties of both the phases play a role in the problem. The governing equations may be written down for the two phases as given below.

$$\text{Liquid phase:} \quad \frac{\partial T}{\partial t} = \alpha_f \frac{\partial^2 T}{\partial x^2}; \quad 0 < x < s(t),\ t > 0 \tag{17.46}$$

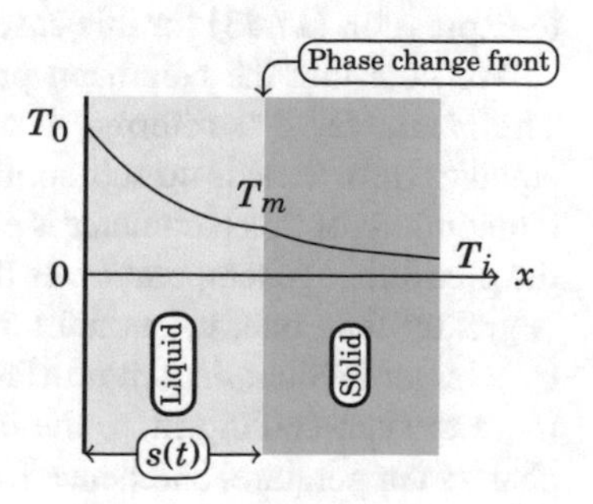

Fig. 17.11 Neumann problem for melting in semi-infinite space

$$\text{Solid phase:} \quad \frac{\partial T}{\partial t} = \alpha_s \frac{\partial^2 T}{\partial x^2}; \quad s(t) < x < \infty, \ t > 0 \tag{17.47}$$

$$\text{Interface condition:} \ -k_f \left.\frac{\partial T}{\partial x}\right|_{x=s(t)} = -k_s \left.\frac{\partial T}{\partial x}\right|_{x=s(t)} + \rho_s h_{sf} \frac{\mathrm{d}s}{\mathrm{d}t} \tag{17.48}$$

The boundary conditions are:

$$T(0, t > 0) = T_0, \quad T(s, t > 0) = T_m, \quad T(x \to \infty, t > 0) = T_i \tag{17.49}$$

Initial condition specifies the entire medium to be at $T = T_i$ at $t = 0$. The solution method is similar to the one used in the solution of the Stefan problem. We use solution of the form (17.37) for both the regions, taking care to use the appropriate material properties in the two phases. Four constants and the interface position as a function of time are to be determined by the application of the above conditions. The interface position is again taken to follow the form $s = m\sqrt{t}$ used in the Stefan problem. Apart from the Stefan number that has already been introduced earlier, the following two new parameters enter this problem:

$$\phi = \frac{T_m - T_i}{T_0 - T_m}, \quad \Gamma = \sqrt{\frac{k_f \rho_f C_{pf}}{k_s \rho_s C_{ps}}} \tag{17.50}$$

The steps leading to the interface position are similar to those in the case of the Stefan problem and will, therefore, be not repeated here. The final expression is obtained as

$$\boxed{\frac{Ste}{\mu\sqrt{\pi}} \left[\frac{1}{\mathrm{e}^{\mu^2} \operatorname{erf} \mu} - \frac{\phi \, \mathrm{e}^{-\mu^2 \left(\frac{\alpha_f}{\alpha_s}\right)}}{\Gamma \operatorname{erfc}\left(\mu \sqrt{\frac{\alpha_f}{\alpha_s}}\right)} \right] = 1} \tag{17.51}$$

We note that $\phi = 0$ corresponds to the Stefan problem and expression (17.51) reduces to expression (17.43) for this case.

We note that the Neumann problem is more general than the Stefan problem. The parameter ϕ is referred to as the sub-cooling parameter. If this parameter is equal to zero, there is no sub-cooling and the medium is initially at the phase change temperature. If this parameter is equal to unity the initial temperature is as far below the phase change temperature as the surface temperature is above it. If the parameter is greater than one, the initial temperature is below the phase change temperature by a larger amount than the surface temperature is above it. It is quite clear that the more the sub-cooling more the energy required to bring the medium to the phase change temperature, and hence lower the melting rate. The thermal diffusivity ratio

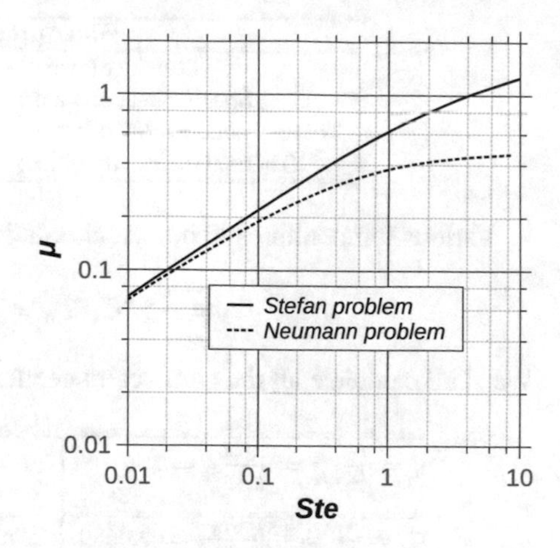

Fig. 17.12 Comparison of cases with and without sub-cooling

also plays an important role since the relative rates at which heat will diffuse in the liquid and solid phases depend on it. The physical parameter Γ also determines the melting rate, being a composite parameter that involves the conductivity, the density and the specific heat ratios for the two phases.

In order to appreciate the differences between the Stefan solution and the Neumann solution, we show in Fig. 17.12 the variation of μ with Ste. The Neumann solution shown is for $\phi = 1$, $\frac{\alpha_f}{\alpha_s} = 1$ and $\Gamma = 1$.

Example 17.6

A very large block of ice is initially at a temperature of $T_i = -5$ °C throughout. At $t = 0$ the surface of the ice block is brought to $T_0 = 5$ °C and held fixed at that temperature thereafter. Determine the time needed for a 2 cm layer of ice to melt.

Solution:
Properties of ice and water are taken from tables of properties and listed below: (We follow the notation used in the text in what follows)

Ice properties:	
Density of ice:	$\rho_s = 916.8$ kg/m^3
Specific heat of ice:	$C_{ps} = 4873$ J/kg K
Thermal conductivity of ice:	$k_s = 2.24$ W/m K
Thermal diffusivity of ice:	$\alpha_s = 5.01 \times 10^{-7}$ m^2/s
Latent heat of melting ice:	$h_{sf} = 3.336 \times 10^5$ J/kg

Water properties:	
Density of water:	$\rho_s = 1000$ kg/m^3
Specific heat of water:	$C_{pf} = 4206$ J/kg K
Thermal conductivity of water:	$k_f = 0.568$ W/m K
Thermal diffusivity of ice:	$\alpha_f = 51.535 \times 10^{-6}$ m^2/s

Various temperatures in our usual notation are

$$T_i = -5\,°\text{C},\ T_m = 0\,°\text{C},\ T_0 = 5\,°\text{C}$$

We shall calculate all the parameters needed in the Neumann solution.

$\phi = \frac{T_0 - T_m}{T_m - T_i} = \frac{5-0}{0-(-5)} = 1$	$\Gamma = \sqrt{\frac{0.568 \times 1000 \times 4206}{2.24 \times 916.8 \times 4873}} = 0.4886$
$\alpha_r = \frac{\alpha_f}{\alpha_s} = \frac{1.535 \times 10^{-6}}{5.01 \times 10^{-7}} = 3.0639$	$Ste = \frac{C_{pf}(T_0 - T_m)}{h_{sf}} = \frac{4206(5-0)}{3.336 \times 10^5} = 0.063$

We now calculate the value of μ using the Neumann solution given by Eq. 17.51. A trial and error solution is required and the correct value of $\mu = 0.1358$ is obtained.

The depth of melt is given as $s = 2$ cm or $s = 0.02$ m. Using the definition of μ, we then have the time to melt 2 cm layer of ice as

$$t = \frac{s^2}{4\alpha_f \mu^2} = \frac{0.02^2}{4 \times 1.535 \times 10^{-6} \times 0.1358^2} = 3532.6 \text{ s} \approx 59 \text{ min}$$

17.3.3 Phase Change in a Finite Domain

Consider a flat slab of a phase change material (PCM) of thickness initially in the solid phase at its melting temperature. We would like to know the time it takes to melt the material by supplying heat symmetrically from the two sides, as shown in Fig. 17.13.

Since the PCM is given to be at its phase change temperature the problem is treatable as a Stefan problem. The time at which $s = \frac{\delta}{2}$ is when the melting is complete and the amount of heat stored in the PCM in this time is the energy storage of the PCM. Let the time required for the above to be satisfied be t_c. From the Stefan

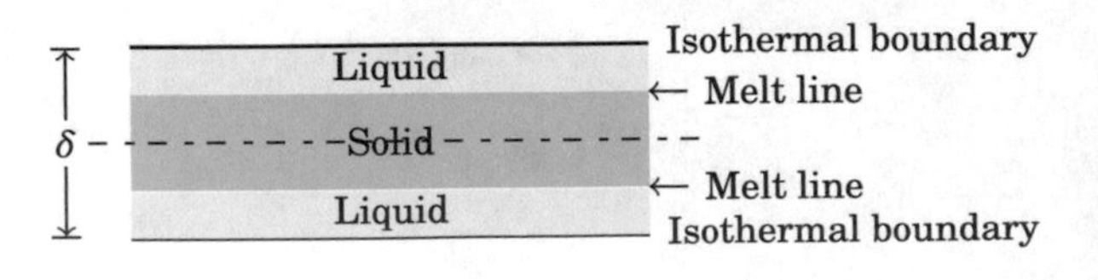

Fig. 17.13 Melting in a finite region

solution, we then have $\frac{\delta}{2} = 2\mu\sqrt{\alpha_f t_c}$ where μ satisfies Eq. 17.43. Thus, we have the charging time given by

$$t_c = \frac{\delta^2}{16\alpha_f \mu^2} \tag{17.52}$$

The total energy stored may be calculated as

$$E_c = -2\int_0^{t_c} k_f \left.\frac{\partial T}{\partial x}\right|_{x=0} \mathrm{d}t \tag{17.53}$$

Using Eq. 17.44, and the definition of θ presented in Eq. 17.42, we may obtain the derivative in the integrand of Eq. 17.53 as

$$\left.\frac{\partial T}{\partial x}\right|_{x=0} = -\frac{(T_0 - T_m)}{\text{erf}\ \mu}\left.\frac{\mathrm{d}\ \text{erf}\ \xi}{\mathrm{d}\xi}\right|_{\xi=0} = -\frac{(T_0 - T_m)}{\text{erf}\ \mu}\frac{1}{\sqrt{\pi\alpha_f t}}$$

Substitute this in Eq. 17.53, perform the indicated integration to get

$$E_c = \frac{(T_0 - T_m)}{\text{erf}\ \mu} \cdot \frac{k_f}{\sqrt{\pi\alpha_f}} \cdot 4\sqrt{t_c}$$

Use expression (17.52) to recast the above as

$$E_c = \frac{(T_0 - T_m)}{\text{erf}\ \mu} \cdot \frac{k_f}{\alpha_f} \cdot \frac{\delta}{\mu\sqrt{\pi}}$$

Noting that $\frac{k_f}{\alpha_f} = \rho_f C_{pf}$, the above may be written in the non-dimensional form

$$\boxed{\frac{E_c}{\rho_f C_{pf}(T_0 - T_m)} = \frac{1}{\sqrt{\pi}} \cdot \frac{\delta}{\mu\ \text{erf}\ \mu}} \tag{17.54}$$

Example 17.7

A 2 cm thick slab of ice is at 0 °C throughout. At $t = 0$ both surfaces of the ice slab are suddenly raised to a temperature of 10 °C and kept at that value thereafter. Determine the time taken for complete melting of the slab and the energy that has entered the slab in this time.

Solution:
We follow the notation used in the text. The properties of ice and water are taken as below

Ice properties:	
Density of ice:	$\rho_s = 916.8$ kg/m^3
Latent heat of melting ice:	$h_{sf} = 3.336 \times 10^5$ J/kg

Water properties:	
Density of water:	$\rho_s = 1000$ kg/m^3
Specific heat of water:	$C_{pf} = 4206$ J/kg K
Thermal diffusivity of ice:	$\alpha_f = 51.535 \times 10^{-6}$ m^2/s

The ice slab thickness is given to be $\delta = 2$ cm $= 0.02$ m. Superheat parameter is zero in this case since $T_i = T_m = 0$ °C. The Stefan solution is appropriate. The Stefan number is calculated as

$$Ste = \frac{C_{pf}(T_0 - T_m)}{h_{sf}} = \frac{4206(10-0)}{3.336 \times 10^5} = 0.126$$

Linear interpolation using the values in Table 17.2 gives

$$\mu = 0.24 + \frac{(0.126 - 0.1197)}{(0.1415 - 0.1197)}(0.26 - 0.24) = 0.246$$

The charge time is then obtained using Eq. 17.52 as

$$t_c = \frac{0.02^2}{16 \times 1.535 \times 10^{-6} \times 0.246^2} = 269.1 \text{ s}$$

The energy that has entered the ice slab in this time is given by Eq. 17.54 as

$$E_c = \frac{\rho_f C_{pf}(T_0 - T_m)}{\sqrt{\pi}} \cdot \frac{\delta}{\mu \operatorname{erf} \mu} = \frac{1000 \times 4206 \times (10-0)}{\sqrt{\pi}} \times \frac{0.02}{0.246 \times \operatorname{erf} 0.246}$$
$$= 7.148 \times 10^6 \text{ J/m}^2$$

17.4 Heat Transfer During Condensation

Heat transfer in condensers involve phase change wherein a vapor turns in to a liquid. Pure substances have a phase change temperature that is fixed once the pressure is fixed. For example, condensers used in power plants operate at a pressure of

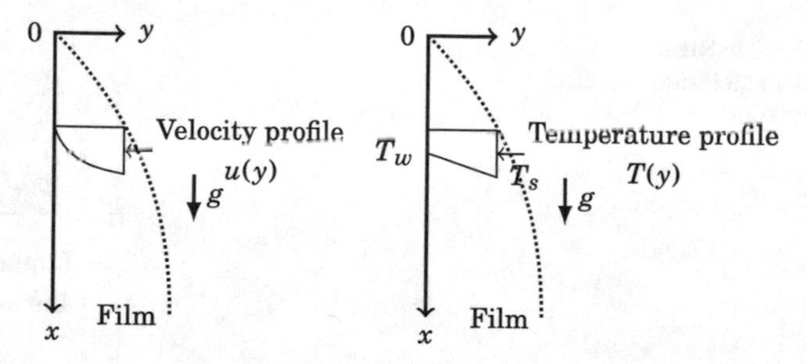

Fig. 17.14 Laminar film condensation over an isothermal vertical surface

~ 0.1 *bar* with the corresponding saturation temperature around 55 °C. Condensation will take place over a surface if it is at a temperature lower than the saturation temperature of the substance. Since there is a large change in density during condensation, the condensate is much heavier and flows or drains away from the surface, in the presence of gravity.

Drop wise condensation takes place if the condensate does not wet the surface. The droplets then run off the surface due to gravitational forces. Heat transfer rates are high since heat transfer takes place directly with the surface over which condensation takes place.

However, condensation may take place with a condensate film forming on the surface, when the condensate wets the surface. Since heat released during condensation has to pass through the film the heat transfer rates are generally lower in the case of film condensation. Film condensation over a vertical surface may be analyzed in a simple way, using the analysis due to Nusselt,[4] as presented below.

17.4.1 *Film Condensation Over An Isothermal Vertical Surface*

Condensation is assumed to take place over an isothermal vertical surface maintained at a temperature $T_w < T_s$, where T_s is the temperature of the vapor that is assumed to be saturated and at *rest*. A film of condensate of thickness $\delta(x)$ forms over the surface, as shown in Fig. 17.14. The flow within the film is assumed to be laminar. Evidently, the film thickness increases with x to account for the condensate that has formed over the length $0 - x$ to flow down.

Consider a fluid element of size $\mathrm{d}x \times \mathrm{d}y$ within the condensate film, located at some distance x from the top edge, as shown in Fig. 17.15. The dimension of this element may be taken as 1 unit in a direction perpendicular to the plane of the figure.

[4]W. Nusselt, Zeitschrift des Verieins deutscher Ingenieure, Vol. 60, No. 27 and 28, pp. 541–546 and pp. 569–575, 1916.

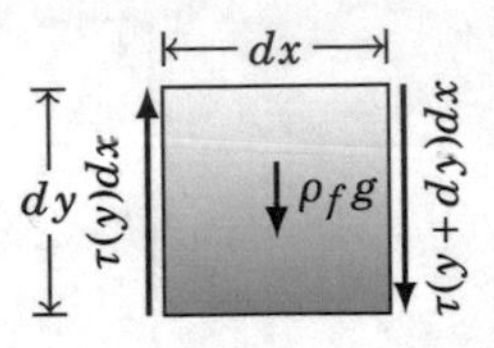

Fig. 17.15 Force balance over a fluid element inside the condensate film

Force balance over the element leads to (as usual, we use Taylor expansion and retain only first order changes)

$$\frac{\partial \tau}{\partial y} = -\rho_f g$$

where ρ_f is the density of condensate. We assume that the density of vapor is negligible in comparison with the density of the liquid. The inertia forces are assumed to be negligible since the velocities involved are very small. The force balance given above is similar to what was observed in the case of fully developed flow in a pipe where viscous forces were balanced by pressure forces. Assuming constant properties, we also have, by Newton's law of viscosity the relation

$$\tau = \mu_f \frac{\partial u}{\partial y}$$

where μ_f is the dynamic viscosity of the condensate. From the above two equations, we get

$$\mu_f \frac{\partial^2 u}{\partial y^2} = -\rho_f g \tag{17.55}$$

The boundary condition on the velocity is obviously the no slip condition at the wall at $y = 0$. At the edge of the condensate layer, the velocity gradient is zero if we *assume* that negligible shear is exerted by the vapor. Thus, the boundary conditions are taken as

$$\text{(a) } u = 0 \text{ at } y = 0 \text{ for all } x, \quad \text{(b) } \frac{\partial u}{\partial y} = 0 \text{ at } y = \delta \text{ for all } x \tag{17.56}$$

Equation 17.55 may be integrated twice with respect to y to get a quadratic velocity profile given by

$$u = A + By - \frac{\rho_f g}{2\mu_f} y^2 \tag{17.57}$$

where A and B are integration constants (these are actually functions of x), to be determined. Wall boundary condition requires that we choose $A = 0$. At the surface of the condensate, we have, using boundary condition (17.56)(b)

$$\left.\frac{\partial u}{\partial y}\right|_{y=\delta} = B - \frac{\rho_f g \delta}{\mu_f} = 0 \text{ or } B = \frac{\rho_f g \delta}{\mu_f}$$

The velocity profile will then read as

$$u = \frac{\rho_f g \delta^2}{\mu_f}\left(\frac{y}{\delta} - \frac{1}{2}\left\{\frac{y}{\delta}\right\}^2\right) = \frac{\rho_f g \delta^2}{\mu_f}\left(\eta - \frac{\eta^2}{2}\right) \tag{17.58}$$

where $\eta = \frac{y}{\delta}$. We may obtain the mean velocity by equating the mass flow rate per unit width $\dot{m} = \rho_f \int_0^\delta u(y)dy$ to $\rho_f U \delta$ where U is the mean velocity. We substitute the velocity profile given by Eq. 17.58, perform the indicated integration to get

$$U = \frac{\rho_f g \delta^2}{\mu_f}\int_0^1 \left(\eta - \frac{\eta^2}{2}\right) \mathrm{d}\eta = \frac{\rho_f g \delta^2}{3\mu_f} \tag{17.59}$$

Thus, the condensate mass flow rate at x is given by

$$\dot{m} = \frac{\rho_f^2 g}{3\mu_f}\delta^3 \tag{17.60}$$

The rate at which the condensate mass flow changes with x is given by

$$\frac{\mathrm{d}\dot{m}}{\mathrm{d}x} = \frac{\rho_f^2 g}{\mu_f}\delta^2\frac{\mathrm{d}\delta}{\mathrm{d}x}$$

The change in the mass flow rate of the condensate is due to the amount of condensation that takes place in an interval $\mathrm{d}x$ around x. Assume that the latent heat for conversion from vapor to liquid is h_{fg}. Then the amount of heat released must be

$$\frac{\mathrm{d}Q}{\mathrm{d}x} = h_{fg}\frac{\mathrm{d}\dot{m}}{\mathrm{d}x} = h_{fg}\frac{\rho_f^2 g}{\mu_f}\delta^2\frac{\mathrm{d}\delta}{\mathrm{d}x} \tag{17.61}$$

We assume that heat transfer across the condensate film is purely by conduction and the temperature variation across the condensate film is linear. Hence, we have

$$\frac{\mathrm{d}Q}{\mathrm{d}x} = q_k = k_f\frac{(T_s - T_w)}{\delta} \tag{17.62}$$

Equating the above two expressions, we get the following first order ordinary differential equation for δ.

$$\delta^3\mathrm{d}\delta = \frac{k_f(T_s - T_w)\mu_f}{\rho_f^2 g h_{fg}}\mathrm{d}x$$

On integration with respect to x and taking $\delta = 0$ at $x = 0$, we get

$$\delta = \left[\frac{4k_f\mu_f(T_s - T_w)x}{\rho_f^2 g h_{fg}}\right]^{\frac{1}{4}} \tag{17.63}$$

Based on Eq. 17.62, we may define a local heat transfer coefficient as

$$h_x = \frac{q_k}{(T_s - T_w)} = \frac{k_f}{\delta} = \left[\frac{k_f^3\rho_f^2 g h_{fg}}{4\mu_f(T_s - T_w)x}\right]^{\frac{1}{4}} \tag{17.64}$$

We may define a mean heat transfer coefficient over length L of the surface in the usual way.

$$\overline{h}_L = \frac{1}{L}\int_0^L h_x \mathrm{d}x$$

Using Eq. 17.64, we then have

$$\overline{h}_L = \frac{4}{3}h_L = \frac{4}{3}\left[\frac{k_f^3\rho_f^2 g h_{fg}}{4\mu_f(T_s - T_w)L}\right]^{\frac{1}{4}} = 0.943\left[\frac{k_f^3\rho_f^2 g h_{fg}}{\mu_f(T_s - T_w)L}\right]^{\frac{1}{4}} \tag{17.65}$$

Example 17.8

Steam at a saturation temperature of $T_s = 50$ °C is condensing over a vertical surface at $T_w = 40$ °C. The length of the surface is $L = 0.3$ m. Laminar film condensation takes place over the entire length of the surface. Determine the film thickness at $x = L$. What is the mean heat transfer coefficient over the length of the surface? Determine the rate at which steam condenses per unit width of the surface. Take acceleration due to gravity of $g = 9.81$ m/s^2. What is the maximum velocity of the condensate?

Solution:
The required properties are taken either from steam tables or table of properties of water.

Steam properties	
Saturation pressure:	$p_s = 0.083$ *bar* absolute
Enthalpy of vaporization:	$h_{fg} = 2.382 \times 10^6$ J/kg
Density of vapor:	$\rho_g = 0.083$ kg/m^3

Properties of water	
Density of water:	$\rho_f = 988$ kg/m^3
Thermal conductivity:	$k_f = 0.643$ W/m°C
Dynamic viscosity:	$\mu_f = 0.000544$ kg/m s

The film thickness at $x = L = 0.3$ m is obtained using Eq. 17.63 as

$$\delta_L = \left[\frac{4 \times 0.643 \times 0.000544(50 - 40) \times 0.3}{988^2 \times 9.81 \times 2.382 \times 10^6}\right]^{\frac{1}{4}} = 0.00012 \text{ m}$$

The mean heat transfer coefficient over the length of the surface may be determined using Eq. 17.65 as

$$\overline{h}_L = 0.943\left[\frac{0.643^3 \times 988^2 \times 9.81 \times 2.382 \times 10^6}{0.000544(50 - 40) \times 0.3}\right]^{\frac{1}{4}} = 7360.9 \text{ W/m}^2\,{}^\circ\text{C}$$

Rate of heat transfer over the length of surface (per unit width basis) is

$$Q_L = \bar{h}_L(T_s - T_w)L = 7360.9 \times (50 - 40) \times 0.3 = 22082.7 \text{ W/m}$$

The rate at which steam condenses is

$$\dot{m} = \frac{Q_L}{h_{fg}} = \frac{22082.7}{2.382 \times 10^6} = 0.0093 \text{ kg/s m}$$

The maximum velocity of the condensate occurs at $x = L$, $y = \delta_L$. From Eq. 17.58, by putting $x = L$ and $\eta = 1$ the maximum velocity is obtained as

$$u = \frac{\rho_f g \delta_L^2}{2\mu_f} = \frac{998 \times 9.81 \times 0.00012^2}{2 \times 0.000544} = 0.13 \text{ m/s}$$

Non-dimensional Form of Equations

Equating the heat transferred to the heat released by the condensing vapor, over length L, we have

$$\dot{m}h_{fg} = \overline{h}_L(T_s - T_w)L$$

where $\dot{m}$ is the mass of steam condensed over length L. This may be rearranged as

$$(T_s - T_w) = \frac{\dot{m}h_{fg}}{\overline{h}_L L}$$

Introduce this in Eq. 17.65 to get

$$\overline{h}_L = 0.943\left[\frac{k_f^3\rho_f^2 g\overline{h}_L}{\mu_f\dot{m}}\right]^{\frac{1}{4}}$$

This may be rearranged as

$$\frac{\overline{h}_L}{k_f}\underbrace{\left(\frac{\mu_f^2}{\rho_f^2 g}\right)^{\frac{1}{3}}}_{L_{ch}} = 0.943^{\frac{4}{3}}\left[\frac{\mu_f}{\dot{m}}\right]^{\frac{1}{3}} = 0.925\underbrace{\left[\frac{\dot{m}}{\mu_f}\right]^{-\frac{1}{3}}}_{Re^{-\frac{1}{3}}} \tag{17.66}$$

The quantity $\left(\frac{\mu_f^2}{\rho_f^2 g}\right)^{\frac{1}{3}}$ provides a characteristic length scale L_{ch} in the problem. Hence, the left-hand side of Eq. 17.66 may be interpreted as a Nusselt number $Nu_{L_{ch}}$ based on this characteristic length scale. The quantity $\frac{\dot{m}}{\mu_f}$ may be interpreted as a Reynolds number Re based on condensate mass flow rate per unit width. Thus, we may write the equation above in the alternate non-dimensional form

$$Nu_{L_{ch}} = 0.925Re^{-\frac{1}{3}} \tag{17.67}$$

The flow within the condensate film becomes turbulent if the Reynolds number is more than a critical value Re_c given by

$$Re_c = 256Pr^{-0.47} \tag{17.68}$$

where the Prandtl number lies between 1 and 10. There is an enhancement of heat transfer as compared to the laminar film case, with the Nusselt number being dependent on Re and Pr. For more details, the reader should refer to the appropriate literature.[5]

[5]K. Stephan, *Heat Transfer in Condensation and Boiling*, (Translated by C. V. Green), Springer Verlag, 1992.

17.4.2 Film Condensation Inside and Outside Tubes

Condensation outside vertical and horizontal tubes are of practical importance. In heat exchange equipment such as condensers it is usual to arrange condensation to take place over a bank of tubes arranged in such a manner that steam flow takes place through the spaces between the tubes. As long as the film thickness is small compared to the diameter of the tube, i.e., $\delta \ll D$ (see Example 17.8), the analysis given for film condensation over a vertical surface is also valid for condensation over inside or outside of vertical tubes, schematically shown in Fig. 17.16a, b.

In the case of horizontal tubes, the condensate film forms as shown schematically in Fig. 17.16c. The condensate formed over the tube runs off vertically due to gravity. One may imagine what will happen if a second tube were to be present below the one shown in the figure. The bottom tube will be inundated by the condensate running off the top tube, and hence we expect the second tube to be less efficient as compared to the top tube.

Nusselt also analyzed condensation over a horizontal tube and has shown that the heat transfer coefficient is given by

$$\overline{h}_h = 0.728\left[\frac{\rho_f^2 g h_{fg} k_f^3}{\mu_f (T_s - T_w) D}\right] \tag{17.69}$$

which may also be written in the alternate form

$$\boxed{\frac{\overline{h}_h}{k_f}\left(\frac{\mu_f^2}{\rho_f^2 g}\right)^{\frac{1}{3}} = 0.959\left(\frac{\dot{m}}{\mu_f}\right)^{-\frac{1}{3}}} \tag{17.70}$$

where $\dot{m}$ is the rate of condensation over a unit length of tube.

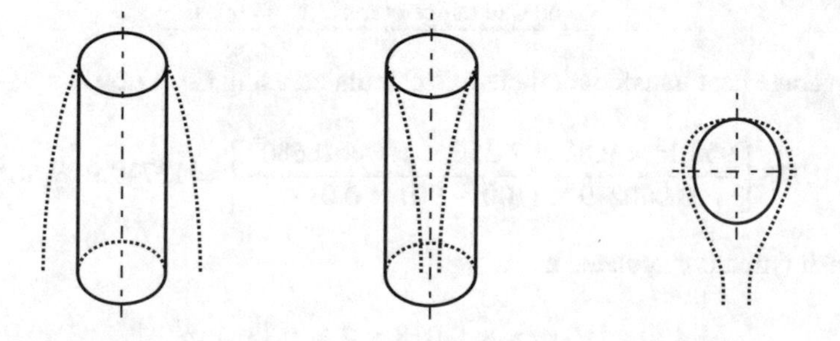

(a) Condensation outside vertical tube (b) Condensation inside vertical tube (c) Condensation outside horizontal tube

Fig. 17.16 Film condensation over tubes

When n tubes are present one above the other, the average heat transfer coefficient $\overline{h}_{h,n}$ for n tubes is given, according to Kern,[6] by

$$\overline{h}_{h,n} = \overline{h}_h n^{-\frac{1}{6}} \tag{17.71}$$

where $\overline{h}_h$ is the average heat transfer coefficient for a single horizontal tube given by Eq. 17.69.

Example 17.9

Steam at atmospheric pressure condenses over a horizontal tube of $D = 18$ mm maintained at a temperature of $T_w = 90$ °C over a length of $L = 2$ m. Determine the rate of condensation based on Nusselt analysis.

Solution:
The required properties are taken either from steam tables or table of properties of water.

Properties of steam	
Saturation pressure:	$p_s = 1.014$ *bar* absolute
Enthalpy of vaporization:	$h_{fg} = 2.256 \times 10^6$ J/kg
Properties of water	
Density of water:	$\rho_f = 958.1$ kg/m^3
Thermal conductivity of water:	$k_f = 0.680$ W/m°C
Dynamic viscosity of water:	$\mu_f = 0.000279$ kg/m s

Geometric parameters specified are

Diameter of tube:	$D = 18$ mm $= 0.018$ m
Tube length:	$L = 2$ m

The temperatures are specified as

Saturation temperature of steam:	$T_s = 100$ °C
Tube wall temperature:	$T_w = 90$ °C

The average heat transfer coefficient is calculated using Eq. 17.69 as

$$\overline{h}_h = 0.728 \left[\frac{958.1^2 \times 9.81 \times 2.256 \times 10^6 \times 0.680^3}{0.000279 \times (100 - 90) \times 0.018} \right] = 13749.9 \text{ W/m}^2\,{}^\circ\text{C}$$

Heat transfer area is calculated as

$$A = \pi D L = \pi \times 0.018 \times 2 = 0.1131 \text{ m}^2$$

[6] D. Q. Kern, AIChE Journal, Vol. 4, No. 2, pp. 157–160, 1958.

The total heat transfer at the tube wall is calculated as

$$Q_w = \overline{h}_h A(T_s - T_w) = 13749.9 \times 0.1131 \times (100 - 90) = 15550.8 \text{ W}$$

This should equal the heat released by condensing steam. Hence, the condensate mass flow rate is

$$\dot{m} = \frac{Q_w}{h_{fg}} = \frac{15550.8}{2.256 \times 10^6} = 0.00689 \text{ kg/s} = 24.8 \text{ kg/h}$$

17.4.3 *Condensation in the Presence of Flowing Vapor*

In most engineering applications the vapor will pass through the heat transfer equipment with a non-zero flow velocity. When the vapor flows downward over a surface over which condensation is taking place, it induces flow of the condensate in the downward direction because of shear force at the interface. In view of this, one expects the condensation rate to improve with respect to that considered previously with stagnant vapor.

Laminar Film Condensation

When the condensate velocity is small, laminar film condensation takes place and the Nusselt analysis may easily be extended to take account of shear at the interface. The velocity profile in the condensate is parabolic and is given by Eq. 17.57. The wall boundary condition remains unchanged while the boundary condition at the surface of the condensate film is changed to

$$\mu_f \left.\frac{\partial u}{\partial y}\right|_{y=\delta} = \tau_\delta \tag{17.72}$$

where τ_δ is the shear stress imposed on the condensate film due to vapor flow. On the vapor side we have a balance between pressure drop and viscous shear such that

$$\frac{\mathrm{d}p}{\mathrm{d}x}\frac{\pi D^2}{4} = \tau_\delta \pi D \quad \text{or} \quad \frac{\mathrm{d}p}{\mathrm{d}x}\frac{D}{4} = \tau_\delta$$

where D is the tube diameter and condensation is assumed to take place on the inside surface. The condensate film thickness is assumed to be negligibly small in comparison with the tube diameter. In terms of the friction factor, we also have

$$\frac{\mathrm{d}p}{\mathrm{d}x} = f\frac{\rho_g U_g^2}{2D}$$

where U_g is the mean velocity of vapor. From these two equations, we have

$$\tau_\delta = f\frac{\rho_g U_g^2}{8} \tag{17.73}$$

The velocity profile in the condensate layer is obtained by using the second boundary condition given by Eq. 17.72 as

$$u = \frac{\rho_f g \delta^2}{\mu_f}\left(\eta - \frac{\eta^2}{2}\right) + \frac{\tau_\delta y}{\mu_f} \tag{17.74}$$

The mean velocity of condensate may now be obtained as

$$U = \int_0^1 \left[\frac{\rho_f g \delta^2}{\mu_f}\left(\eta - \frac{\eta^2}{2}\right) + \frac{\tau_\delta \delta}{\mu_f}\eta\right] d\eta = \frac{\rho_f g \delta^2}{3\mu_f} + \frac{\tau_\delta \delta}{2\mu_f} \tag{17.75}$$

We may follow the steps leading to the results for condensation with stationary vapor to get the following results in this case with flowing vapor:

$$\boxed{\delta^4 + \frac{4}{3}\frac{\tau_\delta \delta^3}{\rho_f g} = \frac{4k_f \mu_f (T_s - T_w)x}{\rho_f^2 g h_{fg}}} \tag{17.76}$$

Once δ is obtained by solving the above equation, the heat transfer is calculated as $q_k = \frac{k_f(T_s - T_w)}{\delta}$.

Example 17.10

Saturated steam at 1 atmosphere pressure passes through a smooth copper tube of inside diameter $D = 18$ mm with a mean velocity of $U_g = 20$ m/s. The copper tube is maintained at a temperature of $T_w = 90$ °C. The tube length is $L = 0.5$ m. What is the condensate mass flow rate at $x = L$ if film condensation takes place at the tube surface. Take in to account the shear exerted by the flow on the condensate film. Express the results in suitable non-dimensional form. What will be your answers in case the vapor is stationary.

Solution:

Step 1 The required properties are taken either from steam tables or table of properties of water. From steam tables we have

Properties of steam	
Saturation tempe:	$T_s = 100\ °C$
Enthalpy of vaporization:	$h_{fg} = 2.257 \times 10^6$ J/kg
Density of vapor:	$\rho_g = 0.5982$ kg/m^3
Viscosity of steam:	$\mu_g = 1.2 \times 10^{-5}$ kg/m s
Properties of water	
Density of water:	$\rho_f = 958.1$ kg/m^3
Thermal conductivity of water:	$k_f = 0.680$ W/m°C
Dynamic viscosity of water:	$\mu_f = 0.000279$ kg/m s
Prandtl number of water:	$Pr_f = 1.73$

Step 2 With steam velocity of $U_g = 20$ m/s the Reynolds number of steam may be calculated as

$$Re_g = \frac{\rho_g D U_g}{\mu_g} = \frac{0.5982 \times 0.018 \times 20}{1.2 \times 10^{-5}} = 17946$$

The steam flow is turbulent.

Step 3 The friction factor is calculated based on Eq. 14.24(a) as

$$f = 0.316 \times 17946^{-0.25} = 0.0273$$

The shear stress imposed by steam flow on the condensate may then be obtained as

$$\tau_\delta = \frac{f \rho_g U_g^2}{8} = \frac{0.0273 \times 0.5982 \times 20^2}{8} = 0.817 \text{ Pa}$$

Step 4 With $x = L = 0.5$ m the equation for δ given by Eq. 17.76 takes the form

$$\delta_L^4 + 0.000111 8 \delta_L^3 = 3.7632 \times 10^{-16}$$

This equation is easily solved to get $\delta_L = 9.59 \times 10^{-5}$ m.

Step 5 The mean flow velocity of the condensate is calculated using Eq. 17.75 as

$$U_l = \frac{958.1 \times 9.81 \times (9.59 \times 10^{-5})^2}{3 \times 2.79 \times 10^{-4}} + \frac{0.817 \times 9.59 \times 10^{-5}}{2 \times 2.79 \times 10^{-4}} = 0.2437 \text{ m/s}$$

The condensate mass flow rate at $x = L$ may then be calculated as

$$\dot{m}_L = \rho_f \pi D \delta_L U_l = 958.1 \times \pi \times 0.018 \times 9.59 \times 10^{-5} \times 0.2427 = 0.00127 \text{ kg/s}$$

Condensate Reynolds number is obtained as

$$Re_l = \frac{\dot{m}_L}{\pi D \mu_f} = \frac{0.00127}{\pi \times 0.018 \times 2.79 \times 10^{-4}} = 80.5$$

Since the condensate Reynolds number is less than $256 \times 1.73^{-0.47} = 197.9$, the condensate flow is laminar. Hence, the use of the quantities derived earlier are appropriate.

Step 6 Heat removed at the wall between $x = 0$ and $x = L$ is then given by

$$Q_L = \dot{m}_L h_{fg} = 0.00127 \times 2.257 \times 10^6 = 2866.4 \text{ W}$$

The mean heat transfer coefficient between $x = 0$ and $x = L$ is then obtained as

$$\overline{h}_L = \frac{Q_L}{\pi D L (T_s - T_w)} = \frac{2866.4}{\pi \times 0.018 \times 0.5(100 - 90)} = 10137.8 \text{ W/m}^2\,°\text{C}$$

Step 7 Now we shall introduce the appropriate non-dimensional variables to represent the results. Characteristic length is given by

$$L_{ch} = \left(\frac{\mu_f^2}{g \rho_f^2}\right)^{\frac{1}{3}} = \left(\frac{(2.79 \times 10^{-4})^2}{9.81 \times 958.1^2}\right)^{\frac{1}{3}} = 2.0523 \times 10^{-5} \text{ m}$$

The average Nusselt number is then calculated as

$$\overline{Nu_L} = \frac{\bar{h}_L L_{ch}}{k_f} = \frac{10137.8 \times 2.0523 \times 10^{-5}}{0.68} = 0.306$$

Non-dimensional shear stress is given by

$$\tau_\delta^* = \frac{\tau_\delta}{\rho_f L_{ch} g} = \frac{0.817}{958.1 \times 2.0523 \times 10^{-5} \times 9.81} = 4.24$$

Thus, the results are depicted as

$$Re_l = 80.5, \ \tau_\delta^* = 4.24, \ \overline{Nu_L} = 0.306$$

Step 8 The reader may calculate the results when the vapor is stationary to get

$$Re_l = 61.6, \ \tau_\delta^* = 0, \ \overline{Nu_L} = 0.234$$

Step 9 Thus, heat transfer is improved by a factor of $\frac{0.306}{0.234} = 1.31$ or some 31% when the vapor is flowing as compared to the stationary vapor case.

Turbulent Film Condensation

When the film Reynolds number is more than the critical value, the flow within the condensate film becomes turbulent. The temperature profile within the film becomes non-linear showing the effect of turbulent fluctuations. The analysis is more complex than in the case of laminar film condensation. Turbulent quantities and their variation across the condensate layer are required to perform the analysis. Leaving out details, we present below a few useful formulae.

In a vertical tube, the condensate film becomes turbulent when the condensate Reynolds number exceeds the critical value. The condensate flow is partly laminar and partly turbulent over the tube height. In the turbulent part of the condensate flow, the results are well correlated by the following relation:

$$\boxed{\overline{Nu}_{L_{ch}} = 0.065(Pr_f \tau_\delta^*)^{\frac{1}{2}}} \tag{17.77}$$

over the following range of parameters:

$$2 \leq Pr_f \leq 3, \quad \text{and} \quad 5 \leq \tau_\delta^* \leq 50 \tag{17.78}$$

The shear stress is calculated based on relations presented earlier. The vapor flow reduces from entry to exit because of condensation. This may be taken into account by basing the mean shear stress on an effective mass flux given by

$$\dot{m}_v = \sqrt{\frac{\dot{m}_{vi}^2 + \dot{m}_{vi}\dot{m}_{ve} + \dot{m}_{ve}^2}{3}} \tag{17.79}$$

where $\dot{m}_{vi}$ and $\dot{m}_{ve}$ are the mass fluxes (i.e., mass flow rate divided by the tube area), respectively, at tube entry and tube exit.

A correlation for local Nusselt number in the form

$$\boxed{Nu_{L_{ch}} = ARe_l^a Pr_f^b (1 + B\tau_\delta^{*\,c})} \tag{17.80}$$

has been proposed. The values of the various constants appearing in Eq. 17.80 are given in Table 17.3. The correlation is valid for non-metallic fluids with $Pr_f \geq 0.1$. The local Nusselt number for laminar and turbulent flows can be obtained by taking

Table 17.3 Constants in Eq. 17.80

τ_δ^*	A	B	a	b	c
0	8.663×10^{-3}	0	0.382	0.569	0
$0 \leq \tau_\delta^* \leq 5$	8.663×10^{-3}	0.145	0.382	0.569	0.541
$5 \leq \tau_\delta^* \leq 10$	2.700×10^{-2}	0.407	0.207	0.500	0.420
$10 \leq \tau_\delta^* \leq 40$	4.294×10^{-2}	0.647	0.096	0.458	0.473

$$Nu_{L_{\text{ch}}} = \left[(1.15 Nu_{L_{\text{ch,lam}}})^4 + Nu_{L_{\text{ch,turb}}}^4\right]^{\frac{1}{4}} \tag{17.81}$$

where $Nu_{L_{\text{ch,lam}}}$ is the Nusselt number in laminar condensate flow and $Nu_{L_{\text{ch,turb}}}$ is the Nusselt number in turbulent condensate flow.

Condensation Over Horizontal Tubes
In many engineering applications, condensation over horizontal cylinders are important. Different flow regimes are possible depending on the the vapor velocity. Many a time finned surfaces are used to augment condensation heat transfer over horizontal tubes. These are beyond the scope of the present discussion. The reader may refer to the appropriate literature for more details regarding these aspects.

17.5 Heat Transfer During Boiling

Change of phase of a liquid to vapor is used in many engineering applications such as in steam power plants, refrigeration systems, and many others. When a liquid is adjacent to a surface at a temperature T_w greater than the saturation temperature T_s, boiling may be expected to take place if the temperature difference $(T_w - T_s)$ is sufficiently large. Boiling may take place by "pool boiling" in a liquid that is taken in a vessel or it may take place in a tube by "flow boiling" in which case the liquid flows within the tube with a specified mass flow rate. The latter is important in most practical applications.

17.5.1 Pool Boiling

Pool boiling experiment is usually conducted by heating a thin tungsten wire completely surrounded by a liquid. The power dissipated may be controlled by controlling the current passing through the wire. Heat transfer from the wire to the liquid is determined by the temperature difference between the wire and the surrounding liquid. When this temperature difference is small heat transfer is by natural convection and there is no change of phase. This corresponds to region I in Fig. 17.17. Note that the

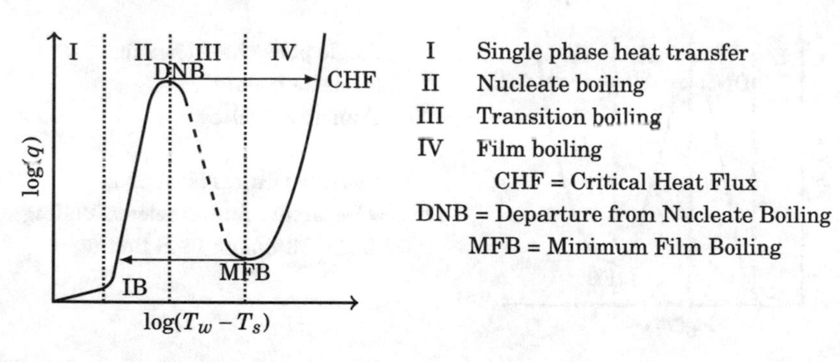

Fig. 17.17 Pool boiling curve with heat flux control

boiling curve uses log – log representation of the data. A straight line on this curve will correspond to a power dependence of heat flux on the temperature difference. As the heat flux is increased, the temperature difference between the wire and the liquid also increases. Boiling requires the wire temperature to be larger than the saturation temperature of the surrounding liquid at its pressure. The temperature difference $(T_w - T_s)$ represents wall superheat. Boiling starts at the point of incipient boiling or IB indicated in the figure. Boiling appears in the form of bubbles of vapor forming on the surface, leaving the surface because of buoyancy forces and moving away from the surface. Formation of vapor bubble is referred to as nucleation and is dependent on the nature of the surface and surface tension of the fluid. The vapor bubble swells in size such that the pressure difference across the bubble balances the forces due to surface tension. Note that the pressure outside the bubble is the saturation pressure, while the pressure inside the vapor bubble is determined by vapor temperature, and hence the wall superheat. The rate at which the bubbles are formed and move away from the surface increases with the surface heat flux, and hence the temperature difference. Heat transfer rates are very high and vary along the line shown in region II. The maximum heat flux that can be sustained is called the critical heat flux (CHF) and is indicated by the peak at the end of region II. The corresponding point is also indicated as point of departure from nucleate boiling or DNB. When the heat flux and hence the temperature difference is increased beyond the critical value, a film of vapor forms on the surface and the boiling curve shifts along the horizontal line from the critical value and moves on to the film boiling line shown as region IV. The dashed line indicated in the figure is inaccessible. If the heat flux is increased further, the temperature difference increases and eventually the wire will melt. The boiling curve was first presented by Nukiyama in 1934.[7]

If we now start cooling the wire by reducing the input power, film boiling proceeds till it reaches the bottommost point in region IV denoted by minimum film boiling point or MFB. On cooling further, it then reduces to the bottom of the region II along the horizontal line, again not going through the inaccessible part of the curve shown by the dashed line.

[7] S. Nukiyama, "Film boiling water on thin wires", Soc. Mech. Engg., Japan 37, 1934

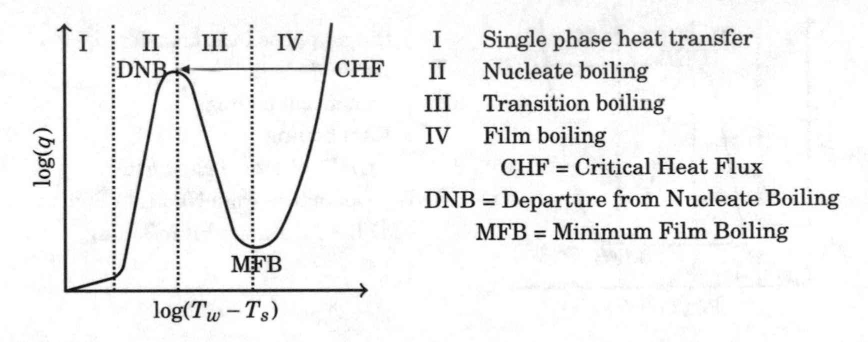

Fig. 17.18 Pool boiling curve with temperature control

The four regimes of heat transfer occur under the following conditions:

- (I) Natural convection for $T_w < T_{IB}$
- (II) Nucleate boiling for $T_{IB} < T_w < T_{DNB}$
- (III) Transition boiling for $T_{DNB} < T_W < T_{MFB}$
- (IV) Film boiling for $T_{MFB} < T_w$

As opposed to the above, the entire boiling curve is accessible when the temperature of the heater is controlled rather than the heat flux. Temperature control is possible by the use of change of phase within a tube immersed in a liquid. On the tube side we may use condensing steam whose temperature may be varied by varying the pressure. The pool boiling curve appears as shown in Fig. 17.18.

Incipient boiling requires that the vapor is formed with the pressure within the bubble large enough to support the diameter of the vapor bubble. Surface tension of saturated water provides the required force to support the bubble. The superheat required depends on the bubble radius r_b and is given by

$$\Delta T = (T_w - T_s) = \frac{2\sigma T_s}{r_b h_{fg} \rho_g} \tag{17.82}$$

where σ is the surface tension of saturated water. For example, if we assume that the vapor bubble diameter is $D_b = 5 \times 10^{-6}$ m, water is the liquid at 1 atmosphere pressure, surface tension $\sigma = 0.059$ N/m, the superheat required is calculated as

$$\Delta T = \frac{2 \times 0.059 \times 373}{5 \times 10^{-6} \times 2.2565 \times 10^6 \times 0.5981} = 6.5 \text{ K}$$

where the saturated steam properties are taken from steam tables. Experimentally observed values are in good agreement with this value.

17.5.2 Some Useful Relations in Pool Boiling

Critical Heat Flux

In this section, we present a few relations that are useful in describing pool boiling heat transfer. The critical heat flux[8] q_{cr} is given by the relation

$$q_{cr} = 0.18\rho_g h_{fg}\left[\frac{\sigma g(\rho_f - \rho_g)}{\rho_g^2}\right]^{\frac{1}{4}} \approx 0.18\rho_g h_{fg}\left[\frac{\sigma g\rho_f}{\rho_g^2}\right]^{\frac{1}{4}} \tag{17.83}$$

where all the symbols have the usual meanings. This relation assumes that the vapor density is negligibly small compared to the liquid density, an approximation that is often made in condensation and boiling heat transfer studies. The constant 0.18 appearing in Eq. 17.83 depends on the shape of the surface and varies from about 0.12–0.2. The value of constant is determined by the dimensionless parameter

$$L' = L\left[\frac{g(\rho_f - \rho_g)}{\sigma}\right]^{\frac{1}{2}} \approx L\left[\frac{g\rho_f}{\sigma}\right]^{\frac{1}{2}} \tag{17.84}$$

where L is the length of side for a square plate and is radius R for a circular plate. For $L' \geq 2.7$, Lienhard and Dhir[9] propose the relation

$$q_{cr} = 0.149\rho_g^{\frac{1}{2}} h_{fg}\left[\sigma g(\rho_f - \rho_g)\right]^{\frac{1}{4}} \tag{17.85}$$

For horizontal cylinders, Sun and Lienhard[10] give the relation

$$q_{cr} = f(L')\frac{\pi\rho_g^{\frac{1}{2}} h_{fg}}{24}\left[\sigma g(\rho_f - \rho_g)\right]^{\frac{1}{4}} \tag{17.86}$$

with

$$f(L') = 0.89 + 2.27\mathrm{e}^{-3.44\sqrt{L'}} \tag{17.87}$$

where L is taken as the radius of the cylinder.

Heat Transfer Correlation in Nucleate Boiling

Rohsenow[11] gives the following relationship between superheat and heat flux q_{nb} in nucleate boiling:

---

[8] N. Zuber, AEC Report, AECV-4439, 1959.

[9] J. H. Lienhard and V. K. Dhir, ASME J. Heat Transfer, Vol. 95, pp. 152–158, 1973.

[10] K. H. Sun and J. H. Lienhard, Int. J Heat and Mass Transfer, Vol. 13, pp. 1425–1429, 1970.

[11] W. M. Rohsenow, ASME J. Heat Transfer, Vol. 74, pp. 969–976, 1952.

$$\Delta T = \frac{Ch_{fg}}{C_{pf}}\left[\left(\frac{q_{nb}}{\mu_f h_{fg}}\right)\left(\frac{\sigma}{g(\rho_f - \rho_g)}\right)^{\frac{1}{2}}\right]^{0.33} Pr_f^{m+1} \tag{17.88}$$

where constant C depends on the fluid surface combination and $m = 0.7$ for fluids other than water and $m = 0$ for water. For example, for water boiling over a polished copper surface, the value of constant C is 0.0128. Note that the above relation may be recast in the alternate non-dimensional form

$$\boxed{Nu_{nb} = \frac{1}{C} Re^{1-n} Pr_f^{-m}} \tag{17.89}$$

where

$$Nu_{nb} = \frac{h_{nb}}{k_f}\left[\frac{\sigma}{g(\rho_f - \rho_g)}\right]^{\frac{1}{2}}, \quad Re = \frac{q_{nb}}{\mu_f h_{fg}}\left[\frac{\sigma}{g(\rho_f - \rho_g)}\right]^{\frac{1}{2}} \frac{\rho_f}{\mu_f} \tag{17.90}$$

and where h_{nb} is the heat transfer coefficient in nucleate boiling region.

Minimum Heat Flux in Film Boiling

The minimum heat flux that occurs at MFB is given by the relation

$$\boxed{q_{MFB} = \frac{q_{cr}}{2\sqrt{\frac{\rho_f+\rho_g}{\rho_g}}}} \tag{17.91}$$

Heat Transfer in Film Boiling

For large horizontal plates, heat transfer coefficient in stable film boiling regime is given by

$$\boxed{h_f = 0.425\left[\frac{k_g^3 \rho_g(\rho_f - \rho_g) g h_e}{\mu_g \Delta T\left(\frac{L_c}{2\pi}\right)}\right]^{\frac{1}{4}}} \tag{17.92}$$

where $L_c = 2\pi\sqrt{\frac{\sigma}{g(\rho_f-\rho_g)}}$ and $h_e = h_{fg} + 0.4C_{pg}\Delta T$ which takes into account the enthalpy of steam due to superheat (C_{pg} is the specific heat of superheated steam). However, for horizontal tubes of diameter D, the heat transfer coefficient in film boiling is given by the relation[12]

[12]L. A. Bromley, Chem. Engr. Progress, Vol. 46, pp. 221–227, 1950.

$$h_f = \left(0.59 + 0.069\frac{L_c}{D}\right)\left[\frac{k_g^3 \rho_g(\rho_f - \rho_g) g h'_e}{\mu_g \Delta T L_c}\right]^{\frac{1}{4}} \tag{17.93}$$

but where

$$h'_e = h_{fg} + 0.34 C_{pg} \Delta T$$

In the film boiling regime, radiation is also important and hence the total heat transfer coefficient is given by $h_{ft} = h_f + \frac{3}{4} h_r$ where h_r is the radiation heat transfer coefficient calculated by assuming the liquid to be a black surface at saturation temperature that surrounds the tube.

Example 17.11

Calculate the salient features in the boiling curve when water at a pressure of 1 atmosphere absolute boils over a polished copper surface. Make use of the various correlations presented above.

Solution:

Step 1 Properties of saturated water and saturated steam at $T_s = 100$ °C are

Saturated water at $T_s = 100$ °C	
Density:	$\rho_f = 958.4$ kg/m^3
Thermal conductivity:	$k_f = 0.680$ W/m°C
Dynamic viscosity:	$\mu_f = 2.79 \times 10^{-4}$ kg/m s
Specific heat:	$C_{pf} = 4216$ J/kg°C
Prandtl number:	$Pr_f = 1.73$
Surface tension:	$\sigma = 0.059$ N/m
Saturated steam at $T_s = 100$ °C	
Density:	$\rho_g = 0.5981$ kg/m^3
Thermal conductivity:	$k_g = 0.0248$ W/m°C
Latent heat of vaporization:	$h_{fg} = 2.257 \times 10^6$ J/kg
Specific heat:	$C_{pg} = 2020$ J/kg°C
Dynamic viscosity:	$\mu_g = 1.2 \times 10^{-5}$ kg/m s

Step 2 The critical heat flux is calculated using Eq. 17.83 as

$$q_{cr} = 0.18 \times 0.5981 \times 2.257 \times 10^6 \left[\frac{0.059 \times 9.81(958.4 - 0.5981)}{0.5981^2}\right]^{\frac{1}{4}}$$

$$= 1.524 \times 10^6 \text{ W/m}^2$$

Step 3 Using Rohsenow correlation (17.88) we calculate the wall superheat parameter under DNB as

$$\Delta T_{DNB} = \frac{0.0182 \times 2.257 \times 10^6}{4216}\left(\frac{1.524 \times 10^6}{2.79 \times 10^{-4} \times 2.257 \times 10^6}\right)^{0.33} \times \left(\frac{0.059}{9.81(958.4 - 0.5981)}\right)^{0.165} \times 1.73 = 17.66\,^\circ\text{C}$$

where the constant C has been chosen as 0.0182 for water boiling over polished copper.

Step 4 Assuming that nucleate boiling initiates when $\Delta T = 6.5$ K, the corresponding heat flux is given by

$$q_{IB} = q_{cr}\left(\frac{\Delta T_{IB}}{\Delta T_{DNB}}\right)^{\frac{1}{0.33}} = 1.524 \times 10^6\left(\frac{6.5}{\Delta 17.66}\right)^{\frac{1}{0.33}} = 73722\ \text{W/m}^2$$

Step 5 The minimum heat flux under film boiling condition is calculated based on Eq. 17.91

$$q_{MFB} = \frac{1.524 \times 10^6}{2\sqrt{\frac{958.4+0.5981}{0.5981}}} = 19029.8\ \text{W/m}^2$$

We shall assume that the heat transfer coefficient in film boiling is given by Eq. 17.92. The heat flux is then given by $q_{MFB} = h_f \Delta T_{MFB}$. We may then solve for ΔT_{MFB} as

$$\Delta T_{MFB} = \frac{\left(\frac{q_{MFB}}{0.425}\right)^{\frac{4}{3}}}{\left(\frac{k_g^3 \rho_g(\rho_f - \rho_g) g h_e}{\mu_g \frac{L_c}{2\pi}}\right)^{\frac{1}{3}}}$$

To begin with let us assume that $h_e = h_e^0 \approx h_{fg} = 2.257 \times 10^6$ J/kg. The characteristic length L_c is given by

$$L_c = 2\pi\sqrt{\frac{0.059}{9.81(958.4 - 0.5981)}} = 0.1157\ \text{m}$$

Then we get

$$\Delta T_{MFB}^0 = \frac{\left(\frac{19029.8}{0.425}\right)^{\frac{4}{3}}}{\left(\frac{0.0248^3 \times 0.5981(958.4 - 0.5981) \times 9.81 \times 2.257 \times 10^6}{1.2 \times 10^{-5} \times \frac{0.1157}{2\pi}}\right)^{\frac{1}{3}}} = 166.3$$

We shall now correct the result by correcting h_e as

$$h_e^1 = h_{fg} + 0.4C_{pg}\Delta T = 2.257 \times 10^6 + 0.4 \times 2020 \times 166.3 = 2.3913 \times 10^6$$

The corrected value of superheat is then obtained as

$$\Delta T_{MFB}^1 = \frac{\Delta T_{MFB}^0}{\left(\frac{h_e^1}{h_e^0}\right)^{0.33}} = \frac{\Delta 166.3}{\left(\frac{2.391\times 10^6}{2.257\times 10^6}\right)^{0.33}} = 163.1\ °\text{C}$$

Step 6 We may also calculate the film boiling heat flux with another superheat value of $\Delta T = 500$ °C. Then we have

$$h_e^1 = 2.257 \times 10^6 + 0.4 \times 2020 \times 500 = 2.661 \times 10^6$$

The corresponding heat flux is then obtained using Eq. 17.92 as

$$q_f = 0.425 \times 500^{\frac{3}{4}} \left[\frac{0.0248^3 \cdot 0.5981(958.4 - 0.5981) \cdot 9.81 \cdot 2.661 \times 10^6}{1.2 \times 10^{-5}\left(\frac{0.1157}{2\pi}\right)}\right]^{\frac{1}{4}}$$

$$= 45296.3\ \text{W/m}^2$$

The calculations made in this example are used to make a plot of the boiling curve as shown in Fig. 17.19. We join the points by straight lines on the log-log plot as shown in the figure. Experimental boiling curves indicate fair agreement with the values calculated here.

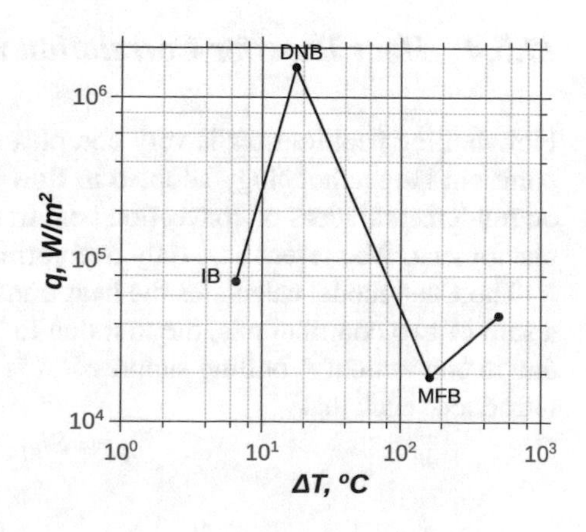

Fig. 17.19 Synthetic pool boiling curve for boiling of water at 1 atmosphere absolute as calculated in Example 17.11

17.5.3 Flow Boiling

In most engineering applications, phase change takes place as the fluid moves through a tube. For example, in boiler tubes, the flow takes place vertically upwards, with the liquid entering the tube at the bottom and the steam being taken off from the top. The flow may be set up by natural circulation due to the lower density of liquid vapor mixture as compared to liquid alone. In general, one may expect the heat transfer rates to be larger than that in the case of boiling in a stationary liquid.

The flow, as it progresses along the tube, undergoes several regime changes, both hydrodynamic as well as thermal. For some distance from the entry the flow is single phase (liquid). This is due to the fact that there is a pressure drop as the fluid (mixture of liquid and vapor) moves through the tube and the local saturation pressure and hence the saturation temperature is more than that at the exit pressure. Subsequently, boiling is initiated at the surface and bubbles of vapor are formed and the flow is termed as bubbly flow. In the slug flow region, the entire section may be filled alternately by pockets of vapor and liquid. Subsequently, the flow becomes what is termed as annular flow wherein a film of liquid exists adjacent to the tube wall and vapor elsewhere. As the flow proceeds further and the quality has increased to a value greater than about 0.6 we have two phase flow with the liquid dispersed as bubbles in an ambiance of vapor.

Correspondingly, several regimes of heat transfer may be identified. Near the tube entry, heat transfer is by single phase convection. Subsequently, there is a region of sub-cooled boiling followed by saturated nucleate boiling. In the annular flow region, heat transfer is by convection through liquid film adjacent to the tube wall. Subsequently, we have a liquid deficient region where heat transfer is to a mixture of vapor and liquid droplets. Finally, we have single phase heat transfer to the vapor. Superheating of the vapor may take place in this region.

17.5.4 Heat Transfer Correlation in Flow Boiling

Flow boiling heat transfer is very complex and hence our intention here is only to point out the methodology adopted in flow boiling heat transfer. Models are based on fundamental ideas of convection heat transfer coupled with the basics of boiling phenomena. We present here only one correlation due to Chen.[13]

The Chen model calculates the heat transfer coefficient in two phase flow h_{tp} as a sum of two contributions, the first due to boiling given by h_{fz} and the second h_f due to convection. A boiling suppression factor S and a two phase multiplier F are introduced such that

$$h_{tp} = Sh_{fz} + Fh_f \tag{17.94}$$

[13]J. C. Chen, Ind. Eng. Chem. and Proc. Des. Dev., Vol. 5, pp. 322–329, 1963.

The boiling heat transfer is based on the work of Foster and Zuber[14] and is given by

$$h_{fz} = 0.00122 \left[\frac{k_f^{0.79} C_{pf}^{0.45} \rho_f^{0.49}}{\sigma^{0.5} \mu_f^{0.29} h_{fg}^{0.24} \rho_g^{0.24}} \right] \Delta T^{0.24} \Delta p^{0.75} \tag{17.95}$$

where $\Delta T = T_w - T_s$, $\Delta p = p_w - p_s$ where the subscript w stands for the conditions at the wall and subscript s indicates the local saturation value. The convective heat transfer coefficient is based on the Dittus–Boelter equation and is given by

$$h_f = 0.023 Re_f^{0.8} Pr_f^{0.4} \frac{k_f}{D} \tag{17.96}$$

where the Reynolds number is defined as

$$Re_f = \frac{\dot{m}(1 - x)D}{\mu_f} \tag{17.97}$$

where x is the local vapor quality. The two phase multiplier F involves the Martinelli parameter X_{tt} that is given by[15]

$$X_{tt} = \left(\frac{1 - x}{x} \right)^{0.9} \left(\frac{\rho_g}{\rho_f} \right)^{0.5} \left(\frac{\mu_f}{\mu_g} \right)^{0.1} \tag{17.98}$$

The factor F itself is given by

$$F = \left(\frac{1}{X_{tt}} + 0.213 \right)^{0.736} \tag{17.99}$$

However F is set to unity if $\frac{1}{X_{tt}} \leq 0.1$. The boiling suppression parameter is given by

$$S = \frac{1}{1 + 2.53 \times 10^{-6} Re_{tp}^{1.17}} \tag{17.100}$$

where the two phase Reynolds number Re_{tp} is related to the liquid phase Reynolds number through the relation

$$Re_{tp} = F^{1.25} Re_f \tag{17.101}$$

---

[14]H. K. Foster and N. Zuber, Dynamics of vapor bubbles and boiling heat transfer, A.1.Ch.E. J1, Vol. 1, pp. 531–535, 1955

[15]Lockhart, R. W., Martinelli, R. C.; Chem. Eng. Prog., Vol. 45, pp. 39–48, 1949

17.6 Mixed Convection

Forced and natural convection processes have been covered individually in the earlier chapters. There are many occasions when these two modes of heat transfer occur simultaneously. For example, when a stream of fluid is forced to flow parallel to a hot surface with a modest velocity, depending on the orientation of the surface, it is possible that density differences will induce flow that is not small compared to the forced flow velocity. In such a case, both forced and natural convection occur simultaneously and the flow regime is said to be a mixed flow regime. In this section, we will look at some elementary cases where the analysis is simple and follow up with cases where correlations based on numerical or experimental studies are presented.

17.6.1 Laminar Mixed Convection For Flow Over A Vertical Isothermal Flat Plate

Consider flow past a vertical isothermal flat plate maintained at a temperature T_w different from the temperature of the free-stream fluid T_∞ as shown in Fig. 17.20. Incoming flow is a parallel stream with an upward uniform velocity of u_∞. Direction of g is downwards as indicated in the figure.

Because of the imposed upward velocity, flow exhibits a forced convection boundary layer adjacent to the plate. Normally, forced convection will be the dominant mode of heat transfer. However, when T_w is different from T_∞ buoyancy forces are also active and hence natural convection may become important if the temperature difference is large enough to induce a sizable body force on the fluid adjacent to the plate surface. If the body force is in the upward direction ($T_w > T_\infty$), natural convection will augment convection and the process is identified as aiding mixed convection. If the body force is in the downward direction ($T_w < T_\infty$), natural convection will diminish convection and the process is identified as opposing mixed convection.

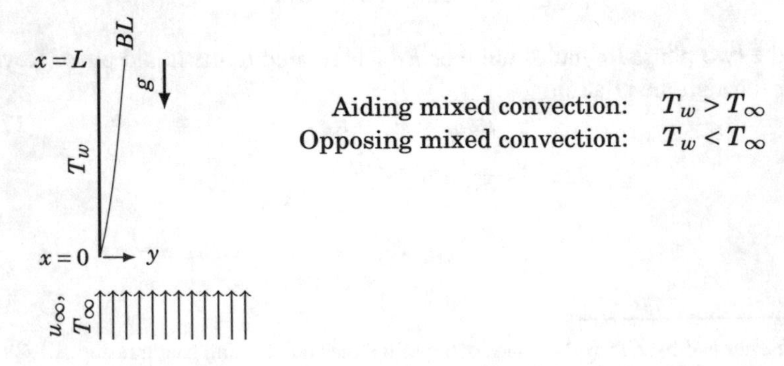

Fig. 17.20 Mixed convection for flow past a vertical isothermal flat plate

Consider steady laminar mixed convection in two dimensions, i.e., all dependent variables depend on x and y only. Under the boundary layer approximation, the Navier–Stokes equations may be simplified as was done in earlier chapters dealing with forced convection boundary layer flow past a flat plate, as well as natural convection boundary layer flow past an isothermal vertical plate. In the present case, the x momentum equation alone is different since the body force term is to be included. Hence, the governing equations may be written as under.

$$\underbrace{\frac{\partial u}{\partial x}+\frac{\partial v}{\partial y}=0}_{\text{Continuity}};\quad \underbrace{u\frac{\partial u}{\partial x}+v\frac{\partial u}{\partial y}=\nu\frac{\partial^2 u}{\partial y^2}+g\beta(T-T_\infty)}_{x\text{ momentum}};\quad \underbrace{u\frac{\partial T}{\partial x}+v\frac{\partial T}{\partial y}=\alpha\frac{\partial^2 T}{\partial y^2}}_{\text{Energy}} \tag{17.102}$$

Introduce the following non-dimensional variables and parameters:

$$\begin{gathered} X=\frac{X}{L},\ Y=\frac{y}{L},\ U=\frac{u}{u_\infty},\ V=\frac{v}{u_\infty},\ \theta=\frac{T-T_\infty}{T_w-T_\infty}, \\ Re_L=\frac{u_\infty L}{\nu},\ Gr_L=g\beta\frac{(T_w-T_\infty)\,L^3}{\nu^2} \end{gathered} \tag{17.103}$$

Invoking the boundary layer assumption the inertia terms in the x momentum equation are both of order $\frac{u_\infty^2}{L}$. Dividing throughout by this factor, it is easily shown that the viscous term is given by $\frac{1}{Re_L}\frac{\partial U}{\partial Y^2}$ and the buoyancy term is given by $\frac{Gr_L}{Re_L^2}\theta$. Hence the x momentum equation in non-dimensional form is given by

$$U\frac{\partial U}{\partial x}+V\frac{\partial U}{\partial Y}=\frac{1}{Re_L}\frac{\partial^2 U}{\partial Y^2}+\frac{Gr_L}{Re_L^2}\theta \tag{17.104}$$

Non-dimensional energy equation may be shown to reduce to

$$U\frac{\partial \theta}{\partial x}+V\frac{\partial \theta}{\partial Y}=\frac{1}{Re_L Pr}\frac{\partial^2 \theta}{\partial Y^2} \tag{17.105}$$

Noting that the derivative of velocity scales as $\sqrt{Re_L}$ and the second derivative of velocity scales as Re_L within the boundary layer, the strength of natural convection vis a vis the forced convection is given by the magnitude of $\frac{Gr_L}{Re_L^2}$ or Ri_L, the Richardson number[16] since $0 \le \theta \le 1$. Hence, it is customary to refer to the Richardson number also as the mixed convection parameter.

Three regimes are possible depending on the magnitude of Ri. As $Ri \rightarrow 0$ forced convection dominates and contribution of natural convection is very small. When $Ri \rightarrow \infty$ natural convection dominates and the contribution of forced convection to heat transfer is not significant. However, when $Ri \sim 1$ both modes of heat transfer are of equal importance and the regime is one of mixed convection.

As far as the boundary conditions are concerned, the no slip velocity condition applies on the plate surface. The velocity and temperature outside the boundary layer remain constant at u_∞ and T_∞, respectively. Thus, we have

$$U = V = 0,\ \theta = 1,\ \text{at } Y = 0,\ U \rightarrow 1,\ \theta \rightarrow 0 \text{ as } Y \rightarrow \infty \text{ for } 0 \le X \le 1 \tag{17.106}$$

Solution of these equations is not presented here. Either perturbation solutions for small and large Ri or full numerical solution are possible. However, we present experimental results available in the heat transfer literature.

Velocity and Temperature Profiles in Mixed Convection Flow Over a Vertical Isothermal Flat Plate

Blasius solution represents the forced convection velocity profile, in the absence of buoyancy. In the absence of forced convection, Ostrach solution represents the velocity profile that is due only to buoyancy. In the case of assisting or aiding mixed convection, the velocity profile is due both to forced flow and buoyancy. Because the governing equations are non-linear, the mixed convection velocity profile is due to complex interaction between the two modes of convection. In the case of aiding flow, we expect the velocity within a part of the boundary layer to be more than that corresponding to forced convection velocity profile. It is indeed so as brought out by Fig. 17.21, where the velocity within the boundary layer is larger than the free-stream velocity. The fluid under consideration is air with a Prandtl number of 0.7. Mixed convection data is based on measured values.

The temperature variation in the boundary layer is again affected by the interplay between natural and forced convection. The temperature gradient at the wall is more for mixed convection as compared to that due to forced convection alone (see Fig. 17.21). In other words, both wall shear stress and wall heat flux are more for aiding mixed convection as compared to pure forced convection. Mixed convection velocity and temperature profiles are based on measurements by Ramachandran et al.[17] These are also in good agreement with those computed by Gururaja Rao using the full Navier–Stokes equations instead of the boundary layer equations.[18]

---

[16]Named after Lewis Fry Richardson 1881–1953, an English mathematician, physicist, meteorologist. He is known for the "Richardson extrapolation" used in numerical mathematics.

[17]Ramachandran, N., Armaly, B. F. and Chen, T. S. Measurements and predictions of laminar mixed convection flow adjacent to a vertical surface, ASME Journal of Heat Transfer, Vol. 13, pp. 299–301, 1985

[18]C. Gururaja Rao, "Conjugate mixed convection with surface radiation from vertical plates and channels", Ph.D. Thesis, IIT Madras, May 2001

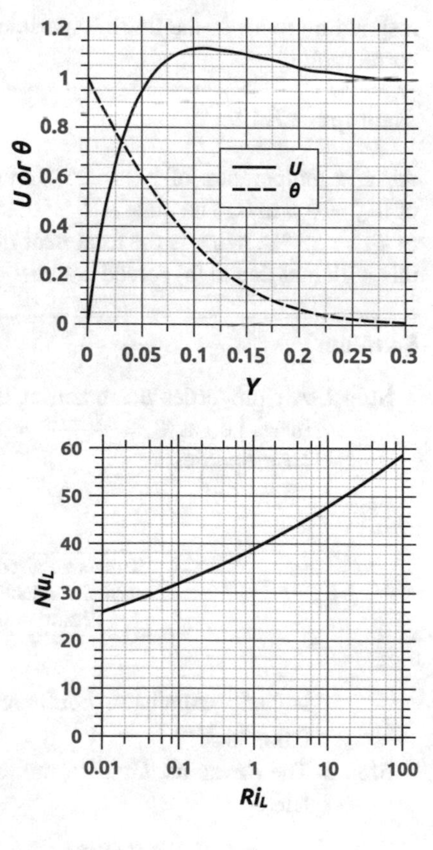

Fig. 17.21 Mixed convection velocity and temperature profiles at the trailing edge of plate for $Ri_L = 0.1884$

Fig. 17.22 Nusselt number variation with Richardson number with air as fluid $(Pr = 0.7)$

Heat Transfer Correlations

From the point of view of applications, it is convenient to have useful correlations for total heat transfer from the plate by mixed convection using non-dimensional parameters. Based on a numerical study, Gururaja Rao has recommended the following relation for the mean Nusselt number $\overline{Nu}_L$ over plate length.

$$\boxed{\overline{Nu}_L = 1.32 Pr^{0.315} Ri_L^{-0.141} Gr_L^{0.23}} \tag{17.107}$$

This relation is valid for $0.1 \leq Pr \leq 100$ and $0.1 \leq Ri_L \leq 100$ with the proviso that the Reynolds and Grashof numbers be so as to assure laminar flow. The above heat transfer correlation has been shown as a plot in Fig. 17.22 for $Pr = 0.7$ corresponding to air as the fluid. Experimental corroboration also exists for the above correlation.[19]

Example 17.12

Air at a temperature of $T_\infty = 25$ °C flows up a vertical flat plate with a velocity of $u_\infty = 0.5$ m/s. The plate is $L = 0.15$ m long and is maintained at a temperature of $T_w = 55$ °C. What is the total heat transfer from one side of the plate? Compare this with that due to (a) forced convection alone and (b) natural convection alone.

Solution:

Step 1 Air properties are taken at the mean of wall and free-stream temperatures, i.e., at $T_m = \frac{T_w + T_\infty}{2} = \frac{55+25}{2} = 40$ °C. We thus have the following air properties:

Kinematic viscosity:	$\nu = 17.07 \times 10^{-6}$ m^2/s
Thermal conductivity:	$k = 0.0274$ W/m°C
Prandtl number:	$Pr = 0.7$

Isobaric expansion coefficient of air is taken as $\beta = \frac{1}{T_\infty} = \frac{1}{273+25} = 0.00336\ K^{-1}$.

Step 2 The Reynolds, Grashof number and Richardson numbers are now calculated.

$$Re_L = \frac{u_\infty L}{\nu} = \frac{0.5 \times 0.15}{17.07 \times 10^{-6}} = 4393.7$$

$$Gr_L = \frac{g\beta(T_w - T_\infty)L^3}{\nu^2} = \frac{9.81 \times 0.00336(55 - 25)0.15^3}{(17.07 \times 10^{-6})^2} = 1.144 \times 10^7$$

Both Reynolds and Grashof numbers fall in the laminar range. The Richardson number is then given by

$$Ri_L = \frac{Gr_L}{Re_L^2} = \frac{1.144 \times 10^{-7}}{4393.7^2} = 0.593$$

Step 3 Using Eq. 17.107 we calculate the average mixed convection Nusselt number as

$$\overline{Nu_L} = 1.32 \times 0.7^{0.315} \times 0.593^{-0.141} \times (1.144 \times 10^7)^{0.23} = 53.38$$

[19] G. Venugopal, "Parameter estimation with steady and transient heat transfer experiments", Ph.D. Thesis, IIT Madras, December 2008

The mean heat transfer coefficient from the plate to air is then given by

$$\overline{h}_L = \frac{\overline{Nu_L}k}{L} = \frac{53.38 \times 0.0274}{0.15} = 9.75\ \text{W/m}^2\,{}^\circ\text{C}$$

Step 4 Heat transfer from one side of the plate is calculated based on a unit width of plate. Area of plate is then numerically the same as the length of the plate. Heat transfer from the plate to air is given by

$$Q = \overline{h}_L L(T_w - T_\infty) = 9.75 \times 0.15 \times (55 - 25) = 43.88\ \text{W}$$

Step 5 If heat transfer were to be by **forced convection** alone, the average Nusselt number is based on the Blasius solution (Eq. 13.54). We have

$$\overline{Nu_L} = 0.664 Re_L^{\frac{1}{2}} Pr^{\frac{1}{3}} = 0.664 \times 4393.7^{\frac{1}{2}} 0.7^{\frac{1}{3}} = 39.06$$

The mean heat transfer coefficient from the plate to air is then given by

$$\overline{h}_L = \frac{\overline{Nu_L}k}{L} = \frac{39.06 \times 0.0274}{0.15} = 7.14\ \text{W/m}^2\,{}^\circ\text{C}$$

Step 6 Heat transfer from one side of the plate is calculated as

$$Q = \overline{h}_L L(T_w - T_\infty) = 7.14 \times 0.15 \times (55 - 25) = 32.11\ \text{W}$$

Step 7 Comment: Mixed convection augments heat loss by $43.88 - 32.11 = 11.77$ W or as a percentage by $\frac{11.77 \times 100}{32.11} = 36.65\%$.

Step 8 Heat transfer by **natural convection** alone may be computed using the Ostrach solution. Using the curve fit, C (Eq. 16.43) is given by

$$C = \left(\frac{0.4Pr}{1 + 2\sqrt{Pr} + 2Pr}\right)^{\frac{1}{4}} = \left(\frac{0.4 \times 0.7}{1 + 2\sqrt{0.7} + 2 \times 0.7}\right)^{\frac{1}{4}} = 0.512$$

Mean Nusselt number is then given by

$$\overline{Nu_L} = C\ Ra_L^{\frac{1}{4}} = C(Gr_L Pr)^{\frac{1}{4}} = 0.512 \times (1.144 \times 10^7 \times 0.7)^{\frac{1}{4}} = 27.22$$

The mean heat transfer coefficient from the plate to air is then given by

$$\overline{h}_L = \frac{\overline{Nu_L}k}{L} = \frac{27.22 \times 0.0274}{0.15} = 4.97\ \text{W/m}^2\,{}^\circ\text{C}$$

Step 9 Heat transfer from one side of the plate is calculated as

$$Q = \overline{h}_L L(T_w - T_\infty) = 4.97 \times 0.15 \times (55 - 25) = 22.38 \text{ W}$$

Step 10 Comment: Mixed convection augments heat loss by $43.88 - 22.38 = 21.50$ W or as a percentage by $\frac{21.50 \times 100}{22.38} = 96.08\%$.

Step 11 General comment: If the heat loss by forced convection and natural convection are added together we get a total heat loss from plate to air of $32.11 + 22.38 = 54.49$ W. This is more than the heat loss by mixed convection and hence the mixed convection heat transfer rate is less than the algebraic sum of forced and natural convection heat transfers occurring in isolation. Essentially, there is an interaction between the two modes of heat transfer.

17.6.2 *Laminar Mixed Convection in a Parallel Plate Channel*

Mixed convection in a vertical channel exhibits some interesting features. For short channels, the two walls act as independent walls over which velocity and temperature boundary layers develop independently as shown in Fig. 17.23. This figure also introduces the nomenclature appropriate to the problem.

In a short channel the two boundary layers do not merge and the problem may be treated using the analysis presented above for mixed convection from a vertical flat plate. Once the two boundary layers meet, the flow tends to become fully developed as $x \to \infty$. In the fully developed limit, the problem is relatively easy since analytical solution is possible to the governing equations.

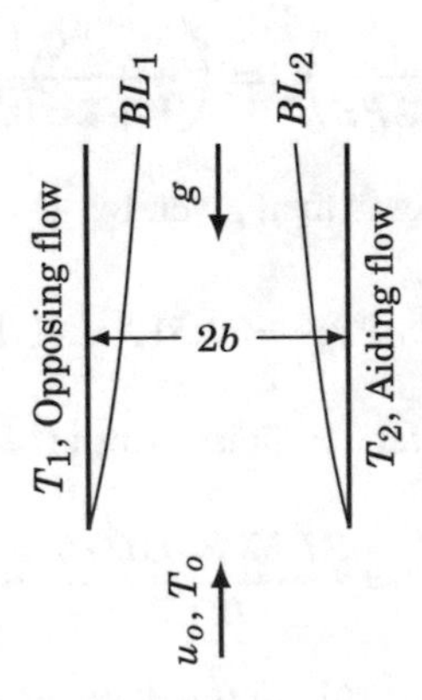

Fig. 17.23 Mixed convection in a short channel with $T_o = T_m = \frac{T_1+T_2}{2}$

In case $T_1 = T_2 = T_w$ and $T_o < T_w$, the channel induces flow similar to that in a chimney that was considered in Chap. 16. In case there is no imposed pressure gradient, chimney type flow may be analyzed using appropriate correlations.

In view of these, we consider the fully developed case in what follows.

17.6.3 *Laminar Mixed Convection in a Vertical Parallel Plate Channel: Fully Developed Solution*

In order to understand laminar mixed convection, we look at a simple case of a parallel plate channel with the two channel walls at different temperatures, both different from the temperature of the fluid as it enters the channel, as shown in Fig. 17.24. The flow and temperature fields are two-dimensional in that both are functions only of x and y. The governing equations are the Navier–Stokes equations in two dimensions and for steady flow all time derivatives are set to zero. Under the Boussinesq approximation natural convection is accounted for by adding buoyancy term in the x momentum equation (see Chap. 16).

Temperature Profile in Fully Developed Mixed Convection

Fluid enters the channel with a uniform velocity of u_o. Assuming that the channel is very long (in principle as $x \to \infty$) the flow becomes fully developed, i.e., $\frac{\partial u}{\partial x} = 0$. Hence, from continuity equation $\frac{\partial v}{\partial y} = 0$ and hence v is independent of y. Since it has to vanish at the two channel walls, it has to be zero for all y. With fully developed temperature, we should have $\frac{\partial T}{\partial x} = 0$. Temperature is a function of y alone. Energy equation will then reduce to

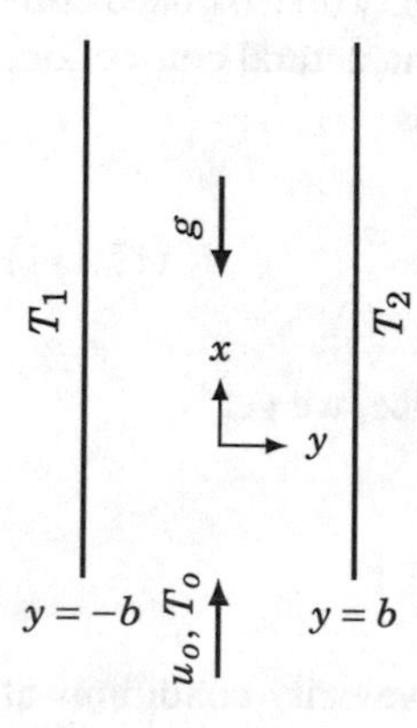

Nomenclature	
$T_2 > T_1 > T_o$	$T_m = \dfrac{T_1 + T_2}{2}$
$X = \dfrac{x}{b}, Y = \dfrac{y}{b}$	$\theta = \dfrac{T(y) - T_m}{T_2 - T_m}, U(Y) = \dfrac{u(y)}{u_o}, P = \dfrac{pb}{\mu u_o}$
$Re_b = \dfrac{u_o b}{\nu}$	$Gr_b = \dfrac{g\beta(T_2 - T_1)b^3}{\nu^2}$

Fig. 17.24 Mixed convection in a parallel plate channel—geometry and nomenclature

$$\frac{d^2T}{dy^2} = 0; \quad \text{with} \quad T(y = -b) = T_1, \quad T(y = b) = T_2 \tag{17.108}$$

This represents simply the conduction equation in one dimension and hence the temperature variation is linear and is given by $T(y) = c_1 + c_2 y$ where c_1 and c_2 are constants. Using the boundary conditions we may show that

$$c_1 = \frac{T_1 + T_2}{2} = T_m, \quad c_2 = \frac{T_2 - T_1}{2b}$$

Hence, the temperature variation of the fluid is given by

$$T(y) = T_m + \frac{T_2 - T_1}{2}\frac{y}{b} \quad \text{or} \quad \theta(Y) = Y \tag{17.109}$$

Note that $-1 \leq Y \leq 1$ and $-1 \leq \theta \leq 1$.

Velocity Profile in Fully Developed Mixed Convection

Now consider the x momentum equation. Under fully developed condition the inertia terms drop off. There is a balance between viscous friction and the combination of imposed pressure drop (forced convection) and body force due to buoyancy (natural convection). It is easily seen that the x momentum equation reduces to

$$0 = g\beta(T - T_o) - \frac{1}{\rho}\frac{dp}{dx} + \nu\frac{d^2u}{dy^2} \tag{17.110}$$

where the body force is due to density difference between local density and density corresponding to the entry temperature T_o. Note that $\frac{dp}{dx}$ is a constant under the fully developed assumption. Since the above equation is linear, velocity due to mixed convection is an algebraic sum of that due to forced convection and natural convection. Equation 17.110 may be recast in the non-dimensional form as

$$\frac{d^2U}{dY^2} + \frac{Gr_b}{Re_b}(\theta + \theta_o) = \frac{dP}{dX} \tag{17.111}$$

where $\theta_o = \frac{T_m - T_o}{T_2 - T_m}$. Introducing Eq. 17.109 and integrating twice, we get

$$U + \frac{Gr_b}{Re_b}\left[\frac{Y^3}{6} + \theta_o\frac{Y^2}{2}\right] = \frac{dP}{dX}\frac{Y^2}{2} + c_3Y + c_4$$

where c_3 and c_4 are constants of integration. Imposing zero velocity conditions at $Y = \pm 1$, the constants of integration may be obtained as

$$c_3 = \frac{Gr_b}{6Re_b} \quad \text{and} \quad c_4 = -\frac{1}{2}\left[\frac{dP}{dX} - \frac{Gr_b}{Re_b}\theta_o\right]$$

With these, the velocity profile may be written as

$$U = \frac{Gr_b}{6Re_b}\left(Y - Y^3\right) - \left[\frac{dP}{dX} - \frac{Gr_b}{Re_b}\theta_o\right]\left(\frac{1 - Y^2}{2}\right)$$

Noting that the total flow across the channel, by definition, is $\int_{Y=-1}^{Y=1} U(Y)dY = 2$ it is easily seen that $\frac{dP}{dX} - \frac{Gr_b}{Re_b}\theta_o = -3$. With this, the velocity profile takes its final form

$$\boxed{U = \frac{Gr_b}{6Re_b}\left(Y - Y^3\right) + \frac{3(1 - Y^2)}{2}} \tag{17.112}$$

It is observed that the first term on the right-hand side determines the importance of natural convection as compared to forced convection. The parameter $\frac{Gr_b}{Re_b}$ is a measure of relative importance of natural convection with respect to forced convection, and hence is termed the mixed convection parameter. It is a normal practice to represent the ratio $\frac{Gr_b}{Re_b^2}$ as Richardson number Ri_b, and hence the mixed convection parameter in the present case is $\frac{Gr_b}{Re_b} = \frac{Gr_b}{Re_b^2} \times Re_b = Ri_b \times Re_b$.

We may now interpret the second term on the right-hand side of Eq. 17.112 as representing the forced convection component of the velocity profile. It is a parabolic profile that is familiar to us from Chap. 12. The first term on the right-hand side is that due to natural convection and does not contribute to net flow across the channel. It is anti-symmetric with respect to mid-plane, and hence carries as much fluid upwards as it does downwards.

We may now look at the effect of natural convection on the mixed convection profile. For this, it is instructive to look at the velocity gradient at $Y = -1$, i.e., at the colder of the two channel walls. Differentiating Eq. 17.112 with respect to Y once and letting $Y = -1$, we get

$$\left.\frac{dU}{dY}\right|_{Y=-1} = -\frac{1}{3}\frac{Gr_b}{Re_b} + 3$$

We note that the gradient is positive if $\frac{Gr_b}{Re_b} < 9$ and negative otherwise. Hence, the velocity becomes negative (downward flow) close to the left wall for $\frac{Gr_b}{Re_b} > 9$. Thus, the effect of natural convection is a reduction in the velocity and possibly flow reversal when its effect is significant. This is clearly brought out by the velocity profiles shown in Figure 17.25 for various values of $\frac{Gr_b}{Re_b}$ both below and above the critical value given above.

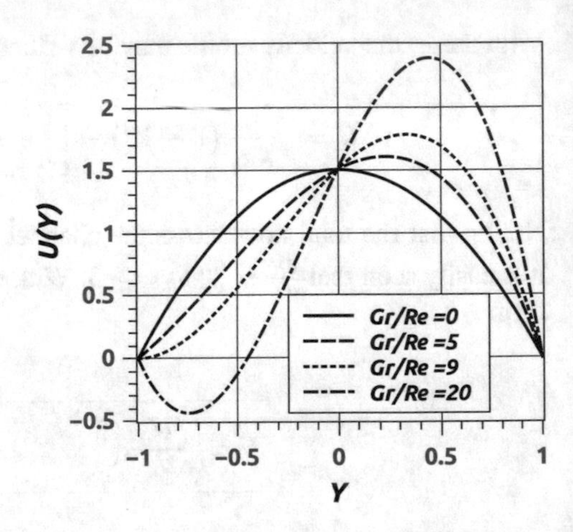

Fig. 17.25 Fully developed mixed convection velocity profiles in a channel

Example 17.13

A very tall vertical channel consists of two parallel plates maintained at temperatures of $T_1 = 25$ °C and $T_2 = 55$ °C. The channel spacing is equal to $2b = 2.5$ cm. Air at $T_o = 40$ °C enters the channel with a uniform vertical velocity of $u_o = 0.5$ m/s. Would you expect flow reversal in this case because of natural convection? What is the axial pressure gradient in this mixed convection flow? What is the maximum velocity in this case? Consider the fully developed part of the flow far downstream of the entry.

Solution:

Step 1 Air properties are taken at the mean temperature $T_m = \frac{T_1+T_2}{2} = \frac{25+55}{2} = 40$ °C.

Density:	$\rho_m = 1.1169\ \text{kg/m}^3$
Dynamic viscosity:	$\mu_m = 19.1 \times 10^{-6}\ \text{kg/m s}$
Kinematic viscosity:	$\nu_m = 17.1 \times 10^{-6}\ \text{m}^2/\text{s}$
Thermal conductivity:	$k_m = 0.0274\ \text{W/m°C}$
Prandtl number:	$Pr = 0.699$
Isobaric coefficient of expansion:	$\beta_o = \frac{1}{T_o} = \frac{1}{273+40} = 3.19 \times 10^{-3}\ K^{-1}$

In addition, we take $g = 9.81\ \text{m/s}^2$, semi-channel spacing $b = 1.25$ cm $=$ 0.0125 m.

Step 2 Calculate the appropriate non-dimensional parameters:
The Reynolds number is calculated as

$$Re_b = \frac{u_o b}{\nu_m} = \frac{0.5 \times 0.0125}{17.1 \times 10^{-6}} = 366.14$$

We note that the flow is laminar. The Grashof number is calculated as

$$Gr_b = \frac{g\beta_m(T_2 - T_1)b^3}{\nu_m^2} = \frac{9.81 \times 3.19 \times 10^{-3}(55 - 25) \times 0.0125^3}{(17.1 \times 10^{-6})^2} = 6302.44$$

Mixed convection parameter is then given by $\frac{Gr_b}{Re_b} = \frac{6302.44}{366.14} = 17.2$. Since the mixed convection parameter is greater than the critical value of 9, flow reversal is expected in the channel.

Step 3 Non-dimensional entry air temperature is calculated as $\theta_o = \frac{T_m - T_o}{T_2 - T_m} = \frac{40-40}{55-40} = 0$. Hence, the non-dimensional pressure drop is given by $\frac{dP}{dX} = -3 + \frac{Gr_b}{Re_b}\theta_o = -3$. Converting this to dimensional variables, we get

$$\frac{dp}{dx} = \frac{\mu_m u_o}{b^2}\frac{dP}{dX} = \frac{19.1 \times 10^{-6} \times 0.5}{0.0125^2}(-3) = -0.183 \text{ Pa/m}$$

Step 4 The maximum velocity occurs when the slope is zero. Differentiating Eq. 17.112 with respect to Y, we get $\frac{dU}{dY} = \frac{Gr_b}{6Re_b}\left(1 - 3Y^2\right) - 3Y$. Setting it to 0, we get a quadratic in Y which may be solved to get

$$Y = -\frac{3Re_b}{Gr_b} \pm \sqrt{\left(\frac{3Re_b}{Gr_b}\right)^2 + \frac{1}{3}}$$

The root that is required for locating the maximum velocity is obtained by taking the plus sign for the second term. We then have

$$Y = -\frac{3}{17.2} \pm \sqrt{\left(\frac{3}{17.2}\right)^2 + \frac{1}{3}} = 0.429$$

The maximum non-dimensional velocity is then given by the use of Eq. 17.112.

$$U_{max} = U(y = 0.429) = \frac{17.2}{6}\left(0.429 - 0.429^3\right) + 1.4\left(1 - 0.429^2\right) = 2.228$$

The maximum velocity is then given by $u_{\max} = U_{\max}u_o = 2.228 \times 0.5 = 1.114$ m/s.

17.7 Heat Transfer in a Particle Bed

A fluid forced through a bed of stationary particles passes along the interstices between the particles along tortuous paths. Hence, the fluid has a long residence time as compared to a fluid that passes through an empty channel. Invariably, this leads to an increase in the heat transfer rate between the fluid and the particles or the fluid and the bounding walls. For a given bed height, the surface area of the bed particles that is exposed to the fluid also is very high and this is yet another reason for the increase in heat transfer between the bed particles and the fluid. If the bed is used as an inert medium that creates disturbance in the flow, heat transfer between the wall of the container and the flowing fluid also goes up as compared to an empty container.

A fixed bed of particles is arranged as shown in Fig. 17.26. The bed of particles is supported by a distributor plate (simply a plate with holes that do not allow the particles to fall through) placed at the bottom of a containment vessel. Height of the bed is H and the particles have a mean diameter of D_p (assuming they are spherical). In case the particles are not spherical, we define an equivalent diameter through the relation

$$D_h = \sqrt{\frac{S_p}{\pi}} \tag{17.113}$$

where S_p is the surface area of the particle. For a spherical particle, the equivalent diameter is also the actual diameter of the particle.

In practice, the particles are added slowly into the container and shaken so that the particles settle down in a near close packed arrangement. In between the particles, interconnected voids exist. These allow a fluid to pass through, along a tortuous path, from entry below the bed to exit at top of bed. The situation when the particles are

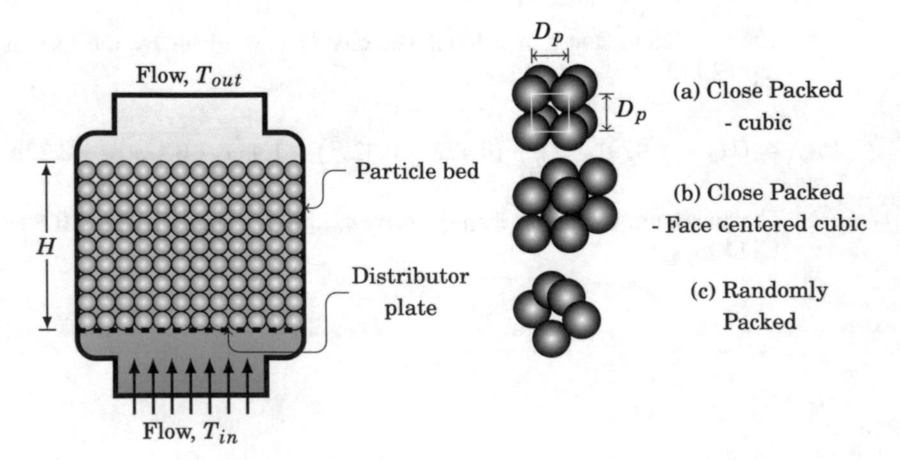

Fig. 17.26 Flow through a fixed bed of particles

close packed is as shown in Fig. 17.26a or b. The packing of particles will normally be random as shown in Fig. 17.26c and the void fraction (defined below) is a strong function of the packing.

In case the cross section area of the containment vessel is A, the voids occupy a volume ϵAH, where ϵ is called the void fraction. For a close packed bed, one can estimate theoretically the void fraction. However, in most applications it is necessary to measure the void fraction by a suitable technique. If the volume flow rate of the fluid through the bed is $\dot{V}$, the superficial velocity U_s is defined as

$$U_s = \frac{\dot{V}}{A} \tag{17.114}$$

Since the area available for flow is only that due to the voids, the actual velocity through the void space is higher, on an average, than the superficial velocity.

Void fraction for a close packed bed of spherical particles:
In the cubic close packed arrangement shown in Fig. 17.26a, we can identify a unit cell in the form of a cube of side D_p. All the particles touch one another and there is exactly one particle per unit cell (eight octants of a sphere at the eight corners of the cube). Thus, the solid volume in a unit cell is the same as the volume of a particle, i.e., V_p. In terms of D_p we have $V_p = \frac{\pi D_p^3}{6}$. Volume of the unit cell is D_p^3. The void volume is then given by $D_p^3 - \frac{\pi D_p^3}{6}$. Divide this by D_p^3 to get the void fraction for close packed spherical particles as $\epsilon = 1 - \frac{\pi}{6} = 0.476$.

In the case of face centered cubic shown in Fig. 17.26b, the void fraction may be shown to be equal to 0.260—the smallest possible value. It is also the same for Hexagonal close packed arrangement (not shown in the figure). For the random packing shown in Fig. 17.26c, the void fraction is close to 0.360.

17.7.1 Flow Characteristics of a Particle Bed

When a fluid passes through the bed, we expect the pressure to drop in the direction of flow. Since the fluid flows past the particles as it passes through the bed, pressure drop is due to fluid friction because of the viscosity of the fluid and also due to form drag (inertia effects) induced because of pressure distribution around the particles. Fully theoretical calculations are difficult and hence we take recourse to experimental correlations to calculate the appropriate friction factor for flow through a bed. Reynolds number is defined based on the particle diameter and the superficial velocity.

$$Re_p = \frac{U_s D_p}{\nu} \tag{17.115}$$

where ν is the kinematic viscosity of the fluid. The friction factor is defined through the relation

$$f = \frac{\Delta p}{\rho U_s^2} \frac{D_p}{H} \tag{17.116}$$

Viscous contribution to the friction factor is proportional to the velocity while that due to inertial effects is proportional to the square of the velocity. Based on empirical evidence, Ergun[20] has proposed the following expression for the friction factor.

$$\text{Ergun equation}: \boxed{f = \frac{(1-\epsilon)}{\epsilon^3}\left[1.75 + \frac{150(1-\epsilon)}{Re_p}\right]} \tag{17.117}$$

When $Re_p << 1$ the viscous term dominates and the friction factor is approximated by the second term in the Ergun equation which is known as the Kozeny–Carman equation. Thus,

$$\text{Kozeny–Carman equation}: \boxed{f = \frac{(1-\epsilon)^2}{\epsilon^3} \frac{150}{Re_p}} \tag{17.118}$$

For $Re_p >> 1$ the inertial effects dominate and the friction factor is approximated by the first term in the Ergun equation known as the Blake–Plummer equation. Thus

$$\text{Blake–Plummer equation}: \boxed{f = \frac{1.75(1-\epsilon)}{\epsilon^3}} \tag{17.119}$$

Ergun equation and the two limiting forms are expected to be valid for small diameter particles with $D_p < 25$ mm. Ergun equation and the asymptotic forms are shown plotted in Fig. 17.27 for a typical porous medium with $\epsilon = 0.4$.

Example 17.14

A pebble bed consists of uniform size spherical bed particles of diameter $D_p = 2$ cm. The void fraction has been experimentally determined to be $\epsilon = 0.36$. Air at a temperature of $T_a = 40$ °C is blown through the bed at the rate of $\dot{m} = 0.5$ kg/m^{2}s. Determine the pressure drop per unit bed height. Discuss the contributions due to viscous and inertial effects.

---

[20] S. Ergun, Fluid flow through packed columns, Chem. Eng. Prog., Vol. 48, pp. 89–94, 1952

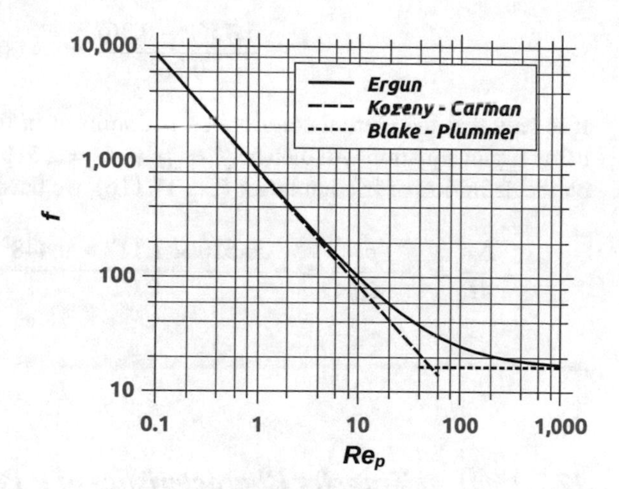

Fig. 17.27 Ergun equation and the asymptotic forms

Solution:

Air properties at 40 °C are given below

$$\rho = 1.117\ kg/m^3; \quad C_p = 1005\ \text{J/kg°C};$$
$$k = 0.0274\ \text{W/m°C}; \quad \nu = 1.71 \times 10^{-5}\ \text{m}^2/\text{s}$$

Other data specified in the problem are:

$$D_p = 2\ \text{cm} = 0.02\ \text{m}; \quad \dot{m} = 0.5\ \text{kg/m}^2 s; \quad \epsilon = 0.36$$

Using the mass flux specified, the superficial velocity is calculated as

$$U_s = \frac{\dot{m}}{\rho} = \frac{0.5}{1.117} = 0.448\ \text{m/s}$$

Reynolds number based on particle diameter of $D_p = 2$ cm $= 0.02$ m is given by

$$Re_p = \frac{U_s D_p}{\nu} = \frac{0.448 \times 0.02}{1.71 \times 10^{-5}} = 523.5$$

Contribution of viscous term to friction factor is given by Eq. 17.118 as

$$f_v = \frac{(1 - 0.36^2)}{0.36^3} \times \frac{150}{523.5} = 2.515$$

Contribution of inertial term to friction factor is given by Eq. 17.119 as

$$f_i = \frac{1.75(1-0.36)}{0.36^3} = 24.005$$

It is seen that the inertial contribution is dominant in this case. The friction factor using Ergun equation will then be $f = f_i + f_v = 2.515 + 24.005 = 26.520$. Based on the definition of friction factor (Eq. 17.116), we have

$$\frac{\Delta p}{H} = -\frac{f\rho U_s^2}{D_p} = -\frac{26.520 \times 1.117 \times 0.448^2}{0.02} = -296.79 \text{ Pa/m}$$

17.7.2 *Heat Transfer Characteristics of a Particle Bed*

Two types of heat transfer situations occur with respect to the operation of a particle bed. The first one involves heat transfer between the bed particles and the fluid flowing through the bed. In this case, the bed particles are at a different temperature compared to the fluid and the wall of the bed may be adiabatic. This is typical of heat storage application, where the bed particles store heat to be transferred to the fluid. The second involves heat transfer from the bed wall to the fluid in the presence of bed particles. In this case, the wall will be at a temperature different from that of the bed and the fluid. Bed particles simply have the role of augmenting heat transfer between the wall and the fluid.

Heat Transfer From Bed to Fluid

Just as the pressure drop incorporates viscous and inertial effects, heat transfer also involves two terms in the heat transfer correlation. Whitaker[21] has proposed the following correlation based on experiments with different types of bed material.

$$Nu_p = \frac{h_p D_p}{k} = \left(\frac{1-\epsilon}{\epsilon}\right)\left(0.5 Re_m^{\frac{1}{2}} + 0.2 Re_m^{\frac{2}{3}}\right) Pr^{\frac{1}{3}} \quad (17.120)$$

where h_p is the heat transfer coefficient for heat transfer from bed material to the gas and $Re_m = \frac{Re_p}{(1-\epsilon)}$ is a modified Reynolds number. Properties of the flowing gas may be taken at a suitable mean temperature. The above correlation is valid for $20 \le Re_m \le 10^4$ and for gases with $0.34 \le Pr \le 0.78$.

---

[21] S. Whitaker, Forced convection heat transfer correlations for flow in pipes, past flat plates, single cylinders, single spheres and for flow in packed beds and tube bundles, AIChE J., vol. 18, pp. 361–371, 1972

Volumetric Heat Transfer Coefficient

The heat transfer coefficient may also be expressed, alternately, as volumetric heat transfer coefficient h_v with units of W/m^3 °C. By definition, we then have, heat transfer per unit volume of bed q_v equals the product of h_v and temperature difference ΔT between the bed particles and the fluid. Thus, we have,

$$q_v = h_v \Delta T = h_p S_v V \Delta T$$

where S_v is the surface area of bed particles per unit volume of bed and V is the bed volume. It is easily seen that $S_v = \frac{6(1-\epsilon)}{D_p}$, and hence the volumetric heat transfer coefficient is given by

$$\boxed{h_v = \frac{6(1-\epsilon)h_p}{D_p}} \tag{17.121}$$

Combining Eqs. 17.120 and 17.121, we get the following relation for the volumetric heat transfer coefficient.

$$\boxed{h_v = \frac{6(1-\epsilon)^2 k}{\epsilon D_p^2}\left(0.5Re_m^{\frac{1}{2}} + 0.2Re_m^{\frac{2}{3}}\right) Pr^{\frac{1}{3}}} \tag{17.122}$$

Other correlations have also been proposed for the volumetric heat transfer coefficient in the literature. For example, Lof and Hawley[22] have proposed the following correlation:

$$h_v = 650\left(\frac{G}{D_p}\right)^{0.7} \tag{17.123}$$

where G is the mass velocity of gas in kg/s m^2, D_p is the particle diameter in m and h_v is in W/m^3 °C.

Example 17.15

Consider a bed consisting of particles of diameter $D_p = 0.02$ m which has been heated to a uniform temperature of $T_b = 80$ °C. Air at a temperature of $T_i = 15$ °C is blown through the bed with a mass velocity of $G = 0.5$ kg/s m^2. Immediately afterwards, it is seen that the air exiting the bed is at $T_e = 65$ °C. Estimate the bed height using the correlations proposed by (a) Whitaker and (b) Lof and Hawley. Porosity of the bed is known to be $\epsilon = 0.5$.

[22] G. O. G. Lof and R. W. Hawley, Unsteady state heat transfer between air and loose solids, Ind. Engg. Chem., vol. 40, pp. 1061–1070, 1948

Solution:

Step 1 Air properties are taken at the mean temperature of $T_m = \frac{T_i + T_o}{2} = \frac{15+65}{2} = 40\ °C$.

Density of air:	$\rho = 1.117\ kg/m^3$
Specific heat of air:	$C_p = 1005\ J/kg°C$
Kinematic viscosity of air:	$\nu = 17.1 \times 10^{-6}\ m^2/s$
Thermal conductivity of air:	$k = 0.0274\ W/m°C$
Prandtl number of air:	$Pr = 0.699$

Step 2 Superficial velocity of air is calculated as (based on unit bed cross section area) $U_s = \frac{G}{\rho} = \frac{0.5}{1.092} = 0.448\ m/s$. Hence, the modified Reynolds number is given by

$$Re_m = \frac{U_s D_p}{\nu(1-\epsilon)} = \frac{0.448 \times 0.02}{17.1 \times 10^{-6}(1-0.5)} = 1047.1$$

The volumetric heat transfer coefficient may then be calculated based on Whitaker correlation (17.122) $h_{v(a)}$ as

$$h_{v(a)} = \frac{6(1-0.5)^2 \times 0.0274}{0.5 \times 0.02^2}\left(0.5 \times 1047.1^{\frac{1}{2}} + 0.2 \times 1047.1^{\frac{2}{3}}\right) 0.0.699^{\frac{1}{3}}$$
$$= 6711.9\ W/m^3\,°C$$

Alternately, the volumetric heat transfer coefficient may be calculated based on Lof and Hawley correlation (17.123) $h_{v(b)}$ as

$$h_{v(b)} = 650\left(\frac{0.5}{0.02}\right)^{0.7} = 6186.9\ W/m^3\,°C$$

Step 3 Modeling heat transfer between bed and air[23]: Consider a bed of height H and unit cross section area as shown in Fig. 17.28. The axis of the bed is vertical and oriented parallel to the z axis. Consider a bed element of thickness Δz as shown. Heat balance requires that heat gained by air as it moves up through the element be equal to the heat lost by the bed particles. Thus

$$GC_pT(z+\Delta z) - GC_pT(z) = h_v(T_b - T)\Delta z$$

[23] Analysis is similar to that of a tubular heat exchanger subject to constant wall temperature

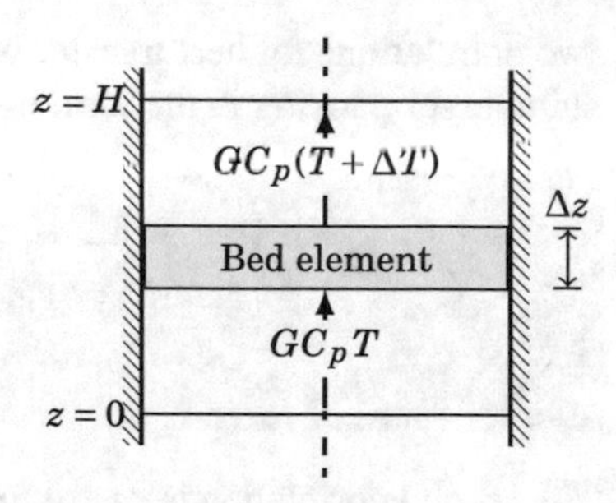

Fig. 17.28 Heat balance for a bed element

Using Taylor expansion and retaining two terms and in the limit $\Delta z \to 0$, the above simplifies to

$$GC_p \frac{\mathrm{d}T}{\mathrm{d}z} = h_v(T_b - T)$$

This equation is in variable separable form and is easily integrated once with respect to z to get $-\ln(T_b - T) = \frac{h_v z}{GC_p} + C$ where C is a constant of integration. Using $T = T_i$ at $z = 0$ we have $-\ln(T_b - T_i) = C$. Hence the air temperature at any z is given by

$$\ln\left(\frac{T_b - T_i}{T_b - T}\right) = \frac{h_v z}{GC_p}$$

We are given that $T = T_e$ at $z = H$.Hence, we have

$$H = \frac{GC_p}{h_v} \ln\left(\frac{T_b - T_i}{T_b - T_e}\right)$$

Step 4 Based on above analysis the bed height may be calculated.

$$\text{Whitaker correlation:} \quad H_{(a)} = \frac{0.5 \times 1005}{6711.9} \ln\left(\frac{80 - 15}{80 - 65}\right) = 0.110\, m$$

$$\text{Lof and Hawley correlation:} \quad H_{(b)} = \frac{0.5 \times 1005}{6186.88} \ln\left(\frac{80 - 15}{80 - 65}\right) = 0.119 \text{ m}$$

From a design point of view the bigger answer may be chosen.

Heat Transfer From Bed Wall to Fluid

A packed bed may be used to improve heat transfer between the wall of the bed and a fluid that is forced through the bed. Based on previous studies, Beek[24] has proposed

[24] J. Beek, Design of packed catalytic reactors, Adv. Chem. Eng., vol. 3, pp. 203–271, 1962

two correlations for heat transfer between the bed wall and the fluid. The first one is suitable for packing in the form of cylinders and is given by

$$\boxed{\frac{h_c D_p}{k} = 2.58 Re_p^{\frac{1}{3}} Pr^{\frac{1}{3}} + 0.094 Re_p^{0.8} Pr^{0.4}} \quad (17.124)$$

The second correlation is useful for particles like spheres and is given by

$$\boxed{\frac{h_c D_p}{k} = 0.203 Re_p^{\frac{1}{3}} Pr^{\frac{1}{3}} + 0.220 Re_p^{0.8} Pr^{0.4}} \quad (17.125)$$

h_c is the convective heat transfer coefficient between the bed wall and the fluid. Both correlations are useful when $40 \le Re_p \le 2000$.

Performance Parameter:
Example 17.16 has shown that a pebble bed heat exchanger improves transfer of heat between the wall of the bed and the fluid passing through the bed. We define Colburn j factor as $j = St_c Pr^{\frac{2}{3}}$. Since heat transfer is proportional to h_c we define a heat transfer parameter $J = j Re_p$. Noting that $St_c = \frac{Nu_c}{Re_p Pr}$, we see that J is proportional to Nu_c and hence the heat transfer.

Heat transfer enhancement with a fixed bed requires higher pressure drop and hence the pumping power. Pumping power P is proportional to the product of friction factor and the cube of velocity, i.e., $f U_s^3$ - product of drag force which is proportional to square of velocity and mass flow rate that is proportional to velocity. We may replace U_s by Re_p, and hence the pumping power factor is proportional to $F = f Re_p^3$.

Performance of the bed as a heat transfer enhancement device is evaluated by taking the ratio of J to $F^{\frac{1}{3}}$, i.e., by taking the ratio of heat transfer to pumping power. Note that the ratio will not involve *explicitly* the Reynolds number. The larger the ratio better is the heat exchange device.

Example 17.16

Air at $T_i = 25$ °C enters a bed of diameter $D_b = 0.05$ m with a mass flux of $G = 2$ kg/m^{2}s. Two types of bed particles are under consideration viz. (1) Spherical particles of diameter $D_p = 0.003$ m, void fraction equal to $\epsilon = 0.4$ and (2) Cylindrical particles of diameter D_{cyl} equal to length L_{cyl} of 0.002 m, void fraction equal to $\epsilon = 0.45$. The bed heights are to be chosen such that the exiting air is at $T_e = 55$ °C when the bed wall is steam jacketed and maintained at $T_w = 100$ °C. Evaluate the two options and decide the better one.

Solution:
Air properties at a mean temperature of $T_m = \frac{T_i+T_e}{2} = \frac{25+55}{2} = 40\ °C$ are read off air tables

$$\rho = 1.117\ \text{kg/m}^3, \quad \mu = 1.91 \times 10^{-5}\ \text{kg/m s}, C_p = 1005\ \text{J/kg K},$$
$$k = 0.0274\ \text{W/m K}, \quad Pr = 0.699$$

Case 1: Bed with spherical particles:
Mass flux through the bed is given as $G = 2\ \text{kg/m}^2\text{s}$. The superficial air velocity is then given by

$$U_s = \frac{G}{\rho} = \frac{2}{1.117} = 1.791\ \text{m/s}$$

The Reynolds number based on particle diameter is then given by

$$Re_p = \frac{\rho U_s D_p}{\mu} = \frac{1.117 \times 1.791 \times \times 0.003}{1.91 \times 10^{-5}} = 314.63$$

Use Eq. 17.125 to obtain the wall to gas heat transfer coefficient as

$$h_c = \left(0.203 Re_p^{\frac{1}{3}} Pr^{\frac{1}{3}} + 0.220 Re_p^{0.8} Pr^{0.4}\right)\frac{k}{D_p} = \left(0.203 \times 314.63^{\frac{1}{3}} \times 0.699^{\frac{1}{3}}\right.$$
$$\left. +0.220 \times 314.63^{0.8} \times 0.699^{0.4}\right)\frac{0.0263}{0.003} = 182.1\ \text{W/m}^2\,°\text{C}$$

The Stanton number St_b is then calculated as

$$St_b = \frac{h_c}{\rho U_s C_p} = \frac{182.1}{1.117 \times 1.791 \times 1005} = 0.0906$$

With all the temperatures specified, the LMTD is calculated as

$$LMTD = \frac{(T_w - T_i) - (T_w - T_e)}{\ln \frac{(T_w - T_i)}{(T_w - T_e)}} = \frac{(100 - 25) - (100 - 55)}{\ln\left(\frac{100-25}{100-55}\right)} = 58.7\ °\text{C}$$

Total heat transferred to air is calculated as

$$Q = G A_b C_p (T_e - T_i) = 2 \times \frac{\pi \times 0.05^2}{4} \times 1005 \times (55 - 25) = 118.4\ \text{W}$$

Noting that $Q = h_c(\pi D_b H) LMTD$, we solve for bed height H as

$$H = \frac{Q}{h_c \pi D_b LMTD} = \frac{118.4}{182.1 \times 0.05 \times 58.7} = 0.0705\ \text{m}$$

Performance parameters are now calculated. The friction factor is calculated using Ergun equation as

$$f = \frac{1-\epsilon}{\epsilon^3}\left(1.75 + 150\frac{1-\epsilon}{Re_p}\right) = \frac{1-0.4}{0.4^3}\left(1.75 + 150\frac{1-0.4}{314.63}\right) = 19.09$$

Colburn j factor is calculated as

$$j = St_b Pr^{\frac{2}{3}} = 0.0906 \times 0.699^{\frac{2}{3}} = 0.0714$$

Performance parameter of the fixed bed heat exahnger is calculated as

$$\frac{J}{F^{\frac{1}{3}}} = \frac{j}{f^{\frac{1}{3}}} = \frac{0.0714}{19.09^{\frac{1}{3}}} = 0.0267$$

Case 2: Bed with cylindrical particles:
Since the particles are not spherical, we make use of equivalent diameter for the particles given by $D_h = \sqrt{\frac{A_s}{\pi}}$, where A_s is the particle surface area given by $A_s = \pi D_{\text{cyl}} L\text{cyl} + 2\frac{\pi D_{\text{cyl}}^2}{4}$. With $D_{\text{cyl}} = L_{\text{cyl}} = 0.002$ m, we have $A_s = 1.5\pi D_{cyl}^2 = 1.5 \times \pi \times 0.002^2 = 1.885 \times 10^{-5}$ m^2. Hence, the equivalent diameter is given by $D_h = \sqrt{\frac{1.885\times10^{-5}}{\pi}} = 0.00245$ m. The rest of the calculations follow the same procedure as was used with the spherical particale case except that Eq. 17.124 is used for obtaining the heat transfer coefficient. Leaving the details to the reader, the results are tabulated below

Results for cylindrical bed particles			
U_s:	1.791 m/s	Re_h:	256.9
Nu_c:	21.45	h_c:	239.98 W/m^2 °C
H:	0.0535 m	St:	0.1194
j:	0.094	f:	12.5
$j/f^{\frac{1}{3}}$:	0.0405		

Comparisons: Bed with cylindrical particles is more compact since it requires a smaller bed height. It is also the better of the two since the heat transfer performance parameter is larger of the two values. Hence, it is recommended that the bed with cylindrical particles be chosen for the application.

Porous media approach:
Flow through a bed of particles may also be approached by treating the bed as a porous medium. A porous medium or matrix is a homogeneous medium with interconnected pores through which a fluid may flow. Such a medium

is characterized by permeability - κ and form coefficient - C such that the pressure drop is given by the equation

$$\boxed{-\frac{\Delta p}{H} = \frac{\mu U_s}{\kappa} + \frac{C}{\sqrt{\kappa}}\rho U_s^2} \tag{17.126}$$

This way of describing the pressure drop is known as the Hazen-Dupuit-Darcy model or as the Forchheimer extended Darcy model. The above is no different from the Ergun model since the first Darcy term that is proportional to velocity accounts for viscous effects while the second Forchheimer term accounts for inertia effects. κ and C are obtained by curve fit to pressure drop data. κ is known as the permeability of the porous medium and has units of m^2. C is a non-dimensional constant.

When the Reynolds number is very small, the viscous term dominates and the velocity and pressure gradient are related by a relation of form

$$\boxed{\frac{\mathrm{d}p}{\mathrm{d}x} = -\frac{\mu U}{\kappa}} \quad \text{or} \quad \boxed{U = -\frac{\kappa}{\mu}\frac{\mathrm{d}p}{\mathrm{d}x}} \tag{17.127}$$

In this limit, the flow is said to be Darcy flow. Velocity is directly proportional to pressure gradient and the equations of motion become very simple. When the Forchheimer term becomes comparable to the Darcy term the flow is said to be non Darcy flow.

In recent times, metal foams have been used in heat transfer applications. These foams consist of an interconnected metal foam with interconnected voids or pores as shown photographically in Fig. 17.29. The foam represented in the figure is made of aluminum and has 10 pores per inch and has a porosity (the same as void fraction) of 0.95, and hence is referred to as high porosity metal foam. The measured pressure drop characteristics shown in Fig. 17.30 is very well represented by a quadratic in U_s with the permeability being given by $\kappa = 2.48 \times 10^{-7}$ m^2 and the form drag coefficient given by $C = 0.473$. The flowing medium is air at atmospheric pressure and 30 °C.

Metallic foams may be used for augmenting heat transfer between the heated channel walls and the air or fluid that passes through it. There are two effects that come into play. The metal foam is highly conducting, and hence the heat from the wall spreads easily into it and is taken away by the fluid—akin to attaching fins to the wall. This effect can be improved by having a good contact between the wall and the foam by brazing it to the wall. The second effect is because of large residence time of the fluid as it flows through the pores and also because of highly disturbed flow within the pores. Compared to an empty channel heat transfer increases many fold.

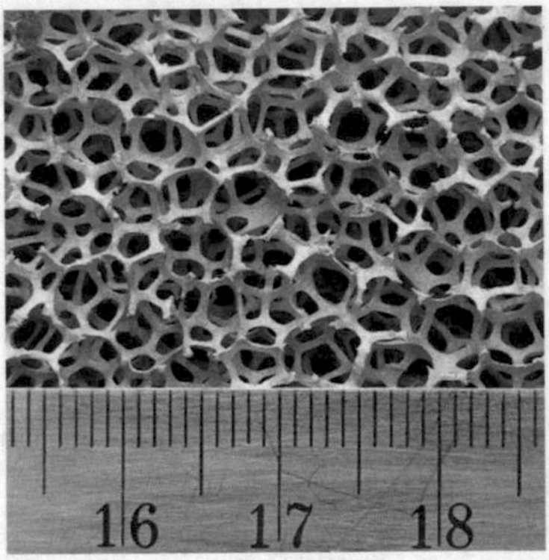

Fig. 17.29 Photograph of aluminum foam with 10 pores per inch

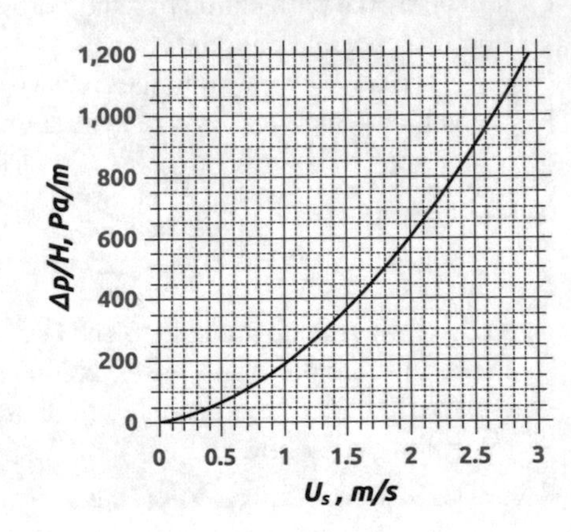

Fig. 17.30 Pressure drop data in a channel filled with aluminum foam

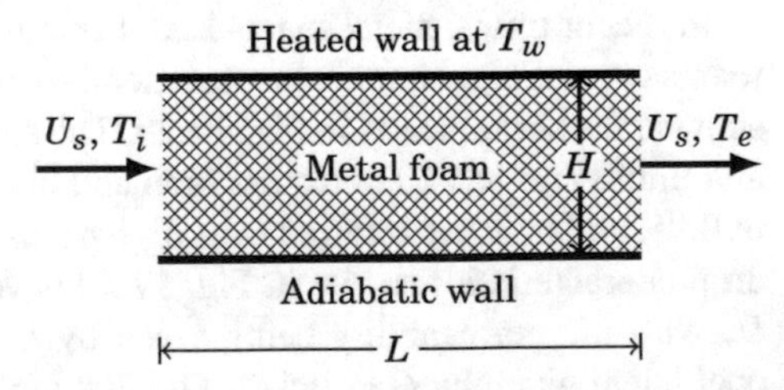

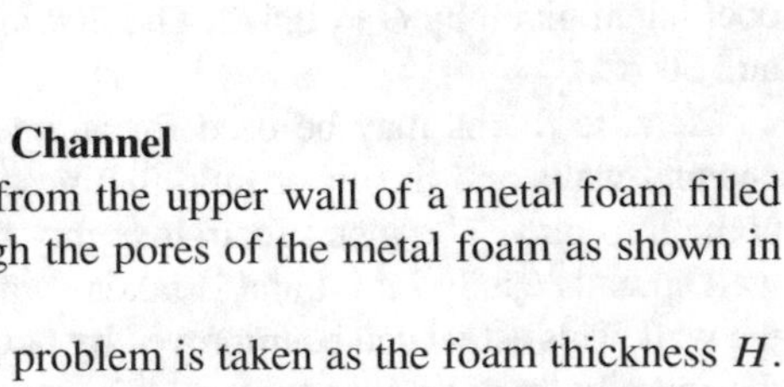

Fig. 17.31 Air flow through a asymmetrically heated metal foam filled channel

Heat Transfer in a Metal Foam Filled Channel

As an example, consider heat transfer from the upper wall of a metal foam filled horizontal channel to air flowing through the pores of the metal foam as shown in Fig. 17.31.[25]

The characteristic length scale in the problem is taken as the foam thickness H. Properties of air are based on the inlet temperature T_i. Apart from the usual non-

[25] Seo Young Kim, Byung Ha Kang, Jin-Ho Kim, Forced convection from aluminum foam materials in an asymmetrically heated channel, International Journal of Heat and Mass Transfer, vol. 44, pp. 1451–1454, 2001.

dimensional parameters, we introduce an additional non-dimensional parameter—the Darcy number Da - defined as $Da = \frac{\kappa}{H^2}$. Experiments performed by Kim et al. on aluminum metal foams of fixed porosity but different pores per inch led to the following correlation.

$$Nu_H = 0.0159 \frac{Re_H^{0.428} Pr^{\frac{1}{3}}}{Da^{0.787}} \tag{17.128}$$

This correlation is valid in the range $1000 \leq Re_H \leq 3000$. It is to be noted that the foam thickness H was constant at 9 mm, length was constant at $L = 188$ mm and the channel width was constant at $W = 90$ mm in all the experiments.

Recently, Kamath[26] has presented an experimental study of heat transfer from an asymmetrically heated foam filled vertical channel to air passing through it. Experiments were done with different foam thicknesses $H = 10 - 30$ mm but with $L = 150$ mm and $W = 250$ mm. The foam material was either aluminum or copper. He has specified the following heat transfer correlation.

$$Nu_H = 9.353 \lambda^{0.0385} \left(\frac{L}{H}\right)^{-0.887} Re_H^{0.446} \tag{17.129}$$

In the above $\lambda = \left(\frac{k_s}{k}\right)^{(1-\epsilon)}$ with k_s the thermal conductivity of the solid constituent of the metal foam. Equation 17.129 is valid for the following range of parameters.

$$1000 \leq Re_H \leq 7300; \quad 2.5 < \frac{L}{H} < 7.5; \quad 1.55 < \lambda < 3.85 \tag{17.130}$$

Example 17.17

Aluminum foam of porosity equal to 0.92 but with three different pores per inch (PPI) are to be compared in situation that was used in the experiment of Kim et al. The permeabilities of three foams of PPI equal to 10, 20, and 40 are, respectively, given by 1.04×10^{-7}, 0.76×10^{-7} and 0.51×10^{-7} m^2. The channel dimensions are those given earlier while discussing the correlation proposed by Kim et al. Air enters the foam filled channel with a velocity of 2 m/s and a temperature of $T_i = 25$ °C. What will be the exit temperature of air if the upper wall is maintained at 65 °C with the three different foams?

[26] P. M. Kamath, Experimental studies on thermal performance of metal foams in a vertical channel, Ph.D. Thesis, IIT Madras, 2012.

Solution:

Step 1 Air properties are taken at air entry temperature viz. $T_i = 25$ °C.

Density of air:	$\rho = 1.165$ kg/m^3
Specific heat capacity of air:	$C_p = 1007$ J/kg K
Thermal conductivity of air:	$k = 0.026$ W/m K
Kinematic viscosity of air:	$\nu = 1.61 \times 10^{-5}$ m^2/s
Prandtl number of air:	$Pr = 0.713$

Step 2 Other data common to all three cases of foam are $H = 0.009$ m; $L = 0.188$ m and $W = 0.09$ m. Area from which heat transfer takes place is also common and given by $A = LW = 0.188 \times 0.09 = 0.01692$ m^2. The Reynolds number is calculated as $Re_H = \frac{U_s H}{\nu} = \frac{2\times0.009}{1.61\times10^{-5}} = 1120.1$ and this remains the same in all the three cases. Detailed calculations are shown for the case of Aluminum foam with 10 PPI.

Step 3 With the specified permeability for this foam, the Darcy number is calculated as $Da = \frac{\kappa}{H^2} = \frac{1.04\times10^{-7}}{0.009^2} = 1.284 \times 10^{-3}$. Using Eq. 17.128, the Nusselt number is calculated as

$$Nu_H = 0.0159 \frac{\times 1120.1^{0.428} 0.713^{\frac{1}{3}}}{\left(1.284 \times 10^{-3}\right)^{0.787}} = 54.09$$

Step 4 The Stanton number is calculated as

$$St = \frac{Nu_H}{Re_H Pr} = \frac{54.09}{1120.1 \times 0.713} = 0.0677$$

We may easily show that the $NTU = \frac{StL}{H}$, and hence we have $NTU = \frac{0.0677\times0.188}{0.009} = 1.415$. Effectiveness may be shown to be $\frac{T_w - T_e}{T_w - T_i} = e^{-NTU}$, and hence the air exit temperature is given by

$$T_e = T_w - (T_w - T_i)e^{-NTU} = 65 - (65 - 25)e^{-1.415} = 55.3\ ^\circ\text{C}$$

Step 5 Calculations may be performed similarly for the other two foams also. The reader may verify the following:

PPI	Nu_H	NTU	T_e °C
10 PPI:	54.09	1.415	55.3
20 PPI:	69.23	1.810	58.5
40 PPI:	94.76	2.479	61.6

17.8 Heat Transfer in High Speed Flows

When a gas flows at a speed greater than about $\frac{1}{3}^{rd}$ the speed of sound in the gas, the incompressible flow assumption breaks down. Density variations cannot be ignored. In case the flow takes place in the vicinity of a solid surface, viscous effects dominate in the boundary layer close to the surface. The velocity gradients within the boundary layer may be such as to give rise to significant heating of the gas by viscous dissipation. The heating of the gas is due to irreversible conversion of kinetic energy to internal energy. The temperature variations may be so large that the gas properties like the viscosity and thermal conductivity may vary significantly within the boundary layer. We will look at these effects in what follows under suitable simplifying assumptions.

17.8.1 Compressible Boundary Layer Flow Parallel to a Flat Plate

Constant Wall Temperature Case

Consider the steady laminar flow of a gas (like air) parallel to a flat plate maintained at a uniform temperature of $T_w \neq T_\infty$. Let $U, T_\infty, \rho_\infty, \mu_\infty, k_\infty$ characterize the conditions in the free stream. The symbols have the usual meaning. We assume the gas to follow the ideal gas relation $p = \rho RT$, have a constant specific heat C_p. However, the gas is viscous and thermally conducting with the viscosity and thermal conductivity varying linearly with temperature. In addition, we shall assume that the Prandtl number is unity. Later we shall look at what happens when the Prandtl number is not equal to unity.

The boundary layer equations for laminar flow of such a gas may be written down as follows:

- Continuity equation

$$\frac{\partial(\rho u)}{\partial x} + \frac{\partial(\rho v)}{\partial y} = 0 \tag{17.131}$$

- Momentum equation

$$\rho\left\{u\frac{\partial u}{\partial x} + v\frac{\partial u}{\partial y}\right\} = \frac{\partial}{\partial y}\left(\mu\frac{\partial u}{\partial y}\right) \tag{17.132}$$

- Energy equation

$$\rho C_p\left\{u\frac{\partial T}{\partial x} + v\frac{\partial T}{\partial y}\right\} = \frac{\partial}{\partial y}\left(k\frac{\partial T}{\partial y}\right) + \mu\left(\frac{\partial u}{\partial y}\right)^2 \tag{17.133}$$

These equations differ from those that were used in the incompressible formulation by the appearance of density, viscosity, and thermal conductivity within the differential sign. This has the effect of coupling the momentum and energy equations. Another difference lies in the fact that the viscous dissipation makes its appearance through the source term $\mu\left(\frac{\partial u}{\partial y}\right)^2$ in the energy equation. It is akin to the heat generation term in the heat equation. A fourth equation is required now in the form of an equation of state that has been introduced already in the previous paragraph. We have made the tacit assumption that the pressure is uniform and is at p_∞, the free-stream value.

The boundary conditions remain the same as those that have been presented earlier in the case of incompressible flow past a flat plate. The no slip condition is satisfied at $y = 0$.

$$u = v = 0 \text{ and } T = T_w \text{ at } y = 0 \text{ for all } x \tag{17.134}$$

The values tend to the free-stream values when $y \to \infty$. Thus

$$u = U,\ T = T_\infty \text{ as } y \to \infty \text{ for all } x \tag{17.135}$$

Transformations

A stream function $\psi(x, y)$ is defined such that the continuity Eq. 17.131 is automatically satisfied. This is easily accomplished by the following:

$$\rho u = \frac{\partial \psi}{\partial y};\ \rho v = -\frac{\partial \psi}{\partial x} \tag{17.136}$$

Linear relationship between viscosity and thermal conductivity with temperature is taken in the form

$$\frac{\mu}{\mu_\infty} = \frac{k}{k_\infty} = \frac{T}{T_\infty} = \frac{\rho_\infty}{\rho} \tag{17.137}$$

The last part follows from the ideal gas equation under the constant pressure assumption. The following non-dimensional scheme is used:

$$x' = \frac{\rho_\infty U x}{\mu_\infty};\ y' = \frac{\rho_\infty U y}{\mu_\infty};\ T' = \frac{T}{T_\infty};\ \eta(x', y') = \frac{1}{\sqrt{x'}} \int_0^{y'} \frac{\rho}{\rho_\infty} dy' \tag{17.138}$$

It is noted that x' and y' are Reynolds numbers based on x and y, respectively, the properties being those in the free stream. Similarity variable η defined above is appropriate for the compressible boundary layer case. It may be shown, by a procedure similar to that presented in Chap. 13 that the momentum equation will reduce to the Blasius equation, if, in addition, we define the Blasius function as

$$f(\eta) = \frac{\psi(x', y')}{\sqrt{\rho_\infty U \mu_\infty x'}} \tag{17.139}$$

The Blasius equation, in the present case, is given by

$$\boxed{2\frac{d^3 f}{d\eta^3} + f\frac{d^2 f}{d\eta^2} = 0} \tag{17.140}$$

with the boundary conditions

$$f(\eta = 0) = \left.\frac{df}{d\eta}\right|_{\eta=0} = 0, \quad \left.\frac{df}{d\eta}\right|_{\eta\to\infty} = 1 \tag{17.141}$$

The solution to the Blasius equation presented earlier in Chap. 13 is valid in the present case also, but with a change in the definition of η. Note that there is an extra factor 2 in the Blasius equation given above as compared to the equation presented in the incompressible case. Now we turn our attention to the energy equation. It may easily be shown to reduce to the following equation:

$$\boxed{\frac{d^2 T'}{d\eta^2} + \frac{f}{2}\frac{dT'}{d\eta} = -Ec\left[\frac{d^2 f}{d\eta^2}\right]^2} \tag{17.142}$$

The accompanying boundary conditions are

$$T'(\eta = 0) = T'_w, \; T'(\eta \to \infty) = 1 \tag{17.143}$$

where $T'_w = \frac{T_w}{T_\infty}$ is the ratio of wall temperature to free-stream temperature. The Prandtl number has been assumed to be unity and the quantity $Ec = \frac{U^2}{C_p T_\infty}$ is a non-dimensional parameter referred to as the Eckert number. Let us seek the solution to Eq. 17.142 in the form

$$T' = A + B\frac{df}{d\eta} + C\left(\frac{df}{d\eta}\right)^2 \tag{17.144}$$

where the constants A, B, and C are to be determined by requiring the above to satisfy Eq. 17.142 and the boundary conditions (17.143). Differentiating expression (17.144) with respect to the similarity variable, we have

$$\frac{\mathrm{d}T'}{\mathrm{d}\eta} = B\frac{\mathrm{d}^2 f}{\mathrm{d}\eta^2} + 2C\frac{\mathrm{d}f}{\mathrm{d}\eta}\frac{\mathrm{d}^2 f}{\mathrm{d}\eta^2}$$

$$\frac{\mathrm{d}^2 T'}{\mathrm{d}\eta^2} = B\frac{\mathrm{d}^3 f}{\mathrm{d}\eta^3} + 2C\left(\frac{\mathrm{d}^2 f}{\mathrm{d}\eta^2}\right)^2 + 2C\frac{\mathrm{d}f}{\mathrm{d}\eta}\frac{\mathrm{d}^3 f}{\mathrm{d}\eta^3}$$

Noting that the Blasius function $f(\eta)$ satisfies Eq. 17.140, substituting the above in Eq. 17.142, we get

$$B\frac{\mathrm{d}^3 f}{\mathrm{d}\eta^3} + 2C\left(\frac{\mathrm{d}^2 f}{\mathrm{d}\eta^2}\right)^2 + 2C\frac{\mathrm{d}f}{\mathrm{d}\eta}\frac{\mathrm{d}^3 f}{\mathrm{d}\eta^3} + \frac{f}{2}\left\{B\frac{\mathrm{d}^2 f}{\mathrm{d}\eta^2} + 2C\frac{\mathrm{d}f}{\mathrm{d}\eta}\frac{\mathrm{d}^2 f}{\mathrm{d}\eta^2}\right\} = -Ec\left[\frac{\mathrm{d}^2 f}{\mathrm{d}\eta^2}\right]^2$$

or

$$B\frac{\mathrm{d}^3 f}{\mathrm{d}\eta^3} + 2C\left(\frac{\mathrm{d}^2 f}{\mathrm{d}\eta^2}\right)^2 + 2C\frac{\mathrm{d}f}{\mathrm{d}\eta}\frac{\mathrm{d}^3 f}{\mathrm{d}\eta^3} - \left(B + 2C\frac{\mathrm{d}f}{\mathrm{d}\eta}\right)\frac{\mathrm{d}^3 f}{\mathrm{d}\eta^3} = -Ec\left[\frac{\mathrm{d}^2 f}{\mathrm{d}\eta^2}\right]^2$$

After cancelation of terms, this simplifies to $C = -\frac{Ec}{2}$. Since $\frac{df}{d\eta} = 0$ at $\eta = 0$, the wall boundary condition requires that $A = T'_w$. Since $\frac{df}{d\eta} \to 1$ as $\eta \to \infty$ the free-stream boundary condition requires that $A + B + C = 1$. Thus, we get $B = 1 + \frac{Ec}{2} - T'_w$. With these, the solution to the energy equation is written down as

$$\boxed{T' = T'_w + \left[1 + \frac{Ec}{2} - T'_w\right]\frac{\mathrm{d}f}{\mathrm{d}\eta} - \frac{Ec}{2}\left[\frac{\mathrm{d}f}{\mathrm{d}\eta}\right]^2} \tag{17.145}$$

Adiabatic wall case

This case is interesting, as we shall see later. The governing equation is 17.142 subject to the boundary conditions

$$\left.\frac{\mathrm{d}T'}{\mathrm{d}\eta}\right|_{\eta=0} = 0, \ T'(\eta \to \infty) = 1 \tag{17.146}$$

The solution again starts with Eq. 17.144 and uses the above procedure (the reader should work out the details) to get

$$T' = 1 + \frac{Ec}{2} - \frac{Ec}{2}\left[\frac{\mathrm{d}f}{\mathrm{d}\eta}\right]^2 \tag{17.147}$$

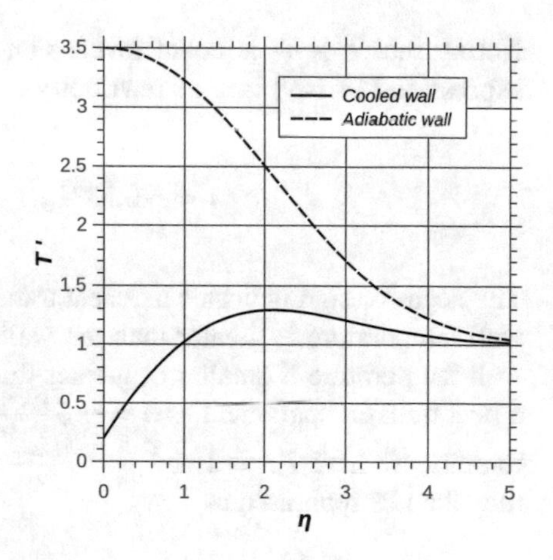

Fig. 17.32 Temperature variation in the boundary layer for two cases **a** Constant temperature wall (cooled wall) with $T_w = 0.2T_\infty$, **b** Adiabatic wall. $Ec = 5$ in both cases

It is seen that the adiabatic wall temperature is given by

$$T'(\eta = 0) = T'_{aw} = 1 + \frac{Ec}{2} \tag{17.148}$$

Figure 17.32 shows the variation of non-dimensional temperature inside the boundary layer for both wall temperature specified and the adiabatic wall cases.

Wall Heat Flux

In the case of isothermal plate, the wall heat flux is obtained by evaluating the conduction heat flux at the wall. By definition, we have

$$q_{w,x} = -\left(k\frac{\partial T}{\partial y}\right)\bigg|_{y=0} = -T_\infty k_\infty \times \frac{k_w}{k_\infty} \times \frac{\rho_\infty U}{\mu_\infty}\frac{\partial \eta}{\partial y'}\bigg|_{y'=0} \times \frac{dT'}{d\eta}\bigg|_{\eta=0}$$

By the definition of the similarity variable given in (17.138), we have

$$\frac{\partial \eta}{\partial y'}\bigg|_{y'=0} = \frac{1}{\sqrt{x'}}\frac{\rho_w}{\rho_\infty} = \sqrt{\frac{\mu_\infty}{\rho_\infty U x}}\frac{T_\infty}{T_w}$$

From Equation 17.145, we have

$$\frac{dT'}{d\eta}\bigg|_{\eta=0} = \left[1 + \frac{Ec}{2} - T'_w\right]\frac{d^2 f}{d\eta^2}\bigg|_{\eta=0} = \left[1 + \frac{Ec}{2} - T'_w\right] \times 0.332$$

With all these and on simplification the wall heat flux is obtained as

$$q_{w,x} = -0.332 k_\infty T_\infty \sqrt{\frac{\rho_\infty U}{\mu_\infty x}}\left[1 + \frac{Ec}{2} - T'_w\right] \tag{17.149}$$

Noting that $1 + \frac{Ec}{2}$ is equal to the non-dimensional adiabatic wall temperature, expression (17.149) may be rewritten as

$$q_{w,x} = -0.332 k_\infty T_\infty \sqrt{\frac{\rho_\infty U}{\mu_\infty x}} \left[T'_{aw} - T'_w\right] \quad (17.150)$$

The negative sign indicates that heat transfer takes place to the wall if the adiabatic wall temperature is greater than the wall temperature, irrespective of whether the wall temperature is smaller or greater than the free-stream temperature. We define a heat transfer coefficient as $h = -\frac{q_w}{T_{aw} - T_w}$. Also, since the Prandtl number is equal to unity, we have $Pr = 1$ or $\frac{C_p \mu}{k} = \frac{C_p \mu_\infty}{k_\infty} = 1$ or $k_\infty = C_p \mu_\infty$. Expression (17.150) may then be rephrased as

$$q_{w,x} = -h(T_{aw} - T_w) = 0.332 C_p \mu_\infty \sqrt{\frac{\rho_\infty U}{\mu_\infty x}}$$
$$= 0.332 \rho_\infty U C_p \sqrt{\frac{\mu_\infty}{\rho_\infty U x}} = 0.332 \frac{\rho_\infty U C_p}{x'}$$

where we have additionally made use of $T_{aw} = T_\infty T'_{aw}$ and $T_w = T_\infty T'_w$. Introducing the familiar Stanton number(note that it is a function of x'), we then get

$$St = \frac{h}{\rho_\infty U C_p} = \frac{0.332}{\sqrt{x'}} \quad (17.151)$$

The above expression is identical to the incompressible case but with the local Reynolds number x' based on the free-stream properties and the heat transfer coefficient based on the difference between the adiabatic wall temperature and the wall temperature.

Reference Temperature Method

The above treatment is elegant but has made the assumption that the Prandtl number is equal to unity. In the case of gases like air, the Prandtl number is close to, but not equal to unity. In order to take this into account, we have to modify the previous treatment. The adiabatic wall temperature is *not* equal to the value given earlier by Eq. 17.148. A correction is applied in the form of a recovery factor r defined such that

$$T_{aw, Pr \neq 1} = T_\infty + r \frac{U^2}{2 C_{p\infty}} \quad (17.152)$$

For engineering purpose, the recovery factor is well approximated, in laminar flow by

$$r = \sqrt{Pr},\ 0.6 \leq Pr \leq 15 \tag{17.153}$$

In order to take care of the variation of gas properties within the boundary layer, a reference temperature T^* defined by the relation given below is made use of.

$$\frac{T^*}{T_\infty} = 1 + 0.032 M_\infty^2 + 0.58 \left[\frac{T_w}{T_\infty} - 1 \right] \tag{17.154}$$

In the above expression M_∞ is the free-stream Mach number defined as the ratio of the free-stream velocity U to the speed of sound under free-stream conditions a_∞. If the ratio of specific heats γ is assumed to be constant, the Mach number is given by

$$M_\infty = \frac{U}{a_\infty} = \frac{U}{\sqrt{\gamma R T_\infty}} \tag{17.155}$$

where R is the gas constant. The following relation replaces Equation 17.151 for heat transfer.

$$St = \frac{0.332}{x'^{*\left(\frac{1}{2}\right)} \, Pr^{*\left(\frac{2}{3}\right)}} \tag{17.156}$$

In the above, the Prandtl and the Reynolds numbers (indicated with *) are evaluated using the gas properties at the reference temperature.

Example 17.18

Air flows past a flat plate at a free-stream Mach number of $M_\infty = 0.8$. The air pressure and temperature in the free-stream are $p_\infty = 1\ atm.$ and $T_\infty = 350$ K, respectively. The plate is maintained at a temperature of $T_w = 300$ K. Determine the heat flux on the plate at a location $x = 0.02$ m downstream of the leading edge. Use the reference temperature method. What will happen if compressibility effects and viscous dissipation are ignored?

Solution:
(Note that all temperatures are in Kelvin in this class of problems)

Step 1 The air properties at the free-stream condition are:

Density:	$\rho_\infty = 0.995$ kg/m^3
Kinematic viscosity:	$\nu_\infty = 20.92 \times 10^{-6}$ m^2/s
Specific heat:	$C_{p\infty} = 1009$ K/kg K
Thermal conductivity:	$k_\infty = 0.030$ W/m K

The Prandtl number of air may be taken as a constant at $Pr = 0.7$.

Step 2 The speed of sound in air at the ambient temperature is based on $\gamma = 1.4$ and $R = 287$ J/kg K.

$$a_\infty = \sqrt{1.4 \times 287 \times 350} = 375 \text{ m/s}$$

The free-stream air velocity is then obtained as

$$U = M_\infty a_\infty = 0.8 \times 375 = 300 \text{ m/s}$$

Step 3 The recovery factor (based on Eq. 17.153) is given by

$$r = \sqrt{Pr} = \sqrt{0.7} = 0.837$$

assuming the flow to be laminar. The adiabatic wall temperature is then given by

$$T_{aw} = T_\infty + r\frac{U^2}{2C_{p\infty}} = 350 + 0.837\frac{300^2}{2 \times 1009} = 387.3 \text{ K}$$

The reference temperature is given by (using Eq. 17.154)

$$\frac{T^*}{T_\infty} = 1 + 0.032 \times 0.8^2 + 0.58\left[\frac{300}{350} - 1\right] = 0.938$$

$$\text{or} \quad T^* = 0.938 \times 350 = 328.2 \text{ K}$$

Step 4 The air properties of interest at the reference temperature are:

Density:	$\rho^* = 1.068 \text{ kg/m}^3$
Dynamic viscosity:	$\mu^* = 19.75 \times 10^{-6} \text{ kg/m s}$

Step 5 We now calculate the Reynolds number using the properties at the reference temperature.

$$x'^* = \frac{\rho^* U x}{\mu^*} = \frac{1.068 \times 300 \times 0.02}{19.75 \times 10^{-6}} = 3.245 \times 10^5$$

The flow is laminar. The Stanton number is then given by (Eq. 17.156)

$$St = \frac{0.332}{\sqrt{(3.245 \times 10^5)}0.7^{\frac{2}{3}}} = 7.39 \times 10^{-4}$$

Step 6 The heat transfer coefficient is then given by

$$h = -St\rho_\infty U C_{p\infty} = -7.39 \times 10^{-4} \times 0.995 \times 300 \times 1009 = -222.67 \text{ W/m}^2\text{K}$$

Step 7 The local wall heat flux is then given by

$$q_{w,x=0.02\ m} = h(T_{aw} - T_w) = -222.67(387.5 - 300) = -19483.6 \text{ W/m}^2$$

Heat transfer is from air to wall.

Step 8 This may be compared with the incompressible boundary layer solution ignoring viscous dissipation. All the properties are based on the free-stream temperature. The various quantities of interest are (the reader may verify these):

Local Reynolds number:	$Re_x = 2.867 \times 10^5$
Stanton number:	$St = 7.864 \times 10^{-4}$
Heat transfer coefficient:	$h = 236.9 \text{ W/m}^2\text{K}$
Local wall heat flux:	$q_{w,x=0.02\ \text{m}} = -11840 \text{ W/m}^2$

There is thus a dramatic difference when compressibility effects are taken in to account, as they should be!

17.9 Current Topics of Interest in Heat Transfer

For the most part, the present book has dealt with conduction—radiation—convection as conventional text books do. Only in the present chapter, we have dealt with a few special topics in heat transfer. Even though some special topics have been considered in enough detail to motivate the reader, it has not been possible to be exhaustive in the coverage of special topics. With changes in technology, new topics keep adding to the list of special topics in heat transfer. In this section, we highlight some of the emerging areas and refer the reader to either more advanced books or the current heat transfer literature.

Typically, heat transfer journals nowadays partition the contents as shown below:

1	**Biological heat and mass transfer**
2	Combustion and reactive flows
3	Conduction
4	**Convection in microtubes and microchannels**
5	**Efficient cooling technology development**
6	**Electronic and photonic cooling**
7	**Experimental techniques for measuring thermal properties**
8	Heat and mass transfer
9	Heat and mass transfer in manufacturing and processing
10	**Heat and mass transfer in porous media**
11	**Heat and mass transfer in thermal and energy systems**
12	Heat exchanger fundamentals
13	Heat transfer applications
14	**Heat transfer enhancement**
15	**Heat transfer in manufacturing**
16	**High efficiency heat pipes**
17	**Jets, wakes, and impingement cooling**
18	Melting and Solidification
19	**Microscale and nanoscale heat transfer**
20	Multiple-phase flow, boiling and condensation
21	Natural and Mixed Convection
22	**New methods of measuring and/or correlating transport-property data**
23	Natural, forced and mixed convection
24	**Phase change heat transfer, melting, solidification and energy storage**
25	Radiation

Entries in **bold font** are new areas

New areas that find place in recent heat transfer literature are interdisciplinary in nature and may involve other branches of science such as physics,chemistry, biology, and material science. In recent times, we see many more examples of cooperative research involving scientists and engineers with background in thermal and other science areas that are involved in the special topics. Recently, much attention is focused on safety of nuclear reactors and thermal hydraulics applicable to reactors is an important area of study. Decay heat removal, loss of coolant analysis, efficient cooling using liquid metals are some of the special heat transfer areas that are important in such applications. Heat transfer in space applications in propulsion, cooling of surfaces subject to extreme heat fluxes such as during reentry are some special areas with tremendous technological importance. Study of energy transfer in the earth's atmosphere plays an important role in the understanding of weather. Radiation transfer in the atmosphere plays an important role and is a subject of considerable interest. Absorption and emission by gases in the atmosphere, scattering of radiation by aerosols are important in this, and hence have become areas of detailed study. Radiation heat transfer in the atmosphere also is important because of its application in remote sensing of the atmosphere, land, and sea surface using satellites that are capable of interrogating in various wavelength regions of the electromagnetic spectrum.

Concluding Remarks

We have come to the end of the book! After discussing some special topics, the book has ended with a list of areas that are currently receiving attention of researchers in heat transfer. When everyone thought that there was nothing new to study in heat transfer, areas opened up because of developments in other areas such as electronic devices, faster computers, concern about the atmosphere, space exploration, nuclear reactor design, Micro-Electro-Mechanical Systems or MEMS, and so on. Study of heat transfer will continue to be relevant.

17.10 Exercises

17.1 Water at a mean temperature of 80 °C flows with a velocity of 0.15 m/s inside a thin walled horizontal tube of 25 mm ID. At the outer surface of the tube, heat is transferred by natural convection to atmospheric air at 15 °C. Determine:
(a) Tube wall temperature, (b) Overall heat transfer coefficient, and (c) Heat loss per meter length of tube.

17.2 Redo Exercise 17.1 with the ambient air blowing normal to the tube axis with a velocity of 5 m/s. All other data is as given there.

17.3 A temperature measurement system uses a copper tube of 1 mm wall thickness and outer diameter of 6 mm. Thermocouple is attached to the wall of the tube. Thermocouple lead wires are such that conduction through them is negligible. The system is exposed to still air at 65 °C. The copper tube is tarnished due to use and has a surface emissivity of 0.65. Determine the temperature indicated by the thermocouple if the copper tube interacts by radiation with a background at 25 °C.

17.4 A thermometer has a spherical shape and is of 3 mm diameter. It is exposed to a moving fluid stream via a heat transfer coefficient of 15 W/m^2K. The thermometer also communicates radiatively to a background at 300 K. If the thermometer indicates a temperature of 333 K what is the fluid temperature? Consider two cases: (a) Emissivity of thermometer surface is 0.18 and (b) Emissivity of thermometer surface is 0.85. How will your answer change in the latter case if the heat transfer coefficient is changed to 45 W/m^2K?

17.5 A thermocouple is attached to the bottom of a cylidrical well that is exposed to an air stream at a temperature of $T_a = 95$ °C moving with a speed of $V = 2$ m/s normal to its axis. The cylinder is attached to a duct wall which is at $T_w = 80$ °C. The well is made of a material of thermal conductivity $k_w = 15$ W/m K, inner diameter $d_i = 4$ *mm* and outer diameter $d_o = 5$ mm. Assume that the cylinder outer surface has a gray emissivity of $\varepsilon = 0.18$.

- What is the reference temperature?
- What is the minimum value for the thermometer error?
- Determine the minimum length of immersion such that the thermometric error is limited to 1 °C.

17.6 A first order system has the following specifications:

- It is a spherical shell of copper with wall thickness 1 mm and OD 6 mm
- Fluid is still air at 30 °C
- Initial shell temperature is 80 °C

  How long should one wait for the cylinder to cool to 50 °C. Assume that the Nusselt number is governed by the correlation $Nu_D = 2 + 0.43 Ra_D^{0.25}$ where the symbols have the usual meanings. The heat transfer coefficient may be calculated at a mean cylinder temperature of 65 °C.

17.7 A first order system consists of a stainless steel shell of wall thickness 0.5 mm and outside diameter 12.5 mm which is initially at a temperature of 60 °C. The ambient fluid is atmospheric air at $10°C$. Convection heat transfer between the shell and ambient air is known to be 12 W/m^2 °C. The surface also loses heat by radiation with the surface emissivity of 0.85 to the ambient at the same temperature as ambient air. Obtain numerically the solution to this problem by the Runge–Kutta method. Determine the time at which the shell temperature is 15 °C.

17.8 List all the quantities that are required to describe heat transfer from a radiating-conducting fin. Identify the number of non-dimensional parameters that are expected based on dimensional arguments. Obtain all these non-dimensional parameters. Discuss the physical significance of each of these.

17.9 A radiating fin is in the form of a flat plate made of a special alloy having a thermal conductivity of 230 W/m K. The exposed surfaces of the fin are specially treated to achieve an emissivity of 0.8 and an absorptivity of 0.1. The fin base is held at a temperature of 127 °C. The fin is irradiated on both sides with an irradiation of 155 W/m^2. Obtain the thickness length combination for the fin that is in accord with the theoretical profile. Determine the net heat dissipated by the fin if it has a width of 0.45 m.

17.10 A radiating fin is in the form of a plate of uniform thickness 0.0035 m that is 0.12 m long. The plate material has a thermal conductivity of 100 W/m K. The plate surfaces have $\varepsilon_{IR} = 0.99$ and $\alpha_S = 0.12$. The fin base is attached to a base structure at 76 °C. Determine the heat loss from the fin if the plate is 0.5 m wide. Consider two cases (a) No irradiation and (b) Irradiation of $G = 400\ \text{W/m}^2$.

17.11 In service the plate surface in Exercise 17.10 deteriorates such that the absorptivity increases to 0.22. Rework the exercise for this condition. Also, determine the proportions of a minimum mass profile using the same amount of material as in Exercise 17.10 with the width remaining the same.

17.12 A tubular radiator consists of the required length of a thin walled tube of 0.025 m diameter. The outside surface has been chemically treated such that $\varepsilon = 0.8$ and $\alpha = 0.16$. The outside surface of the tube is subject to a uniform irradiation of $250\ \text{W/m}^2$. Water flows through the tube at a mass flow rate of 0.1 kg/s, entering at 85 °C and leaving at 45 °C. Determine the tube length.
Hint: The analysis presented in the text needs to be reworked for a duct of circular cross section.

17.13 In order to reduce the tube lengthand the pumping power (observe the answer in Exercise 17.10) fins may be attached to the tubes. Mackay[27] presents analysis of such finned radiators for space applications. Study the book and make a detailed note on how this may be used in the design of finned space radiators.

17.14 Air flows past a flat plate at a free-stream Mach number of 2.5. The pressure and temperature of free-stream air are 1000 Pa and 1000 K. What is the adiabatic wall temperature? If the plate is maintained at a temperature of 450 K determine the heat flux on the plate at a location 0.015 m downstream of the leading edge. Use the reference temperature method to solve the problem.

17.15 Identify all the parameters that are needed in describing the process of solidification. Form as many non-dimensional parameters as you can, using these. Compare these with the non-dimensional parameters used in the text.

17.16 It is clear from the analysis presented in the text that the ratios of solid to liquid properties affect the process of melting or solidification. Using Eq. 17.51 and the defining expressions (17.50) as guides, discuss how the properties affect melting or freezing times. Take ice water system as an example.

17.17 A very long channel of 10 cm spacing contains ice at 0 °C. The ice is allowed to melt by subjecting the top wall of the channel to a constant temperature of 5 °C starting at $t = 0$. The bottom wall of the channel is perfectly insulated. How long does it take to just melt the ice? What is the energy consumed per unit area in this process?

17.18 Ventilation system for a building consists of an underground heat exchanger through which outdoor air is drawn and cooled as shown

[27] D. B. Mackay, *Design of Space Power Plants*, Prentice Hall, 1963.

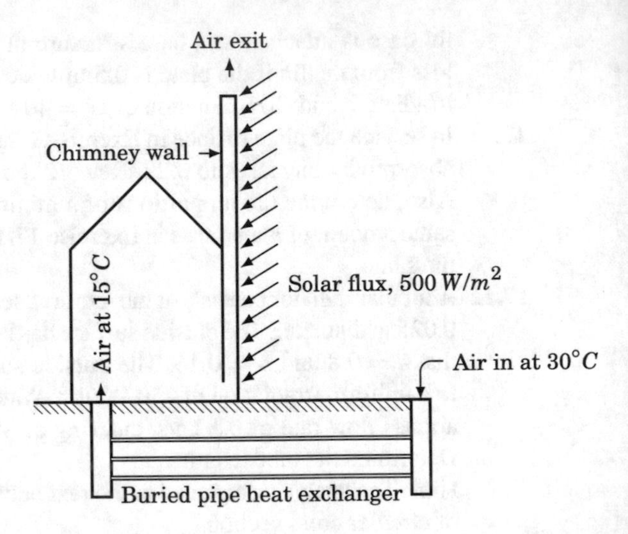

Fig. 17.33 Natural draft ventilation system

in Fig. 17.33. The air then passes into a room that is ventilated by a chimney wall that essentially consists of a double wall with the inner black wall that absorbs all the solar radiation that falls on it after passing through a front glass sheet that is transparent to solar flux. Ventilation is provided by natural draft induced by the chimney wall. Refer to current literature and try to analyze the problem. The goal is to design the chimney wall proportions for a desirable level of air flow through the room.

17.19 Consider a pebble bed heat storage unit. Initially, the entire bed is at 80 °C. The bed consists of spherical pebbles of 1 cm diameter. The length of the bed is 0.1 m. Air enters the bed at 15°C with a mass velocity of 0.5 kg/m^{2}s. The bed material properties may be taken as $\rho = 2800$ kg/m^3, $C_{pm} = 900$ J/kgK. Determine the state of affairs at the end of 5 minutes if the void fraction of the bed is 0.5. (This problem requires the solution and tables provided in S. P. Sukhatme, Solar Energy—Principles of Thermal Collection and Storage, Tata McGraw-Hill, 1984)

17.20 It is desired to heat air at atmospheric pressure from 100°C to 300 °C in a packed bed. The bed is contained in a 10 cm ID pipe filled with random arrangement of solid cylinders 5 mm diameter and 5 mm long. The flow rate of air is 6 kg/hr and the inside surface of the pipe is held at 400 °C. Determine the required bed height.

17.21 Saturated steam at 10^4 Pa and at rest condenses over a vertical tube of diameter equal to 0.025 m. Determine the distance from the top where the condensate film becomes turbulent.

17.22 Stephan (reference has already been cited in this chapter) has presented the details of how turbulent film condensation may be analyzed. Study this material and make a short report. Consider both stationary and moving vapor cases.

17.23 Saturated steam at 0.2 *bar* absolute condenses outside of a tube of 2.5 cm diameter maintained at a uniform temperature 20 °C below the saturation temperature of water. Calculate the average heat transfer coefficient when the tube is horizontal, as well as when it is vertical. The length of the tube is 0.3 m. Condensing steam may be assumed to be at rest.

17.24 A polished copper cylinder of 6 mm diameter is at 300 °C and is immersed in a pool of water at 90 °C. Film boiling of water is expected to take place at the surface of the cylinder. Determine the heat transfer rate immediately after the cylinder is immersed in water.

17.25 In a certain steam condenser steam at 10 kPa is condensed from dry steam to saturated liquid. The cooling water at 1000 kg/hr enters at a temperature of 30 °C and leaves at 37.5 °C. The water side flow takes place through tubes of ID 18 mm and 4 m length each. Determine the steam side heat transfer coefficient. Determine the water side heat transfer coefficient. The tube has an OD of 19 mm and is made of copper. What is the overall heat transfer coefficient based on tube ID?

17.26 Water in a pool boils over a very large horizontal plate in nucleate boiling mode. Water is at its saturation temperature at a pressure of 2 *bar* absolute while the plate is maintained at a temperature 25 °C above the saturation temperature. Determine the heat flux. What is the critical heat flux in this case?

Appendix A
Note on Bessel Functions

A.1 Background

Second-order ordinary differential equations with variable coefficients occur frequently in engineering applications. These equations naturally occur when we deal with problems with variable areas such as when we deal with conduction in cylindrical or spherical coordinate systems. These equations also occur in one-dimensional conduction when the area varies with x. The standard form of the Bessel equation is

$$\frac{d^2y}{dx^2} + \frac{1}{x}\frac{dy}{dx} + \left(1 - \frac{\nu^2}{x^2}\right)y = 0 \tag{A.1}$$

The Bessel equation thus involves a single parameter ν. This equation cannot be solved by elementary methods. The solution proceeds by expressing the solution as a power series.[1] Since Eq. A.1 is of second order, we ought to have two independent solutions. Both of these must come out of the power series solution.

A.1.1 Bessel Equation with Non-integer ν

Let us look for a solution, to begin with, for non-integer value of ν in the form

$$y(x) = \sum_{n=0}^{\infty} a_n x^{n+m} \tag{A.2}$$

[1] The solution method is referred to as the Frobenius method named after Ferdinand Georg Frobenius, 1849–1917, German mathematician.

S. P. Venkateshan, *Heat Transfer*,
https://doi.org/10.1007/978-3-030-58338-5

where m has to be determined as a part of the solution process. Assuming that the power series is convergent for any x, we may differentiate the power series term by term to obtain

$$\frac{\mathrm{d}y}{\mathrm{d}x} = \sum_{n=0}^{\infty} a_n (n+m) x^{n+m-1}; \frac{\mathrm{d}^2 y}{\mathrm{d}x^2} = \sum_{n=0}^{\infty} a_n (n+m)(n+m-1) x^{n+m-2};$$
$$\frac{\nu^2}{x^2} y = \nu^2 \sum_{n=0}^{\infty} a_n x^{n+m-2} \tag{A.3}$$

Substitute Eq. A.3 in Eq. A.1, group terms appropriately to get

$$\sum_{n=0}^{\infty} a_n [(n+m)(n+m-1) + (n+m) - \nu^2] x^{n+m-2} + \sum_{n=0}^{\infty} a_n x^{n+m} = 0 \tag{A.4}$$

Since Eq. A.4 should hold for any x, the coefficient of each power of x must vanish. Hence, for example, for x^0 term, assuming that $a_0 \neq 0$, we should have

$$m(m-1) + m - \nu^2 = m^2 - \nu^2 = 0 \text{ or } m = \pm\nu$$

However, then the coefficient x term is

$$(m+1)m + (m+1) - \nu^2 = (m+1)^2 - \nu^2 \neq 0 \text{ for } m = \pm\nu$$

This simply means that we have to choose $a_1 = 0$. Now we look at the coefficient of x^n in Eq. A.4 given by

$$[(n+m)(n+m-1) + (n+m) - \nu^2] a_n + a_{n-2}$$

and set it to zero. The resulting relationship may be written as

$$a_n = -\frac{a_{n-2}}{(n+m)^2 - \nu^2} \tag{A.5}$$

Case 1: $m = \nu$
Relation given in (A.5) leads to the recurrence relation

$$a_n = -\frac{a_{n-2}}{(n+\nu)^2 - \nu^2} = -\frac{a_{n-2}}{n(n+2\nu)} \tag{A.6}$$

Equation A.6 holds for $n \geq 2$. Repeated use of the recurrence relation will show that

$$a_1 = 0;\ a_3 = -\frac{a_1}{3(3+2\nu)} = 0;\ a_5 = -\frac{a_3}{5(5+2\nu)} = 0 \tag{A.7}$$
$$\text{and, in general } a_{2r-1} = 0 \text{ for } r = 1, 2, \ldots$$

and

$$a_2 = -\frac{a_0}{2(2+2\nu)} = -\frac{a_0}{2^2(2+\nu)};\ a_4 = -\frac{a_2}{4(4+2\nu)} = \frac{a_0}{2^4 2!(1+\nu)(2+\nu)} \text{ and,}$$
$$\text{ingeneral } a_{2r} = (-1)^r \frac{a_0}{2^{2r}(1+\nu)(2+)\cdots(r+\nu)} \text{ for } r = 0, 1, 2, \ldots \tag{A.8}$$

It is customary to choose a_0 as

$$a_0 = \frac{C_1}{2^\nu \nu\Gamma(\nu)} \tag{A.9}$$

where C_1 is an arbitrary constant and $\Gamma(\nu)$ is itself a special function.

A.1.2 Gamma Function: A Short Digression

Gamma function is also referred to as the generalized factorial function and is defined by the following:

$$\Gamma(\nu) = \int_0^\infty x^{\nu-1} e^{-x} \mathrm{d}x \tag{A.10}$$

which, on integration by parts, and for $n > 1$, leads to the recurrence relation

$$\Gamma(\nu + 1) = (\nu)\Gamma(\nu) \tag{A.11}$$

In fact, the Gamma function is defined by Eq. A.10 when the integral exists and by Eq. A.11 otherwise. We easily see that

$$\Gamma(1) = \int_0^\infty x^{1-1} e^{-x} \mathrm{d}x = \int_0^\infty e^{-x} \mathrm{d}x = 1;\ \Gamma(2) = 1\Gamma(1) = 1;$$
$$\Gamma(3) = 2\Gamma(2) = 2 \times 1 \times \Gamma(1) = 2!;\ \text{and, in general}$$
$$\Gamma(p+1) = p! \text{ where p is an integer}$$

The above makes it clear as to why the Gamma function is also referred to as the generalized factorial function. Gamma function may be defined for negative values of ν by the use of the recurrence relation rewritten as

$$\Gamma(\nu) = \frac{\Gamma(\nu + 1)}{\nu} \tag{A.12}$$

For example, we then have

$$\Gamma(-0.5) = \frac{\Gamma(0.5)}{-0.5};\ \Gamma(-1.5) = \frac{\Gamma(-0.5)}{1.5} = \frac{\Gamma(0.5)}{1.5 \times 0.5} \text{ etc.}$$

It is also seen that the values of $\Gamma(\nu)$ for any value of ν may be determined, using the $\Gamma(\nu)$ in the interval $1 < \nu < 2$ by the use of the recurrence relation. This is the reason why the Gamma function is tabulated for ν in this interval. Equation A.12 indicates that

$$\Gamma(0) = \lim_{\nu \to 0} \frac{\Gamma(\nu + 1)}{\nu} \to \pm\infty$$

The $\pm$ sign depends on the direction from which we approach the origin. Use of the recurrence relation then indicates that

$$\Gamma(\nu) \to \pm\infty$$

for $-\nu$ where ν is an integer.

A.1.3 Bessel Function of the First Kind

Now we get back to the solution to the Bessel equation. Using the properties of the Gamma function and the definition of a_0 given by Eq. A.9, Eq. A.8 may be recast as

$$a_{2r} = (-1)^r \frac{C_1}{2^{2r+\nu} r! \Gamma(\nu + r + 1)} \tag{A.13}$$

One solution to the Bessel equation is hence given by

$$y_1 = C_1 \sum_0^\infty (-1)^r \frac{C_1}{2^{2r+\nu} r! \Gamma(\nu + r + 1)} x^{\nu+2r} = C_1 J_\nu(x) \tag{A.14}$$

where $J_\nu(x)$ is the Bessel function of order ν and the first kind.
Case 2: $m = -\nu$
To obtain the second solution, all we have to do is to replace ν by $-\nu$ and get

$$y_2 = C_2 \sum_0^\infty (-1)^r \frac{C_1}{2^{2r-\nu} r! \Gamma(-\nu + r + 1)} x^{-\nu+2r} = C_1 J_{-\nu}(x) \tag{A.15}$$

where $J_{-\nu}(x)$ is the Bessel function of order $-\nu$ and the first kind. The above step is alright as long as ν is not an integer. In case ν is an integer, $J_{-\nu}(x)$ can be easily shown to be the same as $(-1)^\nu J_\nu(x)$. Hence, in case ν is an integer, we need to obtain the second solution by an alternate method.

A.2 Bessel Equation When ν is an Integer: Bessel Function of the Second Kind

We make use of the method of variation of parameters to obtain the solution to the Bessel equation when ν is an integer. Represent the solution to the Bessel equation (A.1) as $y = u(x)J_\nu(x)$. We then have

$$\frac{\mathrm{d}y}{\mathrm{d}x} = u\frac{\mathrm{d}J_\nu}{\mathrm{d}x} + J_\nu\frac{\mathrm{d}u}{\mathrm{d}x}; \quad \frac{\mathrm{d}^2 y}{\mathrm{d}x^2} = u\frac{\mathrm{d}^2 J_\nu}{\mathrm{d}x^2} + 2\frac{\mathrm{d}J_\nu}{\mathrm{d}x}\frac{\mathrm{d}u}{\mathrm{d}x} + J_\nu\frac{\mathrm{d}^2 u}{\mathrm{d}x^2}$$

Substitute these in Eq. A.1, group terms to get

$$J_\nu\frac{\mathrm{d}^2 u}{\mathrm{d}x^2} + 2\frac{\mathrm{d}J_\nu}{\mathrm{d}x}\frac{\mathrm{d}u}{\mathrm{d}x} + \frac{J_\nu}{x}\frac{\mathrm{d}u}{\mathrm{d}x} + u\left[\frac{\mathrm{d}^2 J_\nu}{\mathrm{d}x^2} + \frac{1}{x}\frac{\mathrm{d}J_\nu}{\mathrm{d}x} - \left(1 - \frac{\nu^2}{x^2}\right)J_\nu\right] = 0 \quad \text{(A.16)}$$

The terms in square brackets vanish since J_ν is a solution to the Bessel equation. The other terms may be rearranged in the variable separable form

$$\frac{\dfrac{\mathrm{d}^2 u}{\mathrm{d}x^2}}{\dfrac{\mathrm{d}u}{\mathrm{d}x}} = -\left[2\frac{\dfrac{\mathrm{d}J_\nu}{\mathrm{d}x}}{J_\nu} + \frac{1}{x}\right] \quad \text{(A.17)}$$

This is integrated with respect to x to get

$$\frac{\mathrm{d}u}{\mathrm{d}x} = \frac{C_2}{xJ_\nu^2} \quad \text{(A.18)}$$

where C_2 is a constant of integration. A second integration with respect to x will then give

$$u = C_1 + C_2\int_0^x \frac{\mathrm{d}x}{xJ_\nu^2} \quad \text{(A.19)}$$

where C_1 is a second constant of integration. Finally, the solution is obtained as

$$y = uJ_\nu = C_1 J_\nu + C_2 J_\nu\int_0^x \frac{\mathrm{d}x}{xJ_\nu^2} \quad \text{(A.20)}$$

Thus, we see that both solutions to the Bessel equation have resulted from the use of the method of variation of parameters. The desired second solution is given by

Table A.1 Zeros of $J_0(x)$ and $J_1(x)$

Zeros of $J_0(x)$	Zeros of $J_1(x)$
2.4048	0
5.5201	3.8317
8.6537	7.0156
11.7915	10.1735
14.9309	13.3237

$$Y_\nu(x) = J_\nu \int_0^x \frac{\mathrm{d}x}{x J_\nu^2} \tag{A.21}$$

is called the Bessel function of order ν and the second kind. As an example, it may be shown that the Bessel function of order zero and second kind is given by

$$Y_0(x) = \frac{2}{\pi}\left[J_0(x)\{\ln(x) + \ln 2 - \gamma\} - \frac{\left(\frac{x}{2}\right)^4}{(2!)^2}\left(1 + \frac{1}{2}\right) + \frac{\left(\frac{x}{2}\right)^6}{(3!)^2}\left(1 + \frac{1}{2} + \frac{1}{3}\right) - \cdots + \cdots \right] \tag{A.22}$$

where

$$\gamma = \lim_{n\to\infty} 1 + \frac{1}{2} + \frac{1}{3} + \cdots + \frac{1}{n} - \ln(n) \approx 0.577216\ldots$$

is known as the Euler constant. We make an important observation that $Y_0(x) \to \infty$ as $x \to 0$ and $Y_0(x) \to 0$ as $x \to \infty$. Both functions $J_0(x)$ and $Y_0(x)$ are oscillatory in nature.

Table A.1 gives the first few zeros of $J_0(x)$ and $J_1(x)$. The interval between successive zeros asymptotically approaches π.

Oscillatory behavior, as well as the locations of zeros of $J_0(x)$ and $Y_0(x)$, are brought out from Fig. A.1.

A.3 Asymptotic Behavior of Bessel Functions

It is interesting to look at what happens to Bessel functions when $x \to \infty$. For this purpose, introduce the transformation $y = \frac{u(x)}{\sqrt{x}}$ in Eq. A.1. We have

$$\frac{\mathrm{d}y}{\mathrm{d}x} = \frac{1}{\sqrt{x}}\frac{\mathrm{d}u}{\mathrm{d}x} - \frac{u}{2x^{3/2}}$$

$$\frac{\mathrm{d}^2 y}{\mathrm{d}x^2} = \frac{1}{\sqrt{x}}\frac{\mathrm{d}^2 u}{\mathrm{d}x^2} - \frac{1}{x^{3/2}}\frac{\mathrm{d}u}{\mathrm{d}x} + \frac{3u}{4x^{5/2}}$$

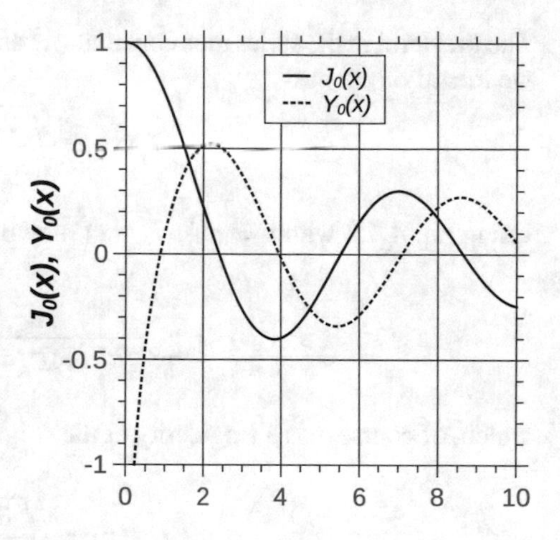

Fig. A.1 Plot of Bessel functions of order zero, first, and second kind

Introduce these in Eq. A.1 and simplify to get

$$\frac{\mathrm{d}^2 u}{\mathrm{d}x^2} - \left(1 - \frac{\nu^2 - \frac{1}{4}}{x^2}\right) u = 0 \tag{A.23}$$

It is seen from Eq. A.23 that when $x \gg \sqrt{\left|\nu^2 - \frac{1}{4}\right|}$, the Bessel equation reduces to the equation

$$\frac{\mathrm{d}^2 u}{\mathrm{d}x^2} + u = 0 \tag{A.24}$$

Also, when $\nu = \pm\frac{1}{2}$ the above is valid for all x! Eq. A.24 has the simple solution

$$u = C_1 \sin x + C_2 \cos x \text{ or } y = C_1 \frac{\sin x}{\sqrt{x}} + C_2 \frac{\cos x}{\sqrt{x}} \tag{A.25}$$

where C_1 and C_2 are constants of integration. Since the circular functions are periodic with a period of π the Bessel functions are also periodic with the period tending asymptotically to π. The Bessel functions exhibit a damping due to $\sqrt{x}$ in the denominator.

Using Eq. A.14, with $\nu = \frac{1}{2}$, $J_{\frac{1}{2}}(x)$ may be obtained as

$$J_{\frac{1}{2}}(x) = \sum_{0}^{\infty} \frac{(-1)^r}{2^{2r+\frac{1}{2}} r! \Gamma(\frac{1}{2} + r + 1)} x^{\frac{1}{2}+2r} \tag{A.26}$$

The leading term is given by

$$\frac{\sqrt{x}}{\sqrt{2}\frac{1}{2}\Gamma\frac{1}{2}} = \sqrt{\frac{2x}{\pi}} \tag{A.27}$$

The first term in the series thus contains $\sqrt{x}$ and hence $J_{\frac{1}{2}}(x) \to 0$ as $x \to 0$. Hence, we identify $J_{\frac{1}{2}}(x)$ as

$$J_{\frac{1}{2}}(x) = \sqrt{\frac{2}{\pi x}} \sin x \tag{A.28}$$

Using Eq. A.15, with $\nu = -\frac{1}{2}$, $J_{-\frac{1}{2}}(x)$ may be obtained as

$$J_{-\frac{1}{2}}(x) = \sum_{0}^{\infty} \frac{(-1)^r}{2^{2r-\frac{1}{2}} r! \Gamma(-\frac{1}{2} + r + 1)} x^{-\frac{1}{2}+2r} \tag{A.29}$$

which, of course, has a singularity at the origin. Hence, we identify $J_{\frac{1}{2}}(x)$ as

$$J_{-\frac{1}{2}}(x) = \sqrt{\frac{2}{\pi x}} \cos x \tag{A.30}$$

We are now able to generalize these observations and note that the Bessel functions of order $\pm\nu$ asymptotically behave as under for $x \gg \sqrt{|\nu^2 - \frac{1}{4}|}$:

$$J_\nu(x) \approx \sqrt{\frac{2}{\pi x}} \cos\left(x - \frac{\pi}{4} - \nu\frac{\pi}{2}\right); \; J_{-\nu}(x) \approx \sqrt{\frac{2}{\pi x}} \cos\left(x - \frac{\pi}{4} + \nu\frac{\pi}{2}\right) \tag{A.31}$$

A.4 Orthogonal Property of Bessel Functions

Bessel functions display an interesting orthogonality property in the interval (0, 1). Consider the Bessel functions $J_n(ax)$ and $J_n(bx)$, where a and b are two different roots of J_n. Then we can show that the following holds:

$$\int_0^1 x J_n(ax) J_n(bx) \mathrm{d}x = 0 \tag{A.32}$$

However, when $a = b$, we have

$$\int_0^1 x J_n^2(ax) \mathrm{d}x = \frac{J_{n+1}^2}{2} \tag{A.33}$$

This property of Bessel functions is useful in arriving at series expansion, known as Fourier Bessel series, of a function $f(x)$ in the interval (0, 1) as

$$\sum_{n=0}^{\infty} C_n J_n(a_n x) = \begin{Bmatrix} f(x) & \text{when it is continuous} \\ \frac{f(x^-)+f(x^+)}{2} & \text{at discontinuous points} \end{Bmatrix} \tag{A.34}$$

where the weights C_n are given by

$$C_n = \frac{2}{J_{n+1}^2(a_n)} \int_0^1 x f(x) J_n(a_n x) \mathrm{d}x \tag{A.35}$$

A.5 Modified Bessel Functions

The reason we have presented a note on Bessel functions in this Appendix is because of the connection they have with steady conduction in a variable area fin. Consider, for example, the case of a trapezoidal fin for which the governing equation has been derived in Chap. 4 and presented as Eq. 4.36. Even though the first two terms are the same as those in the Bessel equation, the third term is different, in that it is *negative*. This difference has crucial implications as far as the nature of the solution is concerned. The fin temperature variation is monotonic, while the solution to the Bessel equation has been seen to be oscillatory. We desire to obtain the solution to the fin problem by finding a suitable transformation that will transform it to the Bessel equation.

Make use of the transformation

$$z = 2ip\sqrt{x} \tag{A.36}$$

where z is a complex number! We may then easily show that

$$\mathrm{d}z = 2ip\frac{1}{2\sqrt{x}}\mathrm{d}x \text{ or } \frac{\mathrm{d}}{\mathrm{d}x} = -\frac{2p^2}{z}\frac{\mathrm{d}}{\mathrm{d}z} \tag{A.37}$$

and

$$\frac{\mathrm{d}^2}{\mathrm{d}x^2} = \frac{\mathrm{d}}{\mathrm{d}x}\left(\frac{\mathrm{d}}{\mathrm{d}x}\right) = -\frac{2p^2}{z}\frac{\mathrm{d}}{\mathrm{d}z}\left(-\frac{2p^2}{z}\frac{\mathrm{d}}{\mathrm{d}z}\right) = \frac{4p^4}{z^2}\left(\frac{\mathrm{d}^2}{\mathrm{d}z^2} - \frac{\mathrm{d}}{\mathrm{d}z}\right) \tag{A.38}$$

Introduce these in Eq. 4.36 to get the equation

$$\frac{\mathrm{d}^2\theta}{\mathrm{d}z^2} + \frac{1}{z}\frac{\mathrm{d}\theta}{\mathrm{d}z} + z = 0 \tag{A.39}$$

The fin equation has thus ended up in the form of Bessel equation with $\nu = 0$. The solution to Eq. A.39 may at once be written down as

$$\theta = C_1 J_0(z) + C_2 Y_0(z) = C_1 J_0(2ip\sqrt{x}) + C_2 Y_0(2ip\sqrt{x}) \tag{A.40}$$

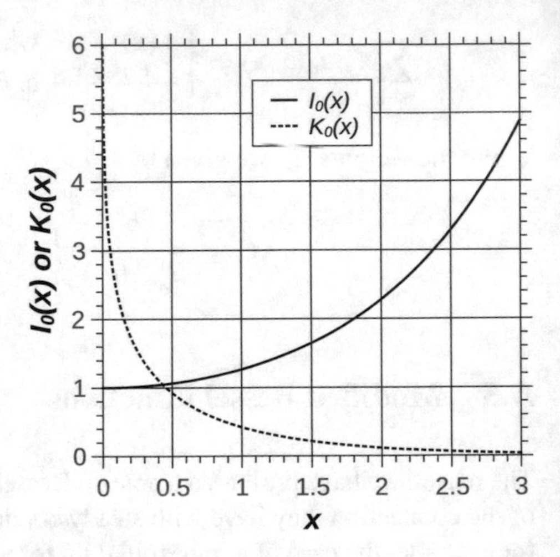

Fig. A.2 Plot of Modified Bessel functions of order zero, first, and second kind

Thus, the solution to the fin equation is obtained by expressing the solution as Bessel functions of order 0, of first and second kind, respectively, and for an imaginary argument! Since the temperature in the fin is a real function, the solution above must give a real function. Indeed it is so, and we, therefore, introduce modified Bessel functions of first and second kind, given by

$$I_0(x) = J_0(ix) \text{ and } K_0(x) = Y_0(ix) \tag{A.41}$$

We may easily show the following, starting from the power series representation of J_0 and Y_0 given earlier.

$$I_0(x) = \sum_0^{\infty} \frac{\left(\frac{x}{2}\right)^{2r}}{r!\Gamma(r+1)} = \sum_0^{\infty} \frac{\left(\frac{x}{2}\right)^{2r}}{r!^2} \tag{A.42}$$

and

$$K_0(x) = -I_0(x)\ln\left(\frac{x}{2}\right) + \sum_0^{\infty} \frac{\left(\frac{x}{2}\right)^{2r}}{r!^2}\phi(r) \tag{A.43}$$

where $\phi(r) = 1 + \frac{1}{2} + \frac{1}{3} + \cdots + \frac{1}{r}$ with $\phi(0) = 0$. In general, we also have

$$J_\nu(ix) = i^\nu I_\nu(x) \tag{A.44}$$

Monotonic behavior of $I_0(x)$ and $K_0(x)$ is apparent from Fig. A.2 that shows a plot of Modified Bessel functions of order zero, first and second kind. Also, $K_0(x)$ has a singularity at the origin. As in the case of the Bessel functions, we may study the asymptotic behavior of modified Bessel functions also. It may be shown, in a manner analogous to that made earlier, that

$$I_{\frac{1}{2}}(x) = \sqrt{\frac{2}{\pi x}}\sinh x; \quad K_{\frac{1}{2}}(x) = \sqrt{\frac{2}{\pi x}}\cosh x \tag{A.45}$$

for all x. It may also be shown that $I_\nu(x) \sim \frac{e^x}{\sqrt{x}}$ and $K_\nu(x) \sim \frac{e^{-x}}{\sqrt{x}}$ for $x \to \infty$.

A.6 General Form of Equation Solvable in Terms of Bessel Functions

A large number of important cases, that lead to solution in terms of Bessel functions, are represented by a general equation of the form

$$x^2\frac{d^2y}{dx^2} + (1-2\alpha)x\frac{dy}{dx} + [\beta^2\gamma^2x^{2\gamma} + \alpha^2 - m^2\gamma^2]y = 0 \tag{A.46}$$

where α, β, γ and m are parameters. The solution to Eq. A.46 is given by

$$y = C_1x^\alpha J_m(\beta x^\gamma) + C_2x^\alpha J_{-m}(\beta x^\gamma), \ \ m \text{ not an integer} \tag{A.47}$$

and

$$y = C_1x^\alpha J_m(\beta x^\gamma) + C_2x^\alpha Y_m(\beta x^\gamma), \ \ m \text{ an integer} \tag{A.48}$$

We shall demonstrate the usefulness of the above by two examples.

Example 1

Obtain the solution for temperature in a radial fin of uniform thickness using the general form of equation given by Eq. A.46.

Solution:
The equation governing the temperature variation in a radial fin of uniform thickness is given by (Eq. 4.52 in Chap. 4)

$$\frac{d^2y}{dx^2} + \frac{1}{x}\frac{dy}{dx} - p^2y = 0$$

where $p = \sqrt{\frac{h}{kt}}$ represents the fin parameter appropriate to this case. We have y in place of θ and x in place of r to avoid confusion. This equation may be recast, after multiplying through by x^2, as

$$x^2\frac{d^2y}{dx^2} + x\frac{dy}{dx} - p^2x^2y = 0 \tag{A.49}$$

We compare term by term Eq. A.49 with Eq. A.46 to arrive at the following:

$$1 - 2\alpha = 1 \text{ or } \alpha = 0;\ \ \beta^2\gamma^2x^{2\gamma} = -p^2x^2$$
$$\text{or } \gamma = 1 \text{ and } \beta^2 = -p^2 \text{ or } \beta = \pm ip;\ m = 0$$

Since the Bessel index is zero, the solution is written after (A.48) as

$$y = C_1x^0J_0(ipx) + C_2Y_0(ipx) \text{ or } \boxed{y = C_1I_0(px) + C_2K_0(px)}$$

Example 2

Express the solution of

$$\frac{d^2y}{dx^2} + a^2x^2y = 0 \tag{A.50}$$

in terms of Bessel functions.

Solution:
The given equation is in non-standard form, and hence is converted to standard form by a suitable transformation. Let $y = x^2Y$. Then, we have

$$\frac{dy}{dx} = x^2\frac{dY}{dx} + 2xY \text{ and } \frac{d^2y}{dx^2} = x^2\frac{d^2Y}{dx^2} + 4x\frac{dY}{dx} + 2Y$$

The given differential equation (A.50) then reduces to the required standard form

$$x^2\frac{d^2Y}{dx^2} + 4x\frac{dY}{dx} + [a^2x^4 + 2]Y = 0 \tag{A.51}$$

Compare Eq. (A.51) with (A.46)to get the following:

$$1 - \alpha = 4 \text{ or } \alpha = -\frac{3}{2};\ \ 2\gamma = 4 \text{ or } \gamma = 2$$
$$\beta^2\gamma^2 = a^2 \text{ or } \beta = \frac{a}{2};\ \ \alpha^2 - m^2\gamma^2 = 2 \text{ or } m = \frac{1}{4}$$

Since the Bessel index is not an integer, we make use of Eq. A.47 to write

$$Y(x) = C_1x^{-\frac{3}{2}}J_{\frac{1}{4}}\left(\frac{a}{2}x^{-\frac{3}{2}}\right) + C_2x^{-\frac{3}{2}}J_{-\frac{1}{4}}\left(\frac{a}{2}x^{-\frac{3}{2}}\right) \tag{A.52}$$

or

$$y(x) = C_1\sqrt{x}J_{\frac{1}{4}}\left(\frac{a}{2}x^{-\frac{3}{2}}\right) + C_2\sqrt{x}J_{-\frac{1}{4}}\left(\frac{a}{2}x^{-\frac{3}{2}}\right) \tag{A.53}$$

A.7 Some Useful Results

Some interesting but useful results are given in this section, without proof.

Useful recurrence relation
Starting from the defining equations, we may easily show that the following recurrence relations are followed by Bessel functions:

$$\begin{aligned}
&\text{(a)} \quad J_{\nu+1}(x) + J_{\nu-1}(x) = \tfrac{2\nu}{x} J_\nu(x)\\
&\text{(b)} \quad Y_{\nu+1}(x) + Y_{\nu-1}(x) = \tfrac{2\nu}{x} Y_\nu(x)\\
&\text{(c)} \quad I_{\nu+1}(x) - I_{\nu-1}(x) = -\tfrac{2\nu}{x} I_\nu(x)\\
&\text{(d)} \quad K_{\nu+1}(x) - K_{\nu-1}(x) = \tfrac{2\nu}{x} K_\nu(x)
\end{aligned} \tag{A.54}$$

This means that, if we have available to us tables of J_0 and J_1, then J_2, J_3, etc., may be calculated by the repeated use of the recurrence relation given by Eq. A.54(a) and so on. Derivative of the Bessel function may be evaluated by term by term differentiation. We can show, by this process that

$$\begin{aligned}
&\text{(a)} \quad \frac{dJ_0(x)}{dx} = -J_1(x); \quad \text{(b)} \quad \frac{dY_0(x)}{dx} = -Y_1(x);\\
&\text{(c)} \quad \frac{dI_0(x)}{dx} = I_1(x); \quad \text{(d)} \quad \frac{dK_0(x)}{dx} = -K_1(x)
\end{aligned} \tag{A.55}$$

We may also show the following general result, by the same procedure:

$$\begin{aligned}
&\text{(a)} \quad \frac{1}{x^{\nu+1}}\frac{d(x^{\nu+1}J_{\nu+1}(x))}{dx} = J_\nu(x); \quad \text{(b)} \quad \frac{1}{x^{\nu+1}}\frac{d(x^{\nu+1}Y_{\nu+1}(x))}{dx} = Y_\nu(x);\\
&\text{(c)} \quad \frac{1}{x^{\nu+1}}\frac{d(x^{\nu+1}I_{\nu+1}(x))}{dx} = I_\nu(x); \quad \text{(d)} \quad \frac{1}{x^{\nu+1}}\frac{d(x^{\nu+1}K_{\nu+1}(x))}{dx} = -K_\nu(x)
\end{aligned} \tag{A.56}$$

We consider the radial fin as an example to demonstrate the use of some of the formulae given above.

Example 3

Consider a radial fin of uniform thickness subject to insulated tip condition. Write down the complete solution to the problem. Also, obtain an expression for heat transfer at the base.

Solution:
We have seen in Example A.1 that the solution to the radial fin problem is given by

$$\theta = C_1 I_0(pr) + C_2 K_0(pr) \tag{A.57}$$

where we have reverted to the notation used in Chap. 4. The boundary conditions are specified as

$$\theta = \theta_b \text{ at } r = r_1 \tag{A.58}$$

and

$$\frac{d\theta}{dr} = 0 \text{ at } r = r_2 \tag{A.59}$$

Base boundary condition (A.58) requires that

$$C_1 I_0(pr_1) + C_2 K_0(pr_1) = \theta_b \tag{A.60}$$

Tip boundary condition (A.59) requires that

$$\left.\frac{d\theta}{dr}\right|_{r=r_2} = C_1 p I_1(pr_2) - C_2 p K_1(pr_2) = 0 \tag{A.61}$$

where the required derivatives follow from Eq. A.55. From Eq. A.61, we have $C_2 = C_1 \frac{I_1(pr_2)}{K_1(pr_2)}$. Substitute this in Eq. A.60 to get

$$C_1 = \theta_b \frac{K_1(pr_2)}{I_0(pr_1)K_1(pr_2) + I_1(pr_2)K_0(pr_1)}$$

This also yields, on back substitution

$$C_2 = \theta_b \frac{I_1(pr_2)}{I_0(pr_1)K_1(pr_2) + I_1(pr_2)K_0(pr_1)}$$

These lead to the fin temperature profile given by

$$\frac{\theta}{\theta_b} = \frac{K_1(pr_2)I_0(pr) + I_1(pr_2)K_0(pr)}{I_0(pr_1)K_1(pr_2) + I_1(pr_2)K_0(pr_1)} \tag{A.62}$$

The heat transfer at the base is given by

$$Q_b = -2\pi r_1 k \frac{d\theta}{dr}\bigg|_{r=r_1} = -2\pi p k \theta_b \frac{K_1(pr_2)I_1(pr_1) - I_1(pr_2)K_1(pr_1)}{I_0(pr_1)K_1(pr_2) + I_1(pr_2)K_0(pr_1)} \quad (A.63)$$

where again the required derivatives follow from Eq. A.55.

A.8 Tables of Bessel Functions and Modified Bessel Functions

We round off this Appendix by presenting a short table of Bessel and modified Bessel functions that will be useful in solving heat transfer problems. In practice, the reader will find it useful to use a spreadsheet program such as Microsoft EXCEL or Linux based LibreOffice Calc which have built in functions that may easily be used directly in calculations (Tables A.2 and A.3).

Table A.2 Table of Bessel functions

x	$J_0(x)$	$Y_0(x)$	$J_1(x)$	$Y_1(x)$
0.0	1	$-\infty$	0	$-\infty$
0.1	0.997502	−1.534239	0.049938	−6.458951
0.2	0.990025	−1.081105	0.099501	−3.323825
0.3	0.977626	−0.807274	0.148319	−2.293105
0.4	0.960398	−0.606025	0.196027	−1.780872
0.5	0.938470	−0.444519	0.242268	−1.471472
0.6	0.912005	−0.308510	0.286701	−1.260391
0.7	0.881201	−0.190665	0.328996	−1.103250
0.8	0.846287	−0.086802	0.368842	−0.978144
0.9	0.807524	0.005628	0.405950	−0.873127
1.0	0.765198	0.088257	0.440051	−0.781213
1.5	0.511828	0.382449	0.557937	−0.412309
2.0	0.223891	0.510376	0.576725	−0.107032
2.5	−0.048384	0.498070	0.497094	0.145918
3.0	−0.260052	0.376850	0.339059	0.324674
3.5	−0.380128	0.189022	0.137378	0.410188
4.0	−0.397150	−0.016941	−0.066043	0.397926
4.5	−0.320543	−0.194705	−0.231060	0.300997
5.0	−0.177597	−0.308518	−0.327579	0.147863
5.5	−0.006844	−0.339481	−0.341438	−0.023758
6.0	0.150645	−0.288195	−0.276684	−0.175010
6.5	0.260095	−0.173242	−0.153841	−0.274091

(continued)

Table A.2 (continued)

x	$J_0(x)$	$Y_0(x)$	$J_1(x)$	$Y_1(x)$
7.0	0.300079	−0.025950	−0.004683	−0.302667
7.5	0.266340	0.117313	0.135248	−0.259129
8.0	0.171651	0.223521	0.234636	−0.158060
8.5	0.041939	0.270205	0.273122	−0.026169
9.0	−0.090334	0.249937	0.245312	0.104315
9.5	−0.193929	0.171211	0.161264	0.203180
10.0	−0.245936	0.055671	0.043473	0.249015

Table A.3 Table of modified Bessel functions

x	$I_0(x)$	$K_0(x)$	$I_1(x)$	$K_1(x)$
0.0	1	∞	0	∞
0.1	1.002502	2.427069	0.050063	9.853845
0.2	1.010025	1.752704	0.100501	4.775973
0.3	1.022627	1.372460	0.151694	3.055992
0.4	1.040402	1.114529	0.204027	2.184354
0.5	1.063483	0.924419	0.257894	1.656441
0.6	1.092045	0.777522	0.313704	1.302835
0.7	1.126303	0.660520	0.371880	1.050284
0.8	1.166515	0.565347	0.432865	0.861782
0.9	1.212985	0.486730	0.497126	0.716534
1.0	1.266066	0.421024	0.565159	0.601907
1.5	1.646723	0.213806	0.981666	0.277388
2.0	2.279585	0.113894	1.590637	0.139866
2.5	3.289839	0.062348	2.516716	0.073891
3.0	4.880793	0.034740	3.953370	0.040156
3.5	7.378203	0.019599	6.205835	0.022239
4.0	11.301922	0.011160	9.759465	0.012483
4.5	17.481172	0.006400	15.389223	0.007078
5.0	27.239872	0.003691	24.335642	0.004045
5.5	42.694645	0.002139	38.588165	0.002326
6.0	67.234407	0.001244	61.341937	0.001344
6.5	106.293	0.000726	97.735011	0.000780
7.0	168.594	0.000425	156.039	0.000454
7.5	268.161	0.000249	249.584	0.000265

(continued)

Table A.3 (continued)

x	$I_0(x)$	$K_0(x)$	$I_1(x)$	$K_1(x)$
8.0	427.564	0.000146	399.873	0.000155
8.5	683.162	0.000086	641.620	0.000091
9.0	1093.588	0.000051	1030.915	0.000054
9.5	1753.481	0.000030	1658.453	0.000032
10.0	2815.717	0.000018	2670.988	0.000019

Appendix B
Note on Legendre Functions

B.1 Background

Second-order ordinary differential equations with variable coefficients occur frequently in engineering applications as indicated in Appendix A. Legendre polynomials make their appearance when we deal with the Laplace equation in spherical coordinates. The Legendre equation is of form

$$(1-x^2)\frac{d^2y}{dx^2} - 2x\frac{dy}{dx} + n(n+1)y = 0 \tag{B.1}$$

where n is a parameter.

B.1.1 Special Simple Case of Legendre Equation

We shall build up the solution to the Legendre equation by considering first a special case, corresponding to $n = 0$. Equation B.1 will then take the variable separable form given by

$$\frac{\frac{\mathrm{d}^2y}{\mathrm{d}x^2}}{\frac{\mathrm{d}y}{\mathrm{d}x}} = \frac{2x}{(1-x^2)} \tag{B.2}$$

This equation may be integrated once to get

$$\ln\left(\frac{\mathrm{d}y}{\mathrm{d}x}\right) = -\ln(1-x^2) + \ln(A_2) \tag{B.3}$$

S. P. Venkateshan, *Heat Transfer*,
https://doi.org/10.1007/978-3-030-58338-5

where A_2 is a constant of integration. The above equation may be rewritten in the form

$$\frac{dy}{dx} = \frac{A_2}{1-x^2} = \frac{A_2}{2}\left[\frac{1}{1+x} + \frac{1}{1-x}\right]$$

A second integration leads immediately to the solution

$$y = \frac{A_2}{2}\ln\left[\frac{1+x}{1-x}\right] + A_1 \tag{B.4}$$

where A_1 is a second constant of integration. Thus, the general solution to the Legendre equation with $n = 0$ may be written down as

$$\begin{aligned} y &= A_1 P_0(x) + A_2 Q_0(x) \\ \text{where}\quad P_0(x) &= 1 \text{ and } \quad Q_0(x) = \tfrac{1}{2}\ln\left[\tfrac{1+x}{1-x}\right] \end{aligned} \tag{B.5}$$

In Eq. B.5 above, $P_0(x)$ is the Legendre function of degree 0 and the first kind while $Q_0(x)$ is the Legendre function of degree 0 and the second kind. We notice that $Q_0(x)$ is singular at the origin.

B.1.2 Legendre Equation with *n* ≥ 1

We make use of the method of series in this case. Let the solution be given by

$$y = \sum_{m=0}^{\infty} C_m x^m \tag{B.6}$$

Term by term differentiation of the above series yields

$$\frac{dy}{dx} = \sum_{m=0}^{\infty} C_m m x^{m-1}; \quad \frac{d^2y}{dx^2} = \sum_{m=0}^{\infty} C_m m(m-1)x^{m-2}$$

Substitute these in Eq. B.1 to get, after minor simplification

$$\sum_{m=0}^{\infty} C_m x^m[-m(m+1) + n(n+1)] + \sum_{m=0}^{\infty} C_m x^{m-2} m(m-1) = 0 \tag{B.7}$$

Equate coefficients of terms containing x^0 to zero and get

$$C_0 n(n+1) + C_2(2 \times 1) = 0 \text{ or } C_2 = -\frac{n(n+1)}{2}C_0$$

Thus, C_2 is related to C_0 which is taken as a non-zero constant. Equate coefficient of x^1 to zero in Eq. B.6 and get

$$C_1[n(n+1)-2]+C_3(3\times 2)=0 \quad \text{or} \quad C_3=-\frac{(n-1)(n+2)}{3!}C_1$$

Thus, C_3 is related to C_1 which again may be taken as a non-zero constant. We may generalize the above and mention that all even coefficients (i.e., C_0, C_2, $C_4\ldots$) are related and yield a series with even powers of x. Similarly, all odd coefficients, i.e., C_1, C_3, $C_5\ldots$) are related and yield a series with odd powers of x. These two series must, therefore, form two independent solutions to the Legendre equation. The required relations between the coefficients of the two series may be represented by the following recurrence relation:

$$C_{s+2}=-\frac{(n-s)(n+s+1)}{(s+2)(s+1)}C_s \tag{B.8}$$

Let us represent the two solutions as $y_1(x)$ and $y_2(x)$. The two series are given by the following relations:

$$\begin{aligned} y_1(x)&=1+\sum_{s=1}^{\infty}(-1)^s\frac{[(n-2s+2)\cdots(n-2)][(n+1)\cdots(n+2s-1)]}{2n!}x^{2s}\\ y_2(x)&=x+\sum_{s=1}^{\infty}(-1)^s\frac{[(n-2s+1)\cdots(n-1)][(n+2)\cdots(n+2s)]}{2n!}x^{2s+1} \end{aligned} \tag{B.9}$$

When n is an even integer, $y_1(x)$ reduces to a polynomial of degree n with only even powers of x and the series $y_2(x)$ diverges. When n is an odd integer, $y_2(x)$ reduces to a polynomial of degree n with only odd powers of x and the series $y_1(x)$ diverges. These polynomials are known as Legendre polynomials and represented by the symbol $P_n(x)$. Thus, we have

$$\begin{aligned} P_n(x)&=C\ y_1(x) \quad \text{if } n \text{ is an even integer}\\ &=C\ y_2(x) \quad \text{if } n \text{ is an odd integer} \end{aligned} \tag{B.10}$$

where C is chosen such that $P_n(1)=1$. A few of these polynomials are given below

$P_0(x)=1$	$P_1(x)=x$
$P_2(x)=\frac{1}{2}(3x^2-1)$	$P_3(x)=\frac{1}{2}(5x^3-3x)$
$P_4(x)=\frac{1}{8}(35x^4-30x^2+3)$	$P_5(x)=\frac{1}{8}(63x^5-70x^3+15x)$

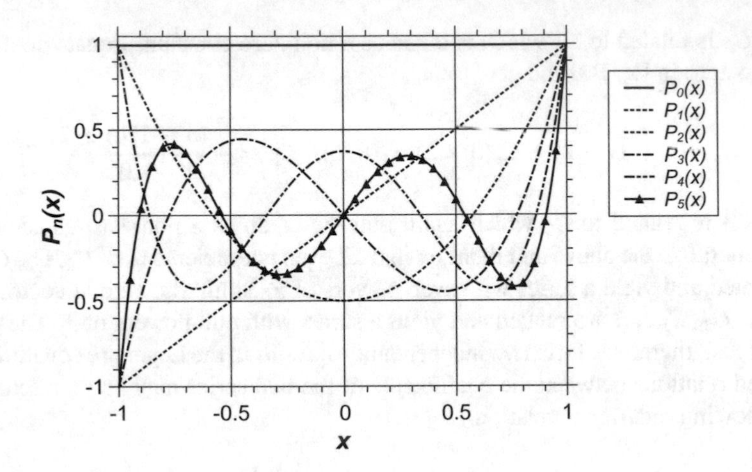

Fig. B.1 Legendre polynomials

B.1.3 Legendre Function of Second Kind

We have already come across the Legendre function of degree n and second kind in the case corresponding to $n = 0$. In fact, the function Q_0 was given by the second of Eq. B.5. Without proof, we give the following:

$$Q_1(x) = Q_0(x)P_1(x) - 1; \quad Q_2(x) = Q_0(x)P_2(x) - \frac{3}{2}x; \quad Q_3(x) = Q_0(x)P_3(x) - \frac{5}{2}x^2 + \frac{2}{3} \tag{B.11}$$

In general, we have

$$Q_n(x) = Q_0(x)P_n(x) - \frac{(2n-1)}{1 \cdot n}P_{n-1}(x) - \frac{(2n-5)}{3 \cdot (n-1)}P_{n-3}(x) - \cdots \tag{B.12}$$

The Legendre functions of second kind are not pursued further. We show plots of a few Legendre polynomials in Fig. B.1. It is clear that $P_n(x)$ are regular at both $x = 1$ and $x = -1$. We also show plots of a few Legendre functions of second kind in Fig. B.2. Clearly, these functions diverge at $x = 1$ and $x = -1$.

B.1.4 Some Useful Relations Involving Legendre Polynomials

In applications, it is useful to have relations among Legendre polynomials that help in calculations. Legendre polynomial for negative values of x are related to $P_n(x)$ according to the relation $P_n(-x) = (-1)^n P_n(x)$ as may be verified. A useful recurrence relation is given by

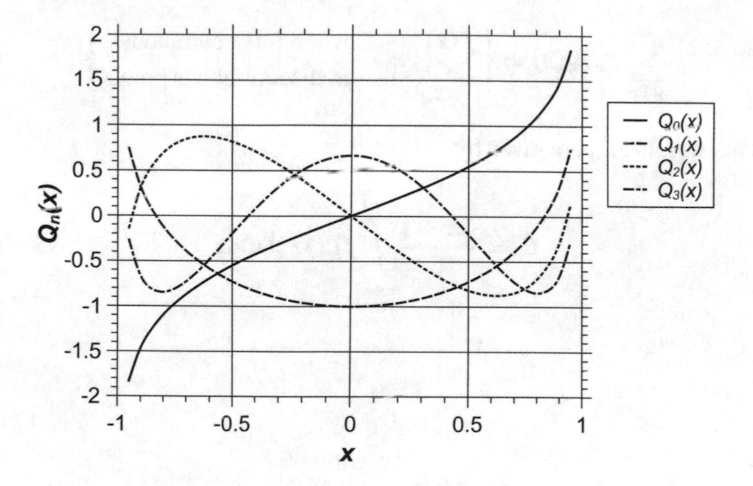

Fig. B.2 Legendre functions of second kind

$$(n+1)P_{n+1}(x) + nP_{n-1}(x) = (2n+1)xP_n(x) \tag{B.13}$$

Rodrigues' formula generates the Legendre polynomial and is given by

$$P_n(x) = \frac{1}{2^n n!}\frac{d^n}{dx^n}(x^2-1)^n \tag{B.14}$$

The first derivatives follow a recurrence formula given by

$$P'_{n+1}(x) - P'_{n-1}(x) = (2n+1)P_n(x) \tag{B.15}$$

where $'$ represents $\frac{\mathrm{d}}{\mathrm{d}x}$.

B.1.5 *Orthogonality property and Fourier Legendre series*

Most importantly the Legendre polynomials have the following orthogonality property over the interval $(-1, +1)$.

$$\int_{-1}^{1} P_n(x)P_m(x)dx = 0 \text{ if } m \neq n; \quad \int_{-1}^{1} P_n(x)P_m(x)\mathrm{d}x = \frac{2}{2n+1} \text{ if } m = n \tag{B.16}$$

The orthogonality property is very useful in that we may represent any piecewise continuous function $f(x)$ in the interval $(-1, +1)$ in terms of Legendre polynomials as Fourier Legendre series given by

$$\sum_{n=0}^{\infty} C_n P_n(x) = \begin{Bmatrix} f(x) & \text{when it is continuous} \\ \frac{f(x^-)+f(x^+)}{2} & \text{at discontinuous points} \end{Bmatrix} \tag{B.17}$$

where the weights C_n are given by

$$C_n = \frac{2n+1}{2} \int_{-1}^{+1} f(x) P_n(x) \mathrm{d}x \tag{B.18}$$

Appendix C
Basics of Complex Variables

C.1 Introduction

Two-dimensional steady conduction problems in cartesian, as well as cylindrical coordinates, may be solved easily by the use of complex variables technique. The complex variables technique is useful as a versatile tool for the solution of Laplace equation. We develop the basic ideas here for use in the text at appropriate places. We make use of an important property of a differentiable function of a complex variable, for arriving at the solution to the Laplace equation.

C.1.1 Definitions

A complex number z is an ordered pair (x, y) defined as below

$$z = x + iy \tag{C.1}$$

x is referred to as the real part and y the imaginary part of the complex number. $i = \sqrt{-1}$ is referred to as pure imaginary number. The magnitude of the complex number z is given by $|z| = \sqrt{x^2 + y^2}$. The angle θ is given by $\theta = \tan^{-1}\left(\frac{y}{x}\right)$. A complex number may also be represented in cylindrical coordinates as

$$z = re^{i\theta} \tag{C.2}$$

with $x = r\cos\theta$, $y = r\sin\theta$ and $r = |z|$. Figure C.1a shows the geometric representation of a complex number. It is represented as a point in the (x, y) plane where the horizontal axis is referred to as the real axis and the vertical axis is referred to as the imaginary axis. The angle made by the line joining the point P with the origin and the real axis is the angle θ. The length of the line joining P and the origin is the

S. P. Venkateshan, *Heat Transfer*,
https://doi.org/10.1007/978-3-030-58338-5

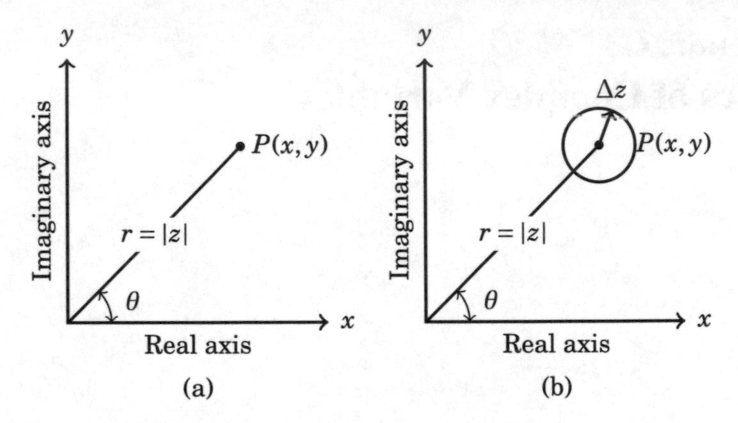

Fig. C.1 **a** Representation of a complex number **b** Neighborhood of z and path Δz

magnitude of the complex number r or $|z|$.

A function $w = f(z)$ is written as a complex number given by

$$w = f(z) = u(x, y) + iv(x, y) \tag{C.3}$$

where $u(x, y)$ and $v(x, y)$[2] are real functions of x and y, referred to, respectively, as the real part and the complex part of w. The function $w = f(z)$ plots as a point in the so-called w plane where the abscissa represents the real part and the ordinate represents the imaginary part of the function.

C.1.2 Conditions for the Existence of Derivative of w

Cartesian coordinates

Let us explore the conditions under which w has a derivative at point P. For this purpose, consider the neighborhood of a point z. From calculus, we know that the derivative, if it exists, is obtained by a limiting process of approaching the point P from its neighborhood, and defining the derivative as

$$\frac{\mathrm{d}f}{\mathrm{d}z} = \lim_{\Delta z \to 0} \frac{f(z + \Delta z) - f(z)}{\Delta z} \tag{C.4}$$

In the present case, the path may be along any radial line, starting from a point on the circle with radius $|\Delta z|$ centered at P. The derivative, if it exists, will have to be the same irrespective of the direction along which the limit is taken. Hence, we choose two such paths, the first, path 1 with $dx = 0$, $dy \to 0$ and the second, path 2 with

[2]It is fairly common to represent w as $\phi(x, y) + i\psi(x, y)$ where $\phi(x, y)$ is referred to as the potential function and $\psi(x, y)$ is referred to as the stream function.

$dx \to 0, dy = 0$, and see what happens.

Path 1

Along this path, we have, $\Delta z = 0 + i\Delta y$ and hence

$$\begin{aligned} \frac{\mathrm{d}f}{\mathrm{d}z} &= \lim_{\Delta x = 0, \Delta y \to 0} \frac{[u(x, y + \Delta y) - u(x, y)] + i[v(x, y + \Delta y) - v(x, y)]}{0 + i\Delta y} \\ &= \frac{1}{i}\frac{\partial u}{\partial y} + \frac{\partial v}{\partial y} = -i\frac{\partial u}{\partial y} + \frac{\partial v}{\partial y} \end{aligned} \tag{C.5}$$

Path 2

Along this path, we have, $\Delta z = \Delta x + i0$ and hence

$$\frac{\mathrm{d}f}{\mathrm{d}z} = \lim_{\Delta x \to 0, \Delta y = 0} \frac{[u(x + \Delta x, y) - u(x, y)] + i[v(x + \Delta x, y) - v(x, y)]}{\Delta x + i0}$$

$$= \frac{\partial u}{\partial x} + i\frac{\partial v}{\partial x} \tag{C.6}$$

If the function $f(z)$ has a derivative at point z, then the two expressions given by Eqs. C.5 and C.6 must be identical. This requires that the real part and imaginary part be identical, and hence we get the relations

$$\text{(a)} \quad \frac{\partial u}{\partial x} = \frac{\partial v}{\partial y}, \quad \text{(b)} \quad \frac{\partial u}{\partial y} = -\frac{\partial v}{\partial x} \tag{C.7}$$

These relations are known as Cauchy-Riemann conditions. Let us assume that u and v are twice differentiable. We take the x derivative of Eq. C.7(a) and y derivative of Eq. C.7(b) to get

$$\frac{\partial^2 u}{\partial x^2} = \frac{\partial^2 v}{\partial y \partial x}; \quad \frac{\partial^2 u}{\partial y^2} = -\frac{\partial^2 v}{\partial x \partial y}$$

Add these two equations to get

$$\frac{\partial^2 u}{\partial x^2} + \frac{\partial^2 u}{\partial y^2} = \frac{\partial^2 v}{\partial y \partial x} - \frac{\partial^2 v}{\partial x \partial y} = 0 \tag{C.8}$$

since the cross derivative is immune to the order of differentiation. Thus, $u(x, y)$ satisfies the Laplace equation. Similarly, one may show that $v(x, y)$ is also a solution of the Laplace equation. Thus, we have the important property that the real part and imaginary part of a function of complex variable are solutions to the Laplace equation, if the complex function has a derivative at the point z.

Cylindrical coordinates

Cauchy–Riemann conditions in cylindrical coordinates may be shown, by a procedure similar to that followed above, to be given by

$$\text{(a)} \quad \frac{\partial u}{\partial r} = \frac{1}{r}\frac{\partial v}{\partial \theta} \quad \text{(b)} \quad \frac{\partial v}{\partial r} = -\frac{1}{r}\frac{\partial u}{\partial \theta} \tag{C.9}$$

where $w = u(r, \theta) + iv(r, \theta)$. Both $u(r, \theta)$ and $v(r, \theta)$ satisfy Laplace equation in cylindrical coordinates given by Eq. 5.144, where T may stand either for u or v.

C.1.3 Complex Potential

A function of a complex variable that has a derivative is referred to as an analytic or regular function of a complex variable. Such a function is also referred to as a complex potential with its real part representing a potential function and the complex part (or the imaginary part) representing a stream or flux function. These terms actually originate from ideal fluid flow theory. As seen earlier, both the potential, as well as the stream function satisfy the Laplace equation.

Consider now the curves represented by $u(x, y) =$ constant. On such a curve, we have $du = 0$ or

$$\mathrm{d}u = \frac{\partial u}{\partial x}\mathrm{d}x + \frac{\partial u}{\partial y}\mathrm{d}y = 0 \quad \text{or} \quad \left.\frac{\mathrm{d}y}{\mathrm{d}x}\right|_u = -\frac{u_x}{u_y} \tag{C.10}$$

where the subscript u means that it is held constant, and subscript x and y represent partial derivatives. Similarly, on curves with $v(x, y) =$ constant, we have

$$\mathrm{d}v = \frac{\partial v}{\partial x}\mathrm{d}x + \frac{\partial v}{\partial y}\mathrm{d}y = 0 \quad \text{or} \quad \left.\frac{\mathrm{d}y}{\mathrm{d}x}\right|_v = -\frac{v_x}{v_y} \tag{C.11}$$

Using the Cauchy–Riemann conditions (Eq. C.7) Eq. C.11 may be rewritten as

$$\left.\frac{dy}{dx}\right|_v = \frac{u_y}{u_x} \tag{C.12}$$

Now consider a possible point of intersection between these two curves. The derivatives given by Eqs. C.10 and C.12 represent the slopes of the tangents to these curves, at the point of intersection. We see that the product of the two slopes is -1 showing that the two curves intersect orthogonally. At once the connection with heat conduction becomes apparent. Isotherms and flux lines are known to be orthogonal. Also, the

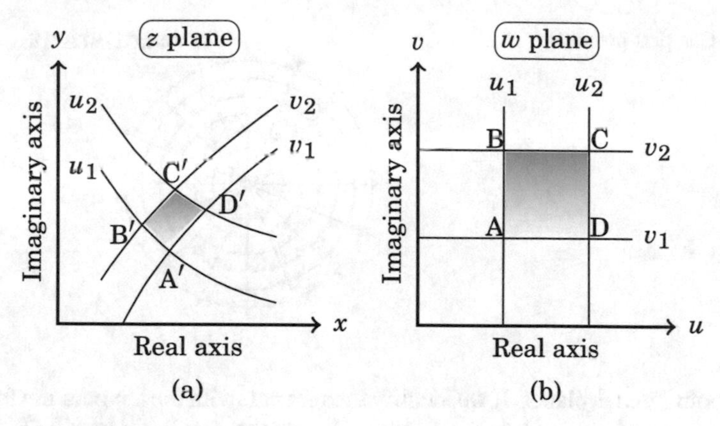

Fig. C.2 Transformation $w = f(z)$ represented in z and w planes

temperature is a solution to the Laplace equation. Hence, temperature function may be identified with the potential function and the heat flux function may be identified with the stream function (or vice versa).

Consider a complex potential w. It may be plotted in the z plane by showing constant u curves and constant v curves as shown in Fig. C.2a. These show up, in the w plane, as lines parallel to the two axes, as shown in Fig. C.2b. The points $ABCD$ form a rectangle when two sets of constant u and constant v lines intersect. In the z plane the intersection points define $A'B'C'D'$, a curvilinear rectangle, as shown. The angles at each of the corners are maintained at 90° in both planes. In fact, we may look upon the function $w = f(z)$ as a transformation that transforms or maps a curvilinear rectangle in the z plane to a rectangle in the w plane.

Even though the transformation that has been indicated in the figure is one to one, it may also be one to many. This is determined by the nature of the complex potential.

C.2 Examples of Complex Potentials

We consider a few examples of complex potentials that will be of use in heat transfer studies and ideal fluid flow. By definition, these will be solutions to the Laplace equation and hence solutions to steady conduction in two dimensions, in the absence of heat generation.

C.2.1 Complex Potential $w = z$

The first example is a simple one given by $w = z = x + iy$. We at once see that $u(x, y) = x$ and $v(x, y) = y$. Constant u lines and constant v lines form rectangular

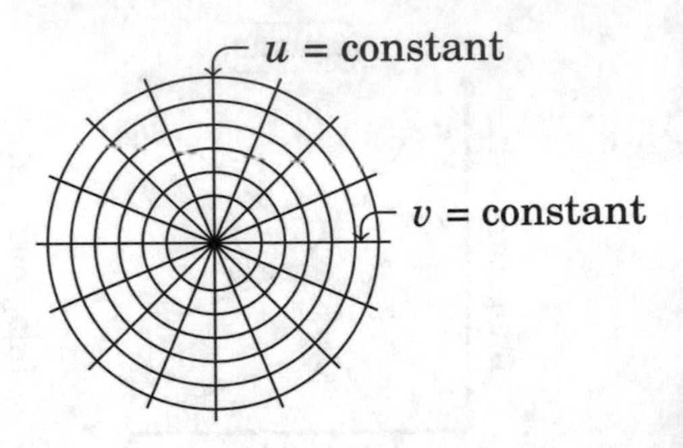

Fig. C.3 Complex potential $w = \mathrm{Ln}(z)$

grids in both z and w planes. If we identify the real part with the temperature function, isotherms are vertical lines in the z plane. Hence, the flux lines must be horizontal lines in the z plane. The solution represented by this complex potential is the solution to one-dimensional heat conduction in a slab. It also represents the parallel flow of an ideal fluid.

C.2.2 Complex Potential $w = Ln(z)$

Consider the complex potential given by

$$w = \mathrm{Ln}(z) \tag{C.13}$$

The symbol Ln means that we consider the principal value of the function $\mathrm{Ln}(z)$ that is defined in the interval $0 \le \theta \le 2\pi$. Using cylindrical coordinates, we have

$$w = \mathrm{Ln}(z) = \mathrm{Ln}(re^{i\theta}) = \ln(r) + i\theta \tag{C.14}$$

We may set $u(r, \theta) = \ln r$ and $v(r, \theta) = \theta$. Constant u lines are concentric circles with center at the origin, while constant v lines are radial lines passing through the origin. The complex potential represents a line source placed at the origin. The solution represents one-dimensional radial conduction in an annulus. The function $w = \mathrm{Ln}(z)$ also is an elementary solution to the Laplace equation in cylindrical coordinates. The isotherms and flux lines appear as shown in Fig. C.3. Alternately, we may consider $u(r, \theta) = \theta$ and $v(r, \theta) = \ln r$. The isotherms will now be radial lines passing through the origin while the heat flux lines will be concentric circles centered at the origin. Figure C.3 represents this case also if we interchange u and v in the figure.

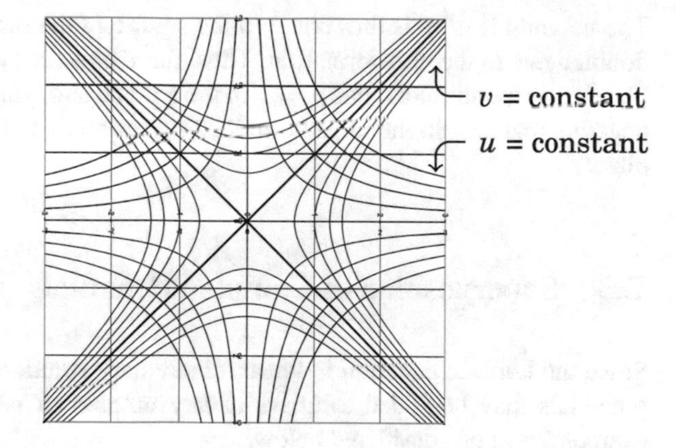

Fig. C.4 Constant u and v lines in the z plane for the complex potential $w = z^2$

C.2.3 Complex Potential $w = e^z$

This complex potential may be written in expanded form as

$$w = \mathrm{e}^z = \mathrm{e}^{x+iy} = \mathrm{e}^x e^{iy} = \mathrm{e}^x[\cos(y) + i\sin(y)] \tag{C.15}$$

Thus, we have

$$u(x, y) = \mathrm{e}^x \cos(y); \quad v(x, y) = \mathrm{e}^x \sin(y) \tag{C.16}$$

It is easily seen that $w = \mathrm{e}^z$ translates to $z = \mathrm{Ln}(w)$ and hence the state of affairs may be visualized by interchanging the z and w planes! This aspect is left to the reader's imagination.

Complex Potential $\mathbf{w = z^2}$

The complex potential $w = z^2$ is an analytic function since the derivative exists and is given by $\frac{\mathrm{d}w}{\mathrm{d}z} = 2z$. The complex potential may be written in expanded form as

$$w = z^2 = (x + iy)^2 = (x^2 - y^2) + i2xy \tag{C.17}$$

Thus, $u(x, y) = x^2 - y^2$ and $v(x, y) = 2xy$. The constant u lines in the w plane map on to rectangular hyperbolas given by $x^2 - y^2 =$ conatant in the z plane, which are asymptotic to the lines $x = \pm y$. The constant v lines in the w plane map on to rectangular hyperbolas given by $xy =$ conatant in the z plane, which are asymptotic to the x and y axes. As illustration, we show a few of these in Fig. C.4. It is clear from the figure that the constant u and v lines are orthogonal at points of intersection of the two sets of curves. In ideal fluid flow, this complex potential represents stagnation point flow. It is interesting to look at the transformation by expressing it in cylindrical coordinates. We have

$$w = z^2 = (re^{i\theta})^2 = r^2 e^{i2\theta} \tag{C.18}$$

The magnitude of w is thus equal to the square of the magnitude of z. The angle is doubled due to the transformation. Thus, the v axis (or the imaginary axis) in the w plane corresponds to the line $y = x$ in the z plane and lying in the first quadrant. The negative real axis in the w plane indeed corresponds to the positive y axis in the z plane.

C.3 Superposition of Complex Potentials

Since the Laplace equation is linear, elementary solutions represented by complex potentials may be added to arrive at any number of new solutions. A couple of examples will be considered below.

C.3.1 Combination of Complex Potentials Considered in Sects. C.2.1 and C.2.2

Consider the two elementary complex potentials given earlier, viz. $w = z$ and $w = Ln(z)$. Let the combination complex potential be given by

$$w = -az + bLn(z) \tag{C.19}$$

where a and b are real constants. The real and complex parts of the complex potential are given by

$$\text{(a)}\quad u(r,\theta) = -ar\cos(\theta) + b\ln(r), \quad \text{(b)}\quad v(r,\theta) = -ar\sin(\theta) + b\theta \tag{C.20}$$

Let us explore the nature of this complex potential. Let us see what would be the curve in the z plane that corresponds to $v = 0$. Consider a point on the real axis such that $\theta = 0$, i.e., any point on the x axis (real axis in the z plane). Indeed, $v(r,\theta) = 0$. This is one possibility. $v(r,\theta)$ is also zero if $r = \frac{b}{a} = x$. Let us see what happens if $\theta = \pm\frac{\pi}{2}$. We then have $r = y$ and $y = \pm\dfrac{b\left(\dfrac{\pi}{2}\right)}{a}$. It is also easily seen that for other values of θ, we have $y = \frac{b\theta}{a}$.

It may be shown easily that the $v = 0$ curve divides the z plane in to two regions with $v = +$ outside and $v = -$ inside as shown in Fig. C.5. In ideal fluid flow the curve is referred to as the Rankine half body. As $x \to -\infty$ the y coordinate on the half body tends to $\pm\pi$. The flow is visualized as flow past a Rankine half body. The flow is uniform far away from the half body.

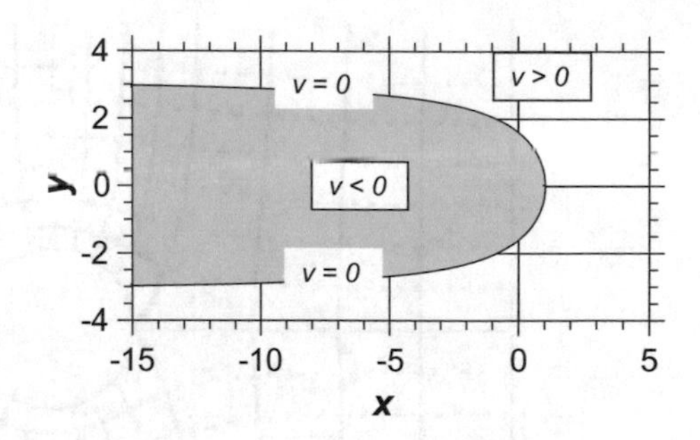

Fig. C.5 Zero v line for the complex potential $w = -az + bLn(z)$ with $a = b = 1$

C.3.2 Flow Past a Cylinder

A second example that we shall consider is the combination of a parallel flow with a dipole (considered in detail in Chap. 5) situated at the origin. The complex potential is given by

$$w = c \underbrace{z}_{\text{Parallel flow}} + \underbrace{\frac{ca^2}{z}}_{\text{Dipole}}, \quad c \neq 0 \tag{C.21}$$

We use the representation of z in cylindrical coordinates to write the stream function as (imaginary part of the complex potential)

$$v(r, \theta) = c \sin(\theta) \left[r - \frac{a^2}{r} \right] \tag{C.22}$$

The potential function is then identified with the real part of the complex potential and is given by

$$u(r, \theta) = c \cos(\theta) \left[r + \frac{a^2}{r} \right] \tag{C.23}$$

The zero streamline consists of either $\sin(\theta) = 0$, i.e., $\theta = 0$ or $r = a$. The flow is thus past a cylinder of radius a centered at the origin. The flow is from left to right and the velocity of the fluid parallel to the real axis is c, far away from the origin. Streamline pattern (broken lines, v=constant) looks like that shown in Fig. C.6. The potential lines (solid lines, u=constnat) form an orthogonal net along with the streamlines.

We also notice that the streamlines are lines parallel to the real axis far away from the cylinder. In the vicinity of the cylinder, the streamlines bend as shown. The indicated pattern also is a possible solution to a heat transfer problem. Imagine half space ($y > 0$) to have a semi-circular cut centered at the origin. If the boundary of the semi-circle and the straight segments along the real axis are maintained at a uniform temperature, the streamlines correspond to the isotherms. The potential lines then correspond to heat flux lines.

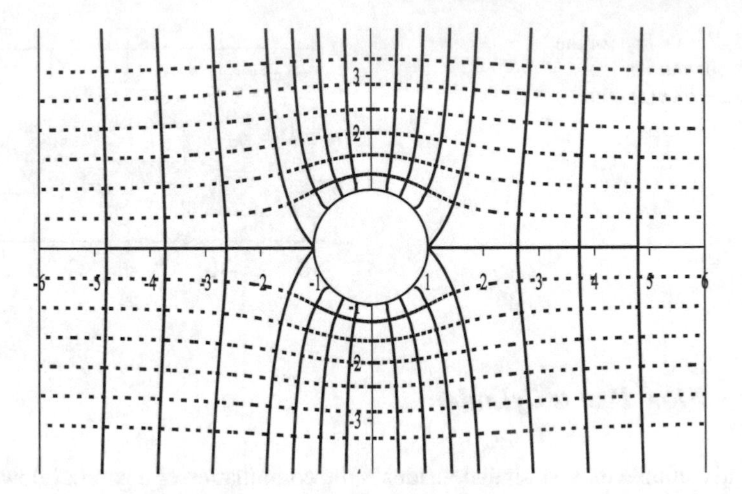

Fig. C.6 Flow past a cylinder. Broken lines are streamlines while solid lines are potential lines

C.3.3 General Comments

Some general comments may now be made. By now, it should be clear to the reader that the constant u and v lines (actually curves) represent an orthogonal net representing either a problem in steady conduction in two dimensions or ideal fluid flow in two dimensions (in a plane). In case we identify v as the stream function, the fluid flow is along constant v lines and there is no flow normal to them. Any streamline may hence represent a solid surface and be assigned arbitrarily a value for the stream function (note that the velocity normal to any solid surface must be zero). In practice, it is customary to take the zero streamline as the surface of the body over which a flow takes place. For example, in the case of the Rankine half body, it is, in fact, the shape of a surface past which ideal fluid flows. The complex potential $w = -az$ represents parallel flow with a constant velocity $-a$, i.e., from right to left. The potential $w = Ln(z)$ represents source flow from a source located at the origin. The two flows combine to define flow past the Rankine half body.

Similarly, any adiabatic surface may be visualized to provide zero flow of heat perpendicular to it, and hence identified with a flux line. Hence, the stream function in fluid flow is analogous to heat flux function in steady conduction. The potential function must, therefore, correspond to the temperature function in the case of conduction. In the case of fluid flow, it is simply referred to as the potential function.

Appendix D
Heisler Charts

D.1 One Term Approximation of the Slab Transient

Three charts are made for the slab transient problem. They are:

1. Plot of mid-plane temperature as a function of Fourier number, with Biot number as a parameter (Fig. D.1). Plot is based on Eq. 6.11 with $\xi = 0$.
2. Correction factor as a function of Biot number with off mid plane location within the slab, as a parameter (Fig. D.2). Plot is based on Eq. 6.11 with $\xi \neq 0$.
3. Heat loss fraction as a function of Fourier number with Biot number as a parameter (Fig. D.3). Plot is based on Eq. 6.13.

D.2 One Term Approximation of the Cylinder Transient

Again three charts are made for the cylinder transient problem. They are:

1. Plot of axial temperature as a function of Fourier number, with Biot number as a parameter (Fig. D.4). Plot is based on Eq. 6.19 with $r = 0$, retaining only the first term.
2. Correction factor as a function of Biot number with off center location within the cylinder, as a parameter (Fig. D.5). Plot is based on Eq. 6.19 with $r \neq 0$.
3. Heat loss fraction as a function of Fourier number, with Biot number as a parameter (Fig. D.6). Plot is based on Eq. 6.26.

S. P. Venkateshan, *Heat Transfer*,
https://doi.org/10.1007/978-3-030-58338-5

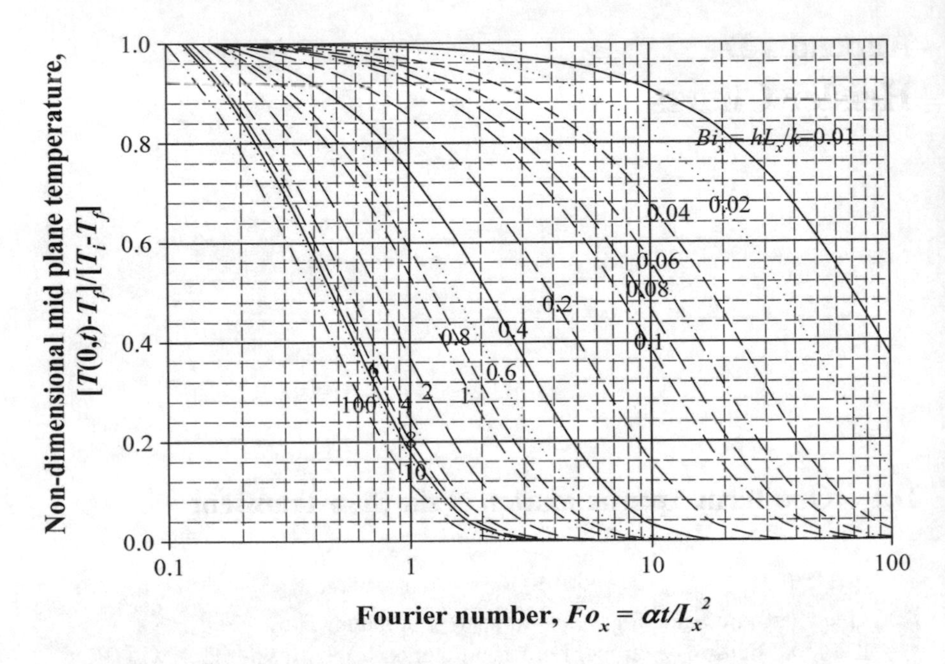

Fig. D.1 Slab mid plane temperature variation with time

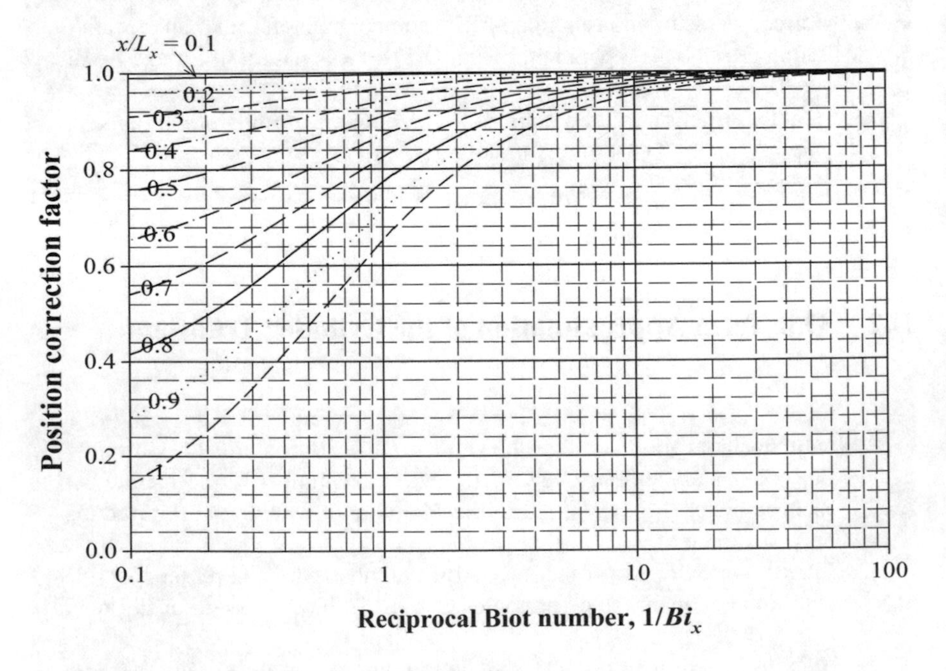

Fig. D.2 Correction factor for non mid-plane location

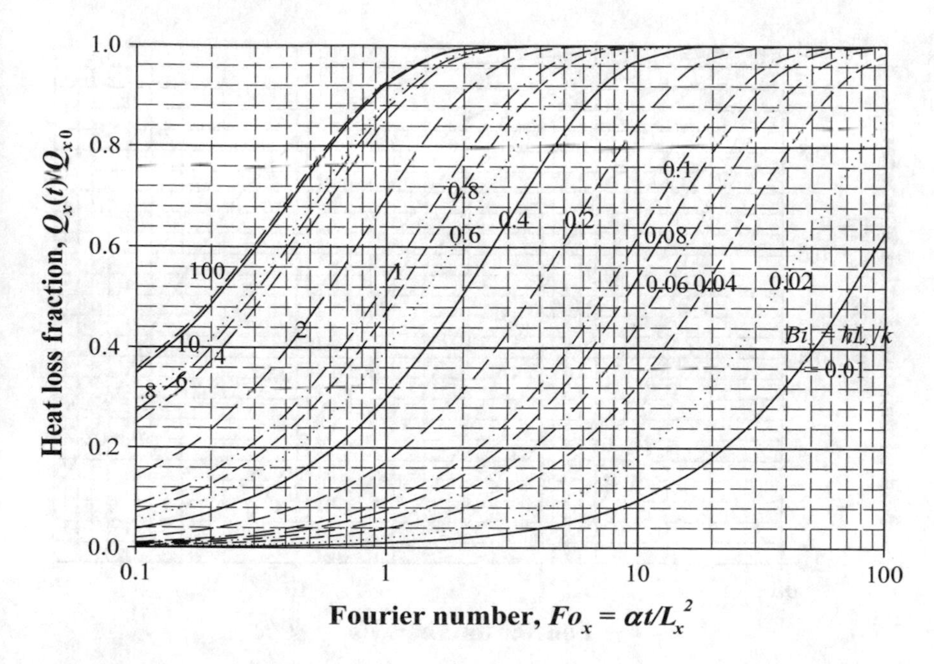

Fig. D.3 Heat loss fraction in the slab transient

D.3 One Term Approximation of the Sphere Transient

Again three charts are made for the cylinder transient problem. They are:

1. Plot of center temperature as a function of Fourier number, with Biot number as a parameter (Fig. D.7). Plot is based on Eq. 6.33 with $r = 0$, retaining only the first term.
2. Correction factor as a function of Biot number with off center location within the sphere, as a parameter (Fig. D.8). Plot is based on Eq. 6.33 with $r \neq 0$.
3. Heat loss fraction as a function of Fourier number, with Biot number as a parameter (Fig. D.9). Plot is based on Eq. 6.39.

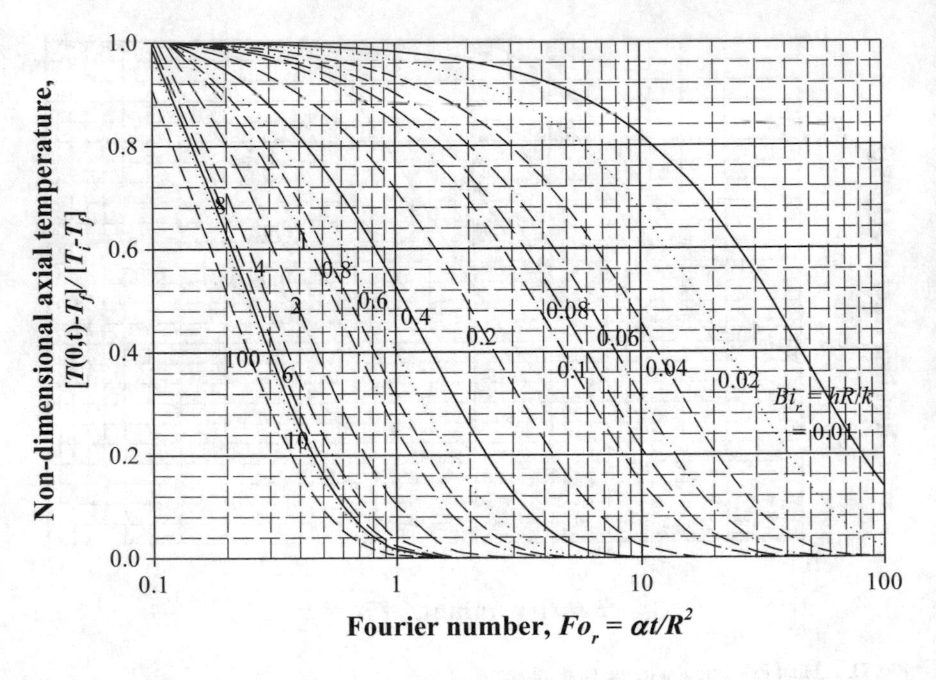

Fig. D.4 Axial temperature variation with time in cylinder transient

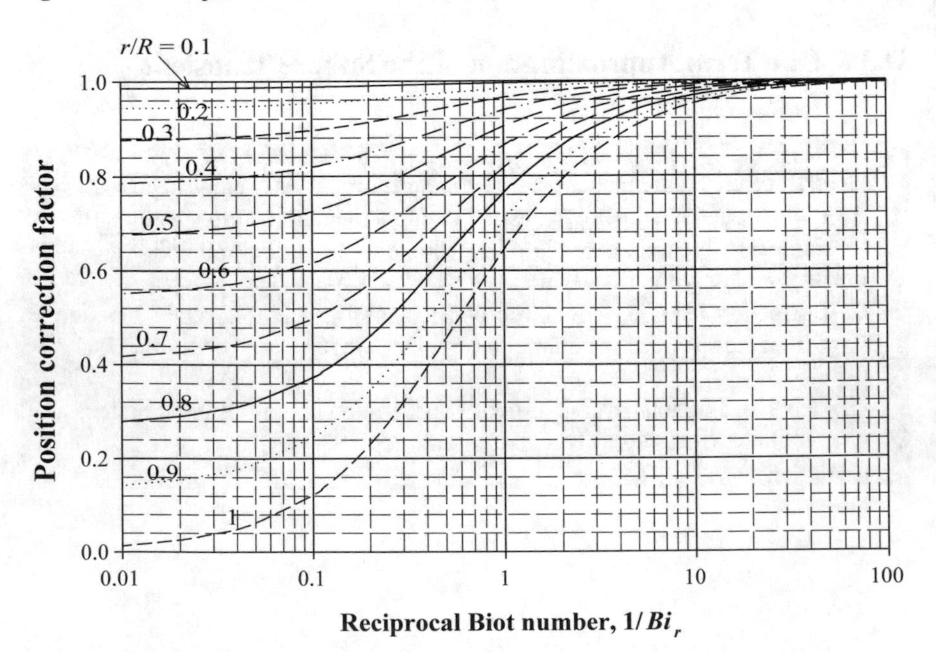

Fig. D.5 Correction factor for non axial location within the cylinder

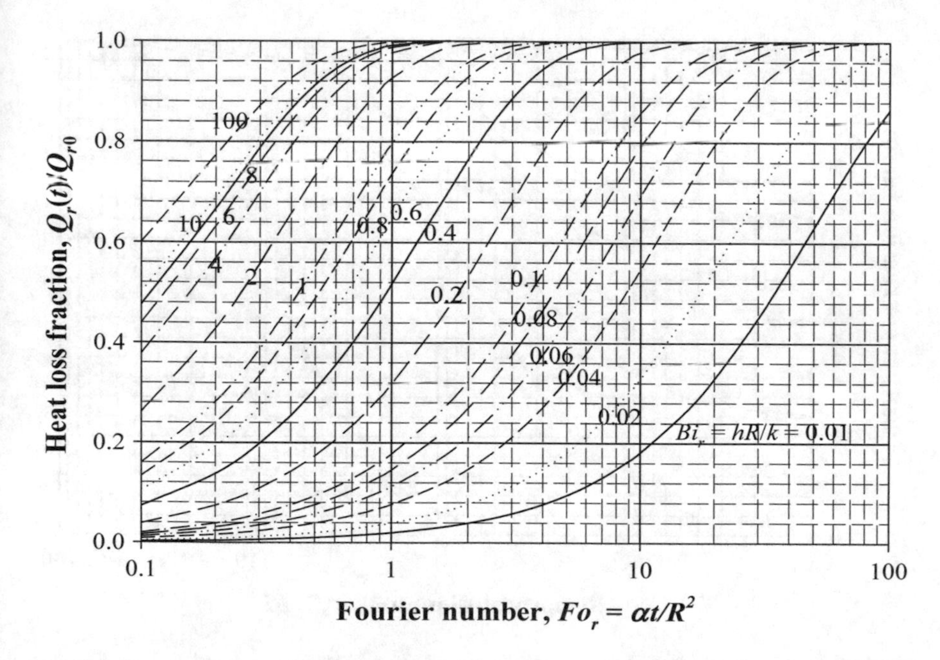

Fig. D.6 Heat loss fraction in the cylinder transient

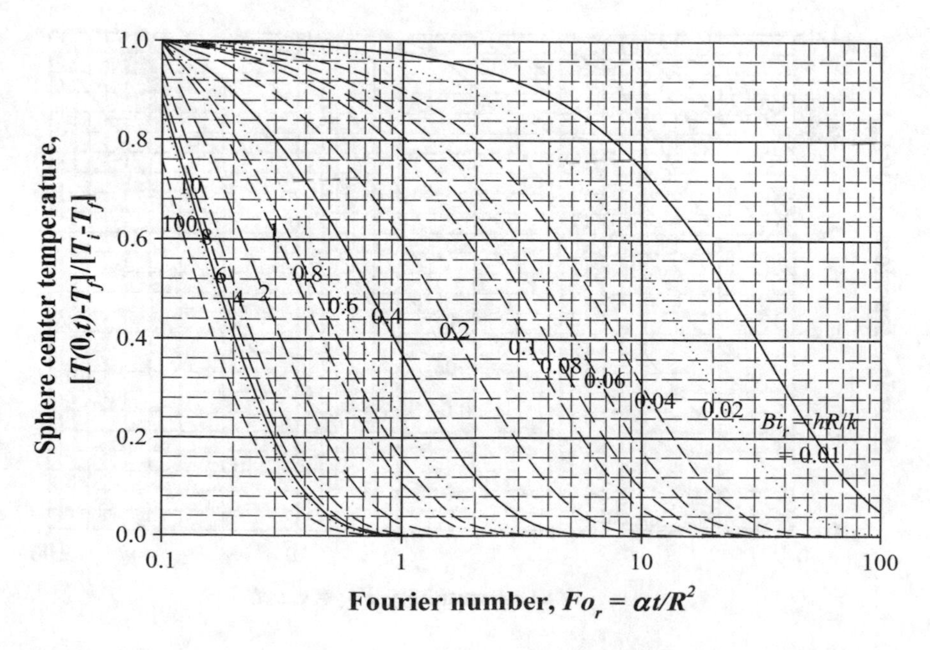

Fig. D.7 Center temperature variation with time in sphere transient

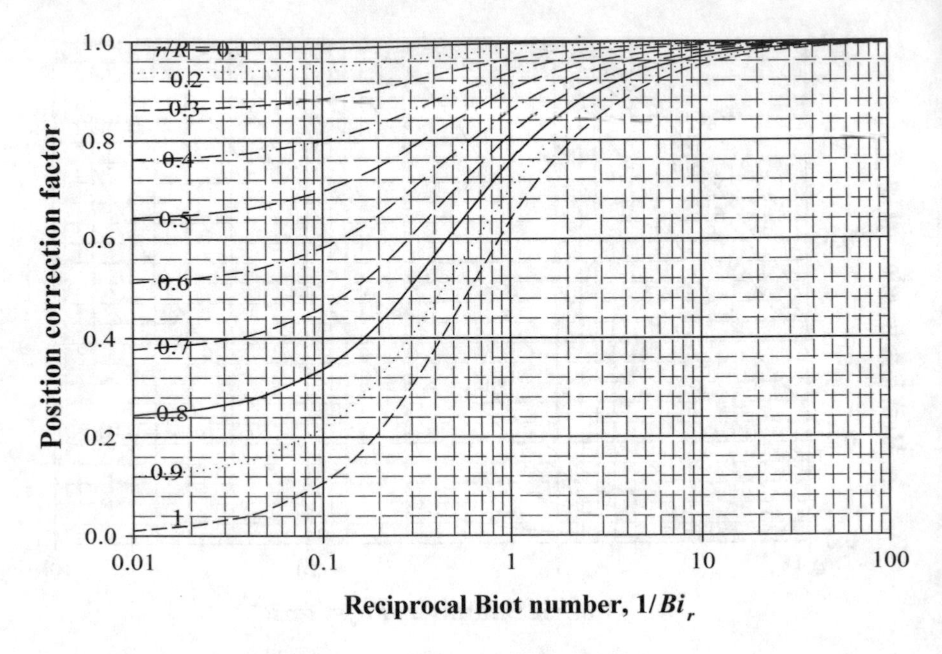

Fig. D.8 Correction factor for non central location within the sphere

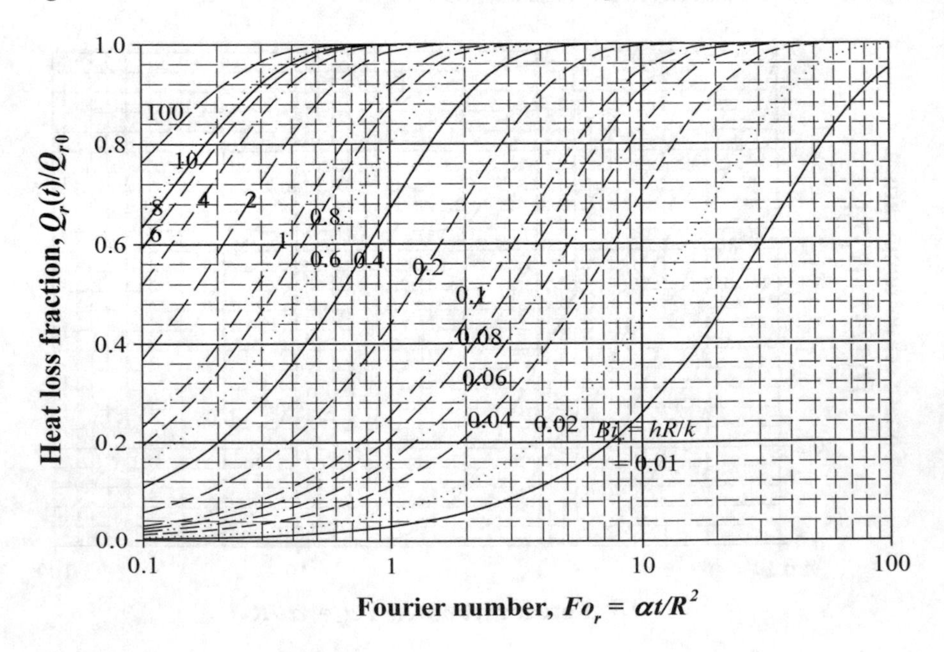

Fig. D.9 Heat loss fraction in the sphere transient

Appendix E
Numerical Solution of Algebraic and Differential Equations

E.1 Introduction

Many heat transfer and fluid flow problems are modeled by either ordinary or partial differential equations. These equations may not be amenable to analytical solution leading to closed form solutions. In such cases, numerical solution is the only alternative. The finite difference or finite volume or finite element method is used to solve the equations after suitably discretizing the computational domain. The governing equations subject to initial and boundary conditions reduce to either ordinary differential equations or a set of algebraic equations. We discuss in brief, some of the more commonly used numerical methods, in solving such equations.

E.2 Solution of Algebraic Equations

E.2.1 Solution of a Single Algebraic Equation

A useful numerical method of solution of s single algebraic equation is the Newton–Raphson method. Consider the problem of obtaining a root of the equation $f(x) = 0$. A graph of the function may be made to roughly get the location(s) of the root(s). Accurate determination of the root(s) assumes that we are close to a root and a guess value x_g is available. In the vicinity of the root, we may use Taylor expansion to get

$$f(x_b) = f(x_g) + \left.\frac{\mathrm{d}f}{\mathrm{d}x}\right|_{x_g} (x_b - x_g) + O(x_b - x_g)^2 \tag{E.1}$$

x_b is assumed to be a better estimate for the root, and hence we set $f(x_b) = 0$ and ignore the square term. Then, we get the relation

S. P. Venkateshan, *Heat Transfer*,
https://doi.org/10.1007/978-3-030-58338-5

$$\boxed{x_b = x_g - \frac{f(x_g}{f'(x_g)}} \tag{E.2}$$

where $'$ indicates differentiation with respect to x. Equation E.2 provides an iteration scheme wherein the x value is updated by its repeated use till successive estimates differ by a predetermined tolerance.

E.2.2 Solution of Several Algebraic Equations

The Newton–Raphson method is easily extended to a set of algebraic equations. We start with a guess set $x_{1g}, x_{2g}, \cdots x_{ng}$for obtaining the solution of a set of equations of the form

$$\begin{aligned} f_1(x_1, x_2, \cdots x_n) &= 0 \\ f_2(x_1, x_2, \cdots x_n) &= 0 \\ \cdots\cdots\cdots&\cdots\cdots\cdots\cdots \\ f_n(x_1, x_2, \cdots x_n) &= 0 \end{aligned} \tag{E.3}$$

Following a procedure similar to the one that lead to Eq. E.2, we obtain the following set of equations for iteration:

$$\begin{aligned} f_{1g} + \tfrac{\partial f_1}{\partial x_1}\Delta x_1 + \tfrac{\partial f_1}{\partial x_2}\Delta x_2 + \cdots + \tfrac{\partial f_1}{\partial x_n}\Delta x_n &= 0 \\ f_{1g} + \tfrac{\partial f_1}{\partial x_1}\Delta x_1 + \tfrac{\partial f_1}{\partial x_2}\Delta x_2 + \cdots + \tfrac{\partial f_1}{\partial x_n}\Delta x_n &= 0 \\ f_{2g} + \tfrac{\partial f_2}{\partial x_1}\Delta x_1 + \tfrac{\partial f_2}{\partial x_2}\Delta x_2 + \cdots + \tfrac{\partial f_2}{\partial x_n}\Delta x_n &= 0 \\ \cdots\cdots\cdots\cdots\cdots\cdots\cdots\cdots&\cdots\cdots \\ f_{ng} + \tfrac{\partial f_n}{\partial x_1}\Delta x_1 + \tfrac{\partial f_n}{\partial x_2}\Delta x_2 + \cdots + \tfrac{\partial f_n}{\partial x_n}\Delta x_n &= 0 \end{aligned} \tag{E.4}$$

where $\Delta x_i = x_{ib} - x_{ig}$ for $i = 1, 2, \ldots, n$ represents the change in the guess value during an iteration and f_{ig} stands for $f_i(x_{1g}, x_{2g}, \cdots x_{ng})$. All the partial derivatives are calculated at $x_{1g}, x_{2g}, \cdots x_{ng}$. We thus see that a set of linear algebraic equations have to be solved for each iteration step. The above equations may be written in the alternate matrix form

$$\underbrace{\begin{bmatrix} \frac{\partial f_1}{\partial x_1} & \frac{\partial f_1}{\partial x_2} & \cdots\cdots & \frac{\partial f_1}{\partial x_n} \\ \frac{\partial f_2}{\partial x_1} & \frac{\partial f_2}{\partial x_2} & \cdots\cdots & \frac{\partial f_2}{\partial x_n} \\ \cdots & \cdots & \cdots\cdots & \cdots \\ \cdots & \cdots & \cdots\cdots & \cdots \\ \frac{\partial f_n}{\partial x_1} & \frac{\partial f_n}{\partial x_2} & \cdots\cdots & \frac{\partial f_n}{\partial x_n} \end{bmatrix}}_{\text{Jacobian matrix}} \begin{bmatrix} \Delta x_1 \\ \Delta x_2 \\ \cdots \\ \cdots \\ \Delta x_n \end{bmatrix} = \begin{bmatrix} -f_{1g} \\ -x_{2g} \\ \cdots \\ \cdots \\ -f_{ng} \end{bmatrix} \tag{E.5}$$

These equations may be solved easily using Kramer's rule if n is not too large, say ≤ 5. For larger n, one may use point by point iterative schemes such as the Gauss or Gauss-Seidel iteration schemes.

E.2.3 Solution of Equations Involving Sparse Matrix—TDMA

We have seen that the ordinary differential equations reduce to a set of linear equations when the derivatives are represented by their finite difference analogs. The resulting matrix equation is of form $\mathbf{AX} = \mathbf{B}$, where $\mathbf{A}$ is an $n \times n$ square matrix and $\mathbf{X}$ and $\mathbf{B}$ are $n \times 1$ column vectors. $\mathbf{X}$ represents the solution vector while $\mathbf{B}$ is referred to as the forcing vector. The case of interest to us here is when the $\mathbf{A}$ is tridiagonal in nature. Tridiagonal matrix algorithm or Thomas algorithm[3] is useful in solving such equations. The matrix $\mathbf{A}$ (diagonal dominant matrix)[4] is normally represented as

$$\begin{bmatrix} a_1 & b_1 & 0 & \cdots & 0 \\ c_2 & a_2 & b_2 & \cdots & 0 \\ \cdots & \ddots & \ddots & \ddots & \cdots \\ \cdots\cdots & c_{n-1} & a_{n-1} & b_{n-1} \\ 0 \;\; 0 & \cdots & c_n & a_n \end{bmatrix} \tag{E.6}$$

The reader should note that only the non-zero elements of the tridiagonal matrix comprising of $3n$ elements a_i, b_i, c_i, $1 \leq i \leq n$ and n elements d_i, $1 \leq i \leq n$ need be stored in the computer memory. It is seen that elementary row operations will be able to transform the matrix $\mathbf{A}$ to upper triangular form, i.e., all the elements below the main diagonal will become zero. This operation will leave the zero elements above the diagonal unchanged. For example, multiplication of elements in the second row by $\frac{a_1}{c_2}$ and subtracting the elements in the first row will make element 2, 1 zero. Similarly, all the elements below the main diagonal may be made to be zero. The same elementary row operations are to be performed on the matrix $\mathbf{B}$ or as is usual, these

[3]Thomas, L. H., Watson Sc Comput. Lab. Rept., Columbia University, New York, 1949.

[4]Diagonal dominance requires $|a_i| > |b_i| + |c_i|$ for every i

row operations are performed on an augmented $n \times n + 1$ matrix obtained by adding the elements of **B** to **A** as the last column. This process will give an upper triangular matrix with the main diagonal and the upper diagonal alone having non-zero terms. The resulting bi-diagonal matrix may be used to obtain the nodal temperatures by going back from the last equation in what is known as back substitution. This process is easy to perform for small n like 3 or 4. However, when n is large, it is easier to use a simple algorithm to achieve the same result. The algorithm that does this is called the TDMA.

The starting point for developing the TDMA is to note that any one of the equations in the set of equations that is being considered is of form

$$a_i X_i = b_i X_{i+1} + c_i X_{i-1} + d_i \tag{E.7}$$

The algorithm should eventually make it possible to back substitute by a recurrence relation of form

$$X_{i-1} = P_{i-1} X_i + Q_{i-1} \tag{E.8}$$

where P_i, Q_i contain the matrix elements a_i, b_i and c_i. We substitute Eq. E.8 in Eq. E.7 to get $a_i X_i = b_i X_{i+1} + c_i(P_{i-1} X_i + Q_{i-1}) + d_i$ which on simplification yields

$$\boxed{X_i = \frac{b_i}{a_i - c_i P_{i-1}} X_{i+1} + \frac{d_i + c_i Q_{i-1}}{a_i - c_i P_{i-1}}} \tag{E.9}$$

which has a form identical with Eq. E.8! Comparing these two equations, we get the following recurrence relations for the P's and Q's:

$$\boxed{\text{(a)}\quad P_i = \frac{b_i}{a_i - c_i P_{i-1}} \qquad \text{(b)}\quad Q_i = \frac{d_i + c_i Q_{i-1}}{a_i - c_i P_{i-1}}} \tag{E.10}$$

Since $c_1 = 0$, we have $P_1 = \frac{b_1}{a_1}$ and $Q_1 = \frac{d_1}{a_1}$. With P_1 and Q_1 thus calculated, the recurrence relations are used to calculate all the P's and Q's. We notice also that since $b_n = 0$, $P_n = 0$. Hence, Eq. E.8 indicates that $X_n = Q_n$. Back substitution may be then continued using (E.8). Thus, the TDMA leads to a simple procedure using recurrence relations that may be programmed easily on a computer.

E.2.4 Point by Point Iteration Methods

When the coefficient matrix is not sparse, it is necessary to use a point by point iteration method such as the Gauss or the Gauss-Seidel iteration methods. The set of equations to be solved are rearranged such that the ith equation is in the form

$$X_i = \frac{\sum_{j=1}^{n} a_{ij} X_j}{a_{ii}}, i \neq j \tag{E.11}$$

In the Gauss iteration method, we start the iterations with a guess set X_{ig}, substitute these on the right hand side of Eq. E.11 and obtain better values X_{ib}. If $|X_{ib} - X_{ig}| \leq \epsilon$, where ϵ is a user defined small number, we stop the iteration. Otherwise the iteration is continued by replacing X_{ig} with X_{ib}.

In the Gauss-Seidel scheme, the values that have already been updated (i.e., for $j < i$) will be used on the right hand side, without waiting to complete one pass through all the i's.

Example E.1

Solve by Gauss and Gauss-Seidel iteration the following two simultaneous equations:

$$4T_1 - T_2 = 90, \quad -T_1 + 4T_2 = 330$$

Compare the numerical solutions with the exact solution and also study the convergence trends of the two iteration schemes.

Solution:
Let T_i^{old}, T_i^{new} for $i = 1, 2$ represent the values of the variables before and after an iteration.
(i) Gauss iteration:
The iteration scheme follows from Eq. E.11 and is written down as

$$\text{(a)} \quad T_1^{new} = \frac{90 + T_2^{old}}{4} \quad \text{(b)} \quad T_2^{\text{new}} = \frac{330 + T_1^{\text{old}}}{4} \tag{E.12}$$

We start with a guess set $T_1^{\text{old}} = 22.5$, $T_2^{\text{old}} = 82.5$. With these, the iteration count is set to 1, and the iteration scheme given by Eq. E.12(a)–(b) is used repeatedly. The following table gives the results of this scheme (Table E.1). We note that the values have converged to $T_1 = 46$ and $T_2 = 94$ to two significant digits after decimals, at the end of seven iterations. In fact, the reader may verify that these represent the exact solution to the two equations.
(ii) Gauss-Seidel iteration:
The iteration scheme following from Eq. E.11 is modified as

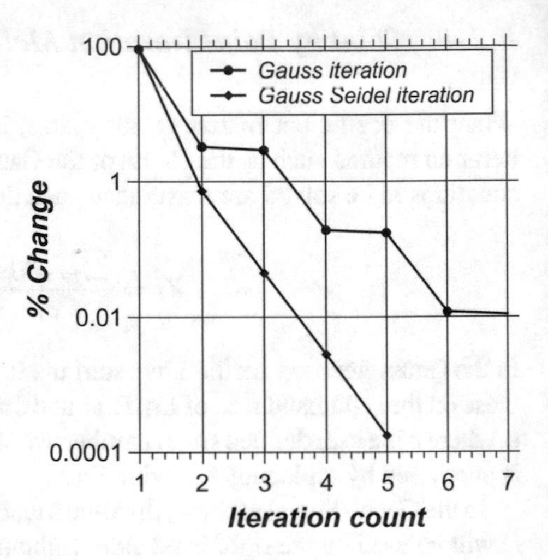

Fig. E.1 Convergence of Gauss and Gauss-Seidel schemes in Example E.1

$$\text{(a)} \quad T_1^{\text{new}} = \frac{90 + T_2^{\text{old}}}{4} \quad \text{(b)} \quad T_2^{\text{new}} = \frac{330 + T_1^{\text{new}}}{4} \tag{E.13}$$

We start with a guess set $T_1^{\text{old}} = 22.5$, $T_2^{\text{old}} = 82.5$. With these, the iteration count is set to 1, and the iteration scheme given by Eq. E.13(a)–(b) is used repeatedly. The following table gives the results of this scheme. We note that the values have converged to $T_1 = 46$ and $T_2 = 94$ to two significant digits after decimals, at the end of four iterations. It is noted that the Gauss-Seidel scheme converges faster than the Gauss iteration scheme. The convergence history for the two schemes are given in Fig. E.1. The change in T_1 as a percentage of its previous value is shown for both the schemes. Gauss iteration shows a staircase type of convergence while the Gauss-Seidel shows a monotonic convergence (Table E.2).

Table E.1 Gauss iteration for Example E.1

Iteration count	1	2	3	4	5	6	7	8
T_1	22.5	43.13	44.53	45.82	45.91	45.99	45.99	46.00
T_2	82.5	88.13	93.28	93.63	93.96	93.98	94.00	94.00

Table E.2 Gauss-Seidel iteration for Example E.1

Iteration count	1	2	3	4	5
T_1	22.5	43.13	45.82	45.99	46.00
T_2	82.5	93.28	93.96	94.00	94.00

E.2.5 Iteration With Over or Under Relaxation

The point by point iteration scheme may be modified by using a suitable relaxation parameter. Sometimes it may be possible to hasten convergence by using over relaxation. Sometimes under relaxation may be used to avoid unpleasant oscillations or divergence in the solution. Let us assume that the two values T_i^{old} and T_i^{new} have been calculated by point by point iteration such as the Gauss scheme. A relaxation parameter is introduced through the equation

$$T_{i,\text{rel}}^{\text{new}} = T_i^{\text{old}} + \omega(T_i^{\text{new}} - T_i^{\text{old}}) \tag{E.14}$$

where the updated value with relaxation is represented by $T_{i,\text{rel}}^{\text{new}}$ and ω is the relaxation parameter. It is noted that $\omega = 1$ in the Gauss and Gauss-Seidel methods. Relaxation parameter $\omega < 1$ represents underrelaxation and $\omega > 1$ represents overrelaxation.

E.3 Solution of Ordinary Differential Equations (ODE)

Ordinary differential equations are often encountered while modeling heat transfer problems (many examples may be found in the present book). It is best to use close form analytical solution when possible. Sometimes it is necessary to solve ODEs by numerical methods. Those methods that are commonly used are summarized here.

ODEs of first order are always initial value problems and the methods are introduced with these in view. Second and higher order ODEs may either be initial or boundary value problems. In the latter case, they pose a challenge which will be addressed in due course. Non-linear ODEs are harder still and need an iterative scheme coupled with one of the numerical methods that will be considered here.

E.3.1 First Order ODE

Simple low order schemes
A general first order ODE is given by

$$\boxed{\frac{\mathrm{d}y}{\mathrm{d}x} = f(x, y) \text{ with } y(x_0) = y_0} \tag{E.15}$$

The simplest method is the first order accurate Euler method. The domain of interest is the region to the right of x_0, i.e., $x > x_0$. This region is covered by taking steps of size $\Delta x = h$, and the points that arise are referred to as nodes and are numbered by the subscript $i = 1, 2, \ldots$. Let us assume that we have arrived at the point x_n by the repeated use of a numerical algorithm and the value of the dependent variable at this point is y_n. We shall look at the step $n \to n+1$. In the Euler method, the following scheme is used:

$$\boxed{y_{n+1} = y_n + f(x_n, y_n)h} \tag{E.16}$$

The reader would recognize that this relation is based on Taylor expansion that retains only the first order term. Improvement of the Euler scheme may be made by using the modified Euler or the Heun method that is second order accurate. In this method, the Euler scheme gives an estimate of y_{n+1} which is represented as y^*_{n+1}, referred to as the predictor. This value is corrected by calculating the slope at x_{n+1}, y^*_{n+1} and replacing the slope by the mean slope at x_n and x_{n+1}.

Thus, we have the Heun scheme given by

$$\begin{aligned}\text{Predictor: } y^*_{n+1} &= y_n + f(x_n, y_n)h \\ \text{Corrector: } y_{n+1} &= y_n + \frac{f(x_n, y_n) + f(x_{n+1}, y^*_{n+1})}{2}h\end{aligned} \tag{E.17}$$

4th Runge Kutta Method (RKM)

A popular single step multi-stage scheme is the 4th Runge–Kutta scheme. In this scheme, the function f is calculated at intermediate points within the step apart from those at the two end points x_n and x_{n+1}. Without proof, we give the algorithm below

$$\begin{aligned}&k_1 = hf(x_n, y_n); \qquad k_2 = hf(x_n + \tfrac{h}{2}, y_n + \tfrac{k_1}{2});\\ &k_3 = hf(x_n + \tfrac{h}{2}, y_n + \tfrac{k_2}{2}); \; k_4 = hf(x_n + h, y_n + k_3)\\ &y_{n+1} = y_n + \tfrac{k_1 + 2k_2 + 2k_3 + k_4}{6}\end{aligned} \tag{E.18}$$

The scheme calculates the $k's$ which are referred to as auxiliary quantities. In the present case, the step x_n to x_{n+1} is covered by two half steps. Four auxiliary quantities are calculated to complete a step. The method is self starting (Euler and Heun methods are also self starting) since we can start the calculation from the initial value and continue the calculation for any number of steps.

Adam Moulton predictor corrector method

This is a 4th order accurate multistep method that uses four nodal values of f. The method is based on analytical integration of an interpolating polynomial. Since more than one nodal value is required to extend the solution by one step, the method is non self starting. In practice, RKM is used to get the first four nodal values and the Adam–Moulton scheme is used to continue the solution from there on. The predictor is based on the solution available at x_n, x_{n-1}, x_{n-2}, and x_{n-3}.

$$y^*_{n+1} = y_n + \frac{h}{24}[55 f_n - 59 f_{n-1} + 37 f_{n-2} - 9 f_{n-3}] \tag{E.19}$$

where f_n indicates the function value evaluated at x_n, y_n and so on. The corrector is based on the solution available at x_{n+1}, x_n, x_{n-1} and x_{n-2}.

$$y_{n+1} = y_n + \frac{h}{24}[9 f^*_{n+1} + 19 f_n - 5 f_{n-1} + f_{n-2}] \tag{E.20}$$

One clear advantage of Adam–Moulton as compared to RKM is that the function values needed are already available and hence needs fewer calculations as compared to the RKM.

Example E.2

A thermal system is governed by the equation $\frac{d\theta}{dt} + 1.5(\theta^4 - 0.5) = 0$, where θ is a suitably defined non-dimensional temperature and t is non-dimensional time. Solve this equation up to $t = 1$ by taking a time step of $t = 0.2$. Use the fourth order RK method and assume that $\theta = 1$ at $t = 0$.

Solution:

We identify the function f in our usual notation as $-1.5(\theta^4 - 0.5)$. The calculations are done using a spreadsheet and the result of the calculation is shown in Table E.3. Even though the calculations have been based on at least 8 significant digits, the k values in the table have been rounded to six digits and the θ values to four digits.

Table E.3 RK table for Example E.2

t	k_1	k_2	k_3	k_4	θ
0	−0.15	−0.069628	−0.110354	−0.037927	1
0.2	−0.054538	−0.031069	−0.040906	−0.020121	0.9087
0.4	−0.023654	−0.014425	−0.017981	−0.009772	0.8722
0.6	−0.010978	−0.006888	−0.008402	−0.004749	0.8559
0.8	−0.005249	−0.003337	−0.004032	−0.002319	0.8482
1	−0.002546	−0.001628	−0.001958	−0.001135	0.8444

E.4 Higher Order ODE

As an example of a higher order ODE, we consider a second-order ODE of form

$$\boxed{\frac{\mathrm{d}^2 y}{\mathrm{d}x^2} = f\left(x, y, \frac{\mathrm{d}y}{\mathrm{d}x}\right)} \tag{E.21}$$

This equation requires two conditions, either two initial conditions or two boundary conditions, to be specified. In the former case, the function and first derivative are specified at an initial point x_0 as

$$\boxed{y = y_0 \text{ and } \frac{\mathrm{d}y}{\mathrm{d}x} = y_0' \text{ at } x = x_0} \tag{E.22}$$

In the latter case, the boundary conditions—one each—is specified at $x = a$ and $x = b$. The boundary conditions may be of three types: (1) First kind:—Function is specified (2)Second kind:—Derivative is specified and (3) Third kind:—A relation between the function and the derivative is specified Both the numerical solution methods described above are essentially initial value solvers. Hence we first consider initial value problem.

E.4.1 Second-Order ODE: Initial Value Problem

The second-order ODE may be written as two first order ODEs as indicated below

$$\frac{dz}{dx} = f(x, y, z), \quad \text{with} \quad \underbrace{\frac{dy}{dx} = z(x, y)}_{\text{Definition of } z(x,y)} \tag{E.23}$$

The initial conditions E.22 become

$$y = y_0 \text{ at } x = x_0, \text{ and } z = y_0' \text{ at } x = x_0 \tag{E.24}$$

The two first order ODEs given by E.23 are solved by the 4th order RKM by introducing auxiliary quantities $k_i, i = 1 - 4$ and $l_i, i = 1 - 4$. Then we have

$$\begin{aligned}
&k_1 = hf(x_n, y_n, z_n), && l_1 = hz_n \\
&k_2 = hf(x_n + \tfrac{h}{2}, y_n + \tfrac{l_1}{2}, z_n + \tfrac{k_1}{2}), && l_2 = h(z_n + \tfrac{k_1}{2}) \\
&k_3 = hf(x_n + \tfrac{h}{2}, y_n + \tfrac{l_2}{2}, z_n + \tfrac{k_2}{2}), && l_3 = h(z_n + \tfrac{k_2}{2}) \\
&k_4 = hf(x_n + h, y_n + l_3, z_n + k_3), && l_4 = h(z_n + k_3) \\
&z_{n+1} = y_n + \tfrac{k_1+2k_2+2k_3+k_4}{6}, && y_{n+1} = y_n + \tfrac{l_1+2l_2+2l_3+l_4}{6}
\end{aligned} \tag{E.25}$$

The algorithm given above may be used till we reach the desired upper value of x.

E.4.2 Second-Order ODE: Boundary Value Problem

The method that is commonly used is called the"shooting method". If the given second-order ODE is linear, we use two solutions, obtained by solving the ODE with assumed initial values, solved as initial value problems. For example, if both boundary conditions are of first kind, we solve the ODE by assuming two different initial slopes, say $z_a^{(1)}$ and $z_a^{(2)}$, but with the specified $y = y_a$ (say) at at $x = a$. The two solutions will give two different values $y_b^{(1)}$ and $y_b^{(2)}$ at $x = b$. Since the ODE is linear, a linear combination of the two solutions also is a solution to the original ODE. Hence, we require that a linear combination of the two values $y_b^{(1)}$ and $y_b^{(2)}$ be combined to satisfy the boundary condition at $x = b$

$$y_b = \alpha y_b^{(1)} + (1-\alpha) y_b^{(2)} \tag{E.26}$$

where α is a constant. We may solve for α to get

$$\alpha = \frac{y_b - y_b^{(2)}}{y_b^{(1)} - y_b^{(2)}} \tag{E.27}$$

The required solution is then given by $y = \alpha y_{(1)} + (1-\alpha) y_{(2)}$. Other kinds of boundary conditions may be treated in a similar fashion.

Example E.3

Solve the fin equation $\frac{d^2\theta}{d\xi^2} - 3\theta = 0$ with the boundary conditions $\theta = 1$ at $\xi = 0$ and $\frac{d\theta}{d\xi} = 0$ at $\xi = 1$ by the shooting method. Use fourth order RKM.

Solution:
The given ordinary differential equation is equivalent to two first order ordinary differential equations

$$\text{(a)} \quad \frac{dz}{d\xi} = -3\theta, \quad \text{(b)} \quad \frac{d\theta}{d\xi} = z$$

We solve these equations twice as initial value problems by choosing $z_0^{(1)} = -1$ and $z_0^{(2)} = -2$, with $h = \Delta\xi = 0.1$. These two solutions yield, respectively, $z_1^{(1)} = 1.82716$ and $z_1^{(2)} = -1.08739$. In order to satisfy the second kind boundary condition at $\xi = 1$, we choose α as

$$\alpha = \frac{z_b - z_1^{(2)}}{z_1^{(1)} - z_1^{(2)}} = \frac{0 - (-1.08739)}{1.82716 - (-1.08739)} = 0.37309$$

The solution is now obtained as $\theta = \alpha\theta_{(1)} + (1-\alpha)\theta_{(2)}$. The results are shown in Table E.4.

The shooting method is also graphically shown in Fig. E.2.

Table E.4 Table for Example E.3

ξ	$\theta^{(1)}$	$z^{(1)}$	$\theta^{(2)}$	$z^{(2)}$	θ	z	$\theta(Exact)$
0	1	−1	1	−2	1	−1.6269	1.0000
0.1	0.9145	−0.7135	0.8140	−1.7286	0.8515	−1.3499	0.8515
0.2	0.8566	−0.4485	0.6526	−1.5091	0.7287	−1.1134	0.7287
0.3	0.8244	−0.1970	0.5107	−1.3351	0.6277	−0.9105	0.6277
0.4	0.8170	0.0486	0.3842	−1.2012	0.5457	−0.7349	0.5457
0.5	0.8341	0.2956	0.2693	−1.1034	0.4800	−0.5814	0.4800
0.6	0.8764	0.5516	0.1624	−1.0388	0.4288	−0.4455	0.4288
0.7	0.9450	0.8241	0.0605	−1.0055	0.3905	−0.3229	0.3905
0.8	1.0420	1.1214	−0.0397	−1.0024	0.3639	−0.2100	0.3639
0.9	1.1704	1.4524	−0.1410	−1.0294	0.3483	−0.1034	0.3483
1	1.3340	1.8272	−0.2466	−1.0874	0.3431	0.0000	0.3431

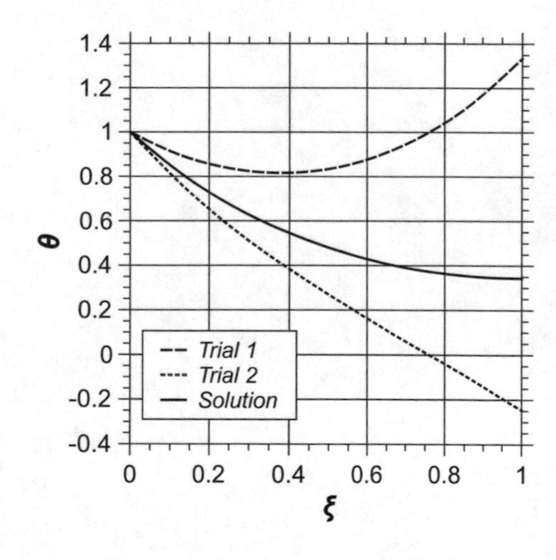

Fig. E.2 Trial solutions and final solution by the "shooting method" in Example E.3

Appendix F
Exponential Integrals

F.1 Introduction

Exponential integrals are special functions that occur in problems involving radiation in participating media. These are defined through integrals over polar angle θ or more commonly in terms of $\mu = \cos\theta$. Exponential integral of order n (where n is an integer) is defined as

$$E_n(t) = \int_0^1 \mu^{n-2} e^{-\frac{t}{\mu}} \mathrm{d}\mu \tag{F.1}$$

where t may take on any value between 0 and ∞. Differentiation of Eq. F.1 with respect to t shows that

$$\frac{\mathrm{d}E_n}{\mathrm{d}t} = \int_0^1 \mu^{n-2} \cdot \frac{-1}{\mu} e^{-\frac{t}{\mu}} \mathrm{d}\mu = -\int_0^1 \mu^{n-3} e^{-\frac{t}{\mu}} \mathrm{d}\mu = -E_{n-1}(t) \tag{F.2}$$

At $t = 0$ we have

$$E_n(0) = \int_0^1 \mu^{n-2} d\mu = \left.\frac{\mu^{n-1}}{n-1}\right|_0^1 = \frac{1}{n-1} \tag{F.3}$$

We also notice that $E_n(t) \to 0$ as $t \to \infty$. Using Eq. F.2, on integration with respect to t, we get

$$E_n(t) = -\int_0^t E_{n-1}(t) + A$$

S. P. Venkateshan, *Heat Transfer*,
https://doi.org/10.1007/978-3-030-58338-5

where A the integration constant is obtained as $A = E_n(0)$. Using Eq. F.3, we then have

$$E_n(t) = E_n(0) - \int_0^t E_{n-1}(t) = \frac{1}{n-1} - \int_0^t E_{n-1}(t) \quad (F.4)$$

The exponential integral may be represented in an alternate form by the use of the transformation $s = \frac{1}{\mu}$. We then have $d\mu = -\frac{ds}{s^2}$, and hence

$$E_n(t) = -\int_\infty^1 \frac{e^{-st}}{s^{n-2}} \frac{ds}{s^2} = \int_1^\infty \frac{e^{-st}}{s} ds \quad (F.5)$$

F.2 Useful Ways of Calculating Exponential Integrals

Exponential integrals may be evaluated by actual numerical integration with efficient routines or may be obtained by using series representations. Expanding the integrand in Eq. F.1 as a power series, integration term by term leads to the following representation of Exponential integral of order 1.

$$E_1(t) = -\gamma - \ln(t) + \sum_{i=1}^{\infty} \frac{(-1)^{i-1} t^i}{i! i} \quad (F.6)$$

where γ is known as the Euler constant and has a numerical value of 0.577216 . . . . Immediately it is apparent that we may use Eq. F.4 to generate Exponential integrals of higher order, by term by term integration. For example, $E_2(t)$ is given by

$$E_2(t) = 1 + t(\gamma + \ln(t) - 1) + \sum_{i=1}^{\infty} \frac{(-1)^i t^{i+1}}{(i+1)! i} \quad (F.7)$$

These series rapidly converge and are useful in the evaluation of the Exponential integrals. Values of Exponential integrals have been generated using the series given above and presented in Table F.1. These are also shown plotted in Figure F.1.

F.2.1 Approximation of $E_3(t)$:

The Exponential integral of order 3 is well approximated by an exponential function of form

$$E_3(t) = \frac{1}{2} e^{-1.8t} \quad (F.8)$$

Table F.1 Table of Exponential integrals

t	$E_1(t)$	$E_2(t)$	$E_3(t)$	t	$E_1(t)$	$E_2(t)$	$E_3(t)$
0	∞	1	0.5000	0.8	0.310596	0.200852	0.1443
0.01	4.037929	0.949671	0.4903	0.85	0.284019	0.185999	0.1347
0.02	3.354707	0.913105	0.4810	0.9	0.260184	0.172404	0.1257
0.03	2.959118	0.881672	0.4720	0.95	0.238737	0.15994	0.1174
0.04	2.681263	0.853539	0.4633	1	0.219384	0.148496	0.1097
0.05	2.467898	0.827835	0.4549	1.2	0.158408	0.111104	0.0839
0.06	2.295307	0.804046	0.4468	1.4	0.116219	0.08389	0.0646
0.07	2.150838	0.781835	0.4388	1.6	0.086308	0.063803	0.0499
0.08	2.026941	0.760961	0.4311	1.8	0.064713	0.048815	0.0387
0.09	1.918744	0.741244	0.4236	2	0.0489	0.037534	0.0301
0.1	1.822924	0.722545	0.4163	2.2	0.037191	0.028983	0.0235
0.15	1.464461	0.641039	0.3823	2.4	0.02844	0.022461	0.0184
0.2	1.22265	0.574201	0.3519	2.6	0.02185	0.017463	0.0144
0.25	1.044282	0.51773	0.3247	2.8	0.016855	0.013615	0.0113
0.3	0.905676	0.469115	0.3000	3	0.013048	0.010642	0.0089
0.35	0.794215	0.426713	0.2777	3.2	0.010133	0.008337	0.0070
0.4	0.70238	0.389368	0.2573	3.4	0.007896	0.006544	0.0056
0.45	0.625331	0.356229	0.2387	3.6	0.00616	0.005146	0.0044
0.5	0.559773	0.326644	0.2216	3.8	0.00482	0.004054	0.0035
0.55	0.503364	0.3001	0.2059	4	0.003779	0.0031982	0.0028
0.6	0.454379	0.276184	0.1916	4.2	0.002968	0.0025268	0.0022
0.65	0.411517	0.25456	0.1783	4.4	0.002336	0.0019988	0.0017
0.7	0.373769	0.234947	0.1661	4.6	0.001841	0.001583	0.0014
0.75	0.34034	0.217111	0.1548	4.8	0.001453	0.001255	0.0011
0.8	0.310596	0.200852	0.1443	5	0.001148	0.000997	0.0009

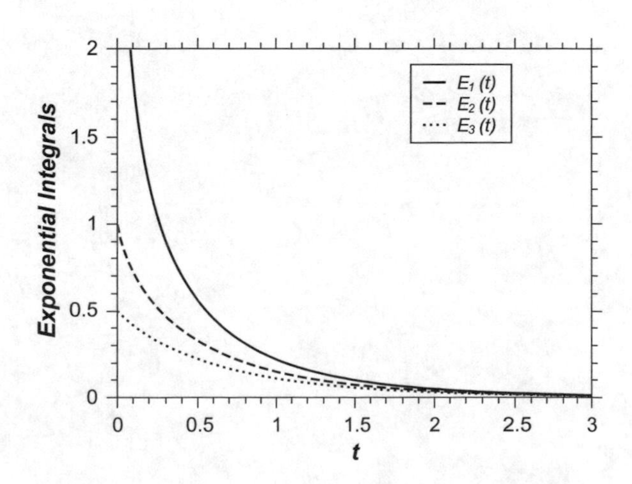

Fig. F.1 The first three exponential integrals

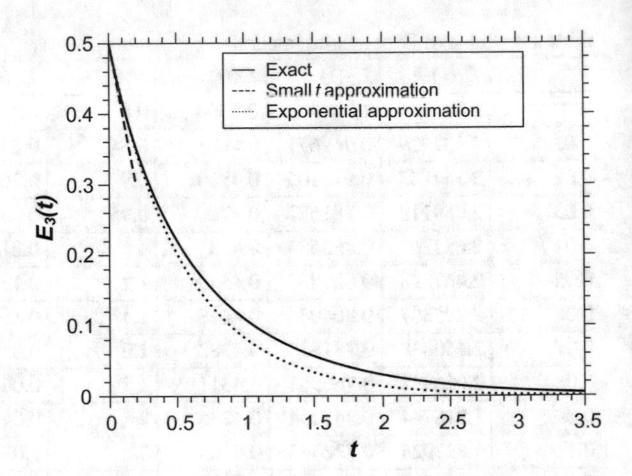

Fig. F.2 $E_3(t)$ and its approximations

For $t \rightarrow 0$ the linear approximation $E_3(t) = \frac{1}{2} - t$ is valid. Figure F.2 shows a plot of $E_3(t)$ and also the two approximations mentioned above. The exponential approximation seems to be a good approximation for all t. However the linear approximation is good only for $t < 0.15$ or so.

Appendix G
Angle Factors and Mean Beam Lengths

G.1 Angle Factors

Enclosures encountered in engineering applications generally involve surfaces oriented in three-dimensional space such that the angle factor determination is more complex than the simple analysis that involved only lengths of sides (based on angle factor algebra), as in the case of two-dimensional enclosures. A catalog of angle factors and the vast literature available on these has been made by Howell.[5]

Common geometries involved in engineering practice involve rectangles and circles. For example, rectangular parallelepiped enclosure or a rectangular box enclosure is a common geometry in engineering practice. Ovens and furnaces are usually of this geometry and our interest is in such an enclosure that is assumed to be evacuated (or contains a radiatively neutral gas). The surfaces of the enclosure are assumed to be perfectly diffuse. Radiation heat transfer among the surfaces is governed by the angle factors that have been introduced in Chap. 10. Two configurations that are relevant to the rectangular box case are:

- Two parallel rectangles of equal size
- Two rectangles that are placed perpendicular to each other and share a common edge

Enclosures involving planar surfaces in the form of circles are met with in the case of cylindrical or conical enclosures. The angle factor that is relevant for this type of enclosure is that between two coaxial circular disks.

We shall look at these cases in what follows.

---

[5]Catalog is accessible at http://www.me.utexas.edu/~howell/index.html.

S. P. Venkateshan, *Heat Transfer*,
https://doi.org/10.1007/978-3-030-58338-5

G.1.1 *Angle Factors Between Rectangles*

Definition of the Geometry

Consider the enclosure as shown in Fig. G.1. The enclosure is a rectangular parallelepiped of width a, depth b, and height c. The non-dimensional ratios shown in the figure are used in the both angle factor and mean beam length calculations (to be discussed later on).

Some general remarks may be made now. Since the opposite sides are parallel rectangles of equal size (there are three such cases) the angle factor between, say the bottom and top, is the same as that between the top and the bottom. However, adjacent sides may not have equal areas and hence there are two angle factor pairs that may or may not be equal.

G.1.2 *Angle Factor Between Equal and Parallel Rectangles*

The angle factor between opposite sides of the parallelepiped pertains to two equal rectangles of size $a \times b$ with spacing c (or two equal rectangles of size $a \times c$ with spacing b or two equal rectangles of size $b \times c$ with spacing a). Each rectangle has area ab and the spacing between the rectangles is c. The angle factor is evaluated by performing the integration of the angle factor defining Eq. 10.10. It is possible to obtain the results in closed analytical form and is available from Howell's catalog. The expression is given by

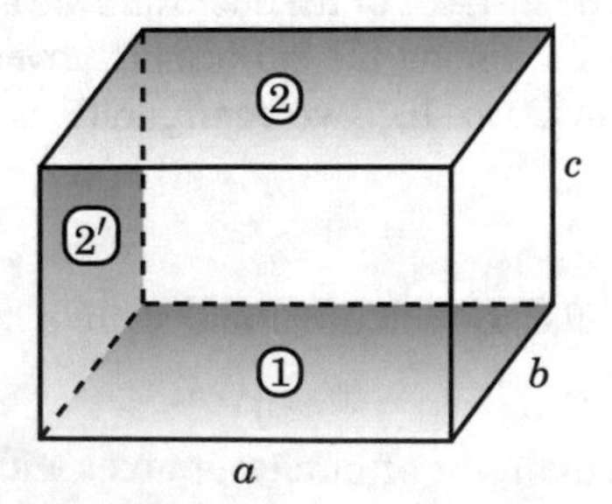

Opposite rectangles:

1 - 2: $X = \dfrac{a}{c}$ $\quad Y = \dfrac{b}{c}$

Adjascent rectangles:

1 - 2′ $Z = \dfrac{a}{b}$ $\quad Y' = \dfrac{c}{b} = \dfrac{1}{Y}$

Fig. G.1 Rectangular parallelepiped enclosure

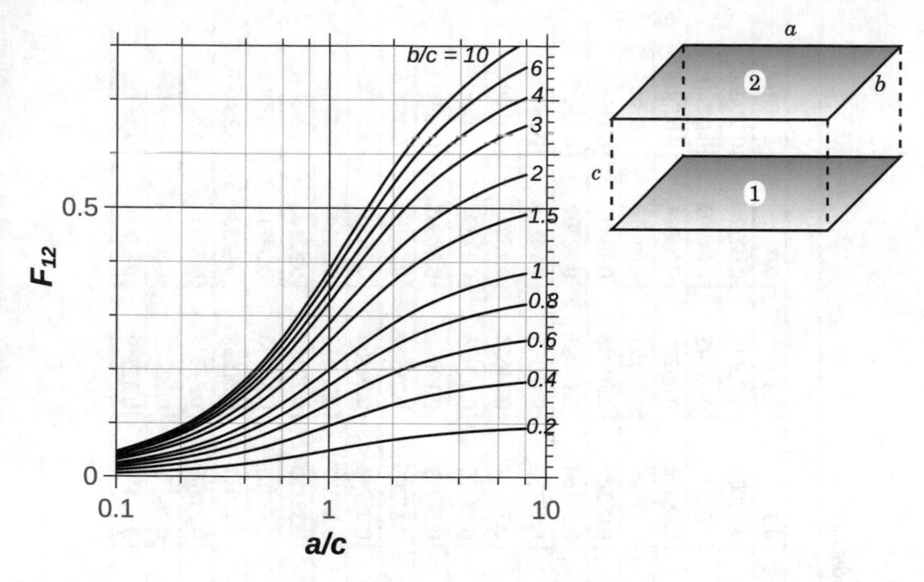

Fig. G.2 Angle factor between parallel rectangles of equal size

$$F_{12} = \frac{2}{\pi XY}\left\{\ln\sqrt{\frac{(1+X^2)(1+Y^2)}{1+X^2+Y^2}} + X\sqrt{1+Y^2}\tan^{-1}\left[\frac{X}{\sqrt{1+Y^2}}\right] + Y\sqrt{1+X^2}\tan^{-1}\left[\frac{Y}{\sqrt{1+X^2}}\right] - X\tan^{-1}X - Y\tan^{-1}Y\right\} \tag{G.1}$$

where the nomenclature used is in accordance with Fig. G.1. The angle factor between two parallel rectangles of equal size F_{12} (either bottom surface is 1 and top surface is 2 or vice versa) depends on two ratios $X = \frac{a}{c}$ and $Y = \frac{b}{c}$. The closed form solution given above is used to generate the angle factor data shown plotted in Fig. G.2. F_{12} is plotted for $0.1 \le X \le 10$ as a family of curves for the indicated values of Y.

The data is also given in Tables G.1 and G.2 for more accurate work, where interpolation may be used to get the desired angle factors for arbitrary values of X and Y.

G.1.3 Angle Factor Between Perpendicular Rectangles

Similarly, the angle factors between perpendicular rectangles that share a common edge are evaluated using the closed form solution (adjacent sides in Fig. G.1). Surfaces are identified as in Fig. G.1 and hence the common edge shared between the

Table G.1 Angle factors for diffuse radiation interchange between equal parallel rectangles (continued)

X	Y									
	0.1	0.2	0.3	0.4	0.5	0.6	0.7	0.8	0.9	1
0.1	0.00316	0.00626	0.00925	0.01207	0.01499	0.01715	0.02002	0.02141	0.02326	0.02492
0.2	0.00626	0.01240	0.01832	0.02392	0.02977	0.03398	0.03977	0.04243	0.04609	0.04941
0.3	0.00925	0.01832	0.02705	0.03532	0.04415	0.05020	0.05898	0.06272	0.06814	0.07305
0.4	0.01207	0.02392	0.03532	0.04614	0.05796	0.06560	0.07746	0.08199	0.08910	0.09554
0.5	0.01471	0.02915	0.04306	0.05625	0.07109	0.08001	0.09505	0.10006	0.10876	0.11665
0.6	0.01715	0.03398	0.05020	0.06560	0.08346	0.09336	0.11166	0.11683	0.12702	0.13627
0.7	0.01938	0.03840	0.05675	0.07417	0.09505	0.10563	0.12722	0.13225	0.14383	0.15435
0.8	0.02141	0.04243	0.06272	0.08199	0.10585	0.11683	0.14174	0.14637	0.15923	0.17092
0.9	0.02326	0.04609	0.06814	0.08910	0.11587	0.12702	0.15525	0.15923	0.17327	0.18605
1	0.02492	0.04941	0.07305	0.09554	0.12516	0.13627	0.16777	0.17092	0.18605	0.19982
1.5	0.03119	0.06186	0.09152	0.11979	0.16216	0.17123	0.21785	0.21527	0.23460	0.25226
2	0.03514	0.06971	0.10318	0.13513	0.18765	0.19342	0.25248	0.24356	0.26564	0.28588
3	0.03965	0.07868	0.11650	0.15268	0.21938	0.21888	0.29577	0.27614	0.30147	0.32474
4	0.04209	0.08353	0.12372	0.16219	0.23798	0.23271	0.32122	0.29387	0.32100	0.34596
5	0.04360	0.08654	0.12820	0.16809	0.25010	0.24130	0.33782	0.30490	0.33314	0.35917
6	0.04463	0.08858	0.13123	0.17209	0.25859	0.24712	0.34947	0.31238	0.34139	0.36813
7	0.04537	0.09005	0.13342	0.17497	0.26486	0.25132	0.35808	0.31778	0.34734	0.37461
8	0.04592	0.09116	0.13507	0.17715	0.26968	0.25449	0.36469	0.32186	0.35184	0.37950
10	0.04671	0.09272	0.13739	0.18021	0.27659	0.25896	0.37419	0.32759	0.35816	0.38638

Table G.2 Angle factors for diffuse radiation interchange between equal parallel rectangles

X	*Y*									
	1.5	2	3	4	5	6	7	8	9	10
0.1	0.03119	0.03514	0.03965	0.04209	0.04360	0.04463	0.04537	0.04592	0.04636	0.04671
0.2	0.06186	0.06971	0.07868	0.08353	0.08654	0.08858	0.09005	0.09116	0.09202	0.09272
0.3	0.09152	0.10318	0.11650	0.12372	0.12820	0.13123	0.13342	0.13507	0.13636	0.13739
0.4	0.11979	0.13513	0.15268	0.16219	0.16809	0.17209	0.17497	0.17715	0.17885	0.18021
0.5	0.14641	0.16527	0.18687	0.19859	0.20586	0.21079	0.21435	0.21703	0.21913	0.22081
0.6	0.17123	0.19342	0.21888	0.23271	0.24130	0.24712	0.25132	0.25449	0.25697	0.25896
0.7	0.19417	0.21951	0.24864	0.26447	0.27431	0.28098	0.28580	0.28943	0.29227	0.29455
0.8	0.21527	0.24356	0.27614	0.29387	0.30490	0.31238	0.31778	0.32186	0.32504	0.32759
0.9	0.23460	0.26564	0.30147	0.32100	0.33314	0.34139	0.34734	0.35184	0.35535	0.35816
1	0.25226	0.28588	0.32474	0.34596	0.35917	0.36813	0.37461	0.37950	0.38332	0.38638
1.5	0.32006	0.36405	0.41541	0.44366	0.46131	0.47331	0.48199	0.48855	0.49367	0.49779
2	0.36405	0.41525	0.47558	0.50899	0.52993	0.54421	0.55454	0.56236	0.56847	0.57338
3	0.41541	0.47558	0.54738	0.58762	0.61300	0.63037	0.64297	0.65251	0.65998	0.66599
4	0.44366	0.50899	0.58762	0.63204	0.66020	0.67954	0.69359	0.70426	0.71261	0.71933
5	0.46131	0.52993	0.61300	0.66020	0.69024	0.71093	0.72599	0.73743	0.74640	0.75363
6	0.47331	0.54421	0.63037	0.67954	0.71093	0.73258	0.74838	0.76039	0.76982	0.77741
7	0.48199	0.55454	0.64297	0.69359	0.72599	0.74838	0.76473	0.77717	0.78695	0.79482
8	0.48855	0.56236	0.65251	0.70426	0.73743	0.76039	0.77717	0.78995	0.80000	0.80811
10	0.49779	0.57338	0.66599	0.71933	0.75363	0.77741	0.79482	0.80811	0.81856	0.82699

two rectangles is the side of length b. The angle factor is dependent on $Z = \frac{a}{b}$ and the ratio $Y' = \frac{c}{b} = \frac{1}{Y}$ and is given by the expression

$$F_{12} = \frac{1}{\pi Y'}\Bigg\{Y' \tan^{-1}\frac{1}{Y'} + Z\tan^{-1}\frac{1}{Z} - \sqrt{Z^2 + Y'^2}\tan^{-1}\left[\frac{1}{\sqrt{Y'^2 + Z^2}}\right] + \frac{1}{4}\ln\left[\frac{(1+Y'^2)(1+Z^2)}{1+Y'^2+Z^2}\right] + \frac{Y'^2}{4}\ln\left[\frac{Y'^2(1+Y'^2+Z^2)}{(1+Y'^2)(Y'^2+Z^2)}\right] + \frac{Z^2}{4}\ln\left[\frac{Z^2(1+Y'^2+Z^2)}{(1+Z^2)(Y'^2+Z^2)}\right]\Bigg\} \tag{G.2}$$

The angle factor data is shown plotted in Fig. G.3 for $0.1 \le \frac{1}{Y} \le 10$ as a family of curves for the indicated values of Z.

The data is also given in Tables G.3 and G.4 for more accurate work, where interpolation may be used to get the desired angle factors for arbitrary values of Z and Y'.

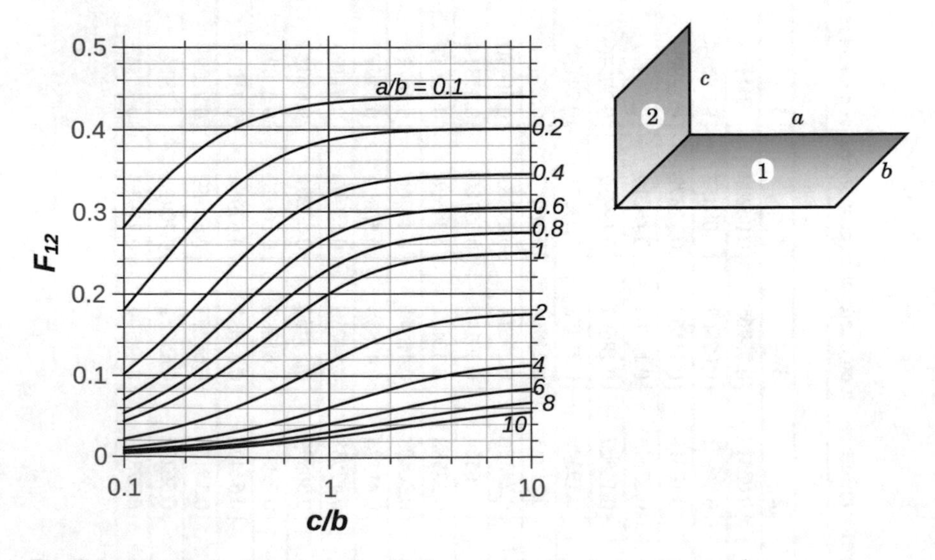

Fig. G.3 Angle factor between perpendicular rectangles sharing a common edge

Table G.3 Angle factors for diffuse radiation interchange between perpendicular rectangles sharing a common edge (continued)

Y'	Z											
	0.1	0.15	0.2	0.25	0.3	0.4	0.5	0.6	0.7	0.8	0.9	1
0.1	0.28189	0.33271	0.36216	0.38072	0.39322	0.40859	0.41738	0.42290	0.42660	0.42920	0.43109	0.43251
0.15	0.22181	0.27643	0.31182	0.33569	0.35245	0.37378	0.38634	0.39434	0.39975	0.40358	0.40638	0.40848
0.2	0.18108	0.23387	0.27104	0.29765	0.31714	0.34295	0.35864	0.36884	0.37581	0.38078	0.38443	0.38719
0.25	0.15229	0.20141	0.23812	0.26571	0.28669	0.31554	0.33371	0.34577	0.35413	0.36014	0.36459	0.36796
0.3	0.13107	0.17622	0.21143	0.23891	0.26048	0.29119	0.31120	0.32477	0.33433	0.34127	0.34644	0.35039
0.4	0.10215	0.14017	0.17147	0.19721	0.21839	0.25032	0.27243	0.28809	0.29945	0.30789	0.31429	0.31924
0.5	0.08348	0.11590	0.14346	0.16686	0.18672	0.21794	0.24064	0.25734	0.26983	0.27931	0.28663	0.29237
0.6	0.07048	0.09859	0.12295	0.14407	0.16239	0.19206	0.21445	0.23147	0.24453	0.25467	0.26263	0.26896
0.7	0.06094	0.08566	0.10737	0.12647	0.14328	0.17111	0.19273	0.20960	0.22284	0.23331	0.24168	0.24842
0.8	0.05365	0.07567	0.09519	0.11254	0.12797	0.15395	0.17457	0.19100	0.20415	0.21474	0.22331	0.23032
0.9	0.04790	0.06773	0.08543	0.10127	0.11548	0.13969	0.15924	0.17509	0.18797	0.19850	0.20715	0.21429
1	0.04325	0.06127	0.07744	0.09199	0.10512	0.12770	0.14619	0.16138	0.17390	0.18425	0.19286	0.20004
1.5	0.02908	0.04139	0.05258	0.06280	0.07216	0.08867	0.10271	0.11471	0.12502	0.13389	0.14156	0.14822
2	0.02188	0.03120	0.03971	0.04753	0.05473	0.06757	0.07865	0.08829	0.09671	0.10411	0.11065	0.11643
3	0.01462	0.02088	0.02661	0.03190	0.03680	0.04559	0.05327	0.06004	0.06605	0.07142	0.07624	0.08059
4	0.01097	0.01568	0.02000	0.02398	0.02768	0.03434	0.04018	0.04536	0.04998	0.05413	0.05789	0.06131
5	0.00878	0.01255	0.01601	0.01921	0.02218	0.02753	0.03223	0.03641	0.04015	0.04352	0.04659	0.04938
6	0.00732	0.01046	0.01335	0.01602	0.01849	0.02296	0.02690	0.03040	0.03354	0.03637	0.03895	0.04130
7	0.00628	0.00897	0.01145	0.01373	0.01586	0.01970	0.02308	0.02609	0.02879	0.03123	0.03345	0.03549
8	0.00549	0.00785	0.01002	0.01202	0.01388	0.01724	0.02020	0.02284	0.02521	0.02735	0.02931	0.03110
10	0.00439	0.00628	0.00802	0.00962	0.01111	0.01380	0.01617	0.01829	0.02019	0.02191	0.02348	0.02492

Table G.4 Angle factors for diffuse radiation interchange between perpendicular rectangles sharing a common edge

Y'	Z									
	1.5	2	3	4	5	6	7	8	9	10
0.1	0.43616	0.43756	0.43860	0.43897	0.43915	0.43925	0.43930	0.43934	0.43937	0.43939
0.15	0.41391	0.41599	0.41755	0.41812	0.41838	0.41853	0.41861	0.41867	0.41871	0.41874
0.2	0.39435	0.39711	0.39919	0.39993	0.40029	0.40048	0.40059	0.40067	0.40072	0.40076
0.25	0.37678	0.38020	0.38279	0.38372	0.38416	0.38440	0.38455	0.38464	0.38471	0.38475
0.3	0.36079	0.36486	0.36795	0.36907	0.36960	0.36988	0.37006	0.37017	0.37025	0.37030
0.4	0.33253	0.33784	0.34191	0.34339	0.34409	0.34447	0.34470	0.34486	0.34496	0.34503
0.5	0.30814	0.31460	0.31961	0.32145	0.32232	0.32280	0.32309	0.32327	0.32340	0.32350
0.6	0.28678	0.29429	0.30019	0.30238	0.30342	0.30399	0.30433	0.30456	0.30471	0.30482
0.7	0.26789	0.27632	0.28306	0.28559	0.28679	0.28745	0.28786	0.28812	0.28830	0.28843
0.8	0.25104	0.26029	0.26781	0.27066	0.27203	0.27278	0.27324	0.27354	0.27375	0.27389
0.9	0.23594	0.24588	0.25412	0.25729	0.25881	0.25966	0.26017	0.26050	0.26074	0.26090
1	0.22232	0.23285	0.24176	0.24522	0.24690	0.24783	0.24840	0.24877	0.24903	0.24921
1.5	0.17077	0.18286	0.19415	0.19889	0.20126	0.20260	0.20343	0.20397	0.20435	0.20462
2	0.13715	0.14930	0.16169	0.16731	0.17024	0.17193	0.17299	0.17370	0.17419	0.17455
3	0.09708	0.10780	0.12019	0.12660	0.13023	0.13245	0.13388	0.13486	0.13555	0.13606
4	0.07458	0.08365	0.09495	0.10136	0.10526	0.10776	0.10944	0.11062	0.11147	0.11210
5	0.06038	0.06809	0.07814	0.08420	0.08810	0.09072	0.09254	0.09385	0.09481	0.09554
6	0.05065	0.05731	0.06622	0.07184	0.07560	0.07821	0.08009	0.08147	0.08251	0.08331
7	0.04359	0.04943	0.05738	0.06254	0.06610	0.06865	0.07053	0.07194	0.07302	0.07386
8	0.03825	0.04343	0.05057	0.05531	0.05865	0.06110	0.06294	0.06435	0.06545	0.06632
10	0.03069	0.03491	0.04082	0.04484	0.04777	0.04998	0.05170	0.05306	0.05414	0.05502

G.1.4 Angle Factor Between Coaxial Disks

The nomenclature for two coaxial disks is shown in Fig. G.4a. The angle factor F_{12} between the bottom disk 1 and top disk 2 depends on the two ratios $R_1 = \frac{r_1}{L}$ and $R_2 = \frac{r_2}{L}$. The angle factor is obtained in the closed form as

$$F_{12} = \frac{1}{2}\left\{X - \sqrt{X^2 - 4\left(\frac{R_2}{R_1}\right)^2}\right\} \quad \text{where} \quad X = 1 + \frac{1 + R_2^2}{R_1^2} \tag{G.3}$$

Based on the closed form solution, data is generated for a range of these two parameters and the results are shown in Fig. G.4b. The data is also given in Table G.5 for more accurate work, where interpolation may be used to get the desired angle factors for arbitrary values of R_1 and R_2.

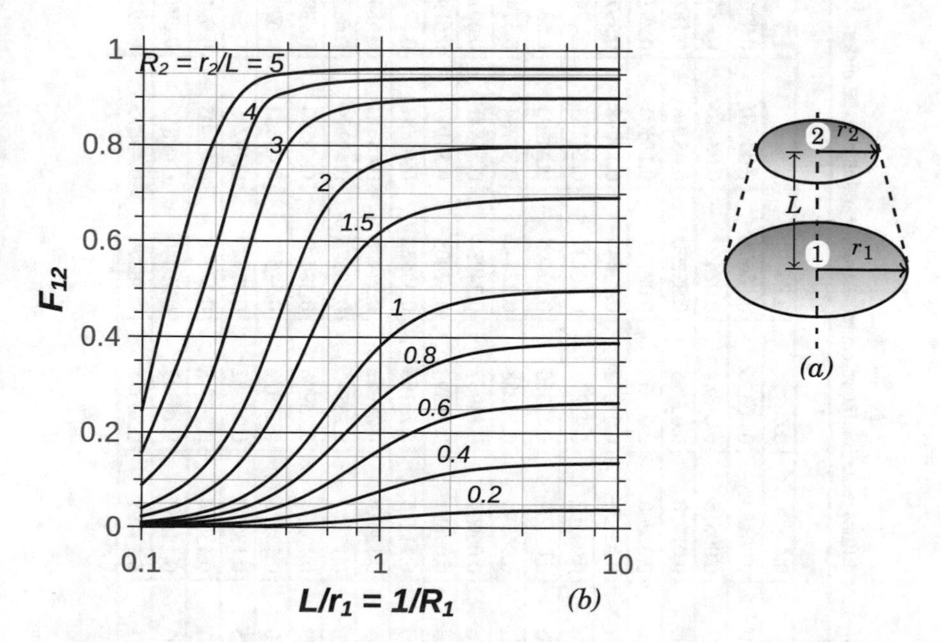

Fig. G.4 **a** Parallel coaxial disk geometry **b** Angle factor between parallel coaxial disks

Table G.5 Angle factors for diffuse radiation interchange between parallel coaxial disks

$\frac{1}{R_1}$	R_2									
	0.2	0.4	0.6	0.8	1	1.5	2	3	4	5
0.1	0.00040	0.00158	0.00356	0.00634	0.00990	0.02227	0.03959	0.08902	0.15812	0.24672
0.2	0.00154	0.00615	0.01384	0.02459	0.03840	0.08623	0.15279	0.33944	0.58388	0.81900
0.3	0.00330	0.01320	0.02965	0.05260	0.08196	0.18242	0.31803	0.64586	0.86472	0.93590
0.4	0.00551	0.02200	0.04930	0.08713	0.13502	0.29352	0.48769	0.80000	0.91153	0.95015
0.5	0.00799	0.03179	0.07092	0.12446	0.19098	0.39754	0.60961	0.84861	0.92482	0.95492
0.6	0.01056	0.04187	0.09278	0.16121	0.24388	0.47901	0.67889	0.86822	0.93063	0.95716
0.7	0.01310	0.05169	0.11360	0.19495	0.28993	0.53614	0.71722	0.87817	0.93375	0.95841
0.8	0.01551	0.06090	0.13258	0.22442	0.32784	0.57481	0.73985	0.88397	0.93564	0.95919
0.9	0.01776	0.06929	0.14937	0.24934	0.35811	0.60123	0.75423	0.88769	0.93688	0.95970
1	0.01980	0.07681	0.16393	0.27005	0.38197	0.61980	0.76393	0.89023	0.93774	0.96006
2	0.03120	0.11586	0.23196	0.35532	0.46887	0.67544	0.79176	0.89770	0.94035	0.96118
3	0.03487	0.12729	0.24941	0.37438	0.48612	0.68493	0.79640	0.89899	0.94081	0.96138
4	0.03636	0.13177	0.25595	0.38125	0.49219	0.68818	0.79798	0.89943	0.94097	0.96145
5	0.03709	0.13393	0.25906	0.38447	0.49500	0.68967	0.79871	0.89964	0.94105	0.96148
6	0.03750	0.13513	0.26077	0.38623	0.49653	0.69048	0.79911	0.89975	0.94109	0.96150
7	0.03775	0.13587	0.26181	0.38729	0.49745	0.69097	0.79935	0.89982	0.94111	0.96151
8	0.03791	0.13634	0.26248	0.38798	0.49805	0.69128	0.79950	0.89986	0.94113	0.96152
9	0.03803	0.13668	0.26295	0.38846	0.49846	0.69150	0.79960	0.89989	0.94114	0.96152
10	0.03811	0.13691	0.26328	0.38879	0.49875	0.69165	0.79968	0.89991	0.94114	0.96152

G.2 Mean Beam Lengths in a Parallelepiped Enclosure

The parallelepiped enclosure geometry is commonly used in ovens and furnaces, that may contain radiatively participating gases. Hence a knowledge of mean beam lengths for surface to surface, surface to gas in this geometry are important. The defining formulae for these are given in Chap. 11. The integrals have been obtained as closed form expressions by Dunkle[6] in the case of rectangles that form either opposite or adjacent sides of a rectangular parallelepiped, corresponding to the optically thin case. He has also presented tables and graphs that may be made use of. We describe here the methodology and give tables in a readily usable form. The entries have been calculated using the closed form expressions given by Dunkle.

G.2.1 Mean Beam Length Between Equal and Parallel Rectangles

The geometry for this case follows the notation used in Fig. G.1. The area of any one of the rectangles is given by $A = ab$.The volume of the parallelepiped is given by $V = abc$. We shall represent the angle factors and mean beam lengths with subscript par. Thus the angle factor $F_{12} = F_{21} = F_{par}$ may be obtained by using Fig. G.2 or by using Tables G.1 and G.2. The mean beam length r_{par} is represented in the form of a non-dimensional ratio given by

$$Z_{par} = \frac{F_{par} A r_{par}}{V} \tag{G.4}$$

Using the expressions for area and volume given earlier, we may rewrite this as

$$Z_{par} = \frac{F_{par} ab r_{par}}{abc} = \frac{F_{par} r_{par}}{c} \tag{G.5}$$

Thus, the mean beam length information may be recast in the alternate non-dimensional form

$$\frac{r_{par}}{c} = \frac{Z_{par}}{F_{par}} \tag{G.6}$$

The closed form expression for Z_{par} is given by

[6]R. V.Dunkle, ASME Journal of Heat Transfer, Vol. 86, pp.75–80, 1964.

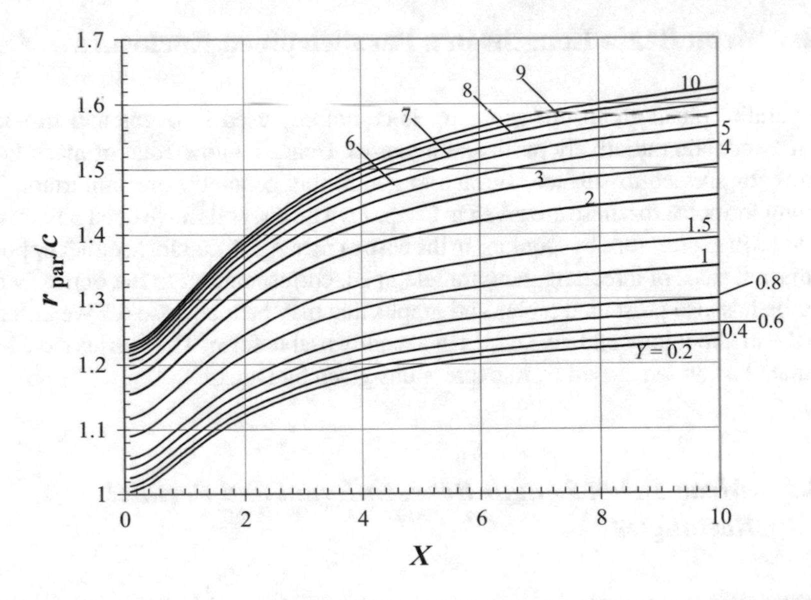

Fig. G.5 Non-dimensional mean beam length for parallel equal rectangles

$$Z_{par} = \frac{4}{\pi}\left\{ \tan^{-1}\left[\frac{XY}{\sqrt{1+X^2+Y^2}}\right] + \frac{1}{X}\ln\left[\frac{Y+\sqrt{1+X^2+Y^2}}{\sqrt{1+X^2}(Y+\sqrt{1+Y^2})}\right] + \frac{1}{Y}\ln\left[\frac{X+\sqrt{1+X^2+Y^2}}{\sqrt{1+Y^2}(X+\sqrt{1+X^2})}\right] + \frac{1}{XY}(\sqrt{1+X^2}+\sqrt{1+Y^2}-1-\sqrt{1+X^2+Y^2}) \right\} \quad (G.7)$$

Evaluating the required angle factors using Eq. G.1 and using Eq. G.7 for evaluating Z_{par} the non-dimensional ratio $\frac{r_{par}}{c}$ may be calculated for the desired values of X and Y. The data thus generated has been shown plotted in Fig. G.5.

The generated data is also given in Tables G.6 and G.7 for ease in interpolation for arbitrary X and Y values.

G.2.2 Mean Beam Length Between Perpendicular Rectangles Sharing a Common Edge

As indicated earlier, the non-dimensional geometric parameters that are appropriate to this geometry are Z and $Y' = \frac{1}{Y}$. The two rectangles are, in general, of different areas, and hence the angle factors, as well as the mean beam lengths, depend on the direction of radiant interaction. We define a non-dimensional mean beam length given by the relation

Table G.6 Non-dimensional mean beam length, $\frac{r_{par}}{c}$ for parallel equal rectangles (continued)

X	Y							
	0.2	0.4	0.6	0.8	1	1.5	2	3
0.1	1.00409	1.01306	1.02588	1.04056	1.05560	1.08993	1.11729	1.15498
0.2	1.00651	1.01547	1.02828	1.04295	1.05799	1.09236	1.119778	1.15758
0.3	1.01038	1.01932	1.03210	1.04676	1.06181	1.09625	1.12377	1.16176
0.4	1.01547	1.02438	1.03714	1.05180	1.06686	1.10140	1.12906	1.16730
0.5	1.02152	1.03041	1.04315	1.05781	1.07289	1.10757	1.13539	1.17395
0.6	1.02828	1.03714	1.04988	1.06454	1.07966	1.11450	1.14253	1.18146
0.7	1.03549	1.04434	1.05707	1.07176	1.08693	1.12196	1.15022	1.18958
0.8	1.04295	1.05180	1.06454	1.07927	1.09449	1.12975	1.15827	1.19808
0.9	1.05049	1.05935	1.07211	1.08689	1.10218	1.13769	1.16649	1.20680
1	1.05799	1.06686	1.07966	1.09449	1.10987	1.14565	1.17475	1.21558
1.5	1.09236	1.10140	1.11450	1.12975	1.14565	1.18300	1.21374	1.25740
2	1.11978	1.12906	1.14253	1.15827	1.17475	1.21374	1.24613	1.29264
3	1.15758	1.16730	1.18146	1.19808	1.21558	1.25740	1.29264	1.34410
4	1.18141	1.19146	1.20613	1.22342	1.24168	1.28560	1.32298	1.37826
5	1.19752	1.20781	1.22287	1.24064	1.25946	1.30494	1.34392	1.40212
6	1.20907	1.21955	1.23489	1.25303	1.27227	1.31893	1.35913	1.41959
7	1.21773	1.22836	1.24392	1.26234	1.28191	1.32949	1.37064	1.43289
8	1.22447	1.23520	1.25094	1.26959	1.28941	1.33772	1.37964	1.44333
10	1.23423	1.24513	1.26114	1.28011	1.30032	1.34972	1.39278	1.45865

Table G.7 Non-dimensional mean beam length, $\frac{r_{par}}{c}$ for parallel equal rectangles

X	*Y*						
	4	5	6	7	8	9	10
0.1	1.17872	1.19477	1.20627	1.21490	1.22160	1.22695	1.23132
0.2	1.18141	1.19752	1.20907	1.21773	1.22447	1.22984	1.23423
0.3	1.18572	1.20194	1.21357	1.22230	1.22907	1.23449	1.23891
0.4	1.19146	1.20781	1.21955	1.22836	1.23520	1.24067	1.24513
0.5	1.19835	1.21488	1.22675	1.23566	1.24258	1.24812	1.25264
0.6	1.20613	1.22287	1.23489	1.24392	1.25094	1.25655	1.26114
0.7	1.21456	1.23153	1.24373	1.25290	1.26002	1.26572	1.27037
0.8	1.22342	1.24064	1.25303	1.26234	1.26959	1.27538	1.28011
0.9	1.23250	1.25000	1.26259	1.27206	1.27944	1.28533	1.29015
1	1.24168	1.25946	1.27227	1.28191	1.28941	1.29541	1.30032
1.5	1.28560	1.30494	1.31893	1.32949	1.33772	1.34432	1.34972
2	1.32298	1.34392	1.35913	1.37064	1.37964	1.38686	1.39278
3	1.37826	1.40212	1.41959	1.43289	1.44333	1.45174	1.45865
4	1.41548	1.44173	1.46110	1.47593	1.48762	1.49707	1.50485
5	1.44173	1.46992	1.49085	1.50696	1.51970	1.53003	1.53857
6	1.46110	1.49085	1.51307	1.53024	1.54388	1.55496	1.56414
7	1.47593	1.50696	1.53024	1.54830	1.56269	1.57443	1.58416
8	1.48762	1.51970	1.54388	1.56269	1.57774	1.59002	1.60024
10	1.50485	1.53857	1.56414	1.58416	1.60024	1.61344	1.62445

Table G.8 Non-dimensional mean beam length for radiation interchange between perpendicular rectangles sharing a common edge (continued)

Y'	Z											
	0.1	0.15	0.2	0.25	0.3	0.4	0.5	0.6	0.7	0.8	0.9	1
0.1	0.20385	0.19304	0.17755	0.16256	0.14919	0.12743	0.11088	0.09801	0.08776	0.07942	0.07251	0.06669
0.15	0.19304	0.19605	0.18896	0.17872	0.16793	0.14799	0.13133	0.11765	0.10637	0.09696	0.08902	0.08225
0.2	0.17755	0.18896	0.18863	0.18315	0.17556	0.15905	0.14369	0.13033	0.11890	0.10912	0.10072	0.09345
0.25	0.16256	0.17872	0.18315	0.18158	0.17698	0.16428	0.15086	0.13842	0.12735	0.11764	0.10913	0.10166
0.3	0.14919	0.16793	0.17556	0.17698	0.17491	0.16580	0.15453	0.14331	0.13292	0.12353	0.11515	0.10769
0.4	0.12743	0.14799	0.15905	0.16428	0.16580	0.16258	0.15540	0.14689	0.13823	0.12996	0.12226	0.11519
0.5	0.11088	0.13133	0.14369	0.15086	0.15453	0.15540	0.15154	0.14554	0.13873	0.13177	0.12501	0.11860
0.6	0.09801	0.11765	0.13033	0.13842	0.14331	0.14689	0.14554	0.14165	0.13651	0.13085	0.12507	0.11942
0.7	0.08776	0.10637	0.11890	0.12735	0.13292	0.13823	0.13873	0.13651	0.13279	0.12830	0.12348	0.11859
0.8	0.07942	0.09696	0.10912	0.11764	0.12353	0.12996	0.13177	0.13085	0.12830	0.12483	0.12088	0.11671
0.9	0.07251	0.08902	0.10072	0.10913	0.11515	0.12226	0.12501	0.12507	0.12348	0.12088	0.11768	0.11416
1	0.06669	0.08225	0.09345	0.10166	0.10769	0.11519	0.11860	0.11942	0.11859	0.11671	0.11416	0.11122
1.5	0.04752	0.05939	0.06833	0.07523	0.08063	0.08821	0.09280	0.09538	0.09657	0.09677	0.09627	0.09527
2	0.03686	0.04638	0.05368	0.05945	0.06408	0.07089	0.07539	0.07832	0.08012	0.08110	0.08147	0.08139
3	0.02542	0.03219	0.03747	0.04174	0.04524	0.05059	0.05437	0.05708	0.05901	0.06035	0.06124	0.06180
4	0.01939	0.02463	0.02875	0.03211	0.03490	0.03923	0.04238	0.04472	0.04646	0.04776	0.04871	0.04940
5	0.01567	0.01994	0.02332	0.02608	0.02839	0.03201	0.03468	0.03670	0.03825	0.03943	0.04034	0.04104
6	0.01315	0.01675	0.01961	0.02195	0.02392	0.02702	0.02934	0.03111	0.03248	0.03355	0.03439	0.03505
7	0.01132	0.01443	0.01691	0.01895	0.02066	0.02338	0.02541	0.02698	0.02821	0.02918	0.02996	0.03057
8	0.00994	0.01268	0.01487	0.01667	0.01819	0.02059	0.02241	0.02382	0.02493	0.02582	0.02653	0.02710
10	0.00799	0.01021	0.01197	0.01344	0.01467	0.01663	0.01813	0.01930	0.02022	0.02097	0.02158	0.02207

Table G.9 Non-dimensional mean beam length for radiation interchange between perpendicular rectangles sharing a common edge

Y'	Z									
	1.5	2	3	4	5	6	7	8	9	10
0.1	0.04752	0.03686	0.02542	0.01939	0.01567	0.01315	0.01132	0.00994	0.00886	0.00799
0.15	0.05939	0.04638	0.03219	0.02463	0.01994	0.01675	0.01443	0.01268	0.01131	0.01021
0.2	0.06833	0.05368	0.03747	0.02875	0.02332	0.01961	0.01691	0.01487	0.01327	0.01197
0.25	0.07523	0.05945	0.04174	0.03211	0.02608	0.02195	0.01895	0.01667	0.01488	0.01344
0.3	0.08063	0.06408	0.04524	0.03490	0.02839	0.02392	0.02066	0.01819	0.01624	0.01457
0.4	0.08821	0.07089	0.05059	0.03923	0.03201	0.02702	0.02338	0.02059	0.01840	0.01653
0.5	0.09280	0.07539	0.05437	0.04238	0.03468	0.02934	0.02541	0.02241	0.02005	0.01813
0.6	0.09538	0.07832	0.05708	0.04472	0.03670	0.03111	0.02698	0.02382	0.02132	0.01930
0.7	0.09657	0.08012	0.05901	0.04646	0.03825	0.03248	0.02821	0.02493	0.02233	0.02022
0.8	0.09677	0.08110	0.06035	0.04776	0.03943	0.03355	0.02918	0.02582	0.02314	0.02097
0.9	0.09627	0.08147	0.06124	0.04871	0.04034	0.03439	0.02996	0.02653	0.02380	0.02158
1	0.09527	0.08139	0.06180	0.04940	0.04104	0.03505	0.03057	0.02710	0.02433	0.02207
1.5	0.08680	0.07723	0.06145	0.05036	0.04246	0.03663	0.03218	0.02867	0.02584	0.02352
2	0.07723	0.07086	0.05869	0.04923	0.04212	0.03670	0.03246	0.02907	0.02631	0.02402
3	0.06145	0.05869	0.05162	0.04505	0.03959	0.03514	0.03152	0.02853	0.02603	0.02392
4	0.05036	0.04923	0.04505	0.04052	0.03641	0.03287	0.02986	0.02730	0.02511	0.02322
5	0.04246	0.04212	0.03959	0.03641	0.03332	0.03052	0.02804	0.02587	0.02397	0.02231
6	0.03663	0.03670	0.03514	0.03287	0.03052	0.02829	0.02625	0.02442	0.02279	0.02133
7	0.03218	0.03246	0.03152	0.02986	0.02804	0.02625	0.02457	0.02303	0.02162	0.02034
8	0.02867	0.02907	0.02853	0.02730	0.02587	0.02442	0.02303	0.02172	0.02050	0.01939
10	0.02352	0.02402	0.02392	0.02322	0.02231	0.02133	0.02034	0.01939	0.01848	0.01762

$$Z_{per} = \frac{F_{12}A_1r_{12}}{abc} \tag{G.8}$$

where (1) and (2) are surface identifiers as shown in Fig. G.1. Z_{per} is obtained in closed form as

$$\begin{aligned} Z_{per} = &\frac{1}{\pi}\left\{\frac{Y'}{Z}\ln\left[\frac{\left(1+\sqrt{1+Y'^2}\right)\sqrt{Y'^2+Z^2}}{Y'\left(1+\sqrt{1+Y'^2+Z^2}\right)}\right] + \frac{Z}{Y'}\ln\left[\frac{\left(1+\sqrt{1+Z^2}\right)\sqrt{Y'^2+Z^2}}{Z\left(1+\sqrt{1+Y'^2+Z^2}\right)}\right]\right. \\ &+ \frac{1}{3Y'Z}\left[(1+Y'^2)^{1.5} + (1+Z^2)^{1.5} + (Y'^2+Z^2)^{1.5} - (1+Y'^2+Z^2)^{1.5}\right] \\ &\left. + \frac{Z}{Y'}\left[\sqrt{1+Y'^2+Z^2} - \sqrt{Y'^2+Z^2} - \sqrt{1+Z^2}\right] + \frac{2}{3}\left[\frac{Y'^2}{Z} + \frac{Z^2}{Y'} - \frac{1}{2Y'Z}\right]\right\} \end{aligned} \tag{G.9}$$

Data generated with the help of the above expression is shown in Tables G.8 and G.9 for ready reference, as well as in a form suitable for interpolation.

Appendix H
Basic Equations of Convection Heat Transfer

H.1 Introduction

Equations of motion for a viscous heat conducting fluid are required while dealing with problems that involve convection heat transfer. The governing equations go by the name of Navier–Stokes equations[7] (or NS equations) and are derived from first principles, by making mass, momentum and energy balance over a suitably chosen control volume. In the case of Cartesian coordinates, the control volume is in the shape of a rectangular parallelepiped, similar to what was used in deriving the heat equation in Chap. 5. The control volumes in the case of cylindrical and spherical coordinates are similar to those that were made use of in Chap. 5. The procedure of deriving the partial differential equations is similar to that used in deriving the heat equation. For the rigorous derivation of the NS equations, the reader may refer to a book such as that by Kays and Crawford.[8]

We shall look at the case of a Newtonian fluid that is incompressible and has constant thermo-physical properties. The last assumption makes the energy equation to decouple from the momentum equations and this is a major advantage while looking for solutions.

H.1.1 NS Equations in Cartesian Coordinates

The flowing fluid has three velocity components u, v, w—these are the components of $\vec{u}$—parallel, respectively, to the three directions x, y, z, i.e.,$\vec{u} = u\hat{i} + v\hat{j} + w\hat{k}$

---

[7]After Claude Louis Marie Henri Navier, 1785–1836, French engineer and Sir George Gabriel Stokes, 1819–1903, British mathematician and physicist

[8]W. M. Kays and M. E. Crawford, *Convective Heat and Mass Transfer*, McGraw Hill International Edition, 1993.

S. P. Venkateshan, *Heat Transfer*,
https://doi.org/10.1007/978-3-030-58338-5

where $\hat{i}, \hat{j}, \hat{k}$ are unit vectors along the three directions. The equation of continuity expresses the fact that the mass (in an incompressible fluid the volume) of a fluid element remains constant as it migrates within the flow domain. The equation of conservation of mass or the equation of continuity is given by

$$\boxed{\frac{\partial u}{\partial x} + \frac{\partial v}{\partial y} + \frac{\partial w}{\partial z} = 0} \quad \text{or} \quad \boxed{\nabla \cdot \vec{u} = 0} \tag{H.1}$$

Let the pressure at any point in the fluid be $p(x, y, z)$. The forces acting on a fluid element are due to changes in pressure across its faces and the shear stresses acting on the faces due to fluid viscosity. These forces result in accelerating the fluid element. The acceleration of the fluid element consists of two parts. The first one is that due to change in velocity at a point with respect to time, if the flow is unsteady. The second one is due to the change in the velocity as the fluid element migrates from one point to another in the flow domain. The total acceleration is the sum of these two and is given in the vector form by

$$\vec{a} = \frac{\partial \vec{u}}{\partial t} + \vec{u} \cdot \nabla \vec{u} \tag{H.2}$$

Note that the acceleration is per unit volume of the fluid. For example, along the x direction the component of acceleration a_x is given by

$$a_x = \frac{\partial u}{\partial t} + u\frac{\partial u}{\partial x} + v\frac{\partial u}{\partial y} + w\frac{\partial u}{\partial z} \tag{H.3}$$

The acceleration given by Eq. H.2 is represented as the total or material derivative such that

$$\vec{a} = \frac{D\vec{u}}{Dt} \tag{H.4}$$

with $\frac{D}{Dt} = \frac{\partial}{\partial t} + \vec{u} \cdot \nabla$. These terms also appear naturally when we perform a momentum balance on a fluid element. The equation of conservation of momentum is written down in the vector form as

$$\boxed{\frac{D\vec{u}}{Dt} = -\frac{1}{\rho}\nabla p + \nu \nabla^2 \vec{u}} \tag{H.5}$$

In scalar form the above equation may be written down as three momentum conservation equations referred to as x momentum, y momentum, and z momentum equations. These are given below

$$\begin{aligned}
&\text{(a)} \quad \frac{\partial u}{\partial t}+u\frac{\partial u}{\partial x}+v\frac{\partial u}{\partial y}+w\frac{\partial u}{\partial z} = -\frac{1}{\rho}\frac{\partial p}{\partial x}+\nu\left(\frac{\partial^2 u}{\partial x^2}+\frac{\partial^2 u}{\partial y^2}+\frac{\partial^2 u}{\partial z^2}\right)\\
&\text{(b)} \quad \frac{\partial v}{\partial t}+u\frac{\partial v}{\partial x}+v\frac{\partial v}{\partial y}+w\frac{\partial v}{\partial z} = -\frac{1}{\rho}\frac{\partial p}{\partial y}+\nu\left(\frac{\partial^2 v}{\partial x^2}+\frac{\partial^2 v}{\partial y^2}+\frac{\partial^2 v}{\partial z^2}\right)\\
&\text{(c)} \quad \frac{\partial w}{\partial t}+u\frac{\partial w}{\partial x}+v\frac{\partial w}{\partial y}+w\frac{\partial w}{\partial z} = -\frac{1}{\rho}\frac{\partial p}{\partial z}+\nu\left(\frac{\partial^2 w}{\partial x^2}+\frac{\partial^2 w}{\partial y^2}+\frac{\partial^2 w}{\partial z^2}\right)
\end{aligned} \tag{H.6}$$

The last equation to be considered is the energy equation which is a scalar equation. We shall assume that the fluid has a constant specific heat c. Also, we shall assume that there is negligible dissipation of kinetic energy to heat because of viscosity. Recall the heat equation that was written down for a stationary medium. All we have to do it to modify the heat equation by adding flux terms due to the movement of the fluid medium. Analogous to momentum the moving fluid transports energy and the required term is $c\vec{u}\cdot\nabla T$. With this addition the heat equation is written, in the scalar form as

$$\boxed{\frac{\partial T}{\partial t}+u\frac{\partial T}{\partial x}+v\frac{\partial T}{\partial y}+w\frac{\partial T}{\partial z} = \alpha\left(\frac{\partial^2 T}{\partial x^2}+\frac{\partial^2 T}{\partial y^2}+\frac{\partial^2 T}{\partial z^2}\right)} \tag{H.7}$$

The NS equations are non-linear because of the inertia terms ($\vec{u}\cdot\nabla\vec{u}$) in the momentum equations. With the constant property assumption the momentum equations may be solved before solving the energy equation which is only a linear equation.

H.1.2 *Ideal Fluid Flow*

Ideal fluid is a fluid that has no viscosity or thermal conductivity. On taking $\nu = 0$, the NS equations reduce to the Euler equations given by

$$\boxed{\frac{D\vec{u}}{Dt} = -\frac{1}{\rho}\nabla p} \tag{H.8}$$

These equations would be of no particular interest were it not for the fact that a fluid *behaves* as an ideal fluid far away from a solid boundary. This fact is exploited in the boundary layer theory that will be discussed in the text in some detail.

For simplicity, we direct our attention to steady flow in two dimensions, confined to the $x - y$ plane. The Euler equations may be written down, in the scalar form as a set of two equations given by

$$\text{(a)} \quad u\frac{\partial u}{\partial x} + v\frac{\partial u}{\partial y} = -\frac{1}{\rho}\frac{\partial p}{\partial x} \quad \text{(b)} \quad u\frac{\partial v}{\partial x} + v\frac{\partial v}{\partial y} = -\frac{1}{\rho}\frac{\partial p}{\partial y} \tag{H.9}$$

The equation of continuity reduces to

$$\frac{\partial u}{\partial x} + \frac{\partial v}{\partial y} = 0 \tag{H.10}$$

A stream function $\psi(x, y)$ defined such that

$$\boxed{u = \frac{\partial \psi}{\partial y}, \quad \text{and} \quad v = -\frac{\partial \psi}{\partial x}} \tag{H.11}$$

satisfies the equation of continuity exactly. Now consider the two momentum equations. Differentiating Eq. H.9(a) with respect to y, we get

$$\underbrace{\frac{\partial u}{\partial y}\frac{\partial u}{\partial x}}_{\text{Term 1}} + u\frac{\partial^2 u}{\partial x \partial y} + \underbrace{\frac{\partial v}{\partial y}\frac{\partial u}{\partial y}}_{\text{Term 3}} + v\frac{\partial^2 u}{\partial y^2} = -\frac{1}{\rho}\frac{\partial^2 p}{\partial x \partial y}$$

Invoking continuity, i.e., $\frac{\partial v}{\partial y} = -\frac{\partial u}{\partial x}$, Terms and 1 and 3 in the above equation cancel each other. Hence, the above equation reduces to

$$u\frac{\partial^2 u}{\partial x \partial y} + v\frac{\partial^2 u}{\partial y^2} = -\frac{1}{\rho}\frac{\partial^2 p}{\partial x \partial y} \tag{H.12}$$

Differentaitng Eq. H.9(b) with respect to x, we get

$$\underbrace{\frac{\partial u}{\partial x}\frac{\partial v}{\partial x}}_{\text{Term 1}} + u\frac{\partial^2 v}{\partial x^2} + \underbrace{\frac{\partial v}{\partial x}\frac{\partial v}{\partial y}}_{\text{Term 3}} + v\frac{\partial^2 v}{\partial x \partial y} = -\frac{1}{\rho}\frac{\partial^2 p}{\partial y \partial x}$$

Equation of continuity is invoked again to cancel terms 1 and 3 to get

$$u\frac{\partial^2 v}{\partial x^2} + v\frac{\partial^2 v}{\partial x \partial y} = -\frac{1}{\rho}\frac{\partial^2 p}{\partial y \partial x} \tag{H.13}$$

Subtract Eq. H.13 from Eq. H.12 to get (after minor manipulation)

$$u\frac{\partial}{\partial x}\left(\frac{\partial u}{\partial y}-\frac{\partial v}{\partial x}\right)+v\frac{\partial}{\partial y}\left(\frac{\partial u}{\partial y}-\frac{\partial v}{\partial x}\right)=0$$

We introduce the notation $\omega=\frac{\partial u}{\partial y}-\frac{\partial v}{\partial x}$ to write the above equation as

$$u\frac{\partial \omega}{\partial x}+v\frac{\partial \omega}{\partial y}=0 \tag{H.14}$$

ω is known as the vorticity. In two-dimensional flow, this represents the rotation of a fluid element. Equation H.14 states simply that vorticity is transported with the fluid without any generation or annihilation. This is a consequence of the fluid being non-viscous or *inviscid*.

Writing the velocities in terms of stream function using Eq. H.11, the above equation simplifies to

$$u\left(\frac{\partial^2 \psi}{\partial y^2}+\frac{\partial^2 \psi}{\partial x^2}\right)+v\left(\frac{\partial^2 \psi}{\partial y^2}+\frac{\partial^2 \psi}{\partial x^2}\right)=0 \tag{H.15}$$

Since there is no restriction on the magnitudes of u and v, the above equation holds only if the stream function satisfies the Laplace equation given by[9]

$$\boxed{\frac{\partial^2 \psi}{\partial x^2}+\frac{\partial^2 \psi}{\partial y^2}=0} \tag{H.16}$$

We may also define a potential function $\phi(x, y)$ such that the velocity components are given by

$$u=\frac{\partial \phi}{\partial x} \text{ and } v=\frac{\partial \phi}{\partial y} \tag{H.17}$$

The vorticity is then given by

$$\frac{\partial u}{\partial y}-\frac{\partial v}{\partial x}=\frac{\partial^2 \phi}{\partial y \partial x}-\frac{\partial^2 \phi}{\partial x \partial y}\equiv 0$$

Thus, the flow is rotation free or *irrotational*. Using the equation of continuity, we also have

[9]Note in general that $\frac{\partial^2 \psi}{\partial x^2}+\frac{\partial^2 \psi}{\partial y^2}=\omega$ since the stream function may be defined even when the flow is that of a viscous fluid.

$$\frac{\partial u}{\partial x} + \frac{\partial v}{\partial y} = \boxed{\frac{\partial^2 \phi}{\partial x^2} + \frac{\partial \phi}{\partial y^2} = 0} \tag{H.18}$$

Thus, the potential function also satisfies the Laplace equation.

Since the stream and potential functions satisfy individually the Laplace equation,, we may consider these as the real and imaginary parts of a complex potential. The complex potential method that is given in Appendix C may also be used for treating potential flows. Actually, typical ideal fluid flow examples have been given there.

H.2 NS Equations in Cylindrical and Spherical Coordinates

If we write the NS equations in the vector form, they are valid representations in all the three coordinate systems. The NS equations are presented (starting page 996 ending with page 999) in scalar forms for easy reference.

H.2.1 NS Equations in Cylindrical Coordinates

The velocity components are represented as u_r, u_θ, u_z, respectively, along the three coordinate directions r, θ, z. The NS equations are given below

- Equation of continuity:

$$\frac{1}{r}\frac{\partial (r u_r)}{\partial r} + \frac{1}{r}\frac{\partial (u_\theta)}{\partial \theta} + \frac{\partial u_z}{\partial z} = 0 \tag{H.19}$$

- r momentum equation:

$$\begin{aligned} \frac{\partial u_r}{\partial t} + u_r \frac{\partial u_r}{\partial r} + \frac{u_\theta}{r}\frac{\partial u_r}{\partial \theta} - \frac{u_\theta^2}{r} + u_z \frac{\partial u_r}{\partial z} = -\frac{1}{\rho}\frac{\partial p}{\partial r} + \\ \nu\left[\frac{\partial}{\partial r}\left(\frac{1}{r}\frac{\partial (r u_r)}{\partial r}\right) + \frac{1}{r^2}\frac{\partial^2 u_r}{\partial \theta^2} - \frac{2}{r^2}\frac{\partial u_\theta}{\partial \theta} = \frac{\partial^2 u_z}{\partial z^2}\right] \end{aligned} \tag{H.20}$$

- θ momentum equation:

$$\frac{\partial u_\theta}{\partial t}+u_r\frac{\partial u_\theta}{\partial r}+\frac{u_\theta}{r}\frac{\partial u_\theta}{\partial \theta}+\frac{u_r u_\theta}{r}+u_z\frac{\partial u_\theta}{\partial z}=-\frac{1}{\rho}\frac{1}{r}\frac{\partial p}{\partial \theta}+ \nu\left[\frac{\partial}{\partial r}\left(\frac{1}{r}\frac{\partial (r u_\theta)}{\partial r}\right)+\frac{1}{r^2}\frac{\partial^2 u_\theta}{\partial \theta^2}+\frac{2}{r^2}\frac{\partial u_r}{\partial \theta}=\frac{\partial^2 u_\theta}{\partial z^2}\right] \tag{H.21}$$

- z momentum equation:

$$\frac{\partial u_z}{\partial t}+u_r\frac{\partial u_z}{\partial r}+\frac{u_\theta}{r}\frac{\partial u_z}{\partial \theta}+u_z\frac{\partial u_z}{\partial z}=-\frac{1}{\rho}\frac{\partial p}{\partial z}+ \nu\left[\frac{1}{r}\frac{\partial}{\partial r}\left(r\frac{\partial u_z}{\partial r}\right)+\frac{1}{r^2}\frac{\partial^2 u_z}{\partial \theta^2}+\frac{\partial^2 u_z}{\partial z^2}\right] \tag{H.22}$$

- Energy equation:

$$\frac{\partial T}{\partial t}+u_r\frac{\partial T}{\partial r}+\frac{u_\theta}{r}\frac{\partial T}{\partial \theta}+u_z\frac{\partial T}{\partial z}=\alpha\left[\frac{1}{r}\frac{\partial}{\partial r}\left(r\frac{\partial T}{\partial r}\right)+\frac{1}{r^2}\frac{\partial^2 T}{\partial \theta^2}+\frac{\partial^2 T}{\partial z^2}\right] \tag{H.23}$$

Inviscid flow in polar coordinates

Steady inviscid flow in a plane may be considered either using cartesian system of coordinates or the polar coordinates (r,θ). The equations in cylindrical coordinates reduce to simple form when we put $u_z\equiv 0$ and $\frac{\partial}{\partial z}\equiv 0$ in Eq. H.19 through H.22.

- Equation of continuity:

$$\frac{\partial (r u_r)}{\partial r}+\frac{\partial (u_\theta)}{\partial \theta}=0 \tag{H.24}$$

- r momentum equation:

$$u_r\frac{\partial u_r}{\partial r}+\frac{u_\theta}{r}\frac{\partial u_r}{\partial \theta}-\frac{u_\theta^2}{r}=-\frac{1}{\rho}\frac{\partial p}{\partial r} \tag{H.25}$$

- θ momentum equation:

$$u_r\frac{\partial u_\theta}{\partial r}+\frac{u_\theta}{r}\frac{\partial u_\theta}{\partial \theta}+\frac{u_r u_\theta}{r}=-\frac{1}{\rho}\frac{1}{r}\frac{\partial p}{\partial \theta} \tag{H.26}$$

We see that a stream function $\psi(r,\theta)$ may be defined as follows:

$$u_r = \frac{1}{r}\frac{\partial \psi}{\partial \theta},\ u_\theta = \frac{\partial \psi}{\partial r} \tag{H.27}$$

We then see that the equation of continuity is identically satisfied. As in the case of cartesian system, $\psi(r, \theta)$ also satisfies the Laplace equation, but in polar form given by

$$\frac{\partial}{\partial r}\left(r\frac{\partial \psi}{\partial r}\right) + \frac{1}{r}\frac{\partial^2 \psi}{\partial \theta^2} = 0 \tag{H.28}$$

The flow is again irrotational and it is possible to define a potential function $\phi(r, \theta)$ such that

$$\vec{u} = \nabla \phi$$

where we use the gradient operator in polar form. We thus have the following:

$$u_r = \frac{\partial \phi}{\partial r},\ u_\theta = \frac{1}{r}\frac{\partial \phi}{\partial \theta} \tag{H.29}$$

Then the potential ϕ also satisfies Laplace equation in polar coordinates given in Eq. H.28.

NS equations for axisymmetric flow

A case that is of much importance in convection heat transfer is axisymmetric flow in cylindrical coordinates. For example, flow inside a circular tube is of this type. The velocity is a function of only r, z and the appropriate equations are obtained by setting $u_\theta \equiv 0$ and $\frac{\partial}{\partial \theta} \equiv 0$ in Eqs. H.19 through H.23. Of course, Eq. H.21 drops off completely. The appropriate equations for axisymmetric flow are given below

- Equation of continuity:

$$\frac{1}{r}\frac{\partial (r u_r)}{\partial r} + \frac{\partial u_z}{\partial z} = 0 \tag{H.30}$$

- r momentum equation:

$$\frac{\partial u_r}{\partial t} + u_r\frac{\partial u_r}{\partial r} + u_z\frac{\partial u_r}{\partial z} = -\frac{1}{\rho}\frac{\partial p}{\partial r} + \nu\left[\frac{\partial}{\partial r}\left(\frac{1}{r}\frac{\partial (r u_r)}{\partial r}\right) + \frac{\partial^2 u_z}{\partial z^2}\right] \tag{H.31}$$

- z momentum equation:

$$\frac{\partial u_z}{\partial t} + u_r\frac{\partial u_z}{\partial r} + u_z\frac{\partial u_z}{\partial z} = -\frac{1}{\rho}\frac{\partial p}{\partial z} + \nu\left[\frac{1}{r}\frac{\partial}{\partial r}\left(r\frac{\partial u_z}{\partial r}\right) + \frac{\partial^2 u_z}{\partial z^2}\right] \tag{H.32}$$

- Energy equation:

$$\frac{\partial T}{\partial t}+u_r\frac{\partial T}{\partial r}+u_z\frac{\partial T}{\partial z}=\alpha\Big[\frac{1}{r}\frac{\partial}{\partial r}\Big(r\frac{\partial T}{\partial r}\Big)+\frac{\partial^2 T}{\partial z^2}\Big] \tag{H.33}$$

H.2.2 NS Equations in Spherical Coordinates

The velocity components are represented as u_r, u_θ, u_ϕ, respectively, along the three coordinate directions r, θ, ϕ. The NS equations are given below

- Equation of continuity:

$$\frac{1}{r^2}\frac{\partial(r^2u_r)}{\partial r}+\frac{1}{r\sin\theta}\frac{\partial(u_\theta\sin\theta)}{\partial\theta}+\frac{1}{r\sin\theta}\frac{\partial u_\phi}{\partial\phi}=0 \tag{H.34}$$

- r momentum equation:

$$\frac{\partial u_r}{\partial t}+u_r\frac{\partial u_r}{\partial r}+\frac{u_\theta}{r}\frac{\partial u_r}{\partial\theta}+\frac{u_\phi}{r\sin\theta}\frac{\partial u_r}{\partial\phi}-\frac{u_\theta^2+u_\phi^2}{r}=-\frac{1}{\rho}\frac{\partial p}{\partial r}$$
$$+\nu\Big[\frac{1}{r^2}\frac{\partial^2(r^2u_r)}{\partial r^2}+\frac{1}{r^2\sin\theta}\frac{\partial}{\partial\theta}\Big(\sin\theta\frac{\partial u_r}{\partial\theta}\Big)+\frac{1}{r^2\sin^2\theta}\frac{\partial^2u_r}{\partial\phi^2}\Big] \tag{H.35}$$

- θ momentum equation:

$$\frac{\partial u_\theta}{\partial t}+u_r\frac{\partial u_\theta}{\partial r}+\frac{u_\theta}{r}\frac{\partial u_\theta}{\partial\theta}+\frac{u_\phi}{r\sin\theta}\frac{\partial u_\theta}{\partial\phi}+\frac{u_ru_\theta}{r}-\frac{u_\phi^2\cot\theta}{r}=-\frac{1}{\rho}\frac{1}{r}\frac{\partial p}{\partial\theta}$$
$$+\nu\Big[\frac{1}{r^2}\frac{\partial}{\partial r}\Big(r^2\frac{\partial u_\theta}{\partial r}\Big)+\frac{1}{r^2\sin^2\theta}\frac{\partial^2u_\theta}{\partial\phi^2}+\frac{2}{r^2}\frac{\partial u_r}{\partial\theta}-\frac{2\cos\theta}{r^2\sin^2\theta}\frac{\partial u_\phi}{\partial\phi}\Big] \tag{H.36}$$

- ϕ momentum equation:

$$\frac{\partial u_\phi}{\partial t}+u_r\frac{\partial u_\phi}{\partial r}+\frac{u_\theta}{r}\frac{\partial u_\phi}{\partial\theta}+\frac{u_\phi}{r\sin\theta}\frac{\partial u_\phi}{\partial\phi}+\frac{u_ru_\phi}{r}+\frac{u_\theta u\phi\cot\theta}{r}=$$
$$-\frac{1}{\rho}\frac{1}{r\sin\theta}\frac{\partial p}{\partial\phi}+\nu\Big[\frac{1}{r^2}\frac{\partial}{\partial r}\Big(r^2\frac{\partial u_\phi}{\partial r}\Big)+\frac{1}{r^2}\Big(\frac{1}{\sin\theta}\frac{\partial}{\partial\theta}(u_\phi\sin\theta)\Big)+$$
$$\frac{1}{r^2\sin^2\theta}\frac{\partial^2u_\phi}{\partial\phi^2}+\frac{2}{r^2\sin\theta}\frac{\partial u_r}{\partial\phi}+\frac{2\cos\theta}{r^2\sin^2\theta}\frac{\partial u_\theta}{\partial\phi}\Big] \tag{H.37}$$

- Energy equation:

$$\frac{\partial T}{\partial t} + u_r \frac{\partial T}{\partial r} + \frac{u_\theta}{r}\frac{\partial T}{\partial \theta} + \frac{u_\phi}{r \sin\theta} =$$
$$\alpha\left[\frac{1}{r^2}\frac{\partial}{\partial r}\left(r^2 \frac{\partial T}{\partial r}\right) + \frac{1}{r^2 \sin^2\theta}\frac{\partial}{\partial \theta}\left(\frac{\sin\theta}{r}\frac{\partial T}{\partial \theta}\right) + \frac{1}{r \sin\theta}\frac{\partial}{\partial \phi}\left(\frac{1}{r \sin\theta}\frac{\partial T}{\partial \phi}\right)\right] \tag{H.38}$$

Appendix I
Useful Tables

See Tables I.1, I.2, I.3, I.4, I.5 and I.6

Table I.1 Physical constants

Atmospheric pressure	101.325	kPa
Avogadro number, N	6.022×10^{23}	$\frac{molecules}{mole}$
Velocity of light in vacuum, c_0	3×10^{8}	$\frac{m}{s}$
Planck constant, h	6.62×10^{-34}	$J\ s$
Stefan-Boltzmann constant, σ	5.67×10^{-8}	$\frac{W}{m^2 K^4}$
Boltzmann constant, k	1.39×10^{-23}	$\frac{J}{K}$
First radiation constant, C_1	3.74×10^{-16}	$W\ m^2$
Second radiation constant, C_2	14387.7	$\mu m\ K$
Standard acceleration due to gravity, g	9.807	$\frac{m}{s^2}$

S. P. Venkateshan, *Heat Transfer*,
https://doi.org/10.1007/978-3-030-58338-5

Table I.2 Units of physical quantities

Quantity	Dimension	Name	Symbol	Unit
Acceleration	LT^{-2}		a	$\frac{m}{s^2}$
Area	L^2		A or S	m^2
Density	ML^{-3}		ρ	$\frac{kg}{m^3}$
Dynamic viscosity	$ML^{-1}T^{-1}$		μ	$\frac{kg}{m\ s}$
Energy	ML^2T^{-2}	*Joule*	E	$\frac{kg\ m^2}{s^2}$ or J
Enthalpy	ML^2T^{-2}		h	$\frac{kg\ m^2}{s^2}$ or J
Force	MLT^{-2}	*Newton*	F	$\frac{kg\ m}{s^2}$ or N
Frequency	T^{-1}	*Hertz*	f	$\frac{1}{s}$ or Hz
Gas constant	$L^2T^{-2}\theta^{-1}$		R_g	$\frac{J}{kg\ K}$ or $\frac{J}{kg\ ^\circ C}$
Heat Flux	MT^{-3}		q	$\frac{J}{s\ m^2}$ or $\frac{W}{m^2}$
Heat transfer coefficient	$MT^{-3}\theta^{-1}$		h	$\frac{W}{m^{2\circ}C}$
Kinematic viscosity	L^2T^{-1}		ν	$\frac{m^2}{s}$
Momentum	MLT^{-1}		p	$\frac{kg\ m}{s}$ or $N\ s$
Power	ML^2T^{-3}	*Watt*	P	W
Pressure or stress	$ML^{-1}T^{-2}$	*Pascal*	p or P	Pa or $\frac{N}{m^2}$
Ratio of specific heats	non-dimensional		γ	No unit
Resistance, fluid due to viscosity	$L^{-1}T^{-1}$		R	$\frac{1}{m\ s}$
Resistance, thermal	$M^{-1}T^3\theta$		R	$\frac{^\circ C\ m^2}{W}$
Specific heat	$L^2T^{-2}\theta^{-1}$		c	$\frac{J}{kg\ K}$ or $\frac{J}{kg\ ^\circ C}$
Thermal conductivity	$MLT^{-3}\theta^{-1}$		k	$\frac{W}{m^\circ C}$
Thermal diffusivity	L^2T^{-1}		α	$\frac{m^2}{s}$
Velocity	LT^{-1}		V	$\frac{m}{s}$
Velocity, sonic	LT^{-1}		a	$\frac{m}{s}$
Wavelength	L		λ	m or μm
Wavenumber	L^{-1}			cm^{-1}
Work	ML^2T^{-2}	*Joule*	W	$\frac{kg\ m^2}{s^2}$ or J

Table I.3 Important dimensionless groups

Biot number	$Bi = \frac{hL}{k}$	Nusselt number	$Nu = \frac{hL}{k}$
Colburn j factor	$j = St \times Pr^{\frac{2}{3}}$	Peclet number	$Pe = Re \times Pr$
Euler number	$Eu = \frac{p}{\rho U^2}$	Prandtl number	$Pr = \frac{\mu C_p}{k}$
Eckert number	$Ec = \frac{U^2}{C_p T}$	Radiation conduction interaction parameter	$N_{RC} = \frac{\varepsilon \sigma T^3 L^2}{kt}$
Fourier number	$Fo = \frac{\alpha t}{2}$	Reynolds number	$Re = \frac{\rho U L}{\mu}$
Graetz number	$Gz = \frac{Re_D Pr}{2} \times \frac{x}{D}$	Richardson number	$\frac{Gr}{Re^2}$
Grashof number	$Gr = \frac{g\beta \Delta T L^3}{\nu^2}$	Stanton number	$St = \frac{h}{\rho U C_p}$
Mach number	$M = \frac{U}{a}$	Stefan number	$Ste = \frac{C_{pf}(T_0 - T_m)}{h_{sf}}$

Table I.4 Thermal properties of metallic solids*

Metal	ρ kg/m^3	C J/kg°C	k W/m°C	$\alpha \times 10^5$ m^2/s
Aluminum	2702	903	237	9.71
Alumel	8574	464	29.2	0.734
Brass	8470	377	116	3.63
Bronze	8830	377	52	1.36
Cast Iron (4% C)	7272	420	51	1.67
Chromel	8730	448	17.3	0.493
Copper	8900	385	21.7	6.33
Constantan	8900	390	22.7	0.654
Duralumin	8933	385	401	11.7
Gold	19300	129	317	12.7
Iron	7870	447	80.3	2.28
Nickel	8900	444	90.7	2.3
Platinum	21450	133	71.6	2.51
Silver	10500	235	429	17.4
Stainless Steel 304	7900	477	14.9	0.395
Tin	7310	227	66.6	4.03
Titanium	4500	522	21.9	0.932
Tungsten	19300	132	174	6.83

*Tabulated values are at room temperature and representative

Table I.5 Thermal properties of non-metallic solids*

Non-metallic solid	ρ kg/m^3	C J/kg°C	k W/m°C	$\alpha \times 10^7$ m^2/s
Alumina (Al_2O_3)	3970	765	36	11.9
Asbestos cement board	1900	1000	0.58	3.05
Asbestos cement roofing	1900	1000	0.27	1.42
Asphalt	2110	920	0.74	3.81
Bakelite	1270	1590	0.233	1.15
Balsa wood	140		0.055	
Cellular polyurethane	24	1590	0.025	6.55
Cement plaster	1860	840	0.72	4.61
Clay	2080	921	1.73	9.03
Common brick	1920	835	0.72	4.49
Concrete	2240	900	1.9	9.42
Concrete blocks	2100	920	1.1	5.69
Crown glass	2500		1.22	
CVD diamond			1000-1800	
Diamond	3500	509	2300	12910
Felt	330		0.05	
Fiberglass	16		0.046	
Fired clay brick	1920	790	0.9	5.93
Granite	2630	775	2.79	13.7
Ice at 0°C	917	2093	2.21	11.5
Magnesia	184		0.05	
Mylar			0.197	
Oak	704		0.17	
Paper	930	1300	0.13	1.08
Paperboard, laminated	480	1380	0.072	1.09
Particle board, high density	1000	1300	0.17	1.31
Particle board, low density	590	1300	0.102	1.33
Plywood	540	1210	0.12	1.84
Polyurethane foam	32		0.025	
Polyvinyl chloride (PVC)	1380	1170	0.15	0.929
Pyrex	2170	716	1.06	6.82
Sand	1520	921	0.317	2.26
Window glass	2800		0.7	
Wool felt	320		0.045	

*Tabulated values are at room temperature and representative

Table I.6 Properties of dry air at atmospheric pressure

T °C	ρ kg/m^3	C_p kJ/kg°C	$\mu \times 10^6$ kg/m · s	$\nu \times 10^6$ m^2/s	$k \times 10^3$ W/m°C	$\alpha \times 10^6$ m^2/s	Pr
0	1.2811	1.004	17.09	13.34	24.2	18.8	0.709
20	1.1934	1.004	18.09	15.16	25.8	21.5	0.704
40	1.1169	1.005	19.07	17.07	27.4	24.4	0.699
60	1.0496	1.007	20.02	19.08	29	27.4	0.696
80	0.9899	1.008	20.95	21.16	30.5	30.5	0.693
100	0.9367	1.011	21.85	23.33	32	33.8	0.691
120	0.8889	1.013	22.74	25.58	33.4	37.1	0.689
140	0.8457	1.016	23.59	27.9	34.9	40.6	0.688
160	0.8065	1.019	24.43	30.29	36.3	44.1	0.687
180	0.7708	1.023	25.25	32.76	37.6	47.7	0.686
200	0.7381	1.026	26.04	35.29	38.9	51.4	0.686
240	0.6803	1.034	27.58	40.54	41.5	59	0.687
280	0.631	1.043	29.05	46.04	43.9	66.8	0.689
320	0.5883	1.051	30.45	51.76	46.3	74.8	0.692
360	0.551	1.06	31.8	57.71	48.5	83	0.695
400	0.5181	1.069	33.09	63.87	50.6	91	0.699
440	0.489	1.078	34.34	70.23	52.7	100	0.703
480	0.4629	1.087	35.55	76.8	54.7	109	0.707
520	0.4395	1.096	36.73	83.58	56.7	118	0.71
560	0.4183	1.105	37.88	90.56	58.6	127	0.714
600	0.3991	1.114	39.01	97.75	60.6	136	0.717

Table I.7 Thermal properties of saturated water

T °C	ρ kg/m^3	C_p kJ/kg°C	$\mu \times 10^6$ kg/m · s	$\nu \times 10^7$ m^2/s	k W/m°C	$\alpha \times 10^6$ m^2/s	Pr
0	999.8	4.217	1752.5	17.53	0.569	1.35	12.988
10	999.7	4.193	1299.2	13	0.586	1.398	9.296
20	998.3	4.182	1001.5	10.03	0.602	1.442	6.957
30	995.7	4.179	797	8.004	0.617	1.483	5.398
40	992.3	4.179	651.3	6.564	0.63	1.519	4.32
50	988	4.181	544	5.506	0.643	1.557	3.537
60	983.2	4.185	463	4.709	0.653	1.587	2.967
70	977.7	4.19	400.5	4.096	0.662	1.616	2.535
80	971.6	4.197	351	3.613	0.669	1.641	2.202
90	965.2	4.205	311.3	3.225	0.675	1.663	1.939

(continued)

Table I.7 (continued)

T °C	ρ kg/m^3	C_p kJ/kg°C	$\mu \times 10^6$ kg/m · s	$\nu \times 10^7$ m^2/s	k W/m°C	$\alpha \times 10^6$ m^2/s	Pr
100	958.1	4.216	279	2.912	0.68	1.683	1.73
110	950.7	4.229	252.2	2.653	0.683	1.699	1.562
120	942.9	4.245	230	2.439	0.685	1.711	1.425
130	934.6	4.263	211	2.258	0.687	1.724	1.309
140	925.8	4.285	195	2.106	0.687	1.732	1.216
150	916.8	4.31	181	1.974	0.686	1.736	1.137
160	907.3	4.339	169	1.863	0.684	1.737	1.072
170	897.3	4.371	158.5	1.766	0.681	1.736	1.017
180	886.9	4.408	149.3	1.683	0.676	1.729	0.974
200	864.7	4.497	133.8	1.547	0.664	1.708	0.906
220	840.3	4.614	121.5	1.446	0.648	1.671	0.865
240	813.6	4.77	111.4	1.369	0.629	1.621	0.845
260	783.9	4.985	103	1.314	0.604	1.546	0.85
280	750.5	5.3	96.1	1.28	0.573	1.441	0.889
300	712.2	5.77	90.1	1.265	0.54	1.314	0.963
320	666.9	6.59	83	1.245	0.503	1.145	1.087
340	610.1	8.27	74.8	1.226	0.46	0.912	1.345
360	528.3	14.99	64.4	1.219	0.401	0.506	2.407

Table I.8 Thermal properties of SAE-20 engine oil

T °C	ρ kg/m^3	C_p kJ/kg°C	$\mu \times 10^6$ kg/m · s	$\nu \times 10^6$ m^2/s	$k \times 10^3$ W/m°C	$\alpha \times 10^7$ m^2/s	Pr
0	902.9	1.726	77.8	861.71	134.5	0.863	9984.9
10	896.2	1.762	33.599	374.9	134.3	0.85	4409.4
20	889.6	1.798	16.396	184.31	134	0.838	2199.3
30	882.9	1.834	8.878	100.55	133.8	0.826	1216.8
40	876.3	1.87	5.247	59.88	133.6	0.815	734.5
50	869.6	1.906	3.335	38.35	133.4	0.805	476.7
60	863	1.942	2.251	26.08	133.1	0.794	328.3
70	856.3	1.978	1.594	18.62	132.9	0.785	237.3
80	849.7	2.014	1.174	13.82	132.7	0.775	178.2
90	843	2.05	0.892	10.58	132.5	0.767	138
100	836.4	2.086	0.694	8.3	132.2	0.758	109.5
110	829.7	2.122	0.551	6.65	132	0.75	88.6
120	823.1	2.158	0.446	5.42	131.8	0.742	73
130	816.4	2.194	0.366	4.49	131.6	0.735	61.1
140	809.8	2.23	0.306	3.78	131.3	0.727	52
150	803.1	2.266	0.261	3.25	131.1	0.721	45.1
160	796.5	2.302	0.227	2.85	130.9	0.714	40
170	789.8	2.338	0.204	2.58	130.7	0.708	36.5

Table I.9 Thermal properties of ethylene glycol

T °C	ρ kg/m^3	C_p kJ/kg°C	$\mu \times 10^2$ kg/m · s	$\nu \times 10^6$ m^2/s	$k \times 10^3$ W/m°C	$\alpha \times 10^7$ m^2/s	Pr
0	1129	2.28	5.89	52.2	304.3	1.18	441.6
10	1123	2.33	3.51	31.24	296.6	1.13	275.5
20	1116	2.38	2.2	19.68	289	1.09	180.8
30	1109	2.429	1.44	13	281.4	1.04	124.4
40	1102	2.479	0.988	8.97	273.8	1	89.4
50	1094	2.529	0.704	6.44	266.3	0.962	66.9
60	1087	2.578	0.521	4.79	258.7	0.923	51.9
70	1079	2.628	0.397	3.68	251.2	0.886	41.6
80	1071	2.677	0.312	2.91	243.6	0.849	34.3
90	1064	2.727	0.251	2.36	236.1	0.814	29
100	1056	2.777	0.206	1.95	228.6	0.78	25
110	1048	2.826	0.172	1.64	221.1	0.747	22
120	1040	2.876	0.146	1.4	213.7	0.714	19.6
130	1032	2.925	0.125	1.21	206.2	0.683	17.7
140	1024	2.975	0.107	1.04	198.7	0.652	16
150	1017	3.025	0.092	0.91	191.2	0.622	14.6
160	1009	3.075	0.079	0.79	183.8	0.592	13.3
170	1002	3.125	0.068	0.68	176.3	0.563	12

Table I.10 Thermal properties of ice

T °C	C_p J/kg · K	k W/m°C	ρ kg/m^3	$\alpha \times 10^6$ m^2/s
0	2040	2.24	917	1.2
−10	1997	2.32	916	1.27
−20	1946	2.43	914	1.37
−30	1886	2.55	913	1.48
−40	1817	2.66	911	1.61

$h_{sf} = 333.6$ kJ/kg at $T = 0$ °C

Table I.11 Emissivities of surfaces at room temperature

Surface	Emissivity	Surface	Emissivity
Aluminum (anodized)	0.77	Glass	0.92
Aluminum (oxidized)	0.11	Gold	0.02
Aluminum (polished)	0.05	Iron oxide	0.56
Aluminum (roughened with emery)	0.17	Paper	0.93

(continued)

Table I.11 (continued)

Surface	Emissivity	Surface	Emissivity
Aluminum foil	0.03	Plaster	0.98
Aluminized Mylar	0.03	Porcelain (glazed)	0.92
Anodized black coating	0.88	Silver-pure, polished	0.020–0.032
Asbestos board	0.94	Steel-galvanized (old)	0.88
Brass (dull)	0.22	Steel-galvanized (new)	0.23
Brass (polished)	0.03	Steel-oxidized	0.79
Brick (dark)	0.9	Steel-polished	0.07
Concrete	0.85	Tile	0.97
Copper (oxidized)	0.87	Tungsten (polished)	0.03
Copper (polished)	0.04	Water	0.95
Fire-clay	0.75		

Table I.12 Emissivity variation with temperature for various surfaces

	Temperature °C					
Material surface	25	100	200	500	600	1200
Aluminum (heavily oxidized)		0.2		0.31		
Aluminum (oxidized)			0.11		0.19	
Aluminum (un-oxidized)	0.02	0.03		0.06		
Aluminum (commercial sheet)		0.09				
Aluminum (highly polished sheet)			0.04		0.06	
Aluminum (highly polished)		0.09				
Aluminum (roughly polished)		0.18				
Brass (polished)			0.03	0.03		
Carborundum						0.92
Cast iron (heavily oxidized)		0.95	0.95			
Cast iron (oxidized)			0.64		0.78	
Cast iron (un-oxidized)		0.21				
Fire brick						0.75–0.80
Iron (oxidized)		0.74		0.84		0.89
Iron (un-oxidized)		0.05				
Mild steel (polished)	0.1					
Refractory (magnesite)						0.75
Silica (glazed)						0.88

(continued)

Table I.12 (continued)

	Temperature °C					
Silica(unglazed)						0.8
Silver (polished)	0.01		0.02		0.03	0.03
Steel (oxidized)	0.8					
Steel (polished sheet)	0.07		0.1		0.14	
Steel (un-oxidized)		0.08				

Index

S. P. Venkateshan, *Heat Transfer*,
https://doi.org/10.1007/978-3-030-58338-5

Printed in the United States
by Baker & Taylor Publisher Services